## Algebraic forms

25. $\displaystyle\int \frac{u\,du}{a+bu} = \frac{u}{b} - \frac{a}{b^2}\ln(a+bu)$

26. $\displaystyle\int \frac{du}{u(a+bu)} = \frac{1}{a}\ln\left|\frac{u}{a+bu}\right|$

27. $\displaystyle\int \frac{u\,du}{(a+bu)^2} = \frac{a}{b^2}\left(\frac{1}{a+bu} + \frac{1}{a}\ln|a+bu|\right)$

28. $\displaystyle\int \frac{du}{u(a+bu)^2} = \frac{1}{a(a+bu)} + \frac{1}{a^2}\ln\left|\frac{u}{a+bu}\right|$

29. $\displaystyle\int \frac{du}{u\sqrt{a+bu}} = \frac{1}{\sqrt{a}}\ln\left|\frac{\sqrt{a+bu}-\sqrt{a}}{\sqrt{a+bu}+\sqrt{a}}\right|$

30. $\displaystyle\int u\sqrt{a+bu}\,du = \frac{2(3\,bu - 2a)\sqrt{(a+bu)^3}}{15b^2}$

31. $\displaystyle\int \frac{\sqrt{a+bu}}{u}\,du = 2\sqrt{a+bu} + a\int \frac{du}{u\sqrt{a+bu}}$

32. $\displaystyle\int \frac{u\,du}{\sqrt{a+bu}} = \frac{2(bu - 2a)}{3b^2}\sqrt{a+bu}$

33. $\displaystyle\int \sqrt{a^2 - u^2}\,du = \frac{u}{2}\sqrt{a^2 - u^2} + \frac{a^2}{2}\arcsin\frac{u}{a}, \quad |u| < a$

34. $\displaystyle\int \sqrt{u^2 \pm a^2}\,du = \frac{u}{2}\sqrt{u^2 \pm a^2} \pm \frac{a^2}{2}\ln|u + \sqrt{u^2 \pm a^2}|$

35. $\displaystyle\int \frac{\sqrt{a^2 \pm u^2}}{u}\,du = \sqrt{a^2 \pm u^2} - a\ln\left|\frac{a + \sqrt{a^2 \pm u^2}}{u}\right|$

36. $\displaystyle\int \frac{\sqrt{u^2 - a^2}}{u}\,du = \sqrt{u^2 - a^2} - a\arccos\frac{a}{u}, \quad 0 < a < |u|$

## Trigonometric forms

37. $\displaystyle\int \tan u\,du = -\ln|\cos u|$

38. $\displaystyle\int \sec u\,du = \ln|\sec u + \tan u|$

39. $\displaystyle\int \sin^2 u\,du = \tfrac{1}{2}u - \tfrac{1}{4}\sin 2u$

40. $\displaystyle\int \sin^n u\,du = -\frac{\sin^{n-1} u \cos u}{n} + \frac{n-1}{n}\int \sin^{n-2} u\,du$

41. $\displaystyle\int \cos^n u\,du = \frac{\cos^{n-1} u \sin u}{n} + \frac{n-1}{n}\int \cos^{n-2} u\,du$

42. $\displaystyle\int \frac{du}{\sin^n u} = -\frac{\cos u}{(n-1)\sin^{n-1} u} + \frac{n-2}{n-1}\int \frac{du}{\sin^{n-2} u}, \quad n \neq 1$

43. $\displaystyle\int \sin mu \sin nu\,du = \frac{\sin(m-n)u}{2(m-n)} - \frac{\sin(m+n)u}{2(m+n)}, \quad m \neq \pm n$

44. $\displaystyle\int \cos mu \cos nu\,du = \frac{\sin(m-n)u}{2(m-n)} + \frac{\sin(m+n)u}{2(m+n)}, \quad m \neq \pm n$

(Continued inside back cover)

# FOURTH EDITION

# CALCULUS

## WITH ANALYTIC GEOMETRY

# FOURTH EDITION

# CALCULUS

## WITH ANALYTIC GEOMETRY

**MURRAY H. PROTTER**
UNIVERSITY OF CALIFORNIA, BERKELEY

**PHILIP E. PROTTER**
PURDUE UNIVERSITY

**JONES** AND **BARTLETT PUBLISHERS**
Boston • Portola Valley

*Sales, editorial and customer service offices:*
Jones and Bartlett Publishers, Inc.
20 Park Plaza
Boston, MA 02116


Photographs for Chapters 5, 16, and 18 from *Living Images* by Gene Shih and
Richard Kessel. © Jones and Bartlett Publishers, 1982.
All other photographs by Rafael Millán.


Library of Congress Cataloging-in-Publication Data
Protter, Murray H.
    Calculus with analytic geometry.
     1. Calculus.    2. Geometry, Analytic.    I. Protter,
Philip E.    II. Title.
QA303.P972  1988    515′.15      87–30983

ISBN: 0-86720-093-6


*Cover art:* Untitled monotype by Robert Siegelman. © 1986.
        Reprinted by permission.
        Courtesy of Gallery NAGA (Boston).
*Book and cover design:* Michael Michaud.
*Production:* Unicorn Production Services, Inc.
*Composition:* H. Charlesworth & Co. Ltd.
*Printing and Binding:* Rand McNally & Company.

Printed in the United States of America.

Printing number (last digit): 10 9 8 7 6 5 4 3 2 1

# PREFACE TO THE FOURTH EDITION

In the teaching of Calculus the selection of topics, the order in which they are treated, and the style in which they are taught continue to change. Once the province of mathematicians, physicists, chemists, and engineers, calculus has also become indispensable to the biological sciences and to many areas of the social sciences. In the fourth edition of *College Calculus with Analytic Geometry* we have changed the selection, the order, and sometimes the treatment of topics to reflect current needs. Our goal remains, as in the three previous editions, to meet the requirements of a majority of students taking either a two or three semester course at a college or university.

A major change in this edition is the addition of two new chapters. Chapter 17, on vector field theory, includes the theorems of Green and Stokes. Our treatment of both theorems is elementary enough for the average student yet sufficiently general for most applications. Chapter 18 treats differential equations; we give techniques for solving both linear and nonlinear first and second order equations. We also include a section on solutions by infinite series, useful for the treatment of numerical solutions of differential equations. This chapter is suitable for students who require only a basic knowledge of differential equations; it also provides a valuable introduction for those who wish to continue studying differential equations at a more advanced level.

We have revised the beginning of the text; in Chapter 1 we present a thorough review of precalculus material. This chapter, together with Appendix 1, which contains all the definitions and formulas from trigonometry which are needed in the course, serves as an introduction for those students who may not have an adequate background for calculus. The well-prepared student, however, can treat Chapter 1 as a review and go on to study limits and the derivative in Chapters 2 and 3.

We have changed and improved the order in which many of the topics are presented. For example, we now develop the formulas for the differentiation of trigonometric functions early in the course, in Chapter 3. Thus, when we introduce integration in Chapter 5, we can provide the integration formulas for trigonometric functions. Such an early introduction of the calculus of trigonometric functions allows us to include many more applications of the derivative in Chapter 4 and correspondingly more applications of the integral in Chapter 6. In the same way we also provide a second set of applications of integration in Chapter 9, after the student has completed the study of logarithms, exponential functions, and inverse functions in Chapter 7 and advanced techniques of integration in Chapter 8.

We have expanded considerably the treatment of infinite series (in Chapter 10), with both more illustrative examples and more tests for convergence.

Chapters 1 through 10, together with Chapter 11 on plane analytic geometry and Chapter 12 on parametric equations and related topics, complete the course in the calculus of functions of one variable.

An emphasis on the wide variety of applications of calculus is present throughout the book: in the discussions, in the examples, and in the problems. We have also added two new optional sections: The first, a section on Newton's method for finding the roots of equations, is in Chapter 4 on the applications of the derivative. This topic is particularly interesting since it combines the techniques of calculus with the use of calculators or computers. The second new section, on applications of the integral to probability problems, which appears in Chapter 9, covers applications of the integral. Here we show that the definition of the integral is the only requirement for solving an array of applied problems which use probability theory.

A stronger emphasis is given in this edition to the graphing of functions as surfaces. We include a treatment of level curves; also, the new laser-drawn graphics illustrate how computers are contributing to a better understanding of calculus.

Other changes include a greatly improved treatment of vectors and vector functions (now in Chapter 14), a simpler and more natural definition of function, improved and simplified notation, and the designation of many sections as optional. The thirteen optional sections, scattered throughout the text, are not necessary for an understanding of subsequent sections, but they provide interesting applications of calculus. The optional sections give the fourth edition considerable flexibility.

The number of problems has been increased by more than one-third; there are now over 6,500. The problems are graded, with many at each of several levels of difficulty. In addition we have added a set of Review Problems at the end of each of the eighteen chapters.

In 1961 Charles B. Morrey, Jr. wrote *University Calculus*, a text designed for an honors course in first year calculus. Two years later, the first edition of Protter/Morrey, *College Calculus with Analytic Geometry* appeared, a companion volume to *University Calculus*, one to be used in the regular first year college calculus course. With succeeding editions many changes were made in *College Calculus with Analytic Geometry*; the present edition differs greatly from the original version, which appeared twenty-five years ago. While this edition is our full responsibility, we wish to acknowledge the debt we owe to the late Charles B. Morrey, Jr. for his invaluable contributions to the original text and his help in preparing the previous revisions.

January 1988

Murray H. Protter
Philip E. Protter

# CONTENTS

# CONTENTS

CONTENTS

CONTENTS

# 1

# INEQUALITIES
# FUNCTIONS
# THE LINE

In this chapter we develop the topics in algebra and geometry needed for learning calculus. The subjects we emphasize are algebraic inequalities, the definition of function, and the basic properties of straight lines in the plane. These are essential tools for mastering the main ideas of calculus.

## 1

### INEQUALITIES

In elementary algebra and plane geometry we study equalities almost exclusively. The solution of linear and quadratic equations, the congruence of triangles, and relationships among trigonometric functions are topics concerned with equality. As we progress in the development of mathematical ideas, we shall see that the study of inequalities is both important and useful. An inequality arises when we are more concerned with the approximate size of a quantity than we are with its true value. Practically all laboratory experiments in science deal with such approximations. Also, since the proofs of the most important theorems in calculus depend on approximations, it is essential that we develop a facility for working with inequalities.

We assume the reader is familiar not only with ordinary numbers which we call the **real number system** but also with the laws of elementary algebra. In this section we shall be concerned with inequalities among real numbers, and we begin by recalling some familiar relationships. Given that $a$ and $b$ are any two real numbers, the symbol

$$a < b$$

means that $a$ **is less than** $b$. We may also write the inequality in the opposite direction,

$$b > a,$$

which is read $b$ **is greater than** $a$.

The rules for handling inequalities are only slightly more complicated than the ones we learned in algebra for manipulating equalities. However, the differences are so important that we state them as four Theorems about Inequalities and they should be studied carefully.

**THEOREM 1**    *If $a < b$ and $b < c$, then $a < c$. In words: if $a$ is less than $b$ and $b$ is less than $c$, then $a$ is less than $c$.*

**THEOREM 2**    *If $c$ is any number and $a < b$, then it is also true that $a + c < b + c$ and $a - c < b - c$. In words: if the same number is added to or subtracted from each side of an inequality, the result is an inequality in the same direction.*

**THEOREM 3**    *If $a < b$ and $c < d$ then $a + c < b + d$. That is, inequalities in the same direction may be added.*

It is important to note that in general inequalities may not be subtracted. For example, $2 < 5$ and $1 < 7$. We can say, by addition, that $3 < 12$, but note that subtraction would state the absurdity that $1$ is less than $-2$.

**THEOREM 4**    *If $a < b$ and $c$ is any positive number, then*

$$ac < bc,$$

*while if $c$ is a negative number, then*

$$ac > bc.$$

*In words: multiplication of both sides of an inequality by the same positive number preserves the direction, while multiplication by a negative number reverses the direction of the inequality.*

Since dividing an inequality by a number $d$ is the same as multiplying it by $1/d$, we see that Theorem 4 applies for division as well as for multiplication.

From the geometric point of view we associate a horizontal axis with the totality of real numbers. The origin may be selected at any convenient point, with positive numbers to the right and negative numbers to the left (Fig. 1). For every real number there will be a corresponding point on the line and, conversely, every point will represent a real number. Then the inequality $a < b$ may be read: $a$ **is to the left of** $b$. This geometric way of looking at inequalities is frequently of help in solving problems. It is also helpful to introduce the notion of an *interval of numbers* or *points*. If $a$ and $b$ are numbers (as shown in Fig. 2), then the **open interval** from $a$ to $b$ is the collection of all numbers which are both larger than $a$ and smaller than $b$.

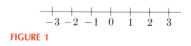

**FIGURE 1**

**FIGURE 2**

That is, an open interval consists of all numbers *between a and b*. A number $x$ is between $a$ and $b$ if *both* inequalities $a < x$ and $x < b$ are true. A compact way of writing this is

$$a < x < b.$$

The **closed interval** from $a$ to $b$ consists of all the points between $a$ and $b$, *including a and b* (Fig. 3). Suppose a number $x$ is either equal to $a$ or larger than $a$, but we don't know which. We write this conveniently as $x \geq a$, which is read: *x is greater than or equal to a*. Similarly, $x \leq b$ is read: *x is less than or equal to b*, and means that $x$ may be either smaller than $b$ or may be $b$ itself. A compact way of designating a closed interval from $a$ to $b$ is to state that it consists of all points $x$ such that

$$a \leq x \leq b.$$

An interval which contains the endpoint $b$ but not $a$ is said to be **half-open on the left**. That is, it consists of all points $x$ such that

$$a < x \leq b.$$

Similarly, an interval containing $a$ but not $b$ is called **half-open on the right**, and we write

$$a \leq x < b.$$

Parentheses and brackets are used as symbols for intervals in the following way:

> $(a, b)$ for the open interval: $a < x < b$,
> $[a, b]$ for the closed interval: $a \leq x \leq b$,
> $(a, b]$ for the interval half-open on the left: $a < x \leq b$,
> $[a, b)$ for the interval half-open on the right: $a \leq x < b$.

We can extend the idea of an interval of points to cover some unusual cases. Suppose we wish to consider *all* numbers larger than 7. This may be thought of as an interval extending to infinity on the right. (See Fig. 4.) Of course, *infinity is not a number*, but we use the symbol $(7, \infty)$ to represent all numbers larger than 7. We could also write: all numbers $x$ such that

$$7 < x < \infty.$$

In a similar way, the symbol $(-\infty, 12)$ will stand for all numbers less than 12. The double inequality

$$-\infty < x < 12$$

is an equivalent way of representing all numbers $x$ less than 12.

The first-degree equation $3x + 7 = 19$ has a unique solution, $x = 4$. The quadratic equation $x^2 - x - 2 = 0$ has two solutions, $x = -1$ and $x = 2$. The trigonometric equation $\sin x = \frac{1}{2}$ has an infinite number of solutions: $x = 30°$, $150°$, $390°$, $510°$, .... *The solution of an equation* involving a single unknown, say $x$, *is the collection of all numbers which make the equation a true statement*. This is called the **solution set** of the equation. Similarly, the solution of an *inequality* involving a single unknown, say $x$, is the collection of all numbers which make the inequality a true statement. For example, the inequality

$$3x - 7 < 8$$

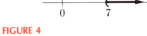

FIGURE 3

FIGURE 4

has as its solution *all* numbers less than 5. To demonstrate this we argue in the following way. If $x$ is a number which satisfies the above inequality we can, by Theorem 2, add 7 to both sides of the inequality and obtain a true statement. That is, we have

$$3x - 7 + 7 < 8 + 7 \qquad \text{or} \qquad 3x < 15.$$

Now, dividing both sides by 3 (Theorem 4), we obtain

$$x < 5$$

and observe that *if* $x$ is a solution, *then* it is less than 5. Strictly speaking, however, we have not *proved* that every number which is less than 5 is a solution. In an actual proof we would begin by supposing that $x$ is any number less than 5; that is,

$$x < 5.$$

We multiply both sides by 3 (Theorem 4) and then subtract 7 (Theorem 2) to get

$$3x - 7 < 8,$$

the original inequality. Since the condition that $x$ is less than 5 implies the original inequality, we have proved the result. The important thing to notice is that the proof consisted of *reversing* the steps of the original argument which led to the solution $x < 5$ in the first place. So long as each of the steps we take is *reversible*, the above procedure is completely satisfactory so far as obtaining solutions is concerned. The step going from $3x - 7 < 8$ to $3x < 15$ is reversible, since these two inequalities are equivalent. Similarly, the inequalities $3x < 15$ and $x < 5$ are equivalent. Finally, note that the solution set consists of all numbers in the interval $(-\infty, 5)$.

Methods for the solution of various types of simple algebraic inequalities are shown in the following examples, which should be studied carefully.

**EXAMPLE 1**     Solve for $x$:

$$-7 - 3x < 5x + 29.$$

**Solution**     Subtract $5x$ from both sides, getting

$$-7 - 8x < 29.$$

Multiply both sides by $-1$, reversing the direction of the inequality, to obtain

$$7 + 8x > -29.$$

Subtracting 7 from both sides yields $8x > -36$, and dividing by 8 gives the solution

$$x > -\tfrac{9}{2},$$

or, stated in interval form: all $x$ in the interval $(-\tfrac{9}{2}, \infty)$.      □

To verify the correctness of the result, it is necessary to perform the above steps in reverse order. However, the observation that each individual step is reversible is sufficient to check the validity of the answer.

**EXAMPLE 2**    Solve for $x(x \neq 0)$:

$$\frac{3}{x} < 5.$$

**Solution**    We have an immediate inclination to multiply both sides by $x$. However, since we don't know in advance whether $x$ is positive or negative, we must proceed cautiously. We do this by considering two cases: (1) $x$ is positive, and (2) $x$ is negative.

**Case 1.**    Suppose $x > 0$. Then multiplying by $x$ preserves the direction of the inequality (Theorem 4), and we get

$$3 < 5x.$$

Dividing by 5, we find that $x > \frac{3}{5}$. This means that we must find all numbers which satisfy *both* of the inequalities

$$x > 0 \qquad \text{and} \qquad x > \tfrac{3}{5}.$$

Clearly, any number greater than $\frac{3}{5}$ is also positive, and the solution in Case 1 consists of all $x$ in the interval $(\frac{3}{5}, \infty)$.

**Case 2.**    $x < 0$. Multiplying by $x$ *reverses* the direction of the inequality. We have

$$3 > 5x,$$

and therefore $\frac{3}{5} > x$. We seek all numbers $x$, such that *both* of the inequalities

$$x < 0 \qquad \text{and} \qquad x < \tfrac{3}{5}$$

hold. The solution in Case 2 is the collection of all $x$ in the interval $(-\infty, 0)$. A way of combining the answers in the two cases is to state that the solution set consists of all numbers $x$ *not* in the closed interval $[0, \frac{3}{5}]$. (See Fig. 5.)

**FIGURE 5**

**EXAMPLE 3**    Solve for $x(x \neq -2)$:

$$\frac{2x - 3}{x + 2} < \frac{1}{3}.$$

**Solution**    As in Example 2, we must consider two cases, according to whether $x + 2$ is positive or negative.

**Case 1.**    $x + 2 > 0$. We multiply by $3(x + 2)$, which is positive, getting

$$6x - 9 < x + 2.$$

Adding $9 - x$ to both sides, we have

$$5x < 11, \qquad \text{from which} \qquad x < \tfrac{11}{5}.$$

Since we have already assumed that $x + 2 > 0$, and since we must have $x < \frac{11}{5}$, we see that $x$ must be larger than $-2$ and smaller than $\frac{11}{5}$. That is, the solution set consists of all $x$ in the interval $(-2, \frac{11}{5})$.

**Case 2.** $x + 2 < 0$. Again multiplying by $3(x + 2)$ and *reversing* the inequality, we obtain

$$6x - 9 > x + 2,$$

or $5x > 11$ and $x > \frac{11}{5}$. In this case, $x$ must be less than $-2$ and greater than $\frac{11}{5}$, which is impossible. Combining the cases, we get as the solution set all numbers in $(-2, \frac{11}{5})$. See Fig. 6.

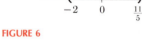

FIGURE 6

## 1  PROBLEMS

In Problems 1 through 18, solve for $x$.

**1** $2x - 3 < 7$

**2** $2x + 4 < x - 5$

**3** $5 - 3x < 14$

**4** $2 - 5x < 3 + 4x$

**5** $2(3x - 6) < 4 - (2 + 5x)$

**6** $x - 4 < \dfrac{2x}{3} + \dfrac{2 - 3x}{5}$

**7** $\dfrac{2}{3} < \dfrac{1}{x}$

**8** $\dfrac{4}{3} < -\dfrac{2}{x}$

**9** $\dfrac{1}{5} < \dfrac{2}{1 - x}$

**10** $\dfrac{3}{4} < \dfrac{2}{x - 2}$

**11** $\dfrac{4}{x} < \dfrac{3}{5}$

**12** $\dfrac{x - 1}{x} < 4$

**13** $\dfrac{x + 3}{x - 2} < 5$

**14** $\dfrac{2}{x} - 3 < \dfrac{4}{x} + 1$

**15** $\dfrac{x + 3}{x - 4} < -2$

**16** $\dfrac{2 - x}{x + 1} < \dfrac{3}{2}$

**17** $\dfrac{x}{3 - x} < 2$

**\*18** $\dfrac{x - 2}{x + 3} < \dfrac{x + 1}{x}$

In Problems 19 through 27, find the values of $x$, if any, for which both inequalities hold.

**19** $2x - 7 < 5 - x$  and  $3 - 4x < \frac{5}{3}$

**20** $3x - 8 < 5(2 - x)$  and  $2(3x + 4) - 4x + 7 < 5 + x$

**21** $\dfrac{2x + 6}{3} - \dfrac{x}{4} < 5$  and  $15 - 3x < 4 + 2x$

**22** $3 - 6x < 2(x + 5)$  and  $7(2 - x) < 3x + 8$

**\*23** $\dfrac{2}{x - 1} < 4$  and  $\dfrac{3}{x - 2} < 7$

**\*24** $\dfrac{x - 2}{x + 1} < 3$  and  $\dfrac{3 - x}{x - 2} < 5$

**\*25** $\dfrac{3}{x} < 5$  and  $x + 2 < 7 - 3x$

**\*26** $\dfrac{2}{x} < 1$  and  $(x - 2)(x + 3) < 0$

**\*27** $\dfrac{3}{x} < 2$  and  $(x - 1)(x + 4) < 0$

**28** Show that Theorem 3 for inequalities may be derived from Theorems 1 and 2.

**29** Given that $a$, $b$, $c$, and $d$ are all positive numbers, and that $a < b$ and $c < d$; show that $ac < bd$.

**\*30** a) State the most general circumstances in which the hypotheses $a < b$ and $c < d$ imply that $ac < bd$.
   b)  Given that $a < b$ and $c < d$, when is it true that $ac > bd$?

**31** If $x$ is a positive number, prove that

$$x + \frac{1}{x} \geq 2.$$

**32** a)  If $x$ and $y$ are positive numbers, show that

$$\left(\frac{1}{x} + \frac{1}{y}\right)(x + y) \geq 4.$$

   b)  If $x$, $y$, and $z$ are positive numbers, show that

$$\left(\frac{1}{x} + \frac{1}{y} + \frac{1}{z}\right)(x + y + z) \geq 9.$$

   c)  If $x$, $y$, $z$, and $w$ are positive numbers, show that

$$\left(\frac{1}{x} + \frac{1}{y} + \frac{1}{z} + \frac{1}{w}\right)(x + y + z + w) \geq 16.$$

   *d)  Generalize the above results to $n$ numbers $x_1, x_2, \ldots, x_n$.

**\*33** If $x$ and $y$ are any numbers different from zero, show that

$$\frac{x^2}{y^2} + \frac{16y^2}{x^2} + 24 \geq \frac{8x}{y} + \frac{32y}{x}.$$

**\*34** Let $x$ and $y$ be positive numbers with $x \geq y$. Show that

$$\frac{x}{y} + 3\frac{y}{x} \geq \frac{y^2}{x^2} + 3.$$

Show that the inequality is reversed if $y \geq x$.

\*Starred problems are those that are unusually difficult.

2

## ABSOLUTE VALUE

If $a$ is any positive number, the **absolute value** of $a$ is defined to be $a$ itself. If $a$ is negative, the absolute value of $a$ is defined to be $-a$. The absolute value of zero is zero. The symbol for the absolute value of $a$ is $|a|$. In other words, we have

**DEFINITION**

$$|a| = a \quad \text{if } a > 0,$$
$$|a| = -a \quad \text{if } a < 0,$$
$$|0| = 0.$$

For example,

$$|7| = 7, \qquad |-13| = 13, \qquad |2 - 5| = |-3| = 3.$$

This rather simple idea has important consequences but, before we can consider them, we must discuss methods of solving equations and inequalities involving absolute values.

**EXAMPLE 1**   Solve for $x$: $|x - 7| = 3$.

**Solution**   This equation, according to the definition of absolute value, expresses the fact that $x - 7$ must be 3 or $-3$, since in either case the absolute value is 3. If $x - 7 = 3$, we have $x = 10$; and if $x - 7 = -3$, then $x = 4$. We see that there are *two* values of $x$ which solve the equation: $x = 4, 10$. ☐

**EXAMPLE 2**   Solve for $x$: $|2x - 6| = |4 - 5x|$.

**Solution**   The two possibilities are

$$2x - 6 = 4 - 5x \qquad \text{and} \qquad 2x - 6 = -(4 - 5x).$$

Solving each of these for $x$, we obtain the two solutions

$$x = \tfrac{10}{7}, \ -\tfrac{2}{3}. \qquad ☐$$

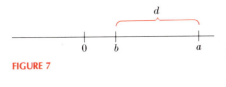

**FIGURE 7**

If $a$ and $b$ are any two numbers, we may represent them as points on the line as shown in Fig. 7. Then the distance between $a$ and $b$, denoted by $d$, is the length of the line segment from $b$ to $a$. This distance is **always positive** (or zero if $a = b$). From the definition of absolute value we see that

$$d = |b - a|.$$

If $x$ is any number, then $|x|$ represents the distance $d$ of $x$ from the origin, i.e., $d = |x - 0| = |x|$. The statement $|x| < 4$ is equivalent to the condition that $x$ is any number in the interval extending from $-4$ to $+4$ (Fig. 8). Or, in terms of the symbol for intervals, $x$ must lie in the interval $(-4, 4)$. Sometimes a

**FIGURE 8**

statement such as $|x| < 4$ is used to denote the interval $(-4, 4)$. In terms of inequalities without absolute value signs, the statement

$$-4 < x < 4$$

is equivalent to $|x| < 4$. In a similar way, the inequality $|x - 3| < 5$ means that $x - 3$ must lie in the interval $(-5, 5)$. We could also write

$$-5 < x - 3 < 5.$$

This consists of two inequalities, *both* of which $x$ must satisfy. If we add 3 to each member of the double inequality above, then

$$-2 < x < 8.$$

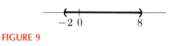

**FIGURE 9**

Therefore $x$ must lie in the interval $(-2, 8)$. See Fig. 9.

**EXAMPLE 3**    Solve for $x$: $|3x - 4| \leq 7$.

**Solution**    We write the inequality in the equivalent form

$$-7 \leq 3x - 4 \leq 7.$$

Now we add 4 to each term:

$$-3 \leq 3x \leq 11,$$

and divide each term by 3:

$$-1 \leq x \leq \tfrac{11}{3}.$$

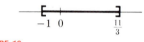

**FIGURE 10**

The solution set consists of all numbers $x$ in the *closed* interval $[-1, \tfrac{11}{3}]$. See Fig. 10.                                  □

**EXAMPLE 4**    Solve for $x$: $|2 - 5x| < 3$.

**Solution**    Here we have

$$-3 < 2 - 5x < 3,$$

and subtracting 2 from each member yields

$$-5 < -5x < 1.$$

Dividing by $-5$ will reverse each of the inequalities, and we get

$$1 > x > -\tfrac{1}{5}.$$

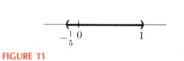

**FIGURE 11**

The solution set consists of those $x$ in the open interval $(-\tfrac{1}{5}, 1)$. See Fig. 11.                                  □

**EXAMPLE 5**    Solve for $x$:

$$\left| \frac{2x - 5}{x - 6} \right| < 3.$$

**Solution**    Proceeding as before, we see that

$$-3 < \frac{2x - 5}{x - 6} < 3.$$

We would like to multiply by $x - 6$, and in order to do this we must distinguish two cases, depending on whether $x - 6$ is positive or negative.

**Case 1.**   $x - 6 > 0$. In this case multiplication by $x - 6$ preserves the direction of the inequalities, and we have

$$-3(x - 6) < 2x - 5 < 3(x - 6).$$

Now the left inequality states that

$$-3x + 18 < 2x - 5,$$

or that

$$\tfrac{23}{5} < x.$$

The right inequality states that

$$2x - 5 < 3x - 18,$$

or that

$$13 < x.$$

In Case 1 we must have $x - 6 > 0$ *and* $\tfrac{23}{5} < x$ *and* $13 < x$. If the third inequality holds, then the other two hold as a consequence. Hence the solution in Case 1 consists of all $x$ in the interval $(13, \infty)$.

**Case 2.**   $x - 6 < 0$. The inequalities *reverse* when we multiply by $x - 6$. We then get

$$-3(x - 6) > 2x - 5 > 3(x - 6).$$

The two inequalities now state that $\tfrac{23}{5} > x$ and $13 > x$. The three inequalities

$$x - 6 < 0 \qquad \text{and} \qquad \tfrac{23}{5} > x \qquad \text{and} \qquad 13 > x$$

all hold if $x < \tfrac{23}{5}$. In Case 2, the solution consists of all numbers in the interval $(-\infty, \tfrac{23}{5})$. We could also describe the solution set by saying that it consists of all numbers not in the closed interval $[\tfrac{23}{5}, 13]$. See Fig. 12.

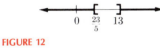

0    $\tfrac{23}{5}$   13

**FIGURE 12**

## 2   PROBLEMS

In Problems 1 through 16, solve for $x$.

**1** $|2x + 1| = 4$

**2** $|x - 3| = 1$

**3** $|4 - 2x| = 3$

**4** $|7 - 5x| = 8$

**5** $|x + \tfrac{1}{3}| = 5$

**6** $|-2x - 4| = 9$

**7** $|3x - 5| = |7x - 2|$

**8** $|2x + 1| = |3x|$

**9** $|x - 6| = |3 - 2x|$

**10** $\left|\dfrac{x + 2}{2x - 5}\right| = 1$

**11** $\dfrac{3}{|x|} = \dfrac{2}{5}$

**12** $\dfrac{4}{|x - 1|} = \dfrac{1}{|2 - x|}$

**13** $\dfrac{2}{|x|} = \dfrac{5}{|x - 3|}$

**14** $\dfrac{|x - 1|}{|4 - x|} = \dfrac{1}{3}$

**15** $\left|\dfrac{x + 2}{x + 5}\right| = 3$

**16** $\left|\dfrac{3x - 1}{4 - 2x}\right| = 6$

In each of Problems 17 through 27, find the values of $x$, if any, for which the following inequalities hold. State answers in terms of intervals.

**17** $|x + 1| < 4$

**18** $|x - 3| < 2$

**19** $|2x + 5| < 3$

**20** $|3 - 7x| \le 6$

**21** $|12 + 5x| \leq 1$

**22** $\dfrac{|x + 1|}{|3x - 2|} < 5$

**23** $\left|\dfrac{3 - 2x}{2 + x}\right| < 4$

**24** $|x + 3| \leq |2x - 6|$

**25** $|3 - 2x| \leq |x + 4|$

**26** $\left|\dfrac{x + 3}{6 - 5x}\right| \leq 2$

**\*27** $|x(x + 1)| < |x + 4|$

**28** Find all solutions of the simultaneous equations $|x - y| = 2$, $|2x + y| = 4$.

**29** Find all solutions of the simultaneous equations $|x + y| = 1$, $|2x - 3y| = 8$.

**30** Find all solutions of the simultaneous equations $|x^2 - 2y^2| = 1$, $|2x - y| = 4$.

---

**3**

## ABSOLUTE VALUE AND INEQUALITIES

If $a$ and $b$ are any two numbers, the reader can easily verify that

$$|ab| = |a| \cdot |b| \qquad \text{and} \qquad \left|\frac{a}{b}\right| = \frac{|a|}{|b|}.$$

In words, *the absolute value of a product is the product of the absolute values, and the absolute value of a quotient is the quotient of the absolute values.*

An important and useful inequality is stated in the following theorem, which we prove.

**THEOREM 5**   *If $a$ and $b$ are any numbers, then*

$$|a + b| \leq |a| + |b|.$$

**Proof**   The statement of the theorem is equivalent to the double inequality

$$-(|a| + |b|) \leq a + b \leq |a| + |b|.$$

However, we know from the definition of absolute value that $|a|$ is $a$ or $-a$ and $|b|$ is $b$ or $-b$. Therefore, we may write $-|a| \leq a \leq |a|$ and $-|b| \leq b \leq |b|$ since one of the equal signs must always hold. Now adding these last two sets of inequalities, we get

$$-|a| - |b| \leq a + b \leq |a| + |b|,$$

which is the desired result.                                    □

**COROLLARY**   *If $a$ and $b$ are any numbers, then*

$$|a - b| \leq |a| + |b|.$$

**Proof**   We write $a - b$ as $a + (-b)$, and apply Theorem 5 to obtain

$$|a - b| = |a + (-b)| \leq |a| + |-b| = |a| + |b|.$$

The final equality holds since, from the definition of absolute value, it is always true that $|-b| = |b|$. □

In calculus we frequently need to know the approximate value of a quantity although we cannot find its exact value. Theorem 5 and its Corollary are invaluable if we wish to make such approximations. We show how this is done by working several examples.

**EXAMPLE 1**     Estimate how large the expression $x^3 - 2$ can become if $x$ is restricted to the interval $[-4, 4]$.

**Solution**     From the Corollary, we have

$$|x^3 - 2| \leq |x^3| + |2|.$$

Since the absolute value of a product is the product of the absolute values, we have $|x^3| = |x \cdot x \cdot x| = |x| \cdot |x| \cdot |x| = |x|^3$. Therefore we get

$$|x^3 - 2| \leq |x|^3 + 2.$$

By hypothesis we stated that $|x|$ is always less than 4, and we conclude that

$$|x^3 - 2| \leq 4^3 + 2 = 66$$

if $x$ is any number in $[-4, 4]$. □

**EXAMPLE 2**     Find a positive number $M$ such that

$$|x^3 - 2x^2 + 3x - 4| \leq M$$

for all values of $x$ in the interval $[-3, 2]$.

**Solution**     From Theorem 5 and the Corollary we can write

$$|x^3 - 2x^2 + 3x - 4| \leq |x^3| + |2x^2| + |3x| + |4|,$$

and from our knowledge of the absolute value of products, we get

$$|x^3| + |2x^2| + |3x| + |4| = |x|^3 + 2|x|^2 + 3|x| + 4.$$

Since $|x|$ can never be larger than 3, it follows that

$$|x|^3 + 2|x|^2 + 3|x| + 4 \leq 27 + 2 \cdot 9 + 3 \cdot 3 + 4 = 58.$$

The positive number $M$ we seek is 58. □

**THEOREM 6**     *If $a$, $b$, $c$, and $d$ are all positive numbers and $a \geq b$, $d \geq c$, then*

$$\frac{a}{c} \geq \frac{b}{d}.$$

**Proof**     Since $c$ is less than or equal to $d$, the *reciprocal* of $c$, the quantity $1/c$, is larger than or equal to the reciprocal of $d$. This follows from Theorem 4 for inequalities, since dividing both sides of the inequality $c \leq d$ by the positive quantity $cd$ gives

$$\frac{1}{c} \geq \frac{1}{d}.$$

Again from Theorem 4, and from the fact that $a \geq b$, we have

$$a\left(\frac{1}{c}\right) \geq b\left(\frac{1}{c}\right) \geq b\left(\frac{1}{d}\right).$$

The first and last terms yield the result.                    □

Theorem 6 may be combined with Theorem 5 to yield estimates for many algebraic expressions. We observe that since these estimates are only approximate, some techniques for making them will give better results than others. We exhibit this fact later in Example 5.

**EXAMPLE 3**    Find a number $M$ such that

$$\left|\frac{x+2}{x-2}\right| \leq M$$

when $x$ is restricted to the interval $[\frac{1}{2}, \frac{3}{2}]$.

**Solution**    We know that

$$\left|\frac{x+2}{x-2}\right| = \frac{|x+2|}{|x-2|}.$$

If we can estimate the *smallest* possible value of the denominator and the *largest* possible value of the numerator then, by Theorem 6, we will have estimated the largest possible value of the entire expression. For the numerator we have

$$|x+2| \leq |x| + |2| \leq \tfrac{3}{2} + 2 = \tfrac{7}{2}.$$

For the denominator we note that the smallest value occurs when $x$ is as close as possible to 2. This occurs when $x = \frac{3}{2}$, and so

$$|x-2| \geq |\tfrac{3}{2} - 2| = \tfrac{1}{2},$$

if $x$ is in the interval $[\frac{1}{2}, \frac{3}{2}]$. We conclude finally that

$$\left|\frac{x+2}{x-2}\right| \leq \frac{\frac{7}{2}}{\frac{1}{2}} = 7.$$                    □

**EXAMPLE 4**    Find an estimate for the largest possible value of

$$\left|\frac{x^2+2}{x+3}\right|$$

when $x$ is restricted to the interval $[-4, 4]$.

**Solution**    The numerator is simple to estimate, since

$$|x^2+2| \leq |x|^2 + 2 \leq 4^2 + 2 = 18.$$

However, we have to find a smallest value for $|x+3|$ if $x$ is in $[-4, 4]$. We see first of all that the expression is not defined for $x = -3$, since then the denominator would be zero, and division by zero is always excluded.

Furthermore, if $x$ is a number near $-3$, then the denominator is near zero, the numerator has a value near 11, and the quotient will be a "large" number. Indeed, by taking $x$ sufficiently close to $-3$ we can make the denominator as small as we like and hence the fraction can be made as large as we like. In this problem there is no largest value of the given expression in $[-4, 4]$. Such questions are discussed further in Chapter 2. □

**EXAMPLE 5**    Find a number $M$ such that

$$\left| \frac{x+2}{x} - 5 \right| \leq M$$

when $x$ is restricted to the interval $(1, 4)$.

**Solution**    **Method 1.**    Since, from elementary algebra,

$$\frac{x+2}{x} - 5 = \frac{-4x+2}{x},$$

we can write

$$\left| \frac{x+2}{x} - 5 \right| = \frac{|-4x+2|}{|x|} \leq \frac{|4||x| + |2|}{|x|}.$$

Since $|x| < 4$, we have $4|x| + 2 < 18$, while the denominator, $|x|$, can never be smaller than 1. We obtain

$$\left| \frac{x+2}{x} - 5 \right| < 18.$$

**Method 2.**    From the Corollary to Theorem 5 it follows that

$$\left| \frac{x+2}{x} - 5 \right| \leq \left| \frac{x+2}{x} \right| + |5| = \frac{|x+2|}{|x|} + 5.$$

Now $|x| + 2 < 6$ and $|x| > 1$. Hence we obtain

$$\left| \frac{x+2}{x} - 5 \right| < \frac{6}{1} + 5 = 11.$$    □

This example shows that some algebraic manipulations lead to better estimates than do others.

## 3    PROBLEMS

In Problems 1 through 12, find a positive number $M$, if there is one, such that the absolute value of the given expression does not exceed $M$ when $x$ is in the interval given.

**1** $x^2 - 3x + 4$;   $x$ in $[-2, 2]$

**2** $x^2 + 4x - 3$;   $x$ in $[-2, 4]$

**3** $x^3 + 2x^2 - 3x - 6$;   $x$ in $[-2, 5]$

**4** $x^4 - 2x^3 + x^2 - 3x - 5$;   $x$ in $[-3, -1]$

**5** $\dfrac{x+2}{x-4}$;   $x$ in $[5, 8]$

6  $\dfrac{x+4}{2x+6};$  $x$ in $(-1, 4)$

7  $\dfrac{x^2 - 6x + 2}{x + 5};$  $x$ in $(-4\frac{1}{2}, 4)$

8  $\dfrac{x^3 - 6x + 5}{x^2 + 5};$  $x$ in $(-2, 3)$

9  $\dfrac{x + 7}{x^2 + 4x + 4};$  $x$ in $(-1, 3)$

10  $\dfrac{x^3 - 3x + 5}{(x^2 - 2x - 3)};$  $x$ in $[0, 4]$

11  $\dfrac{2x + 1}{x(x^2 + 2x - 8)};$  $x$ in $(-3, 1)$

12  $\dfrac{2x^3 - 3x + 1}{1 - x};$  $x$ in $(0.7, 0.9)$

In each of Problems 13 through 18, find in two ways how large the given expression can become in the given interval. Use the methods in Example 5.

13  $\left|\dfrac{x - 3}{x + 1} + 2\right|;$  $x$ in $[1, 3]$

14  $\left|\dfrac{x + 1}{2x - 1} - 3\right|;$  $x$ in $[2, 3]$

15  $\left|\dfrac{x + 2}{3x} + 2x\right|;$  $x$ in $[-2, -1]$

16  $\left|\dfrac{x}{x + 1} + 2x - 1\right|;$  $x$ in $[0, 2]$

17  $\left|\dfrac{x - 1}{x + 2} - \dfrac{x}{x + 1}\right|;$  $x$ in $[0, 3]$

18  $\left|\dfrac{x}{x + 3} - \dfrac{2x}{x + 4}\right|;$  $x$ in $[-2, 2]$

19  Prove for any numbers $a$ and $b$ that $|a| - |b| \le |a - b|$.

20  Prove for any numbers $a$ and $b$ that $||a| - |b|| \le |a - b|$.

21  Given that $a$ and $b$ are positive, $c$ and $d$ are negative, and $a > b$, $c > d$, show that

$$\frac{a}{c} < \frac{b}{d}.$$

22  If $a_1, a_2, a_3$ are any numbers, show that $|a_1 + a_2 + a_3| \le |a_1| + |a_2| + |a_3|$.

23  If $a_1, a_2, a_3$ are any numbers and $|a_1 + a_2| > |a_3|$, show that

$$\frac{a_3}{a_1 + a_2} < 1.$$

Is the same result true if $|a_1 + a_2| > |a_3|$ is replaced by $a_1 + a_2 > a_3$?

---

**4**

## SETS. SET NOTATION. GRAPHS

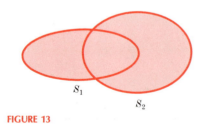

FIGURE 13

A **set** is a collection of objects. The objects may have any character (numbers, points, lines, etc.) so long as we know which objects are in a given set and which are not. If $S$ is a set and $P$ is an object in it, we write $P \in S$ and say that $P$ **belongs to** $S$ or that $P$ **is an element of** $S$. If $S_1$ and $S_2$ are two sets, their **union**, denoted by $S_1 \cup S_2$, consists of all objects each of which is in at least one of the two sets. The **intersection of** $S_1$ **and** $S_2$, denoted by $S_1 \cap S_2$, consists of all objects each of which is in both sets. Schematically, if $S_1$ is the horizontally shaded set (Fig. 13) and $S_2$ the vertically shaded set, then $S_1 \cup S_2$ consists of the entire shaded area and $S_1 \cap S_2$ consists of the doubly shaded area. Similarly, we may form the union and intersection of any number of sets. When we write $S_1 \cup S_2 \cup \cdots \cup S_7$ for the union of the seven sets $S_1, S_2, \ldots, S_7$, this union consists of all elements each of which is in at least one of the 7 sets. The intersection of these 7 sets is written $S_1 \cap S_2 \cap \cdots \cap S_7$. It may happen that two sets $S_1$ and $S_2$ have no elements in common. In such a case we say that their intersection is empty, and we use the term **empty set** for the set which is devoid of members.

Most often we will deal with sets each of which is specified by some property or properties of its elements. For example, we may speak of the set of all even integers. We employ the special symbol

$$\{x : x = 2n \quad \text{and} \quad n \text{ is an integer}\}$$

to represent the set of all even integers. In this notation the letter $x$ stands for a generic element of the set, and the properties which determine membership in the set are listed to the right of the colon. We recall that a **rational number is one which can be expressed as** $a/b$ **where** $a$ **and** $b$ **are integers**. For example, $5/7$, $-3/2$, and $17/3$ are rational numbers. The symbol

$$\{x : x \in (0, 1) \quad \text{and} \quad x \text{ is rational}\}$$

represents the rational numbers in the open interval $(0, 1)$. If a set has only a few elements, we may specify it by listing its members between braces. Thus the symbol $\{-2, 0, 1\}$ denotes the set whose elements are the numbers $-2, 0$, and $1$. Other examples are

$$(0, 2) = \{x : 0 < x < 2\},$$

$$[2, 14) = \{t : 2 \leq t < 14\}.$$

The last set is read as "the set of all numbers $t$ such that $t$ is greater than or equal to two and less than fourteen."

To illustrate the use of the symbols for set union and set intersection, we observe that

$$[1, 3] = [1, 2\tfrac{1}{2}] \cup [2, 3], \qquad (0, 1) = (0, \infty) \cap (-\infty, 1).$$

**Notation.**   The expression "if and only if," a technical one used frequently in mathematics, requires an explanation. Suppose $A$ and $B$ stand for propositions which may be true or false. To say that $A$ *is true* **if** $B$ *is true* means that the truth of $B$ implies the truth of $A$. We often shorten this expression by saying that $B$ *implies* $A$, and we denote this by

$$B \Rightarrow A.$$

The statement $A$ *is true* **only if** $B$ *is true* means that the truth of $A$ implies the truth of $B$. We shorten this expression by saying that $A$ *implies* $B$, and we denote this by

$$A \Rightarrow B.$$

The shorthand statement "$A$ is true if and only if $B$ is true" is equivalent to the *double implication:* the truth of $A$ implies and is implied by the truth of $B$. As a further shorthand notation we use the symbol $\Leftrightarrow$ to represent "if and only if," and we write

$$A \Leftrightarrow B$$

for the two implications above. The term **necessary and sufficient** is sometimes used as a synonym for "if and only if."

In Sections 1–3 we saw that the study of inequalities is greatly illuminated when we represent real numbers by means of points on a straight line. In the study of problems involving two unknowns, we shall see that a corresponding geometric interpretation is both useful and important.

The determination of solution sets of equations and inequalities in two or more unknowns is an important topic in mathematics, especially from the point of view of applications. The geometric representation of a solution set, called the **graph**, is particularly helpful in the study of analytic geometry.

We now introduce some terminology which will be used throughout the text. It is essential that the reader become familiar with these terms as quickly as possible. **The set of all real numbers is denoted by** $R^1$. Any two real numbers $a$ and $b$ form a **pair**. For example, 2 and 5 are a pair of numbers. When the *order* of the pair is prescribed, we say that we have an **ordered pair**. The ordered pair 2 and 5 is different from the ordered pair 5 and 2. We usually designate an ordered pair in parentheses with a comma separating the first and second elements of the pair: $(a, b)$ is used for the ordered pair consisting of the numbers $a$ (first) and $b$ (second).

**DEFINITIONS**  *The set of all ordered pairs of real numbers is called the* **number plane** *and is denoted by* $R^2$. *Each individual ordered pair is called a* **point** *in the number plane. The two elements in a number pair are called its* **coordinates**.

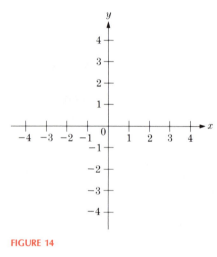

FIGURE 14

The number plane can be represented on a geometric or Euclidean plane. It is important that we keep separate the concept of the number plane, $R^2$, which is an abstract system of ordered pairs, from the concept of the geometric plane, which is the two-dimensional object we studied in Euclidean plane geometry.

In a Euclidean plane we draw a horizontal and a vertical line, denote them as the $x$ and $y$ axes, respectively, and label their point of intersection $O$ (Fig. 14). The point $O$ is called the **origin**. We select a convenient unit of length and, starting from the origin as zero, mark off a number scale on the horizontal axis, positive to the right and negative to the left. Similarly, we insert a scale along the vertical axis, with positive numbers extending upward and negative ones downward. It is not necessary that the units along the horizontal axis have the same length as the units along the vertical axis, although we will usually take them as equal.

We now set up a one-to-one correspondence between the points of the number plane, $R^2$, and the points of the Euclidean (geometric) plane. For each point $P$ in the Euclidean plane we construct perpendiculars from $P$ to the coordinate axes, as shown in Fig. 15. The intersection with the $x$ axis is at the point $Q$, and the intersection with the $y$ axis is at the point $R$. The distance from the origin to $Q$ (measured positive if $Q$ is to the right of $O$ and negative if $Q$ is to the left of $O$) is denoted by $a$. Similarly, the distance $OR$ is denoted by $b$. Then the point in the number plane corresponding to $P$ is $(a, b)$. Conversely, to each point in the number plane there corresponds exactly one point in the Euclidean plane. Because of the one-to-one correspondence between the number plane and the plane of Euclidean geometry, we frequently find it convenient to use geometric terms for the number plane. For example, a "line" in the number plane is used for the set of points corresponding to a line in the geometric plane.

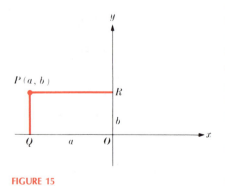

FIGURE 15

The description above suggests the method to be used in plotting (that is, representing geometrically) points of the number plane. When plotting points, we enclose the coordinates in parentheses adjacent to the point, as illustrated in Fig. 16.

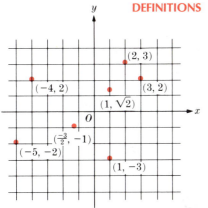

**FIGURE 16**

**DEFINITIONS**

*The* **solution set** *of an equation in two unknowns consists of all points in the number plane,* $R^2$, *whose coordinates satisfy the equation. A geometric representation of the solution set (that is, the actual drawing) is called the* **graph of the equation.**

The solution set of an equation in two unknowns is, of course, a set in $R^2$. It is a simple matter to extend the set notation already introduced to include sets in the number plane. If $S$ denotes the set of all $(x, y)$ in $R^2$ having certain properties, which, for example, we call $A$ and $B$, we describe $S$ as follows:

$$S = \{(x, y): (x, y) \text{ has properties } A \text{ and } B\}.$$

As an illustration, if $S$ is the solution set of the equation

$$2x^2 - 3y^2 = 6,$$

we write

$$S = \{(x, y): 2x^2 - 3y^2 = 6\}.$$

To construct the exact graph of the solution set of some equation is generally impossible, since it would require the plotting of infinitely many points. Usually, in drawing a graph, we select enough points to exhibit the general nature of the graph, plot these points, and then approximate the remaining points by drawing a "smooth curve" through the points already plotted.

It is useful to have a systematic method of choosing points on the graph. We can do this for an equation in two unknowns when we can solve for one of the unknowns in terms of the other. Then, from the formula obtained, we can assign values to the unknown in the formula and obtain values of the unknown for which we solved. The corresponding values are then tabulated as shown in the examples below.

**EXAMPLE 1**     Sketch a graph of the solution set of the equation $2x + 3y = 5$. Use set notation to describe this set.

**Solution**     The solution set $S$ in set notation is

$$S = \{(x, y): 2x + 3y = 5\}.$$

To sketch a graph, we first solve for one of the unknowns, say $x$. We find $x = \frac{5}{2} - \frac{3}{2}y$. Assigning values to $y$, we obtain the following table.

| $x$ | 7 | $\frac{11}{2}$ | 4 | $\frac{5}{2}$ | 1 | $-\frac{1}{2}$ | $-2$ |
|---|---|---|---|---|---|---|---|
| $y$ | $-3$ | $-2$ | $-1$ | 0 | 1 | 2 | 3 |

Plotting these points, we see that they appear to lie on the straight line shown in Fig. 17.

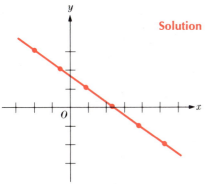

**FIGURE 17**

**EXAMPLE 2**    Sketch a graph of the solution set of the equation

$$\frac{x^2}{25} + \frac{y^2}{16} = 1.$$

Use set notation to describe the solution set $S$.

**Solution**    $S = \{(x, y): (x^2/25) + (y^2/16) = 1\}$. Solving for $y$ in terms of $x$, we obtain

$$y = \pm\tfrac{4}{5}\sqrt{25 - x^2}.$$

Since $x$ enters into the equation only in the term involving $x^2$, we may abbreviate the table as indicated below.

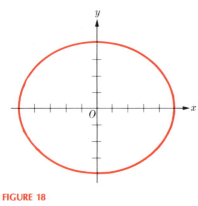

**FIGURE 18**

| $x$ | 0 | $\pm 1$ | $\pm 2$ | $\pm 3$ | $\pm 4$ | $\pm 5$ |
|---|---|---|---|---|---|---|
| $y$ | $\pm 4$ | $\pm\tfrac{4}{5}\sqrt{24}$ | $\pm\tfrac{4}{5}\sqrt{21}$ | $\pm\tfrac{16}{5}$ | $\pm\tfrac{12}{5}$ | 0 |
| $y$ (approx.) | $\pm 4$ | $\pm 3.92$ | $\pm 3.67$ | $\pm 3.2$ | $\pm 2.4$ | 0 |

Plotting these points, we see that they appear to lie on the oval curve of Fig. 18. Note that each column except the first and last really represents four points. What happens if we choose $x > 5$ or $x < -5$?    □

## 4    PROBLEMS

In each of Problems 1 through 6, a set $S$ is given as the union or intersection of sets of real numbers. Describe the sets without using the union or intersection symbol. (*Example:* The set $S = [0, 2] \cup [1, 3)$ can be described as $S = [0, 3)$.)

1    $S = [-1, 2) \cup [1, 3] \cup (2, 7]$

2    $S = [-4, 6) \cap [3, 8]$

3    $S = \{2, 6, 8\} \cup \{1, 6, 8\} \cup \{0, 5, 9\}$

4    $S = [0, 2] \cap [4, 7)$

5    $S = \{\{2, 5, 9\} \cup \{4, 6, 10\}\} \cap \{3, 6, 11\}$

6    $S = [-5, 1] \cap [-3, 8] \cap (0, 2]$

Sketch a graph of the solution set $S$ of each of the following equations.

7    $x - y = 0$

8    $x + 3y = 2$

9    $y - 2x = 4$

10    $3x - 2y = 6$

11    $x = 4$

12    $y = -1$

13    $x + y = 0$

14    $3x + 2y = 6$

15    $y = \tfrac{1}{2}x$

16    $x = \tfrac{1}{2}y^2$

17    $y = x^2 - 4$

18    $x = y^2 - 1$

19    $y = x^2 - 2x - 3$

20    $y = x^2 - 2x + 3$

21    $x^2 + y^2 = 25$

22    $x^2 + y^2 = 9$

23    $4x^2 + y^2 = 36$

24    $x^2 - y^2 = 9$

25    $2x^2 - y^2 = 17$

26    $x^2 - y^2 + 9 = 0$

27    $y = -x^2 + 2x + 3$

28    $y = \tfrac{1}{8}x^3$

29    $y = x^3 - 4x$

30    $y = x^3 - 3x + 2$

31    $2x^2 + 3xy + y^2 = 0$    (*Hint:* Write as a quadratic in $y$.)

32    $x^2 + xy + y^2 = 5$

33    $2x^2 + 3xy + y^2 + 2x + 3y = 10$

34    $x^2 - 5x + 6 = 0$    35    $y = |x|$

36    $y = |x - 1| + 1$    37    $y = 1/(|x| + 1)$

_____  5

## FUNCTIONS. FUNCTIONAL NOTATION

In mathematics and many of the physical sciences, simple formulas occur repeatedly. For example, if $r$ is the radius of a circle and $A$ is its area, then

$$A = \pi r^2.$$

If heat is added to an ideal gas in a container of fixed volume, the pressure $p$ and the temperature $T$ satisfy the relation

$$p = a + cT$$

where $a$ and $c$ are fixed numbers with values depending on the properties of the gas, the units used, and so forth.

The relationships expressed by these formulas are simple examples of the concept of function, to be defined precisely later. However, it is not essential that a function be associated with a particular formula. As an example, consider the cost $C$ in cents of mailing a package which weighs $x$ grams. Suppose postal regulations in some country state that the cost is "6¢ per gram or fraction thereof." We can construct the following table:

| Weight $x$ in grams | $0 < x \le 1$ | $1 < x \le 2$ | $2 < x \le 3$ | $3 < x \le 4$ | $4 < x \le 5$ |
|---|---|---|---|---|---|
| Cost $C$ in cents | 6 | 12 | 18 | 24 | 30 |

TABLE 1

This table could be continued until $x =$ the maximum weight permitted by postal regulations. To each value of $x$ between 0 and the maximum weight there corresponds a precise cost $C$. We have here an example of a function relating $x$ and $C$.

It frequently happens that an experimenter finds by measurement that the numerical value $y$ of some quantity depends in a *unique* way on the measured value $x$ of some other quantity. It is usually the case that no known formula expresses the relationship between $x$ and $y$. All we have is the set of ordered pairs $(x, y)$. In such circumstances, the entire interconnection between $x$ and $y$ is determined by the ordered pairs. Often this correspondence is denoted by a letter such as $f$, which indicates that each value of $y$ is obtained from a particular value of $x$. We write $y = f(x)$ to show the relationship. If $D$ denotes the set of all values of $x$ which occur, and $E$ the set of all values of $y$ which occur, then $f$ indicates that each value in $D$ gives rise to precisely one value in $E$. The value $y$ in $E$ is denoted $f(x)$. (See Fig. 19.)

FIGURE 19

**DEFINITIONS**  *A **function**, called $f$, from a set of numbers $D$ to a set of numbers $E$ is a correspondence that assigns to each number $x$ of $D$ a unique number $y$ of $E$. We write $y = f(x)$ for this correspondence. The set $D$ is called the **domain** of the function $f$ and the set $E$ is called the **range** of the function $f$.*

An example of a function is given by the formula for the area $y$ of a square whose side has length $x$; i.e., $y = x^2$. We may write $f(x) = x^2$, and we see that for each value of $x > 0$ we obtain a number which is the area of a square of side $x$. The domain $D$ of this function, i.e., the set of all possible values of $x$, is the half-infinite interval $(0, \infty)$, and the range $E$, i.e., the set of all possible values for the area $y$, is the half-infinite interval $(0, \infty)$.

The formula $p = 3 + 2T$, a particular case relating the temperature $T$ and the pressure $p$, is an example of a function. Here the letter $T$ replaces the letter $x$ and the letter $p$ replaces the letter $y$. However, we have a function since to each value of $T$ there corresponds exactly one value of $p$. We write $p = f(T)$ for the function $p = 3 + 2T$. The domain of $T$ and the range of $p$ depend on the particular conditions of the gas, the container of the gas, and so forth.

The example of postal rates given in Table 1 also represents a function, although no simple formula is available which expresses the cost of mailing $C$ in terms of the weight $x$. We may still write $C = f(x)$ and use the table to find a value of $C$ for each $x$. For example, if $x = 3\frac{1}{2}$ grams, then $C = 24$ cents. We call this function the **postage function**.

From a geometric point of view a function can be considered as a set of ordered pairs $(x, y)$, where $y$ is identified with $f(x)$. This interpretation is useful in constructing the graph of a function. Consequently, we make the following alternate definition of a function.

**DEFINITIONS**  *A **function** is a set of ordered pairs $(x, y)$ of real numbers in which no two pairs have the same first element. In other words, to each value of $x$ (the first member of the pair) there corresponds exactly one value of $y$ (the second member). The set of all values of $x$ which occur is called the **domain** of the function, and the set of all $y$ which occur is called the **range** of the function.*

When we wish to construct the graph of a function, the definition in terms of ordered pairs is most useful. The special property implied by the term function assures us that every line parallel to the $y$ axis intersects the graph no more than once (Fig. 20). The vertical lines which pass through the graph of a function intersect the $x$ axis. These points of intersection with the $x$ axis form a set called the **projection on the $x$ axis**. This projection is the domain of the function (Fig. 20). Horizontal lines through the graph of the function intersect the $y$ axis and these intersections form a set which we recognize as the range of the function. That is, the range is the **projection on the $y$ axis** of the graph of the function. Note that horizontal lines may intersect the graph of a function many times.

It is important to be able to discuss functions and their properties without actually specifying the particular ones we have in mind. For this purpose we use a symbol, usually a letter of the alphabet, to stand for a function. The letters most often used are $f$, $g$, $F$, $G$, $\phi$, $\Phi$. Sometimes, if a problem concerns many different functions, subscripts are employed, so that, for example, $f_1, f_2$, and $f_3$ would stand for three different functions.

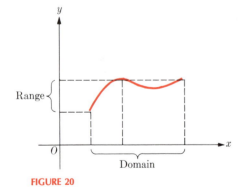

**FIGURE 20**

Several symbols for function are now in common use. One such is $f: x \mapsto y$, where $x$ is a generic element of the domain of $f$ and $y$ is the element of the range which is the **image** of $x$. Another notation for function is $f: D_1 \to D_2$, where $D_1$ is the set forming the domain of $f$ and $D_2$ is the set forming the range.

When specifying a function $f$, we must give its domain and a precise rule for determining the value of $f(x)$ for each $x$ in the domain. For the most part, we shall give functions by means of formulas such as

$$f(x) = x^2 - x + 2.$$

Such a formula, by itself, does not give the domain of $x$. Both in this case and *in general*, we shall take it for granted that *if the domain is not specified, then any value of $x$ may be inserted in the formula so long as the result makes sense. The domain shall consist of the set of all such values $x$.*

In prescribing a function by means of a formula, *the particular letter used is usually of no importance.* The function $F$ determined by the formula

$$F(x) = x^3 - 2x^2 + 5$$

is identical with the function determined by

$$F(t) = t^3 - 2t^2 + 5.$$

The difference is one of notation only.

**EXAMPLE 1**   Suppose that $f$ is the function defined by the equation

$$f(x) = x^2 - 2x - 3.$$

Find $f(0)$, $f(-1)$, $f(-2)$, $f(2)$, $f(3)$, $f(t)$, and $f(f(x))$. Plot a graph of $f$ for the portion of the domain in $-2 \le x \le 3$.

**Solution**   We have

$$f(0) = 0^2 - 2 \cdot 0 - 3 = -3, \qquad f(-1) = (-1)^2 - 2(-1) - 3 = 0,$$
$$f(-2) = (-2)^2 - 2 \cdot (-2) - 3 = 5, \qquad f(2) = 2^2 - 2 \cdot 2 - 3 = -3,$$
$$f(3) = 3^2 - 2 \cdot 3 - 3 = 0, \qquad f(t) = t^2 - 2t - 3.$$

The difficult part is finding $f(f(x))$, and here a clear understanding of the meaning of the symbolism is needed. The formula defining $f$ means that whatever is in the parentheses in $f(\ )$ is substituted in the right side. That is,

$$f(f(x)) = (f(x))^2 - 2 \cdot (f(x)) - 3.$$

However, the right side again has $f(x)$ in it, and we can substitute to get

$$f(f(x)) = (x^2 - 2x - 3)^2 - 2(x^2 - 2x - 3) - 3$$
$$= x^4 - 4x^3 - 4x^2 + 16x + 12.$$

To plot the graph we compute

$$f(1) = 1^2 - 2 \cdot 1 - 3 = -4$$

and assemble all the results above to obtain the following table.

| $x$ | $-2$ | $-1$ | $0$ | $1$ | $2$ | $3$ |
|---|---|---|---|---|---|---|
| $f(x) = y$ | $5$ | $0$ | $-3$ | $-4$ | $-3$ | $0$ |

The graph is plotted in Fig. 21.

Sometimes successions of parentheses become unwieldy, and we may use brackets or braces with the same meaning: $f[f(x)]$ is the same as $f(f(x))$.

If we write

$$g(x) = x^3 + 2x - 6, \quad -2 \le x \le 3,$$

this means that the domain of the function $g$ is the interval $[-2, 3]$. If, in the same formula, the portion $-2 \le x \le 3$ were omitted, we would assume that $g$ is defined by that formula for *all* $x$; the domain would be $(-\infty, \infty)$. This opens up many possibilities. For example, we may define a function $F$ by the following conditions (the function $F$ defined in this way has the interval $[-1, 5]$ for its domain):

$$F(x) = \begin{cases} x^2 + 3x - 2, & -1 \le x < 2, \\ x^3 - 2x + 4, & 2 \le x \le 5. \end{cases}$$

Formulas which define functions may have obvious impossibilities. If these are not explicitly pointed out, the reader should be in a position to find them. For example,

$$f_1(x) = \frac{1}{x - 3}$$

is a function defined for all values of $x$ except $x = 3$, since division by zero is always excluded. If

$$f_2(x) = \sqrt{4 - x^2},$$

then it is clear that the domain of $f_2$ cannot exceed the interval from $-2$ to $+2$, since imaginary numbers are excluded. Any value of $x$ larger than 2 or less than $-2$ is impossible.

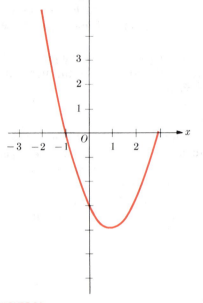

FIGURE 21

**EXAMPLE 2**   Discuss the distinction between the functions

$$F(x) = \frac{x^2 - 4}{x - 2} \quad \text{and} \quad G(x) = x + 2.$$

**Solution**   Since the expression $x^2 - 4$ is factorable into $(x - 2)(x + 2)$, at first glance it appears that the functions are the same. However, the domain of $G$ is all of $R^1$. That is, $x$ can have any value. For the function $F$, however, there is difficulty at $x = 2$. If this value is inserted for $x$, both numerator and denominator are zero. Therefore $F$ and $G$ are identical for all values of $x$ *except* $x = 2$. For $x = 2$ we have $G(2) = 4$. For $x = 2$ the function $F$ is *not defined*. We could define $F(2)$ to be 4 or any other quantity, at our pleasure. But we would have to specify that fact. If we write

$$F(x) = \frac{x^2 - 4}{x - 2} \quad \text{and} \quad F(2) = 4,$$

then this function is identical with $G$. This may seem to be a minor point, but we shall see later that it plays an important part in portions of the calculus.

□

**EXAMPLE 3**   Given that $f(x) = x^2$, show that

$$f(x^2 + y^2) = f[f(x)] + f[f(y)] + 2f(x)f(y).$$

**Solution**
$$f(x^2 + y^2) = (x^2 + y^2)^2 = x^4 + 2x^2y^2 + y^4,$$
$$f[f(x)] = f[x^2] = (x^2)^2 = x^4,$$
$$f[f(y)] = f[y^2] = (y^2)^2 = y^4,$$
$$2f(x)f(y) = 2x^2y^2.$$

Adding the last three lines, we obtain

$$f[f(x)] + f[f(y)] + 2f(x)f(y) = x^4 + y^4 + 2x^2y^2,$$

which we have seen is just $f(x^2 + y^2)$.

□

**EXAMPLE 4**   Given $f(x) = |x|$, plot the graph.

**Solution**   From the definition of absolute value, we have

$$f(x) = \begin{cases} x & \text{for} \quad x \geq 0 \\ -x & \text{for} \quad x < 0. \end{cases}$$

We set up the table of values:

| $x$ | 0 | 1 | 2 | 3 | $-1$ | $-2$ | $-3$ | $-4$ |
|---|---|---|---|---|---|---|---|---|
| $y = f(x)$ | 0 | 1 | 2 | 3 | 1 | 2 | 3 | 4 |

The graph is shown in Fig. 22. Note that the graph has a corner at the origin.

□

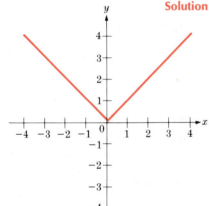

**FIGURE 22**

## 5  PROBLEMS

**1** Given that $f(x) = x^2 + 1$, find $f(-4)$, $f(-3)$, $f(-2)$, $f(-1)$, $f(0)$, $f(1)$, $f(2)$, $f(3)$. Plot the graph of the equation $y = f(x)$ for $-4 \leq x \leq 3$.

**2** Given that $f(x) = x^2 + x + 1$, find $f(-2)$, $f(-1)$, $f(0)$, $f(1)$, $f(2)$, $f(3)$. Plot the graph of the equation $y = f(x)$ for $-2 \leq x \leq 3$.

**3** Given that $f(x) = x^2 + 3x - 2$, find $f(-3)$, $f(-2)$, $f(-1)$, $f(0)$, $f(1)$, $f(2)$, $f(3)$, and $f(a + 2)$. Plot a graph of the equation $y = f(x)$ for $-3 \leq x \leq 3$.

**4** Given that $f(x) = \frac{1}{4}x^3 - x + 3$, find $f(-3)$, $f(-2)$, $f(-1)$,

$f(0)$, $f(1)$, $f(2)$, $f(3)$, and $f(a-1)$. Plot a graph of the equation $y = f(x)$ for $-3 \leq x \leq 3$.

**5** Given that

$$f(x) = \frac{3x + 2}{2x + 3},$$

find $f(-4)$, $f(-3)$, $f(-2)$, $f(-1)$, $f(0)$, $f(1)$, $f(-1000)$, $f(1000)$, and $f[f(x)]$. Is $-\frac{3}{2}$ in the domain of $f$? Plot a graph of $f$ for $x$ on $[-4, 1]$ using the values above and additional values near $-\frac{3}{2}$.

**6** Given the function

$$f(x) = \frac{2x - 3}{3x - 1},$$

find $f(x)$ for $x = -1000, -4, -3, -2, -1, 0, 1, 2, 3, 4,$ 1000. Also find $f[f(x)]$. Is $\frac{1}{3}$ in the domain of $f$? Plot a graph of $f$ for $x$ on $[-4, 4]$ using the values above and additional values near $\frac{1}{3}$.

**7** Given that $f(x) = x^2 - 2$, find $f[f\{f(x)\}]$.

**8** Given that $f(x) = x^3 - 2x + 1$, find $f\{f[f(x)]\}$.

**9** Given the function

$$f(x) = \frac{2x}{x^2 + 1},$$

find $f(x)$ for $x = -1000, -3, -2, -1, 0, 1, 2, 3, 1000$. Show that $f(-x) = -f(x)$ for all values of $x$ in $R^1$. Plot a graph of $f$ for $x$ on $[-3, 3]$.

**10** Given that

$$f(x) = \frac{1 + x}{1 - x},$$

show that

$$\frac{f(x) - f(y)}{1 + f(x)f(y)} = \frac{x - y}{1 + xy}$$

whenever both sides are defined.

**11** Given $f(x) = \sqrt{x + 1}$, find $f(-1), f(0), f(1), f(2), f(3)$. What is the domain of $f$? Plot its graph.

**12** Given $f(x) = |x - 1|$, what is the domain of $f$? Find $f(x)$ for $x = -2, -1, 0, 1, 2, 3, 4$ and plot a graph of $f$ for $x$ on $[-2, 4]$.

**13** Given $f(x) = 3 - |x + 1|$, find $f(x)$ for $x = -4, -3, -2, -1, 0, 1, 2$ and plot a graph of $f$ for $x$ on $[-4, 2]$.

**14** Given $f(x) = |x| + |x - 2|$, what is the domain of $f$? Plot the graph for $x$ on $[-3, 2]$.

**15** Given

$$f(x) = \frac{3}{|x - 1|},$$

what is the domain of $f$? Plot the graph for $x$ on $[-2, 1)$ and $x$ on $(1, 3]$.

**16** Given $f(x) = \sqrt{(x - 1)(x - 3)}$, what is the domain of $f$? Plot a graph of $f$ for $x$ between $-2$ and $6$, using several values close to 1 and 3.

**17** Given $f(x) = \sqrt{2 - 2x - x^2}$, what is the domain of $f$? (*Hint:* Complete the square under the radical.) Plot a graph of $f$.

**18** Given the function

$$f(x) = \frac{x^2 - 2x - 3}{x - 3},$$

find the domain of $f$. Give, by formula, a function defined for all $x$ in $R^1$ which coincides with $f$ wherever $f$ is defined.

**19** Same as Problem 16 for the function

$$f(x) = \frac{(x + 2)(x^2 + 6x - 16)(x - 6)}{(x - 2)(x^2 - 4x - 12)}.$$

In Problems 20 through 26, find the value of

$$\frac{f(x + h) - f(x)}{h}.$$

and simplify, assuming that $h \neq 0$.

**20** $f(x) = x^2$      **21** $f(x) = 2x^2 + x - 3$

**22** $f(x) = x^3$      **23** $f(x) = \dfrac{1}{x}$

**24** $f(x) = \dfrac{1}{x^2}$      **25** $f(x) = \sqrt{x}, \quad x > 0$

**26** $f(x) = \sqrt{x^2 + 3}$

**27** If $f(x) = \sqrt{x}$, show that

$$\frac{f(a) - f(b)}{a - b} = \frac{1}{f(a) + f(b)}.$$

**28** A function $f$ is **linear** if it has the form $f(x) = ax + b$ for some numbers $a$ and $b$. If $f$ and $g$ are linear functions, show that $f + g$ is a linear function. Show that $f[g]$ is a linear function. Under what circumstances is $f \cdot g$ linear?

**29** Given $f(x) = 2x - 1$, $g(x) = x^2 + 2$, $h(x) = 3x + 1$, find $f[g[h(x)]]$.

**30** The volume of a sphere of radius $r$ is given by the formula

$$f(r) = \frac{4}{3}\pi r^3.$$

Show that if the radius is doubled, the volume is multiplied by a factor of 8.

**31** Find a function that gives the volume of a sphere if the radius of the sphere is the square root of the area of a circle of radius $r$.

_____ 6 _____

## RELATIONS. INTERCEPTS. ASYMPTOTES

**DEFINITIONS**   *The solution set of an equation in two unknowns, say x and y, is called a* **relation**. *The* **domain** *of a relation is the set of all numbers $x_0$ such that the vertical line $x = x_0$ intersects the graph of the relation. The* **range** *of a relation is the set of all $y_0$ such that the horizontal line $y = y_0$ intersects the graph.*

Figure 23 shows the graph of a typical relation. We see that the line $x = x_0$ intersects the graph in three points while the line $y = y_0$ intersects the graph in two points. Unlike a function, it is not necessary that in a relation vertical lines such as $x = x_0$ intersect the graph in only one point. If we are given a relation in $x$ and $y$, that is, an equation connecting the variables $x$ and $y$, then we can determine its domain by performing the following two steps: (i) *We solve for y in terms of x.* (We caution that this step is often not possible.) (ii) *In the resulting expression or expressions in x, we determine those values of x for which at least one of the expressions has meaning.* The domain is the totality of such values of $x$. To find the range, we perform steps (i) and (ii) with the roles of $x$ and $y$ interchanged. Several examples illustrate the method.

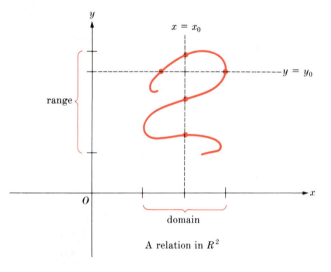

A relation in $R^2$

**FIGURE 23**

**EXAMPLE 1**   Find the domain and the range of the relation defined by

$$y^2 = x^2 - 4x + 3.$$

**Solution**   To find the domain we first solve for $y$ in terms of $x$. Setting $f(x) = x^2 - 4x + 3$, we see that $y^2 = f(x)$ and

$$y = +\sqrt{f(x)}, \qquad y = -\sqrt{f(x)}.$$

Since there is no real number which is the square root of a negative quantity, the domain consists of all $x$ for which $f(x) \geq 0$. To determine when $f(x)$ is nonnegative, we factor the expression $x^2 - 4x + 3$. We obtain

$$f(x) = x^2 - 4x + 3 \equiv (x - 1)(x - 3),$$

and $f$ is positive whenever *both* factors are positive or *both* factors are negative. The factors $x - 1$ and $x - 3$ are both positive for $x > 3$ and both negative for $x < 1$. The domain of the relation is the set

$$(-\infty, 1] \cup [3, +\infty).$$

To find the range we solve for $x$ in terms of $y$. From the relation

$$x^2 - 4x + 3 - y^2 = 0,$$

we obtain

$$x = 2 \pm \sqrt{1 + y^2}.$$

Since $1 + y^2$ is positive for all $y$, the range is $(-\infty, +\infty)$. A graph of the relation is sketched in Fig. 24.    □

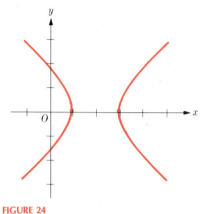

FIGURE 24

EXAMPLE 2    Given the equation $x^2 + xy - 4 = 0$, determine whether or not the solution set $S$ is a function. Plot the graph of $S$. If $S$ is not a function, show how it can be represented as the union of several functions. Find the domain and range.

Solution    Solving the above equation for $y$, we find

$$x^2 + xy - 4 = 0 \quad \Leftrightarrow \quad y = \frac{4 - x^2}{x} = \frac{4}{x} - x.$$

The set $S$ is the solution set of the equation $y = (4/x) - x$. Since there is only one value of $y$ for each value of $x$, $S$ is a function. In fact, $S$ is the function $f$ defined by

$$f(x) = \frac{4 - x^2}{x}.$$

The domain consists of all $x$ in $R^1$ except $x = 0$. To find the range, we solve for $x$ in terms of $y$, getting

$$x = \tfrac{1}{2}(-y \pm \sqrt{y^2 + 16}).$$

Since all possible values of $y$ may be inserted in the right side of the above expression, the range is all of $R^1$. To plot the graph we make the following table:

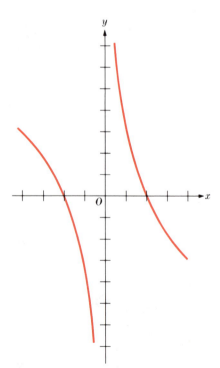

FIGURE 25

| $x$ | $-4$ | $-3$ | $-2$ | $-1$ | $-\frac{1}{2}$ | $\frac{1}{2}$ | $1$ | $2$ | $3$ | $4$ |
|---|---|---|---|---|---|---|---|---|---|---|
| $f(x) = y$ | $3$ | $\frac{5}{3}$ | $0$ | $-3$ | $-\frac{15}{2}$ | $\frac{15}{2}$ | $3$ | $0$ | $-\frac{5}{3}$ | $-3$ |

We draw a smooth curve through the points plotted from the table. Since $f$ is not defined for $x = 0$, we compute $f(-\frac{1}{2})$ and $f(\frac{1}{2})$ to get an indication of the behavior of $f$ near $x = 0$. (See Fig. 25.)    □

**EXAMPLE 3**   Let $S$ be the solution set of the equation $x^4 - 4x^2 + y^2 = 0$. Plot the graph of $S$. If $S$ is not a function, show how it can be represented as the union of several functions.

**Solution**   Solving for $y$, we note that

$$x^4 - 4x^2 + y^2 = 0 \qquad \Leftrightarrow \qquad y = \pm x\sqrt{4 - x^2}.$$

Thus $S$ is a relation which is the union of two functions $f_1$ and $f_2$ defined by $f_1(x) = x\sqrt{4 - x^2}$ and $f_2(x) = -x\sqrt{4 - x^2}$. We obtain the following table:

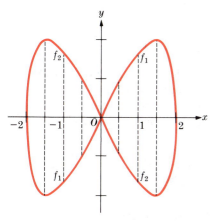

| $x$ | $\pm 2$ | $\pm\frac{3}{2}$ | $\pm 1$ | $\pm\frac{1}{2}$ | $0$ |
|---|---|---|---|---|---|
| $y$ | $0$ | $\pm\frac{3}{4}\sqrt{7}$ | $\pm\sqrt{3}$ | $\pm\frac{1}{4}\sqrt{15}$ | $0$ |
| $y$ approx. | $0$ | $\pm 1.98$ | $\pm 1.73$ | $\pm 0.97$ | $0$ |

The graph is shown in Fig. 26. Note that $f_1$ and $f_2$ are defined *implicitly* by the equation $x^4 - 4x^2 + y^2 = 0$.   □

**FIGURE 26**

We now discuss several facts about graphs which are easily obtained from the equation and which are substantial aids in constructing quick, accurate graphs. Later we shall see how the methods of calculus can be used to get additional information about a graph before it is drawn.

It is useful to know where a graph crosses the $x$ and $y$ axes.

**DEFINITIONS**   *A point at which a graph crosses the x axis is called an x* **intercept**; *a point where it crosses the y axis is called a y* **intercept**.

It may happen that a graph does not intersect a given axis or that it intersects both axes many times. To locate the intercepts, if any, we have the rule:

*to find the x intercepts, set y = 0 and solve for x;*
*to find the y intercepts, set x = 0 and solve for y.*

**EXAMPLE 4**   Find the $x$ and $y$ intercepts of the graph of

$$x^2 - 3y^2 + 6x - 3y = 7.$$

**Solution**   Setting $y = 0$, we get

$$x^2 + 6x - 7 = 0; \qquad \text{thus} \qquad x = -7 \qquad \text{or} \qquad x = 1.$$

The $x$ intercepts are $(-7, 0)$ and $(1, 0)$.
Setting $x = 0$, we get

$$3y^2 + 3y + 7 = 0; \qquad \text{thus} \qquad y = \frac{-3 \pm \sqrt{-75}}{6}.$$

Since the solutions for $y$ are complex numbers, there are no $y$ intercepts and the graph does not intersect the $y$ axis.   □

EXAMPLE 5    Given the curve with equation

$$y(x^2 - 1) = 2,$$

find the intercepts and sketch the graph.

Solution    Setting $x = 0$, we get $y = -2$, which is the $y$ intercept; $y = 0$ yields the impossible statement $0 = 2$, and the curve has no $x$ intercept. We make up a table of values:

| $x$ | 0 | $\pm 1$ | $\pm 2$ | $\pm 3$ | $\pm 4$ | $\pm 5$ |
|---|---|---|---|---|---|---|
| $y$ | $-2$ | * | $\frac{2}{3}$ | $\frac{1}{4}$ | $\frac{2}{15}$ | $\frac{1}{12}$ |

We see that as $x$ increases, the values of $y$ get closer and closer to zero. On the other hand, there is no value of $y$ corresponding to $x = 1$ or $-1$. In order to get a closer look at what happens when $x$ is near 1, we construct an auxiliary table of values for $x$ near 1. We obtain the following set of values:

| $x$ | $\frac{1}{2}$ | $\frac{3}{4}$ | $\frac{7}{8}$ | 0.9 | 1.1 | 1.2 | 1.3 | $1\frac{1}{2}$ |
|---|---|---|---|---|---|---|---|---|
| $y$ | $-\frac{8}{3}$ | $-\frac{32}{7}$ | $-\frac{128}{15}$ | $-\frac{200}{19}$ | $\frac{200}{21}$ | $\frac{50}{11}$ | $\frac{200}{69}$ | $\frac{8}{5}$ |

As $x$ moves closer to 1 from the left, the corresponding values of $y$ become large negative numbers. As $x$ approaches 1 from the right, the values of $y$ become large positive numbers. The curve is sketched in Fig. 27. The portion

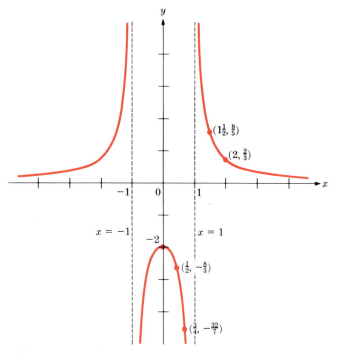

FIGURE 27

to the left of the $y$ axis is a reflection of that to the right of the $y$ axis since negative values of $x$ yield the same values of $y$ as do positive values of $x$.

☐

In Example 5, the vertical line through the point $(1, 0)$ shown in Fig. 27 plays a special role. The curve to the right of the line gets closer and closer to this line as the curve becomes higher and higher. In fact, the distance between the curve and the line tends to zero as the curve continues upward beyond all bound. Such a line is called a **vertical asymptote** to the curve. Similarly, the $x$ axis is called a **horizontal asymptote**, since the distance between the curve and the $x$ axis tends to zero as $x$ increases beyond all bound. A knowledge of the location of the vertical and horizontal asymptotes is of great help in sketching the curve. We now give a rule (which works in many cases) for finding the asymptotes.

**RULE**   *To locate the **vertical asymptotes**, solve the equation for $y$ in terms of $x$. If the result is a quotient of two expressions involving $x$, find all those values of $x$ for which the denominator vanishes and the numerator does not. If $a$ is such a value, the vertical line through the point $(a, 0)$ will be a vertical asymptote. To locate the **horizontal asymptotes**, solve for $x$ in terms of $y$, and find those values of $y$ for which the denominator vanishes (and the numerator does not). If $b$ is such a value, the horizontal line through the point $(0, b)$ is a horizontal asymptote.*

Two examples illustrate the technique.

**EXAMPLE 6**   Find the intercepts, domain, range, and asymptotes, and sketch a graph of the equation $(x^2 - 4)y^2 = 1$.

**Solution**   a) *Intercepts:* The value $y = 0$ yields no $x$ intercept; when $x = 0$, we have $y^2 = -\frac{1}{4}$, and there are no $y$ intercepts.

b) *Domain:* Solving for $y$ in terms of $x$:

$$y = \pm \frac{1}{\sqrt{x^2 - 4}}.$$

The domain is all $x$ with $|x| > 2$.

c) *Range:* Solving for $x$ in terms of $y$: $x = \pm\sqrt{1 + 4y^2}/y$. The range is all $y$ except $y = 0$.

d) *Asymptotes:* To find the vertical asymptotes, we use the expression in part (b) for the domain. Setting the denominator equal to zero, we have the vertical lines $x = 2$, $x = -2$ as asymptotes. To find the horizontal asymptotes, we use the expression in part (c) for the range. Setting the denominator equal to zero, we have the horizontal line $y = 0$ as an asymptote. The graph is sketched in Fig. 28.

☐

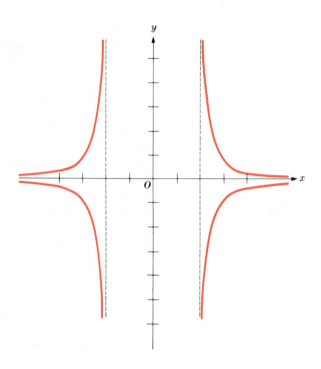

FIGURE 28

**EXAMPLE 7**   Find the intercepts, domain, range, and asymptotes, and sketch the graph of the equation

$$x^2 y = x - 3.$$

**Solution**   a)  *Intercepts:* If $y = 0$, then $x = 3$ and the $x$ intercept is 3. Setting $x = 0$ yields no $y$ intercept.

b)  *Domain:* Solving for $y$ in terms of $x$, we find

$$y = \frac{x - 3}{x^2}.$$

The domain is all $x \neq 0$.

c)  *Range:* To solve for $x$, we write the equation $yx^2 - x + 3 = 0$. The solution of this equation in $x$ obtained by the quadratic formula is

$$x = \frac{1 \pm \sqrt{1 - 12y}}{2y} \quad \text{if } y \neq 0.$$

The original equation shows that $x = 3$ when $y = 0$. The range is all $y \leq \frac{1}{12}$; that is, $(-\infty, \frac{1}{12}]$.

d)  *Asymptotes:* The line $x = 0$ is a vertical asymptote. The line $y = 0$ is a horizontal asymptote. To draw the graph we construct the following table:

| $x$ | $-4$ | $-3$ | $-2$ | $-1$ | $1$ | $2$ | $3$ | $4$ | $5$ |
|-----|------|------|------|------|-----|-----|-----|-----|-----|
| $y$ | $-\frac{7}{16}$ | $-\frac{2}{3}$ | $-\frac{5}{4}$ | $-4$ | $-2$ | $-\frac{1}{4}$ | $0$ | $\frac{1}{16}$ | $\frac{2}{25}$ |

The graph is sketched in Fig. 29.

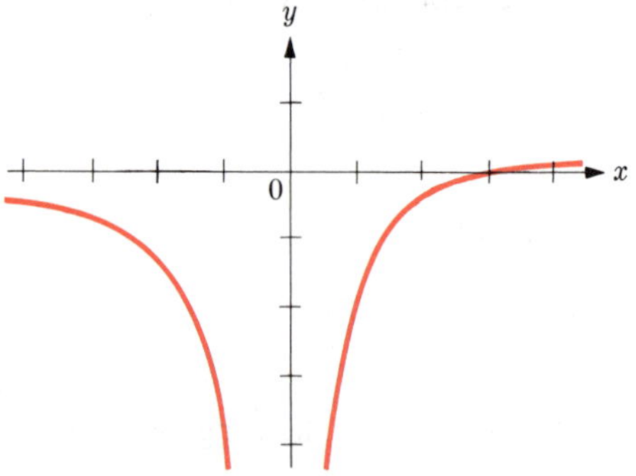

FIGURE 29

## 6　PROBLEMS

In Problems 1 through 33, find in each case the intercepts, domain, range, and vertical and horizontal asymptotes, if any. Also sketch the graph.

**1** $y = 4x^2$　　　　　**2** $y^2 = 2x$

**3** $y^2 = -4x$　　　　**4** $y^2 = 2x - 4$

**5** $y = 3 - x^2$　　　**6** $xy = 6$

**7** $x^2 y = 8$　　　　　**8** $y^2 = x^2 + 5$

**9** $2x^2 - 3y^2 = 6$　　**10** $9x^2 - 4y^2 + 36 = 0$

**11** $2x^2 + 3y^2 = 6$　　**12** $3x^2 + 2y^2 = 18$

**13** $xy^2 = 8$　　　　　**14** $y^2 = 1 - 2x^2$

**15** $x^2 + xy + y^2 = 12$　**16** $y^2 - xy = 2$

**17** $y^2(x + 1) = 4$　　**18** $y(x^2 - 1) = 1$

**19** $x^2(y^2 - 4) = 4$　　**20** $x^2(y - 2) = 2$

**21** $(x^2 + 1)y^2 = 4$　　**22** $y^2(x - 2) + 2 = 0$

**23** $y(x - 1)(x - 3) = 4$　**24** $y^2(x - 1)(x - 3) = 4$

**25** $x(y^2 - 4) = 2y$　　**26** $x^2(y^2 - 4) = 4y$

**27** $y^2 + 2x = x^2 y^2$　　**28** $y^4 = y^2 - x^2$

**29** $y^4 = 4(x^2 - y^2)$　　**30** $x^2 y^2 = x - 2$

**31** $y^2(x^2 - 1) = x + 2$　**32** $(x^2 - 4)y^2 = x^2 - 1$

**33** $5y(x - 1)(x - 3) = 2(5x + 3)$

**\*34** a) A curve may cross an asymptote, as demonstrated in Fig. 29, where we see that the graph crosses the $x$ axis. Construct an equation such that its graph crosses its asymptote twice.

b) Construct an equation which has the $x$ axis as an asymptote and such that the graph crosses the $x$ axis 3 times. State a method for constructing such equations where the number of crossings is $n$, where $n$ is any positive integer.

**35** Find the range and intercepts of the solution set of $|y - 1| = |x - 2|$. Draw the graph.

**36** Discuss the solution set of

$$y = \frac{|x - 2|}{|x + 3|}$$

for intercepts, asymptotes, range, and domain. Draw the graph.

**37** Given $|y^2 - 1| = |x + 2|$. Discuss for domain, range, and intercepts. Draw the graph.

**38** Discuss for intercepts and plot the graph of the equation $x = y + \frac{1}{4}y^3$. Show that if $(x_1, y_1)$ and $(x_2, y_2)$ are on the graph and $y_1 < y_2$, then $x_1 < x_2$. (*Hint:* Consider the cases $0 \le y_1 < y_2$, $y_1 < 0 < y_2$, $y_1 < y_2 \le 0$.) Would you say that the solution set is a function?

**39** Let $n$ be any integer. Define

$$f(x) = n \quad \text{for } n \le x < n + 1, n = 0, \pm 1, \pm 2, \ldots.$$

a) Plot the graph of $f$.
b) Plot the graph of $g(x) = x - f(x)$.
c) Plot the graph of $f(x + \frac{1}{2})$.

----

## 7

## DISTANCE FORMULA. MIDPOINT FORMULA

We consider a rectangular coordinate system with the unit of measurement the same along both coordinate axes. The **distance** between any two points in the plane is the length of the line segment joining them. We shall derive a formula for this distance in terms of the coordinates of the two points.

Let $P_1$, with coordinates $(x_1, y_1)$, and $P_2$, with coordinates $(x_2, y_2)$, be two points, and let $d$ denote the length of the segment between $P_1$ and $P_2$ (Fig. 30). Draw a line parallel to the $y$ axis through $P_1$, and do the same through $P_2$. These lines intersect the $x$ axis at the points $A(x_1, 0)$ and $B(x_2, 0)$. (The symbol $A(x_1, 0)$ is to be read "the point $A$ with coordinates $(x_1, 0)$.") Now draw a line through $P_1$ parallel to the $x$ axis; this line intersects the vertical through $P_2$ and $B$ at a point $C(x_2, y_1)$.

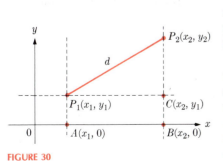

FIGURE 30

The length of the segment between $P_1$ and $C$ (**denoted** $|P_1C|$) is equal to the length $|AB|$. If $x_2$ is to the right of $x_1$ (as shown in the figure), then the length of $|AB|$ is $x_2 - x_1$. If $x_2$ were to the left of $x_1$, the distance would be $x_1 - x_2$. In either case, the length of $|AB|$ is easily written by using absolute value notation: $|AB| = |x_2 - x_1|$. Similarly, the length of $|P_2C| = |y_2 - y_1|$. It is a good exercise for the reader to check the correctness of these facts when the points $P_1$ and $P_2$ are in various quadrants of the plane.

We note that triangle $P_1P_2C$ is a right triangle, and we recall from plane geometry the Pythagorean theorem: "The sum of the squares of the legs of a right triangle is equal to the square of the hypotenuse." Applying the Pythagorean theorem to triangle $P_1P_2C$, we get

$$|P_1P_2|^2 = |P_1C|^2 + |P_2C|^2 \qquad \text{or} \qquad d^2 = |x_2 - x_1|^2 + |y_2 - y_1|^2.$$

This may also be written

$$d^2 = (x_2 - x_1)^2 + (y_2 - y_1)^2$$

and, extracting the square root, we obtain:

**DEFINITION**   **(Distance Formula)**   *The distance $d$ between two points $P_1(x_1, y_1)$ and $P_2(x_2, y_2)$ is defined to be*

$$d = \sqrt{(x_2 - x_1)^2 + (y_2 - y_1)^2}.$$

**Notation.**   The distance $d$ between any two distinct points is always *positive* and, in keeping with this, we shall always use the square-root symbol without any plus or minus sign in front of it to mean the positive square root. If we wish to discuss the negative square root of some number, say 3, we write $-\sqrt{3}$.

**EXAMPLE 1**   Find the distance between the points $P_1(3, 1)$ and $P_2(5, -2)$.

**Solution**   We substitute in the distance formula and get

$$d = \sqrt{(5 - 3)^2 + (-2 - 1)^2} = \sqrt{4 + 9} = \sqrt{13}. \qquad \square$$

**EXAMPLE 2**   The point $P_1(5, -2)$ is 4 units away from a second point $P_2$, whose $y$ coordinate is 1. Locate the point $P_2$.

**Solution**   The point $P_2$ will have coordinates $(x_2, 1)$. From the distance formula we have the equation

$$4 = \sqrt{(x_2 - 5)^2 + (1 - (-2))^2}.$$

To solve for $x_2$, we square both sides and obtain

$$16 = (x_2 - 5)^2 + 9 \qquad \text{or} \qquad x_2 - 5 = \pm\sqrt{7}.$$

There are two possibilities for $x_2$:

$$x_2 = 5 + \sqrt{7}, \qquad 5 - \sqrt{7}.$$

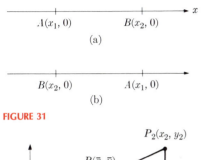

(a)

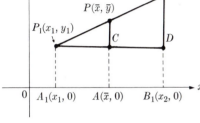

(b)

**FIGURE 31**

**FIGURE 32**

In other words, there are two points $P_2$, one at $(5 + \sqrt{7}, 1)$ and the other at $(5 - \sqrt{7}, 1)$, which have $y$ coordinate 1 and which are 4 units from $P_1$. □

Let $A(x_1, 0)$ and $B(x_2, 0)$ be two points on the $x$ axis. We define the **directed distance from $A$ to $B$** to be $x_2 - x_1$. If $B$ is to the right of $A$, as shown in Fig. 31(a), the directed distance is positive. If $B$ is to the left of $A$, as in Fig. 31(b), the directed distance is negative. If two points $C(0, y_1)$ and $D(0, y_2)$ are on the $y$ axis, the **directed distance from $C$ to $D$** is similarly defined to be $y_2 - y_1$. Directed distances are defined only for pairs of points on a coordinate axis. While ordinary distance between any two points is always positive, note that directed distance may be positive or negative.

Let $P_1(x_1, y_1)$ and $P_2(x_2, y_2)$ be any two points in the plane. We *show how to find the coordinates of the midpoint of the line segment joining $P_1$ and $P_2$*. Let $P$, with coordinates $(\bar{x}, \bar{y})$, be the midpoint. Through $P_1$, $P$, and $P_2$ draw parallels to the $y$ axis, and through $P_1$ a parallel to the $x$ axis, forming the triangles shown in Fig. 32. We recall from plane geometry the statement, "A line parallel to the base of a triangle which bisects one side also bisects the other side." Since $PC$ is parallel to $P_2 D$ and $P$ is the midpoint of $P_1 P_2$, it follows that $C$ is the midpoint of $P_1 D$. We see at once that $A$ is the midpoint of $A_1 B_1$. The coordinates of $A_1$ are $(x_1, 0)$, the coordinates of $A$ are $(\bar{x}, 0)$, and those of $B_1$ are $(x_2, 0)$. Therefore the directed distance $A_1 A$ must be equal to the directed distance $AB_1$. From the definition of directed distance we have

$$\bar{x} - x_1 = x_2 - \bar{x}$$

and, solving for $\bar{x}$, we obtain

$$\bar{x} = \frac{x_1 + x_2}{2}.$$

If we perform the same argument for the $y$ coordinates, the result by analogy is

$$\bar{y} = \frac{y_1 + y_2}{2}.$$

Thus we have established the **Midpoint Formula**. The midpoint $(\bar{x}, \bar{y})$ of the line segment joining $P_1(x_1, y_1)$ and $P_2(x_2, y_2)$ is

$$(\bar{x}, \bar{y}) = \left( \tfrac{1}{2}(x_1 + x_2), \tfrac{1}{2}(y_1 + y_2) \right).$$

**EXAMPLE 3**    Locate the midpoint of the line segment joining the points $P(3, -2)$ and $Q(-4, 5)$.

**Solution**    From the above formula,

$$\bar{x} = \frac{3 - 4}{2} = -\frac{1}{2} \quad \text{and} \quad \bar{y} = \frac{-2 + 5}{2} = \frac{3}{2}. \qquad □$$

**EXAMPLE 4**    Find the length of the line segment joining the point $A(7, -2)$ to the midpoint of the line segment between the points $B(4, 1)$ and $C(3, -5)$.

**Solution**    The midpoint of the line segment between $B$ and $C$ is at

$$\bar{x} = \frac{4+3}{2} = \frac{7}{2}, \qquad \bar{y} = \frac{1-5}{2} = -2.$$

From the distance formula applied to $A$ and this midpoint we have

$$d = \sqrt{(7 - \tfrac{7}{2})^2 + (-2 - (-2))^2} = \tfrac{7}{2}.$$    $\square$

**EXAMPLE 5**    A line segment $AB$ has its midpoint at $C(5, -1)$. Point $A$ has coordinates $(2, 3)$; find the coordinates of $B$.

**Solution**    In the midpoint formula, we know that $\bar{x} = 5$, $\bar{y} = -1$, $x_1 = 2$, $y_1 = 3$, and we have to find $x_2, y_2$. Substituting these values in the midpoint formula, we get

$$5 = \frac{2 + x_2}{2}, \qquad -1 = \frac{3 + y_2}{2},$$

and

$$x_2 = 8, \qquad y_2 = -5.$$    $\square$

## 7   PROBLEMS

In Problems 1 through 4, find the lengths of the sides of the triangles with the given points as vertices.

1.   $A(4, 1)$, $B(2, -1)$, $C(-1, 5)$
2.   $A(-1, 0)$, $B(5, 2)$, $C(3, -2)$
3.   $A(3, -4)$, $B(2, 1)$, $C(6, -2)$
4.   $A(4, 0)$, $B(0, -2)$, $C(5, 7)$

In Problems 5 and 6, locate the midpoints of the line segments joining the given points.

5.   $P_1(2, 5)$, $P_2(7, -4)$      6.   $A(3, 7)$, $B(8, -12)$

In Problems 7 and 8, locate the three points which divide the line segment joining $P_1$ and $P_2$ into four equal parts.

7.   $P_1(7, 1)$, $P_2(-2, 5)$      8.   $P_1(4, 0)$, $P_2(-3, -2)$

9.   The midpoint of a line segment $AB$ is at the point $P(-4, -3)$. The point $A$ has coordinates $(8, -5)$. Find the coordinates of $B$.

10.   The midpoint of a line segment $AB$ is at the point $P(-7, 2)$. The $x$ coordinate of $A$ is at 5, and the $y$ coordinate of $B$ is at $-9$. Find the points $A$ and $B$.

11.   Find the lengths of the medians of the triangle with vertices at $A(4, 1)$, $B(-5, 2)$, $C(3, -7)$.

12.   Same as Problem 11, for $A(-3, 2)$, $B(4, 3)$, $C(-1, -4)$.

13.   Show that the triangle with vertices at $A(1, -2)$, $B(-4, 2)$, $C(1, 6)$ is isosceles.

14.   Same as 13, with $A(2, 3)$, $B(6, 2)$, $C(3, -1)$.

15.   Show that the triangle with vertices at $A(0, 2)$, $B(3, 0)$, $C(4, 8)$ is a right triangle.

16.   Same as 15, with $A(3, 2)$, $B(1, 1)$, $C(-1, 5)$.

17.   Show that the quadrilateral with vertices at $(3, 2)$, $(0, 5)$, $(-3, 2)$, $(0, -1)$ is a square.

18.   Show that the diagonals of the quadrilateral with the following vertices bisect each other: $(-4, -2)$, $(2, -10)$, $(8, -5)$, $(2, 3)$.

19.   Same as 18, for $(5, 3)$, $(2, -3)$, $(-2, 1)$, $(1, 7)$.

20.   The four points $A(1, 1)$, $B(3, 2)$, $C(7, 3)$, $D(0, 9)$ form the vertices of a quadrilateral. Show that the midpoints of the sides are the vertices of a parallelogram.

21.   a) Show how directed distances may be defined in a natural way for pairs of points on a line parallel to a coordinate axis.

     b) Using the definition in (a), carry through the proof of the midpoint formula, using directed distances along the line $P_1D$ in Fig. 32.

22.   The formula for the coordinates of the point $Q(x_0, y_0)$ which divides the line segment from $P_1(x_1, y_1)$ to $P_2(x_2, y_2)$ in the ratio $p$ to $q$ is

$$x_0 = \frac{px_2 + qx_1}{p + q}, \qquad y_0 = \frac{py_2 + qy_1}{p + q}.$$

Derive this formula.

**23** Let $A(x_1, y_1)$, $B(x_2, y_2)$, $C(x_3, y_3)$ be the vertices of any triangle. Show that the length of the line segment joining the midpoints of any two sides of this triangle is $\frac{1}{2}$ the length of the third side.

**24** Find the set of all points $P(x, y)$ which are equidistant from $(-1, 2)$ and $(3, 4)$. (*Hint:* Let $d_1$ be the distance from $P$ to $(-1, 2)$ and $d_2$ the distance from $P$ to $(3, 4)$. Then the condition $d_1 = d_2$ becomes

$$\sqrt{(x+1)^2 + (y-2)^2} = \sqrt{(x-3)^2 + (y-4)^2}.$$

Squaring this equation and simplifying yields the equation of the solution set.) Draw the graph.

**25** Find the set of all points $P(x, y)$ which are at a distance of 2 from $(1, 1)$. Draw the graph.

**26** Find the set of all points $P(x, y)$ which are at a distance of 5 from $(3, 4)$. Draw the graph.

**27** Find the set of all points $P(x, y)$ such that the distance of $P$ from $(4, -1)$ equals the distance of $P$ from the $y$ axis. Draw the graph.

**28** Find the set of all points $P(x, y)$ such that the distance of $P$ from $(1, 2)$ equals the distance of $P$ from the $x$ axis. Draw the graph.

**29** Find the set of all points $P(x, y)$ such that the distance of $P$ from $(1, -1)$ is twice the distance of $P$ from $(4, -1)$. Draw the graph.

**30** Find the set of all points $P(x, y)$ such that the sum of the squares of the distances of $P$ from the coordinate axes is 16. Draw the graph.

---

**8**

## SLOPE OF A LINE. PARALLEL AND PERPENDICULAR LINES

A line $L$, not parallel to the $x$ axis, intersects it. Such a line and the $x$ axis form two angles which are supplementary. To be definite, we denote by $\alpha$ the angle formed by starting on the side of the $x$ axis to the right of $L$ and going counterclockwise until we reach the line $L$. The angle $\alpha$ will have a value between 0 and 180°. Two examples are shown in Fig. 33. The angle $\alpha$ is called the **inclination** of the line $L$. All lines parallel to the $x$ axis are said to make a zero angle with the $x$ axis and therefore have inclination zero. From plane geometry we recall the statement: "If, when two lines are cut by a transversal, corresponding angles are equal, the lines are parallel, and conversely." In Fig. 34, lines $L_1$ and $L_2$ each have inclination $\alpha$. Applying the theorem of plane geometry, with the $x$ axis as the transversal, we conclude that $L_1$ and $L_2$ are

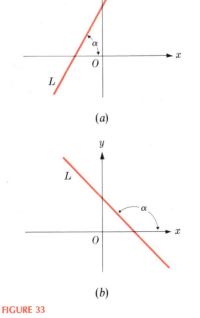

(a)

(b)

FIGURE 33

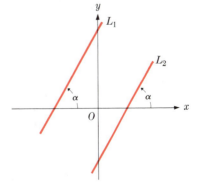

FIGURE 34

parallel. More generally, *all lines with the same inclination are parallel* and, conversely, *all parallel lines have the same inclination.*

The notion of inclination is a simple one which is easy to understand. However, for purposes of analytic geometry and calculus it is cumbersome and difficult to use. For this reason we introduce the notion of the **slope of a line** *L*. The slope is usually denoted by *m* and is defined in terms of the inclination as follows:

DEFINITION 1     **(Trigonometric)**   *If L is a line of inclination* α, *the slope of L, denoted by m, is given by*

$$m = \tan \alpha.$$

Before discussing slope, let us recall some of the properties of the tangent function.* The tangent function is positive for angles between 0° and 90°. It starts at zero for 0° and increases steadily, reaching 1 when α = 45°. It continues to increase until, as the angle approaches 90°, the function increases without bound. The tangent of 90° is not defined. Loosely speaking, we sometimes say that the tangent of 90° is infinite. Between 90° and 180° the tangent function is negative, and its values are obtained with the help of the relation

$$\tan(180° - \alpha) = -\tan \alpha.$$

That is, the tangent of an obtuse angle is the negative of that of the corresponding acute supplementary angle. (See Fig. 35.)

From these facts about the tangent function, we see that any line parallel to the *x* axis has zero slope, while a line with an inclination between 0° and 90° has positive slope. A line parallel to the *y* axis has no slope, strictly speaking, although we sometimes say such a line has infinite slope. If the inclination is an obtuse angle, the slope is negative. The line in Fig. 33(a) has positive slope, and the one in Fig. 33(b) has negative slope.

We can always find the slope of a line if we know two points on it. To see this we consider the line passing through the points $P(2, 1)$ and $Q(5, 3)$, as shown in Fig. 36. To find the slope of this line, draw a parallel to the *x* axis through *P* and a parallel to the *y* axis through *Q*, forming the right triangle *PQR*. The angle α at *P* is equal to the inclination, by corresponding angles of parallel lines. The definition of tangent function in a right triangle is "opposite over adjacent." The slope is therefore given by

$$m = \tan \alpha = \frac{|RQ|}{|PR|}.$$

Since $|RQ| = 3 - 1 = 2$ and $|PR| = 5 - 2 = 3$, the slope *m* is $\frac{2}{3}$.

Suppose that $P(x_1, y_1)$ and $Q(x_2, y_2)$ are any two points in the plane, and we wish to find the slope *m* of the line passing through these two points. The procedure is exactly the same as the one just described. In Fig. 37(a) we

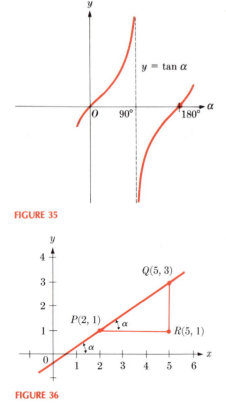

FIGURE 35

FIGURE 36

*We provide a review of the elements of trigonometry in Appendix 1.

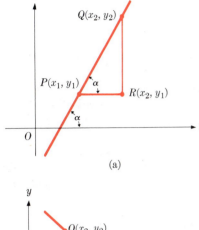

(a)

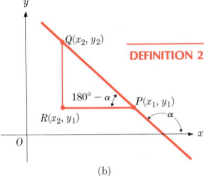

(b)

FIGURE 37

see that

$$m = \tan \alpha = \frac{y_2 - y_1}{x_2 - x_1},$$

and in Fig. 37(b)

$$m = \tan \alpha = -\tan(180 - \alpha) = -\frac{y_2 - y_1}{x_1 - x_2} = \frac{y_2 - y_1}{x_2 - x_1}.$$

A difficulty arises if the points $P$ and $Q$ are on a vertical line, since then $x_1 = x_2$ and the denominator is zero. However, we know that a vertical line has no slope, and we state that the formula holds for all cases except when $x_1 = x_2$. Note that there is no difficulty if $P$ and $Q$ lie on a horizontal line. In that case, $y_1 = y_2$ and the slope is zero, as it should be.

We see that there is an algebraic formula for the slope of a line which is equivalent to the definition in terms of the inclination.

**DEFINITION 2**    **(Algebraic)**    *The* **slope** *$m$ of a line $L$ through the two points $P(x_1, y_1)$ and $Q(x_2, y_2)$ with $x_1 \neq x_2$ is given by the formula*

$$m = \frac{y_2 - y_1}{x_2 - x_1}.$$

If we think of a particle traveling along the line $L$ from $P(x_1, y_1)$ to $Q(x_2, y_2)$, then the particle will *rise* an amount $y_2 - y_1$ while *running* horizontally an amount $x_2 - x_1$. We say, intuitively, that the slope is the "rise" divided by the "run" or, more succinctly,

$$\text{slope} = \frac{\text{rise}}{\text{run}}.$$

Of course, if $y_2 - y_1$ is negative then the "rise" is actually a fall so the expression rise/run is to be taken loosely.

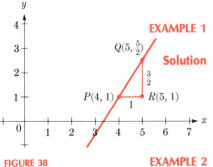

FIGURE 38

**EXAMPLE 1**    Find the slope of the line through the points $(4, -2)$ and $(7, 3)$.

**Solution**    First we note that in the formula for slope it doesn't matter which point we label $(x_1, y_1)$ [the other being labeled $(x_2, y_2)$]. We let $(4, -2)$ be $(x_1, y_1)$ and $(7, 3)$ be $(x_2, y_2)$. This gives

$$m = \frac{3 - (-2)}{7 - 4} = \frac{5}{3}. \qquad \square$$

**EXAMPLE 2**    Through the point $P(4, 1)$ construct a line with slope equal to $\frac{3}{2}$.

**Solution**    Starting at $P$, draw a parallel to the $x$ axis extending to the point $R$ one unit to the right (Fig. 38). Now draw a parallel to the $y$ axis, stopping $\frac{3}{2}$ units above $R$. The coordinates of this point $Q$ are $(5, \frac{5}{2})$. The line through $P$ and $Q$ has slope $\frac{3}{2}$. $\qquad \square$

We have seen that parallel lines always have the same inclination. Therefore, we have the following result.

| | |
|---|---|
| **THEOREM 7** | *Two parallel lines have the same slope. Conversely, two lines with the same slope are parallel.* |

**Proof**  Since parallel lines have the same inclination, the definition of slope in terms of inclination shows they have the same slope. Now suppose two lines have the same slope, say $m$. Then there is exactly one angle between $0°$ and $180°$, call it $\beta$, such that $\tan \beta = m$. To see this observe that the graph of the tangent function in Fig. 34 shows clearly that for any value $m$ in the range of the tangent function, i.e., for any value on the $y$ axis, there is exactly one value of $\beta$ such that $\tan \beta = m$. We conclude that if two lines have the same slope, their inclinations must also be the same, and the lines therefore are parallel. (This works even if the slopes are infinite, since then the lines are perpendicular to the $x$ axis and so parallel.)  □

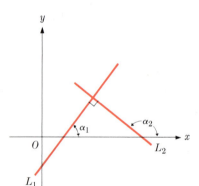

FIGURE 39

When are two lines perpendicular? If a line is parallel to the $x$ axis, it has zero slope and is perpendicular to any line which has infinite slope. Suppose, however, that a line $L_1$ has slope $m_1$ which is not zero (Fig. 39). Its inclination will be $\alpha_1$, also different from zero. Let $L_2$ be perpendicular to $L_1$ and have slope $m_2$ and inclination $\alpha_2$, as shown; it is assumed here that $m_1 > 0$. We recall from plane geometry: "An exterior angle of a triangle is equal to the sum of the remote interior angles." This means that (in Fig. 39)

$$\alpha_2 = 90° + \alpha_1 \qquad \text{and} \qquad \tan \alpha_2 = \tan(90° + \alpha_1).$$

We recall from trigonometry* the basic formulas

$$\sin(A + B) = \sin A \cos B + \cos A \sin B$$

$$\cos(A + B) = \cos A \cos B - \sin A \sin B.$$

We find

$$m_2 = \tan \alpha_2 = \tan(90° + \alpha_1) = \frac{\sin(90° + \alpha_1)}{\cos(90° + \alpha_1)}.$$

Now applying the basic trigonometry formulas to the above equation, we get

$$m_2 = \frac{\sin 90° \cos \alpha_1 + \cos 90° \sin \alpha_1}{\cos 90° \cos \alpha_1 - \sin 90° \sin \alpha_1}.$$

We use the fact that $\sin 90° = 1$, $\cos 90° = 0$ to obtain

$$m_2 = \frac{\cos \alpha_1}{-\sin \alpha_1} = -\cot \alpha_1 = -\frac{1}{\tan \alpha_1}.$$

The last formula, in terms of slopes, states that

$$m_2 = -\frac{1}{m_1}.$$

We have established the next result.

---
*See Appendix 1.

**THEOREM 8**   *Two lines are perpendicular if and only if their slopes are the negative reciprocals of each other.*

**EXAMPLE 3**   Show that the line through $P_1(3, -4)$ and $Q_1(-2, 6)$ is parallel to the line through $P_2(-3, 6)$ and $Q_2(9, -18)$.

**Solution**   The slope of the line through $P_1$ and $Q_1$, according to the formula, is

$$m_1 = \frac{6-(-4)}{-2-3} = \frac{10}{-5} = -2.$$

Similarly, the line through $P_2$ and $Q_2$ has slope

$$m_2 = \frac{-18-6}{9+3} = -2.$$

Since the slopes are the same, the lines must be parallel.   □

**EXAMPLE 4**   Determine whether or not the three points $P(-1, -5)$, $Q(1, 3)$, and $R(7, 12)$ lie on the same straight line.

**Solution**   The line through $P$ and $Q$ has slope

$$m_1 = \frac{3-(-5)}{1-(-1)} = 4.$$

If $R$ were on this line, the line joining $R$ and $Q$ would have to be the very same line; therefore it would have the same slope. The slope of the line through $R$ and $Q$ is

$$m_2 = \frac{12-3}{7-1} = \frac{3}{2},$$

and this is different from $m_1$. Therefore, $P$, $Q$, and $R$ do not lie on the same line.   □

**EXAMPLE 5**   Is the line through the points $P_1(5, -1)$ and $Q_1(-3, 2)$ perpendicular to the line through the points $P_2(-3, 1)$ and $Q_2(0, 9)$?

**Solution**   The line through $P_1$ and $Q_1$ has slope

$$m_1 = \frac{2-(-1)}{-3-5} = -\frac{3}{8}.$$

The line through $P_2$ and $Q_2$ has slope

$$m_2 = \frac{9-1}{0-(-3)} = \frac{8}{3}.$$

The slopes are the negative reciprocals of each other, and the lines are perpendicular.   □

**EXAMPLE 6**   Given the isosceles triangle with vertices at the points $P(-1, 4)$, $Q(0, 1)$, and $R(2, 5)$, show that the median drawn from $P$ is perpendicular to the base $QR$ (Fig. 40).

**Solution**   Let $M$ be the point where the median from $P$ intersects the base $QR$. From the definition of median, $M$ must be the midpoint of the segment $QR$. The coordinates of $M$, from the midpoint formula, are

$$\bar{x} = \frac{0+2}{2} = 1, \qquad \bar{y} = \frac{5+1}{2} = 3.$$

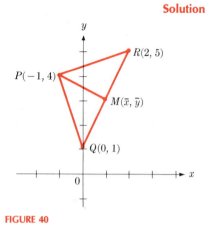

Now we check the slopes of $PM$ and $QR$. The slope of $PM$ is

$$m_1 = \frac{3-4}{1-(-1)} = -\frac{1}{2}.$$

The slope of $QR$ is

$$m_2 = \frac{5-1}{2-0} = 2.$$

**FIGURE 40**     Since

$$m_1 = -\frac{1}{m_2},$$

the median is perpendicular to the base.                           □

---

## 8  PROBLEMS

In Problems 1 through 6, check to see whether the line through the pair of points $P_1, Q_1$ is parallel or perpendicular to the line through the pair of points $P_2, Q_2$.

1. $P_1(-5, 2), Q_1(3, -1)$ and $P_2(4, 2), Q_2(12, -1)$
2. $P_1(3, 1), Q_1(-2, 7)$ and $P_2(5, -3), Q_2(-1, -8)$
3. $P_1(5, 3), Q_1(8, 3)$ and $P_2(7, 4), Q_2(7, -4)$
4. $P_1(-4, 5), Q_1(14, 2)$ and $P_2(6, 0), Q_2(9, -4)$
5. $P_1(12, 8), Q_1(4, 8)$ and $P_2(3, 1), Q_2(-6, 1)$
6. $P_1(7, -1), Q_1(10, 2)$ and $P_2(0, -4), Q_2(1, -5)$

In Problems 7 through 10, determine whether or not the three points all lie on the same straight line.

7. $P(3, 4), Q(8, 5), R(13, 6)$
8. $P(2, -1), Q(5, 3), R(-7, 4)$
9. $P(7, 1), Q(-7, 2), R(4, -5)$
10. $P(-3, 0), Q(4, 1), R(11, 2)$
11. Construct a line passing through the point $(5, -2)$ and having slope $\frac{3}{4}$.
12. Construct a line passing through the point $(-3, 1)$ and having slope $-\frac{2}{3}$.

In Problems 13 through 23, the points $P, Q, R, S$ are the vertices of a quadrilateral. In each case, determine whether the figure is a trapezoid, parallelogram, rhombus, rectangle, square, or none of these.

13. $P(1, 3), Q(2, 5), R(6, 17), S(5, 15)$

14. $P(1, 3), Q(3, 7), R(3, 11), S(1, 7)$
15. $P(2, 3), Q(4, 1), R(0, -8), S(-5, -3)$
16. $P(5, 6), Q(-8, -7), R(-8, -10), S(5, 3)$
17. $P(-1, 2), Q(2, -2), R(\frac{38}{13}, -\frac{8}{13}), S(-\frac{17}{13}, \frac{8}{13})$
18. $P(\frac{4}{3}, \frac{20}{3}), Q(4, 2), R(1, -4), S(5, 12)$
19. $P(3, -1), Q(2, -6), R(1, 7), S(0, -8)$
20. $P(5, 3), Q(8, 3), R(4, 4), S(7, 4)$
21. $P(3, 0), Q(3, 5), R(0, 9), S(0, 4)$
22. $P(2, 1), Q(7, 13), R(-5, 18), S(-10, 6)$
23. $P(6, 2), Q(3, -1), R(7, 5), S(8, -4)$
24. The points $A(3, -2)$, $B(4, 1)$, and $C(-3, 5)$ are the vertices of a triangle. Show that the line through the midpoints of the sides $AB$ and $AC$ is parallel to the base $BC$ of the triangle.
25. The points $A(0, 0)$, $B(a, 0)$, and $C(\frac{1}{2}a, b)$ are the vertices of a triangle. Show that the triangle is isosceles. Prove that the median from $C$ is perpendicular to the base $AB$. How general is this proof?
26. The points $A(0, 0)$, $B(a, 0)$, $C(a + b, c)$, $D(b, c)$ form the vertices of a parallelogram. Prove that the diagonals bisect each other.
27. The points $A(0, 0)$, $B(a, 0)$, $C(b, c)$, $D(e, f)$ are the vertices of a quadrilateral. Show that the line segments joining the midpoints of opposite sides bisect each other. How general is this proof?

**28** Show that in any triangle the length of one side is no larger than the sum of the lengths of the other two sides.

**29** Let $P(-\frac{1}{2}a, 0)$ and $Q(\frac{1}{2}a, 0)$ be two adjacent vertices of a regular hexagon above the side $PQ$. Find the coordinates of the remaining vertices and the length of the diagonal

joining two opposite vertices.

**30** Let $P(-\frac{1}{2}a, 0)$ and $Q(\frac{1}{2}a, 0)$ be two adjacent vertices of a regular octagon situated above the side $PQ$. Find the coordinates of the remaining six vertices and the length of the diagonal joining two opposite vertices.

---

**9**

# THE STRAIGHT LINE

It is easy to verify that the equation $y = 4$ represents all points on a line parallel to the $x$ axis and four units above it. Similarly, the equation $x = -2$ represents all points on a line parallel to the $y$ axis and two units to the left. In general, any line parallel to the $y$ axis has an equation of the form

$$x = a,$$

where $a$ is the number denoting how far the line is to the right or left of the $y$ axis. The equation

$$y = b$$

describes a line parallel to the $x$ axis and $b$ units from it. In this way we obtain the equations of all lines with zero or infinite slope. A line which is not parallel to either axis has a slope $m$ which is different from zero. Suppose that the line passes through a point denoted $P(x_1, y_1)$. To be specific, we consider the slope $m$ to be $-\frac{2}{3}$ and the point $P$ to have coordinates $(4, -3)$. If a point $Q(x, y)$ is on this line, then the slope as calculated from $P$ to $Q$ must be $-\frac{2}{3}$, as shown in Fig. 41. That is,

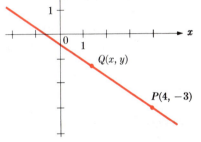

FIGURE 41

$$\frac{y + 3}{x - 4} = -\frac{2}{3} \qquad \text{or} \qquad y + 3 = -\frac{2}{3}(x - 4).$$

*This is the equation of the line passing through the point $(4, -3)$ with slope $-\frac{2}{3}$.* In the general case of a line with slope $m$ passing through $P(x_1, y_1)$, the statement that $Q(x, y)$ is on the line is the same as the statement that the slope $m$ as computed from $P$ to $Q$ (Fig. 42) is

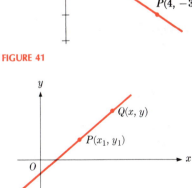

FIGURE 42

$$\frac{y - y_1}{x - x_1} = m \qquad \text{or} \qquad \boxed{y - y_1 = m(x - x_1).}$$

We summarize this general statement as follows:

---

**THEOREM 9**    **(Point-Slope Formula for the Equation of a Line)**    *The equation of the line passing through the point $P(x_1, y_1)$ with slope $m$ is*

$$y - y_1 = m(x - x_1).$$

---

If we are given the coordinates of a point and the numerical value of the slope, substitution in the above formula yields the equation of the line going through the point and having the given slope.

**EXAMPLE 1** Find the equation of the line passing through the point $(-2, 5)$ and having slope $\frac{4}{3}$.

**Solution** Substitution in the above formula gives

$$y - 5 = \tfrac{4}{3}[x - (-2)] \qquad \text{or} \qquad 3y = 4x + 23. \qquad \square$$

We know that two points determine a line. The problem of finding the equation of the line passing through the points $(3, -5)$ and $(-7, 2)$ can be solved in two steps. First we employ the formula for the slope of a line, as given in Section 8, to obtain the slope of the line through the given points. We get

$$m = \frac{2 - (-5)}{-7 - 3} = -\frac{7}{10}.$$

Then, knowing the slope, we use *either* point, together with the slope, in the point-slope formula. This gives [using the point $(3, -5)$]

$$y - (-5) = -\tfrac{7}{10}(x - 3) \qquad \text{or} \qquad 10y = -7x - 29.$$

We verify quite easily that the same equation is obtained if we use the point $(-7, 2)$ instead of $(3, -5)$.

The above process may be transformed into a general formula by applying it to any two points, $P_1(x_1, y_1)$ and $P_2(x_2, y_2)$. The slope of the line through these points is

$$m = \frac{y_2 - y_1}{x_2 - x_1}.$$

Substituting this value for the slope into the point-slope form, we get the **two-point form** for the equation of a line:

$$y - y_1 = \frac{y_2 - y_1}{x_2 - x_1}(x - x_1).$$

Note that this is really not a new formula, but merely the point-slope form with an expression for the slope substituted into it.

Thus we have the following corollary.

**COROLLARY TO THEOREM 9**    (**Two-Point Form for the Equation of a Line**)    *The equation of the line passing through the two points $P_1(x_1, y_1)$ and $P_2(x_2, y_2)$ with $x_1 \neq x_2$ is given by*

$$y - y_1 = \frac{y_2 - y_1}{x_2 - x_1}(x - x_1).$$

Another variation of the point-slope form is obtained by introducing a number called the $y$ intercept. Every line not parallel to the $y$ axis must intersect it; if we denote by $(0, b)$ the point of intersection, **the number $b$ is** called the $y$ **intercept.** * Suppose a line has slope $m$ and $y$ intercept $b$. We substitute in the point-slope form to get

$$y - b = m(x - 0) \qquad \text{or} \qquad \boxed{y = mx + b.}$$

This is called the **slope-intercept form** for the equation of a straight line.

**EXAMPLE 2**      A line has slope 3 and $y$ intercept $-4$. Find its equation.

**Solution**      Substitution in the slope-intercept formula gives

$$y = 3x - 4.$$

The important thing to notice is that the point-slope form is the basic one for the equation of a straight line. The other formulas were derived as simple variations or particular cases.

Examples 1 and 2 led to equations of lines which could be put in the form

$$Ax + By + C = 0,$$

where $A$, $B$, and $C$ are any numbers. This equation is the most general equation of the first degree in $x$ and $y$. We shall establish the theorem:

**THEOREM 10**      *Every equation of the form*

$$Ax + By + C = 0,$$

*so long as $A$ and $B$ are not both zero, is the equation of a straight line.*

**Proof**      We consider two cases, according as $B = 0$ or $B \neq 0$. If $B = 0$, then we must have $A \neq 0$, and the above equation becomes

$$x = -\frac{C}{A},$$

which we know is the equation of a straight line parallel to the $y$ axis and $-C/A$ units from it.

If $B \neq 0$, we divide by $B$ and solve for $y$, getting

$$y = -\frac{A}{B}x - \frac{C}{B}.$$

From the slope-intercept form for the equation of a line, we recognize this as *the equation of a line with slope $-A/B$ and $y$ intercept $-C/B$.*

In the statement of the theorem it is necessary to make the requirement that $A$ and $B$ are not both zero. If both of them vanish and $C$ is zero, the linear equation reduces to the triviality $0 = 0$, which is satisfied by every point $P(x, y)$ in the plane. If $C \neq 0$, then no point $P(x, y)$ satisfies the equation $0 = C$.

---

*The point $(0, b)$ is also called **the $y$ intercept**.

**EXAMPLE 3**    Given the linear equation

$$3x + 2y + 6 = 0,$$

find the slope and $y$ intercept.

**Solution**    Solving for $y$, we have

$$y = -\tfrac{3}{2}x - 3.$$

From this we simply read off that $m = -\tfrac{3}{2}$ and $b = -3$.    □

An equation of the first degree in $x$ and $y$ is called a **linear equation**. When we solve for $y$ in terms of $x$, as in the point-slope form, $y$ becomes a function of $x$. Such a function is called a **linear function**. Any line not parallel to the $y$ axis may be thought of as a function.

**EXAMPLE 4**    Find the equation of the line which is the perpendicular bisector of the line segment joining the points $P(-3, 2)$ and $Q(5, 6)$. (See Fig. 43.)

**Solution**    We give two methods.

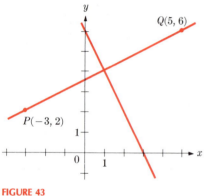

**FIGURE 43**

**Method 1.**    We first find the slope $m$ of the line through $P$ and $Q$. It is

$$m = \frac{6 - 2}{5 + 3} = \frac{1}{2}.$$

The slope of the perpendicular bisector must be $-2$, the negative reciprocal. Next we get the coordinates of the midpoint of the line segment $PQ$. They are

$$\bar{x} = \frac{5 - 3}{2} = 1, \qquad \bar{y} = \frac{6 + 2}{2} = 4.$$

The equation of the line through $(1, 4)$ with slope $-2$ is

$$y - 4 = -2(x - 1) \qquad \Leftrightarrow \qquad 2x + y - 6 = 0,$$

which is the desired equation.

**Method 2.**    We start by noting that any point on the perpendicular bisector must be equidistant from $P$ and $Q$. Let $R(x, y)$ be any such point, and let $d_1$ denote the distance from $R$ to $P$, and $d_2$ the distance from $R$ to $Q$. From the formula for the distance between two points and the condition $d_1 = d_2$, we have

$$\sqrt{(x + 3)^2 + (y - 2)^2} = \sqrt{(x - 5)^2 + (y - 6)^2}.$$

Squaring both sides and multiplying out, we obtain

$$x^2 + 6x + 9 + y^2 - 4y + 4 = x^2 - 10x + 25 + y^2 - 12y + 36.$$

The terms of the second degree cancel and the remaining terms combine to give

$$6x - 4y + 13 = -10x - 12y + 61 \qquad \Leftrightarrow \qquad 16x + 8y - 48 = 0.$$

Dividing by 8, we obtain the same answer as in the first method:

$$2x + y - 6 = 0.$$    □

## 9   PROBLEMS

In Problems 1 through 17, find the equation of the line with the given requirements.

1  Slope 2 and passing through $(-1, 4)$

2  Slope 4 and passing through $(3, -1)$

3  Slope $-\frac{2}{3}$ and passing through $(-2, 5)$

4  Passing through the points $(5, 2)$ and $(-1, -6)$

5  Passing through the points $(2, -3)$ and $(0, -4)$

6  Passing through the points $(1, 4)$ and $(-2, -7)$

7  Slope 0 and passing through $(-2, -7)$

8  Passing through the points $(3, 8)$ and $(3, -7)$

9  Passing through the points $(4, -2)$ and $(-7, -2)$

10  Slope $-\frac{4}{5}$ and $y$ intercept 3

11  Slope 0 and $y$ intercept $-5$

12  Slope $\frac{4}{5}$ and $y$ intercept 8

13  Slope 2 and passing through the midpoint of the line segment connecting $(3, -2)$ and $(4, 7)$

14  Slope $-\frac{3}{2}$ and passing through the midpoint of the line segment connecting $(2, -7)$ and $(5, -1)$

15  Slope $-3$ and $y$ intercept 0

16  Parallel to the $y$ axis and passing through the point $(4, -3)$

17  Parallel to the $x$ axis and passing through the point $(6, -5)$

18  Find the slope and $y$ intercept of the line $2x + 7y + 4 = 0$.

19  Find the slope and $y$ intercept of the line $2x - 3y - 7 = 0$.

20  Find the equation of the line through the point $(1, -4)$ and parallel to the line $x + 5y - 3 = 0$.

21  Find the equation of the line through the point $(-2, -3)$ and parallel to the line $3x - 7y + 4 = 0$.

22  Find the equation of the line through the point $(3, -2)$ and perpendicular to the line $2x + 3y + 4 = 0$

23  Find the equation of the line through the point $(1, 5)$ and perpendicular to the line $5x - 4y + 1 = 0$.

24  Find the equation of the line through the point $(-1, -3)$ and parallel to the line through the points $(3, 2)$ and $(-5, 7)$.

25  Find the equation of the line passing through $(4, -2)$ and parallel to the line through the points $(2, -1)$ and $(5, 7)$.

26  Find the equation of the line through the midpoint of the line segment joining $(2, 1)$ and $(6, -4)$ and through the point which is $\frac{1}{4}$ of the way from $(3, 2)$ to $(7, -6)$.

27  Find the equation of the line passing through $(-5, 3)$ and perpendicular to the line through the points $(7, 0)$ and $(-8, 1)$.

28  Find the equation of the line passing through $(2, -1)$ and perpendicular to the line through the points $(3, 1)$ and $(-2, 5)$.

29  Find the equation of the perpendicular bisector of the line segment joining $(6, 2)$ and $(-1, 3)$.

30  Find the equation of the perpendicular bisector of the line segment joining $(3, -1)$ and $(5, 2)$.

31  The points $P(0, 0)$, $Q(a, 0)$, $R(a, b)$, $S(0, b)$ are the vertices of a rectangle. Show that if the diagonals meet at right angles, then $a = \pm b$ and the rectangle must be a square.

32  Let $P$, $Q$, $R$, $S$ be the vertices of a parallelogram. Show, by analytic geometry, that if the diagonals are equal the parallelogram must be a rectangle.

33  $P$, $Q$, $R$, $S$ are the vertices of a parallelogram. Show, by analytic geometry, that if the diagonals are perpendicular the figure is a rhombus.

34  The points $A(1, 0)$, $B(7, 0)$, $C(3, 4)$ are the vertices of a triangle. Find the equations of the three medians. Show that the three medians intersect in a point.

35  The points $A(0, 0)$, $B(a, 0)$, $C(b, c)$ are vertices of a triangle. Show that the three medians meet in a point.

36  The points $A(1, 0)$, $B(5, 1)$, $C(3, 8)$ are the vertices of a triangle. Find the equation of the perpendicular from each vertex to the opposite side. Show that these three lines intersect in a point.

37  The points $A(0, 0)$, $B(a, 0)$, $C(b, c)$ are the vertices of a triangle. Find the equation of the perpendicular from each vertex to the opposite side. Show that these three lines intersect in a point.

38  Let $a > 0$, $c > 0$ be given numbers. Show that the points $A(0, 0)$, $B(a, 0)$, $C(a + b, c)$, $D(b, c)$ are the vertices of a parallelogram, and show that its area is $ac$.

39  Let $c > 0$ and $d > b$ be given numbers. Show that the points $A(0, 0)$, $B(a, 0)$, $C(d, c)$, $D(b, c)$ are the vertices of a trapezoid and that its area is $\frac{1}{2}c(a + d - b)$.

40  Let $a > 0$, $c > 0$ be given numbers. The points $A(0, 0)$, $B(a, 0)$, $C(b, c)$ are the vertices of a triangle. Suppose that $D$ and $E$ are points on sides $AC$ and $BC$, respectively, with line segment $DE$ parallel to $AB$. Show that $|CD|/|DE| = |CA|/|AB|$.

# CHAPTER 1

## REVIEW PROBLEMS

In Problems 1 through 6, solve for $x$.

**1** $3 + 2x < 4 - x$

**2** $2(6 - x) < 3x + 5$

**3** $\dfrac{2 + x}{x - 1} < 4$

**4** $\dfrac{x + 1}{2 - x} < \dfrac{3}{2}$

**5** $\dfrac{2 + x}{x} < 3$ and $\dfrac{3 + x}{2} < -4$

**6** $\dfrac{2}{1 + x} < 3$ and $2x - 6 < 5$

In Problems 7 through 10, find the values of $x$, if any, for which the inequalities hold.

**7** $\dfrac{|1 + x|}{|2 - x|} < 3$

**8** $|2 + x| < |4 - x|$

**9** $|2x(1 - x)| < 2$

**10** $|x + 2| < |x(x - 1)|$

**11** Find all solutions of the simultaneous equations $|x - y| = 2$, $|2x - y| = 3$.

**12** Write a complete proof of the statement: $|a \cdot b| = |a||b|$ for all real numbers $a, b$.

**13** Write a complete proof of the statement:

$$\frac{|a|}{|b|} = \left|\frac{a}{b}\right|$$

for all real numbers $a, b$ with $b \neq 0$.

In Problems 14 through 17, find an estimate for how large the given quantity can become in the given interval.

**14** $|x(x + 1)|$ in $[-2, 3]$

**15** $|x^2 + 2x - 3|$ in $[-4, 2]$

**16** $\dfrac{|x + 2|}{|x^2 + 2x + 5|}$ in $[0, 2]$

**17** $\dfrac{x^2 + 1}{x^2 + 4}$ in $[3, 7]$

**18** If $a_1, a_2, \ldots, a_n$ are any numbers, show that $|a_1 + a_2 + \cdots + a_n| \leq |a_1| + |a_2| + \cdots + |a_n|$.

**19** In the inequality $|a + b| \leq |a| + |b|$, state all conditions in which the equality sign holds.

**20** Given $f(x) = x^2 - 2x + 4$, find $f(-1)$, $f(0)$, $f(2)$, $f(a)$, and $f(a + 3)$.

**21** Given $f(x) = x^2 + 2$, show that

$$\frac{f(a)f(b) - f(ab)}{f(a) + f(b) - 3} = 2.$$

**22** What is the domain of $f(x) = \sqrt{x^2 - 3x + 2}$? Plot the graph.

**23** Plot the graph of $f(x) = |x + 1| + |x - 1|$.

In Problems 24 through 27, find the intercepts, domain, range, and asymptotes, if any. Sketch the graph.

**24** $y^2(x - 3) = 4$

**25** $y = x^3 + 2$

**26** $y^2 - 3xy + 2x^2 = 0$

**27** $y^4 = x^3$

**28** Draw the graph of

$$f(x) = \frac{x^2}{|x - 1|}.$$

Find the domain, range, intercepts, and asymptotes.

**29** a) Find the midpoint of the line segment $L$ joining $(2, 1)$ and $(5, 6)$.
b) Find the equation of the line through this midpoint and perpendicular to $L$.

**30** Use the formula in Problem 22 of Section 7 to find the point one-third of the way from $(-1, 3)$ to $(4, 1)$.

**31** Describe the set of all points such that the distance from $(2, 1)$ is always equal to the distance from $(-1, 3)$. Sketch this set.

**32** Suppose a line has $x$ intercept $(a, 0)$ and $y$ intercept $(0, b)$. Show that the equation of the line is

$$\frac{x}{a} + \frac{y}{b} = 1.$$

This equation is known as the **two-intercept form** of the straight line.

**33** Find the equation of the line through the point $A(2, -1)$ which is parallel to the line $2x - 3y - 1 = 0$.

**34** Find the equation of the line through the point $B(3, -2)$ which is perpendicular to the line $x + 2y + 3 = 0$.

**35** Given the line $2x + ky - 3 = 0$, find the value of $k$ such that the point $(1, 4)$ is on the line.

**36** Given the vertices of the triangle $A(2, 1)$, $B(4, 0)$, $C(5, 4)$, find the equation of the line passing through $A$ and the point which is one-fourth of the way from $B$ to $C$.

# 2

# LIMITS
# THE DERIVATIVE
# APPLICATIONS

We begin the study of calculus with the process known as
differentiation. To establish this procedure we develop the
important idea of a limit, first in an intuitive and later in a
precise way. We also provide both a geometric and a
physical interpretation of differentiation.

## 1

## LIMITS (INTUITIVE)

The single most important idea in calculus is that of limit. The limit concept is
at the foundation of almost all of mathematical analysis, and an understanding of it is absolutely essential. The reader should not be surprised to find the
discussion rather complex and perhaps difficult. This should not prove
discouraging, however, for once a precise understanding of the limit concept
is achieved, the reward is a good grasp of all the basic processes of calculus.

We begin with an intuitive discussion of limit and postpone a formal
description to Section 6. It may seem strange that it is possible to work with
and apply an idea without defining it precisely, but historically the limit
concept evolved in just this way. In the early development of calculus precise
statements in the modern sense were seldom made. Yet progress was achieved
because there was still a degree of understanding. When imprecise ideas led to
trouble, a later, more careful approach overcame the difficulty, and a more
solid basis for calculus was attained. The same process continues today in
mathematical research.

We start by considering a function $f$ and its graph. We concentrate on a
particular value of $x$, say $x = a$. For example, let the function be

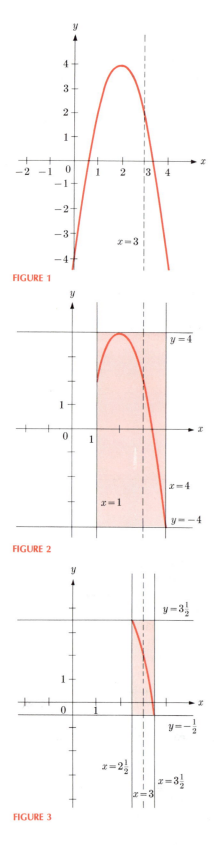

FIGURE 1

FIGURE 2

FIGURE 3

$$y = f(x) = -2x^2 + 8x - 4$$

and take the particular value $x = 3$. A graph of this function is shown in Fig. 1. We are interested in this function not only when $x = 3$ but also when $x$ takes on values in various intervals containing 3.

Suppose that we select the interval from $x = 1$ to $x = 4$. A graph of the function in this region shows that the highest value is attained at $x = 2$ when $y$ is 4, and the lowest value occurs at $x = 4$ when $y$ is $-4$ (Fig. 2). In other words, the graph of the function lies in the rectangle bounded by the lines $x = 1$, $x = 4$ and $y = 4$, $y = -4$. The next step is to select a smaller interval about $x = 3$. Suppose we take the interval from $x = 2\frac{1}{2}$ to $x = 3\frac{1}{2}$ and draw the graph in this region (Fig. 3). The graph of the function is now in the rectangle bounded by the lines $x = 2\frac{1}{2}$, $x = 3\frac{1}{2}$, $y = -\frac{1}{2}$, $y = 3\frac{1}{2}$, as shown. Proceeding further, we take a still smaller interval about $x = 3$, say from $x = 2.9$ to $x = 3.1$. We draw the graph (enlarged) in this region, as shown in Fig. 4. The function values are now situated in the rectangle bounded by the lines $x = 2.9$, $x = 3.1$, $y = 1.58$, $y = 2.38$.

The main point we wish to emphasize concerns the *height* of these rectangles. As the widths of the rectangles become narrower, the heights also shrink in size. If we go further and take an $x$-interval from 2.99 to 3.01, the corresponding rectangle containing the graph of the function is bounded by the lines $x = 2.99$, $x = 3.01$, $y = 1.9598$, $y = 2.0398$. A width of 0.02 unit containing the value 3 leads to a rectangle of height only 0.08 unit. In addition to the heights becoming narrower and narrower as the widths squeeze in to the value $x = 3$, we observe that the heights cluster about the value $y = 2$. The reader may wonder what all this fuss is about, since the formula for the equation yields, by direct substitution, the result that $y = 2$ when $x = 3$. Note, however, that throughout the discussion we never made use of this fact (Fig. 5). Indeed, we carefully avoided any consideration at all of what happens when $x$ *is* 3. We are concerned solely with the behavior of $y$ when $x$ is in some **interval** about the value 3.

For functions we have studied up to this time, the behavior of a function at a point, say at $x = 3$, and its behavior in a sequence of shrinking intervals about this point were never distinguished. Now, however, a striking change occurs as we start to study functions whose behavior cannot be discovered by straight substitution. For example, the function

$$y = f(x) = \frac{\sin x}{x}$$

is defined for all values of $x$ except $x = 0$. A straight substitution at $x = 0$ would tell us that

$$y = f(0) = \frac{0}{0},$$

which is completely meaningless. However, we shall see that by studying a sequence of intervals about $x = 0$ which get smaller and smaller, we find that the corresponding rectangles containing the function above get thinner and thinner, and the heights cluster about a particular value of $y$. Nothing is ever stated about the value of $y$ when $x$ **is** zero. We just study the value of $y$ when $x$ gets closer and closer to zero.

Returning to the first example of the function $f(x) = -2x^2 + 8x - 4$, we see that as $x$ comes closer and closer to the value 3, $f(x)$ gets closer and closer

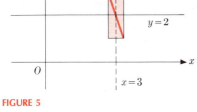

**FIGURE 4**

**FIGURE 5**

to the value 2. We say, "$f(x)$ approaches 2 as $x$ approaches 3." This sentence is abbreviated even further by the statement

$$\lim_{x \to 3} f(x) = 2.$$

Here once again we have the shorthand so typical of mathematics. A complex idea which takes several paragraphs to describe, even in a simple case, is boiled down to a brief symbolic expression. Furthermore, this symbolic expression contains a shorthand symbol for the notion of function, developed previously in Chapter 1, Section 5. As we proceed further, more concepts will be built on this symbolism.

If a function $f$ is defined for values of $x$ about the fixed number $a$, and if, as $x$ tends toward $a$, the values of $f(x)$ get closer and closer to some specific number $L$, then we write

$$\lim_{x \to a} f(x) = L,$$

and we read it, "The limit of $f(x)$ as $x$ approaches $a$ is $L$." Geometrically, this means that the series of rectangles which surround $a$ and which have narrower and narrower widths become smaller and smaller in height and cluster about the point $(a, L)$.

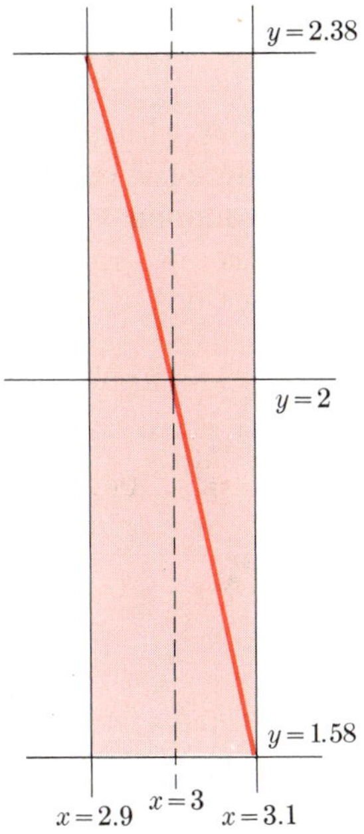

Of course all the above statements containing terms such as "closer," "nearer," "narrower," etc., are quite imprecise and are intended to give only an intuitive idea of what occurs. As mentioned, a precise definition of the expression $\lim_{x \to a} f(x) = L$ is given in Section 6.

Now let us take an example in which the result is quite different from that of the first example. We define the function

$$y = f(x) = \frac{(x - 2)}{(2|x - 2|)},$$

which is well determined for all values of $x$ except $x = 2$. At $x = 2$ the function is not defined, since straight substitution yields $y = \frac{0}{0}$, a meaningless expression. The graph of the function, shown in Fig. 6, is quite simple. If $x$ is larger than 2, then $|x - 2| = x - 2$ and the function has the value $+\frac{1}{2}$. If $x$ is less than 2, then $|x - 2| = -(x - 2)$, and the function is equal to $-\frac{1}{2}$. We wish to study the behavior of the function as $x$ tends to 2. We select an interval containing $x = 2$, say from $x = 1.4$ to $x = 2.3$. We see that the function is contained in the rectangle bounded by the lines $x = 1.4$, $x = 2.3$, $y = -\frac{1}{2}$, $y = +\frac{1}{2}$ (Fig. 7). In fact, no matter how narrow the interval about $x = 2$ becomes, the *height* of the rectangle will always be 1 unit. *There is no limit as $x$ approaches* 2. We say that

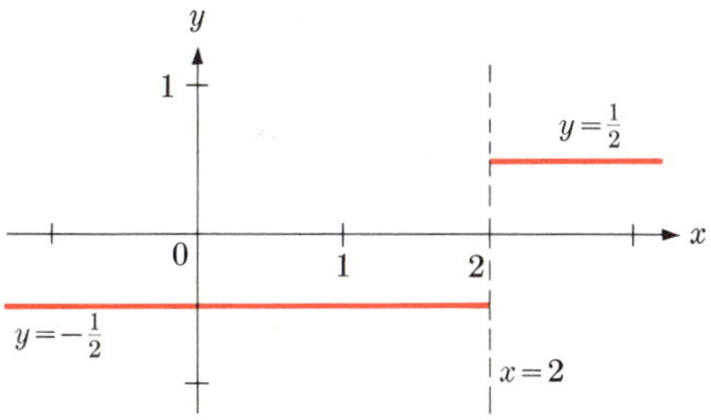

**FIGURE 6**

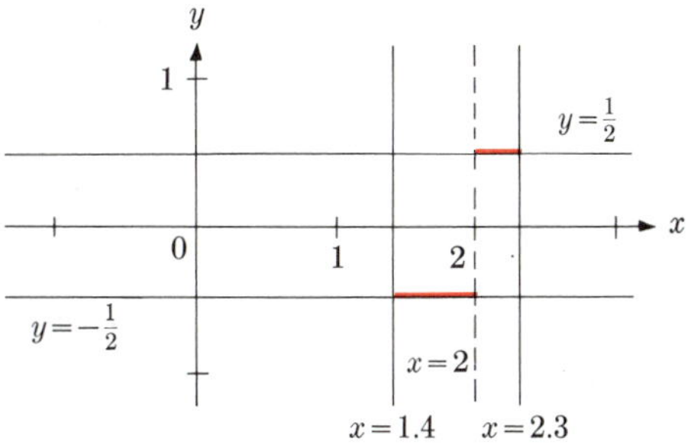

**FIGURE 7**

$$\lim_{x \to 2} \frac{x-2}{2|x-2|} \quad \text{does not exist.}$$

We now examine a number of examples of functions, with a view toward discovering what happens in the neighborhood of a particular value when the function is not defined at that value by straight substitution.

**EXAMPLE 1**    The function

$$f(x) = \frac{2x^2 - x - 3}{x + 1}$$

is defined for all values of $x$ except $x = -1$, since at $x = -1$ both numerator and denominator vanish. Does

$$\lim_{x \to -1} f(x)$$

exist and, if so, what is its value?

**Solution**    To get an idea of what is happening, we construct a table of values:

| $x$ | 0 | 1 | 2 | 3 | $-2$ |
|---|---|---|---|---|---|
| $f(x)$ | $-3$ | $-1$ | 1 | 3 | $-7$ |

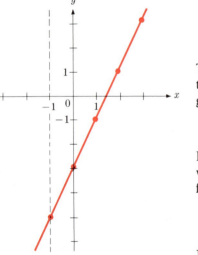

**FIGURE 8**

Then we draw the graph, which appears to be a straight line with a "hole" at the point $(-1, -5)$ (Fig. 8). From the geometric discussion of rectangles given at the beginning of this section we see that

$$\lim_{x \to -1} f(x) = -5.$$

However, we would like a more systematic method of obtaining limits, without relying on pictorial representation and intuition. By means of factoring, we can write $f(x)$ in the form

$$f(x) = \frac{(2x - 3)(x + 1)}{x + 1}.$$

Now if $x \ne -1$, we are allowed to divide both the numerator and the denominator by $(x + 1)$. Then

$$f(x) = 2x - 3, \quad \text{if } x \ne -1.$$

This function tends to $-5$ as $x$ tends to $-1$, since simple substitution now works. We conclude that

$$\lim_{x \to -1} f(x) = -5.$$

Note that we never substituted the value $x = -1$ in the original expression.

$\square$

**EXAMPLE 2**    Find the limit of the function

$$f(x) = \frac{x-4}{3(\sqrt{x}-2)}, \quad x \neq 4,$$

as $x$ tends to 4. (This function is defined for nonnegative values of $x$ except for $x = 4$, which is not in its domain.)

**Solution**   Note that straight substitution of $x = 4$ fails, since $f(4) = 0/0$, which is meaningless. We could proceed graphically, as in Example 1, but instead we introduce a useful algebraic trick. We *rationalize the denominator* by multiplying both numerator and denominator by $\sqrt{x} + 2$. So long as $x \neq 4$, we can write

$$\frac{(x-4)}{3(\sqrt{x}-2)}\frac{(\sqrt{x}+2)}{(\sqrt{x}+2)} = \frac{(x-4)(\sqrt{x}+2)}{3(x-4)},$$

and furthermore the common factor can be canceled if $x \neq 4$. We get

$$f(x) = \frac{\sqrt{x}+2}{3}, \quad \text{if } x \neq 4.$$

The limit of this expression can be found by straight substitution of $x = 4$. We find

$$\lim_{x \to 4} f(x) = \frac{\sqrt{4}+2}{3} = \frac{4}{3}. \qquad \square$$

**EXAMPLE 3**   Find the limit of the function

$$f(x) = \frac{\sqrt{2+x}-1}{x+1}, \quad x \neq -1,$$

as $x$ tends to $-1$. (This function is defined only for values of $x$ larger than or equal to $-2$ except for $x = -1$, which is not in its domain.)

**Solution**   Since straight substitution fails, we employ the device of *rationalizing the numerator*. Multiplication of numerator and denominator by $\sqrt{2+x}+1$ yields

$$f(x) = \frac{(\sqrt{2+x}-1)(\sqrt{2+x}+1)}{(x+1)(\sqrt{2+x}+1)} = \frac{x+1}{(x+1)(\sqrt{2+x}+1)}, \quad x \neq -1,$$

where we have used the fact that $(a-b)(a+b) = a^2 - b^2$ with $a = \sqrt{2+x}$, $b = 1$. We now divide numerator and denominator by $x+1$ and find that

$$f(x) = \frac{1}{\sqrt{2+x}+1} \quad \text{for} \quad x \neq -1, \quad x \geq -2.$$

As $x$ tends to $-1$, $f(x)$ tends to

$$\frac{1}{\sqrt{2-1}+1} = \frac{1}{2}. \qquad \square$$

# 1   PROBLEMS

In Problems 1 through 14, in each case find the limit (if any) approached by $f(x)$ as $x$ tends to $a$. Note all values of $x$ where $f(x)$ is not defined.

**1** $f(x) = \dfrac{x^2 - 4}{x^2 - 5x + 6}, \quad a = 2$

**2** $f(x) = \dfrac{x^2 - 9}{x^2 - 7x + 12}, \quad a = 3$

**3** $f(x) = \dfrac{x^3 + 2x - 1}{x + 2}, \quad a = 3$

**4** $f(x) = \dfrac{x^2 - x - 2}{x^2 - 4x + 4}, \quad a = 2$

**5** $f(x) = \dfrac{x^3 + 1}{x + 1}, \quad a = -1$

**6** $f(x) = \dfrac{x^4 - 1}{x^2 - 1}, \quad a = 1$

**7** $f(x) = \dfrac{\sqrt{x + 2} - 2}{x - 2}, \quad a = 2, x \geq -2$

**8** $f(x) = \dfrac{x^4 - 2x - 3}{x + 1}, \quad a = -1$

**9** $f(x) = \dfrac{\sqrt{2x + 3} - x}{x - 3}, \quad a = 3, x \geq -\dfrac{3}{2}$

**10** $f(x) = \dfrac{\sqrt{x^2 + 3} - 2}{x + 1}, \quad a = -1$

**11** $f(x) = \dfrac{x - 9}{\sqrt{x} - 3}, \quad a = 9, x \geq 0$

**12** $f(x) = \dfrac{x + 3}{\sqrt{x^2 + 7} - 4}, \quad a = -3$

**13** $f(x) = \dfrac{x^3 - 6x^2 + 11x - 6}{x^3 - 7x + 6}, \quad a = 1.$ What is $\lim\limits_{x \to 2} f(x)$? What is $\lim\limits_{x \to 3} f(x)$?

**14** $f(x) = \dfrac{x^3 - 5x^2 + 2x + 8}{x^3 + 2x^2 - 5x - 6}, \quad a = -1.$ What is $\lim\limits_{x \to 2} f(x)$?

In Problems 15 through 28, in each case find the indicated limit.

**15** $\lim\limits_{x \to 3} (x^2 + 4x - 6)$

**16** $\lim\limits_{x \to 2} (2x^4 - 3x^2 + 7x + 5)$

**17** $\lim\limits_{x \to 4} \dfrac{x^2 - 16}{x - 4}$

**18** $\lim\limits_{x \to -2} \dfrac{\sqrt{4 - x^2}}{x + 2}$

**19** $\lim\limits_{x \to 0} \dfrac{\sqrt{5 + x} - \sqrt{5}}{2x}$

**20** $\lim\limits_{h \to 0} \dfrac{\sqrt{9 + 2h} - 3}{h}$

**21** $\lim\limits_{h \to 0} \dfrac{h^2}{\sqrt{4 + 3h^2} - 2}$

**22** $\lim\limits_{h \to 0} \dfrac{\sqrt{3x + h} - \sqrt{3x}}{h}$

**23** $\lim\limits_{h \to 1} \dfrac{\sqrt{b + 2(h - 1)} - \sqrt{b}}{h - 1}$

**24** $\lim\limits_{x \to 2} \dfrac{\sqrt[3]{x} - \sqrt[3]{2}}{x - 2}$

**25** $\lim\limits_{x \to -4} \dfrac{x^6 - 4096}{x + 4}$

**26** $\lim\limits_{x \to -2} \dfrac{x^3 + 4x^2 + x - 6}{x^2 + 6x + 8}$

**27** $\lim\limits_{x \to 8} \dfrac{\sqrt[4]{x} - \sqrt[4]{8}}{\sqrt{x} - \sqrt{8}}$

**28** $\lim\limits_{x \to 0} \dfrac{\sqrt{3 + x} - \sqrt{3 - x}}{\sqrt{4 + x^2} - \sqrt{4 - 2x}}$

**29** Let $f(x) = (1 - |x|)/(1 + x)$ be given with domain all $x \neq -1$. Plot the graph of $f$. Find

$$\lim\limits_{x \to -1} \dfrac{1 - |x|}{1 + x}.$$

**30** Given

$$f(x) = \begin{cases} 1 & \text{for} \quad -\infty < x < 3 \\ 2 & \text{for} \quad 3 \leq x < \infty \end{cases}.$$

Plot the graph. Show that $\lim\limits_{x \to 3} f(x)$ does not exist.

**31** Given $f(x) = (\sin x)/x$ for all $x \neq 0$ where $x$ is measured in radians. Sketch the graph of $f$ for $x$ between $-1$ and $1$ by plotting the values of $f$ for $x = -1, -0.9, -0.8, \ldots, 0.9, 1$. Can any conclusion be suggested about the value of

$$\lim\limits_{x \to 0} \dfrac{\sin x}{x}?$$

**32** Given

$$f(x) = \dfrac{\tan x}{x}$$

where $x$ is measured in radians. Use a calculator to compute $f(0.1), f(0.01), f(0.001),$ and $f(0.0001)$. Then find $f(-0.1), f(-0.01), f(-0.001),$ and $f(-0.0001)$. Make an estimate of

$$\lim\limits_{x \to 0} \dfrac{\tan x}{x}.$$

**33** Same as Problem 32 for

$$f(x) = \dfrac{1 - \cos x}{x}.$$

_____ 2
_____

## COMPUTATION OF LIMITS

In evaluating limits, functional notation plays a convenient and important part. Consider the function

$$f(x) = x^2 + 3,$$

and suppose that we wish to evaluate the limit

$$\lim_{x \to 0} \frac{f(5+x) - f(5)}{x}, \quad x \neq 0.$$

A straight substitution of $x = 0$ leads to the meaningless expression

$$\frac{f(5) - f(5)}{0} = \frac{0}{0},$$

and we must penetrate a little deeper. We have $f(5) = 28$ and $f(5 + x) = (5 + x)^2 + 3 = x^2 + 10x + 28$. Therefore

$$\frac{f(5+x) - f(5)}{x} = \frac{x^2 + 10x + 28 - 28}{x} = x + 10.$$

The cancellation of $x$ from numerator and denominator is most fortunate, since now

$$\lim_{x \to 0} (x + 10)$$

may be found by direct substitution; the answer is 10.

The best way of learning the technique for evaluating such limits is by a careful study of the following four examples.

**EXAMPLE 1**    Find the value of

$$\lim_{h \to 0} \frac{f(4+h) - f(4)}{h}, \quad h \neq 0,$$

where

$$f(x) = \frac{1}{(x+1)^2}, \quad x \neq -1.$$

**Solution**    Direct substitution of $h = 0$ gives

$$\frac{f(4) - f(4)}{0} = \frac{0}{0},$$

which is meaningless. However,

$$f(4) = \frac{1}{25}, \quad f(4+h) = \frac{1}{(4+h+1)^2} = \frac{1}{(5+h)^2}.$$

Therefore

$$\frac{f(4+h)-f(4)}{h} = \frac{\dfrac{1}{(5+h)^2} - \dfrac{1}{25}}{h}.$$

We might try direct substitution at this point, but again we would fail, since the result is again $\frac{0}{0}$. We proceed by finding the lowest common denominator, getting

$$\frac{f(4+h)-f(4)}{h} = \frac{25-(5+h)^2}{25h(5+h)^2} = \frac{-(10h+h^2)}{25h(5+h)^2} = \frac{-h(10+h)}{25h(5+h)^2}.$$

So long as $h \neq 0$, this factor may be canceled in both numerator and denominator. Then

$$\lim_{h\to 0} \frac{-(10+h)}{25(5+h)^2}$$

may be obtained by direct substitution. The answer is $-\frac{2}{125}$.  □

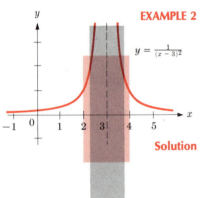

$y = \frac{1}{(x-3)^2}$

No rectangle can contain the graph near $x = 3$.

**FIGURE 9**

**EXAMPLE 2**  Find the value of

$$\lim_{x\to 3} f(x),$$

where

$$f(x) = \frac{1}{(x-3)^2}, \quad x \neq 3.$$

**Solution**  Sketching the graph of this function about $x = 3$, we see that it increases without bound as $x$ tends to 3 (Fig. 9). According to the notion of limit as given in Section 1, we take an interval of $x$ values about 3 and then see in what rectangle the function values are contained. From the figure it is clear that no matter how narrow we make the rectangle, we cannot make it high enough to contain the graph of the function. In such a case we say that

$$\lim_{x\to 3} f(x)$$

does not exist.  □

**EXAMPLE 3**  Find the limit

$$\lim_{t\to 0} \frac{f(2+t)-f(2)}{t^2}$$

where $f(x) = x^2$.

**Solution**  Direct substitution of $t = 0$ fails, and we proceed as before to find that

$$f(2) = 4 \quad \text{and} \quad f(2+t) = (2+t)^2 = 4 + 4t + t^2.$$

Therefore, if $t \neq 0$,

$$\frac{f(2+t)-f(2)}{t^2} = \frac{4+4t+t^2-4}{t^2} = \frac{4+t}{t}.$$

However, this function increases without bound as $t$ tends to zero through positive values, and it decreases without bound as $t$ tends to zero through negative values (Fig. 10). The limit does not exist. ☐

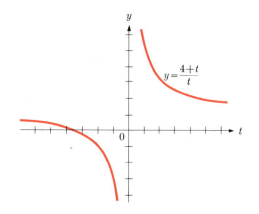

**FIGURE 10**

A convenient symbol used throughout calculus to designate a real number is $\Delta x$, which is read "delta $x$." This expression is to be treated as a single quantity such as $t$ in the above Example 3 or $h$ in Example 1 (and *not* as the product of $\Delta$ and $x$). Later we shall see that such symbols actually help in understanding rather complicated limiting processes.

**EXAMPLE 4**    Find the limit

$$\lim_{\Delta x \to 0} \frac{f(-1 + \Delta x) - f(-1)}{\sqrt{\Delta x}}, \quad \Delta x > 0,$$

where $f(x) = x^3$.

**Solution**    We have $f(-1) = -1$, $f(-1 + \Delta x) = (-1 + \Delta x)^3$ and

$$\frac{f(-1 + \Delta x) - f(-1)}{\sqrt{\Delta x}} = \frac{-1 + 3(\Delta x) - 3(\Delta x)^2 + (\Delta x)^3 - (-1)}{\sqrt{\Delta x}}$$

$$= \frac{(\Delta x)[3 - 3(\Delta x) + (\Delta x)^2]}{\sqrt{\Delta x}}$$

$$= \sqrt{\Delta x}[3 - 3(\Delta x) + (\Delta x)^2].$$

As $\Delta x$ tends to zero, this expression tends to $\sqrt{0}[3 - 3(0) + (0)^2]$, and the answer is zero. ☐

## 2   PROBLEMS

Evaluate the following limits.

**1**   $\displaystyle\lim_{x \to 0} \frac{f(4 + x) - f(4)}{x}, \quad f(x) = 3x^2 - 1$

**2**   $\displaystyle\lim_{h \to 0} \frac{f(-3 + h) - f(-3)}{h}, \quad f(x) = x^2 + 2$

**3**   $\displaystyle\lim_{h \to 0} \frac{f(1 + h) - f(1)}{h}, \quad f(x) = 2x^3 - 2$

**4**   $\displaystyle\lim_{k \to 0} \frac{f(3 + k) - f(3)}{k}, \quad f(x) = \frac{2}{x}, \quad x \neq 0$

**5** $\lim\limits_{h \to 0} \dfrac{f(a+h)-f(a)}{h}, \quad f(x)=2x^2-3$

**6** $\lim\limits_{h \to 0} \dfrac{f(c+h)-f(c)}{h}, \quad f(x)=-\dfrac{2}{x^2}, \quad x \neq 0$

**7** $\lim\limits_{h \to 0} \dfrac{g(2+h)-g(2)}{h}, \quad g(x)=12-x^2$

**8** $\lim\limits_{h \to 0} \dfrac{g(-4+h)-g(-4)}{h}, \quad g(x)=\dfrac{1}{x^2+1}$

**9** $\lim\limits_{h \to 0} \dfrac{F(-2+h)-F(-2)}{h}, \quad F(x)=\dfrac{1}{x-1}, \quad x \neq 1$

**10** $\lim\limits_{h \to 0} \dfrac{F(a+h)-F(a)}{h}, \quad F(x)=\dfrac{1}{\sqrt{x}}, \quad x>0, \quad a>0$

**11** $\lim\limits_{h \to 0} \dfrac{F(x+h)-F(x)}{h}, \quad F(x)=-x^2+5$

**12** $\lim\limits_{k \to 0} \dfrac{G(b+k)-G(b)}{k}, \quad G(x)=x^2-2x+3$

**13** $\lim\limits_{t \to 0} \dfrac{H(3+t)-H(3)}{t}, \quad H(x)=x^3+2x+5$

**14** $\lim\limits_{h \to 0} \dfrac{F(4+h)-F(4)}{h}, \quad F(x)=\dfrac{1}{x^2+4}$

**15** $\lim\limits_{h \to 0} \dfrac{F(2+h)-F(2)}{\sqrt{h}}, \quad F(x)=x^4-2x+5, \quad (h>0)$

**16** $\lim\limits_{h \to 0} \dfrac{L(3+h)-L(3)}{h^{3/2}}, \quad L(x)=\dfrac{1}{x^2+2}-3, \quad (h>0)$

**17** $\lim\limits_{h \to 0} \dfrac{f(x+h)-f(x)}{h}, \quad f(x)=x^{3/2}, \quad x>0$

**18** $\lim\limits_{h \to 0} \dfrac{F(t+h)-F(t)}{h}, \quad F(x)=x^2-\dfrac{1}{x^2}, \quad x \neq 0$

**19** $\lim\limits_{h \to 0} \dfrac{g(c+h)-g(c)}{h}, \quad g(x)=x^2+3x-c$

**20** $\lim\limits_{h \to 0} \dfrac{g(x+3h)-g(x)}{h}, \quad g(x)=3x^2+2x-1$

**21** $\lim\limits_{h \to 0} \dfrac{f(x+h)-f(x)}{h}, \quad f(x)=\dfrac{1}{x^3+1}, \quad x \neq -1$

**22** $\lim\limits_{h \to 0} \dfrac{f(x+h)-f(x)}{h}, \quad f(x)=\dfrac{1}{x^{3/2}}, \quad x>0$

**23** $\lim\limits_{\Delta x \to 0} \dfrac{f(2+\Delta x)-f(2)}{\Delta x}, \quad f(x)=x+\sqrt{x}, \quad x>0$

**24** $\lim\limits_{\Delta x \to 0} \dfrac{f(x+\Delta x)-f(x)}{\Delta x}, \quad f(x)=\sqrt[3]{x}$

**25** $\lim\limits_{\Delta x \to 0} \dfrac{G(x+\Delta x)-G(x)}{\Delta x}, \quad G(x)=\dfrac{x-1}{x+1}, \quad x \neq -1$

**26** $\lim\limits_{h \to 0} \dfrac{f(1+2h)-f(1)}{h}, \quad f(x)=x^3-1$

**27** $\lim\limits_{h \to 0} \dfrac{f(2-3h)-f(2)}{h}, \quad f(x)=\dfrac{x+1}{x+2}$

**28** $\lim\limits_{h \to 0} \dfrac{f(a+h)-f(a)}{h}, \quad f(x)=\dfrac{1}{x+1}-\dfrac{1}{x-1}, \quad x \neq 1, -1$

**29** $\lim\limits_{k \to 0} \dfrac{f(y+k)-f(y)}{2k}, \quad f(t)=\dfrac{t^2-1}{t+2}, \quad t \neq -2$

**30** $\lim\limits_{p \to 0} \dfrac{f(z+p)-f(z)}{3p}, \quad f(z)=\dfrac{\pi z^2+1}{2z}, \quad z \neq 0$

---

**3**

## THE DERIVATIVE

**DEFINITION**  *If f is a function, **the derivative of the function** f, denoted by f' (read "f prime"), is defined by the formula*

$$f'(x)=\lim\limits_{h \to 0} \dfrac{f(x+h)-f(x)}{h}.$$

In this definition, $x$ remains fixed, while $h$ tends to zero. If the limit does not exist for a particular value of $x$, the function has no derivative for that value. We also say that a function $f$ is **differentiable at** $x$ if the above limit exists. A function $f$ is simply said to be **differentiable** if it has a derivative for all $x$ in its domain.

Looking back at the previous section, we see that many of the problems consisted of finding the derivative at a particular value of $x$. We now give a systematic procedure for obtaining derivatives, a method which we call the **three-step rule**. The technique is illustrated by some examples. In the following sections we provide interpretations and applications of the derivative.

**EXAMPLE 1**    Given $f(x) = x^2$. Find the derivative, $f'(x)$, by the three-step rule.

**Solution**    **Step 1:**  Write the formula for the expression $f(x+h) - f(x)$.

We have $f(x) = x^2$ and $f(x+h) = (x+h)^2 = x^2 + 2xh + h^2$. Therefore

$$f(x+h) - f(x) = x^2 + 2xh + h^2 - x^2$$
$$= 2xh + h^2.$$

**Step 2:**  Divide the expression for $f(x+h) - f(x)$ by $h$.

Since $h$ is a factor of this difference, we get

$$\frac{f(x+h) - f(x)}{h} = \frac{2xh + h^2}{h} = 2x + h.$$

**Step 3:**  Take the limit as $h$ tends to zero.

We find

$$\lim_{h \to 0} (2x + h) = 2x.$$

*Answer:*  $f'(x) = 2x.$                                              □

In the three-step rule, the first two steps are purely mechanical and are carried out in a routine manner. It is the third and last step that frequently requires ingenuity and algebraic manipulation.

**EXAMPLE 2**    Given $f(x) = 1/x$, $x \neq 0$. Find $f'(x)$ by the three-step rule.

**Solution**    We have the following.

**Step 1:**   $f(x+h) - f(x) = \dfrac{1}{x+h} - \dfrac{1}{x}.$

**Step 2:**   $\dfrac{f(x+h) - f(x)}{h} = \dfrac{\dfrac{1}{x+h} - \dfrac{1}{x}}{h}.$

It is at this point that the difficulty arises, since letting $h$ tend to zero yields $\frac{0}{0}$. Before applying the third step, we manipulate the expression by finding the lowest common denominator. We obtain

$$\frac{f(x+h) - f(x)}{h} = \frac{x - (x+h)}{hx(x+h)} = \frac{-1}{x(x+h)}.$$

**Step 3:** $\lim\limits_{h \to 0} \dfrac{-1}{x(x+h)} = -\dfrac{1}{x^2}.$

*Answer:* $f'(x) = -\dfrac{1}{x^2}.$ ☐

**EXAMPLE 3** Given $f(x) = \sqrt{x}$ for $x > 0$. Find $f'(x)$ by the three-step rule.

**Solution** In the three-step rule, it is not essential that the letter $h$ be used in computing the limit. Any symbol, such as $t$, $k$, $\Delta x$, and so forth, may be used. We employ $\Delta x$ in solving this problem.

**Step 1:** $f(x + \Delta x) - f(x) = \sqrt{x + \Delta x} - \sqrt{x}.$

**Step 2:** $\dfrac{f(x + \Delta x) - f(x)}{\Delta x} = \dfrac{\sqrt{x + \Delta x} - \sqrt{x}}{\Delta x}.$

Now the trick of *rationalizing the numerator* is used. The result is

$$\dfrac{\sqrt{x + \Delta x} - \sqrt{x}}{\Delta x} \cdot \dfrac{\sqrt{x + \Delta x} + \sqrt{x}}{\sqrt{x + \Delta x} + \sqrt{x}} = \dfrac{(x + \Delta x) - x}{\Delta x(\sqrt{x + \Delta x} + \sqrt{x})}$$

$$= \dfrac{1}{\sqrt{x + \Delta x} + \sqrt{x}}.$$

Now taking the limit, we get

**Step 3:** $\lim\limits_{\Delta x \to 0} \dfrac{1}{\sqrt{x + \Delta x} + \sqrt{x}} = \dfrac{1}{\sqrt{x} + \sqrt{x}} = \dfrac{1}{2\sqrt{x}}.$

*Answer:* $f'(x) = \dfrac{1}{2\sqrt{x}}.$ ☐

Some problems require quite a bit of algebraic manipulation, as the next example shows.

**EXAMPLE 4** Given $f(x) = x/(x + 1)$. Find $f'(x)$ by the three-step rule.

**Solution** We have

**Step 1:** $f(x + h) - f(x) = \dfrac{x + h}{x + h + 1} - \dfrac{x}{x + 1}.$

**Step 2:** $\dfrac{f(x + h) - f(x)}{h} = \dfrac{\dfrac{x + h}{x + h + 1} - \dfrac{x}{x + 1}}{h}.$

At this point we get the lowest common denominator in the numerator so that we can combine the expression into one fraction. We find

$$\frac{f(x+h)-f(x)}{h} = \frac{\dfrac{(x+h)(x+1)-x(x+h+1)}{(x+h+1)(x+1)}}{h}.$$

We now multiply out the numerator and write the above expression as one fraction:

$$\frac{f(x+h)-f(x)}{h} = \frac{x^2+hx+x+h-(x^2+hx+x)}{h(x+h+1)(x+1)}.$$

All terms but one in the numerator cancel, giving

$$\frac{f(x+h)-f(x)}{h} = \frac{h}{h(x+h+1)(x+1)}.$$

Dividing through by $h$, we apply the next step, getting

**Step 3:**   $\displaystyle\lim_{h\to 0}\frac{1}{(x+h+1)(x+1)} = \frac{1}{(x+1)^2}.$

*Answer:*  $f'(x) = \dfrac{1}{(x+1)^2}.$   □

## 3   PROBLEMS

In each of Problems 1 through 33, find the derivative by the three-step rule.

**1** $f(x) = 3x^2$

**2** $f(x) = -2x^2 + 1$

**3** $f(x) = 2x^2 - 3x + 4$

**4** $f(x) = 3x^3$

**5** $f(x) = -x^4$

**6** $f(x) = x^3 - 2x$

**7** $f(x) = \dfrac{1}{x-1}$

**8** $f(x) = \dfrac{1}{x^2}$

**9** $f(x) = \dfrac{1}{x^3}$

**10** $f(x) = \dfrac{2x+3}{3x-2}$

**11** $f(x) = \dfrac{x}{1-x}$

**12** $f(x) = \dfrac{1}{x^2+1}$

**13** $f(x) = \dfrac{x^2-1}{x^2+1}$

**14** $f(x) = \dfrac{x^2-2x}{2}$

**15** $f(x) = \dfrac{1}{\sqrt{x}}$

**16** $f(x) = \dfrac{1}{\sqrt{x+3}}$

**17** $f(x) = x\sqrt{x}$

**18** $f(x) = x\sqrt{x+1}$

**19** $f(x) = \dfrac{x}{\sqrt{x-1}}$

**20** $f(x) = \dfrac{\sqrt{x-1}}{x}$

**21** $f(x) = \sqrt{4-x^2}$

**22** $f(x) = \dfrac{1}{\sqrt{4-x^2}}$

**23** $f(x) = \dfrac{x^2-2x+2}{x^2+x+2}$

**24** $f(x) = 2x - \dfrac{1}{3x} + \dfrac{2}{x^2}$

**25** $f(x) = \sqrt{x^2+x+1}$

**26** $f(x) = x^5 + 2x - 1$

**27** $f(x) = \dfrac{1}{(x+1)^3}$

**28** $f(x) = x^2 + \dfrac{2}{x} + 3x$

**29** $f(x) = \dfrac{1}{x^4}$

**30** $f(x) = \dfrac{1}{(2x+1)^4}$

**31** $f(x) = \dfrac{x}{3x+4}$

**32** $f(x) = |x|, \quad x \neq 0$

**33** $f(x) = |x-5|, \quad x \neq 5$

**34** We define $f(x) = [x]$ as follows: $f(x)$ is the largest integer that is less than or equal to $x$. For example, $f(3\frac{1}{2}) = 3$, $f(7) = 7$, $f(8.2) = 8$. We call $f$ the "greatest integer function." Find the derivative of $f(x)$ when $x$ is not an integer. What happens if $x$ is an integer?

## 4

## GEOMETRIC INTERPRETATION OF THE DERIVATIVE

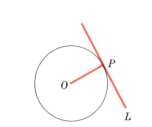

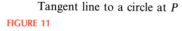

Tangent line to a circle at $P$

FIGURE 11

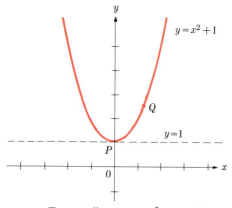

Tangent line to $y = x^2 + 1$ at $P$

FIGURE 12

It is easy to define a *tangent line* to a circle at a point $P$ on the circle. A radius from the center $O$ to $P$ is drawn and then, at $P$, a line $L$ perpendicular to this radius is constructed (Fig. 11). **The line $L$ is called the tangent to the circle at the point $P$.** We next draw the curve given by the equation $y = x^2 + 1$, as shown in Fig. 12. How do we define a tangent line to this curve at a point $P$?

If the point $P$ is $(0, 1)$, the lowest point on the curve, it seems natural to take the line $y = 1$, as shown, as the tangent line. However, at a point such as $Q$, it is not evident what line is tangent to the curve. Intuitively, the definition should state that a tangent line should "touch" the curve at one point only. But this "definition" is not good enough, since there may be several such lines and, in fact, a vertical line through $Q$ has only one point in common with the curve. While the definition of tangent line to a circle is quite elementary, the definition for other kinds of curves requires the sophisticated notion of limit.

Let a curve be given by a function $y = f(x)$, as shown in Fig. 13. Draw a line, denoted $L_h$, through $P_0$ and $P$, two points on the curve. Any line through two points on a curve is called a **secant line**. The line $L_h$ is such a secant line. Suppose that the coordinates of $P_0$ are $(x_0, y_0)$. This means that $y_0 = f(x_0)$. The coordinates of $P$ are taken to be $(x_0 + h, f(x_0 + h))$, where $h$ is some positive or negative number. Figure 13 is drawn with $h$ positive. If $P$ were to the left of $P_0$, then $h$ would be negative. The slope of the line $L_h$ is obtained from the usual formula for slope when two points are known:

$$\text{slope of } L_h = \frac{f(x_0 + h) - f(x_0)}{(x_0 + h) - x_0} = \frac{f(x_0 + h) - f(x_0)}{h}.$$

As $h$ tends to zero, we see geometrically that the point $P$ slides along the curve $y = f(x)$ and tends to the point $P_0$. The secant line $L_h$ rotates about $P_0$. Intuitively it seems that $L_h$ approaches a limiting line as $h \to 0$, and this limiting line should be the tangent line to the curve at $P_0$. Figure 14 shows several positions for $L_h$ and the limiting position as $h \to 0$. On the other hand,

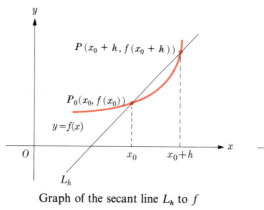

Graph of the secant line $L_h$ to $f$

FIGURE 13

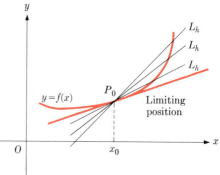

Secant lines tending to a limit line as $h$ tends to $O$

FIGURE 14

if the function $f$ possesses a derivative, then

$$\lim_{h \to 0} \frac{f(x_0 + h) - f(x_0)}{h} = f'(x_0).$$

In other words, *the slopes of the lines $L_h$ tend to a limiting slope which is the derivative of the function $f$ at $P_0$.* Suppose that we now construct, at $P_0$, the line passing through $P_0$ and having slope $f'(x_0)$. This can be done by the point-slope formula for the straight line.

**DEFINITION**   *The **tangent line** to the curve with equation $y = f(x)$ at $P_0(x_0, f(x_0))$ is the line through $P_0(x_0, f(x_0))$ with slope $f'(x_0)$.*

We recall that the point-slope form for the equation of a straight line is

$$y - y_0 = m(x - x_0).$$

If in this formula we take $y_0 = f(x_0)$ and $m = f'(x_0)$, we obtain the equation of the line tangent to the curve at the point $P_0$.

The equation of the **tangent line at** $(x_0, f(x_0))$ is $y - y_0 = f'(x_0)(x - x_0)$.

**EXAMPLE 1**   Find the equation of the line which is tangent to the curve that has equation $y = x^2 + 1$ at the point on the curve where $x = 2$.

**Solution**   In this problem $f(x) = x^2 + 1$, $x_0 = 2$. Therefore $y_0 = f(x_0) = f(2) = 5$. To find $f'(x_0)$, we apply the three-step rule to $f(x) = x^2 + 1$.

**Step 1:**   $f(x + h) - f(x) = x^2 + 2xh + h^2 + 1 - x^2 - 1.$

**Step 2:**   $\dfrac{f(x + h) - f(x)}{h} = \dfrac{2xh + h^2}{h} = 2x + h.$

**Step 3:**   Taking the limit, we get $f'(x) = 2x$.

Then $f'(x_0) = f'(2) = 4$ and $m = 4$. The desired equation is

$$y - 5 = 4(x - 2).$$                                                        □

**DEFINITION**   *The **normal line** to the curve with equation $y = f(x)$ at $P_0(x_0, f(x_0))$ is the line through $P_0$ which is perpendicular to the tangent line at $P_0$.*

**EXAMPLE 2**   In the problem of Example 1, find the equation of the line normal to the curve at the point where $x = 3$.

**Solution**   Since $x_0 = 3$ and $f(x) = x^2 + 1$, we have $y_0 = f(x_0) = 10$. The tangent line at $x_0 = 3$ has slope $f'(x_0) = f'(3)$. We have already found that $f'(x) = 2x$ for any value of $x$. Therefore the tangent line at $x = 3$ has slope 6. The normal line must have a slope which is the negative reciprocal of the slope of the tangent line, namely $-\frac{1}{6}$. The equation of the normal line at $(3, 10)$ is

$$y - 10 = -\tfrac{1}{6}(x - 3).$$                                            □

We have just seen that the derivative is merely the slope of the tangent line. In fact, the tangent line is defined in this way. The intuitive notion of the nature of a line tangent to a curve is of great help in plotting graphs. If the derivative $f'(x)$ is positive, the slope is positive, showing that the curve must be rising as we go from left to right. Similarly, if $f'(x) < 0$, the curve must be falling as we go from left to right. These statements will be proved rigorously in Chapter 4, Section 3. The fact that $f'(x) = 0$ means that the tangent line is horizontal, and the curve *may* have either a low point or a high point at such a location. (However, other possibilities exist, as we shall see later in Chapter 4, Section 3.)

**EXAMPLE 3**    Find the intervals in which the function $y = f(x) = x^3 - 3x + 2$ is increasing and those in which it is decreasing. Sketch the graph.

**Solution**    We first find the derivative by the three-step rule.

**Step 1:**  $f(x+h) - f(x) = (x+h)^3 - 3(x+h) + 2 - (x^3 - 3x + 2)$

$$= x^3 + 3x^2h + 3xh^2 + h^3 - 3x - 3h + 2$$

$$- x^3 + 3x - 2$$

$$= 3x^2h + 3xh^2 + h^3 - 3h.$$

**Step 2:**  $\dfrac{f(x+h) - f(x)}{h} = 3x^2 - 3 + 3xh + h^2.$

**Step 3:**   Taking the limit as $h \to 0$, we get $f'(x) = 3x^2 - 3$. We first find the places at which $f'(x) = 0$. That is, we solve the equation $3x^2 - 3 = 0$ to get $x = 1, -1$. An examination of the function $f'(x) = 3x^2 - 3 = 3(x-1)(x+1)$ shows that $f'(x) > 0$ for $x < -1$; also we see that $f'(x) < 0$ for $-1 < x < 1$ and $f'(x) > 0$ for $x > 1$. We conclude that

$$f(x) \text{ is increasing for } x < -1,$$

$$f(x) \text{ is decreasing for } -1 < x < 1,$$

$$f(x) \text{ is increasing for } x > 1.$$

A table of values gives the following points.

| $x$ | $-2$ | $-1$ | $0$ | $1$ | $2$ |
|---|---|---|---|---|---|
| $y = f(x)$ | $0$ | $4$ | $2$ | $0$ | $4$ |

The graph is sketched in Fig. 15.                                   □

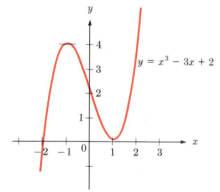

$y = x^3 - 3x + 2$

Graph of $f(x) = x^3 - 3x + 2$

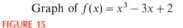

**FIGURE 15**

The fact that the curve reached a low point at exactly $(1, 0)$ and a high point at exactly $(-1, 4)$ was discovered from studying the derivative. The location of high and low points could not easily be obtained merely from plotting points and sketching the graph. A low point such as $(1, 0)$ is a minimum point only if we restrict our attention to a small interval about this point. Such a point is called a **relative minimum**. It is clearly not the minimum point on the entire curve, since the curve falls below the $x$ axis as we go to the left beyond $x = -2$. Similarly, the point $(-1, 4)$ is called a **relative maximum**. At a relative maximum or minimum it is always true that if $f$ has a derivative, then $f'(x) = 0$.

**Notation.** We saw earlier that the symbol $\Delta x$ can be used instead of $h$ in calculating the derivative of a function $f$. Writing $y = f(x)$, we **define** the symbol $\Delta y$ by the formula

$$\Delta y = f(x + \Delta x) - f(x).$$

Then Step 2 in the three-step rule states that

$$\frac{\Delta y}{\Delta x} = \frac{f(x + \Delta x) - f(x)}{\Delta x}.$$

The derivative consists of taking the limit of the above expression as $\Delta x \to 0$. The symbol we have been using is $f'(x)$. However, it is helpful to use a symbol related to the quantity $\dfrac{\Delta y}{\Delta x}$. It turns out to be most useful to write

$$f'(x) = \frac{dy}{dx}$$

where, at this time, $dy$ and $dx$ *will not be defined separately*. We simply use the ratio $\dfrac{dy}{dx}$ as an alternate expression for $f'(x)$. That is, we write

$$\lim_{\Delta x \to 0} \frac{\Delta y}{\Delta x} = \frac{dy}{dx} = f'(x).$$

## 4   PROBLEMS

In Problems 1 through 12, find the equation of the tangent line and the normal line at the point corresponding to the given value of $x_0$.

1  $y = x^2 + x - 1$, $\quad x_0 = 1$

2  $y = x^2 + 3x$, $\quad x_0 = -1$

3  $y = x^3 - 2x^2 + 3$, $\quad x_0 = 2$

4  $y = x^3 - x + 2$, $\quad x_0 = -2$

5  $y = -x^3 + 3x + 1$, $\quad x_0 = 2$

6  $y = \dfrac{x + 1}{x - 2}$, $\quad x_0 = 3$

7  $y = \sqrt{2x}$, $\quad x_0 = 2$

8  $y = \sqrt{3 - x}$, $\quad x_0 = -1$

9  $y = \dfrac{1}{\sqrt{x - 2}}$, $\quad x_0 = 6$

10  $y = 2x + 3\sqrt{x}$, $\quad x_0 = 4$

11  $y = 2x + \dfrac{2}{x}$, $\quad x_0 = 1$

12  $y = x^4 - 2x + 5$, $\quad x_0 = -1$

In Problems 13 through 20, find the intervals in which $f$ is increasing and those in which it is decreasing. Plot the graph of $y = f(x)$. Note relative maxima and minima.

13  $f(x) = x^2 - 4x + 5$

14  $f(x) = 1 + 2x - x^2$

15  $f(x) = \frac{1}{3}x^3 - \frac{1}{2}x^2 - 2x$

16  $f(x) = \frac{1}{3}x^3 - x^2 - 3x + 2$

17  $f(x) = 1 + 3x^2 - x^3$

18  $f(x) = x^3 + 3x - 2$

19  $f(x) = 3x^4 - 8x^3 - 6x^2 + 24x + 2$

20  $f(x) = 3x^5 - 25x^3 + 60x + 10$

In Problems 21 through 23, a function $y = f(x)$ is given. Compute $\Delta y = f(x + \Delta x) - f(x)$ and $\Delta y/\Delta x$ for the given values of $x$ and $\Delta x$.

**21** $y = f(x) = x^2 - 3x + 4$, $x = 1$, $\Delta x = 0.1$

**22** $y = f(x) = \dfrac{x - 3}{x + 3}$, $x = 2$, $\Delta x = -0.2$

**23** $y = f(x) = \dfrac{x^3 - 9}{x^2 + 1}$, $x = 0$, $\Delta x = 0.01$

In Problems 24 through 30, find $dy/dx$ and evaluate the derivative at the given value of $x$.

**24** $y = x^4 + 2x - 1$, $x = -2$

**25** $y = \dfrac{x^2 - 1}{x^2 + 1}$, $x = 1$

**26** $y = \sqrt[3]{x}$, $x = 8$

**27** $y = \dfrac{x}{x^2 + 1}$, $x = 2$

**28** $y = \sqrt{x^2 + x + 2}$, $x = 1$

**29** $y = \dfrac{x^3}{x + 2}$, $x = 2$

**30** $y = x^5 + 3$, $x = -1$

---

**5**

## INSTANTANEOUS VELOCITY AND SPEED: ACCELERATION

If the needle of the speedometer in a car is pointing to 60, we say that the car is traveling at 60 kilometers per hour\*. An analysis of this statement requires a definition of the phrase "60 km per hour." As a start, we might say that if the car were to continue in exactly the same manner, then in one hour it would have traveled exactly 60 kilometers. But such a statement avoids the problem completely, since the needle pointing to 60 does so without any knowledge of the way the car behaved some minutes ago, or how it intends to behave in the future.

Before any further discussion of this question we introduce two simplifying assumptions.

1) Motion will be assumed to take place along a *straight line*, although the object in motion may go in either direction. One direction will arbitrarily be selected as positive, and the other direction will then be negative.

2) The object in motion is idealized to be a point or particle. Such a simplification is necessary, since a mathematical treatment of the exact motion of every portion of a complicated object such as an automobile would defy analysis.

Suppose that the path of a particle is along a horizontal line $L$, with distance to the right designated as positive and that to the left as negative (Fig. 16). We let $t$ denote time in some convenient unit such as seconds or minutes, and we measure the distance $s$ of the object from the point $O$ in units such as centimeters, meters, or kilometers. If the particle in motion is at the point $s_1$ at time $t_1$ and at the point $s_2$ at time $t_2$, then it took $t_2 - t_1$ units of time to travel the distance $s_2 - s_1$. We define the **average velocity** of the particle over this time interval to be

$$\frac{s_2 - s_1}{t_2 - t_1}.$$

If $s$ is measured in meters and $t$ in seconds, the units of this average velocity would be denoted as **meters per second** (m/sec). If a car travels 40 kilometers

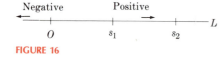

Negative          Positive

$O$         $s_1$         $s_2$

**FIGURE 16**

---

\*The metric system will be used throughout.

in one hour, the average velocity of the car is 40 kilometers per hour (km/hr). But we still know very little about how fast the car is traveling at various times during the hour. The needle of a speedometer pointing to 60 is an instantaneous event, and we need a notion of velocity at a given instant of time along the path. The development of this notion proceeds in the following way.

We mark off two points along a straight stretch of road, accurately measure the distance between them, and provide a timing device to indicate the exact moment the car passes each location. The average velocity is then computed. We believe that if the distance between the two points is small, the car's motion cannot change too radically as it travels from one point to the other; we therefore obtain a good indication of the velocity of the car at any time during this interval. For example, if the points are 100 meters apart and the average velocity is 60 km/hr, the variation in the motion along such a short stretch of road is probably not too great. If we desire greater accuracy, we can shorten the interval to ten meters, improve the measurement of the distance, and increase the precision of the timing device. Now we again argue that, in such a short distance, the motion of the car varies so slightly that the average velocity closely approximates the velocity indicated on the speedometer. On the other hand, some variation may still be possible, and so we cut the interval again, this time to one meter, say. The continuation of this process shows that instantaneous velocity is the result of computing the average velocity over a sequence of smaller and smaller intervals.

To make the above discussion precise, we first note that the distance $s$ traveled along the straight line is a function of the time $t$; that is, $s = f(t)$. Then the **average velocity** between the times $t_1$ and $t_2$ is simply expressed as

$$\frac{f(t_2) - f(t_1)}{t_2 - t_1} = \frac{\text{distance}}{\text{time}}.$$

**DEFINITION**    *The **instantaneous velocity** at time $t_1$ is defined to be*

$$\lim_{t_2 \to t_1} \frac{f(t_2) - f(t_1)}{t_2 - t_1}.$$

We see that the instantaneous velocity is the limit of the average velocities as the time over which these averages are taken tends to zero. Since $t_2$ must be different from $t_1$, we can write $t_2 = t_1 + \Delta t$, where $\Delta t$ may be *positive or negative*. The above expression for instantaneous velocity then becomes

$$\lim_{t_1 + \Delta t \to t_1} \frac{f(t_1 + \Delta t) - f(t_1)}{t_1 + \Delta t - t_1} \qquad \text{or} \qquad \lim_{\Delta t \to 0} \frac{f(t_1 + \Delta t) - f(t_1)}{\Delta t}.$$

We note that this is precisely the derivative of $f(t)$ evaluated at $t_1$. In other words,

> the instantaneous velocity of a particle moving in a straight line according to the law $s = f(t)$, describing the motion, is $f'(t)$.

We also use the symbol $\dfrac{ds}{dt}$ to describe this derivative. The quantity $ds/dt$ may be positive or negative, according to whether the particle is moving along the line in the positive or negative direction. The **speed** of the particle is defined to be $|ds/dt| = |f'(t)|$. The speed is merely the magnitude of the velocity and is always positive or zero. It may seem strange to introduce a special term specifically for the absolute value of the velocity. However, the notion of speed will be especially useful later when we study motion along curved paths. The speed tells us how fast the particle is moving but gives no information about its direction.

**EXAMPLE 1**   A projectile shot directly upward with a speed of 100 meters/sec moves according to the law

$$s = 100t - 5t^2,$$

where $s$ is the height in meters above the starting point, and $t$ is the time in seconds after it is thrown. Find the velocity of the projectile after 2 seconds. Is it still rising or is it falling? For how many seconds does it continue to rise? How high does the projectile go?

**Solution**   We assume that the motion of the projectile (particle) is in a straight vertical line. If we let $f(t) = 100t - 5t^2$, the velocity is given by $ds/dt$. Applying the three-step rule, we obtain

$$\frac{ds}{dt} = 100 - 10t.$$

Let $v$ denote the velocity at any time $t$; then

$$v = \frac{ds}{dt} = 100 - 10t.$$

If $t = 2$, then $v = 100 - 10(2) = 80$ m/sec. Since $v$ is positive, the projectile is still rising. The velocity is zero when $100 - 10t = 0$, or $t = 10$. When $t$ is larger than 10, the velocity is negative and the motion is downward, i.e., the projectile is falling. When $t = 10$, we see that $f(10) = 100(10) - 5(10)^2 = 500$ meters, and this is the highest point the projectile reaches. When $t = 20$ the particle has returned to its starting point. (See Fig. 17.)   □

In motion along a straight line, whenever a particle changes direction the velocity goes from positive to negative or from negative to positive. *Therefore at the instant the particle reverses direction the velocity must be zero.*

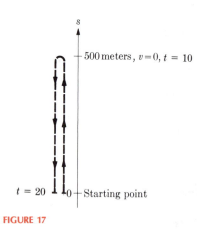

**FIGURE 17**

**EXAMPLE 2**   A particle moves along a horizontal line (positive to the right) according to the law

$$s = t^3 - 3t^2 - 9t + 5.$$

During which intervals of time is the particle moving to the right, and during which is it moving to the left?

**Solution**   It is moving to the right whenever the velocity is positive and to the left when the velocity is negative. The velocity $v$ is just the derivative of $f(t) = t^3 - 3t^2 - 9t + 5$. Applying the three-step rule, we have

$$v = \frac{ds}{dt} = 3t^2 - 6t - 9 = 3(t+1)(t-3).$$

If $t = -1, 3$, the velocity is zero. By examining the signs of the factors $t + 1$ and $t - 3$, we find that:

> If $t < -1$, $v$ is positive and the motion is to the right.

> If $-1 < t < 3$, $v$ is negative and the motion is to the left.

> If $t > 3$, $v$ is positive and the motion is to the right.

Motion to the left and motion to the right are separated by points of zero velocity, i.e., at $t = -1, 3$ (Fig. 18). Although the motion is along the line $L$, its schematic behavior is shown by the path above this line. Zero velocity occurs at points $P, Q$. The point $P$ corresponds to $t = -1$, $s = 10$, $v = 0$; $Q$ corresponds to $t = 3$, $s = -22$, $v = 0$.  □

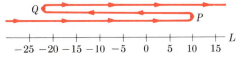

Showing the direction of motion of a particle moving
according to the law $s = t^3 - 3t^2 - 9t + 5$

**FIGURE 18**

*Acceleration is a measure of the change in velocity.* If a particle moves along a straight line with constant velocity, the acceleration is zero. In an automobile race, the cars pass the starting position traveling at a uniform velocity—say 30 km/hr. Within ten seconds one of the cars is traveling at 120 km/hr. The **average acceleration** of this car is

$$\frac{120 - 30}{10} = 9 \ (\text{km/hr})/\text{sec}.$$

The units appear rather strange, since velocity is in kilometers per hour and the time in the denominator is in seconds. We can convert 120 km/hr to $33\frac{1}{3}$ m/sec if we multiply 120 by 1000/3600. Similarly, 30 km/hr $= 8\frac{1}{3}$ m/sec. With this change the average acceleration is expressed as

$$\frac{33\frac{1}{3} - 8\frac{1}{3}}{10} = 2.5 \ (\text{m/sec})/\text{sec}.$$

**DEFINITION**   If the velocity $v$ of the particle is varying according to the law $v = F(t)$, where $t$ is the time, then the **instantaneous acceleration** $a$, or simply the **acceleration**, is defined as the limit of the average acceleration:

$$a = \lim_{h \to 0} \frac{F(t+h) - F(t)}{h} = F'(t).$$

This is in complete analogy with the definition of instantaneous velocity as the limit of the average velocity.

If a particle is moving (along a straight line) so that the distance $s$ is given by $s = f(t)$, then the velocity $v$ is found by taking the derivative $f'(t)$. The derivative $f'$ is the law which describes the velocity. The derivative of this derivative is the acceleration. We call it the **second derivative** and use the notation $f''(t)$. That is, if $s = f(t)$, then

$$v = f'(t) \qquad \text{and} \qquad a = \frac{dv}{dt} = f''(t).$$

We could continue this process and take third derivatives, fourth derivatives, etc., and indeed we sometimes do. The notation for third derivative is $f'''(t)$ or $f^{(3)}(t)$. [Beyond the third derivative, primes become unwieldy, and a numeral in parentheses is used. For example, the seventh derivative is denoted $f^{(7)}(t)$.] We should mention that from the point of view of physics, the first and second derivatives are the most important—especially in the study of the motion of objects. Derivatives higher than the second are of occasional interest only.

**EXAMPLE 3**    A particle moves along a straight line according to the law

$$s = 132 + 108t - 16t^2 + 3t^3,$$

$s$ being the distance in meters and $t$ the time in seconds. Find the velocity and acceleration at any time $t$. What is the velocity when $t = 2$? What is the acceleration when $t = 1$ and when $t = 3$?

**Solution**    The velocity is found by taking the derivative, which we do according to the three-step rule. Then

$$v = 108 - 32t + 9t^2.$$

The acceleration is obtained by taking the derivative of this function, which we also do by the three-step rule. We have

$$a = -32 + 18t.$$

When $t = 2$, we substitute in the equation for $v$ to get $v = 108 - 64 + 36 = 80$ m/sec. When $t = 1$, we see that $a = -32 + 18 = -14$ (m/sec)/sec, and when $t = 3$, $a = -32 + 54 = 22$ (m/sec)/sec.

Since the acceleration is negative when $t = 1$, we conclude that the particle is slowing down, while $a$ is positive when $t = 3$ and therefore the particle is speeding up.    □

## 5  PROBLEMS

In Problems 1 through 8, a particle is moving along a horizontal line (positive to the right), according to the stated law. Determine whether the particle is moving to the right or to the left at the given time.

**1**  $s = t^2 - 2t + 7, \quad t = 3$

**2**  $s = t^2 + t - 5, \quad t = 1$

**3**  $s = t^3 - 2t^2 + 7t - 6, \quad t = -1$

**4**  $s = t^3 - 5t^2 + 2t + 4, \quad t = 3$

**5**  $s = t^3 - 4t^2 - 2t - 1, \quad t = -2$

**6**  $s = t^4 + 2t - 1, \quad t = 1$

**7**  $s = \dfrac{1}{1 + t^2}, \quad t = 2$     **8**  $s = \dfrac{t - 1}{t + 1}, \quad t = 3$

In Problems 9 through 16, calculate the velocity and acceleration of the given laws of motion. Determine the times, if any, when the velocity is zero.

**9**  $s = t^2 - 4t + 2$     **10**  $s = 6 - 2t - 16t^2$

11  $s = \frac{1}{3}t^3 - 2t^2 + 3t - 5$     12  $s = t^3 + 2t^2 - t - 1$

13  $s = bt^2 + ct + d$   ($b, c, d$ constants)

14  $s = bt^3 + ct^2 + dt + e$   ($b, c, d, e$ constants)

15  $s = \dfrac{t^2 - 1}{t + 2}$     16  $s = \dfrac{t}{t^2 + 4}$

In Problems 17 through 26, motion is along a horizontal line, positive to the right. When is the particle moving to the right and when is it moving to the left? Discuss the acceleration.

17  $s = t^2 + 3t - 1$

18  $s = t^2 - 2t + 3$

19  $s = 8 - 4t + t^2$

20  $s = t^3 + 3t^2 - 9t + 4$

21  $s = 2t^3 - 3t^2 - 12t + 8$

22  $s = t^3 - 3t^2 + 3t$

23  $s = \dfrac{t}{1 + t^2}$     24  $s = \dfrac{1 + t}{4 + t^2}$

25  $s = \dfrac{t + 1}{t - 1}$     26  $s = \dfrac{t^2 + 4}{t^2 - 1}$

In Problems 27 and 28, find expressions for the velocity and acceleration.

27  $s = bt^4 + ct + d$   ($b, c, d$ constants)

28  $s = bt^4 + ct^2 + dt + e$   ($b, c, d, e$ constants)

29  A stone is thrown upward from the top of a building 50 meters high with a velocity of 15 meters/sec. The height $s$ at any time $t$ is given by the formula $s = 50 + 15t - 5t^2$. Find the maximum height the stone will reach and how long it takes to reach this height. When will the stone hit the ground?

30  A projectile hurled straight up from the earth travels according to the law $s = x_0 + v_0 t - 5t^2$, where $x_0$ is the height above the ground at the start of the motion, $v_0$ is the initial velocity, $t$ is the time traveled, and $s$ is the height at time $t$. Distance is measured in meters, time in seconds.

A man on a tower throws a ball upward. He observes that it passes him on the way down exactly 3 seconds later and that it hits the ground in one more second. How high is the tower? What is the initial velocity?

31  A projectile hurled upward from the surface of the moon follows the law $s = v_0 t - \frac{1}{2}gt^2$ where $v_0$ is the initial velocity and $g$ is the gravitational constant for the moon. A stone thrown up with an initial velocity of 20 meters/sec is observed hitting the surface 25 seconds later. What is the value of $g$, the gravitational constant?

---

## 6

## DEFINITION OF LIMIT (OPTIONAL)*

In Section 1 we introduced the notion of limit in an informal way. We spoke of intervals as being "small," numbers as being "close," quantities "approaching" zero, and so forth. However, these nonmathematical words vary widely in meaning from person to person and cannot be the basis for a mathematical structure. Therefore we give the following statement as a precise definition of limit:

DEFINITION    *Given a function $f$ and numbers $a$ and $L$, we say that $f(x)$ **tends to $L$ as a limit** as $x$ **tends to** $a$ if for each positive number $\varepsilon$ there is a positive number $\delta$ such that $f(x)$ is defined and*

$$|f(x) - L| < \varepsilon \qquad whenever \qquad 0 < |x - a| < \delta.$$

*In abbreviated notation, we write*

$$f(x) \to L \qquad as \qquad x \to a$$

*for this definition of limit.*

---

*This section may be omitted without loss of continuity. The problems at the end of this section are above average in difficulty.

We now elaborate on the meaning of the above definition. First of all, the definition implies that there can be at most one limit $L$. (We omit the proof of this fact.) Next we recall that

$$|x - a| < \delta$$

is the same as the two inequalities

$$a - \delta < x < a + \delta.$$

This double inequality states that $x$ must lie in an interval of length $2\delta$ having $a$ as its center (Fig. 19). In the definition, the part of the inequality which states that $0 < |x - a|$ merely means that $x$ is not allowed to be equal to $a$ itself. This is done for convenience. The inequality

$$|f(x) - L| < \varepsilon$$

is equivalent to

$$L - \varepsilon < f(x) < L + \varepsilon,$$

which asserts that the function $f$ lies above the line $y = L - \varepsilon$ and below the line $y = L + \varepsilon$ (Fig. 20).

The definition itself may be interpreted as a test. If I am given any positive number whatsoever (call it $\varepsilon$), the test consists of finding a number $\delta$ such that $f(x)$ lies between the values $L - \varepsilon$ and $L + \varepsilon$ if $x$ is in the interval $(a - \delta, a + \delta)$ and $x \neq a$. If such a $\delta$ can be found for *every* positive number $\varepsilon$, then we say that $f(x)$ has the limit $L$ as $x$ approaches $a$. Note that the value of delta will be different for different epsilons. Also the test must be performed for *every* positive epsilon, which means in general that it is an extremely difficult thing to check.

The geometric explanation, one similar to that given in Section 1, states that if an $\varepsilon$ is given, a $\delta$ can be found such that the graph of the function $f$ lies in the rectangle bounded by the lines $x = a - \delta, x = a + \delta, y = L - \varepsilon, y = L + \varepsilon$ (Fig. 21). Nothing at all is said about the value of $f$ when $x$ is $a$.

It is good to get some practice in finding the $\delta$ which corresponds to a given $\varepsilon$; the $\delta$ can actually be found in very simple cases. To consider an easy case, we let $f(x) = 3x - 2$ and take $a = 5$. We know intuitively from our earlier work that

$$\lim_{x \to 5} f(x) = 3(5) - 2 = 13.$$

We wish to show that, given an $\varepsilon$, we can find a $\delta$ such that

$$|3x - 2 - 13| < \varepsilon \qquad \text{whenever} \qquad 0 < |x - 5| < \delta.$$

But $|3x - 15| = |3(x - 5)|$. If someone gives us an $\varepsilon$, we simply take $\delta = \varepsilon/3$. Then, if $|x - 5| < \delta = \varepsilon/3$, we find (by multiplying through by 3) that $3|x - 5| < \varepsilon$, which is the same as $|3x - 15| < \varepsilon$, as is required.

FIGURE 19

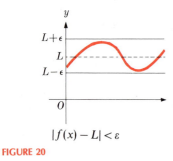

$|f(x) - L| < \varepsilon$

FIGURE 20

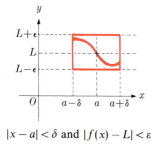

$|x - a| < \delta$ and $|f(x) - L| < \varepsilon$

FIGURE 21

**EXAMPLE 1**    Draw a graph of the function

$$f(x) = \frac{1}{x + 1}, \quad x \neq -1.$$

Find a $\delta$ such that $|f(x) - \tfrac{1}{2}| < 0.01$ if $0 < |x - 1| < \delta$.

**Solution**

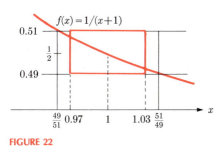

**FIGURE 22**

Part of the graph is sketched in Fig. 22. In the definition of limit we have $L = \frac{1}{2}$ and $a = 1$. We must find an interval of the $x$ axis about $x = 1$ such that the graph lies in the proper rectangle. The function decreases steadily as we go to the right and therefore, when we erect vertical lines where the lines $y = 0.51$, $y = 0.49$ intersect the curve, the largest possible interval on the $x$ axis is obtained. When we solve the following equation for $x$,

$$\frac{1}{x+1} = 0.51,$$

we get $x = \frac{49}{51}$, and similarly, solving $1/(x+1) = 0.49$ gives $x = \frac{51}{49}$. In Fig. 22 these values are shown with units greatly exaggerated. Fortunately, once we find a $\delta$, then any *smaller* $\delta$ will also be valid; for if the function lies in a rectangle, it certainly lies in a similar rectangle which is of the same height but narrower. So we may take $\delta = 0.03$, since $\frac{49}{51} < 0.97$ and $\frac{51}{49} > 1.03$. ☐

**EXAMPLE 2**    Draw a graph of

$$f(x) = \frac{2(x-4)}{\sqrt{x}-2}, \quad x \geq 0, \quad x \neq 4,$$

and find a $\delta$ such that

$$|f(x) - 8| < 0.01 \qquad \text{whenever} \qquad 0 < |x-4| < \delta.$$

**Solution**    The function $f$ is not defined at $x = 4$, but for $x \neq 4$ we can multiply numerator and denominator by $\sqrt{x} + 2$ to obtain

$$f(x) = \frac{2(x-4)(\sqrt{x}+2)}{(x-4)} = 2\sqrt{x} + 4 \quad \text{for } x \neq 4.$$

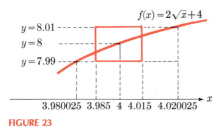

**FIGURE 23**

The graph of this function is shown in Fig. 23. We construct the lines $y = 7.99$ and $y = 8.01$, shown with greatly enlarged units in the figure. The intersections of these lines with the function are found by solving the equations

$$2\sqrt{x} + 4 = 7.99 \qquad \text{and} \qquad 2\sqrt{x} + 4 = 8.01.$$

We get $x = 3.980025$ and $x = 4.020025$. Since the function steadily increases to the right, an adequate selection for $\delta$ is 0.015. In other words, it is true that

$$|f(x) - 8| < 0.01 \qquad \text{whenever} \qquad 0 < |x-4| < 0.015. \qquad ☐$$

*Remark.* To establish the existence of a limit we must find a $\delta$ for each positive $\varepsilon$. It is not necessary that we find the largest possible $\delta$ and, if work can be avoided by selecting a $\delta$ smaller than the largest possible one, we usually do so.

**EXAMPLE 3**    Show directly from the definition that

$$\lim_{x \to 1} \frac{x}{x+1} = \frac{1}{2},$$

and draw the graph of the function $f(x) = x/(x+1)$.

**Solution**    We have $L = \frac{1}{2}$ and $a = 1$. We must show that for every $\varepsilon > 0$ we can find a $\delta > 0$ such that

$$\left| \frac{x}{x+1} - \frac{1}{2} \right| < \varepsilon \qquad \text{whenever} \qquad 0 < |x - 1| < \delta.$$

In order to get an idea of the appearance of the function we sketch the graph (Fig. 24). From the graph we see that the function is a steadily increasing one. We verify this fact by writing the identity

$$\frac{x}{x+1} = 1 - \frac{1}{x+1},$$

and noting that as $x$ gets larger, $1/(x+1)$ gets smaller and, as a result, $1 - [1/(x+1)]$ increases.

Let us first suppose that $\varepsilon < \frac{1}{2}$. Then $L + \varepsilon < 1$ and $L - \varepsilon > 0$, since $L = \frac{1}{2}$. Next, to see where the lines $y = \frac{1}{2} - \varepsilon$ and $y = \frac{1}{2} + \varepsilon$ intersect the curve, we solve the equations

$$\frac{x}{x+1} = \frac{1}{2} - \varepsilon \qquad \text{and} \qquad \frac{x}{x+1} = \frac{1}{2} + \varepsilon.$$

The first equation gives

$$x = (\tfrac{1}{2} - \varepsilon)(x+1) \quad \Leftrightarrow \quad (\tfrac{1}{2} + \varepsilon)x = \tfrac{1}{2} - \varepsilon \quad \Leftrightarrow \quad x = \frac{\tfrac{1}{2} - \varepsilon}{\tfrac{1}{2} + \varepsilon} \equiv x_1.$$

Similarly, the second equation yields

$$x = \frac{\tfrac{1}{2} + \varepsilon}{\tfrac{1}{2} - \varepsilon} \equiv x_2.$$

We select as $\delta$ the smaller of the distances between 1 and $x_1$ and between 1 and $x_2$. The reader can check the fact that $1 - x_1$ is smaller than $x_2 - 1$. Therefore

$$\delta = 1 - x_1 = 1 - \frac{\tfrac{1}{2} - \varepsilon}{\tfrac{1}{2} + \varepsilon} = \frac{2\varepsilon}{\tfrac{1}{2} + \varepsilon} = \frac{4\varepsilon}{1 + 2\varepsilon}. \qquad \square$$

*Remark.* We restricted $\varepsilon$ to a value smaller than $\frac{1}{2}$ and then found a $\delta$ for every positive $\varepsilon < \frac{1}{2}$. While it is true that the basic definition states that a $\delta$ has to be found for *every* $\varepsilon$, in actuality this is not so. Once we have found a $\delta$ for a specific $\varepsilon$, we can use the same $\delta$ for *all* larger $\varepsilon$. Geometrically, this means that once the function is known to lie in a particular rectangle, clearly it lies in every rectangle which has the same sides and contains the particular rectangle.

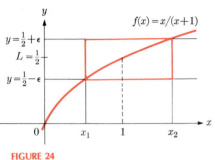

$y = \frac{1}{2} + \epsilon$

$L = \frac{1}{2}$

$y = \frac{1}{2} - \epsilon$

$f(x) = x/(x+1)$

$x_1 \qquad 1 \qquad x_2$

**FIGURE 24**

## 6  PROBLEMS

In Problems 1 through 17, the numbers $a$, $L$, and $\varepsilon$ are given. Determine a number $\delta$ so that $|f(x) - L| < \varepsilon$ for all $x$ such that $0 < |x - a| < \delta$. Draw a graph.

**1** $f(x) = 2x + 3$, $a = 1$, $L = 5$, $\varepsilon = 0.001$

**2** $f(x) = 1 - 2x$, $a = -1$, $L = 3$, $\varepsilon = 0.01$

**3** $f(x) = (x^2 - 9)/(x + 3)$, $a = -3$, $L = -6$, $\varepsilon = 0.005$

**4** $f(x) = \sqrt{x}$, $a = 1$, $L = 1$, $\varepsilon = 0.01$

**5** $f(x) = \sqrt[3]{x}$, $a = 1$, $L = 1$, $\varepsilon = 0.01$

**6** $f(x) = \sqrt[3]{x}$, $a = 0$, $L = 0$, $\varepsilon = 0.1$

7 $f(x) = \sqrt{2x}$,   $a = 2, L = 2, \varepsilon = 0.02$

8 $f(x) = 1/x$,   $a = 2, L = \frac{1}{2}, \varepsilon = 0.002$

9 $f(x) = 2/\sqrt{x}$,   $a = 4, L = 1, \varepsilon = 0.1$

10 $f(x) = 3/(x + 2)$,   $a = 1, L = 1, \varepsilon = 0.001$

11 $f(x) = 1/x$,   $a = -1, L = -1, \varepsilon = 0.01$

12 $f(x) = (x - 1)/(x + 1)$,   $a = 0, L = -1, \varepsilon = 0.01$

13 $f(x) = (\sqrt{x} - 1)/(x - 1)$,   $a = 1, L = \frac{1}{2}, \varepsilon = 0.01$

14 $f(x) = (\sqrt{2x} - 2)/(x - 2)$,   $a = 2, L = \frac{1}{2}, \varepsilon = 0.01$

15 $f(x) = x^2$,   $a = 1, L = 1, \varepsilon = 0.01$

16 $f(x) = x^3 - 6$,   $a = 1, L = -5, \varepsilon = 0.1$

*17 $f(x) = x^3 + 3x$,   $a = -1, L = -4, \varepsilon = 0.5$

In Problems 18 through 25, show that

$$\lim_{x \to a} f(x) = L$$

directly by finding the $\delta$ corresponding to every positive $\varepsilon$. (Use the method of Example 3.)

18 $f(x) = x$,   $a = 5, L = 5$          19 $f(x) = 5$,   $a = 7, L = 5$

20 $f(x) = 2x$,   $a = \pi/2, L = \pi$

21 $f(x) = (x^2 - 4)/(x - 2)$,   $a = 2, L = 4$

22 $f(x) = \sqrt{x}$,   $a = 2, L = \sqrt{2}$

23 $f(x) = \sqrt[3]{x}$,   $a = 3, L = \sqrt[3]{3}$

24 $f(x) = 1/(x + 1)$,   $a = 2, L = \frac{1}{3}$

25 $f(x) = 1/(x + 2)$,   $a = -3, L = -1$

_____ 7 _____

## THEOREMS ON LIMITS

In Sections 1 through 5 we discussed limits without any attempt at rigorous or exact mathematical statements. In the process of doing this, we performed all sorts of algebraic manipulations. The skeptical reader realizes that each of these needs justification, even though on the surface many appear obvious. The first step in such a justification requires the precise definition of limit which we have just given in Section 6. The next step should be the statements and the proofs of the theorems which allow us to manipulate limits. However, we shall restrict ourselves to statements of the theorems only, since many of the proofs are beyond the scope of a course in elementary calculus. The statements themselves will help the reader understand the kind of work that has to be done.

The first theorem we state asserts that a function cannot approach more than one limit. While this fact appears fairly obvious the proof, which we skip, requires the rigorous definition of limit we gave in Section 6.

**THEOREM 1**    **(Uniqueness of Limits)**    *Suppose that $f(x) \to L_1$ as $x \to a$, and $f(x) \to L_2$ as $x \to a$. Then $L_1 = L_2$.*

As an illustration of Theorem 1, suppose that

$$f(x) = \frac{x}{x^2 + 1} \qquad \text{and} \qquad a = 2.$$

Then we already know from our intuitive discussion that

$$\lim_{x \to 2} f(x) = \frac{2}{4 + 1} = \frac{2}{5}.$$

Theorem 1 tells us that $\frac{2}{5}$ is the *only* number that $f(x)$ can approach as $x$ tends to 2.

---

**THEOREM 2**  **(Limit of a Constant)**  *If $c$ is a constant and $f(x) = c$ for all values of $x$, then for any number $a$*

$$\lim_{x \to a} f(x) = c.$$

---

Theorem 2 is established by applying the definition of limit to the particular function $f(x) = c$. Geometrically, the function $f(x) = c$ represents a line parallel to the $x$ axis and $c$ units from it. For example, if $c = -4$, then Theorem 2 states that

$$\lim_{x \to a} (-4) = -4,$$

and this limit holds for *every* number $a$. We have

$$\lim_{x \to 3} (-4) = -4, \qquad \lim_{x \to 8} (-4) = -4,$$

and so on.

---

**THEOREM 3**  **(Obvious Limit)**  *If $a$ is a real number and $f(x) = x$ for all $x$, then*

$$\lim_{x \to a} f(x) = a.$$

---

Theorem 3 says that for the special function $f(x) = x$, the limit is always obtained by *direct substitution* of the value $a$ for $x$. In other words,

$$\lim_{x \to 2} x = 2, \qquad \lim_{x \to -3} x = -3, \qquad \lim_{x \to 5} x = 5,$$

and so on.

---

**THEOREM 4**  **(Limit of Equal Functions)**  *Suppose there is a number $h > 0$ such that $f(x) = g(x)$ for all $x$ for which $0 < |x - a| < h$. Suppose also that*

$$\lim_{x \to a} g(x) = L.$$

*Then*

$$\lim_{x \to a} f(x) = L.$$

---

This theorem is useful whenever the limit of $f(x)$ cannot be found by direct substitution but where there is a "simplified function" $g$ which is identical with $f$ except at the point $a$ and is such that the limit of $g$ as $x \to a$ is found easily. For example, suppose

$$f(x) = \frac{(x + 2)(x - 2)}{x - 2}, \quad g(x) = x + 2.$$

We see that $f(x)$ and $g(x)$ are identical except at $x = 2$, where $f$ is not defined and $g$ has the value 4. Then Theorem 4 says that since

THEOREMS ON LIMITS

$$\lim_{x \to 2} g(x) = 4, \quad \text{then} \quad \lim_{x \to 2} f(x) = 4.$$

Another illustration is given in Example 2 of Section 1. The functions $f$ and $g$ are

$$f(x) = \frac{x - 4}{3(\sqrt{x} - 2)} \quad \text{and} \quad g(x) = \frac{\sqrt{x} + 2}{3}.$$

By rationalizing the denominator we see that these functions are identical except when $x = 4$. Therefore

$$\lim_{x \to 4} f(x) = \lim_{x \to 4} g(x) = \tfrac{4}{3}.$$

---

**THEOREM 5**    **(The Limit of a Sum Is the Sum of the Limits)**   *If $f$ and $g$ are two functions, with*

$$\lim_{x \to a} f(x) = L_1 \quad \text{and} \quad \lim_{x \to a} g(x) = L_2,$$

*then*

$$\lim_{x \to a} (f(x) + g(x)) = L_1 + L_2.$$

---

The hypothesis states that $|f(x) - L_1|$ can be made "small" if $x$ is "close to" $a$; the same is true about $|g(x) - L_2|$. The conclusion asserts that $|f(x) + g(x) - L_1 - L_2|$ can be made "small" if $x$ is "close to" $a$.

Once Theorem 5 is established, we can use it over and over to add the limits of any number of functions. For example, if

$$\lim_{x \to a} f(x) = L_1, \quad \lim_{x \to a} g(x) = L_2, \quad \text{and} \quad \lim_{x \to a} h(x) = L_3,$$

then

$$\lim_{x \to a} (f(x) + g(x) + h(x)) = L_1 + L_2 + L_3.$$

To demonstrate this, we first apply Theorem 5 to $f$ and $g$, designating $f(x) + g(x) = F(x)$. Then we apply Theorem 5 again to $F(x)$ and $h(x)$. This technique of combining and using the same theorem over and over occurs frequently in the study of limits. In fact, we may use Theorem 5 any number of times to get the following general result.

---

**COROLLARY**    *The limit of the sum of any ( finite) number of functions is the sum of the limits of each of the functions.*

---

**THEOREM 6**    **(The Limit of a Product Is the Product of the Limits)**   *If $f$ and $g$ are two functions, with*

$$\lim_{x \to a} f(x) = L_1 \quad \text{and} \quad \lim_{x \to a} g(x) = L_2,$$

*then*

$$\lim_{x \to a} [f(x) \cdot g(x)] = L_1 \cdot L_2.$$

In analogy with the discussion following Theorem 5, we note the following corollary.

**COROLLARY**  *The limit of the product of any finite number of functions is the product of the limits.*

**EXAMPLE 1**  Given

$$\lim_{x \to a} f(x) = L_1, \qquad \lim_{x \to a} g(x) = L_2, \qquad \text{and} \qquad \lim_{x \to a} h(x) = L_3,$$

find the value of

$$\lim_{x \to a} [f(x) \cdot g(x) + h(x)].$$

**Solution**  We define $F(x) = f(x) \cdot g(x)$. From Theorem 6 we know that

$$\lim_{x \to a} f(x) \cdot g(x) = L_1 \cdot L_2.$$

That is,

$$\lim_{x \to a} F(x) = L_1 \cdot L_2.$$

Applying Theorem 5 to $F(x) + h(x)$, we now find that

$$\lim_{x \to a} (F(x) + h(x)) = L_1 \cdot L_2 + L_3. \qquad \square$$

**THEOREM 7**  **(The Limit of a Quotient Is the Quotient of the Limits)**  *If f and g are two functions, with*

$$\lim_{x \to a} f(x) = L_1, \qquad \lim_{x \to a} g(x) = L_2, \qquad \text{and} \qquad L_2 \neq 0,$$

*then*

$$\lim_{x \to a} \frac{f(x)}{g(x)} = \frac{L_1}{L_2}.$$

It is necessary to assume that $L_2 \neq 0$ if the expression $L_1/L_2$ is to have a meaning.

**EXAMPLE 2**  Given $F(x) = x^2$, show that

$$\lim_{x \to a} F(x) = a^2.$$

**Solution**  It is possible to obtain this result by appealing to the definition of limit. However, we can also prove the statement simply by correct application of the theorems on limits. Using Theorem 3, we have

$$\lim_{x \to a} x = a.$$

Let $f(x) = x$ and $g(x) = x$. Then

$$\lim_{x \to a} f(x) = a \qquad \text{and} \qquad \lim_{x \to a} g(x) = a.$$

We can apply Theorem 6, with $L_1 = a$, $L_2 = a$, to get

$$\lim_{x \to a} f(x) \cdot g(x) = a \cdot a = a^2,$$

which says that

$$\lim_{x \to a} F(x) = a^2. \qquad \square$$

**EXAMPLE 3**    Given $F(x) = x^2/(3x - 2)$, show that

$$\lim_{x \to a} \frac{x^2}{3x - 2} = \frac{a^2}{3a - 2} \quad \text{if } a \neq \tfrac{2}{3}.$$

**Solution**    From Example 2 we know that

$$\lim_{x \to a} x^2 = a^2,$$

and from Theorems 2 and 3, we know that

$$\lim_{x \to a} 3 = 3, \qquad \lim_{x \to a} (-2) = -2, \qquad \lim_{x \to a} x = a.$$

From Theorem 6 with $f(x) = 3$ and $g(x) = x$, we see that

$$\lim_{x \to a} 3x = 3a.$$

An application of Theorems 5 and 2 yields

$$\lim_{x \to a} (3x - 2) = \lim_{x \to a} [3x + (-2)] = 3a + (-2) = 3a - 2.$$

Finally, using Theorem 7 with $f(x) = x^2$, $g(x) = 3x - 2$, $L_1 = a^2$, and $L_2 = 3a - 2$, we conclude that

$$\lim_{x \to a} \frac{x^2}{3x - 2} = \frac{a^2}{3a - 2}. \qquad \square$$

**THEOREM 8**    **(The Sandwich Theorem)**    *Suppose that* $f(x), g(x), h(x)$ *have the property that*

$$f(x) \le g(x) \le h(x) \quad \text{for } 0 < |x - a| < \delta.$$

*Suppose also that*

$$\lim_{x \to a} f(x) = L, \qquad \lim_{x \to a} h(x) = L.$$

*Then*

$$\lim_{x \to a} g(x) = L.$$

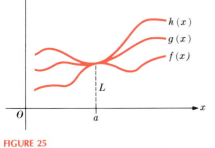

*y*

$h(x)$

$g(x)$

$f(x)$

$L$

$0$    $a$    *x*

FIGURE 25

The situation is shown clearly in Fig. 25. The function $g(x)$ is squeezed between $f$ and $h$ and so must tend to $L$ as $x$ tends to $a$. We provide an application of Theorem 8 in Example 6 below.

In working the problems at the end of this section the reader should give the reason for each step. An abbreviation for the *name*, rather than the number of the theorem being used, helps in the development of a clear understanding of the processes. We illustrate this procedure in the next three examples.

**EXAMPLE 4**    Find the value of

$$\lim_{x \to 2} \frac{x^3 + 3}{2x^2 + 5}$$

and justify each step.

**Solution**    **Step 1:**    $\lim_{x \to 2} 2 = 2,$    $\lim_{x \to 2} 3 = 3,$    $\lim_{x \to 2} 5 = 5$    (lim. const.)

**Step 2:**    $\lim_{x \to 2} x = 2$    (obv. lim.)

**Step 3:**    $\lim_{x \to 2} 2x^2 = \lim_{x \to 2} (2 \cdot x \cdot x) = 2 \cdot 2 \cdot 2 = 8,$

$\lim_{x \to 2} x^3 = \lim_{x \to 2} (x \cdot x \cdot x) = 8$    (lim. prod.)

**Step 4:**    $\lim_{x \to 2} (x^3 + 3) = 8 + 3 = 11,$

$\lim_{x \to 2} (2x^2 + 5) = 8 + 5 = 13$    (lim. sum)

**Step 5:**    $\lim_{x \to 2} \frac{x^3 + 3}{2x^2 + 5} = \frac{11}{13}$    (lim. quot.)

□

---

**THEOREM 9**    **(Limit of a Composite Function)**    *Suppose f and g are functions, a and b are numbers, and*

$$\lim_{x \to b} f(x) = f(b) \qquad and \qquad \lim_{x \to a} g(x) = b.$$

*Then*

$$\lim_{x \to a} f[g(x)] = f(b).$$

---

**THEOREM 10**    *If n is a positive integer and a > 0, then*

$$\lim_{x \to a} \sqrt[n]{x} = \sqrt[n]{a}.$$

The next result can be obtained by combining Theorems 9 and 10.

---

**THEOREM 11**    *If n is a positive integer, L > 0, and*

$$\lim_{x \to a} f(x) = L, \qquad then \qquad \lim_{x \to a} \sqrt[n]{f(x)} = \sqrt[n]{L}.$$

---

**EXAMPLE 5**    Evaluate

$$\lim_{x \to 2} \sqrt{\frac{x^4 - 16}{x^3 - 8}}.$$

**Solution**   Straight substitution shows that the function is undefined when $x = 2$. If we denote the quantity under the radical by $f(x)$, we find by factoring that for $x \neq 2$:

$$f(x) = \frac{(x-2)(x^3 + 2x^2 + 4x + 8)}{(x-2)(x^2 + 2x + 4)}.$$

Setting

$$g(x) = \frac{x^3 + 2x^2 + 4x + 8}{x^2 + 2x + 4},$$

we observe that $g(x)$ is defined for all values of $x$ and $f(x) = g(x)$ except when $x = 2$.

**Step 1:**  $\lim_{x \to 2} 2 = 2,$     $\lim_{x \to 2} 4 = 4,$     $\lim_{x \to 2} 8 = 8$                    (lim. const.)

**Step 2:**  $\lim_{x \to 2} x = 2$                                                                    (obv. lim.)

**Step 3:**  $\lim_{x \to 2} x^3 = 8,$     $\lim_{x \to 2} 2x^2 = 8,$     $\lim_{x \to 2} 4x = 8,$

$\lim_{x \to 2} x^2 = 4,$     $\lim_{x \to 2} 2x = 4$                                      (lim. prod.)

**Step 4:**  $\lim_{x \to 2} (x^3 + 2x^2 + 4x + 8) = 32,$

$\lim_{x \to 2} (x^2 + 2x + 4) = 12$                                          (lim. sum)

**Step 5:**  $\lim_{x \to 2} \dfrac{x^3 + 2x^2 + 4x + 8}{x^2 + 2x + 4} = \dfrac{32}{12}$                    (lim. quot.)

**Step 6:**  $\lim_{x \to 2} g(x) = \lim_{x \to 2} f(x) = \dfrac{32}{12} = \dfrac{8}{3}$                    (lim. equal fcts.)

**Step 7:**  $\lim_{x \to 2} \sqrt{\dfrac{x^4 - 16}{x^3 - 8}} = \sqrt{\dfrac{8}{3}}$                    (lim. $\sqrt[n]{f(x)}$ )

□

**EXAMPLE 6**   Let

$$g(x) = \left| x \sin \frac{1}{x} \right|$$

and show that

$$\lim_{x \to 0} g(x) = 0.$$

**Solution**   Straight substitution shows that $g(x)$ is undefined when $x = 0$. Since $g(x) \geq 0$ always, and since

$$\left| \sin \frac{1}{x} \right| \leq 1$$

for all $x \neq 0$, if we define the functions $f(x)$ and $h(x)$ as $f(x) = 0$ for all $x$ and $h(x) = |x|$ for all $x$, we have

$$f(x) \leq g(x) \leq h(x) \quad \text{for all } x \neq 0.$$

We then proceed:

**Step 1:** $\lim\limits_{x \to 0} x = 0$                                                    (obv. lim.)

**Step 2:** $\lim\limits_{x \to 0} h(x) = \lim\limits_{x \to 0} |x| = 0$                      (lim. comp. function)

**Step 3:** $\lim\limits_{x \to 0} f(x) = 0$                                                  (lim. const.)

**Step 4:** $\lim\limits_{x \to 0} g(x) = 0$                                                  (Sandwich Theorem)

$\square$

It is important to know the eleven theorems of this section and have a clear understanding of how they are used. However, as a practical approach to the evaluation of limits of algebraic expressions, straight substitution always works *except when there is a zero in the denominator*. In such cases either there is no limit or, if the numerator is also zero, Theorems 4 and 8 can sometimes be used. In Chapter 10 we will study additional techniques for evaluating limits.

## 7   PROBLEMS

In Problems 1 through 20, evaluate the limits by following the methods given in Examples 4 and 5. Give the reason for each step as in those examples.

**1** $\lim\limits_{x \to 2} (x^2 + 4x - 3)$

**2** $\lim\limits_{x \to 3} (x^3 + 2x^2 - 7x + 4)$

**3** $\lim\limits_{x \to -1} \dfrac{2x + 1}{x^2 + 3x + 4}$

**4** $\lim\limits_{x \to 1} \dfrac{x^2 + 6x - 5}{x^3 + 2x + 7}$

**5** $\lim\limits_{x \to 2} \dfrac{x^3 - 8}{x - 2}$

**6** $\lim\limits_{x \to -2} \dfrac{x^3 + 8}{x + 2}$

**7** $\lim\limits_{x \to -2} \dfrac{x^2 + x - 2}{x^2 - 4}$

**8** $\lim\limits_{x \to -2} \dfrac{x^2 - 4}{x^3 + 8}$

**9** $\lim\limits_{t \to 2} \sqrt{\dfrac{2t + 5}{3t - 2}}$

**10** $\lim\limits_{r \to 1} \sqrt{\dfrac{2r^2 + 3r - 1}{r^2 + 1}}$

**11** $\lim\limits_{y \to 2} \sqrt{\dfrac{y^2 - 4}{y^2 - 3y + 2}}$

**12** $\lim\limits_{x \to 3} \sqrt[3]{\dfrac{x^3 - 27}{x^2 - 2x - 3}}$

**13** $\lim\limits_{h \to 2} \sqrt{\dfrac{h^3 - 8}{h^2 - 4}}$

**14** $\lim\limits_{h \to 0} \dfrac{\sqrt{1 + h} - 1}{h}$

**15** $\lim\limits_{h \to 0} \dfrac{\sqrt{x + h} - \sqrt{x}}{h}, \quad x > 0$

**16** $\lim\limits_{h \to 0} \dfrac{1}{h}\left( \dfrac{1}{x + h} - \dfrac{1}{x} \right)$

**17** $\lim\limits_{h \to 0} \dfrac{1}{h}\left( \dfrac{1}{\sqrt{1 + h}} - 1 \right)$

**18** $\lim\limits_{h \to 0} \dfrac{(1 + h)^{3/2} - 1}{h}$

**19** $\lim\limits_{x \to 2} \dfrac{x^5 - 32}{x - 2}$

**20** $\lim\limits_{x \to 3} \dfrac{x^6 - 729}{x + 3}$

In each of Problems 21 and 22, use the Sandwich Theorem to evaluate the limit.

**21** $\lim\limits_{x \to 0} \dfrac{|x|}{\sqrt{x^4 + 3x^2 + 6}}$

**22** $\lim\limits_{x \to 0} x^2 \left( \sin \dfrac{1}{x} \right)^2$

***23** Prove Theorem 1 by employing the definition of limit.

***24** Prove Theorem 2 by employing the definition of limit.

***25** Prove Theorem 3 by employing the definition of limit.

**\*26** Show that

$$\lim_{x \to 0} x \sin \frac{1}{x} = 0$$

by using the definition of limit in Section 6.

**27** Given

$$\lim_{x \to a} f(x) = L_1, \qquad \lim_{x \to a} g(x) = L_2,$$

$$\lim_{x \to a} h(x) = L_3, \qquad \lim_{x \to a} p(x) = L_4,$$

state the theorems which justify the following statements:

a) $\lim_{x \to a} \dfrac{f(x) + g(x)}{h(x)} = \dfrac{L_1 + L_2}{L_3}, \quad$ if $L_3 \neq 0$.

b) $\lim_{x \to a} \dfrac{f(x)g(x) - h(x)}{g(x) + p(x)} = \dfrac{L_1 L_2 - L_3}{L_2 + L_4}, \quad$ if $L_2 + L_4 \neq 0$.

c) $\lim_{x \to a} \left( \dfrac{f(x) - g(x) + h(x)p(x)}{g(x) + h(x)} \right)^{2/3}$

$$= \left( \dfrac{L_1 - L_2 + L_3 L_4}{L_2 + L_3} \right)^{2/3}, \quad \text{if } L_2 + L_3 \neq 0.$$

**28** Use the definition of limit to show that if $a$ is any number then

$$\lim_{x \to a} x^2 = a^2.$$

**29** Let $P(x) = a_n x^n + a_{n-1} x^{n-1} + \cdots + a_1 x + a_0$ be any polynomial. Show that

$$\lim_{x \to b} P(x) = P(b).$$

**30** a) Suppose that $\lim_{x \to a} f(x) = L$. Show that $\lim_{x \to a} |f(x)| = |L|$.

b) Suppose that $\lim_{x \to a} |f(x)| = |L|$. Show that it is not necessarily true that $\lim_{x \to a} f(x) = L$. Give an example.

---

**8**

# CONTINUITY. ONE-SIDED LIMITS

In Section 6 we analyzed the meaning of

$$\lim_{x \to a} f(x) = L.$$

In doing so we made a point of ignoring the actual value of the function $f$ when $x = a$. In fact, for many expressions the function was not even defined at $x = a$. Suppose we have a function $f$ which is defined at $x = a$, and suppose the limit $L$, which $f$ approaches when $x$ tends to $a$, is the value of $f$ when $x$ is $a$—just the quantity we call $f(a)$. When this happens, we say that the function is **continuous** at $x = a$.

DEFINITIONS    **"The function $f$ is continuous at the number $a$"** *means that*

> i) $f(a)$ *is defined, and*
>
> ii) $\lim_{x \to a} f(x) = f(a)$.

*When a function is not continuous at $a$, we say that it is* **discontinuous at** $a$. *We say that a function is* **continuous** *if it is continuous at each point of its domain.*

Most, but not all, of the functions we have been studying are continuous everywhere. For example, the functions

$$f(x) = 3x - 2, \qquad g(x) = 5 - 2x + 4x^2, \qquad \text{and} \qquad h(x) = \frac{3x + 1}{2 + x^2}$$

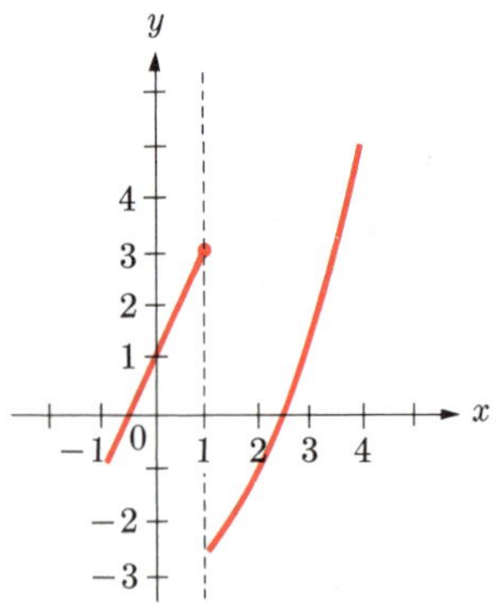

**FIGURE 26**

are defined for all values of $x$ and are continuous for all values of $x$. On the other hand, we shall show that the function

$$F(x) = \begin{cases} 2x + 1, & -\infty < x \le 1 \\ \frac{1}{2}x^2 - 3, & 1 < x < \infty \end{cases}$$

is discontinuous at $x = 1$. The graph of part of $F$ is shown in Fig. 26. We define the functions

$$G(x) = 2x + 1, \qquad H(x) = \tfrac{1}{2}x^2 - 3,$$

in each case the domain being all of $R^1$; that is, $G$ and $H$ are given by the above formulas for all values of $x$. Now we apply Theorem 4 on the limit of equal functions. We observe that $F$ and $G$ are identical for all values of $x$ in the interval $-\infty < x < 1$; therefore the limit as $x$ tends to 1 from the left (that is, as $x$ tends to 1 through values less than 1) of the functions $F$ and $G$ must be the same. Since $G$ is continuous everywhere, we see that $F(1) = 2 \cdot 1 + 1 = 3$ is equal to the limit of $G(x)$ as $x$ tends to 1 from the left. We employ a similar argument for $F$ and $H$; these functions are identical for $1 < x < \infty$. Also, $H$ is continuous for all values of $x$. Therefore the limit as $x$ tends to 1 from the right (that is, as $x$ tends to 1 through values larger than 1) of the functions $F$ and $H$ must be the same. We have $H(1) = -\frac{5}{2}$ and so $F(x)$ tends to $-\frac{5}{2}$ as $x$ tends to 1 through values larger than 1. These statements are intuitively clear in the graph shown in Fig. 26. The definition of continuity for $F$ is not satisfied at $x = 1$; this assertion is justified in Theorem 12 on page 85.

In the definition of limit, when we state that

$$f(x) \to L \qquad \text{as} \qquad x \to a,$$

we must consider values of $x$ larger than $a$ and values smaller than $a$ as $x$ tends to $a$. However, in the above discussion on the discontinuity of $F$, we saw that it is important to consider the limit of $F$ as the number 1 is approached through values entirely on one side of 1. This leads to the following symbol for "one-sided limit." *When $f(x)$ tends to $L$ as $x$ tends to $a$ through values larger than $a$, we write*

$$f(x) \to L \qquad \text{as} \qquad x \to a^+.$$

The plus sign after the letter $a$ indicates that $x$ tends to $a$ through values larger than $a$. We also say that $L$ is the **one-sided limit from the right of** $f$.

The definition of **one-sided limit from the left** is entirely analogous to that above, with the change that $x$ is smaller than $a$ instead of larger than $a$. We use the notation

$$f(x) \to L \qquad \text{as} \qquad x \to a^-,$$

the minus sign after the letter $a$ indicating that $x$ tends to $a$ through values smaller than $a$. Additional notations for one-sided limits are

$$\lim_{x \to a^+} f(x) = L, \qquad \lim_{x \to a^-} f(x) = L.$$

All the theorems on limits stated in Section 7 carry over to one-sided limits with the exception of Theorem 9 on composite functions. In that theorem we must assume that the limit of $f$ exists, although one-sided limits of $g$ are allowed. In addition, we have the following extension of Theorem 10.

**THEOREM 10′**    *If $n$ is a positive integer, then*

$$\lim_{x \to 0^+} \sqrt[n]{x} = 0.$$

We note that the restriction to one-sided limits is necessary in Theorem 10′ since $\sqrt[n]{x}$ is not defined for negative values of $x$ when $n$ is an even integer. An extended theorem similar to Theorem 11 holds for

$$\lim_{x \to a^+} \sqrt[n]{f(x)} \qquad \text{if} \qquad f(x) \to 0^+ \qquad \text{as} \qquad x \to a^+.$$

The next theorem is an immediate consequence of the definitions of one-sided limits.

**THEOREM 12**    *If $f$ is a function and $a$ and $L$ are numbers, then*

$$\lim_{x \to a} f(x) = L$$

*if and only if*

$$\lim_{x \to a^+} f(x) = L \qquad \text{and} \qquad \lim_{x \to a^-} f(x) = L.$$

Theorem 12 asserts that the limit of $f$ exists if and only if *both* one-sided limits exist *and they have the same value.* For this reason, we sometimes say that the ordinary limit is the **two-sided limit**.

Using Theorem 12, we can justify the conclusion that the function $F$ given on page 84 is not continuous at $x = 1$. Both one-sided limits of $F$ exist, but they have different values. Therefore the two-sided limit (ordinary limit) does not exist and the function cannot be continuous.

**EXAMPLE 1**    Find the value of the one-sided limit

$$\lim_{x \to 3^-} \frac{9 - x^2}{\sqrt{3 + 2x - x^2}}$$

and give a reason for each step.

**Solution**    We factor the expression, getting

$$\frac{9 - x^2}{\sqrt{3 + 2x - x^2}} = \frac{(3 + x)(3 - x)}{\sqrt{(1 + x)(3 - x)}}$$

$$= \frac{(3 + x)\sqrt{3 - x}}{\sqrt{1 + x}} \quad \text{for } -1 < x < 3.$$

From the theorems on limit of a constant and obvious limit, applied to one-sided limits, we have

$$\lim_{x \to 3^-} 1 = 1, \qquad \lim_{x \to 3^-} 3 = 3, \qquad \lim_{x \to 3^-} x = 3.$$

Therefore, from the limit of a sum (one-sided case), we find

$$\lim_{x \to 3^-} (3 + x) = 6, \qquad \lim_{x \to 3^-} (1 + x) = 4, \qquad \lim_{x \to 3^-} (3 - x) = 0.$$

Also,

$$\lim_{x \to 3^-} \sqrt{3-x} = 0, \qquad \lim_{x \to 3^-} \sqrt{1+x} = 2, \qquad \text{(lim. } \sqrt[n]{f(x)}\text{)}$$

$$\lim_{x \to 3^-} \frac{(3+x)\sqrt{3-x}}{\sqrt{1+x}} = \frac{6 \cdot 0}{2} = 0, \quad \text{(lim. prod.; lim. quot.)}$$

$$\lim_{x \to 3^-} \frac{9-x^2}{\sqrt{3+2x-x^2}} = 0. \qquad \text{(lim. equal func.)}$$

☐

**EXAMPLE 2**   For what values is the function

$$f(x) = \frac{x^2 - 9}{x + 3}$$

continuous?

**Solution**   This function is defined for all numbers except $x = -3$. Straight substitution at $-3$ yields $\frac{0}{0}$ and $f$ is undefined. Therefore $f$ is not continuous at $-3$. For all other numbers in the domain of $f$, the function has the same value as the function $g(x) = x - 3$, and so is continuous. In fact,

$$\lim_{x \to -3} f(x) = -6.$$

☐

*Remark.*   The important distinction between the function $f$ in the above example and the function $g(x) = x - 3$ is found in the *domain* of these functions. The domain of $g$ is all of $R^1$, while the domain of $f$ is all of $R^1$ with the exception of $-3$. The functions coincide wherever they are both defined, but the functions are not identical. These fine points occur frequently and they cannot be overlooked.

**EXAMPLE 3**   Find the values of $x$ where the function

$$h(x) = \begin{cases} 2x^2, & -1 \le x < 1 \\ 3 - x, & 1 \le x < 2 \end{cases}$$

is continuous.

**Solution**   The domain of $h$ is the half-open interval $[-1, 2)$. Clearly $h$ is continuous except possibly at $-1$ and at $1$. Taking one-sided limits, we find that

$$\lim_{x \to 1^-} 2x^2 = 2 \qquad \text{and} \qquad \lim_{x \to 1^+} (3-x) = 2.$$

Since both one-sided limits are equal and since $h(1) = 2$, we conclude that $h$ is continuous at $x = 1$. (See Fig. 27.)

At $x = -1$, the function $h$ has the value 2. Also, the limit as $x$ tends to $-1$ from the right exists and the limiting value is 2. Under these circumstances, we say that $h$ is **continuous on the right at** $x = -1$.   ☐

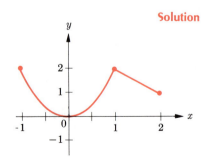

**FIGURE 27**

*Remark.*   If a function $f$ is defined on a closed interval $[a, b]$, then continuity of $f$ may be defined at every interior point. The function $f$ is **continuous on the right at the endpoint** $a$ if and only if

$$\lim_{x \to a^+} f(x) = f(a).$$

Similarly, $f$ is **continuous on the left at** $b$ if and only if

$$\lim_{x \to b^-} f(x) = f(b).$$

DEFINITION

*A function $f$ is **continuous on the closed interval** $[a, b]$ if and only if $f$ is continuous at every interior point, continuous on the right at $a$, and continuous on the left at $b$.*

To say that a function has a derivative at the value $a$ means that a tangent line to the curve can be drawn at the point $(a, f(a))$. In fact, the value of the derivative, defined as

$$\lim_{h \to 0} \frac{f(a + h) - f(a)}{h},$$

is simply the slope of this tangent line. As we have seen in Example 3, a continuous function may have corners, and it is hard to imagine how a tangent to a curve can be constructed at a corner; in general, it cannot. Not every continuous function possesses a tangent line at each point. An example is the function

$$F(x) = \begin{cases} 2x^2, & -1 \le x < 1 \\ 2(x - 2)^2, & 1 \le x < 2 \end{cases}$$

As Fig. 28 exhibits, $F(x)$ is continuous at $x = 1$, but does not possess a tangent line at $(1, 2)$. The two dotted lines in the graph show that the requirement that the tangent line to a curve is unique must be violated.

If a function has a derivative, however, it is continuous, as we prove next.

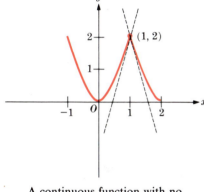

A continuous function with no derivative at $x = 1$

FIGURE 28

THEOREM 13

*If the function $f$ possesses a derivative at the value $a$, it is continuous at $a$.*

Proof

We recall that two things must be shown:

i) $f(a)$ is defined     and     ii) $\lim_{x \to a} f(x) = f(a)$.

If a function has a derivative at $a$, then

$$\lim_{h \to 0} \frac{f(a + h) - f(a)}{h} = f'(a)$$

exists. This statement would have no meaning if $f(a)$ did not; therefore having a derivative at $a$ means that $f(a)$ must have a meaning. As for (ii), we see that it has the same meaning as

$$\lim_{h \to 0} f(a + h) = f(a).$$

We can write

$$f(a + h) - f(a) = \left( \frac{f(a + h) - f(a)}{h} \right) \cdot h.$$

On the right, the term in parentheses approaches $f'(a)$, a finite number, as $h$ tends to zero. The second part, namely $h$, tends to zero. Therefore, as $h$ tends to zero, the right side—according to the theorem on the limit of a product— tends to zero. We conclude that

$$f(a + h) - f(a) \to 0 \quad \text{as } h \to 0,$$

and (ii) is established; $f(x)$ is continuous at $a$.  □

## 8  PROBLEMS

In each of Problems 1 through 12, evaluate the one-sided limit, giving a reason for each step.

**1** $\lim\limits_{x \to 1^+} \dfrac{x - 1}{\sqrt{x^2 - 1}}$

**2** $\lim\limits_{x \to 1^+} (x + \sqrt{x^2 - 1})$

**3** $\lim\limits_{x \to 2^-} \dfrac{4 - x^2}{\sqrt{2 + x - x^2}}$

**4** $\lim\limits_{x \to 2^+} \dfrac{x^2 - 3x + 2}{\sqrt{x^2 - 4}}$

**5** $\lim\limits_{x \to 2^+} \dfrac{x^4 - 16}{\sqrt{x - 2}}$

**6** $\lim\limits_{x \to 2^+} \sqrt{\dfrac{x^4 - 16}{x - 2}}$

**7** $\lim\limits_{x \to 0^-} \dfrac{\sqrt[3]{x^2(2 - x)}}{\sqrt[4]{x(3 - x)}}$

**8** $\lim\limits_{x \to 1^-} |x - 1| + |x|$

**9** $\lim\limits_{x \to 0^+} x\sqrt{1 + \dfrac{1}{x^2}}$

**10** $\lim\limits_{x \to 0^-} x\sqrt{1 + \dfrac{1}{x^2}}$

**11** $\lim\limits_{x \to 4^-} (x - \sqrt{16 - x^2})$

**12** $\lim\limits_{x \to 2^-} \dfrac{\sqrt{4 - x^2}}{\sqrt{6 - 5x + x^2}}$

**13** Evaluate

$$\lim_{x \to 1^-} |x - 1| \quad \text{and} \quad \lim_{x \to 1^+} |x - 1|.$$

Does (the two-sided limit) $\lim_{x \to 1} |x - 1|$ exist? Draw the graph.

**14** Evaluate

$$\lim_{x \to 0^+} \frac{x}{|x|} \quad \text{and} \quad \lim_{x \to 0^-} \frac{x}{|x|}.$$

Does

$$\lim_{x \to 0} \frac{x}{|x|}$$

exist? Give a reason for your answer.

In each of Problems 15 through 30, a function is defined in a certain domain. Determine those values of $x$ at which the function is continuous. Sketch the graph.

**15** $f(x) = \dfrac{1}{x^2 + 3}$ for $-4 < x < 7$

**16** $f(x) = \dfrac{x + 3}{x^2 - 2x - 15}$ for $-7 < x < 4, x \neq -3$

$f(-3) = 2$

**17** $f(x) = \dfrac{x^2 - 4}{x^2 - 8x + 12}, \quad 0 < x < 5, x \neq 2$

$f(2) = 1$

**18** $f(x) = \dfrac{x + 2}{x^2 - 2x - 8}$ for $-1 < x < 3$

**19** $f(x) = \dfrac{x^3 - 1}{x^2 + x - 2}, \quad 0 < x < 2, x \neq 1$

$f(1) = 1$

**20** $f(x) = \begin{cases} x - 4, & -1 < x \leq 2 \\ x^2 - 6, & 2 < x < 5 \end{cases}$

**21** $f(x) = \begin{cases} x^2 - 6, & -\infty < x < -1 \\ -5, & -1 \leq x \leq 10 \\ x - 15, & 10 < x < \infty \end{cases}$

**22** $f(x) = \begin{cases} \dfrac{x^2 - 1}{x^4 - 1}, & -1 < x < 2, \quad x \neq 1 \\ x^2 + 3x - 2, & 2 \leq x < 5 \end{cases}$

$f(1) = \frac{1}{2}$

**23** $f(x) = |x + 3|$ for all values of $x$

**24** $f(x) = |2x - 5|$ for all values of $x$

**25** $f(x) = \dfrac{x - 2}{|x - 2|}$ for all $x$ except $x = 2$

$f(2) = 0$

**26** $f(x) = \dfrac{3 + |x - 2|}{1 + x^2}$ for all values of $x$

**27** $f(x) = \dfrac{x^2 - x - 6}{x - 3}$ for all values of $x$ except $x = 3$

$f(3) = 5$

***28** $f(x) = \sin \dfrac{\pi}{x}$ for $-1 < x < 1$, $x \neq 0$

$f(0) = 0$

***29** $f(x) = x \sin \dfrac{\pi}{x}$ for $-1 < x < 1$, $x \neq 0$

$f(0) = 0$

**30** $f(x) = |x| + |1 - x|$ for $-1 < x < 3$

**31** Given

$$f(x) = \begin{cases} x - 2n & \text{for} \quad 2n \leq x < 2n + 1, \quad n = 0, 1, 2, \ldots \\ 2n - x & \text{for} \quad 2n - 1 \leq x < 2n, \quad n = 1, 2, \ldots, \end{cases}$$

where is $f$ continuous? Sketch the graph for $x$ in the interval $[0, 6)$.

**32** a) Given $f(x) = |x - 1| + |2 - x|$, show that $f$ has the constant value 1 on the interval $1 \leq x \leq 2$. For what values of $x$ is $f$ continuous?

b) Write a formula for a continuous function $f$ which has the constant value 2 on the interval $0 \leq x \leq 3$ and is not constant outside this interval.

---

**9**

## LIMITS AT INFINITY; INFINITE LIMITS (OPTIONAL)*

The graph of the function $f(x) = x^2/(1 + x^2)$ is shown in Fig. 29. The values of this function are always less than 1; in particular, we note the following:

$$f(1) = \frac{1}{2}, \quad f(10) = \frac{100}{101}, \quad f(100) = \frac{10,000}{10,001}, \quad f(1000) = \frac{10^6}{10^6 + 1}, \ldots.$$

As $x$ gets larger and larger, the graph of $f$ gets closer and closer to the line $y = 1$, as shown in Fig. 29. We use the term "$x$ tends to infinity" when referring to values which increase or decrease without bound. If $x$ is becoming larger through positive values only, we write $x \to +\infty$; if through negative values only, we write $x \to -\infty$. The symbol $x \to \infty$ means that $|x|$ increases without bound.† In the above example, we would write

$$f(x) \to 1 \quad \text{as} \quad x \to \infty.$$

More generally, we formulate the following definition.

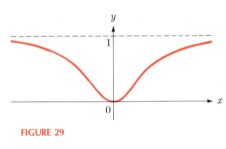

FIGURE 29

DEFINITION   **We say that** $f(x) \to c$ **as** $x \to \infty$, *if for each $\varepsilon > 0$ there is a number $A > 0$, such that $|f(x) - c| < \varepsilon$ for all $x$ for which $|x| > A$.*

---

If we want to define

$$f(x) \to c \quad \text{as} \quad x \to +\infty$$

(i.e., $f(x)$ tends to $c$ as $x$ tends to infinity through positive values only), the only change in the definition would be to write $x > A$ instead of $|x| > A$.

---

*This section may be omitted without loss of continuity.

†Some texts use the symbols $\infty$, $-\infty$, $+\infty$ corresponding to our use in this section of $+\infty$, $-\infty$, $\infty$, respectively.

Further, the definition of

$$f(x) \to c \quad \text{as} \quad x \to -\infty$$

would require $x < -A$ instead of $|x| > A$.

The theorems on uniqueness of limit and on limit of a constant, equal functions, sum, product, and quotient, and the Sandwich Theorem remain unchanged. In place of the theorem on limits of functions of the form $f(x) = x$, we have the following theorem on obvious limits.

**THEOREM 14**

$$\lim_{x \to \infty} \frac{1}{x} = 0, \qquad \lim_{x \to +\infty} \frac{1}{x} = 0, \qquad \lim_{x \to -\infty} \frac{1}{x} = 0.$$

**EXAMPLE 1**   Evaluate

$$\lim_{x \to +\infty} \frac{3x - 2}{5x + 4},$$

giving the reason for each step.

**Solution**   We employ a standard trick in working with infinite limits by dividing each term in the numerator and denominator by $x$:

$$\frac{3x - 2}{5x + 4} = \frac{3 - (2/x)}{5 + (4/x)}, \quad x \neq 0, \quad -\frac{4}{5}.$$

For the numerator, we find that

$$\lim_{x \to +\infty} \left(3 - \frac{2}{x}\right) = \lim_{x \to +\infty} 3 + \lim_{x \to +\infty} \left(-\frac{2}{x}\right),$$

since the limit of a sum is the sum of the limits. We now reason that

$$\lim_{x \to +\infty} 3 = 3 \quad \text{(using limit of a constant)}$$

and

$$\lim_{x \to +\infty} \left(-\frac{2}{x}\right) = -2 \lim_{x \to +\infty} \frac{1}{x} = 0 \quad \text{(using limit of a product and Theorem 14).}$$

Therefore

$$\lim_{x \to +\infty} \left(3 - \frac{2}{x}\right) = 3.$$

In exactly the same way,

$$\lim_{x \to +\infty} \left(5 + \frac{4}{x}\right) = 5.$$

Now we can say that

$$\lim_{x \to +\infty} \frac{3 - (2/x)}{5 + (4/x)} = \frac{\lim\limits_{x \to +\infty} (3 - (2/x))}{\lim\limits_{x \to +\infty} (5 + (4/x))},$$

since the limit of a quotient is the quotient of the limits. We conclude that (limit of equal functions)

$$\lim_{x \to +\infty} \frac{3x-2}{5x+4} = \frac{3}{5}.$$

**EXAMPLE 2**    Evaluate

$$\lim_{x \to +\infty} \frac{\sqrt{x^2-1}}{2x+1},$$

giving a reason for each step.

**Solution**    We first divide numerator and denominator by $x$, getting

$$\frac{\sqrt{x^2-1}}{2x+1} = \frac{\sqrt{1-(1/x^2)}}{2+(1/x)}.$$

Since

$$\frac{1}{x^2} = \frac{1}{x} \cdot \frac{1}{x},$$

we have

$$\lim_{x \to +\infty} \frac{1}{x^2} = \lim_{x \to +\infty} \frac{1}{x} \cdot \lim_{x \to +\infty} \frac{1}{x} = 0,$$

using the limit of a product theorem and Theorem 14. Then

$$\lim_{x \to +\infty} \left(1 - \frac{1}{x^2}\right) = \lim_{x \to +\infty} 1 - \lim_{x \to +\infty} \frac{1}{x^2} = 1 \quad \text{(using limit of a sum).}$$

Further, by using Theorem 11, we find that

$$\lim_{x \to +\infty} \sqrt{1 - \frac{1}{x^2}} = 1.$$

We observe that

$$\lim_{x \to +\infty} \left(2 + \frac{1}{x}\right) = 2,$$

as in Example 1, and when we apply the quotient rule and the theorem on the limit of equal functions, we have

$$\lim_{x \to +\infty} \frac{\sqrt{x^2-1}}{2x+1} = \frac{1}{2}.$$

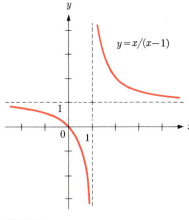

**FIGURE 30**

The graph of the function $f(x) = x/(x-1)$ is shown in Fig. 30. We note the following values of this function:

$$f(2) = 2, \quad f(\tfrac{3}{2}) = 3, \quad f(1.1) = 11, \quad f(1.01) = 101, \quad f(1.001) = 1{,}001, \dots;$$

$$f(0) = 0, \quad f(\tfrac{1}{2}) = -1, \quad f(0.9) = -9, \quad f(0.99) = -99, \quad f(0.999) = -999, \dots.$$

This type of behavior implies that the function grows without bound as $x$ gets closer and closer to 1. We use the expression

$$f(x) \to \infty \quad \text{as} \quad x \to 1,$$

which we define in the following way.

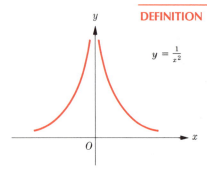

$y = \frac{1}{x^2}$

FIGURE 31

**DEFINITION**  *We say that $f(x)$ **becomes infinite as** $x \to a$, and write $f(x) \to \infty$ as $x \to a$, if for each number $A > 0$ there is a $\delta > 0$ such that $|f(x)| > A$, for all $x$ for which $0 < |x - a| < \delta$.*

This definition does not state whether the function goes off to infinity "upward" or "downward." In other words, we may have $f(x) \to +\infty$ or $f(x) \to -\infty$ as special cases of $f(x) \to \infty$. For example, if $f(x) \to +\infty$ as $x \to a$, we say that "$f$ **becomes positively infinite as** $x$ **tends to** $a$." The only change required in the above definition is the replacement of $|f(x)| > A$ by $f(x) > A$. Figure 31 shows the function $g(x) = 1/x^2$, which becomes positively infinite as $x \to 0$.

Similarly, a function may become **negatively infinite as** $x$ **tends to** $a$. In this case, we replace $|f(x)| > A$ by $f(x) < -A$ in the above definition.

We also consider functions which become infinite as a value $a$ is approached from one side. In Fig. 30 it is clear that $f(x) \to +\infty$ as $x \to 1^+$ and $f(x) \to -\infty$ as $x \to 1^-$. We now give a precise definition of one-sided limit from the right.

**DEFINITION**  *We say that $f(x)$ **becomes positively infinite as** $x$ **tends to** $a$ **from the right** if for each number $A$ there is a $\delta > 0$ such that $f(x) > A$ for all $x$ for which $0 < |x - a| < \delta$ and $x$ is larger than $a$. For this one-sided limit, we use the notations*

$$f(x) \to +\infty \quad \text{as} \quad x \to a^+ \qquad \text{and} \qquad \lim_{x \to a^+} f(x) = +\infty.$$

Analogous definitions and notations are used for functions which become negatively infinite and for one-sided limits from the left.

Figure 31 shows a function which becomes positively infinite as $x$ tends to zero both from the left and from the right. There is an analog to the "two-sided limit" theorem for functions which become infinite. We can conclude, for example, that $g(x) = 1/x^2$ becomes positively infinite as $x \to 0$.

Some functions have infinite limits from one side only. For example, the function

$$h(x) = \frac{1}{\sqrt{x - 2}}$$

becomes positively infinite as $x \to 2^+$, but $h$ is not defined for $x$ less than 2. Therefore the limit from the left cannot exist and consequently there can be no (two-sided) limit as $x$ tends to 2.

The rules for operating with limits are somewhat tricky when one of the quantities becomes infinite. We must remember that $\infty$ *is not a number* and cannot be treated as such. There are some simple facts about infinite limits which the reader will easily recognize. For example, in a case such as

$$f(x) \to +\infty \quad \text{as} \quad x \to a,$$

$$g(x) \to c \quad \text{as} \quad x \to a,$$

where $c$ is any number, then

$$f(x) + g(x) \to +\infty \quad \text{as} \quad x \to a.$$

If $c \neq 0$, we have

$$f(x) \cdot g(x) \to +\infty \quad \text{when} \quad c > 0$$

and

$$f(x) \cdot g(x) \to -\infty \quad \text{when} \quad c < 0.$$

*On the other hand, if $c = 0$, further investigation is necessary.* To see this, we let $f(x) = 1/(x-3)^2$ and $g(x) = (x-3)^2$. Then $f(x) \to +\infty$ as $x \to 3$ and $g(x) \to 0$ as $x \to 3$. However,

$$f(x) \cdot g(x) \equiv 1,$$

and so, $f \cdot g \to 1$ as $x \to 3$. If we replace $f(x)$ by the function $A/(x-3)^2$, where $A$ is *any number*, the product $f \cdot g$ tends to $A$. The reader can easily construct examples in which $f \cdot g \to +\infty$ and $f \cdot g \to 0$ as $x \to 3$. (See Problems 25 and 26 at the end of this section.)

For cases in which $f(x) \to +\infty$ and $g(x) \to -\infty$ as $x \to a$, nothing can be said about the behavior of

$$f(x) + g(x)$$

without a closer examination of the particular functions.

## 9   PROBLEMS

In each of Problems 1 through 15, evaluate the limit or show that the function tends to $\infty$, $+\infty$, or $-\infty$. Give a reason for each step.

**1** $\displaystyle\lim_{x \to \infty} \frac{2x + 4}{3x + 1}$

**2** $\displaystyle\lim_{x \to \infty} \frac{x^2 + 2x + 5}{2x^2 - 6x + 1}$

**3** $\displaystyle\lim_{x \to \infty} \frac{2x^2 + 7x + 5}{x^3 + 2x + 1}$

**4** $\displaystyle\lim_{x \to +\infty} \frac{\sqrt{2x^2 + 1}}{x + 3}$

**5** $\displaystyle\lim_{x \to -\infty} \frac{\sqrt{x^2 + 4}}{x + 2}$

**6** $\displaystyle\lim_{x \to +\infty} (\sqrt{x^2 + a^2} - x)$

**7** $\displaystyle\lim_{x \to 1} \frac{x^2}{1 - x^2}$

**8** $\displaystyle\lim_{x \to 2} \frac{x}{4 - x^2}$

**9** $\displaystyle\lim_{x \to 0} \frac{\sqrt{1 + x}}{x}$

**10** $\displaystyle\lim_{x \to \infty} \frac{x^2 + 1}{x}$

**11** $\displaystyle\lim_{x \to 2^+} \frac{\sqrt{x^2 - 4}}{x - 2}$

**12** $\displaystyle\lim_{x \to +\infty} (\sqrt{x^2 + 2x} - x)$

**13** $\displaystyle\lim_{x \to -\infty} (\sqrt{x^2 + 2x} - x)$

**14** $\displaystyle\lim_{x \to 2^-} \frac{\sqrt{4 - x^2}}{\sqrt{6 - 5x + x^2}}$

**15** $\displaystyle\lim_{x \to 0} \frac{x + 1}{|x|}$

In each of Problems 16 through 19, determine those values of $x$ at which the function is continuous. Sketch the graph.

**16** $\begin{cases} f(x) = \dfrac{1}{x + 3} & \text{for } -7 < x < -1, \quad x \neq -3 \\ f(-3) = 2 \end{cases}$

**17** $\begin{cases} f(x) = \dfrac{x + 3}{x^2 - 9}, & -6 < x < 6, \quad x \neq -3, 3 \\ f(-3) = 2, \quad f(3) = 1 \end{cases}$

**18** $f(x) = \begin{cases} \dfrac{2x}{x^2 - 4} & \text{for } 0 < x < 2 \\ 3x - 5 & \text{for } 2 \leq x \leq 5 \\ x^2 + 6 & \text{for } 5 < x < 7 \end{cases}$

**19** $\begin{cases} f(x) = \dfrac{x + 1}{|x - 2|} & \text{for all } x \text{ except } x = 2 \\ f(2) = 3 \end{cases}$

In each of Problems 20 through 23, sketch the graph of the given function. Then decide what limit, if any, is approached.

**\*20** $\displaystyle\lim_{x\to 0^+} f(x)$   where   $f(x) = x \sin \dfrac{\pi}{x}$

**\*21** $\displaystyle\lim_{x\to +\infty} f(x)$   where   $f(x) = \dfrac{1}{x} \sin \dfrac{\pi}{x}$

**\*22** $\displaystyle\lim_{x\to 0^+} f(x)$   where   $f(x) = \dfrac{1}{x} \sin \dfrac{\pi}{x}$

**\*23** $\displaystyle\lim_{x\to 0} f(x)$   where   $f(x) = \sin \dfrac{\pi}{x}$

**\*24** Given the function $f$ defined by the conditions

$$f(x) = \begin{cases} 2x & \text{if } x \text{ is a rational number,} \\ 0 & \text{if } x \text{ is an irrational number.} \end{cases}$$

Find $\displaystyle\lim_{x\to 0^+} f(x)$.

**25** Find functions $f$ and $g$ such that $f(x) \to +\infty$ and $g(x) \to 0$ as $x \to a$ and $f(x) \cdot g(x) \to +\infty$ as $x \to a$. Also find functions $f$ and $g$ such that $f(x) \to +\infty$, $g(x) \to 0$ as $x \to a$ and $f(x) \cdot g(x) \to 0$ as $x \to a$. Is it possible that $f(x) \to +\infty$, $g(x) \to 0$, and $f(x) \cdot g(x) \to A$ as $x \to a$, where $A$ is negative? Justify the answer.

**26** Find functions $f$ and $g$ such that $f(x) \to +\infty$ and $g(x) \to$ $-\infty$ as $x \to a$ and $f(x) + g(x) \to +\infty$ as $x \to a$. Find functions $f$ and $g$ such that $f(x) \to +\infty$, $g(x) \to -\infty$ as $x \to a$ and $f(x) + g(x) \to -\infty$. If $A$ is any given number, find functions $f$ and $g$ such that $f(x) \to +\infty$, $g(x) \to -\infty$ as $x \to a$, and $f(x) + g(x) \to A$ as $x \to a$.

**\*27** Let $P(x) = a_n x^n + a_{n-1} x^{n-1} + \cdots + a_1 x + a_0$ and $Q(x) = b_m x^m + b_{m-1} x^{m-1} + \cdots + b_1 x + b_0$ be two polynomials with $a_n \neq 0$ and $b_m \neq 0$. Find the value of

$$\lim_{x\to +\infty} \frac{P(x)}{Q(x)}$$

by considering three cases: (i) $m > n$, (ii) $m = n$, (iii) $m < n$.

**\*28** Let $P$ and $Q$ be defined as in Problem 27. Find the value of

$$\lim_{x\to -\infty} \frac{P(x)}{Q(x)}.$$

**\*29** Suppose that $P(x)$ and $Q(x)$ in Problem 27 are written in the form $P(x) = a_n(x - c_1)(x - c_2)(x - c_3) \cdots (x - c_n)$, $Q(x) = b_m(x - d_1)(x - d_2) \cdots (x - d_m)$. Describe the various possible results in evaluating the

$$\text{limit of } \frac{P(x)}{Q(x)} \quad \text{as } x \to d_i$$

where $i$ is an integer between 1 and $m$.

---

# CHAPTER 2

## REVIEW PROBLEMS

In Problems 1 through 4, find the limit (if any) approached by $f(x)$ as $x$ tends to $a$.

**1** $f(x) = \dfrac{x^2 + x - 2}{x^2 - 4x + 3}, \quad a = 1$

**2** $f(x) = \dfrac{x^2 - 2x - 3}{x^2 - 7x + 12}, \quad a = 3$

**3** $f(x) = \dfrac{\sqrt{x + 2} - 3}{x^2 - 7x}, \quad a = 7$

**4** $f(x) = \dfrac{x^6 - 1}{x^2 + x}, \quad a = -1$

In Problems 5 through 14, use the three-step rule to find the derivative.

**5** $f(x) = 2x^2 - x + 1$

**6** $f(x) = -x^2 + 3x - 2$

**7** $f(x) = x^3 - 2x^2 + 1$

**8** $f(x) = 2x^3 + x^2 - 3x + 1$

**9** $f(x) = \dfrac{2x + 1}{x - 1}$

**10** $f(x) = \dfrac{3x}{x + 4}$

**11** $f(x) = x^{-2} + 2x$

**12** $f(x) = \dfrac{1}{\sqrt{2x - 3}}$

**13** $f(x) = x^4 + x^2 - 2$

**14** $f(x) = x^5$

In Problems 15 through 18, find the equation of the line tangent to the given curve at the given value of $a$.

**15** $y = 2x^2 + 3x - 1, \quad a = -1$

**16** $y = \dfrac{2x - 3}{x + 4}, \quad a = 1$

**17** $y = \sqrt{x^2 + 5}, \quad a = 2$

**18** $y = \dfrac{x^2}{x + 2}, \quad a = -4$

In Problems 19 through 24, a particle is moving along a straight line according to the given law. Find the times, if any, when the velocity and the acceleration are zero.

**19** $s = t^2 + 2t - 3$

**20** $s = 2t^2 - 3t + 5$

**21** $s = t^3 + 2t^2 - t + 1$     **22** $s = 2t^3 + t^2 - t + 4$

**23** $s = \dfrac{t+2}{t-1}$          **24** $s = \dfrac{t^2}{t+1}$

**25** A man in a stationary balloon 500 meters high hurls a stone straight down with an initial velocity of 2 meters per second. How long does it take for the stone to strike the earth?

In Problems 26 through 29, evaluate each of the limits, citing the appropriate theorem to justify each step in the evaluation.

**26** $\lim\limits_{x \to 3} (x^2 - 2x + 5)$     **27** $\lim\limits_{x \to -1} \dfrac{x^2 - x - 2}{x^2 + 4x + 3}$

**28** $\lim\limits_{x \to 4} \dfrac{2 - \sqrt{x}}{4 - x}$          **29** $\lim\limits_{x \to 2} \sqrt{x^3 - 2x + 2}$

**30** Suppose that $\lim\limits_{x \to a} f(x) = L$ and $L$ is negative.

Show that $\lim\limits_{x \to a} \sqrt[n]{f(x)} = \sqrt[n]{L}$ if $n$ is an odd integer.

In Problems 31 through 34, evaluate the one-sided limit.

**31** $\lim\limits_{x \to 0^+} (3 + \sqrt{x^2 + 2x})$     **32** $\lim\limits_{x \to 2^-} \dfrac{1}{x^2 + 2}$

**33** $\lim\limits_{x \to 2^+} (|x| + |2 - x|)$     **34** $\lim\limits_{x \to 2^-} (|x| + |2 - x|)$

In Problems 35 through 39, determine the values of $x$ for which the given function is continuous.

**35** $f(x) = \begin{cases} x + 2, & -1 \le x \le 2 \\ 6 - x, & 2 < x \le 8 \end{cases}$

**36** $f(x) = \begin{cases} x^2 + 2, & -1 \le x < 1 \\ 4 - x, & 1 \le x < 2 \end{cases}$

**37** $f(x) = \begin{cases} |x| - |x + 1|, & -1 \le x \le 1 \\ -|2 - x|, & 1 < x \le 5 \end{cases}$

**38** $f(x) = \begin{cases} |x|, & -5 < x < 2 \\ |x - 4|, & 2 \le x < 5 \end{cases}$

**39** $f(x) = \dfrac{x^2 + x - 2}{x^2 - x - 6}$, for all $x$ except $x = -2, 3$; $f(-2) = 1$, $f(3) = 2$.

# 3

# DIFFERENTIATION

The rules of differentiation are essential for a working knowledge of calculus. In particular, the Chain Rule and its applications are especially useful. We also broaden our ability to differentiate by finding the derivatives of trigonometric functions.

## 1

### THEOREMS ON DIFFERENTIATION

The process of finding the derivative is called **differentiation**. The reader will recall that use of the three-step rule for differentiation, as shown in Chapter 2, Section 3, frequently involves a lengthy and complicated process. There are certain simple expressions and combinations of expressions which occur repeatedly; it pays to be able to differentiate these at sight. In this section we establish some theorems which help us in the process of differentiation so that we do not always have to use the three-step rule.

**THEOREM 1** *If $f(x) = c$, a constant, for all $x$, then $f'(x) = 0$ for all $x$.* **(The derivative of a constant is zero.)**

**Proof** We prove this by the three-step rule. Since $f$ has the value $c$ for all values of $x$, we have

**Step 1:** $f(x+h) - f(x) = c - c = 0.$

**Step 2:** $\dfrac{f(x+h) - f(x)}{h} = \dfrac{0}{h} = 0.$

**Step 3:** $\lim_{h \to 0} 0 = 0.$

That is, $f'(x) = 0.$ □

Expressions such as

$$3x + 7, \quad x^2 - 6x + 5, \quad x^3 + 2x^2 - 7$$

are examples of *polynomials* in $x$. In general, an expression of the form

$$a_n x^n + a_{n-1} x^{n-1} + a_{n-2} x^{n-2} + \cdots + a_1 x + a_0,$$

where $n$ is a positive integer and the coefficients $a_n, a_{n-1}, a_{n-2}, \ldots, a_1, a_0$ are numbers, is called a polynomial in $x$. The **degree** of a polynomial is the highest exponent which appears. In the above examples the polynomials are of degree 1, 2, and 3, respectively.

Combinations of the form

$$\frac{2x - 6}{x^2 + 1}, \quad \frac{3x^3 - 2x + 5}{7x - 6}, \quad \frac{x^2 + 5x - 2}{x^5 - 3x^2 + 1}$$

are called *rational functions*. More generally, any function which can be written as the ratio of two polynomials is, by definition, a **rational function**. That is, it must be expressible as

$$\frac{P(x)}{Q(x)}$$

where $P(x)$ and $Q(x)$ are polynomials.

---

**THEOREM 2**    *If $n$ is a positive integer and $f(x) = x^n$, then $f'(x) = nx^{n-1}$.*    √

---

**Proof**    We use the three-step rule. Recalling that $\Delta f$ is a shorthand expression for $f(x + h) - f(x)$, we have

**Step 1:**    $\Delta f = (x + h)^n - x^n$.

At this point we make use of the binomial theorem which states that

$$(x + h)^n = x^n + nx^{n-1}h + \frac{n(n-1)}{2!}x^{n-2}h^2$$

$$+ \frac{n(n-1)(n-2)}{3!}x^{n-3}h^3 + \cdots + h^n,$$

and so we find by subtracting $x^n$ from both sides

$$\Delta f = nx^{n-1}h + \frac{n(n-1)}{2!}x^{n-2}h^2 + \frac{n(n-1)(n-2)}{3!}x^{n-3}h^3 + \cdots + h^n.$$

Note that every term on the right side of the above equation has $h$ as a factor.

**Step 2:**    $\dfrac{\Delta f}{h} = nx^{n-1} + \dfrac{n(n-1)}{2!}x^{n-2}h + \dfrac{n(n-1)(n-2)}{3!}x^{n-3}h^2 + \cdots + h^{n-1}$.

Now we observe that every term on the right *except* the first has $h$ in it. Since the limit of a sum is the sum of the limits, we take the limit of each of the terms on the right as $h \to 0$ and add the result. However, each term on the right, except the first, tends to zero. The term $nx^{n-1}$ is completely unaffected as $h \to 0$, since it has no $h$ in it at all.

**Step 3:** $f'(x) = \lim\limits_{h \to 0} \dfrac{\Delta f}{h} = nx^{n-1}.$     □

Theorem 2 asserts that whenever we see an expression such as $x^8$ or $x^{17}$, the derivative may be found by inspection. The derivative of $x^8$ is $8x^7$, and the derivative of $x^{17}$ is $17x^{16}$.

**THEOREM 3**    *If $f(x)$ has a derivative $f'(x)$, then the derivative of $g(x) = cf(x)$ is $cf'(x)$, where c is any constant.* **(The derivative of a constant times a function is the constant times the derivative of the function.)**

**Proof**    We use the three-step rule.

**Step 1:**   $\Delta g = g(x + h) - g(x) = cf(x + h) - cf(x) = c\Delta f.$

**Step 2:**   $\dfrac{\Delta g}{h} = c\dfrac{\Delta f}{h}.$

**Step 3:**   $g'(x) = \lim\limits_{h \to 0} c\dfrac{\Delta f}{h} = c \lim\limits_{h \to 0} \dfrac{\Delta f}{h} = cf'(x).$

The justification for Step 3 comes from the theorem that the limit of a product is the product of the limits.     □

Theorems 2 and 3 can be combined to find derivatives of expressions such as $4x^7$, $-2x^5$, and so forth. We know from Theorem 2 that $x^7$ has the derivative $7x^6$ and therefore, from Theorem 3, $4x^7$ has the derivative $4(7x^6) = 28x^6$. Similarly, the derivative of $-2x^5$ is $-10x^4$.

**THEOREM 4**    *If $f(x)$ and $g(x)$ have derivatives and $F(x) = f(x) + g(x)$, then $F'(x) = f'(x) + g'(x)$.* **(The derivative of the sum is the sum of the derivatives.)**

**Proof**    We use the three-step rule.

**Step 1:**   $\Delta F = F(x + h) - F(x) = f(x + h) + g(x + h) - f(x) - g(x)$
$$= \Delta f + \Delta g.$$

**Step 2:**   $\dfrac{\Delta F}{h} = \dfrac{\Delta f}{h} + \dfrac{\Delta g}{h}.$

Since the limit of a sum is the sum of the limits, we see that the right side tends to $f'(x) + g'(x)$. Therefore,

**Step 3:**   $F'(x) = f'(x) + g'(x).$     □

While Theorem 4 was proved for the sum of two functions, the same proof works for the sum of any (finite) number of functions. Theorems 2, 3, and 4 combine to enable us to differentiate any polynomial.

**EXAMPLE 1**   Find the derivative of

$$f(x) = 3x^5 + 2x^4 - 7x^2 + 2x + 5.$$

**Solution**   The quantity $x^5$ has the derivative $5x^4$, as we know from Theorem 2. Using this fact and Theorem 3, we find the derivative of $3x^5$ to be $15x^4$. Similarly, the derivative of

$$2x^4 \text{ is } 8x^3, \quad -7x^2 \text{ is } -14x, \quad 2x \text{ is } 2, \quad 5 \text{ is } 0.$$

Theorem 4 tells us that the derivative of the sum of these terms is the sum of the derivatives. We obtain

$$f'(x) = 15x^4 + 8x^3 - 14x + 2. \qquad \square$$

For every polynomial we now know how to compute its derivative at any point. Therefore, by Theorem 13 of Chapter 2 (page 87) we can state that **every polynomial is a continuous function**.

---

**THEOREM 5**   *If $u(x)$ and $v(x)$ are any two functions which have a derivative and if $f(x) = u(x) \cdot v(x)$, then*

$$f'(x) = u(x) \cdot v'(x) + v(x) \cdot u'(x).$$

**(The derivative of the product of two functions is the first times the derivative of the second plus the second times the derivative of the first.)**

---

**Proof**   We use the three-step rule, as usual.

**Step 1:**   $\Delta f = f(x + h) - f(x) = u(x + h) \cdot v(x + h) - u(x) \cdot v(x).$

At this point we use a trick of a type we have used before. We write the quantity $f(x + h)$ in a more complicated way, as follows:

$$f(x + h) = [u(x) + u(x + h) - u(x)] \cdot [v(x) + v(x + h) - v(x)].$$

With the symbols $\Delta u$ and $\Delta v$ this becomes

$$f(x + h) = (u + \Delta u)(v + \Delta v).$$

Therefore

$$\Delta f = f(x + h) - f(x) = (u + \Delta u)(v + \Delta v) - uv.$$

Multiplying out the right side, we obtain

$$\Delta f = u\Delta v + v\Delta u + \Delta u \cdot \Delta v.$$

**Step 2:**   $\dfrac{\Delta f}{h} = u\dfrac{\Delta v}{h} + v\dfrac{\Delta u}{h} + \Delta u \cdot \dfrac{\Delta v}{h}.$

We now examine each term on the right side of the equation as $h$ tends to zero. The first term tends to $u(x)v'(x)$, since

$$\frac{\Delta v}{h} \to v'(x) \qquad \text{as} \qquad h \to 0.$$

The second term tends to $v(x)u'(x)$. The last term has two parts, $\Delta v/h$ and $\Delta u$.

The quantity $\Delta v / h$ tends to $v'(x)$. However, the function $u(x)$ is continuous (according to the theorem which says that if a function has a derivative it is continuous), and so $u(x + h) - u(x) = \Delta u \to 0$ as $h \to 0$. This means that the last term on the right tends to zero as $h \to 0$. We conclude

**Step 3:**   $f'(x) = \lim\limits_{h \to 0} \dfrac{\Delta f}{h} = u(x)v'(x) + v(x)u'(x).$       □

**EXAMPLE 2**   Find the derivative of

$$f(x) = (x^2 - 3x + 2)(x^3 + 2x^2 - 6x).$$

What is the value of $f'(3)$? the value of $f'(2)$?

**Solution**   One way to do this would be to multiply the two expressions in parentheses, thus obtaining a polynomial of the 5th degree, and then to find the derivative as in Example 1. However, Theorem 5 gives us an idea for a simpler way. We write

$$u(x) = x^2 - 3x + 2 \qquad \text{and} \qquad v(x) = x^3 + 2x^2 - 6x.$$

This means that $f(x) = u(x)v(x)$. We can now readily find $u'(x)$ and $v'(x)$, using the method in Example 1: $u'(x) = 2x - 3$ and $v'(x) = 3x^2 + 4x - 6$. Then

$$f'(x) = u(x)v'(x) + v(x)u'(x)$$
$$= (x^2 - 3x + 2)(3x^2 + 4x - 6) + (x^3 + 2x^2 - 6x)(2x - 3),$$
$$f'(3) = (9 - 9 + 2)(27 + 12 - 6) + (27 + 18 - 18)(6 - 3)$$
$$= 66 + 81 = 147,$$
$$f'(2) = (4 - 6 + 2)(12 + 8 - 6) + (8 + 8 - 12)(4 - 3)$$
$$= 0 + 4 = 4. \qquad\qquad □$$

**THEOREM 6**   *If $u(x)$ and $v(x)$ are any two functions which have a derivative and if*

$$f(x) = \frac{u(x)}{v(x)}$$

*with $v(x) \neq 0$, then*

$$f'(x) = \frac{v(x)u'(x) - u(x)v'(x)}{[v(x)]^2}.$$

**(The derivative of a quotient of two functions is the denominator times the derivative of the numerator minus the numerator times the derivative of the denominator, all divided by the square of the denominator.)**

**Proof**   We proceed by the three-step rule.

**Step 1:**   $\Delta f = f(x + h) - f(x) = \dfrac{u(x + h)}{v(x + h)} - \dfrac{u(x)}{v(x)}.$

We may write $u(x + h) = u(x) + u(x + h) - u(x) = u + \Delta u$. Similarly, $v(x + h) = v(x) + v(x + h) - v(x) = v + \Delta v$. Therefore

$$\Delta f = \frac{u + \Delta u}{v + \Delta v} - \frac{u}{v} = \frac{v(u + \Delta u) - u(v + \Delta v)}{v(v + \Delta v)} = \frac{v\Delta u - u\Delta v}{v(v + \Delta v)}.$$

**Step 2:**  $\dfrac{\Delta f}{h} = \dfrac{v\dfrac{\Delta u}{h} - u\dfrac{\Delta v}{h}}{v(v + \Delta v)}.$

**Step 3:**  $\lim\limits_{h \to 0} \dfrac{\Delta f}{h} = f'(x) = \dfrac{vu'(x) - uv'(x)}{v^2}.$  $\square$

**EXAMPLE 3**  Find the derivative of

$$f(x) = \frac{2x^2 - 3x}{x^2 + 3}.$$

**Solution**  We recognize that the function $f$ is of the form one function divided by another. We set $u = 2x^2 - 3x$ and $v = x^2 + 3$. Then $u'(x) = 4x - 3$ and $v'(x) = 2x$. Now, by Theorem 6,

$$f'(x) = \frac{(x^2 + 3)(4x - 3) - (2x^2 - 3x) \cdot 2x}{(x^2 + 3)^2} = \frac{3x^2 + 12x - 9}{(x^2 + 3)^2}.$$

The last step was merely an algebraic simplification and, strictly speaking, not a part of the differentiation process.  $\square$

We have seen that for $n$ a positive integer, the function $g(x) = x^n$ has derivative $g'(x) = nx^{n-1}$. We now prove that this formula holds when $n$ is a negative integer.

**THEOREM 7**  *If $n$ is a positive integer and $f(x) = x^{-n}$, with $x \neq 0$, then*  ✓

$$f'(x) = -nx^{-n-1}.$$

**Proof**  A negative exponent simply means that we may write $f(x)$ in the form

$$f(x) = \frac{1}{x^n}.$$

We can apply Theorem 6: Let $u(x) = 1$ and $v(x) = x^n$. Then

$$f(x) = \frac{u(x)}{v(x)}.$$

From Theorem 1, $u'(x) = 0$ and, from Theorem 2, $v'(x) = nx^{n-1}$. We substitute in the formula of Theorem 6 to get

$$f'(x) = \frac{x^n \cdot 0 - 1 \cdot nx^{n-1}}{(x^n)^2},$$

which gives us

$$f'(x) = \frac{-nx^{n-1}}{x^{2n}} = -nx^{-n-1}. \qquad \Box$$

Combining Theorems 2 and 7, we see that **if $k$ is any positive or negative integer and**

**if $f(x) = x^k$, then $f'(x) = kx^{k-1}$.**

This formula is called the **Power Rule**.

## 1   PROBLEMS

In Problems 1 through 30, differentiate the functions by the methods given in this section. $1{-}28$

**1** $x^2 + 3x - 4$

**2** $2x^3 + 5x^2 - 3x + 7$

**3** $x^{14} + 10x^2 + 7x - 4$

**4** $x^8 + 2x^6 - 4x^3 + 6x + 9$

**5** $x^{-2} - 4x^{-5} + 3x^{-8}$

**6** $2x^{-1} - 3x^{-5} + 2x^2 - 7$

**7** $x^3 + 2x^2 - 3x + 5 - 2x^{-1} + 4x^{-2}$

**8** $x^5 + 12x - 1 + 3x^{-4} - 5x^{-7}$

**9** $x^2 + 2x - \dfrac{1}{x^2}$

**10** $x^3 - 3x - \dfrac{2}{x^4}$

**11** $\dfrac{3}{4x^7}$

**12** $2x^4 - 3x + \dfrac{5}{8x^3}$

**13** $(x^2 + 2x)(3x + 1)$

**14** $(2x^3 + 6x)(7x - 5)$

**15** $(x^3 + 6x^2 - 2x + 1)(x^2 + 3x - 5)$

**16** $(x^4 + 2x - 3)(x^6 - 7x^5 + 8x^3 + 9x^2 + 1)$

**17** $\dfrac{2x^2 - 3x + 4}{x}$

**18** $\dfrac{x^3 - 3x + 5}{x^2}$

**19** $\dfrac{2x^3 - 3x^2 + 4x - 2}{x^3}$

**20** $\dfrac{x^3 - 4x^2 + 3x - 2}{x^4}$

**21** $\dfrac{x - 1}{x + 1}$

**22** $\dfrac{2x + 3}{3x - 2}$

**23** $\dfrac{3x - 2}{2x + 3}$

**24** $\dfrac{5x - 2}{4x + 3}$

**25** $\dfrac{x}{x^2 + 1}$

**26** $\dfrac{x^2 + 1}{x^2 - 1}$

**27** $\dfrac{x + 1}{x^2 + 2x + 2}$

**28** $\dfrac{2x^2 - 3x + 4}{x^2 - 2x + 3}$

**29** $\dfrac{x + 1}{2x + 3}(2x - 5)$

**30** $\dfrac{x^2 - x + 6}{x^2 + 1}(x^2 + x + 1)$

**\*31** Using Theorem 5, find a formula for the derivative of $f(x) = [u(x)]^2$ given that $u$ is a differentiable function. Apply this to the functions

a) $(x^2 + 2x - 1)^2$,

b) $(x^3 + 7x^2 - 8x - 6)^2$,

c) $(x^7 - 2x + 3x^{-2})^2$.

**32** Given that $f(x) = u(x) \cdot v(x) \cdot w(x)$, with $u$, $v$, and $w$ differentiable functions, find a formula for $f'(x)$. (*Hint:* Apply Theorem 5 twice.) Using this formula, find the derivative of the following functions:

a) $(x^2 + 2x - 6)(3x - 2)(x^2 + 5)$,

b) $(2x^2 + x^{-2})(x^2 - 3)(4x + 1)$,

c) $(2x + 3)^2(x^2 + 1)$,

d) $(x^2 + x + 1)^3$.

**33** Let $u$ be differentiable, and let $f(x) = [u(x)]^3$. Use the result of Problem 32 to show that $f'(x) = 3[u(x)]^2 u'(x)$.

**34** Given that

$$f(x) = \frac{u(x)}{v(x)} w(x),$$

find a formula for $f'(x)$. (*Hint:* Apply Theorems 5 and 6.) Using this formula, find the derivative of the functions

a) $\dfrac{x + 2}{3x + 1}(x - 6)$,     b) $\dfrac{x^2 - 1}{2x + 6}(x^2 + 5)$.

**35** Let $P(x) = a_n x^n + a_{n-1} x^{n-1} + \cdots + a_1 x + a_0$ and $Q(x) = b_m x^m + b_{m-1} x^{m-1} + \cdots + b_1 x + b_0$ be two polynomials. Let $y = P(x)/Q(x)$. Find a formula for $dy/dx$.

**36** Let $P(x) = a_n x^n + a_{n-1} x^{n-1} + \cdots + a_1 x + a_0$ be a polynomial of degree $n$. Denote by $P^{(k)}(x)$ the $k$th derivative of $P$. Prove that $P^{(n+1)}(x) \equiv 0$.

**\*37** Suppose $f(x) = u(x) \cdot v(x)$ where $u$ and $v$ are given functions. We denote the $k$th derivative of $u$ by $u^{(k)}(x)$, and similarly for $v$ and $f$. Show that the "binomial formula" holds:

$$f^{(k)}(x) = u^{(k)}(x)v + \frac{k}{1!} u^{(k-1)}(x)v'(x)$$

$$+ \frac{k(k-1)}{2!} u^{(k-2)}(x)v^{(2)}(x) + \cdots + u(x)v^{(k)}(x).$$

**38** Suppose that $f(x) = u(x)/v(x)$ and $v(x) \neq 0$. Find a formula for $f''(x)$.

---

## 2

## THE CHAIN RULE. APPLICATIONS. THE POWER FUNCTION

The Chain Rule is one of the most important and useful tools in differentiation. The proof of the theorem establishing this rule is somewhat difficult to follow but if it is studied carefully, the reader will achieve a deeper understanding of differentiation, which will make the effort well worth while. The examples should be gone over thoroughly, as a means of studying the applications of this rule.

**THEOREM 8**    **(Fundamental Lemma of Differentiation)**    *Suppose that F has a derivative at a value u so that F'(u) exists. We define the function*

$$G(h) = \begin{cases} \dfrac{F(u+h) - F(u)}{h} - F'(u), & \text{if } h \neq 0, \\ 0, & \text{if } h = 0. \end{cases}$$

*Then* (a) *G is continuous at h = 0, and* (b) *the formula*

$$F(u+h) - F(u) = [F'(u) + G(h)]h \tag{1}$$

*holds.*

**Proof**    From the definition of derivative, we know that

$$\lim_{h \to 0} \frac{F(u+h) - F(u)}{h} = F'(u)$$

or, equivalently,

$$\lim_{h \to 0} \left[ \frac{F(u+h) - F(u)}{h} - F'(u) \right] = 0.$$

Hence $G(h) \to 0$ as $h \to 0$. Therefore $G$ is continuous at 0, and (a) holds. To establish (b), we observe that for $h \neq 0$, the formula (1) is a restatement of the definition of $G$. For $h = 0$, both sides of (1) are zero.    □

**THEOREM 9**      **(Chain Rule)** *Suppose that f, g, and u are functions with $f(x) = g[u(x)]$, and suppose that g and u are differentiable. Then f is differentiable and the following formula holds:*

$$f'(x) = g'[u(x)]u'(x).$$

**Proof**     We use the three-step rule to find the derivative of $f$.

**Step 1:**    $\Delta f = f(x+h) - f(x) = g[u(x+h)] - g[u(x)].$

We now write $g[u(x+h)]$ in the more complicated form:

$$g[u(x+h)] = g[u(x) + u(x+h) - u(x)]$$

$$= g[u + \Delta u].$$

Therefore,

$$\Delta f = g(u + \Delta u) - g(u).$$

We now apply Theorem 8 to the right side, where $g$ is used instead of $F$ and $\Delta u$ instead of $h$. Then Step 1 can be written

$$\Delta f = [g'(u) + G(\Delta u)]\Delta u.$$

**Step 2:**    $\dfrac{\Delta f}{h} = [g'(u) + G(\Delta u)]\dfrac{\Delta u}{h}.$

**Step 3:**    Since $\Delta u/h \to u'(x)$ as $h \to 0$, since $\Delta u \to 0$ as $h \to 0$, and since $G(\Delta u) \to 0$ as $\Delta u \to 0$, we conclude that

$$\lim_{h \to 0} \frac{\Delta f}{h} = f'(x) = g'(u)u'(x).$$

This is exactly the formula given in the statement of the theorem.      □

Before illustrating the Chain Rule by examples, we establish one of the most important special cases.

**COROLLARY**     *If $f(x) = [u(x)]^n$ and n is an integer, then*

$$f'(x) = n[u(x)]^{n-1}u'(x).$$

**Proof**     In the Chain Rule we take $g(u) = u^n$. Then $f = g[u(x)]$ means that $f(x) = [u(x)]^n$. We obtain

$$f'(x) = g'(u)u'(x) = nu^{n-1}u'(x). \qquad □$$

We can also write schematically

$$f(x) = (\text{expression in } x)^n,$$

$$f'(x) = n(\text{expression in } x)^{n-1}(\text{derivative of expression in } x).$$

*Remark.*    If we write $y = f(x)$ then, in Theorem 9, $y$ is a function of $u$ in the form $y = g(u)$. Also, $u$ is a function of $x$ which is expressed as $u = u(x)$. The symbol $\dfrac{dy}{dx}$ may be used for $f'(x)$, the symbol $\dfrac{dy}{du}$ may be used for $g'(u)$, and the symbol $\dfrac{du}{dx}$ may be used for $u'(x)$. The Chain Rule then takes the particularly simple form

$$\frac{dy}{dx} = \frac{dy}{du} \cdot \frac{du}{dx}.$$

We recall that the individual symbols $dx$, $du$, and $dy$ (called **differentials**) have not as yet been defined. In Chapter 4 we shall give a precise meaning to each of these quantities, and show that the above formula is obtained by dividing and multiplying by $du$.

**EXAMPLE 1**   Find $f'(x)$, given that $f(x) = (x^2 + 3x - 2)^4$.

**Solution**   From the Corollary, with $n = 4$ and the expression in $x$ being $x^2 + 3x - 2$, we find

$$f'(x) = 4(x^2 + 3x - 2)^3 \cdot (2x + 3),$$

since the derivative of $x^2 + 3x - 2$ is $2x + 3$.   □

**EXAMPLE 2**   Find $dy/dx$, given that

$$y = \frac{1}{x^3 + 3x^2 - 6x + 4}.$$

**Solution**   We write $y = (x^3 + 3x^2 - 6x + 4)^{-1}$ and apply the Corollary with $n = -1$:

$$\frac{dy}{dx} = -1(x^3 + 3x^2 - 6x + 4)^{-2} \cdot (3x^2 + 6x - 6)$$

$$= \frac{-3(x^2 + 2x - 2)}{(x^3 + 3x^2 - 6x + 4)^2}.$$   □

We note that this example could also have been worked by using the formula for the derivative of a quotient.

**EXAMPLE 3**   Find $dy/dx$, given that

$$y = \left(\frac{3x - 2}{2x + 1}\right)^7.$$

**Solution**   This requires a combination of formulas. First, from the Corollary, we see that

$$\frac{dy}{dx} = 7\left(\frac{3x - 2}{2x + 1}\right)^6 \cdot \left(\text{derivative of } \frac{3x - 2}{2x + 1}\right).$$

To find the derivative of $(3x - 2)/(2x + 1)$ we apply the quotient formula to get

$$\frac{dy}{dx} = 7\left(\frac{3x - 2}{2x + 1}\right)^6 \frac{(2x + 1)(3) - (3x - 2)(2)}{(2x + 1)^2}$$

$$= \frac{49(3x - 2)^6}{(2x + 1)^8}.$$

**EXAMPLE 4**  Find $f'(x)$, given that

$$f(x) = (x^2 + 2x - 3)^{16}(2x + 5)^{13}.$$

**Solution**  We first observe that if we let $u(x) = (x^2 + 2x - 3)^{16}$ and $v(x) = (2x + 5)^{13}$, we may use Theorem 5, on the product of two functions:

$$f'(x) = u(x)v'(x) + v(x)u'(x).$$

To find $u'(x)$ we employ the Chain Rule:

$$u'(x) = 16(x^2 + 2x - 3)^{15}(2x + 2);$$

and similarly,

$$v'(x) = 13(2x + 5)^{12}(2).$$

Substituting in the formula for $f'(x)$, we get

$$f'(x) = (x^2 + 2x - 3)^{16}26(2x + 5)^{12}$$

$$+ (2x + 5)^{13}32(x^2 + 2x - 3)^{15}(x + 1)$$

$$= (x^2 + 2x - 3)^{15}(2x + 5)^{12}[26(x^2 + 2x - 3) + 32(2x + 5)(x + 1)].$$

The function

$$f(x) = x^n,$$

where $n$ is some number, is called the **power function**. That is, $x$ is raised to an exponent called the **power**. If $n$ is an *even positive integer*, the graph of the function appears as in Fig. 1. If $n$ is an odd positive integer the curve is below the $x$ axis for negative values of $x$, as shown in Fig. 2.

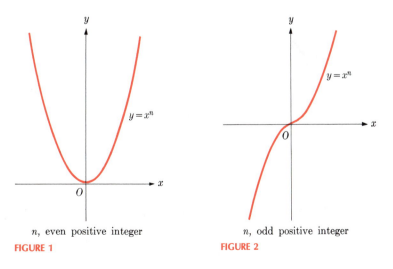

$y = x^n$  (Figure 1)    $y = x^n$  (Figure 2)

*n*, even positive integer        *n*, odd positive integer

FIGURE 1                          FIGURE 2

For negative exponents, the $y$ axis is a vertical asymptote and the $x$ axis is a horizontal asymptote. The general behavior of these functions is shown in Fig. 3.

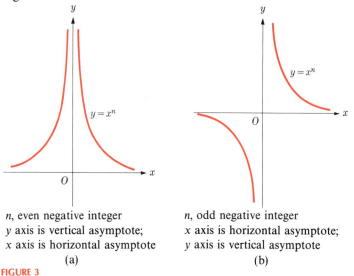

$n$, even negative integer
$y$ axis is vertical asymptote;
$x$ axis is horizontal asymptote
(a)

$n$, odd negative integer
$x$ axis is horizontal asymptote;
$y$ axis is vertical asymptote
(b)

FIGURE 3

If $n$ is of the form $1/q$, where $q$ is a positive integer, the graph of the function has a completely different character. For example, the function

$$y = x^{1/2} = \sqrt{x}$$

is not even defined for negative values of $x$. In fact, if $q$ is any even number, the function

$$y = x^{1/q} = \sqrt[q]{x}$$

is defined only for $x$ positive or zero. On the other hand, if $q$ is odd, we know that because the odd root of a negative number can be found (for example, $\sqrt[5]{-32} = -2$), the function is defined for all values of $x$. The graphs of $\sqrt[q]{x}$ are shown in Fig. 4.

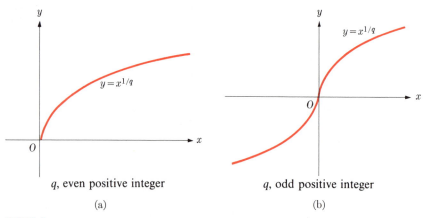

$q$, even positive integer

(a)

$q$, odd positive integer

(b)

FIGURE 4

We follow the same rule for the derivative of $x^{1/q}$ as for the derivative of the power function when the exponent is an integer. The following theorem establishes this fact.

**THEOREM 10**    *If $f(x) = x^{1/q}$ and $q$ is an integer, then*

$$f'(x) = \frac{1}{q} x^{(1/q)-1}.$$

This theorem is a special case of Theorem 10 in Chapter 7, Section 4. The result established there shows that if $s$ is *any real number* then the derivative of $f(x) = x^s$ is $f'(x) = sx^{s-1}$. In the meantime we shall use the result of Theorem 10 above and the corollaries below. For a direct proof of Theorem 10 see Problem 56 at the end of this section and the hint given there.

A **rational number** is one which can be written as one integer over another. "The number $r$ is rational" means that $r = p/q$, where $p$ and $q$ are integers, with $q \neq 0$. Theorem 10 can be combined with the Chain Rule to give the general rule for differentiating the power function when the exponent is any rational number.

**COROLLARY 1**    *If $f(x) = x^r$ and $r$ is any rational number, then $f'(x) = rx^{r-1}$.*

**Proof**    Since $r$ is rational, we can write

$$f(x) = x^{p/q} = (x^{1/q})^p,$$

where $p$ and $q$ are integers. By the special case of the Chain Rule, we have

$$f'(x) = p(x^{1/q})^{p-1} \cdot (\text{derivative of } x^{1/q});$$

and so, by Theorem 10,

$$f'(x) = p(x^{1/q})^{p-1} \frac{1}{q} x^{(1/q)-1} = \frac{p}{q} x^{(p/q)-(1/q)+(1/q)-1} = rx^{r-1}. \qquad \square$$

**COROLLARY 2**    *If $f(x) = [u(x)]^r$ and $r$ is rational, then $f'(x) = r[u(x)]^{r-1} u'(x)$.*

**Proof**    This is an immediate consequence of Corollary 1 and the Chain Rule. $\qquad \square$

**EXAMPLE 5**    Given that $f(t) = \sqrt[3]{t^3 + 3t + 1}$, find $f'(t)$.

**Solution**    We write $f(t) = (t^3 + 3t + 1)^{1/3}$ and use Corollary 2:

$$f'(t) = \tfrac{1}{3}(t^3 + 3t + 1)^{-2/3}(3t^2 + 3) = \frac{t^2 + 1}{(t^3 + 3t + 1)^{2/3}}. \qquad \square$$

**EXAMPLE 6**    Given that $f(x) = (x+1)^3 (2x-1)^{4/3}$, find $f'(x)$.

**Solution**    We use the rule for differentiating a product, setting $u(x) = (x+1)^3$ and $v(x) = (2x-1)^{4/3}$, so that $f(x) = u(x)v(x)$. Then

$$f'(x) = (x+1)^3 v'(x) + (2x-1)^{4/3} u'(x).$$

By the Chain Rule for powers we have

$$u'(x) = 3(x+1)^2 \cdot 1 \qquad \text{and} \qquad v'(x) = \tfrac{4}{3}(2x-1)^{1/3} \cdot 2.$$

Therefore

$$f'(x) = \tfrac{8}{3}(x+1)^3(2x-1)^{1/3} + 3(2x-1)^{4/3}(x+1)^2$$
$$= (x+1)^2(2x-1)^{1/3}[\tfrac{8}{3}(x+1) + 3(2x-1)]$$
$$= \tfrac{1}{3}(x+1)^2(2x-1)^{1/3}(26x-1).$$

☐

**EXAMPLE 7**　　Given the function

$$f(s) = \frac{s}{\sqrt{s^2-1}},$$

find $f'(s)$.

**Solution**　　We write

$$f(s) = \frac{s}{(s^2-1)^{1/2}}$$

and use Theorem 6 for the derivative of a quotient:

$$f'(s) = \frac{(s^2-1)^{1/2}\cdot 1 - s\tfrac{1}{2}(s^2-1)^{-1/2}(2s)}{s^2-1}.$$

This expression may be simplified in several ways. One easy method is to multiply through both the numerator and the denominator by $(s^2-1)^{1/2}$. This gives

$$f'(s) = \frac{(s^2-1)^1 - s^2(s^2-1)^0}{(s^2-1)^{3/2}}.$$

Since $(s^2-1)^0 = 1$, we get

$$f'(s) = \frac{-1}{(s^2-1)^{3/2}}.$$

☐

## 2　PROBLEMS

In each of Problems 1 through 36, find the derivative.

**1** $(2x+3)^9$

**2** $(x^2+4x-3)^6$

**3** $(2-5x)^4$

**4** $(x^3+4x^2-3x+7)^5$

**5** $(2x+1)^{-4}$

**6** $(x^2-3x+2)^{-5}$

**7** $(x^3+2x-3+x^{-2})^4$

**8** $(x^4+5x-6x^{-1})^3$

**9** $(x^2+1)^2(x^3-2x)^2$

**10** $(x^3+2x-6)^3(x^2-4x+5)^7$

**11** $(x^2-x^{-1}+1)(x^3+2x-6)^7$

**12** $\dfrac{(3x-2)^2}{(2x-6)^2}$

**13** $\dfrac{(x^2+1)^3}{(x^2+2)^2}$

**14** $\dfrac{(2x-6)^4}{(x+1)^7}$

**15** $\dfrac{(x^2+2)^2}{(x^2+x^{-1})^3}$

**16** $\dfrac{(2x-6)^{-1}}{(x^2+3)^{-2}}$

**17** $x^{5/3}+2x^{4/3}-3x^{-1/3}$

**18** $x^{3/2}+2x^{3/5}-4x^{4/7}$

**19** $x^{-2/3}+x^{-4/3}-2x^{4/7}$

**20** $2x^{2/3}+5x^{-1/4}-\tfrac{2}{5}x^{5/2}$

**21** $\tfrac{3}{4}x^{4/3}+2x^{1/2}-2x^{-1}$

**22** $\tfrac{3}{8}x^{8/3}+\tfrac{1}{2}x^{-1}-4x^{1/4}$

**23** $(2x+3)^{10/3}$

**24** $(3x-2)^{4/3}$

**25** $(x^2+2x+3)^{3/2}$

**26** $(x^3+3x^2+6x+5)^{2/3}$

**27** $(2x^3-3x^2+3x-1)^{-1/3}$

**28** $(x^4+2x^2+1)^{-7/4}$

**29** $\sqrt[3]{(x^3+3x+2)^2}$

**30** $\dfrac{1}{\sqrt[4]{x^2+2x+3}}$

**31** $\dfrac{(x^2+x-3)\sqrt{x^2+x-3}}{5}$

**32** $\dfrac{1}{2(x^2+2x+3)\sqrt[3]{(x^2+2x+3)^2}}$

**33** $(2x + 3)^4(3x - 2)^{7/3}$

**34** $(3 - 2x)^{5/2}(2 - 3x)^{1/3}$

**35** $(2x - 1)^{5/2}(7x - 3)^{3/7}$

**36** $(2 - 3x)^{4/3}(5x + 2)^3$

In Problems 37 through 40, find the value of $f'(x)$ for the given value of $x$.

**37** $f(x) = (x^2 + 2x + 3)^{2/3}, \quad x = -1$

**38** $f(x) = \dfrac{2x}{\sqrt{4x^2 + 1}}, \quad x = 2$

**39** $f(x) = (x^2 + 1)^3(2x + 4)^{1/3}, \quad x = 2$

**40** $f(x) = (x^2 + 6)^2\sqrt[3]{5x + 12}, \quad x = 3$

**41** Find $g'(t)$, given $g(t) = \dfrac{\sqrt[3]{2 + 6t}}{t}$

**42** Find $\dfrac{dy}{ds}$, given $y = \dfrac{\sqrt{s^2 + 1}}{s}$

**43** Find $f'(y)$, given $f(y) = \dfrac{2y + 3}{\sqrt{y^2 + 3y + 4}}$

**44** Find $\dfrac{dy}{dr}$, given $y = \dfrac{\sqrt{1 + 4r^2}}{r^2}$

In Problems 45 through 48, find the equations of the tangent line and normal line to the curve $y = f(x)$, at the point having the given value of $x$.

**45** $f(x) = (2x - 3)^{5/2}, \quad x = 2$

**46** $f(x) = \dfrac{1}{\sqrt[3]{2 - 3x}}, \quad x = 1$

**47** $f(x) = x\sqrt{25 - x^2}, \quad x = 4$

**48** $f(x) = x\sqrt[3]{2 - x}, \quad x = 1$

**49** Suppose that $v$ is a function of $x$ so that $v = v(x)$, that $u$ is a function of $v$, and that $g$ is a function of $u$ where $u$, $v$, and $g$ are differentiable. Writing $f(x) = g\{u[v(x)]\}$, show that

$$f'(x) = g'u'v'$$

or, in differential notation with $y = f(x)$, establish the formula

$$\frac{dy}{dx} = \frac{dy}{du} \cdot \frac{du}{dv} \cdot \frac{dv}{dx}.$$

**50** Let $P(u) = a_n u^n + a_{n-1} u^{n-1} + \cdots + a_1 u + a_0$ be a polynomial in $u$. Suppose that $u$ is a differentiable function of $x$. Writing $f(x) = P[u(x)]$, find $f'(x)$ in terms of $u$ and $u'$.

**51** Let $u$ and $v$ be differentiable functions of $x$. Suppose that

$$f(x) = g(u + v), \quad F(x) = G(u \cdot v), \quad \text{and} \quad H(x) = h\left(\frac{u}{v}\right),$$

where $g$, $G$, and $h$ are differentiable functions. Find formulas for $f'(x)$, $F'(x)$, and $H'(x)$.

**52** Find a formula for the derivative of

$$f(x) = [u(x)]^m [v(x)]^n$$

and apply it to the functions

a) $f(x) = (2x + 1)^6(x^3 - 6)^{4/3}$,

b) $f(x) = (x^2 + 2x - 1)^{3/2}(x^4 - 3)^3$.

**53** Find a formula for the derivative of

$$f(x) = [u(x)]^r [v(x)]^s [w(x)]^t,$$

where $r$, $s$, and $t$ are rational numbers. Apply it to the functions

a) $f(x) = (x^2 + 1)^{3/2}(2x - 5)^{1/3}(x^2 + 4)^2$,

b) $f(x) = (x - 1)^3(3 - 2x)^2(x^2 - 5)^{4/3}$.

**54** Use the three-step rule to find the derivative of $f(x) = x^{1/2}$.

**\*55** Use the three-step rule to find the derivative of $f(x) = x^{1/3}$. [*Hint:* Use the formula

$$a^3 - b^3 = (a - b)(a^2 + ab + b^2).]$$

**\*56** Use the three-step rule to find the derivative of $f(x) = x^{1/q}$ where $q$ is a positive integer. [*Hint:* Use the formula

$$a^m - b^m = (a - b)(a^{m-1} + a^{m-2}b + a^{m-3}b^2 + \cdots$$
$$+ ab^{m-2} + b^{m-1})$$

and set $a = (x + h)^{1/q}, b = x^{1/q}, m = q$. Then "rationalize the numerator" in the formula obtained in Step 2 of the three-step rule by multiplying numerator and denominator by $a^{q-1} + a^{q-2}b + \cdots + ab^{q-2} + b^{q-1}.]$

---

**3**

## IMPLICIT DIFFERENTIATION

Before discussing implicit differentiation, we shall introduce additional symbolism which will be particularly helpful. Suppose that $f$ is a function of $x$, such as

$$f(x) = x^4 - 2x^2 + 3.$$

If $y$ is used as an element in the range of the function, we can write

$$y = f(x) = x^4 - 2x^2 + 3.$$

Sometimes, instead of $f(x)$, it is useful to write $y(x)$, *which means the same thing*. The derivative of $f$ is denoted by $f'$, and, similarly, we may denote this derivative as $y'$ or as $y'(x)$. The symbol $\frac{dy}{dx}$ is interchangeable with $y'$ and $y'(x)$.

Thus we could write

$$y = y(x) = x^4 - 2x^2 + 3, \qquad \text{and} \qquad y' = y'(x) = 4x^3 - 4x.$$

In Chapter 1, Section 6, we discussed relations in the plane. The graph of an equation such as

$$x^2 + y^2 - 4 = 0$$

is an example of such a relation. The solution set $S$ defined by

$$S = \{(x, y): x^2 + y^2 - 4 = 0\}$$

is a circle, as shown in Fig. 5. Although this relation is not a function, we may describe $S$ by means of two functions. Defining

$$S_1 = \{(x, y): y = \sqrt{4 - x^2}\}$$

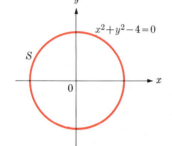

and

$$S_2 = \{(x, y): y = -\sqrt{4 - x^2}\},$$

we see that $S = S_1 \cup S_2$. The functions describing $S_1$ and $S_2$ are

$$y_1 = \sqrt{4 - x^2} \qquad \text{and} \qquad y_2 = -\sqrt{4 - x^2}.$$

See Figs. 6 and 7.

Examples of relations which may be defined by decomposition into one or more functions are

$$x^2 - 2xy + 3y^2 - 7 = 0, \quad x^3 y^2 - 4 = 0, \quad y^3 - x - 5 = 0.$$

Note that all these expressions are of the general form

$$f(x, y) = 0.$$

When this situation occurs we say that the functions, if any, are **defined implicitly** by the equation. We say "if any" because the equation may not define any function at all. For example, there are no values of $x$ and $y$ which satisfy an equation such as

$$x^2 + y^2 + 9 = 0.$$

We could easily compose other examples.

Returning to the equation

$$x^2 + y^2 - 4 = 0,$$

we see that this consists of two functions, each function having for its domain the interval $-2 \le x \le 2$. Inside this interval the functions possess derivatives which can readily be found by differentiating the expressions

$$\sqrt{4 - x^2} \qquad \text{and} \qquad -\sqrt{4 - x^2}$$

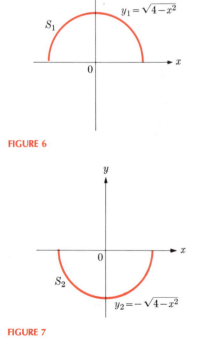

**FIGURE 5**

**FIGURE 6**

**FIGURE 7**

which we have denoted $y_1$ and $y_2$, respectively.

However, it is possible to proceed directly, without solving for $y$ in terms of $x$. In the equation

$$x^2 + y^2 - 4 = 0,$$

we first differentiate $x^2$; its derivative is simply $2x$. The derivative of $y^2$ is more complicated, since $y = y(x)$ and we must use the Chain Rule. The derivative of $y^2$ is $2yy'$. The derivative of $-4$ is zero. We have now found the derivative of the left side, and we know that the right side, being zero, has derivative zero. We conclude that

$$2x + 2yy' = 0.$$

This may be solved for $y'$ to yield (when $y \neq 0$),

$$y' = -\frac{x}{y}.$$

We now ask ourselves which function of $x$ has been differentiated. The answer is *both*. If $y_1$ is substituted for $y$ in the right side of the expression for $y'$, we have the derivative of $y_1$; if $y_2$ is substituted, we have the derivative of $y_2$. Further examples illustrate the method.

**EXAMPLE 1**      Assuming that $y$ is a differentiable function defined implicitly by the equation

$$y^3 + 3xy + x^3 - 5 = 0,$$

find $dy/dx$ in terms of $x$ and $y$.

**Solution**      The derivative of $y^3$ is $3y^2(dy/dx)$. The term $3xy$ must be treated as a product. The derivative of $3xy$ is $3x(dy/dx) + 3y$. The derivative of $x^3$ is $3x^2$. The derivative of $-5$ is $0$. Therefore

$$3y^2 \frac{dy}{dx} + 3x \frac{dy}{dx} + 3y + 3x^2 = 0.$$

We can now solve for $\dfrac{dy}{dx}$ (assuming $y^2 + x \neq 0$):

$$\frac{dy}{dx} = -\frac{y + x^2}{y^2 + x}. \qquad\qquad \square$$

*Remark.* If it turns out that the implicit relation is determined by several functions, then the answer gives the appropriate derivative according to which $y(x)$ is put into the right side of the equation.

**EXAMPLE 2**      Find the derivative of

$$4x^2 + 9y^2 = 36$$

implicitly, and check the result by solving for $y$ and differentiating explicitly.

**Solution**      The implicit method gives

$$8x + 18yy' = 0 \qquad \text{and} \qquad y' = -\frac{4x}{9y}, \quad y \neq 0.$$

On the other hand, using the explicit method, we find that

$$y_1 = +\tfrac{1}{3}\sqrt{36 - 4x^2}, \qquad y_2 = -\tfrac{1}{3}\sqrt{36 - 4x^2},$$

$$y_1'(x) = \tfrac{1}{3} \cdot \tfrac{1}{2}(36 - 4x^2)^{-1/2}(-8x) = -\frac{4x}{3\sqrt{36 - 4x^2}},$$

and $y_2'(x)$ is the same expression preceded by a plus sign.

From the implicit method,

$$y_1' = -\frac{4x}{9y_1} = -\frac{4x}{9 \cdot \tfrac{1}{3}\sqrt{36 - 4x^2}} = -\frac{4x}{3\sqrt{36 - 4x^2}},$$

which is identical with the explicit result; a similar result is obtained for $y_2(x)$. ☐

**EXAMPLE 3**  Assuming that $y$ is a differentiable function, find the derivative when $y^5 + 3x^2 y^3 - 7x^6 - 8 = 0$.

**Solution**  In this problem we have no choice as to whether we shall use the explicit or the implicit method, since it is impossible to solve for $y$ in terms of $x$ or for $x$ in terms of $y$. We obtain

$$5y^4 \frac{dy}{dx} + 3x^2 \cdot 3y^2 \frac{dy}{dx} + 3y^3 \cdot 2x - 42x^5 = 0,$$

and solving for $dy/dx$ we find

$$\frac{dy}{dx} = \frac{42x^5 - 6xy^3}{5y^4 + 9x^2 y^2}.$$

☐

## 3  PROBLEMS

In each of Problems 1 through 12, find $y'$ in terms of $x$ and $y$ by implicit differentiation, assuming that $y$ is a differentiable function.

1  $x^2 - 2y^2 + 5 = 0$
2  $4x^2 + 8y^3 - 5 = 0$
3  $x^4 + 2y^4 - 4 = 0$
4  $2x^3 + 3y^3 + 6 = 0$
5  $x^2 - 3xy + y^2 = 6$
6  $x^2 + 2xy - 2y^2 + 3x - y - 9 = 0$
7  $x^3 + 6xy + 5y^3 = 3$
8  $x^3 + 2x^2 y - xy^2 + 2y^3 = 4$
9  $2x^3 - 3x^2 y + 2xy^2 - y^3 = 2$
10  $x^4 + 2x^2 y^2 + xy^3 + 2y^4 = 6$
11  $x^6 + 2x^3 y - xy^7 = 10$
12  $x^5 - 2x^3 y^2 + 3xy^4 - y^5 = 5$

In Problems 13 through 23, find $dy/dx$ in terms of $x$ and $y$ by implicit differentiation; then solve for $y$ in terms of $x$, and show that each solution and its derivative satisfy the equation obtained.

13  $x^2 - 2xy = 3$
14  $y^2 = 4x$
15  $xy - 4 = 0$
16  $x^{1/2} + y^{1/2} = 1$
17  $x^{2/3} + y^{2/3} = a^{2/3}, \quad a = \text{const}$
18  $y^2 - 2x = 4$
19  $y^2 + 2x = 8$
20  $2x^2 + 3y^2 = 5$
21  $2x^2 - y^2 = 1$
22  $x^2 - xy + y^2 = 1$
23  $2x^2 - 3xy - 4y^2 = 5$

In Problems 24 through 29, find $y'$ by implicit differentiation; then solve for $y$ in terms of $x$ and select the particular function passing through the given point. Find the general form for $y'$ for this function and evaluate $y'$ at the given point.

24  $x^2 + y^2 = 25 \quad (4, 3)$
25  $y^2 = 2x \quad (2, -2)$
26  $xy = 3 \quad (1, 3)$
27  $x^2 + y^2 = 10 \quad (3, -1)$
28  $3x^2 - 2xy - y^2 = 3 \quad (1, 0)$

**29** $2x^2 - 3xy + 2y^2 = 2$   $(-1, -\frac{3}{2})$

**30** Find the derivative, $dy/dx$, of the relation $|x - 3y| - 4 = 0$. Sketch the graph.

**31** Find the derivative, $y'$, of the relation $|y^2 - 3xy| - 1 = 0$. Sketch the graph.

**32** Given the relation $|x^2 - y^2| - 10(x^2 + y^2) = 0$, find $dy/dx$. For what values of $x$ is there a corresponding value of $dy/dx$?

**33** Let $r$ be a rational number of the form $p/q$ where $p$ and $q$ are integers. Let $y = x^r$. By raising both sides of this equation to the $q$th power, use implicit differentiation to obtain a proof of Theorem 10 of Section 2. What facts about implicit functions are used in this proof?

In Problems 34 through 37, find the equation of the tangent line to the given curve at the point or points given.

**34** $x^2 + y^2 - 4 = 0$,   $(1, \sqrt{3})$

**35** $x^2 + xy^3 - 20 = 0$,   $x = 2$

**36** $x^3 - 2xy + y^3 + 3 = 0$,   $(-2, 1)$

**37** $2xy + 2x - y - 6 = 0$,   $x = 1$

---

**4**

---

## TRIGONOMETRIC LIMITS

In Chapter 2, Sections 1 and 2, we learned several techniques for evaluating the limits of various algebraic expressions. We shall now employ a new device for finding the limit in a special case. We consider the function

$$f(\theta) = \frac{\sin \theta}{\theta}$$

and suppose that $\theta$ is measured in *radians*.* This function is defined for all values of $\theta$ except $\theta = 0$. Since $\sin 0 = 0$, straight substitution gives $f(0) = 0/0$, a meaningless expression. Nevertheless, we can find the value of

$$\lim_{\theta \to 0} \frac{\sin \theta}{\theta}.$$

We shall start by supposing that $\theta$ is positive and less than $\pi/2$. Figure 8 shows a typical value of $\theta$. We draw an arc $\overset{\frown}{IP}$ of a circle of radius 1 and construct the perpendiculars $PQ$ and $TI$, as shown. From this figure, it is clear that

$$\text{area } \triangle IOP < \text{area sector } IOP < \text{area } \triangle IOT.$$

As usual, we use the symbol $|AB|$ to denote the length (always positive) of a line segment $AB$. The area of a circle of radius $r$ is $\pi r^2$ which we write as $2\pi(\frac{1}{2}r^2)$. Therefore, the area of a sector $IOP = \theta(\frac{1}{2}r^2)$, $\theta$ measured in radians. Since $|OI| = |OP| = 1$, we have

$$\text{area } \triangle IOP = \tfrac{1}{2}|PQ| \cdot |OI| = \tfrac{1}{2}|PQ|;$$

$$\text{area sector } IOP = \tfrac{1}{2}\theta r^2 = \tfrac{1}{2}\theta \ (\theta \text{ in radians});$$

$$\text{area } \triangle IOT = \tfrac{1}{2}|IT| \cdot |OI| = \tfrac{1}{2}|IT|.$$

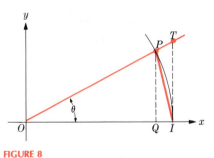

**FIGURE 8**

---

*We provide a review of the elements of trigonometry in Appendix 1.

On the other hand, we know that

$$\sin \theta = \frac{|PQ|}{|OP|} = |PQ|; \qquad \tan \theta = \frac{|IT|}{|OI|} = |IT|.$$

Therefore the above area inequalities are the same as the inequalities

$$\tfrac{1}{2} \sin \theta < \tfrac{1}{2} \theta < \tfrac{1}{2} \tan \theta.$$

We multiply by 2 and divide by $\sin \theta$. (This keeps the same direction for the inequalities, since everything is positive.) We obtain

$$1 < \frac{\theta}{\sin \theta} < \frac{1}{\cos \theta}.$$

Now we take reciprocals; this has the effect of reversing the direction. (Recall that if $a$ and $b$ are any positive numbers and $a < b$, then $1/a > 1/b$.) We conclude that

$$1 > \frac{\sin \theta}{\theta} > \cos \theta.$$

As $\theta$ tends to zero, we know that $\cos \theta \to 1$. Since $\sin \theta/\theta$ is always between 1 and a number tending to 1, we know from the Sandwich Theorem that it also must approach 1 as $\theta$ approaches zero. In the proof we considered $\theta$ positive, but now we note that

$$f(\theta) = \frac{\sin \theta}{\theta}$$

is an *even* function of $\theta$. That is, if we replace $\theta$ by $-\theta$, we have

$$f(-\theta) = \frac{\sin (-\theta)}{-\theta} = \frac{-\sin \theta}{-\theta} = \frac{\sin \theta}{\theta} = f(\theta).$$

In general, a function $f(x)$ is said to be **even** if $f(-x) = f(x)$ for all $x$; it is said to be **odd** if $f(-x) = -f(x)$ for all $x$. Therefore, as $\theta \to 0$ through negative values, the result must be identical with the one obtained when $\theta \to 0$ through positive values. The above development gives the following result.

**THEOREM 11**   *If $\theta$ is measured in radians, then*

$$\lim_{\theta \to 0} \frac{\sin \theta}{\theta} = 1.$$

**COROLLARY**   *We have*

$$\lim_{\theta \to 0} \frac{1 - \cos \theta}{\theta} = 0.$$

**Proof**   This result may be obtained from Theorem 1 by using the identity

$$\frac{1-\cos\theta}{\theta} = \frac{(1-\cos\theta)(1+\cos\theta)}{\theta(1+\cos\theta)} = \frac{1-\cos^2\theta}{\theta(1+\cos\theta)} = \frac{\sin^2\theta}{\theta(1+\cos\theta)}$$

$$= \frac{\sin\theta}{\theta}\cdot\frac{\sin\theta}{1+\cos\theta}$$

We know that

$$\lim_{\theta\to 0}\frac{\sin\theta}{\theta} = 1 \quad \text{and} \quad \lim_{\theta\to 0}\frac{\sin\theta}{1+\cos\theta} = 0,$$

since the second limit may be obtained by straight substitution. From the fact that the limit of a product is the product of the limits, we obtain the formula of the corollary.                                                                                □

**EXAMPLE 1**    Find the value of

$$\lim_{\theta\to 0}\frac{\sin 2\theta}{\theta},$$

$\theta$ measured in radians.

**Solution**    We write

$$\frac{\sin 2\theta}{\theta} = \frac{2\sin(2\theta)}{(2\theta)}.$$

Therefore

$$\lim_{\theta\to 0}\frac{\sin 2\theta}{\theta} = 2\lim_{(2\theta)\to 0}\frac{\sin(2\theta)}{(2\theta)} = 2\cdot 1.$$

We made use of the fact that when $\theta \to 0$, so does $2\theta$.                    □

**EXAMPLE 2**    Find the value of

$$\lim_{\alpha\to 0}\frac{\sin\alpha}{\alpha},$$

$\alpha$ measured in degrees.

**Solution**    The functions $\sin\alpha°$ and $\sin(\pi\alpha/180)$ have the same values, with $\alpha$ the measure in degrees. We have

$$\lim_{\alpha\to 0}\frac{\sin\alpha}{\alpha} = \lim_{\alpha\to 0}\frac{\pi}{180}\frac{\sin\dfrac{\pi\alpha}{180}}{\dfrac{\pi\alpha}{180}} = \frac{\pi}{180}\lim_{\frac{\pi\alpha}{180}\to 0}\frac{\sin\dfrac{\pi\alpha}{180}}{\dfrac{\pi\alpha}{180}} = \frac{\pi}{180}.$$                    □

Example 2 exhibits one of the reasons radian measure is used throughout calculus. The limit of $\sin\theta/\theta$ is 1 if $\theta$ is measured in radians and $\pi/180$ if $\theta$ is measured in degrees. In the next section we shall see why the first is highly preferable to the second.

_____ 5

## THE DIFFERENTIATION OF TRIGONOMETRIC FUNCTIONS

With the help of the limits obtained in the preceding section, we can find the derivatives of all the trigonometric functions. *We shall suppose throughout that all angles are measured in radians.*

**THEOREM 12**    *If* $f(x) = \sin x$, *then* $f'(x) = \cos x$.

**Proof**    We apply the three-step rule.

**Step 1:**    $f(x + h) - f(x) = \sin(x + h) - \sin x$.

**Step 2:**    $\dfrac{f(x + h) - f(x)}{h} = \dfrac{\sin(x + h) - \sin x}{h}$.

As usual, it is at this point that we must employ ingenuity. We first use the fact that $\sin(x + h) = \sin x \cos h + \cos x \sin h$ and write

$$\frac{f(x + h) - f(x)}{h} = \frac{\sin x \cos h + \cos x \sin h - \sin x}{h}$$

$$= \cos x \frac{\sin h}{h} + \sin x \frac{\cos h - 1}{h}.$$

Then, taking the limit as $h \to 0$, we use Theorem 11 and its Corollary to perform Step 3 and conclude that

$$f'(x) = \lim_{h \to 0} \frac{f(x + h) - f(x)}{h} = \cos x. \qquad \square$$

*The simplicity of the result—that the derivative of* $\sin x$ *is* $\cos x$*—is due to the use of radians.* If, for example, degrees were used, then $\lim_{h \to 0} (\sin h/h)$ would be $\pi/180$, and all the derivative formulas for trigonometric functions would have an extra, inconvenient factor in them. Later we shall see that a similar situation occurs in the study of logarithms.

Suppose that $u$ is a function of $x$; that is, $u = u(x)$. Then, by using the Chain Rule, we easily find the formula for the derivative of $\sin u$. If we write

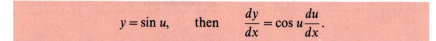

$$y = \sin u, \qquad \text{then} \qquad \frac{dy}{dx} = \cos u \frac{du}{dx}.$$

**EXAMPLE 1**    Find the derivative of $y = \sin(3x^2 - 6x + 1)$.

**Solution**    Taking $u = 3x^2 - 6x + 1$, we obtain from the formula for the derivative of the sine function and the Chain Rule

$$\frac{dy}{dx} = \cos(3x^2 - 6x + 1) \cdot (6x - 6) = 6(x - 1)\cos(3x^2 - 6x + 1). \qquad \square$$

**THEOREM 13**   *If $f(x) = \cos x$,   then $f'(x) = -\sin x$.*

**Proof**   We can obtain this result by the three-step rule. However, by writing

$$f(x) = \cos x = \sin\left(\frac{\pi}{2} - x\right),$$

we may apply the Chain Rule and obtain the result more simply. We write

$$u(x) = \frac{\pi}{2} - x.$$

Then

$$f(x) = \sin u \qquad \text{and} \qquad f'(x) = \cos u \, \frac{du}{dx}.$$

But $du/dx = -1$ and so

$$f'(x) = \cos u(-1) = -\cos\left(\frac{\pi}{2} - x\right) = -\sin x. \qquad \square$$

The Chain Rule again shows that if

$$y = \cos u, \qquad \text{then} \qquad \frac{dy}{dx} = -\sin u \, \frac{du}{dx}.$$

With the help of the derivatives of the sine and cosine functions, we may find the derivatives of the remaining trigonometric functions. The following example shows the method.

**EXAMPLE 2**   Find the derivative of $y = \tan x$.

**Solution**   Since $\tan x = \sin x/\cos x$, we write

$$y = \frac{\sin x}{\cos x}$$

and use the formula for the derivative of a quotient. We have

$$\frac{dy}{dx} = \frac{\cos x(\cos x) - \sin x(-\sin x)}{\cos^2 x}.$$

This yields

$$\frac{dy}{dx} = \frac{1}{\cos^2 x} = \sec^2 x. \qquad \square$$

The formulas for differentiating trigonometric functions are summarized in the following table (in each case $u$ is a function of $x$):

$$\text{If } y = \sin u, \quad \text{then } \frac{dy}{dx} = \cos u \frac{du}{dx}$$

$$y = \cos u, \quad \text{then } \frac{dy}{dx} = -\sin u \frac{du}{dx}$$

$$y = \tan u, \quad \text{then } \frac{dy}{dx} = \sec^2 u \frac{du}{dx}$$

$$y = \cot u, \quad \text{then } \frac{dy}{dx} = -\csc^2 u \frac{du}{dx}$$

$$y = \sec u, \quad \text{then } \frac{dy}{dx} = \sec u \tan u \frac{du}{dx}$$

$$y = \csc u, \quad \text{then } \frac{dy}{dx} = -\csc u \cot u \frac{du}{dx}.$$

The general rules for differentiating polynomials and rational functions, together with the formulas for products and quotients, open up vast possibilities for differentiating complicated expressions. The next examples show some of the ramifications.

**EXAMPLE 3**   Given that $f(x) = \sin^3 x$, find $f'(x)$.

**Solution**   $f(x) = (\sin x)^3$, and setting $u = \sin x$ we have by the Chain Rule

$$f(x) = u^3, \qquad f'(x) = 3u^2 u'(x).$$

Since $u'(x) = \cos x$, we obtain

$$f'(x) = 3 \sin^2 x \cos x. \qquad \square$$

**EXAMPLE 4**   Given that $y = \tan^4(2x + 1)$, find $\dfrac{dy}{dx}$.

**Solution**   We use the Chain Rule twice. We first set $u = \tan(2x + 1)$. Then by the Chain Rule,

$$y = u^4 \qquad \text{and} \qquad \frac{dy}{dx} = 4u^3 \frac{du}{dx}.$$

To find $\dfrac{du}{dx}$ we use the Chain Rule again. Setting $v = 2x + 1$, we have

$$u = \tan v \qquad \text{and} \qquad \frac{du}{dx} = \sec^2 v \frac{dv}{dx}.$$

We easily see that $dv/dx = 2$, and therefore

$$\frac{dy}{dx} = 4[\tan(2x + 1)]^3 \sec^2(2x + 1) \cdot 2$$

$$= 8 \tan^3(2x + 1) \sec^2(2x + 1). \qquad \square$$

**EXAMPLE 5**   Given that $f(x) = x^3 \sec^2 3x$, find $f'(x)$.

**Solution**   $f(x) = x^3[\sec(3x)]^2$. We see that this is in the form of a product, and write $u = x^3$, $v = [\sec(3x)]^2$. Then

$$f'(x) = u\frac{dv}{dx} + v\frac{du}{dx}.$$

We have $\dfrac{du}{dx} = 3x^2$ and to obtain $\dfrac{dv}{dx}$ we use the Chain Rule. Setting $w = 3x$ we have

$$v = [\sec w]^2, \qquad \frac{dv}{dx} = 2\sec w \sec w \tan w\frac{dw}{dx}.$$

Thus we find

$$f'(x) = x^3 \cdot 6\sec^2(3x)\tan(3x) + \sec^2(3x) \cdot 3x^2$$

$$= 3x^2\sec^2 3x(2x\tan 3x + 1).$$

## 5   PROBLEMS

In each of Problems 1 through 18, evaluate the limit (all angles in radians).

**1** $\displaystyle\lim_{h\to 0}\frac{\sin 5h}{h}$

**2** $\displaystyle\lim_{\theta\to 1}\frac{\sin 2(\theta - 1)}{\theta - 1}$

**3** $\displaystyle\lim_{h\to 0}\frac{\sin 2h}{\sin 5h}$

**4** $\displaystyle\lim_{h\to 0}\frac{h}{\tan h}$

**5** $\displaystyle\lim_{\theta\to 0}\frac{1 - \cos\theta}{4\theta^2}$

**6** $\displaystyle\lim_{\theta\to\pi/2}\frac{\theta - \dfrac{\pi}{2}}{\cos\left(\theta - \dfrac{\pi}{2}\right)}$

**7** $\displaystyle\lim_{\phi\to 0}\frac{1}{\phi\cot\phi}$

**8** $\displaystyle\lim_{\theta\to\pi/2}\frac{\cos\theta}{(\pi/2) - \theta}$

**9** $\displaystyle\lim_{x\to 0}\frac{x}{\csc^2 x}$

**10** $\displaystyle\lim_{x\to 0}\frac{\cot^2 x}{4x^2}$

**11** $\displaystyle\lim_{x\to 0}\frac{x^3}{\tan^3 2x}$

**12** $\displaystyle\lim_{x\to 0}\frac{4x^3}{1 - \cos^2\frac{1}{2}x}$

**13** $\displaystyle\lim_{\theta\to 0}\frac{\sin^2\theta}{\theta^2}$

**14** $\displaystyle\lim_{\theta\to 0}\frac{\sin^2\theta}{\theta}$

**15** $\displaystyle\lim_{\theta\to 0}\frac{\cos\theta\sin\theta}{\theta}$

**16** $\displaystyle\lim_{\theta\to 0^+}\frac{\sin\theta}{\sqrt\theta}$

**17** $\displaystyle\lim_{\theta\to\frac{\pi}{2}}\frac{\cos\theta}{\theta - \dfrac{\pi}{2}}$

**18** $\displaystyle\lim_{\theta\to\frac{\pi}{2}}\frac{\sin\theta\cos\theta}{\theta - \dfrac{\pi}{2}}$

In Problems 19 through 45, differentiate the trigonometric functions.

**19** $f(x) = \tan 3x$

**20** $g(y) = 2\sin y\cos y$

**21** $f(s) = \sec(s^2)$

**22** $\phi(t) = \sin^2 3t$

**23** $f(y) = \csc^3\frac{1}{3}y$

**24** $g(x) = \sin 2x - \frac{1}{3}(\sin^3 2x)$

**25** $f(x) = 2x + \cot 2x$

**26** $h(u) = u\sin u + \cos u$

**27** $x(s) = \sqrt{\tan 2s}$

**28** $h(t) = \sec^2 2t - \tan^2 2t$

**29** $f(x) = x^2\tan^2(x/2)$

**30** $g(x) = (\tan^3 4x)/3 + \tan 4x$

**31** $f(x) = \frac{1}{3}(\sec^3 2x) - \sec 2x$

**32** $f(x) = \sqrt{1 + \cos x}$

**33** $f(x) = (\sin 2x)/x^2$

**34** $f(x) = (\sin 2x)/x^5$

**35** $f(x) = (\cos^2 3x)/(1 + x^2)$

**36** $f(x) = \sin^3 2x\tan^2 3x$

**37** $f(x) = \cos^2 3x\csc^3 2x$

**38** $f(x) = (\sin 2x)/(1 + \cos 2x)$

**39** $f(x) = (\cot^2 ax)/(1 + x^2)$

**40** $f(x) = (1 + \tan 3x)^{2/3}$

**41** $f(x) = (1 + \sin^2 2x)^{1/2}$

**42** $f(x) = \sin nx\sin^n x$

**43** $f(x) = \sin(\cos x)$

**44** $f(x) = \dfrac{\cot(x/2)}{\sqrt{1 - \cot^2(x/2)}}$

**45** $f(x) = \dfrac{(1 + \cos^2 3x)^{1/3}}{(1 + x^2)^{1/2}}$

**46** If $f(x) = \cos x$ prove, by using the three-step rule, that $f'(x) = -\sin x$.

**47** Prove that if $f(x) = \cot x$, then $f'(x) = -\csc^2 x$.

**48** Prove that if $f(x) = \sec x$, then $f'(x) = \sec x\tan x$.

**49** Prove that if $f(x) = \csc x$, then $f'(x) = -\csc x\cot x$.

**50**   If $f(x) = \sin \lambda x$, show that $f''(x) + \lambda^2 f(x) = 0$.

In each of Problems 51 through 56, find the equation of the line tangent to the given curve at the given value of $x_0$.

**51**   $y = \sin x, \quad x_0 = \dfrac{\pi}{4}$

**52**   $y = \cos 3x, \quad x_0 = \dfrac{\pi}{9}$

**53**   $y = \tan x + \sin 2x, \quad x_0 = \dfrac{\pi}{4}$

**54**   $y = \sec \frac{1}{3}x, \quad x_0 = \pi$

**55**   $y = \sin x \cos 2x, \quad x_0 = \dfrac{\pi}{6}$

**56**   $y = \sin x \tan 3x, \quad x_0 = \dfrac{2\pi}{3}$

# CHAPTER 3

## REVIEW PROBLEMS

In Problems 1 through 10 find $\dfrac{dy}{dx}$.

**1**   $y = x^3 - 2x^2 + 7x - 6$

**2**   $y = x^5 + 4x^2 - 5$

**3**   $y = (x^2 + 2x - 1)(x^4 + 7x - 5)$

**4**   $y = (x^3 + 4x^2 - 5x)(x^7 + 2x^5 + 3)$

**5**   $y = \dfrac{x^2 + x - 1}{x^3 + 2x + 5}$     **6**   $y = \dfrac{x^3 + 7x - 3}{x^2 + 2x + 4}$

**7**   $y = (x^3 + 12x + 5)^7$     **8**   $y = (x^4 - 2x^2 + 3x - 5)^7$

**9**   $y = \dfrac{(x^2 - 2x + 1)(x^3 + x + 5)}{x + 6}$

**10**   $y = \dfrac{(x^3 - 7x - 5)(x^2 + 6x - 2)}{x^2 - 3x + 8}$

In Problems 11 through 22 find the derivative of the given function.

**11**   $f(x) = x^{3/2} - 2x^{1/3} + 7$

**12**   $f(x) = x^{4/5} - 2x^{-3} + 7x$

**13**   $g(x) = (x^2 + 2x - 1)^{5/3}$

**14**   $g(x) = (x^3 + x^{1/2} - 3)^{2/3}$

**15**   $h(x) = (x^2 + 2x^{1/3} - 4)^{-3}$

**16**   $h(t) = (t^3 + 2t^{-1} + \sqrt{t})^3$

**17**   $f(s) = \sqrt{\dfrac{s-1}{s+1}}$     **18**   $K(x) = \sqrt{\dfrac{3x-2}{2x+3}}$

**19**   $\phi(t) = \sqrt[3]{\dfrac{3t+4}{3t-2}}$     **20**   $\psi(t) = \sqrt[3]{\dfrac{t^2-1}{t^2+1}}$

**21**   $f(y) = \dfrac{y^3 - 3y^2 - 5y + 2}{5y^3 - \sqrt[3]{y^2}}$

**22**   $g(z) = \frac{1}{4}z\sqrt[3]{z} - \frac{1}{7}z^2 - \sqrt[3]{z} + \frac{1}{10}z^3\sqrt[3]{z}$

**23**   Let $u$ and $v$ be differentiable functions of $x$. Given that $f(x) = g(3u - 2v)$, find $f'(x)$. (*Hint:* Use the Chain Rule by setting $z = 3u - 2v$.)

**24**   Let $u$ and $v$ be differentiable functions of $x$. Given that $f(x) = g(2u + 1) \cdot h(3v - 2)$, find $f'(x)$. (*Hint:* Let $s = 2u + 1$ and $t = 3v - 2$. Then use the Chain Rule.)

In Problems 25 through 32 find $\dfrac{dy}{dx}$.

**25**   $x^3 - 2y^3 + 7 = 0$

**26**   $x^4 - 4y^3 + 3x - 2y - 4 = 0$

**27**   $x^2 + 3x^2y^3 - 2xy + 2 = 0$

**28**   $x^3 + xy^2 - 3x + 2y + 4 = 0$

**29**   $x^{1/3} - y^{1/3} + x - y = 2$

**30**   $x^{2/3} + y^{2/3} - 3x + 2y = 4$

**31**   $xy \tan(xy) + x^2y^2 = 3$

**32**   $\sin(x - y) = y \sin x$

**33**   Find $dy/dx$ given that $|2x - 4| - 3y = 0$. At what point or points does $dy/dx$ fail to exist? Sketch the graph.

Problems 34 through 44 find the derivative.

**34**   $f(x) = \sin 3x - 2 \cos 4x$

**35**   $f(x) = \tan 2x - \cos \frac{1}{3}x$

**36**   $f(t) = \sin^3 2t$

**37**   $f(t) = \cos^{4/3} 2t$

**38**   $g(x) = \tan^3(3x^2 + 1)$

**39**   $h(x) = \sec^2 2x + \cot 4x$

**40**   $F(x) = \csc^3 4x + \sin^{-3} 2x$

**41**   $G(x) = 2 \sec^4 3x - 7 \tan^{4/3} 5x$

**42**   $K(s) = 2s \cos 3s + 4s \sin s$

**43**   $f(x) = \sin^2 3x \cos^4 2x$

**44**   $g(t) = 2t \cos t \sin \frac{1}{3}t + t^2 \tan^3 2t$

In each of Problems 45 through 50 find the equation of the tangent line and the equation of the normal line to the given curve at the given value of $x_0$.

**45** $y = (x^2 + 1)^3$, $x_0 = 0$

**46** $y = (x^2 + 1)^2(x^3 - 2)^4$, $x_0 = 1$

**47** $y = \dfrac{(x^2 + 2)(x^3 - 1)}{x + 3}$, $x_0 = -1$

**48** $y = 2 \sin \frac{1}{2}x \cos 2x$, $x_0 = \dfrac{\pi}{3}$

**49** $y = \tan^2 2x \sec x$, $x_0 = \dfrac{\pi}{6}$

**50** $y = \dfrac{\sin 2x \cos x}{\tan 3x}$, $x_0 = \dfrac{\pi}{6}$

**51** Given that $f(x) = \cos[2 \sin(3x)]$, find $f'(x)$.

**52** Given that $g(x) = \tan[2 \cos(\frac{1}{3}x)]$, find $g'(x)$.

# 4

# APPLICATIONS OF THE DERIVATIVE

Applications of the derivative abound. In this chapter we first develop the theoretical underpinnings for applications of the derivative. Then we show how the derivative is used in solving a variety of numerical, geometrical, and physical problems.

## 1

## TOOLS FOR APPLICATIONS OF THE DERIVATIVE

As we learned earlier, a **closed interval** along the $x$ axis is an interval which includes its endpoints, an **open interval** is one which excludes the endpoints, and a **half-open** interval contains one endpoint but not the other.

The basic theorem of this section is the **Extreme Value Theorem**, a result which is usually proved in a more advanced mathematics course. We shall show how this theorem can be used to obtain important applications for the differentiation process we studied in Chapter 3.

**THEOREM 1**    **(Extreme Value Theorem)**    *If $f$ is a continuous function defined on the closed interval $[a, b]$, there is (at least) one point in $[a, b]$ (call it $x_1$) where $f$ has a largest value, and there is (at least) one point (call it $x_2$) where $f$ has a smallest value.*

This theorem is fairly clear intuitively if we think of a continuous function as one with no breaks or gaps. As we move along the curve from the point corresponding to $x = a$ to the point corresponding to $x = b$, there must be a place where the curve has a high point (called the **maximum value**) and there must also be a place where it has a low point (called the **minimum value**). In spite of the simplicity of the situation from the intuitive point of

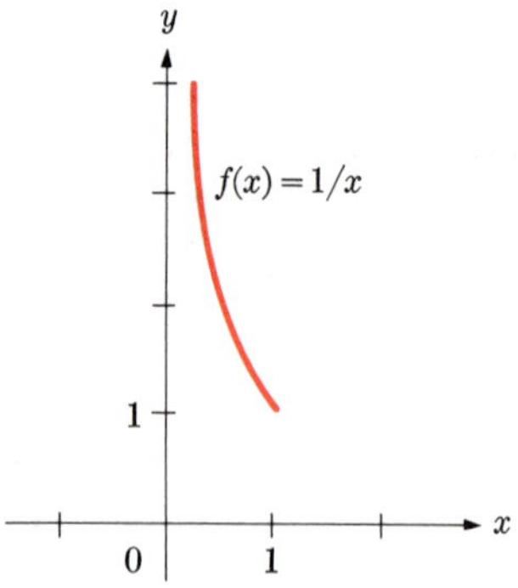

$f(x) = 1/x$

**FIGURE 1**

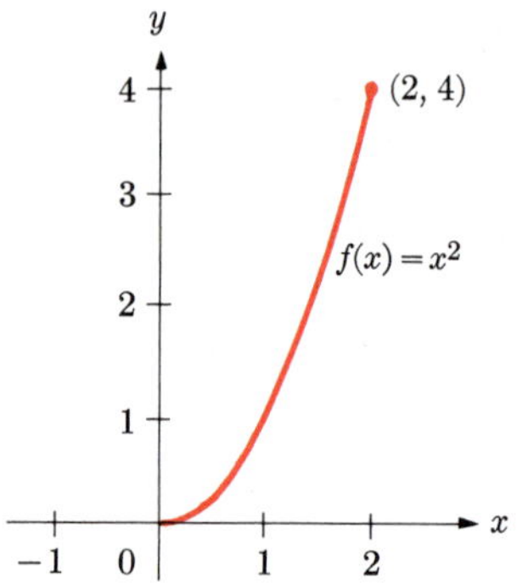

$f(x) = x^2$

(2, 4)

**FIGURE 2**

view, the proof of Theorem 1 is hard. Consequently we shall restrict ourselves to a discussion of the meaning of this theorem.

A good way to examine a theorem critically is to see what happens if some of the hypotheses are altered. In Theorem 1 there are two principal hypotheses: (1) the interval $[a, b]$ is closed, and (2) the function $f$ is continuous. We shall show by example that if we tamper with either hypothesis, the conclusion of the theorem may be false.

Suppose the assumption that the interval is closed is replaced by the assumption that the interval is open. The function $f(x) = 1/x$ is continuous on the open interval $0 < x < 1$ (Fig. 1), and it has no maximum value in this open interval. A more subtle example is given by the function $f(x) = x^2$ defined in the open interval $0 < x < 2$ (Fig. 2). This function has no maximum or minimum value on the *open* interval but does have a maximum of 4 and a minimum of 0 on the *closed* interval. This situation comes about because $f(x) = x^2$ is continuous on the closed interval $0 \le x \le 2$. In the first example the function $1/x$ is continuous on $0 < x < 1$ but not on the closed interval $0 \le x \le 1$. Even making the interval half-open is not good enough, since the function $1/x$ is continuous on the half-open interval $0 < x \le 1$ and still has no maximum value there. The second example, $f(x) = x^2$, is continuous on the half-open interval $0 < x \le 2$ and does not have a minimum in this half-open interval.

The second hypothesis, that of continuity, is also essential. The function

$$f(x) = \frac{1}{x - 1}, \quad 0 \le x \le 2, \quad x \ne 1$$

$$f(1) = 2$$

is continuous except at $x = 1$ (Fig. 3). As the graph clearly shows, the function has no maximum and no minimum on the closed interval $[0, 2]$. A second, more sophisticated, example is the function

$$f(x) = \begin{cases} x^2 + 1, & 0 \le x < 1 \\ x - 1, & 1 \le x \le 2 \end{cases}$$

whose graph is shown in Fig. 4. This function is continuous except at $x = 1$, where its minimum occurs with the value zero. There is no point on the closed

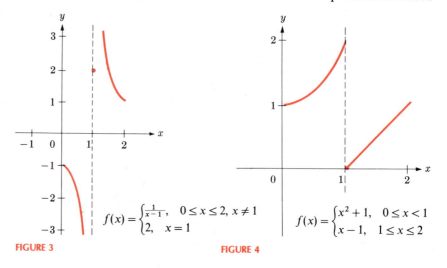

$$f(x) = \begin{cases} \frac{1}{x-1}, & 0 \le x \le 2, \; x \ne 1 \\ 2, & x = 1 \end{cases}$$

**FIGURE 3**

$$f(x) = \begin{cases} x^2 + 1, & 0 \le x < 1 \\ x - 1, & 1 \le x \le 2 \end{cases}$$

**FIGURE 4**

interval $[0, 2]$ where it has a maximum value. Note that the function gets close to the value 2 as $x \to 1$ from the left. But there is no point where it actually *has* the value 2.

If $f$ is continuous on $[a, b]$, with its maximum value $M$ at $x_1$ and its minimum value $m$ at $x_2$, then Theorem 1 shows that

$$m \leq f(x) \leq M$$

for all $x$ in $[a, b]$. We could equally well write $f(x_1)$ instead of $M$ and $f(x_2)$ instead of $m$. In words, Theorem 1 states that

> *a function which is continuous on a closed interval takes on its maximum and minimum values.*

We shall now prove a theorem concerning the value of the derivative at a maximum or minimum point.

**THEOREM 2**   *Suppose that $f$ is continuous on an interval and takes on its maximum (or minimum) at some point $x_0$ which is in the interior of the interval. If $f'(x_0)$ exists, then*

$$f'(x_0) = 0.$$

**Proof**   We prove the theorem for the case where $f(x_0)$ is a maximum. The proof for a minimum is similar. If $f(x_0)$ is the maximum value, then

$$f(x_0 + h) \leq f(x_0)$$

for every possible $h$, both positive and negative. The only restriction is that $x_0 + h$ must be in the interval in order for $f(x_0 + h)$ to have a meaning. We can also write (see Fig. 5)

$$f(x_0 + h) - f(x_0) \leq 0. \tag{1}$$

If $h$ is positive we may divide by $h$ to get

$$\frac{f(x_0 + h) - f(x_0)}{h} \leq 0, \quad h > 0.$$

Taking the one-sided limit as $h \to 0^+$, we conclude* that

$$\lim_{h \to 0^+} \frac{f(x_0 + h) - f(x_0)}{h} \leq 0. \tag{2}$$

If $h$ is negative, *the inequality* (1) *reverses* when we divide by $h$, so that

$$\frac{f(x_0 + h) - f(x_0)}{h} \geq 0, \quad h < 0.$$

Taking the one-sided limit as $h \to 0^-$, we conclude* that

$$\lim_{h \to 0^-} \frac{f(x_0 + h) - f(x_0)}{h} \geq 0. \tag{3}$$

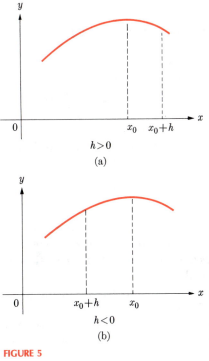

$h > 0$

(a)

$h < 0$

(b)

**FIGURE 5**

*We are employing the following (intuitively clear) theorem on limits: *if $F(x) \leq 0$ for all $x$ and if $F(x) \to L$ as $x \to a$, then $L \leq 0$.* The same result holds for one-sided limits; also the analogous result holds for functions $F(x) \geq 0$, in which case $L \geq 0$.

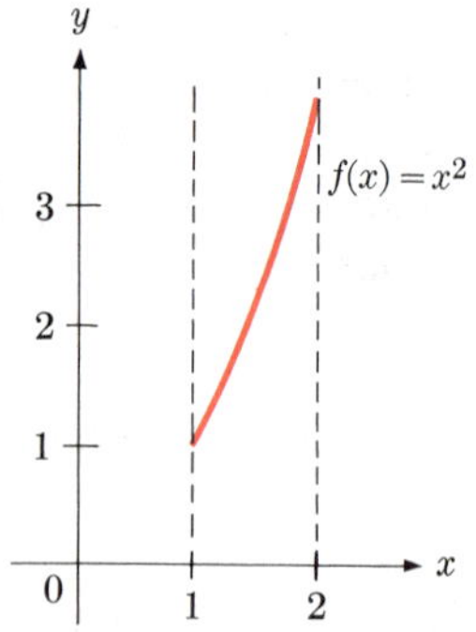

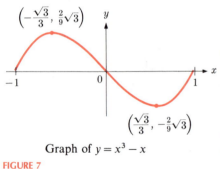

FIGURE 6

Since the ordinary limit (two-sided limit) exists as $h$ tends to zero, the one-sided limits exist and are equal (see Theorem 11 of Chapter 2). Examining (2) and (3), we see that they can only be equal if $f'(x_0) = 0$, which is what we wished to prove. $\qquad\square$

*Discussion.* The important hypotheses in Theorem 2 are (1) that $x_0$ is an interior point, and (2) that $f$ has a derivative at $x_0$. If the first hypothesis is neglected the theorem is false, as is shown by the example $f(x) = x^2$ on the interval $1 \le x \le 2$ (Fig. 6). The maximum occurs at $x = 2$ (not an interior point) and the minimum occurs at $x = 1$ (not an interior point). The derivative of $f(x) = x^2$ is $f'(x) = 2x$, and this is different from zero throughout the interval $[1, 2]$. The fact that $x_0$ is **interior** to the interval is important; whether the interval is open, closed, or half-open is irrelevant.

Using the methods described in Chapter 2, Section 4, we now show that the function

$$f(x) = x^3 - x, \quad -1 < x < 1$$

has a maximum at the point

$$\left( -\frac{\sqrt{3}}{3}, \frac{2}{9}\sqrt{3} \right)$$

and a minimum at

$$\left( \frac{\sqrt{3}}{3}, -\frac{2}{9}\sqrt{3} \right).$$

(see Fig. 7). To find these points we first get the derivative of $f$:

$$f'(x) = 3x^2 - 1.$$

Then, setting this equal to zero and solving for $x$, we obtain $3x^2 - 1 = 0$, $x = \frac{1}{3}\sqrt{3}, -\frac{1}{3}\sqrt{3}$. Substitution of the $x$ values into the original expression yields the desired points. We shall prove in Section 3 that these points are the true maximum and minimum points of the function.

The second hypothesis in Theorem 2, namely, that $f$ has a derivative at $x_0$, is also essential, as is shown by the following example. Let

$$f(x) = \begin{cases} -\frac{1}{3}x^2 + 2x - \frac{5}{3}, & 1 \le x \le 2 \\ -x^2 + 2x + 1, & 2 < x \le 3. \end{cases}$$

This function is continuous on the interval $1 \le x \le 3$, since both expressions approach the same value at $x = 2$. As Fig. 8 shows, however, there is a "corner" at $x = 2$, and the derivative does not exist at this point. Nevertheless, the point $P(2, 1)$ is exactly where the maximum occurs.

The derivative may not exist by virtue of being infinite at a maximum or minimum point. The relation $x^{2/3} + y^{2/3} = 1$ for $-1 \le x \le 1$, has the graph shown in Fig. 9. The portion of the graph above the $x$ axis represents a function. The maximum occurs at $x = 0$ and at $x = 0$ we have $y = 1$. We can compute the derivative implicitly and obtain

$$\frac{2}{3}x^{-1/3} + \frac{2}{3}y^{-1/3}\frac{dy}{dx} = 0 \quad \text{or} \quad \frac{dy}{dx} = -\frac{y^{1/3}}{x^{1/3}},$$

$\left( -\frac{\sqrt{3}}{3}, \frac{2}{9}\sqrt{3} \right)$

$\left( \frac{\sqrt{3}}{3}, -\frac{2}{9}\sqrt{3} \right)$

Graph of $y = x^3 - x$

FIGURE 7

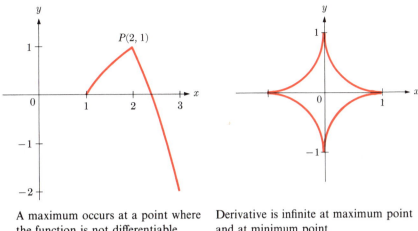

A maximum occurs at a point where
the function is not differentiable

**FIGURE 8**

Derivative is infinite at maximum point
and at minimum point

**FIGURE 9**

and we see that this tends to infinity as $x$ tends to zero. Therefore there is no
derivative at the maximum point.

## 1   PROBLEMS

**1** Show that the continuous function

$$f(x) = \begin{cases} x, & 0 \leq x < 1 \\ 2 - x, & 1 \leq x \leq 2 \\ x - 2, & 2 < x \leq 3 \end{cases}$$

has its maximum value at two different points. Also show
that the minimum value is taken on at two points. Sketch.

**2** Construct by formula (as in Problem 1) a continuous
function on the interval $[-2, 5]$ which assumes its maximum value at two different points and its minimum value
at three different points. Sketch the graph.

**3** Construct a continuous function (by formula) defined on
the interval $[1, 6]$ which assumes its maximum value at
every point of the subinterval $[2, 4.5]$. Sketch the graph.

**4** Construct a continuous function (by formula) defined on
the interval $[0, 5]$ which assumes its minimum at $x = 1$ and
at every point of the interval $[2, 3]$ but at no other points.
Sketch the graph.

**5** Construct a continuous function (by formula) which is
defined on $[-1, 4]$, which has its maximum at $x = 0$, its
minimum at $x = 2$ and is such that $f'$ does not exist at
$x = 0$ and at $x = 2$. Sketch the graph.

**6** The function

$$f(x) = \frac{x^2 - 1}{x^2 + 1}$$

is continuous for all values of $x$ from $-\infty$ to $+\infty$. Sketch

the graph. Show that $f$ has the range $[-1, 1)$, and that the
function never assumes its maximum value. Why does
Theorem 1 not apply?

**7** After studying the function in Problem 6, construct a
continuous function defined for $-\infty < x < \infty$ with range
$(-2, 3]$ and which never assumes its minimum value.

**8** Given the continuous function

$$f(x) = \begin{cases} x^2/(x^2 + 1), & 0 \leq x < +\infty \\ -x^2/(x^2 + 1), & -\infty < x < 0. \end{cases}$$

Show that $f$ never assumes its maximum or minimum
value. Sketch.

In Problems 9 through 14, use the method described in
Chapter 2, Section 4. These methods will be justified in Section 3
of this chapter.

**9** Given the function $f(x) = 3x^2 + 6x + 1$ on the interval
$-2 \leq x \leq 0$, find the minimum point.

**10** Given the function $f(x) = 3x^2 + 6x + 1$ on the interval
$0 \leq x \leq 3$, find the minimum point.

**11** Find the maximum value of

$$f(x) = \begin{cases} x + 1, & 0 \leq x \leq 1 \\ -x^2 + 3, & 1 < x \leq 3 \end{cases}$$

on the interval $0 \leq x \leq 3$. Where is the minimum value?

**12** Find the maximum point of $f(x) = \frac{1}{3}x^3 - x + 2$ on the
interval $[-2, 2]$. What is the maximum point on the
interval $[-3, 3]$?

**13** Given the function

$$f(x) = \begin{cases} -x^2 + 4x - 2, & 1 \leq x \leq 2 \\ -\frac{1}{2}x^2 + 2x, & 2 < x \leq 3, \end{cases}$$

find the maximum point. Does the function have a derivative at this point? If so, find the value of the derivative.

**14** Examine the relation

$$x^{1/3} + y^{1/3} = 2, \quad -8 \leq x \leq 8$$

with regard to maximum and minimum points.

**15** Given the function

$$f(x) = \frac{1}{x^2 + 1}, \quad -\infty < x < \infty,$$

show that $f$ has a maximum value, but no minimum value. Sketch the graph.

**16** Draw the graph of $y = \sin x$, $-\infty < x < \infty$, and identify the infinite collection of maximum and minimum points.

**\*17** The function $y = \arctan x$ has domain $(-\infty, +\infty)$ and range $(-\pi/2, \pi/2)$. Sketch the graph. Does the function have maximum and minimum values?

**\*18** Prove the following result: If $F(x) \geq 0$ for all $x$ on an interval $[a, b]$ containing the point $c$ and if $F(x) \rightarrow L$ as $x \rightarrow c$, then $L \geq 0$. (*Hint:* Assume that $L$ is negative and use the definition of limit to show there must be a value of $x$ near $c$ where $F(x) < 0$, thereby reaching a contradiction. (See the footnote on page 127.))

**19** Write out a proof of Theorem 2 (page 127) for the case where $f$ takes on its minimum value.

**20** Suppose that a function $f$ defined on the closed interval $[a, b]$ assumes its maximum and minimum values on this interval. Is $f$ continuous on $[a, b]$? Justify your answer.

**\*21** A function $f$ defined on $[a, b]$ has a **left-hand derivative** at $b$ if

$$\lim_{h \to 0^-} \frac{f(b + h) - f(b)}{h}$$

exists. The **right-hand derivative** at $a$ is defined similarly. Show that the Extreme Value Theorem holds for functions which are differentiable on $(a, b)$ and have a right-hand derivative at $a$ and a left-hand derivative at $b$.

---

**2**

---

## FURTHER TOOLS: ROLLE'S THEOREM; MEAN VALUE THEOREM

Theorems 1 and 2 of Section 1 are the basis for the next result, known as Rolle's Theorem. While Rolle's Theorem is quite special, it has the advantage of being intuitively clear, especially when viewed geometrically. Also, the very next theorem, known as the Mean Value Theorem, is then an easy consequence even though it does not appear to be as elementary as Rolle's Theorem.

**THEOREM 3**    **(Rolle's Theorem)**    *Suppose that $f$ is continuous for $a \leq x \leq b$ and that $f'(x)$ exists for each $x$ between $a$ and $b$. If*

$$f(a) = f(b) = 0,$$

*then there must be (at least) one point, call it $x_0$, between $a$ and $b$ such that*

$$f'(x_0) = 0.$$

---

**Proof**    There are three possibilities:

**Case 1.**    (The trivial case.) $f(x) = 0$ for all $x$ between $a$ and $b$; then $f'(x) = 0$ for all $x$, and $x_0$ can be chosen to be any value between $a$ and $b$.

**Case 2.** $f(x)$ is positive somewhere between $a$ and $b$. Then the maximum of $f$ is positive, and we choose $x_0$ (Theorem 1) to be a place where this maximum occurs. (See Fig. 10a.) According to Theorem 2 of the previous section, $f'(x_0) = 0$, since $x_0$ must be interior to the interval.

**Case 3.** $f(x)$ is negative somewhere between $a$ and $b$. Then the minimum of $f$ is negative, and we choose $x_0$ to be a place where this minimum occurs. (See Fig. 10b.) According to Theorem 2 of the previous section, $f'(x_0) = 0$.

Since every function which is zero at $a$ and $b$ must fall into one of the three cases, the theorem is proved.

Figure 10(c) shows that a function may fall into both Case 2 and Case 3, and Fig. 10(d) illustrates the possibility of several choices for $x_0$ even though it satisfies only Case 2.

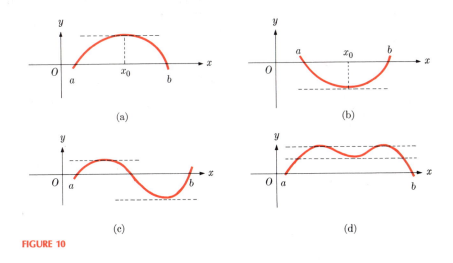

(a)        (b)
(c)        (d)

**FIGURE 10**

*We can state Rolle's Theorem in a simple way: If a differentiable curve crosses the x axis twice there must be a point between successive crossings at which the line tangent to the curve is parallel to the x axis.*

We have been building up to the proof of the theorem known as the **Mean Value Theorem**, one which is used over and over again throughout the branch of mathematics known as analysis. Every mathematician who works in analysis encounters it often, uses it, and feels at home with it. We shall see that, important though it is, the Mean Value Theorem is merely a variation of Rolle's Theorem.

**THEOREM 4**    **(Mean Value Theorem)**    *Suppose that $f$ is continuous for $a \le x \le b$ and that $f'(x)$ exists for each $x$ between $a$ and $b$. Then there is an $x_0$ between $a$ and $b$ (that is, $a < x_0 < b$) such that*

$$f'(x_0) = \frac{f(b) - f(a)}{b - a}.$$

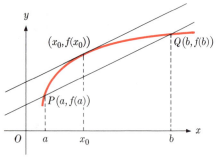

**FIGURE 11**

Before proving this theorem we shall discuss its meaning from a geometric standpoint. Figure 11 shows a typical function $f$ between the points $a$ and $b$. The point $P$ has coordinates $(a, f(a))$ and $Q$ has coordinates $(b, f(b))$. We construct the straight line through $PQ$ and calculate its slope. We know that the slope $m$ is the difference of the $y$ values over the difference of the $x$ values, that is,

$$m = \frac{f(b) - f(a)}{b - a}.$$

This is exactly the same expression that occurs in the statement of the Mean Value Theorem. The theorem says there is a point $(x_0, f(x_0))$ on the curve where the slope has the value $m$; that is, the tangent line at $(x_0, f(x_0))$ is parallel to the line through $PQ$. Glancing at the figure, we see that there must be such a point. In fact, by means of the following device, we can see that there must be one: We look at the figure and tilt it so that the line through $PQ$ appears horizontally as the $x$ axis. Then the Mean Value Theorem resembles Rolle's Theorem.

**Proof of Theorem 4**   The equation of the line through $PQ$ (according to the two-point formula for the equation of a straight line) is

$$y - f(a) = \frac{f(b) - f(a)}{b - a}(x - a).$$

We construct the function

$$F(x) = f(x) - \frac{f(b) - f(a)}{b - a}(x - a) - f(a).$$

By straight substitution with $x = a$ and then $x = b$, we find

$$F(a) = f(a) - \frac{f(b) - f(a)}{b - a}(a - a) - f(a) = 0,$$

$$F(b) = f(b) - \frac{f(b) - f(a)}{b - a}(b - a) - f(a) = 0.$$

Therefore $F(x)$ satisfies all the hypotheses of Rolle's Theorem. There must be a value $x_0$ such that $F'(x_0) = 0$. But (by differentiation) we see that

$$F'(x) = f'(x) - \frac{f(b) - f(a)}{b - a}.$$

This implies that

$$f'(x_0) = \frac{f(b) - f(a)}{b - a},$$

which is what we wished to prove.                                                                $\square$

**EXAMPLE 1**   Given that

$$f(x) = \frac{x + 2}{x + 1} \quad \text{and} \quad a = 1, b = 2,$$

find all values $x_0$ in the interval $1 < x < 2$ such that

$$f'(x_0) = \frac{f(b) - f(a)}{b - a}.$$

**Solution**   We differentiate $f$, getting

$$f'(x) = \frac{(x + 1) \cdot 1 - (x + 2) \cdot 1}{(x + 1)^2} = \frac{-1}{(x + 1)^2},$$

and

$$f(1) = \frac{3}{2}, \qquad f(2) = \frac{4}{3}, \qquad \frac{f(b) - f(a)}{b - a} = -\frac{1}{6}.$$

We solve the equation $f'(x_0) = -\frac{1}{6}$, which yields

$$\frac{-1}{(x_0 + 1)^2} = -\frac{1}{6} \qquad \text{or} \qquad x_0^2 + 2x_0 - 5 = 0,$$

and

$$x_0 = \frac{-2 \pm \sqrt{24}}{2} = -1 \pm \sqrt{6}.$$

The value $x_0 = -1 + \sqrt{6}$ is in the interval $(1, 2)$, while the value $x_0 = -1 - \sqrt{6}$ is rejected since it is outside this interval.   □

**EXAMPLE 2**   Given that $f(x) = x^3 - 2x^2 + 3x - 2$ and $a = 0$, $b = 2$, find all possible values for $x_0$ in the interval $0 < x < 2$ such that

$$f'(x_0) = \frac{f(b) - f(a)}{b - a}.$$

**Solution**   We differentiate $f$, getting

$$f'(x) = 3x^2 - 4x + 3. \quad \text{Also,} \quad f(0) = -2, \qquad f(2) = 4,$$

so that

$$\frac{f(b) - f(a)}{b - a} = \frac{4 + 2}{2} = 3.$$

We solve the equation

$$3x_0^2 - 4x_0 + 3 = 3,$$

which has two solutions: $x_0 = 0$ and $x_0 = \frac{4}{3}$. But $x_0 = 0$ is not *between* 0 and 2, and therefore the only answer is $x_0 = \frac{4}{3}$.   □

**EXAMPLE 3**   Given the function

$$f(x) = \frac{x^2 - 4x + 3}{x - 2}$$

and

$$a = 1, b = 3,$$

discuss the validity of the Mean Value Theorem.

**Solution**  If we proceed formally, we see that $f(1) = 0$ and $f(3) = 0$, so we must find an $x_0$ in the interval $(1, 3)$ such that $f'(x_0) = 0$. Computing the derivative by the quotient rule, we write

$$f'(x) = \frac{(x-2)(2x-4) - (x^2 - 4x + 3)}{(x-2)^2} = \frac{x^2 - 4x + 5}{(x-2)^2}.$$

Setting this equal to zero, we obtain

$$x_0^2 - 4x_0 + 5 = 0$$

and

$$x_0 = \frac{4 \pm \sqrt{16 - 20}}{2} = \frac{4 \pm \sqrt{-4}}{2},$$

which is impossible! We look once again at the function $f$ and see that it becomes infinite at $x = 2$. Therefore the hypotheses of the Mean Value Theorem are not satisfied, and the theorem is not applicable. There is no value $x_0$. $\square$

## 2  PROBLEMS

In each of Problems 1 through 12, find all numbers $x_0$ between $a$ and $b$ which satisfy the equation

$$f'(x_0) = \frac{f(b) - f(a)}{b - a}.$$

**1** $f(x) = x^2 - 3x - 2$, $\quad a = 0, b = 2$

**2** $f(x) = x^3 - 5x^2 + 4x - 2$, $\quad a = 1, b = 3$

**3** $f(x) = x^3 - 7x^2 + 5x$, $\quad a = 1, b = 5$

**4** $f(x) = x^3 + 2x^2 - x$, $\quad a = -3, b = 2$

**5** $f(x) = \dfrac{x - 2}{x + 2}$, $\quad a = 0, b = 1$

**6** $f(x) = \dfrac{2x + 3}{3x - 2}$, $\quad a = 1, b = 5$

**7** $f(x) = \sqrt{25 - x^2}$, $\quad a = -3, b = 4$

**8** $f(x) = \sqrt{x^2 + 81}$, $\quad a = 12, b = 40$

**9** $f(x) = x^4 - 2x^3 + x^2 - 2x$, $\quad a = -1, b = 2$

**10** $f(x) = x^4 + x^3 - 3x^2 + 2x$, $\quad a = -2, b = -1$

**11** $f(x) = \dfrac{x^2 - 3x - 4}{x + 5}$, $\quad a = -1, b = 4$

**12** $f(x) = \dfrac{x^2 + 6x + 5}{x - 6}$, $\quad a = 1, b = 5$

**13** Given $f(x) = x^3 - 3x^2 - 6x + 8$, find all values of $x$ such that $f(x) = 0$. Then find the values of $x$ between each two zeros of $f$ such that $f'(x) = 0$, thus verifying Rolle's Theorem.

**14** Same as Problem 13 for $f(x) = x^3 + 2x^2 - 11x - 12$.

**15** Given that $f(x) = (2x + 3)/(3x + 2)$, $a = -1$, $b = 0$, show that there is no number $x_0$ between $a$ and $b$ which satisfies the Mean Value Theorem. Sketch the graph.

**16** Given that $f(x) = 2x^{2/3}$, $a = -1$, $b = 1$, show that there is no number $x_0$ between $a$ and $b$ which satisfies the Mean Value Theorem. Sketch the graph.

**17** Given that

$$f(x) = 1/(x - 1)^2, \quad a = -1, b = 2.$$

Is there a number $x_0$ between $a$ and $b$ which satisfies the equation

$$(b - a)f'(x_0) = f(b) - f(a)?$$

Can the Mean Value Theorem be used?

**18** Given that $f$ is a quadratic function of $x$, and $a$ and $b$ are any numbers, show that $x_0 = \frac{1}{2}(a + b)$ is the value that satisfies the Mean Value Theorem. (*Hint:* Assume that $f(x) = cx^2 + dx + e$, where $c$, $d$, and $e$ are any constants.)

**\*19** Given that $f$ is a general equation of the third degree. That is, suppose $f(x) = \alpha x^3 + \beta x^2 + \gamma x + \delta$, with $\alpha \neq 0$. Show that there are at most two values of $x_0$ which satisfy the Mean Value Theorem for any given numbers $a$ and $b$. Prove that these are given by the formula

$$x_0 = -\frac{\beta}{3\alpha} \pm \frac{1}{3\alpha}\sqrt{\beta^2 + 3\alpha^2(a^2 + ab + b^2) + 3\alpha\beta(a + b)}.$$

**20** Given the function $f$ defined for $0 \le x \le 2$ by the formula

$$f(x) = \begin{cases} x & \text{for} \quad 0 \le x < 1 \\ \frac{1}{2}x^2 + \frac{1}{2} & \text{for} \quad 1 \le x \le 2. \end{cases}$$

Does the Mean Value Theorem hold for $0 \le x \le 2$? Justify your answer.

**21** Same as Problem 18 for the function

$$f(x) = \begin{cases} 2x^3 - 1 & \text{for} \quad 0 \le x < 1 \\ 6x - 5 & \text{for} \quad 1 \le x \le 2. \end{cases}$$

**22** Same as Problem 18 for the function

$$f(x) = \begin{cases} x^2 + 2x - 1 & \text{for} \quad 0 \le x < 1 \\ x^4 + 2 & \text{for} \quad 1 \le x \le 2. \end{cases}$$

**\*23** Use the Mean Value Theorem to show that for any two real numbers $x$ and $y$,

$$|\sin x - \sin y| \le |x - y|.$$

**\*24** Use the Mean Value Theorem to show that if $f(x) = x^2 + 2x + 3$, then for any values $a$ and $b$ in the interval $(1, 3)$,

$$|f(b) - f(a)| \le 8|b - a|.$$

**\*25** Use the Mean Value Theorem to show that if $f$ is continuous for $a \le x \le b$ and $f'(x)$ exists for each $x$ between $a$ and $b$, and if $|f'(x)| \le K$ for all $x$ between $a$ and $b$, where $K$ is a positive constant, then

$$|f(x) - f(y)| \le K|x - y|$$

for all $x$, $y$ between $a$ and $b$.

---

## 3

## APPLICATIONS TO GRAPHS OF FUNCTIONS

Knowing how the derivative of a function behaves helps us in obtaining an accurate idea of the graph of a function. We shall see that the derivative tells us when a graph is rising and falling, as well as when it has peaks and troughs. In this and the following section we establish some rules (in the form of theorems) which are useful for graphing functions; we also give examples to illustrate the technique.

**DEFINITION**   *A function $f$ is said to be **increasing on the interval** $I$ if $f(x_2) > f(x_1)$ whenever $x_2 > x_1$, so long as both $x_1$ and $x_2$ are in $I$. It is **decreasing** if $f(x_2) < f(x_1)$ whenever $x_2 > x_1$ (Fig. 12). The interval $I$ may contain one endpoint, both endpoints, or neither endpoint.*

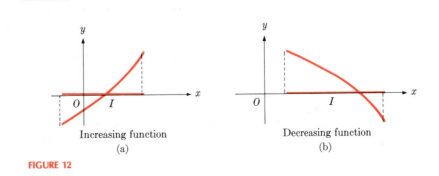

Increasing function
(a)

Decreasing function
(b)

**FIGURE 12**

**DEFINITION**　*A function $f$ is said to have a **relative maximum** at $x_0$ if there is some interval with $x_0$ as an interior point, such that $f(x_0)$ is the true maximum of $f$ in this interval. Similarly, it has a **relative minimum** at $x_1$ if there is some interval with $x_1$ as an interior point, such that $f(x_1)$ is the true minimum of $f$ in this interval.*

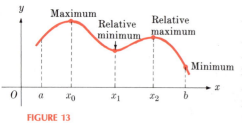

**FIGURE 13**

The latter definition needs some explanation, as provided in Fig. 13, which shows a typical situation. The function $f$ has its maximum on $[a, b]$ at $x_0$, a point which is a relative maximum. However, $f$ has a relative maximum at $x_2$, although $f(x_2)$ is not the maximum of the function on $[a, b]$. The function $f$ has a relative minimum at $x_1$. The true minimum on $[a, b]$ occurs at $b$. However, $f$ does not have a relative minimum at $b$, since the definition for relative minimum is not satisfied there. In Section 5 we investigate further the distinction between true maxima and minima and relative maxima and minima.

**THEOREM 5**　*If $f$ is continuous on an interval $I$ and if $f'(x) > 0$ for each $x$ in the interior of $I$, then $f$ is increasing on $I$.*

**Proof**　We apply the Mean Value Theorem to two points $x_1$, $x_2$ in $I$. We find that

$$f(x_2) - f(x_1) = f'(x_0)(x_2 - x_1),$$

where $x_0$ is between $x_1$ and $x_2$ and hence interior to $I$. For $x_2 > x_1$ and $f'$ positive, we obtain $f'(x_0)(x_2 - x_1) > 0$, and so $f(x_2) - f(x_1) > 0$. This means, by definition, that $f$ is increasing. □

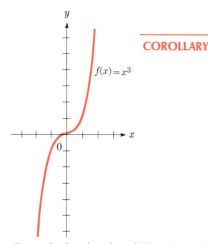

$f(x) = x^3$

Example showing that $f'(0) = 0$, yet $f$ does not have a relative minimum or relative maximum at $x = 0$

**FIGURE 14**

**COROLLARY**　*If $f$ is continuous on $I$ and $f'(x) < 0$ for each $x$ interior to $I$, then $f$ is decreasing on $I$.*

Theorem 5 is useful for graphing; we can find out whether the graph of a function is rising or falling by observing whether the derivative is positive or negative. This theorem (together with Theorem 2 of Section 1, which says that at a relative maximum or minimum the derivative of a function is zero) helps us find the peaks and troughs.

There is a complication in using Theorem 2 because, although the derivative of a function is zero at a maximum or minimum point (assuming the derivative exists there), *the converse may not hold true.* For example, the function $f(x) = x^3$ (Fig. 14) has derivative $f'(x) = 3x^2$, which is zero at $x = 0$. But $(0, 0)$ is neither a maximum nor a minimum point of the function. Thus the knowledge that the derivative of a function vanishes at a certain point is not enough to guarantee that the function has a maximum or minimum there.

**DEFINITIONS**　*A **critical value** of a function $f$ is a value of $x$ where $f'(x) = 0$. A **critical point** of a function $f$ is the point $(x, f(x))$ on the graph corresponding to the critical value $x$.*

A critical point occurs at any relative maximum or minimum point of a function which has a derivative at that point. However, the function $f(x) = x^3$ has a critical point at $(0, 0)$, but this point is neither a maximum nor a minimum of the function.

Suppose a function $f$ and its derivative $f'$ are continuous. If the equation $f'(x) = 0$ has only a finite number of solutions, we may use Theorem 5 and its corollary as an aid in graphing. To do so, we let $x_1, x_2, \ldots, x_n$ be *all* the values of $x$ for which $f'(x) = 0$, arranged in order of increasing size. In each of the intervals $(-\infty, x_1)$, $(x_1, x_2)$, $(x_2, x_3)$, $\ldots$, $(x_{n-1}, x_n)$, $(x_n, +\infty)$, the quantity $f'(x)$ must remain either *positive throughout* or *negative throughout*. This fact is a result of Theorem 6, stated below without proof. As a result, to determine the sign of $f'(x)$ in any one of the intervals, we need only find its sign at a single interior point in that interval.

**THEOREM 6**    *Suppose $f$ is continuous on an interval $I$ and $f$ is not zero at any point of $I$. Then $f$ is either positive on all of $I$ or negative on all of $I$.*

This theorem is intuitively clear when we graph a continuous function. If $f$ starts out positive, (i.e., above the $x$ axis) and then becomes negative, (i.e., goes below the $x$ axis) it can do so only by crossing the axis. The place where it crosses the $x$ axis is a point where $f$ is zero.

Theorem 6 is usually proved in a course in advanced calculus. It is a consequence of the Intermediate Value Theorem, stated in Chapter 5 (Theorem 11, page 206).

We give an example showing how the determination of the zeros of $f'$ can be used as an aid in graphing.

**EXAMPLE 1**    Study the derivative of the function $f(x) = \frac{1}{3}x^3 - x^2 - 3x + 3$ and use the resulting knowledge to sketch the graph of the function.

**Solution**    The derivative is $f'(x) = x^2 - 2x - 3 = (x+1)(x-3)$. The derivative is zero for $x_1 = -1$ and $x_2 = 3$. We consider the intervals $(-\infty, -1)$, $(-1, 3)$, and $(3, +\infty)$. We pick a single point in each interval, selecting the point $-3$ in $(-\infty, -1)$, the point $0$ in $(-1, 3)$, and the point $5$ in $(3, +\infty)$. We determine the sign of $f'(x)$ at each of these points and the value of $f$ at each of them, as shown in the table below. With these aids we are ready to graph the function (Fig. 15).

| $x$ | $-3$ | $-1$ | $0$ | $3$ | $5$ |
|---|---|---|---|---|---|
| $f'(x)$ | $+$ | $0$ | $-$ | $0$ | $+$ |
| $f(x)$ | $-6$ | $4\frac{2}{3}$ | $3$ | $-6$ | $4\frac{2}{3}$ |

The critical value $x = -1$ corresponds to a relative maximum, and the critical value $x = 3$ corresponds to a relative minimum.    □

The theorems and definitions presented in this section give us a method for finding relative maxima and minima. We state the method in the form of a test.

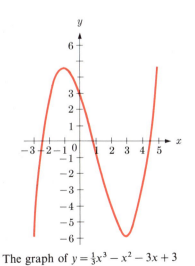

The graph of $y = \frac{1}{3}x^3 - x^2 - 3x + 3$

FIGURE 15

**THEOREM 7**  **(First Derivative Test)**  i) *If $f$ is increasing ($f' > 0$) on some interval to the left of $x_0$ with $x_0$ as endpoint of this interval, and if $f$ is decreasing ($f' < 0$) on some interval to the right of $x_0$ (with $x_0$ as endpoint), then $f$ has a relative maximum at $x_0$ if it is continuous there.*

ii) *If $f$ is decreasing ($f' < 0$) in some interval to the left of $x_0$ with $x_0$ as endpoint of this interval, and if $f$ is increasing ($f' > 0$) in some interval to the right of $x_0$ (with $x_0$ as endpoint), then $f$ has a relative minimum at $x_0$ if it is continuous there.*

*Remarks.* If $f$ has a derivative at $x_0$, then the derivative will be zero. However, we include the possibility that there may be a cusp or a corner. Figure 16 shows various possibilities.

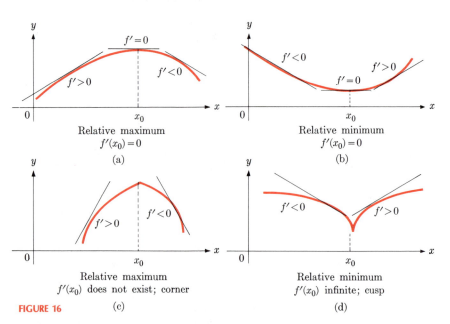

FIGURE 16

**Proof of Theorem 7**   To prove part (i), we denote by $I'$ the interval to the left (including $x_0$) and by $I''$ the interval to the right of $x_0$; then we combine them into one interval and call this interval $I$ (Fig. 17). That is, $I' \cup I'' = I$. Then $x_0$ is an interior point of $I$. Since $f$ is increasing on $I'$ and decreasing on $I''$ and both contain the point $x_0$, we must have $f(x_0) \geq f(x)$ for all $x$ in $I$.

The proof of (ii) is the same.                              □

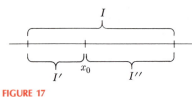

FIGURE 17

**EXAMPLE 2**   Discuss the function $f$ defined by

$$f(x) = x^{5/3} + 5x^{2/3}$$

for relative maxima and minima, and determine the intervals in which $f$ is increasing and those in which $f$ is decreasing. Sketch the graph.

**Solution**   The derivative is

$$f'(x) = \tfrac{5}{3}x^{2/3} + \tfrac{10}{3}x^{-1/3} = \tfrac{5}{3}x^{-1/3}(x + 2).$$

The critical value is $x = -2$. The derivative is not defined at $x = 0$. We construct the following table (* means undefined):

| $x$ | $-5$ | $-2$ | $-1$ | $0$ | $1$ |
|-----|------|------|------|-----|-----|
| $f'(x)$ | $+$ | $0$ | $-$ | $*$ | $+$ |
| $f(x)$ | $0$ | $3(2)^{2/3}$ | $4$ | $0$ | $6$ |

We conclude that

$$f \text{ increases for } x \leq -2;$$

$$f \text{ decreases for } -2 \leq x \leq 0;$$

$$f \text{ increases for } x \geq 0.$$

We now apply the First Derivative Test and conclude that there is a relative maximum at $x = -2$ and a relative minimum at $x = 0$. The graph is sketched in Fig. 18. □

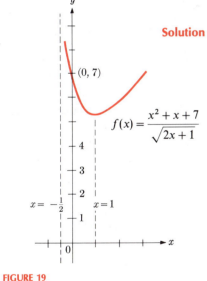

The graph of $y = x^{5/3} + 5x^{2/3}$

**FIGURE 18**

**EXAMPLE 3**   Discuss the function $f$ defined by

$$f(x) = \frac{x^2 + x + 7}{\sqrt{2x + 1}}, \quad x > -\frac{1}{2},$$

for relative maxima and minima, and determine where $f$ is increasing and where it is decreasing. Sketch the graph.

**Solution**   The derivative is

$$f'(x) = \frac{(2x + 1)^{1/2}(2x + 1) - (x^2 + x + 7)\frac{1}{2}(2x + 1)^{-1/2}(2)}{(2x + 1)}.$$

We simplify by multiplying numerator and denominator through by $(2x + 1)^{1/2}$ and obtain

$$f'(x) = \frac{(2x + 1)^2 - (x^2 + x + 7)}{(2x + 1)^{3/2}} = \frac{3(x + 2)(x - 1)}{(2x + 1)^{3/2}}.$$

We set this equal to zero and find that $x = 1$ is a critical value. The apparent solution $x = -2$ is excluded, since the function is defined only for $x > -\frac{1}{2}$. We easily discover that

$$f(x) \text{ decreases for } -\frac{1}{2} < x \leq 1; \quad f(x) \text{ increases for } x \geq 1.$$

The First Derivative Test now applies, and we find that the value $x = 1$ yields a relative minimum. The graph is sketched in Fig. 19. □

$(0, 7)$

$$f(x) = \frac{x^2 + x + 7}{\sqrt{2x + 1}}$$

$x = -\frac{1}{2} \qquad x = 1$

**FIGURE 19**

## 3   PROBLEMS

In Problems 1 through 36, discuss each function for relative maxima and minima, and determine those intervals in which the function is increasing and those in which it is decreasing. Sketch the graphs.

**1**  $x^2 + 3x + 4$

**2**  $x^2 - 4x + 5$

**3**  $-2x^2 + x - 6$

**4**  $-3x^2 - 3x + 2$

**5**  $2x^3 + 3x^2 - 12x$

**6**  $x^3 - 6x^2 + 9x + 5$

**7**  $x^3 + 2x^2 - 3x - 2$

**8**  $x^3 - 3x^2 + 6x - 3$

**9**  $x^3 + 6x^2 + 12x - 5$

**10**  $x^3 + 3x^2 + 6x - 3$

**11**  $-x^3 + 2x^2 - x + 1$

**12**  $-x^3 - 2x^2 + 3x - 6$

**13**  $x^4 - \frac{4}{3}x^3 - 4x^2 + \frac{2}{3}$

**14**  $x^4 + \frac{4}{3}x^3 - 12x^2$

**15** $x^4 + 2x^3$

**16** $x^4 + 4x^3 + 6x^2$

**17** $x + \dfrac{1}{x}$

**18** $\dfrac{2x}{x^2 + 1}$

**19** $\cos x$

**20** $\sin x$

**21** $\sin x \cos x$

**22** $\cos^2 x$

**23** $\dfrac{x - 1}{x + 1}$

**24** $\dfrac{3x - 2}{2x + 3}$

**25** $\dfrac{x^2}{\sqrt{x + 1}}, \quad x > -1$

**26** $3x^{1/2} - x^{3/2}$

**27** $x^{2/3}(x + 3)^{1/3}$

**28** $x/(x + 1)$

**29** $x\sqrt{3 - x}, \quad x \le 3$

**30** $x^2\sqrt{5 - x}, \quad x \le 5$

**31** $x\sqrt{2 - x^2}, \quad |x| \le \sqrt{2}$

**32** $\dfrac{x^2 - 2x + 1}{x + 1}$

**33** $\dfrac{x^2 - 3x - 4}{x - 2}$

**34** $|x^2 - 2x| + 1$

**35** $x^2 + |2x + 2|$

**36** $\dfrac{x - 2}{|x^2 - 2x| + 1}$

In each of Problems 37 through 44, find the relative maxima and minima of the given function $f$ for $0 \le x \le 2\pi$. Determine where the function is increasing, where decreasing, and sketch the graph.

**37** $f(x) = \cos x + \sin x$

**38** $f(x) = \cos x - \sin x$

**39** $f(x) = x - \sin x$

**40** $f(x) = 2x + \cos x$

**41** $f(x) = 2 \cos x + \sin 2x$

**42** $f(x) = 2 \cos x + \cos 2x$

**43** $f(x) = 3x + \sin 3x$

**44** $f(x) = -4x - \cos 3x$

In each of Problems 45 through 48, find the relative maximum and minimum points of the graph.

**45** $y = 3 \sin x + \cos x$

**46** $y = \sin 3x - 3 \sin x$

**47** $y = \sqrt{3} \sin 2x - \cos 2x$

**48** $y = 4 \sin^3 x - 3 \sin x$

**49** The function $f(x) = (\tan x)^{1/2}$ is defined for $0 \le x < \pi/2$. Show that $f'(x) > 0$ for $0 < x < \pi/2$. Is it true that $f''(x) > 0$ for $0 < x < \pi/2$?

**50** Find all the relative maximum and minimum points of the function

$$f(x) = \sin \frac{\pi}{x}$$

defined for $0 < x \le 1$. (There are infinitely many.)

**\*51** Given

$$f(x) = x^p \sin \frac{\pi}{x}$$

for $0 < x \le 1$. Show that if $p$ is any positive number, there are always infinitely many relative maxima and minima. If these points are labeled $(x_1, y_1)$, $(x_2, y_2)$, ... in order of decreasing $x_i$, then it follows that $x_i \to 0$, $y_i \to 0$ as $i \to \infty$.

**\*52** Given the function $f(x) = ax^3 + bx^2 + cx + d$ which is the most general equation of the third degree. Figure 15 exhibits the shape of this function when $a = \frac{1}{3}$, $b = -1$, $c = -3$, $d = 3$. Figure 14 exhibits the shape when $a = 1$, $b = 0$, $c = 0$, $d = 0$. Show that there are basically four possible shapes for a third-degree equation and sketch the remaining two shapes. (Assume $a \ne 0$.)

**\*53** The general equation of the fourth degree has the form $f(x) = ax^4 + bx^3 + cx^2 + dx + e$. Assuming that $a \ne 0$, decide how many different general shapes the graph of such an equation can exhibit. Sketch one of each type.

---

**4**

## APPLICATIONS USING THE SECOND DERIVATIVE

The more information we have about a function the more accurately we can construct its graph. In Section 3 we saw that a knowledge of the first derivative helps us decide when the graph is increasing and when it is decreasing. Furthermore, we learned that the First Derivative Test enables us to locate the relative maxima and minima. The second derivative yields additional facts which are helpful in determining the nature of the graph. The first of these facts is a test for relative maxima and minima which employs the second derivative.

**DEFINITION**   *If, at each point of an interval, the graph of a function f always remains above the line tangent to the curve at this point, we say that the curve is* **concave upward** *on the interval (see Fig. 20). If the curve always remains below its tangent line, we say it is* **concave downward** *(see Fig. 21).*

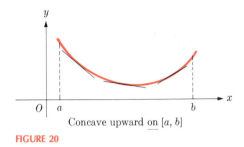

Concave upward on $[a, b]$

**FIGURE 20**

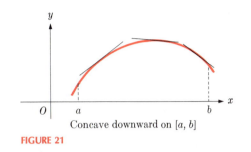

Concave downward on $[a, b]$

**FIGURE 21**

Before stating and proving the basic theorem of this section we recall the equation of the line tangent to a curve. At any value $x_0$ the function $f$ has value $f(x_0)$ and the slope of the curve at this point is $f'(x_0)$. The equation of a line through the point $(x_0, f(x_0))$ with slope $f'(x_0)$ is, according to the point-slope formula,

$$y - f(x_0) = f'(x_0)(x - x_0) \qquad \text{or} \qquad y = f(x_0) + f'(x_0)(x - x_0).$$

This is the tangent line.

**THEOREM 8**   **(Concavity Test)**   *Assume that f has a second derivative on an interval I.*

a) *If $f''(x) > 0$ for all $x$ interior to $I$, then the curve is concave upward on $I$.*

b) *If $f''(x) < 0$ for all $x$ interior to $I$, then the curve is concave downward on $I$.*

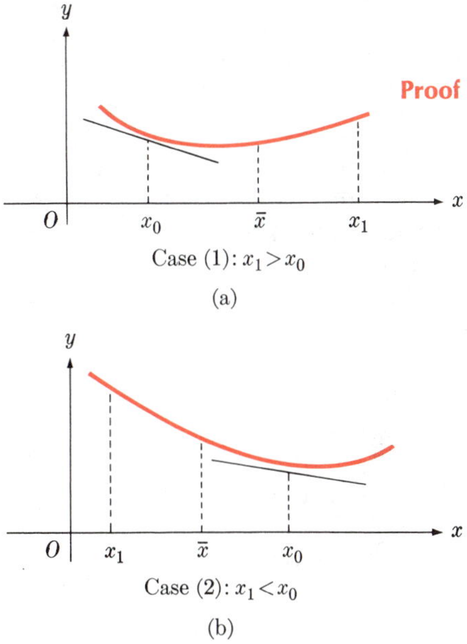

**FIGURE 22**

**Proof**   To prove (a), we must show that the curve lies above the tangent line at any point. Let $x_0$ be any (fixed) point in $I$ (see Fig. 22). The tangent line at $x_0$ has equation

$$y = f(x_0) + f'(x_0)(x - x_0),$$

and we therefore have to establish (for all $x$ in $I$) the inequality

$$f(x) \geq f(x_0) + f'(x_0)(x - x_0).$$

If $x = x_0$, this becomes $f(x_0) \geq f(x_0)$, which is true. If $x$ is some other value, say $x_1$, then we can apply the Mean Value Theorem to the function $f$ between the two points $x_0$ and $x_1$. This theorem says that there is a value $\bar{x}$ such that

$$f'(\bar{x}) = \frac{f(x_1) - f(x_0)}{x_1 - x_0},$$

where $\bar{x}$ is between $x_0$ and $x_1$. We have two possibilities: (1) $x_1 > x_0$, and (2) $x_1 < x_0$. In Case (1) we write the Mean Value Theorem in the form

$$f(x_1) = f(x_0) + f'(\bar{x})(x_1 - x_0). \tag{1}$$

Case (1): $x_1 > x_0$

(a)

Case (2): $x_1 < x_0$

(b)

Since $f'' > 0$ by hypothesis, we know from Theorem 5 that $f'$ is increasing on $I$. If $x_1 > x_0$, then $x_1 > \bar{x} > x_0$ and

$$f'(\bar{x}) > f'(x_0).$$

We multiply this last inequality through by the positive quantity $(x_1 - x_0)$ to get

$$f'(\bar{x})(x_1 - x_0) > f'(x_0)(x_1 - x_0),$$

and therefore, by substitution in Eq. (1), we obtain

$$f(x_1) > f(x_0) + f'(x_0)(x_1 - x_0),$$

which is what we wished to show.

In Case (2), since $x_0 > \bar{x}$, we must have $f'(x_0) > f'(\bar{x})$. But now we multiply through by the negative number $(x_1 - x_0)$, which reverses the inequality, giving

$$f'(\bar{x})(x_1 - x_0) > f'(x_0)(x_1 - x_0)$$

as before, and therefore, using (1)

$$f(x_1) > f(x_0) + f'(x_0)(x_1 - x_0).$$

The proof of part (b) of Theorem 8 is similar.                         □

Theorem 8 shows that knowledge of the second derivative gives us an even clearer picture of the appearance of the curve. In addition, it gives us the following useful test, known as the **Second Derivative Test**, for relative maxima and minima.

**THEOREM 9**     **(Second Derivative Test)**   *Assume that $f$ has a second derivative, that $f''$ is continuous, and that $x_0$ is a critical value $(f'(x_0) = 0)$. Then*

a) *If $f''(x_0) > 0$, $f$ has a relative minimum at $x_0$.*

b) *If $f''(x_0) < 0$, $f$ has a relative maximum at $x_0$.*

c) *If $f''(x_0) = 0$, the test fails.*

**Proof**     To prove part (a), we see from Theorem 8 that the curve is concave upward and must lie above the tangent line at $x_0$. But this line is horizontal, since $f'(x_0) = 0$. Therefore $f(x_0)$ must be a minimum value. The proof of (b) is the same. Part (c) is added for the sake of completeness.                         □

**EXAMPLE 1**     Discuss the function

$$f(x) = x^3 - \tfrac{21}{4}x^2 + 9x - 4$$

for relative maxima and minima. Sketch the graph.

**Solution**     The derivative is

$$f'(x) = 3x^2 - \tfrac{21}{2}x + 9 = 3(x^2 - \tfrac{7}{2}x + 3),$$

and the critical values are solutions of

$$x^2 - \tfrac{7}{2}x + 3 = 0 \qquad \text{or} \qquad x = \tfrac{3}{2}, 2.$$

We apply the Second Derivative Test.

$$f''(x) = 6x - \tfrac{21}{2}; \quad f''(\tfrac{3}{2}) = -\tfrac{3}{2} < 0, \quad f''(2) = \tfrac{3}{2} > 0.$$

Therefore $x = \tfrac{3}{2}$ corresponds to a relative maximum, and $x = 2$ corresponds to a relative minimum. Also, the second derivative is positive if $x \geq \tfrac{7}{4}$ and negative if $x \leq \tfrac{7}{4}$. That is, the curve is concave upward for $x \geq \tfrac{7}{4}$ and concave downward for $x \leq \tfrac{7}{4}$. We construct the following table.

| $x$ | $0$ | $\tfrac{3}{2}$ | $\tfrac{7}{4}$ | $2$ | $3$ |
|---|---|---|---|---|---|
| $f$ | $-4$ | $\tfrac{17}{16}$ | $1\tfrac{1}{32}$ | $1$ | $\tfrac{11}{4}$ |
| $f'$ | $+$ | $0$ | $-$ | $0$ | $+$ |
| $f''$ | $-$ | $-$ | $0$ | $+$ | $+$ |

This problem indicates the difficulty that may arise when a relative maximum and a relative minimum are close together. Without knowledge of the derivatives given in the table, we would normally plot the points at $x = 0$, 1, 2, 3, etc. This would miss the maximum at $x = \tfrac{3}{2}$ and give a misleading picture of the curve. We know that the curve is concave downward to the left of $x = \tfrac{7}{4}$ and concave upward to the right of $\tfrac{7}{4}$. These facts prevail indefinitely far to the left and right. The graph is sketched in Fig. 23. In this example the point $x = \tfrac{7}{4}$ is important, as it separates a concave downward region from a concave upward region. □

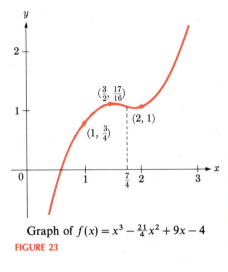

Graph of $f(x) = x^3 - \tfrac{21}{4}x^2 + 9x - 4$

**FIGURE 23**

**DEFINITION**   *A point on a curve is a* **point of inflection** *if $f''(x_0) = 0$ at this point and if the graph is concave upward on one side and concave downward on the other.*

We make two remarks which the reader should keep in mind: The first derivative may or may not vanish at a point of inflection. In Example 1 the value $x = \tfrac{7}{4}$ corresponds to a point of inflection, and $f'(\tfrac{7}{4}) = -\tfrac{3}{16}$. On the other hand, the function $f(x) = (x - 2)^3 + 1$ has a point of inflection (see Fig. 24) at $x = 2$, since $f''(x) = 6(x - 2)$ and $f''(2) = 0$. However, we notice that $f'(x) = 3(x - 2)^2$ and $f'(2) = 0$, also.

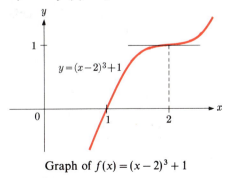

Graph of $f(x) = (x - 2)^3 + 1$

**FIGURE 24**

The second remark: It is not enough to know that $f''(x_0) = 0$ to guarantee that $x_0$ corresponds to a point of inflection. We must also know that $f''(x) > 0$ on one side and that $f''(x) < 0$ on the other. An example which

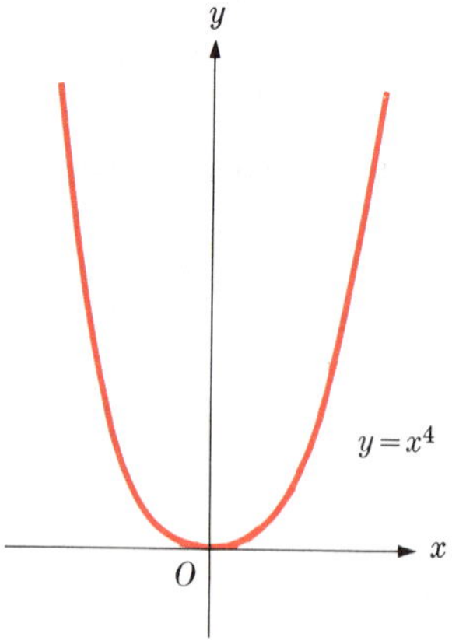

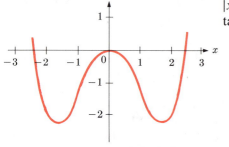

$y = x^4$

FIGURE 25

shows this difficulty is the function $f(x) = x^4$ (see Fig. 25). In this case, $f''(x) = 12x^2$ and $f''(0) = 0$. However, $f''(x)$ is always positive for $x \neq 0$ and, in fact, $x = 0$ is the value for a minimum. (Part (c) of Theorem 9 for the Second Derivative Test is the rule which holds here.) The First Derivative Test, however, works to give a minimum at $x = 0$.

**EXAMPLE 2**  Discuss the function

$$f(x) = \tfrac{1}{4}x^4 - \tfrac{3}{2}x^2$$

for relative maxima, relative minima, and points of inflection. Sketch the graph.

**Solution**  The first two derivatives are

$$f'(x) = x^3 - 3x \quad \text{and} \quad f''(x) = 3x^2 - 3.$$

The critical values are solutions of $x^3 - 3x = 0$, and so we get $x = 0$, $\sqrt{3}$, $-\sqrt{3}$. The Second Derivative Test tells us that

$$x = 0 \text{ is at a relative maximum;}$$

$$x = \sqrt{3}, \ -\sqrt{3} \text{ are at relative minima.}$$

The possible points of inflection are obtained from solutions of $3x^2 - 3 = 0$; that is, $x = +1, -1$. Since $f''(x)$ is negative for $-1 < x < 1$ and positive for $|x| > 1$, both $x = 1$ and $x = -1$ yield points of inflection. We construct the table:

| $x$ | $-2$ | $-\sqrt{3}$ | $-1$ | $0$ | $1$ | $\sqrt{3}$ | $2$ |
|-----|------|-------------|------|-----|-----|------------|-----|
| $f$ | $-2$ | $-\tfrac{9}{4}$ | $-\tfrac{5}{4}$ | $0$ | $-\tfrac{5}{4}$ | $-\tfrac{9}{4}$ | $-2$ |
| $f'$ | $-$ | $0$ | $+$ | $0$ | $-$ | $0$ | $+$ |
| $f''$ | $+$ | $+$ | $0$ | $-$ | $0$ | $+$ | $+$ |

The graph is shown in Fig. 26.    □

Graph of $f(x) = \tfrac{1}{4}x^4 - \tfrac{3}{2}x^2$

FIGURE 26

**EXAMPLE 3**  Discuss the function

$$f(x) = x^{4/3} + 4x^{1/3}$$

for relative maxima, relative minima, and points of inflection. Sketch the graph.

**Solution**  We have

$$f'(x) = \tfrac{4}{3}x^{1/3} + \tfrac{4}{3}x^{-2/3} = \tfrac{4}{3}x^{-2/3}(x + 1);$$

$$f''(x) = \tfrac{4}{9}x^{-2/3} - \tfrac{8}{9}x^{-5/3} = \tfrac{4}{9}x^{-5/3}(x - 2).$$

The first and second derivatives do not exist at $x = 0$. The value $x = -1$ is a critical value and $x = 2$ corresponds to a possible point of inflection. We construct the table:

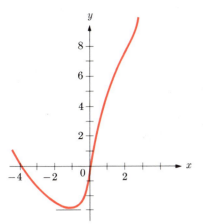

| $x$ | $-4$ | $-1$ | $-\frac{1}{2}$ | $0$ | $1$ | $2$ | $3$ |
|---|---|---|---|---|---|---|---|
| $f$ | $0$ | $-3$ | $-\frac{7}{4}\sqrt[3]{4}$ | $0$ | $5$ | $6\sqrt[3]{2}$ | $7\sqrt[3]{3}$ |
| $f'$ | $-$ | $0$ | $+$ | $*$ | $+$ | $+$ | $+$ |
| $f''$ | $+$ | $+$ | $+$ | $*$ | $-$ | $0$ | $+$ |

From the table we see that $x = -1$ yields a relative minimum and that $x = 2$ yields a point of inflection. Furthermore, $x = 0$ yields a point where the curve changes from concave upward to concave downward. The graph is shown in Fig. 27.     □

Graph of $f(x) = x^{4/3} + 4x^{1/3}$

**FIGURE 27**

**EXAMPLE 4**   Discuss the function $f(x) = \sin 2x$,   $-\pi < x < \pi$, for relative maxima, relative minima, and points of inflection. Sketch the graph.

**Solution**   We have

$$f'(x) = 2\cos 2x, \qquad f''(x) = -4\sin 2x.$$

The critical points are solutions of $2\cos 2x = 0$. These occur when

$$2x = \pm\frac{\pi}{2}, \qquad \pm\frac{3\pi}{2} \qquad \text{or} \qquad x = \pm\frac{\pi}{4}, \qquad \pm\frac{3\pi}{4}.$$

Using the Second Derivative Test, we find

$$f''\left(\frac{\pi}{4}\right) < 0, \quad f''\left(\frac{3\pi}{4}\right) > 0, \quad f''\left(-\frac{\pi}{4}\right) > 0, \quad f''\left(-\frac{3\pi}{4}\right) < 0.$$

Thus relative maxima occur at

$$x = \frac{\pi}{4}, \qquad -\frac{3\pi}{4}$$

and relative minima at

$$x = \frac{3\pi}{4}, \qquad -\frac{\pi}{4}.$$

To obtain the points of inflection, we set $f''(x) = 0$ and find that these occur when

$$2x = 0, \qquad \pm\pi \qquad \text{or} \qquad x = 0, \qquad \pm\frac{\pi}{2}.$$

The points of inflection are

$$(0, 0), \qquad \left(\frac{\pi}{2}, 0\right), \qquad \left(-\frac{\pi}{2}, 0\right).$$

The graph is shown* in Fig. 28.

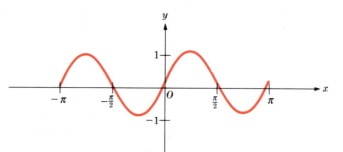

## 4   PROBLEMS

In Problems 1 through 35, discuss each of the functions for relative maxima and minima, concavity, and points of inflection. Sketch the graphs.

1 $f(x) = x^2 - 3x + 4$

2 $f(x) = -2x^2 + 6x + 5$

3 $f(x) = x^3 - 27x + 4$

4 $f(x) = x^3 + 3x^2 + 6x - 4$

5 $f(x) = x^4 + 4x^3$

6 $f(x) = x^4 - 4x^3 - 2x^2 + 12x - 8$

7 $f(x) = x^2 + \dfrac{1}{x^2}$

8 $f(x) = 2x^2 - \dfrac{1}{x^2}$

9 $f(x) = \dfrac{2x}{x^2 + 1}$

10 $f(x) = x\sqrt{x + 3}$

11 $f(x) = x^3 - \frac{3}{2}x^2 - 6x + 2$

12 $f(x) = x^3 + x^2 - x - 1$

13 $f(x) = x^3 - 4x^2 + 4x - 1$

14 $f(x) = x^3 + 3x^2 - 3x - 5$

15 $f(x) = x^3 - x^2 + x - 1$

16 $f(x) = x^4 + \frac{4}{3}x^3 - 4x^2 - \frac{4}{3}$

17 $f(x) = (x + 2)(x - 2)^3$

18 $f(x) = x^4 - 3x^3 + 3x^2$

19 $f(x) = x^4 + 5x^3 + 6x^2$

20 $f(x) = x - 3 + \dfrac{2}{x + 1}$

21 $f(x) = \dfrac{4x}{x^2 + 4}$

22 $f(x) = 5x^{2/3} - x^{5/3}$

23 $f(x) = x\sqrt{8 - x^2}, \quad |x| \le \sqrt{8}$

24 $f(x) = x^{2/3}(x + 2)^{-1}$

25 $f(x) = x^2\sqrt{5 + x}, \quad x \ge -5$

26 $f(x) = x^2\sqrt{3 - x^2}, \quad |x| \le \sqrt{3}$

27 $f(x) = x^{1/3}(x + 2)^{-2/3}$

28 $f(x) = |x^2 - 6x| + 2$

29 $f(x) = 2\sin\frac{1}{2}x, \quad -\pi < x < \pi$

30 $f(x) = \cos 3x, \quad -\dfrac{\pi}{2} < x < \dfrac{\pi}{2}$

31 $f(x) = 4\sin 2x \cos 2x, \quad -\pi < x < \pi$

32 $f(x) = 2\cos^2\frac{1}{3}x, \quad -\dfrac{\pi}{2} < x < \dfrac{\pi}{2}$

33 $f(x) = \tan x, \quad -\dfrac{\pi}{2} < x < \dfrac{\pi}{2}$

34 $f(x) = \sin x + \cos x, \ -2\pi < x < 2\pi.$

35 $f(x) = \sqrt{3}\sin x + \cos x, \quad -\pi < x < \pi.$

In Problems 36 and 37, find the relative maxima and minima.

36 $f(x) = x^3 + |2x^2 - 2x| + 1$

37 $f(x) = \dfrac{x - 1}{|x^2 - 4x| + 2}$

---

*We provide a review on graphing of trigonometric functions in Section 2 of Appendix 1.

*38 Given the function $f(x) = x^n$, where $n$ is a positive integer. Show that $f$ has a minimum at $x = 0$ if $n$ is an even integer and that $f$ has a point of inflection at $x = 0$ if $n$ is an odd integer.

*39 Using the results of Problem 38 as a guide, devise a "Higher Derivative Test" for relative maxima and minima when both $f'(x_0) = 0$ and $f''(x_0) = 0$. Describe the situation if the first $k$ derivatives of $f$ vanish at $x_0$ but the derivative of order $k + 1$ is different from zero.

40 If $f$ is a given function, devise a definition of a point of inflection at a point at which the second derivative does not exist. Construct an example of a function which has such a point of inflection.

41 Show that if the function $f(x) = a_0 x^3 + a_1 x^2 + a_2 x + a_3$ has exactly one critical point, then it is a point of inflection.

42 Show that the function

$$f(x) = \frac{ax + b}{cx + d}$$

has no relative maxima or minima, where $a, b, c, d$ are any numbers with $ad - bc \neq 0$.

---

## 5

# THE MAXIMUM AND MINIMUM VALUES OF A FUNCTION ON AN INTERVAL

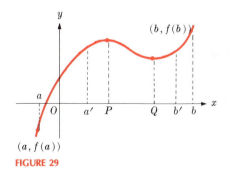

FIGURE 29

In the preceding sections we studied ways of finding relative maxima and minima of a function. One of these relative maxima may be the true maximum of the function, or it may not be. As Fig. 29 shows, the function exhibited has a relative maximum at $P$ and a relative minimum at $Q$. As we go off to the right, however, the function gets larger than the value at $P$, and as we go off to the left, the function gets smaller than the value at $Q$. Suppose that we look at the above function only on some interval $[a, b]$, as shown in Fig. 29. Then the maximum occurs at $b$ and is $f(b)$, while the minimum occurs at $a$ and is $f(a)$. If $[a, b']$ were the interval of interest, as shown, then the maximum of $f$ on $[a, b']$ is at $P$, while the minimum is at $a$. Further, on the interval $[a', b']$ the maximum is at $P$, the minimum at $Q$.

---

DEFINITIONS  *Let $f$ be a continuous function defined on the closed interval $a \leq x \leq b$. The maximum value $f$ takes on in this interval is called the **absolute maximum** of $f$ on $[a, b]$. The minimum value $f$ takes on in the interval $[a, b]$ is called the **absolute minimum** of $f$ on $[a, b]$.*

---

To obtain the maximum and minimum of a continuous function on an interval $[a, b]$, we employ the following procedure (which is based on Theorem 5 and the First Derivative Test):

---

RULE  a) *Find the relative maximum and minimum values of $f$.*
b) *Find the value of the function at each of the endpoints.*

*The largest of the values of* (a) *and* (b) *is the absolute maximum. The smallest value is the absolute minimum.*

**EXAMPLE**    Given the function

$$f(x) = \tfrac{1}{3}x^3 + \tfrac{1}{2}x^2 - 2x,$$

find the absolute maximum and absolute minimum on the interval $[-3, 4]$.

**Solution**    We follow the Rule and find relative maxima and minima:

$$f'(x) = x^2 + x - 2,$$

and the critical values are $x = -2, 1$. The Second Derivative Test gives us

$$f''(x) = 2x + 1; \qquad f''(-2) = -3,$$

making $\tfrac{10}{3}$ a relative maximum value, and

$$f''(1) = 3,$$

which makes $-\tfrac{7}{6}$ a relative minimum value. That is, the point $(1, -\tfrac{7}{6})$ is a relative minimum, and $(-2, \tfrac{10}{3})$ is a relative maximum. Now we go on to part (b) of the Rule: $f(-3) = \tfrac{3}{2}$ and $f(4) = \tfrac{64}{3}$. Since $\tfrac{64}{3}$ is larger than $\tfrac{10}{3}$, the absolute maximum of the function occurs at $x = 4$, the right endpoint. Since $-\tfrac{7}{6}$ is smaller than $\tfrac{3}{2}$, the absolute minimum of the function is at $x = 1$, the relative minimum. Figure 30 shows the various points.    □

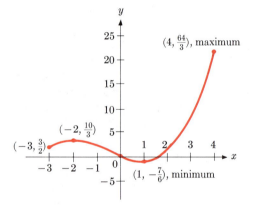

FIGURE 30

## 5  PROBLEMS

In Problems 1 through 14, find the absolute maximum and absolute minimum of the given function on the interval indicated.

1  $f(x) = x^2 + 2x - 4, \quad -4 \le x \le 3$

2  $f(x) = -2x^2 + x - 3, \quad -1 \le x \le 6$

3  $f(x) = 2x^3 - x^2 + 2, \quad -2 \le x \le 1$

4  $f(x) = x^3 + 2x^2 + 18x, \quad -1 \le x \le 2$

5  $f(x) = x^3 + 3x^2 - 9x + 4, \quad -5 \le x \le 6$

6  $f(x) = x^4 - 2x^2 + 1, \quad -2 \le x \le 1$

7  $f(x) = x^4 - 2x^2 + 1, \quad 0 \le x \le 5$

8  $f(x) = x^4 - 2x^2 + 1, \quad -5 \le x \le 5$

9  $f(x) = \dfrac{x}{x+1}, \quad -\tfrac{1}{2} \le x \le 1$

10  $f(x) = \dfrac{x+3}{x-2}, \quad -3 \le x \le 1$

11  $f(x) = \tan x, \quad -\dfrac{\pi}{3} \le x \le \dfrac{\pi}{3}$

12  $f(x) = \sin 3x, \quad 0 \le x \le \dfrac{\pi}{6}$

**13** $f(x) = \sqrt{3} \sin x - \cos x, \quad -\dfrac{\pi}{3} \le x \le \dfrac{\pi}{4}$

**14** $f(x) = \cos^2 2x, \quad -\dfrac{\pi}{2} \le x \le \dfrac{2\pi}{3}$

**15** Give an example which shows that the Rule of this section may fail if the function considered is not continuous.

In each of Problems 16 through 24, find the absolute maxima and minima of the functions on the interval indicated, or show that there are none.

**16** $f(x) = x^2, \quad -2 \le x < 1$

**17** $f(x) = x^3 - 2x^2 + x - 1, \quad -3 < x < 3$

**18** $f(x) = \dfrac{x-1}{x+1}, \quad -2 \le x \le 2$

**19** $f(x) = \dfrac{x^2}{x-2}, \quad 1 \le x \le 3$

**20** $f(x) = \dfrac{x^2+2}{x+2}, \quad -2 < x \le 0$

**21** $f(x) = \dfrac{1}{x+1}, \quad 2 \le x < \infty$

**22** $f(x) = \dfrac{x-1}{x+1}, \quad -\infty < x \le -2$

**23** $f(x) = \sec x, \quad \dfrac{\pi}{3} \le x \le \dfrac{5\pi}{6}$

**24** $f(x) = \tan 2x, \quad -\dfrac{\pi}{6} \le x \le \dfrac{\pi}{6}$

**25** Given $f(x) = x^3 + ax^2 + bx + c$ defined on an interval $[d_1, d_2]$. Show that if $a^2 < 3b$ then the absolute maximum and minimum of $f$ must occur at the endpoints.

**26** Let $r$ be a rational number. Let $f(x) = (1 + x)^r - (1 + rx)$ be defined for $-1 \le x \le a$. Locate the absolute maximum and absolute minimum values of $f$. Show that if $r \ge 1$, then the minimum must occur at $x = 0$.

## 6

# APPLICATIONS OF MAXIMA AND MINIMA

Until now we have encountered polynomials, rational functions, trigonometric functions, and so on. Once we know these functions, we set out to find the various relative maxima and minima, maxima at endpoints, and such other properties as concavity and points of inflection.

In applications, however, we usually encounter problems in which we must discover the function itself before we can discuss its properties. Having found the function, we then use the methods described in the previous sections of this chapter to determine the function's characteristics. In order to be equal to the task, the reader should memorize the formulas for areas and volumes of simple geometric shapes, as listed below. In many of the problems these formulas turn out to be likely candidates for functions.

### Elementary Geometric Formulas

i)    Circle of radius $r$. Circumference $= 2\pi r$. Area $= \pi r^2$.
ii)   Circular sector. Area $= \frac{1}{2} r^2 \alpha$, $\alpha$ being the central angle, measured in radians.
iii)  Trapezoid of height $h$ and bases $b$ and $B$. Area $= \frac{1}{2} h(b + B)$.
iv)   Right circular cylinder of height $h$, radius of base $r$. Volume $= \pi r^2 h$. Lateral surface area $= 2\pi rh$.
v)    Right circular cone of height $h$, radius of base $r$. Volume $= \frac{1}{3} \pi r^2 h$. Lateral surface area $= \pi rL$, where $L = \sqrt{r^2 + h^2}$.
vi)   Sphere of radius $r$. Volume $= \frac{4}{3} \pi r^3$. Surface area $= 4\pi r^2$.

If a manufacturer wants to make a tin can at the lowest possible cost, he faces the problem of finding a *minimum*. Making the strongest possible bridge of a certain size, type and span presents a problem of finding a *maximum*. Whenever we use words such as largest, most, least, smallest, best, etc., we can translate them into mathematical language in terms of maxima and minima. If we have a specific formula for the function in question (a situation which is very often impossible in actual practice), then we may be able to use the methods of the calculus to find the required maximum or minimum.

We start by giving several examples of the operating technique and then outline the procedure by giving a set of rules consisting of five steps.

**EXAMPLE 1**    A man has a stone wall alongside a field. He has 1200 meters of fencing material and he wishes to make a rectangular pen, using the wall as one side. What should the dimensions of the pen be in order to enclose the largest possible area?

**Solution**    Here we see that the problem is to find the largest area; clearly this is a maximum problem. If we can find the area as a function of something or other and then differentiate it, we may be able to find the maximum of this function and so get the answer. Let us draw a figure (Fig. 31), and call the area of the pen by some letter, say $A$. The length and width of the pen are both unknown. However, if $x$ is the width, then the length must be $1200 - 2x$, since there are 1200 meters of fencing to be used. Now we can get an expression for the area, namely, length times width or

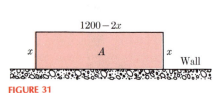

**FIGURE 31**

$$A = A(x) = x(1200 - 2x) = 1200x - 2x^2.$$

We note that $x$ must lie between 0 and 600, since otherwise one side would have negative length. That is, $0 \leq x \leq 600$. The derivative gives us

$$A'(x) = 1200 - 4x,$$

and this vanishes when $x = 300$.

The Second Derivative Test tells us that

$$A'' = -4,$$

and so a relative maximum occurs at $x = 300$. Actually, it is the absolute maximum, since $A(x)$ is a quadratic function which is concave downward everywhere. In fact, $A = 0$ when $x = 0$ and $x = 600$, so the endpoints give a minimum. We conclude that the width of the pen must be 300 meters and the length $1200 - 600 = 600$ meters. $\square$

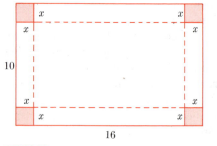

**FIGURE 32**

**EXAMPLE 2**    A rectangular box with an open top is to be made in the following way. A piece of tin 10 cm by 16 cm has a small square cut from each corner (shaded portion in Fig. 32) and then the edges (dashed lines) are folded vertically. What should be the size of the squares cut out if the box is to have as large a volume as possible?

**Solution**    We see that if we cut out exceptionally small squares the box will have practically no height at all and so will have a small volume. Similarly, if the squares are too large the base of the box will be tiny and again the volume will be small. We seek a formula for the volume. We let $x$ be the edge of one of the squares cut out. Then:

$16 - 2x$ will be the length of the base of the box,

$10 - 2x$ will be the width of the base of the box,

$x$ will be the height of the box.

The volume, of course, is length times width times height. Therefore, calling the volume $V$, we have

$$V = V(x) = (16 - 2x)(10 - 2x)x \qquad \text{and} \qquad V(x) = 4(40x - 13x^2 + x^3),$$

which gives volume as a function of $x$. Now we must note that there are restrictions on $x$. First of all, $x$ must be positive in order to make any sense at all. Secondly, $x$ must be less than 5. For, once it gets to 5, the width of the base is zero and we have no box at all. We write

$$0 < x < 5.$$

With this in mind we calculate the derivative:

$$V'(x) = 4(40 - 26x + 3x^2),$$

and by solving

$$40 - 26x + 3x^2 = 0,$$

we obtain

$$x = 2, 6\tfrac{2}{3}.$$

But $6\tfrac{2}{3}$ is outside the interval of interest and we reject it. We note that

$$V''(x) = (-26 + 6x) \qquad \text{and} \qquad V''(2) < 0,$$

which gives a relative maximum at $x = 2$. The endpoints of the interval $[0, 5]$ both give zero volume. Therefore according to the Rule in Section 5, $x = 2$ cm yields the absolute maximum. The volume of the box is $(2)(6)(12) = 144 \text{ cm}^3$.

□

EXAMPLE 3    The sum of one number and three times a second number is 60. Among the possible numbers which satisfy this condition, find the pair whose product is as large as possible.

Solution    We start by letting $x$ be one of the numbers and $y$ the other. We want to find the maximum of the product, which we shall call $P$. So we write

$$P = xy.$$

We have expressed $P$ in terms not of one quantity but of two. We can overcome this difficulty by recalling that the first sentence of the problem asserts that

$$x + 3y = 60.$$

We can eliminate either $x$ or $y$; if we eliminate $y$, we obtain

$$P = x\tfrac{1}{3}(60 - x) = 20x - \tfrac{1}{3}x^2.$$

Since $P$ is a function of $x$, we can differentiate. We write

$$P' = P'(x) = 20 - \tfrac{2}{3}x.$$

Setting this equal to zero, we find that $x = 30$ and $y = 10$. It is easy to verify that this value of $x$ gives the maximum value to $P$. The result is the same if we eliminate $x$, express $P$ as a function of $y$, and differentiate.  □

Using these three examples as a guide, we now list the steps to take in attacking this type of problem in maxima and minima.

**Step 1:** Draw a figure when appropriate.

**Step 2:** Assign a letter to each of the quantities mentioned in the problem.

**Step 3:** Select the quantity which is to be made a maximum or minimum and express it as a function of the other quantities.

**Step 4:** Use the information in the problem to eliminate all quantities but one so as to have a function of one variable. Determine the possible domain of this function.

**Step 5:** Use the methods of Sections 4 and 5 to get the maximum or minimum.

With this five-step procedure in mind, we now study two more examples.

**EXAMPLE 4**  Find the dimensions of the right circular cylinder of maximum volume which can be inscribed in a sphere of radius 12.

**Solution**  When a cylinder is inscribed in a sphere we mean that the upper and lower bases of the cylinder have their bounding circles on the surface of the sphere. The axis of the cylinder is along a diameter of the sphere. We first draw a cross section of the inscribed cylinder, as in Fig. 33 (Step 1). The cylinder might be short and fat or tall and thin. We label the radius of the base $r$ and the height of the cylinder $h$; the volume is denoted by $V$ (Step 2). Then (Step 3),

$$V = \pi r^2 h.$$

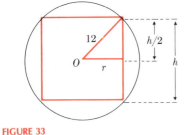

**FIGURE 33**

We have now expressed the volume $V$ in terms of the two quantities $r$ and $h$, so we must eliminate one (Step 4). Referring to Fig. 33, we use the Pythagorean Theorem to get

$$r^2 + \frac{h^2}{4} = 144.$$

Solving for $h$, we get

$$h = 2\sqrt{144 - r^2},$$

and substituting this value of $h$ in the formula for $V$, we find

$$V = 2\pi r^2 \sqrt{144 - r^2}, \quad 0 < r < 12.$$

By differentiating (Step 5), we obtain

$$V'(r) = 4\pi r \sqrt{144 - r^2} + 2\pi r^2 \cdot \tfrac{1}{2}(144 - r^2)^{-1/2}(-2r)$$

$$= 4\pi r \sqrt{144 - r^2} - \frac{2\pi r^3}{\sqrt{144 - r^2}}$$

$$= \frac{2\pi r(288 - 2r^2 - r^2)}{\sqrt{144 - r^2}}.$$

Setting this equal to zero, we find that $r = 0$ or $r = \pm 4\sqrt{6}$. We reject $r = 0$ and the negative value of $r$. Since the Second Derivative Test is rather messy, we reason as follows: $V$ is continuous for $0 \le r \le 12$; when $r = 0$ and $r = 12$, we get $V = 0$. The volume is positive for $r$ in the interval $(0, 12)$. $V$ takes on its maximum on $[0, 12]$ at some $r_0$ which must be interior to the interval. Thus $V'(r_0) = 0$. But $V'(r) = 0$ for only one value of $r$ in $(0, 12)$, so that value must be $4\sqrt{6}$. Therefore the critical value $r_0 = 4\sqrt{6}$ must give the maximum. From the relation between $r$ and $h$ we find that $h = 8\sqrt{3}$.    □

**EXAMPLE 5**    A lighthouse is at point $A$, 4 km offshore from the nearest point $O$ of a straight beach; a store is at point $B$, 4 km down the beach from $O$. If the lighthouse keeper can row 4 km/hr and walk 5 km/hr, how should she proceed in order to get from the lighthouse to the store in the least possible time?

**Solution**    Clearly she will row to some point on the beach between $O$ (the nearest point) and the store at $B$ and then walk the rest of the way. We draw a figure such as Fig. 34 (Step 1), where $C$ is the spot on the beach where she lands and $x$ is the distance from $O$ to $C$. We denote by $T$ the time of the trip from $A$ to $B$ (Step 2); $T$ is the quantity we wish to make a minimum. To get a formula for $T$, we use the fact that *rate times time equals distance*. We then have

$$T = \text{time of trip along } AC + \text{time of trip along } CB$$

$$= \frac{\text{distance } AC}{4} + \frac{\text{distance } CB}{5}.$$

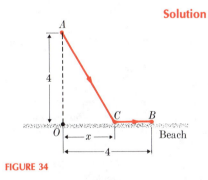

**FIGURE 34**

To express this in terms of one quantity is not hard, since

$$CB = 4 - x,$$

and, using right triangle $AOC$,

$$AC = \sqrt{16 + x^2}.$$

Therefore (Step 3),

$$T(x) = \tfrac{1}{4}\sqrt{16 + x^2} + \tfrac{1}{5}(4 - x).$$

The only part of Step 4 we need is the determination of the domain of the function $T$. Since $C$ is between $O$ and $B$ we have the restriction

$$0 \le x \le 4.$$

Taking the derivative (Step 5), we obtain

$$T'(x) = \tfrac{1}{8}(16 + x^2)^{-1/2}(2x) - \tfrac{1}{5}$$

and, setting this equal to zero, we find that

$$\frac{x}{4\sqrt{16+x^2}} = \frac{1}{5}.$$

We solve for $x$:

$$x = \pm \frac{16}{3}.$$

But these values of $x$ are *outside* the interval $[0, 4]$. The conclusion is that there are no relative maxima or minima in the interval $0 \le x \le 4$. Therefore the absolute minimum must occur at one of the endpoints. We have

$$T(0) = \tfrac{9}{5}$$

and

$$T(4) = \sqrt{2}.$$

Since $\sqrt{2} < \tfrac{9}{5}$, the fastest method for the lighthouse keeper is to row directly to the store $B$ and do no walking. ☐

## 6  PROBLEMS

**1** A rectangle has a perimeter of 120 meters. What length and width yield the maximum area? What is the result when the perimeter is $L$ units?

**2** Find the dimensions of the rectangle of maximum area that can be inscribed in a circle of radius 6. What is the result for a circle of radius $R$?

**3** Find the dimensions of the right circular cylinder of maximum volume which can be inscribed in a sphere of radius $R$.

**4** A horizontal gutter is to be made from a long piece of sheet iron 8 cm wide by turning up equal widths along the edges into a vertical position. How many centimeters should be turned up at each side to yield the maximum carrying capacity?

**5** The difference between two numbers is 20. Select the numbers so that the product is as small as possible.

**6** A box with a square base is to have an open top. The area of the material in the box is to be $100\,\text{cm}^2$. What should the dimensions be in order to make the volume as large as possible? What is the result for an area of $S$ square centimeters?

**7** A window is in the shape of a rectangle surmounted by a semicircle. Find the dimensions when the perimeter is 12 meters and the area is as large as possible.

**8** Find the radius and central angle (in radians) of the circular sector of maximum area having a perimeter of 16 cm.

**9** The top and bottom margins of a page are each $1\frac{1}{2}$ cm and the side margins are each 1 cm. If the area of the printed material per page is fixed at $30\,\text{cm}^2$, what are the dimensions of the page of least area?

**10** A closed right circular cylinder (i.e., top and bottom included) has a surface area of $100\,\text{cm}^2$. What should the radius and altitude be in order to provide the largest possible volume? Find the result if the surface area is $S\,\text{cm}^2$.

**11** A right circular cone has a volume of $120\,\text{cm}^3$. What shape should it be in order to have the smallest lateral surface area? Find the result if the volume is $V\,\text{cm}^3$.

**12** At midnight, ship $B$ was 90 km due south of ship $A$. Ship $A$ sailed east at 15 km/hr and ship $B$ sailed north at 20 km/hr. At what time were they closest together?

**13** Suppose that in Example 5 on page 153, the lighthouse is 5 km from shore, the store is 6 km down the beach from $O$, the lighthouse keeper can row 2 km/hr, and she can walk 4 km/hr. Where should she land in order to get from the lighthouse to the store in the shortest time?

**14** Find the dimensions of the rectangle having the largest area which can be inscribed in the ellipse

$$\frac{x^2}{a^2} + \frac{y^2}{b^2} = 1.$$

**15** Find the coordinates of the point or points on the curve $y = 2x^2$ which are closest to the point $(9, 0)$.

**16** Find the coordinates of the point or points on the curve $x^2 - y^2 = 16$ which are nearest to the point $(0, 6)$.

**17** (a) A right triangle has hypotenuse of length 13 and one leg of length 5. Find the dimensions of the rectangle of largest area which has one side along the hypotenuse and the ends of the opposite side on the legs of this triangle. (b) What is the result for a hypotenuse of length $H$ with an altitude to it of length $h$?

**18** A trough is to be made from a long strip of sheet metal 12 cm wide by turning up strips 4 cm wide on each side so that they make the same angle with the bottom of the trough (trapezoidal cross section). Find the width across the top such that the trough will have maximum carrying capacity.

**19** The sum of three positive numbers is 30. The first plus twice the second plus three times the third add up to 60. Select the numbers so that the product of all three is as large as possible.

**20** The stiffness of a given length of beam is proportional to the product of the width and the cube of the depth. Find the shape of the stiffest beam which can be cut from a cylindrical log (of the given length) with cross-sectional diameter of 4 cm.

**21** (a) A manufacturer makes aluminium cups of a given volume ($16 \text{ cm}^3$) in the form of right circular cylinders open at the top. Find the dimensions which use the least material. (b) What is the result for a given volume $V$?

**22** In Problem 21, suppose that the material for the bottom is $1\frac{1}{2}$ times as expensive as the material for the sides. Find the dimensions which give the lowest cost.

**23** Find the shortest segment with ends on the positive $x$ and $y$ axes which passes through the point $(1, 8)$.

**24** The product of two positive numbers is 16. Determine them so that the square of one plus the cube of the other is as small as possible.

**25** Find the dimensions of the cylinder of greatest lateral area which can be inscribed in a sphere of given radius $R$.

**26** A piece of wire of length $L$ is cut into two parts, one of which is bent into the shape of a square and the other into the shape of a circle. (a) How should the wire be cut so that the sum of the enclosed areas is a minimum? (b) How should it be cut to get the maximum enclosed areas?

**27** Solve Problem 26 if one part is bent into the shape of an equilateral triangle and the other into a circle.

**28** Solve Problem 26 if one part is bent into the shape of an equilateral triangle and the other into a square.

**29** Find the dimensions of the right circular cone of maximum volume which can be inscribed in a sphere of given radius $R$.

**30** Find the dimensions of the right circular cone of minimum volume which can be circumscribed about a sphere of radius $R$.

**31** A right circular cone is to have a volume $V$. Find the dimensions so that the lateral surface area is as small as possible.

**32** A silo is to be built in the form of a right circular cylinder surmounted by a hemisphere. If the cost of the material per square meter is the same for floor, walls, and top, find the most economical proportions for a given capacity $V$.

**33** Work Problem 32, given that the floor costs twice as much per square meter as the sides and the hemispherical top costs three times as much per square meter as the sides.

**34** A tank is to have a given volume $V$ and is to be made in the form of a right circular cylinder with hemispheres attached at each end. The material for the ends costs twice as much per square meter as that for the sides. Find the most economical proportions.

**35** Find the length of the longest rod which can be carried horizontally around a corner from a corridor 8 m wide into one 4 m wide. (*Hint:* Observe that this length is the minimum value of certain lengths.)

**36** Suppose the velocity of light is $V_1$ in air and $V_2$ in water. A ray of light traveling from a point $P_1$ above the surface of the liquid to a point $P_2$ below the surface will travel by the path which requires the least time. Show that the ray will cross the surface at the point $Q$ in the vertical plane through $P_1$ and $P_2$ so placed that

$$\frac{\sin \theta_1}{V_1} = \frac{\sin \theta_2}{V_2},$$

where $\theta_1$ and $\theta_2$ are the angles shown in Fig. 35.

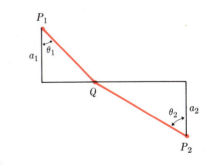

**FIGURE 35**

**37** A manufacturer makes widgets to sell so that an order of 1000 costs \$3 per hundred. For each additional 100 in the order, the price is reduced by 6¢ per hundred. What size order will yield the maximum dollar value?

_____ 7

## THE DIFFERENTIAL. APPROXIMATION

Until now we have been dealing with functions whose domain is a set of real numbers (i.e., a set in $R^1$) and whose range is also a set of real numbers (a set in $R^1$). Symbolically, if $D$ is the set of numbers comprising the domain of a function $f$ and $S$ is the set of numbers comprising the range, we may write $f: D \to S$. In the case of functions given by formulas, we write, for example, $f(x) = x^2 - 2x + 7$ to indicate the particular function.

We now enlarge the notion of function by considering domains which are ordered pairs of numbers instead of numbers in $R^1$. That is, we consider the case in which an element in the domain is a point in $R^2$. The range is the same as before: a collection of real numbers (a set in $R^1$). These functions are defined as **functions on $R^2$**, and we give the following precise statement.

**DEFINITIONS**   *Let D be a set of ordered pairs of real numbers $(x, y)$, i.e., D is a set in $R^2$. Let E be a set of real numbers. A **function** f **from** D **to** E is a correspondence that assigns to each element $(x, y)$ of D a unique number z where z is a number in E. The set D is the **domain** of f and the set E is the **range** of f.*

We recognize the above definition as a straightforward extension of the definition of function given in Chapter 1, Section 5, page 20. We also have the following equivalent definition, which extends the one given on page 20.

**DEFINITIONS**   *Consider a collection of ordered pairs $(A, w)$, in which the elements A are themselves ordered pairs of real numbers and the elements w are real numbers. If no two members of the collection have the same item A as a first element—i.e., if it can never happen that there are two members $(A_1, w_1)$ and $(A_1, w_2)$ with $w_1 \neq w_2$—then we call this collection a **function on $R^2$**. The totality of possible ordered pairs A is called the **domain** of the function. The totality of possible values for w is called the **range** of the function.*

We now revise our terminology and call a function as defined on page 20 a **function on $R^1$**. We use the same letters $f$, $F$, $\phi$, $g$, $H$, etc. to represent functions on $R^2$ that we used for functions on $R^1$. However, the formulas used for describing *particular* functions on $R^2$ appear quite different from those we have used until now. If the pair $(x, y)$ is used to represent a typical element in the domain of a function $F$ on $R^2$, a particular function may be written

$$F(x, y) = x^2 - 2y^2 + 3x + 4y.$$

If the letter $z$ is used as a typical element in the range of a function $F$, we write $z = F(x, y)$.

Employing classical terminology, we call a function on $R^2$ a **function of two variables**, while a function on $R^1$ is called a **function of one variable**. These terms, although not quite precise, have considerable intuitive value, and we shall use them from time to time.*

_____

*It is not difficult to extend the definition of function to the case in which the domain consists of ordered triples, ordered quadruples, etc., of numbers and in which the range also consists of ordered pairs, triples, etc., of numbers.

The letters used for elements of the domain in $R^2$ are not restricted to $x$ and $y$. Simple examples of functions of two variables are

$$z = u^2 + 2v^2 + 3v^4, \qquad w = \frac{x^2 + y^2 - 2}{1 + x^2}, \qquad t = \frac{x^2 + u^2 + 1}{4 - 2u^2 - x^2}.$$

The domain in each of these formulas is not described explicitly. In each such case we shall always assume that *the domain consists of all possible values which may be substituted in the right side of the formula.*

We are now ready to define a function which we call *the differential.* Suppose that $f$ is a function of one variable.

**DEFINITION**     *We designate by df **the differential of** f which is a function of two variables given by the formula*

$$df(x, h) = f'(x) \cdot h.$$

In this definition, the symbol $df$ is considered to be a single entity and does *not* mean $d$ times $f$. Since $df$ is the symbol for a function on $R^2$, we recognize $df(x, h)$ as the symbol for the value of the function when $x$ and $h$ are substituted into the right side of the above formula.

**EXAMPLE 1**     Given that $f(x) = x^2(1 - x^2)^{1/2}$, find the formula for $df$. Also, find the value of $df(\frac{1}{2}, 3)$.

**Solution**     By differentiation, we have

$$f'(x) = 2x(1 - x^2)^{1/2} - x^3(1 - x^2)^{-1/2}.$$

Therefore

$$df(x, h) = [2x(1 - x^2)^{1/2} - x^3(1 - x^2)^{-1/2}] \cdot h$$

$$= (1 - x^2)^{-1/2}[2x(1 - x^2) - x^3]h$$

$$= \frac{x(2 - 3x^2)}{\sqrt{1 - x^2}} h.$$

Substituting $x = \frac{1}{2}$ and $h = 3$ in this expression, we get

$$df(\tfrac{1}{2}, 3) = \tfrac{5}{4}\sqrt{3}. \qquad \square$$

Earlier (in Chapter 2, Section 2) we introduced the symbol $\Delta f$ by defining

$$\Delta f = f(x + h) - f(x).$$

We did not say so at the time, but the reader can now see that $\Delta f$ is a function on $R^2$. That is, we may write $\Delta f = \Delta f(x, h)$.

It will help our understanding to compare the functions $df$ and $\Delta f$. We can do this geometrically, as shown in Fig. 36, where $P$ represents a point on the graph of $y = f(x)$. If $h$ has any value,* then the point $(x + h, f(x + h))$ is

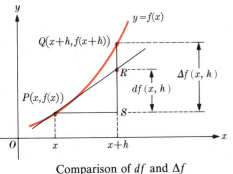

Comparison of $df$ and $\Delta f$

**FIGURE 36**

_____

*Figure 36 is drawn with $h$ positive. If $h$ is negative, the point $x + h$ lies to the left of $x$. The discussion is unchanged.

on the graph and is denoted by $Q$. We see that

$$\Delta f(x, h) = \text{height } \overline{SQ}.$$

The tangent at $P$ intersects the line through $Q$ and $S$ at $R$, and we know that $\overline{SR}/\overline{PS}$ is the slope of this line, or simply $f'(x)$. Since $\overline{PS} = h$, we have

$$df(x, h) = \text{height } \overline{SR}.$$

This tells us that the height $\overline{RQ}$ is just the difference between $\Delta f$ and $df$. That is,

$$\Delta f - df = \text{height } \overline{RQ}.$$

Also we see from the graph that as $h$ tends to zero (thus implying that the point $Q$ slides along the curve to $P$), the difference between $\Delta f$ and $df$ tends to zero. In fact, we have

$$\lim_{h \to 0} \frac{\Delta f - df}{h} = 0.$$

The proof of this statement is easy. We know from the definition of derivative that

$$\frac{\Delta f}{h} \to f'(x) \quad \text{as } h \to 0.$$

On the other hand, $df/h$ **is** the derivative $f'(x)$. Therefore

$$\lim_{h \to 0} \frac{\Delta f - df}{h} = \lim_{h \to 0} \frac{\Delta f}{h} - \lim_{h \to 0} \frac{df}{h} = f'(x) - f'(x) = 0.$$

If $h$ is small, $\Delta f$ and $df$ must be close together. The function $\Delta f$ measures the *change* in $f$ as we go from one point to another. The differential $df$ can be used as a good approximation for $\Delta f$ if the points are near each other. We write

$$df \approx \Delta f$$

for this approximation.

EXAMPLE 2   Given that $f(x) = 1/x$, compute $df$ and $\Delta f$ when $x = 1$ and $h$ is 0.1, 0.01, 0.001. Do this to four significant figures.

Solution   $f(1) = 1$; $f'(x) = -1/x^2$ and $f'(1) = -1$. For $h = 0.1$,

$$f(1.1) = 0.909091 \qquad \text{and} \qquad \Delta f = 0.909091 - 1 = -0.09091.$$

On the other hand, $df = f'(1) \cdot h = -1 \cdot (0.1) = -0.10000$. In this way we obtain the table:

| $h$ | $df$ | $\Delta f$ |
|-----|------|------------|
| 0.1 | $-0.10000$ | $-0.09091$ |
| 0.01 | $-0.010000$ | $-0.009901$ |
| 0.001 | $-0.0010000$ | $-0.0009990$ |

The point of this example—as the reader can readily see if she has worked along with it—is that the quantities $df$ require practically no computing, while finding $\Delta f$ is quite a bit of work. Since the error made in using $df$ instead of $\Delta f$ is small, the saving in computation time may make it worthwhile, especially in very complicated problems.

**EXAMPLE 3**  Use the differential to compute $\sqrt[5]{33}$ approximately.

**Solution**  To work this problem we need an idea; it is simply that we know $\sqrt[5]{32}$ *exactly* and 33 is "close to" 32. The function we are talking about is

$$f(x) = \sqrt[5]{x} = x^{1/5}.$$

If we think of $x = 32$ and $h = 1$, then $f(33)$ is what we want to find. We know that

$$\Delta f = f(33) - f(32),$$

while

$$df = f'(32) \cdot (1).$$

The derivative of $f$ is

$$f'(x) = \tfrac{1}{5} x^{-4/5},$$

and

$$f'(32) = \tfrac{1}{5} \cdot \tfrac{1}{16} = \tfrac{1}{80}, \qquad df = \tfrac{1}{80}.$$

Using the fact that $df$ is almost the same as $\Delta f$, we write

$$f(33) = f(32) + \Delta f \approx f(32) + df.$$

(The symbol $\approx$ means "nearly equal.") Therefore

$$\sqrt[5]{33} = f(33) \approx 2 + \tfrac{1}{80} = 2.0125,$$

approximately. The actual error in this computation is less than $\tfrac{1}{6400}$.  $\square$

Of course the fifth root of 33 can be found easily with a hand calculator. The point of Example 3 and other problems of this type is to illustrate the power of the differential in making approximations without intricate computations. There are many problems, especially where symbols rather than numbers are manipulated, in which differentials are crucial in simplifying approximate formulas.

A measurement of a length of pipe is made, and it is found that there is an error of $\tfrac{1}{2}$ cm. Is this error large or small? If the length of the pipe is about 8 cm we would say that the error is pretty large. If the total length measured were $\tfrac{1}{2}$ km, we would consider the error extremely small. The error is the difference between the planned length and the actual length.

If a quantity is being measured and the true value is $a$ but there is an error of an amount $h$, we define the **proportional error** as

$$\frac{h}{a}$$

and the **percentage error** as

$$\frac{h}{a} \cdot 100\%.$$

In the above example of the $\frac{1}{2}$-cm error, when $a = 8$ cm and $h = \frac{1}{2}$ cm, the proportional error is $\frac{1}{16}$, and the percentage error is $\frac{1}{16} \cdot 100 = 6\frac{1}{4}\%$. If $a = \frac{1}{2}$ km and $h = \frac{1}{2}$ cm, then the proportional error is $1/100{,}000$ and the percentage error is $0.0001\%$—one ten-thousandth of one percent.

Suppose that $f$ is a function and we wish to find $f(a)$, the value $a$ to be determined by measurement. This measurement is not precise (measurements never are) and there is an error $h$ in measuring the value $a$. The error in the function $f$ is $\Delta f = f(a + h) - f(a)$ and the *proportional error* in the function is simply

$$\frac{\Delta f}{f} = \frac{f(a + h) - f(a)}{f(a)}.$$

However, in such computations, we conveniently use the **approximation to the proportional error:**

$$\frac{df}{f} = \frac{f'(a) \cdot h}{f(a)}.$$

The **approximate percentage error** is then found by multiplying by 100.

EXAMPLE 4    The radius of a sphere is found by measurement to be 3 cm, but there is a possible error of $\pm 0.03$ cm in the measurement. Find, approximately, the error and percentage error in the value of the surface area of the sphere that might occur because of the error in the radius.

Solution    The surface area $S$ is given by the formula $S = 4\pi r^2$, and for $r = 3$ the area is $36\pi$ cm$^2$. The error is approximated by $dS = S'(r)h = 8\pi rh$. The quantity $h = \pm 0.03$, and so the approximate error is

$$dS = 8\pi(3)(\pm 0.03) = \pm 0.72\pi \text{ cm}^2.$$

The approximate proportional error is

$$\frac{dS}{S} = \frac{\pm 0.72\pi}{36\pi} = \pm 0.02,$$

and the approximate percentage error is $\pm 2\%$. $\qquad\square$

## 7  PROBLEMS

In Problems 1 through 10, find $df(x, h)$.

**1** $f(x) = x^4 + 3x^2 - 2x$     **2** $f(x) = \dfrac{x^2 + 2}{x^2 + 4}$

**3** $f(x) = x^2\sqrt{x^4 + 1}$     **4** $f(x) = \dfrac{(x + 1)^2}{x^2 + 4x + 1}$

**5** $f(x) = (x^2 + 2)^{1/2}(2x + 1)^{1/3}$

**6** $f(x) = (x^2 + 1)^{1/4}(x + 2)^{1/5}$

**7** $f(x) = \sin 3x$     **8** $f(x) = x \cos^3 2x$

**9** $f(x) = \tan \frac{1}{2}x + \sec 2x$     **10** $f(x) = \sec^3 4x + \csc x$

In Problems 11 through 20, find $df$ and $\Delta f$ and evaluate them for the quantities given.

**11** $f(x) = x^2 + x - 1$,   $x = 1$, $h = 0.01$

**12** $f(x) = x^2 - 2x - 3$,   $x = -1$, $h = -0.02$

**13** $f(x) = x^3 + 3x^2 - 6x - 3$,   $x = 2$, $h = 0.01$

**14** $f(x) = x^3 - 2x^2 + 3x + 4$,   $x = -1$, $h = 0.02$

**15** $f(x) = 1/x$,   $x = 2$, $h = 0.05$

**16** $f(x) = x^{1/2}$,   $x = 1$, $h = -0.1$

**17** $f(x) = x^{-1/2}$,   $x = 1$, $h = 0.1$

**18** $f(x) = \dfrac{x}{1+x}, \quad x = 0, h = 0.1$

**19** $f(x) = x \sin 2x, \quad x = \dfrac{\pi}{6}, h = 0.5$

**20** $f(x) = \cos^2 x + 2 \sin 3x, \quad x = \dfrac{\pi}{3}, h = 0.2$

In Problems 21 through 28, calculate (approximately) the given quantity by means of the differential. Then use a hand calculator to compute the quantity to 2 decimal places and compare the results.

**21** $\sqrt{65}$          **22** $\sqrt[3]{124}$

**23** $(0.98)^{-1}$      **24** $\sqrt[4]{80}$

**25** $(31)^{1/5}$       **26** $(17)^{-1/4}$

**27** $\sqrt{0.0024}$     **28** $\sqrt{82} + \sqrt[4]{82}$

**29** The diameter of a sphere is to be measured and its volume

computed. The diameter is 9 cm, with a possible error of $\pm 0.05$ cm. Find (approximately) the maximum possible percentage error in the volume.

**30** The radius of a sphere is to be measured and its volume computed. If the diameter can be accurately measured to within 0.1%, find (approximately) the maximum percentage error in the determination of the volume.

**31** A coat of paint of thickness $t$ cm is applied evenly to the faces of a cube of edge $a$ cm. Use differentials to find approximately the number of cubic centimeters of paint used. Compare this with the exact amount used by computing volumes before and after painting.

**32** Work Problem 31 with the cube replaced by a sphere of radius $R$.

**33** State precisely the definition of a function when the domain consists of elements of $R^3$ and the range is a set in $R^1$. Do the same when the domain is a set in $R^3$ and the range is a set in $R^2$.

---

## 8

## DIFFERENTIAL NOTATION

Let $f$ be a function of one variable. Writing $y = f(x)$, we recall that in Chapter 2, Section 4 the symbol $\dfrac{dy}{dx}$ was introduced as an alternate notation for the derivative $f'(x)$. At this point we identify $dx$ with the number $h$ used in the definition of the differential, and we identify $dy$ with the differential $df$. That is, setting

$$dx = h \quad \text{and} \quad dy = df,$$

we obtain the equivalent formula for the differential

$$dy = f'(x) \, dx.$$

We call $dx$ the **differential of** $x$ and $dy$ **the differential of** $y$. It is important to observe that $dx$ is an *independent variable* and $dy$, which is defined by the above formula, is not. As expected, the ratio $dy/dx$ is the derivative whenever $dx \neq 0$, and the use of this notation for the derivative is justified in terms of differentials.

Suppose that $y = f(x)$ is a function and that $x = g(t)$ is a second function with the range of $g$ in the domain of $f$. Then we may consider $y$ as a function of $t$ by writing

$$y = f[g(t)].$$

To get the derivative of $f[g(t)]$ we apply the Chain Rule to obtain

$$f'[g(t)]g'(t).$$

We also have the following formulas for differentials:

$$dy = f'(x)\, dx \qquad \text{and} \qquad dx = g'(t)\, dt.$$

By direct substitution, we find

$$dy = f'(x)g'(t)\, dt.$$

The Chain Rule may now be expressed in terms of differentials. If $dx \neq 0$ and $dt \neq 0$, then

$$\frac{dy}{dt} = \frac{dy}{dx} \cdot \frac{dx}{dt}.$$

Differentials may be multiplied, divided (whenever different from zero), added, and subtracted. One differential divided by another may be thought of as a derivative. We can now write all the elementary rules for derivatives as differentials. Since they become derivative formulas merely by division by $dx$, there is really not much new in them.

If $c$ is a constant, $dc = 0$ and $d(cu) = c\, du$. Also,

$$d(u + v) = du + dv,$$

$$d(u \cdot v) = u\, dv + v\, du,$$

$$d\!\left(\frac{u}{v}\right) = \frac{v\, du - u\, dv}{v^2},$$

$$d(u^n) = nu^{n-1}\, du.$$

**EXAMPLE 1**   Given that $y = (x^2 + 2x + 1)^3$ and $x = 3t^2 + 2t - 1$, find $dy/dt$.

**Solution**   Here we have

$$\frac{dy}{dt} = \frac{dy}{dx} \cdot \frac{dx}{dt}$$

and

$$\frac{dy}{dx} = 3(x^2 + 2x + 1)^2(2x + 2), \qquad \frac{dx}{dt} = 6t + 2.$$

Therefore

$$\frac{dy}{dt} = 6(x + 1)(x^2 + 2x + 1)^2 \cdot 2(3t + 1).$$

Since $y$ is a function of $x$ and $x$ is a function of $t$, we see that $y$ is actually a function of $t$. By replacing $x$ with $3t^2 + 2t - 1$ in the expression for $y$, we could simply differentiate $y$ with respect to $t$. If we substitute for $x$ in the above expression for $\dfrac{dy}{dt}$, we get

$$\frac{dy}{dt} = 6(3t^2 + 2t)[(3t^2 + 2t - 1)^2 + 2(3t^2 + 2t - 1) + 1]^2 \cdot 2(3t + 1)$$

and $x$ is eliminated from the final result.                                  □

**EXAMPLE 2**  Given that $y = \sqrt{x^2 + 1}/(2x - 3)$, find $dy$.

**Solution**  One way would be to find the derivative and multiply by $dx$. But we may also use the formula for the differential of a quotient, and we do it here for practice:

$$dy = \frac{(2x - 3)d(\sqrt{x^2 + 1}) - \sqrt{x^2 + 1} \cdot d(2x - 3)}{(2x - 3)^2}.$$

We see now that $d(\sqrt{x^2 + 1}) = \frac{1}{2}(x^2 + 1)^{-1/2}2x \, dx$ and $d(2x - 3) = 2 \, dx$. We obtain

$$dy = \frac{x(2x - 3)(x^2 + 1)^{-1/2} \, dx - 2\sqrt{x^2 + 1} \, dx}{(2x - 3)^2}.$$

We multiply both the numerator and the denominator by $\sqrt{x^2 + 1}$ to get

$$dy = \frac{-3x - 2}{\sqrt{x^2 + 1}(2x - 3)^2} \, dx. \qquad \square$$

The method of differentials is particularly helpful in *implicit differentiation*. Suppose two functions, each defined on $R^1$, satisfy a relation. We may take differentials if one of the functions is expressible in terms of the other. However, the actual process of solving for one quantity in terms of the other need not be carried out. An example will illustrate the idea.

**EXAMPLE 3**  Suppose that $u$ and $v$ are functions defined on $R^1$ and that they satisfy the relation

$$u^2 + 2uv^2 + v^3 - 6 = 0.$$

Find $dv/du$.

**Solution**  Take differentials:

$$2u \, du + 2u \cdot 2v \, dv + 2v^2 \, du + 3v^2 \, dv = 0.$$

Divide through by $du$ and solve:

$$\frac{dv}{du} = -\frac{2(u + v^2)}{4uv + 3v^2},$$

if $du \neq 0$ and if $4uv + 3v^2 \neq 0$. $\qquad \square$

If $y = f(x)$, there are now two symbols for the derivative: $f'(x)$ and $dy/dx$. These are the most prevalent symbols, commonly used in texts and papers on various related subjects. Another symbol, not quite so common but nevertheless used often, is $D_x f$. The only notation we learned so far for the **second derivative** is $f''(x)$. The expression

$$\frac{d^2 y}{dx^2},$$

which is read: $d$ second $y$ by $dx$ second, is a classical one for the second derivative. The numerator, $d^2 y$, and the denominator, $dx^2$, *have absolutely no meaning by themselves*. (In elementary calculus there is no such thing as the differential of a differential.) We just use $d^2y/dx^2$ as an equivalent for $f''(x)$. Similarly, third, fourth, and fifth derivatives are written

$$\frac{d^3y}{dx^3}, \quad \frac{d^4y}{dx^4}, \quad \frac{d^5y}{dx^5},$$

and so on. In each case, the expression is to be thought of not as a fraction dividing two quantities but as an inseparable symbol representing the appropriate derivative.

## 8    PROBLEMS

In Problems 1 through 18, find the differential $dy$.

**1** $y = (x^3 - 3x^2 + 2)^5$     **2** $y = (x^2 + 1)^{1/3}$

**3** $y = (x^3 + 4)^{-5}$     **4** $y = \dfrac{x+1}{x^2+1}$

**5** $y = x^2\sqrt{2x+3}$     **6** $y = (x-1)\sqrt{x+1}$

**7** $y = (x+1)^2(2x-1)^3$     **8** $y = (x+2)^{2/3}(x-1)^{1/3}$

**9** $y = \dfrac{2x}{x^2+1}$     **10** $y = \dfrac{x}{\sqrt{x^2+1}}$

**11** $y = \dfrac{x^{2/3}}{(x+1)^{2/3}}$     **12** $y = \dfrac{\sqrt{x^2+1}}{x}$

**13** $y = \sqrt{\dfrac{x+1}{x-1}}$     **14** $y = \dfrac{x}{\sqrt{x^2+x+1}}$

**15** $y = x^2 \sin 2x$

**16** $y = \cos 3x + \tan 2x$

**17** $y = x \tan^2 \frac{1}{2}x$

**18** $y = \sec(x+4) + \cot^2 2x$

In Problems 19 through 26, find the derivative $dy/dx$ by the method of differentials.

**19** $2x^2 + xy - y^2 + 2x - 3y + 5 = 0$

**20** $x^3 + x^2y - 2y^3 = 0$     **21** $\sqrt{x} + \sqrt{y} = 2$

**22** $x^{2/3} + y^{2/3} = a^{2/3}$, $a$ const.

**23** $2x^3 - xy^2 - y^3 + 2x - y = 0$

**24** $x^2 \cos^2 y + x^2 y^3 - 1 = 0$

**25** $xy \cos^2 x + 2x^2 \cos y - 3 = 0$

**26** $x^3 + 2x^2y - 3x \cos x \sin y - 4 = 0$

**27** Find $dy/dt$, given that $y = (x^2 + 2x + 5)^{3/4}$ and $x = \sqrt{t^2 - 2t + 1}$.

**28** Find $dz/dr$, given that $z = (2u+1)/(u^2+u)$ and $u = (r^2 + 5)^4$.

**29** Find $ds/dt$, given that $s = x^2 + 3x - 6$, $x = r^3 - 8r + 5$, $r = \sqrt{t^3 + 5}$.

**30** Find $dz/ds$, given that

$$z = \frac{y+1}{y-1}, \qquad y = \frac{x^3 - 8x + 1}{\sqrt{x+1}}, \qquad x = s^3 - 8s + 5.$$

**31** Find $dy/dt$, given that $x^3 + 2xy - y^3 + 8 = 0$ and $x = t^3 - 2t + 1$.

**32** Find $dy/dt$, given that

$$x^4 + 2x^2y - y^4 - 3y = 0, \qquad x = \sqrt{2t+3}.$$

**33** Find $dz/dt$, given that

$$z^3 + 2y^3 - 3z^2 = 1, \qquad y = x^2 + 2x - 6, \qquad x = (t+1)^4.$$

**34** Find $dy/dx$, given that

$$x^3 + t^2 - 2\sqrt{t} - 4 = 0, \qquad y^4 - t^3 + 2y - 7 = 0.$$

**\*35** Find $d^2y/dx^2$, given that

$$x^3 - 4y^4 + 7x - 6y - 3 = 0.$$

**36** a) Given $y = f(x)$, $x = g(t)$, $t = h(s)$. State appropriate conditions which make the formula

$$\frac{dy}{ds} = \frac{dy}{dx} \cdot \frac{dx}{dt} \cdot \frac{dt}{ds}$$

valid.

b) Extend the above formula to the case of $n$ functions $f_1, f_2, \ldots, f_n$.

**\*37** Given the relations

$$x^2 + 2xt + y^3 - t^2 + 3xy - 2t^3 = 1,$$

$$2x^3 - y^3 + 2xy^2 + 3yt^3 + x = 3.$$

Use differentials to find a formula (in terms of $x$, $y$, $t$) for the derivative $dx/dt$. (Hint: Observe that theoretically $y$ may be eliminated between the two equations and $x$ expressed in terms of $t$.)

9

## NEWTON'S METHOD (OPTIONAL)

In many calculus problems as well as in problems throughout mathematics we often seek solutions of an equation $f(x) = 0$. The values of $x$ which satisfy such an equation are called its **roots**. These numbers are also designated the **zeros** *of the function* $f$. We already had practice in finding roots of polynomial equations and equations involving trigonometric functions when we were finding relative maxima and minima and points of inflection. In most problems the roots of an equation cannot be found exactly. This is particularly evident if the equation contains a polynomial of high degree or if it involves combinations of trigonometric functions. Therefore it is important to have a method which approximates a root to any desired degree of accuracy. **Newton's Method** is a simple and powerful technique for accomplishing this task.

If we plot the graph of a function $y = f(x)$ which we suppose is smooth, then each place where the curve crosses the $x$ axis is a root of $f(x) = 0$, i.e., a zero of $f$. If for some value $a$ we have $f(a) < 0$ and if for some value $b$ we have $f(b) > 0$, then we know there is at least one root of $f(x) = 0$ between $a$ and $b$. Suppose we choose some value of $x$ between $a$ and $b$, designated $x_1$, as a first guess for the root $r$ of $f(x) = 0$, as shown in Fig. 37. To get a closer approximation to the root $r$ we draw the line $L$ tangent to the curve at the point $(x_1, f(x_1))$. This line intersects the $x$ axis at a point which we designate $x_2$. If the situation is actually as it appears in Fig. 37, then $x_2$ will be a better approximation to $r$ than $x_1$ is. In many circumstances this process is entirely correct. To determine $x_2$, the point where $L$ intersects the $x$ axis, we use the point-slope formula for the equation of $L$. We find

$$y - f(x_1) = f'(x_1)(x - x_1).$$

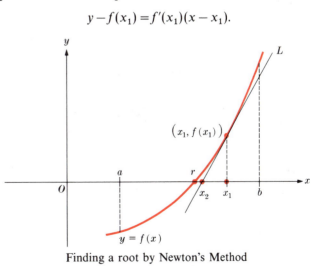

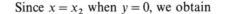

Finding a root by Newton's Method

FIGURE 37

Since $x = x_2$ when $y = 0$, we obtain

$$-f(x_1) = f'(x_1)(x_2 - x_1)$$

and, supposing that $f'(x_1) \neq 0$, we can solve for $x_2$ to get

$$x_2 = x_1 - \frac{f(x_1)}{f'(x_1)}.$$

In other words, given the function $f(x)$ and knowing a first approximation $x_1$ to the root $r$, the above formula tells us how to find $x_2$, a better approximation to $r$. Once we have $x_2$ we can repeat the process. Drawing the line $L_2$ tangent to the curve at $(x_2, f(x_2))$, we denote the point where $L_2$ intersects the $x$ axis by $x_3$. Then $x_3$ is a better approximation to $r$ than is $x_2$. Fig. 38 shows the process. Then if $f'(x_2) \neq 0$ the formula for finding $x_3$ is identical to that for $x_2$:

$$x_3 = x_2 - \frac{f(x_2)}{f'(x_2)}.$$

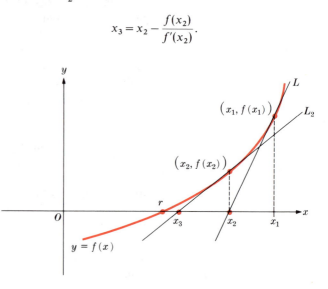

FIGURE 38

The continuation of the procedure is now clear and the general term, known as **Newton's Formula**,

$$x_{n+1} = x_n - \frac{f(x_n)}{f'(x_n)},$$

is valid so long as $f'(x)$ is never zero in the interval $a \leq x \leq b$. It can be shown that if the first guess $x_1$ is sufficiently close to $r$ then the above process yields a sequence $x_1, x_2, \ldots, x_n$, which tends to $r$ as $n \to \infty$. We work an example before stating a precise theorem.

**EXAMPLE 1**   Find the cube root of 3 with an accuracy of three decimal places.

**Solution**   The key to solving the problem is the observation that the cube root of 3 is the positive root of the equation $x^3 - 3 = 0$. We first draw the graph of $y = f(x) \equiv x^3 - 3$ by setting up a table of values:

| $x$ | $-2$ | $-1$ | $0$ | $1$ | $2$ | $3$ |
|---|---|---|---|---|---|---|
| $f(x)$ | $-11$ | $-4$ | $-3$ | $-2$ | $5$ | $24$ |

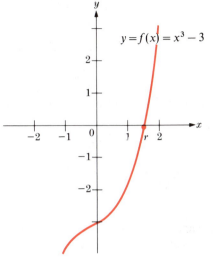

$y = f(x) = x^3 - 3$

**FIGURE 39**

The graph is shown in Fig. 39, and we note that there is a root between 1 and 2. We also observe that $f'(x) = 3x^2$ is never zero between 1 and 2. We choose any number in this interval, say 1.5 as a first guess for $r$. Then by Newton's Formula

$$x_2 = 1.5 - \frac{f(1.5)}{f'(1.5)}.$$

We find $f(1.5) = (1.5)^3 - 3 = 0.375$ and $f'(1.5) = 6.75$. Therefore

$$x_2 = 1.5 - 0.05556 = 1.44444.$$

Continuing the process, we obtain for $x_3$,

$$x_3 = 1.44444 - \frac{f(1.44444)}{f'(1.44444)} = 1.44444 - 0.00219$$

$$= 1.44225.$$

We carry out one more step:

$$x_4 = 1.44225 - \frac{f(1.44225)}{f'(1.44225)} = 1.44225 - 0.00000268$$

$$= 1.44224.$$

We see that the third step in Newton's Method leaves the first four decimal places unchanged and we conclude that 1.442 is the cube root of 3 with an accuracy of three decimal places. □

Generally, if we require an approximation to a fixed number of decimal places the computation should be made with *two* more places than is actually needed. With a hand calculator the additional work is usually not lengthy. Then when the iterations yield no change in any of the digits we drop the two extra digits to get the final result.

Newton's Method is quite fast. In most polynomial equations and combinations of polynomial and trigonometric equations each successive step *doubles* the number of digits of accuracy from the preceding step. Thus accuracy improves quickly with successive iterations.

If the first guess $x_1$ is poor then it may happen that Newton's Formula will give a value of $x_2$ *that is a worse approximation to $r$ than $x_1$ is*. Figure 40 shows how this can happen. If the slope $f'(x_1)$ is very small, then the tangent line $L$ may intersect the $x$ axis at a point which is farther away from $r$ than $x_1$ is. In such a case it usually happens that the successive values $x_1, x_2, x_3, \ldots$ become dispersed rather than clustering about a specific root $r$. To guarantee that the sequence $x_1, x_2, \ldots, x_n$ actually converges to a root, we require not only that $f'(x)$ does not vanish but also that $f''(x)$ is not too large in the interval $a \leq x \leq b$. A typical theorem on the convergence of Newton's Method, which we state without proof, is given in the following result.

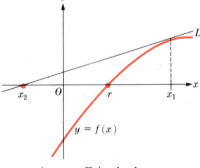

$y = f(x)$

$x_1$ is not sufficiently close to $r$

**FIGURE 40**

**THEOREM 10**   *Suppose there is a number $M > 1$ such that $|f'(x)| \geq M$ and $|f''(x)| \leq 2/M$ for $x$ on $I = \{x : a \leq x \leq b\}$. If $f(a)$ and $f(b)$ have opposite signs and $x_1$ is any point of $I$ with $|f(x_1)| \leq 1$, then the sequence $x_1, x_2, \ldots, x_n, \ldots$ determined by Newton's Formula converges to a number $r$ such that $f(r) = 0$. That is, $r$ is a root of the equation $f(x) = 0$.*

We illustrate Newton's Method with a second example.

**EXAMPLE 2**   Find the positive root of the equation $2x^4 + 2x^3 - 3x^2 - 5x - 5 = 0$ with an accuracy of two decimal places.

**Solution**   We first draw the graph of $y = f(x) \equiv 2x^4 + 2x^3 - 3x^2 - 5x - 5$ by constructing a table of values:

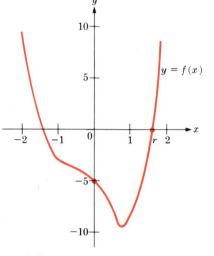

| $x$ | $-2$ | $-1$ | $0$ | $1$ | $2$ |
|---|---|---|---|---|---|
| $y = f(x)$ | $9$ | $-3$ | $-5$ | $-9$ | $21$ |

From the graph in Fig. 41 we observe that the positive root $r$ lies between 1 and 2, and that it seems to be somewhat closer to 2 than to 1. We choose $x_1 = 1.6$ as a first guess. With $f'(x) = 8x^3 + 6x^2 - 6x - 5$ we apply Newton's Formula and obtain

$$x_2 = 1.6 - \frac{f(1.6)}{f'(1.6)} = 1.6 - \frac{0.6192}{33.528} = 1.6 - 0.0185 = 1.5815.$$

We calculate the next approximation:

$$x_3 = 1.5815 - \frac{f(1.5815)}{f'(1.5815)}$$

and we find $f(1.5815) = 0.114$. We next calculate $f'(1.5815)$, getting 32.1623. Hence

$$x_3 = 1.5815 - \frac{0.114}{32.1623} = 1.5815 - 0.0035$$

$$= 1.5880,$$

**FIGURE 41**

and the root is $r = 1.59$ accurately to two decimal places. In this problem we can simplify some of the computation by using the value of $f'$ obtained in finding $x_2$ rather than by computing $f'(x_2)$ as we did above. If we use $f'(1.6)$ instead of $f'(1.5815)$ in finding $x_3$ we get

$$x_3 = 1.5815 - \frac{0.114}{33.528} = 1.5815 - 0.0034.$$

The answer accurate to two decimal places is the same. In general, if $f'$ is large then it won't change much through successive steps and it saves time to use the same value of $f'$ throughout. The resulting sequence will converge to the root only slightly more slowly than would be the case if the precise Newton Formula were used. Occasionally, the reduction in computation makes the simplification worthwhile.   □

## 9  PROBLEMS

In each of Problems 1 through 6, find the appropriate root with an accuracy of three decimal places. Use the methods of this section and then check your results with a calculator.

**1** $\sqrt{13}$        **2** $\sqrt[3]{25}$

**3** $\sqrt{67}$        **4** $\sqrt[4]{14}$

**5** $\sqrt[4]{23}$        **6** $\sqrt[5]{30}$

**7** Find the positive root of $2x^4 - 3x^2 - 5 = 0$ with an accuracy of three decimal places.

**8** Find the positive root of $3x^3 + 6x^2 - 7x - 14 = 0$ with an accuracy of three decimal places.

**9** Find the negative root of $x^4 - 2x^3 - 3x^2 - 2x - 4 = 0$ with an accuracy of two decimal places.

**10** Find the root of $\cos x - x = 0, 0 \le x \le \dfrac{\pi}{2}$, with an accuracy of two decimal places.

**11** Suppose we wish to find the positive square root of a number $c$. We take any reasonable guess and call it $d_1$. We divide $d_1$ into $c$, getting a quotient $q_1$. We average $d_1$ and $q_1$; that is, we form $d_2 = \frac{1}{2}(d_1 + q_1)$. We then divide $d_2$ into $c$, getting a quotient $q_2$. Repeat the process. Show that $d_1, d_2, \ldots, d_n, \ldots$ converges to $\sqrt{c}$ by showing that we have described Newton's Method for the equation $x^2 - c = 0$.

**12** Find all the roots of $2x^3 - x^2 - 5x + 3 = 0$ with an accuracy of two decimal places.

**13** Find all the roots of $3x^3 + x^2 - 11x + 6 = 0$ with an accuracy of two decimal places.

**14** a) Find the positive root of $3x^3 + 16x^2 - 8x - 16 = 0$ with an accuracy of four decimal places.
b) Find the same root as in (a) by the modified Newton's Method in which $f'(x_1)$ is used in all iterations, as described in Example 2. Compare the speed of convergence of (b) with that of (a).

**15** Find the positive roots of $12x^5 + 12x^4 - 23x^3 + 10x + 10 = 0$ with an accuracy of two decimal places.

**16** Find the positive roots of $10x^5 + 20x^4 - 33x^3 - 66x^2 + 27x + 54 = 0$ with an accuracy of two decimal places.

---

## 10

## RELATED RATES

In Chapter 2, Section 5, we discussed motion along a straight line and developed the ideas of velocity and speed. We learned that if a particle moves in a straight line so that the distance traveled, $s$, depends on the time $t$, according to some law $s = f(t)$, the velocity is obtained by finding the derivative $f'(t)$. We can also write

$$f'(t) = \frac{ds}{dt}.$$

The velocity may be thought of as the *rate of change* of distance with respect to time. If there are several particles, each moving in a straight line according to some law, then we can talk about the rate of change of each of the particles. Suppose the motion of these particles is related in some way. (One may go up as the other goes down, as in a lever, for example.) Then we say that we have a problem in **related rates**.

We are not limited to particles moving in a straight line. If a tank is being filled with water, the level of the surface is rising with time. We talk about the rate of change of the depth of the water. If the depth is denoted by $h$, then $dh/dt$ is the rate of change of the depth. Similarly, the volume $V$ is increasing; $dV/dt$ measures the rate of this increase. Any quantity which grows or diminishes with time is a candidate for a problem in related rates.

In the problems we shall consider, it is important to remember that **every**

**quantity is a function of time**. Therefore we can take derivatives of each quantity with respect to $t$, the time. This derivative is called the **rate of change**. We first give two examples and then map out general rules of procedure.

**EXAMPLE 1**  One airplane flew over an airport at the rate of 300 km/hr. Ten minutes later another airplane flew over the airport at 240 km/hr. If the first airplane was flying west and the second flying south (both at the same altitude), determine the rate at which they were separating 20 minutes after the second plane flew over the airport. (We assume the airplanes are traveling at constant speed.)

**Solution**  Draw a diagram, as shown in Fig. 42. Let $s$ be the distance between the airplanes at time $t$, let $x$ be the distance the westbound airplane has traveled at time $t$, and let $y$ be the distance the southbound airplane has traveled at time $t$. We know that $dx/dt = 5$ km/min, $dy/dt = 4$ km/min. We wish to find $ds/dt$ when $t = 20$ min, measured from the time the second airplane passes over $O$. We have the relation

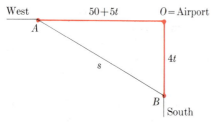

West          $50+5t$          $O$ = Airport

$A$

$s$                    $4t$

$B$

South

One airplane flying west and the other flying south

**FIGURE 42**

$$s^2 = x^2 + y^2.$$

Take differentials:

$$2s\,ds = 2x\,dx + 2y\,dy.$$

Divide by $dt$:

$$s\frac{ds}{dt} = x\frac{dx}{dt} + y\frac{dy}{dt}.$$

When $t = 20$, the second airplane has traveled $(20)(4)$ km and $y = 80$. The first airplane, which started 10 minutes earlier, has traveled $(30)(5)$ km and $x = 150$. Therefore $s = 170$, and substituting all these values we get

$$170 \cdot \frac{ds}{dt} = 150 \cdot 5 + 80 \cdot 4 \qquad \Leftrightarrow \qquad \frac{ds}{dt} = \frac{107}{17} \text{ km/min.} \qquad \square$$

We begin the solution of such a problem by using a letter to denote each of the quantities which change. In the above example we use $x$, $y$, and $s$; each of these is a function of $t$. Some relation is found among the letters ($s^2 = x^2 + y^2$, as we saw), and differentials by implicit methods are used to get derivatives with respect to $t$. Then the numerical value for each quantity is substituted to obtain the answer. Let us work another example.

**EXAMPLE 2**  Water is flowing at the rate of 5 cubic meters/min into a tank (Fig. 43) in the form of a cone of altitude 20 meters and base radius 10 meters and with its

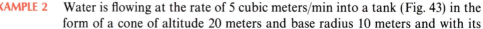

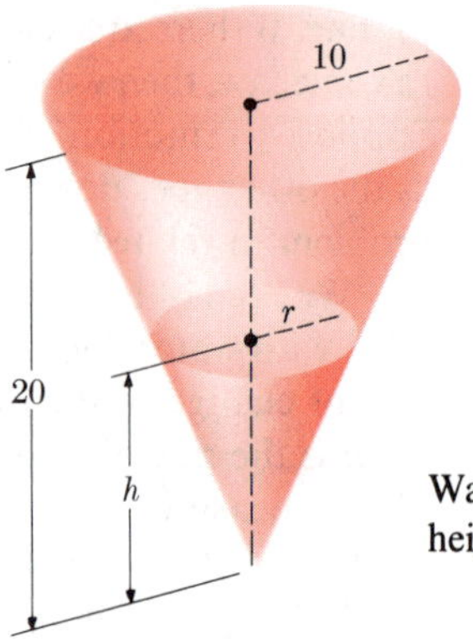

10

$r$

20

$h$

Water flowing into tank is at height $h$

**FIGURE 43**

vertex in the downward direction. How fast is the water level rising when the water is 8 meters deep?

**Solution**  Let $h$ be the depth, $r$ the radius of the surface, and $V$ the volume of the water at an arbitrary time $t$. We wish to find $dh/dt$. We know $dV/dt = 5$. The volume of water is given by

$$V = \tfrac{1}{3}\pi r^2 h,$$

where all quantities depend on $t$. By similar triangles,

$$\frac{r}{h} = \frac{10}{20} \qquad \text{or} \qquad r = \tfrac{1}{2}h,$$

and so

$$V = \frac{1}{3}\pi\frac{h^3}{4} = \frac{1}{12}\pi h^3.$$

We take differentials:

$$dV = \tfrac{1}{4}\pi h^2\, dh,$$

and, dividing by $dt$:

$$\frac{dV}{dt} = \frac{1}{4}\pi h^2 \frac{dh}{dt}.$$

We want to find $dh/dt$ when $h = 8$. and since $dV/dt = 5$, we obtain

$$\frac{dh}{dt} = \frac{5}{16\pi} \text{ m/min.} \qquad \square$$

The major trap to avoid in this problem is the premature use of the fact that $h = 8$ at the instant we want to find $dh/dt$. The height $h$ *changes* with time and must be denoted by a letter. If $h = 8$ is put in the diagram, it usually leads to disaster. *All quantities which change with time must be denoted by letters*.

The rules of procedure illustrated in Examples 1 and 2 are now outlined in the form of four steps.

> **Step 1:**  Draw a diagram. Label any numerical quantities which remain fixed throughout the problem (such as the dimensions of the cone in Example 2).
>
> **Step 2:**  Denote all quantities which change with time by letters. A relation (or relations) is found among the quantities which vary; these relations must hold for all time.
>
> **Step 3:**  Take differentials of the relation (or relations) found in Step 2. Divide by $dt$ to obtain a relation among the derivatives.
>
> **Step 4:**  Insert the special numerical values of all quantities to get the desired result.

We apply these rules in the following example.

**EXAMPLE 3**  An airplane at an altitude of 3000 meters, flying horizontally at 300 km/hr, passes directly over an observer. Find the rate at which it is approaching the observer when it is 5000 meters away.

**Solution** We draw Fig. 44. The airplane is at point $A$ flying toward point $B$. The distance $OB$ is labeled as 3000 (Step 1). Note that this distance does not change in the course of the problem. Let $x$ be the distance of the airplane $A$ from the point $B$ directly over the observer, $O$, and $s$ be the distance from the airplane to the observer. We want to find $ds/dt$ when $s = 5000$. We are given $dx/dt = -300$ km/hr. The negative sign is used because $x$ is decreasing. A change of units gives us $dx/dt = -\dfrac{(300)(1000)}{3600}$ m/sec $= -83\frac{1}{3}$ m/sec.

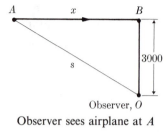

Observer sees airplane at $A$

**FIGURE 44**

We have (Step 2)

$$s^2 = x^2 + (3000)^2,$$

and, applying differentials (Step 3),

$$2s\, ds = 2x\, dx.$$

Dividing by $dt$, we get

$$s\frac{ds}{dt} = x\frac{dx}{dt}.$$

We now insert the values (Step 4), noting that $x = 4000$ when $s = 5000$:

$$5000\frac{ds}{dt} = 4000(-83\tfrac{1}{3}) \quad \Leftrightarrow \quad \frac{ds}{dt} = -66\tfrac{2}{3}\text{ m/sec} = -240\text{ km/hr.} \quad \square$$

## 10  PROBLEMS

**1** Water is flowing into a vertical cylindrical tank of radius 2 meters at the rate of 8 cubic meters per minute. How fast is the water level rising?

**2** A launch whose deck is 7 meters below the level of a wharf is being pulled toward the wharf by a rope attached to a ring on the deck. If a winch pulls in the rope at the rate of 15 m/min, how fast is the launch moving through the water when there are 25 m of rope out?

**3** Two automobiles start from a point $A$ at the same time. One travels west at 60 km/hr and the other travels north at 35 km/hr. How fast is the distance between them increasing 3 hr later?

**4** At noon of a certain day, ship $A$ is 60 km due north of ship $B$. If $A$ sails east at 12 km/hr and $B$ sails north at 9 km/hr, determine how rapidly the distance between them is changing 2 hr later. Is it increasing or decreasing?

**5** At a given instant the legs of a right triangle are 16 cm and 12 cm, respectively. The first leg decreases at $\frac{1}{2}$ cm/min and the second increases at 2 cm/min. At what rate is the area increasing after 2 min?

**6** A rhombus with an acute angle of 45° has sides which are growing at the rate of 2 cm/min, while always retaining the same shape. How fast is the area changing when the sides are 10 cm?

**7** At a certain instant a small balloon is released (from ground level) at a point 75 m away from an observer (on ground level). If the balloon goes straight up at a rate of 4 m/sec, how rapidly will it be receding from the observer 30 sec later?

**8** A street light is 5 meters above a sidewalk. A man 2 meters tall walks away from the point under the light at the rate of 2 m/sec. How fast is his shadow lengthening when he is 7 m away from the point under the light?

**9** A balloon is being inflated at the rate of 15 m³/min. At what rate is the diameter increasing after 5 min? Assume that the diameter is zero at time zero.

**10** A baseball diamond is 90 ft on a side. (It is really a square.) A man runs from first base to second base at 25 ft/sec. At what rate is his distance from third base decreasing when he is 30 ft from first base? At what rate is his distance from home plate increasing at the same instant?

**11** A woman starts walking eastward at 2 m/sec from a point $A$. Ten minutes later a man starts walking west at the rate of 2 m/sec from a point $B$, 3000 meters north of $A$. How fast are they separating 10 min after the man starts?

**12** A point moves along the curve $y = \sqrt{x^2 + 1}$ in such a way that $dx/dt = 4$. Find $dy/dt$ when $x = 3$.

13  A point moves along the upper half of the curve $y^2 = 2x + 1$ in such a way that $dx/dt = \sqrt{2x+1}$. Find $dy/dt$ when $x = 4$.

14  A point moves along the curve $y = \sin^2 3x$ in such a way that $dx/dt = 3$. Find $dy/dt$ when $x = \pi/4$.

15  The variables $x$, $y$, and $z$ are all functions of $t$ and satisfy the relation $x^3 - 2xy + y^2 + 2xz - 2xz^2 + 3 = 0$. Find $dz/dt$ when $x = 1$, $y = 2$, if $dx/dt = 3$ and $dy/dt = 4$ for all times $t$.

16  The variables $x$ and $y$ are functions of $t$ and satisfy the relation $\sin 2x - 2 \tan 3y + 1 = 0$. Find $\dfrac{dy}{dt}$ when $x = \pi/4$ and $dx/dt = 2$.

17  A ladder 5 meters long leans against a vertical wall of a house. If the bottom of the ladder is pulled horizontally away from the house at 4 m/sec, how fast is the top of the ladder sliding down when the bottom is 3 meters from the wall?

18  A light is on the ground 40 m from a building. A man 2 m tall walks from the light toward the building at 2 m/sec. How rapidly is his shadow on the building growing shorter when he is 20 m from the building?

19  A trough is 10 m long and its ends are isosceles triangles with altitude 2 m and base 2 m, their vertices being at the bottom. If water is let into the trough at the rate of $3 \text{ m}^3/\text{min}$, how fast is the water level rising when it is 1 m deep?

20  A trough 10 m long has as its ends isosceles trapezoids, altitude 2 m, lower base 2 m, upper base 3 m. If water is let in at the rate of $3 \text{ m}^3/\text{min}$, how fast is the water level rising when the water is 1 m deep?

21  A swimming pool is 9 m wide, 15 m long, 1 m deep at the shallow end, and 3 m deep at the deep end, the bottom being an inclined plane. If water is pumped into the pool at the rate of $3 \text{ m}^3/\text{min}$, how fast is the water level rising when it is 2 m deep at the deep end?

22  Sand is issuing from a spout at the rate of $3 \text{ m}^3/\text{min}$ and falling on a conical pile whose diameter at the base is always three times the altitude. At what rate is the altitude increasing when the altitude is 4 m?

23  Water is leaking out of a conical tank (vertex down) at the rate of $0.5 \text{ m}^3/\text{min}$. The tank is 30 m across at the top and 10 m deep. If the water level is rising at the rate of $1\frac{1}{2}$ m/min, at what rate is water being poured into the tank from the top?

24  In Example 2 on page 170, find the rate at which the uncovered surface of the conical tank is decreasing at the instant in question.

25  A boat is anchored in such a way that its deck is 25 m above the level of the anchor. If the boat drifts directly away from the point above the anchor at the rate of 5 m/min, how fast does the anchor rope slip over the edge of the deck when there are 65 m of rope out? (Assume that the rope forms a straight line from deck to anchor.)

26  A man lifts a bucket of cement to a scaffold 30 m above his head by means of a rope which passes over a pulley on the scaffold. The rope is 60 m long. If he keeps his end of the rope horizontal and walks away from beneath the pulley at 4 m/sec, how fast is the bucket rising when he is $22\frac{1}{2}$ m away?

27  Water is flowing into a tank in the form of a hemisphere of radius 10 m with flat side up (Fig. 45) at the rate of $4 \text{ m}^3/\text{min}$. At any instant let $h$ denote the depth of the water, $r$ the radius of the surface, and $V$ the volume of the water. Assuming $dV = \pi r^2 \, dh$, find how fast the water level is rising when $h = 5$ m.

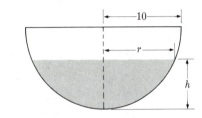

**FIGURE 45**

28  A bridge is 10 m above a canal. A motorboat going 3 m/sec passes under the center of the bridge at the same instant that a woman walking 2 m/sec reaches that point. How rapidly are they separating 3 sec later?

29  If in Problem 28 the woman reaches the center of the bridge 5 sec before the boat passes under it, find the rate at which the distance between them is changing 4 sec after the woman crosses.

30  When a gas expands or contracts adiabatically, it obeys the law $pv^\gamma = K$ where $p$ is the pressure, $v$ the volume, and $\gamma$ and $K$ are constants. At a certain instant a container of air is under pressure of 40 gram/cm$^2$, the volume is 32 cm$^3$, and the volume is increasing at the rate of 5 cm$^3$ per second. Assuming that $\gamma = 1.4$, find how rapidly the pressure is decreasing at that instant.

31  A light is at the top of a pole which is $h$ meters high. A ball is dropped from a height $\frac{1}{2}h$ at a point which is at a horizontal distance $d$ meters from the pole. Assume that the ball falls according to the law $s = gt^2$, where $t$ is the time in seconds, $s$ is the distance in meters, and $g$ is a constant. Find how fast the tip of the shadow of the ball is moving along the ground $t_0$ seconds after it is dropped.

32  Ohm's law for a certain electrical circuit states that $E = IR$ where $E$ is the voltage in volts, $I$ the current in amperes, and $R$ the resistance in ohms. If the circuit heats up and the voltage is kept constant, the resistance increases at the rate of 0.5 ohms per second. Find the rate at which the current decreases when $I = 2$ amps and $E$ is kept constant at 10 volts.

## 11

### ANTIDERIVATIVES

Until now we have studied the problem of finding the derivative of a function. We learned techniques for differentiating polynomials, rational functions, and trigonometric functions. Now we take up the inverse problem: we are given a function $f$ and we seek a function $F$ such that $f$ is the derivative of $F$. That is, we want to find an $F$ such that $F'(x) = f(x)$. For example, suppose that $f(x) = x^3$. It is not difficult to guess that if we choose $F(x) = \frac{1}{4}x^4$ then $F'(x) = f(x)$. However, $\frac{1}{4}x^4$ is not the only answer to the problem. The functions $\frac{1}{4}x^4 + 8$, $\frac{1}{4}x^4 + 5$, and $\frac{1}{4}x^4 - 2$ all have derivatives equal to $x^3$. The answer to the problem we posed is not unique. In Chapters 5 and 8 we will develop systematic methods for finding antiderivatives.

**DEFINITION**    *A given function $f$ has an* **antiderivative** *$F$ if $F'(x) = f(x)$ for all $x$ in the domain of $f$.*

The next example shows that finding the antiderivative is more complex than the process of differentiation.

**EXAMPLE 1**    Given the function $f(x) = 4x^3 + 2x - 5$, find its antiderivative.

**Solution**    To find an antiderivative, we simply guess. We see that $4x^3$ is the derivative of $x^4$; $2x$ is the derivative of $x^2$; $-5$ is the derivative of $-5x$. Hence, a possible antiderivative $F(x)$ is given by

$$F(x) = x^4 + x^2 - 5x + 7,$$

which we verify by computing $F'(x)$. Another antiderivative is $F(x) = x^4 + x^2 - 5x + 4$; still another is $F(x) = x^4 + x^2 - 5x - 3$. We conclude that there is no single solution to the problem of finding the function if the derivative is given. That is, *the problem of finding the antiderivative does not have a unique solution.* It is easy to see that any function $F$ of the form

$$F(x) = x^4 + x^2 - 5x + c$$

is a solution to the above problem, where $c$ is a number which may have any value whatsoever.* If, in addition to being given the derivative $F'$, we are given one more fact about the function $F$, we can usually determine the quantity $c$. In the above illustration, if we also know that $F(2) = 4$, we can write

$$F(2) = 4 = 16 + 4 - 10 + c,$$

and conclude that $c = -6$. The function $F$ is then

$$F(x) = x^4 + x^2 - 5x - 6. \qquad \square$$

The process described in Example 1 can be formalized into the following rule.

---

*We shall see in Chapter 5 that all solutions are of this form.

**POLYNOMIAL ANTIDERIVATIVE RULE**    *Let $p(x)$ be the polynomial*

$$p(x) = a_n x^n + a_{n-1} x^{n-1} + \cdots + a_1 x + a_0.$$

*Then the antiderivative of $p(x)$, denoted $P(x)$, is*

$$P(x) = a_n \frac{x^{n+1}}{n+1} + a_{n-1} \frac{x^n}{n} + \cdots a_1 \frac{x^2}{2} + a_0 x + c,$$

*where $c$ is a number which may have any value.*

The validity of the Polynomial Antiderivative Rule is easily checked by differentiating $P(x)$ and noting that $P'(x) = p(x)$. An important special case is $p(x) = x^n$. We have

If $p(x) = x^n$, then the antiderivative $P(x)$ is given by

$$P(x) = \frac{x^{n+1}}{n+1} + c,$$

where $c$ is an arbitrary constant.

We also observe that if $r$ is any *rational* number, the derivative of $x^r$ is $rx^{r-1}$. Hence the Polynomial Antiderivative Rule also works when $x$ is raised to any rational exponent, positive or negative.

**EXAMPLE 2**    Find an antiderivative of

$$f(x) = x^{1/7} - x^{2/3} + \sin x.$$

**Solution**    An antiderivative of $x^{1/7}$ is $\frac{7}{8}x^{8/7}$; an antiderivative of $-x^{2/3}$ is $-\frac{3}{5}x^{5/3}$; and we easily guess that an antiderivative of $\sin x$ is $-\cos x$. Adding these antiderivatives, we get $F(x)$, an antiderivative of $f$:

$$F(x) = \tfrac{7}{8}x^{8/7} - \tfrac{3}{5}x^{5/3} - \cos x.$$

This $F(x)$ is, however, only one antiderivative of $f$. Since the derivative of a constant is zero, we can always add a constant to $F(x)$ to get another antiderivative. To indicate the freedom in the choice of a constant we write the antiderivative of $f$ in the form

$$\tfrac{7}{8}x^{8/7} - \tfrac{3}{5}x^{5/3} - \cos x + c,$$

where $c$ is an arbitrary constant.    □

*Remark.*    In Chapter 3 we developed systematic methods for finding the derivatives of algebraic functions. So far, we can only find antiderivatives by guessing at the result and then verifying the correctness of the guess by differentiating the answer.

Let us work another example by the guessing method.

**EXAMPLE 3**    Suppose that $F$ is a function whose derivative is

$$\tfrac{1}{2}x^3(x+1)^{-1/2} + 3x^2(x+1)^{1/2}.$$

Given that $F(3) = 5$, find the function $F$.

**Solution**    We observe that if $u(x) = x^3$, then $u'(x) = 3x^2$, and if $v(x) = (x + 1)^{1/2}$, then $v'(x) = \frac{1}{2}(x + 1)^{-1/2}$. Guessing that the expression for $F'(x)$ is in the form $uv' + u'v$, we observe that $F(x) = u(x) \cdot v(x)$. Therefore

$$F(x) = x^3(1 + x)^{1/2} + c.$$

Setting $f(3) = 5$, we find

$$5 = 27 \cdot \sqrt{4} + c \qquad \Leftrightarrow \qquad c = -49.$$

We conclude that

$$f(x) = x^3(1 + x)^{1/2} - 49.$$    □

The next example employs the definitions of velocity and acceleration as given in Chapter 2, Section 5.

**EXAMPLE 4**    A ball is thrown upward, and we know that the acceleration is $-10$(m/sec)/sec. (a) What is the velocity at any instant of time if the ball was launched with a velocity of 60 m/sec? (b) Find the height of the ball above the launching point at any time $t$

**Solution**    Since the acceleration $a$ is given, finding the velocity $v$ is a problem in antiderivatives. We write $a(t) = v'(t)$ and, using the Polynomial Rule, we have

$$a = -10, \qquad v = -10t + c,$$

where $t$ is the time (starting at $t = 0$ when the ball is thrown). We have $v = 60$ when $t = 0$, which means that $60 = -10 \cdot 0 + c$, or $c = 60$. We obtain the formula

$$v = -10t + 60.$$

To get the distance traveled, we again have a problem in antiderivatives. We find that $v = s'(t)$ and again using the Polynomial Rule, we obtain

$$s = -5t^2 + 60t + k.$$

To evaluate $k$, we note that at time $t = 0$ the distance traveled is also zero. This makes $k = 0$, and we have the formula

$$s = -5t^2 + 60t.$$    □

Suppose $F(x)$ is the antiderivative of $f(x)$, and we write $y = F(x)$. Then the equation $F'(x) = f(x)$ can be written in the form

$$\frac{dy}{dx} = f(x). \tag{1}$$

Here $f(x)$ is a known function and we are seeking $y$ as a function of $x$. An equation with derivatives, such as (1) above, is called a **differential equation**. We provide a systematic study of differential equations in Chapter 18. Typically, in solving a differential equation such as (1), i.e., in finding the antiderivative, there is usually an additional condition imposed in order to determine the constant $c$ which arises. We call this added condition an **initial condition** for $y = y(x)$. The next example describes the method.

**EXAMPLE 5**   Solve the differential equation

$$\frac{dy}{dx} = 6x^2 + 3x - 5$$

with the initial condition $y = 4$ when $x = 0$.

**Solution**   Since $6x^2 + 3x - 5$ is a polynomial, we obtain the antiderivative

$$y = 2x^3 + \frac{3x^2}{2} - 5x + c,$$

where $c$ is any constant. The initial condition yields $4 = 0 + 0 + 0 + c$ and the solution is

$$y = 2x^3 + \frac{3x^2}{2} - 5x + 4. \qquad \square$$

## 11   PROBLEMS

In each of Problems 1 through 16, find the antiderivative $F$. Use the additional condition to determine the solution uniquely.

**1** $f(x) = x^4 - 2x^3 + 7x - 6$;   $F(1) = 2$

**2** $f(x) = -2x^5 + 7x^3 - 5x^2 + 2x$;   $F(2) = 1$

**3** $f(x) = (x + 2)^7$;   $F(-2) = 1$

**4** $f(x) = 3x^2(x + 1)^2 + 2x(x + 1)^3$;   $F(3) = -1$

**5** $f(x) = 2(2x + 1)^3(x + 2) + 6(2x + 1)^2(x + 2)^2$;
$F(-\frac{1}{2}) = 0$

**6** $f(x) = (x + 3)^{-2}[2(x + 3) - (2x + 1)]$;   $F(1) = 1$

**7** $f(x) = x^{1/5} + 6x^{2/3} + x^2$;   $F(0) = 3$

**8** $f(x) = x^{3/4} + x^3 + \cos x$;   $F(0) = 2$

**9** $f(x) = x^{1/4} + \sec x \tan x$;   $F\left(\frac{\pi}{4}\right) = 1$

**10** $f(x) = \dfrac{x^2 + x^3(x^2 + 1)}{x^4}$;   $F(1) = 2$

**11** $f(x) = 2x \cos x - x^2 \sin x$;   $F\left(\frac{\pi}{2}\right) = 3$

**12** $f(x) = \dfrac{1}{2\sqrt{x}} \sec 2x + 2\sqrt{x} \sec 2x \tan 2x$;   $F\left(\frac{\pi}{3}\right) = 1$

**13** $f(x) = \cos 2x$;   $F\left(\frac{\pi}{6}\right) = \sqrt{3}$

**14** $f(x) = -\sin 4x$;   $F\left(\frac{\pi}{8}\right) = 2$

**15** $f(x) = \sin^2 x \cos x$;   $F\left(\frac{\pi}{4}\right) = \sqrt{2}$

**16** $f(x) = \sec^2 2x$;   $F\left(\frac{\pi}{3}\right) = 1$

In Problems 17 through 22, find the distance $s$ traveled along a straight line by a particle if the velocity $v = v(t)$ follows the law given. The additional fact required for the result is also given.

**17** $v = t^2 + t - 2$ and $s = 2$ when $t = -1$

**18** $v = -t^2 + 2t + 1$ and $s = 0$ when $t = 2$

**19** $v = 6 - 2t - 3t^2$ and $s = 0$ when $t = 0$

**20** $v = 5 - 7t^2$ and $s = 5$ when $t = 0$

**21** $v = \cos^2 2t \sin 2t$ and $s = 2$ when $t = 0$

**22** $v = -\sin \frac{1}{2}t$ and $s = 1$ when $t = 0$

**23** A ball is thrown upward from the ground with a launching velocity of 80 m/sec. Its acceleration is constant and equal to $-10$(m/sec)/sec. How long does the ball continue to rise? How long does the ball stay in the air?

**24** A woman driving an automobile in a straight line at a speed of 80 m/sec applies the brakes at a certain instant (which we take to be $t = 0$). If the brakes furnish a constant acceleration of $-20$(m/sec)/sec (actually a deceleration), how far will she go before she stops?

In each of Problems 25 through 30, solve the differential equation with the given initial condition.

**25** $\dfrac{dy}{dx} = 3x^2 - 5x + 2$;   $y(0) = 6$

**26** $\dfrac{dy}{dx} = 4x^7 - 6x + 3$;   $y(0) = 7$

**27** $\dfrac{dy}{dx} = 6 \sin x$;   $y(0) = -3$

**28** $\dfrac{dy}{dx} = 14x^2 - 3x + 7$;   $y(1) = 12$

**29** $\dfrac{dy}{dx} = 27x^3 - 3x^2 + 4x + 6; \quad y(-1) = 2$

**30** $\dfrac{dy}{dx} = 4 \cos x - 3 \sin x; \quad y(\pi) = \pi$

***31** Suppose that $f'(x) = 0$ for $a \leq x \leq b$. Show that $f(x)$ is constant. (*Hint:* Use the Mean Value Theorem.)

***32** If $F_1(x)$ and $F_2(x)$ are two solutions of the differential equation

$$\frac{dy}{dx} = f(x),$$

show that $F_1(x) - F_2(x)$ is constant. (*Hint:* Use the result of Problem 31.)

***33** Let $F(x)$ be a solution of the differential equation

$$\frac{dy}{dx} = f(x)$$

where $F(0) = a$. If $f(x)$ is continuous show that the solution is unique. (*Hint:* Use the result of Problem 32.)

**34** Find a solution $y = y(x)$ of the differential equation

$$\frac{d^2y}{dx^2} = 3x - 2.$$

**35** Find the solution $y = y(x)$ of the differential equation

$$\frac{d^2y}{dx^2} = 6x - 3$$

subject to the conditions $y(0) = 6$ and $y'(0) = 7$.

**36** Find the solution $y = y(x)$ of the differential equation

$$\frac{d^2y}{dx^2} = 2 \sin 3x + 4 \cos 2x$$

subject to the conditions $y(0) = 3$ and $y'(0) = 5$.

**37** Show that $y(x) = \sin x + \cos x$ is a solution of

$$\frac{d^2y}{dx^2} + y = 0.$$

# CHAPTER 4

## REVIEW PROBLEMS

In Problems 1 through 10 find the relative maxima and minima and sketch the graph.

**1** $f(x) = -x^2 + 2x + 3$     **2** $f(x) = 2x^2 + x - 5$

**3** $f(x) = x^3 + 4x^2 - 2x$     **4** $f(x) = x^3 - 3x^2 + 2x + 4$

**5** $f(x) = 3x^4 - 8x^3 - 6x^2 + 24x + 2$

**6** $f(x) = x^4 - 14x^2 + 24x + 4$

**7** $f(x) = \sin 2x, \quad -\pi \leq x \leq \pi$

**8** $f(x) = \cos 3x, \quad -\pi \leq x \leq \pi$

**9** $f(x) = \dfrac{x^2 - 2}{x^2 + 3}$     **10** $f(x) = \dfrac{x - 1}{x^2 + 2}$

In Problems 11 through 20 find the relative maxima and minima and the points of inflection. Find the intervals of concavity both upward and downward.

**11** $f(x) = -3x^2 + 5$     **12** $f(x) = x^3 + 4x$

**13** $f(x) = 2x^3 - x^2 + 4$     **14** $f(x) = x^3 + 4x^2 - 3x - 1$

**15** $f(x) = x^4 - 14x^2 - 24x + 4$

**16** $f(x) = 3x^4 - 4x^3 - 18x^2 + 36x + 5$

**17** $f(x) = 2 \sin 3x, \quad -0 \leq x \leq 2\pi$

**18** $f(x) = \tan 2x, \quad -\pi \leq x \leq \pi$

**19** $f(x) = \dfrac{x - 1}{x + 1}$     **20** $f(x) = \dfrac{x^2 - 4}{x^2 + 4}$

In Problems 21 through 30 find the absolute maximum and absolute minimum values of the function on the interval given.

**21** $f(x) = x^3 + 2x, \quad -1 \leq x \leq 4$

**22** $f(x) = -2x^3 + x^2, \quad -2 \leq x \leq 3$

**23** $f(x) = x^4 - 2x^2, \quad -2 \leq x \leq 2$

**24** $f(x) = x^3 + 2x^2 - 3x - 4, \quad -1 \leq x \leq 4$

**25** $f(x) = \begin{cases} -2x + 5, & -2 \leq x \leq -1 \\ x^2 + 3x + 9, & -1 < x \leq 4 \end{cases}$

**26** $f(x) = \begin{cases} x^2 - 3x + 1, & -4 \leq x < 1 \\ -x^2 + 2x - 2, & 1 \leq x \leq 3 \end{cases}$

**27** $f(x) = \begin{cases} 2x + 5, & -2 \leq x < -1 \\ x^2 - 3x + 1, & -1 \leq x \leq 2 \end{cases}$

**28** $f(x) = \begin{cases} \sin x, & 0 \leq x \leq \pi/2 \\ 1 - \cos x, & \dfrac{\pi}{2} < x \leq \pi \end{cases}$

**29** $f(x) = 3x^4 - 16x^3 - 12x^2 + 96x + 2, \quad -3 \leq x \leq 4$

**30** $f(x) = \dfrac{3x}{x^2 + 2}, \quad -1 \leq x \leq 2$

**31** The product of two numbers is 51. Find the numbers if the sum of three times one number and the other is as large as possible. Find them if the same sum is as small as possible.

**32** A rectangular box with square base has the top and bottom made of one material while the sides are made of a different material which costs twice as much. The volume is to be 100 cm³. Find the dimensions so that the cost is as small as possible.

**33** Find the point on the curve $y = x^2 - 2x$ which is closest to the point $(1, -3)$.

**34** A window is designed in the shape of a rectangle surmounted by an equilateral triangle. If the perimeter is 300 cm, find the dimensions of the window which provides the most light.

In Problems 35 through 38 calculate the differential.

**35** $f(x) = \dfrac{x^2 + 2x - 3}{x^2 + 7x + 9}$     **36** $f(x) = (x^2 + 2)^{1/3} - 2x^{7/4}$

**37** $f(x) = \sec^2 2x$     **38** $f(x) = 2 \cot \frac{1}{2}x + \sin 4x$

**39** Find $dy/dx$ given that $x^2 + z^2 - 2xz - 4 = 0$ and $y^3 + 2y^2z + 2z^4 - 3 = 0$.

In Problems 40 through 42 use Newton's Method to find the value accurately to three decimal places.

**40** $\sqrt[3]{24}$     **41** $\sqrt[3]{-11}$     **42** $\sqrt[4]{41}$

**43** To find the cube root of a number $c$, show that Newton's Method leads to the formula

$$x_{i+1} = \left(\frac{2}{3}\right)x_i + \frac{c}{3x_i^2}.$$

**44** To find the $n$th root of a number $c$, show that Newton's Method leads to the formula

$$x_{i+1} = \frac{n-1}{n}x_i + \frac{c}{nx_i^{n-1}}.$$

**45** A man at point $A$ is walking east at 2 m/sec. A second man who is 60 meters northeast of $A$ starts walking north at the same instant at the rate of 3 m/sec. How fast are they separating 1 minute later?

**46** Two buildings, one 20 m tall and the other 40 m tall, are 20 m apart. A ladder stretches from the roof of one to the roof of the other. At a certain instant a man beginning at the smaller building climbs the ladder at the rate of 2 m/sec. Starting at the same time, a woman runs along the ground from the smaller to the larger building at the rate of 3 m/sec. How fast is the distance between the two changing after 4 seconds?

In each of Problems 47 through 51 find $y = y(x)$ from the information given.

**47** $\dfrac{dy}{dx} = 2 - 3x + x^2, \quad y(0) = -2$

**48** $\dfrac{dy}{dx} = x^5 + 2x^{3/2} + 1, \quad y(1) = 2$

**49** $\dfrac{dy}{dx} = \sin 2x + \cos x, \quad y(\pi/4) = 1$

**50** $\dfrac{dy}{dx} = 2 \sin x \cos x, \quad y(0) = 0$

**51** $\dfrac{dy}{dx} = \cos 3x + \sec^2 2x, \quad y\left(\frac{\pi}{6}\right) = 3$

# 5

# THE DEFINITE INTEGRAL

We have seen that derivatives are useful for solving geometric and physical problems. The definition of the integral involves a limiting process which is quite different from that of the derivative. In this chapter we develop the basic properties of integrals, and in the Fundamental Theorem of Calculus we show that integrals and derivatives are inverses just as multiplication and division are. In Chapter 6 we provide applications of the integral, applications that show us how to calculate areas, volumes, and the length of curves, especially for irregularly shaped regions.

## 1

### AREA

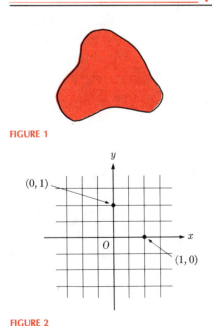

**FIGURE 1**

**FIGURE 2**

The area of a region is a measure of its size. A rectangle of length $l$ and width $w$ has area $A = l \cdot w$. The area of a triangle of base $b$ and altitude $h$ is just $\frac{1}{2}bh$. Similarly, we may obtain the area of a polygon simply by decomposing it into triangles and adding the areas of the component parts. Moreover, we readily see that the area of a polygon is independent of the way in which the polygon is cut up into triangles.

Suppose we have an irregularly shaped region in the plane, such as the one shown in Fig. 1. How do we go about defining its area? The process is rather complicated and depends intrinsically on the idea of limit, which we have previously learned. In this section the procedure for defining area will be given, although the proofs of some of the statements will be omitted.

We start with the notion of a **square grid**. The entire plane is divided into squares by constructing equally spaced lines parallel to the coordinate axes. Figure 2 shows a square grid with squares $\frac{1}{2}$ unit on each side, and Fig. 3 shows a square grid with squares $\frac{1}{4}$ unit on each side. We always select the coordinate axes themselves as lines of the grid, and then the size of the

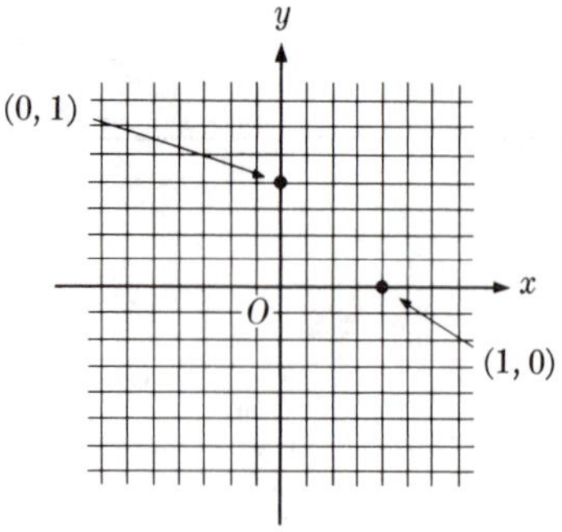

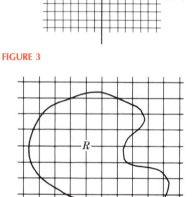

**FIGURE 3**

**FIGURE 4**

squares determines the location of all other grid lines. We may have a grid with squares of any size, but for convenience we consider only grids of size $\frac{1}{2}$, $\frac{1}{4}$, $\frac{1}{8}$, $\frac{1}{16}$, ..., $1/2^n$, ..., where $n$ is any positive integer. We obtain a sequence of grids with the property that the lines of any grid in the sequence are also lines in all the grids farther along in the sequence.

We want to *define* the area of an irregularly shaped region $R$, such as the one shown in Fig. 4. We begin by constructing a square grid (one of our sequence of grids) in the plane. All the squares of this grid can be divided into three types: (1) squares which are completely inside the region $R$; (2) squares which have some points in the region and some points not in the region; (3) squares which are completely outside the region.

If the grid we select is the 5th in the sequence ($n = 5$), then the length of the side of each square is $1/2^5 = \frac{1}{32}$ and the area of each square is

$$\frac{1}{32} \cdot \frac{1}{32} = \frac{1}{1024} = \frac{1}{2^{10}}$$

square units. We define for the square grid with $n = 5$:

$A_5^L$ = sum of the areas of all squares of type (1)

$A_5^U$ = sum of the areas of all squares of type (1) plus type (2).

(The $L$ stands for *lower* and the $U$ stands for *upper*.) According to the way we have defined these quantities, if the region we started with is to have something we call "area," then we would expect that $A_5^L$ would be less than this area, while $A_5^U$ would be greater. We have the inequality

$$A_5^L \leq A_5^U,$$

which comes from the very definition. The quantity $A_5^U$ contains everything in $A_5^L$ and more.

We have described a typical step in a process which consists of a sequence of steps. The next step consists of constructing a grid with $n = 6$. Each square is of side $\frac{1}{64}$, the area being $\frac{1}{64} \cdot \frac{1}{64} = \frac{1}{4096}$ square units. Each square of the $n = 5$ grid is divided into exactly 4 subsquares to form the $n = 6$ grid. We now define the quantities:

$A_6^L$ = sum of areas of all squares of type (1) of the $n = 6$ grid;

$A_6^U$ = sum of areas of all squares of type (1) plus type (2) of the $n = 6$ grid.

As before, the definition tells us that

$$A_6^L \leq A_6^U.$$

But now we can see that there is a relation between $A_5^L$ and $A_6^L$. Any square in $A_5^L$ is the same as 4 squares of $A_6^L$, with the *same* total area. However, $A_6^L$ may have some squares of type (1) near the boundary which are part of type (2) regions for the $n = 5$ grid. Figure 5 shows that the lower left corner is of type (1) for $n = 6$, while the large square is of type (2) for $n = 5$. It can be proved (although our discussion has not done so) that for *any* region, we have

$$A_5^L \leq A_6^L.$$

Similarly, $A_5^U$ may have a type (2) region (see Fig. 6) which, when considered in the $n = 6$ grid, gives three type (2) regions and one type (3) region. This

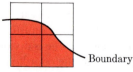

Large square of type (2) for $n = 5$. For $n = 6$, 3 squares of type (2) and one square of type (1)

**FIGURE 5**

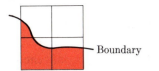

Large square of type (2) for $n = 5$. For $n = 6$, 3 squares of type (2) and one square of type (3)

**FIGURE 6**

means that for such a square the contribution to $A_6^U$ (three smaller squares) is less than the corresponding contribution to $A_5^U$ (the larger square). *It can be shown in a general way that*

$$A_6^U \leq A_5^U.$$

The inequalities may also be combined to give

$$A_5^L \leq A_6^L \leq A_6^U \leq A_5^U.$$

When the above process is performed for $n = 1, 2, 3, \ldots$ we get a **sequence** of numbers

$$A_1^L \leq A_2^L \leq A_3^L \leq A_4^L \leq \cdots \leq A_n^L \leq \cdots.$$

and a corresponding **sequence**

$$A_1^U \geq A_2^U \geq A_3^U \geq A_4^U \geq \cdots \geq A_n^U \geq \cdots.$$

The situation is shown in Fig. 7. The sequence $A_1^L, A_2^L, \ldots, A_n^L, \ldots$ is steadily increasing but all the terms remain below the sequence $A_1^U, A_2^U, \ldots, A_n^U$, whose terms are steadily decreasing. It is intuitively clear from Fig. 7 that the terms $A_1^L, A_2^L, \ldots, A_n^L, \ldots$ tend to a limiting value denoted by $A^-$ and called the **inner area** of the region. Since the region is designated by $R$, we also write $A^-(R)$.

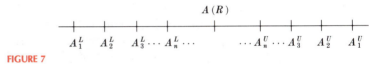

$A(R)$

$A_1^L \quad A_2^L \quad A_3^L \cdots A_n^L \cdots \qquad \cdots A_n^U \cdots A_3^U \quad A_2^U \quad A_1^U$

**FIGURE 7**

In the same way $A_n^U$ tends to a limit which we denote by $A^+$, and this number is called the **outer area** of the region. We also write $A^+(R)$.

It is always true that

$$A^-(R) \leq A^+(R),$$

but it is not always true that these numbers are equal. However, the sets $R$ for which they are unequal are weird, and for the kinds of regions we shall be concerned with it can be shown that

$$A^-(R) = A^+(R).$$

*This common value is then* **defined to be the area of the region,** *and we shall denote it* $A(R)$.

Note that the above discussion did not *prove* the validity of the inequalities for $A_n^L$ and $A_n^U$. The arguments may or may not have appeared convincing, but arguing "pictorially" is not a proof. The development given here is intended only to establish the reasonableness of the results.

The fundamental unit for determining areas is the square. A completely analogous development could be made with rectangles instead of squares. A sequence of *rectangular grids* is then used, the lengths and widths of the rectangles tending to zero as the $n$th term in the sequence tends to infinity. We would naturally expect, and it can be proved, that both the rectangular and the square grids yield the same number for the area of $R$. If a region $R$ is formed as the sum of two nonoverlapping regions $R_1$ and $R_2$, we can easily

prove the expected fact that

$$A(R) = A(R_1) + A(R_2).$$

Finally, it can be shown that the value of $A(R)$ is not influenced by the location of the coordinate axes.

To justify the statement above that there are limiting values $A^-(R)$ and $A^+(R)$ for the two sequences shown in Fig. 7, we require a new axiom known as the **Axiom of Continuity**. While the statement of the axiom is simple and appears to be one we should be able to prove, it turns out that there are certain number systems for which the conclusion of the axiom is not valid.

---

**AXIOM C**  **(Axiom of Continuity)**  *Suppose that a sequence $a_1$, $a_2$, ..., $a_n$, ... has the properties (1) $a_{n+1} \geq a_n$ for all n, and (2) there is a number M such that $a_n \leq M$ for all n. Then there is a number $b \leq M$ such that*

$$\lim_{n \to \infty} a_n = b \qquad and \qquad a_n \leq b$$

*for all n.*

---

The Axiom of Continuity is also known as the **Completeness Axiom** for the real numbers.

Figure 8 shows the situation. The numbers $a_n$ move steadily to the right, and yet they can never get beyond $M$. It is reasonable to have an axiom which states that there must be some number $b$ (perhaps $M$ itself) toward which the $a_n$ cluster. Axiom C is usually stated in the form: **Every bounded, nondecreasing sequence of numbers tends to a limit**. A similar statement holds if the sequence is bounded and nonincreasing.

**FIGURE 8**

By choosing $a_n = A_n^L$ and applying Axiom C, we find that the limiting value $b$ is $A^-(R)$. Similarly, by choosing $a_n = A_n^U$ and noting that the sequence is nonincreasing, we get $b = A^+(R)$.

We have seen that the development of the derivative is helped greatly by the use of special symbols for the derivative and the differential. We now introduce symbols for sums which will not only help in deriving formulas for the area but will prove to be useful in other topics.

The sum of seven terms,

$$a_1 + a_2 + a_3 + a_4 + a_5 + a_6 + a_7,$$

can be written in abbreviated form using a special symbol. The Greek letter sigma and certain subscripts are combined in the following way:

$$\sum_{i=1}^{7} a_i \quad \text{means} \quad a_1 + a_2 + a_3 + a_4 + a_5 + a_6 + a_7.$$

We read it: "the sum from 1 to 7 of $a$ sub $i$." The lower number (1 in this case) indicates where the sum starts, and the upper number (7 in this case) indicates where it ends. The expression

$$\sum_{i=3}^{8} b_i \quad \text{means} \quad b_3 + b_4 + b_5 + b_6 + b_7 + b_8.$$

The symbol $i$ is a "dummy" symbol (called the **index of summation**), since

$$\sum_{k=3}^{8} b_k \quad \text{means} \quad b_3 + b_4 + b_5 + b_6 + b_7 + b_8,$$

which is exactly the same thing.

Carrying the process one step further, we have

$$\sum_{i=1}^{5} (2i + 1) = (2 \cdot 1 + 1) + (2 \cdot 2 + 1) + (2 \cdot 3 + 1) + (2 \cdot 4 + 1)$$
$$+ (2 \cdot 5 + 1) = 3 + 5 + 7 + 9 + 11,$$

and

$$\sum_{k=2}^{5} [(k + 1)^3 - k^3] = [(2 + 1)^3 - 2^3] + [(3 + 1)^3 - 3^3]$$
$$+ [(4 + 1)^3 - 4^3] + [(5 + 1)^3 - 5^3] = -2^3 + 6^3.$$

The **summation notation**, as it is called, is of great help in manipulating sums. As an example, consider

$$\sum_{i=1}^{6} 2i = 2 \cdot 1 + 2 \cdot 2 + 2 \cdot 3 + 2 \cdot 4 + 2 \cdot 5 + 2 \cdot 6$$
$$= 2(1 + 2 + 3 + 4 + 5 + 6)$$
$$= 2 \sum_{i=1}^{6} i.$$

More generally, we have the rule that for any sum

$$\sum_{k=1}^{n} ca_k = c \sum_{k=1}^{n} a_k, \quad c = \text{const},$$

where the symbol means that the sum starts with $a_1$ and ends with $a_n$. The number at the bottom is called the **lower limit** of the sum and the number at the top the **upper limit**. We also have

$$\sum_{k=1}^{n} (a_k + b_k) = \sum_{k=1}^{n} a_k + \sum_{k=1}^{n} b_k.$$

Sums of the form

$$\sum_{k=1}^{n} (b_{k+1} - b_k) \tag{1}$$

occur frequently in practice. When we write out the individual terms we get

$$(b_2 - b_1) + (b_3 - b_2) + (b_4 - b_3) + \cdots + (b_n - b_{n-1}) + (b_{n+1} - b_n),$$

and we make the important observation that almost all the terms cancel. In fact, the above sum reduces to

$$-b_1 + b_{n+1}.$$

For this reason we designate any sum of the form (1) a **telescoping sum**.

Sometimes a sum may be put in a telescoping form even when at first glance it appears quite different. For example, the sum

$$\sum_{k=1}^{n} \frac{1}{k^2 + k}$$

is actually a telescoping sum. To see this we write

$$\frac{1}{k^2 + k} = \frac{1}{k(k+1)} = \frac{A}{k} + \frac{B}{k+1},$$

where $A$ and $B$ are numbers to be determined. Multiplying through by $k(k+1)$ we find

$$1 = A(k+1) + Bk.$$

This equation is an identity only if $A = 1$ and $A + B = 0$, and thus $B = -1$. Finally,

$$\frac{1}{k^2 + k} = \frac{1}{k} - \frac{1}{k+1},$$

and the sum

$$\sum_{k=1}^{n} \frac{1}{k^2 + k} = \sum_{k=1}^{n} \left( \frac{1}{k} - \frac{1}{k+1} \right)$$

is in a telescoping form.

**EXAMPLE**    Evaluate the sum

$$\sum_{k=1}^{100} \frac{1}{k^2 + k}.$$

**Solution**    From the above discussion, we find

$$\sum_{k=1}^{100} \frac{1}{k^2 + k} = \sum_{k=1}^{100} \left( \frac{1}{k} - \frac{1}{k+1} \right) = 1 - \frac{1}{101} = \frac{100}{101}.$$

Observe that calculating the sum

$$\sum_{k=1}^{100} \frac{1}{k^2 + k} = \frac{1}{2} + \frac{1}{6} + \frac{1}{12} + \frac{1}{20} + \cdots + \frac{1}{10100}$$

is long and tedious.                                                     □

## 1    PROBLEMS

In Problems 1 through 8, find in each case the value of a given sum.

**1** $\displaystyle\sum_{i=1}^{12} i$

**2** $\displaystyle\sum_{i=3}^{7} (2i - 3)$

**3** $\displaystyle\sum_{i=-4}^{3} (i + 2)$

**4** $\displaystyle\sum_{k=3}^{7} b_k,$   given that $b_j = 2^j$ for every $j$.

**5** $\displaystyle\sum_{j=1}^{5} \frac{1}{j}$

**6** $\displaystyle\sum_{k=-4}^{-1} \frac{1}{1 + k^2}$

**7** $\displaystyle\sum_{j=1}^{10} (2^{j+1} - 2^j)$

**8** $\displaystyle\sum_{k=-1}^{2} \cos k\pi$

**9** Show that $\sum\limits_{k=1}^{8} (3k+2) = \sum\limits_{k=0}^{7} (3k+5)$.

**10** Find the value of $\sum\limits_{k=1}^{6} (a_{k+1} - a_k)$, given that $a_k = 10^k$.

**11** Find the value of $\sum\limits_{k=2}^{7} (a_k - a_{k+1})$, given that $a_k = \dfrac{1}{k}$.

In each of Problems 12 through 15, find the value of the given sum.

**12** $\sum\limits_{k=4}^{50} \dfrac{2}{k^2+k}$        **13** $\sum\limits_{k=1}^{70} \dfrac{-6}{k^2+2k}$

**14** $\sum\limits_{k=10}^{25} \dfrac{3}{k^2+3k}$        **15** $\sum\limits_{k=3}^{50} \dfrac{1}{k^2-k}$

**16** a) If $c$ is any real number, show that

$$\sum_{k=1}^{n} ca_k = c \sum_{k=1}^{n} a_k.$$

b) Show that $\sum\limits_{k=1}^{n} c = cn$.

**17** Let $p$ and $q$ be positive integers with $q \geq p$. Show that

$$\sum_{i=p}^{q} (-1)^i$$

always has the value 0, 1 or $-1$.

**18** Let $R$ be the triangle with vertices at $(0,0)$, $(1,0)$, $(0,1)$. Consider square meshes in the plane with side lengths $1/2^n$, $n = 1, 2, \dots$ . These give rise to $A_n^L$, $A_n^U$, $n = 1, 2, \dots$ for the region $R$. Compute the value of $A_3^L$ and $A_3^U$.

**19** Same as Problem 18 with $R$ the trapezoid with vertices at $(0,0)$, $(1,1)$, $(2,1)$, $(3,0)$.

**20** a) Let $R$ be the square with vertices at $(0,0)$, $(1,0)$, $(1,1)$, $(0,1)$. Show that for all $n$, $A_n^L = A_n^U = 1$.

b) Let $S$ be the square with vertices at $(\frac{1}{3},0)$, $(1\frac{1}{3},0)$, $(1\frac{1}{3},1)$, $(\frac{1}{3},1)$ and use the same meshes as in Part (a). Show that $A_n^L < A_n^U$ for all $n$ and that $A_n^L \to 1$, $A_n^U \to 1$ as $n \to \infty$.

**21** Given

$$a_n = 2 + \frac{n}{n+1}, \quad n = 1, 2, \dots .$$

Show that $a_{n+1} > a_n$ for all $n$. Find the number $b$ in Axiom C.

---

**2**

## THE DEFINITE INTEGRAL

The definite integral and the derivative are the two basic concepts of the calculus. We are now in a position to develop a precise definition of the integral and show its relationship to the antiderivative which we discussed in Section 11 of Chapter 4.

Suppose that $f$ is a function defined on an interval $[a, b]$. We make a *subdivision* of this interval by introducing $n - 1$ intermediate points; call them $x_1, x_2, x_3, \dots, x_{n-1}$ (Fig. 9). These points are not necessarily equally spaced. Letting $x_0 = a$ and $x_n = b$, we see that there are exactly $n$ subintervals. Since it is convenient to have a symbol for such a subdivision, we shall use the Greek letter $\Delta$ for this purpose. That is, $\Delta$ stands for the process of introducing $n - 1$ points between $a$ and $b$. The subdivision can be made in an infinite variety of ways for each value of $n$ and can also be made for every positive integer $n$. However, we will simply use the symbol $\Delta$ to indicate such a subdivision without attempting to describe the way in which it is made. We now introduce the important symbol $\Delta_i x$

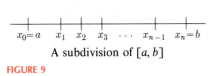

A subdivision of $[a, b]$

**FIGURE 9**

$$\Delta_i x = x_i - x_{i-1} = \text{length of } i\text{th subinterval in the subdivision } \Delta.$$

There are exactly $n$ subintervals, and among them there is always one which is largest. (Of course there may be several of equal size.) We introduce

$\|\Delta\| = $ length of largest subinterval in subdivision $\Delta$.

This quantity $\|\Delta\|$ is called the **norm** of the subdivision. When we want to know how "fine" a subdivision is, the norm $\|\Delta\|$, the maximum distance between points, is a measure of this "fineness." For example, if $a = 2$, $b = 5$, and $\|\Delta\| = 0.05$, we know that $n$ has to be at least 60. This reasoning holds because it takes 60 intervals of length 0.05 to add up to $b - a = 3$. Of course, $\|\Delta\|$ could be 0.05 and $n = 100$, with many points quite close together and only one or two intervals actually of length 0.05.

In each of the subintervals of a subdivision $\Delta$ we select a point. *Such a selection may be made in any way whatsoever.* Let $\xi_1$ be the point selected in $[x_0, x_1]$; let $\xi_2$ be the point in $[x_1, x_2]$, and, in general, let $\xi_k$ be the point in $[x_{k-1}, x_k]$. See Fig. 10.

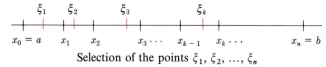

Selection of the points $\xi_1, \xi_2, ..., \xi_n$

**FIGURE 10**

We now form the sum

$$f(\xi_1)(x_1 - x_0) + f(\xi_2)(x_2 - x_1) + \cdots$$
$$+ f(\xi_k)(x_k - x_{k-1}) + \cdots + f(\xi_n)(x_n - x_{n-1}).$$

We can abbreviate this to

$$f(\xi_1)\Delta_1 x + f(\xi_2)\Delta_2 x + \cdots + f(\xi_k)\Delta_k x + \cdots + f(\xi_n)\Delta_n x,$$

and even further, by using the summation notation, to

$$\sum_{k=1}^{n} f(\xi_k)\Delta_k x.$$

This sum is called a **Riemann Sum**.

DEFINITION    *A function $f$ is said to be integrable on the interval $[a, b]$ if there is a number $A$ with the following property; for each $\varepsilon > 0$ there is a $\delta > 0$ such that*

$$\left| \sum_{k=1}^{n} f(\xi_k)\Delta_k x - A \right| < \varepsilon,$$

*for every subdivision $\Delta$ with $\|\Delta\| < \delta$ and for any choices of the $\xi_k$ in $[x_{k-1}, x_k]$.*

A simpler but less precise way of writing the above definition nevertheless gives the idea of what is happening:

$$\lim_{\|\Delta\| \to 0} \sum_{k=1}^{n} f(\xi_k)\Delta_k x = A.$$

The lack of precision comes about because, for any given value of the norm, $\|\Delta\|$, the $\Delta_k x$ can vary greatly and the $\xi_k$ can wander throughout the subinterval. This limiting process is quite different from the one we studied earlier.

It is natural to ask if there are several possible values for the number $A$ in any particular case. *It can be shown that if there is a number $A$ which satisfies the definition, then it is unique.* There cannot be two different values.

**DEFINITIONS**    *The number $A$ in the above definition is called* **the definite integral of $f$ from $a$ to $b$** *and is denoted by*

$$\int_a^b f(x)\,dx.$$

*The function $f$ is the* **integrand**, *and the numbers $a$ and $b$ are the* **lower** *and* **upper limits of integration**, *respectively. The letter $x$ is the* **variable of integration**. *This is a "dummy" variable and may be replaced by any other letter (although to prevent confusion we would avoid using, a, b, d, or f for the variable of integration).*

When does a function $f$ satisfy the conditions of the definition? To find out in any particular case, we would have to investigate all possible subdivisions with all possible choices for the $\xi_i$ and be sure that the same number $A$ is approached in every instance. The task is a hopeless one. It is imperative that we have some theorems which tell us when a function is integrable. We have seen that not all functions possess derivatives. If a function has a "corner" or a discontinuity at a point, it has no tangent line and therefore is not differentiable at that point. Similarly, we wish to know which of the various types of functions are integrable. Is a function with a corner integrable? Does a function with a "jump" discontinuity satisfy the definition of integrability?

In order to gain further insight into the process of integration we use a method of selecting special sums of the type which appear in the definition of integral. Suppose that $f$ is a continuous function on a closed interval $[a, b]$. The first theorem in Chapter 4 states that there is a place in $[a, b]$ where $f$ has a largest value and another place where $f$ has a smallest value (Extreme Value Theorem). Starting with this fact, we make a subdivision $\Delta$ of the interval $[a, b]$. That is,

$$\Delta \text{ is } \{a = x_0 < x_1 < x_2 < \cdots < x_{n-1} < x_n = b\}.$$

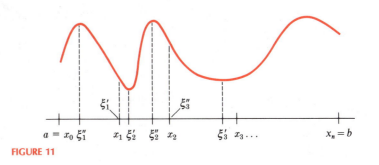

**FIGURE 11**

Since $f$ is continuous on $[x_0, x_1]$, there is a point on this subinterval, call it $\xi'_1$, where $f$ has a minimum value and a point, call it $\xi''_1$, where $f$ has a maximum value. On $[x_1, x_2]$ $f$ is continuous. Let $\xi'_2$ and $\xi''_2$ be the places where $f$ has minimum and maximum values, respectively. See Fig. 11. Continue this process. The values $\xi'_k$, $\xi''_k$ are respectively the places where $f$ has minimum and maximum values on the interval $[x_{k-1}, x_k]$. In this way corresponding to the subdivision $\Delta$ we get *two* sums:

$$\sum_{k=1}^{n} f(\xi'_k)\, \Delta_k x \qquad \text{and} \qquad \sum_{k=1}^{n} f(\xi''_k)\, \Delta_k x.$$

We abbreviate this even further by writing

$$\underline{S}(\Delta) = \sum_{k=1}^{n} f(\xi'_k)\, \Delta_k x, \qquad \overline{S}(\Delta) = \sum_{k=1}^{n} f(\xi''_k)\, \Delta_k x.$$

From our method of selection we know that $\underline{S}(\Delta) \le \overline{S}(\Delta)$, always. But we know even more: since in each subinterval the function $f$ is always between its minimum and maximum, then for *any* choice $\xi_k$ in a particular subdivision $\Delta$ we always have

$$\underline{S}(\Delta) \le \sum_{k=1}^{n} f(\xi_k)\, \Delta_k x \le \overline{S}(\Delta). \tag{1}$$

That is, the Riemann Sum is always between $\underline{S}(\Delta)$ and $\overline{S}(\Delta)$.

If we can show that $\underline{S}(\Delta)$ and $\overline{S}(\Delta)$ tend to the same limit, $A$, for all possible $\Delta$, with $\|\Delta\| \to 0$, then the same will be true for every possible choice of the $\xi_i$ and therefore for all subdivisions. The function $f$ will satisfy the definition of integrability.

**EXAMPLE**   Given that $f(x) = x^2$ and given the subdivision

$$\Delta: \{a = x_0 = -1, x_1 = -0.6, x_2 = -0.2, x_3 = 0.3, x_4 = 0.7, x_5 = b = 1\},$$

find $\underline{S}(\Delta)$ and $\overline{S}(\Delta)$.

**Solution**   We construct the graph shown in Fig. 12, and the corresponding table:

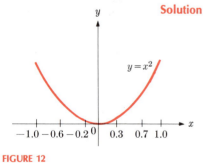

FIGURE 12

| | $i$ | 1 | 2 | 3 | 4 | 5 |
|---|---|---|---|---|---|---|
| $\Delta_i x = x_i - x_{i-1}$ | $\Delta_i x$ | 0.4 | 0.4 | 0.5 | 0.4 | 0.3 |
| Place where minimum occurs | $\xi'_i$ | −0.6 | −0.2 | 0 | 0.3 | 0.7 |
| Minimum value | $f(\xi'_i)$ | 0.36 | 0.04 | 0 | 0.09 | 0.49 |
| Place where maximum occurs | $\xi''_i$ | −1 | −0.6 | 0.3 | 0.7 | 1 |
| Maximum value | $f(\xi''_i)$ | 1 | 0.36 | 0.09 | 0.49 | 1 |

Thus we obtain

$$\underline{S}(\Delta) = \sum_{i=1}^{5} f(\xi_i') \, \Delta_i x$$

$$= 0.36(0.4) + 0.04(0.4) + 0(0.5) + 0.09(0.4) + 0.49(0.3)$$

$$= 0.343;$$

$$\overline{S}(\Delta) = \sum_{i=1}^{5} f(\xi_i'') \, \Delta_i x$$

$$= 1(0.4) + 0.36(0.4) + 0.09(0.5) + (0.49)(0.4) + 1(0.3)$$

$$= 1.085. \qquad \square$$

We now sketch a proof of a theorem which indicates when certain functions are integrable.

**THEOREM 1**   *If $f(x)$ is defined and increasing (or at least nondecreasing) on the closed interval $a \le x \le b$, then it is integrable there.*

**Proof**   Let $\Delta$ be any subdivision. On each subinterval $[x_{i-1}, x_i]$ the minimum must occur at the left endpoint and the maximum must occur at the right endpoint. That is, $\xi_i' = x_{i-1}$ and $\xi_i'' = x_i$. We construct the sums

$$\underline{S}(\Delta) = \sum_{i=1}^{n} f(x_{i-1}) \, \Delta_i x, \qquad \overline{S}(\Delta) = \sum_{i=1}^{n} f(x_i) \, \Delta_i x.$$

We know that $\underline{S}(\Delta) \le \overline{S}(\Delta)$. We now subtract:

$$\overline{S}(\Delta) - \underline{S}(\Delta) = \sum_{i=1}^{n} [f(x_i) - f(x_{i-1})] \Delta_i x.$$

Each $\Delta_i x$ is positive and the quantities $f(x_i) - f(x_{i-1})$ are all positive (or at least nonnegative). If we replace each $\Delta_i x$ by its largest possible value, the right side becomes larger, and we get the inequality

$$\overline{S}(\Delta) - \underline{S}(\Delta) \le \sum_{i=1}^{n} [f(x_i) - f(x_{i-1})] \|\Delta\|.$$

The quantity $\|\Delta\|$, by definition, stands for the largest $\Delta_i x$. We now note that the terms on the right form a telescoping sum and we have:

$$\overline{S}(\Delta) - \underline{S}(\Delta) \le [f(x_n) - f(x_0)] \|\Delta\| = [f(b) - f(a)] \|\Delta\|.$$

Then, given any $\varepsilon > 0$, if we select a subdivision with

$$\|\Delta\| < \delta = \frac{\varepsilon}{f(b) - f(a)}, \quad \text{we have } \overline{S}(\Delta) - \underline{S}(\Delta) < \varepsilon.$$

From the definition of area as given in Section 1 and inequality (1) on page 190, it is possible to show that there is a limit $A$ to which both $\overline{S}(\Delta)$ and $\underline{S}(\Delta)$ tend. Furthermore, every intermediate sum also tends to $A$. The function $f$ is integrable. $\qquad \square$

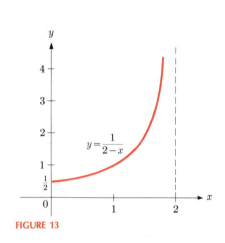

**FIGURE 13**

Theorem 1 illustrates how careful we must be in stating the hypotheses of a theorem. To make the sloppy statement that an increasing function in an interval is integrable is to make a false statement. For example, consider the function

$$f(x) = \frac{1}{2-x}$$

in the interval $0 \le x < 2$. (See Fig. 13.) This function is increasing everywhere from 0 to 2, and it tends to infinity as $x \to 2$. This function is not integrable on $[0, 2]$. On the other hand, Theorem 1 tells us that the function is integrable on any interval $[0, c]$ if $c < 2$.

In Theorem 1, the hypothesis that $f$ is increasing is a fairly restrictive one. Actually functions are integrable under much more general hypotheses. The following theorem, which we state without proof, is an example of the type of result which establishes the integrability of a large class of functions— namely, those which are continuous on a *closed* interval.

**THEOREM 2**    *If $f$ is continuous on $[a, b]$, then it is integrable on $[a, b]$.*

The proof of this theorem depends on the notion of *uniform continuity*, a concept usually taken up in more advanced courses in analysis. However, the idea behind the proof can be discussed here. As in the discussion of the preceding theorem, we consider

$$\bar{S}(\Delta) - \underline{S}(\Delta) = \sum_{i=1}^{n} \left[ f(\xi_i'') - f(\xi_i') \right] \Delta_i x.$$

We apply the "triangle inequality" for absolute values: $(|a + b| \le |a| + |b|)$ and, since $|\Delta_i x| = \Delta_i x$, we can write

$$\bar{S}(\Delta) - \underline{S}(\Delta) \le \sum_{i=1}^{n} |f(\xi_i'') - f(\xi_i')| \cdot \Delta_i x.$$

**Uniform continuity on an interval** $[a, b]$ means that, given any $\varepsilon > 0$, there is a $\delta > 0$ such that

$$|f(x') - f(x'')| < \varepsilon$$

for *any* two points in $[a, b]$ such that $|x' - x''| < \delta$. This concept is used to establish the inequality

$$\bar{S}(\Delta) - \underline{S}(\Delta) \le \sum_{i=1}^{n} \varepsilon \Delta_i x = \varepsilon \sum_{i=1}^{n} \Delta_i x = \varepsilon(b - a),$$

from which it follows that $f$ is integrable.

## 2  PROBLEMS

In Problems 1 through 10, a function $f$ and a subdivision are given. Sketch the graph of the function and find the value of $\underline{S}(\Delta)$ and $\bar{S}(\Delta)$. Use knowledge of $f'(x)$ when necessary.

**1** $f(x) = x^2 + 1$;   $\Delta: a = x_0 = 1$, $x_1 = 1.2$, $x_2 = 1.5$, $x_3 = 1.6$, $x_4 = 2$, $x_5 = 2.5$, $x_6 = b = 3$.

**2** $f(x) = 1/(1 + x)$;   $\Delta: a = x_0 = 0$, $x_1 = 0.2$, $x_2 = 0.4$, $x_3 = 0.6$, $x_4 = 0.8$, $x_5 = b = 1$.

**3** $f(x) = x^2 - x + 1$;   $\Delta: a = x_0 = 0$, $x_1 = 0.1$, $x_2 = 0.3$, $x_3 = 0.5$, $x_4 = 0.7$, $x_5 = 0.8$, $x_6 = 0.9$, $x_7 = b = 1$.

4 $f(x) = 1/(1 + x^2)$;  $\Delta: a = x_0 = -1, x_1 = -0.9, x_2 = -0.5,$
$x_3 = -0.2, x_4 = 0, x_5 = 0.2, x_6 = 0.3, x_7 = 0.7, x_8 = b = 1.$

5 $f(x) = x^3 + x$;  $\Delta: a = x_0 = 0, x_1 = 0.1, x_2 = 0.2, x_3 = 0.3,$
$x_4 = 0.5, x_5 = b = 0.6.$

6 $f(x) = x^2 + 2x - 1$;  $\Delta: a = x_0 = -2, x_1 = -1.8, x_2 = -1.6, x_3 = -1.4, x_4 = -1.2, x_5 = b = -1.$

7 $f(x) = (x^2 - 2)/(x + 1)$;  $\Delta: a = x_0 = 0, x_1 = 0.5, x_2 = 1,$
$x_3 = 1.5, x_4 = 2, x_5 = 2.5, x_6 = 3, x_7 = b = 3.5.$

8 $f(x) = x/(x^2 + 1)$;  $\Delta: a = x_0 = 0, x_1 = 1, x_2 = 2, x_3 = 3,$
$x_4 = 4, x_5 = 4.5, x_6 = 5, x_7 = b = 6.$

9 $f(x) = x^3 - 3x + 1$;  $\Delta: a = x_0 = -2, x_1 = -1.5, x_2 = -1,$
$x_3 = -0.5, x_4 = 0, x_5 = 0.5, x_6 = 1, x_7 = 1.5, x_8 = b = 2.$

10 $f(x) = x^3/(x^3 + 1)$;  $\Delta: a = x_0 = 0, x_1 = 1, x_2 = 2, x_3 = 3,$
$x_4 = b = 4.$

11 Prove that if a function $f$ is decreasing (or nonincreasing) on a closed interval $[a, b]$ it is integrable.

12 Suppose that a function $f$ is differentiable on a closed interval $[a, b]$. If $f$ has only a finite number of maximum and minimum points on $[a, b]$, use the results of Theorem 1 and Problem 11 to show that $f$ is integrable there.

*13 Show that every polynomial function is integrable on any interval $[a, b]$. (*Hint:* Use the result in Problem 12.)

14 Why is the "postage" function (see Chapter 1, Section 5) integrable even though it is not continuous? Find the value of the integral from $x = 0$ to $x = 4$.

15 Suppose that $f$ and $f'$ are continuous on $[a, b]$ and that $|f'(x)| \le M$ on $[a, b]$. Use the Mean Value Theorem to show that for any subdivision $\Delta$,

$$\bar{S}(\Delta) - \underline{S}(\Delta) \le M(b - a) \cdot \|\Delta\|.$$

16 In what intervals is the function $f(x) = -1/x$ integrable?

17 Let $\Delta$ be a subdivision of the interval $[0, 1]$ into $n$ equal parts. Compute $\underline{S}(\Delta)$ for $f(x) = 1/x^2$. What happens to $\underline{S}(\Delta)$ as $n \to \infty$? What can be said about $\bar{S}(\Delta)$ and therefore about the integrability of $f$?

*18 Use mathematical induction to show that

$$\sum_{k=1}^{n} k = \tfrac{1}{2} n(n + 1).$$

*19 Use mathematical induction to show that

$$\sum_{k=1}^{n} k^2 = \tfrac{1}{6} n(n + 1)(2n + 1).$$

*20 Given $f(x) = x^2$,  $0 \le x \le 1$. Divide $[0, 1]$ into $n$ equal parts and form a Riemann Sum for $f$. Use the results of Problems 18 and 19 to find the integral of $f$ on $[0, 1]$ by letting $n \to \infty$.

---

## 3

## THE DEFINITE INTEGRAL AND AREA

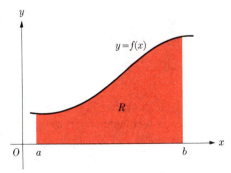

FIGURE 14

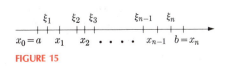

FIGURE 15

Suppose that $y = f(x)$ is a function which happens to lie above the $x$ axis, as shown in Fig. 14. At two points on the $x$ axis, $a$ and $b$, vertical lines are drawn. We shall concern ourselves with the problem of calculating the area of the region (denoted $R$) bounded by the $x$ axis, the vertical lines through $a$ and $b$, and the graph of the function $f$. The fact that three sides of the region are straight lines simplifies the problem. At this stage we assume that the only kind of region for which we can calculate the area is a rectangle. The problem seems fairly hopeless until we realize that we can get an *approximate* idea of how large the area is in the following way.

We make a subdivision of the interval $[a, b]$

$$\Delta = \{a = x_0 < x_1 < x_2 < \cdots < x_n = b\}$$

as shown in Fig. 15. In each subinterval $[x_{i-1}, x_i]$ we choose a point $\xi_i$ and form the Riemann Sum

$$f(\xi_1)\Delta_1 x + f(\xi_2)\Delta_2 x + \cdots + f(\xi_n)\Delta_n x = \sum_{i=1}^{n} f(\xi_i)\Delta_i x$$

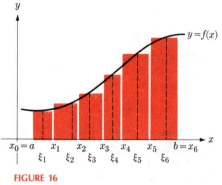

**FIGURE 16**

where we recall that $\Delta_i x = x_i - x_{i-1}$. As we note in Fig. 16 each term $f(\xi_i)\,\Delta_i x$ in the above sum is the area of the shaded rectangle of width $\Delta_i x$ and height $f(\xi_i)$. We observe that the sum of all these shaded areas is an approximation to the area of $R$.

Figure 16 shows an approximation for six intervals of subdivision. It is natural to expect that if we take 60 intervals of subdivision, each of them much narrower than the ones shown, the approximation to the area would be much better. And if we take 600 subintervals, it would be better still. We recognize that if the function $f(x)$ represents a smooth curve then in the limit as $n \to \infty$ we get

$$\int_a^b f(x)\,dx,$$

which measures the area of the region $R$. We formalize the above discussion into the next theorem.

**THEOREM 3**   *If $f(x)$ is a continuous nonnegative function on $[a, b]$, then the area $A$ of the region bounded by the x axis, the lines $x = a$, $x = b$ and the curve $y = f(x)$ is given by*

$$A = \int_a^b f(x)\,dx.$$

The concepts of definite integral and area under a curve have an intimate connection with the idea of derivative. Consider the problem of finding the area under the curve $y = f(x)$ between the points $a$ and $X$, where $X$ is between $a$ and $b$, as shown in Fig. 17. We would write this area as the definite integral

$$\int_a^X f(x)\,dx.$$

As $X$ changes, this area assumes different values. If $X$ is $a$, the value is zero. Let's call the area $A$ and, since $A$ is a function of $X$, we write

$$A = F(X).$$

We now select an $h > 0$ and compute $F(X + h)$. That is, we calculate the area under the curve between $a$ and $X + h$. This is just the integral

$$\int_a^{X+h} f(x)\,dx.$$

The difference $F(X + h) - F(X)$ is the area of the shaded region shown in Fig. 17. If $h$ is "small," the shaded region is almost a rectangle and, if we let $Y$ be some average value of the function $f(x)$ between $X$ and $X + h$, we could say that

$$F(X + h) - F(X) = Y \cdot h \qquad \text{or} \qquad Y = \frac{F(X + h) - F(X)}{h}.$$

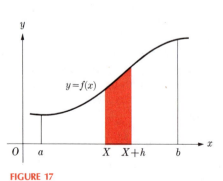

**FIGURE 17**

If $h \to 0$, we know from our earlier experience that the limit approached is the derivative $F'(X)$. On the other hand, "geometrically" we see that as $h \to 0$, the average height $Y$ must approach the height of the function $f$ at the point $X$, namely $f(X)$. We conclude by this "argument" that

$$F'(X) = f(X),$$

which states that the function $f$ is the derivative of the area function $F$. We recognize $F$ as the antiderivative of the function $f$.

The above informal treatment of the relationship between the derivative and integral will be made precise when we discuss the Fundamental Theorem of Calculus in Section 5.

**EXAMPLE**   Approximate, by rectangles, the area under the curve $f(x) = x^2$ if $a = 0, b = 1$, the number of rectangles is 5, the subintervals are all equal in size, and the $\xi_i$ are all taken at the midpoints of the subintervals. Draw a graph. Compute the exact area by the antiderivative method.

**Solution**   Each $\Delta_i x = 0.2$, and $\xi_1 = 0.1$, $\xi_2 = 0.3$, $\xi_3 = 0.5$, $\xi_4 = 0.7$, $\xi_5 = 0.9$. We have (Fig. 18)

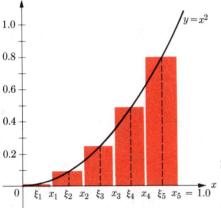

$$f(\xi_1)\,\Delta_1 x = 0.002$$
$$f(\xi_2)\,\Delta_2 x = 0.018$$
$$f(\xi_3)\,\Delta_3 x = 0.050$$
$$f(\xi_4)\,\Delta_4 x = 0.098$$
$$f(\xi_5)\,\Delta_5 x = 0.162$$
$$\text{Sum} = 0.330$$

Since $F'(x) = f(x) = x^2$, we see from our knowledge of antiderivatives that

$$F(x) = \tfrac{1}{3}x^3 + c,$$

**FIGURE 18**

which gives the area under the curve at any point $x$. We must, however, use the fact that we are starting at the point $a$. This is equivalent to the statement that $F(a) = 0$. In our case $a = 0$, and so $F(0) = 0$. Substituting in the above expression for $F(x)$, we get $c = 0$. The area we seek extends from $a = 0$ to $b = 1$; substituting $x = 1$ in the formula for $F(x) = \tfrac{1}{3}x^3$, we obtain

$$F(1) = \tfrac{1}{3} = \text{area under curve.} \qquad \square$$

## 3   PROBLEMS

In Problems 1 through 4, approximate by rectangles the area under the curve $y = f(x)$ from the data given. Then compute each term to three decimal places as in the above example. Sketch a graph.

**1**  $f(x) = 2x + 3$, $n = 5$, $a = -1$, $b = 0$.

| $i$ | 0 | 1 | 2 | 3 | 4 | 5 |
|-----|-----|-----|-----|-----|-----|-----|
| $x_i$ | $-1$ | $-0.8$ | $-0.6$ | $-0.4$ | $-0.2$ | 0 |
| $\xi_i$ | | $-1.0$ | $-0.8$ | $-0.5$ | $-0.2$ | 0 |

(In this problem, compute the exact area.)

**2**  $f(x) = x^2$, $n = 5$, $a = -1$, $b = 0$

| $i$ | 0 | 1 | 2 | 3 | 4 | 5 |
|-----|-----|-----|-----|-----|-----|-----|
| $x_i$ | $-1$ | $-0.8$ | $-0.6$ | $-0.4$ | $-0.2$ | 0 |
| $\xi_i$ | | $-1.00$ | $-0.75$ | $-0.50$ | $-0.25$ | 0 |

**3**  $f(x) = x^2$, $n = 7$, $a = -2.0$, $b = 0$

| $i$ | 0 | 1 | 2 | 3 | 4 | 5 | 6 | 7 |
|-----|-----|-----|-----|-----|-----|-----|-----|-----|
| $x_i$ | $-2$ | $-1.8$ | $-1.6$ | $-1.4$ | $-1.2$ | $-0.8$ | $-0.4$ | 0 |
| $\xi_i$ | | $-2.0$ | $-1.6$ | $-1.5$ | $-1.3$ | $-1.0$ | $-0.7$ | $-0.1$ |

**4** $f(x) = 2x - x^2$, $n = 5$, $a = 0$, $b = 2$

| $i$ | 0 | 1 | 2 | 3 | 4 | 5 |
|---|---|---|---|---|---|---|
| $x_i$ | 0 | 0.4 | 0.8 | 1.2 | 1.6 | 2.0 |
| $\xi_i$ | | 0 | 0.5 | 1.0 | 1.5 | 2.0 |

In Problems 5 through 15, find the exact area under the curve $y = f(x)$ from $a$ to $b$ by finding $F(x)$, the area function, and then $F(b)$.

**5** $f(x) = x^2$,   $a = -1$, $b = 0$

**6** $f(x) = x^2$,   $a = -2$, $b = 0$

**7** $f(x) = 2x - x^2$,   $a = 0$, $b = 2$

**8** $f(x) = x^2 - 2x + 2$,   $a = 1$, $b = 3$

**9** $f(x) = 3x - x^2$,   $a = 1$, $b = 2$

**10** $f(x) = x^2 + x + 1$,   $a = -2$, $b = 1$

**11** $f(x) = 6 + x - x^2$,   $a = -2$, $b = -1$

**12** $f(x) = \cos x$,   $a = 0$, $b = \pi/2$

**13** $f(x) = \sin x$,   $a = 0$, $b = \pi$

**14** $f(x) = \dfrac{1}{2}\sin 2x$,   $a = \dfrac{\pi}{6}$, $b = \dfrac{\pi}{2}$

**15** $f(x) = 2\cos 2x$,   $a = -\dfrac{\pi}{4}$, $b = \dfrac{\pi}{6}$

In Problems 16 through 19, approximate the area by rectangles with the number $n$ given. Take the $\xi_i$ at the midpoints.

**16** $f(x) = x + 1$,   $a = 1$, $b = 3$, $n = 1$

**17** $f(x) = x^2$,   $a = -2$, $b = -1$, $n = 5$

**18** $f(x) = \dfrac{1}{1 + x}$,   $a = 0$, $b = 1$, $n = 5$

**19** $f(x) = \dfrac{1}{1 + x^2}$,   $a = 0$, $b = 1$, $n = 5$

**20** Compute the exact area in Problems 16 and 17 by using the antiderivative method.

---

**4**

## PROPERTIES OF THE DEFINITE INTEGRAL

Later we shall develop techniques for evaluating integrals without recourse to the definition. This is analogous to what we did for derivatives: first we showed how to find the derivative directly from the definition, and then we developed methods for differentiation which bypassed the definition entirely. The properties we shall now discuss will be useful in learning methods for integrating various kinds of functions.

The simplest properties are given in the two following theorems.

**THEOREM 4**   *If $c$ is any number and $f$ is integrable on $[a, b]$, then the function $cf(x)$ is integrable on $[a, b]$ and*

$$\int_a^b cf(x)\, dx = c \int_a^b f(x)\, dx.$$

**THEOREM 5**   *If $f(x)$ and $g(x)$ are integrable on $[a, b]$, then $f(x) + g(x)$ is integrable on $[a, b]$ and*

$$\int_a^b [f(x) + g(x)]\, dx = \int_a^b f(x)\, dx + \int_a^b g(x)\, dx.$$

We prove Theorem 5.

**Proof** If $M$ and $N$ are any numbers, we know that $|M + N| \leq |M| + |N|$. Let $\Delta$ be any subdivision and take

$$M = \sum_{i=1}^{n} f(\xi_i)\,\Delta_i x - \int_a^b f(x)\,dx, \qquad N = \sum_{i=1}^{n} g(\xi_i)\,\Delta_i x - \int_a^b g(x)\,dx.$$

From the definition of integrability, we have $|M| < \varepsilon$ and $|N| < \varepsilon$ if $\|\Delta\|$ is sufficiently small. But

$$M + N = \sum_{i=1}^{n} [f(\xi_i) + g(\xi_i)]\,\Delta_i x - \left[ \int_a^b f(x)\,dx + \int_a^b g(x)\,dx \right].$$

We conclude that $|M + N| < 2\varepsilon$. This means that $M + N$ tends to zero as $\|\Delta\|$ tends to zero and, therefore, the function $f + g$ is integrable and its integral is equal to the sum of the integrals of $f$ and $g$. $\qquad\square$

The proof of Theorem 4 is similar.

---

**THEOREM 6** *If $f(x)$ is integrable on an interval $[a, b]$, then $f$ is bounded there.*

This theorem is proved in more advanced courses in mathematics. The theorem means that if a function is integrable there must be two numbers $m$ and $M$ ($m$ may be negative and $M$ may be very large, positive) such that for all values $x$ in $[a, b]$, $f(x)$ lies between $m$ and $M$. That is,

$$m \leq f(x) \leq M.$$

Theorem 5 is of considerable importance in developing the theory of integration. For example, the function $f(x) = 1/(1 - x)$ is unbounded on any interval which contains $x = 1$ as an endpoint or an interior point. Then Theorem 6 tells us that $f$ cannot be integrated over such an interval.

The next theorem gives a specific method for estimating the value of an integral.

---

**THEOREM 7** *If $f(x)$ is integrable on an interval $[a, b]$ and $m$ and $M$ are numbers such that*

$$m \leq f(x) \leq M \qquad \text{for } a \leq x \leq b,$$

*then it follows that*

$$m(b - a) \leq \int_a^b f(x)\,dx \leq M(b - a).$$

---

**Proof** Let $\Delta$ be a subdivision of $[a, b]$. We know that the Riemann Sum

$$\sum_{i=1}^{n} f(\xi_i)\,\Delta_i x$$

tends to the integral if $x_{i-1} \leq \xi_i \leq x_i$ and if $\|\Delta\| \to 0$ as $n \to \infty$. By hypothesis, $m \leq f(\xi_i) \leq M$ for each $i$. Multiplying through by $\Delta_i x$ yields

$$m\,\Delta_i x \leq f(\xi_i)\,\Delta_i x \leq M\,\Delta_i x.$$

**THE DEFINITE INTEGRAL**

Summing these inequalities from 1 to $n$, we get

$$\sum_{i=1}^{n} m \, \Delta_i x \leq \sum_{i=1}^{n} f(\xi_i) \, \Delta_i x \leq \sum_{i=1}^{n} M \, \Delta_i x.$$

On the left and the right we can factor out the $m$ and $M$, respectively, to find

$$m \sum_{i=1}^{n} \Delta_i x \leq \sum_{i=1}^{n} f(\xi_i) \, \Delta_i x \leq M \sum_{i=1}^{n} \Delta_i x.$$

But

$$\sum_{i=1}^{n} \Delta_i x$$

"telescopes" to give $b - a$. Therefore

$$m(b - a) \leq \sum_{i=1}^{n} f(\xi_i) \, \Delta_i x \leq M(b - a).$$

The term in the middle tends to the integral and, since the inequalities hold in the limit, the result follows. □

**THEOREM 8**  *If $f$ and $g$ are integrable on the interval $[a, b]$ and $f(x) \leq g(x)$ for each $x$ in $[a, b]$, then*

$$\int_{a}^{b} f(x) \, dx \leq \int_{a}^{b} g(x) \, dx.$$

**Proof**  The proof furnishes an excellent illustration of the way in which mathematics is composed of building blocks. We show how earlier theorems form the blocks which produce the result. We define $F(x) = g(x) - f(x)$. Then by hypothesis $F(x) \geq 0$ on $[a, b]$. By Theorem 4, if $f(x)$ is integrable, so is $-f(x), (c = -1)$. According to Theorem 5, if $g(x)$ and $-f(x)$ are integrable, so is $g(x) + (-f(x)) = F(x)$. We recall that Theorem 7 tells us that

$$m(b - a) \leq \int_{a}^{b} F(x) \, dx$$

and, since $F(x) \geq 0$ on $[a, b]$, we can select $m = 0$. Therefore

$$\int_{a}^{b} [g(x) - f(x)] \, dx \geq 0,$$

or

$$\int_{a}^{b} g(x) \, dx - \int_{a}^{b} f(x) \, dx \geq 0 \quad \text{(by Theorem 5)},$$

or

$$\int_{a}^{b} g(x) \, dx \geq \int_{a}^{b} f(x) \, dx,$$

which is what we wanted to prove. □

**THEOREM 9**    *If $f(x)$ is continuous on the interval $[a, b]$ and $c$ is any number in $[a, b]$, then*

$$\int_a^b f(x)\, dx = \int_a^c f(x)\, dx + \int_c^b f(x)\, dx.$$

**Sketch of proof**    Consider a subdivision $\Delta$ of the interval $[a, b]$. This subdivision may or may not contain $c$ as one of its points $x_i$. If it does not, we insert $c$ into it and obtain a new subdivision $\Delta'$, with one more point and one more interval. The new subdivision has a norm $\|\Delta'\|$ which is smaller than or the same as the norm $\|\Delta\|$; at least it is not larger. Then the part of the subdivision of $\Delta'$ from $a$ to $c$ gives a Riemann Sum of the form

$$\sum_{i=1}^{n'} f(\xi_i)\, \Delta_i x,$$

where $n'$ is the number of subintervals in $[a, c]$; similarly, if $n''$ is the number of subintervals from $c$ to $b$ there will be a Riemann Sum of the form

$$\sum_{i=1}^{n''} f(\xi_i)\, \Delta_i x$$

for the interval $[c, b]$. The first sum tends to $\int_a^c f(x)\, dx$, the second sum tends to $\int_c^b f(x)\, dx$, and the two added together tend to $\int_a^b f(x)\, dx$. This is exactly the statement of the result of the theorem.    $\square$

Two useful definitions,

$$\int_a^a f(x)\, dx = 0 \qquad \text{and} \qquad \int_a^b f(x)\, dx = -\int_b^a f(x)\, dx,$$

imply that the relation

$$\int_a^b f(x)\, dx = \int_a^c f(x)\, dx + \int_c^b f(x)\, dx$$

holds for any value of $c$ whether or not it is between $a$ and $b$. If $c = b$ the result is clear, while if $c > b$ or $c < a$, the appropriate reversal of sign reduces the result to the one stated in Theorem 9.

## 4   PROBLEMS

In Problems 1 through 18, in each case study the function in the given interval and apply Theorem 7 to find largest and smallest values the integral can have.

1   $\displaystyle\int_1^4 2x\, dx$

2   $\displaystyle\int_0^3 x^2\, dx$

3   $\displaystyle\int_{-1}^2 x^3\, dx$

4   $\displaystyle\int_0^5 \frac{2}{1 + 2x}\, dx$

5   $\displaystyle\int_3^4 \frac{x}{1 + x}\, dx$

6   $\displaystyle\int_{-2}^2 \frac{1}{1 + x^2}\, dx$

7   $\displaystyle\int_0^4 \frac{x^2}{1 + x^2}\, dx$

8   $\displaystyle\int_{-3}^{-1} \frac{1}{1 - x}\, dx$

9   $\displaystyle\int_2^5 \sqrt{1 + x}\, dx$

10   $\displaystyle\int_0^3 \sqrt{x^2 + 1}\, dx$

11   $\displaystyle\int_0^4 \frac{2\sqrt{x}}{1 + x}\, dx$

12   $\displaystyle\int_1^2 \frac{1 - x^2}{1 + x^2}\, dx$

13   $\displaystyle\int_{-2}^3 \frac{2 + x}{\sqrt{1 + x^2}}\, dx$

14   $\displaystyle\int_0^{\pi/2} \sin x\, dx$

**15** $\displaystyle\int_0^{\pi/4} \tan x \, dx$

**16** $\displaystyle\int_0^{\pi} \cos 2x \, dx$

**17** $\displaystyle\int_0^{\pi/2} \sin^2 \tfrac{1}{3}x \, dx$

**18** $\displaystyle\int_0^{\pi/4} \sec x \, dx$

**19** Prove Theorem 3.

**20** Assuming that $\int_a^b x^n \, dx$ is integrable for every nonnegative integer $n$, use Theorems 3 and 4 to write out a proof that every polynomial is integrable on every finite interval $[a, b]$.

**21** Given the function

$$f(x) = \begin{cases} x, & -1 \leq x \leq 0, \\ x^2 - x + 1, & 0 < x \leq 1, \end{cases}$$

show, by combining Theorems 3 and 4, that $f$ is integrable on $[-1, 1]$.

In each of Problems 22 through 29, decide on what interval or intervals the given function is integrable. Use Theorem 6.

**22** $f(x) = \dfrac{1}{x^2 - 3x + 2}$

**23** $f(x) = \tan x$

**24** $f(x) = \cos x$

**25** $f(x) = \sec x$

**26** $f(x) = \dfrac{1+x}{x^2 + 2x + 5}$

**27** $f(x) = -\dfrac{1}{\sqrt{x^2 + x + 4}}$

**28** $f(x) = \dfrac{x^2 + 1}{x^3 - 2x^2 - x + 2}$

**29** $f(x) = \dfrac{x^3 + 1}{x}$

**30** Write out the details of the proof that

$$\int_a^b f(x) \, dx = \int_a^c f(x) \, dx + \int_c^b f(x) \, dx$$

when $c$ is not between $a$ and $b$.

**31** Let $c_1, c_2, \ldots, c_k$ be any real numbers. Show that

$$\int_{c_1}^{c_k} f(x) \, dx = \sum_{i=1}^{k-1} \int_{c_i}^{c_{i+1}} f(x) \, dx.$$

**32** Let $a_0 < a_1 < \cdots < a_k$ be any $k+1$ real numbers. Let $P_i(x)$, $i = 1, 2, \ldots, k$ be any polynomial functions. Define $F(x) = P_i(x)$ for $a_{i-1} \leq x < a_i$. Show that $F$ is integrable on the interval $[a_0, a_k]$. (See Problems 20 and 21.)

---

**5**

# EVALUATION OF DEFINITE INTEGRALS. FUNDAMENTAL THEOREM OF CALCULUS

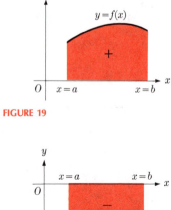

FIGURE 19

FIGURE 20

It is apparent from the work in Section 3, that the definite integral is intimately connected with area. In fact, Theorem 3 states that *if $f(x) \geq 0$ on an interval $[a, b]$ and if it is integrable there, then*

$$\int_a^b f(x) \, dx$$

*is the area bounded by the lines $x = a$, $x = b$, the $x$ axis, and the curve of $y = f(x)$* (shaded region in Fig. 19). If $f(x)$ is entirely *below* the $x$ axis, then the sums

$$\sum_{i=1}^n f(\xi_i) \, \Delta_i x$$

are always *negative*. If $f$ is integrable on $[a, b]$, this integral will have a *negative value*. The value of

$$\int_a^b f(x) \, dx$$

will be *the negative of the area* bounded by the lines $x = a$, $x = b$, the $x$ axis, and the curve of $y = f(x)$ (shaded region in Fig. 20). If a function $f$ is partly

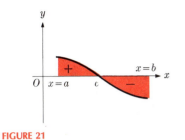

**FIGURE 21**

positive and partly negative on $[a, b]$ so that there is a point $c$ where it crosses the $x$ axis (see Fig. 21), the value of

$$\int_a^b f(x)\, dx$$

is the difference between the area above the $x$ axis and the area below it.

> To obtain the true area of a region such as the shaded one shown in Fig. 21, we must first find the point $c$ where the function $f$ crosses the $x$ axis and compute
>
> $$\int_a^c f(x)\, dx \qquad \text{and} \qquad \int_c^b f(x)\, dx$$
>
> separately. The value of the first integral is positive and that of the second negative. We add the absolute value of the second integral to the value of the first, and the resulting quantity is the total area.

The next theorem we shall establish, known as the **Fundamental Theorem of Calculus**, is basic for actually finding the value of a definite integral. In Section 3, we saw in an intuitive way the relationship between derivative and integral. The integral was interpreted as an area (we considered positive functions only) and the derivative of the integral turned out to be the integrand. Now we establish the relation between derivative and integral in an analytic manner.

**THEOREM 10**     **(Fundamental Theorem of Calculus, first form)**     *Suppose that F is continuous and that F′ is integrable on the interval $[a, b]$. Then*

$$\int_a^b F'(x)\, dx = F(b) - F(a).$$

**Proof**     We make a subdivision

$$\Delta: \{a = x_0 < x_1 < x_2 < \cdots < x_{n-1} < x_n = b\}$$

of the interval $[a, b]$. We assemble the following three facts:

1)  $F(b) - F(a) = \displaystyle\sum_{i=1}^{n} [F(x_i) - F(x_{i-1})].$

    This is true since the right side is a telescoping sum.

2)  From the Mean Value Theorem (Chapter 4, Theorem 4, page 131) we know that for any two points $\bar{x}$, $\bar{\bar{x}}$, we have $F(\bar{\bar{x}}) - F(\bar{x}) = F'(\xi)(\bar{\bar{x}} - \bar{x})$, where $\xi$ is some number between $\bar{x}$ and $\bar{\bar{x}}$.

3)  From the definition of integral, if $\varepsilon$ is any positive number, then there is a number $\delta > 0$ such that if the norm of the subdivision, $\|\Delta\|$, is less than $\delta$, then the Riemann Sum is close to the value of the integral, that is,

$$\left| \sum_{i=1}^{n} F'(\xi_i)\, \Delta_i x - \int_a^b F'(x)\, dx \right| < \varepsilon.$$

We apply the Mean Value Theorem (fact (2)) to $F(x_i) - F(x_{i-1})$, calling the intermediate point $\xi_i$, to obtain

$$F(x_i) - F(x_{i-1}) = F'(\xi_i)(x_i - x_{i-1}).$$

We use fact (1) and the customary abbreviation $\Delta_i x = x_i - x_{i-1}$, to write

$$F(b) - F(a) = \sum_{i=1}^{n} F'(\xi_i)\,\Delta_i x.$$

Fact (3) allows us to substitute $F(b) - F(a)$ for the sum and so obtain

$$\left| F(b) - F(a) - \int_a^b F'(x)\,dx \right| < \varepsilon.$$

We now make use of a typical mathematical argument which seems to give us something for nothing. Since the above inequality must hold for *every* possible $\varepsilon > 0$, and since the left side of the inequality has nothing to do with $\varepsilon$, the part within the absolute-value sign must be *equal* to zero. (If it were not, we would simply choose an $\varepsilon$ smaller than that value to get a contradiction.) Thus the proof is complete. □

The Fundamental Theorem states that if we are given any integrable function $f$ we can compute the definite integral over $[a, b]$ by finding *any* antiderivative of $f$ and then evaluating this antiderivative at $b$, evaluating it at $a$, and then subtracting. A function $f$ has more than one antiderivative; in fact it has infinitely many of them. However, as the following Corollary shows, any two antiderivatives differ by a constant.

**COROLLARY**   *If $F_1(x)$ and $F_2(x)$ are antiderivatives of the same function $f$ on the interval $[a, b]$, then*

$$F_1(x) = F_2(x) + \text{const} \quad on \; [a, b].$$

**Proof**   Since $F_1(x)$ and $F_2(x)$ are antiderivatives of the same function, the function $F(x) = F_1(x) - F_2(x)$ has a derivative which is identically zero. It follows from the Mean Value Theorem that $F(x)$ must be constant. □

As we saw in Chapter 4, Section 11, to get antiderivatives of functions we simply work backward from our knowledge of derivatives. If

$$f(x) = x^n, \quad n \text{ rational}, \neq -1,$$

then

$$\text{antiderivative of } f = \frac{x^{n+1}}{n+1} + \text{const},$$

since the derivative of

$$\frac{x^{n+1}}{n+1}$$

is exactly $x^n$. We illustrate the procedure with some examples.

**EXAMPLE 1**   Evaluate $\int_1^2 x^2\,dx$.

**Solution**   The antiderivative of $x^2$ is

$$\frac{x^3}{3} + C, \quad \text{where } C \text{ is a constant.}$$

We evaluate this at $x = 2$ and at $x = 1$ and then subtract:

$$\frac{x^3}{3} + C \quad \text{at} \quad x = 2 \quad \text{is } \frac{8}{3} + C;$$

$$\frac{x^3}{3} + C \quad \text{at} \quad x = 1 \quad \text{is } \frac{1}{3} + C.$$

Difference $= \frac{8}{3} - \frac{1}{3} = \frac{7}{3}$. Then $\int_1^2 x^2\, dx = \frac{7}{3}$, which is the desired evaluation. $\square$

> Observe that the constant in the antiderivative canceled. This is always the case when we evaluate definite integrals; therefore, from now on, the constant of integration will be omitted in evaluating definite integrals.

**EXAMPLE 2**   Evaluate $\int_0^4 (x^3 - 2)\, dx$.

**Solution**   An antiderivative of $x^3 - 2$ is $\frac{1}{4}x^4 - 2x$.

$$\frac{1}{4}x^4 - 2x \quad \text{at} \quad x = 4 \quad \text{is} \quad 64 - 8 = 56;$$

$$\frac{1}{4}x^4 - 2x \quad \text{at} \quad x = 0 \quad \text{is} \quad 0.$$

Therefore

$$\int_0^4 (x^3 - 2)\, dx = 56. \qquad \square$$

There is a convenient and simple notation which reduces the work of evaluating integrals. We write

$$F(x)\big]_a^b \quad \text{or} \quad [F(x)]_a^b \quad \text{for } F(b) - F(a).$$

The way in which the notation works is exhibited in the following example.

**EXAMPLE 3**   Evaluate $\int_{-1}^2 (2x^3 - 3x^2 + x - 1)\, dx$.

**Solution**

$$\int_{-1}^2 (2x^3 - 3x^2 + x - 1)\, dx = \left[\frac{x^4}{2} - x^3 + \frac{x^2}{2} - x\right]_{-1}^2$$

$$= (8 - 8 + 2 - 2) - \left(\frac{1}{2} + 1 + \frac{1}{2} + 1\right)$$

$$= -3. \qquad \square$$

**EXAMPLE 4**   Evaluate

$$\int_0^{\pi/4} \sin 2x\, dx.$$

**Solution**   An antiderivative of $\sin 2x$ is $-\frac{1}{2}\cos 2x$ since the derivative of $-\frac{1}{2}\cos 2x$ is precisely sin 2x. We find

$$\int_0^{\pi/4} \sin 2x \, dx = -\frac{1}{2}\cos 2x \bigg]_0^{\pi/4} = -\frac{1}{2}\cos\frac{\pi}{2} + \frac{1}{2}\cos 0 = 0 + \frac{1}{2} = \frac{1}{2}.$$

□

If the integrand is a complicated function, it may be difficult or even impossible to get an expression for the antiderivative. In such cases we may be interested in obtaining an *approximate value* for the definite integral. The very definition of a definite integral yields useful methods for finding an approximate value. One such process, which we shall describe, is called the **midpoint rule**. We first subdivide the interval of integration into *n equal* parts so that each subinterval has length $(b-a)/n$. That is, we choose $\Delta_i x = (b-a)/n$ for every $i$. In forming the Riemann Sum, we select $\xi_i$ to be the *midpoint* of the $i$th subinterval. The Riemann Sum then becomes

$$\sum_{i=1}^n f(\xi_i)\,\Delta_i x = \frac{b-a}{n}\sum_{i=1}^n f(\xi_i).$$

We use this last sum as an approximation of $\int_a^b f(x)\,dx$.

**EXAMPLE 5**     Using the midpoint rule, compute

$$\int_1^2 \frac{1}{x}\,dx$$

approximately, with $n = 5$

**Solution**     Each $\Delta_i x = 0.2$, since there are five subintervals from 1 to 2. Then $x_0 = 1.0$, $x_1 = 1.2$, $x_2 = 1.4$, $x_3 = 1.6$, $x_4 = 1.8$, $x_5 = 2.0$. Using midpoints, we find that $\xi_1 = 1.1$, $\xi_2 = 1.3$, $\xi_3 = 1.5$, $\xi_4 = 1.7$, $\xi_5 = 1.9$. See Fig. 22. We obtain

$$f(\xi_1)\,\Delta_1 x = \frac{1}{1.1}(0.2) = 0.18182,$$

$$f(\xi_2)\,\Delta_2 x = \frac{1}{1.3}(0.2) = 0.15385,$$

$$f(\xi_3)\,\Delta_3 x = \frac{1}{1.5}(0.2) = 0.13333,$$

$$f(\xi_4)\,\Delta_4 x = \frac{1}{1.7}(0.2) = 0.11765,$$

$$f(\xi_5)\,\Delta_5 x = \frac{1}{1.9}(0.2) = 0.10526,$$

$$\sum_{i=1}^5 f(\xi_i)\,\Delta_i x = 0.69191.$$

The true value of the integral, correct to three decimal places, is $0.693^+$. □

In Chapter 9 we discuss more accurate methods for finding the approximate value of integrals.

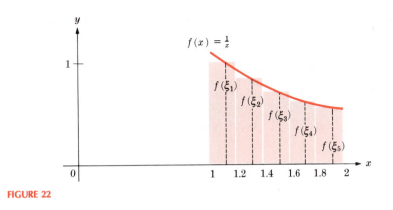

**FIGURE 22**

## 5    PROBLEMS

In Problems 1 through 18, in each case evaluate the given definite integral.

**1** $\displaystyle\int_{1}^{2} (x^2 + x - 5)\, dx$      **2** $\displaystyle\int_{1}^{3} (x^3 + 4x^2 - 6)\, dx$

**3** $\displaystyle\int_{-1}^{0} (3x^4 + 2x^2 - 5)\, dx$      **4** $\displaystyle\int_{-2}^{2} (x^5 + 4x^3 - 3x)\, dx$

**5** $\displaystyle\int_{1}^{3} (x^{3/2} + 2x^{1/2})\, dx$      **6** $\displaystyle\int_{0}^{2} (2x^3 - 4x^{2/3} + 1)\, dx$

**7** $\displaystyle\int_{1}^{9} \frac{x^2 + x + 1}{\sqrt{x}}\, dx$      **8** $\displaystyle\int_{1}^{2} (2t + 1)^2\, dt$

**9** $\displaystyle\int_{0}^{2} (x + 1)(2x + 6)\, dx$      **10** $\displaystyle\int_{1}^{2} (2x + 1)^2 (x - 6)^2\, dx$

**11** $\displaystyle\int_{-1}^{1} \frac{(x^2 + 6x - 2)^2}{9}\, dx$      **12** $\displaystyle\int_{1}^{3} \frac{t^2 + 2t - 1}{\sqrt[3]{t}}\, dt$

**13** $\displaystyle\int_{0}^{x} t^2\, dt$      **14** $\displaystyle\int_{1}^{x} \frac{1}{\sqrt{t}}\, dt, \quad x > 0$

**15** $\displaystyle\int_{0}^{\pi/6} \cos 2x\, dx$      **16** $\displaystyle\int_{0}^{\pi/4} \sec^2 x\, dx$

**17** $\displaystyle\int_{\pi/6}^{\pi/3} \sin 3x\, dx$      **18** $\displaystyle\int_{0}^{\pi/2} \sin x \cos x\, dx$

In Problems 19 through 32, compute the integrals approximately, using the midpoint rule. Compute each term $f(\xi_i)\,\Delta_i x$ to four decimals and round off the result to three decimal places. *Also* compute the exact value in each problem by the rule for integrating.

**19** $\displaystyle\int_{-2}^{-1} x^2\, dx, \quad n = 5$      **20** $\displaystyle\int_{0}^{1} x^3\, dx, \quad n = 5$

**21** $\displaystyle\int_{1}^{2} \frac{1}{x^2}\, dx, \quad n = 5$      **22** $\displaystyle\int_{0.5}^{1} \frac{1}{x^3}\, dx, \quad n = 5$

**23** $\displaystyle\int_{1}^{3} (2x + 1)\, dx, \quad n = 1$      **24** $\displaystyle\int_{-1}^{0} (2 - 3y)\, dy, \quad n = 1$

**25** $\displaystyle\int_{a}^{b} (At + B)\, dt, \quad n = 1$      **26** $\displaystyle\int_{1}^{4} \sqrt{u}\, du, \quad n = 6$

**27** $\displaystyle\int_{1}^{2} (x + 1)(2x + 1)\, dx, \quad n = 4$

**28** $\displaystyle\int_{1}^{3} \frac{x^2 + 2}{x^2}\, dx, \quad n = 2$

**29** $\displaystyle\int_{0}^{\pi/2} \sin x\, dx, \quad n = 4$

**30** $\displaystyle\int_{\pi/4}^{\pi/2} \csc^2 x\, dx, \quad n = 2$

**31** $\displaystyle\int_{\pi/4}^{\pi/2} \cos 4x\, dx, \quad n = 3$

**32** $\displaystyle\int_{0}^{\pi/2} \cos^2 x \sin x\, dx, \quad n = 2$

In Problems 33 through 38, use the midpoint rule to compute the integrals approximately as in Problems 19–32. The answers given are the exact values rounded off to the number of decimals indicated. They are *not* necessarily the answers which the reader should obtain by following the procedure given.

**33** $\displaystyle\int_{0}^{1} \frac{dx}{x + 1}, \quad n = 5; \quad \text{answer, } 0.693^{+}$

**34** $\displaystyle\int_{0}^{1} \frac{dx}{x^2 + 1}, \quad n = 5; \quad \text{answer, } \frac{\pi}{4} = 0.7854^{-}$

**35** $\displaystyle\int_{0}^{1} \sqrt{1 - x^2}\, dx, \quad n = 5; \quad \text{answer, } \frac{\pi}{4} = 0.7854^{-}$

**36** $\displaystyle\int_{0}^{0.5} \frac{dx}{\sqrt{1 - x^2}}, \quad n = 5; \quad \text{answer, } \frac{\pi}{6} = 0.5236^{-}$

37 $\displaystyle\int_0^\pi \sin x\,dx$,   $n=2$;   answer, 2

38 $\displaystyle\int_{-\pi/2}^{\pi/2} \cos x\,dx$,   $n=3$;   answer, 2

*39 Show that if $f$ is integrable then

$$\lim_{n\to\infty} \frac{1}{n}\sum_{k=1}^{n} f\left(\frac{k}{n}\right) = \int_0^1 f(x)\,dx.$$

*40 Let $m$ be a nonnegative integer and set $f(x) = x^m$. Use the formula in Problem 39 to prove that

$$\lim_{n\to\infty} \frac{1^m + 2^m + \cdots + n^m}{n^{m+1}} = \frac{1}{1+m}.$$

*41 Use the formula in Problem 39 to prove that

$$\lim_{n\to\infty} \frac{(n+1)+(n+2)+\cdots+(n+n)}{n^2} = \frac{3}{2}.$$

## 6

## MEAN VALUE THEOREM FOR INTEGRALS

In Chapter 4 we took up the Extreme Value Theorem (Theorem 1), which says that *a function which is continuous on a closed interval takes on its maximum and minimum values there.* A companion to this theorem is the **Intermediate Value Theorem** given below.

**THEOREM 11**   **(Intermediate Value Theorem)**   *Suppose that $f$ is continuous on an interval $[a, b]$ and that $f(a) = A$, $f(b) = B$. If $C$ is any number between $A$ and $B$, there is a number $c$ between $a$ and $b$ such that $f(c) = C$.*

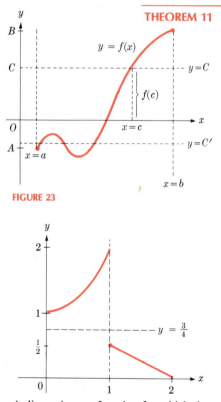

**FIGURE 23**

A discontinuous function for which the Intermediate Value Theorem does not hold

**FIGURE 24**

This theorem is usually proved in advanced analysis courses. However, we shall discuss the plausibility of the result. Figure 23 shows a typical function defined on $[a, b]$ with $f(a) = A$, $f(b) = B$. We give a geometrical interpretation of Theorem 11: If $C$ is any point on the $y$ axis between $A$ and $B$ and we draw the line $y = C$, then this line must intersect the curve representing $y = f(x)$. Furthermore, the $x$ value of the point of intersection is in the interval $(a, b)$. Figure 23 shows the intersection which occurs when $x = c$. That is, we have $f(c) = C$. Intuitively, the theorem asserts that if the function is continuous and if it extends from a point below the line $y = C$ to a point above the line $y = C$, it must cross this line. Note that for some values of $C$ (say $C'$, as shown in Fig. 23) there may be several possible values of $c$. The theorem says that there is always *at least* one.

If the function is not continuous the theorem is false, as the following example shows. Define

$$f(x) = \begin{cases} x^2 + 1, & 0 \le x < 1, \\ \tfrac{1}{2}(2-x), & 1 \le x \le 2. \end{cases}$$

The graph of this function is shown in Fig. 24. If we select $a = 0$, $b = 2$, then $f(0) = 1$, $f(2) = 0$, and $A = 1$, $B = 0$. We see that if $C$ is chosen so that $\tfrac{1}{2} < C < 1$, there is no value of $c$ such that $f(c) = C$. The function never assumes any value between $\tfrac{1}{2}$ and 1. The discontinuity at $x = 1$ causes the trouble.

Another example in which Theorem 11 fails is given by the function (Fig. 25)

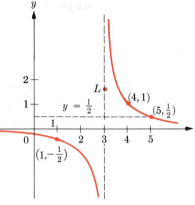

**FIGURE 25**

$$f(x) = \frac{1}{x-3}, \quad x \neq 3, \, f(3) = L,$$

where $L$ is any number. As the graph shows, $f$ is not continuous at $x = 3$. Any interval $[a, b]$ which contains 3 will have values $C$ for which we cannot find numbers $c$ such that $f(c) = C$. If $a = 1$, $b = 4$, then $f(1) = -\frac{1}{2}$, $f(4) = 1$, and a value of $C = \frac{1}{2}$ which is between $A = -\frac{1}{2}$, $B = 1$ is taken on only when

$$\frac{1}{x-3} = \frac{1}{2} \qquad \text{or} \qquad x = 5.$$

Then $f(5) = \frac{1}{2}$, but 5 is *not* in the interval $[1, 4]$, and Theorem 11 is not applicable.

**EXAMPLE 1**    Given the function $f(x) = (x-1)/(x^2+1)$, $a = 0$, $b = 2$. Select a value of $C$ between the $A$ and $B$ of the Intermediate Value Theorem and verify the validity of the result.

**Solution**    $f(0) = -1$, $f(2) = \frac{1}{5}$. We select $C$ between $-1$ and $\frac{1}{5}$, say $-\frac{1}{2}$. Then we solve for $x$:

$$\frac{x-1}{x^2+1} = -\frac{1}{2}.$$

This yields

$$x = -1 \pm \sqrt{2}.$$

The value $-1 + \sqrt{2}$ is in the interval $[0, 2]$ and is the value for which

$$f(-1 + \sqrt{2}) = -\frac{1}{2}.$$

The number $-1 - \sqrt{2}$ is rejected, since it falls outside $[0, 2]$.    □

The Intermediate Value Theorem is used in establishing the **Mean Value Theorem for Integrals**.

---

**THEOREM 12**    **(Mean Value Theorem for Integrals)**   *If $f$ is continuous on a closed interval with endpoints $a$ and $b$, there is a number $\xi$ between $a$ and $b$ (i.e., in $[a, b]$) such that*

$$\int_a^b f(x)\, dx = f(\xi)(b - a).$$

---

We shall first prove this theorem and then discuss it geometrically.

**Proof**    We know by the Extreme Value Theorem (Chapter 4, Theorem 1) that there are two numbers $m$ and $M$ such that $m \leq f(x) \leq M$ for all $x$ in $[a, b]$. Furthermore, in Theorem 7 of this chapter we showed that

$$m(b - a) \leq \int_a^b f(x)\, dx \leq M(b - a).$$

Since the integral is between two numbers, there must be some number which gives the exact value. That is, there is a number $D$ between $m$ and $M$ such that

$$\int_a^b f(x)\,dx = D(b-a).$$

By the Extreme Value Theorem, we know that there are values $x_0$ and $x_1$ such that $f(x_0) = m$ and $f(x_1) = M$ with $x_0$ and $x_1$ in $[a, b]$. Now we bring into play the Intermediate Value Theorem, which states that there is a number $\xi$ between $x_0$ and $x_1$ such that

$$f(\xi) = D.$$

In other words,

$$\int_a^b f(x)\,dx = f(\xi)(b-a). \qquad \square$$

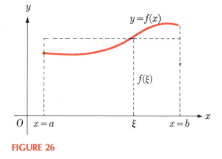

**FIGURE 26**

In Fig. 26 we exhibit a geometric interpretation for a nonnegative function $f$. Then $\int_a^b f(x)\,dx$ is the area bounded by the lines $x = a$, $x = b$, the $x$ axis, and the curve of $y = f(x)$.

In effect, Theorem 12 states that there is a point $\xi$ (between $a$ and $b$) such that the area determined by the integral is equal to that of the rectangle of height $f(\xi)$ and width $b - a$, namely, $f(\xi)(b-a)$. If $f(x)$ is considered to be made of rubber and if the low parts are pushed up and the high parts are pushed down, always keeping the area under the curve the same, the result would be a rectangle of height $f(\xi)$ and width $b - a$.

Finding the number $\xi$ is usually very difficult. However, Theorem 12 proves its worth (particularly in the development of mathematical theory) by demonstrating that such a value $\xi$ does exist. It may seem strange that a use can be found for the knowledge of the existence of something without knowledge of its value or even of how to obtain its value. But that is exactly the state of affairs, as is shown in the proof of Theorem 13 on page 209.

In some cases the value of $\xi$ can be determined, and the following example shows one way to do it.

**EXAMPLE 2**    Find a value of $\xi$ such that

$$\int_1^3 f(x)\,dx = f(\xi)(3-1),$$

where $f(x) = x^2 - 2x + 1$.

**Solution**    We find the value of the integral

$$\int_1^3 (x^2 - 2x + 1)\,dx = \tfrac{1}{3}x^3 - x^2 + x \Big]_1^3$$

$$= (9 - 9 + 3) - (\tfrac{1}{3} - 1 + 1) = \tfrac{8}{3}.$$

Therefore $f(\xi) \cdot (2) = \tfrac{8}{3}$ and $f(\xi) = \tfrac{4}{3}$. We solve for $\xi$:

$$\xi^2 - 2\xi + 1 = \tfrac{4}{3},$$

and the result is

$$\xi = 1 \pm \tfrac{2}{3}\sqrt{3}.$$

Of these numbers, the value $1 + \frac{2}{3}\sqrt{3}$ is in the interval $[1, 3]$. We conclude that

$$\int_1^3 f(x)\,dx = f(1 + \tfrac{2}{3}\sqrt{3})(3 - 1). \qquad \Box$$

The next theorem, **The Fundamental Theorem of Calculus** (as Theorem 10 is also called) tells us that under most circumstances, differentiation and integration are inverse processes. That is, if we integrate a function and then take the derivative of the result we get back the original function.

**THEOREM 13**  **(Fundamental Theorem of Calculus, second form)**  *Suppose that $f$ is continuous on an interval $[a, b]$ and $c$ is some number in this interval. Define the function $F$ by*

$$F(x) = \int_c^x f(t)\,dt$$

*for each $x$ in the interval $(a, b)$. Then*

$$F'(x) = f(x).$$

**Proof**  We know by Theorem 9 (see also Problem 30 of Section 4) that

$$\int_c^{x+h} f(t)\,dt = \int_c^x f(t)\,dt + \int_x^{x+h} f(t)\,dt,$$

so long as $h$ is a number sufficiently small that $x + h$ does not fall outside the interval $[a, b]$. The way in which we defined $F$ tells us that this relationship is nothing but

$$F(x + h) = F(x) + \int_x^{x+h} f(t)\,dt.$$

We apply the Mean Value Theorem for Integrals (Theorem 12) to the integral $\int_x^{x+h} f(t)\,dt$, getting

$$\int_x^{x+h} f(t)\,dt = f(\xi) \cdot h,$$

where $\xi$ is some value between $x$ and $x + h$. We now have

$$\frac{F(x + h) - F(x)}{h} = f(\xi).$$

The hypothesis that $f$ is continuous states (by the definition of continuity) that given $\varepsilon > 0$ there is a $\delta > 0$ such that

$$|f(\xi) - f(x)| < \varepsilon$$

whenever $|\xi - x| < \delta$. Note that as $h \to 0$, then $\xi \to x$, since $\xi$ is always between $x$ and $x + h$. Therefore, if $h$ is sufficiently close to zero,* we may write

———————————

*In fact, if $-\delta < h < \delta$ and $h \neq 0$.

$$\left| \frac{F(x+h) - F(x)}{h} - f(x) \right| < \varepsilon.$$

Since, by definition,

$$\frac{F(x+h) - F(x)}{h} \to F'(x) \quad \text{as } h \to 0,$$

we conclude that

$$F'(x) = f(x). \qquad \square$$

In the proof of Theorem 13, we employed the Mean Value Theorem for Integrals (Theorem 12) without ever finding the value of $\xi$ which occurs. This is a common practice in the proof of theorems in mathematical analysis.

A geometric interpretation of the above result for nonnegative functions $f$ was given in Section 3. We shall repeat it here. From Fig. 27, where $F(x)$ is the shaded area on the left, $F(x+h)$ is the sum of the shaded areas, and $F(x+h) - F(x)$ is the shaded area on the right, we find that

$$\frac{F(x+h) - F(x)}{h}$$

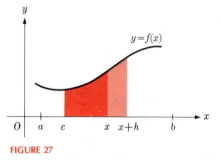

**FIGURE 27**

is the *average* height of $f(x)$ in the area on the right. As $h \to 0$ this average height tends to the height at $x$, namely $f(x)$. On the other hand, as $h \to 0$, the expression above tends to $F'(x)$.

The real meaning of the Fundamental Theorem of Calculus is that **differentiation and integration are inverse processes**. This is meant in the same sense that addition and subtraction are inverse processes, as are multiplication and division (excluding division by zero). What about squaring and taking the square root? These are inverse to each other if an additional condition is tacked on: only positive numbers are allowed. An analogous situation prevails for derivatives and integrals; they are inverse if some additional condition is added (such as continuity of $f$ in Theorem 13). In more advanced courses theorems similar to Theorem 10 and Theorem 13 are established, which show under much less restrictive hypotheses that differentiation and integration are inverse processes.

**EXAMPLE 3**    Find the value of

$$\int_0^3 (x^3 + 3x^2 + x)^2 (3x^2 + 6x + 1) \, dx.$$

**Solution**    One way to proceed would be to multiply $(x^3 + 3x^2 + x)$ by itself and then multiply the result by $(3x^2 + 6x + 1)$. We get a polynomial of degree 8. Then using the Polynomial Antiderivative Rule (page 175), we would be able to evaluate the integral.

A more effective method, one we shall be using often, starts with the observation that the function

$$G(x) = \tfrac{1}{3}(x^3 + 3x^2 + x)^3$$

has the derivative (using the Chain Rule)

$$G'(x) = (x^3 + 3x^2 + x)^2 (3x^2 + 6x + 1),$$

which is the integrand, precisely. The Fundamental Theorem of Calculus tells us that

$$\int_0^3 G'(x) = G(3) - G(0)$$

$$= \tfrac{1}{3}(27 + 27 + 3)^3 - 0 = \tfrac{1}{3}(57)^3 = 61,731. \qquad \square$$

In the next section we present a systematic method for evaluating integrals of the type given in Example 3.

**EXAMPLE 4**    Evaluate

$$\int_{-2}^2 |x + 1| \, dx$$

**Solution**    From the definition of absolute value, we have

$$f(x) \equiv |x + 1| = \begin{cases} x + 1 & \text{if } x + 1 > 0 \\ -(x + 1) & \text{if } x + 1 < 0 \end{cases}$$

A graph of $f$ is shown in Fig. 28.

We find

$$\int_{-2}^2 |x + 1| \, dx = \int_{-2}^{-1} -(x + 1) \, dx + \int_{-1}^2 (x + 1) \, dx.$$

An antiderivative of $-(x + 1)$ is $F_1(x) \equiv -\tfrac{1}{2}(x + 1)^2$, and an antiderivative of $(x + 1)$ is $F_2(x) \equiv \tfrac{1}{2}(x + 1)^2$. Thus

$$\int_{-2}^2 |x + 1| \, dx = F_1(-1) - F_1(-2) + F_2(2) - F_2(-1) = 0 + \tfrac{1}{2} + \tfrac{9}{2} - 0 = 5.$$

$\square$

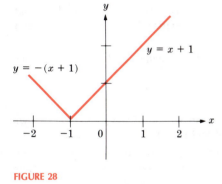

**FIGURE 28**

# 6   PROBLEMS

In each of Problems 1 through 8, a function $f$ and numbers $a$, $b$, and $C$ are given. Verify the Intermediate Value Theorem by finding the value of $c$, or show that this theorem does not apply.

**1** $f(x) = x^2 + x + 3$,   $a = 1$, $b = 3$, $C = 9$

**2** $f(x) = 2 + 4x - x^2$,   $a = 0$, $b = 2$, $C = 4$

**3** $f(x) = (x + 2)/(x^2 + 2)$,   $a = 0$, $b = 3$, $C = 2$

**4** $f(x) = (x^2 + 1)/(x - 2)$,   $a = 3$, $b = 5$, $C = 12$

**5** $f(x) = (x - 1)/(x + 1)$,   $a = -2$, $b = 1$, $C = 1$

**6** $f(x) = 3x^2 + x - 6$,   $a = 0$, $b = 2$, $C = -4$

**7** $f(x) = \begin{cases} 2x, & -1 \le x < 1 \\ 3 - x, & 1 \le x < 5 \end{cases}$,   $a = 0$, $b = 2$, $C = \tfrac{1}{2}$

**8** $f(x) = \begin{cases} 1 - x^2, & -10 \le x < 0 \\ x - 3, & 0 \le x < 5 \end{cases}$,   $a = -1$, $b = 2$, $C = -\tfrac{1}{2}$

In Problems 9 through 15, verify the Mean Value Theorem for Integrals by finding appropriate values of $\xi$. Sketch graphs.

**9** $\int_2^5 f(x) \, dx$,   with $f(x) = 3x - 2$

**10** $\int_{-1}^2 f(x) \, dx$,   with $f(x) = 2 + 4x$

**11** $\int_2^7 f(x) \, dx$,   with $f(x) = x^2 - 2x + 1$

**12** $\displaystyle\int_0^1 f(x)\,dx,$ with $f(x)=(x+1)^2$

**13** $\displaystyle\int_{-1}^1 f(x)\,dx,$ with $f(x)=3-2x+4x^2$

**14** $\displaystyle\int_0^2 f(x)\,dx,$ with $f(x)=4x^3-3x^2-4x$

**15** $\displaystyle\int_{-2}^2 f(x)\,dx,$ with $f(x)=x^5-4x^3+3x$

**16** Given that $f(x)=(x^2+2)^3$, verify that $f'(x)=6x(x^2+2)^2$. Use the Fundamental Theorem of Calculus to evaluate $\int_0^2 6x(x^2+2)^2\,dx$.

**17** Given $f(x)=(x^2+5)^{1/2}$, verify that $f'(x)=x(x^2+5)^{-1/2}$. Use the Fundamental Theorem of Calculus to evaluate

$$\int_1^3 \frac{x\,dx}{\sqrt{x^2+5}}.$$

**18** Given $f(x)=6(x+2)^{3/2}(x^2+1)^{4/3}$, verify that

$$f'(x)=(x+2)^{1/2}(x^2+1)^{1/3}(25x^2+32x+9).$$

Evaluate

$$\int_0^2 (x+2)^{1/2}(x^2+1)^{1/3}(25x^2+32x+9)\,dx.$$

**19** Given $f(x)=(x+3)/(x-2)$, verify that $f'(x)=-5/(x-2)^2$. Evaluate

$$\int_{-1}^1 \frac{dx}{(x-2)^2}.$$

**20** Given $f(x)=\sin^3 x$, verify that $f'(x)=3\sin^2 x\cos x$. Evaluate

$$\int_0^{\pi/3} \sin^2 x\cos x\,dx.$$

**21** Given $f(x)=\cos 4x$, verify that $f'(x)=-4\sin 4x$. Evaluate

$$\int_0^{\pi/8} \sin 4x\,dx.$$

**22** Given $f(x)=\tan 2x$, verify that $f'(x)=2\sec^2 2x$. Evaluate

$$\int_0^{\pi/6} \sec^2 2x\,dx.$$

**23** Given $f(x)=\sec \tfrac{1}{2}x$, verify that $f'(x)=\tfrac{1}{2}\sec\tfrac{1}{2}x\tan\tfrac{1}{2}x$. Evaluate

$$\int_{\pi/3}^{\pi/2} \sec\tfrac{1}{2}x\tan\tfrac{1}{2}x\,dx.$$

**24** Evaluate $\displaystyle\int_0^2 (x^3+2x+1)^4(3x^2+2)\,dx.$

**25** Evaluate $\displaystyle\int_0^1 (x^3-2x)^{49}(3x^2-2)\,dx.$

**26** Evaluate $\displaystyle\int_{-2}^3 |x|\,dx.$

**27** Evaluate $\displaystyle\int_{-1}^3 |x-1|\,dx.$

**28** Evaluate $\displaystyle\int_0^{\pi/2} \sin^4 x\cos x\,dx.$

**29** If a function is **odd**, that is, if $f(-x)=-f(x)$ for all $x$, what can be said about $\int_{-b}^b f(x)\,dx$? (*Hint:* Draw a sketch.)

**30** If a function is **even**, that is, if $f(-x)=f(x)$ for all $x$, what can be said about the relation between $\int_{-b}^b f(x)\,dx$ and $\int_0^b f(x)\,dx$?

**31** Let $f$ be a given function. Define $F$ and $G$ by the formulas

$$F(x)=\int_c^x f(t)\,dt, \qquad G(x)=\int_d^x f(t)\,dt,$$

where $c$ and $d$ are given numbers. Show that $F$ and $G$ differ by a constant and find the value of this constant.

**32** Suppose $f(x)$ is a function partly above and partly below the $x$ axis. Interpret the Mean Value Theorem for Integrals in terms of area. (*Hint:* See discussion after Theorem 12.)

**33** Given

$$f(x)=\begin{cases} x, & 0\le x\le 1 \\ 1-x, & 1<x\le 2. \end{cases}$$

Then

$$f'(x)=\begin{cases} 1, & 0<x<1 \\ -1, & 1<x<2. \end{cases}$$

Does the Fundamental Theorem of Calculus (second form) hold for

$$\int_c^x f'(x)\,dx$$

for each $x$ in the interval $(0,2)$?

_____ 7

## THE INDEFINITE INTEGRAL. INTEGRATION BY SUBSTITUTION

To evaluate the definite integral of a function, we must first obtain an antiderivative of the function and then substitute the appropriate limits of integration into this antiderivative. Every function has an infinite number of antiderivatives, since a constant of any size may be added to one antiderivative to yield another. A common notation used for all the antiderivatives of a function $f$ is

$$\int f(x)\,dx,$$

in which there are no upper and lower limits. This is called the **indefinite integral** of $f$. For example, the formula for the antiderivative of $x^n$, $n$ rational, $\neq -1$, is written

$$\int x^n\,dx = \frac{x^{n+1}}{n+1} + C, \quad n \neq -1,$$

where the constant $C$ is the **constant of integration**.

The following seemingly trivial variation of the above formula has many useful consequences. We write

$$\int u^n\,du = \frac{u^{n+1}}{n+1} + C, \quad n \neq -1.$$

True, this *is* the same formula. However, according to the Chain Rule, we know that

$$d\left(\frac{u^{n+1}}{n+1}\right) = u^n\,du$$

holds if $u$ is *any function* of $x$. Schematically we write

$$\int (\text{expression in } x)^n\,d(same \text{ expression in } x) = \frac{(\text{expression in } x)^{n+1}}{n+1} + C.$$

We shall illustrate the significance of this formula with two examples.

**EXAMPLE 1**    Find $\int (3x+2)^7\,dx$.

**Solution**    The integrand could be multiplied out and then each term could be evaluated separately. Instead, we observe that $(3x+2)^7$ is an expression in $x$ raised to the 7th power. So we write

$$u = 3x + 2, \qquad du = 3\,dx \quad \text{or} \quad dx = \tfrac{1}{3}\,du.$$

Substituting, we get

$$\int (3x+2)^7\,dx = \int u^7 \cdot \tfrac{1}{3}\,du.$$

A constant (*and only a constant*) may be moved in and out of integrals at will. We obtain

$$\int u^7 \frac{1}{3} \, du = \frac{1}{3} \int u^7 \, du = \frac{1}{3} \left( \frac{u^8}{8} + C \right).$$

Since $C$ represents *any* constant, so does $\frac{1}{3}C$, and we write, after substituting the value of $u$ in terms of $x$,

$$\int (3x + 2)^7 \, dx = \tfrac{1}{24}(3x + 2)^8 + C. \qquad \square$$

**EXAMPLE 2** Find $\int (x^2 + 1)^{5/2} x \, dx$.

**Solution** Let $u = x^2 + 1$. Then $du = 2x \, dx$ and, after substitution, we obtain

$$\int (x^2 + 1)^{5/2} x \, dx = \int u^{5/2} \frac{1}{2} \, du = \frac{1}{2} \int u^{5/2} \, du$$

$$= \frac{1}{2} \left( \frac{2u^{7/2}}{7} + C \right) = \frac{1}{7}(x^2 + 1)^{7/2} + C,$$

which is our answer. $\qquad \square$

*Remark.* The integrand in Example 2 was rigged so that the result is readily obtainable by the method we are learning. For example, the integral $\int (x^2 + 1)^{5/2} \, dx$, which appears "simpler" than the one in Example 2, cannot be evaluated by this method. For, if we let $u = x^2 + 1$, then $du = 2x \, dx$ and, substituting, we get

$$\int (x^2 + 1)^{5/2} \, dx = \int u^{5/2} \frac{1}{2x} \, du.$$

There is no way to get rid of the $x$ in the denominator. Later we shall study other methods which will enable us to perform this type of integration.

The technique illustrated in Examples 1 and 2, known as the **Method of Substitution**, may be used for evaluating definite integrals as well. However, *great care* must be exercised in evaluating the limits. There are two possible methods of procedure, and we illustrate each with an example.

**EXAMPLE 3** Evaluate $\int_1^2 (x + 1)(x^2 + 2x + 2)^{1/3} \, dx$.

**Solution** We make the substitution $u = x^2 + 2x + 2$. Then

$$du = (2x + 2) \, dx = 2(x + 1) \, dx.$$

Substituting, we find for the *indefinite integral*

$$\int (x + 1)(x^2 + 2x + 2)^{1/3} \, dx = \int u^{1/3} \frac{1}{2} \, du = \frac{1}{2} \frac{3u^{4/3}}{4} + C.$$

**Method I** involves substituting back for $u$ in terms of $x$, thereby obtaining for the *definite integral*

$$\int_1^2 (x + 1)(x^2 + 2x + 2)^{1/3} \, dx = \tfrac{3}{8}(x^2 + 2x + 2)^{4/3}\Big]_1^2$$

$$= \tfrac{3}{8}(10^{4/3} - 5^{4/3}). \qquad \square$$

The next example exhibits the second method.

**EXAMPLE 4**     Evaluate

$$\int_{-2}^{0} x\sqrt{2x^2 + 1}\ dx.$$

**Solution**     Let $u = 2x^2 + 1$, $du = 4x\ dx$.

**Method II** involves changing the limits of the definite integral to values of $u$ instead of values of $x$. Since $u = 2x^2 + 1$, we easily see that when $x = -2$, then $u = 9$, and when $x = 0$, $u = 1$. Therefore we can write

$$\int_{-2}^{0} x\sqrt{2x^2 + 1}\ dx = \int_{9}^{1} u^{1/2} \cdot \frac{1}{4}\ du = \frac{1}{4} \cdot \frac{2u^{3/2}}{3}\bigg]_{9}^{1}$$

$$= \frac{1}{6}(1^{3/2} - 9^{3/2}) = -\frac{13}{3}. \qquad \square$$

Most of the time Method II is simpler and shorter. Occasionally there are problems in which it pays to go back to the original variables before evaluating the limits of integration. The justification of Method II depends on the formula

$$\int_{a}^{b} f[u(x)]u'(x)\ dx = \int_{u(a)}^{u(b)} f(u)\ du, \qquad (1)$$

which the next theorem establishes. Formula (1) is called the **Change of Variables Formula**.

---

**THEOREM 14**     *Suppose that $u$ and $u'$ are continuous functions defined on $[a, b]$ and that $f$ is continuous on an interval $[c, d]$. Suppose, also, that the range\* of $u$ is in $(c, d)$. Then the above substitution formula (1) holds.*

---

**Proof**     We define functions $F$ and $G$ by the formulas

$$F(u) = \int_{x_0}^{u} f(t)\ dt, \qquad (2)$$

where $x_0$ is any fixed number in $[c, d]$, and

$$G(x) = F[u(x)].$$

Then, from the Fundamental Theorem of Calculus, second form (Theorem 13), we see that

$$F'(u) = f(u) \quad \text{for each } u \text{ in } (c, d).$$

Also, the Chain Rule applied to $G$ gives

$$G'(x) = F'[u(x)]u'(x) = f[u(x)]u'(x).$$

We integrate this last equation between the limits $a$ and $b$, getting

$$\int_{a}^{b} G'(x)\ dx = \int_{a}^{b} f[u(x)]u'(x)\ dx.$$

---

\*We omit the technical discussion which occurs if for some $x$, $u(x) = c$ or $u(x) = d$.

Now using the Fundamental Theorem of Calculus, first form (Theorem 10), we find

$$\int_a^b f[u(x)]u'(x)\,dx = G(b) - G(a)$$

$$= F[u(b)] - F[u(a)].$$

However,

$$F[u(b)] - F[u(a)] = \int_{x_0}^{u(b)} f(t)\,dt - \int_{x_0}^{u(a)} f(t)\,dt = \int_{u(a)}^{u(b)} f(t)\,dt.$$

Therefore

$$\int_{u(a)}^{u(b)} f(t)\,dt = \int_a^b f[u(x)]u'(x)\,dx,$$

and formula (1) is established. □

Every differentiation formula automatically gives us a corresponding integration formula. From our knowledge of the derivatives of the trigonometric functions, we obtain at once the following indefinite integral formulas:

$$\int \cos u\,du = \sin u + C,$$

$$\int \sin u\,du = -\cos u + C,$$

$$\int \sec^2 u\,du = \tan u + C,$$

$$\int \csc^2 u\,du = -\cot u + C,$$

$$\int \sec u \tan u\,du = \sec u + C,$$

$$\int \csc u \cot u\,du = -\csc u + C.$$

In the above formulas $u$ may be any function of $x$, and $du$ is the differential of $u$. These formulas may be combined with the methods of integration of the power function to help us integrate a wide variety of functions. We illustrate the technique with some examples.

**EXAMPLE 5**    Find $\int \sin 4x\,dx$.

**Solution**    We let $u = 4x$ and $du = 4\,dx$. Then the above integral becomes

$$\int \sin u \cdot \tfrac{1}{4}\,du = \tfrac{1}{4} \int \sin u\,du = -\tfrac{1}{4} \cos u + C = -\tfrac{1}{4} \cos 4x + C. \quad \square$$

**EXAMPLE 6**    Find $\int \tan^2(2x - 3)\,dx$.

Solution   This does not seem to fit any of the formulas for integration until we recall* the trigonometric identity $1 + \tan^2 \theta = \sec^2 \theta$. Then the above integral becomes

$$\int [\sec^2 (2x - 3) - 1] \, dx = \int \sec^2 (2x - 3) \, dx - \int dx.$$

We now let $u = 2x - 3$, $du = 2 \, dx$, and the first integral on the right becomes

$$\int \sec^2 u \cdot \tfrac{1}{2} \, du = \tfrac{1}{2} \int \sec^2 u \, du = \tfrac{1}{2} \tan u + C = \tfrac{1}{2} \tan (2x - 3) + C.$$

The second integral gives us merely $x + C$. We now have two arbitrary constants of integration. Their sum is therefore an arbitrary constant and, adding the constants of integration, we have

$$\int \tan^2 (2x - 3) \, dx = \tfrac{1}{2} \tan (2x - 3) - x + C. \qquad \square$$

EXAMPLE 7   Evaluate

$$\int_0^{\pi/4} \sin^3 x \, dx.$$

Solution   We again make use of trigonometric identities and write

$$\sin^3 x = \sin x \sin^2 x = \sin x (1 - \cos^2 x) = \sin x - \sin x \cos^2 x.$$

Therefore

$$\int_0^{\pi/4} \sin^3 x \, dx = \int_0^{\pi/4} \sin x \, dx - \int_0^{\pi/4} \sin x \cos^2 x \, dx.$$

The first integral is easy and gives us $-\cos x]_0^{\pi/4} = -\tfrac{1}{2}\sqrt{2} + 1$. The second integral is a little harder but still manageable, if we let $u = \cos x$. Then $du = -\sin x \, dx$. Also, $u = 1$ when $x = 0$ and $u = \tfrac{1}{2}\sqrt{2}$ when $x = \pi/4$. Therefore

$$\int_0^{\pi/4} \sin x \cos^2 x \, dx = -\int_1^{\frac{1}{2}\sqrt{2}} u^2 \, du = -\tfrac{1}{3} u^3 \Big]_1^{\frac{1}{2}\sqrt{2}} = -\tfrac{1}{12}\sqrt{2} + \tfrac{1}{3}.$$

The final result is

$$\int_0^{\pi/4} \sin^3 x \, dx = \tfrac{2}{3} - \tfrac{5}{12}\sqrt{2}. \qquad \square$$

## 7   PROBLEMS

In each of Problems 1 through 40, find the indefinite integral and check the result by differentiation.

1. $\displaystyle \int (3x - 2)^5 \, dx$

2. $\displaystyle \int (2x + 5)^4 \, dx$

3. $\displaystyle \int (2 - x)^{-4} \, dx$

4. $\displaystyle \int (4 - 6x)^{-7} \, dx$

5. $\displaystyle \int (2y + 1)^{3/2} \, dy$

6. $\displaystyle \int (2x^2 + 3)^4 x \, dx$

*The basic trigonometric formulas may be found in Appendix 1 at the end of the book.

**7** $\displaystyle\int (2x^2 + 3)^{7/3} x\, dx$

**8** $\displaystyle\int (4 - x^2)^3 x\, dx$

**9** $\displaystyle\int (3 - 2x^2)^{-2/3} x\, dx$

**10** $\displaystyle\int \frac{x\, dx}{\sqrt{x^2 - 1}}$

**11** $\displaystyle\int \frac{x\, dx}{(3x^2 + 2)^2}$

**12** $\displaystyle\int x\sqrt{2x^2 - 1}\, dx$

**13** $\displaystyle\int x^2 \sqrt{x^3 + 1}\, dx$

**14** $\displaystyle\int u^3 \sqrt{u^4 + 1}\, du$

**15** $\displaystyle\int \frac{(x^2 + 2x)\, dx}{\sqrt[3]{x^3 + 3x^2 + 1}}$

**16** $\displaystyle\int \left(1 + \frac{1}{t}\right)^2 \frac{dt}{t^2}$

**17** $\displaystyle\int \sin 6x\, dx$

**18** $\displaystyle\int \cos 4v\, dv$

**19** $\displaystyle\int \csc^2\left(\frac{2x + 1}{3}\right) dx$

**20** $\displaystyle\int \sec(x - 1)\tan(x - 1)\, dx$

**21** $\displaystyle\int \frac{\sin 3x}{\cos^3 3x}\, dx$

**22** $\displaystyle\int \frac{\sec^2 4x}{\tan^7 4x}\, dx$

**23** $\displaystyle\int (x^2 + 1)^3 x^3\, dx$

**24** $\displaystyle\int (2x^2 - 3)^{4/3} x^3\, dx$

**25** $\displaystyle\int (x^3 + 1)^{7/5} x^5\, dx$

**26** $\displaystyle\int (x^2 - 2x + 1)^{4/3}\, dx$

**27** $\displaystyle\int \cot^2(2x - 6)\, dx$

**28** $\displaystyle\int \cos^3 2x\, dx$

**29** $\displaystyle\int \sin^5 x\, dx$

**30** $\displaystyle\int \cos^5 x\, dx$

**31** $\displaystyle\int x \sin(x^2)\, dx$

**32** $\displaystyle\int x^2 \sec^2(4x^3)\, dx$

**33** $\displaystyle\int \frac{\sin x\, dx}{\cos^5 x}$

**34** $\displaystyle\int \cos^6 2x \sin 2x\, dx$

**35** $\displaystyle\int \tan^5 x \sec^2 x\, dx$

**36** $\displaystyle\int \csc^7 2x \cot 2x\, dx$

**37** $\displaystyle\int \frac{\csc^2(3 - 2x)}{\cot^4(3 - 2x)}\, dx$

**38** $\displaystyle\int \cos^2 \tfrac{1}{2}x\, dx$ (*Hint:* Use half-angle formula in Appendix 1.)

**39** $\displaystyle\int \sin^2 3x\, dx$

**40** $\displaystyle\int \sin^4 x\, dx$

In Problems 41 through 48, evaluate the integrals by Method I.

**41** $\displaystyle\int_0^3 \sqrt[3]{(3t - 1)^2}\, dt$

**42** $\displaystyle\int_{-2}^1 \sqrt{2 - x}\, dx$

**43** $\displaystyle\int_{-3}^{-1} \frac{dx}{(x - 1)^2}$

**44** $\displaystyle\int_0^1 x\sqrt{x^2 + 1}\, dx$

**45** $\displaystyle\int_0^{\pi/6} \sin 2x\, dx$

**46** $\displaystyle\int_0^{\pi/3} \sin^{2/3} 3x \cos 3x\, dx$

**47** $\displaystyle\int_0^{\pi/2} \tan^3 \tfrac{1}{2}x \sec^2 \tfrac{1}{2}x\, dx$

**48** $\displaystyle\int_0^{\pi/2} \cos^2\left(2x - \frac{\pi}{4}\right)\sin\left(2x - \frac{\pi}{4}\right) dx$

In Problems 49 through 56, evaluate each integral by Method II.

**49** $\displaystyle\int_0^2 \frac{(x^2 + 1)\, dx}{\sqrt{x^3 + 3x + 1}}$

**50** $\displaystyle\int_1^2 (2x + 1)\sqrt{x^2 + x + 1}\, dx$

**51** $\displaystyle\int_1^2 \frac{(x^{1/3} + 2)^4\, dx}{\sqrt[3]{x^2}}$

**52** $\displaystyle\int_1^3 \left(x + \frac{1}{x}\right)^{3/2}\left(\frac{x^2 - 1}{x^2}\right) dx$

**53** $\displaystyle\int_{\pi/6}^{\pi/3} \cos 2x\, dx$

**54** $\displaystyle\int_0^{2\pi/3} \cos^{4/3} \tfrac{1}{2}x \sin \tfrac{1}{2}x\, dx$

**55** $\displaystyle\int_0^{\pi/6} \tan^2 x \sec^4 x\, dx$

**56** $\displaystyle\int_{\pi/6}^{\pi/4} \sec^{5/3} x \sin x\, dx$

**57** Verify that a straightforward evaluation by Method II of the integral

$$\int_{-1}^3 \left(x - \frac{3}{x}\right)^5 \left(1 + \frac{3}{x^2}\right) dx$$

yields the value zero. What is wrong with the result?

# CHAPTER 5

## REVIEW PROBLEMS

In Problems 1 through 5 find the value of the given sum.

**1** $\displaystyle\sum_{i=1}^4 3^{i-1}$

**2** $\displaystyle\sum_{i=2}^5 (4 - 3i)$

**3** $\displaystyle\sum_{i=2}^8 3^i - 3^{i+1}$

**4** $\displaystyle\sum_{i=1}^4 \sin \tfrac{1}{2}i\pi$

**5** $\displaystyle\sum_{k=10}^{200} \frac{1}{k^2 + 3k}$

**6** Given

$$a_n = 3 + \frac{n^2}{n + 1}, \quad n = 1, 2, \ldots .$$

Does Axiom C apply? Justify.

**7** Give an example of a region in which $A_3^L < A_3^U$ but $A_n^L = A_n^U$ for all $n > 3$. (*Hint:* See Problem 18 on page 187.)

In Problems 8 through 11 a function $f$ and a subdivision $\Delta$ are given. Find the value of $\underline{S}(\Delta)$ and $\overline{S}(\Delta)$.

**8** $f(x) = x^2 + 2x - 3$;   $\Delta: a = x_0 = 0$,   $x_1 = 0.2$,   $x_2 = 0.3$, $x_3 = 0.5$, $x_4 = 0.8$, $x_5 = b = 1$.

**9** $f(x) = \sin \pi x$;   $\Delta: a = x_0 = 0$, $x_1 = \frac{1}{6}$, $x_2 = \frac{1}{4}$, $x_3 = \frac{1}{3}$, $x_4 = \frac{1}{2}$, $x_5 = \frac{3}{4}$, $x_6 = b = 1$.

**10** $f(x) = \dfrac{x}{x+2}$;   $\Delta: a = x_0 = -1.5$,   $x_1 = -1$,   $x_2 = -\frac{1}{2}$, $x_3 = 0$, $x_4 = \frac{1}{2}$, $x_5 = b = 2$.

**11** $f(x) = \tan \pi x$;   $\Delta: a = x_0 = -\frac{1}{3}$,   $x_1 = -\frac{1}{4}$,   $x_2 = -\frac{1}{6}$, $x_3 = 0$, $x_4 = \frac{1}{6}$, $x_5 = b = \frac{1}{4}$.

**12** Suppose that $f(x) \leq M$ on an interval $[a, b]$. Let $\Delta$ be a subdivision of $[a, b]$. Show that the Riemann Sum satisfies the inequality

$$\sum_{i=1}^n f(\xi_i) \Delta_i x \leq M(b - a).$$

**13** Given

$$f(x) = \begin{Bmatrix} x, & 0 \leq x < 1 \\ \frac{1}{2}(x - 1), & 1 \leq x \leq 2 \end{Bmatrix}.$$

Let $\Delta$ be any subdivision of $[0, 2]$. Show that $\overline{S}(\Delta) > \underline{S}(\Delta)$. Also show that $\overline{S}(\Delta)$ is strictly larger than any Riemann Sum taken over the same subdivision.

**14** Are there any functions other than constants for which $\overline{S}(\Delta) = \underline{S}(\Delta)$ for some subdivision?

**15** a) Exhibit a function other than a constant for which

$$\sum_{i=1}^n f(\xi_i) \Delta_i x = \overline{S}(\Delta)$$

for some subdivision.

b) Same as (a) with $\overline{S}(\Delta)$ replaced by $\underline{S}(\Delta)$.

In Problems 16 through 25 find upper and lower bounds for the given integral or show there are none. Do not attempt to integrate the expression.

**16** $\displaystyle\int_0^2 (x^2 + 1)\, dx$     **17** $\displaystyle\int_1^2 \frac{x^2}{2 + x^2}\, dx$

**18** $\displaystyle\int_0^{\pi/4} \cos^2 x\, dx$     **19** $\displaystyle\int_{-\pi/4}^{\pi/3} \tan^3 x\, dx$

**20** $\displaystyle\int_1^4 \frac{3x}{\sqrt{x^2 + 1}}\, dx$     **21** $\displaystyle\int_2^4 \frac{x^2}{\sqrt{x^2 - 1}}\, dx$

**22** $\displaystyle\int_{-4}^{-1} \frac{x}{x + 1}\, dx$     **23** $\displaystyle\int_{-2}^2 \frac{\sqrt{x^2 + 5}}{2x}\, dx$

**24** $\displaystyle\int_0^{\pi/3} \csc^2 x\, dx$     **25** $\displaystyle\int_0^{\pi/2} \sec^3 \tfrac{1}{2}x\, dx$

In Problems 26 through 31 evaluate the given definite integral.

**26** $\displaystyle\int_0^2 (2x + 3)^{5/2}\, dx$     **27** $\displaystyle\int_{-1}^1 \frac{dx}{(4 - 2x)^{1/3}}$

**28** $\displaystyle\int_{\pi/6}^{\pi/2} \sin^{4/3} 2x \cos 2x\, dx$

**29** $\displaystyle\int_2^3 (x^2 + 2x - 3)^{1/4}(x + 1)\, dx$

**30** $\displaystyle\int_0^{\pi/2} \tan \tfrac{1}{2}x \sec^4 \tfrac{1}{2}x\, dx$

**31** $\displaystyle\int_{\pi/4}^{\pi/2} \cot^3 x \csc^2 x\, dx$

**32** Find the value of

$$\int_1^3 \frac{dx}{(x^2 + 1)^{3/2}}$$

by differentiating $f(x) = \sqrt{x^2 + 1}$ twice and using the Fundamental Theorem of Calculus.

**33** Find the value of

$$\int_0^{\sqrt{\pi/2}} x \sin(x^2) \cos(x^2)\, dx.$$

**34** Given

$$F(x) = \int_c^x \frac{dt}{\sqrt[3]{t^2 - 1}},$$

for what values of $x$ and $c$ is the formula

$$F'(x) = \frac{1}{\sqrt[3]{x^2 - 1}}$$

valid?

**35** Give an example of a function $f$ defined on $[0, 1]$ such that $f$ is not continuous and yet the Mean Value Theorem for Integrals is valid.

**36** Give an example of a function $f$ defined on $[0, 1]$ which has exactly one point of discontinuity and for which the Mean Value Theorem for Integrals is invalid.

**37** Verify the Mean Value Theorem for Integrals for the integral

$$\int_0^{\pi/2} \sin x\, dx$$

by finding (approximately) the appropriate value of $\xi$.

In Problems 38 through 49 find the indefinite integral.

**38** $\displaystyle\int (3x^3 + 4)^7 x^2\, dx$     **39** $\displaystyle\int \left(2 + \frac{3}{t^4}\right) \frac{dt}{t^5}$

**40** $\displaystyle\int \sec^2(2x+5)\,dx$

**41** $\displaystyle\int \cos^3(x^2+2x-6)\sin(x^2+2x-6)(x+1)\,dx$

**42** $\displaystyle\int \cot^4 3x\,\csc^4 3x\,dx$

**43** $\displaystyle\int \sec^4(2x^3)x^2\,dx$

**44** $\displaystyle\int \sec(2x+4)\tan(2x+4)\,dx$

**45** $\displaystyle\int \csc(x^2+1)\cot(x^2+1)x\,dx$

**46** $\displaystyle\int \sec^2(3x-5)\tan(3x-5)\,dx$

**47** $\displaystyle\int \sec^{5/3}(2x)\tan(2x)\,dx$

**48** $\displaystyle\int \csc^{2/3}(1-x)\cot(1-x)\,dx$

**49** $\displaystyle\int \cot(x^3)\csc(x^3)x^2\,dx$

# 6

# APPLICATIONS OF THE INTEGRAL

The integral is useful for calculating areas of regions, lengths of curves, and volumes of figures which have complicated shapes. In fact, one of the principal motivations for developing the integral is the goal of measuring lengths and areas in cases where the elementary methods of Euclidean geometry fail. In this chapter we show how to use the integral in making such calculations.

## 1

## AREA BETWEEN CURVES

Suppose that $f$ and $g$ are two continuous functions defined on the interval $[a, b]$ and, furthermore, suppose that

$$f(x) \geq g(x), \quad a \leq x \leq b.$$

Figure 1 shows a typical situation. Let $R$ denote the region bounded by the lines $x = a$, $x = b$, and the two curves. The area $A$ of the region $R$ is given by

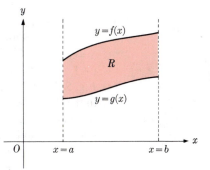

**FIGURE 1**

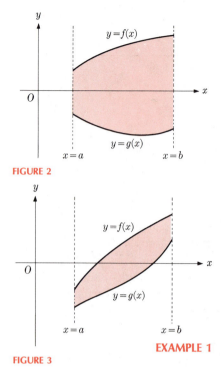

**FIGURE 2**

**FIGURE 3**

$$A = \int_a^b f(x)\,dx - \int_a^b g(x)\,dx = \int_a^b [f(x) - g(x)]\,dx.$$

The interesting point about this formula is that *it holds whether the curves are above or below the x axis.* Figures 2 and 3 show other possible situations. A moment's thought about Fig. 2 discloses the fact that $\int_a^b f(x)\,dx$ is the area below $f(x)$ and above the x axis. Also, $-\int_a^b g(x)\,dx$ is the area between the x axis and $g(x)$. Adding these, we get for the total area $A$ between the curves

$$A = \int_a^b [f(x) - g(x)]\,dx.$$

Figure 3 appears slightly more complicated, but the reader can easily satisfy himself that the same formula holds.

When we apply the formula for the area between two curves, the main problem lies in verifying the fact that $f(x)$ is larger than $g(x)$ *throughout the interval*. Once this verification is made, we perform the integration in the usual way.

**EXAMPLE 1**   Find the area of the region $R$ bounded by the lines $x = 1$, $x = 2$, $y = 3x$, and the curve $y = x^2$.

**Solution**   We draw a sketch, as shown in Fig. 4, which makes it obvious that $y = 3x$ is above $y = x^2$ on the interval $[1, 2]$. This can be shown analytically first by finding where the curves intersect and then by noting that between intersection points one curve must *always remain above* the other; $y = 3x$ and $y = x^2$ meet when $3x = x^2$ or $x = 0$, $3$. Between $x = 0$ and $x = 3$, say at $x = 1$, we know that $3x > x^2$, since $3 > 1$. This means that $y = 3x$ is above $y = x^2$ *throughout* the interval $[0, 3]$.

The area is given by

$$A = \int_1^2 (3x - x^2)\,dx = \tfrac{3}{2}x^2 - \tfrac{1}{3}x^3]_1^2 = (6 - \tfrac{8}{3}) - (\tfrac{3}{2} - \tfrac{1}{3}) = \tfrac{13}{6}. \qquad \square$$

In the next example we show how to find the area when the two curves intersect at several points.

**EXAMPLE 2**   Find the area bounded by the curves $y = x$ and $y = x^3$.

**Solution**   To determine the nature of the region bounded by the curves, we begin by finding the points of intersection of the two curves. Solving the equation $x = x^3$, we get $x = 0, 1, -1$; therefore the curves intersect at $(0, 0)$, $(1, 1)$, and $(-1, -1)$. After plotting a few points, we see that the graph appears as in Fig. 5. There are two regions, one in the first quadrant (denoted $S_1$) and one in the third (denoted $S_2$). We label their areas $A_1$ and $A_2$, respectively. For $S_1$ the line $y = x$ is above the curve $y = x^3$, since for $x = \tfrac{1}{2}$ (a typical value between 0 and 1), $\tfrac{1}{2} > (\tfrac{1}{2})^3$. We obtain

$$A_1 = \int_0^1 (x - x^3)\,dx = \tfrac{1}{2}x^2 - \tfrac{1}{4}x^4]_0^1 = (\tfrac{1}{2} - \tfrac{1}{4}) - 0 = \tfrac{1}{4}.$$

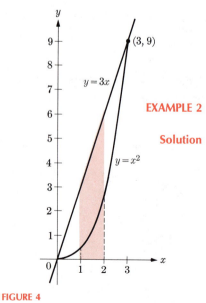

**FIGURE 4**

**AREA BETWEEN CURVES**

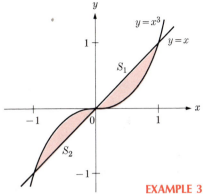

FIGURE 5

To find $A_2$ we note that $y = x^3$ is above $y = x$ for $S_2$ [since $(-\frac{1}{2})^3 > -\frac{1}{2}$], and so

$$A_2 = \int_{-1}^{0} (x^3 - x)\, dx = \frac{1}{4}x^4 - \frac{1}{2}x^2]_{-1}^{0} = 0 - (\frac{1}{4} - \frac{1}{2}) = \frac{1}{4}.$$

The total area between the curves is $A_1 + A_2 = \frac{1}{2}$.  ☐

We work the next example by two methods and, in the process, a number of ideas will be developed which will show how the scope of the method for finding the area between curves may be expanded considerably.

**EXAMPLE 3**  Find the area bounded by the curve $y^2 = x$ and the line $x = 4$.

**Solution**  **Method I.** A sketch of the region $R$ is shown in Fig. 6. The first difficulty is that the relation $y^2 = x$ does not express $y$ as a function of $x$. The upper boundary of $R$ is given by the function $y = \sqrt{x}$ and the lower boundary by the function $y = -\sqrt{x}$. Applying the method used in the previous examples, we obtain

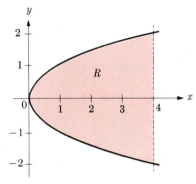

FIGURE 6

$$A = \int_{0}^{4} [\sqrt{x} - (-\sqrt{x})]\, dx$$

$$= \int_{0}^{4} (\sqrt{x} + \sqrt{x})\, dx$$

$$= 2\int_{0}^{4} x^{1/2}\, dx = \frac{4}{3}x^{3/2}]_{0}^{4} = \frac{32}{3}.$$

**Method II.** To solve the problem by this method, we change our point of view. We observe that the equation $x = y^2$ expresses $x$ as a function of $y$. Therefore we may integrate along the $y$ axis. The region bounded by the curve $x = y^2$, the lines $y = \pm 2$, and the $y$ axis is shown as the shaded region in Fig. 7. Its area is

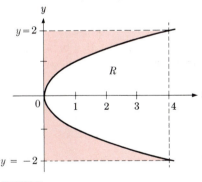

FIGURE 7

$$\int_{-2}^{2} y^2\, dy = \frac{1}{3}y^3]_{-2}^{2} = \frac{8}{3} - (-\frac{8}{3}) = \frac{16}{3}.$$

The area of the rectangle bounded by the four lines $y = -2$, $y = 2$, $x = 0$, and $x = 4$ is 16. The area of the region $R$ is obtained by subtraction:

$$A = 16 - \frac{16}{3} = \frac{32}{3}.$$  ☐

We could have performed the integration of this second method in one step by finding the area between the curves $x = 4$ and $x = y^2$. Integration along the $y$ axis then yields

$$\int_{-2}^{2} (4 - y^2)\, dy = 4y - \frac{1}{3}y^3]_{-2}^{2}$$

$$= (8 - \frac{8}{3}) - (-8 + \frac{8}{3}) = \frac{32}{3}.$$

The technique of looking at the same problem from several points of view is frequently helpful. One way of looking at a problem often yields an insight which other ways miss. If in Example 3 we note that the region is *symmetric* with respect to the $x$ axis, the problem becomes simpler. In that case we integrate the upper half,

$$\int_0^4 \sqrt{x}\,dx,$$

and then double the result to get the answer.

**EXAMPLE 4**　Find the area bounded by $y^2 = x - 1$ and $y = x - 3$.

**Solution**　The curves are sketched in Fig. 8. They intersect where

$$y^2 + 1 = y + 3,$$

or $y = -1, 2$. The corresponding values of $x$ are 2, 5.

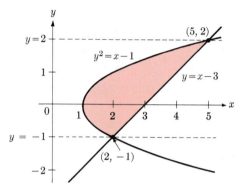

**FIGURE 8**

There are two ways of proceeding. In the first we write the equations

$$x = y^2 + 1 \qquad \text{and} \qquad x = y + 3$$

and observe that in each case $x$ is a function of $y$. Then the area between the curves is obtained by integrating along the $y$ axis from $y = -1$ to $y = 2$. We get

$$A = \int_{-1}^{2} [(y + 3) - (y^2 + 1)]\,dy$$

$$= \int_{-1}^{2} (-y^2 + y + 2)\,dy$$

$$= -\tfrac{1}{3}y^3 + \tfrac{1}{2}y^2 + 2y]_{-1}^{2}$$

$$= (-\tfrac{8}{3} + 2 + 4) - (\tfrac{1}{3} + \tfrac{1}{2} - 2) = \tfrac{9}{2}.$$

The second method employs integration along the $x$ axis. From the figure, we see that the upper portion of the region is bounded by

$$y = \sqrt{x - 1}.$$

The lower boundary is made up of two parts:

$$y = -\sqrt{x - 1} \qquad \text{if } x \text{ is between 1 and 2;}$$

$$y = x - 3 \qquad \text{if } x \text{ is between 2 and 5.}$$

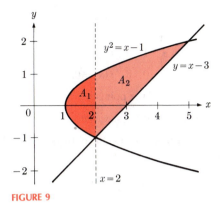

**FIGURE 9**

Figure 9 shows how we solve the problem in two stages. The area $A$ is divided into two parts $A_1$ and $A_2$, as shown.

$$A_1 = \int_1^2 \left[\sqrt{x-1} - (-\sqrt{x-1})\right] dx$$

$$= 2\int_1^2 \sqrt{x-1}\, dx;$$

$$A_2 = \int_2^5 \left[\sqrt{x-1} - (x-3)\right] dx.$$

These integrations can be performed to yield $A_1 + A_2 = \frac{9}{2}$. ☐

It is easy to imagine a complicated region in which the only way to find the area is to decompose it into a number of simpler subregions.

## 1  PROBLEMS

In Problems 1 through 31, in each case sketch the bounded region formed by the given curves and find its area.

1  $x = 2$, $x = 4$, $y = x^2$, $y = -x$

2  $x = 2$, $x = -2$, $y = x^2$, $y = 16 - x^2$

3  $x = 0$, $x = 3$, $y = x^2$, $y = 12x$

4  $y = 0$, $y = \frac{3}{4}$, $x = y^2$, $x = y$

5  $x = -2$, $x = 0$, $y = x^3$, $y = -x$

6  $x = 0$, $x = 3$, $y = \sqrt{x+1}$, $y = \frac{1}{2}x$

7  $x = 1$, $x = 2$, $x^2 y = 2$, $x + y = 4$

8  $x = 0$, $x = \frac{1}{2}$, $y = x^2$, $y = x^4$

9  $y = x^2$, $y = x^4$

10  $x = 0$, $x = 3$, $y = 0$, $y = 1/\sqrt{x+1}$

11  $x = 1$, $x = 2$, $y^2 = x$

12  $y = 0$, $y = 2$, $x = 0$, $x = y^2$

13  $y = -1$, $y = 1$, $x - y + 1 = 0$, $x = 1 - y^2$

14  $y = 1$, $x = y$, $xy^2 = 4$

15  $x = y^2$, $x = 18 - y^2$

16  $y = x^2$, $y^2 = x$

17  $y = \sqrt{x}$, $y = x^3$

18  $y^2 = 4x + 1$, $x + y = 1$

19  $y = x^2 + 4x + 2$, $2x - y + 5 = 0$

20  $y = 1$, $y = 2$, $x = 0$, $x = y/\sqrt{9 - y^2}$

21  $y^2 = 2 - x$, $y = 2x + 2$

22  $y = x^2$, $y = 3x$, $y = 1$, $y = 2$

23  $y = 2x$, $4y = x$, $y = 2/x^2$, $x > 0$

24  $x = 1$, $x = 2$, $x^2 y = 1$, $y = 0$

25  $x = 1$, $x = 2$, $y = \frac{1}{2}x$, $x^2 y = 1$. (Find area of all bounded regions.)

26  $y = \cos x$, $x = 0$, $y = x - \dfrac{\pi}{2}$

27  $y = \sin x$, $y = x + 1 - \dfrac{\pi}{2}$, $x = 0$

28  $x = \frac{1}{2}$, $x = 2$, $y = 0$, $y = |x - 1|^3$

29  $x = 0$, $y = x + 1 - \dfrac{\pi}{4}$, $y = \tan^2 x$

30  $x = -\pi$, $x = \dfrac{\pi}{2}$, $y = |\sin x|$, $y = -x^2 - 1$

31  $y = |\sin 2x|$, $y = \dfrac{4}{\pi}x$

32  Find the area enclosed by the curves

$$y = x^2 - x, \qquad y = x - x^2.$$

33  The curve $|x| + |y| = 1$ encloses a square. Find its area by integration.

In Problems 34 through 36, find the areas of the triangles having the given vertices, using the method of this section.

34  $(0, 0)$, $(4, 1)$, $(2, 4)$

35  $(-2, -1)$, $(2, 2)$, $(3, -2)$

36  $(0, 0)$, $(a, 0)$, $(b, c)$

37  A trapezoid connects the four points $(0, 0)$, $(B, 0)$, $(a, h)$, $(a + b, h)$. Use the methods of this section to obtain the formula for the area $A = \frac{1}{2}h(B + b)$.

38  Find the area of the region bounded by $x = 1$, $y = 0$, $y = 1/x^2$, and the line $x = a$, where $a$ is some number greater than 1. The result will depend on the number $a$. What happens to the value of the area as $a \to +\infty$?

39  Same as Problem 38, with the equation $y = 1/x^2$ replaced by $y = 1/x^{1/2}$.

**40** Same as Problem 38, with the equation $y = 1/x^2$ replaced by $y = 1/x^p$, where $p$ is a fixed positive number larger than 1. What happens for $p$ between 0 and 1 ($p \neq 0, 1$)?

**41** Let $f(x)$ be integrable on $[a, b]$. Define

$$g(x) = \tfrac{1}{2}[|f(x)| + f(x)], \qquad h(x) = \tfrac{1}{2}[|f(x)| - f(x)].$$

Show that

$$\int_a^b f(x)\, dx = \int_a^b g(x)\, dx - \int_a^b h(x)\, dx,$$

$$\int_a^b |f(x)|\, dx = \int_a^b g(x)\, dx + \int_a^b h(x)\, dx.$$

**42** Let $f(x)$ be continuous on $[a, b]$ with $f(a) = f(b) = 0$. Define

$$\operatorname{sign}[f(x)] = \begin{cases} 1 & \text{if } f(x) > 0 \\ -1 & \text{if } f(x) < 0 \\ 0 & \text{if } f(x) = 0. \end{cases}$$

Suppose that $f$ is zero only at the points

$$a = x_0 < x_1 < x_2 < \cdots < x_n = b.$$

Find an expression in terms of the lengths of the subintervals for the value of

$$\int_a^b \operatorname{sign}[f(x)]\, dx.$$

---

## 2

# VOLUME OF A SOLID OF REVOLUTION. DISK METHOD

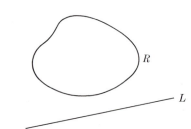

**FIGURE 10**

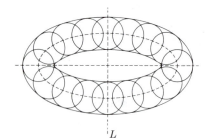

**FIGURE 11**

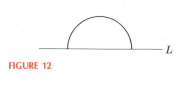

**FIGURE 12**

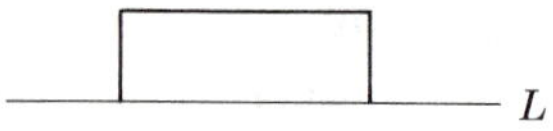

**FIGURE 13**

In Chapter 5 we developed a method for measuring the area of plane regions. The theory rests on two ideas: (1) the formula for the area of a rectangle, and (2) a method of approximating any region by a combination of rectangular regions.

It is possible to extend the above process to the measurement of volumes of solids. First, we recall that the formula for the volume of a rectangular parallelepiped is length times width times height; next we must find a method for approximating a solid by a combination of rectangular parallelepipeds. Instead of discussing this topic in its most general form, we take up the problem of measuring the volume of solids which have special shapes.

Let $R$ be a region in the $xy$ plane (Fig. 10), and $L$ a line which does not intersect it (although $L$ may touch the boundary of $R$). If the region $R$ is revolved about the line $L$, a solid results which is called a **solid of revolution**. If the region $R$ is a circle, the resulting solid will have the shape of a doughnut (called a **torus**), as seen in Fig. 11. A semicircle revolved about its diameter, as shown in Fig. 12, generates a sphere. A rectangle revolved about one of the edges, as in Fig. 13, yields a right circular cylinder.

In Chapter 16 we develop general methods for obtaining the volume of a solid by integration. In this section we shall discuss a method for obtaining the volume of a solid of revolution by the techniques of integration we have already studied. The basis of this method depends on two assumptions. The first is that a right circular cylinder (Fig. 14) with radius of base $a$ and altitude $h$ has a volume $V$ given by the formula

$$V = \pi a^2 h.$$

The second assumption is that any solid of revolution may be approximated by a combination of right circular cylinders.

Let $R$ be a region in the plane bounded by the curve $y = f(x)$, the lines $x = a$ and $x = b$, and the $x$ axis, as shown in Fig. 15. The following theorem

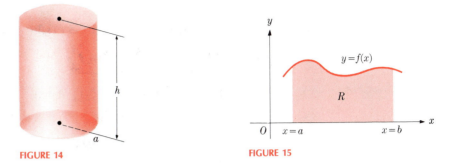

FIGURE 14

FIGURE 15

gives the formula for finding the volume of the solid obtained by revolving the region $R$ about the $x$ axis.

THEOREM 1    **(Disk Method)**    *Suppose that $f$ is a continuous, nonnegative function on the interval $[a, b]$. Then the solid $S$ obtained by revolving the region*

$$R = \{(x, y): a \leq x \leq b, 0 \leq y \leq f(x)\}$$

*about the $x$ axis has a volume $V$ given by the formula*

$$V = \pi \int_a^b [f(x)]^2 \, dx.$$

Proof    We first divide the interval $[a, b]$ into $n$ subintervals by the subdivision $\{a = x_0 < x_1 < \cdots < x_{n-1} < x_n = b\}$. For each integer $i$ between 1 and $n$ let $\xi_i$ be the value of $x$ at which $f$ takes on its minimum value on the $i$th interval $[x_{i-1}, x_i]$. Similarly, let $\eta_i$ be the value of $x$ at which $f$ takes on its maximum value on $[x_{i-1}, x_i]$. Denote by $r_i$ and $R_i$ the rectangles

$$r_i = \{(x, y): x_{i-1} \leq x \leq x_i, 0 \leq y \leq f(\xi_i)\},$$

$$R_i = \{(x, y): x_{i-1} \leq x \leq x_i, 0 \leq y \leq f(\eta_i)\}.$$

Figure 16 shows some of the smaller rectangles and Fig. 17 some of the larger ones. When the rectangle $r_i$ is revolved about the $x$ axis, a right circular cylinder is obtained which we denote by $s_i$. Similarly, revolving $R_i$ yields a cylinder which we denote by $S_i$. The volumes of these cylinders are designated $V(s_i)$ and $V(S_i)$, respectively. (Figure 18 shows one quarter of a disk.) We

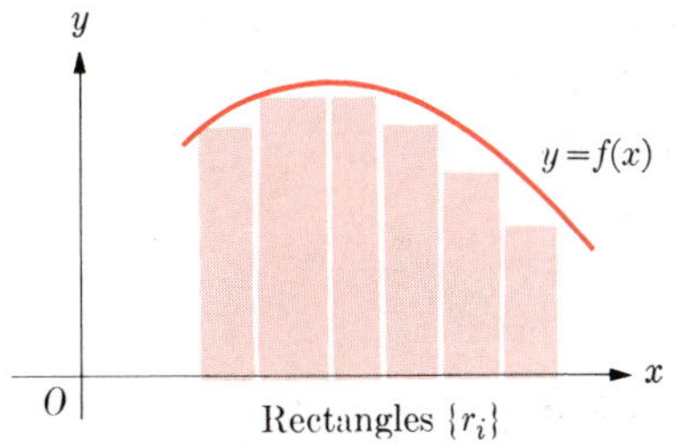

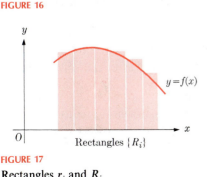

FIGURE 16

FIGURE 17

Rectangles $r_i$ and $R_i$

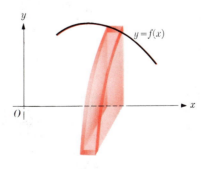

FIGURE 18

have

$$V(s_1) + V(s_2) + \cdots + V(s_n) \leq V \leq V(S_1) + V(S_2) + \cdots + V(S_n).$$

We recall that $\Delta_i x = x_i - x_{i-1}$ and from the formula for the volume of a right circular cylinder, we obtain

$$\pi[f(\xi_1)]^2 \Delta_1 x + \pi[f(\xi_2)]^2 \Delta_2 x + \cdots$$
$$+ \pi[f(\xi_n)]^2 \Delta_n x \leq V \leq \pi[f(\eta_1)]^2 \Delta_1 x + \cdots$$
$$+ \pi[f(\eta_n)]^2 \Delta_n x.$$

From the properties of integrals as described in Chapter 5, Section 2, we can conclude that the left and right sides of the above inequalities tend to

$$\pi \int_a^b [f(x)]^2 \, dx$$

as the norm of the subdivision, $\|\Delta\|$, tends to zero.*                    □

EXAMPLE 1    Find the volume of the solid generated by revolving about the $x$ axis the region bounded by the $x$ axis and one arch of the curve $y = \sin x$.

Solution    The region $R$ to be revolved is shown in Fig. 19. We describe it in set notation by writing $R = \{(x, y) : 0 \leq x \leq \pi, 0 \leq y \leq \sin x\}$. According to the formula of Theorem 1, we have

$$V = \pi \int_0^\pi \sin^2 x \, dx = \frac{\pi}{2} \int_0^\pi (1 - \cos 2x) \, dx = \frac{\pi}{2}\left[ x - \frac{1}{2} \sin 2x \right]_0^\pi = \frac{\pi^2}{2}. \quad □$$

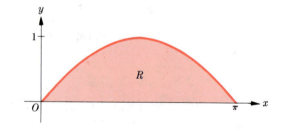

FIGURE 19

EXAMPLE 2    The curve $y = \sqrt{x}$, the line $x = 2$, and the $x$ axis form the sides of a bounded region $R$. Find the volume of the solid generated by revolving $R$ about the $x$ axis.

Solution    We draw the region $R$ which may be described in set notation by $R = \{(x, y) : 0 \leq x \leq 2, 0 \leq y \leq \sqrt{x}\}$. Figure 20 is a sketch of the solid of revo-

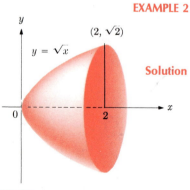

FIGURE 20

*A precise definition of volume would require a discussion analogous to the one for area. An **inner volume** $V^-$ and an **outer volume** $V^+$ may be obtained in a way similar to the way in which inner and outer area are obtained. Then we must show that for regions of the type we shall consider here, we have $V^- = V^+$, and this common value, called the **volume**, is given by the integral.

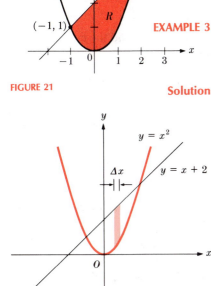

**FIGURE 21**

lution. We have

$$V = \pi \int_0^2 (\sqrt{x})^2 \, dx = \frac{\pi}{2} \left[ x^2 \right]_0^2 = 2\pi.$$

The solid is called a **paraboloid**.

**EXAMPLE 3**   The line $y = x + 2$ and the parabola $y = x^2$ contain a bounded region $R$ between them. (See Fig. 21.) Find the volume $V$ of the solid $S$ generated by revolving $R$ about the $x$ axis.

**Solution**   To describe $R$ in set notation, we find the points of intersection of the line and the parabola. Solving simultaneously, we have

$$x^2 = x + 2 \qquad \text{and} \qquad x = -1, 2.$$

The points of intersection are $(-1, 1)$ and $(2, 4)$. Then $R$ is described by

$$R = \{(x, y): -1 \le x \le 2, x^2 \le y \le x + 2\}.$$

We first find the volume $V_1$ of the solid $S_1$ generated by revolving the region $R_1$ defined by

$$R_1 = \{(x, y): -1 \le x \le 2, 0 \le y \le x^2\}$$

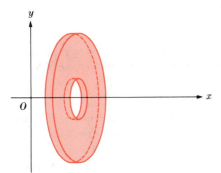

**FIGURE 22**

about the $x$ axis. Then we subtract $V_1$ from the volume $V_2$ of the solid $S_2$ generated by revolving about the $x$ axis the region $R_2$ defined by

$$R_2 = \{(x, y): -1 \le x \le 2, 0 \le y \le x + 2\}.$$

We get

$$V = V_2 - V_1 = \pi \int_{-1}^2 (x + 2)^2 \, dx - \pi \int_{-1}^2 x^4 \, dx$$

$$= \pi \left[ \tfrac{1}{3}(x + 2)^3 - \tfrac{1}{5}x^5 \right]_{-1}^2 = \pi \left[ \tfrac{64}{3} - \tfrac{32}{5} - \left( \tfrac{1}{3} + \tfrac{1}{5} \right) \right] = \tfrac{72}{5} \pi.$$

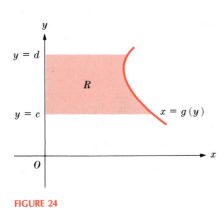

**FIGURE 23**

*Remark.*   In Example 3 the volume $V_1$ obtained by revolving the region $R_1$ about the $x$ axis is subtracted from the volume $V_2$ obtained by revolving the region $R_2$ about the $x$ axis. This procedure can be combined into a single integration. We consider a narrow rectangle of width $\Delta x$ stretching between the curve $y = x^2$ and the curve $y = x + 2$, as shown in Fig. 22. When this rectangle is revolved about the $x$ axis we get a large thin disk with a small disk cut out. (See Fig. 23.) This technique is called the **Washer Method**; the integral in this case is

$$\pi \int_{-1}^2 [(x + 2)^2 - x^4] \, dx,$$

which, of course yields the same result as the one obtained in Example 3.

Suppose a function $g$ is given as a function of $y$, so that $x = g(y)$. We consider a region $R$ defined by the lines $x = 0$, $y = c$, $y = d$, and the curve $x = g(y)$. (See Fig. 24.) If this region is *revolved about the $y$ axis*, then the formula analogous to that in Theorem 1 holds. We merely interchange $x$ and $y$ and obtain the following result for the volume of revolution.

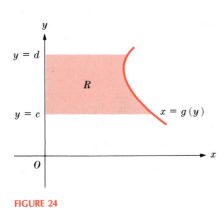

**FIGURE 24**

**COROLLARY**   (**Disk Method**) *Suppose that* $x = g(y)$ *is a continuous nonnegative function for* $c \leq y \leq d$. *Then the solid* $S$ *obtained by revolving the region* $R = \{(x,\ y): 0 \leq x \leq g(y),\ c \leq y \leq d\}$ *about the* $y$ *axis has volume*

$$V = \pi \int_c^d [g(y)]^2\, dy.$$

**FIGURE 25**

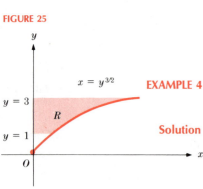

**EXAMPLE 4**   The region $R$ bounded by the lines $x = 0$, $y = 1$, $y = 3$, and the curve $x = y^{3/2}$ is revolved about the $y$ axis. Find the volume generated.

**Solution**   The region $R$ is shown in Fig. 25. The formula for the volume of revolution by the disk method is

$$V = \pi \int_1^3 (y^{3/2})^2\, dy = \pi \int_1^3 y^3\, dy = \pi \frac{y^4}{4} \bigg]_1^3 = 20\pi.$$ □

## 2   PROBLEMS

In Problems 1 through 22 a solid $S$ is generated by revolving a *bounded* region $R$ about the $x$ axis. In each case find the volume of $S$. Describe the region $R$ in set notation whenever it is not prescribed in such terms. Draw $R$ and sketch $S$ in each case.

1   $R = \{(x, y): 0 \leq x \leq 4, 0 \leq y \leq 2\sqrt{5x}\}$.

2   $R = \{(x, y): 0 \leq x \leq b,\ 0 \leq y \leq 2\sqrt{ax}\}$; $a$, $b$ are positive constants.

3   $R$ is bounded by the $x$ axis, the $y$ axis, and the line $2x + 3y = 1$.

4   $R$ is bounded by the $x$ axis, the $y$ axis, and the line $(x/h) + (y/r) = 1$, where $h$ and $r$ are positive constants. Note that the solid generated is a right circular cone of altitude $h$ and radius of base $r$.

5   $R = \{(x, y): -a \leq x \leq a, 0 \leq y \leq \sqrt{a^2 - x^2}\}$. Note that the result is the formula for the volume of a sphere of radius $a$.

6   $R = \{(x, y): a - h \leq x \leq a, 0 \leq y \leq \sqrt{a^2 - x^2}\}$, where $h$ and $a$ are positive constants with $h < a$.

7   $R$ is bounded by $y = 0$ and $y = 4 - x^2$.

8   $R = \{(x, y): 0 \leq x \leq 4 - y^2, 0 \leq y \leq 2\}$.

9   $R$ is bounded by the lines $x = \pi/3$, $y = 0$, and the curve $y = \tan x$.

10   $R = \{(x, y): 0 \leq x \leq 3, 0 \leq y \leq x/\sqrt{4 - x}\}$.

11   $R$ is bounded by the $x$ axis, the $y$ axis, the line $x = \pi/3$, and the curve $y = \cos x$.

12   $R$ is bounded by the $x$ axis, the $y$ axis, the line $x = \pi/4$, and the curve $y = \sec x$.

13   $R$ is bounded by $y = 2$, $x = 0$, $x = y^2$.

14   $R = \{(x, y): y \leq x \leq 4/y, 1 \leq y \leq 2\}$.

15   $R = \{(x, y): 0 \leq x \leq \sqrt{y^2 + 4}, 0 \leq y \leq 2\}$.

16   $R$ is bounded by $x = 0$ and $x + y^2 - 4y = 0$.

17   $R$ is bounded below by $y = 0$ and above by the curves $x = y^2$ and $x = 8 - y^2$.

18   $R = \{(x, y): 0 \leq x \leq 1, x^2 \leq y \leq \sqrt{x}\}$.

19   $R$ is bounded by $y = \sqrt{x}$ and $y = x^3$.

20   $R$ is bounded by $x + y = 5$ and $xy = 4$.

21   $R$ is bounded by $y = x + 2$ and $y^2 - 3y = 2x$.

22   $R = \left\{(x, y): 1 \leq y \leq 2, \dfrac{y}{3} \leq x \leq \sqrt{y}\right\}$.

In Problems 23 through 30 a solid $S$ is generated by revolving a bounded region $R$ about the $y$ axis. Find the volume of $S$.

23   $R$ is bounded by $x = 0$, $y = 2$, $y = 3$, and $x = y^{2/3}$.

24   $R$ is bounded by the $x$ axis, the $y$ axis, and the line $x = 3 - y$.

25   Find the volume of a right circular cone of height $h$ and radius of base $a$ by revolving about the $y$ axis the region $R$ bounded by $x = 0$, $y = 0$, and $x = a - (a/h)y$.

26   The region $R$ in the first quadrant bounded by $x = 0$, $y = 0$, $x = \sqrt{4 - y}$.

27   The region $R$ bounded by the lines $x = 0$, $y = \pi/6$, $y = \pi/4$, and the curve $x = \cos y$.

28   The region $R$ bounded by the lines $x = 0$, $y = \pi/4$, $y = \pi/3$, and the curve $x = \tan y$.

29   The region $R$ bounded by the curves $x = y^2$, $x = \sqrt{y}$.

30   The curves $x = 4y - y^2$ and the line $x = 3y$ form two bounded regions in the first quadrant. Find the volume generated by both regions.

31   Suppose that $f$ is positive and increasing on the closed interval $[a, b]$. Give a detailed proof of Theorem 1 of this section for such a function. (*Hint:* Study the proof of Theorem 1 in Chapter 5.)

**3**

## VOLUME OF A SOLID OF REVOLUTION. SHELL METHOD

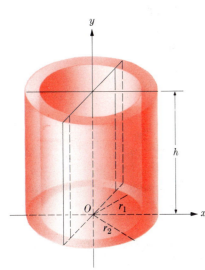

A **cylindrical shell** is the solid contained between two concentric cylinders, as in Fig. 26. A cross section of such a shell is shown in Fig. 27. The volume $V$ of a cylindrical shell with inner radius $r_1$, outer radius $r_2$, and height $h$ is

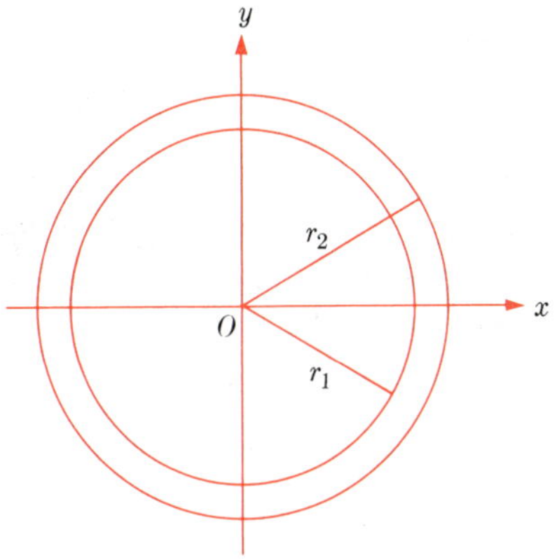

$$V = \pi r_2^2 h - \pi r_1^2 h.$$

Since $r_2^2 - r_1^2 = (r_2 + r_1)(r_2 - r_1)$ we may write the above formula in the form

$$V = \pi(r_2 + r_1)(r_2 - r_1)h = 2\pi\left(\frac{r_2 + r_1}{2}\right)(r_2 - r_1)h.$$

If we define $\bar{r}$ and $\Delta r$ as the average and difference of $r_1$ and $r_2$, respectively, so that

$$\bar{r} = \frac{r_2 + r_1}{2}, \qquad \Delta r = r_2 - r_1,$$

then the formula for the volume of a cylindrical shell becomes

$$V = 2\pi\bar{r}h\,\Delta r.$$

Note that $2\pi\bar{r}$ is the length of the circumference of a circle whose radius is the *average* of $r_1$ and $r_2$, while $\Delta r$ is the *thickness* of the shell.

A solid of revolution may be approximated by a combination of cylindrical shells. To see this, suppose that $R$ is a region in the first quadrant bounded by the lines $x = a$, $x = b$, $y = 0$, and the curve $y = f(x)$. A solid $S$ is formed when $R$ is revolved about the $y$ axis. (See Fig. 28.) To find the volume of $S$, first make a subdivision of the interval $[a, b]$:

$$\{a = x_0 < x_1 < \cdots < x_{n-1} < x_n = b\}.$$

Next let $f(\xi_i)$ and $f(\eta_i)$ be the minimum and maximum values, respectively, of $f$ on the interval $[x_{i-1}, x_i]$. As in the disk method, form the rectangles (see Fig. 29)

$$r_i = \{(x, y): x_{i-1} \le x \le x_i, 0 \le y \le f(\xi_i)\},$$

$$R_i = \{(x, y): x_{i-1} \le x \le x_i, 0 \le y \le f(\eta_i)\}.$$

**FIGURE 26**

**FIGURE 27**

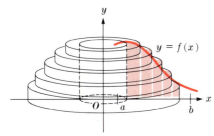

**FIGURE 28**

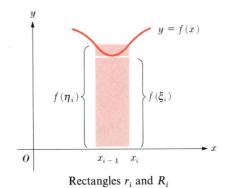

**FIGURE 29**                    Rectangles $r_i$ and $R_i$

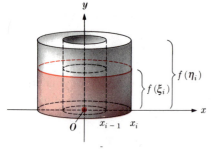

Cylindrical shells $t_i$ and $T_i$

**FIGURE 30**

Since the region $R$ is revolved about the $y$ axis, the rectangles $r_i$ and $R_i$ will generate cylindrical shells $t_i$ and $T_i$ (see Fig. 30). We have for the volumes of the cylindrical shells

$$V(t_i) = 2\pi\left(\frac{x_i + x_{i-1}}{2}\right)\Delta_i x f(\xi_i), \qquad V(T_i) = 2\pi\left(\frac{x_i + x_{i-1}}{2}\right)\Delta_i x f(\eta_i).$$

The volume $V$ of the solid $S$ will be between the quantities

$$\sum_{i=1}^{n} V(t_i) \qquad \text{and} \qquad \sum_{i=1}^{n} V(T_i).$$

We thus obtain

$$\sum_{i=1}^{n} V(t_i) = 2\pi \sum_{i=1}^{n} \bar{x}_i f(\xi_i)\,\Delta_i x \leq V \leq 2\pi \sum_{i=1}^{n} \bar{x}_i f(\eta_i)\,\Delta_i x = \sum_{i=1}^{n} V(T_i),$$

where $\bar{x}_i = \frac{1}{2}(x_i + x_{i-1})$. As the norm of the subdivision, $\|\Delta\|$, tends to zero, we get the following result.

**THEOREM 2**   (*Shell Method*) *Suppose that $f$ is a continuous, nonnegative function on the interval $[a, b]$ with $a \geq 0$. Then the solid $S$ obtained by revolving the region $R = \{(x, y): a \leq x \leq b, 0 \leq y \leq f(x)\}$ about the $y$ axis has volume $V$ given by*

$$V = 2\pi \int_a^b xf(x)\,dx.$$

**FIGURE 31**

When a region $R$ in the first quadrant bounded by the lines $y = c$, $y = d$, $x = 0$, and the curve $x = g(y)$ is revolved about the $x$ axis (see Fig. 31), the volume $V$ of the resulting solid is obtained by substituting $y$ for $x$ in the formula we just found. The result is given as a corollary.

**COROLLARY**   (*Shell Method*) *Suppose that $x = g(y)$ is a continuous nonnegative function for $c \leq y \leq d$ with $c \geq 0$. Then the solid $S$ obtained by revolving the region $R = \{(x, y): 0 \leq x \leq g(y), c \leq y \leq d\}$ about the $x$ axis has volume*

$$V = 2\pi \int_c^d yg(y)\,dy.$$

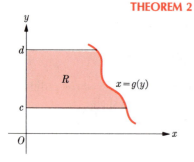

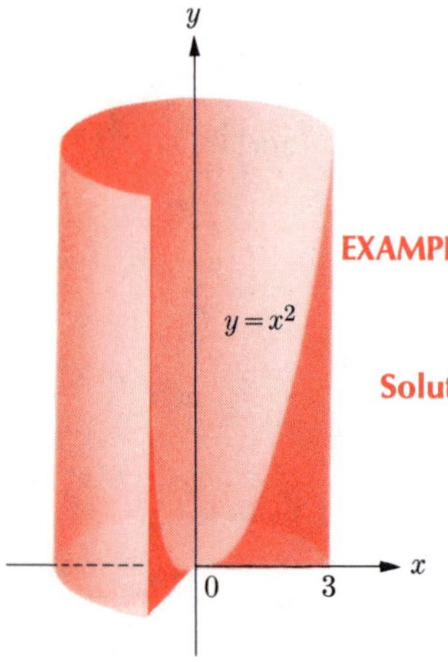

**EXAMPLE 1**   The region $R$ bounded by $y = x^2$ and the lines $y = 0$ and $x = 3$ is revolved about the $y$ axis. Find the volume of the solid generated.

**Solution**   In set notation the region is described by writing

$$R = \{(x, y): 0 \leq x \leq 3, 0 \leq y \leq x^2\}.$$

Figure 32 shows the situation. According to the shell method we have

$$V = 2\pi \int_0^3 x \cdot x^2\,dx = \frac{2\pi}{4}[x^4]_0^3 = \frac{81\pi}{2}.\qquad \square$$

**FIGURE 32**

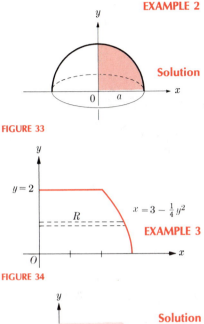

FIGURE 33

FIGURE 34

FIGURE 35

**EXAMPLE 2** The region bounded by the positive $x$ axis, the positive $y$ axis, and the curve $y = \sqrt{a^2 - x^2}$ is revolved about the $y$ axis. Find the volume of the solid generated.

**Solution** The solid generated is a hemisphere of radius $a$, as shown in Fig. 33. Applying the method of shells, we have

$$V = 2\pi \int_0^a x\sqrt{a^2 - x^2}\, dx.$$

To integrate the above expression, we let $u = a^2 - x^2$, $du = -2x\, dx$ and obtain

$$V = -\frac{2\pi}{2}\int_{a^2}^0 \sqrt{u}\, du = -\pi\left[\frac{1}{3}u^{3/2}\right]_{a^2}^0 = \frac{2}{3}\pi a^3.$$

**EXAMPLE 3** The region $R$, bounded by the $x$ axis, the $y$ axis, the line $y = 2$, and the parabola

$$y^2 = 12 - 4x,$$

is revolved about the $x$ axis. Find the volume of the solid generated.

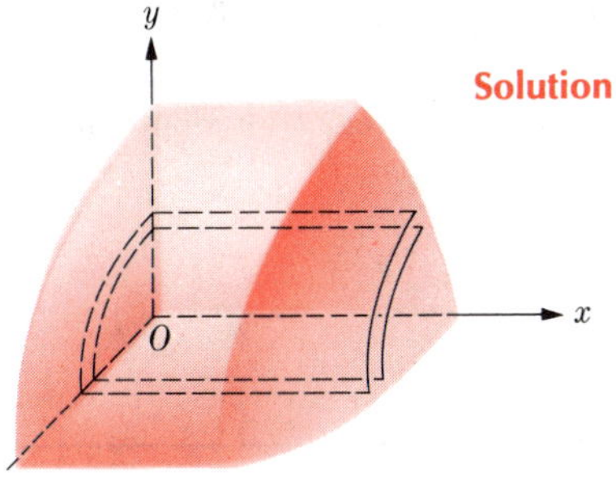

**Solution** The region $R$ described in set notation as

$$R = \{(x, y): 0 \le x \le 3 - \tfrac{1}{4}y^2, 0 \le y \le 2\}$$

is shown in Fig. 34. One-quarter of the solid generated is illustrated in Fig. 35. We apply the shell method, which leads to an integration along the $y$ axis. We have $g(y) = 3 - \tfrac{1}{4}y^2$, and therefore

$$V = 2\pi \int_0^2 y(3 - \tfrac{1}{4}y^2)\, dy = 2\pi[\tfrac{3}{2}y^2 - \tfrac{1}{16}y^4]_0^2 = 10\pi.$$

## 3   PROBLEMS

In Problems 1 through 18, the bounded region $R$ is revolved about the given axis. Describe the region $R$ in set notation whenever it is not prescribed in such terms. Use the method of shells to find the volume of the solid generated.

**1** $R = \{(x, y): 0 \le x \le y^2, 0 \le y \le 2\}$; $R$ is revolved about the $y$ axis.

**2** $R$ is bounded by $y = x^3$, $y = 0$, and $x = 1$; $R$ is revolved about the $y$ axis.

**3** $R$ is the same as in Problem 2, but the region is revolved about the $x$ axis.

**4** $R$ is bounded by the lines $y = 0$, $x = \pi/6$, $x = \pi/2$, and the curve $y = (\sin x)/x$; $R$ is revolved about the $y$ axis.

**5** $R$ is bounded by the lines $y = 0$, $x = 1/4$, $x = 1/3$, and the curve $y = (\cos \pi x)/x$; $R$ is revolved about the $y$ axis.

**6** $R$ is bounded by the lines $x = 0$, $y = 0$, $y = 2$, and the curve $x = \sqrt{y^2 + 4}$; $R$ is revolved about the $y$ axis.

**7** $R$ is the same as in Problem 6, but the region is revolved about the $x$ axis.

**8** $R$ is bounded by $y = 0$, $x = 3$, $y = x/\sqrt{4 - x}$; $R$ is revolved about the $y$ axis.

**9** $R = \{(x, y): y^2 \le x \le 8 - y^2, 0 \le y \le 2\}$; $R$ is revolved about the $y$ axis.

**10** $R$ is bounded by $x = 0$ and $x + y^2 - 4y = 0$; $R$ is revolved about the $y$ axis.

**11** $R$ is bounded by $y = \sqrt{x}$ and $y = x^3$; $R$ is revolved about the $y$ axis.

**12** $R$ is bounded by the lines $y = 0$, $x = 1/3$, $x = 1/2$, and $y = \sec^2(\pi x/2)/x$; $R$ is revolved about the $y$ axis.

**13** $R$ is bounded by the lines $x = 0$, $y = \tfrac{1}{8}$, $y = \tfrac{1}{6}$, and $x = \tan^2(2\pi y)/y$; $R$ is revolved about the $x$ axis.

**14** $R$ is bounded by $y = x^2$, $y = x^3$; $R$ is revolved about the $y$ axis.

**15** Same as Problem 14 but $R$ is revolved about the $x$ axis.

**16** Let $m$ and $n$ be positive integers, $m \ne n$. $R$ is the region in the first quadrant bounded by $y = x^m$, $y = x^n$; $R$ is revolved about the $y$ axis.

**17** Same as Problem 16 but $R$ is revolved about the $x$ axis.

**18** $R$ is bounded by

$$y = 0, \ x = \frac{\pi}{6}, x = \frac{\pi}{4} \quad \text{and} \quad y = (\csc^2 x)/x;$$

$R$ is revolved about the $y$ axis.

**19** The region $R$ bounded by the $x$ axis, the $y$ axis, and the line $x + y = 1$ is revolved about the line $y = -1$. Find the volume of the solid generated.

**20** The region $R$ bounded by $y = 0$ and $y = 4 - x^2$ is revolved about the line $y = -1$. Find the volume of the solid generated.

**21** Find the volume of the solid generated when the region of Problem 20 is revolved about the line $x = -2$.

**22** The region $R$ bounded by the lines $y = 0$, $x = 1$, $x = 2$, and the curve $y = x^2 + 1$ is revolved about the line $x = -1$. Use the shell method to find the volume generated.

**23** The same region $R$ as in Problem 22 is revolved about the line $x = -1$. Use the disk method to find the volume generated.

**24** The region $R$ bounded by the curves $y = x^2$, $y = x^3$ is revolved about the line $y = -2$. Find the volume generated.

**25** The region $R$ bounded by the curves $y = x^2$, $y = x^5$ is revolved about the line $x = 3$. Find the volume generated.

**26** The region $R$ bounded by the lines $x = 0$, $y = 1$, $y = 2$, and the curve $y = \sqrt{x}$ is revolved about the line $x = 4$. Find the volume generated.

**27** The same region $R$ as in Problem 26 is revolved about the line $y = 5$. Find the volume generated.

**28** Let $a$, $b$, $c$ be positive numbers. Suppose that $f$ is a positive function for $a \le x \le b$. Find a formula for the volume of the solid generated when the region $R = \{(x, y): \ a \le x \le b, \ 0 \le y \le f(x)\}$ is revolved about the line $x = -c$.

**29** Same as Problem 28, except that the region $R$ is revolved about the line $y = -c$.

**30** Same as Problem 28, except that

$$R = \{(x, y): 0 \le x \le g(y), \ \alpha \le y \le \beta\}$$

and the line is $x = -\gamma$ where $\alpha$, $\beta$, $\gamma$ are positive numbers.

**31** Same as Problem 30, except that the line is $y = -\gamma$.

**32** Find by the shell method the volume of the torus generated by revolving about the $x$ axis the region interior to the circle $x^2 + (y - b)^2 = a^2$, $b > a > 0$.

---

## 4

## ARC LENGTH

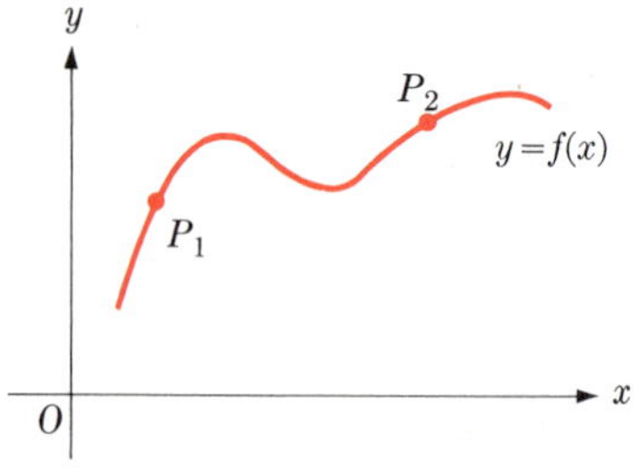

We frequently draw the graph of a function $y = f(x)$ and refer to it as the curve representing the function. When picturing such a graph, we automatically associate a length with any part of it (e.g., the portion going from $P_1$ to $P_2$ in Fig. 36). We now raise three questions:

1. What are the kinds of curves with which we shall associate a length?
2. How do we define length?
3. Once we have defined length, how do we measure it?

These questions are easily answered for straight lines. Every straight line segment has a length given by the distance formula

$$d = \sqrt{(x_1 - x_2)^2 + (y_1 - y_2)^2},$$

where $(x_1, y_1)$ and $(x_2, y_2)$ are the coordinates of the endpoints of the segment.

We get an intuitive idea of the length of a curve in the following way: place a string along the curve which is lying flat in the plane. Remove the string and make it taut. The length of this straight string is measured and the value is taken as the length of the curve. A first step in a *precise* discussion of the length of curves, of which a circular arc is a special case, is the definition of what we call an *arc*.

FIGURE 36

ARC LENGTH

---

**DEFINITION**   *If a graph is given by a function such as $y = f(x)$, $a \leq x \leq b$, and if $f$ is continuous on this interval, then the graph of $f$ is called an* **arc**.

---

The first of the three questions we raised may now be answered by stating that we shall discuss the lengths of only those curves which are arcs.

Turning to the second question—that of defining length—we shall proceed by making use of the one type of arc whose length we know how to measure: the line segment. Let $C$ be an arc in the plane, as shown in Fig. 37.

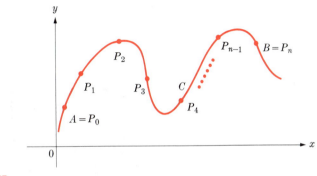

**FIGURE 37**

Suppose that we wish to define the length of such an arc between the points $A$ and $B$. *The definition of length requires a limiting process.* As a first step we mark off a number of points on the arc (in order) between $A$ and $B$ and label them $P_1, P_2, ..., P_{n-1}$. Setting $P_0 = A$ and $P_n = B$, we call this *a subdivision of C* and we write

$$\Delta = \{A = P_0, P_1, ..., P_n = B\}.$$

We next draw a straight line segment from each point $P_i$ to the next point $P_{i+1}$, as in Fig. 38. From the distance formula for the length of a line segment, we compute the lengths

$$|P_0 P_1|, |P_1 P_2|, |P_2 P_3|, ..., |P_{n-1} P_n|,$$

and add them. We can write the sum concisely in the form

$$\sum_{i=1}^{n} |P_{i-1} P_i|.$$

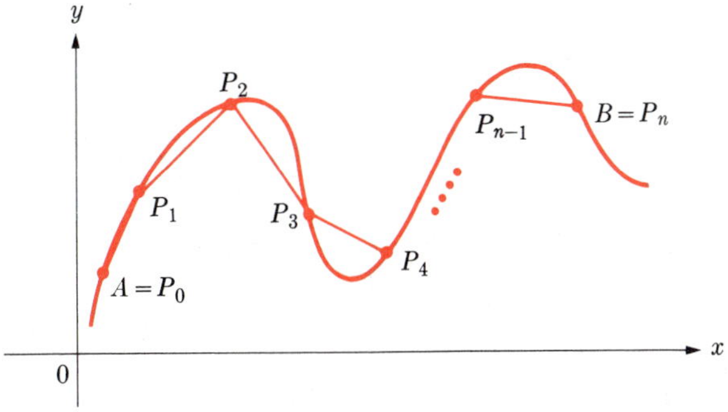

**FIGURE 38**

If the points $P_i$ are "close together," we feel intuitively that the total length of the line segments will be "close to" the as-yet-undefined length of the curve. Therefore, if a sequence of subdivisions is made with an increasing number of points in succeeding subdivisions, we would expect that in the limit the length of the arc would be attained. This is indeed the case. In any subdivision of $C$ we denote the length of the longest line segment connecting successive points by $\|\Delta\|$ and call this the **norm of the subdivision**. We recall that the norm of a subdivision of a section of the $x$ axis was used in the definition of integral.

DEFINITION     **An arc $C$ from $A$ to $B$ has length** *if there is a number $L$ with the following property: For each $\varepsilon > 0$ there is a $\delta > 0$ such that*

$$\left| \sum_{i=1}^{n} |P_{i-1}P_i| - L \right| < \varepsilon$$

*for every subdivision $\Delta = \{A = P_0, P_1, P_2, \ldots, P_{n-1}, P_n = B\}$ with $\|\Delta\| < \delta$. The number $L$ is called the* **length** *of the arc $C$.*

This number $L$ is unique if it exists at all.

Any arc which has a length is called **rectifiable**. According to the definition, *the length $L$ is the limit of the lengths of inscribed line segments, as the maximum distance between successive points of these inscribed line segments tends to zero.*

We now turn to the third question—that of actually finding ways of computing the length of an arc. Suppose that the arc $C$ which has length $s$ is given in the form $y = f(x)$, with $f$ possessing a continuous derivative and with the endpoints of the arc at $A(a, f(a))$, $B(b, f(b))$, as shown in Fig. 39. We introduce subdivision points $P_1, P_2, \ldots, P_{n-1}$, with coordinates $P_i(x_i, y_i)$. The length of the inscribed line segments is given by the formula

$$\sum_{i=1}^{n} |P_{i-1}P_i| = \sum_{i=1}^{n} \sqrt{(x_i - x_{i-1})^2 + (y_i - y_{i-1})^2},$$

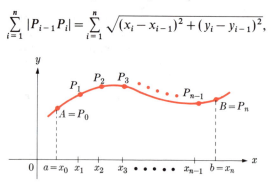

FIGURE 39

where we have denoted $a = x_0$, $f(a) = y_0$, $b = x_n$, and $f(b) = y_n$. At this point we recall the Mean Value Theorem (Chapter 4, Section 2), which states that if a function $f$ has a continuous derivative in an interval $(c, d)$, then there is a value $\xi$ between $c$ and $d$ such that

$$\frac{f(d) - f(c)}{d - c} = f'(\xi), \quad c < \xi < d.$$

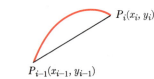

FIGURE 40

This theorem can be used in the present situation. A typical line segment is shown in Fig. 40, and we remember that $y_{i-1} = f(x_{i-1})$, $y_i = f(x_i)$, and therefore there is a value $\xi_i$ between $x_{i-1}$ and $x_i$ such that

$$\frac{f(x_i) - f(x_{i-1})}{x_i - x_{i-1}} = f'(\xi_i), \quad x_{i-1} < \xi_i < x_i,$$

or

$$y_i - y_{i-1} = f'(\xi_i)(x_i - x_{i-1}).$$

We find such an equation for every $i$ from 1 to $n$, and we substitute in the formula for the length of the inscribed line segments, getting

$$\sum_{i=1}^{n} |P_{i-1}P_i| = \sum_{i=1}^{n} \sqrt{(x_i - x_{i-1})^2 + [f'(\xi_i)]^2(x_i - x_{i-1})^2}$$

$$= \sum_{i=1}^{n} \sqrt{1 + [f'(\xi_i)]^2}(x_i - x_{i-1}).$$

We now set $\Delta_i x = x_i - x_{i-1}$ and, recalling the definition of an integral, we see that the limit of

$$\sum_{i=1}^{n} \sqrt{1 + [f'(\xi_i)]^2}\, \Delta_i x$$

is nothing but

$$s = \int_a^b \sqrt{1 + [f'(x)]^2}\, dx. \tag{1}$$

This is the desired formula for the length of an arc, and we have the answer to the third question raised at the beginning of the section. The next theorem formalizes this result.

THEOREM 3    **(Length of an Arc)** *Suppose $f$ is a function with a continuous derivative on an interval containing $a \le x \le b$. Then the length of the arc $C = \{(x, y): a \le x \le b, y = f(x)\}$ is given by Formula (1) above.*

EXAMPLE 1    Find the length of the arc $y = x^{3/2}$ from the point $A(1, 1)$ to $B(2, 2\sqrt{2})$. (See Fig. 41.)

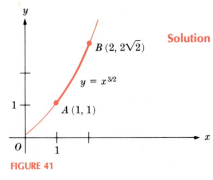

FIGURE 41

Solution    We have $dy/dx = \frac{3}{2}x^{1/2}$ and $(dy/dx)^2 = \frac{9}{4}x$. According to Formula (1), the length of arc, $s$, is given by

$$s = \int_1^2 \sqrt{1 + \tfrac{9}{4}x}\, dx.$$

To integrate this we set $u = 1 + \frac{9}{4}x$, $du = \frac{9}{4}dx$, and obtain

$$s = \int_{13/4}^{11/2} u^{1/2} \cdot \tfrac{4}{9}\, du,$$

new limits having been inserted because $u = \frac{13}{4}$ when $x = 1$ and $u = \frac{11}{2}$ when $x = 2$. Therefore

$$s = [\tfrac{4}{9} \cdot \tfrac{2}{3} u^{3/2}]_{13/4}^{11/2} = \tfrac{2}{27}(11\sqrt{22} - \tfrac{13}{2}\sqrt{13}). \qquad \square$$

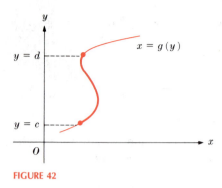

**FIGURE 42**

If a function is given in the form $x = g(y)$ so that $y$ is the independent and $x$ the dependent variable, we obtain the formula for arc length simply by interchanging the $x$ and $y$ variables. We have in this case

$$s = \int_c^d \sqrt{1 + [g'(y)]^2}\, dy, \tag{2}$$

where $s$ is the length of the arc along the curve $x = g(y)$ from $y = c$ to $y = d$. (See Fig. 42.) We observe that arc length is *always positive* so that we always choose the positive square root in Formulas (1) and (2).

Differentials are helpful in remembering the formulas for arc length. We recall that when $y = f(x)$, then $dy = f'(x)\, dx$, and when $x = g(y)$, then $dx = g'(y)\, dy$. From the Fundamental Theorem of Calculus, which states that differentiation and integration are inverse processes, the above formulas for $s$ yield, when $y = f(x)$,

$$\frac{ds}{dx} = \sqrt{1 + [f'(x)]^2} = \sqrt{1 + \left(\frac{dy}{dx}\right)^2},$$

and when $x = g(y)$,

$$\frac{ds}{dy} = \sqrt{1 + [g'(y)]^2} = \sqrt{1 + \left(\frac{dx}{dy}\right)^2}.$$

We square both sides in these formulas and multiply through by $(dx)^2$ in the first and $(dy)^2$ in the second. We get

$$(ds)^2 = \left[1 + \left(\frac{dy}{dx}\right)^2\right](dx)^2 = (dx)^2 + (dy)^2$$

and

$$(ds)^2 = \left[1 + \left(\frac{dx}{dy}\right)^2\right](dy)^2 = (dx)^2 + (dy)^2.$$

In other words, the differential of arc length always satisfies the formula

$$(ds)^2 = (dx)^2 + (dy)^2.$$

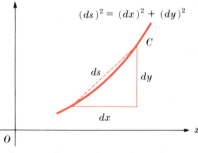

$(ds)^2 = (dx)^2 + (dy)^2$

*ds* approximates the element of arc *C*

**FIGURE 43**

This formula is the basic one for arc length and is easily remembered from Fig. 43. Starting here we divide by $(dx)^2$ or $(dy)^2$, whichever is appropriate, take the positive square root, and then integrate to get the right formula for $s$.

**EXAMPLE 2**    Given $x = y^{2/3} + 4$. Find the length of the arc from $y = 1$ to $y = 8$.

**Solution**    We have

$$\frac{dx}{dy} = \frac{2}{3} y^{-1/3} \qquad \text{and} \qquad \left(\frac{dx}{dy}\right)^2 = \frac{4}{9} y^{-2/3}.$$

Therefore

$$s = \int_1^8 \sqrt{1 + \frac{4}{9} y^{-2/3}}\, dy = \int_1^8 \sqrt{\frac{9y^{2/3} + 4}{y^{2/3}}}\, dy$$

$$= \int_1^8 \sqrt{9y^{2/3} + 4}\, \frac{1}{y^{1/3}}\, dy.$$

We make the substitution $u = 9y^{2/3} + 4$, $du = 6y^{-1/3}\, dy$. Then $u = 13$ when

$y = 1$ and $u = 40$ when $y = 8$. Hence

$$s = \int_{13}^{40} \sqrt{u} \tfrac{1}{6} \, du = \tfrac{1}{6} \int_{13}^{40} u^{1/2} \, du = \tfrac{1}{6} \cdot \tfrac{2}{3} u^{3/2} \Big]_{13}^{40}$$

$$= \tfrac{1}{9} [(40)^{3/2} - (13)^{3/2}].$$

☐

## 4   PROBLEMS

In Problems 1 through 14, find the length of arc in each case.

**1**   $y = \dfrac{1}{6}x^3 + \dfrac{1}{2x}, \quad 1 \le x \le 3$

**2**   $y = 2x^{3/2}, \quad 0 \le x \le 2$

**3**   $y = x^{2/3}, \quad -8 \le x \le -2$

**4**   $f(x) = (9 - x^{2/3})^{3/2}, \quad 1 \le x \le 2$

**5**   $x = \tfrac{1}{6}y^4 + \tfrac{3}{16}y^{-2}, \quad 1 \le y \le 4$

**6**   $x = (a^{2/3} - y^{2/3})^{3/2}, \quad -a \le y \le a$

**7**   $x = \tfrac{1}{3}(y - 3)\sqrt{y}, \quad 0 \le y \le 3$

**8**   $y = \tfrac{1}{2}(3x)^{2/3}, \quad 1 \le x \le 2$

**9**   $y = \tfrac{1}{8}x^4 + \tfrac{1}{4}x^{-2}, \quad 1 \le x \le 2$

**10**   $x = \tfrac{3}{5}y^{5/3} - \tfrac{3}{4}y^{1/3}, \quad 1 \le y \le 2$

**11**   $y = \tfrac{1}{4}x^5 + \tfrac{1}{15}x^{-3}, \quad 2 \le x \le 3$

**12**   $x = 2y^{1/2} - \tfrac{1}{6}y^{3/2}, \quad 1 \le y \le 2$

**13**   $y = \tfrac{1}{8}x^8 + \tfrac{1}{24}x^{-6}, \quad 2 \le x \le 3$

**14**   $x = \tfrac{1}{6}y^{4/5} - \tfrac{25}{16}y^{6/5}, \quad 1 \le y \le 2$

In Problems 15 through 22, in each case set up the integral for arc length but do not attempt to evaluate the integral.

**15**   $y = \sqrt{x}, \quad 1 \le x \le 4$

**16**   $y = x^3, \quad 1 \le x \le 2$

**17**   $x = y^4 - 2y^2 + 3, \quad 1 \le y \le 2$

**18**   $y = \tfrac{1}{3}(x - 3)\sqrt{x}, \quad 1 \le x \le 2$

**19**   $y = \cos 2x, \quad 0 \le x \le \dfrac{\pi}{2}$

**20**   $y = \tfrac{4}{5}\sqrt{25 - x^2}, \quad 0 \le x \le 4$

**21**   $x = \tan \dfrac{1}{2}y, \quad 0 \le y \le \dfrac{\pi}{4}$

**22**   $x = \sec y, \quad \dfrac{\pi}{6} \le y \le \dfrac{\pi}{3}$.

In Problems 23 through 26, in each case subdivide the given interval into the number of equal subintervals indicated by the integer $n$. Compute the quantity

$$\sum_{i=1}^{n} \sqrt{1 + [f'(\xi_i)]^2}\,(x_i - x_{i-1}),$$

by taking the value $\xi_i$ at the midpoint of the $i$th subinterval, and in this way approximate the length of the arc.

**23**   $y = x^3/3, \, 0 \le x \le 2, \, n = 4$

**24**   $y = \tfrac{2}{5}x^{5/2}, \, 0 \le x \le 2, \, n = 4$

**25**   $y = \tfrac{4}{5}\sqrt{25 - x^2}, \quad 0 \le x \le 2, \, n = 4$

**26**   $y = \dfrac{2}{\sqrt{x^2 + 1}}, \quad 0 \le x \le 2, \, n = 4$

In Problems 27 through 30, follow the directions for Problems 23 through 26, except that the quantity

$$\sum_{i=1}^{n} \sqrt{1 + [g'(\eta_i)]^2}\,(y_i - y_{i-1})$$

is to be computed, with $\eta_i$ the midpoint of $[y_{i-1}, y_i]$.

**27**   $x = y^2, \quad 0 \le y \le 1, \, n = 4$

**28**   $x = \sin y, \quad 0 \le y \le 2\pi, \, n = 4$

**29**   $x = 2y^{3/2}, \quad 1 \le y \le 3, \, n = 3$

**30**   $x = \tan y, \quad -\dfrac{\pi}{3} \le y \le \dfrac{\pi}{3}, \, n = 2$

**31**   Find the total length of the loop of the curve $6y^2 = x(x - 2)^2$ between $x = 0$ and $x = 2$.

**\*32**   Find the total length of the loop of the curve $3y^2 = x^2(2x + 1)$ between $x = 0$ and $x = -\tfrac{1}{2}$.

**33**   Given the curve $ay^2 = x(x - b)^2$ which is similar to that in Problem 31. Find three pairs of values for $a$ and $b$ so that the length of the loop from $x = 0$ to $x = b$ can be found by evaluating an elementary integral.

**34**   Let $f$ be any function which has arc length on $[a, b]$. Let $s(x)$ be the length of the arc from the point $a$ to $x$. Show that $s$ is an increasing function of $x$ for $x \in [a, b]$.

**35**   Suppose that $f$ is continuous on $[a, b]$ and that $f$ has a derivative except at one point where the derivative has a jump. Show how to find the length of the arc from $a$ to $b$.

**36**   Find the length of the arc $y = x^{4/5}$ from $x = 1$ to $x = 2$.

**37**   Let $f(x)$, $0 \le x < \infty$ be a positive function with a continuous derivative. Let $l(X)$ be the length of the curve between 0 and $X$. Show that $l(X) \to +\infty$ as $X \to +\infty$.

## 5

## AREA OF A SURFACE OF REVOLUTION

Right circular cone

FIGURE 44

When an arc situated above the $x$ axis in the $xy$ plane is revolved about the $x$ axis, a **surface of revolution** is generated. We shall define the area of such a surface and show how to calculate it by integration. In defining the area of a plane region, we know that the fundamental quantity is the area of a rectangle. For the volume of a solid of revolution, we saw that the fundamental quantity is the volume of a right circular cylinder. *For determining the area of a surface of revolution, the fundamental quantity is the lateral area of a right circular cone.*

A right circular cone is shown in Fig. 44. We denote the radius of the base $r$, the slant height $l$, and the area of the lateral surface $S$. A line from the vertex of the cone to a point on the boundary of the circular base is called a **generator** of the cone. The surface area $S$ is given by the formula

$$S = \pi r l,$$

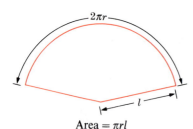

Area = $\pi r l$

FIGURE 45

which may be seen intuitively by cutting the cone along a generator, $l$, and flattening the surface onto a plane. The result is a circular sector, as shown in Fig. 45, whose area is exactly $\pi r l$.

If a cone is cut by a plane parallel to the base, the portion of the cone between the base and the disk cut off by the parallel plane is called a **frustum** of a cone. Such a frustum is shown in Fig. 46.

The smaller radius is $r_1$, the larger is $r_2$, and the slant height is $l$. The surface area $F$ of such a frustum is

$$F = \pi r_2(l_1 + l) - \pi r_1 l_1.$$

Because of similar triangles (see Fig. 46), we have the proportion

$$\frac{l_1}{r_1} = \frac{l}{r_2 - r_1} \qquad \text{or} \qquad l_1 = \frac{lr}{r_2 - r_1}.$$

We substitute this value of $l_1$ in the formula for $F$ and, after some algebra we obtain for the **lateral surface area of a frustum of a cone**

$$F = \pi(r_1 + r_2)l.$$

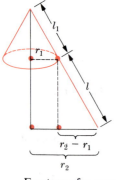

Frustum of a cone

FIGURE 46

We write this formula in the more suggestive form

$$F = 2\pi \bar{r} l,$$

where $\bar{r} = \frac{1}{2}(r_1 + r_2)$ is the average of the radii $r_1$ and $r_2$.

Suppose that an arc $C$, situated entirely above the $x$ axis, is given in the form

$$y = f(x), \quad a \le x \le b.$$

We revolve $C$ about the $x$ axis, obtaining a surface of revolution. We let $\{a = x_0 < x_1 < x_2 < \cdots < x_n = b\}$ be a subdivision of $[a, b]$, and we construct the inscribed polygon to $C$ exactly as we did in defining arc length. When this inscribed polygon is revolved about the $x$ axis, we obtain a collection of frustums of cones, as in Fig. 47. The sum of the lateral surface areas of these frustums will be an approximate measure of the (as yet undefined) area of the surface $S$ generated by the arc $C$. We argue intuitively that the "closer" the inscribed polygon is to the curve $C$, the more accurate will be the approximation to the surface area.

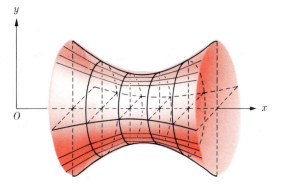

**FIGURE 47**

**DEFINITION**   *Assume that $C: y = f(x)$,   $a \leq x \leq b$ is situated in the upper half-plane (that is, $f(x) \geq 0$). Let $y_i = f(x_i)$, $i = 1, 2, \ldots, n$; define $S$ as the surface of revolution obtained by revolving $C$ about the $x$ axis. Then the* **area** *$A(S)$ is defined as the limit approached by the sums*

$$\sum_{i=1}^{n} 2\pi \bar{y}_i \sqrt{(x_i - x_{i-1})^2 + (y_i - y_{i-1})^2},$$

*where*

$$\bar{y}_i = \frac{y_i + y_{i-1}}{2},$$

*as the norms of the subdivisions tend to zero, provided the limit exists. Each term in the above sum is the lateral surface area of the inscribed frustum of a cone (Fig. 48).*

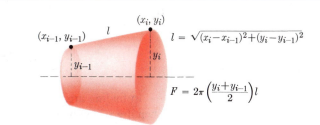

**FIGURE 48**

We now state without proof the formula for evaluating the area $A(S)$. *If an arc*

$$C: y = f(x), \quad a \leq x \leq b$$

*is in the upper half-plane, then the area $A(S)$ of the surface generated by* *revolving C about the x axis is given by the formula*

$$A(S) = 2\pi \int_a^b y\sqrt{(dx)^2 + (dy)^2}. \tag{1}$$

Since $y = f(x)$ and $dy = f'(x)\,dx$, we can write the above formula in the useful form

$$A(S) = 2\pi \int_a^b f(x)\sqrt{1 + [f'(x)]^2}\,dx.$$

We recall from the section on arc length that

$$ds = \sqrt{(dx)^2 + (dy)^2}.$$

The above formula (1) can be written in a simpler, more useful form. For revolution about the $x$ axis, we write

$$A(S) = 2\pi \int_a^b y\,ds, \tag{2}$$

while for revolution about the $y$ axis, we write

$$A(S) = 2\pi \int_a^b x\,ds. \tag{3}$$

Then *any particular form for x, y, and ds is usable in Formulas (2) and (3).* For example, if $C: y = f(x)$ is revolved about the $y$ axis, we get the formula

$$A(S) = 2\pi \int_a^b x\sqrt{1 + [f'(x)]^2}\,dx.$$

Also, if $C$ is given in the form $x = g(y)$, $c \le y \le d$, then $ds = \sqrt{1 + [g'(y)]^2}\,dy$ and if $C$ is revolved about the $y$ axis, then

$$A(S) = 2\pi \int_c^d x\sqrt{1 + [g'(y)]^2}\,dy.$$

Hence

$$A(S) = 2\pi \int_c^d g(y)\sqrt{1 + [g'(y)]^2}\,dy.$$

In finding areas of revolution, it pays to remember formulas (2) and (3). Then if $y = f(x)$ or $x = g(y)$ we make the appropriate substitution depending on whether we are revolving the curve about the $x$ axis or $y$ axis.

We observe that the area of a surface of revolution is always positive.

EXAMPLE 1    Find the surface area of a zone of a sphere obtained by revolving about the $x$ axis the part of the semicircle

$$\{(x, y): b \le x \le c, y = \sqrt{a^2 - x^2}\}, \quad (\text{where } -a < b < c < a).$$

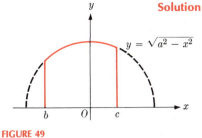

**FIGURE 49**

**Solution**   We draw the graph (Fig. 49) and use the formula

$$A = 2\pi \int_b^c y \, ds,$$

with $ds = \sqrt{1 + (dy/dx)^2} \, dx$ and $y = \sqrt{a^2 - x^2}$. We have

$$\frac{dy}{dx} = -\frac{x}{\sqrt{a^2 - x^2}} \quad \text{and} \quad ds = \frac{a}{\sqrt{a^2 - x^2}} \, dx.$$

Therefore

$$A = 2\pi \int_b^c \sqrt{a^2 - x^2} \cdot \frac{a}{\sqrt{a^2 - x^2}} \, dx = 2\pi a(c - b). \qquad \square$$

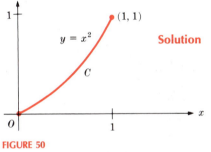

**FIGURE 50**

**EXAMPLE 2**   Find the area of the surface generated by revolving about the $y$ axis the arc $C$ given by $C = \{(x, y): 0 \le x \le 1, \, y = x^2\}$. (See Fig. 50.)

**Solution**   We use the formula

$$A = 2\pi \int_0^1 x \, ds,$$

with $ds = \sqrt{1 + (dy/dx)^2} \, dx = \sqrt{1 + 4x^2} \, dx$. Therefore

$$A = 2\pi \int_0^1 x\sqrt{1 + 4x^2} \, dx.$$

We make the substitution $u = 1 + 4x^2$, $du = 8x \, dx$. Then

$$A = 2\pi \int \frac{1}{8}\sqrt{u} \, du = \frac{\pi}{6} u^{3/2} = \frac{\pi}{6}[(1 + 4x^2)^{3/2}]_0^1 = \frac{\pi}{6}(5\sqrt{5} - 1). \qquad \square$$

**EXAMPLE 3**   Find the area of the surface generated by revolving about the $y$ axis the arc $C$ given by $x = 2\sqrt{y}$, $1 \le y \le 2$. (See Fig. 51.)

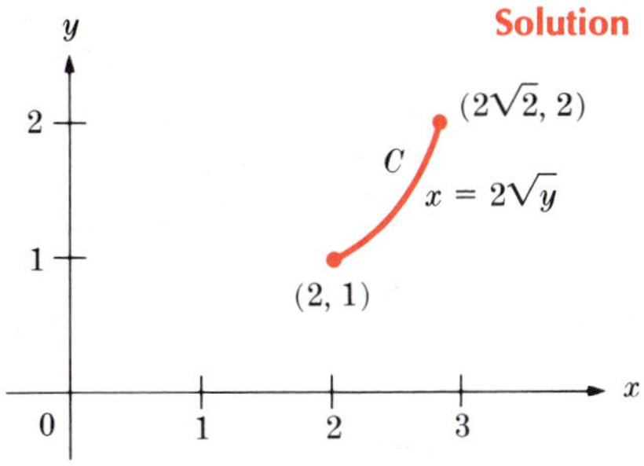

**Solution**   We use the formula

$$A(S) = 2\pi \int_1^2 x \, ds$$

with $ds = \sqrt{1 + [g'(y)]^2} \, dy$. We have $(dx/dy)^2 = y^{-1}$ and

$$A(S) = 2\pi \int_1^2 2\sqrt{y}\sqrt{1 + y^{-1}} \, dy$$

$$= 4\pi \int_1^2 \sqrt{y}\sqrt{\frac{y + 1}{y}} \, dy = 4\pi \int_1^2 \sqrt{y + 1} \, dy$$

$$= 4\pi \cdot \frac{2}{3}(1 + y)^{3/2} \Big]_1^2 = \frac{8\pi}{3}[3^{3/2} - 2^{3/2}]. \qquad \square$$

**FIGURE 51**

## 5  PROBLEMS

In Problems 1 through 12, find the area of each surface $S$ obtained by revolving the given arc $C$ about the $x$ axis.

1  $C = \{(x, y): 0 \leq x \leq 2, y = \frac{1}{3}x^3\}$

2  $C = \{(x, y): 1 \leq x \leq 2, y = \frac{1}{6}x^3 + (1/2x)\}$

3  $C = \{(x, y): 1 \leq x \leq 3, y = \sqrt{x}\}$

4  $C = \{(x, y): 0 \leq x \leq 6, y^2 = 6x\}$

5  $C = \{(x, y): 0 \leq x \leq 2, y^2 = 2 - x\}$

6  $C = \{(x, y): 18y^2 = x(6 - x)^2, 0 \leq x \leq 6\}$

7  $C = \{(x, y): 1 \leq x \leq 2, y = (x^4/8) + (1/4x^2)\}$

8  $C = \{(x, y): -a \leq x \leq a, x^{2/3} + y^{2/3} = a^{2/3}, y \geq 0\}$

9  $C = \{(x, y): 0 \leq x \leq 1, y = \frac{3}{2}x^{2/3}\}$

10  $C = \{(x, y): 0 \leq x \leq 1, y = \frac{5}{3}x^{3/5}\}$

11  $C = \{(x, y): 0 \leq x \leq 1/2, x^2 = 1 - y^2\}$

12  $C = \left\{(x, y): 1 \leq x \leq 2, y = \frac{1}{10}x^5 + \frac{1}{6x^3}\right\}$

In Problems 13 through 17, find the area of each surface $S$ obtained by revolving the given arc $C$ about the $y$ axis.

13  $C = \{(x, y): y = \frac{2}{3}x^{3/2}, 0 \leq x \leq 2\}$

14  $C = \left\{(x, y): x = \frac{1}{6}y^3 + \frac{1}{2y}, 1 \leq y \leq 3\right\}$

15  $C = \{(x, y): x = \frac{5}{4}y^{4/5}, 1 \leq y \leq 2\}$

16  $C = \left\{(x, y): x = \frac{1}{\sqrt{8}}y(1 - y^2)^{1/2}, 0 \leq y \leq 1\right\}$

17  $C = \{(x, y): x^2 = 1 - y, 0 \leq y \leq 1\}$

18  a) Let $C$ be an arc in the first quadrant given by $y = f(x)$, $a \leq x \leq b$. Develop a formula for the area of the surface $S$ generated when $C$ is revolved about the line $y = -3$. Do the same for the line $y = -k$, where $k > 0$.
b) The arc $C = \{(x, y): y = \frac{3}{2}x^{2/3}, 1 \leq x \leq 8\}$ is revolved about the line $y = -1$. Find the surface area generated.

19  a) An arc $C$ in the first quadrant is given by $y = f(x)$, $a \leq x \leq b$. Find the formula for the area of the surface $S$ generated when $C$ is revolved about the line $x = -h$, where $h > 0$.
b) The arc $C = \{(x, y): y = \frac{2}{3}x^{3/2}, 0 \leq x \leq 4\}$ is revolved about the line $x = -2$. Find the area of the surface generated.

## CHAPTER 6

## REVIEW PROBLEMS

In Problems 1 through 10, find the area bounded by the given curves.

1  $y = 2 - x, y = 4 - x^2$

2  $y = 2 - x^4, y = x^2$

3  $x = 2, y = 3, x^2y = 1$

4  $x = 3 - y^2, x = 1$

5  $xy^2 = 2, x + y = 1, x = 4$

6  $y = x^5, y = x^4$

7  $C = \left\{(x, y): y = \sin 2x, \frac{\pi}{6} \leq x \leq \frac{\pi}{3}\right\}, x = \frac{\pi}{6}, x = \frac{\pi}{3}, y = 0$

8  $y = \sec^2 x, x = 0, y = 0, x = \frac{\pi}{4}$

9  $y = |2 - x|, y = 5$

10  $x = |3 - y|, x = 4$

In Problems 11 through 18, find the volume of the solid obtained by revolving the given region $R$ about the axis stated. Use either the disk or the shell method.

11  $R$ is bounded by the curve $y = 1 - x + x^2$ and the lines $x = 0, x = 2, y = 0$; about the $x$ axis.

12  $R$ is bounded by $x = 2 + y^4, x = 0, y = 0, y = 2$; about the $y$ axis.

13  $R = \{(x, y): x^5 \leq y \leq x^{2/3}, 0 \leq x \leq 1\}$; about the $x$ axis.

14  $R = \{(x, y): x^5 \leq y \leq x^{2/3}, 0 \leq x \leq 1\}$; about the $y$ axis.

15  $R = \{(x, y): 0 \leq x \leq 9 - y^2, 0 \leq y \leq 3\}$; about the $x$ axis.

16  $R = \{(x, y): 0 \leq x \leq 9 - y^2, 0 \leq y \leq 3\}$; about the $y$ axis.

17  $R = \left\{(x, y): 0 \leq y \leq \sin 2x, 0 \leq x \leq \frac{\pi}{6}\right\}$; about the $x$ axis.

18  $R = \left\{(x, y): 0 \leq x \leq \csc 2y, \frac{\pi}{4} \leq y \leq \frac{\pi}{3}\right\}$; about the $y$ axis.

19  a) Suppose a region $R = \{(x, y): a \leq x \leq b, f_1(x) \leq y \leq f_2(x)\}$ in the first quadrant is revolved about the line $y = -k, k > 0$. Find a formula for the volume generated.
b) Find the volume generated when the region $R = \{(x, y): 0 \leq x \leq 2, 0 \leq y \leq 4 - x^2\}$ is revolved about the line $y = -3$.

**20** a) Suppose $R = \{(x, y): a \leq x \leq b, f_1(x) \leq y \leq f_2(x)\}$ in the first quadrant is revolved about the line $x = -h$, $h > 0$. Find a formula for the volume generated.

b) Find the volume generated when the region $R = \{(x, y): 0 \leq x \leq 2, 0 \leq y \leq 4 - x^2\}$ is revolved about the line $x = -2$.

In Problems 21 through 24, find the length of arc for each curve $C$.

**21** $C = \{(x, y): 1 \leq x \leq 4, y = x^3 + (\frac{1}{12})x^{-1}\}$

**22** $C = \{(x, y): 0 \leq x \leq 1, y = x^{2/3} + 1\}$

**23** $C = \{(x, y): 0 \leq x \leq 3, y = |1 - x^{3/2}|\}$

**24** $C = \{(x, y): 1 \leq x \leq 3, y = |\frac{1}{2}x^{2/3} - 4|\}$

In Problems 25 through 28, find the area of each surface $S$ obtained by revolving the given arc $C$ about the line stated.

**25** $C = \{(x, y): 1 \leq x \leq 32, y = \frac{5}{7}x^{7/5}\}$;  $x$ axis

**26** $C = \{(x, y): 1 \leq x \leq 32, y = \frac{5}{6}x^{6/5}\}$;  $y$ axis

**27** $C = \{(x, y): 0 \leq x \leq b, y = ax\}$;  $y = -a$, where $a, b$ are any positive constants. (Surface area of a frustum of a cone.)

**28** $C = \{(x, y): 0 \leq x \leq b, y = ax^2\}$;  $x = -c$, where $a, b, c$ are any positive constants.

# 7

# LOGARITHMS, THE EXPONENTIAL FUNCTION, INVERSE FUNCTIONS

Logarithms were invented as an aid to computation. Since addition and subtraction are simpler than multiplication and division, the purpose behind logarithms is the reduction of problems in multiplication and division to the easier ones of addition and subtraction. With the advent of computers and calculators there is now little or no need for using logarithms in making computations. However, it turns out that logarithms play an important theoretical role in calculus and other branches of mathematics. Moreover, there are extensive uses for the logarithm and related functions in various branches of technology. In this chapter we show how the logarithm function can be defined by means of an integral, and we study the exponential function which is the inverse of the logarithm.

## 1

### THE LOGARITHM FUNCTION

The reader has undoubtedly learned the basic ideas of logarithms in elementary courses in algebra and trigonometry. The underlying principle occurs in the *law of exponents*, which states that

$$b^{\alpha} \cdot b^{\beta} = b^{\alpha + \beta},$$

where $\alpha$, $\beta$, and $b$ are any numbers. If the exponents $\alpha$ and $\beta$ are integers, the meaning of the above equation is clear from the very definition of an exponent as a symbol for repeated multiplication. Similarly, if the exponents are rational and $b$ is positive, the notion of the $n$th root of a number allows us to derive the above law of exponents. However, it is not clear how we would go about defining the number

$$3^{\sqrt{2}}.$$

The definition of numbers with irrational exponents requires an additional effort. The customary introduction of logarithms proceeds on the *assumption* that all such numbers are known. In particular, if $b$ and $N$ are any numbers, with $b > 0$, $b \neq 1$, it is assumed that the equation

$$b^x = N$$

may always be solved uniquely for $x$. The value of $x$ which solves this equation is written

$$x = \log_b N.$$

The usual laws of logarithms, such as

$$\log_b M + \log_b N = \log_b (MN),$$

$$\log_b M - \log_b N = \log_b \left( \frac{M}{N} \right),$$

follow immediately from the law of exponents.

Rather than follow the traditional method of elementary mathematics to define logarithms, we shall use the help of the calculus. First we recall the formula

$$\int t^n \, dt = \frac{t^{n+1}}{n+1} + C, \quad n \neq -1.$$

This integration formula fails for $n = -1$. Nevertheless, if we plot the graph of the integrand $y = 1/t$ for positive values of $t$, the expression

$$\int_1^x \frac{1}{t} \, dt, \quad x > 0,$$

has meaning, as it is simply the area under the curve between the points 1 and $x$. (See Fig. 1.) Its value will depend on $x$ and, in fact, it is a function of $x$ so long as $x$ is positive. We define a new function in the following way.

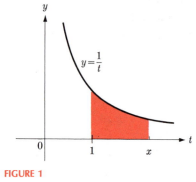

$y = \frac{1}{t}$

**FIGURE 1**

**DEFINITION**   *For values of $x > 0$, we define*

$$\ln x = \int_1^x \frac{1}{t} \, dt.$$

*Remark.*   For the present, read ln $x$ as "ell-en $x$" or "ell-en of $x$."

The integral of the function $1/t$ has a useful property which we state in the form of a lemma.

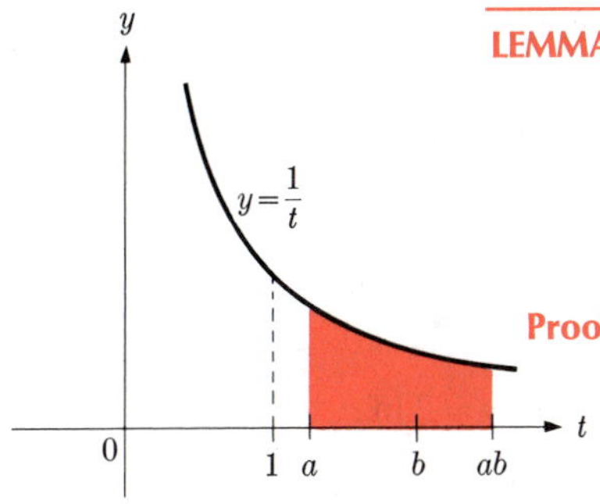

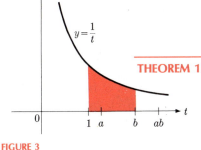

**FIGURE 2**

**FIGURE 3**

**LEMMA**  *If a and b are any positive numbers, then*

$$\int_a^{a \cdot b} \frac{1}{t} \, dt = \int_1^b \frac{1}{t} \, dt.$$

**Proof**  We let $u = (1/a)t$ in the integral on the left; then $du = (1/a) \, dt$. After making this change, we have for the limits of integration: $t = a$ corresponds to $u = 1$, and $t = ab$ corresponds to $u = b$. Therefore

$$\int_a^{a \cdot b} \frac{dt}{t} = \int_1^b \frac{a \, du}{au} = \int_1^b \frac{du}{u} = \int_1^b \frac{dt}{t},$$

the last equality being valid because the letter used for the variable of integration is irrelevant.  □

Geometrically, the preceding lemma states that the shaded areas shown in Figs. 2 and 3 are equal. With the aid of this lemma the following basic theorem is easy to establish.

**THEOREM 1**  *If a and b are any positive numbers, then*

*i) $\ln (a \cdot b) = \ln a + \ln b$,*
*ii) $\ln (a/b) = \ln a - \ln b$,*
*iii) $\ln 1 = 0$,*
*iv) $\ln a^r = r \ln a$, if r is any rational number.*

**Proof**  To prove (i), we have from the definition of the ln function that

$$\ln (a \cdot b) = \int_1^{a \cdot b} \frac{dt}{t}.$$

This integral may be written in the form

$$\int_1^{a \cdot b} \frac{dt}{t} = \int_1^a \frac{dt}{t} + \int_a^{a \cdot b} \frac{dt}{t}.$$

At this point we use the lemma to replace the second integral on the right and get

$$\ln (a \cdot b) = \int_1^a \frac{dt}{t} + \int_1^b \frac{dt}{t} = \ln a + \ln b.$$

To prove (ii), we write $a = b \cdot (a/b)$ and apply (i) to the product of $b$ and $a/b$. We then have

$$\ln a = \ln \left( b \cdot \frac{a}{b} \right) = \ln b + \ln \left( \frac{a}{b} \right),$$

and when we transfer $\ln b$ to the left side, we see that

$$\ln \frac{a}{b} = \ln a - \ln b.$$

As for (iii), we apply (ii) with $b = a$. This says that

$$\ln 1 = \ln \frac{a}{a} = \ln a - \ln a = 0.$$

To establish (iv) we proceed in stages. If the exponent is a positive integer, we use (i) and mathematical induction. First, if $b = a$, we observe that (i) states that

$$\ln a^2 = \ln (a \cdot a) = \ln a + \ln a = 2 \ln a.$$

Then, using a step-by-step method, we obtain

$$\ln a^n = n \ln a, \quad \text{if } n \text{ is a positive integer.}$$

To obtain the result for negative integers, we write $a^{-n} = 1/a^n$ and use (ii) to get

$$\ln (a^{-n}) = \ln \left( \frac{1}{a^n} \right) = \ln 1 - \ln a^n.$$

Since $\ln 1 = 0$ and $\ln a^n = n \ln a$, we find that

$$\ln (a^{-n}) = -n \ln a.$$

If $r$ is any rational number, then by the definition of rational number, $r = p/q$, where $p$ and $q$ are integers (we take $q$ as positive). We define $u = a^{1/q}$; then $u^q = a$ and $\ln a = q \ln u$, since $q$ is an integer. Also we have

$$u^p = a^{p/q} = a^r.$$

Therefore

$$\ln (a^r) = \ln (u^p) = p \ln u = \frac{p}{q} q \ln u = \frac{p}{q} \ln a;$$

we conclude

$$\ln (a^r) = r \ln a. \qquad \square$$

The **domain** of the ln function consists of all positive numbers. The next theorem establishes some of the basic properties of this function and also tells us that the **range** of the ln function consists of all real numbers.

| THEOREM 2 | *If $f(x) = \ln x$, then* |
|---|---|

i) $f'(x) = 1/x$ *for* $x > 0$, *and* $f$ *is increasing,*
ii) $\frac{1}{2} \leq \ln 2 \leq 1$,
iii) $\ln x \to +\infty$ *as* $x \to +\infty$,
iv) $\ln x \to -\infty$ *as* $x \to 0^+$,
v) *the range of* $f$ *consists of all real numbers.*

**Proof** Item (i) is simply the statement that differentiation and integration are inverse processes (Fundamental Theorem of Calculus). The derivative is positive, and a function with a positive derivative is increasing, a fact which we have already learned. To establish (ii), we make use of our knowledge of

THE LOGARITHM FUNCTION

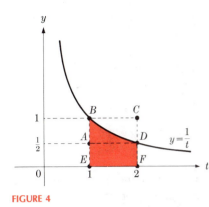

**FIGURE 4**

integrals as areas. The value of

$$\ln 2 = \int_1^2 \frac{1}{t}\, dt$$

is shown as the shaded area in Fig. 4. From Theorem 7 of Chapter 5, page 197, which gives upper and lower bounds of integrals, we have

$$\text{area } ADFE \le \int_1^2 \frac{1}{t}\, dt \le \text{area } BCFE \qquad \text{or} \qquad \tfrac{1}{2} \le \ln 2 \le 1.$$

To prove (iii), we must show that $\ln x$ increases without bound as $x$ does. Let $n$ be any positive integer. If $x > 2^n$, then

$$\ln x > \ln(2^n),$$

since we know from (i) that $\ln x$ is an increasing function. But

$$\ln x > \ln 2^n = n \ln 2 \ge \tfrac{1}{2} n \quad \text{(by (ii))}.$$

Since $\tfrac{1}{2} n \to +\infty$ as $n \to +\infty$, we have

$$\ln x \to +\infty \quad \text{as } x \to +\infty.$$

To obtain (iv), we use a device similar to that used for (iii). If $n$ is a positive integer, $(1/2^n) \to 0$ as $n \to +\infty$. If $0 < x < 1/2^n$, we know (by (i)) that

$$\ln x < \ln\left(\frac{1}{2^n}\right) = -n \ln 2.$$

Since $n \ln 2 > \tfrac{1}{2} n$, we can say that

$$\ln x < -\tfrac{1}{2} n.$$

Since $x < 1/2^n$, we know that $x \to 0$ as $n \to +\infty$. As $n \to +\infty$, the quantity $-\tfrac{1}{2} n$ tends to $-\infty$, and so $\ln x \to -\infty$ as $x \to 0$. Note that $x$ must tend to zero through positive values, since $\ln x$ is not defined for $x \le 0$.

Items (iii) and (iv) show that the range of $\ln x$ extends from $-\infty$ to $+\infty$. The Intermediate Value Theorem then allows us to conclude that no numbers in the range are omitted, thus proving (v). □

A graph of the function $\ln x$ is given in Fig. 5.

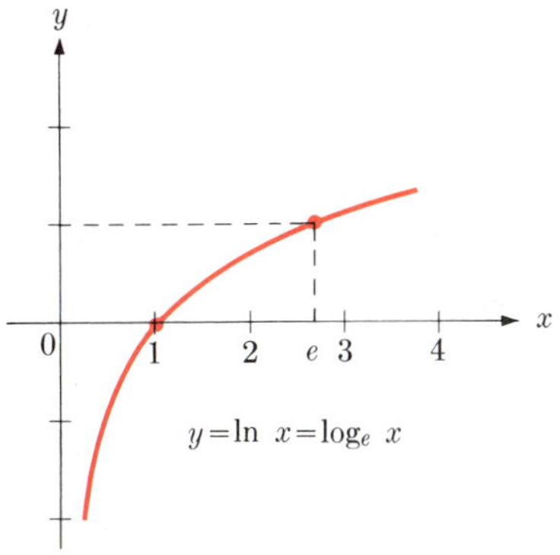

*Remark.* Suppose that $u$ is any positive function of $x$. Then from the Chain Rule, we get the important formula

$$\frac{d}{dx}\ln u = \frac{1}{u}\frac{du}{dx}.$$

**FIGURE 5**

**EXAMPLE 1** Given that $\ln 2 = 0.69315$ and $\ln 3 = 1.09861$, find $\ln 24$ and $\ln 0.1875$ without using a calculator.

**Solution** We write $24 = 2^3 \cdot 3$ and use the properties of Theorem 1 to obtain

$$\ln 24 = 3 \ln 2 + \ln 3 = 3(0.69315) + 1.09861 = 3.17806.$$

Observe that $0.1875 = \tfrac{3}{16} = 3 \cdot 2^{-4}$, and so

$$\ln 0.1875 = \ln 3 - 4 \ln 2 = -1.67399.$$ □

**EXAMPLE 2** Use Theorem 7 of Chapter 5 giving upper and lower bounds for integrals to obtain an upper and lower bound for ln 5.

**Solution** From Fig. 6, we see that a lower bound is given by the shaded area $ABCD$; it is $\frac{4}{5}$. An upper bound is given by the area $ABEF$; it is 4. In other words,

$$\tfrac{4}{5} \le \ln 5 \le 4.$$

A refined upper bound is given by the area of the trapezoid $ABCF$; it is

$$\tfrac{1}{2}(4)(\tfrac{1}{5} + 1) = \tfrac{12}{5}.$$

A more precise lower bound may be obtained by dividing the interval from 1 to 5 into sections as shown in Fig. 7 and computing the areas separately:

$$\text{Area} = 1 \cdot (\tfrac{1}{2}) + 1 \cdot (\tfrac{1}{3}) + 1 \cdot (\tfrac{1}{4}) + 1 \cdot (\tfrac{1}{5}) = \tfrac{77}{60}.$$

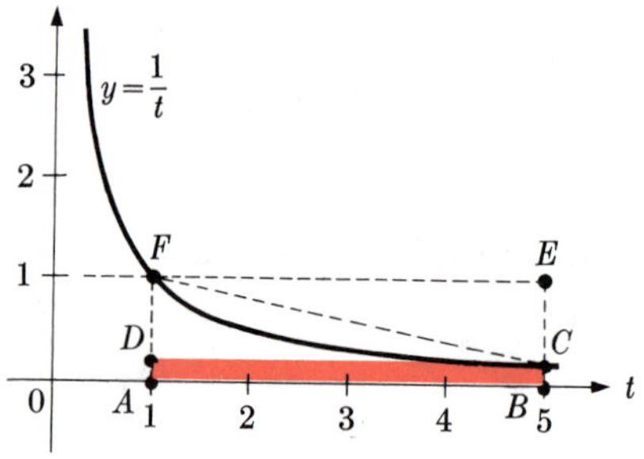

FIGURE 6 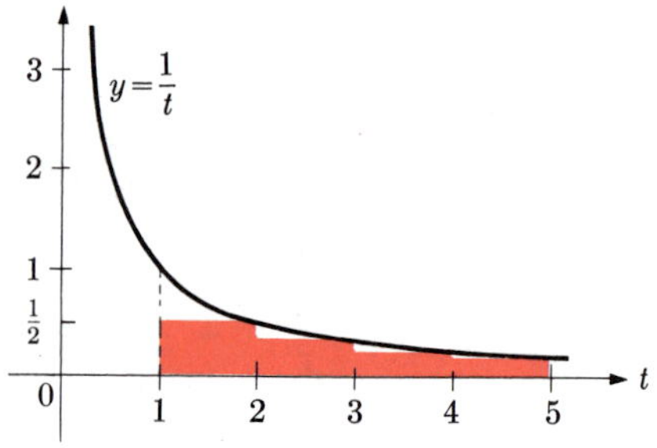                    FIGURE 7

The estimates for ln 5 are

$$\tfrac{77}{60} \le \ln 5 \le \tfrac{12}{5}.$$

**EXAMPLE 3** Given that $F(x) = \ln(x^2 + 5)$, find $F'(x)$.

**Solution** The Chain Rule tells us that

$$\frac{d}{dx} \ln u = \frac{1}{u} \cdot \frac{du}{dx}.$$

Therefore, letting $u = x^2 + 5$, we obtain

$$F'(x) = \frac{1}{x^2 + 5} \cdot 2x = \frac{2x}{x^2 + 5}.$$

**EXAMPLE 4** Given that $G(x) = \ln \sqrt{(1 + x)/(1 - x)}$ for $-1 < x < 1$, find $G'(x)$.

**Solution** The use of the properties of logarithms simplifies the work of differentiation. We first write

$$G(x) = \frac{1}{2} \ln \frac{1 + x}{1 - x} = \frac{1}{2} \ln(1 + x) - \frac{1}{2} \ln(1 - x).$$

Then

$$G'(x) = \frac{1}{2} \frac{1}{1 + x} - \frac{1}{2} \frac{1}{1 - x} \cdot (-1) = \frac{1}{2} \left( \frac{1}{1 + x} + \frac{1}{1 - x} \right) = \frac{1}{1 - x^2}.$$

## 1  PROBLEMS

In Problems 1 through 6, compute the required quantities, using the fact that ln 2 = 0.69315, ln 3 = 1.09861, and ln 10 = 2.30259. Check the result with a calculator.

**1** ln 4, ln 5, ln 6, ln 8, ln 9

**2** ln 12, ln 15, ln 16, ln 18, ln 20

**3** ln 24, ln 25, ln 27, ln 30, ln 32

**4** ln 1.5, ln 2.7, ln 0.25, ln 1.25

**5** $\ln\sqrt{30}$, $\ln\sqrt[3]{270}$, $\ln\sqrt{2/3}$, $\ln\sqrt[4]{7.2}$

**6** $\ln\sqrt{0.6}$, $\ln\sqrt[3]{15}$, $\ln\sqrt[5]{1.8}$, $\ln\sqrt{2.4}$

In Problems 7 through 16, use the definition of the function ln x and the properties of the area of elementary figures to obtain in each case an upper and lower bound for the given quantity (method of Example 2). Compare the estimates with the exact value obtained on a calculator.

**7** ln 4.5            **8** ln 7

**9** ln 14            **10** ln 3.2

**11** ln 1.05         **12** $\ln\frac{1}{2}$

**13** ln 0.15         **14** ln 0.001

**15** ln 0.95         **16** ln 0.995

In Problems 17 through 34, in each case perform the differentiation.

**17** $f(x) = \ln(2x + 3)$         **18** $f(x) = \ln(x^4)$

**19** $f(x) = \ln(x^2 + 2x + 3)$   **20** $f(x) = \ln(x^3 - 3x + 1)$

**21** $g(x) = 2\ln\sin x$          **22** $h(x) = \ln\sec 2x$

**23** $f(x) = \ln\cos^2 x$         **24** $F(x) = \ln(\sec x + \tan x)$

**25** $F(x) = (\ln x)^4$           **26** $F(x) = \ln^2(x^3)$

**27** $G(x) = \ln(\ln x)$, $(x > 1)$  **28** $G(x) = x\ln x$

**29** $H(x) = \ln\sqrt{1 - x^2}$   **30** $f(x) = \ln(x + \sqrt{x^2 + a^2})$

**31** $H(x) = \ln(x^2\sqrt{x^2 + 1})$

**32** $H(x) = \ln[(x^2 + 1)/(x^2 - 1)]$

**33** $T(x) = \ln\sqrt{(1 + \cos x)/(1 - \cos x)}$

**34** $f(x) = (\ln x)^2/(1 + x^2)$

In Problems 35 through 43, use the definition of the function ln x and the values of ln 2, ln 3, and ln 10 which are given at the beginning of the problems to evaluate the following integrals. Check the results with a calculator.

**35** $\int_1^5 \frac{1}{x}\,dx$

**36** $\int_4^{10} \frac{dx}{x}$

**37** $\int_3^{15} \frac{du}{u}$

**38** $\int_5^8 \frac{du}{u}$

**39** $\int_1^{1/4} \frac{du}{u}$

**40** $\int_{1/3}^1 \frac{du}{u}$

**41** $\int_{1/2}^{12} \frac{du}{u}$

**42** $\int_{1/8}^{1/5} \frac{du}{u}$

**43** $\int_{0.01}^{10} \frac{du}{u}$

**44** Using a method similar to that of the proof of the lemma of this section, show that

$$\int_a^{a\cdot b} \frac{du}{u^2} = \frac{1}{a}\int_1^b \frac{du}{u^2}, \quad a, b \text{ positive.}$$

**45** Extend the result given in Problem 44 to show that if n is any integer, then

$$\int_a^{a\cdot b} \frac{du}{u^n} = \frac{1}{a^{n-1}}\int_1^b \frac{du}{u^n}.$$

**46** Find the equation of the line tangent to the curve $y = 2\ln(x^2 - 1)$ at the point where $x = 2$.

**47** Find the equation of the line tangent to the curve

$$y = \ln\frac{x}{x + 1}$$

at the point where $x = 3$.

**48** Find the area of the region bounded by the curve $y = x/(x^2 + 1)$ and the lines $x = 1$, $x = 3$, $y = 0$.

**49** Find the area of the region

$$R = \{(x, y): 1 \le x \le 2, x - 1 \le y \le 5x^2/(2x^3 - 1)\}.$$

**50** Show that if k is any integer larger than 1, then

$$\frac{1}{2} + \frac{1}{3} + \cdots + \frac{1}{k} < \ln k < 1 + \frac{1}{2} + \cdots + \frac{1}{k - 1}.$$

(*Hint:* See Example 2.)

**51** Find the domain and range of the function $f(x) = \ln(\ln x)$. Show that

$$x\frac{d}{dx}[xf'(x)] = -(\ln x)^{-2}.$$

**52** a) Show that $\ln(1 + x) = \int_0^x 1/(1 + u)\,du$ for $x > -1$.
b) If $x > 0$, show that $\ln(1 + x) < x$.
c) If $x > 0$, show that $x - \frac{1}{2}x^2 < \ln(1 + x)$.
(*Hint:* Use the inequality $1 - u < 1/(1 + u)$ for $u > 0$.)

**\*53** Let $f(x)$ be a differentiable function with the property $f(xy) = f(x) + f(y)$.
a) Show that

$$f'(x) = \frac{k}{x}$$

for $x > 0$, where k is a constant. (*Hint:* Consider

$$f\left(x\frac{1}{a}\right),$$

where a is an arbitrary positive constant. Then differentiate, using the Chain Rule.)
b) Show that $k = f'(1)$.
c) Conclude that $f(x) = k\ln x$.

## 2

### RELATIONS AND INVERSE FUNCTIONS

In Section 6 of Chapter 1 we discussed relations in connection with drawing the graphs of equations in two variables. In order to describe inverse functions accurately we first recall some of the terminology we developed earlier.

The solution set of an equation in two unknowns, say $x$ and $y$, is called a **relation in** $R^2$. The **domain** of the relation is the set of all numbers $x_0$ such that the vertical line $x = x_0$ intersects the graph of the relation. The **range** of the relation is the set of all $y_0$ such that the horizontal line $y = y_0$ intersects the graph. See Fig. 23 of Chapter 1 on page 25.

The distinction between a function and a relation is simply that for a function, to each element of the domain there corresponds exactly one element of the range. No such requirement is imposed for a relation. However, as is the case for functions, the *domain* of a relation is the projection of the graph of the relation on the $x$ axis and the *range* of the relation is the projection on the $y$ axis.

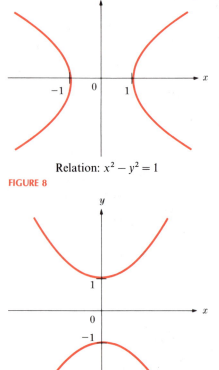

Relation: $x^2 - y^2 = 1$

**FIGURE 8**

Inverse relation: $y^2 - x^2 = 1$

**FIGURE 9**

**DEFINITION** *Suppose that S is a relation. That is, S is made up of points $(x, y)$ in $R^2$. We define the **inverse relation of** S to consist of all points $(x, y)$ such that $(y, x)$ is a point of S.*

The inverse relation merely interchanges the role of domain and range. For example, if $(3, 5)$ is in the relation $S$, then $(5, 3)$ is in the inverse relation. More generally, the domain of the inverse relation is the range of the original relation and the range of the inverse is the domain of the given relation. Also, the inverse of the inverse of a given relation is just the given relation.

A relation is usually defined by an equation involving $x$ and $y$. According to the definition above, the inverse relation is the solution set of the equation we get when we interchange $x$ and $y$ in the equation. We give four examples.

i) If a relation is defined by

$$x^2 - y^2 = 1,$$

the inverse relation is the solution set of

$$y^2 - x^2 = 1.$$

The domain of the relation shown in Fig. 8 is $\{-\infty < x \leq -1\} \cup \{1 \leq x < +\infty\}$. The range of the relation is $-\infty < y < +\infty$. Consequently, the domain of the inverse relation, shown in Fig. 9, is $-\infty < x < +\infty$, and the range is the set $\{-\infty < y \leq -1\} \cup \{1 \leq y < +\infty\}$.

ii) If a relation is defined by

$$y = x^2,$$

the inverse relation is the solution set of

$$y^2 = x.$$

The domain of the relation shown in Fig. 10 is $-\infty < x < +\infty$. The range of the relation is $0 \le y < +\infty$. Consequently, the domain of the inverse relation, shown in Fig. 11, is $0 \le x < +\infty$ and the range is $-\infty < y < +\infty$.

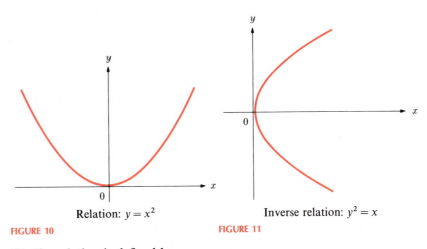

Relation: $y = x^2$

FIGURE 10

Inverse relation: $y^2 = x$

FIGURE 11

iii) If a relation is defined by

$$y = (x - 1)^3,$$

the inverse relation is the solution set of

$$(y - 1)^3 = x \qquad \Leftrightarrow \qquad y = 1 + x^{1/3}.$$

The domain of the relation shown in Fig. 12 is $-\infty < x < +\infty$. The range of the relation is $-\infty < y < +\infty$. Therefore, the domain of the inverse relation, shown in Fig. 13, is $-\infty < x < +\infty$ and the range is $-\infty < y < +\infty$.

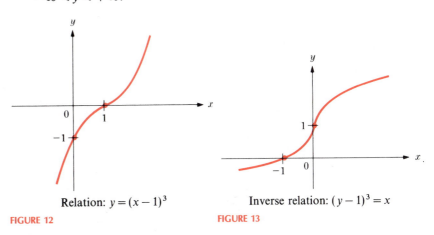

Relation: $y = (x - 1)^3$

FIGURE 12

Inverse relation: $(y - 1)^3 = x$

FIGURE 13

iv) If a relation is defined by

$$y^2 = x^3,$$

the inverse relation is the solution set of

$$y^3 = x^2 \qquad \Leftrightarrow \qquad y = x^{2/3}.$$

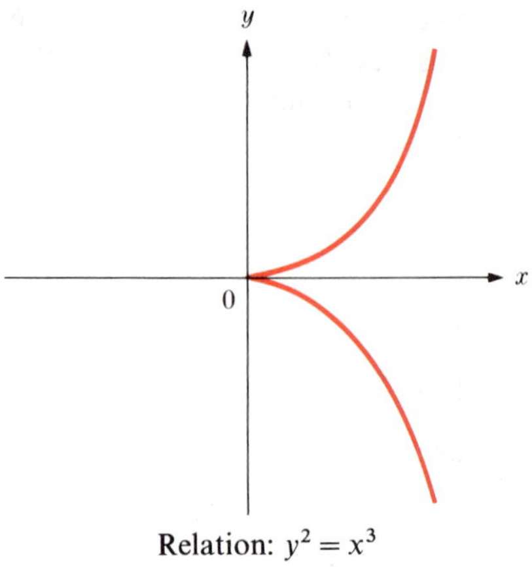

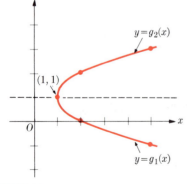

Relation: $y^2 = x^3$

**FIGURE 14**

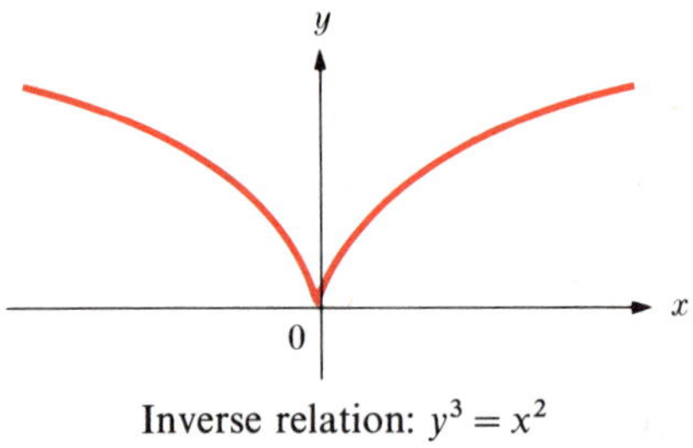

Inverse relation: $y^3 = x^2$

**FIGURE 15**

The domain of the relation, shown in Fig. 14, is $0 \le x < +\infty$. The range of the relation is $-\infty < y < +\infty$. Hence the domain of the inverse relation, shown in Fig. 15, is $-\infty < x < +\infty$, and the range is $0 \le y < +\infty$.

In example (i), neither the original nor the inverse relation is a function. In example (ii), the given relation is a function and the inverse relation is not. In example (iii), both the given relation and its inverse are functions. Finally, in example (iv), the given relation is not a function, but its inverse is. Therefore we see that there is no simple interconnection between inverse relations and functions.

Since a function is a special case of a relation, no additional effort is needed to define the **inverse** of a function. As we did for a relation, *to find the inverse of a function we interchange the variables x and y in the equation which defines the function.* As we saw in example (ii) above, the inverse of a function need not be a function, although it may be one, as example (iii) shows. In general, a function $f$ which has the equation $y = f(x)$ has its inverse given by

$$x = f(y).$$

The inverse of a function is a relation which may consist of several functions. These functions are called the **branches** of the inverse relation. If we can solve the equation $x = f(y)$ for $y$ in terms of $x$, then we can obtain explicit expressions for the branches of the inverse of $f$.

**EXAMPLE 1**    Given the function $f(x) = x^2 - 2x + 2$, plot the inverse relation of $f$. Show that the inverse consists of two branches and find explicit formulas for these branches.

**Solution**    We write $y = x^2 - 2x + 2$ and then the inverse relation is the solution set of the equation

$$x = y^2 - 2y + 2.$$

The graph of the inverse is the parabola sketched in Fig. 16. Solving the above quadratic equation for $y$, we get

$$y = 1 \pm \sqrt{x - 1}.$$

Thus the inverse relation has two branches

$$y = 1 - \sqrt{x - 1} \equiv g_1(x), \quad 1 \le x < +\infty,$$

$$y = 1 + \sqrt{x - 1} \equiv g_2(x), \quad 1 \le x < +\infty. \qquad \square$$

$y = g_2(x)$

$(1, 1)$

$y = g_1(x)$

**FIGURE 16**

Suppose we are given a function $y = f(x)$. There are many times when it is impossible to solve the equation $x = f(y)$ explicitly for $y$ in terms of $x$. However, it is not always necessary to do this in order to determine whether or not the inverse relation is a function. The next theorem, stated without proof, gives a procedure for deciding when the inverse of a function is also a function.

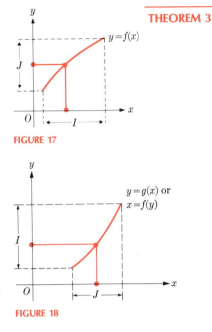

**THEOREM 3**  **(Inverse Function Theorem)**  *Let f be a continuous, increasing function which has an interval I for domain and has range J (Fig. 17). Then f has an inverse g which is a continuous, increasing function with domain J and range I (Fig. 18). We have, moreover,*

$$g[f(x)] = x \quad for \ x \ on \ I, \tag{1}$$

$$f[g(x)] = x \quad for \ x \ on \ J. \tag{2}$$

*If f is decreasing on I instead of increasing, the same result holds with g decreasing on J. If $f'(x) \neq 0$ on the interior of I, then g is differentiable on the interior of J.*

**FIGURE 17**

**FIGURE 18**

Theorem 3 can be given a graphic interpretation. We may think of a function $f$ as an **operator** which carries one object (an element of the domain) into another (the image element of the range). The function $f$ "operates" in that it carries elements of $R^1$ into elements of $R^1$. The term **mapping** is also used as a synonym for operation. We also say that $f$ **maps** elements of $R^1$ (the domain) into elements of $R^1$ (the range). We may interpret formulas (1) and (2) of Theorem 3 by saying that $g$ is the operator which reverses the action of $f$. It is the inverse operator or inverse mapping. Whatever action $f$ performs, $g$ undoes it.

> Whenever we can differentiate a function $f$, then Theorem 3 may be used to decide whether or not the inverse of $f$ is a function. We merely compute $f'$ and if this quantity is always positive or always negative on the domain of $f$, then the inverse of $f$ must be a function.

The next examples show how this idea may be applied to analyze the inverse of a function.

**EXAMPLE 2**    Given the function

$$y = f(x) = \frac{x}{1 + x^2} \quad \text{with domain } 1 < x < +\infty.$$

Determine the range of this function and decide whether or not the inverse of $f$ is a function.

**Solution**    We first compute the derivative of $f$:

$$y' = \frac{1 - x^2}{(1 + x^2)^2}.$$

Since the derivative is negative for $1 < x < +\infty$, the function is decreasing (see Fig. 19). We have $f(1) = \frac{1}{2}$ and $\lim_{x \to \infty} f(x) = 0$. Therefore the range of the function is the interval $(0, \frac{1}{2})$. By the Inverse Function Theorem we know that $f$ has a continuous inverse function $g$ which has domain $(0, \frac{1}{2})$ and range $(1, \infty)$ and is decreasing; $g$ is determined by the equation

$$x = \frac{y}{1 + y^2} \qquad \text{or} \qquad xy^2 - y + x = 0.$$

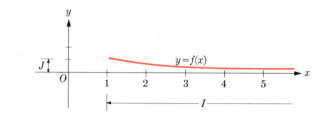

FIGURE 19

Solving this quadratic equation for $y$ in terms of $x$, we get

$$y = \frac{1 \pm \sqrt{1 - 4x^2}}{2x}, \quad 0 < x < \frac{1}{2}.$$

If we choose the plus sign in the above expression then $y$ is in the range $1 < y < +\infty$, while if we choose the minus sign, we see that $y \leq 1$. Therefore the inverse of $f$ is

$$y = g(x) = \frac{1 + \sqrt{1 - 4x^2}}{2x} \quad \text{for } 0 < x < \frac{1}{2}.$$

See Fig. 20.

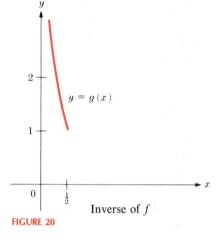

FIGURE 20

Inverse of $f$

In case $f$ is neither always increasing nor always decreasing, its inverse may be analyzed as follows: First find the intervals $I_1, I_2, \ldots$ on which $f$ is increasing or decreasing. If $f_1, f_2, \ldots$ denote the parts of $f$ for $x$ on $I_1, I_2, \ldots$, respectively, then the inverse of each of these functions is a function. The various inverse functions which we call $g_1, g_2, \ldots$ may be identified and plotted provided the equation $x = f(y)$ can be solved for $y$. The next example shows a systematic method for obtaining the branches of the inverse relation.

**EXAMPLE 3**    Given the function

$$f(x) = x^3 - 3x + 2,$$

discuss and plot the inverse relation of $f$, indicating the various functions $g_1$, $g_2$, ....

**Solution**    The graph of $f$ is shown in Fig. 21. The graph of the inverse is determined by the equation

$$x = f(y) = y^3 - 3y + 2.$$

We find that

$$f'(x) = 3x^2 - 3,$$

and the slope is zero at $x = \pm 1$. To determine where $f$ is increasing and decreasing, we construct the following table.

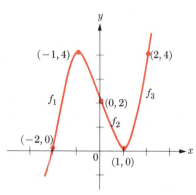

FIGURE 21

| $x$ | $-2$ | $-1$ | $0$ | $1$ | $2$ |
|---|---|---|---|---|---|
| $f'(x)$ | $+$ | $0$ | $-$ | $0$ | $+$ |
| $f(x)$ | $0$ | $4$ | $2$ | $0$ | $4$ |

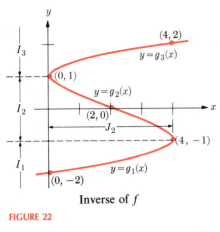

**FIGURE 22**

Using the table and the graph in Fig. 21 as a guide, we conclude that

$f$ is increasing on $I_1 = (-\infty, -1]$, and $J_1 = (-\infty, 4]$,

$f$ is decreasing on $I_2 = [-1, 1]$, and $J_2 = [0, 4]$,

$f$ is increasing on $I_3 = [1, \infty)$, and $J_3 = [0, \infty)$.

The inverse relation of $f$ is plotted in Fig. 22. □

## 2　PROBLEMS

In each of Problems 1 through 12, a relation between $x$ and $y$ is given. Find the domain and range of the relation. Sketch the graph.

**1** $y^2 = x^3$　　　　　　　**2** $y^3 = x^2$

**3** $y^2 = x^3 + 8$　　　　　　**4** $y^4 = x^3$

**5** $y^4 = x^2$　　　　　　　**6** $x^2 + 2y^2 = 1$

**7** $x^2 - 2y^2 = 1$　　　　　**8** $2x^2 + 9y^2 = 1$

**9** $y^2 - 4x^2 = 1$　　　　　**10** $y^2 = (x + 1)^2$

**11** $y^2 = (x + 1)^3$　　　　**12** $x^3 + y^3 = 1$

For each of the functions $f$ given in Problems 13 through 22, determine whether or not there is an inverse function $g$. If there is, determine the domain of $g$ and find an expression for $g(x)$, if possible. Draw a graph.

**13** $f(x) = 3x + 2, \ -\infty < x < \infty$

**14** $f(x) = x^2 + 2x - 3, \ 0 \le x < \infty$

**15** $f(x) = x^3 + 4x - 5, \ -\infty < x < \infty$

**16** $f(x) = \dfrac{x}{x + 1}, \ -1 < x < \infty$

**17** $f(x) = 3 - 2x - x^2, \ -\infty < x < 0$

**18** $f(x) = 3 - 2x - x^2, \ -5 < x < 1$

**19** $f(x) = 3 - 2x - x^2, \ 0 < x < \infty$

**20** $f(x) = \dfrac{x^2}{(x + 1)^2}, \ -2 < x < 2, \ x \ne -1$

**21** $f(x) = \dfrac{1}{x^2}, \ \ 0 < x \le 1$　　**22** $f(x) = \dfrac{2 + x}{2 - x}, \ \ 0 \le x < 2$

For each of the functions $f$ given in Problems 23 through 34, find the domains for which the Inverse Function Theorem applies. Find the inverse functions $g$ and give their domains, if possible.

**23** $f(x) = x^2 + 4x - 1$　　　**24** $f(x) = 6x - x^2$

**25** $f(x) = 2x^2 + 3x + 4$　　**26** $f(x) = 6 + 5x - 2x^2$

**27** $f(x) = \dfrac{x}{x + 1}$　　　　　**28** $f(x) = \dfrac{2x}{1 + x^2}$

**29** $f(x) = \dfrac{2x - 1}{x + 2}$　　　　**30** $f(x) = x + \dfrac{1}{x}$

**31** $f(x) = x^3 + 3x^2 - 9x + 7$　**32** $f(x) = 2x^3 + 3x^2 - 3x$

**33** $f(x) = x^3 + 3x^2 + 6x - 3$

**34** $f(x) = x^4 + \frac{4}{3}x^3 - 4x^2 + \frac{2}{3}$

**35** Assume that $f$ and $g$ in Theorem 3 are differentiable. By using the Chain Rule and implicit differentiation in the equation $f[g(x)] = x$, show that at each point of the domain of $f$, the derivative of $f$ is the reciprocal of the derivative of $g$.

**36** Assume that $f$ and $g$ in Theorem 3 have second derivatives. If $f$ and $g$ are increasing, show that $f''$ and $g''$ have opposite signs at each point. If $f$ and $g$ are decreasing, show that $f''$ and $g''$ have the same sign at each point.

**37** a) Suppose that $f$ is a differentiable function on an interval $I$ and that $g$ is the inverse function of $f$. Show that if $f'$ is zero for some value $x_0$ in $I$, then $g$ cannot be differentiable at $x_0$.

b) Give an example of a differentiable function $f$ on an interval $I$ which has an inverse on $I$ and is such that $f'(x_0) = 0$ for some $x_0$ in $I$.

**38** Suppose that $f_1$ and $f_2$ are increasing functions on an interval $I$. (a) Is $f_1 + f_2$ increasing on $I$? (b) Is $f_1 \cdot f_2$ increasing on $I$? (c) Is $f_1(f_2(x))$ increasing on $I$?

**39** Define the function $f$ on $[0, \infty)$ by the formulas

$$f(x) = \begin{cases} x & \text{for } 0 \le x \le 1 \\ f(n) + \dfrac{1}{n}(x - n) & \text{for } n < x \le n + 1, \quad n = 1, 2, \dots \end{cases}$$

Find the inverse function $g$. Describe its domain and range. Is $g$ differentiable?

_____3_____

## THE EXPONENTIAL FUNCTION. THE NUMBER $e$

Theorems 1 and 2 of Section 1 exhibit the principal properties of the ln function. Since the derivative of ln $x$ is always positive, we may apply the Inverse Function Theorem (Theorem 3), and so define the *inverse function* of ln $x$.

**DEFINITION**   **The exponential function**, *denoted by* exp, *is defined to be the inverse of the function* ln.

*Remark.* Since the function $y = \ln x$ has domain $0 < x < +\infty$ and range $-\infty < y < +\infty$, it is clear that the domain of exp $x$ consists of all real numbers, and the range consists of all positive numbers.

The basic properties of exp $x$ are given in the next theorem.

**THEOREM 4**   *If* $f(x) = \exp x$, *then*

> i) $f$ *is increasing for all values of* $x$,
> ii) $\exp(x_1 + x_2) = (\exp x_1)(\exp x_2)$,
> iii) $\exp(x_1 - x_2) = (\exp x_1)/(\exp x_2)$,
> iv) $\exp rx = (\exp x)^r$ *if* $r$ *is any rational number*,
> v) $f(x) \to +\infty$ *as* $x \to +\infty$,
> vi) $f(x) \to 0$ *as* $x \to -\infty$.

**Proof**   To establish (i), let $y_1 = \exp x_1$ and $y_2 = \exp x_2$, with $x_2 > x_1$. We wish to show that $y_2 > y_1$. From the definition of inverse functions we know that

$$x_1 = \ln y_1 \qquad \text{and} \qquad x_2 = \ln y_2.$$

Since ln is an increasing function, the only way that the inequality $x_2 > x_1$ can hold is if $y_2 > y_1$.

Item (ii) is proved by the same technique—going back to the ln function. Let

$$y_1 = \exp x_1 \qquad \text{and} \qquad y_2 = \exp x_2.$$

Then $x_1 = \ln y_1$, $x_2 = \ln y_2$, and $x_1 + x_2 = \ln y_1 + \ln y_2 = \ln(y_1 y_2)$. This last statement asserts that

$$y_1 y_2 = \exp(x_1 + x_2)$$

or

$$(\exp x_1)(\exp x_2) = \exp(x_1 + x_2).$$

The argument for (iii) is analogous to that for (ii).

We prove part (iv) by mathematical induction. We know that

$$\exp(2x) = \exp(x + x) = (\exp x)(\exp x) = (\exp x)^2,$$

and therefore that

$$\exp(nx) = (\exp x)^n, \quad \text{for } n \text{ a positive integer.}$$

The remaining argument parallels the proof of (iv) in Theorem 1, page 251. Parts (v) and (vi) are derived by appealing to the analogous result for the function ln. To state that ln and exp are inverse functions is equivalent to writing the compact formulas (see Eqs. (1) and (2) in Theorem 3, page 259)

$$\ln(\exp x) = x \text{ for all } x \qquad \text{and} \qquad \exp(\ln x) = x, \quad \text{if } x > 0. \qquad \square$$

Figure 23 shows the functions ln and exp. *They are the reflections of each other in the line $y = x$, since an interchange of the roles of $x$ and $y$ sends one function into the other.*

If $a$ is any positive number and $r$ is any rational number, the formula

$$\exp(\ln x) = x$$

becomes for $x = a^r$ the formula

$$\exp(\ln a^r) = a^r.$$

But $\ln a^r = r \ln a$, and so we have

$$\exp(r \ln a) = a^r, \quad \text{if } a > 0 \text{ and } r \text{ is rational.}$$

The right side of this expression has meaning if $r$ is rational. The left side has meaning if $r$ is *any real number*. This gives us a clue for defining $a^r$ when $r$ is any real number.

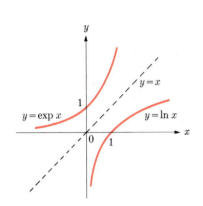

FIGURE 23

**DEFINITION**   *If $a > 0$, we define*

$$a^x = \exp(x \ln a), \quad \textit{for any real number } x.$$

*Remark.*   The procedure used in defining $a^x$ for any real number $x$ is typical of a technique that is used in many branches of mathematics. The process is one that starts with an equation which holds under certain restrictions. If one side of the equation has meaning without these restrictions while the other side is *undefined* if the restrictions are relaxed, the equation itself may be used to define the meaningless side.*

The above definition assures us that $a^x$ has a meaning for every real number $x$ so long as $a$ is positive. Moreover, the range of $a^x$ is the set of all positive numbers if $a \neq 1$. Figure 24 shows the graph of $a^x$ for two typical values of $a$.

Now that numbers of the form $a^x$ have meaning for every real number $x$, we are in a position to define logarithm in the familiar way.

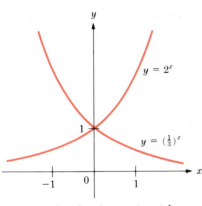

Graphs of $a^x$ for $a = 2$ and $\frac{1}{3}$

FIGURE 24

---

*An example of this is given by the equation $x^2 = b$, where $b$ is a number. The equation determines the square root of a number $b$ if it is nonnegative; but the right side has meaning if $b$ is *any real number*. This equation may be used as the *definition* of imaginary numbers.

**DEFINITION**

*If $b > 0$, $b \neq 1$, and $N > 0$, we define $\log_b N$ as that number $y$ such that $b^y = N$. In words:*

*The logarithm of a positive number $N$ to the positive base $b$ is that power to which $b$ must be raised to obtain $N$.*

We now see why this definition of logarithm was not possible earlier. Since we needed the exp and ln functions to define $b^y$, we could not develop logarithms in a logically consistent way without them.

The familiar properties of exponents and logarithms to any base can now be established from the definition of $a^x$ and from the properties of the ln and exp functions as given in Theorems 1, 2, and 4. Some of these properties are:

i)   $a^x \cdot a^y = a^{x+y}$; $a^x/a^y = a^{x-y}$; $(a^x)^y = a^{xy}$; $(ab)^x = a^x \cdot b^x$;
ii)  $\log_b (x \cdot y) = \log_b x + \log_b y$;
iii) $\log_b (x/y) = \log_b x - \log_b y$;
iv)  $\log_b (x^y) = y \log_b x$.

**DEFINITION**

*We define a number which we designate $e$ by the formula*

$$e = \exp 1.$$

The importance of this number is exhibited in the next two theorems.

**THEOREM 5**   *If $x > 0$, then $\ln x = \log_e x$.*

**Proof**   Since, by definition, $e = \exp 1$, we have $\ln e = \ln \exp 1$. However, because ln and exp are inverse functions, we know that $\ln \exp 1 = 1$. Therefore $\ln e = 1$. If we set $y = \log_e x$, then $x = e^y$; further, $\ln x = \ln e^y = y \ln e = y$.    □

Theorem 5 states that the function we have been calling "ell-en of $x$" is simply a logarithm function to the base $e$.

**DEFINITIONS**

*Logarithms to the base $e$ are called **natural logarithms** (also Napierian logarithms). Logarithms to the base 10 are called **common logarithms**.*

**THEOREM 6**   *We have*

i)  *$\exp x = e^x$ for all $x$,*
ii) *$e$ lies between 2 and 4.*

**Proof**   For the first part, we know by definition of a number to an exponent, as given on page 263 that

$$e^x = \exp(x \ln e).$$

But we saw in the proof of Theorem 5 that $\ln e = 1$. This yields $e^x = \exp x$. To establish (ii), we first recall that in Theorem 2, part (ii), we found that

$$\tfrac{1}{2} \le \ln 2 \le 1.$$

The inequality

$$\ln 2 = \log_e 2 \le 1 = \ln e$$

is interpreted from the definition of logarithm as $2 \le e^1 = e$. From the relation $\ln 4 = \ln 2 + \ln 2 \ge \tfrac{1}{2} + \tfrac{1}{2} = 1$, we see that $\log_e 4 \ge 1$, and therefore

$$4 \ge e^1 = e. \qquad \square$$

The actual numerical value of $e$ is $2.71828^+$, correct to five decimal places.

> *Notation.*   The notations $\ln x$ and $\log x$ are sometimes replaced by $\log x$. *Therefore, whenever no base is indicated in the logarithm function, it will always be understood that the base is $e$.*

**EXAMPLE 1**   Find the value of $\log_2 \tfrac{1}{8}$.

**Solution**   Letting $x = \log_2 \tfrac{1}{8}$, we see from the definition of logarithm that

$$2^x = \tfrac{1}{8}.$$

Writing $\tfrac{1}{8}$ as $2^{-3}$, we obtain $x = -3$. $\qquad \square$

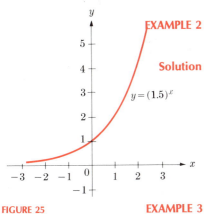

**EXAMPLE 2**   Sketch the graph of $y = (1.5)^x$.

**Solution**   We construct a table of values, as shown. Observe that $y \to +\infty$ as $x \to +\infty$, and $y \to 0$ as $x \to -\infty$. The curve is sketched in Fig. 25.

| $x$ | 0 | 1 | 2 | 3 | 4 | $-1$ | $-2$ | $-3$ | $-4$ |
|---|---|---|---|---|---|---|---|---|---|
| $y$ | 1 | 1.5 | 2.25 | 3.375 | 5.0625 | 0.667 | 0.444 | 0.296 | 0.198 |

$\square$

**FIGURE 25**

**EXAMPLE 3**   Sketch the graph of $y = \log_3 x$.

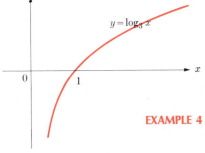

**Solution**   One way to get a table of values is to note that the equation $y = \log_3 x$ is equivalent to $3^y = x$. Letting $y$ take on a succession of values, we construct the table and then sketch the curve, as shown in Fig. 26.

| $y$ | 0 | 1 | 2 | 3 | 4 | $-1$ | $-2$ | $-3$ | $-4$ |
|---|---|---|---|---|---|---|---|---|---|
| $x = 3^y$ | 1 | 3 | 9 | 27 | 81 | $\tfrac{1}{3}$ | $\tfrac{1}{9}$ | $\tfrac{1}{27}$ | $\tfrac{1}{81}$ |

$\square$

**EXAMPLE 4**   Prove that if $x > 0$, $b > 0$, and $b \ne 1$, *then*

$$\log_b x = \frac{\ln x}{\ln b}.$$

**FIGURE 26**

**Solution**    We set $y = \log_b x$ and write the equivalent expression $b^y = x$. Taking the natural logarithm, we have

$$\ln(b^y) = \ln x$$

and, consequently,

$$y \ln b = \ln x \qquad \Leftrightarrow \qquad y = \frac{\ln x}{\ln b}. \qquad \square$$

We now investigate properties of the number $e$. From the definition of the natural logarithm, we have

$$\log_e x = \int_1^x \frac{1}{t} \, dt,$$

and from the Fundamental Theorem of Calculus we see that the derivative of $\log_e x$ is $1/x$. On the other hand, we may find the derivative of $f(x) = \log_e x$ from the very definition of derivative. That is, we use the three-step rule.

**Step 1:**    $\Delta f = f(x + h) - f(x) = \log_e(x + h) - \log_e x.$

**Step 2:**    $\dfrac{1}{h} \Delta f = \dfrac{\log_e(x + h) - \log_e x}{h}.$

We use the properties of logarithms to write this last expression in the form

$$\frac{1}{h} \Delta f = \frac{1}{h} \log_e \frac{x + h}{x} = \frac{1}{h} \log_e \left( 1 + \frac{h}{x} \right).$$

Before letting $h \to 0$ we multiply and divide by $x$, getting

$$\frac{1}{h} \Delta f = \frac{1}{x} \cdot \frac{x}{h} \log_e \left( 1 + \frac{h}{x} \right) = \frac{1}{x} \cdot \log_e \left( 1 + \frac{h}{x} \right)^{x/h}.$$

Now we are ready to let $h$ tend to zero.

**Step 3:**    We have

$$\lim_{h \to 0} \frac{1}{h} \Delta f = f'(x) = \frac{1}{x} \lim_{h \to 0} \log_e \left( 1 + \frac{h}{x} \right)^{x/h}.$$

Since the answer *must be* $1/x$, we conclude that

$$\lim_{h \to 0} \log_e \left( 1 + \frac{h}{x} \right)^{x/h} = 1.$$

Since $\log_e e = 1$, this answer is possible only if

$$\lim_{h \to 0} \left( 1 + \frac{h}{x} \right)^{x/h} = e.$$

Letting $t = h/x$, we rewrite this last expression in the form:

$$e = \lim_{t \to 0} (1 + t)^{1/t}. \qquad (1)$$

The above development is formalized in the next statement.

**COROLLARY**    *The number $e$ is obtainable as the $\lim\limits_{t \to 0} (1+t)^{1/t}$.*

By simple substitution, we can get an idea of how the function

$$G(t) = (1+t)^{1/t}$$

behaves as $t \to 0$. We have

$$G(1) = 2,\ G(\tfrac{1}{2}) = \tfrac{9}{4},\ G(\tfrac{1}{3}) = \tfrac{64}{27},\ G(\tfrac{1}{4}) = \tfrac{625}{256} = 2.4^+,\ G(\tfrac{1}{10}) = 2.59^+.$$

Therefore we can obtain an approximation to the number $e$ to any desired accuracy by computing the function $G(t)$ for smaller and smaller values of $t$.

**EXAMPLE 5**    (*Continuously compounded interest rates*)   Suppose $D$ dollars deposited in a bank earns interest at an annual percentage rate of 10%. However, the interest is compounded monthly. Then after one month $\tfrac{1}{12}$ of 10% interest is earned on $D$ dollars. That is, the account has in it

$$D + \frac{0.10}{12}D = D\left(1 + \frac{0.10}{12}\right).$$

For the second month interest is paid on

$$D\left(1 + \frac{0.10}{12}\right)$$

dollars and so after two months the account has in it

$$D\left(1 + \frac{0.10}{12}\right) + D\left(1 + \frac{0.10}{12}\right)\left(\frac{0.10}{12}\right) = D\left(1 + \frac{0.10}{12}\right)^2.$$

Thus after a full year the original investment of $D$ dollars becomes

$$D\left(1 + \frac{0.10}{12}\right)^{12}$$

dollars. If the money were to be compounded daily instead of monthly, then it is not difficult to see that after one year $D$ dollars would become

$$D\left(1 + \frac{0.10}{365}\right)^{365}$$

at the end of the year. In general, if the interest rate is $r$% and the interest is compounded $n$ times per year, then after $t$ years the amount of money accumulated is

$$D\left(1 + \frac{r}{100n}\right)^{nt}.$$

*Continuous compounding*, by definition, occurs when $n$, the number of compoundings per year, tends to infinity. To obtain continuous compounding we calculate

$$\lim_{n \to \infty} D\left(1 + \frac{r}{100n}\right)^{nt} = \lim_{n \to \infty} D\left[\left(1 + \frac{r}{100n}\right)^{100n/r}\right]^{rt/100}.$$

We set

$$h = \frac{r}{100n}.$$

Then the above limit becomes

$$\lim_{h \to 0} D[(1+h)^{1/h}]^{rt/100} = De^{rt/100}.$$

We conclude that if $D$ dollars are invested and interest is paid at $r\%$, compounded continuously, then after $t$ years the total money accumulated is

$$De^{rt/100} \text{ dollars.}$$

## 3  PROBLEMS

In Problems 1 through 14, sketch the curves.

**1** $y = 4^x$         **2** $y = 3^x$

**3** $y = (\frac{1}{2})^x$      **4** $y = (\frac{1}{3})^{2x}$

**5** $y = 2^{-x}$       **6** $y = 3^{-2x}$

**7** $y = 2^{x+1}$      **8** $y = 2^{3-x}$

**9** $y = \log_2 x$     **10** $y = \log_4 x$

**11** $y = \log_{1/2} x$    **12** $y = \log_{1/5} x$

**13** $y = \log_3 2x$    **14** $y = \log_2(x+1)$

In Problems 15 through 18, find in each case the value of the given expression.

**15** a) $\log_3 81$    b) $\log_4 16$    c) $\log_2 \frac{1}{32}$

**16** a) $\log_3 \frac{1}{27}$    b) $\log_5 125$    c) $\log_4 \frac{1}{64}$

**17** a) $\log_2 1$    b) $\log_7 \frac{1}{49}$    c) $\log_a a$

**18** a) $\log_{1/2} 8$    b) $\log_{1/6} 216$    c) $\log_{1/4} \frac{1}{16}$

**19** From the definition of the function $a^x$, prove that $a^x \cdot a^y = a^{x+y}$ for any real numbers $x$ and $y$ and $a > 0$.

**20** a) Prove that $(2^{\sqrt{3}})^{\sqrt{3}} = 8$.
b) Given that $x$ and $y$ are any real numbers, prove that $(a^x)^y = a^{xy}$.

**21** From the definition of the ln function, show that $\ln 2.4 \le 1$ and, consequently, that $2.4 \le e$.

**22** From the definition of the ln function, show that $\ln 3.5 \ge 1$ and, consequently, that $3.5 \ge e$.

Using the results of Example 4, express the quantities in Problems 23 and 24 in terms of natural logarithms.

**23** $\log_3 8$, $\log_7 12$, $\log_2 15$

**24** $\log_{1/3} 28$, $\log_{1/3} x^2$, $\log_8 12$

**25** Compute values of the function $G(t) = (1+t)^{1/t}$ for $t = \frac{1}{5}$, $\frac{1}{6}$, $\frac{1}{7}$, $\frac{1}{8}$. Use a calculator to find $G(\frac{1}{20})$, $G(\frac{1}{50})$.

**26** Compute $G(2)$, $G(3)$, $G(4)$, $G(5)$. Sketch the graph of $y = G(x)$.

**27** Suppose \$250 is placed in a bank account with an annual percentage rate of 8%. Suppose interest is compounded daily. Use continuous compounding to estimate the interest after 6 months. Calculate the actual interest after 6 months, assuming (as many banks do) a year has 360 days. (Use a hand calculator.)

**28** Suppose \$300 is invested in a money market account which guarantees a rate of return of 9%, compounded continuously. What is the minimum amount of money that will accumulate after one year?

**29** Economists like to measure quantities in "uniform" or fixed dollars. Suppose the inflation rate since 1967 has averaged 6% per year, compounded continuously. If \$550 is put into a cookie jar in 1967, how much is this money worth in 1980 in terms of 1967 dollars?

**\*30** Referring to the inflation described in Problem 29, if a car costs \$23,000 in 1988, what would its price be in 1967 dollars?

**31** Show that the function $H(t) = [1 + (1/t)]^t$ is an increasing function of $t$.

**32** Let $H(t) = (1 + 1/t)^t$ and consider $H(n)$ for large integer values of $n$. The binomial formula yields

$$H(n) = \left(1 + \frac{1}{n}\right)^n$$

$$= 1 + \frac{n}{1!}\left(\frac{1}{n}\right) + \frac{n(n-1)}{2!}\left(\frac{1}{n}\right)^2 + \cdots + \left(\frac{1}{n}\right)^n.$$

Use this result to show that

$$H(n) \le 1 + 1 + \frac{1}{2!} + \cdots + \frac{1}{n!}.$$

Since $\lim_{n \to \infty} H(n) = e$, get an approximate value for $e$ to three decimal places.

**33** For $0 \le u \le 1$, integrate the obvious inequality $0 \le e^u \le e$ between 0 and $x$ to obtain the inequality $0 \le e^x - 1 \le ex$. Repeat this $n$ times and obtain the inequality

$$0 \le e^x - 1 - x - \frac{x^2}{2!} - \cdots - \frac{x^n}{n!} \le \frac{x^{n+1}}{(n+1)!}e.$$

In Problems 34 through 39, without using a table or calculator find the value of the following, given that $\ln 2 = 0.693$, $\ln 3 = 1.099$, $\ln 10 = 2.303$.

**34** $\log_3 16$        **35** $\log_2 30$

**36** $\log_{1/5} 6$      **37** $\log_{10} 24$

**38** $\log_6 18$       **39** $\log_{15} 9$

**40** Suppose that $a$ is a number such that $0 < a < 1$. Show that

$$\lim_{x \to +\infty} a^x = 0.$$

**41** Find the value of $\lim\limits_{x \to 0+} x \log x$.

**42** Find the value of $\lim\limits_{n \to \infty} \dfrac{\ln n}{n}$.

**43** Suppose that $f$ is a continuous function for $-\infty < x < \infty$ and $f(a+b) = f(a)f(b)$ for all $a$, $b$. Show that either $f(0) = 0$ or $f(0) = 1$. If $f(0) = 1$ show that $f$ can never

vanish. If also $f(1) \neq 1$, show that $\lim_{x \to +\infty} f(x) = 0$ or $+\infty$, in which case $\lim_{x \to -\infty} f(x) = +\infty$ or 0.

**44** Determine the domain and range of the function

$$f(x) = \ln\left(\frac{x-1}{x+1}\right).$$

Find the inverse function (or functions) and determine its domain and range.

---

**4**

# DIFFERENTIATION OF EXPONENTIAL FUNCTIONS; LOGARITHMIC DIFFERENTIATION

In Section 1 we saw that the derivative of $\ln x$ is $1/x$. In Section 3 we showed that $\ln x$ is $\log_e x$ and, therefore, we have the formula

$$\frac{d}{dx}(\log_e x) = \frac{1}{x}, \quad x > 0.$$

The following theorem is convenient when $x$ may have negative values.

**THEOREM 7**    *If $f(x) = \log_e |x|$, then $f'(x) = 1/x$, $x \neq 0$.*

**Proof**    If $x$ is positive, $\log_e |x| = \log_e x$, and the result is nothing new. If $x$ is negative, then $-x$ is positive and the Chain Rule yields

$$\frac{d}{dx}\log_e(-x) = \frac{1}{-x}\frac{d(-x)}{dx} = \frac{1}{x}.$$

Therefore, whatever $x$ is, (so long as $x \neq 0$) we have

$$\frac{d}{dx}\log_e |x| = \frac{1}{x}. \qquad \square$$

The derivative of the exponential function is obtained by using the fact that it is the inverse of the natural logarithm.

**THEOREM 8**    *If $f(x) = e^x$, then $f'(x) = e^x$.*

**Proof**    By letting $y = e^x = \exp x$, we may write $\ln y = x$. Then, by the Chain Rule, we have

$$\frac{1}{y}\frac{dy}{dx} = 1 \quad \text{or} \quad f'(x) = \frac{dy}{dx} = y = e^x. \qquad \square$$

We note the interesting fact that the exponential function is its own derivative, which is the content of Theorem 8.

The problem of finding the derivative of $a^x$ can be reduced to that of obtaining the derivative of the exponential function, as the proof of the next theorem shows.

---

**THEOREM 9**     *If $f(x) = a^x$, then $f'(x) = a^x \log_e a$.*

---

**Proof**   We set $y = a^x$. Then, taking the natural logarithm of both sides, we have $\log_e y = x \log_e a$. By differentiating, we find

$$\frac{1}{y}\frac{dy}{dx} = \log_e a \qquad \text{and} \qquad f'(x) = \frac{dy}{dx} = y \log_e a = a^x \log_e a. \qquad \square$$

We summarize the formulas in the following table:

$$\ln u = \log_e u, \qquad a^u = e^{u \ln a},$$

$$\exp u = e^u, \qquad \log_b x = \frac{\ln x}{\ln b},$$

$$d \ln u = \frac{du}{u}, \qquad d \log_a u = \frac{du}{u \ln a},$$

$$de^u = e^u \, du, \qquad da^u = a^u (\ln a) \, du.$$

**EXAMPLE 1**   Given that $f(x) = \ln |\sin x|$, find $f'(x)$.

**Solution**   We have

$$d \ln |\sin x| = \frac{1}{\sin x} d(\sin x) = \frac{\cos x \, dx}{\sin x}.$$

Therefore

$$f'(x) = \cot x. \qquad \square$$

**EXAMPLE 2**   Given that $f(x) = x^2 e^{-x^2}$, find $f'(x)$.

**Solution**   We have

$$d(x^2 e^{-x^2}) = x^2 d(e^{-x^2}) + e^{-x^2} d(x^2)$$

$$= x^2 e^{-x^2} d(-x^2) + 2x e^{-x^2} dx = e^{-x^2}(-2x^3 + 2x) \, dx.$$

Therefore

$$f'(x) = 2x e^{-x^2}(1 - x^2). \qquad \square$$

**EXAMPLE 3**   Given that $f(x) = \ln |\sec x + \tan x|$, find $f'(x)$.

**Solution**    We have

$$f'(x) = \frac{1}{\sec x + \tan x} \cdot \frac{d}{dx}(\sec x + \tan x)$$

$$= \frac{1}{\sec x + \tan x}(\sec x \tan x + \sec^2 x) = \sec x. \qquad \square$$

*Remark.*   Since each differentiation formula carries with it a corresponding integration formula, we see that the result

$$\frac{d}{dx}\ln|\sec x + \tan x| = \sec x$$

gives the interesting integration formula

$$\int \sec x\, dx = \ln|\sec x + \tan x| + C.$$

This follows, of course, from the fact that differentiation and integration are inverse processes. In a similar way, we can easily deduce that

$$\int \csc x\, dx = -\ln|\csc x + \cot x| + C.$$

The function $x^n$ is now defined for $n$ any *real* number (see definition on page 263). The next theorem shows how we can extend to the case for any real number the differentiation formula for $x^n$ stated previously for the case $n$ a rational number on page 109.

**THEOREM 10**    *If $n$ is any real number and $f$ is defined by $f(x) = x^n$, then $f'(x) = nx^{n-1}$, $x > 0$.*

**Proof**    We let $y = x^n$. From the definition on page 263, we have

$$y = x^n = \exp(n \log x) = e^{n \log x}.$$

From the derivative formulas, we get

$$\frac{dy}{dx} = e^{n \log x}\frac{d}{dx}(n \log x) = e^{n \log x} n\left(\frac{1}{x}\right).$$

Therefore

$$\frac{dy}{dx} = x^n \cdot n\frac{1}{x} = nx^{n-1}. \qquad \square$$

We learned in Theorem 9 how to differentiate functions of the form $a^u$, where $a$ is constant and $u$ is variable, and Theorem 10 yields the formula for $u^a$, where $u$ is a variable and $a$ is any constant. Sometimes we get functions of the form $u(x)^{v(x)}$, where $u$ and $v$ are both functions of $x$. If $u$ and $v$ can each be differentiated by known methods it is possible to develop a formula for the derivative of $u^v$. The following examples illustrate the procedure, called **logarithmic differentiation**.

**EXAMPLE 4**    Given that $f(x) = x^x$, find $f'(x)$.

**Solution**    Write $y = x^x$, and take logarithms of both sides of this equation. Then $\log y = x \log x$. Differentiating this last equation implicitly, we find

$$\frac{1}{y}\frac{dy}{dx} = x \cdot \frac{1}{x} + \log x \cdot 1 = 1 + \log x.$$

Therefore

$$f'(x) = \frac{dy}{dx} = y(1 + \log x) = x^x(1 + \log x). \qquad \square$$

**EXAMPLE 5**    Given that

$$f(x) = \left| \frac{x^2 \sqrt[3]{3x+2}}{(2x-3)^3} \right|, \quad x \neq \tfrac{3}{2}, 0, -\tfrac{2}{3},$$

find $f'(x)$.

**Solution**    Taking logarithms of both sides, we have

$$\log f(x) = 2 \log |x| + \tfrac{1}{3} \log |3x + 2| - 3 \log |2x - 3|.$$

Therefore

$$\frac{1}{f} f' = \frac{2}{x} + \frac{1}{3x+2} - \frac{6}{2x-3}$$

and

$$f'(x) = \left| \frac{x^2 \sqrt[3]{3x+2}}{(2x-3)^3} \right| \left( \frac{2}{x} + \frac{1}{3x+2} - \frac{6}{2x-3} \right). \qquad \square$$

**EXAMPLE 6**    Perform the integration:

$$\int (x-1)e^{-x^2+2x} \, dx.$$

**Solution**    Set $u = -x^2 + 2x$. Then $du = (-2x + 2)\, dx$ and

$$\int (x-1)e^{-x^2+2x} \, dx = -\tfrac{1}{2} \int e^u \, du = -\tfrac{1}{2} e^u = -\tfrac{1}{2} e^{-x^2+2x} + C. \qquad \square$$

## 4    PROBLEMS

In each of Problems 1 through 27, find the derivative.

**1** $f(x) = e^{-3x}$

**2** $f(x) = e^{x^3 + 2x^2 - 6}$

**3** $f(x) = x^3 e^{4x}$

**4** $f(x) = x^7 e^{-2x^2}$

**5** $H(x) = e^{3x} \ln |x|$

**6** $f(x) = xe^{-x} \ln(x^3)$

**7** $g(x) = e^{\tan x}$

**8** $F(x) = \exp(\sin x)$

**9** $f(x) = 3^{5x}$

**10** $f(x) = 4^{-2x}$

**11** $f(x) = (x^3 + 3)2^{-7x}$

**12** $F(x) = 2^{5x} \cdot 3^{4x^2}$

**13** $F(x) = x^{\sin x}$

**14** $F(x) = x^{\ln x}$

**15** $G(x) = x^{\sqrt{x}}$

**16** $G(x) = (\sin x)^x$

**17** $H(x) = (\ln x)^x$

**18** $H(x) = (3/x)^x$

**19** $f(x) = \dfrac{x^2\sqrt{2x+3}}{(x^2+1)^4}$     **20** $f(x) = |x\sqrt{2x-1}\,\sqrt[3]{3x+3}|$

**21** $g(x) = \sqrt{(x+2)(x+3)/(x+1)}$

**22** $g(x) = \left|\dfrac{(x+2)^2(2x-3)^{1/2}}{\sqrt[3]{3x-2}}\right|$

**23** $f(x) = \dfrac{x^3\sin 2x}{(1+x^2)^{1/2}}$     **24** $f(x) = \dfrac{x^4\cos^2 x}{1+x^3}$

**25** $f(x) = \dfrac{x^{1/3}\tan 2x}{(2x-1)^7}$     **26** $h(x) = \dfrac{(1+x^2)^{2/3}\sec 4x}{1+\sin^2 x}$

**27** $f(x) = \dfrac{\exp(x^2)}{\sqrt{x^2+1}\,\sqrt[4]{2x^2+3}}$

**28** By differentiating the right side, verify the integration formula

$$\int \csc x\, dx = -\log|\cot x + \csc x| + C.$$

**29** Integrate $\displaystyle\int x\sec(3x^2)\,dx$.

**30** Integrate $\displaystyle\int \dfrac{dx}{x^2\cos(1/x)}$.

**31** Integrate $\displaystyle\int \dfrac{\cos^2 x\, dx}{\sin x}$.

In each of Problems 32 through 40, perform the integration.

**32** $\displaystyle\int e^{-3x}\,dx$     **33** $\displaystyle\int xe^{-1+x^2}\,dx$

**34** $\displaystyle\int e^{-\sin x}(\cos x)\,dx$     **35** $\displaystyle\int \dfrac{1}{x}e^{-2\log_3 x}\,dx$

**36** $\displaystyle\int e^{\cos 3x}\sin 3x\,dx$     **37** $\displaystyle\int 2^x\,dx$

**38** $\displaystyle\int 4^{2x}\,dx$     **39** $\displaystyle\int e^{-\tan 4x}\sec^2 4x\,dx$

**40** $\displaystyle\int x3^{-x^2}\,dx$

**41** If $n$ is any positive integer and $y = \ln x$, show that

$$\frac{d^n y}{dx^n} = (-1)^{n-1}\frac{(n-1)!}{x^n}.$$

**42** If $n$ is any positive integer and $y = xe^x$, show that

$$\frac{d^n y}{dx^n} = y\left(1+\frac{n}{x}\right).$$

**43** Suppose that $F(x) = [f(x)]^{g(x)}$ where $f$ and $g$ are differentiable functions and $f(x) > 0$ for all $x$. Show that

$$F'(x) = gf^g\left[\frac{f'}{f} + \frac{g'}{g}\ln f\right].$$

**44** The symbol $x^{x^x}$ is interpreted as $x^{(x^x)}$.
a) Find $f'(x)$ if $f(x) = x^{x^x}$.
b) Find $g'(x)$ if $g(x) = x^{x^{x^x}}$.

---

**5**

# THE INVERSE TRIGONOMETRIC FUNCTIONS

A procedure for constructing graphs of the elementary trigonometric functions is given in Section 2 of Appendix 1. We reproduce in Fig. 27 the graph of $f(x) = \sin x$. The inverse of the function $y = \sin x$ is the *relation* defined by

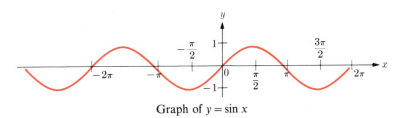

Graph of $y = \sin x$

**FIGURE 27**

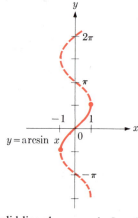

$x = \sin y$

FIGURE 28

interchanging $x$ and $y$. That is, the inverse is given by

$$x = \sin y.$$

A graph of this relation is sketched in Fig. 28, with the curve repeating indefinitely in both upward and downward directions. Since the *range* of the sine function is the interval $[-1, 1]$, this interval is the *domain* of the inverse of the sine function. For each number $x$ on $[-1, 1]$, there are infinitely many values of $y$ such that $\sin y = x$. For example, if $x = \frac{1}{2}$, then

$$y = \frac{\pi}{6} + 2n\pi, \qquad y = \frac{5\pi}{6} + 2n\pi, \quad \text{where } n = 0, \pm 1, \pm 2, \ldots.$$

Referring to Fig. 27, we see that $\sin x$ increases as $x$ goes from $-\pi/2$ to $\pi/2$, decreases from $\pi/2$ to $3\pi/2$, increases from $3\pi/2$ to $5\pi/2$, decreases from $5\pi/2$ to $7\pi/2$, and so on. We define the *increasing* function $f_1$ by restricting the domain of the sine function to the interval $[-\pi/2, \pi/2]$:

$$f_1(x) = \sin x, \quad -\frac{\pi}{2} \le x \le \frac{\pi}{2}.$$

The graph of $f_1$ is shown in Fig. 29.

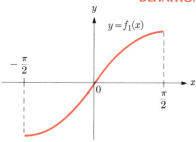

$y = f_1(x)$

Restriction of $y = \sin x$ to the interval

$$-\frac{\pi}{2} \le x \le \frac{\pi}{2}$$

FIGURE 29

**DEFINITION** We define the **arcsin function** as the inverse of $f_1$; that is,

$$y = \arcsin x \quad \Leftrightarrow \quad x = \sin y \text{ and } -\frac{\pi}{2} \le y \le \frac{\pi}{2}.$$

The common notation $\sin^{-1} x$ will sometimes be used for arcsin $x$. This alternate notation will also be used for the other inverse trigonometric functions. When this symbol is used, *the $-1$ is not an exponent*. It is a synonym for the "arc" notation.

The graph of the arcsin function is indicated by the solid line in Fig. 30. Since arcsin is the inverse of $f_1$, formulas (1) and (2) of Section 2 (page 259) become

$$\arcsin(\sin x) = x \quad \text{for } -\frac{\pi}{2} \le x \le \frac{\pi}{2},$$

$$\sin(\arcsin x) = x \quad \text{for } -1 \le x \le 1.$$

$y = \arcsin x$

Solid line shows arcsin function

FIGURE 30

The graph of the function $f(x) = \cos x$ is shown in Fig. 31. The inverse of the function $y = \cos x$ is the relation defined by the equation

$$x = \cos y.$$

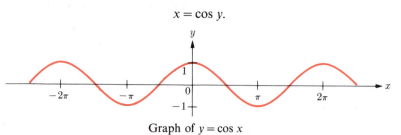

Graph of $y = \cos x$

FIGURE 31

The graph of this relation is shown in Fig. 32, with part of the curve drawn solid and the remainder dashed. By restricting cos $x$ to the interval $[0, \pi]$, we obtain a decreasing function of $x$.

---

**DEFINITION**    We define (see Fig. 33)

$$f_2(x) = \cos x, \quad 0 \leq x \leq \pi,$$

and so obtain the **arccos function** as the inverse of $f_2$; that is,

> $y = \arccos x$   if $x = \cos y$   and   $0 \leq y \leq \pi.$

The graph of this function is indicated by the solid line in Fig. 32.

**EXAMPLE 1**    Sketch the graph of

$$y = 2 \arccos\left(\frac{x}{2}\right).$$

**Solution**    We have

$$y = 2 \arccos\left(\frac{x}{2}\right) \quad \Leftrightarrow \quad \left(\frac{y}{2}\right) = \arccos\left(\frac{x}{2}\right)$$

$$\Leftrightarrow \quad \left(\frac{x}{2}\right) = \cos\left(\frac{y}{2}\right) \quad \text{and} \quad 0 \leq \frac{y}{2} \leq \pi$$

$$\Leftrightarrow \quad x = 2\cos\left(\frac{y}{2}\right) \quad \text{and} \quad 0 \leq y \leq 2\pi.$$

The graph is shown in Fig. 34.                    □

The graph of the function $f(x) = \tan x$ is shown in Fig. 35. The inverse of the function $y = \tan x$ is the relation defined by the equation

$$x = \tan y.$$

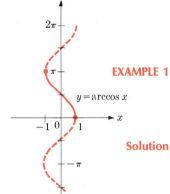

Solid line shows arccos function
**FIGURE 32**

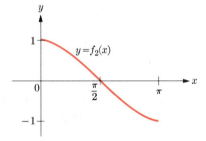

Restriction of $y = \cos x$ to the interval $0 \leq x \leq \pi$
**FIGURE 33**

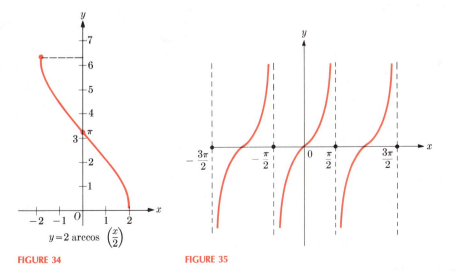

$y = 2\arccos\left(\frac{x}{2}\right)$

**FIGURE 34**                    **FIGURE 35**

The graph of this relation is shown in Fig. 36. Repeating the method used for the sine and cosine functions, we observe that $y = \tan x$ is an increasing function in the interval $(-\pi/2, \pi/2)$.

DEFINITION     The **arctan function** is defined by

$$y = \arctan x \quad \text{if } x = \tan y \quad \text{and} \quad -\frac{\pi}{2} < y < \frac{\pi}{2}.$$

This function is shown as the solid curve in Fig. 36.

Using exactly the same procedure as for the sine, cosine, and tangent functions, we defined arccot $x$ as the number $y$ for which $0 < y < \pi$ and cot $y = x$ (Fig. 37).

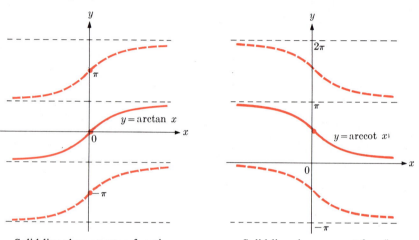

Solid line shows arctan function

FIGURE 36

Solid line shows arccot function

FIGURE 37

For $x \geq 1$, arcsec $x$ is defined as the number $y$ such that $0 \leq y < \pi/2$ and sec $y = x$; for $x \leq -1$, arcsec $x$ is defined as the number $y$ such that $-\pi \leq y < -\pi/2$ and sec $y = x$ (Fig. 38).

For $x \geq 1$, arccsc $x$ is defined as the number $y$ such that $0 < y \leq \pi/2$ and csc $y = x$; for $x \leq -1$, arccsc $x$ is defined as the number $y$ such that $-\pi < y \leq -\pi/2$ and csc $y = x$ (Fig. 39).

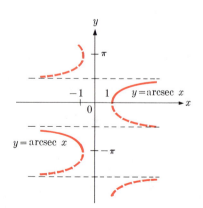

Solid line shows arcsec function

FIGURE 38

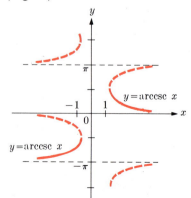

FIGURE 39     Solid line shows arccsc function

To sum up:

a) the values (i.e. the range) of arcsin $x$ and arctan $x$ are between $-\pi/2$ and $\pi/2$;

b) the values (i.e. the range) of arccos $x$ and arccot $x$ are between $0$ and $\pi$;

c) the values (i.e. the range) of arcsec $x$ and arccsc $x$ are between $0$ and $\pi/2$ if $x \geq 1$ and between $-\pi$ and $-\pi/2$ if $x \leq -1$.

**EXAMPLE 2**    Evaluate (a) arcsin $\frac{1}{2}$,    (b) arccot $(-1)$,    (c) arctan $(-\sqrt{3})$,    (d) arcsec $(-2)$.

**Solution**    (a) We set $y = \arcsin \frac{1}{2}$, and this gives us $\sin y = \frac{1}{2}$ or $y = \pi/6$. In the same way we have (b) arccot $(-1) = 3\pi/4$ and (c) arctan $(-\sqrt{3}) = -\pi/3$. As for (d), by setting $y = \operatorname{arcsec}(-2)$, we obtain $\sec y = -2$, and $y$ must be in the third quadrant. Therefore $y = -2\pi/3$ ($\cos y = -\frac{1}{2}$, $y = -120°$).    □

The following theorems tell us how to find the derivatives of the inverse trigonometric functions.

**THEOREM 11**    *Let $y = \arcsin x$. Then*

$$\frac{dy}{dx} = \frac{1}{\sqrt{1-x^2}} \quad \text{if } -1 < x < 1.$$

**Proof**    From the definition of inverse trigonometric functions, we have

$$\sin y = x, \quad -\frac{\pi}{2} \leq y \leq \frac{\pi}{2}.$$

From the Inverse Function Theorem we know that $dy/dx$ exists for $-1 < x < 1$. By the Chain Rule for derivatives, we get

$$\cos y \cdot \frac{dy}{dx} = 1,$$

and we note that $\cos y > 0$ so long as $y$ is in the interval $(-\pi/2, \pi/2)$. Therefore, from the relation

$$\sin^2 y + \cos^2 y = 1,$$

we find that

$$\cos y = +\sqrt{1 - \sin^2 y}, \quad -\frac{\pi}{2} < y < \frac{\pi}{2},$$

or

$$\cos y = \sqrt{1 - x^2}.$$

We conclude that

$$\frac{dy}{dx} = \frac{1}{\cos y} = \frac{1}{\sqrt{1-x^2}}.$$
   □

**THEOREM 12**  *If $y = \arccos x$, then*

$$\frac{dy}{dx} = -\frac{1}{\sqrt{1-x^2}}, \quad if -1 < x < 1.$$

**Proof**  We have

$$\cos y = x, \quad 0 \le y \le \pi,$$

and by the Chain Rule for derivatives

$$-\sin y \frac{dy}{dx} = 1.$$

We see that $\sin y > 0$ for $0 < y < \pi$, and so

$$\sin y = +\sqrt{1 - \cos^2 y} = \sqrt{1-x^2}, \quad -1 < x < 1,$$

from which we conclude that

$$\frac{dy}{dx} = -\frac{1}{\sqrt{1-x^2}}, \quad -1 < x < 1. \qquad \square$$

The next two formulas for derivatives are established in a similar way.

**THEOREM 13**  *If $y = \arctan x$, then*

$$\frac{dy}{dx} = \frac{1}{1+x^2}.$$

**THEOREM 14**  *If $y = \text{arccot } x$, then*

$$\frac{dy}{dx} = -\frac{1}{1+x^2}.$$

We prove the following theorem.

**THEOREM 15**  *If $y = \text{arcsec } x$, then*

$$\frac{dy}{dx} = \frac{1}{x\sqrt{x^2-1}}, \quad if \ |x| > 1.$$

**Proof**  From the definition of the inverse secant function, we may write

$$\sec y = x, \quad 0 \le y < \frac{\pi}{2} \quad \text{or} \quad -\pi \le y < -\frac{\pi}{2}.$$

We note that $\tan y > 0$ in the first and third quadrants, and therefore from

$\tan^2 y + 1 = \sec^2 y$ we obtain

$$\tan y = +\sqrt{\sec^2 y - 1} = \sqrt{x^2 - 1}, \quad |x| > 1.$$

By the Chain Rule for derivatives we see that

$$\sec^2 y \frac{dy}{dx} = \frac{1}{2}(x^2 - 1)^{-1/2}(2x) = \frac{x}{\sqrt{x^2 - 1}}.$$

Now we solve for $dy/dx$ and set $\sec^2 y = x^2$ to obtain

$$\frac{dy}{dx} = \frac{1}{x\sqrt{x^2 - 1}}. \qquad \square$$

THEOREM 16    *If* $y = \text{arccsc } x$, *then*

$$\frac{dy}{dx} = \frac{-1}{x\sqrt{x^2 - 1}}, \quad |x| > 1.$$

The proof of this result follows the same lines as the proof of Theorem 15.

If $u$ is any function of $x$, we may use the Chain Rule to find derivatives of the inverse trigonometric functions of $u$. We state the basic formulas in terms of differentials:

$$d \arcsin u = \frac{du}{\sqrt{1 - u^2}}, \qquad d \arccos u = \frac{-du}{\sqrt{1 - u^2}},$$

$$d \arctan u = \frac{du}{1 + u^2}, \qquad d \arccot u = \frac{-du}{1 + u^2},$$

$$d \text{arcsec } u = \frac{du}{u\sqrt{u^2 - 1}}, \qquad d \text{arccsc } u = \frac{-du}{u\sqrt{u^2 - 1}}.$$

EXAMPLE 3    Given that $y = \arcsin\left(\frac{5}{3}x\right)$, find $dy/dx$.

Solution    We let

$$u = \tfrac{5}{3}x \qquad \text{and} \qquad du = \tfrac{5}{3}dx.$$

Therefore

$$\frac{dy}{dx} = \frac{\frac{5}{3}}{\sqrt{1 - (\frac{5}{3}x)^2}} = \frac{5}{\sqrt{9 - 25x^2}}. \qquad \square$$

EXAMPLE 4    Given that $f(x) = x \arctan(x^2)$, find $f'(x)$.

Solution    This is in the form of the product of $x$ and $\arctan(x^2)$. From the product formula we have

$$f'(x) = x \cdot \frac{d}{dx}(\arctan(x^2)) + \arctan(x^2) \cdot 1.$$

Since

$$\frac{d}{dx}(\arctan(x^2)) = \frac{2x}{1+(x^2)^2} = \frac{2x}{1+x^4},$$

we conclude that

$$f'(x) = \frac{2x^2}{1+x^4} + \arctan(x^2). \qquad \Box$$

## 5   PROBLEMS

In Problems 1 through 8, in each case plot the graph of the equation.

**1**  $y = 2\sin 3x, \quad -\pi \le x \le \pi$

**2**  $y = 3\cos\frac{1}{2}x, \quad 0 \le x \le \pi$

**3**  $y = 2\sin^2 x, \quad -\frac{\pi}{2} \le x \le \frac{\pi}{2}$

**4**  $y = 3\tan 2x, \quad -\frac{\pi}{4} < x < \frac{\pi}{4}$

**5**  $y = 2\arccos x$   **6**  $y = 3\arcsin(x/2)$

**7**  $y = \frac{1}{2}\arcsin 2x$   **8**  $y = \frac{1}{2}\arctan 2x$

In Problems 9 through 12, find the value of each of the expressions. Do not use a calculator.

**9** a)  $\arcsin(\sqrt{3}/2)$   **10** a)  $\arccos(-\frac{1}{2})$

 b)  $\arccos(-1/\sqrt{2})$   b)  $\text{arcsec}(-2/\sqrt{3})$

 c)  $\arctan(-1)$   c)  $\text{arccsc}(-2)$

**11** a)  $\text{arccot}\sqrt{3}$   **12** a)  $\arcsin(-\frac{1}{2})$

 b)  $\text{arcsec}\,2$   b)  $\text{arccot}(-\sqrt{3})$

 c)  $\arctan(-1/\sqrt{3})$   c)  $\text{arccsc}(-2/\sqrt{3})$

In Problems 13 through 39, in each case find the derivative.

**13**  $f(x) = \arctan 2x$   **14**  $g(s) = \text{arccot}(s^2)$

**15**  $\phi(y) = \arccos\sqrt{y}$   **16**  $f(t) = \text{arcsec}(1/t)$

**17**  $y = x\arccos x$   **18**  $f(u) = u^2\arcsin 2u$

**19**  $f(x) = (\text{arccot}\,3x)/(1+x^2)$

**20**  $f(x) = \arctan(x/\sqrt{1-x^2})$

**21**  $f(x) = \arcsin(x/\sqrt{1+x^2})$

**22**  $g(x) = \text{arccot}[(1-x)/(1+x)]$

**23**  $f(x) = \arctan[(x-3)/(1+3x)]$

**24**  $f(x) = (\arccos 2x)/\sqrt{1+4x^2}$

**25**  $f(x) = x\arcsin 2x + \frac{1}{2}\sqrt{1-4x^2}$

**26**  $f(x) = \arctan(3\tan x)$

**27**  $f(x) = \sqrt{x^2-4} - 2\arctan[\frac{1}{2}\sqrt{x^2-4}]$

**28**  $f(x) = \sec^{-1}(\sqrt{1+x^2}/x)$

**29**  $f(x) = \arctan[(3\sin x)/(4+5\cos x)]$

**30**  $y = (\ln x)(\arcsin 2x)$

**31**  $y = e^{-3x}\arccos 3x$

**32**  $f(x) = e^{\arcsin 2x}$   **33**  $y = \ln|\arctan 4x|$

**34**  $g(x) = \arccos(e^{-x})$

**35**  $\phi(x) = x^2 e^{-x}\arcsin x$

**36**  $h(x) = (\text{arcsec}\,x)^2\ln|2x|$

**37**  $f(x) = (\arccos 4x)^{1/2}e^{2x}$

**38**  $y = \text{arccsc}(e^{-3x})$

**39**  $f(t) = t^3\ln|2+t|\arcsin(1+t)$

**40** Given that $f(x) = \arcsin x + \arccos x$, find $f'(x)$. What can you conclude about $f(x)$?

**41** Given that $f(x) = \text{arcsec}\,x + \text{arccsc}\,x$, find $f'(x)$. What can you conclude about $f(x)$?

**42** Prove Theorem 13.   **43** Prove Theorem 14.

**44** Prove Theorem 16, showing the required restrictions on the domain of the arccsc function.

**45** Given $f(x) = \arcsin(\cos x)$ for $0 \le x < \pi$, show that $f''(x) = 0$ for all $x$. Hence $f$ must be a linear function of the form $ax + b$. Find the values of $a$ and $b$.

**46** Same as Problem 45 for $g(x) = \text{arccot}(\tan x)$, $-(\pi/2) < x < (\pi/2)$.

**47** Prove that

$$\arctan x + \arctan c = \arctan\left(\frac{x+c}{1-xc}\right) \qquad (1)$$

provided that $0 < c < 1$ and $0 \le x \le 1$ where $c$ is a constant. (*Hint:* Show that the derivative of the left side equals the derivative of the right side and that the two sides of (1) are equal for one particular value of $x$.)

**48** Prove that if $x \ge 1$, then $\text{arcsec}(1/x) = \arccos x$. (Same hint as for Problem 47.)

_____ 6

## INTEGRATIONS YIELDING INVERSE TRIGONOMETRIC FUNCTIONS

According to the Fundamental Theorem of Calculus, differentiation and integration are inverse processes. Therefore the theorems on derivatives in Section 5 bring with them the following integration formulas.

$$\int \frac{du}{\sqrt{1 - u^2}} = \arcsin u + C,$$

$$\int \frac{du}{1 + u^2} = \arctan u + C,$$

$$\int \frac{du}{u\sqrt{u^2 - 1}} = \text{arcsec } u + C.$$

**EXAMPLE 1**    Find

$$\int \frac{dx}{\sqrt{1 - 4x^2}}.$$

**Solution**    Let $u = 2x$ and $du = 2\, dx$. Then

$$\int \frac{dx}{\sqrt{1 - 4x^2}} = \int \frac{\frac{1}{2}\, du}{\sqrt{1 - u^2}} = \frac{1}{2} \arcsin u + C = \frac{1}{2}\arcsin 2x + C. \qquad \square$$

**EXAMPLE 2**    Find

$$\int \frac{dx}{9 + x^2}.$$

**Solution**    We write

$$\int \frac{dx}{9 + x^2} = \int \frac{dx}{9[1 + (x/3)^2]} = \frac{1}{9} \int \frac{dx}{1 + (x/3)^2}.$$

Making the substitution $u = \frac{1}{3}x$, $du = \frac{1}{3}\, dx$, we obtain

$$\frac{1}{9} \int \frac{3\, du}{1 + u^2} = \frac{1}{3} \arctan u + C = \frac{1}{3}\arctan \frac{1}{3}x + C. \qquad \square$$

The reader may wonder why three of the formulas for integration were omitted. For example, we did not include the formula

$$\int \frac{du}{\sqrt{1 - u^2}} = -\arccos u + C.$$

This is not a new formula because it can be verified that for all values of $u$ between $-1$ and $+1$ we have

$$\arccos u = \frac{\pi}{2} - \arcsin u.$$

(See Problem 40 of Section 5.) Therefore the above integration formula is the same as the one leading to arcsin $u$. The other two formulas present similar situations.

## 6  PROBLEMS

In Problems 1 through 20, in each case perform the integration.

**1** $\displaystyle\int \frac{dx}{\sqrt{1-16x^2}}$

**2** $\displaystyle\int \frac{dx}{1+25x^2}$

**3** $\displaystyle\int \frac{dx}{x\sqrt{9x^2-1}}$

**4** $\displaystyle\int \frac{dx}{\sqrt{9-16x^2}}$

**5** $\displaystyle\int \frac{dx}{1+7x^2}$

**6** $\displaystyle\int \frac{dx}{16+9x^2}$

**7** $\displaystyle\int \frac{dx}{x\sqrt{x^2-4}}$

**8** $\displaystyle\int \frac{dx}{\sqrt{12-5x^2}}$

**9** $\displaystyle\int \frac{(x^2+1)\,dx}{4+x^2}$

**10** $\displaystyle\int \frac{(2x-3)\,dx}{\sqrt{1-4x^2}}$

**11** $\displaystyle\int \frac{dx}{x\sqrt{2x^2-10}}$

**12** $\displaystyle\int \frac{dx}{7+(\frac{1}{3}x)^2}$

**13** $\displaystyle\int \frac{e^x\,dx}{\sqrt{1-e^{2x}}}$

**14** $\displaystyle\int \frac{e^{-2x}\,dx}{\sqrt{1-e^{-4x}}}$

**15** $\displaystyle\int \frac{dx}{x\sqrt{1-(\ln x)^2}}$

**16** $\displaystyle\int \frac{dx}{x(1+(\ln x)^2)}$

**17** $\displaystyle\int \frac{dx}{\sqrt{4e^{2x}-1}}$

**18** $\displaystyle\int \frac{dx}{x\ln|x|\sqrt{9(\ln|x|)^2-1}}$

**19** $\displaystyle\int \frac{e^x\,dx}{36+4e^{2x}}$

**20** $\displaystyle\int \frac{e^{(1/2)x}\,dx}{9+16e^x}$

In Problems 21 through 24, in each case find the value of the integral.

**21** $\displaystyle\int_{-1/2}^{(1/2)\sqrt{3}} \frac{dx}{\sqrt{1-x^2}}$

**22** $\displaystyle\int_{-1}^{\sqrt{3}} \frac{dx}{1+x^2}$

**23** $\displaystyle\int_{(2/3)\sqrt{3}}^{2} \frac{dx}{x\sqrt{x^2-1}}$

**24** $\displaystyle\int_{-2}^{-\sqrt{2}} \frac{dx}{x\sqrt{x^2-1}}$

**25** Show that arctan $x +$ arccot $x =$ constant. Find the value of the constant.

**\*26** Find the area of the region

$$R = \{(x, y): 0 \le x \le 1,\ x^2 \le y \le \sqrt{4-3x^2}\}.$$

(*Hint:* Let $x = (2\sin\theta)/\sqrt{3}$.)

**27** Find the area of the region

$$R = \{(x, y): 0 \le x \le \tfrac{1}{2}\sqrt{2},\ x^2 \le y \le (1+2x^2)^{-1}\}.$$

**28** Define the function

$$F(x; a, b) = \int_0^x \frac{1}{\sqrt{a-bu^2}}\,du, \quad a > 0, b > 0.$$

Show that

$$F(x; a, b) = \frac{1}{\sqrt{b}} F\left(\sqrt{\frac{b}{a}}\,x;\, 1, 1\right)$$

and observe that $F(x; 1, 1) = \arcsin x$.

**29** Define the function

$$G(x; a, b) = \int_0^x \frac{1}{a+bu^2}\,du, \quad a > 0, b > 0.$$

Show that

$$G(x; a, b) = \frac{1}{\sqrt{ab}} G\left(\sqrt{\frac{b}{a}}\,x;\, 1, 1\right)$$

and observe that $G(x; 1, 1) = \arctan x$.

## 7

## APPLICATIONS

In Chapter 4 we discussed various applications of differentiation. Now we are able to differentiate many more types of functions than we could at that time—including logarithmic, exponential, and inverse trigonometric functions. Although the general methods we shall now discuss are similar to those described in Chapter 4, the degree of permissible complication has expanded greatly. The following examples illustrate some of the techniques we can use.

**EXAMPLE 1**    Find the equation of the line tangent to the curve $y = xe^{2x}$ at the point where $x = 2$. Also find the minimum point of the curve.

**Solution**    We have

$$\frac{dy}{dx} = 2xe^{2x} + e^{2x}$$

and, at $x = 2$,

$$\frac{dy}{dx} = 4e^4 + e^4 = 5e^4.$$

Further, if $x = 2$, then $y = 2e^4$. The tangent line has the equation

$$y - 2e^4 = 5e^4(x - 2) \quad \Leftrightarrow \quad 5e^4x - y = 8e^4.$$

To find the minimum point, we set

$$\frac{dy}{dx} = 0$$

and solve for $x$:

$$2xe^{2x} + e^{2x} = 0 \quad \Leftrightarrow \quad x = -\tfrac{1}{2}.$$

We employ the second derivative test for maxima and minima:

$$\frac{d^2y}{dx^2} = 4e^{2x} + 4xe^{2x},$$

and $d^2y/dx^2 > 0$ when $x = -\tfrac{1}{2}$. Therefore $(-\tfrac{1}{2}, -\tfrac{1}{2}e^{-1})$ is a minimum point. The graph is shown in Fig. 40 (with the scale exaggerated).    □

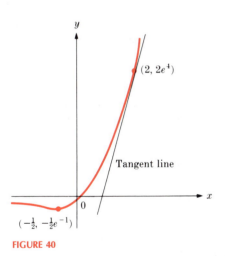

$(2, 2e^4)$

Tangent line

$(-\tfrac{1}{2}, -\tfrac{1}{2}e^{-1})$

**FIGURE 40**

**EXAMPLE 2**    A gutter is to be made out of a long sheet of metal 12 cm wide by turning up strips of width 4 cm along each side so that they make equal angles $\theta$ with the vertical, as in Fig. 41. For what value of $\theta$ will the carrying capacity be greatest?

**Solution**    The carrying capacity is proportional to the area $A$ of a cross section. If we find the maximum value for $A$, the capacity will also be a maximum. From Fig. 41 we see that the area of each of the triangles is $8 \sin \theta \cos \theta$. The area of the rectangle is $16 \cos \theta$.

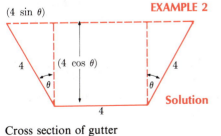

$(4 \sin \theta)$

$(4 \cos \theta)$

$4$      $4$

$\theta$      $\theta$

$4$

Cross section of gutter

**FIGURE 41**

Therefore, for $A = A(\theta)$, we have the formula

$$A = 16 \cos \theta + 16 \sin \theta \cos \theta, \quad 0 \leq \theta < \frac{\pi}{2},$$

$$\frac{dA}{d\theta} = -16 \sin \theta - 16 \sin^2 \theta + 16 \cos^2 \theta = 16(1 - \sin \theta - 2 \sin^2 \theta).$$

Setting the derivative equal to zero, we get

$$0 = 1 - \sin \theta - 2 \sin^2 \theta = (1 + \sin \theta)(1 - 2 \sin \theta).$$

The equation $\sin \theta + 1 = 0$ gives no root for $0 \leq \theta < (\pi/2)$, while

$$1 - 2 \sin \theta = 0 \quad \text{yields} \quad \theta = \frac{\pi}{6}.$$

The corresponding value of $\cos \theta$ is $\frac{1}{2}\sqrt{3}$, and we find that

$$A\left(\frac{\pi}{6}\right) = 12\sqrt{3}.$$

To get the true maximum we examine the endpoints. Since $A(0) = 16$ and $A(\theta) \to 0$ as $\theta \to (\pi/2)$, the maximum carrying capacity occurs at $\pi/6$ rather than at an endpoint, and $A = 12\sqrt{3} \approx 20.8$. ☐

**EXAMPLE 3**    A balloon leaving the ground 1200 meters from an observer rises straight up at the rate of 200 meters per minute. How fast is the angle of elevation of the observer's line of sight increasing when the balloon is at an altitude of 1600 meters?

**Solution**    As shown in Fig. 42, the angle of elevation is $\theta$ and the rate of change of this angle is $d\theta/dt$, where $t$ denotes the time in minutes. Letting $y = y(t)$ be the height of the balloon at time $t$, we have the formula

$$\tan \theta = \frac{y}{1200} \quad \Leftrightarrow \quad \theta = \arctan\left(\frac{y}{1200}\right).$$

The rate at which the balloon is rising is $dy/dt = 200$. Differentiating with respect to $t$, we obtain

$$\frac{d\theta}{dt} = \frac{1}{1 + (y/1200)^2} \frac{d}{dt}\left(\frac{y}{1200}\right) = \frac{1200}{(1200)^2 + y^2} \frac{dy}{dt}.$$

Inserting $y = 1600$ and $dy/dt = 200$, we find $d\theta/dt = 0.06$ radians/min. ☐

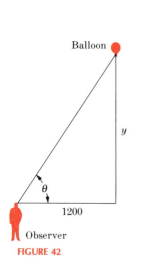

Balloon

$y$

$\theta$

1200

Observer

**FIGURE 42**

Radioactive substances such as radium, cesium, radon, and certain forms of carbon all decay in a similar way. If $t$ denotes the time and $F(t)$ is the amount of a given radioactive substance at time $t$, then the rate of decay, $F'(t)$, satisfies the equation

$$F'(t) = -cF(t),$$

where $c$ is a positive constant whose value depends on the particular substance. It is a simple matter to verify that the function

$$F(t) = Ae^{-ct}$$

satisfies the above equation. Here we observe that the number $A$ is the amount of substance present at time $t = 0$, since

$$F(0) = Ae^0 = A.$$

**EXAMPLE 4**    The half-life of a radioactive substance is the length of time it takes for the original amount to decay so that one half of it remains. It is known that radium has a half-life of 1690 years. Find the value of $c$ in the above formula for radioactive decay.

**Solution**    Since $F(t) = Ae^{-ct}$ we know that when $t = 1690$, then $F(1690) = (\frac{1}{2})A$. Therefore

$$\tfrac{1}{2}A = Ae^{-c(1690)} \qquad \Leftrightarrow \qquad \tfrac{1}{2} = e^{-1690c},$$

or

$$e^{1690c} = 2.$$

We take logarithms of both sides, getting

$$1690 \ln e = \ln 2,$$

and using a table of natural logarithms or a calculator to determine that $\ln 2 = 0.693$, we find

$$c = \frac{\ln 2}{1690} = \frac{0.693}{1690} = 0.000410. \qquad \square$$

In a proper environment, bacteria grow at a rate proportional to the number present. If we denote by $F(t)$ the number of bacteria present at time $t$, then the formula

$$F'(t) = cF(t),$$

where $c$, a positive constant, is an *approximation* to the growth rate of bacteria. We easily verify that

$$F(t) = Ae^{ct}$$

satisfies the above equation, with $A$ the number of bacteria present at time $t = 0$, i.e., $A = F(0)$.

**EXAMPLE 5**    Suppose there are 1000 bacteria present in a dish at time $t = 0$, and it is known that the number of bacteria present always doubles every 90 minutes. How long will it take to have 5000 bacteria in the dish?

**Solution**    We have the formula

$$F(t) = 1000e^{ct},$$

and $F(t) = 2000$ when $t = 90$. Hence $2 = e^{90c}$, and taking logarithms we get

$$90c \ln e = \ln 2 \qquad \Leftrightarrow \qquad c = \frac{\ln 2}{90}.$$

Since $\ln 2 = 0.693$ we have $c = 0.0077$. Therefore

$$F(t) = 1000e^{0.0077t}.$$

We are seeking the value of $t$ which yields $F(t) = 5000$. Thus

$$5000 = 1000e^{0.0077t} \quad \text{or} \quad 5 = e^{0.0077t}.$$

Taking logarithms we find

$$0.0077t \ln e = \ln 5.$$

Since (from a calculator or a table of natural logarithms) $\ln 5 = 1.609$, we have

$$t = \frac{\ln 5}{0.0077} = \frac{1.609}{0.0077} = 209.$$

There will be 5000 bacteria after 209 minutes. □

## 7  PROBLEMS

In Problems 1 through 14, find in each case the equations of the lines tangent and normal to the given curve at the point on the curve corresponding to the given value of $x$.

1  $y = \arctan x, \quad x = 1$

2  $y = 2^x, \quad x = -1$

3  $y = \text{arcsec } x, \quad x = -2$

4  $y = \arcsin x, \quad x = -\frac{1}{2}$

5  $y = \log x, \quad x = e$

6  $y = x^2 e^{-x}, \quad x = 1$

7  $y = x \log x, \quad x = e$

8  $y = x^{-2} e^{2x}, \quad x = -1$

9  $y = 2 \arctan 3x, \quad x = 0$

10  $y = 3^{-x}, \quad x = -1$

11  $y = x^2 \log(1 + x), \quad x = 1$

12  $y = 3 \arcsin 2x, \quad x = \frac{1}{4}$

13  $y = e^{-x} \ln(2 + x), \quad x = 0$

14  $y = (1 + x^2) \arctan x, \quad x = 1$

In Problems 15 through 24, find in each case the point or points of intersection of the curves given; then find the tangent of the acute angle between the curves at these intersection points. *Hint*: If $\tan \theta_1$ and $\tan \theta_2$ are the slopes of the two lines, recall from trigonometry (see Appendix 1) the formula

$$\tan(\theta_2 - \theta_1) = \frac{\tan \theta_2 - \tan \theta_1}{1 + \tan \theta_2 \tan \theta_1}.$$

15  $y = \arcsin x, \quad y = \arccos x$

16  $y = \arctan x, \quad y = \arcsin(x/2)$

17  $y = e^{2x}, \quad y = 2e^x$

18  $y = \log x, \quad y = \log(x^2)$

19  $y = \text{arccot } x, \quad y = \frac{1}{2} \arcsin x$

20  $y = xe^x, \quad y = x^2 e^x$

21  $y = xe^{-x}, \quad y = x^2 e^{-x}$

22  $y = e^{1+x}, \quad y = e^{2x}$

23  $y = \ln(1 + x), \quad y = \ln(3 - x^2), \quad -1 < x < \sqrt{3}$

24  $y = 3^{-x}, \quad y = 3^{1+x}$

In Problems 25 through 39, find in each case the relative maxima and minima, the points of inflection, the intervals in which $f$ is increasing, those in which $f$ is decreasing, those in which the graph is concave upward, and those in which it is concave downward. Sketch the graph.

25  $f(x) = e^{-x}$

26  $f(x) = xe^{-x}$

27  $f(x) = x^2 e^{-x}$

28  $f(x) = e^{-x^2}$

29  $f(x) = 2xe^{-(1/2)x^2}$

30  $f(x) = x^2 e^{-x^2}$

31  $f(x) = \log(1 + x)$

32  $f(x) = x \log x$

33  $f(x) = 2x^2 \log x$

34  $f(x) = 2x(\log x)^2$

35  $f(x) = e^{-x} \sin x$

36  $f(x) = e^{-x} \cos x$

37  $f(x) = 3^{-x}$

38  $f(x) = x2^{-x}$

39  $f(x) = \tan^2 x, \quad -\frac{\pi}{2} < x < \frac{\pi}{2}$

40  Find the minimum value of $x^x$ for $x > 0$.

41  A revolving light 3 km from a straight shoreline makes 2 revolutions/min. Find the speed of the spot of light along the shore when it is 2 km away from the point on the shore nearest the light.

42  A wall 8 m high is $\frac{27}{8}$ m from a building. Find the length of the shortest ladder which will clear the wall and rest with one end on the ground and the other end on the building. Also find the angle which this ladder makes with the horizontal (see Fig. 43).

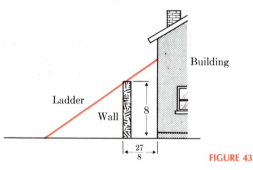

Building

Ladder

Wall

8

$\frac{27}{8}$

**FIGURE 43**

43  An airplane at an altitude of 4400 m is flying horizontally directly away from an observer. At the instant when the angle of elevation is 45°, or $\pi/4$ radians, the angle is decreasing at the rate of 0.05 radian/sec. How fast is the airplane flying at that instant?

44  A cone whose generators make an angle $\theta$ with its axis is inscribed in a sphere of radius $R$. For what value of $\theta$ will the lateral area of the cone be greatest?

45 A man is running along a sidewalk at the rate of 5 m/sec. A searchlight on the ground 30 m from the walk is kept trained on him. At what rate is the searchlight revolving when the man is 20 m away from the point on the sidewalk nearest the light?

46 A tablet 7 m high is placed on a wall with its base 9 m above the level of an observer's eye. How far from the wall should an observer stand in order that the angle subtended at his eye by the tablet be a maximum?

47 A steel girder 27 m long is moved horizontally along a corridor 8 m wide and around a corner into a hall at right angles to the corridor. How wide must the hall be to permit this? Neglect the width of the girder (see Fig. 44).

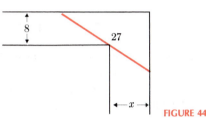

FIGURE 44

48 A man on a wharf is pulling in a small boat. His hands are 20 m above the level of the point on the boat where the rope is tied. If he is pulling in the rope at 2 m/sec, how fast is the angle that the rope makes with the horizontal increasing when there are 52 m of rope out?

49 Find the angle of the sector which should be removed from a circular piece of canvas of radius 12 m so that the conical tent made from the remaining piece will have the greatest volume. Repeat the problem for a circular piece of canvas of radius $R$ m.

50 The end $B$ of a piston rod moves back and forth on a line through $O$; $B$ is attached to a rod $AB$ of length 10 cm which, in turn, is attached to a crank $OA$ of length 4 cm, which revolves about $O$. Find an expression for the velocity of $B$ if $OA$ makes 10 rev/sec (see Fig. 45).

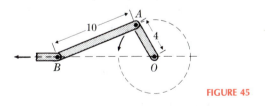

FIGURE 45

51 A hemispherical dome is 50 m in diameter. Just at sunset a ball is dropped from a point on the line joining the top of the dome with the sun. Assuming that the ball drops according to the law $u = 10t^2$ ($u$ in m, $t$ in sec), obtain an expression for the speed of its shadow along the dome in terms of $t$. Find its speed when $t = 1$ and the limit of its speed as $t \to 0$.

52 A weight is drawn along a level table by means of a rope, attached to a point $P$, which passes over a pulley whose axle is fixed horizontally at the level of the table and perpendicular to the line of motion of $P$. If the pulley, which has a radius of 1 m, turns at the rate of $\frac{1}{2}$ rev/sec, find an expression for the rate of change of the angle which the rope makes with the table. Assume that $P$ is at the level of the table.

53 The hour hand of a clock in a tower is $\frac{3}{4}$ meter long and the minute hand is 1 meter long. At what rate in meters per hour are the ends of the hands approaching each other at 3 o'clock?

54 A radioactive substance disintegrates at a rate proportional to the amount of substance present. Let $F(t)$ be the amount at time $t$. Then $F(t)$ is determined by the relation

$$F'(t) = -cF(t)$$

where $c$ is a positive constant depending on the substance. Hence, as in Example 4, $F(t)$ is given by the formula $F(t) = Ae^{-ct}$, where $A$ is the amount of substance at time $t = 0$. Find the half-life of the substance; that is, find the value of $t$ such that $F$ has the value $\frac{1}{2}A$. (Answer in terms of $c$.)

55 Bacteria grow at a rate proportional to the amount present. Using Example 5 as a guide, show that the number of bacteria $B(t)$ follows the law

$$B(t) = Ae^{ct},$$

where $c$ is a positive constant and $A$ is the number at time $t = 0$. Find the length of time required for the population to double. (Answer in terms of $c$.)

56 (See Example 5 on page 285.) When a bank provides interest at 6% annually "compounded daily," then it provides each day 0.06/360 of the principal $P$ as interest. (Many banks consider a year to be 360 days.) In $n$ days the total amount $A$ becomes $A = P(1 + 0.06/360)^n$. Actually banks compound interest "continuously" by approximating the above formula with

$$A = Pe^{0.06n/360}.$$

a) Justify the approximation by finding the value of $A$ by the two methods for $n = 1, 2, 3$ days.

b) At 6% interest compounded continuously, how long does it take for an initial deposit to double?

c) If an amount of money is deposited so that it doubles in 10 years, what must the interest rate be if the compounding is continuous?

57 The current $I$ (in amperes) in a battery (and other electrical networks) obeys the law $LI'(t) = -RI(t)$ where $L$ is the inductance (henrys) and $R$ is the resistance (ohms). Express $I$ as a function of $t$ if $L$ and $R$ are constant and $I = 2$ when $t = 0$.

## 8

## THE HYPERBOLIC FUNCTIONS

Exponential functions occur often in many applications, especially those in physics and engineering. Certain combinations of them occur so often that they have been tabulated and given special names: the **hyperbolic functions**. These functions satisfy relations which bear a great similarity to the basic formulas which the trigonometric functions satisfy. Furthermore, the differentiation and integration properties of hyperbolic functions are quite similar to those of the trigonometric functions. However, one of the most important properties of the trigonometric functions—that of periodicity—is not shared by any of the hyperbolic functions.

We provide in Appendix 3 at the end of the book two sections giving the definitions, elementary properties, and differentiation and integration formulas of hyperbolic and inverse hyperbolic functions. Graphs of the various functions may be found there. Readers not interested in hyperbolic functions may skip this subject without loss of continuity in the development of either the theory or the techniques of calculus.

## CHAPTER 7

## REVIEW PROBLEMS

1 Given $f(x) = \ln(4x + 5)$, find $f'(x)$.

2 Given $y = \ln(2x^3 - 6x + 7)$, find $dy/dx$.

3 Given $f(x) = \ln(\ln(2x + 10))$, find $f'(x)$ and $f''(x)$.

4 Given $y = (x^2 + 1)\ln(x + 3)$, find

$$\frac{dy}{dx} \quad \text{and} \quad \frac{d^2y}{dx^2}.$$

5 Find the line tangent to the curve $y = x^2 \ln(1 + x)$ at the point $(1, \ln 2)$. Find the normal to the curve at this point.

6 Integrate: $\displaystyle\int \frac{dx}{2x + 3}$.     7 Integrate: $\displaystyle\int \frac{x\,dx}{3x^2 + 4}$.

8 Find $f'(x)$ if $f(x) = \ln(\ln(\ln(9 + 2x)))$.

9 Find the area of the region bounded by the curve

$$y = \frac{x^3}{x^4 + 2}$$

and the lines $x = 1$, $x = 4$, $y = 0$.

10 Integrate: $\displaystyle\int \frac{\sin x\,dx}{3\cos x + 4}$.

11 Given the function $f(x) = 1 + x^2$, $0 \leq x < \infty$, show that $f$ has an inverse $g$ and find the formula for $g$. Show by substitution that $f[g(x)] = x$ and $g[f(x)] = x$.

12 Given the function

$$f(x) = \frac{x}{x^2 + 4}, \quad -2 < x < 2,$$

show that $f$ has an inverse $g$ and find the formula for $g$. Show by substitution that $f[g(x)] = x$ and $g[f(x)] = x$.

13 Given the function

$$f(x) = \frac{x}{4x^2 + 9}, \quad -\infty < x < \infty,$$

find the intervals where $f$ is increasing and those where $f$ is decreasing. In each of these intervals, find the inverse function.

**14** Given the relation $2x^2 + y^2 = 12$, find the domain, range, and inverse relation. Is the inverse relation a function?

**15** Given the relation $y^4 - 2x^2 = 4$, find the domain, range, and the inverse relation. Is the inverse relation a function?

**16** Given the relation $2x^2 + y^3 = 10$, find the domain, range and the inverse relation. Is the inverse relation a function?

**17** Given the function $f(x) = x^3 + 4$, $-\infty < x < \infty$, show that $f$ has an inverse function $g$. Find $f'$ and $g'$. What are the domains of $f'$ and $g'$?

**18** Suppose $f$ and $g$ are inverse functions and $f' > 0$, $g' > 0$. Show that if $g''' < 0$ then $f''' > 0$. (*Hint:* Differentiate the formula $f[g(x)] = x$ by the Chain Rule three times.)

In Problems 19 through 26, sketch the curves.

**19** $y = \ln(2x + 1)$     **20** $y = \ln(4 - x)$

**21** $y = e^{2x+1}$     **22** $y = e^{3-x}$

**23** $y = 2^{3+2x}$     **24** $y = \log_3(x^2 + 1)$

**25** $y = xe^{4x}$     **26** $y = xe^{-2x}$

**27** Since

$$e = \lim_{t \to 0}(1 + t)^{1/t},$$

it can be shown that

$$\lim_{t \to 0}(1 + t)^{x/t} = e^x.$$

Use a calculator to approximate $\sqrt{e}$ by letting $t = 0, 1, 0.01$, and $0.001$ in the above formula with $x = \frac{1}{2}$.

**28** Since $2^a = e^{a \ln 2}$, use the formula in Problem 27 for $e^x$ with $x = \frac{1}{2}\ln 2$ to approximate $\sqrt{2}$ by setting $t = 0.1, 0.01$, and $0.001$; use a calculator and take $\ln 2 = 0.693$.

In each of Problems 29 through 36, find the derivative.

**29** $f(x) = e^{3x^2 - 2x + 5}$     **30** $f(x) = (x^2 + 2x - 4)e^{2x+5}$

**31** $g(x) = (\ln|2 + x|)e^{4x+1}$

**32** $g(x) = (\arcsin x)\ln|2 + x|$

**33** $F(x) = \dfrac{(x^2 + 1)^{3/2}(2x + 5)^{2/3}}{(x + 1)(x + 2)^{3/4}}$

**34** $F(x) = \dfrac{(x^2 + 2x)\sin(3x + 1)}{(\cos 2x)\ln|x^2 + 1|}$

**35** $h(x) = (\sin x)(\ln|x|)e^x$

**36** $h(x) = e^{2\sin x}\tan(x^2 + 1)$

**37** Find $f'(x)$ and $f''(x)$, given $f(x) = (x^2 + 1)e^{-1/x}$.

**38** Find $f'(x)$ and $f''(x)$, given $f(x) = e^{x+1}\ln(\ln|10 + x|)$.

In each of Problems 39 through 44, perform the integration.

**39** $\displaystyle\int xe^{2-x^2}\,dx$     **40** $\displaystyle\int e^{4\sin(x^2)}x\cos(x)^2\,dx$

**41** $\displaystyle\int 3^{(1/2)x}\,dx$     **42** $\displaystyle\int \dfrac{\ln|x + 1|\,dx}{3x + 3}$

**43** $\displaystyle\int \tan x\ln(\cos x)\,dx$     **44** $\displaystyle\int \cot 2x\ln(\sin 2x)\,dx$

In each of Problems 45 through 52, find the derivative.

**45** $f(x) = \arcsin(x^2 - 1)$     **46** $f(x) = \arctan(2 - x)$

**47** $f(x) = \dfrac{\arccos x}{x}$     **48** $f(x) = (x^2 + 1)\operatorname{arcsec} x$

**49** $g(x) = (\ln x)\arcsin x$     **50** $g(x) = \dfrac{\arcsin 2x}{\ln(1 + x^2)}$

**51** $F(x) = e^{-3\sin x}\ln x$     **52** $F(x) = x^2\ln(1 + x)e^{x^2}$

In each of Problems 53 through 58, perform the integration.

**53** $\displaystyle\int \dfrac{dx}{\sqrt{16 - 9x^2}}$     **54** $\displaystyle\int \dfrac{dx}{25 + 4x^2}$

**55** $\displaystyle\int \dfrac{dx}{\sqrt{4 - 9x^2}}$     **56** $\displaystyle\int \dfrac{dx}{16 + 9x^2}$

**57** $\displaystyle\int \dfrac{e^{2x}\,dx}{\sqrt{4 - e^{4x}}}$     **58** $\displaystyle\int \dfrac{dx}{x\sqrt{16 - 9(\ln x)^2}}$

**59** Find the equation of the line tangent to the curve $y = \arcsin(1 + 2x)$ at the point on the curve where $x = -\frac{1}{4}$.

**60** Find the equation of the line tangent to the curve and the equation of the line normal to the curve $y = x\ln(1 + x)$ at the point on the curve where $x = 2$.

**61** Find the relative maxima and minima of the curve $y = (x + 2)e^{-x+1}$.

**62** Find the relative maxima and minima of the curve $y = \arcsin(1 - x + x^2)$.

**63** Radioactive carbon decays so that its half-life is approximately 5700 years. Suppose that at time $t = 0$ there is 1 gram of radioactive carbon. How long will it take before there is 0.2 gram left?

**64** Suppose that in a certain environment rabbits multiply in proportion to the number of rabbits present. If it takes 2 months for any given population to double, find how many rabbits there will be at the end of one year starting with 100 rabbits.

**65** It is observed that a certain strain of bacteria increase sixfold in 15 hours. How long would it take for an initial quantity of 10,000 of these bacteria to double?

# 8

# TECHNIQUES OF INTEGRATION

We saw in Theorem 2 of Chapter 5 that any function which is continuous on a closed interval can be integrated. This fact, while it is of theoretical value, gives no hint as to ways of actually finding the integral of a function. For this reason we learned in Section 6 of Chapter 5 how to integrate by the method of substitution simple algebraic and trigonometric expressions. We also showed in Chapter 7 how to integrate logarithmic, exponential, and inverse trigonometric functions.

In this chapter we expand the techniques we have already studied in order to integrate a vast array of functions. In particular, we will study the method known as "integration by parts," a technique of important theoretical as well as practical value.

## 1

### INTEGRATION BY SUBSTITUTION

The formula

$$\int u^n \, du = \frac{u^{n+1}}{n+1} + C, \quad n \neq -1$$

may be used to integrate many expressions which do not have the same appearance. For example, the specific integral

$$\int (4 - 7x)^{15} \, dx,$$

while not of the above form, may be transformed so that it becomes identical to it. We set

$$u = 4 - 7x, \qquad du = -7 \, dx$$

and the integral becomes one we can integrate:

$$-\tfrac{1}{7}\int u^{15}\,du = -\tfrac{1}{112}u^{16} + C = -\tfrac{1}{112}(4-7x)^{16} + C.$$

By making the change of variable from $x$ to $u = 4 - 7x$, we change the form of the integrand so that the integral can be calculated quite easily. We recognize this technique as the method of substitution.

It is easy to find functions for which there is no way of performing the integration. For example, it is possible to show that the integrals

$$\int \frac{dx}{\sqrt{1+x^3}}, \qquad \int e^{-x^2}\,dx, \qquad \int \frac{\sin x}{x}\,dx$$

cannot be integrated in terms of any of the functions we have studied so far (polynomials, rational functions, trigonometric functions, logarithms, etc.).

Appearances may be deceiving, for while

$$\int e^{-x^2}\,dx$$

cannot be integrated, the slightly more "complicated" integral

$$\int xe^{-x^2}\,dx$$

easily can be. (See Example 1 below.)

In this chapter we shall learn a number of methods for performing the integrations in many feasible cases. For ready reference, we collect the formulas for integration which we have already learned:

1. $\displaystyle\int u^n\,du = \frac{u^{n+1}}{n+1} + C, \quad n \neq -1$  2. $\displaystyle\int \frac{du}{u} = \ln|u| + C$

3. $\displaystyle\int e^u\,du = e^u + C$  4. $\displaystyle\int a^u\,du = \frac{a^u}{\ln a} + C, \quad a > 0$

The trigonometric formulas are:

5. $\displaystyle\int \sin u\,du = -\cos u + C$  6. $\displaystyle\int \cos u\,du = \sin u + C$

7. $\displaystyle\int \sec^2 u\,du = \tan u + C$  8. $\displaystyle\int \csc^2 u\,du = -\cot u + C$

9. $\displaystyle\int \sec u \tan u\,du = \sec u + C$  10. $\displaystyle\int \csc u \cot u\,du = -\csc u + C$

11. $\displaystyle\int \sec u\,du = \ln|\sec u + \tan u| + C$

12. $\displaystyle\int \csc u\,du = -\ln|\csc u + \cot u| + C$

The inverse trigonometric formulas ($a > 0$) are:

$$13. \int \frac{du}{\sqrt{a^2 - u^2}} = \arcsin \frac{u}{a} + C \qquad 14. \int \frac{du}{a^2 + u^2} = \frac{1}{a} \arctan \frac{u}{a} + C$$

$$15. \int \frac{du}{u\sqrt{u^2 - a^2}} = \frac{1}{a} \operatorname{arcsec} \frac{u}{a} + C, \quad |u| > a$$

We recall that as a consequence of the Fundamental Theorem of Calculus every one of the above integration formulas may be verified merely by differentiating the right side and then verifying that the result is the integrand on the left. For example, in Formula 11 above, we may verify that the derivative of $\ln |\sec u + \tan u|$ is $\sec u$.

It is easy to check that Formulas 13 through 15 are derived by using the substitution $v = u/a$ in the simpler differentiation formulas

$$d \arcsin v = \frac{dv}{\sqrt{1 - v^2}}, \quad d \arctan v = \frac{dv}{1 + v^2}, \quad d \operatorname{arcsec} v = \frac{dv}{v\sqrt{v^2 - 1}}.$$

We illustrate the method of substitution with twelve examples.

**EXAMPLE 1**　Find $\int xe^{x^2} \, dx$.

**Solution**　Try $u = x^2$. Then $du = 2x \, dx$, and therefore

$$\int xe^{x^2} \, dx = \frac{1}{2} \int e^u \, du = \frac{1}{2} e^u + C = \frac{1}{2} e^{x^2} + C. \qquad \square$$

**EXAMPLE 2**　Find

$$\int \frac{\cos 3x \, dx}{1 + \sin 3x}.$$

**Solution**　Try $u = 1 + \sin 3x$. Then $du = 3 \cos 3x \, dx$, and

$$\int \frac{\cos 3x \, dx}{1 + \sin 3x} = \frac{1}{3} \int \frac{du}{u} = \frac{1}{3} \ln |u| + C = \frac{1}{3} \ln |1 + \sin 3x| + C. \qquad \square$$

**EXAMPLE 3**　Find

$$\int \frac{dt}{\sqrt{4 - 9t^2}}.$$

**Solution**　Try $u = 3t$, $du = 3 \, dt$. Then

$$\int \frac{dt}{\sqrt{4 - 9t^2}} = \frac{1}{3} \int \frac{du}{\sqrt{4 - u^2}}.$$

This expression is Formula 13 with $a = 2$. Therefore

$$\int \frac{dt}{\sqrt{4 - 9t^2}} = \frac{1}{3} \arcsin \frac{u}{2} + C = \frac{1}{3} \arcsin \frac{3t}{2} + C. \qquad \square$$

**EXAMPLE 4**    Find

$$\int \frac{t\,dt}{\sqrt{4-9t^2}}.$$

**Solution**    Try $u = 4 - 9t^2$. Then $du = -18t\,dt$, and therefore

$$\int \frac{t\,dt}{\sqrt{4-9t^2}} = -\frac{1}{18}\int \frac{du}{u^{1/2}} = -\frac{1}{9}u^{1/2} + C = -\frac{1}{9}\sqrt{4-9t^2} + C. \qquad \square$$

*Remark.* The method of substitution proceeds by trial and error. If a particular substitution does not reduce the integral to a known formula, we should feel no hesitation about abandoning that substitution and trying one of an entirely different nature. Example 4, although it is similar in appearance to Example 3, integrates according to a different substitution.

**EXAMPLE 5**    Find

$$\int \frac{e^y\,dy}{(1+e^y)^3}.$$

**Solution**    Try $u = 1 + e^y$. Then $du = e^y\,dy$, and

$$\int \frac{e^y\,dy}{(1+e^y)^3} = \int u^{-3}\,du = -\frac{1}{2u^2} + C = -\frac{1}{2(1+e^y)^2} + C. \qquad \square$$

*Remarks.* When an expression such as $e^{x^2}$ or $\sin(\ln x)$ occurs, it is generally worthwhile to set $u = x^2$ in the first case or $u = \ln x$ in the second case. Such a substitution may reduce the integral to a more recognizable form. It is also fairly easy to spot integrals which are reducible to the forms given in Formulas 13 through 15. But, as Examples 2, 4, and 5 show, it may happen that Formulas 1 and 2 are applicable. Sometimes we must make several substitutions in succession, as the next example shows.

**EXAMPLE 6**    Find

$$\int \frac{dx}{x[9 + 4(\ln x)^2]}.$$

**Solution**    Try $u = \ln x$. Then $du = (1/x)\,dx$, and

$$\int \frac{dx}{x[9 + 4(\ln x)^2]} = \int \frac{du}{9 + 4u^2}.$$

Let $v = 2u$. Then $dv = 2\,du$, and therefore the integral becomes

$$\frac{1}{2}\int \frac{dv}{9 + v^2} = \frac{1}{2}\cdot\frac{1}{3}\arctan\frac{v}{3} + C$$

$$= \frac{1}{6}\arctan\frac{2u}{3} + C = \frac{1}{6}\arctan\frac{2\ln x}{3} + C. \qquad \square$$

Substitution provides us with a short method for evaluating definite integrals. We illustrate the process in the following example.

**EXAMPLE 7**    Evaluate

$$\int_0^{\pi/4} \tan^2 x \sec^2 x \, dx.$$

**Solution**    Try $u = \tan x$, $du = \sec^2 x \, dx$. Note that $u(0) = 0$, $u(\pi/4) = 1$. Therefore

$$\int_0^{\pi/4} \tan^2 x \sec^2 x \, dx = \int_0^1 u^2 \, du = \tfrac{1}{3}u^3 \Big]_0^1 = \tfrac{1}{3}. \qquad \square$$

There are handbooks that contain extensive tables of integrals. Therefore, when we are confronted with a problem in integration, we should first look through a table to seek the answer. However, the integrals that usually occur in practice are not precisely in a form in which the answer can be read directly from a table. The seven examples above show how an integral may be manipulated and reduced so that the answer can be obtained from one of the Formulas 1 through 15, formulas which constitute our "table of integrals."

We now enlarge the table by adding several formulas. These are integrals of algebraic expressions in which the result may be expressed in terms of the logarithm function. The verification of the formulas below follows the customary method of differentiation of the result. (See Example 8 which follows, and Problems 17 through 20 at the end of this section.) We assume that $a > 0$ in the following formulas.

16. $$\int \frac{du}{\sqrt{u^2 + a^2}} = \ln\left(u + \sqrt{u^2 + a^2}\right) + C$$

17. $$\int \frac{du}{\sqrt{u^2 - a^2}} = \ln\left|u + \sqrt{u^2 - a^2}\right| + C \text{ if } |u| > a$$

18. $$\int \frac{du}{a^2 - u^2} = \begin{cases} \dfrac{1}{2a}\ln\left(\dfrac{u + a}{u - a}\right) + C \text{ if } |u| > a \\[2ex] \dfrac{1}{2a}\ln\left(\dfrac{a + u}{a - u}\right) + C \text{ if } |u| < a \end{cases}$$

19. $$\int \frac{du}{u\sqrt{a^2 - u^2}} = -\frac{1}{a}\ln\left|\frac{a + \sqrt{a^2 - u^2}}{u}\right| + C \text{ for } 0 < |u| < a$$

20. $$\int \frac{du}{u\sqrt{a^2 + u^2}} = -\frac{1}{a}\ln\left|\frac{a + \sqrt{a^2 + u^2}}{u}\right| + C \text{ for } u \neq 0$$

**EXAMPLE 8**    Use differentiation to verify Formula 16 above.

**Solution**   We have

$$\frac{d}{du} \ln (u + \sqrt{u^2 + a^2}) = \frac{1}{u + \sqrt{u^2 + a^2}} \left[ 1 + \frac{1}{2}(u^2 + a^2)^{-1/2} \cdot 2u \right]$$

$$= \frac{1}{u + \sqrt{u^2 + a^2}} \left[ 1 + \frac{u}{\sqrt{u^2 + a^2}} \right]$$

$$= \frac{1}{u + \sqrt{u^2 + a^2}} \cdot \frac{\sqrt{u^2 + a^2} + u}{\sqrt{u^2 + a^2}}$$

$$= \frac{1}{\sqrt{u^2 + a^2}}. \qquad \square$$

If the indicated substitution is simple enough, we may shorten the integration process by rewriting the integral so that it is expressed in one of the standard forms. The following examples exhibit the method.

**EXAMPLE 9**   Find

$$\int \frac{\sin \theta \, d\theta}{\sqrt{1 + \cos \theta}}.$$

**Solution**   We see that

$$d(1 + \cos \theta) = -\sin \theta \, d\theta,$$

and so

$$\int \frac{\sin \theta \, d\theta}{\sqrt{1 + \cos \theta}} = - \int (1 + \cos \theta)^{-1/2} \, d(1 + \cos \theta).$$

This is Formula 1 with $n = -\frac{1}{2}$, and we obtain

$$\int \frac{\sin \theta \, d\theta}{\sqrt{1 + \cos \theta}} = -2(1 + \cos \theta)^{1/2} + C. \qquad \square$$

**EXAMPLE 10**   Find

$$\int \frac{dx}{x \ln x}.$$

**Solution**   Since $d(\ln x) = (1/x) \, dx$, we obtain

$$\int \frac{dx}{x \ln x} = \int \frac{d(\ln x)}{\ln x} = \ln |\ln x| + C. \qquad \square$$

When the integrand consists of a **rational function** (one polynomial divided by another polynomial) a simplification can be achieved when *the degree of the numerator is greater than or equal to the degree of the denominator,* simply by performing the division. The next two examples illustrate this technique.

**EXAMPLE 11**   Find

$$\int \frac{x^2 - 2x - 1}{x + 2} \, dx.$$

**Solution**   We first divide $x^2 - 2x - 1$ by $x + 2$, using ordinary division of one polynomial by another:

$$
\begin{array}{r}
x - 4 \\
x+2\overline{)\,x^2 - 2x - 1} \\
\underline{x^2 + 2x} \\
-4x - 1 \\
\underline{-4x - 8} \\
+7
\end{array}
$$

Then

$$\int \frac{x^2 - 2x - 1}{x + 2} \, dx = \int \left( x - 4 + \frac{7}{x + 2} \right) dx$$

$$= \frac{1}{2} x^2 - 4x + 7 \int \frac{d(x + 2)}{x + 2}$$

$$= \frac{1}{2} x^2 - 4x + 7 \ln |x + 2| + C. \qquad \square$$

**EXAMPLE 12**   Find

$$\int \frac{x + 5}{x - 1} \, dx.$$

**Solution**   Since the degree of the numerator is equal to the degree of the denominator, we may use ordinary division of one polynomial by another to obtain

$$\frac{x + 5}{x - 1} = 1 + \frac{6}{x - 1}.$$

Therefore

$$\int \frac{x + 5}{x - 1} \, dx = \int \left( 1 + \frac{6}{x - 1} \right) dx = x + 6 \ln |x - 1| + C. \qquad \square$$

## 1  PROBLEMS

In Problems 1 through 16, find the indefinite integrals.

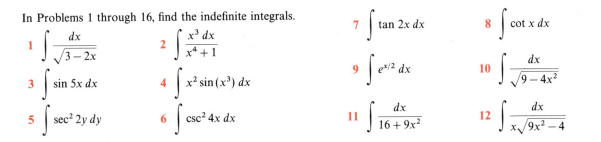

**1** $\int \dfrac{dx}{\sqrt{3 - 2x}}$    **2** $\int \dfrac{x^3 \, dx}{x^4 + 1}$

**3** $\int \sin 5x \, dx$    **4** $\int x^2 \sin(x^3) \, dx$

**5** $\int \sec^2 2y \, dy$    **6** $\int \csc^2 4x \, dx$

**7** $\int \tan 2x \, dx$    **8** $\int \cot x \, dx$

**9** $\int e^{x/2} \, dx$    **10** $\int \dfrac{dx}{\sqrt{9 - 4x^2}}$

**11** $\int \dfrac{dx}{16 + 9x^2}$    **12** $\int \dfrac{dx}{x\sqrt{9x^2 - 4}}$

13 $\int \dfrac{\sin \sqrt{x} \, dx}{\sqrt{x}}$  14 $\int x \sec^2(x^2) \, dx$

15 $\int x e^{x^2+1} \, dx$  16 $\int x^2 e^{2x^3} \, dx$

17 Verify by differentiation Formula 17.

18 Verify by differentiation Formula 18.

19 Verify by differentiation Formula 19.

20 Verify by differentiation Formula 20.

In Problems 21 through 38, in each case find the indefinite integral.

21 $\int \sin^2 x \cos x \, dx$  22 $\int \tan 2x \sec 2x \, dx$

23 $\int \tan 2x \sec^2 2x \, dx$  24 $\int \dfrac{\cos(\ln x) \, dx}{x}$

25 $\int \dfrac{\ln x \, dx}{x[1+(\ln x)^2]}$  26 $\int \dfrac{2 \, dx}{x\sqrt{4x^2-9}}$

27 $\int \dfrac{\cos 2x \, dx}{\sin 2x}$  28 $\int \left(x - \dfrac{1}{x}\right)^2 dx$

29 $\int (\ln x)^3 \dfrac{dx}{x}$  30 $\int \dfrac{dx}{x \ln x}$

31 $\int \dfrac{e^x \, dx}{(e^x+2)^2}$  32 $\int \sec^2 3x \sec 3x \tan 3x \, dx$

33 $\int \sec^7 2x \tan 2x \, dx$  34 $\int \dfrac{x^2}{x^2+1} \, dx$

35 $\int \dfrac{e^x \, dx}{e^x+1}$  36 $\int \dfrac{\sqrt{x} \, dx}{4+x^3}$

37 $\int \dfrac{dy}{1+e^y}$  38 $\int \dfrac{x^5 \, dx}{1+x^2}$

In Problems 39 through 50, evaluate the definite integrals.

39 $\int_0^2 \dfrac{(2x+1) \, dx}{\sqrt{x^2+x+1}}$  40 $\int_1^3 \dfrac{\sqrt[3]{\ln x} \, dx}{x}$

41 $\int_0^{\pi/3} \sec^3 x \tan x \, dx$  42 $\int_0^{\pi/4} \tan^3 x \sec^2 x \, dx$

43 $\int_0^{1/\sqrt{2}} \dfrac{x \, dx}{\sqrt{1-x^4}}$  44 $\int_0^{\sqrt{2}} \dfrac{3x \, dx}{4+x^4}$

45 $\int_0^{\pi/2} \sin^3 x \cos x \, dx$  46 $\int_0^{\pi/2} \dfrac{\cos x \, dx}{1+\sin^2 x}$

47 $\int_0^{\pi/3} \dfrac{\sin x \, dx}{\cos^3 x}$  48 $\int_{\pi/8}^{\pi/4} \cot 2x \csc^2 2x \, dx$

49 $\int_0^1 \dfrac{\sqrt{1+e^{-2x}}}{e^{-3x}} \, dx$  50 $\int_0^{1/2} \dfrac{3 \arcsin x}{\sqrt{1-x^2}} \, dx$

In Problems 51 through 85, find the indefinite integrals.

51 $\int \dfrac{(x+1) \, dx}{x^2+2x+2}$  52 $\int \csc^2\left(\dfrac{x}{2}\right) dx$

53 $\int \dfrac{\sec^2 x \, dx}{1+\tan x}$  54 $\int \dfrac{e^x \, dx}{1+e^{2x}}$

55 $\int \cot 2x \, dx$  56 $\int \dfrac{dx}{x\sqrt{\ln x}}$

57 $\int \dfrac{x^2+3x-2}{x+1} \, dx$  58 $\int \dfrac{x^3+2x^2-x+1}{x+2} \, dx$

59 $\int \dfrac{x}{2x-1} \, dx$  60 $\int \dfrac{x+1}{2x+3} \, dx$

61 $\int \dfrac{dx}{4+3x^2}$  62 $\int x e^{-x^2} \, dx$

63 $\int \dfrac{x \, dx}{(x^2+1)^2}$  64 $\int \cos 2x e^{\sin 2x} \, dx$

65 $\int x 2^{x^2} \, dx$  66 $\int \cot^2 \tfrac{1}{3}x \csc^2 \tfrac{1}{3}x \, dx$

67 $\int \dfrac{x^2+2x-1}{3-x} \, dx$  68 $\int \dfrac{x^4+x^3-2}{x+5} \, dx$

69 $\int \dfrac{x^2+2x+1}{x^2+1} \, dx$  70 $\int \dfrac{3x-2}{4x+7} \, dx$

71 $\int \dfrac{2x^2+6x-2}{x^2+2x-2} \, dx$  72 $\int \dfrac{4-5x}{2-3x} \, dx$

73 $\int e^{\tan x} \sec^2 x \, dx$  74 $\int \dfrac{e^{\arctan x}}{1+x^2} \, dx$

75 $\int \dfrac{e^x \, dx}{\sqrt{3-e^{2x}}}$  76 $\int \cot 2x \csc^3 2x \, dx$

77 $\int x^{-1/2} \sec \sqrt{x} \tan \sqrt{x} \, dx$

78 $\int \dfrac{\arctan 3x}{1+9x^2} \, dx$  79 $\int \dfrac{\sin 2x \, dx}{2+\cos 2x}$

80 $\int \dfrac{2x+1}{x^2+9} \, dx$  81 $\int \dfrac{x-2}{\sqrt{5-x^2}} \, dx$

82 $\int \dfrac{x^2+4}{x\sqrt{x^2-7}} \, dx$  83 $\int \dfrac{(2-\sin x) \, dx}{(2x+\cos x)^3}$

84 $\int \dfrac{1+e^{2x}}{e^x} \, dx$  85 $\int \dfrac{\sin 2x \, dx}{3+\cos^2 2x}$

In Problems 86 through 95, in each case evaluate the definite integral.

86 $\int_0^2 \dfrac{2x+3}{x^2+3x+5} \, dx$  87 $\int_{\pi/3}^{\pi} \dfrac{\cos(x/2)}{\sin(x/2)} \, dx$

88 $\displaystyle\int_1^2 \frac{x^2+1}{x+1}\,dx$

89 $\displaystyle\int_{-1}^1 xe^{-3x^2}\,dx$

92 $\displaystyle\int_{1/2}^1 \frac{\arctan(2x)}{1+4x^2}\,dx$

93 $\displaystyle\int_0^{\pi^3/27} \frac{\sin(x^{1/3})}{x^{2/3}\cos(x^{1/3})}\,dx$

90 $\displaystyle\int_0^1 x^2 3^{(x^3)}\,dx$

91 $\displaystyle\int_{\pi/6}^{\pi/4} \tan(3x/2)\sec^2(3x/2)\,dx$

94 $\displaystyle\int_0^1 \frac{3x+2}{x^2+4}\,dx$

95 $\displaystyle\int_{-1}^2 \frac{e^x}{2+e^x}\,dx$

---

2

## INTEGRATION BY PARTS

The formula for the differential of a product is

$$d(uv) = u\,dv + v\,du.$$

By integrating both sides, we obtain $uv = \int u\,dv + \int v\,du$ or the formula for integration by parts:

### INTEGRATION BY PARTS

$$\int u\,dv = uv - \int v\,du. \qquad (1)$$

This formula is useful not only for evaluating integrals but also for investigating many theoretical questions.

*Every* integral can be written in the form $\int u\,dv$. *If it happens that $\int v\,du$ is easily calculated, then the integration-by-parts formula allows us to calculate $\int u\,dv$.* The next eight examples show how this method is applied.

EXAMPLE 1    Find $\int xe^x\,dx$.

Solution    We wish to write this integral in the form $\int u\,dv$. To do so let

$$u = x \qquad \text{and} \qquad dv = e^x\,dx.$$

Then

$$du = dx \qquad \text{and} \qquad v = e^x.$$

The integration-by-parts formula (1) yields

$$\int xe^x\,dx = xe^x - \int e^x\,dx = xe^x - e^x + C = e^x(x-1) + C.$$

**EXAMPLE 2**    Find $\int \ln x \, dx$.

**Solution**    To express this integral in the form $\int u \, dv$, let

$$u = \ln x \qquad \text{and} \qquad dv = dx.$$

Then

$$du = \frac{1}{x} \, dx \qquad \text{and} \qquad v = x.$$

Integrating by parts, we get

$$\int \ln x \, dx = x \ln x - \int x \cdot \frac{1}{x} \, dx = x \ln x - x + C. \qquad \square$$

Sometimes integration by parts must be performed several times in succession, as the following example shows.

**EXAMPLE 3**    Find $\int x^2 e^x \, dx$.

**Solution**    We let

$$u = x^2 \qquad \text{and} \qquad dv = e^x \, dx.$$

Then

$$du = 2x \, dx \qquad \text{and} \qquad v = e^x.$$

Therefore

$$\int x^2 e^x \, dx = x^2 e^x - 2 \int x e^x \, dx.$$

The last integral on the right may be found by a second integration by parts, as we saw in Example 1. Taking the result given there, we obtain

$$\int x^2 e^x \, dx = x^2 e^x - 2e^x(x - 1) + C = e^x(x^2 - 2x + 2) + C. \qquad \square$$

Integration by parts may be used to evaluate definite integrals. Here is the appropriate formula.

**INTEGRATION BY PARTS FOR DEFINITE INTEGRALS**

$$\int_a^b u \, dv = \left[ uv \right]_a^b - \int_a^b v \, du.$$

**EXAMPLE 4**    Evaluate $\int_0^{\pi/2} x \cos x \, dx$.

**Solution**    We let

$$u = x \qquad \text{and} \qquad dv = \cos x \, dx.$$

Then

$$du = dx \quad \text{and} \quad v = \sin x.$$

We find that

$$\int_0^{\pi/2} x \cos x \, dx = \left[ x \sin x \right]_0^{\pi/2} - \int_0^{\pi/2} \sin x \, dx$$

$$= \frac{\pi}{2} + \left[ \cos x \right]_0^{\pi/2} = \frac{\pi}{2} - 1.$$

*Remark.* In the four examples above, we always begin with a selection of $u$ and $dv$. Then we compute $du$ and $v$ and these choices work. In many problems the correct choices for $u$ and $dv$ are not immediately evident. To illustrate this, we see that in Example 4 we might easily have chosen $u = \cos x$ and $dv = x \, dx$. However, these selections lead to an integral which is more complicated than the original one. We advise the reader to use *trial and error as astutely as possible* in selecting $u$ and $dv$. There are no hard and fast rules. Moreover, the reader should quickly abandon a selection which appears unpromising, and then go on to another choice.

The next example exhibits an interesting technique which uses integration by parts twice.

**EXAMPLE 5**   Find $\int e^x \sin x \, dx$.

**Solution**   We try integration by parts by letting

$$u = e^x, \quad dv = \sin x \, dx, \quad du = e^x \, dx, \quad \text{and} \quad v = -\cos x.$$

Then

$$\int e^x \sin x \, dx = -e^x \cos x + \int e^x \cos x \, dx.$$

The integral on the right appears to be no simpler than the one with which we started. However, we now apply integration by parts to the integral on the right by letting

$$\bar{u} = e^x, \quad d\bar{v} = \cos x \, dx, \quad d\bar{u} = e^x \, dx, \quad \text{and} \quad \bar{v} = \sin x.$$

We then obtain

$$\int e^x \cos x \, dx = e^x \sin x - \int e^x \sin x \, dx.$$

Substituting this expression in the right side of the above expression for $\int e^x \sin x \, dx$, we get

$$\int e^x \sin x \, dx = -e^x \cos x + e^x \sin x - \int e^x \sin x \, dx.$$

Moving the integral on the right side of the equation to the left side and dividing by 2, we get

$$\int e^x \sin x \, dx = \tfrac{1}{2} e^x (\sin x - \cos x) + C.$$

*Remark.* If, in the second integration by parts, we had let $\bar{u} = \cos x$, $d\bar{v} = e^x\, dx$, everything would have canceled out, as this step would have reversed the first integration by parts. The device of Example 5 is rather special, but it is spectacular when it works.

As another illustration of the power of integration by parts, we shall show how certain classes of integrals may be simplified.

**EXAMPLE 6**    Find $\int \sin^n x\, dx$, where $n$ is a positive integer.

**Solution**    The trick here is to write $\sin^n x = \sin^{n-1} x \sin x$. Then we let $u = \sin^{n-1} x$ and $dv = \sin x\, dx$. Therefore

$$du = (n-1)\sin^{n-2} x \cos x\, dx, \quad v = -\cos x.$$

Consequently

$$\int \sin^n x\, dx = -\cos x \sin^{n-1} x + (n-1) \int \sin^{n-2} x \cos^2 x\, dx.$$

We now let $\cos^2 x = 1 - \sin^2 x$ in the integral on the right, and we find that

$$\int \sin^n x\, dx = -\cos x \sin^{n-1} x + (n-1)\int \sin^{n-2} x\, dx - (n-1)\int \sin^n x\, dx.$$

Moving the last integral on the right side of the equation to the left side and dividing by $n$, we obtain

$$\int \sin^n x\, dx = -\frac{\cos x \sin^{n-1} x}{n} + \frac{n-1}{n}\int \sin^{n-2} x\, dx.$$

Applying the same method again to the integral on the right, we get

$$\int \sin^n x\, dx = -\frac{\cos x \sin^{n-1} x}{n}$$
$$+ \frac{n-1}{n}\left( -\frac{\cos x \sin^{n-3} x}{n-2} + \frac{n-3}{n-2}\int \sin^{n-4} x\, dx \right).$$

We continue this process until all of the integrals are evaluated.    □

*Remark.* The result in Example 6 illustrates a procedure known as obtaining a *Reduction Formula*. We begin with an expression such as $\sin x$ raised to a high exponent, which we denote as $n$. Then integration by parts expresses the integral in terms of one of the same form but in which the exponent has been reduced to a lower power. We then continue the process until the exponent is so small that the integration can be performed directly. In Example 6 the Reduction Formula is

$$\int \sin^n x\, dx = -\frac{\cos x \sin^{n-1} x}{n} + \frac{n-1}{n}\int \sin^{n-2} x\, dx.$$

For example, if $n = 6$ the above Reduction Formula when applied three times in succession yields

$$\int \sin^6 x\, dx = -\frac{\cos x \sin^5 x}{6} - \frac{5\cos x \sin^3 x}{24} - \frac{5\cos x \sin x}{16} + \frac{5}{16}x + C.$$

As an illustration, let $n = 5$. We have

$$\int \sin^5 x \, dx = -\frac{\cos x \sin^4 x}{5} + \frac{4}{5}\left(-\frac{\cos x \sin^2 x}{3} + \frac{2}{3}\int \sin x \, dx\right)$$

$$= -\frac{\cos x \sin^4 x}{5} - \frac{4 \cos x \sin^2 x}{15} - \frac{8}{15}\cos x + C$$

$$= -\frac{\cos x}{5}\left(\sin^4 x + \frac{4}{3}\sin^2 x + \frac{8}{3}\right) + C.$$

**EXAMPLE 7**    Find $\int \sec^3 x \, dx$.

**Solution**    Writing $\sec^3 x = \sec x \cdot \sec^2 x$, we let $u = \sec x$, $dv = \sec^2 x \, dx$. Then $du = \tan x \sec x \, dx$ and $v = \tan x$. Therefore

$$\int \sec^3 x \, dx = \sec x \tan x - \int \tan^2 x \sec x \, dx$$

$$= \sec x \tan x - \int (\sec^2 x - 1)\sec x \, dx$$

$$= \sec x \tan x - \int \sec^3 x \, dx + \int \sec x \, dx.$$

Transferring the term $-\int \sec^3 x \, dx$ to the left side and dividing by 2, we find

$$\int \sec^3 x \, dx = \tfrac{1}{2}\sec x \tan x + \tfrac{1}{2}\int \sec x \, dx.$$

Now using Formula 11 on page 292 for $\int \sec x \, dx$, we obtain

$$\int \sec^3 x \, dx = \tfrac{1}{2}\sec x \tan x + \tfrac{1}{2}\ln |\sec x + \tan x| + C. \qquad \square$$

**EXAMPLE 8**    Given that $z$ and $w$ are functions on $R^1$, with $z(0) = 0$, $z'(0) = 1$, $w(0) = 0$, and $w'(0) = 3$. Show that

$$\int_0^a z(x)w''(x) \, dx = z(a)w'(a) - z'(a)w(a) + \int_0^a w(x)z''(x) \, dx.$$

**Solution**    We let

$$u = z(x) \qquad \text{and} \qquad dv = w''(x) \, dx.$$

Hence

$$du = z'(x) \, dx \qquad \text{and} \qquad v = w'(x).$$

Then we can write

$$\int_0^a z(x)w''(x) \, dx = [z(x)w'(x)]_0^a - \int_0^a w'(x)z'(x) \, dx$$

$$= z(a)w'(a) - \int_0^a w'(x)z'(x) \, dx.$$

Continuing with the integral on the right side, we let

$$u = z'(x), \quad dv = w'(x)\,dx, \quad du = z''(x)\,dx, \quad \text{and} \quad v = w(x),$$

which yields

$$\int_0^a z(x)w''(x)\,dx = z(a)w'(a) - \left\{ [z'(x)w(x)]_0^a - \int_0^a w(x)z''(x)\,dx \right\}$$

$$= z(a)w'(a) - z'(a)w(a) + \int_0^a w(x)z''(x)\,dx. \qquad \square$$

## 2   PROBLEMS

In Problems 1 through 33, find the indefinite integrals.

**1** $\displaystyle\int x \ln x\,dx$

**2** $\displaystyle\int x \sin x\,dx$

**3** $\displaystyle\int x^2 \sin x\,dx$

**4** $\displaystyle\int x^2 \ln x\,dx$

**5** $\displaystyle\int (\ln x)^2\,dx$

**6** $\displaystyle\int x^3 e^{2x}\,dx$

**7** $\displaystyle\int \arctan x\,dx$

**8** $\displaystyle\int \arcsin x\,dx$

**9** $\displaystyle\int x \operatorname{arcsec} x\,dx$

**10** $\displaystyle\int x^2 \arctan x\,dx$

**11** $\displaystyle\int x^2 \arcsin x\,dx$

**12** $\displaystyle\int x^3 \operatorname{arcsec} x\,dx$

**13** $\displaystyle\int x \csc^2 \tfrac{1}{2}x\,dx$

**14** $\displaystyle\int (2x\,dx)/(\cos^2 2x)$

**15** $\displaystyle\int 9x \tan^2 3x\,dx$

**16** $\displaystyle\int x^m \ln x\,dx, \quad m \neq -1$

**17** $\displaystyle\int 6x^2 \arcsin 2x\,dx$

**18** $\displaystyle\int (x^2 - 1) \ln x\,dx$

**19** $\displaystyle\int (x^2 - 2x + 2) \ln x\,dx$  **20** $\displaystyle\int (4x^3 - 3x^2 + 4) \ln x\,dx$

**21** $\displaystyle\int \sin \sqrt{2x}\,dx$   (*Hint:* Let $2x = z^2$.)

**22** $\displaystyle\int \sqrt{3x}\,\arcsin \sqrt{3x}\,dx$  **23** $\displaystyle\int 2x^3 e^{x^2}\,dx$

**24** $\displaystyle\int x^4 \arctan(x^2)\,dx$  **25** $\displaystyle\int e^x \sin 2x\,dx$

**26** $\displaystyle\int e^{3x} \sin 2x\,dx$  **27** $\displaystyle\int e^{-x} \cos 3x\,dx$

**28** $\displaystyle\int e^{ax} \cos bx\,dx$  **29** $\displaystyle\int e^{ax} \sin bx\,dx$

**30** $\displaystyle\int x^m (\ln x)^2\,dx$

**31** $\displaystyle\int \sin 4x \sin 2x\,dx$

**32** $\displaystyle\int \cos x \sin 3x\,dx$

**33** $\displaystyle\int \sin^6 x\,dx$

In Problems 34 through 41, evaluate the definite integrals.

**34** $\displaystyle\int_1^2 x^3 \ln x\,dx$

**35** $\displaystyle\int_0^{(1/2)\pi^2} \cos \sqrt{2x}\,dx$

**36** $\displaystyle\int_0^{1/2} 3x^2 \arcsin 2x\,dx$  **37** $\displaystyle\int_0^{\pi/3} 3x^2 \arctan 2x\,dx$

**38** $\displaystyle\int_0^{\pi/2} \cos^4 x\,dx$

**39** $\displaystyle\int_0^{\pi/4} e^{3x} \sin 4x\,dx$

**40** $\displaystyle\int_1^2 (\ln x)^3\,dx$

**41** $\displaystyle\int_0^{\pi/3} x \sec^2 x\,dx$

**42** Find a formula for $\int \cos^n x\,dx$ in terms of $\int \cos^{n-4} x\,dx$.

**43** Find a formula for $\int \sin^n x\,dx$ in terms of $\int \sin^{n-6} x\,dx$.

**44** Given that $z$ and $w$ are functions of $x$ with $z(0) = 1$, $z'(0) = 2$, $z''(0) = 3$, $z(1) = 2$, $z'(1) = -1$, $z''(1) = 2$, $w(0) = -1$, $w'(0) = 2$, $w''(0) = 0$, $w(1) = -1$, $w'(1) = -2$, and $w''(1) = -3$. Express $\int_0^1 zw'''\,dx$ in terms of $\int_0^1 wz'''\,dx$.

**45** Given that $f''(x) = -af(x)$ and that $g''(x) = bg(x)$, where $a$ and $b$ are constants, find the indefinite integral of $\int f(x)g''(x)\,dx$.

**46** Derive the formula

$$\int (\log x)^n\,dx = x(\log x)^n - n \int (\log x)^{n-1}\,dx.$$

Apply the formula to compute $\int (\log x)^3\,dx$.

**47** Derive the formula

$$\int x^p (\log x)^q\,dx = \frac{x^{p+1}(\log x)^q}{p+1} - \frac{q}{p+1} \int x^p (\log x)^{q-1}\,dx,$$

$p \neq -1$.

Apply the formula to compute $\int x^5 (\log x)^2\,dx$.

**48** Use integration by parts to evaluate $\int_1^2 x^4 \arctan x \, dx$.

**49** Let $F(x)$ be a polynomial of the form

$$F(x) = b_0 + b_1 x + \cdots + b_n x^n.$$

Derive the formula

$$\int F(x) e^{ax} \, dx = e^{ax} \left[ \sum_{k=0}^{n} (-1)^{k+1} a^{-k-1} F^{(k)}(x) \right],$$

where $F^{(0)}(x) = F(x)$ and $F^{(k)}(x)$ is the $k$th derivative of $F$ for $k \geq 1$.

**50** Derive the formula

$$\int \tan^n x \, dx = \frac{\tan^{n-1} x}{n-1} - \int \tan^{n-2} x \, dx.$$

Apply the formula to compute $\int \tan^6 x \, dx$.

**51** Define the function $F$ by the formula

$$F(x) = \int_2^x \frac{dy}{\ln y}, \quad 2 \leq x < \infty.$$

Use repeated integration by parts to show that

$$F(x) = a_0 + a_1 \frac{x}{\ln x} + a_2 \frac{x}{\ln^2 x} + \cdots + a_n \frac{x}{\ln^n x}$$
$$+ n! \int_2^x \frac{dy}{\ln^{n+1} y}.$$

Find the values of the constants $a_0, a_1, \ldots, a_n$.

**52** Use integration by parts to derive the formula

$$\int \sin(\log x) \, dx = \tfrac{1}{2} x \sin(\log x) - \tfrac{1}{2} x \cos(\log x) + C.$$

Also verify the result by differentiation of the right side.

---

## 3

# INTEGRATION OF TRIGONOMETRIC EXPRESSIONS

There are special techniques for integrating products and powers of products of trigonometric functions. The methods involve using the elementary trigonometric identities given in Appendix 1 to reduce the integrand to one of the simple forms we know how to handle. Although we recognize that the variety of trigonometric integrands is unlimited, in this section we show how to integrate three of the most common types of integrals:

**A** $\displaystyle\int \sin^m u \cos^n u \, du$

**B** $\displaystyle\int \tan^m u \sec^n u \, du$

**C** $\displaystyle\int \cot^m u \csc^n u \, du$

---

**INTEGRATIONS OF TYPE A**     $\int \sin^m u \cos^n u \, du$.

---

Under Type A there are two cases:

**Case 1.** **Either $m$ or $n$ is an odd, positive integer**. If $m$ is odd, we factor out $\sin u \, du$ and change the remaining even power of sine to powers of cosine by the trigonometric identity

$$\sin^2 u + \cos^2 u = 1.$$

If $n$ is odd, we factor out $\cos u \, du$ and change the remaining even power of cosine to powers of sine by the same identity. Three examples illustrate the technique.

**EXAMPLE 1**    Find $\int \sin^5 x \, dx$.

**Solution**    Since $m = 5$ is an odd positive integer and $n = 0$, we have

$$\int \sin^5 x \, dx = \int \sin^4 x \sin x \, dx$$

$$= \int (\sin^2 x)^2 \sin x \, dx$$

$$= \int (1 - \cos^2 x)^2 \sin x \, dx$$

$$= \int (1 - 2\cos^2 x + \cos^4 x) \sin x \, dx.$$

We set $u = \cos x$, $du = -\sin x \, dx$ and obtain

$$\int \sin^5 x \, dx = \int (1 - 2u^2 + u^4)(-du)$$

$$= -u + \frac{2u^3}{3} - \frac{u^5}{5}$$

$$= -\cos x + \frac{2\cos^3 x}{3} - \frac{\cos^5 x}{5} + C. \qquad \square$$

**EXAMPLE 2**    Find $\int \sin^3 x \cos^{-5} x \, dx$.

**Solution**    Since $m(=3)$ is odd and positive, we have

$$\int \sin^3 x \cos^{-5} x \, dx = \int \sin^2 x \cos^{-5} x \sin x \, dx$$

$$= \int (1 - \cos^2 x) \cos^{-5} x \sin x \, dx$$

$$= -\int \cos^{-5} x \, d(\cos x) + \int \cos^{-3} x \, d(\cos x)$$

$$= \frac{1}{4\cos^4 x} - \frac{1}{2\cos^2 x} + C. \qquad \square$$

**EXAMPLE 3**    Find $\int \sin^4 2x \cos^5 2x \, dx$.

**Solution**   Since $n(=5)$ is odd and positive, we write

$$\int \sin^4 2x \cos^5 2x\, dx = \int \sin^4 2x \cos^4 2x \cos 2x\, dx$$

$$= \int \sin^4 2x(1 - \sin^2 2x)^2 \cos 2x\, dx$$

$$= \tfrac{1}{2} \int (\sin^4 2x - 2\sin^6 2x + \sin^8 2x)\, d(\sin 2x)$$

$$= \tfrac{1}{10}\sin^5 2x - \tfrac{1}{7}\sin^7 2x + \tfrac{1}{18}\sin^9 2x + C. \qquad \square$$

**Case 2.**   **Both $m$ and $n$ are even and positive integers or zero.** In this case, the *half-angle formulas* (see Appendix 1) are used to lower the *degree* of the expression. These formulas (which are worth memorizing) are

$$\sin^2 u = \frac{1 - \cos 2u}{2}, \qquad \cos^2 u = \frac{1 + \cos 2u}{2}.$$

The method of reduction is shown in the next two examples.

**EXAMPLE 4**   Find $\int \sin^2 x \cos^2 x\, dx$.

**Solution**   By the half-angle formulas, we have

$$\int \sin^2 x \cos^2 x\, dx = \frac{1}{4}\int (1 - \cos 2x)(1 + \cos 2x)\, dx$$

$$= \frac{1}{4}\int (1 - \cos^2 2x)\, dx = \frac{x}{4} - \frac{1}{4}\int \cos^2 2x\, dx.$$

To this last integral we again apply the half-angle formula, and get

$$\int \sin^2 x \cos^2 x\, dx = \frac{x}{4} - \frac{1}{4}\int \frac{1 + \cos 4x}{2}\, dx$$

$$= \frac{x}{4} - \frac{x}{8} - \frac{1}{32}\sin 4x + C = \frac{1}{8}\left(x - \frac{1}{4}\sin 4x\right) + C. \qquad \square$$

**EXAMPLE 5**   Find $\int \sin^4 3u\, du$.

**Solution**   This is Case 2, with $m = 4$, $n = 0$. We have

$$\int \sin^4 3u\, du = \frac{1}{4}\int (1 - \cos 6u)^2\, du$$

$$= \frac{1}{4}\int (1 - 2\cos 6u + \cos^2 6u)\, du$$

$$= \frac{u}{4} - \frac{1}{12}\sin 6u + \frac{1}{4}\int \frac{1 + \cos 12u}{2}\, du$$

$$= \frac{3u}{8} - \frac{1}{12}\sin 6u + \frac{1}{96}\sin 12u + C. \qquad \square$$

We recognize that Cases 1 and 2 do not cover all possible integrals of Type A. For example, if both $m$ and $n$ are negative, the methods just described are not applicable, and other techniques must be employed. However, when both $m$ and $n$ are negative, an integral of Type A can often be changed to one of Type B or Type C with different values of $m$ and $n$. Then the methods described below for Types B and C may be tried.

**INTEGRATION OF TYPE B**    $\int \tan^m u \sec^n u \, du$.

Under Type B there are also two cases:

**Case 1.**    *$n$ is an even positive integer*. We factor out $\sec^2 u \, du$ and change the remaining secants to tangents, using the trigonometric identity

$$\sec^2 u = 1 + \tan^2 u.$$

**EXAMPLE 6**    Find

$$\int \frac{\sec^4 u \, du}{\sqrt{\tan u}}.$$

**Solution**    We write

$$\int (\tan u)^{-1/2} \sec^4 u \, du = \int (\tan u)^{-1/2} (1 + \tan^2 u) \sec^2 u \, du$$

$$= \int (\tan u)^{-1/2} d(\tan u) + \int (\tan u)^{3/2} d(\tan u)$$

$$= 2(\tan u)^{1/2} + \tfrac{2}{5}(\tan u)^{5/2} + C. \qquad \square$$

**Case 2.**    *$m$ is an odd positive integer*. We factor out $\sec u \tan u \, du$ and change the remaining even power of the tangents to secants, again using the identity $\tan^2 u = \sec^2 u - 1$.

**EXAMPLE 7**    Find

$$\int \frac{\tan^3 x \, dx}{\sqrt[3]{\sec x}}.$$

**Solution**    We write $(\sec x)^{-1/3} \tan^3 x = (\sec x)^{-4/3} \sec x \tan^2 x \tan x$ and get

$$\int (\sec x)^{-1/3} \tan^3 x \, dx = \int (\sec x)^{-4/3} \tan^2 x (\sec x \tan x \, dx)$$

$$= \int (\sec x)^{-4/3} (\sec^2 x - 1) d(\sec x)$$

$$= \int [(\sec x)^{2/3} - (\sec x)^{-4/3}] d(\sec x)$$

$$= \tfrac{3}{5}(\sec x)^{5/3} + 3(\sec x)^{-1/3} + C. \qquad \square$$

> The breakdown into cases of integrals of Type C is identical with that of Type B. We simply employ the formula $\csc^2 u = 1 + \cot^2 u$ whenever necessary.

We saw in Section 2 that certain trigonometric integrals can be evaluated by integration by parts. Example 6 (page 302) in that section gives an alternate way of integrating an expression of the form $\int \sin^n x \, dx$ where $n$ is *any* positive integer. Also, Example 7 (page 303) shows how to integrate an expression of Type B where $n$ is an odd positive integer and $m = 0$, a case not covered by the above discussion.

## 3   PROBLEMS

In Problems 1 through 14, in each case evaluate the definite integral.

**1** $\displaystyle\int_{-\pi/3}^{\pi/6} \cos^2 x \sin x \, dx$

**2** $\displaystyle\int_{0}^{\pi/6} \cos^3 3x \, dx$

**3** $\displaystyle\int_{0}^{\pi/4} \frac{\sin^3 x}{\sqrt{\cos x}} \, dx$

**4** $\displaystyle\int_{-\pi/6}^{\pi/6} \tan^2 x \sec^2 x \, dx$

**5** $\displaystyle\int_{\pi/6}^{\pi/3} \cot^2 2x \csc^2 2x \, dx$

**6** $\displaystyle\int_{0}^{\pi/3} \tan^3 x \sec x \, dx$

**7** $\displaystyle\int_{0}^{\pi/2} \cos^2 x \, dx$

**8** $\displaystyle\int_{0}^{\pi/4} \sin^2 2x \cos^2 2x \, dx$

**9** $\displaystyle\int_{0}^{\pi/2} \cos^4 x \, dx$

**10** $\displaystyle\int_{0}^{\pi/3} \sin^6 x \, dx$

**11** $\displaystyle\int_{0}^{\pi/2} \cos^3 \tfrac{1}{2}x \sin \tfrac{1}{2}x \, dx$

**12** $\displaystyle\int_{0}^{\pi/4} \cos^2 3x \sin^3 3x \, dx$

**13** $\displaystyle\int_{\pi/6}^{\pi/3} \sin^3 x \cos^{-4} x \, dx$

**14** $\displaystyle\int_{0}^{\pi/4} \sin^5 x \cos^{-1/3} x \, dx$

In Problems 15 through 40, find the indefinite integrals.

**15** $\displaystyle\int \sin^5 (x/2) \, dx$

**16** $\displaystyle\int \sin^3 2x \cos^2 2x \, dx$

**17** $\displaystyle\int \frac{\cos^3 x}{\sin x} \, dx$

**18** $\displaystyle\int \frac{\sin^5 3x}{\cos 3x} \, dx$

**19** $\displaystyle\int \tan 2x \sec^2 2x \, dx$

**20** $\displaystyle\int \tan (x/3) \sec^3 (x/3) \, dx$

**21** $\displaystyle\int \sec^4 2x \, dx$

**22** $\displaystyle\int \csc^6 x \, dx$

**23** $\displaystyle\int \cot^5 x \csc^3 x \, dx$

**24** $\displaystyle\int \tan^2 2x \sec^4 2x \, dx$

**25** $\displaystyle\int \tan^2 x \, dx$

**26** $\displaystyle\int \cot^3 (x/2) \csc^2 (x/2) \, dx$

**27** $\displaystyle\int \cot^4 x \, dx$

**28** $\displaystyle\int \cot^3 4x \csc^4 4x \, dx$

**29** $\displaystyle\int x \sin^3 (x^2) \, dx$

**30** $\displaystyle\int \frac{\sec^4 x}{\tan^2 x} \, dx$

**31** $\displaystyle\int \frac{\cot x}{\csc^3 x} \, dx$

**32** $\displaystyle\int \frac{\sec^2 \sqrt{x}}{\sqrt{x} \tan \sqrt{x}} \, dx$

**33** $\displaystyle\int \frac{\sin^2 x}{\cos^4 x} \, dx$

**34** $\displaystyle\int \tan^2 2x \cos^2 2x \, dx$

**35** $\displaystyle\int \tan x \cos^4 x \, dx$

**36** $\displaystyle\int \frac{\sin^4 x}{\cos^2 x} \, dx$

**37** $\displaystyle\int \cos^{-2} 3x \cot^{-2} 3x \, dx$

**38** $\displaystyle\int \tan^{3/5} x \sec^4 x \, dx$

**39** $\displaystyle\int \cot^{-3} 2x \csc^2 2x \, dx$

**40** $\displaystyle\int \sin^4 x \cos^4 x \, dx$

**41** Verify the formula

$$\int_0^{\pi/2} \sin^{2m+1} \theta \cos^{2n+1} \theta \, d\theta = \frac{1}{2} \frac{m! \, n!}{(m+n+1)!}$$

for $m = n = 0$, $m = n = 1$, and $m = 2$, $n = 1$. (Recall that $0! = 1$.)

**42** Find the area bounded by the curve $y = \sin^3 x$, $0 \le x \le \pi$, and the $x$ axis.

**43** Find the area of the region

$$R = \{(x, y): 0 \le x \le \pi/4, \, 0 \le y \le (\cos x)^3 (\sin x)^{5/2}\}.$$

**44** Use the formula $\sin 2A = 2 \sin A \cos A$ to integrate an expression of Type A with $m = n = -1$. Do the same for $m = n = -2$. How general is this method?

**45** We have
a) $\int \sin x \cos x \, dx = \int \sin x \, d(\sin x) = \tfrac{1}{2} \sin^2 x + C$,
b) $\int \sin x \cos x \, dx = -\int \cos x \, d(\cos x) = -\tfrac{1}{2} \cos^2 x + C$.
Why are both results correct?

_____4

## TRIGONOMETRIC SUBSTITUTION

The method of substitution described in Section 1 and the method for evaluating trigonometric integrals can be combined to yield integration formulas for another class of integrals. If the integrand contains expressions of the form $\sqrt{a^2 - x^2}$, $\sqrt{a^2 + x^2}$, or $\sqrt{x^2 - a^2}$, it is frequently possible to transform the integral into one of the forms discussed in Section 3 by means of a **trigonometric substitution**. The way the method works is shown in the following three examples.

> **A)** If an expression of the form $\sqrt{a^2 - x^2}$ occurs, make the substitution $x = a \sin \theta$ (where we take $\theta = \arcsin(x/a)$).

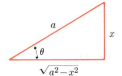

In performing the substitution, the reader should sketch a right triangle, as shown in Fig. 1. The figure contains all the essential ingredients of the process, and it is evident that $\sqrt{a^2 - x^2} = a \cos \theta$.

**FIGURE 1**

**EXAMPLE 1**  Evaluate

$$\int_{-1}^{\sqrt{3}} \sqrt{4 - x^2} \; dx.$$

**Solution**  Let $x = 2 \sin \theta$. Then $dx = 2 \cos \theta \, d\theta$ and, according to Fig. 2 we have $\sqrt{4 - x^2} = 2 \cos \theta$. Therefore

$$\int_{-1}^{\sqrt{3}} \sqrt{4 - x^2} \; dx = 4 \int_{-\pi/6}^{\pi/3} \cos^2 \theta \; d\theta = 2 \int_{-\pi/6}^{\pi/3} (1 + \cos 2\theta) \; d\theta$$

$$= 2[\theta + \tfrac{1}{2} \sin 2\theta]_{-\pi/6}^{\pi/3} = \pi + \sqrt{3}. \qquad \square$$

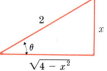

We note that Fig. 2 is only an aid in determining the substitution. In making the change $x = 2 \sin \theta$ or, equivalently, $\theta = \arcsin(x/2)$, the range allowed for $\theta$ is $-\pi/2 \le \theta \le \pi/2$.

**FIGURE 2**

> **B)** If an expression of the form $\sqrt{a^2 + x^2}$ occurs, make the substitution $x = a \tan \theta$ (or $\theta = \arctan(x/a)$).

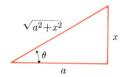

A sketch of this substitution is shown in Fig. 3, from which we see that $\sqrt{a^2 + x^2} = a \sec \theta$.

**FIGURE 3**

**EXAMPLE 2**  Find $\int x^3 \sqrt{7 + x^2} \; dx$.

**Solution**  Let $x = \sqrt{7} \tan \theta$. According to Fig. 3, with $a = \sqrt{7}$, we have $\sqrt{7 + x^2} = \sqrt{7} \sec \theta$. Also $dx = \sqrt{7} \sec^2 \theta \; d\theta$. Therefore

$$\int x^3 \sqrt{7+x^2}\, dx = \int 7\sqrt{7}\tan^3\theta \sqrt{7}\sec\theta \cdot \sqrt{7}\sec^2\theta\, d\theta$$

$$= 49\sqrt{7}\int \tan^3\theta \sec^3\theta\, d\theta.$$

This last integral is of Type B of Section 3, with $m$ odd and positive. We obtain

$$\int x^3 \sqrt{7+x^2}\, dx = 49\sqrt{7}\int (\sec^2\theta - 1)\sec^2\theta \cdot (\tan\theta\sec\theta)\, d\theta$$

$$= 49\sqrt{7}(\tfrac{1}{5}\sec^5\theta - \tfrac{1}{3}\sec^3\theta) + C.$$

Referring to Fig. 3 again with $a = \sqrt{7}$, we now find that $\sec\theta = (1/\sqrt{7})\sqrt{7+x^2}$, and so

$$\int x^3 \sqrt{7+x^2}\, dx = \tfrac{1}{5}(7+x^2)^{5/2} - \tfrac{7}{3}(7+x^2)^{3/2} + C. \qquad \square$$

> C) If an expression of the form $\sqrt{x^2 - a^2}$ occurs, make the substitution $x = a\sec\theta$.

A sketch of this substitution is shown in Fig. 4. It follows that $\sqrt{x^2 - a^2} = a\tan\theta$ when we take $\theta = \operatorname{arcsec}(x/a)$.

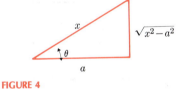

**FIGURE 4**

**EXAMPLE 3**    Evaluate

$$\int_3^6 \frac{\sqrt{x^2 - 9}}{x}\, dx.$$

**Solution**    We make the substitution $x = 3\sec\theta$, $dx = 3\tan\theta\sec\theta\, d\theta$. Further, referring to Fig. 4 with $a = 3$, we see that $\sqrt{x^2 - 9} = 3\tan\theta$. When $x = 3$, we have $\sec\theta = 1$ and $\theta = 0$; when $x = 6$, we have $\sec\theta = 2$ and $\theta = \pi/3$. Therefore

$$\int_3^6 \frac{\sqrt{x^2 - 9}}{x}\, dx = \int_0^{\pi/3} \frac{3\tan\theta}{3\sec\theta} \cdot 3\tan\theta\sec\theta\, d\theta$$

$$= 3\int_0^{\pi/3} \tan^2\theta\, d\theta$$

$$= 3\int_0^{\pi/3} (\sec^2\theta - 1)\, d\theta$$

$$= 3[\tan\theta - \theta]_0^{\pi/3}$$

$$= 3\left(\sqrt{3} - \frac{\pi}{3}\right). \qquad \square$$

*Remarks.*    When there is an expression of the form $(a^2 - x^2)$, $(a^2 + x^2)$, or $(x^2 - a^2)$, it sometimes helps to make a substitution of Type A, B, or C, respectively. We can frequently perform an integration when the integrand contains the square root of a quadratic expression. If a third, fourth, or higher-degree polynomial occurs under the radical sign, it can be shown that there is no general method for carrying out the integration. In most cases whether or not an integration can be performed is determined by trial and error.

We summarize trigonometric substitutions in the following table:

| Given expression | Trigonometric substitution |
|---|---|
| $\sqrt{a^2 - x^2}$ | $x = a \sin \theta$ |
| $\sqrt{a^2 + x^2}$ | $x = a \tan \theta$ |
| $\sqrt{x^2 - a^2}$ | $x = a \sec \theta$ |

## 4    PROBLEMS

In Problems 1 through 10, evaluate the definite integrals.

**1** $\displaystyle\int_0^1 \frac{dx}{\sqrt{4 - x^2}}$

**2** $\displaystyle\int_0^2 \frac{x^2\, dx}{x^2 + 4}$

**3** $\displaystyle\int_{-6}^{-2\sqrt{3}} \frac{dx}{x\sqrt{x^2 - 9}}$

**4** $\displaystyle\int_0^2 \frac{x^3\, dx}{\sqrt{16 - x^2}}$

**5** $\displaystyle\int_0^{2\sqrt{3}} \frac{x^3\, dx}{\sqrt{x^2 + 4}}$

**6** $\displaystyle\int_0^1 \frac{x^2\, dx}{\sqrt{4 - x^2}}$

**7** $\displaystyle\int_0^{5/6} \frac{dx}{\sqrt{25 - 9x^2}}$

**8** $\displaystyle\int_0^1 \frac{x^2\, dx}{4x^2 + 9}$

**9** $\displaystyle\int_0^1 \frac{x^3\, dx}{(9 - x^2)^{3/2}}$

**10** $\displaystyle\int_0^1 \frac{x^3\, dx}{(9 + x^2)^{3/2}}$

In Problems 11 through 28, find the indefinite integrals.

**11** $\displaystyle\int \frac{2\, dx}{x\sqrt{x^2 - 5}}$

**12** $\displaystyle\int \frac{\sqrt{2x^2 - 5}}{x}\, dx$

**13** $\displaystyle\int \frac{x^3\, dx}{\sqrt{9 - x^2}}$

**14** $\displaystyle\int \frac{x^3\, dx}{\sqrt{3x^2 - 5}}$

**15** $\displaystyle\int \frac{x^3\, dx}{\sqrt{2x^2 + 7}}$

**16** $\displaystyle\int t^3\sqrt{a^2 t^2 - b^2}\, dt$

**17** $\displaystyle\int u^3\sqrt{a^2 u^2 + b^2}\, du$

**18** $\displaystyle\int u^3\sqrt{a^2 - b^2 u^2}\, du$

**19** $\displaystyle\int \frac{\sqrt{x^2 - a^2}}{x}\, dx$

**20** $\displaystyle\int \frac{dx}{x^2\sqrt{a^2 - x^2}}$

**21** $\displaystyle\int \frac{dx}{x^2\sqrt{x^2 + a^2}}$

**22** $\displaystyle\int \frac{dx}{(a^2 - x^2)^{3/2}}$

**23** $\displaystyle\int \frac{dx}{x^4\sqrt{a^2 - x^2}}$

**24** $\displaystyle\int \frac{dx}{(x^2 + a^2)^2}$

**25** $\displaystyle\int x^3(9 - x^2)^{3/2}\, dx$

**26** $\displaystyle\int x^3(4 + x^2)^{3/2}\, dx$

**27** $\displaystyle\int x(9 - 4x^2)^{5/2}\, dx$

**28** $\displaystyle\int x(16 - 9x^2)^{n/2}\, dx$,    $n$ a positive integer

Problems 29 through 36, evaluate the definite integrals.

**29** $\displaystyle\int_0^4 \frac{dx}{(16 + x^2)^{3/2}}$

**30** $\displaystyle\int_{2/\sqrt{3}}^2 \frac{dx}{(x^2 - 1)^{3/2}}$

**31** $\displaystyle\int_0^1 \frac{x^2\, dx}{(4 - x^2)^{3/2}}$

**32** $\displaystyle\int_{2\sqrt{3}}^6 \frac{x^3\, dx}{(x^2 - 1)^{3/2}}$

**33** $\displaystyle\int_{\sqrt{3}}^{3\sqrt{3}} \frac{dx}{x^2\sqrt{x^2 + 9}}$

**34** $\displaystyle\int_{-2}^{2\sqrt{3}} x^3\sqrt{x^2 + 4}\, dx$

**35** $\displaystyle\int_0^{\sqrt{5}} x^2\sqrt{5 - x^2}\, dx$

**36** $\displaystyle\int_1^3 \frac{dx}{x^4\sqrt{x^2 + 3}}$

**37** State the domain for $\theta$ in the substitutions (a) $x = 3 \tan \theta$ of type (B); and (b) $x = 3 \sec \theta$ of type (C).

**38** Suppose that an integrand contains an expression of the form

$$\sqrt{bx^2 + cx + d}.$$

Complete the square and describe the conditions on $b$, $c$, and $d$ which are required to reduce the expression to one of the three forms:

$$A\sqrt{a^2 - u^2}, \quad A\sqrt{a^2 + u^2}, \quad A\sqrt{u^2 - a^2},$$

where $A$ is some positive constant.

**39** Given the integral $\int \dfrac{\sqrt{1-x^2}}{x^4}\,dx$,

evaluate it two ways: (a) by trigonometric substitution and (b) by setting $x = 1/t$. Compare the results.

**40** Evaluate $\int \dfrac{1}{\sqrt{1-e^{-x}}}\,dx$.

(*Hint:* Let $u = -e^{(1/2)x}$.)

**41** Given the integral

$$\int \frac{dx}{x^n\sqrt{1-x^2}}.$$

For what values of $n$ can the integral be evaluated by trigonometric substitution and the methods of Section 3 applied? For what values of $n$ can the integral be evaluated by the substitution $x = 1/t^a$?

---

## 5

## INTEGRANDS INVOLVING QUADRATIC FUNCTIONS

When an integrand involves a quadratic function of the form $ax^2 + bx + c$, the integration is usually simplified by *completing the square*. To complete the square of a quadratic, we write the identity

$$ax^2 + bx + c = a\left(x^2 + \frac{b}{a}x\right) + c = a\left(x + \frac{b}{2a}\right)^2 + c - \frac{b^2}{4a^2}.$$

Then the substitution

$$y = x + \frac{b}{2a}$$

changes the quadratic term in the integrand to a form resembling those considered in the four previous sections. The next three examples show this procedure.

**EXAMPLE 1**   Find

$$\int \frac{(2x-3)\,dx}{x^2 + 2x + 2}.$$

**Solution**   We complete the square (with $a = 1$, $b = 2$, $c = 2$), obtaining

$$x^2 + 2x + 2 = (x^2 + 2x) + 2 = (x+1)^2 + 2 - 1 = (x+1)^2 + 1.$$

Let $y = x + 1$, $dy = dx$, and write

$$\int \frac{(2x-3)\,dx}{x^2 + 2x + 2} = \int \frac{2y-5}{y^2+1}\,dy = \int \frac{2y\,dy}{y^2+1} - 5\int \frac{dy}{y^2+1}$$

$$= \log(y^2 + 1) - 5\arctan y + C$$

$$= \log(x^2 + 2x + 2) - 5\arctan(x+1) + C.$$

**EXAMPLE 2**   Evaluate

$$\int \frac{dx}{\sqrt{2x - x^2}}.$$

**Solution**  To complete the square, we note that $a = 1$, $b = 2$, $c = 0$. Then we find

$$2x - x^2 = -(x^2 - 2x + 1) + 1 = -(x - 1)^2 + 1.$$

We let $y = x - 1$, $dy = dx$, and so

$$\int \frac{dx}{\sqrt{2x - x^2}} = \int \frac{dy}{\sqrt{1 - y^2}} = \arcsin y + C = \arcsin(x - 1) + C. \qquad \square$$

**EXAMPLE 3**  Find

$$\int \frac{2x^3\, dx}{2x^2 - 4x + 3}.$$

**Solution**  We perform the ordinary long division of $2x^3$ divided by $2x^2 - 4x + 3$, getting

$$\int \frac{2x^3\, dx}{2x^2 - 4x + 3} = \int \left( x + 2 + \frac{5x - 6}{2x^2 - 4x + 3} \right) dx$$

$$= \frac{1}{2}x^2 + 2x + \int \frac{5x - 6}{2x^2 - 4x + 3}\, dx.$$

We complete the square to obtain

$$2x^2 - 4x + 3 = 2(x - 1)^2 + 1.$$

Therefore the substitution $y = x - 1$ yields

$$\int \frac{2x^3\, dx}{2x^2 - 4x + 3}$$

$$= \frac{1}{2}x^2 + 2x + \int \frac{5y - 1}{2y^2 + 1}\, dy$$

$$= \frac{1}{2}x^2 + 2x + \frac{5}{4} \int \frac{d(2y^2 + 1)}{2y^2 + 1} - \frac{1}{2} \int \frac{dy}{y^2 + \frac{1}{2}}$$

$$= \frac{1}{2}x^2 + 2x + \frac{5}{4} \ln(2y^2 + 1) - \frac{1}{2}\sqrt{2} \arctan(y\sqrt{2}) + C$$

$$= \frac{1}{2}x^2 + 2x + \frac{5}{4} \ln(2x^2 - 4x + 3) - \frac{1}{2}\sqrt{2} \arctan \sqrt{2}(x - 1) + C. \qquad \square$$

## 5  PROBLEMS

In Problems 1 through 24, find the indefinite integrals.

**1** $\displaystyle\int \frac{dx}{x^2 + 4x + 5}$

**2** $\displaystyle\int \frac{dx}{\sqrt{5 + 4x - x^2}}$

**3** $\displaystyle\int \frac{x^2\, dx}{x^2 + 2x + 5}$

**4** $\displaystyle\int \frac{x^3 - 2x}{x^2 + x + 3}\, dx$

**5** $\displaystyle\int \frac{dx}{(x + 2)\sqrt{x^2 + 4x + 3}}$

**6** $\displaystyle\int \frac{dx}{x^2 - x + 1}$

**7** $\displaystyle\int \frac{dx}{\sqrt{5 - 2x + x^2}}$

**8** $\displaystyle\int \frac{dx}{(x - 1)\sqrt{x^2 - 2x - 3}}$

**9** $\displaystyle\int \frac{x + 3}{x^2 + 2x + 5}\, dx$

**10** $\displaystyle\int \frac{(2x - 5)\, dx}{\sqrt{4x - x^2}}$

**11** $\displaystyle\int \frac{x^4\, dx}{x^2 + 3x + 1}$

**12** $\displaystyle\int \frac{4x^3}{2x^2 + 6x + 1}\, dx$

13 $\int \dfrac{(3x+4)\,dx}{\sqrt{2x+x^2}}$

14 $\int \dfrac{\sqrt{x^2+2x}}{x+1}\,dx$

15 $\int \dfrac{4x+7}{(x^2-2x+3)^2}\,dx$

16 $\int \dfrac{(5x-3)\,dx}{(x^2+4x+7)^2}$

17 $\int \dfrac{4x+5}{(2x^2+3x+4)^2}\,dx$

18 $\int \dfrac{(3x-2)\,dx}{(2x^2-5x+8)^2}$

19 $\int \dfrac{(4x+5)\,dx}{(x^2-2x+2)^{3/2}}$

20 $\int \dfrac{(x+2)\,dx}{(3+2x-x^2)^{3/2}}$

21 $\int \dfrac{(2x-3)\,dx}{(x^2+2x-3)^{3/2}}$

22 $\int \dfrac{dx}{(x^2-2x+5)^2}$

23 $\int \dfrac{(x+1)\,dx}{(2-x-x^2)^{5/2}}$

24 $\int \dfrac{(2x-1)\,dx}{(3-2x-x^2)^{5/2}}$

In Problems 25 through 34, in each case find the value of the definite integral.

25 $\int_0^2 \dfrac{dx}{x^2+3x+4}$

26 $\int_1^2 \dfrac{x^2\,dx}{x^2+4x+5}$

27 $\int_{-1}^1 \dfrac{dx}{x^2-4x+5}$

28 $\int_2^4 \dfrac{dx}{\sqrt{8x-x^2}}$

29 $\int_0^1 \dfrac{2x+3}{(x^2-2x+5)^2}\,dx$

30 $\int_0^1 \dfrac{dx}{(x^2+4x+13)^{3/2}}$

31 $\int_1^2 \dfrac{dx}{x^2-2x+5}$

32 $\int_0^2 \dfrac{x\,dx}{x^2-4x+7}$

33 $\int_0^1 \dfrac{dx}{\sqrt{5-2x-x^2}}$

34 $\int_0^1 \dfrac{x\,dx}{\sqrt{6-4x-x^2}}$

35 Find the indefinite integral of

$$\int \dfrac{dx}{x^2-1}.$$

Is it possible to use this formula to evaluate the definite integral from $-2$ to $+2$? Explain. Find all possible intervals in which the above integral can be evaluated as a definite integral.

In Problems 36 through 41, we state general integration formulas. Verify the validity of each by differentiating the right side. Specify conditions on $a$, $b$, and $c$ which ensure the correctness of the formulas.

36 $\int \dfrac{dx}{\sqrt{ax^2+bx+c}}$

$$= \dfrac{1}{\sqrt{a}}\log\left(\sqrt{ax^2+bx+c}+\sqrt{a}\,x+\dfrac{b}{2\sqrt{a}}\right)+C$$

37 $\int \dfrac{dx}{\sqrt{ax^2+bx+c}} = \dfrac{1}{\sqrt{-a}}\arcsin\left(\dfrac{-2ax-b}{\sqrt{b^2-4ac}}\right)+C$

38 $\int \dfrac{dx}{(ax^2+bx+c)^{3/2}} = 2\dfrac{2ax+b}{(4ac-b^2)\sqrt{ax^2+bx+c}}+C$

39 $\int \dfrac{dx}{x\sqrt{ax^2+bx+c}}$

$$= -\dfrac{1}{\sqrt{c}}\log\left(\dfrac{\sqrt{ax^2+bx+c}+\sqrt{c}}{x}+\dfrac{b}{2\sqrt{c}}\right)+C$$

40 $\int \dfrac{dx}{x\sqrt{ax^2+bx+c}} = \dfrac{1}{\sqrt{-c}}\arcsin\left(\dfrac{bx+2c}{x\sqrt{b^2-4ac}}\right)+C$

41 $\int \dfrac{dx}{(ax^2+bx+c)^{5/2}} = \dfrac{2}{3}\cdot\dfrac{2ax+b}{(4ac-b^2)\sqrt{ax^2+bx+c}}$

$$\times\left(\dfrac{1}{\sqrt{ax^2+bx+c}}+\dfrac{8a}{4ac-b^2}\right)+C$$

---

**6**

## INTEGRATION OF RATIONAL FUNCTIONS; METHOD OF PARTIAL FRACTIONS

If $P(x)$ and $Q(x)$ are polynomials, an expression of the form

$$\dfrac{P(x)}{Q(x)}$$

is called a **rational function**. The integration of expressions of the form

$$\int \dfrac{P(x)}{Q(x)}\,dx$$

can, in theory, always be performed. In practice, however, the actual calculation of the integral depends on whether or not the denominator, $Q(x)$, can be factored. A theorem which is proved in more advanced courses states that every polynomial can be factored into a product of linear factors. That is, when $Q(x)$ is a polynomial of degree $r$, it may be written as a product of $r$ linear factors:

$$Q(x) = a(x - \alpha_1)(x - \alpha_2) \cdots (x - \alpha_r).$$

In this decomposition some of the numbers $\alpha_1, \alpha_2, \ldots, \alpha_r$ may be complex. In order to be sure that we do not become involved with complex quantities, we shall use the following theorem, which is stated without proof.

THEOREM 1    *Every polynomial (with real coefficients) may be decomposed into a product of linear and quadratic factors in such a way that each of the factors has real coefficients.*

For example, the polynomial

$$Q(x) = x^3 - 2x^2 + x - 2$$

can be decomposed into the linear factors

$$Q(x) = (x - 2)(x - i)(x + i).$$

Two of these factors are complex, i.e., $i = \sqrt{-1}$, the square root of $-1$. However, the decomposition

$$Q(x) = (x - 2)(x^2 + 1)$$

into a linear and a quadratic factor has only real quantities.

The basic method of integrating the rational function $P/Q$ consists of three steps:

**Step 1:    If the degree of $P$ is larger than or equal to the degree of $Q$, apply long division.**

Then $P/Q$ will equal a polynomial (quotient) plus a rational function (remainder divided by the divisor) in which the degree of the numerator is definitely less than the degree of the denominator. From now on, *we shall always suppose that this simplification has already been performed.*

**Step 2:    Factor $Q(x)$ into a product of linear and quadratic factors.**

**Step 3:    Write $P(x)/Q(x)$ as a sum of simpler rational functions, each of which can be integrated by methods we already learned.**

We assume the reader is already familiar with the elementary algebraic techniques needed to perform Steps 1 and 2. Finding the factors may be very difficult in specific cases, but we shall assume that this can always be done.

Before discussing Step 3 in detail, let us look at some examples of simplifications leading to integrations which can be performed. It can be shown that each of the following equations is an identity in $x$ for properly chosen constants $A$, $B$, $C$, $D$, etc.

$$\frac{2x^2 - 3x + 5}{(x+2)(x-1)(x-3)} = \frac{A}{x+2} + \frac{B}{x-1} + \frac{C}{x-3},$$

$$\frac{x^4 - 2x^2 + 3x + 4}{(x-1)^3(x^2+2x+2)} = \frac{D}{x-1} + \frac{E}{(x-1)^2} + \frac{F}{(x-1)^3} + \frac{Gx+H}{x^2+2x+2},$$

$$\frac{2x^4 + 3x^3 - x - 1}{(x-1)(x^2+2x+2)^2} = \frac{J}{x-1} + \frac{Kx+L}{x^2+2x+2} + \frac{Mx+N}{(x^2+2x+2)^2}.$$

Note that every term on the right side in the above identities is one we can integrate. For example,

$$\int \frac{A}{x+2}\, dx = A \ln|x+2| + C.$$

The integrals

$$\int \frac{dx}{(x-1)^2}, \qquad \int \frac{dx}{(x-1)^3}$$

are routine. An expression of the form

$$\int \frac{Gx+H}{x^2+2x+2}\, dx$$

is easily integrated if we complete the square in the denominator and proceed as described in Section 5, Example 1. The integration of

$$\int \frac{Mx+N}{(x^2+2x+2)^2}\, dx$$

is performed by completing the square in the denominator as in Section 5 and then making a trigonometric substitution of the type described in Section 4.

We now turn to the general problem of decomposing a rational function $P/Q$ into simpler expressions.

The decomposition of a rational function into the sum of simpler expressions is known as the **method of partial fractions**. We divide the method into four cases, depending on the way the denominator factors. We state the results without proof.

**Case 1.   The denominator $Q(x)$ can be factored into linear factors, all different.**

If we can make the decomposition

$$Q(x) = (x - a_1)(x - a_2) \cdots (x - a_r),$$

*with no two of the $a_i$ the same,* then we can decompose $P/Q$ so that

$$\frac{P(x)}{Q(x)} = \frac{A_1}{x - a_1} + \frac{A_2}{x - a_2} + \frac{A_3}{x - a_3} + \cdots + \frac{A_r}{x - a_r},$$

where $A_1$, $A_2$, ..., $A_r$ are properly chosen constants.

We shall show by example how the constants may be found.

**EXAMPLE 1**  Decompose

$$(x^2 + 2x + 3)/(x^3 - x)$$

into partial fractions and integrate.

**Solution**  $Q(x) = x^3 - x = (x - 0)(x - 1)(x + 1)$. We write

$$\frac{x^2 + 2x + 3}{x(x - 1)(x + 1)} = \frac{A_1}{x} + \frac{A_2}{x - 1} + \frac{A_3}{x + 1},$$

which is an identity for all $x(x \neq 0, 1, -1)$ if and only if

$$x^2 + 2x + 3 = A_1(x - 1)(x + 1) + A_2x(x + 1) + A_3x(x - 1). \qquad (1)$$

Multiplying out the right side, we obtain

$$x^2 + 2x + 3 = (A_1 + A_2 + A_3)x^2 + (A_2 - A_3)x - A_1. \qquad (2)$$

If these two polynomials are to be *identical*, every coefficient on the left must equal every coefficient on the right. Then the polynomials are the same for *all values* of $x$. Generally, it is better to work with (1) than with (2). We proceed as follows:

If in (1) we set $x = 0$:    $3 = -A_1$ and so $A_1 = -3$.

If in (1) we set $x = 1$:    $6 = 2A_2$  and so $A_2 = 3$.

If in (1) we set $x = -1$:   $2 = 2A_3$  and so $A_3 = 1$.

We conclude that

$$\int \frac{(x^2 + 2x + 3)\, dx}{x(x - 1)(x + 1)} = \int \left( -\frac{3}{x} + \frac{3}{x - 1} + \frac{1}{x + 1} \right) dx$$

$$= -3 \ln |x| + 3 \ln |x - 1| + \ln |x + 1| + C$$

$$= \ln \left| \frac{(x - 1)^3(x + 1)}{x^3} \right| + C. \qquad \square$$

In the example above, we made use of the following theorem about polynomials.

**THEOREM 2**  *Suppose that two polynomials*

$$S(x) = a_0 + a_1 x + a_2 x^2 + \cdots + a_n x^n,$$

$$T(x) = b_0 + b_1 x + \cdots + b_n x^n$$

*are equal for all except possibly a finite number of values of* $x$. *Then* $a_i = b_i$ *for all* $i = 0, 1, 2, \ldots, n$.

**Proof**  Since polynomial functions are continuous, $S(x)$ and $T(x)$ must be equal for *all* values of $x$. We form the expression

$$S(x) - T(x) = (a_0 - b_0) + (a_1 - b_1)x + \cdots + (a_n - b_n)x^n \equiv 0.$$

Setting $x = 0$, we obtain $S(0) - T(0) = a_0 - b_0 = 0$, or $a_0 = b_0$. Then

$$S(x) - T(x) = (a_1 - b_1)x + (a_2 - b_2)x^2 + \cdots + (a_n - b_n)x^n \equiv 0.$$

Dividing through by $x$ and again setting $x = 0$, we obtain $a_1 - b_1 = 0$. Continuing in this way, we get $a_i = b_i$ for each $i$ from 0 to $n$. ☐

> **Case 2.   The denominator $Q(x)$ can be factored into linear factors, some of which are repeated.**

If we can make the decomposition

$$Q(x) = (x - a_1)^{s_1}(x - a_2)^{s_2} \cdots (x - a_r)^{s_r}$$

where the exponents $s_1$, $s_2$, ..., $s_r$ are positive integers, then the partial fraction decomposition introduces a number of different types of denominators. For example, a factor such as $(x - 1)^4$ gives rise to the four terms

$$\frac{A_1}{x - 1} + \frac{A_2}{(x - 1)^2} + \frac{A_3}{(x - 1)^3} + \frac{A_4}{(x - 1)^4},$$

where $A_1$, ..., $A_4$ are properly chosen constants. In general, a factor such as $(x - a)^q$ gives rise to the terms ($q$ in number)

$$\frac{A_1}{x - a} + \frac{A_2}{(x - a)^2} + \frac{A_3}{(x - a)^3} + \cdots + \frac{A_q}{(x - a)^q}.$$

We illustrate the technique with an example.

**EXAMPLE 2**    Find

$$\int \frac{x + 5}{x^3 - 3x + 2}\, dx.$$

**Solution**    We factor the denominator, obtaining

$$x^3 - 3x + 2 = (x - 1)^2(x + 2).$$

This falls under Case 2 with $a_1 = 1$, $a_2 = -2$, $s_1 = 2$, $s_2 = 1$, and we write

$$\frac{x + 5}{(x - 1)^2(x + 2)} = \frac{A_1}{x - 1} + \frac{A_2}{(x - 1)^2} + \frac{A_3}{x + 2}.$$

Multiplying through by $(x - 1)^2(x + 2)$, we obtain

$$x + 5 = A_1(x - 1)(x + 2) + A_2(x + 2) + A_3(x - 1)^2.$$

We let

$$x = 1: \qquad 6 = 3A_2 \qquad\qquad\quad \Leftrightarrow \quad A_2 = 2;$$
$$x = -2: \quad 3 = 9A_3 \qquad\qquad\quad \Leftrightarrow \quad A_3 = \tfrac{1}{3};$$
$$x = 0: \qquad 5 = -2A_1 + 2A_2 + A_3 \quad \Leftrightarrow \quad A_1 = -\tfrac{1}{3}.$$

Therefore

$$\int \frac{(x + 5)\, dx}{(x - 1)^2(x + 2)} = -\frac{1}{3}\ln|x - 1| - \frac{2}{x - 1} + \frac{1}{3}\ln|x + 2| + C. \qquad ☐$$

> **Case 3.   The denominator $Q(x)$ can be factored into linear and quadratic factors, and none of the quadratic factors is repeated.**

If, for example, the denominator is

$$Q(x) = (x - a_1)(x - a_2)(x - a_3)(x^2 + b_1 x + c_1)(x^2 + b_2 x + c_2),$$

then

$$\frac{P(x)}{Q(x)} = \frac{A_1}{x - a_1} + \frac{A_2}{x - a_2} + \frac{A_3}{x - a_3} + \frac{A_4 x + A_5}{x^2 + b_1 x + c_1} + \frac{A_6 x + A_7}{x^2 + b_2 x + c_2}.$$

In other words, each unrepeated quadratic factor gives rise to a term of the form

$$\frac{Ax + B}{x^2 + bx + c}.$$

We illustrate Case 3 with an example.

**EXAMPLE 3**    Find

$$\int \frac{3x^2 + x - 2}{(x - 1)(x^2 + 1)} \, dx.$$

**Solution**    According to Case 3, we have

$$\frac{3x^2 + x - 2}{(x - 1)(x^2 + 1)} = \frac{A_1}{x - 1} + \frac{A_2 x + A_3}{x^2 + 1}$$

or

$$3x^2 + x - 2 = A_1(x^2 + 1) + (A_2 x + A_3)(x - 1).$$

To find $A_1, A_2$, and $A_3$ we substitute any values of $x$ in the above identity. When

$$x = 1: \qquad 2 = 2A_1 \qquad\qquad\qquad \Leftrightarrow \quad A_1 = 1;$$
$$x = 0: \quad -2 = A_1 - A_3 \qquad\qquad \Leftrightarrow \quad A_3 = 3;$$
$$x = 2: \quad 12 = 5A_1 + (2A_2 + A_3) \quad \Leftrightarrow \quad A_2 = 2.$$

Therefore, after carrying out the appropriate integrations, we find

$$\int \frac{3x^2 + x - 2}{(x - 1)(x^2 + 1)} \, dx = \ln |x - 1| + \ln (x^2 + 1) + 3 \arctan x + C. \qquad \square$$

*Remarks.* (1) In working a problem, we set $x$ equal to convenient values, values which make the equations for the $A$'s easy to solve. However, the method works no matter what values we choose for $x$. (2) It is possible to have a combination of cases. For example, we decompose the expression

$$\frac{P(x)}{(x - 1)(x - 2)^2(3x^2 + 4)}$$

as follows:

$$\frac{P(x)}{(x - 1)(x - 2)^2(3x^2 + 4)} = \frac{A_1}{(x - 1)} + \frac{A_2}{(x - 2)} + \frac{A_3}{(x - 2)^2} + \frac{A_4 x + A_5}{(3x^2 + 4)}.$$

The above decomposition is a combination of Cases 2 and 3.

> **Case 4.** The denominator $Q(x)$ can be factored into linear and quadratic factors, and some of the quadratic factors are repeated.

If the denominator contains a factor such as $(x^2 + 3x + 5)^3$, it will give rise to the three terms

$$\frac{A_1 x + A_2}{x^2 + 3x + 5} + \frac{A_3 x + A_4}{(x^2 + 3x + 5)^2} + \frac{A_5 x + A_6}{(x^2 + 3x + 5)^3}.$$

In general, a factor of the form $(x^2 + bx + c)^q$ will give rise to the $q$ terms

$$\frac{A_1 x + A_2}{x^2 + bx + c} + \frac{A_3 x + A_4}{(x^2 + bx + c)^2} + \cdots + \frac{A_{2q-1} x + A_{2q}}{(x^2 + bx + c)^q}.$$

We illustrate with an example.

**EXAMPLE 4**   Find

$$\int \frac{2x^3 + 3x^2 + x - 1}{(x + 1)(x^2 + 2x + 2)^2}\, dx.$$

**Solution**   This is Case 4 since $x^2 + 2x + 2$ cannot be factored into real linear factors. Thus we have

$$\frac{2x^3 + 3x^2 + x - 1}{(x + 1)(x^2 + 2x + 2)^2} = \frac{A_1}{x + 1} + \frac{A_2 x + A_3}{x^2 + 2x + 2} + \frac{A_4 x + A_5}{(x^2 + 2x + 2)^2}$$

or

$$2x^3 + 3x^2 + x - 1 = A_1(x^2 + 2x + 2)^2$$
$$+ (A_2 x + A_3)(x^2 + 2x + 2)(x + 1)$$
$$+ (A_4 x + A_5)(x + 1).$$

When

$$x = -1: \quad A_1 = -1;$$
$$x = 0: \quad -1 = 4A_1 + 2A_3 + A_5 \quad \text{and} \quad 2A_3 + A_5 = 3;$$
$$x = 1: \quad 5 = 25A_1 + (A_2 + A_3) \cdot (10) + (A_4 + A_5)(2) \quad \text{and}$$
$$5A_2 + 5A_3 + A_4 + A_5 = 15;$$
$$x = 2: \quad 20A_2 + 10A_3 + 2A_4 + A_5 = 43;$$
$$x = -2: \quad 4A_2 - 2A_3 + 2A_4 - A_5 = -3.$$

Solving the four equations for the four unknowns, $A_2, A_3, A_4,$ and $A_5$, we get (after quite a bit of algebra)

$$A_2 = 1, \quad A_3 = 3, \quad A_4 = -2, \quad A_5 = -3.$$

Therefore

$$\int \frac{2x^3 + 3x^2 + x - 1}{(x + 1)(x^2 + 2x + 2)^2}\, dx = -\ln|x + 1| + \int \frac{(x + 3)\, dx}{x^2 + 2x + 2} + \int \frac{(-2x - 3)\, dx}{(x^2 + 2x + 2)^2}.$$

We complete the square and set $u = x + 1$, $du = dx$, obtaining

$$\int \frac{(x+3)\,dx}{x^2 + 2x + 2} = \int \frac{u+2}{u^2 + 1}\,du = \frac{1}{2}\ln|x^2 + 2x + 2| + 2\arctan(x+1),$$

$$\int \frac{-2x - 3}{(x^2 + 2x + 2)^2}\,dx = \int \frac{-2u - 1}{(u^2 + 1)^2}\,du = \frac{1}{u^2 + 1} - \int \frac{du}{(u^2 + 1)^2}$$

$$= \frac{1}{x^2 + 2x + 2} - \frac{1}{2}\arctan(x+1) - \frac{1}{2}\frac{x+1}{x^2 + 2x + 2}.$$

The last two terms on the right are obtained by using a trigonometric substitution in the integral immediately above. Combining all the integrals, we conclude that

$$\int \frac{2x^3 + 3x^2 + x - 1}{(x+1)(x^2 + 2x + 2)^2}\,dx = -\ln|x+1| + \frac{1}{2}\ln(x^2 + 2x + 2)$$

$$+ \frac{3}{2}\arctan(x+1) - \frac{1}{2}\frac{x-1}{x^2 + 2x + 2} + C.$$

□

# 6  PROBLEMS

In Problems 1 through 38, find the indefinite integrals.

1. $\displaystyle\int \frac{x^2 + 3x + 4}{x - 2}\,dx$

2. $\displaystyle\int \frac{x^3 + x^2 - x - 3}{x + 2}\,dx$

3. $\displaystyle\int \frac{x^3 - x^2 + 2x + 3}{x^2 + 3x + 2}\,dx$

4. $\displaystyle\int \frac{2x^3 + 3x^2 - 4}{x^2 - 4x + 3}\,dx$

5. $\displaystyle\int \frac{x^2 + 2x + 3}{x^2 - 3x + 2}\,dx$

6. $\displaystyle\int \frac{x^4 + 1}{x^3 - x}\,dx$

7. $\displaystyle\int \frac{x^2 - 2x - 1}{x^2 - 4x + 4}\,dx$

8. $\displaystyle\int \frac{x^2 + 2x + 3}{(x+1)(x-1)(x-2)}\,dx$

9. $\displaystyle\int \frac{3x^2 - 3x - 2}{x^3 - x^2 - x + 1}\,dx$

10. $\displaystyle\int \frac{-x^2 + 5x + 3}{x^3 - 3x - 2}\,dx$

11. $\displaystyle\int \frac{x^3}{x^2 - 6x + 9}\,dx$

12. $\displaystyle\int \frac{x^3 + 1}{x^5 - x^4}\,dx$

13. $\displaystyle\int \frac{3x - 2}{(x+2)(x+1)(x-1)}\,dx$

14. $\displaystyle\int \frac{x^3 + 2}{x^2 + 4}\,dx$

15. $\displaystyle\int \frac{2x^2 + 3x - 1}{(x+3)(x+2)(x-1)}\,dx$

16. $\displaystyle\int \frac{x^2 - 2}{(x+1)(x-1)^2}\,dx$

17. $\displaystyle\int \frac{x^2 + 3x + 3}{(x+1)(x^2 + 1)}\,dx$

18. $\displaystyle\int \frac{x^2 - 2x - 3}{(x-1)(x^2 + 2x + 2)}\,dx$

19. $\displaystyle\int \frac{x - 3}{(x+1)^2(x-2)}\,dx$

20. $\displaystyle\int \frac{x^2 + 1}{(x-1)^3}\,dx$

21. $\displaystyle\int \frac{2x + 3}{(x+2)(x-1)^2}\,dx$

22. $\displaystyle\int \frac{2x^2 - 1}{(x+1)^2(x-3)}\,dx$

23. $\displaystyle\int \frac{x^3 - 3x + 4}{(x+1)(x-1)^3}\,dx$

24. $\displaystyle\int \frac{x^3 + 1}{(x^2 - 1)^2}\,dx$

25. $\displaystyle\int \frac{x^3 + 3x^2 - 2x + 1}{x^4 + 5x^2 + 4}\,dx$

26. $\displaystyle\int \frac{x^2}{x^4 - 5x^2 + 4}\,dx$

27. $\displaystyle\int \frac{x^2 - x + 1}{x^4 - 5x^3 + 5x^2 + 5x - 6}\,dx$

28. $\displaystyle\int \frac{3x\,dx}{x^5 + 2x^4 - 10x^3 - 20x^2 + 9x + 18}$

29. $\displaystyle\int \frac{x^2 - 2x + 3}{(x-1)^2(x^2 + 4)}\,dx$

30. $\displaystyle\int \frac{x^3 - 2x^2 + 3x - 4}{(x-1)^2(x^2 + 2x + 2)}\,dx$

31. $\displaystyle\int \frac{x^3 + x^2 - 2x - 3}{(x+1)^2(x-2)^2}\,dx$

32. $\displaystyle\int \frac{x^2 - 3x + 5}{x^4 - 8x^2 + 16}\,dx$

33. $\displaystyle\int \frac{4x^3 + 8x^2 - 12}{(x^2 + 4)^2}\,dx$

34. $\displaystyle\int \frac{x^2 + 3x + 5}{x^3 + 8}\,dx$

35. $\displaystyle\int \frac{x^2 + 2x - 1}{x^3 - 27}\,dx$

36. $\displaystyle\int \frac{x^3 - x^2 + 2x + 3}{(x^2 + 2x + 2)^2}\,dx$

37. $\displaystyle\int \frac{x^4 + 1}{(x^2 + 4)^3}\,dx$

38. $\displaystyle\int \frac{2x^5 - 6}{(x^2 + 1)^4}\,dx$

39. Let $f$ and $g$ each be polynomials of degree $n$. If $f^{(k)}(0) = g^{(k)}(0)$, $k = 0, 1, 2, \ldots, n$, show that $f(x) \equiv g(x)$ for all $x$.

**40** Verify that $x - \alpha$ is a factor of the third-degree polynomial $x^3 + cx^2 + bx - \alpha^3 - c\alpha^2 - b\alpha$, where $c$, $b$, and $\alpha$ are real numbers. Then find conditions on $c$, $b$, and $\alpha$ such that the polynomial is factorable into real, linear factors.

**41** Suppose that $Q(x)$ is a polynomial with factors $(x - a_1)$, $(x - a_2)$, ..., $(x - a_r)$, and that at least one of the factors is repeated. Show that in general it is *not* possible to obtain a partial fraction expansion

$$\frac{1}{Q(x)} = \frac{A_1}{x - a_1} + \frac{A_2}{x - a_2} + \cdots + \frac{A_r}{x - a_r}.$$

That is, show that Case 1 is not applicable.

**42** Derive the formula

$$\int \frac{x \, dx}{(ax + b)(cx + d)}$$

$$= \frac{1}{bc - ad}\left[\frac{b}{a}\log(ax + b) - \frac{d}{c}\log(cx + d)\right], \quad bc - ad \neq 0.$$

What is the result if $bc - ad = 0$?

**43** Derive the formula

$$\int \frac{x \, dx}{(ax + b)^2(cx + d)}$$

$$= \frac{1}{bc - ad}\left[-\frac{b}{a(ax + b)} - \frac{d}{bc - ad}\log\frac{cx + d}{ax + b}\right],$$

$bc - ad \neq 0$. What is the result if $bc - ad = 0$?

**44** Write the partial fraction expansion formula for the most general rational function

$$\frac{P(x)}{Q(x)}$$

in which the degree of $P$ is less than the degree of $Q$. That is, assume $Q$ has $r$ simple, distinct roots, $s$ multiple roots of multiplicities $t_1, t_2, ..., t_s$; also that it has $p$ distinct, simple quadratic factors, and $q$ multiple quadratic factors of multiplicities $z_1, z_2, ..., z_q$.

---

**7**

## TWO RATIONALIZING SUBSTITUTIONS (OPTIONAL)

I. Whenever an integrand contains a single irrational expression of the form

$$(ax + b)^{p/q}, \quad p \text{ and } q \text{ integers},$$

the substitution

$$u = (ax + b)^{1/q} \quad \Leftrightarrow \quad x = \frac{u^q - b}{a}, \quad dx = \frac{q}{a}u^{q-1}\, du$$

will convert the given integrand into a rational function of $u$.

**EXAMPLE 1**   Find

$$\int \frac{\sqrt[3]{x + 1}}{x}\, dx.$$

**Solution**   Let

$$u = (x + 1)^{1/3} \quad \Leftrightarrow \quad x = u^3 - 1, \quad dx = 3u^2 \, du.$$

Then

$$\int \frac{\sqrt[3]{x + 1}\, dx}{x} = \int \frac{u \cdot 3u^2 \, du}{u^3 - 1} = \int 3 \, du + \int \frac{3 \, du}{u^3 - 1}.$$

To evaluate the second integral, we use partial fractions and write

$$\frac{3}{u^3 - 1} = \frac{3}{(u-1)(u^2+u+1)} = \frac{A_1}{u-1} + \frac{A_2 u + A_3}{u^2+u+1}$$

$$\Leftrightarrow 3 = A_1(u^2 + u + 1) + (A_2 u + A_3)(u - 1).$$

When

$$u = 1: \qquad 3 = 3A_1 \qquad\qquad \text{and} \quad A_1 = 1;$$

$$u = 0: \qquad 3 = A_1 - A_3 \qquad\quad \text{and} \quad A_3 = -2;$$

$$u = -1: \quad 3 = A_1 + 2A_2 - 2A_3 \quad \text{and} \quad A_2 = -1.$$

Therefore

$$\int \frac{3\, du}{u^3 - 1} = \int \frac{du}{u-1} - \int \frac{u+2}{u^2+u+1}\, du$$

$$= \ln|u-1| - \frac{1}{2}\ln(u^2+u+1) - \sqrt{3}\arctan\left(\frac{2u+1}{\sqrt{3}}\right) + C.$$

We finally obtain

$$\int \frac{\sqrt[3]{x+1}}{x}\, dx = 3(x+1)^{1/3} + \ln|(x+1)^{1/3} - 1|$$

$$- \frac{1}{2}\ln|(x+1)^{2/3} + (x+1)^{1/3} + 1|$$

$$- \sqrt{3}\arctan\left[\frac{2(x+1)^{1/3} + 1}{\sqrt{3}}\right] + C. \qquad \Box$$

---

**II.** If a *single* irrational expression of one of the forms

$$\sqrt{a^2 - x^2}, \quad \sqrt{x^2 + a^2}, \quad \sqrt{x^2 - a^2}$$

appears in the integrand, and if $x^q$ ($q$ being an *odd* integer, positive or negative) also appears in the integrand, an appropriate substitution will transform the integrand into a rational function. The correct substitution is, respectively, one of the following:

$$u = (a^2 - x^2)^{1/2} \quad \text{or} \quad u = (x^2 + a^2)^{1/2} \quad \text{or} \quad u = (x^2 - a^2)^{1/2},$$

depending on which expression is involved. The integrand will become a rational function of $u$.

---

**EXAMPLE 2**  Find

$$\int \frac{\sqrt{a^2 - x^2}}{x^3}\, dx.$$

**Solution**  This is of the form II, with $q = -3$. We let

$$u = (a^2 - x^2)^{1/2} \quad \Leftrightarrow \quad x^2 = a^2 - u^2, \quad x\, dx = -u\, du.$$

Therefore

$$\int \frac{\sqrt{a^2-x^2}}{x^3}\,dx = \int \frac{\sqrt{a^2-x^2}\,x\,dx}{x^4}$$

$$= -\int \frac{u^2\,du}{(a^2-u^2)^2}$$

$$= -\int \frac{u^2\,du}{(u+a)^2(u-a)^2}.$$

We proceed by partial fractions and obtain

$$\frac{-u^2}{(u+a)^2(u-a)^2} = \frac{A_1}{u+a} + \frac{A_2}{(u+a)^2} + \frac{A_3}{u-a} + \frac{A_4}{(u-a)^2}$$

and

$$-u^2 = A_1(u+a)(u-a)^2 + A_2(u-a)^2 + A_3(u-a)(u+a)^2 + A_4(u+a)^2.$$

We find

$$u = a: \quad -a^2 = A_4 4a^2 \quad \text{and} \quad A_4 = -\frac{1}{4};$$

$$u = -a: \quad -a^2 = A_2 4a^2 \quad \text{and} \quad A_2 = -\frac{1}{4};$$

$$u = 0: \quad 0 = a^3 A_1 + a^2 A_2 - a^3 A_3 + a^2 A_4 \quad \text{and} \quad A_1 - A_3 = \frac{1}{2a}.$$

$$u = 2a: \quad -4a^2 = 3a^3 A_1 + a^2 A_2 + 9a^3 A_3 + 9a^2 A_4 \quad \text{and} \quad A_1 + 3A_3 = -\frac{1}{2a}.$$

We get

$$A_1 = \frac{1}{4a}, \qquad A_3 = -\frac{1}{4a},$$

which yields

$$\int \frac{\sqrt{a^2-x^2}}{x^3}\,dx = \frac{1}{4(u+a)} + \frac{1}{4(u-a)} + \frac{1}{4a}\ln\left|\frac{u+a}{u-a}\right| + C$$

$$= -\frac{\sqrt{a^2-x^2}}{2x^2} + \frac{1}{2a}\ln\left|a + \sqrt{a^2-x^2}\right| - \frac{1}{2a}\ln|x| + C.$$

$\square$

## 7 PROBLEMS

In Problems 1 through 33, find the indefinite integrals.

**1** $\displaystyle\int \frac{(2x+3)\,dx}{\sqrt{x+2}}$

**2** $\displaystyle\int \frac{3x-2}{\sqrt{2x-3}}\,dx$

**5** $\displaystyle\int \frac{x-2}{(3x-1)^{2/3}}\,dx$

**6** $\displaystyle\int \frac{2x-1}{(x-2)^{1/3}}\,dx$

**3** $\displaystyle\int x\sqrt{x+1}\,dx$

**4** $\displaystyle\int \frac{2x+1}{(x+2)^{2/3}}\,dx$

**7** $\displaystyle\int \frac{x^2\,dx}{\sqrt[3]{2x+1}}$

**8** $\displaystyle\int (x+2)\sqrt{x-1}\,dx$

9  $\int \dfrac{\sqrt{x+4}}{x}\, dx$

10  $\int \dfrac{\sqrt{2x+3}}{x+1}\, dx$

11  $\int \dfrac{\sqrt{x+2}}{\sqrt{x-1}}\, dx$

12  $\int \dfrac{2\sqrt{x+1}-3}{3\sqrt{x+1}-2}\, dx$

13  $\int \dfrac{x^3}{\sqrt{x^2-4}}\, dx$

14  $\int \dfrac{x^3-x}{\sqrt{9-x^2}}\, dx$

15  $\int \dfrac{x^5+2x^3}{\sqrt{x^2+4}}\, dx$

16  $\int x^3\sqrt{x^2+1}\, dx$

17  $\int (x^3-x)\sqrt{16-x^2}\, dx$

18  $\int x^3\sqrt{x^2-a^2}\, dx$

19  $\int \dfrac{dx}{x\sqrt{a^2-x^2}}$

20  $\int \dfrac{dx}{x\sqrt{x^2+4}}$

21  $\int \dfrac{(x+1)\, dx}{(2x+1)^{9/2}}$

22  $\int \dfrac{(x-3)\, dx}{(2x+3)^{7/4}}$

23  $\int \dfrac{(x^2+x+1)\, dx}{(x+5)^{7/3}}$

24  $\int \dfrac{(1-3x^3)\, dx}{(5x-1)^{8/5}}$

25  $\int \dfrac{\sqrt{a^2-x^2}}{x}\, dx$

26  $\int \dfrac{\sqrt{a^2+x^2}}{x}\, dx$

27  $\int \dfrac{\sqrt{x^2-a^2}}{x}\, dx$

28  $\int \dfrac{\sqrt{4-x}}{x}\, dx$

29  $\int \dfrac{dx}{x^{1/2}+x^{2/3}}$

30  $\int \sqrt{2+\sqrt{x}}\, dx$

31  $\int \dfrac{dx}{x^3\sqrt{x^2+4}}$

32  $\int \dfrac{\sqrt{4-x^2}}{x^3}\, dx$

33  $\int \dfrac{\sqrt{x^2+9}}{x^3}\, dx$

*34  Derive the formula

$$\int x^{m-1}(ax+b)^{p/q}\, dx$$

$$= \frac{q}{a(qm+p)}\left[ x^{m-1}(ax+b)^{(p/q)+1}\right.$$

$$\left. -(m-1)b\int x^{m-2}(ax+b)^{p/q}\, dx\right].$$

35  Given an integral of the form

$$\int \left(x+\frac{b}{2a}\right)^q F(\sqrt{ax^2+bx+c})\, dx,$$

where $q$ is an odd integer and $F$ is a polynomial function. Show how this integral can be transformed into an integral of a rational function.

36  Use the substitution $u=\tan(\theta/2)$ to transform the integral

$$\int \frac{d\theta}{5-4\cos\theta}$$

into the integral of a rational function.
*Hint:* Use the schematic triangle in Fig. 5 to read off

$$\sin\frac{1}{2}\theta = \frac{u}{\sqrt{1+u^2}}, \qquad \cos\frac{1}{2}\theta = \frac{1}{\sqrt{1+u^2}},$$

and also employ the relations

$$\sin\theta = 2\sin\frac{1}{2}\theta\cos\frac{1}{2}\theta = \frac{2u}{1+u^2},$$

$$\cos\theta = \cos^2\frac{\theta}{2}-\sin^2\frac{\theta}{2} = \frac{1-u^2}{1+u^2}.$$

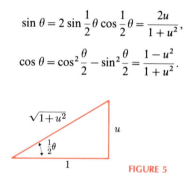

FIGURE 5

In Problems 37 through 44, use the substitution of Problem 36 to compute the integral.

37  $\int \dfrac{dx}{\tan x - \sin x}$

38  $\int \dfrac{dx}{3\cos x + 4\sin x}$

39  $\int \dfrac{d\theta}{\cos\theta + \cot\theta}$

40  $\int \dfrac{dx}{a\cos x + b\sin x}$

41  $\int \dfrac{d\theta}{4+5\sin\theta}$

42  $\int \cos^2 x\, dx$

43  $\int \dfrac{d\theta}{3-2\sin\theta}$

44  $\int \dfrac{d\theta}{2+3\cos\theta}$

# CHAPTER 8

## REVIEW PROBLEMS

In Problems 1 through 25, find the indefinite integral.

1  $\int \dfrac{dx}{(4x+1)^{1/2}}$

2  $\int \dfrac{dx}{(3-2x)^{4/3}}$

3  $\int (2x+5)^{7/2}\, dx$

4  $\int \dfrac{\cos\sqrt{2x}\, dx}{\sqrt{x}}$

**5** $\displaystyle\int x^3 e^{x^4+3}\, dx$

**6** $\displaystyle\int \frac{\sin(2\ln x)\, dx}{x}$

**7** $\displaystyle\int \frac{e^{2x}\, dx}{e^{2x}+4}$

**8** $\displaystyle\int \frac{e^x\, dx}{e^{2x}+9}$

**9** $\displaystyle\int \frac{x+2}{\sqrt{x^2+4x+2}}\, dx$

**10** $\displaystyle\int \frac{x\arcsin(x^2)}{\sqrt{1-x^4}}\, dx$

**11** $\displaystyle\int x\ln(1+x)\, dx$

**12** $\displaystyle\int (1+2x)\ln(1+x)\, dx$

**13** $\displaystyle\int x\arcsin(x^2)\, dx$

**14** $\displaystyle\int (1+x)^3 \ln x\, dx$

**15** $\displaystyle\int x\arctan(x^2)\, dx$

**16** $\displaystyle\int x^5 \arcsin(x^2)\, dx$

**17** $\displaystyle\int x^3 \sec^2(x^2)\, dx$

**18** $\displaystyle\int e^{5x}\sin 5x\, dx$

**19** $\displaystyle\int \sin^3 2x \cos^3 2x\, dx$

**20** $\displaystyle\int \sin^5 4x\, dx$

**21** $\displaystyle\int \tan^3 2x \sec 2x\, dx$

**22** $\displaystyle\int \sin^3 \tfrac{1}{2}x \sqrt{\cos \tfrac{1}{2}x}\, dx$

**23** $\displaystyle\int \frac{\sec^4 3x}{\tan^{2/3} 3x}\, dx$

**24** $\displaystyle\int \tan^6 2x\, dx$

**25** $\displaystyle\int x^3 \sin^3 x^2\, dx$

**26** Find the area bounded by the curve

$$y = \cos^3 x, \quad -\frac{\pi}{2} \le x \le \frac{\pi}{2}$$

and the $x$ axis.

**27** Find the area bounded by the curve $y = 2x\ln x$ and the lines $x = 2$, $x = 3$, $y = 0$.

**28** Find a formula for $\int x^{m+2k+1}\arctan(x^m)\, dx$, $m$ and $k$ positive integers.

In Problems 29 through 38, find the indefinite integral.

**29** $\displaystyle\int \frac{dx}{16+25x^2}$

**30** $\displaystyle\int \frac{x^3}{4+9x^2}\, dx$

**31** $\displaystyle\int \frac{dx}{\sqrt{5-12x^2}}$

**32** $\displaystyle\int \frac{dx}{\sqrt{5+12x^2}}$

**33** $\displaystyle\int \frac{x^2\, dx}{\sqrt{9+4x^2}}$

**34** $\displaystyle\int \frac{x\, dx}{\sqrt{x^2+2x+3}}$

**35** $\displaystyle\int \frac{dx}{3x^2+2x+4}$

**36** $\displaystyle\int \frac{x^2\, dx}{2x^2+x+6}$

**37** $\displaystyle\int \frac{dx}{x\sqrt{4-(\ln x)^2}}$

**38** $\displaystyle\int \frac{e^x\, dx}{\sqrt{1-e^{-2x}}}$

In Problems 39 through 50, use partial fractions to perform the integrations.

**39** $\displaystyle\int \frac{x^3+x^2-3}{x-3}\, dx$

**40** $\displaystyle\int \frac{x^2-2x+5}{x+2}\, dx$

**41** $\displaystyle\int \frac{x^2+2x+2}{(x-2)(x+1)}\, dx$

**42** $\displaystyle\int \frac{x^3+x+1}{(x+2)(x-3)}\, dx$

**43** $\displaystyle\int \frac{x^2}{(x-1)(x+2)^2}\, dx$

**44** $\displaystyle\int \frac{x^3}{(x+1)(x-3)^2(x-2)^2}\, dx$

**45** $\displaystyle\int \frac{x}{x^2+x-2}\, dx$

**46** $\displaystyle\int \frac{x}{x^2+x+2}\, dx$

**47** $\displaystyle\int \frac{x+1}{(x-1)(x^2+2x+5)}\, dx$

**48** $\displaystyle\int \frac{x^2+2}{(x+1)(x^2+2x+5)^2}\, dx$

**49** $\displaystyle\int \frac{1}{x^3+4x^2+x-6}\, dx$

**50** $\displaystyle\int \frac{(x^2+5)\, dx}{x^5-3x^4-5x^3+15x^2+4x-12}$

In Problems 51 through 56, use the substitutions in Section 7 to integrate the expression.

**51** $\displaystyle\int \frac{2x+1}{\sqrt{3x+4}}\, dx$

**52** $\displaystyle\int \frac{dx}{x\sqrt{4x^2+9}}$

**53** $\displaystyle\int \frac{x^5\, dx}{\sqrt{x^2-9}}$

**54** $\displaystyle\int \frac{2x+1}{(x+4)^{5/3}}\, dx$

**55** $\displaystyle\int \frac{dx}{3\sin x+2\cos x}$

**56** $\displaystyle\int \frac{dx}{3+4\sin x}$

# 9

# FURTHER APPLICATIONS OF INTEGRATION

In Chapter 6 we showed how the integral can be used to solve a variety of geometric problems. Techniques were developed to calculate areas, volumes, and arc length. In Chapters 7 and 8 we expanded enormously the class of functions we can integrate. In this chapter we use the processes of integration of Chapter 8 to solve new geometric and physical problems.

## 1

### CENTER OF MASS

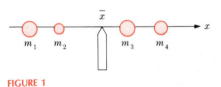

FIGURE 1

Consider a line segment made of some stiff but very light material and four masses, $m_1, m_2, m_3, m_4$ located at various points along the line. By definition, the *center of mass* of this configuration is the point $\bar{x}$ on the line where, if a knife edge were placed, the system would balance. See Fig. 1.

A familiar example in determining the center of mass occurs when two children of unequal weight are on a seesaw. See Fig. 2. The system is in balance when the fulcrum is moved closer to the heavier child by just the right amount. Then the fulcrum is located at the center of mass of the system consisting of the board and the two weights (the children).

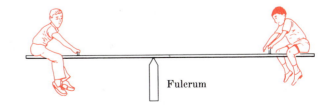

FIGURE 2

329

If a system is composed of $n$ masses $m_1, m_2, \ldots, m_n$ situated on a straight line, we first introduce a coordinate system on the line with the origin at some convenient point. See Fig. 3. Let $x_i$ be the distance of $m_i$ from the origin $O$ measured positive if $m_i$ is to the right of $O$ and negative if $m_i$ is to the left. We define the *moment of $m_i$ with respect to $O$* as the product $m_i x_i$. For the system of $n$ masses on a weightless line we define the **center of mass** $\bar{x}$ to be

$$\bar{x} = \frac{m_1 x_1 + m_2 x_2 + \cdots + m_n x_n}{m_1 + m_2 + \cdots + m_n}.$$

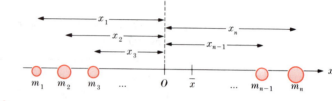

**FIGURE 3**

The center of mass is the point where the system will balance if a knife-edge is put at $\bar{x}$ and the line itself is considered weightless. Furthermore, each mass is assumed to occupy exactly one point. An important fact is that the location of $\bar{x}$ does not depend on the position of the origin $O$. (See Problem 5 at the end of this section.)

**EXAMPLE 1**   Weights of 2, 3, 5, and 4 grams are located on the $x$ axis at the points $(4, 0)$, $(2, 0)$, $(-6, 0)$, and $(-4, 0)$, respectively. Find the center of mass of the system.

**Solution**
$$\bar{x} = \frac{2(4) + 3(2) + 5(-6) + 4(-4)}{2 + 3 + 5 + 4} = -\frac{16}{7}. \qquad \square$$

Suppose that a number of masses, say five, are located at various points in the $xy$ plane. We wish to find the center of mass of this system. From the point of view of mechanics, we imagine the masses supported by a weightless tray and assume that each mass occupies a single point. The center of mass is the point at which the tray will balance when supported by a sharp nail (Fig. 4). To locate this point, we suppose that the mass $m_i$ is situated at the point $(x_i, y_i)$.

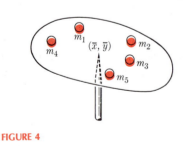

**FIGURE 4**

We define the following quantities:

**DEFINITIONS**   The moment of $m_i$ **with respect to the** $y$ **axis** *is the product* $m_i x_i$. **The moment of** $m_i$ **with respect to the** $x$ **axis** *is the product* $m_i y_i$.

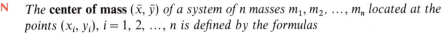 

**DEFINITION**

*The* **center of mass** $(\bar{x}, \bar{y})$ *of a system of n masses* $m_1, m_2, \ldots, m_n$ *located at the points* $(x_i, y_i)$, $i = 1, 2, \ldots, n$ *is defined by the formulas*

$$\bar{x} = \frac{m_1 x_1 + m_2 x_2 + \cdots + m_n x_n}{m_1 + m_2 + \cdots + m_n},$$

$$\bar{y} = \frac{m_1 y_1 + m_2 y_2 + \cdots + m_n y_n}{m_1 + m_2 + \cdots + m_n}.$$

*Remark.* The center of mass is also called the **center of gravity** as well as the **centroid**.

**EXAMPLE 2**

Find the center of mass of the system consisting of masses of 4, 2, and 3 grams located at the points $(1, 3)$, $(-2, 1)$, and $(4, -2)$, respectively.

**Solution**

$$\bar{x} = \frac{4(1) + 2(-2) + 3(4)}{4 + 2 + 3} = \frac{4}{3},$$

$$\bar{y} = \frac{4(3) + 2(1) + 3(-2)}{4 + 2 + 3} = \frac{8}{9}.$$

An important fact is that the center of mass is independent of the location of the coordinate axes. (See Problems 14 and 21 at the end of this section.)

A thin piece of metal of uniform density has its center of mass at the place at which it will balance horizontally when supported on the point of a nail. We idealize the situation by imagining the metal as two-dimensional and think of it as a region located in the $xy$ plane. If the metal is uniform (and we always assume that it is), then the density is constant, and the total mass of the piece of metal is proportional to the area of the region. (See Fig. 5.) For example, a rectangular piece of length $l$ and width $w$ made up of material of constant density $\rho$ has total mass proportional to

$$\rho l w.$$

It is intuitively clear that the *center of mass of a rectangular region is at the center of the rectangle* (Fig. 6). In addition, we have the following definitions:

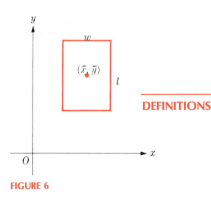

Thin piece of uniform material with center of mass $(\bar{x}, \bar{y})$

**FIGURE 5**

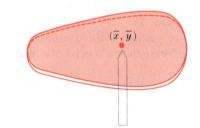

**FIGURE 6**

**DEFINITIONS**

**The moment of a rectangle with respect to the** $y$ **axis** *is*

$$\rho l w \cdot \bar{x},$$

*where* $\bar{x}$ *is the x coordinate of the center of mass of the rectangle. Similarly,* **the moment of a rectangle with respect to the** $x$ **axis** *is*

$$\rho l w \cdot \bar{y},$$

*where* $\bar{y}$ *is the y coordinate of the center of mass of the rectangle.*

With these facts it is possible to find the center of mass of regions composed of a combination of rectangles. Each rectangle may be treated as if its mass were concentrated at the center of the rectangle. We illustrate with the following example.

**EXAMPLE 3**    A region is made up of a combination of rectangles of uniform density $\rho$, as shown in Fig. 7. Find the center of mass.

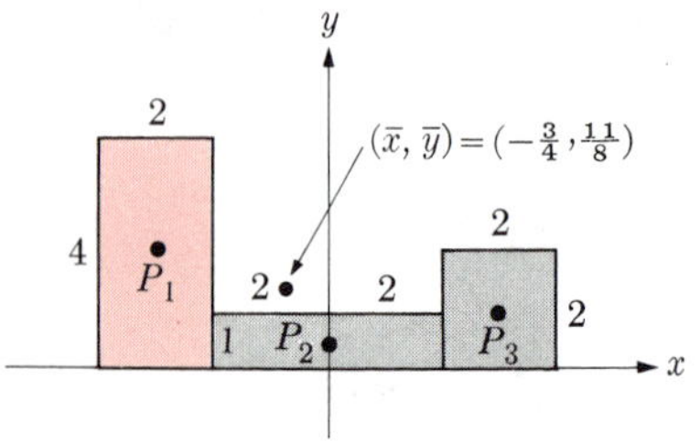

**Solution**    The centers of mass of the three rectangles are at $P_1(-3, 2)$, $P_2(0, \frac{1}{2})$, and $P_3(3, 1)$. The total masses of the rectangles are $8\rho$, $4\rho$, and $4\rho$, respectively. We may treat the region as a system of three point masses located at the centers of mass. We obtain

$$\bar{x} = \frac{[8(-3) + 4(0) + 4(3)]\rho}{16\rho} = -\frac{3}{4},$$

$$\bar{y} = \frac{[8(2) + 4(\frac{1}{2}) + 4(1)]\rho}{16\rho} = \frac{11}{8}.$$

Rectangles of uniform density

**FIGURE 7**

*Remarks.*    The numerator of the expression for $\bar{x}$ is the sum of the moments of the rectangles with respect to the $y$ axis and, correspondingly, the one for $\bar{y}$ is the sum of the moments of the rectangles with respect to the $x$ axis. In each case the denominator is the total mass.

So long as the region is of uniform density, the actual value of the density plays no part in the location of the center of mass. The above example shows how the factor $\rho$ cancels in the computation of $\bar{x}$ and $\bar{y}$. Therefore, *we always assume that the density has the value* 1, in which case the total mass of a region is its area.

As Fig. 7 shows, it is possible for the center of mass of a region to lie outside the region!

## 1  PROBLEMS

In Problems 1 through 4, the masses $m_i$ are located at the points $P_i$ on the $x$ axis; find the center of mass.

1    $m_1 = 3, m_2 = 4, m_3 = 1$; $P_1(2, 0), P_2(6, 0), P_3(-4, 0)$

2    $m_1 = 7$, $m_2 = 1$, $m_3 = 5$, $m_4 = 12$; $P_1(-4, 0)$, $P_2(1, 0)$, $P_3(5, 0), P_4(-7, 0)$

3    $m_1 = 2, m_2 = 1, m_3 = 5, m_4 = 6, m_5 = 1$; $P_1(0, 0), P_2(1, 0),$ $P_3(-1, 0), P_4(2, 0), P_5(-2, 0)$

4    $m_1 = 1, m_2 = 2, m_3 = 1, m_4 = 1, m_5 = 2$; $P_1(1, 0), P_2(2, 0),$ $P_3(3, 0), P_4(4, 0), P_5(5, 0)$

5    (a) Find the center of mass $\bar{x}$ of the system $m_1 = 7, m_2 = 5,$ $m_3 = 4, m_4 = 2$, located on the $x$ axis at the points $P_1(5, 0),$ $P_2(-5, 0)$, $P_3(4, 0), P_4(-2, 0)$. (b) Show that when the origin is shifted a distance $h$ to the right, the location of the center of mass with respect to the masses does not change.

In Problems 6 through 13, the masses $m_i$ are located in the $xy$ plane at the points $P_i$; find the centers of mass.

6    $m_1 = 1, m_2 = 3, m_3 = 2$; $P_1(2, 1), P_2(-1, 3), P_3(1, 2)$

7    $m_1 = 3, m_2 = 1, m_3 = 4$; $P_1(1, 0), P_2(0, 3), P_3(1, 2)$

8    $m_1 = 4, m_2 = 5, m_3 = 1, m_4 = 6$; $P_1(1, 1), P_2(5, 0), P_3(-4, 0),$ $P_4(0, 5)$

9    $m_1 = 2, m_2 = 8, m_3 = 5, m_4 = 2$; $P_1(0, 0), P_2(0, 4), P_3(5, 1),$ $P_4(-1, -1)$

10    $m_1 = 1, m_2 = 3, m_3 = 2, m_4 = 5, m_5 = 2$; $P_1(1, 2), P_2(-1, 2),$ $P_3(3, -1), P_4(0, 5), P_5(-2, -3)$

11    $m_1 = 3, m_2 = 3, m_3 = 1, m_4 = 2, m_5 = 6$; $P_1(2, 1), P_2(0, 4),$ $P_3(-4, 0), P_4(0, -2), P_5(-4, -4)$

12    $m_1 = 2$, $m_2 = 4$, $m_3 = 3$, $m_4 = 6$, $m_5 = 5$; $P_1(1, -3),$ $P_2(-2, 1), P_3(4, -3), P_4(-2, 5), P_5(1, 1)$

13    $m_1 = 1, m_2 = 3, m_3 = 2, m_4 = 6, m_5 = 4, m_6 = 1$; $P_1(1, 2),$ $P_2(-1, -2), P_3(4, -3), P_4(-3, 4), P_5(0, 4), P_6(-3, 0)$

**14** (a) Find the center of mass $(\bar{x}, \bar{y})$ of the system $m_1 = 3$, $m_2 = 4$, $m_3 = 2$, located at the points $P_1(1, 1)$, $P_2(3, 0)$, $P_3(-1, 1)$. (b) Show that if the coordinate axes are translated so that the origin is at the point $(h, k)$, the location of the center of mass with respect to the masses does not change. Can the computation of the center of mass be simplified in this manner?

**15** Five equal masses are located at the vertices of a regular pentagon of side length 2. Find the center of mass. (Use a table or a calculator for the values of trigonometric functions.)

**16** Eight masses of equal size are equally distributed on the rim of a circle. Show that the center of mass is at the center of the circle.

In Problems 17 through 20, introduce a convenient rectangular coordinate system and find the coordinates of the center of mass in that coordinate system.

**17** A plane region of uniform density has the shape shown in Fig. 8. Find the center of mass.

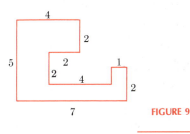

FIGURE 8

**18** A plane region of uniform density has the shape shown in Fig. 9. Find the center of mass.

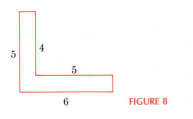

FIGURE 9

**19** A plane region of uniform density has the shape shown in Fig. 10. Find the center of mass.

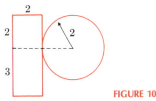

FIGURE 10

**20** A plane region is composed of the regions $R_1$, $R_2$, and $R_3$, as shown in Fig. 11. The density of $R_2$ is twice that of $R_1$, and the density of $R_3$ is three times that of $R_2$. Find the center of mass.

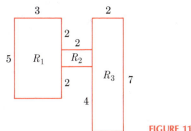

FIGURE 11

**21** Suppose that $n$ masses $m_1, m_2, \ldots, m_n$ are located in the plane at the points $(x_1, y_1)$, $(x_2, y_2)$, $\ldots$, $(x_n, y_n)$, respectively. Denote the center of mass of this system by $(\bar{x}, \bar{y})$. If we translate coordinates by the formulas: $x' = x + h$, $y' = y + k$, show that in the new coordinate system the center of mass is located at $(\bar{x} + h, \bar{y} + k)$. That is, prove that the geometric location of the center of mass is unchanged by a translation of coordinates.

**22** A parallelogram has one side of length 2 and another of length 1 with an acute angle of $\pi/3$ between the two sides. Four equal masses are located at the vertices. Locate the center of mass.

**23** An isosceles triangle has two sides of length $a$ and one of length $b$. Two masses each of size $m$ are at the vertices of the base and one of size $3m$ is at the third vertex. Locate the center of mass.

**2**

## CENTER OF MASS OF A PLANE REGION

In the preceding section we showed how to find the center of mass of a plane region composed of a collection of rectangles, each of uniform density. We now establish methods for finding the center of mass $(\bar{x}, \bar{y})$ of more general plane figures. *We always assume that the density is constant.* It is convenient to take the density equal to 1, since then the total mass of a region is numerically equal to its area. The results rest on the following two principles, which we state without proof.

**PRINCIPLE 1**     *The center of mass of a plane region lies on any axis of symmetry of that region.*

**PRINCIPLE 2**     *Suppose that a plane region R with center of mass $(\bar{x}, \bar{y})$ is divided into regions $R_1, R_2, \ldots, R_n$ (no two of which overlap), having areas $A_1, A_2, \ldots, A_n$, and centers of mass at $(\bar{x}_1, \bar{y}_1), (\bar{x}_2, \bar{y}_2), \ldots, (\bar{x}_n, \bar{y}_n)$. Then*

$$\bar{x} = \frac{A_1\bar{x}_1 + A_2\bar{x}_2 + \cdots + A_n\bar{x}_n}{A_1 + A_2 + \cdots + A_n}, \qquad \bar{y} = \frac{A_1\bar{y}_1 + A_2\bar{y}_2 + \cdots + A_n\bar{y}_n}{A_1 + A_2 + \cdots + A_n}.$$

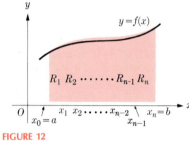

**FIGURE 12**

Let $R$ be the region bounded by the graph of a nonnegative function $y = f(x)$ and the lines $x = a$, $x = b$, and $y = 0$, as shown in Fig. 12. We make a subdivision of the interval $[a, b]$ into $n$ parts:

$$\{a = x_0 < x_1 < x_2 < \cdots < x_n = b\}.$$

We erect vertical lines at the subdivision points, thereby dividing $R$ into $n$ nonoverlapping regions $R_1, R_2, \ldots, R_n$. According to Principle 2, the center of mass of $R$ can be found if the center of mass and the area of each of the regions $R_1, R_2, \ldots, R_n$ is known. The procedure is similar to that used in defining integrals. For each $i$ we select a point $\xi_i$ in the interval $[x_{i-1}, x_i]$ and erect a rectangle of height $f(\xi_i)$ and width $\Delta_i x = x_i - x_{i-1}$, as shown in Fig. 13. This rectangle approximates the area $R_i$, and the center of mass of the rectangle approximates the center of mass of $R_i$. According to Principle 1, the center of mass $(\bar{x}_i, \bar{y}_i)$ of the rectangle is at the point

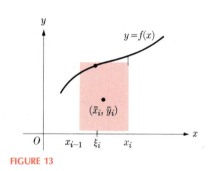

**FIGURE 13**

$$\bar{x}_i = \tfrac{1}{2}(x_{i-1} + x_i), \qquad \bar{y}_i = \tfrac{1}{2}f(\xi_i).$$

The area of the rectangle is $f(\xi_i)\,\Delta_i x$. Therefore the center of mass $(\bar{x}, \bar{y})$ of $R$ is approximated (according to Principle 2) by the expressions

$$\frac{(f(\xi_1)\,\Delta_1 x)\bar{x}_1 + (f(\xi_2)\,\Delta_2 x)\bar{x}_2 + \cdots + (f(\xi_n)\,\Delta_n x)\bar{x}_n}{f(\xi_1)\,\Delta_1 x + f(\xi_2)\,\Delta_2 x + \cdots + f(\xi_n)\,\Delta_n x},$$

$$\frac{(f(\xi_1)\,\Delta_1 x)\bar{y}_1 + (f(\xi_2)\,\Delta_2 x)\bar{y}_2 + \cdots + (f(\xi_n)\,\Delta_n x)\bar{y}_n}{f(\xi_1)\,\Delta_1 x + f(\xi_2)\,\Delta_2 x + \cdots + f(\xi_n)\,\Delta_n x}.$$

We made the above computation for one subdivision. We now make a sequence of such subdivisions, with the norms of the subdivisions tending to zero. The denominators in the above expressions form a Riemann sum for the function $f$:

$$\sum_{i=1}^{n} f(\xi_i)\,\Delta_i x,$$

and as the norm of the subdivision tends to zero we know that this sum tends to

$$\int_a^b f(x)\,dx. \tag{1}$$

The numerators are the sums

$$\sum_{i=1}^{n} \bar{x}_i f(\xi_i)\,\Delta_i x, \qquad \sum_{i=1}^{n} \bar{y}_i f(\xi_i)\,\Delta_i x = \tfrac{1}{2}\sum_{i=1}^{n} [f(\xi_i)]^2\,\Delta_i x.$$

It can be shown, in the same way that the Riemann sum for the area tends to the integral (1), that the above two sums are Riemann sums which tend to

$$\int_a^b xf(x)\,dx \qquad \text{and} \qquad \frac{1}{2}\int_a^b [f(x)]^2\,dx,$$

respectively. We thus have the following result for the center of mass of a plane region.

**THEOREM 1**  *Let R be a region bounded by the nonnegative continuous function $y = f(x)$ and the lines $x = a$, $x = b$, $y = 0$. Then the center of mass $(\bar{x}, \bar{y})$ of R is given by the formula*

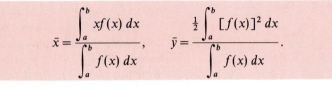

$$\bar{x} = \frac{\displaystyle\int_a^b xf(x)\,dx}{\displaystyle\int_a^b f(x)\,dx}, \qquad \bar{y} = \frac{\frac{1}{2}\displaystyle\int_a^b [f(x)]^2\,dx}{\displaystyle\int_a^b f(x)\,dx}.$$

Observe that the denominator in each case is the area of the region $R$.

**EXAMPLE 1**  Compute the center of mass of the region

$$R = \{(x, y): 1 \le x \le 3, 0 \le y \le x^2\}.$$

(See Fig. 14.)

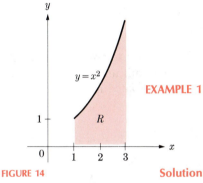

FIGURE 14

**Solution**  The region is of the type discussed above with $f(x) = x^2$, and therefore

$$\bar{x} = \frac{\displaystyle\int_1^3 x^3\,dx}{A}, \qquad \bar{y} = \frac{\frac{1}{2}\displaystyle\int_1^3 x^4\,dx}{A},$$

where $A$ is the area of $R$. We have

$$A = \int_1^3 x^2\,dx = \tfrac{26}{3},$$

and so

$$\bar{x} = \tfrac{30}{13}, \qquad \bar{y} = \tfrac{363}{130}.$$

Rather than memorize the formulas in Theorem 1 for the center of mass of a region of a particular type, the reader might learn the *method* for deriving such formulas. The steps in the method are as follows: (1) Make a subdivision of the interval on the $x$ axis. The region $R$ is approximated by erecting rectangles at the subdivision points. (2) Find the center of mass of each rectangle by Principle 1. (3) Use Principle 2 to obtain an approximate expression for $(\bar{x}, \bar{y})$, the center of mass of $R$. (4) Proceed to the limit by writing the appropriate integrals. (5) Evaluate the integrals.

The next example illustrates this technique, and shows us how the center of mass for general regions may be calculated.

EXAMPLE 2    Find the center of mass of the region $R$ bounded by the curves $y = x^2$ and $y = x + 2$.

Solution    By solving the equations simultaneously, we find that the curves intersect at $(-1, 1)$ and $(2, 4)$. (See Fig. 15.) Therefore, in set notation, we may write $R = \{(x, y): -1 \le x \le 2, x^2 \le y \le x + 2\}$. We make a subdivision of the interval $[-1, 2]$ and erect rectangles at the subdivision points. (A typical one is shown in Fig. 15.) The center of mass $(\bar{x}_i, \bar{y}_i)$ of this rectangle is at its center, which is

$$\bar{x}_i = \tfrac{1}{2}(x_{i-1} + x_i),$$

$$\bar{y}_i = \tfrac{1}{2}[(\xi_i + 2) + \xi_i^2],$$

where $\xi_i$ is some number in the interval $[x_{i-1}, x_i]$.

Since we shall always be dealing with a typical rectangle, it pays to abbreviate the notation in the following way. We drop the subscript $i$ and instead of $\tfrac{1}{2}(x_{i-1} + x_i)$, we write $x$. Since $\xi_i$ is between $x_{i-1}$ and $x_i$, we also use $x$ for this quantity. We then have in the simplified notation

$$\bar{x}_i = x, \qquad \bar{y}_i = \tfrac{1}{2}[(x + 2) + x^2].$$

Although it may appear that such sloppiness will lead to errors, we need not worry, since we shall eventually proceed to the limit and replace sums by integrals. The area of a typical rectangle (in abbreviated notation) is

$$[(x + 2) - x^2]\,\Delta x.$$

Using Principle 2, we obtain as approximate expressions for $\bar{x}$ and $\bar{y}$

$$\frac{\sum \{[(x+2)-x^2]\Delta x\}x}{\sum [(x+2)-x^2]\Delta x}, \qquad \frac{\sum \{[(x+2)-x^2]\Delta x\}\tfrac{1}{2}[(x+2)+x^2]}{\sum [(x+2)-x^2]\Delta x}.$$

These sums extend from 1 to $n$. Proceeding to the limit, we find that

$$\bar{x} = \frac{\displaystyle\int_{-1}^{2} x[(x+2)-x^2]\,dx}{A},$$

$$\bar{y} = \frac{\tfrac{1}{2}\displaystyle\int_{-1}^{2} [(x+2)+x^2][(x+2)-x^2]\,dx}{A},$$

where $A$ is the area of the region $R$. We get $A$ in the usual way from the formula

$$A = \int_{-1}^{2} [(x+2)-x^2]\,dx.$$

Evaluating the various integrals, we obtain

$$\bar{x} = \tfrac{1}{2}, \qquad \bar{y} = \tfrac{8}{5}$$

for the solution of Example 2.    □

Example 2 illustrates how a general formula can be obtained for the center of mass of a region $R$ bounded by the lines $x = a$, $x = b$ and the curves

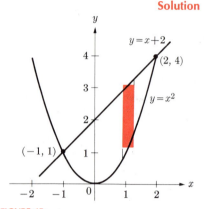

FIGURE 15

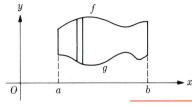

FIGURE 16

$y = f(x)$, $y = g(x)$ with $f(x) \geq g(x)$. See Fig. 16. Following the procedure outlined in Example 2, we construct a typical rectangle, as shown, and locate its center of mass. Continuing the analogy with Example 2 we use $f(x)$ instead of $x + 2$ and $g(x)$ instead of $x^2$. The lines $x = 1$, $x = 2$ are replaced by $x = a$, $x = b$. We state the final result:

**THEOREM 2**  *Suppose $f(x)$ and $g(x)$, $a \leq x \leq b$ are continuous with $f(x) \geq g(x)$. The center of mass $(\bar{x}, \bar{y})$ of the region $R = \{(x, y) : a \leq x \leq b, g(x) \leq y \leq f(x)\}$ is given by*

$$\bar{x} = \frac{\displaystyle\int_a^b x[f(x) - g(x)]\,dx}{\displaystyle\int_a^b [f(x) - g(x)]\,dx}, \quad \bar{y} = \frac{\frac{1}{2}\displaystyle\int_a^b [f(x) + g(x)][f(x) - g(x)]\,dx}{\displaystyle\int_a^b [f(x) - g(x)]\,dx}. \quad (2)$$

Observe that the numerator of $\bar{y}$ may be written

$$\tfrac{1}{2}\textstyle\int_a^b ([f(x)]^2 - [g(x)]^2)\,dx.$$

Suppose a region $R$ is situated between the curves $x = F(y)$, $x = G(y)$ (with $F(y) \geq G(y)$) and the lines $y = c$, $y = d$, as shown in Fig. 17. Then the center of mass $(\bar{x}, \bar{y})$ is obtained by formulas analogous to Formula (2) in Theorem 2 with $x$ and $y$ interchanged. In this case, we have

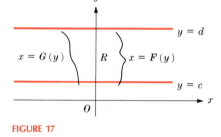

FIGURE 17

$$\bar{x} = \frac{\frac{1}{2}\displaystyle\int_c^d [F(y) + G(y)][F(y) - G(y)]\,dy}{\displaystyle\int_c^d [F(y) - G(y)]\,dy}, \quad \bar{y} = \frac{\displaystyle\int_c^d y[F(y) - G(y)]\,dy}{\displaystyle\int_c^d [F(y) - G(y)]\,dy}. \quad (3)$$

The next example illustrates the use of these formulas.

**EXAMPLE 3**  Find the center of mass of the triangular region with vertices at $(0, 0)$, $(a, 0)$, and $(b, c)$.

**Solution**  We first concentrate on finding $\bar{y}$. It is convenient to subdivide the interval $[0, c]$ along the $y$ axis. A typical rectangle is shown shaded in Fig. 18. We suppose that its height is $y$. Then the $y$ coordinate of its center of mass is simply $y$. Let $l$ be the length and $\Delta y$ the width of this rectangle. By similar triangles, we have the proportion

$$\frac{l}{a} = \frac{c - y}{c};$$

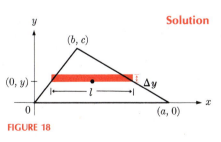

FIGURE 18

therefore the area of a typical rectangle is

$$l\,\Delta y = \frac{a}{c}(c - y)\,\Delta y.$$

According to Principle 2, $\bar{y}$ is approximated by

$$\frac{\sum (a/c)(c - y)\,\Delta y \cdot y}{\sum (a/c)(c - y)\,\Delta y}.$$

Proceeding to the limit, we conclude that

$$\bar{y} = \frac{(a/c) \displaystyle\int_0^c (c - y)\, y \, dy}{\frac{1}{2}ac} = \frac{\frac{1}{6}ac^2}{\frac{1}{2}ac} = \frac{1}{3}c.$$

The center of mass is one-third of the distance from the base to the vertex. Since any side could have been chosen as base, we deduce that the center of mass of any triangle is at the point of intersection of its medians.

Alternately, we may use the formula for $\bar{y}$ given in (2). The equation of the line through $(b, c)$ and $(a, 0)$ is $x = (b - a)\, y/a + a$, and the equation of the line through $(b, c)$ and $(0, 0)$ is $x = by/c$. Thus

$$F(y) = \frac{1}{a}(b - a)\, y + a \qquad \text{and} \qquad G(y) = by/c.$$

Hence

$$\bar{y} = \frac{\displaystyle\int_0^c y\left[\frac{1}{c}(b - a)\, y + a - \frac{b}{c}y\right] dy}{\displaystyle\int_0^c \left[\frac{1}{c}(b - a)\, y + a - \frac{b}{c}y\right] dy}.$$

A calculation yields $\bar{y} = \frac{1}{3}c$, as before.                    □

## 2   PROBLEMS

In Problems 1 through 10, find the center of mass in each case.

**1**  $R$ is bounded by $y = \sqrt{4x}$, the $x$ axis, and the line $x = 6$.

**2**  $R$ is bounded by $y = x^2$ and $y = x$.

**3**  $R$ is bounded by one arch of $y = \sin x$ and the $x$ axis.

**4**  $R$ is bounded by $y = \sin x$, and the lines $x = 0$, $x = \pi/2$, $y = 0$.

**5**  $R$ is the semicircular region bounded by $y = \sqrt{a^2 - x^2}$, $-a \le x \le a$, and the $x$ axis.

**6**  $R$ is the quarter-circle bounded by $y = \sqrt{a^2 - x^2}$ and the lines $x = 0$, $y = 0$.

**7**  $R$ is the region to the right of $y = 3x$, and bounded by $y = 3x$, $y = x^2$, $y = 1$, and $y = 2$.

**8**  $R$ is bounded by $y = \cos x$ and the lines $x = \pi/6$, $x = \pi/3$, $y = 0$.

**9**  $R$ is bounded by $y = e^x$ and the lines $x = 1$, $x = 2$, $y = 0$.

**10**  $R$ is bounded by $y = e^x$, and the lines $y = \frac{1}{2}x$, $x = 1$, $x = 2$.

In Problems 11 through 16, find the center of mass of each of the regions $R$.

**11**  $R = \{(x, y): 0 \le x \le \pi/4, 0 \le y \le \sec^2 x\}$

**12**  $R = \{(x, y): 2 \le x \le 4, 0 \le y \le \ln x\}$

**13**  $R = \{(x, y): 0 \le x \le 1, x^3 \le y \le x\}$

**14**  $R = \{(x, y): 0 \le x \le \frac{1}{4}, x^2 \le y \le x - x^2\}$

**15**  $R = \{(x, y): 1 \le x \le 2, 0 \le y \le e^{2x}\}$

**16**  $R = \{(x, y): 0 \le x \le 1, 0 \le y \le xe^x\}$

**17**  Use the result of Example 3 of this section to find the center of mass of the trapezoid with vertices $(0, 0)$, $(a, 0)$, $(b, c)$, $(d, c)$, where

$$0 < b < d < a.$$

**18**  Complete the details of the proof of Formula (2) on page 337.

In Problems 19 through 22, use Formula (2) on page 337 to compute the center of mass of the region $R$.

**19**  $R = \left\{(x, y): 0 \le x \le \dfrac{\pi}{3}, \sin x \le y \le \sin 2x\right\}$

**20**  $R = \{(x, y): 0 \le x \le 1, x^3 \le y \le x^2\}$

**21**  $R = \{(x, y): 1 \le x \le 2, \frac{1}{2}(x - 1) \le y \le \ln x\}$

**22**  $R$ is the region in the first quadrant bounded by the curves $y = x^m$, $y = x^n$, $m > n > 0$.

In Problems 23 through 26, use Formula (3) on page 337 to compute the center of mass of the region $R$.

**23** $R$ is bounded by the curves $x = e^{(1/2)y}$ and the lines $x = 0$, $y = 0$, $y = 1$.

**24** $R = \left\{(x, y): 0 \le x \le \cos y, \ 0 \le y \le \dfrac{\pi}{2}\right\}$

**25** $R = \{(x, y): -2 \le x \le \ln y, \ 1 \le y \le 2\}$

**26** $R = \left\{(x, y): \sin y \le x \le \cos y, \ 0 \le y \le \dfrac{\pi}{4}\right\}$

**27** Let $R_1$ consist of the union of the region described in Problem 3 and the rectangle

$$R' = \{(x, y): 0 \le x \le \pi, \ -2 \le y \le 0\}.$$

Use Principle 2 to find the center of mass of $R_1$.

**28** Let $R_1$ consist of the union of the region described in Problem 2 and the region

$$R' = \{(x, y): 0 \le x \le 1, \ -x \le y \le 0\}.$$

Find the center of mass of $R_1$.

---

**3**

# CENTER OF MASS OF A SOLID OF REVOLUTION

The methods we have developed enable us now to find the centers of mass of certain solids of revolution. As before, we shall assume that the object is made of uniform material of constant density. For convenience we shall select the density equal to 1, so that the total mass of an object is numerically the same as its volume.

**PRINCIPLE 3**    *A solid of revolution will have its center of mass on the axis of revolution. If the axis of revolution is taken as the $x$ axis, then only the coordinate $\bar{x}$ has to be determined.*

The principle we use in locating this value is the analog of Principle 2 of the last section.

**PRINCIPLE 4**    *Suppose that a solid $F$ is divided into solids $F_1, F_2, \ldots, F_n$ (no two of which overlap), having volumes $V_1, V_2, \ldots, V_n$. Let $\bar{x}_1, \bar{x}_2, \ldots, \bar{x}_n$ be the $x$ values of the centers of mass of $F_1, F_2, \ldots, F_n$, respectively. Then*

$$\bar{x} = \frac{V_1 \bar{x}_1 + V_2 \bar{x}_2 + \cdots + V_n \bar{x}_n}{V_1 + V_2 + \cdots + V_n}.$$

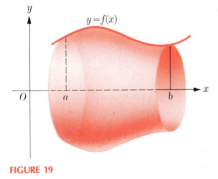

FIGURE 19

Let $R$ be a region bounded by the nonnegative function $y = f(x)$ and the lines $x = a$, $x = b$, and $y = 0$. That is,

$$R = \{(x, y): a \le x \le b, 0 \le y \le f(x)\}.$$

We revolve this region about the $x$ axis, obtaining a solid of revolution $F$ (Fig. 19). The center of mass of $F$ will be on the $x$ axis, according to Principle 3, and we shall show how to locate it. We make a subdivision of the interval $[a, b]$: $\{a = x_0 < x_1 \cdots < x_n = b\}$, and slice the solid $F$ into domains $F_1, F_2, \ldots, F_n$, by planes through the subdivision points. We approximate each $F_i$ by a disk of thickness $\Delta_i x = x_i - x_{i-1}$ and radius $f(\xi_i)$, where $x_{i-1} \le \xi_i \le x_i$.

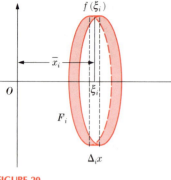

FIGURE 20

Then the volume of the disk $F_i$ is

$$\pi[f(\xi_i)]^2 \Delta_i x,$$

and the center of mass $\bar{x}_i$ of the disk $F_i$ is on the $x$ axis at $\frac{1}{2}(x_{i-1}+x_i)$. We use the abbreviated notation described in Section 2 and replace $\xi_i$ by $x_i$ and $\frac{1}{2}(x_{i-1}+x_i)$ by $x_i$. Then from Principle 4 we obtain the approximation to $\bar{x}$ (see Fig. 20)

$$\frac{\pi \sum \{[f(x_i)]^2 \Delta_i x\} x_i}{\pi \sum [f(x_i)]^2 \Delta_i x}.$$

We recognize the denominator as a Riemann sum for the volume of revolution of the solid $F$. The numerator is also a Riemann sum and proceeding to the limit we get the following result.

**THEOREM 3**   *Let $f(x) \geq 0$, $a \leq x \leq b$ be a continuous function and suppose the region $R = \{(x, y): a \leq x \leq b, 0 \leq y \leq f(x)\}$ is revolved about the $x$ axis. The center of mass of the resulting solid $F$ is on the $x$ axis at the point*

$$\bar{x} = \frac{\displaystyle\int_a^b x[f(x)]^2\,dx}{\displaystyle\int_a^b [f(x)]^2\,dx}.$$

Note that the constant $\pi$, which appears in both numerator and denominator, cancels.

**EXAMPLE**   Find the center of mass of the hemisphere of radius $a$, center at the origin, and axis along the positive $x$ axis.

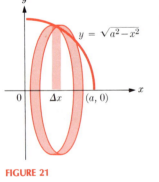

FIGURE 21

**Solution**   The solid is generated by revolving about the $x$ axis the quadrant $Q$ of a disk given by $Q = \{(x, y): 0 \leq x \leq a, 0 \leq y \leq \sqrt{a^2 - x^2}\}$ (Fig. 21). We divide the solid into thin slabs by planes through subdivision points. According to the above description of the method, we then have

$$\bar{x} = \frac{\displaystyle\int_0^a x(a^2 - x^2)\,dx}{\displaystyle\int_0^a (a^2 - x^2)\,dx} = \frac{3a}{8}. \qquad \square$$

### 3   PROBLEMS

In each of Problems 1 through 14, a region $R$ in the plane is given. Find the center of mass of the solid generated by revolving $R$ about the $x$ axis.

**1**   $R$ is bounded by $y = 0$, $x = h$, $y = (a/h)x$. (Note that the solid is a cone of altitude $h$ and radius of base $a$.)

**2**   $R = \{(x, y): 0 \leq x \leq b, 0 \leq y \leq \sqrt{2px}\}$;   $b$, $p$ are constants.

**3**   $R = \{(x, y): 0 \leq x \leq a, 0 \leq y \leq (b/a)\sqrt{a^2 - x^2}\}$

**4**   $R$ is bounded by $x = 0$, $x = \pi/3$, $y = 0$, $y = \sec x$.

**5**   $R$ is bounded by $y = 0$, $x = 3$, $y = x/\sqrt{4 - x}$.

**6**   $R = \{(x, y): 1 \leq x \leq 2, 0 \leq y \leq \ln x\}$

**7**   $R = \{(x, y): 0 \le x \le \pi, 0 \le y \le \sin x\}$

**8**   $R = \{(x, y): 0 \le x \le 1, 0 \le y \le \frac{1}{2}(e^x + e^{-x})\}$

**9**   $R$ is bounded by the curves $y = x^2$ and $y^2 = x$.

**10**   $R$ is bounded by $x + y = 5$ and $xy = 4$.

**11**   $R$ is bounded by $y = x^2$ and $y = x + 2$.

**12**   $R = \{(x, y): 0 \le x \le \dfrac{\pi}{4}, 0 \le y \le \cos 2x\}$

**13**   $R = \left\{(x, y): 0 \le x \le \dfrac{\pi}{4}, 0 \le y \le \tan x\right\}$

**14**   $R = \{(x, y): 0 \le x \le 1, x^m \le y \le x^n\}, \quad m > n > 0$

In each of Problems 15 through 20, find the center of mass of the solid generated by revolving $R$ about the $y$ axis.

**15**   $R$ is bounded by $y = 1, y = 3, x = 0, x = y^2$.

**16**   $R = \{(x, y): 0 \le x \le \sqrt{y^2 + 1}, 0 \le y \le 1\}$

**17**   $R$ is bounded by $x = 0$ and $x + y^2 - 6y = 0$.

**18**   $R$ is bounded by $x = 1/(y + 1), y = 0, y = 3, x = 0$.

**19**   $R = \left\{(x, y): 0 \le x \le \cos y, 0 \le y \le \dfrac{\pi}{2}\right\}$

**20**   $R = \{(x, y): 0 \le x \le e^{2y}, 1 \le y \le 2\}$

**21**   A cylindrical hole of radius 4 cm is bored through a solid hemisphere of radius 5 cm in such a way that the axis of the hole coincides with that of the hemisphere. Locate the center of mass of the remaining solid.

**22**   Using Principle 4 and the result of Example 1, find the center of mass of a solid in the form of a right circular cylinder of altitude $h$ and radius $a$, capped by a hemisphere of the same radius.

**23**   Given     $R_1 = \{(x, y): 0 \le x \le 1, 0 \le y \le x^2\},$

$$R_2 = \{(x, y): 0 \le x \le \sqrt{y}, 0 \le y \le 1\}.$$

Suppose $F_1$ is the solid obtained by revolving $R_1$ about the $x$ axis and $F_2$ is the solid obtained by revolving $R_2$ about the $y$ axis. Find the center of mass of $F_1 \cup F_2$.

**24**   Using Principle 4 and the results of Example 1 and Problem 1, find the center of mass of a solid "top" in the form of a cone of altitude $h$ and base of radius $a$, with vertex downward, surmounted by a hemisphere of radius $a$.

**25**   The region $R = \{(x, y): 0 \le x < \infty, 0 \le y \le e^{-x}\}$ is revolved about the $x$ axis. Find the center of mass of the solid generated. (*Note:* $\lim\limits_{h \to +\infty} he^{-bh} = 0$ when $b > 0$.)

**26**   Let $S$ be a pyramid with a square base of side 2 and an altitude of 3. Suppose that the vertex of the pyramid is directly above the center of the square base. Use Principles 3 and 4 to find the center of mass of $S$.

---

**4**

---

# CENTERS OF MASS OF WIRES AND SURFACES

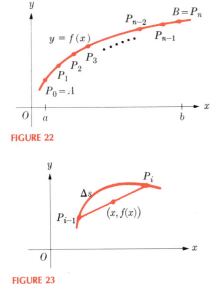

FIGURE 22

FIGURE 23

Continuing the development of the problem of finding the centers of mass of various kinds of objects, we now show how to locate the center of mass of a thin wire. As usual, we assume that the material is uniform and the density constant and equal to 1; we idealize the problem further by supposing that the wire is one-dimensional. The total mass will then be numerically equal to the length of the curve.

Suppose that the wire is described by the equation $y = f(x)$ between the points $A(a, f(a))$ and $B(b, f(b))$. We make a decomposition of the arc (Fig. 22)

$$\{A = P_0, P_1, P_2, \ldots, P_{n-1}, P_n = B\}$$

and approximate each subarc by a straight-line segment (Fig. 23). *The center of mass of a straight-line segment is at its midpoint.* We use the analog of Principle 2 to approximate the center of mass $(\bar{x}, \bar{y})$ by

$$\frac{\sum (\Delta s) \cdot x}{\sum (\Delta s)}, \quad \frac{\sum (\Delta s) f(x)}{\sum (\Delta s)},$$

respectively, where $\Delta s$ is the abbreviated notation for the length of the arc $P_{i-1}P_i$. (See Fig. 23.) We observe that the denominators in the above expressions are a Riemann sum approximation to the length of the arc $\widehat{AB}$. Proceeding to the limit we obtain the next result.

**THEOREM 4**  *Let $f(x)$, $a \le x \le b$ be a continuous function which has finite length. The center of mass $(\bar{x}, \bar{y})$ is given by the formulas*

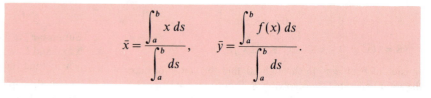

$$\bar{x} = \frac{\displaystyle\int_a^b x\,ds}{\displaystyle\int_a^b ds}, \qquad \bar{y} = \frac{\displaystyle\int_a^b f(x)\,ds}{\displaystyle\int_a^b ds}.$$

**EXAMPLE 1**  Find the center of mass of the arc of the circle $x^2 + y^2 = a^2$ which is in the first quadrant.

**Solution**  The line $y = x$ is a line of symmetry, and therefore $\bar{x} = \bar{y}$. We have

$$y = \sqrt{a^2 - x^2} \qquad \text{and} \qquad \frac{dy}{dx} = \frac{-x}{\sqrt{a^2 - x^2}}.$$

Hence

$$ds = \sqrt{1 + (dy/dx)^2}\,dx = \frac{a}{\sqrt{a^2 - x^2}}\,dx,$$

and therefore from Theorem 4

$$\bar{x} = \frac{\displaystyle\int_0^a x\,ds}{\displaystyle\int_0^a ds} = \frac{\displaystyle\int_0^a x(a/\sqrt{a^2 - x^2})\,dx}{\tfrac{1}{2}\pi a} = \frac{2a}{\pi} = \bar{y}.$$

In Fig. 24, note that the center of mass is not on the arc.

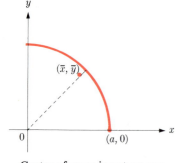

Center of mass is not on arc

**FIGURE 24**

**EXAMPLE 2**  Find the center of mass of the arc in Fig. 25:

$$y = \frac{1}{6}x^3 + \frac{1}{2x}, \quad 1 \le x \le 2.$$

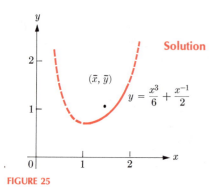

**FIGURE 25**

**Solution**  If we subdivide the interval $1 \le x \le 2$, a typical section will have mass $\Delta s$, and its center of mass will be at $(x, \frac{1}{6}x^3 + 1/2x)$. As an *approximation* to $(\bar{x}, \bar{y})$, we obtain

$$\frac{\sum(\Delta s) \cdot x}{\sum \Delta s}, \qquad \frac{\sum(\Delta s)\left(\dfrac{x^3}{6} + \dfrac{1}{2x}\right)}{\sum(\Delta s)}.$$

Therefore, proceeding to the limit in this approximation or, what is the same thing, using the formulas in Theorem 4, we find

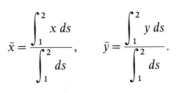

$$\bar{x} = \frac{\displaystyle\int_1^2 x\,ds}{\displaystyle\int_1^2 ds}, \qquad \bar{y} = \frac{\displaystyle\int_1^2 y\,ds}{\displaystyle\int_1^2 ds}.$$

We also have

$$\frac{dy}{dx} = \frac{1}{2}x^2 - \frac{1}{2x^2}$$

and

$$ds = \sqrt{1 + (dy/dx)^2}\,dx = \frac{1}{2}\left(x^2 + \frac{1}{x^2}\right)dx,$$

$$\bar{x} = \frac{\frac{1}{2}\displaystyle\int_1^2 [x^3 + (1/x)]\,dx}{\frac{1}{2}\displaystyle\int_1^2 [x^2 + (1/x^2)]\,dx}, \qquad \bar{y} = \frac{\frac{1}{2}\displaystyle\int_1^2 [\frac{1}{6}x^3 + (1/2x)][x^2 + (1/x^2)]\,dx}{\frac{1}{2}\displaystyle\int_1^2 [x^2 + (1/x^2)]\,dx}.$$

Computing the various integrals, we find

$$\bar{x} = \tfrac{45}{34} + \tfrac{6}{17}\ln 2, \qquad \bar{y} = \tfrac{141}{136}. \qquad \square$$

When a thin wire is revolved about an axis, a shell of revolution is obtained. If we consider this as a two-dimensional object, the center of mass may sometimes be found by simple integrations. As usual, the density is assumed constant and equal to 1. The center of mass will always lie on the axis of revolution.

Let $f(x)$, $a \le x \le b$ be a nonnegative function which represents a wire and suppose a shell is formed by revolving this wire about the $x$ axis. To locate $\bar{x}$, the center of mass, we first observe that the surface area of a thin slice of the shell is given by

$$2\pi f(x)\,\Delta s.$$

See Fig. 26. Then an approximation for $\bar{x}$, according to Principle 4, is the quantity

$$\frac{2\pi \sum (f(x)\,\Delta s)x}{2\pi \sum f(x)\,\Delta s}.$$

We note that the denominator is an approximation to the surface area of the shell and the numerator is an approximation to the first moment of the surface area. Proceeding to the limit we have the next result.

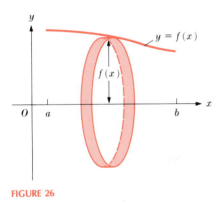

**FIGURE 26**

**THEOREM 5**　　*Let $f(x)$, $a \le x \le b$ be a nonnegative, continuous function which is revolved about the $x$ axis. Then the center of mass of the surface generated is on the $x$ axis at the point*

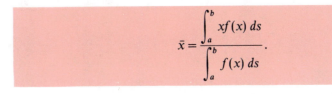

$$\bar{x} = \frac{\displaystyle\int_a^b xf(x)\,ds}{\displaystyle\int_a^b f(x)\,ds}.$$

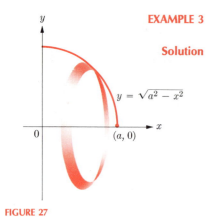

FIGURE 27

**EXAMPLE 3**  Find the center of mass of the surface of a hemisphere of radius $a$.

**Solution**  The hemisphere is obtained by revolving about the $x$ axis the quarter circle $y = \sqrt{a^2 - x^2} \equiv f(x)$, $0 \leq x \leq a$, and the center of mass is on the $x$ axis. (See Fig. 27.) To apply Theorem 5, we compute

$$\frac{dy}{dx} = \frac{-x}{\sqrt{a^2 - x^2}}, \qquad ds = \sqrt{1 + \left(\frac{dy}{dx}\right)^2}\, dx = \frac{a}{\sqrt{a^2 - x^2}}\, dx$$

and

$$\bar{x} = \frac{\displaystyle\int_0^a \left[ x\sqrt{a^2 - x^2}\, a/\sqrt{a^2 - x^2} \right] dx}{\displaystyle\int_0^a \frac{a\sqrt{a^2 - x^2}}{\sqrt{a^2 - x^2}}\, dx}.$$

Therefore

$$\bar{x} = \frac{1}{a^2}\int_0^a x\sqrt{a^2 - x^2}\, \frac{a}{\sqrt{a^2 - x^2}}\, dx = \frac{1}{2}a. \qquad \square$$

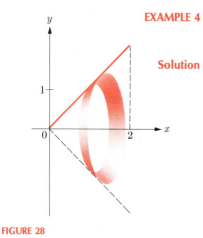

FIGURE 28

**EXAMPLE 4**  The line segment $y = x$, $0 \leq x \leq 2$, is revolved about the $x$ axis, generating the surface of a cone. Find the center of mass of this surface (Fig. 28).

**Solution**  With $f(x) = x$, we have the formula for $\bar{x}$:

$$\bar{x} = \frac{2\pi \displaystyle\int_0^2 x f(x)\, ds}{2\pi \displaystyle\int_0^2 f(x)\, ds}.$$

Therefore

$$\bar{x} = \frac{\displaystyle\int_0^2 x^2 \sqrt{2}\, dx}{2\sqrt{2}} = \frac{1}{2}\left[\frac{1}{3}x^3\right]_0^2 = \frac{4}{3}. \qquad \square$$

*Remark.*  If a function $x = g(y)$, $c \leq y \leq d$, represents a wire in the first quadrant which is revolved about the $y$ axis, then the center of mass of the surface of revolution is on the $y$ axis at the point

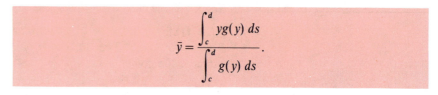

$$\bar{y} = \frac{\displaystyle\int_c^d y g(y)\, ds}{\displaystyle\int_c^d g(y)\, ds}.$$

## 4   PROBLEMS

In Problems 1 through 6, in each case find the center of mass of the given arc $C$.

1   $C = \{(x, y): 1 \le x \le 2, y = \frac{1}{8}x^4 + \frac{1}{4}x^{-2}\}$

2   $C = \{(x, y): -a \le x \le a, y = \sqrt{a^2 - x^2}\}$

3   $C = \{(x, y): 0 \le x \le a, y^{2/3} = a^{2/3} - x^{2/3}, y \ge 0\}$

4   $C = \{(x, y): 0 \le x \le 1, y = \frac{1}{2}(e^x + e^{-x})\}$

5   $C = \{(x, y): 0 \le x \le 2, y = x^2\}$

6   $C = \{(x, y): x = \frac{1}{4}y^2 - \frac{1}{2}\ln y, 1 \le y \le 2\}$

7   Find $\bar{y}$ only for the arc $C = \{(x, y): 1 \le x \le 2, y = e^x\}$.

8   Find the center of mass of the arc of the circle

$$C = \{(x, y): 0 \le x \le \tfrac{1}{2}a, y = \sqrt{a^2 - x^2}\}.$$

In each of Problems 9 through 14, an arc $C$ is revolved about the $x$ axis. Find the center of mass of the surface generated.

9   $C = \{(x, y): 0 \le x \le 4, y^2 = 4x, y \ge 0\}$

10   $C = \{(x, y): 0 \le x \le h, y = ax/h, a > 0\}$

11   $C = \{(x, y): 0 \le x \le 3, y^2 = 3 - x, y \ge 0\}$

12   $C = \{(x, y): 0 \le x \le 1, y = \frac{1}{2}(e^x + e^{-x})\}$

13   $C = \{(x, y): 0 \le x \le 3, 4x^2 + 9y^2 = 36, y \ge 0\}$

14   $C = \{(x, y): 1 \le x \le 2, y = \frac{1}{4}x^2 - \frac{1}{2}\ln x\}$

15   Let $C$ be an arc given by $y = f(x)$, $a \le x \le b$. If $C$ is revolved about the $y$ axis, find a formula for the value of $\bar{y}$ similar to that in Theorem 5.

In each of Problems 16 through 21, an arc $C$ is revolved about the $y$ axis. Find the center of mass of the surface generated.

16   $C = \{(x, y): 0 \le x \le 2, y = x^2\}$

17   $C = \{(x, y): 1 \le x \le 2, y = \frac{1}{6}x^3 + \frac{1}{2}x^{-1}\}$

18   $C = \left\{(x, y): 0 \le x \le a, y = \dfrac{a}{2}(e^{x/a} + e^{-x/a})\right\}$

19   $C = \{(x, y): x^2 = 3 - y, 1 \le y \le 2\}$

20   $C = \{(x, y): 4x^2 + y^2 = 4, 0 \le y \le 2\}$

21   $C = \{(x, y): x = \frac{1}{4}y^2 - \frac{1}{2}\ln y, 2 \le y \le 3\}$

## 5

## THEOREMS OF PAPPUS (OPTIONAL)

The two theorems given (without proof) in this section are useful tools for finding volumes of solids of revolution and areas of surfaces of revolution. They provide interesting applications of the material on finding the center of mass.

THEOREM 6   **(First Theorem of Pappus)**   *If a region R lies on one side of a line l in its plane, the volume V of the solid generated by revolving R about l is equal to the product of the area A of R and the length of the path described by the center of mass of R. That is,*

$$V = 2\pi A d$$

*where d is the distance of the center of mass of R to the line l.*

THEOREM 7 — **(Second Theorem of Pappus)** *If an arc C in a plane lies on one side of a line l in the plane, the area S of the surface generated by revolving C about l is the product of the length of C and the length of the path described by the center of mass of C. That is,*

$$S = 2\pi L d$$

*where L is the length of C and d is the distance from the center of mass of C to the line l.*

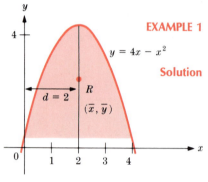

FIGURE 29

EXAMPLE 1 — Find the volume of the solid generated by revolving about the $y$ axis the region $R = \{(x, y): 0 \leq x \leq 4, 0 \leq y \leq 4x - x^2\}$. (See Fig. 29.)

Solution — Since $y = 4x - x^2$ is a curve which has the line $x = 2$ as a line of symmetry, we have $\bar{x} = 2$. The area of the region $R$ is

$$A = \int_0^4 (4x - x^2)\, dx = \tfrac{32}{3}.$$

Therefore (by Theorem 6) the volume of the solid is $2\pi A d$ with $d = 2$. We find

$$V = \frac{32}{3} \cdot 2\pi \cdot 2 = \frac{128\pi}{3}.$$   □

The above example illustrates how a volume of revolution can be found by computing the area of the region being revolved and its center of mass.

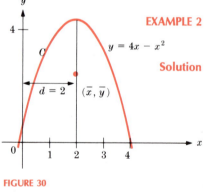

FIGURE 30

EXAMPLE 2 — The arc $C = \{(x, y): 0 \leq x \leq 4, y = 4x - x^2\}$ is revolved about the $y$ axis. (See Fig. 30.) Find the surface area generated.

Solution — By symmetry we have $\bar{x} = 2$. The length of the arc is

$$L = \int_0^4 \sqrt{1 + (4 - 2x)^2}\, dx = \frac{8\sqrt{17} + 2\ln(4 + \sqrt{17})}{4}.$$

By Theorem 7, the total surface area $S = 2\pi L d$ with $d = 2$ is

$$S = \frac{2\pi \cdot 2[8\sqrt{17} + 2\ln(4 + \sqrt{17})]}{4} = \pi[8\sqrt{17} + 2\ln(4 + \sqrt{17})].$$   □

Example 2 shows how a surface of revolution can be computed by finding the length of the arc being revolved and its center of mass.

## 5  PROBLEMS

In each of Problems 1 through 4, the region $R$ is revolved about the axis stated. Use the First Theorem of Pappus to find the volume generated.

1  $R = \{(x, y): 0 \leq x \leq 2, 0 \leq y \leq \sqrt{x}\}$; $R$ is revolved about the $x$ axis.

2  $R = \{(x, y): 0 \leq x \leq \pi, 0 \leq y \leq \sin x\}$; $R$ is revolved about the $y$ axis.

3  $R$ is the same as in Problem 2 but is revolved about the $x$ axis.

4  $R = \{(x, y): 0 \leq x \leq h, 0 \leq y \leq ax/h\}$; $R$ is revolved about the $x$ axis. (Use the result of Problem 1 of Section 3.)

5  The region of the Example in Section 3, is revolved about the $x$ axis. Find the volume generated.

**6** The region of the Example in Section 3, is revolved about the line $y = -3$. Find the volume generated.

**7** $R = \{(x, y): 0 \le x \le 1, x^3 \le y \le x\}$. $R$ is revolved about the $x$ axis. Find the volume generated.

**8** Find the volume and the surface area of a sphere of radius $a$ by using the Theorems of Pappus.

**9** Find the volume and surface area of the torus obtained by revolving about the $y$ axis the circle $(x - b)^2 + y^2 = a^2$, where $a < b$.

**10** Use the First Theorem of Pappus to find the volume of the solid generated by revolving about the $x$ axis the figure bounded by $y = x^2$ and $y = 8 - x^2$.

**11** Use the First Theorem of Pappus to find the volume of a right circular cylinder of height $h$ and radius of base $r$.

**12** Use the Second Theorem of Pappus to find the lateral surface area of a right circular cylinder of height $h$ and radius of base $r$.

---

## 6

## IMPROPER INTEGRALS

Thus far, integrals have been defined when the interval $[a, b]$ is of finite length and the function $f$ being integrated is bounded. We now extend the process of integration in two ways: (1) we allow the length of the interval of integration to become infinite, and (2) we allow the function $f$ to be unbounded. Before defining these two new types of integrals, which we denote *improper integrals*, we illustrate the concepts with several examples.

The integral of $f(x) = e^{-x}$ on an interval $[0, a]$ is obtained easily. We have

$$\int_0^a e^{-x} \, dx = -e^{-x}\big]_0^a = -e^{-a} + 1.$$

For large values of $a$, the quantity $e^{-a}$ is small, and we know (see Theorem 4, Chapter 7, page 262) that $e^{-a}$ tends to zero as $a$ tends to plus infinity. (See the sketch of $f(x) = e^{-x}$ in Fig. 31.) Letting $a$ tend to plus infinity in the above integration, we obtain

$$\lim_{a \to +\infty} \int_0^a e^{-x} \, dx = \lim_{a \to +\infty} (-e^{-a} + 1) = 0 + 1 = 1.$$

In other words, the area of the shaded region in Fig. 31 tends to the value 1 as $a$ tends to plus infinity. It is natural to define

$$\int_0^{+\infty} e^{-x} \, dx = \lim_{a \to +\infty} \int_0^a e^{-x} \, dx$$

and assign the value 1 to the area of the entire region under the curve $y = e^{-x}$ situated to the right of the $y$ axis.

As a second illustration, let us consider the integral

$$\int_1^a \frac{1}{x} \, dx = [\ln x]_1^a = \ln a,$$

where $a$ is any number larger than 1. We recall that $\ln a$ increases without

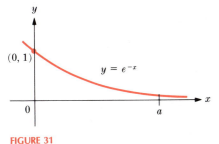

FIGURE 31

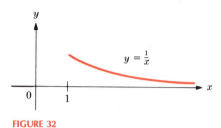

**FIGURE 32**

bound as $a$ increases without bound (see Theorem 2 in Chapter 7). Since

$$\lim_{a \to +\infty} (\ln a)$$

does not exist, we conclude that

$$\lim_{a \to +\infty} \int_1^a \frac{1}{x} dx \quad \text{does not exist.}$$

There is no way of assigning a value to the area of the shaded region under the entire curve in Fig. 32. We say that $\int_1^{+\infty} (1/x) \, dx$ does not exist.

**DEFINITIONS**    *Suppose that $f$ is continuous for all $x \geq x_0$. If*

$$\lim_{a \to +\infty} \int_{x_0}^a f(x) \, dx \quad \text{exists, we say that} \quad \int_{x_0}^{+\infty} f(x) \, dx$$

*is **convergent** to the value given by the above limit. If the limit does not exist, we say that $\int_{x_0}^{+\infty} f(x) \, dx$ is **divergent**, and the integral is not defined.*

**EXAMPLE 1**    Determine whether the following integral is convergent or divergent, and if it is convergent, determine its value:

$$\int_0^{+\infty} \frac{dx}{1+x^2}.$$

**Solution**
$$\int_0^a \frac{dx}{1+x^2} = \arctan a, \quad \lim_{a \to +\infty} (\arctan a) = \frac{\pi}{2},$$

and therefore

$$\int_0^{+\infty} \frac{dx}{1+x^2} = \frac{\pi}{2}.$$

**EXAMPLE 2**    Determine whether the following integral is convergent or divergent; if it is convergent, determine its value:

$$\int_0^{+\infty} \frac{x^2}{1+x^2} \, dx.$$

**Solution**    By division, we have

$$\int_0^a \frac{x^2}{1+x^2} \, dx = \int_0^a \left(1 - \frac{1}{1+x^2}\right) dx$$

$$= [x - \arctan x]_0^a = a - \arctan a.$$

Therefore

$$\lim_{a \to +\infty} \int_0^a \frac{x^2}{1+x^2} \, dx = \lim_{a \to +\infty} [a - \arctan a].$$

The limit on the right does not exist and the integral is divergent.

The above definitions and examples extend the definition of integral to *unbounded intervals*. That is, a function $f$ which is continuous on an interval [b, +∞) or (−∞, b] can be integrated provided a certain limiting process can be performed. It may happen, however, that the *interval of integration is of finite length*, but that the function $f$ to be integrated is unbounded somewhere on the interval of integration. For example, the integral

$$\int_1^3 \frac{1}{\sqrt{x-1}}\,dx$$

is not defined, since

$$\frac{1}{\sqrt{x-1}} \to +\infty \quad \text{as} \quad x \to 1.$$

(We recall Theorem 5 of Chapter 5, which states that if a function is integrable on an interval, then it is bounded on that interval.)

Since the function $1/\sqrt{x-1}$ is not defined for values of $x$ less than 1, we clearly intend $x \to 1$ in the above expression to mean that $x$ tends to 1 through values larger than 1. We state this fact with more precision by writing $x \to 1^+$.

The expression

$$\int_a^3 \frac{1}{\sqrt{x-1}}\,dx$$

**is** integrable for every number $a > 1$, since the integrand is bounded and continuous on the interval $[a, 3]$. We can perform the integration by the simple substitution, $u = x - 1$:

$$\int_a^3 \frac{1}{\sqrt{x-1}}\,dx = 2\sqrt{x-1}\,\Big]_a^3 = 2\sqrt{2} - 2\sqrt{a-1}.$$

Furthermore, we can perform a limiting process:

$$\lim_{a \to 1^+} \int_a^3 \frac{1}{\sqrt{x-1}}\,dx = \lim_{a \to 1^+} [2\sqrt{2} - 2\sqrt{a-1}] = 2\sqrt{2}.$$

We assign the value $2\sqrt{2}$ to the area of the shaded region in Fig. 33.

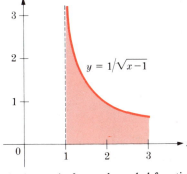

The integral of an unbounded function may have a finite value

**DEFINITIONS**    *Suppose that $f(x)$ is continuous for $a < x \le b$ but is not continuous at a. Then $\int_a^b f(x)\,dx$ is **convergent** if*

$$\lim_{\varepsilon \to 0} \int_{a+\varepsilon}^b f(x)\,dx$$

*exists, where $\varepsilon$ approaches zero through positive values. The value of $\int_a^b f(x)\,dx$ is the value of the limit. If the limit does not exist, the integral is said to be **divergent**. A similar definition is given for the case in which $f(x)$ is continuous for $a \le x < b$ but not at b.*

**EXAMPLE 3**    Determine whether

$$\int_0^1 \frac{dx}{1-x}$$

is convergent and, if so, find its value.

**Solution** The function $f(x) = 1/(1-x)$ tends to infinity as $x \to 1^-$. We let $\varepsilon$ be a positive number and consider the integral

$$\int_0^{1-\varepsilon} \frac{dx}{1-x}.$$

To integrate we set $u = 1 - x$ and $du = -dx$. Then we have

$$\int_0^{1-\varepsilon} \frac{dx}{1-x} = -\int_1^\varepsilon \frac{du}{u} = -\ln u \Big]_1^\varepsilon = -\ln \varepsilon.$$

However, $-\ln \varepsilon \to +\infty$ as $\varepsilon \to 0^+$, and the integral is divergent. □

**EXAMPLE 4** Determine whether

$$\int_0^3 \frac{dx}{(3-x)^{2/3}}$$

is convergent and, if so, find its value.

**Solution** The function $f(x) = (3-x)^{-2/3}$ tends to infinity as $x \to 3^-$. We choose $\varepsilon > 0$ and let $u = 3 - x$, $du = -dx$. Then we find

$$\int_0^{3-\varepsilon} \frac{dx}{(3-x)^{2/3}} = -\int_3^\varepsilon \frac{du}{u^{2/3}} = -3u^{1/3} \Big]_3^\varepsilon = 3\sqrt[3]{3} - 3\sqrt[3]{\varepsilon}.$$

The limit as $\varepsilon \to 0^+$ exists. We obtain

$$\int_0^3 \frac{dx}{(3-x)^{2/3}} = 3\sqrt[3]{3}$$

for the value of the convergent integral. □

It may happen that the integrand $f(x)$ is unbounded somewhere in the *interior of the interval of integration*. In such cases, the next definition is employed.

**DEFINITION** *Suppose that $f(x)$ is continuous for $a \le x \le b$, except for $x = c$, where $a < c < b$. Then we say that $\int_a^b f(x)\,dx$ is **convergent** if both of the integrals $\int_a^c f(x)\,dx$ and $\int_c^b f(x)\,dx$ are convergent. We define*

$$\int_a^b f(x)\,dx = \int_a^c f(x)\,dx + \int_c^b f(x)\,dx.$$

**EXAMPLE 5** Determine whether

$$\int_{-1}^1 \frac{1}{x^2}\,dx$$

is convergent and, if so, find its value.

**Solution** $f(x) = 1/x^2$ tends to infinity as $x \to 0$. We consider

$$\int_{-1}^{-\varepsilon} \frac{1}{x^2}\,dx \qquad \text{and} \qquad \int_\varepsilon^1 \frac{1}{x^2}\,dx, \quad \varepsilon > 0.$$

The second of these yields, upon integration,

$$\int_\varepsilon^1 \frac{1}{x^2}\, dx = \left[ -\frac{1}{x} \right]_\varepsilon^1 = \frac{1}{\varepsilon} - 1.$$

The limit as $\varepsilon \to 0$ does not exist, and the integral is divergent.   □

*Remark.*   If we had not noted that $f(x) = 1/x^2$ tends to infinity as $x \to 0$, and if we had gone naively ahead with the integration, we would have obtained

$$\int_{-1}^1 \frac{1}{x^2}\, dx = \left[ -\frac{1}{x} \right]_{-1}^1 = -1 - \left( -\frac{1}{-1} \right) = -2.$$

This result is obviously wrong, since $f(x) = 1/x^2$ is positive everywhere and could never yield a negative value for the integral.

**EXAMPLE 6**   Determine whether

$$\int_{-2}^3 x^{-1/3}\, dx$$

is convergent and, if so, find its value.

**Solution**   $f(x) = (1/x^{1/3})$ is unbounded as $x \to 0$. Therefore we evaluate

$$\int_\varepsilon^3 x^{-1/3}\, dx = \tfrac{3}{2} x^{2/3}]_\varepsilon^3 = \tfrac{3}{2}\sqrt[3]{9} - \tfrac{3}{2}\sqrt[3]{\varepsilon^2}, \quad \varepsilon > 0,$$

getting (as $\varepsilon \to 0$)

$$\int_0^3 x^{-1/3}\, dx = \tfrac{3}{2}\sqrt[3]{9}.$$

Similarly,

$$\int_{-2}^{-\varepsilon} x^{-1/3}\, dx = \tfrac{3}{2} x^{2/3}]_{-2}^{-\varepsilon} = \tfrac{3}{2}\sqrt[3]{\varepsilon^2} - \tfrac{3}{2}\sqrt[3]{4}, \quad \varepsilon > 0,$$

and (letting $\varepsilon \to 0$)

$$\int_{-2}^0 x^{-1/3}\, dx = -\tfrac{3}{2}\sqrt[3]{4}.$$

The integral is convergent, and

$$\int_{-2}^3 x^{-1/3}\, dx = \tfrac{3}{2}(\sqrt[3]{9} - \sqrt[3]{4}).$$   □

*Remarks.*   The extended integrals defined in this section are called **improper integrals**. Improper integrals may be defined, in more general cases, where $f$ may become infinite at several points in the interval of integration, which may itself be infinite. In such a case, we then divide that interval into smaller intervals of two types: In the first type, the interval is finite and $f$ becomes infinite at one end while, in the second type, $f$ remains continuous but the interval is infinite in one direction only. Then the original improper integral converges if and only if each of the component parts converges. The **value** is the sum of the values of the integrals taken over the smaller intervals.

*Operating Method.* In the evaluation of integrals the reader should, as a routine procedure, check to see if the integrand becomes infinite at either of the endpoints of the interval of integration or at any interior point. If, for example, the integrand of the integral $\int_a^b f(x)\,dx$ becomes infinite at the upper endpoint $b$, then $b$ is replaced by $b-\varepsilon$ and the integration of $\int_a^{b-\varepsilon} f(x)\,dx$ is carried out (if possible). Then we allow $\varepsilon$ to tend to zero. Similarly, if the integrand becomes infinite at $a$, the integration $\int_{a+\varepsilon}^b f(x)\,dx$ is performed and then $\varepsilon$ is allowed to tend to zero.

## 6  PROBLEMS

In Problems 1 through 38, determine whether or not each of the improper integrals is convergent, and compute its value if it is.

1. $\displaystyle\int_1^{+\infty} \frac{dx}{x^2}$

2. $\displaystyle\int_1^{+\infty} \frac{dx}{x^3}$

3. $\displaystyle\int_0^{+\infty} \frac{1}{(x+1)^{3/2}}\,dx$

4. $\displaystyle\int_1^{+\infty} \frac{1}{x^p}\,dx,\quad p>1$

5. $\displaystyle\int_1^{+\infty} \frac{1}{x}\,dx$

6. $\displaystyle\int_1^{+\infty} \frac{1}{x^p}\,dx,\quad p<1$

7. $\displaystyle\int_0^1 \frac{1}{x^p}\,dx,\quad p<1$

8. $\displaystyle\int_0^1 \frac{1}{x}\,dx$

9. $\displaystyle\int_0^1 \frac{1}{x^p}\,dx,\quad p>1$

10. $\displaystyle\int_0^{+\infty} \frac{dx}{(x^2+1)^2}$

11. $\displaystyle\int_0^1 \frac{dx}{\sqrt{1-x}}$

12. $\displaystyle\int_0^2 \frac{dx}{\sqrt{4-x^2}}$

13. $\displaystyle\int_3^4 \frac{dx}{\sqrt{x^2-9}}$

14. $\displaystyle\int_{-\infty}^0 \frac{dx}{9+4x^2}$

15. $\displaystyle\int_0^{5/4} \frac{dx}{\sqrt{25-16x^2}}$

16. $\displaystyle\int_0^1 \frac{x^2}{x^3-1}\,dx$

17. $\displaystyle\int_0^1 \frac{x}{x^3-1}\,dx$

18. $\displaystyle\int_0^1 \frac{x}{x^2-1}\,dx$

19. $\displaystyle\int_2^{+\infty} \frac{x}{x^3-1}\,dx$

20. $\displaystyle\int_2^{+\infty} \frac{x^2}{x^3-1}\,dx$

21. $\displaystyle\int_0^{+\infty} xe^{-x}\,dx^{\dagger}$

22. $\displaystyle\int_{-\infty}^0 x^2 e^x\,dx^{\dagger}$

23. $\displaystyle\int_{-8}^1 x^{-2/3}\,dx$

24. $\displaystyle\int_{-1}^1 x^{-3}\,dx$

25. $\displaystyle\int_{-2}^0 \frac{dx}{\sqrt[3]{x+1}}$

26. $\displaystyle\int_0^{+\infty} \frac{e^{-\sqrt{x}}}{\sqrt{x}}\,dx$

27. $\displaystyle\int_0^{+\infty} xe^{-x^2}\,dx$

28. $\displaystyle\int_0^{+\infty} x^p e^{-x^{p+1}}\,dx,\quad p>0$

29. $\displaystyle\int_0^{\pi/2} \cot\theta\,d\theta$

30. $\displaystyle\int_0^{\pi/4} \frac{\sec^2 x\,dx}{\sqrt{\tan x}}$

31. $\displaystyle\int_0^4 \frac{x\,dx}{\sqrt{16-x^2}}$

32. $\displaystyle\int_0^{+\infty} \frac{dx}{\sqrt{x}(x+4)}$

33. $\displaystyle\int_{-1}^1 \sqrt{1+x^{-2/3}}\,dx$

34. $\displaystyle\int_0^4 \frac{(2-x)\,dx}{\sqrt{4x-x^2}}$

35. $\displaystyle\int_0^4 \frac{dx}{\sqrt{4x-x^2}}$

36. $\displaystyle\int_{-4}^{+\infty} \frac{dx}{x\sqrt{x+4}}$

37. $\displaystyle\int_{-\infty}^{+\infty} x^2 e^{-x^3}\,dx$

38. $\displaystyle\int_{-\infty}^{+\infty} x^3 e^{-x^4}\,dx$

In Problems 39 through 42, use the substitution $t = \tan(\theta/2)$, if necessary, to evaluate each integral.

39. $\displaystyle\int_0^{\pi} \frac{\sin\theta\,d\theta}{\sqrt{1+\cos\theta}}$

40. $\displaystyle\int_0^{\pi} \frac{d\theta}{5+4\cos\theta}$

41. $\displaystyle\int_0^{\pi} \frac{d\theta}{2-\sin\theta}$

42. $\displaystyle\int_0^{\pi} \frac{\cos\theta\,d\theta}{5+4\cos\theta}$

43. The region bounded by $f(x) = 1/x$, the $x$ axis, and the line $x = 1$, and situated to the right of $x = 1$, is revolved about the $x$ axis. Evaluate the improper integral and assign a value to the volume of the solid generated. Note that the plane region does not have a finite area.

44. Rework Problem 43, with $f(x) = 1/x^p$, $p > 0$. For what values of $p$ does the solid have a finite value?

45. Let $f(x) = 3^{-n}$ for $n-1 \le x < n$, $n = 1, 2, \ldots$ . Evaluate $\int_0^{+\infty} f(x)\,dx$.

*46. Let $f(x) = (x-n+1)/n$ for $n-1 \le x < n$, $n = 1, 2, \ldots$ . Decide whether $\int_0^{+\infty} f(x)\,dx$ is convergent or divergent.

47. Let $f(x)$ be continuous and positive for $0 \le x < \infty$. Suppose there is a number $B$ such that $\int_0^a f(x)\,dx \le B$ for all $a > 0$. Show that $\int_0^{+\infty} f(x)\,dx$ is convergent and that its value must be less than $B$.

48. Suppose that $f$ and $g$ are continuous, positive functions for $0 \le x < \infty$ and that $\int_0^{+\infty} g(x)\,dx$ is divergent. If $f(x) \ge g(x)$ for all $x$, show that $\int_0^{+\infty} f(x)\,dx$ is divergent. (Assume the contrary and use the result of Problem 47.)

---

$^{\dagger}$ $\displaystyle\lim_{h\to+\infty} h^n e^{-bh} = 0$ for any $n$ if $b > 0$.

_____ 7 _____

## WORK (OPTIONAL)*

We shall consider only *the motion of objects along a straight line in one direction*. It is convenient to suppose that the motion is along the $x$ axis or the $y$ axis in the positive direction. The restriction to motions along a line is essential, since the study of motion along curved paths requires types of derivatives and integrals of a nature more complicated than those we have taken up so far.

The term *work* is a technical one which requires for its definition the notion of **force**. Using mass, distance, and time as the basic *undefined terms*, we know from Newton's law that

$$\text{force} = \text{mass} \times \text{acceleration}.$$

Force is measured in units of dynes, newtons, etc. To take the simplest possible case, suppose that an object (assumed to occupy a single point) is moving along the $x$ axis (to the right) and suppose that this object is subject to a *constant force* (to the right) of $B$ newtons. Let $d$ be the distance it moves, measured in meters.

---

**DEFINITION**   *The* **work** *$W$ done when a force of $B$ newtons moves an object a distance of $d$ meters is*

$$W = B \cdot d.$$

---

The units of $W$ are measured in joules or newton-meters. When the force is in dynes and the distance in centimeters, the work is measured in ergs. The definition above is fine so long as the force exerted is constant, but what is to be the definition when the force is variable (the most common situation)? As a start toward the definition we state the following principle:

---

**PRINCIPLE 1**   *If a body is moved from position $A_0$ to position $A_1$, then from $A_1$ to $A_2$, then from $A_2$ to $A_3$, etc., and finally from $A_{n-1}$ to $A_n$, the total work done in moving the body from $A_0$ to $A_n$ is the sum of the amounts of work done in moving the body from one position to the next.*

---

If the force exerted in moving a body from one position to another is constant, the work is known; otherwise it is not defined as yet. If the positions are sufficiently close together and if the force acting is continuous, then the force is *almost constant* on each short interval. Suppose that the body is moving along the $x$ axis from $a$ to $b$ ($b > a$) and suppose that there is a law

---

*Readers not interested in the physical applications may skip Sections 7 and 8 without loss of continuity in the development of either the theory or the techniques of calculus.

$F(x)$ which gives the amount of force (exerted to the right) at point $x$. According to Principle 1, if we subdivide the interval $[a, b]$ into $n$ parts: $\{a = x_0 < x_1 < \cdots < x_n = b\}$, and if we *assume that the force is constant* in each subinterval, we can compute the total work in moving the body from point $a$ to point $b$. Suppose that on the subinterval $[x_{i-1}, x_i]$ the force has a constant value equal to $F(\xi_i)$, where $\xi_i$ is some number in the interval $[x_{i-1}, x_i]$. Then, writing $\Delta_i x = x_i - x_{i-1}$, the *total work done* is

$$\sum_{i=1}^{n} F(\xi_i)\, \Delta_i x.$$

This expression which is a Riemann sum, is suggestive of the way we define the integral and leads to the following definition of work.

**DEFINITION**     **The total work** $W$ *done in moving an object along the x axis from a to b, if the force exerted at point x obeys the law $F = F(x)$, is*

$$W = \int_a^b F(x)\, dx.$$

**EXAMPLE 1**     Suppose that an object is moved along the $x$ axis from $x = 2$ to $x = 5$ (units in centimeters), and suppose that the force exerted obeys the law $F(x) = x^2 + x$ dynes. Find the total work done.

**Solution**
$$W = \int_2^5 (x^2 + x)\, dx = \tfrac{1}{3}x^3 + \tfrac{1}{2}x^2\big]_2^5 = 49\tfrac{1}{2} \text{ ergs.} \qquad \square$$

There are situations in which the laws of physics help us get expressions for the force $F(x)$ so that the total work can be calculated. One such case is **Hooke's Law**. Suppose a coil spring is stretched or compressed a distance $x$. Then there will be a force $F$ which tends to restore the spring to its original, natural position. Hooke's Law states that

$$F = kx,$$

where $k$ is a constant. That is, the restoring force of a spring is proportional to the distance the spring is compressed or stretched. The constant $k$ will vary from spring to spring. We illustrate with an example.

**EXAMPLE 2**     A spring has a natural length of 12 cm. When it is stretched $x$ cm Hooke's Law states that it pulls back with a force equal to $kx$, where $k$ is a constant. The value of $k$ depends on the material, thickness of the wire, temperature, how tightly coiled the spring is, etc. If 10 dynes of force are required to hold it stretched $\tfrac{1}{2}$ cm, how much work is done in stretching it from its natural length to a length of 16 cm?

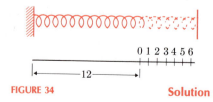

0 1 2 3 4 5 6

|←——12——→|

**FIGURE 34**

**Solution**     We locate the spring along the $x$ axis and place the origin at the position where we start stretching (Fig. 34). The fact that the force is 10 dynes when $x = \tfrac{1}{2}$ is used to find $k$. We have $10 = k \cdot \tfrac{1}{2}$ or $k = 20$. From Hooke's Law, we

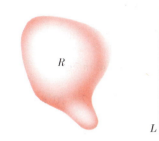

FIGURE 35

may write $F = 20x$. From the definition of work,

$$W = \int_0^4 20x \, dx = 10x^2]_0^4 = 160 \text{ ergs.} \qquad \square$$

Another type of work problem requires a somewhat different principle. Figure 35 represents a substance which occupies a large region $R$. Suppose the material in this region (sand, water, etc.) is to be moved to some location $L$. The following principle tells how to find the work done in moving this mass. It is supposed that each particle in $R$ moves until it reaches the location $L$.

**PRINCIPLE 2**  *If a mass $M$ is moved to a position $L$, the total work done may be obtained by (i) subdividing the mass into smaller masses*

$$M_1 + M_2 + \cdots + M_n = M,$$

*by (ii) finding the work done in moving each of the smaller masses to the position $L$, and by (iii) adding the results.*

Principle 2 is used as a guide for setting up an integral for the total work done in the same way that Principle 1 served as a guide for defining work. We illustrate the method in the following example.

**EXAMPLE 3**  A tank having the shape of a right circular cylinder of altitude 8 meters and radius of base 5 meters is full of water. Find the amount of work done in pumping all the water in the tank up to a level 6 meters above the top of the tank.

**Solution**  Figure 36 shows the tank. Since the motion is in a vertical direction we set up the $x$ axis so that it is vertically downward, as shown. According to Principle 2, we divide the tank into $n$ portions, in the form of slabs or disks. That is, we divide the $x$ axis into $n$ parts,

$$0 = x_0 < x_1 < x_2 < \cdots < x_n = 8,$$

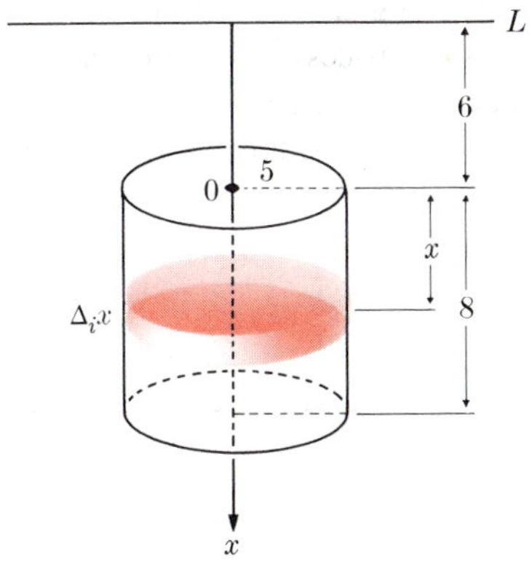

and consider the $i$th disk to be of thickness $\Delta_i x = x_i - x_{i-1}$. Then the total work done is the work required to move each disk up to line $L$, 6 meters above the top. The only force exerted on the water is that due to gravity. By definition, *the weight of any object is the measure of the force of gravity.* Since water has a density of 1000 kg/m$^3$ and the effect $g$ of gravity is approximately $g = 10$ m/sec$^2$, raising one cubic meter of water requires a force of 10,000 newtons. (A newton is 1 kg-m/sec$^2$.)

The volume of the disk of width $\Delta_i x$ is $\pi \cdot 5^2 \cdot \Delta_i x = 25\pi \, \Delta_i x$. The force required to raise this volume is its weight $(10,000)(25\pi \, \Delta_i x)$. Since the force is *constant*, the work done is merely the product of the force and the distance that it moves. Here is the real crux of the problem: Each disk moves a distance which is different from that of any other disk. In fact, even within one disk different particles are raised different distances.

A disk with midpoint at location $x$ meters below the top of the tank moves *approximately* $(6 + x)$ meters. Since $x$ is between $x_{i-1}$ and $x_i$, the work done moving one disk is approximately $10,000(25\pi \, \Delta_i x)(6 + x)$. We now recall that Principle 2 says that we may add the work done in moving each of these

FIGURE 36

disks to find the total work, and so (approximately)

$$W = \sum_{i=1}^{n} 10{,}000(25\pi \, \Delta_i x)(6 + x), \quad x \text{ between } x_{i-1} \text{ and } x_i.$$

When we proceed to the limit, this formula suggests the integral

$$W = \int_0^8 10{,}000(25\pi)(6 + x)\, dx = 250{,}000\pi[6x + \tfrac{1}{2}x^2]_0^8 = 2 \times 10^7 \pi \text{ joules.}$$

Such an integral is used to define the work done in moving a large mass. Its evaluation yields the total work performed.

GENERAL PROCEDURE

1. *A physical law is given in the simplest case. (Work is defined when the force is constant and the object occupies a single point.)*
2. *A principle is stated which says that a complicated situation may be decomposed into a large number of simple ones; each of these may be solved by the simple physical law (1), and then the results may be added.*
3. *The sum obtained in (2) is a Riemann sum of the form used in defining an integral.*
4. *We proceed to the limit, obtain the integral, and use this expression as the* **definition** *of the physical law in the complicated case.*

Now it is plain that the reader cannot be expected to be able to master this procedure immediately when confronted with an unfamiliar physical law. However, practice in carrying out the process in familiar situations makes possible an understanding of the physics, the definition, and the integral concept. Invention is a modification of experience. Once the reader has a backlog of experience in solving problems involving familiar situations, he or she is capable of attacking an unfamiliar physical process right from the start.

With this in mind, the reader will appreciate the benefits of mastering the exercises in this and the following section. Additional material of the same kind is given in subsequent chapters.

## 7  PROBLEMS

In Problems 1 through 14, a particle is moving along the $x$ axis from $a$ to $b$ according to the force law, $F$, which is given. Find the work done.

1  $F(x) = x^3 - 2x^2 + 6x + 4, \quad a = 0, b = 3$

2  $F(x) = x^4 + 2x + 7, \quad a = -1, b = 2$

3  $F(x) = x\sqrt{1 + x^2}, \quad a = 1, b = 3$

4  $F(x) = (x^2 + 2x + 1)^3(x + 1), \quad a = 0, b = 1$

5  $F(x) = (x + 3)^{-4}, \quad a = -1, b = 1$

6  $F(x) = x^3 + x, \quad a = -2, b = 2$. Interpret the result physically in this problem.

7  $F(x) = e^{2x}, \quad a = 0, b = 3.$

8  $F(x) = \dfrac{1}{1 + x^2}, \quad a = 0, b = 1$

9  $F(x) = \sin 3x, \quad a = 0, b = \pi/3$

10  $F(x) = x \cos x, \quad a = 0, b = \pi/2$

11  $F(x) = \dfrac{x}{\sqrt{1 - x^2}}, \quad a = -\dfrac{1}{2}, b = \dfrac{1}{2}$

12  $F(x) = \dfrac{x}{x^3 + 1}, \quad a = 0, b = 2$

13  $F(x) = xe^{-x^2}, \quad a = 1, b = 3$

14  $F(x) = xe^{-2x}, \quad a = 0, b = 2$

In Problems 15 through 19, assume that each spring obeys Hooke's Law ($F = kx$) as stated in Example 2. Find the work done in each case.

15  Natural length, 8 cm; 20 dynes force stretches spring $\frac{1}{2}$ cm. Find the work done in stretching the spring from 8 to 11 cm.

16  Natural length 6 cm; 500 dynes stretches spring $\frac{1}{4}$ cm. Find the work done in stretching it 1 cm.

17  Natural length 10 cm; 30 dynes stretches it to $11\frac{1}{2}$ cm. Find the work done in stretching it (a) from 10 to 12 cm; (b) from 12 to 14 cm.

18  Natural length 6 cm; 12,000 dynes compresses it $\frac{1}{2}$ cm. Find the work done in compressing it from 6 cm to 5 cm. (Hooke's Law works for compression as well as for extension.)

19  Natural length 6 cm; 1200 dynes compresses it $\frac{1}{2}$ cm. Find the work done in compressing it from 6 cm to $4\frac{1}{2}$ cm. What is the work required to bring the spring to 9 cm from its compressed state of $4\frac{1}{2}$ cm?

20  A cable 100 meters long and weighing 5 kg/m is hanging from a winch. Find the work done in winding it up.

21  A boat is anchored so that the anchor is 100 meters directly below the post about which the anchor chain is wound. The anchor weighs 3000 kg and the chain weighs 20 kg/m. How much work is done in bringing up the anchor?

22  A tank full of water is in the form of a right circular cylinder of altitude 5 m and radius of base 3 m. How much work is done in pumping the water up to a level 10 m above the top of the tank?

23  A swimming pool full of water is in the form of a rectangular parallelepiped 5 m deep, 15 m wide, and 25 m long. Find the work required to pump the water up to a level 1 m above the surface of the pool.

24  The base of one cylindrical tank of radius 6 m and altitude 10 m is 3 m above the top of another tank of the same size and shape. The lower tank is full of oil (900 kg/m$^3$) and the upper tank is empty. How much work is done in pumping all the oil from the lower tank into the upper through a pipe which enters the upper tank through its base?

25  A trough full of water is 10 m long, and its cross section is in the shape of an isosceles triangle 2 m wide across the top and 2 m high. How much work is done in pumping all the water out of the trough over one of its ends?

26  A trough full of water is 12 m long and its cross section is in the shape of an isosceles trapezoid 1 m high, 2 m wide at the bottom, and 3 m wide at the top. Find the work done in pumping the water to a point 2 m above the top of the trough.

27  A trough full of water is 6 m long and its cross section is in the shape of a semicircle with the diameter at the top 2 m wide. How much work is required to pump the water out over one of its ends?

28  A steam shovel is excavating sand. Each load weighs 500 kg when excavated. The shovel lifts each load to a height of 15 m in $\frac{1}{2}$ min, then dumps it. A leak in the shovel lets sand drop out while the shovel is being raised. The rate at which sand leaks out is 160 kg/min; find the amount of work done by the shovel in raising one load.

29  Two unlike charges of amount $e_1$ and $e_2$ electrostatic units attract each other with a force $e_1e_2/r^2$, where $r$ is the distance between them (assuming appropriate units). A positive charge $e_1$ of 7 units is held fixed at a certain point $P$. (See Fig. 37.) How much work is done in moving a negative charge $e_2$ of 1 unit from a point 4 units away from the positive charge to a point 20 units away from that charge along a straight line pointing directly away from the positive charge?

Charge $= +7$

$P$   4   Charge $= -1$   20   $r$   **FIGURE 37**

30  The base of a tank of radius 5 m and altitude 8 m is directly above a tank of the same radius but of altitude 6 m. The lower tank is full of water. How much work is done in pumping the water from the lower tank to the upper tank if the pipe enters the upper tank halfway up its side?

31  A metal bar of length $L$ and cross section $A$ obeys Hooke's Law when stretched. If it is stretched $x$ units, the force $F$ required is given by

$$F = \frac{EA}{L}x,$$

where $E$ is a constant known as *Young's modulus* or the *modulus of elasticity*. The quantity $E$ depends only on the type of material used, not on the shape or size of the bar. Given that a steel bar 12 cm long and of uniform cross section 4 cm$^2$ is stretched $\frac{1}{2}$ cm, find the work done. (Give answer in terms of $E$.)

_____8

## FLUID PRESSURE (OPTIONAL)*

The definition of pressure, like the definition of work, depends on the notion of force. In the simplest case, suppose that a fluid (liquid or gas) is in a container and, at some point in the container, a small flat plate is inserted. The mere presence of the fluid exerts a force on this plate both from the weight of the fluid and from the action of the molecules. The **pressure** is defined as the *force per unit area* of the fluid on the plate.

When a liquid is in an open container so that there is a *free surface* at the top, the pressure at any depth of the liquid is given by a particularly simple law. Suppose that the liquid weighs $w$ kg/m³ (i.e., its density is $w$). **The pressure $p$ at a depth $h$ meters below the surface is**

$$p = w \cdot h.$$

This is a remarkable formula in many ways; it says that the size of the vessel is irrelevant. At a depth of 3 meters in a swimming pool the pressure is the same as it is at a depth of 3 meters in an ocean (assuming that the pool is filled with salt water). The pressure is determined by depth alone, all other dimensions of the vessel being of no consequence.

The following principle is also needed in the study of liquid pressure:

PRINCIPLE 1    *At any point in a liquid the pressure is the same in all directions.*

This means that a plate below the surface has the same pressure on it whether it is located vertically, horizontally, or at an angle. For example, any skin diver knows that the pressure on his eardrums depends only on how deep he is and not on the angle at which his head happens to be tilted.

The problem we shall concern ourselves with is the determination of the *total force* on a plate situated in a fluid. In the simplest case, if the plate has area $A$ m² and if it is at a depth where the pressure is $p$ kg/m², then the **total force $F$** is given by

$$F = p \cdot A.$$

If the density of the liquid is $w$ and the plate is $h$ ft below the surface, we may also write

$$F = w \cdot h \cdot A,$$

since $p = w \cdot h$.

If a large plate is located in a fluid either vertically or at an angle with the horizontal, then different portions of it are at different depths. The elementary rule given above for finding the force will not work, since we do not know

_____
*See the footnote at the beginning of Section 7.

what depth $h$ to use. To handle problems of this sort we use the following principle:

---

**PRINCIPLE 2**    *If a flat plate is divided into several parts, then the total force on the entire plate is the sum of the forces on each of the parts.*

---

Note the similarity between this principle and those in the previous section on work. We shall now show how Principles 1 and 2 lead to an integral which will define the total force on a plate in a liquid. A specific example will be worked first.

**EXAMPLE 1**    The face of a dam adjacent to the water is vertical and rectangular in shape with width 30 m and height 10 m. Find the total force exerted by the liquid on the face of the dam when the water surface is level with the top of the dam.

**Solution**    The depth of water varies from 0 to 10 m and the pressure changes with depth. We divide the dam into horizontal strips $0 = h_0 < h_1 < h_2 < \cdots < h_n = 10$. (See Fig. 38.) If the number of subdivisions is large, then the pressure on a strip such as the $i$th one will be *almost* constant, since all of its points are at approximately the same depth. The width of the strip is

$$\Delta_i h = h_i - h_{i-1}.$$

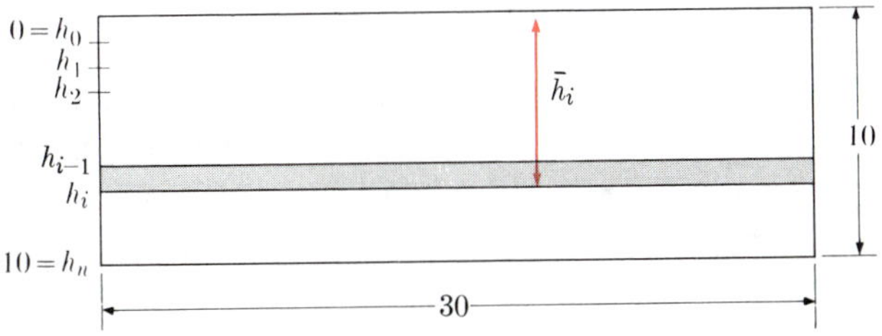

**FIGURE 38**

If $\bar{h}_i$ is an average value between $h_i$ and $h_{i-1}$, then the force on the $i$th strip is the area $(= 30\,\Delta_i h)$ times the pressure $(= w\bar{h}_i)$. We have

$$\text{Force on the } i\text{th strip} = 30w\bar{h}_i\,\Delta_i h.$$

According to Principle 2, we obtain the total force by adding the forces on the individual strips:

$$\text{Total force} = \sum_{i=1}^{n} 30w\bar{h}_i\,\Delta_i h.$$

This expression is a Riemann sum suggestive of the integral

$$\int_0^{10} 30wh\,dh,$$

which is the *definition of the* **total force** on the dam. Therefore, by integration,

$$\text{Total force} = 30w\tfrac{1}{2}h^2\big]_0^{10} = 1500w.$$

The technique for working Example 1 follows the outline given at the end of the previous section. The four steps in the general procedure are:

**GENERAL PROCEDURE**

1. *In the simple case of constant pressure, force = pressure × area. (We know additionally that, for a liquid, pressure = weight × depth.)*
2. *A principle is given which says that the total force is the sum of the forces on the various parts.*
3. *The sum obtained is of the type used in defining integrals.*
4. *We proceed to the limit to obtain an integral defining the total force.*

The preceding example suggests the formula

$$F = \int_a^b w p(h)\, dh,$$

where $w$ is the density of the fluid, $p(h)$ is the pressure as a function of the height $h$, and $a$, $b$ are the lower and upper limits, respectively, of the region in which the pressure is exerted.

Another example illustrates the procedure in a slightly more complicated case.

**EXAMPLE 2**    The face of a dam adjacent to the water has the shape of an isosceles trapezoid of altitude 20 m, upper base 50 m, and lower base 40 m. Find the total force exerted by the water on the dam when the water is 15 m deep.

**Solution**    Figure 39 shows the dam and a typical very narrow strip along which the pressure is almost constant. The width of the strip is $\Delta_i h = h_i - h_{i-1}$. The force on this strip is

$$\text{Force on } i\text{th strip} = w \cdot \bar{h}_i \cdot \text{area},$$

where $\bar{h}_i$ is between $h_{i-1}$ and $h_i$. The main problem is to find the area of the strip. For practical purposes we suppose it to be a rectangle. Then the area is (length) × ($\Delta_i h$). But what is the length $l_i$ as a function of $h$? Here the shape of the dam plays a role. In order to find $l_i$ *as a function of the depth* $h$, we use the method of similar triangles (Fig. 40). From triangles *ABC* and *DEC* we have

$$\frac{l_i - 40}{15 - \bar{h}_i} = \frac{10}{20} \qquad \text{or} \qquad l_i = \frac{95}{2} - \frac{1}{2}\bar{h}_i,$$

which is the length of the $i$th strip. Its area is

$$\left(\tfrac{95}{2} - \tfrac{1}{2}\bar{h}_i\right) \Delta_i h,$$

and the total force on the $i$th strip is (approximately)

$$w\bar{h}_i\left(\tfrac{95}{2} - \tfrac{1}{2}\bar{h}_i\right) \Delta_i h.$$

The total force on the dam, according to Principle 2, is (approximately)

$$\tfrac{1}{2} \sum_{i=1}^n w\bar{h}_i(95 - \bar{h}_i)\, \Delta_i h.$$

We proceed to the limit and obtain

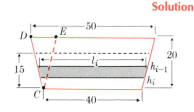

**FIGURE 39**

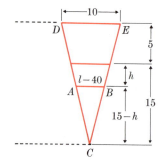

**FIGURE 40**

$$\text{Total force} = \int_0^{15} \frac{1}{2} wh(95 - h)\, dh = w\left[\frac{95}{4}h^2 - \frac{1}{6}h^3\right]_0^{15} = \frac{19{,}125}{4}w.$$

□

The essential step in Example 2 is finding the area of the $i$th strip in terms of the depth $h$, which is necessary in order that the sum we obtain be of the type used in approximating an integral.

## 8   PROBLEMS

In Problems 1 through 10, the face of a dam adjacent to the water is vertical and has the shape indicated. Find the total force on the dam due to fluid pressure.

**1** A rectangle 200 m wide, 15 m high; water 10 m deep.

**2** A rectangle 150 m wide, 12 m high; water 8 m deep.

**3** An isosceles triangle 30 m wide at the top, 20 m high in the center; dam full of water.

**4** An isosceles triangle 30 m wide at the top, 15 m high in the center; water 10 m deep in center.

**5** An isosceles triangle 250 m across the top, 100 m high in the center; water 60 m deep in center.

**6** A right triangle, one leg horizontal (at top) and 10 m wide, one leg vertical and 20 m high; dam full of water.

**7** An isosceles trapezoid 90 m across the top, 60 m wide at the bottom, 20 m high; dam full of water.

**8** An isosceles trapezoid 60 m across the top, 30 m wide at the bottom, 10 m high; water 8 m deep.

**9** A region bounded by the curve

$$y = 10(1 - \cos x), \quad -\frac{\pi}{2} \le x \le \frac{\pi}{2}$$

and the line $y = 10$; water 8 m deep.

**10** A region bounded by the curve

$$y = 15(1 - \sin 2x), \quad 0 \le x \le \frac{\pi}{2}$$

and the line $y = 15$; water 12 m deep.

**11** A square of side 5 m is submerged in water so that its diagonal is perpendicular to the surface. Find the total force on the square (due to the liquid on one side) if one vertex is at the surface.

**12** The vertical ends of a trough are in the shapes of semicircles of diameter 2 m with diameter horizontal. Find the total force on one end when the trough is full of water.

**13** An oil tank is in the shape of a right circular cylinder of diameter 4 m with axis horizontal. Find the total force on one end when the tank is half full of oil weighing 900 kg/m³.

In Problems 14 through 20, it is assumed that in a dam the face of a gate of the shape indicated is vertical. If the gate is adjacent to the water, find the total force on the gate in each case.

**14** A rectangle 4 m wide, 3 m high, with upper edge 20 m below the surface of the water.

**15** An isosceles triangle 4 m wide at the top, 3 m high, with upper edge 15 m below the water surface.

**16** An isosceles triangle 2 m wide at the bottom, 2 m high, with top vertex 10 m below the surface of the water.

**17** An isosceles trapezoid 3 m wide at the top, 4 m wide at the bottom, and 3 m high, with upper base 20 m below the water surface.

**18** An isosceles right triangle, length of leg 5 m, with one leg vertical and the horizontal leg at the bottom. The horizontal leg is 30 m below the water surface.

**19** A region bounded by the semicircle $y = -\sqrt{16 - x^2}$, $-4 \le x \le 4$ and the line $y = 0$. The top of the region is 10 m below the surface of the water.

**20** A region bounded by the curve $y = 5(1 - \cos \frac{1}{2}x)$, $-\pi \le x \le \pi$ and the line $y = 5$. The top of the region is 8 m below the surface of the water.

In Problems 21 through 23, it is assumed that the face of the dam adjacent to the water is a plane figure of the given shape and is inclined at an angle of 30° from the vertical. Find the total force on that face in each case.

**21** A rectangle 50 m wide with slant height 30 m; dam full of water.

**22** An isosceles trapezoid 60 m wide at the top, 40 m wide at the bottom, slant height 20 m; dam full of water.

**23** An isosceles trapezoid 200 m wide at the top, 120 m wide at the bottom, slant height 50 m; water $20\sqrt{3}$ m deep.

**24** A swimming pool is 25 m wide, 40 m long, 2 m deep at one end, and 8 m deep at the other, the bottom being an inclined plane. Find the total force on the bottom.

25  Consider a gas confined to a right circular cylinder closed at one end by a movable piston. Assume that the pressure of the gas is $p$ newtons/m$^2$ when the volume is $v$ m$^3$. The pressure depends on the volume and we write $p = p(v)$. Show that the *work done* by the gas in expanding from a volume $v_1$ to a volume $v_2$ may be defined by

$$\int_{v_1}^{v_2} p(v)\, dv \text{ joules.}$$

(*Hint:* Let the area of the cross section be $A$ m$^2$, and let $x$ be the variable distance from the piston to the fixed end of the cylinder.)

26  Suppose that $v$ and $p$ are related by the equation $p \cdot v^{1.4} =$ const. Find the work done by such a gas as it expands from a volume of 5 m$^3$ to 200 m$^3$ if the pressure is 2500 nt/m$^2$ when $v = 5$ m$^3$. (See Problem 25.)

27  Suppose that $v$ and $p$ are related by the equation $p \cdot v^{1.4} =$ const. If $p = 60$ nt/m$^2$ when $v = 10$ m$^3$, (a) find $v$ when $p = 15$ nt/m$^2$, and (b) find the work done by the gas as it expands until the pressure reaches this value (15 nt/m$^2$). (See Problem 25.)

## 9

## NUMERICAL INTEGRATION

We saw that many geometrical and physical quantities can be found by evaluating definite integrals. Unfortunately, in most cases of integration which occur in practice, the integrals cannot be found exactly. For example, integrals which have a simple appearance, such as

$$\int \sqrt{1 + x^3}\, dx, \qquad \int \frac{\sin x}{x}\, dx, \qquad \int e^{-x^2}\, dx$$

cannot be integrated precisely; in fact it can be shown that there is no method by which the antiderivatives of the above integrals can be expressed in terms of the usual elementary functions. For this reason it is important to have techniques for finding the approximate value of a definite integral. We already know that one such approximation is the Riemann sum. In fact, the limit of the Riemann sum defines the integral. Using the symbol "$\approx$" to indicate that two quantities are approximately equal, we can write

$$\int_a^b f(x)\, dx \approx \sum_{i=1}^{n} f(\xi_i)\, \Delta_i x, \quad x_{i-1} \le \xi_i \le x_i, \quad \Delta_i x = x_i - x_{i-1},$$

whenever $f$ is a continuous function and the norm $\|\Delta\|$ of the subdivision is sufficiently small. Therefore whenever we need to find the numerical value of a particular definite integral we can use the above Riemann sum as an approximation. However, there are better methods, which yield closer approximations with fewer computations. No one method is best; how good a method is depends both on the function being integrated and the interval along which the integration is performed. The first such method we establish is known as the **Trapezoidal Rule**.

THEOREM 8      **(The Trapezoidal Rule)**    *Suppose that $f(x)$ is continuous for $a \leq x \leq b$, and that $\{a = x_0 < x_1 < x_2 < \cdots < x_{n-1} < x_n = b\}$ is a subdivision of the interval $[a, b]$ into $n$ equal subintervals of length $h = (b - a)/n$. Then*

$$\int_a^b f(x)\, dx \approx \frac{h}{2}[f(x_0) + 2f(x_1) + 2f(x_2) + \cdots + 2f(x_{n-1}) + f(x_n)].$$

**Proof**    We consider the interval $[a, b]$ subdivided as in the statement of the theorem and approximate the integral by a Riemann sum, using for each $\xi_i$ the left endpoint of the subinterval $[x_{i-1}, x_i]$. We find

$$\int_a^b f(x)\, dx \approx \sum_{i=1}^n f(x_{i-1})h = h[f(x_0) + f(x_1) + \cdots + f(x_{n-1})].$$

We can also approximate the integral by a Riemann sum in which we take for $\xi_i$ the right endpoint of each subinterval. We obtain

$$\int_a^b f(x)\, dx \approx \sum_{i=1}^n f(x_i)h = h[f(x_1) + f(x_2) + \cdots + f(x_n)].$$

Each of the two above approximations tends to the integral as $n \to \infty$. Consequently, if we take the average of these two sums we get

$$\int_a^b f(x)\, dx \approx \frac{h}{2}[f(x_0) + 2f(x_1) + \cdots + 2f(x_{n-1}) + f(x_n)],$$

and this approximation tends to the integral as $n \to \infty$. $\quad\square$

In case $f(x) \geq 0$ we have a geometrical interpretation of the Trapezoidal Rule. As shown in Fig. 41 we approximate the area under $y = f(x)$ by inscribed trapezoids. The area of a typical trapezoid is $\frac{1}{2}h[f(x_{i-1}) + f(x_i)]$. Since each base of a trapezoid occurs twice except for the first and last, the Trapezoidal Rule follows.

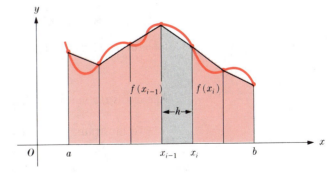

**FIGURE 41**

EXAMPLE 1    Compute the approximate value of

$$\int_0^1 \frac{dx}{1 + x^2}$$

using the Trapezoidal Rule with $n = 5$. Use four decimal places in the computation and check the result by integrating and finding the exact value.

**Solution**     We have $(b-a)/n = (1-0)/5 = 0.2 = h$. With $f(x) = 1/(1+x^2)$, we use a calculator to form the table:

| $x_i$ | 0 | 0.2 | 0.4 | 0.6 | 0.8 | 1.0 |
|---|---|---|---|---|---|---|
| $f(x_i)$ | 1.0000 | 0.9615 | 0.8620 | 0.7353 | 0.6097 | 0.5000 |

Then

$$\frac{h}{2}[f(0) + 2f(0.2) + 2f(0.4) + 2f(0.6) + 2f(0.8) + f(1.0)] = 0.7837.$$

The exact value is obtained from

$$\int_0^1 \frac{dx}{1+x^2} = \arctan x \Big]_0^1 = \frac{\pi}{4} = 0.7854^-.$$

The error in the approximation is 0.0017 or 17 parts in ten thousand.    □

Those readers accustomed to working with computers will observe that the numerical techniques we discuss here are well adapted for programming on a computer or a sophisticated hand calculator. In fact, there exist many programs for evaluating definite integrals, some of which have been incorporated into the more technically oriented hand calculators.

In most problems involving numerical integration there is no way of checking the accuracy of the approximation against the true value of the integral. However, if the function $f$ to be integrated is sufficiently smooth that its second derivative can be calculated, then *it is possible to find out how close the approximation is to the true value.* The absolute value of the difference between the true value and the Trapezoidal approximation is called the estimated error and denoted by $E_T$. We state without proof the next result.

---

**THEOREM 9**     **(Estimate of Error in the Trapezoidal Rule)** *Suppose that $f$ has two derivatives for $a \le x \le b$, and $M$ is a number such that $|f''(x)| \le M$ for $a \le x \le b$. Then the error $E_T$ in using the Trapezoidal Rule to approximate the integral $\int_a^b f(x)\, dx$ satisfies*

$$E_T \le \tfrac{1}{12} M(b-a)h^2,$$

*where $n$ is the number of equal subintervals and $h = (b-a)/n$.*

---

To illustrate the use of Theorem 9 we use the formula in Example 1. We have

$$f(x) = \frac{1}{1+x^2} \qquad \text{and} \qquad f''(x) = \frac{6x^2 - 2}{(1+x^2)^3}.$$

Hence, for $0 \le x \le 1$, we may use the estimate $|f''(x)| \le \frac{4}{1} = M$. Therefore

$$E_T = (\tfrac{1}{12})4(0.2)^2 = 0.01333.$$

Thus before knowing the exact value of the integral in Example 1 we can deduce that the Trapezoidal approximation is within 14 parts in a thousand. It is clear that as the number of subintervals increases the value of $E_T$

decreases since the value of $h^2$ diminishes. Taking $h$ very small leads to accurate approximations of the integral.

We now establish a second formula for approximating definite integrals, one which complements the Trapezoidal Rule. Known as **Simpson's Rule** it frequently yields more accurate approximations with fewer computations than is required by the Trapezoidal Rule. In order to derive Simpson's Rule we first establish the Prismoidal Formula.

---

**THEOREM 10**     **(Prismoidal Formula)**   *If $f(x)$ is a polynomial of degree three or less, then we have the exact formula*

$$\int_a^{a+2h} f(x)\,dx = \frac{h}{3}[f(a) + 4f(a+h) + f(a+2h)].$$

---

**Proof**    Let $f(x)$ be any polynomial of degree three or less, say

$$f(x) = a + bx + cx^2 + dx^3.$$

When we make the substitution $x = a + h + u$, we have

$$f(a+h+u) = g(u),$$

where $g(u)$ is still a polynomial of the same degree as $f$. We write

$$g(u) = A + Bu + Cu^2 + Du^3.$$

Then

$$\int_a^{a+2h} f(x)\,dx = \int_{-h}^{h} g(u)\,du = Au + \frac{Bu^2}{2} + \frac{Cu^3}{3} + \frac{Du^4}{4}\bigg]_{-h}^{h}$$

$$= 2Ah + \frac{2Ch^3}{3}.$$

From the expression for $g(u)$, we have

$$g(0) = A, \qquad g(h) + g(-h) = 2A + 2Ch^2,$$

and therefore

$$C = \frac{g(h) + g(-h) - 2g(0)}{2h^2}.$$

We conclude that

$$\int_a^{a+2h} f(x)\,dx = 2g(0)h + \frac{h}{3}[g(h) + g(-h) - 2g(0)].$$

We now have $g(0) = f(a+h), g(h) = f(a+2h), g(-h) = f(a)$ and, substituting these values in the last equation, we obtain the Prismoidal Formula.    □

**EXAMPLE 2**    Evaluate $\int_1^3 x^3\,dx$, using the Prismoidal Formula. Check by integration.

**Solution**    $a = 1, a + 2h = 3, h = 1, a + h = 2$. Therefore

$$\int_1^3 x^3\,dx = \tfrac{1}{3}(1^3 + 4 \cdot 2^3 + 3^3) = 20.$$

On the other hand,

$$\int_1^3 x^3 \, dx = [\tfrac{1}{4}x^4]_1^3 = \tfrac{81}{4} - \tfrac{1}{4} = 20.$$ □

We now establish the second method, known as **Simpson's Rule**.

**THEOREM 11**     **(Simpson's Rule)**   *Suppose that $f(x)$ is continuous for $a \leq x \leq b$ and $\{a = x_0 < x_1 < x_2 < \cdots < x_{2n-1} < x_{2n} = b\}$ is a subdivision of $[a, b]$ into $2n$ intervals of length $h = (b-a)/2n$. Then*

$$\int_a^b f(x) \, dx \approx \frac{h}{3}[f(x_0) + 4f(x_1) + 2f(x_2) + 4f(x_3) + 2f(x_4) + \cdots$$

$$+ 2f(x_{2n-2}) + 4f(x_{2n-1}) + f(x_{2n})].$$

**Sketch of proof**   Applying the Prismoidal Formula to the first two intervals, we obtain

$$\frac{h}{3}[f(x_0) + 4f(x_1) + f(x_2)].$$

Adding to this the result of applying the Prismoidal Formula to the third and fourth intervals, we get

$$\frac{h}{3}[f(x_0) + 4f(x_1) + 2f(x_2) + 4f(x_3) + f(x_4)].$$

Simpson's Rule results from a continuation of the process to include all $n$ pairs of intervals. □

**EXAMPLE 3**   Compute the approximate value of

$$\int_0^1 \frac{dx}{1 + x^4}$$

using Simpson's Rule with $n = 2$. Do the calculations with five decimal places.

**Solution**   We have $(b - a)/2n = (1 - 0)/4 = 0.25 = h$. With $f(x) = 1/(1 + x^4)$ we use a calculator to form the table:

| $x_i$ | 0 | 0.25 | 0.5 | 0.75 | 1 |
|-------|---|------|-----|------|---|
| $f(x_i)$ | 1 | 0.99611 | 0.94118 | 0.75964 | 0.50000 |

Hence

$$\frac{0.25}{3}[f(0) + 4f(0.25) + 2f(0.5) + 4f(0.75) + f(1)]$$

$$= 0.08333[1 + 3.98444 + 1.88236 + 3.03856 + 0.50000]$$

$$= 0.87110.$$ □

*Remarks.* The integral in Example 3 can be computed exactly since the integrand is a rational function, and the method of partial fractions states that every rational function can be integrated. While theoretically this technique can be used, the required step of decomposing $1 + x^4$ into linear and quadratic factors is not easy. We first observe that $1 + x^4$ has no linear factors, since the only possibilities are $1 + x$ and $1 - x$, and neither of these is a factor. Thus $1 + x^4$ must be decomposable into two quadratic factors. After some trial and error it can be shown that

$$1 + x^4 = (1 - \sqrt{2}x + x^2)(1 + \sqrt{2}x + x^2).$$

Then we get the partial fraction expansion

$$\frac{1}{1 + x^4} = \frac{Ax + B}{1 + \sqrt{2}x + x^2} + \frac{Cx + D}{1 - \sqrt{2}x + x^2}.$$

We next have to solve four equations in four unknowns to find the values of $A$, $B$, $C$, $D$. Then the expressions on the right above can be integrated by known methods.

The above discussion shows that many times even when a definite integral can be found exactly by analytical methods it may be simpler and quicker to get an approximate answer by the use of numerical integration. However, it is important to have an estimate for the error in Simpson's Rule similar to the one we provided in the Trapezoidal Rule. We denote by $E_S$ the absolute value of the difference between the numerical approximation in Simpson's Rule and the actual value of the integral. We state without proof the following result.

**THEOREM 12**   **(Estimate of the Error in Simpson's Rule)** *Suppose that $f(x)$ has four derivatives for $a \le x \le b$, and $M$ is a number such that $|f^{(iv)}(x)| \le M$ for $a \le x \le b$. Then the error $E_S$ in using Simpson's Rule to approximate the integral*

$$\int_a^b f(x)\, dx$$

*satisfies*

$$E_S \le M(b - a)h^4,$$

*where $2n$ is the number of equal subintervals and $h = (b - a)/2n$.*

To show how Theorem 12 works, we estimate the error in the approximation of Example 3. First we compute $f^{(iv)}(x)$ with $f(x) = 1/(1 + x^4)$. After a lengthy calculation, we obtain

$$f^{(iv)}(x) = 24 \frac{35x^{12} - 155x^8 + 65x^4 - 1}{(1 + x^4)^5}.$$

An estimate shows that $|f^{(iv)}(x)| \le (24)(64) = 1536$ for $0 \le x \le 1$. Hence

$$E_S \le \tfrac{1}{180}(1536) \cdot 1 \cdot (\tfrac{1}{4})^4 = 0.03333.$$

In other words, the error in the approximation for the integral obtained in Example 3 is less than 1 part in 30.

# 9   PROBLEMS

In Problems 1 through 10, use the Trapezoidal Rule with the given values of $n$ to compute the approximate values of the given integrals. Keep the number of decimal places indicated. Compute the exact values by integration. Also, compute an upper bound for the error $E_T$ in using the Trapezoidal Rule.

1   $\int_1^4 x^2 \, dx$,   $n = 6$, 2 dec

2   $\int_0^3 x\sqrt{16 + x^2} \, dx$,   $n = 6$, 2 dec

3   $\int_1^2 dx/x$,   $n = 5$, 4 dec

4   $\int_0^{1/2} dx/\sqrt{1 - x^2}$,   $n = 5$, 4 dec

5   $\int_0^1 dx/\sqrt{1 + x^2}$,   $n = 5$, 4 dec

6   $\int_0^{\pi/2} \sin x \, dx$,   $n = 10$, 4 dec

7   $\int_0^2 \dfrac{dx}{4 + x^2}$,   $n = 5$, 3 dec

8   $\int_0^\pi \cos x \, dx$,   $n = 8$, 2 dec

9   $\int_2^4 \dfrac{dx}{1 - x^2}$,   $n = 6$, 2 dec

10   $\int_0^2 e^{-x} \, dx$,   $n = 6$, 4 dec

In Problems 11 through 14, compute the exact value of each integral by the Prismoidal Formula.

11   $\int_1^3 x^2 \, dx$

12   $\int_1^3 (x^3 - 2x^2 + 3x - 4) \, dx$

13   $\int_{-2}^0 (x^3 + x^2 - 3x + 4) \, dx$

14   $\int_{-1}^3 (2x^3 + 3x^2 - 4x - 5) \, dx$

In Problems 15 through 22, use Simpson's Rule with the given values of $2n$ to compute the approximate values of the given integrals. Keep the number of decimal places indicated. Compute the exact values by integration. Also, compute an upper bound for the error $E_S$ in using Simpson's Rule.

15   $\int_1^2 dx/x$,   $2n = 4$, 5 dec

16   $\int_0^{1/2} dx/\sqrt{1 - x^2}$,   $2n = 4$, 6 dec

17   $\int_0^1 dx/\sqrt{1 + x^2}$,   $2n = 4$, 5 dec

18   $\int_0^{\pi/2} \sin x \, dx$,   $2n = 10$, 5 dec

19   $\int_0^1 dx/(x^2 + x + 1)$,   $2n = 4$, 5 dec

20   $\int_0^2 x \, dx/(x^2 + 1)$,   $2n = 10$, 5 dec

21   $\int_0^\pi \cos x \, dx$,   $2n = 8$, 4 dec

22   $\int_0^2 \dfrac{dx}{1 + x^2}$,   $2n = 10$, 3 dec

In Problems 23 through 28, compute approximate values for the integrals, using Simpson's Rule. Also, estimate the value of $E_S$.

23   $\int_0^1 dx/(1 + x^3)$,   $2n = 4$, 5 dec

24   $\int_0^1 dx/\sqrt{1 + x^3}$,   $2n = 4$, 5 dec

25   $\int_0^1 \sqrt[3]{1 + x^2} \, dx$,   $2n = 4$, 5 dec

26   $\int_0^1 e^{-x^2} \, dx$,   $2n = 4$, 5 dec

27   $\int_0^1 x \, dx/\sqrt{1 + x^3}$,   $2n = 4$, 5 dec

28   $\int_0^{\pi/2} dx/\sqrt{1 - \frac{1}{2}\sin^2 x}$,   $2n = 4$, 5 dec

*29 Compute the exact value of the integral

$$\int_0^1 \frac{dx}{1+x^4}$$

by the method of partial fractions as described in the remarks after Example 3. Compare this result with the approximation by Simpson's Rule.

*30 Compute the approximate value of the integral

$$\int_2^3 \frac{dx}{x^6-1}$$

by the Trapezoidal Rule with $n = 6$, keeping 4 decimal places. Calculate the error $E_T$. Then compute the exact value by the method of partial fractions.

31 Let $f(x)$ be continuous for $a \le x \le b$ and let $\{a = x_0 < x_1 < \cdots < x_n = b\}$ be a subdivision of $[a, b]$ into $n$ subintervals of length $h = (b - a)/n$. Let $\bar{x}_i = \frac{1}{2}(x_{i-1} + x_i)$ be the midpoint of each subinterval. The **Midpoint Rule** gives the approximation

$$\int_a^b f(x)\, dx \approx h[f(\bar{x}_1) + f(\bar{x}_2) + \cdots + f(\bar{x}_n)].$$

If $|f''(x)| \le M$ for $a \le x \le b$, then the error $E_M$ satisfies

$$E_M \le \tfrac{1}{24} M(b - a)h^2.$$

Rework Problems 1 and 9 using the Midpoint Rule and compute the maximum error $E_M$ in each case.

---

## 10

## APPLICATIONS TO PROBABILITY THEORY (OPTIONAL)

Probability theory is the mathematical description of the rules of chance. In calculus we study independent and dependent variables, but in probability we study *random variables*. Typically a random variable describes the outcome of an experiment in which chance is involved. For example, imagine an experiment where 10 coins are simultaneously tossed into the air and for each coin there is an equal possibility that it will come up heads or tails. It is impossible to say beforehand how many heads will come up. The experiment of tossing the coins is a random one. Let $X$ denote the number of heads which come up in such a tossing. The possible values of $X$ are 0, 1, 2, ..., 10. We say $X$ is a random variable and its *range* is the set of integers 0, 1, ..., 10. The value of $X$ in each case is the result of a random experiment.

EXAMPLE 1    An experiment consists of tossing two dice. $X$ is the number of dots showing on the top faces of the dice. What is the range of $X$? What is the range of $X$ if 3 dice are tossed? What is the range of $X$ if $n$ dice are tossed?

Solution    The possible values for each die are 1, 2, ..., 6. If two dice are tossed, a table can represent the possible outcomes:

$$(1, 1), (1, 2), (1, 3), \ldots, (1, 6)$$
$$(2, 1), (2, 2), (2, 3), \ldots, (2, 6)$$
$$\vdots \qquad\qquad\qquad\qquad \vdots$$
$$(6, 1), (6, 2), \quad \ldots, \quad \ldots, (6, 6)$$

TABLE 1. Possible outcomes from tossing two dice

In Table 1, the left component of each ordered pair represents the face of the first die; the right component represents the face of the second die. We are interested in the sum of the two faces; an inspection of Table 1 shows that the smallest sum is 2; the largest is 12; and that every integer between 2 and 12 is included. Thus the range of $X$ is $\{2, 3, 4, \ldots, 12\}$. If three dice are tossed we can make a table as follows:

$$(1, 1, 1), (1, 1, 2), \ldots, (1, 1, 6)$$
$$(1, 2, 1), (1, 2, 2), \ldots, (1, 2, 6)$$
$$\vdots \qquad\qquad\qquad \vdots$$
$$(1, 6, 1), (1, 6, 2), \ldots, (1, 6, 6)$$
$$(2, 1, 1), (2, 1, 2), \ldots, (2, 1, 6)$$
$$\vdots \qquad\qquad\qquad \vdots$$
$$(3, 1, 1), (3, 1, 2), \ldots, (3, 1, 6)$$
$$\vdots \qquad\qquad\qquad \vdots$$
$$(6, 6, 1), (6, 6, 1), \ldots, (6, 6, 6)$$

**TABLE 2.** Possible outcomes from tossing three dice

Note that Table 2 has six columns but eighteen rows. We easily see that the range of $X$ is $\{3, 4, 5, \ldots, 18\}$. Analogous reasoning allows us to conclude that if we toss $n$ dice, the range of $X$ is $\{n, n+1, n+2, \ldots, 6n\}$.  □

We give a second example of a random variable, one involving light bulbs. An electric utility installs 50 light bulbs along a street, and the company is interested in the *average life* of these bulbs. Suppose we let $X$ denote the average life of the 50 bulbs. Since we don't know in advance how long it takes for the filament of each bulb to burn out, the value of $X$ may be any number larger than zero. Thus the range of $X$ is from 0 to $\infty$, denoted $[0, \infty)$. Each time the company installs a set of 50 bulbs and it waits until all of them burn out, it can compute the average lifetime of the bulbs and hence a value of $X$. The quantity $X$ is a random variable.

While we cannot know in advance what value a random variable $X$ will take on, we can study the *probability* that $X$ will take on a given value.

> The probability, usually denoted $P$, can have any value between 0 and 1, with the value 1 corresponding to certainty that $X$ will have a specific value and the value 0 corresponding to certainty that $X$ will not have a specific value.

When the range of $X$ is all numbers between 0 and $\infty$, that is, $[0, \infty)$, we can also consider the probability that $X$ will fall between two values, say $a$ and $b$. We write this statement in the form

$$P(a < X < b)$$

and this value will, of course, always be a number between 0 and 1. In the light bulb case we have the obvious statement $P(0 \leq X < \infty) = 1$.

Let $x$ denote an ordinary variable with values between $-\infty$ and $\infty$ and let $X$ be a random variable with range $-\infty$ to $\infty$. If $x_1$ and $x_2$ are values of $X$ with $x_1 < x_2$, then $P(-\infty < X \leq x_1)$ and $P(-\infty < X \leq x_2)$ are numbers between 0 and 1. It is not hard to see that $P(-\infty < X \leq x_1) \leq P(-\infty < X \leq x_2)$, since the probability that the result of an experiment will fall in a small interval must be less than the probability that the result will fall in a larger interval containing the smaller one.

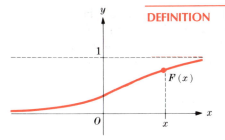

$F(x)$ is probability $X$ is less than $x$

**FIGURE 42**

**DEFINITION**

*Let $X$ be a random variable. Then the **distribution function** of $X$, usually denoted by $F$, is defined by $F(x) = P(-\infty < X \le x)$, where $x$ is an ordinary real variable.*

That is, $F(x)$ is just the probability that the result of a random experiment will fall in the interval $-\infty < X \le x$. (See Fig. 42.)

Clearly $F(x)$ is always a number between 0 and 1. In fact, suppose a random variable can take on only nonnegative values. Then $F(x) = 0$ for $x$ negative. Also, in the coin-tossing experiment with the 10 coins we described, $X$ is always between 0 and 10. Thus $F(10) = 1$ and, in fact $F(x) = 1$ for all $x > 10$. We have, clearly, for *every* random variable $X$

$$\lim_{x \to -\infty} F(x) = 0 \qquad \text{and} \qquad \lim_{x \to +\infty} F(x) = 1.$$

Suppose $F(x)$ is a differentiable function and we define $f(x) = F'(x)$. Then $f(x) \ge 0$ for all $x$ since $F(x)$ always increases (or at least doesn't decrease) as $x$ increases. From the Fundamental Theorem of Calculus we have

$$F(x) = \int_{-\infty}^{x} f(u)\, du. \tag{1}$$

The function $f$ in (1) is called the **probability density function**, and it plays an important part in the study of probability theory.

Suppose $X$ is a random variable which can have values only in an interval $(a, b)$. Then it follows from the definition above that $F(x) = 0$ for all $x \le a$. Consequently, $f(x) = 0$ for all $x \le a$. Hence in this case we can write

$$F(x) = \int_{a}^{x} f(u)\, du.$$

Also, $F(x) = 1$ for all $x \ge b$.

The probability of the outcome of an experiment can be expressed in terms of the probability density function. To compute $P(a < X \le b)$, we can write (see Fig. 43)

$$P(a < X \le b) = P(-\infty < X \le b) - P(-\infty < X \le a)$$

$$= F(b) - F(a)$$

$$= \int_{-\infty}^{b} f(u)\, du - \int_{-\infty}^{a} f(u)\, du = \int_{a}^{b} f(u)\, du.$$

$P(-\infty < X \le b)$

$P(-\infty < X \le a)$

$a$    $b$

$P(a < X \le b)$

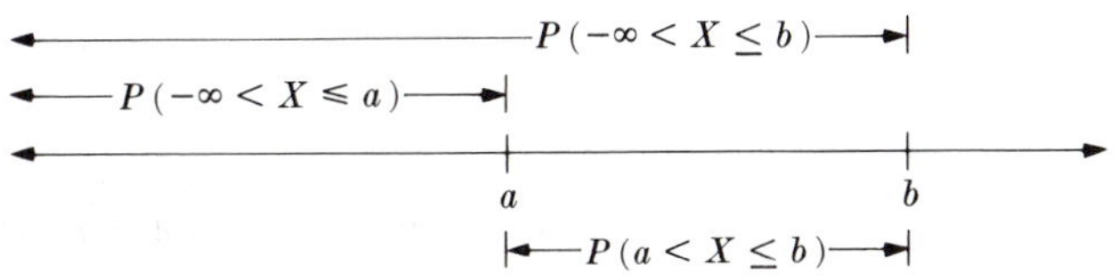

**FIGURE 43**

An important concept of probability is the expected value of a random variable. Intuitively, the expected value of a random variable, denoted $E(X)$, is the "average" value of $X$. For example, in the coin-tossing experiment with 10 coins which we described above, we imagine that the average value for $X$ is

5. In fact, if the coin is what is known as a *fair coin*, i.e., one in which heads and tails are equally likely, then it is true that $E(X) = 5$ in this case.

We now derive a useful formula for $E(X)$. Suppose an experiment can have 10 possible outcomes, $x_1, x_2, \ldots, x_9, x_{10}$, and suppose that $x_1, x_2, \ldots, x_9$ are all between 0 and 2 while $x_{10} = 1000$. Then the average

$$\frac{x_1 + x_2 + \cdots + x_{10}}{10}$$

would be "very large," while intuitively we expect that $X$ would be between 0 and 2 most of the time. Taking the simple average does not yield a good measure for $E(X)$, the expected value, when the likelihood that $x_{10} = 1000$ would occur is small. What we do instead is take each value for $X$ and multiply it by the probability that it will happen. That is, denoting by $P(X = x_i)$ the probability that $X$ has the value $x_i$, we compute

$$E(X) = x_1 P(X = x_1) + \cdots + x_n P(x = x_n)$$

$$= \sum_{i=1}^{n} x_i P(X = x_i).$$

We now extend the definition of $E(X)$ to the case where the random variable $X$ has a range which consists of a portion of, or all of, the real axis. We do this by an approximation method similar to the definition of the integral. To see this suppose the range of $X$ is an interval $[a, b]$. We introduce a subdivision $a = x_0 < x_1 < \cdots < x_{n-1} < x_n = b$ (see Fig. 44) and we define an approximation to the expected value

$$E(X) \approx x_0 P(x_0 < X \le x_1) + \cdots + x_{n-1} P(x_{n-1} < X \le x_n)$$

$$\approx \sum_{i=0}^{n-1} x_i P(x_i < X \le x_{i+1}).$$

FIGURE 44

We recall that $F(d) - F(c) = P(c < X \le d)$, and so

$$E(X) \approx \sum_{i=0}^{n-1} x_i [F(x_{i+1}) - F(x_i)].$$

Assuming that $F$ is differentiable, we may apply the Mean Value Theorem to the right side and obtain

$$E(X) \approx \sum_{i=0}^{n-1} x_i F'(\xi_i)(x_{i+1} - x_i),$$

where $\xi_i$ is some number in the interval $[x_i, x_{i+1}]$. We defined the probability distribution function by $f(x) = F'(x)$, and so

$$E(X) \approx \sum_{i=0}^{n-1} x_i f(\xi_i) \Delta_i x, \quad \Delta_i x = x_{i+1} - x_i.$$

If $\Delta_i x$ is very small then $x_i$ is close to $\xi_i$ and we can write

$$E(X) \approx \sum_{i=0}^{n-1} \xi_i f(\xi_i) \Delta_i x, \quad x_i \le \xi_i \le x_{i+1}.$$

If the function $xf(x)$ is continuous then the sums on the right side are Riemann sums and as the length of $\Delta_i x$ tends to 0, the sums tend to $\int_a^b uf(u)\,du$. This motivates the next definition.

**DEFINITION** *If $X$ is a random variable taking values in an interval $[a, b]$, and if its probability density function $f$ exists and is continuous, then $E(X)$, the* **expected value of** $X$, *is defined to be*

$$E(X) = \int_a^b uf(u)\,du.$$

**EXAMPLE 2** Let $X$ denote the lifetime of a lightbulb. Such an $X$ is known to have an *exponential distribution with parameter* $\alpha$ $(\alpha > 0)$. That is, its probability density function is given by

$$f(x) = \begin{cases} \alpha e^{-\alpha x} & x \geq 0 \\ 0 & x < 0. \end{cases}$$

Figure 45 gives a graph of the probability density function.

The distribution function, $F$, is then

$$F(x) = \int_{-\infty}^x f(u)\,du = \begin{cases} 0 & x \leq 0 \\ \int_0^x f(u)\,du, & x > 0 \end{cases}$$

$$= \begin{cases} 0 & x \leq 0 \\ 1 - e^{-\alpha x}, & x > 0. \end{cases}$$

Note that $\int_{-\infty}^\infty f(u)\,du = \lim_{x \to \infty}(1 - e^{-\alpha x}) = 1$, so $f$ is a true probability density function.

The expected value of $X$ is given by:

$$E(X) = \int_0^\infty x\alpha e^{-\alpha x}\,dx = \frac{1}{\alpha},$$

as a straightforward integration by parts reveals. The quantity $E(X) = 1/\alpha$ is printed on each box of lightbulbs sold in the United States; it varies from 750 hours to 3000 hours. ☐

**EXAMPLE 3** Let $X$ be a random variable with probability density function

$$f(x) = \frac{1}{\sqrt{2\pi}} e^{-x^2/2}.$$

Such a random variable is said to have a *normal distribution* (it is also called a *Gaussian distribution*). This is the most important distribution, since it can be proved that an enormous class of random variables that arise in practice have this distribution, or one closely related to it. Figure 46 gives a sketch of the graph of $f$.

The distribution function,

$$F(x) = \int_{-\infty}^x \frac{1}{\sqrt{2\pi}} e^{-x^2/2}\,dx,$$

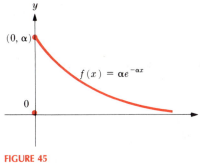

FIGURE 45

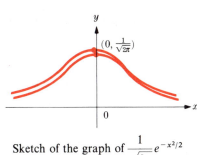

Sketch of the graph of $\dfrac{1}{\sqrt{2\pi}} e^{-x^2/2}$

FIGURE 46

cannot be expressed in terms of elementary functions. To compute $E(X)$, we note that

$$E(X) = \int_{-\infty}^{\infty} xf(x)\,dx = \frac{1}{\sqrt{2\pi}}\int_{-\infty}^{\infty} xe^{-x^2/2}\,dx$$

$$= \frac{1}{\sqrt{2\pi}}\left(\lim_{a \to -\infty}\int_{a}^{0} xe^{-x^2/2}\,dx + \lim_{b \to +\infty}\int_{0}^{b} xe^{-x^2/2}\,dx\right)$$

$$= \frac{1}{\sqrt{2\pi}}\left(\lim_{a \to -\infty}(e^{-a^2/2}-1) + \lim_{b \to +\infty}(1-e^{-b^2/2})\right)$$

$$= 0.$$

It is not obvious that

$$f(x) = \frac{1}{\sqrt{2\pi}}e^{-x^2/2}$$

is indeed a probability density function. We need to show that $\int_{-\infty}^{\infty} e^{-x^2/2}\,dx = \sqrt{2\pi}$. This can be done by a device using methods not yet covered. (See Example 4 of Section 5, Chapter 16.)    □

**EXAMPLE 4**    A random variable $X$ can be the net gain from entering a sweepstakes, or lottery. Suppose the cost of entry is 22¢, the prize is \$1,000,000, and 10,000,000 people enter. Then

$$X = \begin{cases} 1,000,000 - 0.22 & \text{if lottery is won} \\ -0.22 & \text{if lottery is lost.} \end{cases}$$

What is $E(X)$, the expected winnings?

**Solution**    The range of $X$ is just the two numbers: 0.22 and 999,999.78. To compute $E(X)$ we need to know the probabilities of winning and losing. Assuming each entry has an equal chance of winning, the probability of winning the lottery is one in ten million, or $10^{-7}$. Therefore

$$E(X) = (999,999.78)10^{-7} - (0.22)(1 - 10^{-7})$$

$$= 0.099999978 - 0.219999978$$

$$= -0.12.$$

The expected winnings from such a sweepstakes is actually the loss of 12¢.    □

**EXAMPLE 5**    A random variable $X$ is said to have a *Cauchy distribution* if its probability density function is given by

$$f(x) = \frac{1}{\pi(1+x^2)}, \quad -\infty < x < \infty.$$

Such an $f$ is a true probability density function, since

$$F(x) = \int_{-\infty}^{x} f(u)\,du = \int_{-\infty}^{x} \frac{1}{\pi(1+u^2)}\,du = \frac{1}{2} + \frac{1}{\pi}\arctan x,$$

and $\lim_{x \to +\infty} \arctan x = \pi/2$. A graph of $f$ is given in Fig. 47. Note that it resembles the graph of the normal probability density function.

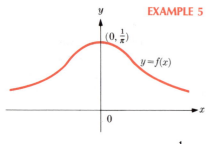

Sketch of the graph of $f(x) = \dfrac{1}{\pi(1+x^2)}$

**FIGURE 47**

If $X$ has the Cauchy distribution, then $E(X)$ does not exist (!). We can see this by looking at

$$\frac{1}{\pi}\int_0^\infty u\frac{1}{1+u^2}\,du,$$

which certainly must exist (in the sense that it is finite) if

$$\frac{1}{\pi}\int_{-\infty}^\infty u\frac{1}{1+u^2}\,du$$

is to be defined. But

$$\frac{1}{\pi}\int_0^\infty \frac{u}{1+u^2}\,du = \lim_{b\to\infty}\frac{1}{\pi}\int_0^b \frac{u}{1+u^2}\,du$$

$$= \lim_{b\to\infty}\ln(1+u^2)\Big]_0^b$$

$$= \lim_{b\to\infty}\ln(1+b^2) = +\infty. \qquad \square$$

**EXAMPLE 6**　A random variable $X$ is said to have a *Gamma distribution* if the probability density function of $f$ is given by:

$$f(x) = \begin{cases} \dfrac{\alpha}{(r-1)!}(\alpha x)^{r-1}e^{-\alpha x}, & x>0 \\ 0, & x\le 0. \end{cases}$$

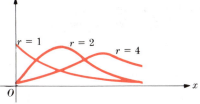

Graphs of the probability density function for a Gamma distribution with different values of $r$ and $\alpha = 1$

**FIGURE 48**

Each integer $r \ge 1$ and $\alpha > 0$ yields a different Gamma distribution. In the study of statistics one often knows from theoretical considerations that a random variable $X$ has a Gamma distribution, but the actual parameters $\alpha$ and $r$ are unknown. A statistician will then try to infer from experimental data what the true parameters $\alpha$ and $r$ actually are. 　　□

## 10　PROBLEMS

In each of Problems 1 through 5, determine the range of the random variable.

**1** A store manager is curious as to how many coffee beans are in each one-pound bag of coffee. He selects bags at random and counts the number of beans. $X$ is the number of beans in a bag he selects.

**2** A store manager is curious as to how long it takes to grind a one-pound bag of coffee beans. She selects bags of coffee beans at random, grinds them, and records how much time it took to grind them. $X$ is the amount of time it took to grind the beans.

**3** A sociologist is studying how many times a caller will let an unanswered telephone ring. She selects a phone at random and lets it ring. $X$ is the number of rings.

**4** A botanist is curious how long a certain flowering plant remains in bloom. He selects plants at random and records the amount of time the flower of the plant is in bloom. $X$ is the blooming time of one of these plants.

**5** A civil engineer is measuring the elasticity factor of soil under an interstate highway in Indiana. She takes a soil sample and performs an experiment which gives her an elasticity factor between 0 and 1. $X$ is the elasticity factor for one of these soil samples.

**6** Five thousand people enter a sweepstakes with one prize of $100,000. How much would be a fair price to pay to enter the sweepstakes?

**7** A contest has a grand prize of $25,000, a second prize of $10,000, and ten third prizes of $100. If winners are selected by lot, and if you and 49,999 other people enter, what are your expected winnings?

**8** A random variable $X$ has a **uniform distribution** on $[a, b]$ if its probability density function is

$$f(x) = \begin{cases} 0 & x \leq a \\ \dfrac{1}{b-a} & a < x < b \\ 0 & x \geq b. \end{cases}$$

a) Show that $f$ really is a probability density function.
b) Find the distribution function $F$ of $X$.
c) Calculate $E(X)$.

**9** Verify that

$$E(X) = \frac{1}{\alpha}$$

if $X$ has an exponential distribution with parameter $\alpha$.

**10** A random variable $X$ is said to have a *normal distribution with parameters $\mu$ and $\sigma^2$* (written $N(\mu, \sigma^2)$) if its probability density function is given by

$$f(x) = \frac{1}{\sqrt{2\pi}\sigma} e^{-(x-\mu)^2/2\sigma^2}.$$

Show that $E(X) = \mu$ for such an $X$.

**11** Let $X$ have an $N(\mu, \sigma^2)$ distribution as defined in Problem 10. Show that the maximum of its probability density function occurs at $x = \mu$.

**12** Suppose $X$ has a uniform distribution over $[-3, 3]$ (see Problem 8). Compute
a) $P(X \leq 2)$           b) $P(|X| \leq 2)$
c) $P(X = 2)$

**13** Suppose $X$ has a Cauchy distribution (see Example 5). Compute $P(X > 3)$.

**14** Let $X$ be a random variable having a Gamma distribution with parameters $\alpha$ and $r$. Compute $E(X)$.

**\*15** Let $X$ be a random variable with distribution function $F_X$. Let $Y$ be another random variable given by $Y = X^2$. Show that $F_Y(y) = F_X(\sqrt{y}) - F_X(-\sqrt{y})$.

**\*16** Use Problem 15 to find the probability density function of $Y = X^2$, if $X$ is distributed $N(0, 1)$.

**17** Suppose a random variable $X$ has a probability density function

$$f(x) = \begin{cases} 2x & 0 < x < 1 \\ 0 & -\infty < x \leq 0;\ 1 \leq x < \infty. \end{cases}$$

Find the following:
a) $P(X < -1)$           b) $P(X < \frac{1}{3})$
c) $P(\frac{1}{3} < X < \frac{2}{3})$     d) $P(X > \frac{3}{4})$

**\*18** Let $X$ be a uniform random variable on $[0, 1]$. Let $Y = \sin(2\pi X)$. Find the probability density function of $Y$.

**\*19** Let $X$ be a random variable with distribution function $F$, and probability density function $f$. Let $Y = F(X)$ define another random variable. Show that $Y$ has a uniform distribution on $[0, 1]$ (see Problem 8).

**20** Suppose $X$ has a normal distribution as defined in Example 3. Use Simpson's Rule to find $P(0 < X \leq 1)$ and give an estimate of the error. (*Hint:* See Problem 26 in Section 9.) Use $2n = 6$.

_____ **CHAPTER 9**

## REVIEW PROBLEMS

In each of Problems 1 through 4, point masses $m_i$ are located at points $P_i$ in the plane. Find the center of mass.

**1** $m_1 = 2,\ m_2 = 1,\ m_3 = 5,\ m_4 = 4;\ P_1(1, 0),\ P_2(0, 3),\ P_3(1, 1),$
$P_4(5, 2).$

**2** $m_1 = 3,\ m_2 = 2,\ m_3 = 4;\ P_1(0, 0),\ P_2(-1, 2),\ P_3(4, -2).$

**3** $m_1 = 1,\ m_2 = 5,\ m_3 = 4,\ m_4 = 12;\ P_1(1, -2),\ P_2(-1, 2),$
$P_3(4, 1),\ P_4(0, 0).$

**4** $m_1 = 4,\ m_2 = 1,\ m_3 = 1,\ m_4 = 1;\ P_1(5, 5),\ P_2(0, 1),\ P_3(1, 0),$
$P_4(-1, -1).$

**5** A plane region of uniform density has the shape shown in Fig. 49. Find its center of mass.

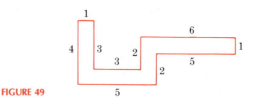

**FIGURE 49**

In Problems 6 through 12, the region $R$ has uniform density. Find the center of mass.

**6** $R$ is bounded by $y = 2x^2$ and $y = x$.

**7** $R$ is the region bounded by $y = \cos x$, $y = 0$, $x = -\pi/4$,
$x = \pi/4$.

**8** $R$ is the region bounded by $y = \cos x$, $y = 0$, $x = -\pi/4$, $x = \pi/3$.

**9** $R$ is bounded by $y = e^{-x}$, $x = 1$, $x = 2$, $y = 0$.

**10** $R$ is bounded by $y = e^{2x}$, $y = 0$, $x = 0$, $x = 1$.

**11** $R$ is bounded by $x = 2y - y^2$ and $x = 0$.

**12** $R$ is bounded by $y = 2x$, $x + y = 6$, $y = 0$.

**13** The center of mass of the region bounded by $y^2 = 4x$ and the line $x = a$ is at $(2, 0)$. Find the value of $a$.

**14** Find the center of mass of the half-ellipse

$$0 \le \frac{x^2}{4} + \frac{y^2}{9} \le 1, \quad y \ge 0.$$

In each of Problems 15 through 19, a region $R$ in the plane is revolved about the $x$ axis. Find the center of mass of the solid generated.

**15** $R = \left\{ (x, y) : \frac{\pi}{4} \le x \le \frac{\pi}{2}, 0 \le y \le \cos x \right\}.$

**16** $R$ is bounded by $2x + y = 5$ and $xy = 2$.

**17** $R$ is bounded by $y = 3/x$, $x = 2$, $x = 3$, $y = 0$.

**18** $R$ is bounded by $y = e^{3x}$, $y = 0$, $x = 0$, $x = 1$.

**19** $R$ is bounded by $y = \tan x$, $y = 0$, $x = \pi/6$, $x = \pi/4$.

**20** The region bounded by $y = \cos x$, $x = 0$, $x = \pi/3$, $y = 0$ is revolved about the $y$ axis. Find the center of mass.

**21** Find the center of mass of the arc given by

$$x^{2/3} = 4 - y^{2/3}, \quad 0 \le y \le 8, x \ge 0.$$

**22** Find the center of mass of the arc given by

$$y = \tfrac{1}{4}x^2 - \tfrac{1}{2} \ln |x|, \quad \tfrac{1}{2} \le x \le 1.$$

In Problems 23 through 31, determine whether or not the given integral is convergent. When it is convergent, compute its value.

**23** $\displaystyle\int_0^{+\infty} \frac{dx}{x^2}$

**24** $\displaystyle\int_0^4 \frac{dx}{\sqrt{x}}$

**25** $\displaystyle\int_{-1}^1 \frac{dx}{x^{1/3}}$

**26** $\displaystyle\int_{-1}^1 \frac{dx}{x^{3/2}}$

**27** $\displaystyle\int_{-1}^{+\infty} \frac{dx}{x^{3/2}}$

**28** $\displaystyle\int_{-\infty}^{-1} \frac{dx}{x^2}$

**29** $\displaystyle\int_0^{+\infty} x^3 e^{-x^2} dx$

**30** $\displaystyle\int_1^{+\infty} x^5 e^{-x^2} dx$

**31** $\displaystyle\int_0^{\pi/2} \tan x \, dx$

**32** A particle moves along the $x$ axis from $x = 1$ to $x = 4$ according to the force law $F(x) = x^2 - 3x + 1$. Find the work done.

**33** A hemispherical tank 2 meters in radius is full of water. Find the work done in pumping the water out the top.

**34** Find the work done in lifting a 1 kg weight from the surface of the Earth to a height of 10 km.

**35** A force of 9 kg is needed to stretch a spring from its natural length of 6 cm to 8 cm. Find the work done in stretching the spring to a length of 10 cm.

In Problems 36 through 41, use the Trapezoidal Rule to compute an approximate value of the given integral with an accuracy of two decimal places. Use the value of $n$ indicated. Calculate an upper bound for the error $E_T$ in each case. A hand calculator should be used.

**36** $\displaystyle\int_1^2 \frac{dx}{1 + x^3}$, $n = 4$

**37** $\displaystyle\int_2^4 \frac{dx}{\sqrt{1 + x^3}}$, $n = 6$

**38** $\displaystyle\int_0^2 e^{-x^2} dx$, $n = 5$

**39** $\displaystyle\int_0^2 \sqrt{1 + x^4} \, dx$, $n = 6$

**40** $\displaystyle\int_0^{2\pi} \cos 2x \, dx$, $n = 10$

**41** $\displaystyle\int_{-2}^1 (x^3 + 2x - 6) \, dx$, $n = 4$

In Problems 42 through 47, use Simpson's Rule to compute an approximate value of the given integral with an accuracy of two decimal places. Use the value of $2n$ indicated. Calculate the maximum error $E_S$. A hand calculator should be used.

**42** $\displaystyle\int_1^3 \frac{dx}{1 + x^2}$, $2n = 6$

**43** $\displaystyle\int_0^2 \frac{dx}{\sqrt{2 + x^2}}$, $2n = 8$

**44** $\displaystyle\int_0^{1/2} \frac{dx}{\sqrt{1 - x^3}}$, $2n = 6$

**45** $\displaystyle\int_0^2 (x^3 + 2x^2 - x + 1) \, dx$, $2n = 10$

**46** $\displaystyle\int_0^2 e^{-x^2} dx$, $2n = 10$

**47** $\displaystyle\int_0^{\pi/2} \sin 2x \, dx$, $2n = 4$

In each of Problems 48 through 51, use the Midpoint Rule (see Problem 31, page 369) to calculate the integral approximately with an accuracy of two decimal places. Use the given value of $n$. Calculate the maximum error $E_M$.

**48** $\displaystyle\int_1^2 \frac{dx}{1 + x^3}$, $n = 4$ (compare with Problem 36, above).

**49** $\displaystyle\int_0^2 e^{-x^2} dx$, $n = 5$ (compare with Problem 38, above).

**50** $\displaystyle\int_2^4 \frac{dx}{\sqrt{1 + x^3}}$, $n = 6$ (compare with Problem 37, above).

**51** $\displaystyle\int_0^{\pi/2} \sin 2x \, dx$, $n = 4$ (compare with Problem 47, above).

**52** Let $X$ be a random variable with a normal distribution. Use Simpson's Rule with $2n = 10$ to give an estimate for $P(-1 < X \le 1)$.

**53** Let $X$ be a random variable with a Gamma distribution (see Example 6 in Section 10) with $r = 10$ and $\alpha = 1/2$. Use the Trapezoidal Rule with $n = 5$ to estimate $P(X > 1)$. (*Hint:* First estimate $P(X \le 1)$.)

# 10

# INFINITE SERIES

The trigonometric, logarithmic, and exponential functions as well as many related functions can be expressed as infinite series. These series help us determine many of the basic properties of the functions they represent and are frequently used for the numerical calculation of the values of functions. In this chapter we develop the fundamental properties of infinite series and we show how these properties are related to the functions we have studied thus far.

## 1

### INDETERMINATE FORMS

If $f(x)$ and $F(x)$ both approach 0 as $x$ tends to a value $a$, the quotient

$$\frac{f(x)}{F(x)}$$

may approach a limit, may become infinite, or may fail to have any limit. We saw in the definition of derivative that it is the evaluation of just such expressions that leads to the usual differentiation formulas. We are aware that the expression

$$\frac{f(a)}{F(a)} = \frac{0}{0}$$

is in itself a meaningless one, and we use the term **indeterminate form** for the ratio 0/0.

If $f(x)$ and $F(x)$ both tend to infinity as $x$ tends to a number $a$, the ratio $f(x)/F(x)$ may or may not tend to a limit. We use the same term, **indeterminate form**, for the expression $\infty/\infty$, obtained by direct substitution of $x = a$ into the quotient $f(x)/F(x)$.

We recall the Mean Value Theorem, which we established (Chapter 4, page 131).

---

**THEOREM 1**  **(Mean Value Theorem)**  *Suppose that $f$ is continuous for*

$$a \leq x \leq b$$

*and that $f'(x)$ exists for each $x$ between $a$ and $b$. Then there is an $x_0$ between $a$ and $b$ (that is, $a < x_0 < b$) such that*

$$\frac{f(b) - f(a)}{b - a} = f'(x_0).$$

---

*Remark.*  Rolle's Theorem (page 130) is the special case $f(a) = f(b) = 0$.

The evaluation of indeterminate forms requires an extension of the Mean Value Theorem which we now prove.

---

**THEOREM 2**  **(Generalized Mean Value Theorem)**  *Suppose that $f$ and $F$ are continuous for $a \leq x \leq b$, and $f'(x)$ and $F'(x)$ exist for $a < x < b$ with $F'(x) \neq 0$ there. Then $F(b) - F(a) \neq 0$ and there is a number $\xi$ with*

$$a < \xi < b$$

*such that*

$$\frac{f(b) - f(a)}{F(b) - F(a)} = \frac{f'(\xi)}{F'(\xi)}. \tag{1}$$

---

**Proof**  The fact that $F(b) - F(a) \neq 0$ is obtained by applying the Mean Value Theorem (Theorem 1) to $F$. For then, $F(b) - F(a) = F'(x_0)(b - a)$ for some $x_0$ such that $a < x_0 < b$. By hypothesis, the right side is different from zero.

For the proof of the main part of the theorem, we define the function $\phi(x)$ by the formula

$$\phi(x) = f(x) - f(a) - \frac{f(b) - f(a)}{F(b) - F(a)}[F(x) - F(a)].$$

We compute $\phi(a)$, $\phi(b)$, and $\phi'(x)$, getting

$$\phi(a) = f(a) - f(a) - \frac{f(b) - f(a)}{F(b) - F(a)}[F(a) - F(a)] = 0,$$

$$\phi(b) = f(b) - f(a) - \frac{f(b) - f(a)}{F(b) - F(a)}[F(b) - F(a)] = 0,$$

$$\phi'(x) = f'(x) - \frac{f(b) - f(a)}{F(b) - F(a)}F'(x).$$

Applying the Mean Value Theorem (i.e., in the special form of Rolle's Theorem) to $\phi(x)$ in the interval $[a, b]$, we find

$$\frac{\phi(b) - \phi(a)}{b - a} = 0 = \phi'(\xi) = f'(\xi) - \frac{f(b) - f(a)}{F(b) - F(a)} F'(\xi)$$

for some $\xi$ between $a$ and $b$. Dividing by $F'(\xi)$, we obtain Formula (1). $\quad\square$

The next theorem, known as **l'Hôpital's Rule**, is useful in the evaluation of indeterminate forms. Let $I$ be an interval in which $a$ is an interior point. The interval $I$ with $a$ removed from it is called a **deleted interval about** $a$.

**THEOREM 3**    (**l'Hôpital's Rule**)  *Suppose that*

$$\lim_{x \to a} f(x) = 0, \qquad \lim_{x \to a} F(x) = 0, \qquad and \qquad \lim_{x \to a} \frac{f'(x)}{F'(x)} = L,$$

*and that the hypotheses of Theorem 2 hold in some deleted interval about* $a$. *Then*

$$\lim_{x \to a} \frac{f(x)}{F(x)} = \lim_{x \to a} \frac{f'(x)}{F'(x)} = L.$$

**Proof**    Since $f$ and $F$ are not defined at $a$, we define $f(a) = 0$ and $F(a) = 0$. For some $h > 0$ we apply Theorem 2 (the Generalized Mean Value Theorem) in the interval $a < x < a + h$. Then

$$\frac{f(a + h)}{F(a + h)} = \frac{f(a + h) - f(a)}{F(a + h) - F(a)} = \frac{f'(\xi)}{F'(\xi)}, \quad a < \xi < a + h.$$

As $h$ tends to $0$, $\xi$ tends to $a$, and so

$$\lim_{h \to 0} \frac{f(a + h)}{F(a + h)} = \lim_{\xi \to a} \frac{f'(\xi)}{F'(\xi)} = L.$$

A similar proof is valid for $x$ in the interval $a - h < x < a$. $\quad\square$

**EXAMPLE 1**    Evaluate

$$\lim_{x \to 3} \frac{x^3 - 2x^2 - 2x - 3}{x^2 - 9}.$$

**Solution**    We set $f(x) = x^3 - 2x^2 - 2x - 3$ and $F(x) = x^2 - 9$. We see at once that $f(3) = 0$ and $F(3) = 0$, and we have an indeterminate form. We calculate

$$f'(x) = 3x^2 - 4x - 2, \qquad F'(x) = 2x.$$

By Theorem 3 (l'Hôpital's Rule):

$$\lim_{x \to 3} \frac{f(x)}{F(x)} = \lim_{x \to 3} \frac{f'(x)}{F'(x)} = \frac{3(9) - 4(3) - 2}{2(3)} = \frac{13}{6}. \quad\square$$

*Remarks.* It is essential that $f(x)$ and $F(x)$ *both* tend to zero as $x$ tends to $a$ before applying l'Hôpital's Rule. If either or both functions tend to finite limits $\neq 0$, or if one tends to zero and the other does not, then the limit of the quotient is found by the method of direct substitution as given in Chapter 2.

It may happen that $f'(x)/F'(x)$ is an indeterminate form as $x \to a$. Then l'Hôpital's Rule may be applied again, and the limit $f''(x)/F''(x)$ may exist as $x$ tends to $a$. In fact, for some problems l'Hôpital's Rule may be required a number of times before the limit is actually determined. Example 3 below exhibits this point.

**EXAMPLE 2**   Evaluate

$$\lim_{x \to a} \frac{x^p - a^p}{x^q - a^q}, \quad a > 0.$$

**Solution**   We set $f(x) = x^p - a^p$, $F(x) = x^q - a^q$. Then $f(a) = 0$, $F(a) = 0$. We compute $f'(x) = px^{p-1}$, $F'(x) = qx^{q-1}$. Therefore

$$\lim_{x \to a} \frac{f(x)}{F(x)} = \lim_{x \to a} \frac{f'(x)}{F'(x)} = \lim_{x \to a} \frac{px^{p-1}}{qx^{q-1}} = \frac{p}{q} a^{p-q}. \qquad \square$$

**EXAMPLE 3**   Evaluate

$$\lim_{x \to 0} \frac{x - \sin x}{x^3}.$$

**Solution**   We set $f(x) = x - \sin x$, $F(x) = x^3$. Since $f(0) = 0$, $F(0) = 0$, we apply l'Hôpital's Rule and get

$$\lim_{x \to 0} \frac{f(x)}{F(x)} = \lim_{x \to 0} \frac{1 - \cos x}{3x^2}.$$

But we note that $f'(0) = 0$, $F'(0) = 0$, and so we apply l'Hôpital's Rule again:

$$f''(x) = \sin x, \quad F''(x) = 6x.$$

Therefore

$$\lim_{x \to 0} \frac{f(x)}{F(x)} = \lim_{x \to 0} \frac{f''(x)}{F''(x)}.$$

Again we have an indeterminate form: $f''(0) = 0$, $F''(0) = 0$. We continue, to obtain $f'''(x) = \cos x$, $F'''(x) = 6$. Now we find that.

$$\lim_{x \to 0} \frac{f(x)}{F(x)} = \lim_{x \to 0} \frac{f'''(x)}{F'''(x)} = \frac{\cos 0}{6} = \frac{1}{6}. \qquad \square$$

L'Hôpital's Rule can be extended to the case where *both* $f(x) \to \infty$ and $F(x) \to \infty$ as $x \to a$. The proof of the next theorem, which we omit, is analogous to the proof of Theorem 3.

THEOREM 4    **(l'Hôpital's Rule)**    *Suppose that f and F are differentiable in a deleted interval about a. Suppose further that*

$$\lim_{x \to a} f(x) = \infty, \qquad \lim_{x \to a} F(x) = \infty, \qquad and \qquad \lim_{x \to a} \frac{f'(x)}{F'(x)} = L.$$

*Then*

$$\lim_{x \to a} \frac{f(x)}{F(x)} = \lim_{x \to a} \frac{f'(x)}{F'(x)} = L.$$

*Remark.*    Theorems 3 and 4 hold for one-sided limits as well as for ordinary limits. In many problems a one-sided limit is required even though this statement is not made explicitly. The next example illustrates such a situation.

EXAMPLE 4    Evaluate

$$\lim_{x \to 0} \frac{\ln x}{\ln (2e^x - 2)}.$$

Solution    First note that $x$ must tend to zero through positive values since otherwise the logarithm function is not defined. We set

$$f(x) = \ln x, \qquad F(x) = \ln (2e^x - 2).$$

Then $f(x) \to -\infty$ and $F(x) \to -\infty$ as $x \to 0^+$. Therefore

$$\lim_{x \to 0^+} \frac{f(x)}{F(x)} = \lim_{x \to 0^+} \frac{1/x}{e^x/(e^x - 1)} = \lim_{x \to 0^+} \frac{e^x - 1}{xe^x}$$

$$= \lim_{x \to 0^+} \frac{1 - e^{-x}}{x}.$$

We still have an indeterminate form, and we take derivatives again. We obtain

$$\lim_{x \to 0^+} \frac{\ln x}{\ln (2e^x - 2)} = \lim_{x \to 0^+} \frac{e^{-x}}{1} = 1. \qquad \square$$

*Remark.*    Theorems 3 and 4 are valid when $a = +\infty$ or $-\infty$. That is, if we have an indeterminate expression for $f(+\infty)/F(+\infty)$, then

$$\lim_{x \to +\infty} \frac{f(x)}{F(x)} = \lim_{x \to +\infty} \frac{f'(x)}{F'(x)},$$

and a similar statement holds when $x \to -\infty$. The next example exhibits this type of indeterminate form.

**EXAMPLE 5**  Evaluate

$$\lim_{x \to +\infty} \frac{8x}{e^x}.$$

**Solution**

$$\lim_{x \to +\infty} \frac{8x}{e^x} = \lim_{x \to +\infty} \frac{8}{e^x} = 0. \qquad \square$$

*Remarks.* Indeterminate forms of the type $0 \cdot \infty$ or $\infty - \infty$ can often be evaluated by transforming the expression into a quotient of the form $0/0$ or $\infty/\infty$. Limits involving exponential expressions may often be evaluated by taking logarithms. Of course, algebraic or trigonometric reductions may be made at any step. The next examples illustrate the procedure.

**EXAMPLE 6**  Evaluate

$$\lim_{x \to \pi/2} (\sec x - \tan x).$$

**Solution**  We employ trigonometric reduction to change $\infty - \infty$ into a standard form and then differentiate. We find

$$\lim_{x \to \pi/2} (\sec x - \tan x) = \lim_{x \to \pi/2} \left( \frac{1}{\cos x} - \frac{\sin x}{\cos x} \right) = \lim_{x \to \pi/2} \frac{1 - \sin x}{\cos x}$$

$$= \lim_{x \to \pi/2} \frac{-\cos x}{-\sin x} = 0. \qquad \square$$

**EXAMPLE 7**  Evaluate

$$\lim_{x \to 0} (1 + x)^{1/x}.$$

**Solution**  We have $1^\infty$, which is indeterminate. Set $y = (1 + x)^{1/x}$ and take logarithms. Then

$$\ln y = \ln (1 + x)^{1/x} = \frac{\ln (1 + x)}{x}.$$

By l'Hôpital's Rule,

$$\lim_{x \to 0} \frac{\ln (1 + x)}{x} = \lim_{x \to 0} \frac{1}{1 + x} = 1.$$

Therefore, $\lim_{x \to 0} \ln y = 1$, and since $e^{\ln y} = y = e^1$, we conclude that

$$\lim_{x \to 0} y = \lim_{x \to 0} (1 + x)^{1/x} = e. \qquad \square$$

**EXAMPLE 8**  Evaluate

$$\lim_{x \to 2} \frac{\ln (x - 1)}{(x - 2)^2}.$$

**Solution**  We have $\ln (2 - 1) = \ln (1) = 0$ and $\lim_{x \to 2} (x - 2)^2 = 0$, so we have the indeterminate form $0/0$. Applying l'Hôpital's Rule, we get

$$\lim_{x \to 2} \frac{\ln(x-1)}{(x-2)^2} = \lim_{x \to 2} \frac{\frac{1}{(x-1)}}{2(x-2)} = \lim_{x \to 2} \frac{1}{2(x-1)(x-2)}.$$

When we examine closely the limit as $x$ tends to 2, we see that

$$\lim_{x \to 2^+} \frac{1}{2(x-1)(x-2)} = +\infty \qquad \text{and} \qquad \lim_{x \to 2^-} \frac{1}{2(x-1)(x-2)} = -\infty.$$

In order for a limit to exist, the same value must be approached as $x$ tends to 2 from the right and from the left. We conclude that the limit does not exist.

□

# 1   PROBLEMS

In each of Problems 1 through 50, find the limit (finite or infinite) when it exists.

**1** $\displaystyle\lim_{x \to -2} \frac{2x^2 + 5x + 2}{x^2 - 4}$

**2** $\displaystyle\lim_{x \to 2} \frac{x^3 - x^2 - x - 2}{x^3 - 8}$

**3** $\displaystyle\lim_{x \to 1} \frac{x^3 - 3x + 2}{x^3 - x^2 - x + 1}$

**4** $\displaystyle\lim_{x \to 2} \frac{x^4 - 3x^2 - 4}{x^3 + 2x^2 - 4x - 8}$

**5** $\displaystyle\lim_{x \to +\infty} \frac{2x^3 - x^2 + 3x + 1}{3x^3 + 2x^2 - x - 1}$

**6** $\displaystyle\lim_{x \to 4} \frac{x^3 - 8x^2 + 2x + 1}{x^4 - x^2 + 2x - 3}$

**7** $\displaystyle\lim_{x \to +\infty} \frac{x^3 - 3x + 1}{2x^4 - x^2 + 2}$

**8** $\displaystyle\lim_{x \to +\infty} \frac{x^4 - 2x^2 - 1}{2x^3 - 3x^2 + 3}$

**9** $\displaystyle\lim_{x \to 0} \frac{\tan 3x}{\sin x}$

**10** $\displaystyle\lim_{x \to 0} \frac{\sin 7x}{x}$

**11** $\displaystyle\lim_{x \to 0} \frac{e^{2x} - 2x - 1}{1 - \cos x}$

**12** $\displaystyle\lim_{x \to 0} \frac{e^{3x} - 1}{1 - \cos x}$

**13** $\displaystyle\lim_{x \to 0} \frac{\ln x}{e^x}$

**14** $\displaystyle\lim_{x \to +\infty} \frac{\ln x}{x^h}, \quad h > 0$

**15** $\displaystyle\lim_{x \to 0} \frac{\ln(1 + 2x)}{3x}$

**16** $\displaystyle\lim_{x \to 0} \frac{3^x - 2^x}{x}$

**17** $\displaystyle\lim_{x \to 0} \frac{3^x - 2^x}{x^2}$

**18** $\displaystyle\lim_{x \to 0} \frac{3^x - 2^x}{\sqrt{x}}$

**19** $\displaystyle\lim_{x \to \pi/2} \frac{1 - \sin x}{\cos x}$

**20** $\displaystyle\lim_{x \to 2} \frac{\sqrt{2x} - 2}{\ln(x-1)}$

**21** $\displaystyle\lim_{x \to \pi/2} \frac{\ln \sin x}{1 - \sin x}$

**22** $\displaystyle\lim_{x \to \pi/2} \frac{\cos x}{\sin^2 x}$

**23** $\displaystyle\lim_{x \to +\infty} \frac{x^3}{e^x}$

**24** $\displaystyle\lim_{x \to \pi/2} \frac{\tan x}{\ln \cos x}$

**25** $\displaystyle\lim_{x \to +\infty} \frac{\sin x}{x}$

**26** $\displaystyle\lim_{x \to \pi/2} \frac{\sin x}{x}$

**27** $\displaystyle\lim_{x \to 0} \frac{x - \arctan x}{x - \sin x}$

**28** $\displaystyle\lim_{x \to 0} \sqrt{x} \ln x$

**29** $\displaystyle\lim_{x \to 0} x \cot x$

**30** $\displaystyle\lim_{x \to \pi/2} (x - \pi/2) \sec x$

**31** $\displaystyle\lim_{x \to +\infty} \frac{\arctan x}{x}$

**32** $\displaystyle\lim_{\theta \to 0} \left( \csc \theta - \frac{1}{\theta} \right)$

**33** $\displaystyle\lim_{x \to 0} \left( \cot^2 x - \frac{1}{x^2} \right)$

**34** $\displaystyle\lim_{x \to 0} x^x$

**35** $\displaystyle\lim_{x \to 0} x^{4x}$

**36** $\displaystyle\lim_{x \to +\infty} \left( 1 + \frac{k}{x} \right)^x$

**37** $\displaystyle\lim_{x \to 0} x^{(x^2)}$

**38** $\displaystyle\lim_{x \to 0} (\cot x)^x$

**39** $\displaystyle\lim_{x \to 0} x^{(1/\ln x)}$

**40** $\displaystyle\lim_{x \to +\infty} \frac{x^p}{e^x}, \quad p > 0$

**41** $\displaystyle\lim_{x \to 0} \frac{\arcsin 2x}{\arcsin x}$

**42** $\displaystyle\lim_{x \to 0} \frac{x - \arctan x}{x \sin x}$

**43** $\displaystyle\lim_{x \to 0} \frac{e^x - e^{-x}}{x^3}$

**44** $\displaystyle\lim_{x \to 0} \frac{e^x + e^{-x}}{x^2}$

**45** $\displaystyle\lim_{x \to +\infty} \frac{x^n}{e^x}, \quad n \text{ a positive integer}$

**46** $\displaystyle\lim_{x \to +\infty} \frac{e^x}{x^n}, \quad n \text{ a positive integer}$

**\*47** $\displaystyle\lim_{x \to +\infty} \frac{e^x}{\sin x}$

**\*48** $\displaystyle\lim_{x \to +\infty} \frac{\sin x}{e^x}$

**\*49** $\displaystyle\lim_{x \to 0^+} (\sin x)^{\tan x}$

**50** $\displaystyle\lim_{x \to 0^+} \frac{\ln x}{x^h}, \quad h \text{ a positive real number}$

**51** Let $f(x) = \int_0^x e^{t^2} \, dt$, $g(x) = e^{x^2}$. Find

$$\lim_{x \to +\infty} \frac{f(x)}{g(x)}.$$

*52 In statistics it is often necessary to estimate the value of $\int_x^\infty e^{-u^2/2}\,du$ when $x$ is large. Show that if $e^{-x^2/2}$ is used as an estimate, it is "conservative." That is, $e^{-x^2/2}$ is probably larger than the quantity being estimated.

53 a) Prove that

$$\lim_{x \to 0^+} x^3 e^{1/x} = +\infty.$$

b) Prove that for every positive integer $n$,

$$\lim_{x \to 0^+} x^n e^{1/x} = +\infty.$$

c) Find the result if in part (b) we have $x \to 0^-$ instead of $x \to 0^+$.

54 By direct methods, find the value of

$$\lim_{x \to +\infty} (x \sin x)/(x^2 + 1).$$

What happens if l'Hôpital's Rule is used? Explain.

55 Prove the following form of l'Hôpital's Rule: If

$$\lim_{x \to +\infty} f(x) = 0, \qquad \lim_{x \to +\infty} g(x) = 0$$

and

$$\lim_{x \to +\infty} \frac{f'(x)}{g'(x)} = L,$$

then

$$\lim_{x \to +\infty} \frac{f(x)}{g(x)} = L.$$

(*Hint:* Consider $\lim_{x \to 0^+} f(1/x)/g(1/x)$.)

56 Suppose that

$$\lim_{x \to 0} f(x) = \lim_{x \to 0} f'(x) = \lim_{x \to 0} f''(x) = \lim_{x \to 0} f'''(x) = 0,$$

and that

$$\lim_{x \to 0} \frac{x^2 f'''(x)}{f''(x)} = 2.$$

Find $\lim_{x \to 0} x^2 f'(x)/f(x)$.

57 If the second derivative $f''$ of a function $f$ exists at a value $x_0$, show that

$$\lim_{h \to 0} \frac{f(x_0 + h) - 2f(x_0) + f(x_0 - h)}{h^2} = f''(x_0).$$

58 Let $P(x)$ and $Q(x)$ be polynomials of degree $m$ and $n$, respectively. Analyze

$$\lim_{x \to +\infty} \frac{P(x)}{Q(x)}$$

according as $m > n$ or $m = n$ or $m < n$.

---

## 2

## SEQUENCES

The numbers

$$3, 8, 17, -12, 15$$

form a sequence of numbers. Since this set contains both a first and a last element, the sequence is termed **finite**. The numbers

$$a_1, a_2, a_3, \ldots, a_{25}, a_{26}, a_{27}$$

form a sequence with 27 elements. The subscripts used here to identify the location of each element are more than a convenience; they provide a way of associating a number with each of 27 positive integers. The process of determining one number when another is given reminds us of the idea of function.

DEFINITION        *A **sequence** is a function the domain of which is a portion of, or all of, the positive integers, and the range of which is any collection of real numbers.*

In writing a sequence, the subscripts form the domain and the members of the sequence make up the range. When the domain of a sequence is all the positive integers or an infinite portion of them, we say the sequence is an **infinite sequence**.

A simple example of an infinite sequence is

$$1, \frac{1}{2}, \frac{1}{3}, \frac{1}{4}, \frac{1}{5}, \dots, \frac{1}{n}, \dots.$$

In this sequence

$$a_1 = 1, \; a_2 = \frac{1}{2}, \; a_3 = \frac{1}{3}, \dots, \; a_n = \frac{1}{n}, \dots.$$

When we draw a horizontal axis we see graphically that the successive terms in the sequence come closer and closer to zero, and yet no term in the sequence actually is zero (Fig. 1). Intuitively, it appears that the further along one gets in this sequence, the more closely the terms approach zero.

**FIGURE 1**

A second example is the sequence

$$\frac{1}{2}, \frac{2}{3}, \frac{3}{4}, \frac{4}{5}, \frac{5}{6}, \dots, \frac{n}{n+1}, \dots,$$

where

$$a_1 = \frac{1}{1+1}, \; a_2 = \frac{2}{2+1}, \; a_3 = \frac{3}{3+1}, \dots, \; a_n = \frac{n}{n+1}, \dots.$$

Graphically it is readily seen that these terms approach 1 as $n$ gets larger, although no individual element in the sequence actually has the value 1 (Fig. 2). We write $a_n \to 1$ as $n \to \infty$ in this case and say that the limit of the sequence is 1. When the natural numbers tend to infinity, we follow the general custom and write $n \to \infty$ rather than $n \to +\infty$.

**FIGURE 2**

**DEFINITION**     *Given the infinite sequence $a_1, a_2, \dots, a_n, \dots$,* **we say that $a_n \to c$ as $n \to \infty$** *if for each $\varepsilon > 0$ there is a positive integer $N$ such that $|a_n - c| < \varepsilon$ for all $n > N$.*

We can visualize this definition by first marking off the quantity $c$ on a number scale (taking $c = 0$ in Fig. 1, and $c = 1$ in Fig. 2), as in Fig. 3. The definition asserts that given any positive number, $\varepsilon$, then, after a certain stage in the sequence is reached, all the terms lie in the interval $(c - \varepsilon, c + \varepsilon)$. That is,

$$c - \varepsilon < a_n < c + \varepsilon$$

for all $n$ larger than some particular integer $N$. The first few (or few million) terms may be scattered anywhere. But if $c$ is to be the limit, then eventually all the terms must be in this interval; the quantity $\varepsilon$ may have any value. If $\varepsilon$ is quite "small," then $N$, the place in the sequence where the terms must begin to be in the interval about $c$, may be required to be very "large."

**FIGURE 3**

Another interesting sequence is given by

$$1, -\tfrac{1}{2}, +\tfrac{1}{3}, -\tfrac{1}{4}, +\tfrac{1}{5}, \ldots,$$

where

$$a_1 = 1, \; a_2 = -\frac{1}{2}, \; a_3 = \frac{1}{3}, \; a_4 = -\frac{1}{4}, \; \ldots, \; a_n = \frac{(-1)^{n+1}}{n}, \; \ldots.$$

The "general term" is worth examining, since $(-1)^{n+1}$ is just equal to $+1$ when $n$ is odd and to $-1$ when $n$ is even. This comes out right, since all the terms with even denominators are negative and all the terms with odd denominators are positive. The sequence tends to zero, but the terms oscillate about the value zero, as shown in Fig. 4.

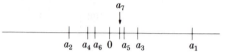

**FIGURE 4**

Suppose that we have a sequence of numbers in which each term in the sequence is larger than the preceding term. We usually think that there are only two possibilities: either (1) the terms increase without bound, i.e., they go off to $+\infty$ as does, for example, the sequence of odd integers

$$1, 3, 5, 7, 9, \ldots,$$

or (2) they cluster about a point which is a limit of the sequence as does, for example, the sequence

$$\tfrac{1}{2}, \tfrac{3}{4}, \tfrac{7}{8}, \tfrac{15}{16}, \tfrac{31}{32}, \ldots,$$

in which the successive terms are

$$a_1 = \frac{1}{2}, \; a_2 = \frac{2^2 - 1}{2^2}, \; a_3 = \frac{2^3 - 1}{2^3}, \; a_4 = \frac{2^4 - 1}{2^4}, \; a_n = \frac{2^n - 1}{2^n}, \; \ldots,$$

and the limit is 1. These facts cannot be proved on the basis of the axioms which form the foundation of arithmetic and elementary algebra. We require an additional axiom known as the **Axiom of Continuity** or **Axiom C** which we now provide.

AXIOM C    **(Axiom of Continuity)**    *Every bounded nondecreasing sequence of real numbers tends to a limit.*

The example above with $a_n = (2^n - 1)/2^n$ illustrates Axiom C. The sequence has the limit $b = 1$ and an upper bound $M$ for the sequence is any number greater than or equal to 1, say $M = 3/2$. Fig. 5 shows the situation.

The evaluation of limits of sequences is quite similar to that of limits of functions. For example, we know that

$$\frac{1}{n} \to 0 \quad \text{as } n \to \infty$$

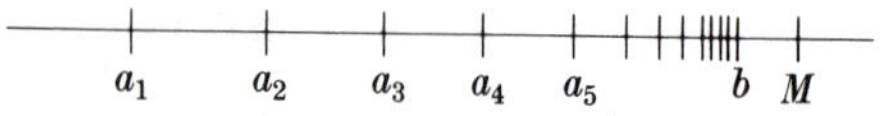

**FIGURE 5**

in much the same way that

$$\frac{1}{x} \to 0 \quad \text{as } x \to \infty.$$

**EXAMPLE 1**    Evaluate the limit

$$\lim_{n \to \infty} \frac{3n^2 - 2n + 1}{4n^2 + 1}.$$

**Solution**    When we divide both numerator and denominator by $n^2$ (a customary device in evaluating the limit of sequences), we have

$$\frac{3n^2 - 2n + 1}{4n^2 + 1} = \frac{3 - (2/n) + (1/n^2)}{4 + (1/n^2)}.$$

The theorems on sums, products, quotients, etc., for limits apply equally well to sequences, and so we get

$$\lim_{n \to \infty} \frac{3 - (2/n) + (1/n^2)}{4 + (1/n^2)} = \frac{\lim_{n \to \infty} [3 - (2/n) + (1/n^2)]}{\lim_{n \to \infty} [4 + (1/n^2)]}$$

$$= \frac{\lim_{n \to \infty} 3 - \lim_{n \to \infty} (2/n) + \lim_{n \to \infty} (1/n^2)}{\lim_{n \to \infty} 4 + \lim_{n \to \infty} (1/n^2)}.$$

We now see that

$$\lim_{n \to \infty} 3 = 3; \qquad \lim_{n \to \infty} \frac{2}{n} = 2 \lim_{n \to \infty} \frac{1}{n} = 0;$$

$$\lim_{n \to \infty} \frac{1}{n^2} = \lim_{n \to \infty} \frac{1}{n} \cdot \lim_{n \to \infty} \frac{1}{n} = 0 \cdot 0 = 0,$$

and so on. This gives us

$$\lim_{n \to \infty} \frac{3n^2 - 2n + 1}{4n^2 + 1} = \frac{3}{4}. \qquad \square$$

An alternative approach to the evaluation of limits is the use of l'Hôpital's Rule. Unfortunately l'Hôpital's Rule at $\infty$ is only defined for functions whose domain contains an interval of the form $(a, \infty)$, for some $a$, and a sequence $a_n$, $a_{n+1}$, $a_{n+2}$, ... is only defined on the positive integers larger than $n$. However if we can find a function $f(x)$ such that $f(n) = a_n$ for each $n$, then we may be able to use the following theorem, which we state without proof.

**THEOREM 5**    *Let $a_1, a_2, \ldots, a_n, \ldots$ be an infinite sequence, and let $f(x)$ be a function defined on $[1, \infty)$ such that $f(n) = a_n$. If $\lim_{x \to \infty} f(x) = L$, then $\lim_{n \to \infty} a_n = L$.*

**EXAMPLE 2**    Find the limit of the sequence $\{a_n\}$, $n = 1, 2, \ldots$, where $a_n$ is given by

$$a_n = \frac{n^{1/3} + n^{2/3}}{1 + n^{1/2} + n^{3/2}}.$$

**Solution**   We observe that the function

$$f(x) = \frac{x^{1/3} + x^{2/3}}{1 + x^{1/2} + x^{3/2}}$$

is defined on $[1, \infty)$ and is such that $f(n) = a_n$. Since

$$f(x) = \frac{g(x)}{h(x)}$$

with $g(x) = x^{1/3} + x^{2/3}$ and $h(x) = 1 + x^{1/2} + x^{3/2}$, and since $\lim_{x \to \infty} g(x) = \infty$ and $\lim_{x \to \infty} h(x) = \infty$, to compute $\lim_{x \to \infty} f(x)$ we can use l'Hôpital's Rule. We have

$$\lim_{x \to \infty} f(x) = \lim_{x \to \infty} \frac{g'(x)}{h'(x)} = \lim_{x \to \infty} \frac{\frac{1}{3}x^{-2/3} + \frac{2}{3}x^{-1/3}}{\frac{1}{2}x^{-1/2} + \frac{3}{2}x^{1/2}}$$

which tends to $0/\infty$; thus the limit is 0. Hence

$$\lim_{n \to \infty} \frac{n^{1/3} + n^{2/3}}{1 + n^{1/2} + n^{3/2}} = 0.$$

☐

What does it mean to say that a sequence does not approach a limit as $n$ tends to infinity? The definition of the limit of a sequence contains a test for deciding when a limit is approached and when it is not. There are many ways in which a sequence may fail to approach a limit, some of which we illustrate with examples.

**EXAMPLE 3**   **(Arithmetic Progression)**   A sequence is called an **arithmetic progression** if it has the property that the difference between successive terms always has the same value. That is, there is a number $d$, called the **common difference**, such that $a_{n+1} - a_n = d$ for all $n$. The sequence

$$1, 4, 7, 10, \ldots, 3n - 2, \ldots,$$

in which $d = 3$, is an example of an arithmetic progression. When $d$ is positive, the $n$th term of such a sequence tends to $+\infty$, and when $d$ is negative, it tends to $-\infty$.

A sequence such as

$$\tfrac{1}{2}, \tfrac{3}{4}, \tfrac{1}{4}, \tfrac{7}{8}, \tfrac{1}{8}, \tfrac{15}{16}, \tfrac{1}{16}, \tfrac{31}{32}, \tfrac{1}{32}, \ldots,$$

in which

$$a_1 = \tfrac{1}{2}, \ a_2 = \tfrac{3}{4}, \ a_3 = \tfrac{1}{4}, \ a_4 = \tfrac{7}{8}, \ a_5 = \tfrac{1}{8}, \ldots,$$

has as its formula for the general term

$$a_{2n-1} = \frac{1}{2^n} \quad \text{(for odd-numbered terms),}$$

and

$$a_{2n} = \frac{2^{n+1} - 1}{2^{n+1}} \quad \text{(for even-numbered terms).}$$

SEQUENCES

This sequence does not tend to a limit, since there are *two* numbers, namely 0 and 1, toward which the terms cluster (Fig. 6). In order for a sequence to approach a limit, there must be *exactly one number* about which the terms cluster (uniqueness of limits). □

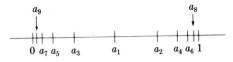

FIGURE 6

EXAMPLE 4 **(Geometric Progression)** A **geometric progression** is a sequence in which there is some number $r$, called the **common ratio**, with the property that

$$\frac{a_{n+1}}{a_n} = r \quad \text{for all } n.$$

Examples of geometric progressions are

$$\frac{3}{2}, \frac{3}{4}, \frac{3}{8}, \frac{3}{16}, \ldots, \frac{3}{2^n}, \ldots \quad \left(r = \frac{1}{2}\right);$$

and

$$6, 18, 54, 162, 486, \ldots, 2 \cdot 3^n, \ldots \quad (r = 3).$$

If $a, a_1, a_2, \ldots, a_n, \ldots$ is a geometric progression with common ratio $r$, it is not hard to show that $a_n = a_0 r^n$, $n = 1, 2, \ldots$. □

The following facts about geometric progressions can be proved.

THEOREM 6 *If $-1 < r < 1$ in a geometric progression, the limit of the sequence is zero. If $r > 1$, the sequence does not tend to a limit. If $r = 1$, the terms are all identical and the limit is this common term. If $r = -1$, the sequence is of the form*

$$a, -a, a, -a, a, -a, \ldots,$$

*which has no limit except in the case $a = 0$.*

## 2   PROBLEMS

In each of Problems 1 through 20, evaluate the limit of the sequence (or show that $a_n \to +\infty$, or $-\infty$).

**1** $\displaystyle\lim_{n \to \infty} \frac{3}{\sqrt{n}}$

**2** $\displaystyle\lim_{n \to \infty} \frac{2n - 3}{5n + 2}$

**3** $\displaystyle\lim_{n \to \infty} \frac{3n - 1}{4 - 2n}$

**4** $\displaystyle\lim_{n \to \infty} \frac{5n - 6}{n^2 + 3n}$

**5** $\displaystyle\lim_{n \to \infty} \frac{2n^2 + 3n - 1}{n + 4}$

**6** $\displaystyle\lim_{n \to \infty} \frac{2n^2 + 3n - 5}{3n^2}$

**7** $\displaystyle\lim_{n \to \infty} \frac{\sqrt[3]{n + 1}}{n}$

**8** $\displaystyle\lim_{n \to \infty} \frac{1}{\sqrt[3]{n^2 + 2}}$

**9** $\displaystyle\lim_{n \to \infty} \frac{2n^2 + n + 1}{n^2 + 1}$

**10** $\displaystyle\lim_{n \to \infty} \frac{2n^2 + 3n + 4}{n^3}$

**11** $\displaystyle\lim_{n \to \infty} \frac{n + 1}{n^2 + 3}$

**12** $\displaystyle\lim_{n \to \infty} (\sqrt{2n + 1} - \sqrt{2n - 1})$

**13** $\displaystyle\lim_{n \to \infty} (\sqrt{n^2 + 1} - n)$

**14** $\displaystyle\lim_{n \to \infty} (\sqrt{n^2 + n + 1} - n)$

**15** $\lim\limits_{n\to\infty} \dfrac{\sqrt[2]{n} + \sqrt[3]{n} + 1}{3\sqrt[2]{2n} - \sqrt[4]{n} + 5}$

**16** $\lim\limits_{n\to\infty} \dfrac{\sqrt[2]{n^3 + 2n} + \sqrt[4]{n}}{\sqrt[3]{n^2 + 1} + \sqrt[5]{n}}$

**17** $\lim\limits_{n\to\infty} \dfrac{n}{e^n}$

**18** $\lim\limits_{n\to\infty} \dfrac{2^n}{e^n}$

**19** $\lim\limits_{n\to\infty} \dfrac{n}{\ln(n^2)}$

**20** $\lim\limits_{n\to\infty} \dfrac{\sin(1/n)}{1/n}$

**21** Prove that an arithmetic progression with $d \neq 0$ cannot tend to a limit.

**22** Prove that a geometric progression with $0 < r < 1$ must tend to the limit zero.

**23** Show that

$$0 < \frac{n}{3^n} < \left(\frac{2}{3}\right)^n$$

for every $n \geq 1$ and, therefore, that

$$\lim_{n\to\infty} \frac{n}{3^n} = 0.$$

**24** Show that

$$\lim_{n\to\infty} \frac{n}{2^n} = 0.$$

**25** Does the sequence

$$\frac{1}{2}, \frac{2}{3}, \frac{3}{4}, \frac{4}{5}, \dots, \frac{n}{n+1}, \dots$$

tend to a limit?

**26** Does the sequence

$$\frac{1}{.1}, \frac{1}{.01}, \frac{1}{.001}, \frac{1}{.0001}, \dots$$

tend to a limit?

**27** Does the sequence

$$\frac{1}{2}, \frac{3}{2}, \frac{2}{3}, \frac{4}{3}, \dots, \frac{n-1}{n}, \frac{n+1}{n}, \dots$$

tend to a limit?

**28** Does the sequence

$$2, \frac{3}{2}, 2, \frac{7}{4}, 2, \frac{15}{8}, 2, \frac{31}{16}, \dots, 2, \frac{2^{n+1} - 1}{2^n}, \dots$$

tend to a limit?

**29** Does the sequence

$$\frac{1}{2}, -\frac{1}{4}, \frac{1}{8}, -\frac{1}{16}, \frac{1}{32}, -\frac{1}{64}, \dots, \frac{(-1)^{n+1}}{2^n}, \dots$$

tend to a limit?

**30** Does the sequence

$$1, -1, 1, -1, 1, -1, \dots$$

tend to a limit?

**\*31** Consider all positive solutions of the equation $\sin x = \frac{1}{2}$ arranged in order of size. Denote these numbers $r_1, r_2, r_3,$ .... Show that

$$\lim_{n\to\infty} \left(\frac{1}{r_n}\right) = 0.$$

**\*32** Let $P(x) = a_r x^r + a_{r-1} x^{r-1} + \cdots + a_1 x + a_0$ and $Q(x) = b_s x^s + b_{s-1} x^{s-1} + \cdots + b_1 x + b_0$ be two polynomials with $a_r \neq 0$, $b_s \neq 0$. Find the value of

$$\lim_{n\to\infty} \frac{P(n)}{Q(n)}$$

according as $r > s$, $r = s$, and $r < s$.

**\*33** Show that every nonincreasing sequence of numbers which is bounded from below must tend to a limit.

---

## 3

## CONVERGENT AND DIVERGENT SERIES

Let $u_1, u_2, \dots, u_{24}$ be a finite sequence. An expression such as

$$u_1 + u_2 + u_3 + \cdots + u_{24}$$

is called a *finite series*. The **sum** of such a series is obtained by adding the 24 terms. We now extend the notion of a finite series by considering an expression of the form

$$u_1 + u_2 + u_3 + \cdots + u_n + \cdots$$

which is nonterminating and which we call an **infinite series**.*

Our first task is to give a meaning, if possible, to an infinite succession of additions.

**DEFINITION** | *Given the infinite series* $u_1 + u_2 + u_3 + \cdots + u_n + \cdots$, *the quantity*

$$s_k = u_1 + u_2 + \cdots + u_k$$

*is called the $k$th **partial sum** of the series. That is,*

$$s_1 = u_1, \qquad s_2 = u_1 + u_2, \qquad s_3 = u_1 + u_2 + u_3,$$

*etc. Each partial sum is obtained simply by a **finite** number of additions.*

**DEFINITION** | *Given the series*

$$u_1 + u_2 + u_3 + \cdots + u_n + \cdots \tag{1}$$

*with the sequence of partial sums*

$$s_1, s_2, s_3, \ldots, s_n, \ldots, \tag{2}$$

*we define the **sum of the series** (1) to be*

$$\lim_{n \to \infty} s_n$$

*whenever the limit exists.*

Using the $\Sigma$ notation for sum, we can also write

$$\sum_{n=1}^{\infty} u_n = \lim_{n \to \infty} s_n.$$

If the limit (2) does not exist, then the sum (1) *is not defined.*

**DEFINITIONS** | *If the limit (2) exists, the series $\Sigma_{n=1}^{\infty} u_n$ is said to **converge** to that limit; otherwise the series is said to **diverge**.*

*Remark.* The expression $\Sigma_{n=1}^{\infty} u_n$ is a shorthand notation for the formal series expression (1). However, the symbol $\Sigma_{n=1}^{\infty} u_n$ is also used as a synonym for the numerical value of the series when it converges. There will be no difficulty in recognizing which meaning we are employing in any particular case.

The sequence of terms

$$a, ar, ar^2, ar^3, \ldots, ar^{n-1}, ar^n, \ldots$$

---

*The definition given here is informal. A more formal definition is as follows: An **infinite series** is an ordered pair ($\{u_n\}, \{s_n\}$) of infinite sequences in which $s_k = u_1 + \cdots + u_k$ for each $k$. The infinite series ($\{u_n\}, \{s_n\}$) is denoted by $u_1 + u_2 + \cdots + u_n + \cdots$ or $\Sigma_{n=1}^{\infty} u_n$. When no confusion can arise we also denote by $\Sigma_{n=1}^{\infty} u_n$ the limit of the sequence $\{s_n\}$ when it exists.

forms a *geometric progression*. Each term (except the first) is obtained by multiplication of the preceding term by $r$, the *common ratio*. The partial sums of the **geometric series**

$$a + ar + ar^2 + ar^3 + \cdots + ar^n + \cdots$$

are

$$s_1 = a,$$

$$s_2 = a + ar,$$

$$s_3 = a + ar + ar^2,$$

$$s_4 = a + ar + ar^2 + ar^3,$$

and, in general,

$$s_n = a(1 + r + r^2 + \cdots + r^{n-1}).$$

For example, with $a = 2$ and $r = \frac{1}{2}$,

$$s_n = 2\left(1 + \frac{1}{2} + \frac{1}{4} + \cdots + \frac{1}{2^{n-1}}\right).$$

The identity

$$(1 + r + r^2 + \cdots + r^{n-1})(1 - r) = 1 - r^n,$$

which may be verified by straightforward multiplication, leads to the formula

$$s_n = a\frac{1 - r^n}{1 - r}$$

for the $n$th partial sum. The example $a = 2$, $r = \frac{1}{2}$ gives

$$s_n = 2\frac{1 - 2^{-n}}{\frac{1}{2}} = 4 - \frac{1}{2^{n-2}}.$$

In general, we may write

$$s_n = a\frac{1 - r^n}{1 - r} = \frac{a}{1 - r} - \frac{a}{1 - r}r^n, \quad r \neq 1. \tag{3}$$

The next theorem is a direct consequence of Formula (3).

**THEOREM 7**    *A geometric series*

$$a + ar + ar^2 + \cdots + ar^n + \cdots$$

*converges if $-1 < r < 1$ and diverges if $|r| \geq 1$. In the convergent case we have*

$$\sum_{n=1}^{\infty} ar^{n-1} = \frac{a}{1 - r}. \tag{4}$$

**Proof**    From (3) we see that $r^n \to 0$ if $|r| < 1$, yielding (4); also, $r^n \to \infty$ if $|r| > 1$. For $r = 1$, the partial sum $s_n$ is $na$, and $s_n$ does not tend to a limit as $n \to \infty$. If $r = -1$, the partial sum $s_n$ is $a$ if $n$ is odd and $0$ if $n$ is even; the partial sums oscillate between $0$ and $a$, and the series diverges.    □

The geometric series is particularly simple, since we only need to find the value of $r$ to determine whether or not the series converges. In the case of a general series it is much more difficult to decide if it converges or diverges. Nevertheless, several important results can be obtained easily. The next theorem is useful in that it exhibits a limitation on the behavior of the terms of a convergent series.

---

**THEOREM 8**   *If the series*

$$\sum_{k=1}^{\infty} u_k = u_1 + u_2 + u_3 + \cdots + u_n + \cdots$$

*converges, then*

$$\lim_{n \to \infty} u_n = 0.$$

---

**Proof**   Writing

$$s_n = u_1 + u_2 + \cdots + u_n,$$

$$s_{n-1} = u_1 + u_2 + \cdots + u_{n-1},$$

we have, by subtraction, $u_n = s_n - s_{n-1}$. Letting $c$ denote the sum of the series, we see that $s_n \to c$ as $n \to \infty$; also, $s_{n-1} \to c$ as $n \to \infty$. Therefore

$$\lim_{n \to \infty} u_n = \lim_{n \to \infty} (s_n - s_{n-1}) = \lim_{n \to \infty} s_n - \lim_{n \to \infty} s_{n-1} = c - c = 0. \qquad \square$$

*Remark.*   The converse of Theorem 8 is not necessarily true. Example 3 of the next section shows that it is possible both for $u_n$ to tend to 0 and for the series to diverge.

The following corollary, a restatement of Theorem 8, is useful in establishing the divergence of infinite series.

---

**COROLLARY**   *If $u_n$ does not tend to zero as $n \to \infty$, then the series $\Sigma_{n=1}^{\infty} u_n$ is divergent.*

Convergent series may be added, subtracted, and multiplied by constants, as the next theorem shows.

---

**THEOREM 9**   *If $\Sigma_{n=1}^{\infty} u_n$ and $\Sigma_{n=1}^{\infty} v_n$ both converge and c is any number, then the series*

$$\sum_{n=1}^{\infty} (cu_n), \qquad \sum_{n=1}^{\infty} (u_n + v_n), \qquad \sum_{n=1}^{\infty} (u_n - v_n)$$

*all converge, and*

$$\sum_{n=1}^{\infty} (cu_n) = c \sum_{n=1}^{\infty} u_n,$$

$$\sum_{n=1}^{\infty} (u_n \pm v_n) = \sum_{n=1}^{\infty} u_n \pm \sum_{n=1}^{\infty} v_n.$$

**Proof**    For each $n$, we have the following equalities for the partial sums:

$$\sum_{j=1}^{n} (cu_j) = c \sum_{j=1}^{n} u_j;$$

$$\sum_{j=1}^{n} (u_j \pm v_j) = \sum_{j=1}^{n} u_j \pm \sum_{j=1}^{n} v_j.$$

The results follow from the theorems on limits as given in Chapter 2, Section 7.    □

**EXAMPLE**    Express the repeating decimal $A = 0.151515 \ldots$ as the ratio of two integers. (A **rational number** is any number expressible as the ratio of two integers.)

**Solution**    We write $A$ in the form of a geometric series:

$$A = 0.15(1 + 0.01 + (0.01)^2 + (0.01)^3 + \cdots),$$

in which $a = 0.15$ and $r = 0.01$. This series is convergent and has sum

$$s = \frac{0.15}{1 - 0.01} = \frac{0.15}{0.99} = \frac{5}{33} = A.$$

□

## 3    PROBLEMS

In Problems 1 through 13, express each repeating decimal as the ratio of two integers.

**1**  0.717171 …

**2**  0.464646 …

**3**  0.013013013 …

**4**  2.718718718 …

**5**  0.000141414 …

**6**  32.46513513513 …

**7**  3.614361436143 …

**8**  42.000100010001 …

**9**  0.454545 …

**10**  0.636363 …

**11**  0.46666 …

**12**  1.13333 …

**13**  0.714285714285 …

**14**  Find the sum of the geometric series if $a = 3$, $r = -\frac{1}{3}$.

**15**  Find the sum of the geometric series if $a = -2$, $r = \frac{1}{4}$.

**16**  The first term of a geometric series is 3 and the fifth term is $\frac{16}{27}$. Find the sum of the infinite series.

**17**  The fourth term of a geometric series is $-1$, and the seventh term is $\frac{1}{8}$. Find the sum of the infinite series.

**18**  A ball is dropped from a height of 6 meters. Upon each bounce the ball rises to one-third of its previous height. Find the total distance traveled by the ball.

In Problems 19 through 27, write the first five terms of the series given. Use the corollary to Theorem 8 to show that the series is divergent.

**19**  $\displaystyle\sum_{n=1}^{\infty} \frac{n^2}{n+1}$

**20**  $\displaystyle\sum_{n=1}^{\infty} \frac{2n}{3n+5}$

**21**  $\displaystyle\sum_{n=1}^{\infty} \frac{n^2 - 2n + 3}{2n^2 + n + 1}$

**22**  $\displaystyle\sum_{n=1}^{\infty} (-1)^{n+1} \frac{e^n}{n^3}$

**23**  $\displaystyle\sum_{n=1}^{\infty} \frac{n^2 + n + 2}{\ln(n+1)}$

**24**  $\displaystyle\sum_{n=1}^{\infty} \frac{n}{n+1}$

**25**  $\displaystyle\sum_{n=1}^{\infty} \frac{n}{e^{-n}}$

**26**  $\displaystyle\sum_{n=1}^{\infty} \frac{3^n}{2^n}$

**27**  $\displaystyle\sum_{n=1}^{\infty} \frac{n \cos(n\pi)}{2n+3}$

In Problems 28 through 34, *assume* that the series

$$\sum_{n=1}^{\infty} \frac{1}{n^2}, \quad \sum_{n=1}^{\infty} \frac{1}{n^3}, \quad \sum_{n=1}^{\infty} \frac{1}{n^4}$$

all converge. In each case use Theorem 9 to show that the given series is convergent.

**28**  $\displaystyle\sum_{n=1}^{\infty} \frac{3n+2}{n^3}$

**29**  $\displaystyle\sum_{n=1}^{\infty} \frac{n-2}{n^3}$

**30** $\displaystyle\sum_{n=1}^{\infty} \frac{3n^2 + 4}{n^4}$

**31** $\displaystyle\sum_{n=1}^{\infty} \frac{3n^2 - 2n + 4}{n^4}$

**32** $\displaystyle\sum_{n=1}^{\infty} \frac{n^2 + 1}{n^4}$

**33** $\displaystyle\sum_{n=1}^{\infty} \frac{2(n-1)^2}{3n^5}$

**34** $\displaystyle\sum_{n=1}^{\infty} \frac{\ln(2^n)}{n^3}$

**35** Suppose that the series $\sum_{k=1}^{\infty} u_k$ converges. Show that any series obtained from this one by deleting a finite number of terms also converges. (*Hint:* Since $s_n = \sum_{k=1}^{n} u_k$ converges to

a limit, find the value to which $S_n$, the partial sums of the deleted series, must tend.)

**36** Show that any number of the form

$$0.a_1 a_2 a_3 a_1 a_2 a_3 a_1 a_2 a_3 \ldots,$$

where $a_1$, $a_2$, and $a_3$ are digits between 0 and 9, is expressible as the ratio of two integers and therefore is a rational number.

**37** Show that any repeating decimal is a rational number.

---

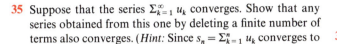

# SERIES OF POSITIVE TERMS. COMPARISON AND LIMIT COMPARISON TESTS. THE INTEGRAL TEST

Except in very special cases, it is not possible to tell if a series converges by finding whether or not $s_n$, the $n$th partial sum, tends to a limit. (The geometric series, however, is one of the special cases where it **is** possible.) In this section, we present some *indirect* tests for convergence and divergence which apply only to series with positive (or at least nonnegative) terms. That is, we assume throughout this section that $u_n \geq 0$ for $n = 1, 2, \ldots$. Tests for series with terms which may be positive or negative will be discussed in the following sections.

**THEOREM 10** *Suppose that $u_n \geq 0$, $n = 1, 2, \ldots$ and $s_n = \sum_{k=1}^{n} u_k$ is the $n$th partial sum. Then, either (a), there is a number $M$ such that all the $s_n \leq M$, in which case the series $\sum_{k=1}^{\infty} u_k$ converges to a value $s \leq M$, or else (b), $s_n \to +\infty$ and the series diverges.*

**Proof** By subtraction, we have

$$u_n = s_n - s_{n-1} \geq 0,$$

and so the $s_n$ form an increasing (or at least nondecreasing) sequence. If all $s_n \leq M$, then by Axiom C (page 388), which says that a nondecreasing bounded sequence has a limit, we conclude that $s_n \to s \leq M$. Thus part (a) of the theorem is established. If there is no such $M$, then for each number $E$, no matter how large, there must be an $s_n > E$; and all $s_m$ with $m > n$ are greater than or equal to $s_n$. This is another way of saying $s_n \to +\infty$. $\square$

The next theorem is one of the most useful tests for deciding convergence and divergence of series.

**THEOREM 11**    **(Comparison Test)**  *Suppose that all $u_n \geq 0$. (a) If $\Sigma_{n=1}^{\infty} a_n$ is a convergent series and $u_n \leq a_n$ for all $n$, then $\Sigma_{n=1}^{\infty} u_n$ is convergent and*

$$\sum_{n=1}^{\infty} u_n \leq \sum_{n=1}^{\infty} a_n.$$

b) *If $\Sigma_{n=1}^{\infty} a_n$ is a divergent series of nonnegative terms and $u_n \geq a_n$ for all $n$, then $\Sigma_{n=1}^{\infty} u_n$ diverges.*

**Proof**    Let

$$s_n = u_1 + u_2 \cdots + u_n, \qquad S_n = a_1 + a_2 + \cdots + a_n$$

be the $n$th partial sums. The $s_n$, $S_n$ are both nondecreasing sequences. In case (a), we let $S$ be the limit of $S_n$ and, since $s_n \leq S_n \leq S$ for every $n$, we apply Theorem 10 to conclude that $s_n$ converges. In case (b), we have $S_n \to +\infty$ and $s_n \geq S_n$ for every $n$. Hence, $s_n \to +\infty$.    □

*Remarks.*    (i) In order to apply the Comparison Test, the reader must show either (a), that the terms $u_n$ of the given series are $\leq a_n$ where $\Sigma_{n=1}^{\infty} a_n$ is a *known* convergent series or (b), that each $u_n \geq a_n$ where $\Sigma_{n=1}^{\infty} a_n$ is a *known* divergent series. *In all other cases, no conclusion can be drawn.* (ii) Since any *finite* number of terms at the beginning of a series does not affect convergence or divergence, the comparison between $u_n$ and $a_n$ in Theorem 11 is not required for all $n$. It is required for all $n$ *except a finite number.*

For the Comparison Test to be useful, we must have at hand as large a number as possible of series (of positive terms) about whose convergence and divergence we are fully informed. Then, when confronted with a new series of positive terms, we shall have available a collection of series for comparison purposes. So far, the only series which we have shown to be convergent are the geometric series with $r < 1$, and the only series which we have shown to be divergent are those in which $u_n$ does not tend to zero. We now study the convergence and divergence of a few special types of series in order to obtain material which can be used for the Comparison Test.

**DEFINITION**    *If $n$ is a positive integer, we define $n!$ (read $n$ factorial) $= 1 \cdot 2 \cdots n$; it is convenient to define $0! = 1$.*

For example, $5! = 1 \cdot 2 \cdot 3 \cdot 4 \cdot 5 = 120$, etc. We see that

$$(n + 1)! = (n + 1) \cdot n!, \quad n \geq 0.$$

**EXAMPLE 1**    Test the series

$$\sum_{n=1}^{\infty} \frac{1}{n!}$$

for convergence or divergence.

**Solution**    Writing the first few terms, we obtain

$$\frac{1}{1!} = \frac{1}{1}, \quad \frac{1}{2!} = \frac{1}{1 \cdot 2}, \quad \frac{1}{3!} = \frac{1}{1 \cdot 2 \cdot 3}, \quad \frac{1}{4!} = \frac{1}{1 \cdot 2 \cdot 3 \cdot 4}.$$

Since each factor except 1 and 2 in $n!$ is larger than 2, we have the inequalities (which may be rigorously established by mathematical induction):

$$n! \geq 2^{n-1} \quad \text{and therefore} \quad \frac{1}{n!} \leq \frac{1}{2^{n-1}}.$$

The series $\sum_{n=1}^{\infty} a_n$ with $a_n = 1/2^{n-1}$ is a geometric series with $r = \frac{1}{2}$, and is therefore convergent. Hence, by the Comparison Test, $\sum_{n=1}^{\infty} 1/n!$ converges. $\square$

The next theorem gives us an entire collection of series useful for comparison purposes.

**THEOREM 12**    **(The $p$-series)**    *The series*

$$\sum_{n=1}^{\infty} \frac{1}{n^p},$$

*known as the p-series, is convergent if $p > 1$ and divergent if $p \leq 1$.*

The proof of this theorem is deferred until later in the section. We note that it is not necessary that $p$ be an integer.

**EXAMPLE 2**    Test the series

$$\sum_{n=1}^{\infty} \frac{1}{n(n+1)}$$

for convergence or divergence.

**Solution**    For each $n$ we have

$$\frac{1}{n(n+1)} \leq \frac{1}{n^2}.$$

Since $\sum_{n=1}^{\infty} 1/n^2$ is a $p$-series with $p = 2$ and so converges, we are in a position to use the Comparison Test. Therefore, the series

$$\sum_{n=1}^{\infty} \frac{1}{n(n+1)}$$

converges. $\square$

**EXAMPLE 3**    Test the series

$$\sum_{n=1}^{\infty} \frac{1}{n+10}$$

for convergence or divergence.

**Solution 1**   Writing out a few terms, we have

$$\tfrac{1}{11} + \tfrac{1}{12} + \tfrac{1}{13} + \tfrac{1}{14} \dots,$$

and we see that the series is just like $\Sigma_{n=1}^{\infty} 1/n$, except that the first ten terms are missing. According to Remark (ii) after Theorem 11, we may compare the given series with the $p$-series for $p = 1$. The comparison establishes divergence.

**Solution 2**   We have, for every $n \geq 1$,

$$n + 10 \leq 11n, \quad \text{and so} \quad \frac{1}{n+10} \geq \frac{1}{11n}.$$

The series

$$\sum_{n=1}^{\infty} \frac{1}{11n} = \frac{1}{11} \sum_{n=1}^{\infty} \frac{1}{n}$$

is divergent ($p$-series with $p = 1$) and, therefore, the given series diverges. $\square$

A useful variation of the Comparison Test is given in the next theorem.

**THEOREM 13**   **(Limit Comparison Test)**   *Suppose that all $a_n \geq 0$ and all $b_n \geq 0$, $n = 1, 2, 3, \dots$, and suppose that*

$$\lim_{n \to \infty} \frac{a_n}{b_n} = L > 0.$$

*Then either $\Sigma_{n=1}^{\infty} a_n$ and $\Sigma_{n=1}^{\infty} b_n$ both converge, or they both diverge.*

**Proof**   Since

$$\lim_{n \to \infty} \frac{a_n}{b_n} = L > 0,$$

there must exist an $N$ such that if $n \geq N$, then

$$\frac{L}{2} < \frac{a_n}{b_n} < \frac{3L}{2}.$$

This is equivalent to

$$\left(\frac{L}{2}\right) b_n < a_n < \left(\frac{3L}{2}\right) b_n,$$

since $b_n \geq 0$, for each $n \geq N$. To show that both series converge together we first apply the Comparison Test (Theorem 11) to the left half of the above inequality. We find that if $\Sigma_{n=N}^{\infty} a_n < \infty$ then

$$\sum_{n=N}^{\infty} \left(\frac{L}{2}\right) b_n < \infty,$$

and since $L/2$ is a constant we have $\Sigma_{n=N}^{\infty} b_n < \infty$. But $\Sigma_{n=1}^{\infty} a_n < \infty$ if and only if $\Sigma_{n=N}^{\infty} a_n < \infty$, since the first part of a series does not affect convergence.

The same is true for $\Sigma_{n=1}^{\infty} b_n$, and $\Sigma_{n=N}^{\infty} b_n$. Thus if $\Sigma_{n=1}^{\infty} a_n < \infty$, then $\Sigma_{n=1}^{\infty} b_n < \infty$. Also using the right half of the inequality above, if $\Sigma_{n=1}^{\infty} b_n$ converges, then so does

$$\sum_{n=1}^{\infty} \left(\frac{3L}{2}\right) b_n,$$

and again by the Comparison Test (Theorem 11) we have that $\Sigma_{n=1}^{\infty} a_n$ converges. We may apply the same arguments (part (b) of Theorem 11) to show the equivalence of the divergence of the two series.  □

**EXAMPLE 4**   Test the series

$$\sum_{n=1}^{\infty} \frac{2n^2 + n + 1}{10n^3 + n}$$

for convergence or divergence.

**Solution**   For large $n$ we see that $2n^2$ is the dominant term in $2n^2 + n + 1$ and $10n^3$ is the dominant term in $10n^3 + n$. Hence

$$\frac{2n^2 + n + 1}{10n^3 + n}$$

behaves like

$$\frac{2n^2}{10n^3} = \frac{1}{5n},$$

so we try the Limit Comparison Test using the $p$-series, with $p = 1$ for comparison. Let

$$a_n = \frac{2n^2 + n + 1}{10n^3 + n}, \qquad \text{and} \qquad b_n = \frac{1}{n}.$$

Then

$$\frac{a_n}{b_n} = \frac{2n^3 + n^2 + n}{10n^3 + n}.$$

We then observe that

$$\frac{a_n}{b_n} = \frac{\frac{1}{n^3} a_n}{\frac{1}{n^3} b_n} = \frac{2 + \frac{1}{n} + \frac{1}{n^2}}{10 + \frac{1}{n^2}},$$

which tends to $\frac{1}{5}$ as $n$ tends to $\infty$. Since $\frac{1}{5} > 0$, the series converges or diverges together with the $p$-series, with $p = 1$. Since this $p$-series diverges by Theorem 12, we conclude the series

$$\sum_{n=1}^{\infty} \frac{2n^2 + n + 1}{10n^3 + n}$$

also diverges.  □

The next example illustrates how useful the Limit Comparison Test can be in complicated series, if we combine it with l'Hôpital's Rule.

**EXAMPLE 5**   Test the series

$$\sum_{n=1}^{\infty} \frac{1}{\sqrt[3]{n^2+7}}$$

for convergence or divergence.

**Solution**   For large $n$ we observe that $(n^2+7)^{1/3}$ behaves like $n^{2/3}$, so we try the Limit Comparison Test with the $p$-series with $p=\frac{2}{3}$. Letting

$$a_n = \frac{1}{\sqrt[3]{n^2+7}}, \quad \text{and} \quad b_n = \frac{1}{n^{2/3}},$$

we have

$$\frac{a_n}{b_n} = \frac{n^{2/3}}{\sqrt[3]{n^2+7}} = \frac{n^{2/3}}{(n^2+7)^{1/3}}.$$

To calculate the limit we use l'Hôpital's Rule with $f(x) = x^{2/3}$ and $g(x) = (x^2+7)^{1/3}$. We then have

$$\lim_{x \to \infty} \frac{f(x)}{g(x)} = \lim_{x \to \infty} \left( \frac{x^2}{x^2+7} \right)^{1/3} = \left( \lim_{x \to \infty} \frac{x^2}{x^2+7} \right)^{1/3}$$

$$= \left( \lim_{x \to \infty} \frac{2x}{2x} \right)^{1/3} = 1.$$

Since $1 > 0$, by the Limit Comparison Test the series converges or diverges with the $p$-series, with $p=\frac{2}{3}$. By Theorem 12 this $p$-series diverges, and we conclude the given series diverges.  $\square$

The next theorem yields another test used frequently in conjunction with the Comparison Test and the Limit Comparison Test.

**THEOREM 14**   **(Integral Test)**   *Assume that $f$ is a continuous, nonnegative, and nonincreasing function defined for all $x \geq 1$. That is, suppose that*

$$f(x) \geq 0, \ (nonnegative)$$

*and*

$$f(x) \geq f(y) \quad for \ x \leq y \ (nonincreasing).$$

*Suppose that $\sum_{n=1}^{\infty} u_n$ is a series with*

$$u_n = f(n) \quad for \ each \ n \geq 1.$$

*Then (a) $\sum_{n=1}^{\infty} u_n$ is convergent if the improper integral $\int_1^{\infty} f(x) \, dx$ is convergent and, conversely (b), the improper integral converges if the series does.*

**Proof**   a) Suppose first that the improper integral is convergent. Then, since $f(x) \geq f(j)$ for $x \leq j$, it follows that

$$\int_{j-1}^{j} f(x) \, dx \geq f(j), \tag{1}$$

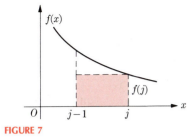

FIGURE 7

a fact verified by noting in Fig. 7 that $f(j)$ is the shaded area and the integral is the area under the curve from $j-1$ to $j$. We define

$$a_j = \int_{j-1}^{j} f(x)\, dx,$$

and then (see Fig. 8)

$$a_1 + \int_{1}^{n} f(x)\, dx = a_1 + a_2 + \cdots + a_n,$$

where we have set $a_1 = f(1) = u_1$. By hypothesis, $\int_{1}^{\infty} f(x)\, dx$ is finite, and so the series $\Sigma_{n=1}^{\infty} a_n$ is convergent. Since $f(j) = u_j$, the inequality $a_j \geq u_j$ is a restatement of (1), and now the Comparison Test applies to yield the result.

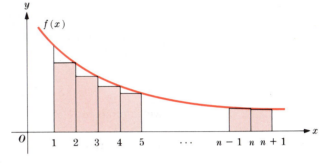

FIGURE 8

b) Suppose next that $\Sigma_{n=1}^{\infty} u_n$ converges. Let

$$v_n = \int_{n}^{n+1} f(x)\, dx \leq f(n) = u_n, \quad n = 1, 2, \ldots;$$

the inequality holds since $f(x) \leq f(n)$ for $n \leq x$ (Figs. 9 and 10). Because $f(x) \geq 0$, we have each $v_n \geq 0$, and so $\Sigma_{n=1}^{\infty} v_n$ converges to some number $S$. That is,

$$\sum_{k=1}^{n} v_k = \int_{1}^{n+1} f(x)\, dx \leq S$$

for every $n$. Let $\varepsilon$ be any positive number. There is an $N$ such that

$$S - \varepsilon < \int_{1}^{n+1} f(x)\, dx \leq S \quad \text{for all } n \geq N.$$

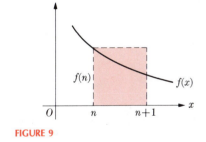

FIGURE 9

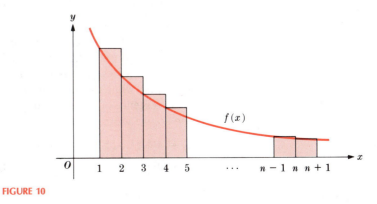

FIGURE 10

Since $f(x) \geq 0$, we see that

$$S - \varepsilon < \int_1^{n+1} f(x)\,dx \leq \int_1^X f(x)\,dx \leq S \quad \text{if } X \geq N + 1.$$

Hence the improper integral converges. $\qquad\square$

We now employ the Integral Test to establish the convergence and divergence of the $p$-series.

**Proof of Theorem 12**    We define the function $f(x) = 1/x^p$, which satisfies all the conditions of the Integral Test if $p > 0$. We have

$$\int_1^\infty \frac{dx}{x^p} = \lim_{t \to +\infty} \int_1^t \frac{dx}{x^p} = \lim_{t \to +\infty} \left( \frac{t^{1-p} - 1}{1 - p} \right) \quad \text{for } p \neq 1.$$

The limit exists for $p > 1$ and fails to exist for $p < 1$. Therefore, by Theorem 14, the $p$-series converges for $p > 1$ and diverges for $p < 1$. As for the case $p = 1$, we have

$$\int_1^t \frac{dx}{x} = \ln t,$$

which tends to $+\infty$ as $t \to +\infty$. Thus Theorem 12 is established. $\qquad\square$

**EXAMPLE 6**    Test the series

$$\sum_{n=1}^\infty \frac{1}{(n+1)\ln(n+1)}$$

for convergence or divergence.

**Solution**    Let

$$f(x) = \frac{1}{(x+1)\ln(x+1)}$$

and note that all conditions for the Integral Test are fulfilled. We let $u = x + 1$ and obtain

$$\int_1^t \frac{dx}{(x+1)\ln(x+1)} = \int_2^{t+1} \frac{du}{u\ln u} = \int_2^{t+1} \frac{d(\ln u)}{\ln u}$$
$$= \ln[\ln(t+1)] - \ln(\ln 2).$$

The expression on the right diverges as $t \to \infty$ and, therefore, the given series is divergent. $\qquad\square$

The particular $p$-series with $p = 1$, known as the **harmonic series**, is interesting, as it is on the borderline between convergence and divergence of the various $p$-series. It is an example of a series in which the general term $u_n$ tends to zero while the series diverges. (See the Remark on page 395). We can prove divergence of the harmonic series *without recourse to the Integral Test*. To do so, we write

$$\tfrac{1}{1} + \tfrac{1}{2} + (\tfrac{1}{3} + \tfrac{1}{4}) + (\tfrac{1}{5} + \tfrac{1}{6} + \tfrac{1}{7} + \tfrac{1}{8})$$

$$+ (\tfrac{1}{9} + \tfrac{1}{10} + \tfrac{1}{11} + \tfrac{1}{12} + \tfrac{1}{13} + \tfrac{1}{14} + \tfrac{1}{15} + \tfrac{1}{16}) + (\text{next 16 terms}) + \cdots. \qquad (2)$$

We have the simple inequalities

$$\tfrac{1}{3} + \tfrac{1}{4} > \tfrac{1}{4} + \tfrac{1}{4} = \tfrac{1}{2},$$

$$\tfrac{1}{5} + \tfrac{1}{6} + \tfrac{1}{7} + \tfrac{1}{8} > \tfrac{1}{8} + \tfrac{1}{8} + \tfrac{1}{8} + \tfrac{1}{8} = \tfrac{1}{2},$$

$$\tfrac{1}{9} + \tfrac{1}{10} + \tfrac{1}{11} + \tfrac{1}{12} + \tfrac{1}{13} + \tfrac{1}{14} + \tfrac{1}{15} + \tfrac{1}{16}$$

$$> \tfrac{1}{16} + \tfrac{1}{16} + \tfrac{1}{16} + \tfrac{1}{16} + \tfrac{1}{16} + \tfrac{1}{16} + \tfrac{1}{16} + \tfrac{1}{16} = \tfrac{1}{2},$$

and so forth.

In other words, each set of terms in a set of parentheses in series (2) is larger than $\tfrac{1}{2}$. By taking a sufficient number of parentheses, we can make the partial sum $s_n$ of the harmonic series as large as we please. Therefore the series diverges.

**EXAMPLE 7**    Show that the series

$$\sum_{n=1}^{\infty} \frac{1}{3n - 2}$$

diverges.

**Solution**    We may use the Integral Test, or observe that

$$\sum_{n=1}^{\infty} \frac{1}{3n - 2} = \frac{1}{3} \sum_{n=1}^{\infty} \frac{1}{(n - \tfrac{2}{3})} \qquad \text{and} \qquad \frac{1}{n - \tfrac{2}{3}} \geq \frac{1}{n} \quad \text{for all } n \geq 1.$$

The Comparison Test shows divergence. The divergence may also be shown directly by a recombination of the terms, as we did for the harmonic series. □

Often it is useful to combine two tests, as the next example illustrates.

**EXAMPLE 8**    Show that the series $\sum_{n=1}^{\infty} e^{-n^2}$ converges.

**Solution**    We would like to use the Integral Test and if we show that $\int_1^{\infty} e^{-x^2}\, dx$ converges, then the series will. However, we cannot perform the integration. We observe, on the other hand, that $e^{-n^2} \leq n e^{-n^2}$ for $n \geq 1$, and so by the Comparison Test it suffices to show that $\sum_{n=1}^{\infty} n e^{-n^2}$ is convergent. We have

$$\int_1^{\infty} x e^{-x^2}\, dx = \lim_{t \to \infty} \int_1^t x e^{-x^2}\, dx$$

$$= \lim_{t \to \infty} \left[ -\frac{1}{2} e^{-x^2} \right]_1^t$$

$$= \lim_{t \to \infty} \left[ \frac{1}{2e} - \frac{1}{2e^{t^2}} \right] = \frac{1}{2e},$$

and thus the series $\sum_{n=1}^{\infty} n e^{-n^2}$ is convergent by the Integral Test. Therefore $\sum_{n=1}^{\infty} e^{-n^2}$ converges. □

## 4   PROBLEMS

In each of Problems 1 through 38, test for convergence or divergence.

**1** $\displaystyle\sum_{n=1}^{\infty} \frac{1}{n\sqrt{n}}$

**2** $\displaystyle\sum_{n=1}^{\infty} \frac{1}{\sqrt{n}}$

**3** $\displaystyle\sum_{n=1}^{\infty} \frac{1}{(n+1)(n+2)}$

**4** $\displaystyle\sum_{n=1}^{\infty} \frac{n+1}{n\sqrt{n}}$

**5** $\displaystyle\sum_{n=1}^{\infty} \frac{2n+3}{n^2+3n+2}$

**6** $\displaystyle\sum_{n=1}^{\infty} \frac{1}{n\cdot 2^n}$

**7** $\displaystyle\sum_{n=1}^{\infty} \frac{n-1}{n^3}$

**8** $\displaystyle\sum_{n=1}^{\infty} \frac{n^2+3n-6}{n^4}$

**9** $\displaystyle\sum_{n=1}^{\infty} \frac{1}{\sqrt{n(n+1)}}$

**10** $\displaystyle\sum_{n=1}^{\infty} \frac{1}{2n+3}$

**11** $\displaystyle\sum_{n=1}^{\infty} \frac{1}{n+100}$

**12** $\displaystyle\sum_{n=1}^{\infty} \frac{2}{2^n+3}$

**13** $\displaystyle\sum_{n=1}^{\infty} \frac{1}{n^5+n^3+1}$

**14** $\displaystyle\sum_{n=1}^{\infty} \frac{\sqrt{n}}{n^2+7}$

**15** $\displaystyle\sum_{n=1}^{\infty} \frac{1}{n^{3n}}$

**16** $\displaystyle\sum_{n=1}^{\infty} \frac{n^4+3n^3+2n}{(n^3+n^2)^2}$

**17** $\displaystyle\sum_{n=1}^{\infty} \frac{n+\ln n}{n^3+3n+6}$

**18** $\displaystyle\sum_{n=1}^{\infty} \frac{\ln n}{n^3}$

**19** $\displaystyle\sum_{n=1}^{\infty} \frac{1}{n^2+2}$

**20** $\displaystyle\sum_{n=1}^{\infty} \frac{2^n}{1000}$

**21** $\displaystyle\sum_{n=1}^{\infty} \frac{2n+5}{n^3}$

**22** $\displaystyle\sum_{n=1}^{\infty} \frac{n-1}{n^2}$

**23** $\displaystyle\sum_{n=1}^{\infty} \frac{\ln n}{n}$

**24** $\displaystyle\sum_{n=1}^{\infty} \frac{1}{(n+1)\left[\ln(n+1)\right]^2}$

**25** $\displaystyle\sum_{n=1}^{\infty} \frac{\ln(n+1)}{(n+1)^3}$

**26** $\displaystyle\sum_{n=1}^{\infty} \frac{n!}{1\cdot 3\cdot 5\cdot 7\cdots(2n-1)}$

**27** $\displaystyle\sum_{n=1}^{\infty} \frac{n}{e^n}$

**28** $\displaystyle\sum_{n=1}^{\infty} \frac{1}{\sqrt{n^2+1}}$

**29** $\displaystyle\sum_{n=1}^{\infty} \frac{n+1}{n\cdot 2^n}$

**30** $\displaystyle\sum_{n=1}^{\infty} \frac{n^4}{e^n}$

**31** $\displaystyle\sum_{n=1}^{\infty} \frac{n}{2^n}$

**32** $\displaystyle\sum_{n=1}^{\infty} \frac{n^2}{2^n}$

**33** $\displaystyle\sum_{n=1}^{\infty} \frac{\ln n}{n^2}$

**34** $\displaystyle\sum_{n=1}^{\infty} \frac{\ln n}{n\sqrt{n}}$

**35** $\displaystyle\sum_{n=1}^{\infty} \frac{n^4}{n!}$

**36** $\displaystyle\sum_{n=1}^{\infty} \frac{(2n)!}{(3n)!}$

**\*37** $\displaystyle\sum_{n=1}^{\infty} \sin\left(\frac{1}{n^2}\right)$

**\*38** $\displaystyle\sum_{n=1}^{\infty} \tan\left(\frac{1}{n}\right)$

**\*39** Let $a_n \geq 0$ and suppose $\Sigma_{n=1}^{\infty} a_n$ is a convergent series. Show that $\Sigma_{n=1}^{\infty} \sin(a_n)$ is a convergent series.

**40** For what values of $p$ does the series

$$\sum_{n=2}^{\infty} \frac{1}{n(\ln n)^p}$$

converge?

**41** For what values of $q$ does the series

$$\sum_{n=1}^{\infty} \frac{\ln n}{n^q}$$

converge?

**42** For what values of $p$ and $q$ does the series

$$\sum_{n=1}^{\infty} \frac{(\ln n)^p}{n^q}$$

converge?

**43** State and prove as general an Integral Test as you can in which $f$ is not nonincreasing. (*Hint:* The behavior of $f$ for small values of $x$ can be disregarded. Also, the precise behavior of $f$ between two successive values of $n$ is not crucial.)

**44** Let $f(x) = \sin^2 \pi x$. Show that the integral $\int_1^{\infty} \sin^2 \pi x\, dx$ diverges. However, $\Sigma_{n=1}^{\infty} f(n)$ converges. Explain.

**45** Suppose that the series $\Sigma_{n=1}^{\infty} a_n$ is convergent with $a_n \geq 0$ for all $n$. Show that $\Sigma_{n=1}^{\infty} a_n^2$ is a convergent series.

**\*46** Let $a_n \geq 0$ for all $n$ and consider $r_n = \sqrt[n]{a_n}$. Show that if $\lim_{n\to\infty} r_n = r < 1$, then $\Sigma_{n=1}^{\infty} a_n$ converges. If $r > 1$, the series diverges. (*Hint:* Show that for sufficiently large $n$, a comparison with a geometric series can be made.)

**\*47** (**The Polynomial Test**)   Given the polynomials

$$P(x) = a_n x^n + a_{n-1}x^{n-1} + \cdots + a_1 x + a_0, \quad a_n > 0,$$
$$Q(x) = b_m x^m + b_{m-1}x^{m-1} + \cdots + b_1 x + b_0, \quad b_m > 0.$$

Show that the infinite series

$$\sum_{r=r_0}^{\infty} \frac{P(r)}{Q(r)}$$

converges if $m > n + 1$ and $r_0$ is a sufficiently large integer.

**\*48** Show that

$$\sum_{n=2}^{\infty} \frac{1}{(\ln(n))^{\ln(n)}}$$

is convergent.

_____5_____

# SERIES OF POSITIVE AND NEGATIVE TERMS. THE ALTERNATING SERIES TEST. THE RATIO TEST. THE ROOT TEST

In this section we establish four theorems which serve as important tests for the convergence and divergence of series whose terms are not necessarily positive.

**THEOREM 15**    *If $\sum_{n=1}^{\infty} |u_n|$ converges, then $\sum_{n=1}^{\infty} u_n$ converges, and*

$$\left| \sum_{n=1}^{\infty} u_n \right| \le \sum_{n=1}^{\infty} |u_n|.$$

**Proof**    We define numbers $v_1, v_2, \ldots, v_n, \ldots$ by the relations

$$v_n = \begin{cases} u_n & \text{if } u_n \text{ is nonnegative,} \\ 0 & \text{if } u_n \text{ is negative.} \end{cases}$$

In others words, the series $\sum_{n=1}^{\infty} v_n$ consists of all the nonnegative entries in $\sum_{n=1}^{\infty} u_n$. Similarly, we define the sequence $w_n$ by

$$w_n = \begin{cases} 0 & \text{if } u_n \text{ is nonnegative,} \\ -u_n & \text{if } u_n \text{ is negative.} \end{cases}$$

The $w_n$ are all positive and for each $n$

$$v_n + w_n = |u_n|, \qquad v_n - w_n = u_n. \tag{1}$$

Since $v_n \le |u_n|$ and $w_n \le |u_n|$, and since, by hypothesis, $\sum_{n=1}^{\infty} |u_n|$ converges, we apply the Comparison Test to conclude that the series

$$\sum_{n=1}^{\infty} v_n, \quad \sum_{n=1}^{\infty} w_n$$

converge. Consequently, (see Theorem 9) the series

$$\sum_{n=1}^{\infty} v_n - \sum_{n=1}^{\infty} w_n = \sum_{n=1}^{\infty} (v_n - w_n) = \sum_{n=1}^{\infty} u_n$$

converges. Also,

$$\left| \sum_{n=1}^{\infty} u_n \right| = \left| \left( \sum_{n=1}^{\infty} v_n \right) - \left( \sum_{n=1}^{\infty} w_n \right) \right|$$

$$\le \sum_{n=1}^{\infty} v_n + \sum_{n=1}^{\infty} w_n = \sum_{n=1}^{\infty} |u_n|. \qquad \square$$

*Remark.*    Theorem 15 shows that if the series of absolute values $\sum_{n=1}^{\infty} |u_n|$ is convergent, then the series itself also converges. The converse is not necessarily true. We give an example on page 409 in which $\sum_{n=1}^{\infty} u_n$ converges while $\sum_{n=1}^{\infty} |u_n|$ diverges.

**DEFINITIONS**

*A series $\sum_{n=1}^{\infty} u_n$ which is such that $\sum_{n=1}^{\infty} |u_n|$ converges is said to be **absolutely convergent**. However, if $\sum_{n=1}^{\infty} u_n$ converges and $\sum_{n=1}^{\infty} |u_n|$ diverges, then the series $\sum_{n=1}^{\infty} u_n$ is said to be **conditionally convergent**.*

The next theorem yields a test for series whose terms are alternately positive and negative. Since the hypotheses are rather stringent, the test can be used only under special circumstances.

**THEOREM 16**

**(Alternating Series Theorem)**  *Suppose that the numbers $u_1$, $u_2$, ..., $u_n$, ... satisfy the hypotheses:*

i)      *the $u_n$ are alternately positive and negative,*
ii)     *$|u_{n+1}| < |u_n|$ for every n, and*
iii)    *$\lim_{n \to \infty} u_n = 0$.*

*Then $\sum_{n=1}^{\infty} u_n$ is convergent. Furthermore, if the sum is denoted by s, then s lies between the partial sums $s_n$ and $s_{n+1}$ for each n.*

**Proof**

Assume that $u_1$ is positive. (If it is not, we can consider the series beginning with $u_2$, since discarding a finite number of terms does not affect convergence.) Therefore, all $u_k$ with odd subscripts are positive and all $u_k$ with even subscripts are negative. We state this fact in the form

$$u_{2n-1} > 0, \qquad u_{2n} < 0$$

for each $n$. We now write

$$s_{2n} = (u_1 + u_2) + (u_3 + u_4) + (u_5 + u_6) + \cdots + (u_{2n-1} + u_{2n}).$$

Since $|u_{2k}| < u_{2k-1}$ for each $k$, we know that each quantity in the parentheses is positive. Hence $s_{2n}$ increases for all $n$. On the other hand,

$$s_{2n} = u_1 + (u_2 + u_3) + (u_4 + u_5) + \cdots + (u_{2n-2} + u_{2n-1}) + u_{2n}.$$

Each quantity in parentheses in the above expression is negative, and so is $u_{2n}$. Therefore $s_{2n} < u_1$ for all $n$. We conclude that $s_{2n}$ is an increasing sequence bounded by the number $u_1$. It must tend to a limit (Axiom C).

We apply similar reasoning to $s_{2n-1}$. We have

$$s_{2n-1} + u_{2n} = s_{2n}$$

and, since $u_{2n} < 0$, we have $s_{2n-1} > s_{2n}$ for every $n$. Therefore for $n > 1$, $s_{2n-1}$ is bounded from below by $s_2 = u_1 + u_2$. Also,

$$s_{2n+1} = s_{2n-1} + (u_{2n} + u_{2n+1}).$$

Since the quantity in parentheses on the right is negative,

$$s_{2n-1} > s_{2n+1};$$

in other words, the partial sums with odd subscripts form a decreasing bounded sequence (Fig. 11). The limits approached by $s_{2n}$ and $s_{2n-1}$ must be

**FIGURE 11**

the same, since by hypothesis (iii),

$$u_{2n} = s_{2n} - s_{2n-1} \to 0 \text{ as } n \to \infty.$$

If $s$ is the limit, we see that every even sum is less than or equal to $s$, while every odd sum is greater than or equal to $s$. □

---

**COROLLARY**  *Suppose the numbers $u_1, u_2, \ldots, u_n, \ldots$ satisfy the hypotheses of Theorem 16 with $u_1 > 0$. Let $s_n = \sum_{k=1}^{n} u_k$ and $s = \sum_{k=1}^{\infty} u_k$. Then $|s - s_n| < |u_{n+1}|$.*

**Proof**  From Theorem 16 we know that $s$ lies between $s_{2n}$ and $s_{2n+1}$, and also between $s_{2n+1}$ and $s_{2n+2}$. Therefore $|s - s_{2n}| < |s_{2n+1} - s_{2n}| = |u_{2n+1}|$. Similarly, $|s - s_{2n+1}| < |s_{2n+1} - s_{2n+2}| = |u_{2n+2}|$. Thus for $n$ both even and odd, the inequality $|s - s_n| < |u_{n+1}|$ holds. □

**EXAMPLE 1**  Test the series

$$\sum_{n=1}^{\infty} \frac{(-1)^{n+1}}{n}$$

for convergence or divergence. If it is convergent, determine whether it is conditionally convergent or absolutely convergent.

**Solution**  Set $u_n = (-1)^{n+1}/n$ and observe that the three hypotheses of Theorem 16 hold; i.e., the terms alternate in sign, $1/(n+1) < 1/n$ for each $n$, and $\lim_{n \to \infty} (-1)^n/n = 0$. Therefore the given series converges. However, the series $\sum_{n=1}^{\infty} |u_n|$ is the harmonic series

$$\sum_{n=1}^{\infty} \frac{1}{n},$$

which is divergent. Therefore the original series is conditionally convergent. □

For any series which satisfies the hypotheses of the Alternating Series Theorem, we can approximate numerically the value of the sum $s$. The next example illustrates the technique.

**EXAMPLE 2**  Test the series

$$\sum_{n=1}^{\infty} (-1)^n \frac{1}{n!}$$

for convergence or divergence. If it is convergent, approximate the sum of the series to three decimal places.

**Solution**  Here

$$u_n = (-1)^n \frac{1}{n!} \qquad \text{and} \qquad |u_n| = \frac{1}{n!}.$$

Clearly $|u_{n+1}| < |u_n|$ and $\lim_{n \to \infty} u_n = 0$, so the series is convergent. If we sum the first six terms, we have

$$-1 + \frac{1}{2!} - \frac{1}{3!} + \frac{1}{4!} - \frac{1}{5!} + \frac{1}{6!} = -1 + \frac{1}{2} - \frac{1}{6} + \frac{1}{24} - \frac{1}{120} + \frac{1}{720} = -0.631944,$$

and we know by the Corollary of the Alternating Series Theorem that the sum of the infinite series, $s$, lies between this sum, $s_6$, and $s_7$. Therefore $|s - s_6| \leqslant |s_7|$, and the accuracy of $s_6$ as an estimate of $s$ is better than $|s_7|$, which in this case is

$$\frac{1}{7!} = \frac{1}{5040} \approx 0.000198.$$

Since this number is less than .0005, we conclude that this estimate is accurate to three decimal places.    $\square$

The next test is one of the most useful for determining absolute convergence of series.

**THEOREM 17**    **(Ratio Test)**    *Suppose that in the series $\Sigma_{n=1}^{\infty} u_n$ every $u_n \neq 0$ and that*

$$\lim_{n \to \infty} \left| \frac{u_{n+1}}{u_n} \right| = \rho \quad \text{or} \quad \left| \frac{u_{n+1}}{u_n} \right| \to +\infty \text{ as } n \to \infty.$$

*Then*

i)    *if $\rho < 1$, the series $\Sigma_{n=1}^{\infty} u_n$ converges absolutely;*
ii)    *if $\rho > 1$, or if $|u_{n+1}/u_n| \to +\infty$, the series diverges;*
iii)    *if $\rho = 1$, the test gives no information.*

**Proof**    (i) Suppose that $\rho < 1$. Choose any $\rho'$ such that $\rho < \rho' < 1$. Then, since

$$\lim_{n \to \infty} \left| \frac{u_{n+1}}{u_n} \right| = \rho,$$

there must be a sufficiently large $N$ for which

$$\left| \frac{u_{n+1}}{u_n} \right| < \rho', \quad \text{for all } n \geq N.$$

Then we obtain

$$|u_{N+1}| < \rho'|u_N|, \quad |u_{N+2}| < \rho'|u_{N+1}|, \quad |u_{N+3}| < \rho'|u_{N+2}|, \text{ etc.}$$

By substitution we find

$$|u_{N+2}| < \rho'^2|u_N|, \quad |u_{N+3}| < \rho'^3|u_N|, \quad |u_{N+4}| < \rho'^4|u_N|, \text{ etc.}$$

and, in general,

$$|u_{N+k}| < (\rho')^k|u_N| \quad \text{for } k = 1, 2, \dots. \tag{2}$$

The series

$$\sum_{k=1}^{\infty} |u_N|(\rho')^k = |u_N| \sum_{k=1}^{\infty} (\rho')^k$$

is a geometric series with ratio less than 1 and hence convergent. From (2) and the Comparison Test, we conclude that

$$\sum_{k=1}^{\infty} |u_{N+k}| \tag{3}$$

converges. Since (3) differs from the series

$$\sum_{n=1}^{\infty} |u_n|$$

in only a finite number of terms (the $N$ terms $|u_1|$, $|u_2|$, ..., $|u_N|$, to be exact), statement (i) of the theorem is established.

ii) Suppose that $\rho > 1$ or $|u_{n+1}/u_n| \to +\infty$. There is an $N$ such that

$$\frac{|u_{n+1}|}{|u_n|} > 1 \quad \text{for all } n \geq N.$$

By induction $|u_n| > |u_N|$ for all $n > N$. Therefore $u_n$ does not tend to zero, and the series diverges (Corollary to Theorem 8).

To establish (iii), we exhibit two cases in which $\rho = 1$, one of them corresponding to a divergent series, the other to a convergent series. We show that for all $p$, the $p$-series.

$$\sum_{n=1}^{\infty} \frac{1}{n^p}$$

has $\rho = 1$. Setting $u_n = 1/n^p$, we have

$$\left|\frac{u_{n+1}}{u_n}\right| = \frac{n^p}{(n+1)^p} = 1 \Big/ \left(1 + \frac{1}{n}\right)^p.$$

Since for any $p$

$$\lim_{n \to \infty} 1 \Big/ \left(1 + \frac{1}{n}\right)^p = 1 = \rho,$$

we see that if $p > 1$ the series converges (and $\rho = 1$), while if $p \leq 1$, the series diverges (and $\rho = 1$). $\qquad\square$

*Remarks.* A good working procedure for a reader is to try first the Ratio Test for convergence or divergence. If the limit $\rho$ turns out to be 1, some other test must then be tried. The Integral Test is one possibility. (We observe that the Integral Test established convergence and divergence for the $p$-series, while the Ratio Test fails.) When the terms have alternating signs, the Alternating Series Theorem is suggested. In addition, we may also try comparison tests.

In the statement of Theorem 17, it may appear at first glance that all possible situations for $\rho$ have been considered. That is not the case, since it may happen that

$$\left|\frac{u_{n+1}}{u_n}\right|$$

does not tend to any limit and does not tend to $+\infty$. In such circumstances, more sophisticated ratio tests are available. They are, however, beyond the scope of this course.

**EXAMPLE 3**   Test for absolute convergence:

$$\sum_{n=1}^{\infty} \frac{2^n}{n!}.$$

**Solution**   Applying the Ratio Test, we have

$$u_{n+1} = \frac{2^{n+1}}{(n+1)!}, \quad u_n = \frac{2^n}{n!}, \quad \text{and} \quad \left|\frac{u_{n+1}}{u_n}\right| = \frac{2^{n+1}}{(n+1)!} \cdot \frac{n!}{2^n} = \frac{2}{n+1}.$$

Therefore

$$\lim_{n \to \infty} \left|\frac{u_{n+1}}{u_n}\right| = 0 = \rho.$$

The series converges absolutely.  □

**EXAMPLE 4**   Test for absolute and conditional convergence:

$$\sum_{n=1}^{\infty} \frac{(-1)^n n}{2^n}.$$

**Solution**   We try the Ratio Test:

$$u_{n+1} = (-1)^{n+1} \frac{n+1}{2^{n+1}}, \qquad u_n = \frac{(-1)^n n}{2^n},$$

and therefore

$$\left|\frac{u_{n+1}}{u_n}\right| = \frac{n+1}{2^{n+1}} \cdot \frac{2^n}{n} = \frac{1}{2}\left(\frac{n+1}{n}\right) = \frac{1}{2}\left(\frac{1+1/n}{1}\right).$$

Hence

$$\lim_{n \to \infty} \left|\frac{u_{n+1}}{u_n}\right| = \frac{1}{2} = \rho.$$

The series converges absolutely.  □

**EXAMPLE 5**   Test for absolute convergence:

$$\sum_{n=1}^{\infty} \frac{(2n)!}{n^{100}}.$$

**Solution**   We try the Ratio Test:

$$u_n = \frac{(2n)!}{n^{100}} \quad \text{and} \quad u_{n+1} = \frac{[2(n+1)]!}{(n+1)^{100}}.$$

Therefore

$$\left|\frac{u_{n+1}}{u_n}\right| = \frac{(2n+2)!}{(n+1)^{100}} \cdot \frac{n^{100}}{(2n)!} = (2n+1)(2n+2)\left(\frac{n}{1+n}\right)^{100}.$$

$$= (2n+1)(2n+2)\left(\frac{1}{1+1/n}\right)^{100}.$$

Hence

$$\lim_{n \to \infty} \left| \frac{u_{n+1}}{u_n} \right| = +\infty,$$

and the series is divergent.                                                                        □

---

**THEOREM 18**   **(Root Test)**   *Let $\Sigma_{n=1}^{\infty} u_n$ be a series such that either*

$$\lim_{n \to \infty} (|u_n|)^{1/n} = r \qquad or \qquad (|u_n|)^{1/n} \to +\infty \ as \ n \to \infty.$$

*Then*

i)         *if $r < 1$, the series $\Sigma_{n=1}^{\infty} u_n$ converges absolutely;*
ii)        *if $r > 1$, or if $(|u_n|)^{1/n} \to +\infty$, the series diverges;*
iii)       *if $r = 1$, the test gives no information.*

---

**Proof**   (i) Suppose $r < 1$: Choose $\varepsilon > 0$ so small that $r + \varepsilon < 1$ as well. Since $(|u_n|)^{1/n} \to r$, we have $(|u_n|)^{1/n} < r + \varepsilon$, for all $n \geq N$, for some $N$ sufficiently large. Therefore $|u_n| < (r + \varepsilon)^n$ for all $n \geq N$. But $\Sigma_{n=1}^{\infty} (r + \varepsilon)^n$ converges since it is a geometric series with ratio $r + \varepsilon < 1$. The Comparison Test then gives that $\Sigma_{n=N}^{\infty} |u_n|$ converges, and hence $\Sigma_{n=1}^{\infty} u_n$ is absolutely convergent.

ii) Suppose $r > 1$ or $(|u_n|)^{1/n} \to \infty$. We choose $\varepsilon > 0$ sufficiently small that $r - \varepsilon > 1$. ($r$ can be chosen arbitrarily—but larger than 1—if $(|u_n|)^{1/n} \to +\infty$.) Since $(|u_n|)^{1/n} \to r$ or $+\infty$, we have that $r - \varepsilon < (|u_n|)^{1/n}$ for all $n \geq N$, for some $N$ sufficiently large. Therefore $(r - \varepsilon)^n < |u_n|$ for all $n \geq N$. Since $r - \varepsilon > 1$, we conclude that $\lim_{n \to \infty} |u_n| \neq 0$, and hence both $\Sigma_{n=1}^{\infty} |u_n|$ and $\Sigma_{n=1}^{\infty} u_n$ are divergent series, by the Corollary to Theorem 8.                       □

**EXAMPLE 6**   Test the series

$$\sum_{k=2}^{\infty} \frac{1}{(\ln(k))^k}$$

for convergence.

**Solution**   Here

$$u_k = \frac{1}{(\ln(k))^k},$$

and since it is positive we do not need to use its absolute value. Then

$$(u_k)^{1/k} = \frac{1}{\ln k} \to 0,$$

and the series converges by the Root Test.                                                          □

**EXAMPLE 7**   Test the series

$$\sum_{n=1}^{\infty} \frac{(-2)^n}{n^3}$$

for convergence.

**Solution**    Since

$$u_n = \frac{(-2)^n}{n^3},$$

we have

$$(|u_n|)^{1/n} = \left(\frac{2^n}{n^3}\right)^{1/n} = \frac{2}{(n^3)^{1/n}} = 2\left[\left(\frac{1}{n}\right)^{1/n}\right]^3. \qquad (4)$$

To calculate

$$\lim_{n \to \infty} \left(\frac{1}{n}\right)^{1/n},$$

we resort to l'Hôpital's Rule. While $n$ is an integer-valued variable, we can let $y$ be a real-valued variable, and we observe that

$$\lim_{y \to \infty} \left(\frac{1}{y}\right)^{1/y} = \lim_{n \to \infty} \left(\frac{1}{n}\right)^{1/n}.$$

Letting

$$x = \frac{1}{y},$$

we have

$$\lim_{x \to 0} (x)^x = \lim_{y \to \infty} \left(\frac{1}{y}\right)^{1/y}.$$

Therefore

$$\lim_{x \to 0} x^x = \lim_{x \to 0} e^{\ln(x^x)} = \lim_{x \to 0} e^{x \ln x}$$
$$= e^{\lim_{x \to 0} (x \ln x)}.$$

We now use l'Hôpital's Rule to calculate $\lim_{x \to 0} x \ln x$. To do this we rewrite $x \ln x$ as a quotient. Doing this, we have

$$\lim_{x \to 0} x \ln x = \lim_{x \to 0} \frac{\ln x}{\dfrac{1}{x}};$$

and using l'Hôpital's Rule, we find

$$= \lim_{x \to 0} \frac{\dfrac{1}{x}}{\dfrac{-1}{x^2}} = \lim_{x \to 0} -x = 0.$$

Therefore $\lim_{x \to 0} x^x = e^{\lim_{x \to 0}(x \ln x)} = e^0 = 1$, and returning to (4) we have

$$\lim_{n \to \infty} (|u_n|)^{1/n} = \lim_{n \to \infty} 2\left[\left(\frac{1}{n}\right)^{1/n}\right]^3 = 2.$$

Therefore the series diverges by the Root Test.    $\square$

We note that the Ratio Test could have been used in Example 7.

## 5   PROBLEMS

In each of Problems 1 through 45, test for convergence or divergence. If the series is convergent, determine whether it is absolutely or conditionally convergent.

**1** $\displaystyle\sum_{n=1}^{\infty} \frac{n!}{10^n}$

**2** $\displaystyle\sum_{n=1}^{\infty} \frac{10^n}{n!}$

**3** $\displaystyle\sum_{n=1}^{\infty} \frac{(-1)^{n-1} n!}{10^n}$

**4** $\displaystyle\sum_{n=1}^{\infty} \frac{(-1)^{n-1} 10^n}{n!}$

**5** $\displaystyle\sum_{n=1}^{\infty} n\left(\frac{3}{4}\right)^n$

**6** $\displaystyle\sum_{n=1}^{\infty} n^2\left(\frac{3}{4}\right)^n$

**7** $\displaystyle\sum_{n=1}^{\infty} \frac{(-1)^n}{\sqrt{n}}$

**8** $\displaystyle\sum_{n=1}^{\infty} \frac{(-1)^n}{n^p}, \quad 0 < p < 1$

**9** $\displaystyle\sum_{n=1}^{\infty} \frac{(-1)^n}{n\sqrt{n}}$

**10** $\displaystyle\sum_{n=1}^{\infty} \frac{(-1)^{n+1}(n-1)}{n^2+1}$

**11** $\displaystyle\sum_{n=1}^{\infty} \frac{(-1)^n n^2}{2^n}$

**12** $\displaystyle\sum_{n=1}^{\infty} \frac{(-1)^{n+1}(n-1)^2}{n^3}$

**13** $\displaystyle\sum_{n=1}^{\infty} \frac{(-1)^{n-1}(4/3)^n}{n^2}$

**14** $\displaystyle\sum_{n=1}^{\infty} \frac{(-1)^n(3/2)^2}{n^4}$

**15** $\displaystyle\sum_{n=1}^{\infty} \frac{(-5)^{n-1}}{n \cdot n!}$

**16** $\displaystyle\sum_{n=1}^{\infty} \frac{(-2)^{n-1}(n+1)}{(2n)!}$

**17** $\displaystyle\sum_{n=1}^{\infty} \frac{(-1)^{n-1} n!}{1 \cdot 3 \cdot 5 \cdots (2n-1)}$

**18** $\displaystyle\sum_{n=1}^{\infty} \frac{(-1)^{n-1}(n!)^2 2^n}{(2n)!}$

**19** $\displaystyle\sum_{n=1}^{\infty} \frac{(-1)^{n-1}(n+1)}{n\sqrt{n}}$

**20** $\displaystyle\sum_{n=1}^{\infty} \frac{(n!)^2 5^n}{(2n)!}$

**21** $\displaystyle\sum_{n=1}^{\infty} \frac{(-1)^n 2 \cdot 4 \cdot 6 \cdots (2n)}{1 \cdot 4 \cdot 7 \cdots (3n-2)}$

**22** $\displaystyle\sum_{n=1}^{\infty} \frac{(-1)^{n+1} 3^{n+1}}{2^{4n}}$

**23** $\displaystyle\sum_{n=1}^{\infty} \frac{(-1)^{n-1} n}{n+1}$

**24** $\displaystyle\sum_{n=1}^{\infty} \frac{(-1)^n(n-2)}{n^{7/4}}$

**25** $\displaystyle\sum_{n=1}^{\infty} \frac{(-1)^n(6n^2 - 9n + 4)}{n^3}$

**26** $\displaystyle\sum_{n=1}^{\infty} \frac{(-1)^{n+1}}{(n+1)\ln(n+1)}$

**27** $\displaystyle\sum_{n=1}^{\infty} \frac{(-1)^{n+1}\ln(n+1)}{n+1}$

**28** $\displaystyle\sum_{n=1}^{\infty} \frac{(-1)^{n-1}\ln n}{n^2}$

**29** $\displaystyle\sum_{n=1}^{\infty} (-1)^n e^{-n}$

**30** $\displaystyle\sum_{n=1}^{\infty} \frac{n!}{e^n}$

**31** $\displaystyle\sum_{n=1}^{\infty} \frac{3^{n+1}}{5^n(n+3)}$

**32** $\displaystyle\sum_{n=1}^{\infty} \frac{\sin(\pi n)}{n}$

**33** $\displaystyle\sum_{n=1}^{\infty} \frac{10^n}{n^2+3}$

**34** $\displaystyle\sum_{n=1}^{\infty} \frac{(-1)^{n+1}\arctan n}{n^2}$

**35** $\displaystyle\sum_{n=1}^{\infty} \frac{n^n}{10^n}$

**36** $\displaystyle\sum_{k=1}^{\infty} \frac{\ln k}{k}$

**37** $\displaystyle\sum_{n=1}^{\infty} \left(\frac{n}{2n+1}\right)^n$

**38** $\displaystyle\sum_{n=1}^{\infty} \frac{2^{n-1}}{(n+1)(n^2+3)}$

**\*39** $\displaystyle\sum_{n=1}^{\infty} \frac{2^n n!}{n^n}$

**40** $\displaystyle\sum_{n=1}^{\infty} \frac{(n!)^2}{(2n)!}$

**41** $\displaystyle\sum_{n=1}^{\infty} \frac{(-1)^n}{n^{(1+(1/n))}}$

**42** $\displaystyle\sum_{n=1}^{\infty} n\left(\frac{3}{4}\right)^n$

**43** $\displaystyle\sum_{n=1}^{\infty} \frac{(-1)^n n 3^{2n}}{5^{n+1}}$

**44** $\displaystyle\sum_{n=1}^{\infty} \frac{n!}{n^n}$

**45** $\displaystyle\sum_{n=2}^{\infty} \frac{1}{(\ln n)^n}$

**46** For what values of $p$ does the series

$$\sum_{n=1}^{\infty} \frac{n!}{n^p}$$

converge?

**47** For what values of $p$ does the series

$$\sum_{n=1}^{\infty} \frac{n^p}{n!}$$

converge?

**48** Given a series $\sum_{n=1}^{\infty} u_n$ in which, for each $n \geq 1$, $u_{4n}$ and $u_{4n+1}$ are positive while $u_{4n+2}$ and $u_{4n+3}$ are negative. After determining appropriate additional hypotheses on the $\{u_n\}$, state and prove a theorem analogous to the Alternating Series Theorem. Generalize your result if possible.

**49** Given that $\sum_{n=1}^{\infty} u_n$ is a convergent series of positive terms, show that $\sum_{n=1}^{\infty} u_n^p$ is convergent for every $p > 1$.

**50** Given that $\sum_{n=1}^{\infty} u_n$ is a divergent series of positive terms, show that $\sum_{n=1}^{\infty} u_n^p$ diverges for $0 < p < 1$.

**\*51** Consider the series $\sum_{n=1}^{\infty} (-1)^{n-1}/n$ and let $A$ be any real number. Show that by rearranging the terms of the series, the sum will be a number in the interval $(A - 1, A + 1)$.

**52** Suppose that $\rho$ is a positive number less than 1 and that there is a positive integer $N$ such that $|u_{n+1}/u_n| < \rho$ for all integers $n > N$. Prove that $\sum_{n=1}^{\infty} |u_n|$ converges even though $\lim_{n \to \infty} |u_{n+1}/u_n|$ may not exist.

**53** Let $a, b$ be two numbers larger than 1 with $a \neq b$. Show that if $u_n = (1/a^n)$ when $n$ is odd and $u_n = (1/b^n)$ when $n$ is even, then the series $\sum_{n=1}^{\infty} u_n$ converges, although $|u_{n+1}/u_n|$ does not tend to a limit as $n \to \infty$.

**54** Use the Root Test to show that the series

$$\sum_{n=1}^{\infty} \frac{a^n}{2 + b^n}$$

converges if $0 < a < b$.

*55 Let $\Sigma_{n=1}^{\infty} a_n$ be an absolutely convergent series. Show that

$$\sum_{n=1}^{\infty} \frac{\sqrt{|a_n|}}{n^p}$$

converges for $p > \frac{1}{2}$. What happens if

$$a_n = n^{-1}(\log(n+1))^{-3/2}$$

and $p = \frac{1}{2}$?

In each of Problems 56 through 61, use the Alternating Series Theorem and a hand calculator to approximate the sum of the series to three decimal places.

**56** $\displaystyle\sum_{n=1}^{\infty} (-1)^{n-1} \frac{1}{(2n-1)!}$          **57** $\displaystyle\sum_{n=0}^{\infty} (-1)^n \frac{1}{(2n)!}$

**58** $\displaystyle\sum_{n=1}^{\infty} \frac{\cos(\pi(n+1))}{n^3}$          **59** $\displaystyle\sum_{n=0}^{\infty} \sin(\pi n) e^{-n}$

**60** $\displaystyle\sum_{n=1}^{\infty} (-1)^{n-1} \frac{1}{n^5}$          **61** $\displaystyle\sum_{n=1}^{\infty} \cos(\pi n) \frac{1}{n}$

---

## 6

## POWER SERIES

A **power series** is a series of the form

$$c_0 + c_1(x-a) + c_2(x-a)^2 + \cdots + c_n(x-a)^n + \cdots,$$

in which $a$ and the $c_i$, $i = 0, 1, 2, \ldots$, are constants. If a particular value is given to $x$, we then obtain an infinite series of numbers of the type we have been considering. The special case $a = 0$ occurs frequently, in which case the series becomes

$$c_0 + c_1 x + c_2 x^2 + c_3 x^3 + \cdots + c_n x^n + \cdots.$$

Most often, we use the $\Sigma$-notation, writing

$$\sum_{n=0}^{\infty} c_n(x-a)^n \qquad \text{and} \qquad \sum_{n=0}^{\infty} c_n x^n.$$

If a power series converges for certain values of $x$, we may define a function of $x$ by setting

$$f(x) = \sum_{n=0}^{\infty} c_n(x-a)^n \qquad \text{or} \qquad g(x) = \sum_{n=0}^{\infty} c_n x^n$$

for those values of $x$. We shall see that all the elementary functions we have studied can be represented by convergent power series.

The Ratio Test is particularly useful in determining when a power series converges. We begin with several examples.

**EXAMPLE 1**    Find the values of $x$ for which the series

$$\sum_{n=1}^{\infty} \frac{1}{n} x^n$$

converges.

**Solution**    We apply the Ratio Test, noting that

$$u_n = \frac{1}{n} x^n, \qquad u_{n+1} = \frac{1}{n+1} x^{n+1}.$$

Then

$$\left| \frac{u_{n+1}}{u_n} \right| = \frac{|x|^{n+1}}{n+1} \cdot \frac{n}{|x|^n} = |x| \frac{n}{n+1}.$$

It is important to observe that $x$ remains *unaffected* as $n \to \infty$. Hence

$$\lim_{n \to \infty} \left| \frac{u_{n+1}}{u_n} \right| = \lim_{n \to \infty} |x| \frac{n}{n+1} = |x| \lim_{n \to \infty} \frac{1}{1 + 1/n} = |x|.$$

That is, $\rho = |x|$ in the Ratio Test.

We conclude: (a) the series converges if $|x| < 1$; (b) the series diverges if $|x| > 1$; (c) if $|x| = 1$, the Ratio Test gives no information. The last case corresponds to $x = \pm 1$, and we may try other methods for these two series, which are

$$\sum_{n=1}^{\infty} \frac{1}{n} \quad (\text{if } x = 1) \qquad \text{and} \qquad \sum_{n=1}^{\infty} \frac{(-1)^n}{n} \quad (\text{if } x = -1).$$

The first series above is the divergent harmonic series. The second series converges by the Alternating Series Theorem. Therefore the given series converges for $-1 \le x < 1$.   □

**EXAMPLE 2**   Find the values of $x$ for which the series

$$\sum_{n=1}^{\infty} \frac{(-1)^n (x+1)^n}{2^n n^2}$$

converges.

**Solution**   We apply the Ratio Test:

$$u_n = \frac{(-1)^n (x+1)^n}{2^n n^2}, \qquad u_{n+1} = \frac{(-1)^{n+1} (x+1)^{n+1}}{2^{n+1} (n+1)^2}.$$

Therefore

$$\left| \frac{u_{n+1}}{u_n} \right| = \frac{|x+1|^{n+1}}{2^{n+1}(n+1)^2} \cdot \frac{2^n n^2}{|x+1|^n} = \frac{1}{2}|x+1| \left( \frac{n}{n+1} \right)^2$$

and

$$\lim_{n \to \infty} \frac{|u_{n+1}|}{|u_n|} = \frac{1}{2}|x+1| \lim_{n \to \infty} \left( \frac{1}{1 + 1/n} \right)^2 = \frac{|x+1|}{2}.$$

According to the Ratio Test: (a) the series converges if $\frac{1}{2}|x+1| < 1$; (b) the series diverges if $\frac{1}{2}|x+1| > 1$; and (c) if $|x+1| = 2$, the test fails.

The inequality $\frac{1}{2}|x+1| < 1$ may be written

$$-2 < x+1 < 2 \qquad \Leftrightarrow \qquad -3 < x < 1,$$

and the series converges in this interval, while it diverges for $x$ outside this interval. The values $x = -3$ and $x = 1$ remain for consideration. The corresponding series are

$$\sum_{n=1}^{\infty} \frac{(-1)^n (-2)^n}{2^n n^2} = \sum_{n=1}^{\infty} \frac{1}{n^2} \qquad \text{and} \qquad \sum_{n=1}^{\infty} \frac{(-1)^n 2^n}{2^n n^2} = \sum_{n=1}^{\infty} \frac{(-1)^n}{n^2}.$$

Both series converge absolutely by the *p*-series test. The original series converges for $x$ in the interval $-3 \le x \le 1$. ☐

**EXAMPLE 3**    Find the values of $x$ for which the series

$$\sum_{n=0}^{\infty} \frac{(-1)^n x^n}{n!}$$

converges.

**Solution**    We apply the Ratio Test:

$$u_n = \frac{(-1)^n x^n}{n!}, \qquad u_{n+1} = \frac{(-1)^{n+1} x^{n+1}}{(n+1)!},$$

and

$$\left| \frac{u_{n+1}}{u_n} \right| = \frac{|x|^{n+1}}{(n+1)!} \cdot \frac{n!}{|x|^n} = |x| \frac{1}{n+1}.$$

Hence $\rho = 0$, regardless of the value of $|x|$. The series converges for all values of $x$; that is, $-\infty < x < \infty$. ☐

**EXAMPLE 4**    Find the values of $x$ for which the series

$$\sum_{n=0}^{\infty} \frac{(-1)^n n! x^n}{10^n}$$

converges.

**Solution**    We apply the Ratio Test:

$$u_n = \frac{(-1)^n n! x^n}{10^n}, \qquad u_{n+1} = \frac{(-1)^{n+1} (n+1)! x^{n+1}}{10^{n+1}}$$

and,

$$\left| \frac{u_{n+1}}{u_n} \right| = \frac{|x|^{n+1}(n+1)!}{10^{n+1}} \cdot \frac{10^n}{|x|^n n!} = |x| \cdot \frac{n+1}{10}.$$

Therefore, if $x \ne 0$, $|u_{n+1}/u_n| \to +\infty$ and the series diverges. The series converges only for $x = 0$. ☐

The convergence properties of the most general power series are illustrated in the examples above. However, the proof of the theorem which states this fact (given in Theorem 20 on page 419) is beyond the scope of this text. In all the examples above we see that it always happened that $|u_{n+1}/u_n|$ tended to a limit or to $+\infty$. The examples are deceptive, since there are cases in which $|u_{n+1}/u_n|$ may neither tend to a limit nor tend to $+\infty$.

We next establish a simple lemma which is needed in the proof of Theorem 19 which gives a comparison test for the convergence of power series.

**LEMMA**    *If the series $\sum_{n=0}^{\infty} u_n$ converges, there is a number M such that $|u_n| \le M$ for every n.*

**Proof**    By Theorem 8 we know that $\lim_{n \to \infty} u_n = 0$. From the definition of a limit, there must be a number $N$ such that

$$|u_n| < 1 \quad \text{for all } n > N$$

(by taking $\varepsilon = 1$ in the definition of limit). We define $M$ to be the largest of the numbers

$$|u_0|, |u_1|, |u_2|, \ldots, |u_N|, 1,$$

and the result is established.    □

---

**THEOREM 19**    *If the series $\sum_{n=0}^{\infty} a_n x^n$ converges for some $x_1 \neq 0$, then the series converges absolutely for all $x$ for which $|x| < |x_1|$, and there is a number $M$ such that*

$$|a_n x^n| \leq M \left| \frac{x}{x_1} \right|^n \quad \text{for all } n.$$

---

**Proof**    Since the series $\sum_{n=0}^{\infty} a_n x_1^n$ converges, we know from the Lemma above that there is a number $M$ such that

$$|a_n x_1^n| \leq M \quad \text{for all } n.$$

Then

$$|a_n x^n| = \left| a_n x_1^n \frac{x^n}{x_1^n} \right| = |a_n x_1^n| \cdot \left| \frac{x}{x_1} \right|^n \leq M \left| \frac{x}{x_1} \right|^n.$$

The series

$$\sum_{n=0}^{\infty} M \left| \frac{x}{x_1} \right|^n$$

is a geometric series with ratio less than 1, and so convergent. Hence, by the Comparison Test, the series

$$\sum_{n=0}^{\infty} a_n x^n$$

converges absolutely.    □

*Remark.*    Theorem 19 may be established for series of the form

$$\sum_{n=0}^{\infty} a_n (x - a)^n$$

in an analogous manner.

---

**THEOREM 20**    *Let $\sum_{n=0}^{\infty} a_n (x - a)^n$ be any given power series. Then either*

   i) *the series converges only for $x = a$; or*
   ii) *the series converges for all values of $x$; or*
   iii) *there is a number $R > 0$ such that the series converges for all $x$ for which $|x - a| < R$ and diverges for all $x$ for which $|x - a| > R$.*

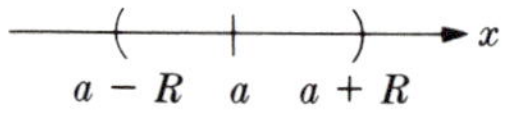

Radius of convergence $R$ about the point $a$, corresponding to the series $\Sigma_{n=0}^{\infty} a_n(x-a)^n$.

FIGURE 12

We omit the proof. The consequence (iii) in Theorem 20 states that there is an *interval of convergence* $-R < x - a < R$ or, equivalently, $a - R < x < a + R$. Nothing is stated about what happens when $x = a - R$ or $a + R$. (See Fig. 12.) These *endpoint* problems must be settled on a case-by-case basis. The alternatives (i) and (ii) correspond to $R = 0$ and $R = +\infty$, respectively. When $0 < R < \infty$, the number $R$ is called the **radius of convergence**.

## 6    PROBLEMS

In Problems 1 through 41, find the values of $x$ for which the following power series converge. Include a discussion of the endpoints.

1. $\displaystyle\sum_{n=0}^{\infty} x^n$

2. $\displaystyle\sum_{n=0}^{\infty} (-1)^n x^n$

3. $\displaystyle\sum_{n=0}^{\infty} (2x)^n$

4. $\displaystyle\sum_{n=0}^{\infty} (\tfrac{1}{4}x)^n$

5. $\displaystyle\sum_{n=0}^{\infty} (-1)^n (n+1) x^n$

6. $\displaystyle\sum_{n=1}^{\infty} \frac{(x-1)^n}{3^n n^2}$

7. $\displaystyle\sum_{n=1}^{\infty} \frac{(x-1)^n}{2^n n^3}$

8. $\displaystyle\sum_{n=1}^{\infty} \frac{(-1)^{n+1}(x-2)^n}{n\sqrt{n}}$

9. $\displaystyle\sum_{n=1}^{\infty} \frac{(x+2)^n}{\sqrt{n}}$

10. $\displaystyle\sum_{n=0}^{\infty} \frac{(10x)^n}{n!}$

11. $\displaystyle\sum_{n=0}^{\infty} \frac{x^n}{(2n)!}$

12. $\displaystyle\sum_{n=0}^{\infty} \frac{n!(x+1)^n}{5^n}$

13. $\displaystyle\sum_{n=0}^{\infty} \frac{(-1)^n(3/2)^n x^n}{n+1}$

14. $\displaystyle\sum_{n=0}^{\infty} \frac{(2n)! x^n}{n!}$

15. $\displaystyle\sum_{n=1}^{\infty} n^2 (x-1)^n$

16. $\displaystyle\sum_{n=1}^{\infty} \frac{n(x+2)^n}{2^n}$

17. $\displaystyle\sum_{n=1}^{\infty} \frac{(-1)^{n-1}(x+4)^n}{3^n \cdot n^2}$

18. $\displaystyle\sum_{n=1}^{\infty} \frac{n!(x-3)^n}{1 \cdot 3 \cdot 5 \cdots (2n-1)}$

19. $\displaystyle\sum_{n=1}^{\infty} \frac{(-1)^{n+1}(n!)^2(x-2)^n}{2^n(2n)!}$

20. $\displaystyle\sum_{n=1}^{\infty} \frac{n!(x-1)^n}{4^n \cdot 1 \cdot 3 \cdot 5 \cdots (2n-1)}$

21. $\displaystyle\sum_{n=1}^{\infty} \frac{(-1)^{n-1} n!(3/2)^n x^n}{1 \cdot 3 \cdot 5 \cdots (2n-1)}$

22. $\displaystyle\sum_{n=0}^{\infty} \frac{(-1)^n 3^{n+1} x^n}{2^{3n}}$

23. $\displaystyle\sum_{n=1}^{\infty} \frac{(n-2)x^n}{n^2}$

24. $\displaystyle\sum_{n=0}^{\infty} \frac{(6n^2 + 3n + 1)x^n}{2^n(n+1)^3}$

25. $\displaystyle\sum_{n=1}^{\infty} \frac{(-1)^{n-1} x^n}{(n+1)\ln(n+1)}$

26. $\displaystyle\sum_{n=1}^{\infty} \frac{\ln(n+1)3^n(x-1)^n}{n+1}$

27. $\displaystyle\sum_{n=1}^{\infty} \frac{(-1)^{n-1}(\ln n)2^n x^n}{3^n n^2}$

28. $\displaystyle\sum_{n=1}^{\infty} \frac{x^n}{(2n)!}$

29. $\displaystyle\sum_{n=1}^{\infty} \frac{2x^n}{(3n)!}$

30. $\displaystyle\sum_{n=1}^{\infty} \frac{(2x)^n}{(3n)!}$

31. $\displaystyle\sum_{n=1}^{\infty} \frac{(-1)^n x^n}{2\sqrt{n}}$

32. $\displaystyle\sum_{n=1}^{\infty} \frac{nx^n}{n+1}$

33. $\displaystyle\sum_{n=1}^{\infty} \frac{n^2 x^n}{(n+1)^2}$

34. $\displaystyle\sum_{n=1}^{\infty} \frac{(2x)^n}{3^n}$

35. $\displaystyle\sum_{n=1}^{\infty} \frac{(2x)^n}{3^{n+2}}$

36. $\displaystyle\sum_{n=1}^{\infty} \frac{(2+x)^n}{3^n}$

37. $\displaystyle\sum_{n=1}^{\infty} \frac{(x+1)^n}{n!}$

38. $\displaystyle\sum_{n=1}^{\infty} \frac{(x+1)^{2n}}{n!}$

39. $\displaystyle\sum_{n=1}^{\infty} \frac{3^{n^2} x^n}{e^n}$

40. $\displaystyle\sum_{n=1}^{\infty} \frac{x^{n^2}}{n!}$

41. $\displaystyle\sum_{n=1}^{\infty} \frac{(x+1)^{n^2}}{n!}$

42. Prove Theorem 19 for series of the form

$$\sum_{n=0}^{\infty} a_n(x-a)^n.$$

43. a) Find the interval of convergence of the series

$$\sum_{n=1}^{\infty} \frac{1 \cdot 3 \cdot 5 \cdots (2n-1)}{2 \cdot 4 \cdot 6 \cdots (2n)} x^n.$$

   b) Show that the series in (a) is identical with the series

$$\sum_{n=1}^{\infty} \frac{(2n)!}{2^{2n}(n!)^2} x^n.$$

44. Find the interval of convergence of the series

$$\sum_{n=1}^{\infty} \frac{1 \cdot 3 \cdot 5 \cdots (2n-1)(x-2)^n}{2^n \cdot 1 \cdot 4 \cdot 7 \cdots (3n-2)}.$$

45. Find the interval of convergence of the **binomial series**

$$1 + \sum_{n=1}^{\infty} \frac{m(m-1)\cdots(m-n+1)}{n!} x^n; \quad m \text{ fixed.}$$

*46 Suppose that the series $\sum_{n=0}^{\infty} a_n x^n$ has an interval of convergence $(-R, R)$. (a) Show that the series $\sum_{n=0}^{\infty} a_n x^{n+1}/(n+1)$ has the same interval of convergence. b) Show that the series $\sum_{n=1}^{\infty} n a_n x^{n-1}$ has the same interval of convergence.

47 Given the series $\sum_{n=0}^{\infty} a_n x^n$. Suppose that $\lim_{n \to \infty} \sqrt[n]{|a_n|} = r$. Show that the series converges for

$$-\frac{1}{r} < x < \frac{1}{r}.$$

48 Given the two series $\sum_{n=0}^{\infty} a_n x^n$ and $\sum_{n=0}^{\infty} b_n x^n$. Suppose there is an $N$ such that $|a_n| \le |b_n|$ for all $n \ge N$. Show that the interval of convergence of the first series is at least as large as the interval of convergence of the second series.

49 Suppose the series $\sum_{n=0}^{\infty} a_n x^n$ converges for $-R < x < R$. Show that the series $\sum_{n=0}^{\infty} a_n x^{kn}$ for some fixed positive integer $k$ converges for $-\sqrt[k]{R} < x < \sqrt[k]{R}$.

---

## 7

## TAYLOR'S SERIES

Suppose that a power series

$$\sum_{n=0}^{\infty} a_n(x-a)^n$$

converges in some interval $-R < x - a < R$ $(R > 0)$. Then the sum of the series has a value for each $x$ in this interval and so defines a function of $x$. We can therefore write

$$f(x) = a_0 + a_1(x-a) + a_2(x-a)^2 + a_3(x-a)^3 + \cdots, \qquad (1)$$

$$a - R < x < a + R.$$

We ask the question: What is the relationship between the coefficients $a_0, a_1, a_2, a_3, \ldots, a_n, \ldots$ and the function $f$?

We shall proceed naively, as if the right side of (1) were a polynomial. Setting $x = a$, we find at once that

$$f(a) = a_0.$$

We differentiate (1) (as if the right side were a polynomial) and get

$$f'(x) = a_1 + 2a_2(x-a) + 3a_3(x-a)^2 + 4a_4(x-a)^3 + \cdots.$$

For $x = a$, we find that

$$f'(a) = a_1.$$

We continue both differentiating and then setting $x = a$, to obtain

$$f''(x) = 2a_2 + 3 \cdot 2a_3(x-a) + 4 \cdot 3a_4(x-a)^2 + 5 \cdot 4a_5(x-a)^3 + \cdots,$$

$$f''(a) = 2a_2 \qquad \text{or} \qquad a_2 = \frac{f''(a)}{2!},$$

$$f'''(x) = 3 \cdot 2a_3 + 4 \cdot 3 \cdot 2a_4(x-a) + 5 \cdot 4 \cdot 3a_5(x-a)^2$$
$$+ 6 \cdot 5 \cdot 4a_6(x-a)^3 + \cdots,$$

$$f'''(a) = 3 \cdot 2a_3 \qquad \text{or} \qquad a_3 = \frac{f'''(a)}{3!},$$

and so forth. The pattern is now clear. The general formula for the coefficients $a_0, a_1, a_2, \ldots, a_n, \ldots$ is

$$a_n = \frac{f^{(n)}(a)}{n!}.$$

In Section 9, it will be shown that all of the above steps are legitimate so long as the series is convergent in some positive interval. Substituting the formulas for the coefficients $a_n$ into the power series, we obtain

$$f(x) = \sum_{n=0}^{\infty} \frac{f^{(n)}(a)}{n!}(x-a)^n. \qquad (2)$$

**DEFINITION**   *The right side of Eq. (2) is called the* **Taylor series for** $f$ **about the point** $a$ *or the* **expansion of** $f$ **into a power series about** $a$.

For the special case $a = 0$, the Taylor series is

$$f(x) = \sum_{n=0}^{\infty} \frac{f^{(n)}(0)}{n!}x^n. \qquad (3)$$

The right side of (3) is called the **Maclaurin series** *for* $f$.

**EXAMPLE 1**   Assuming that $f(x) = \sin x$ is given by its Maclaurin series, expand $\sin x$ into such a series.

**Solution**   We have

$$\begin{aligned}
f(x) &= \sin x, & f(0) &= 0, \\
f'(x) &= \cos x, & f'(0) &= 1, \\
f''(x) &= -\sin x, & f''(0) &= 0, \\
f^{(3)}(x) &= -\cos x, & f^{(3)}(0) &= -1, \\
f^{(4)}(x) &= \sin x, & f^{(4)}(0) &= 0.
\end{aligned}$$

It is clear that $f^{(5)} = f'$, $f^6 = f''$, etc., so that the sequence $0, 1, 0, -1, 0, 1, 0, -1, \ldots$ repeats itself indefinitely. Therefore, from (3) we obtain

$$\sin x = x - \frac{x^3}{3!} + \frac{x^5}{5!} - \frac{x^7}{7!} + \frac{x^9}{9!} - \cdots$$

or

$$\sin x = \sum_{k=0}^{\infty} \frac{(-1)^k x^{2k+1}}{(2k+1)!}. \qquad (4)$$

□

*Remark.*   It may be verified (by the Ratio Test, for example) that the series (4) *converges for all values of* $x$.

**EXAMPLE 2**    Assuming that $f(x) = \cos x$ is given by its Maclaurin series, expand $\cos x$ into such a series.

**Solution**    We have

$$f(x) = \cos x \qquad\qquad f(0) = 1$$

$$f'(x) = -\sin x \qquad\qquad f'(0) = 0$$

$$f''(x) = -\cos x \qquad\qquad f''(0) = -1$$

$$f^{(3)}(x) = \sin x \qquad\qquad f^{(3)}(0) = 0$$

$$f^{(4)}(x) = \cos x \qquad\qquad f^{(4)}(0) = 1$$

Proceeding as in Example 1 we obtain:

$$\cos x = 1 - \frac{x^2}{2!} + \frac{x^4}{4!} - \frac{x^6}{6!} + \cdots + (-1)^n \frac{x^{2n}}{(2n)!} + \cdots$$

or

$$\cos x = \sum_{k=0}^{\infty} (-1)^k \frac{x^{2k}}{(2k)!}.$$

**EXAMPLE 3**    Expand the function

$$f(x) = \frac{1}{x}$$

into a Taylor series about $x = 1$, assuming that such an expansion is valid.

**Solution**    We have

$$f(x) = x^{-1}, \qquad\qquad\qquad f(1) = 1,$$

$$f'(x) = (-1)x^{-2}, \qquad\qquad\qquad f'(1) = -1,$$

$$f''(x) = (-1)(-2)x^{-3}, \qquad\qquad f''(1) = (-1)^2 \cdot 2!,$$

$$f^{(3)}(x) = (-1)(-2)(-3)x^{-4}, \qquad f^{(3)}(1) = (-1)^3 \cdot 3!,$$

$$f^{(n)}(x) = (-1)(-2)\cdots(-n)x^{-n-1}, \qquad f^{(n)}(1) = (-1)^n \cdot n!.$$

Therefore from (2) with $a = 1$, we obtain

$$f(x) = \frac{1}{x} = \sum_{n=0}^{\infty} (-1)^n (x-1)^n. \tag{5}$$

*Remark.*    The series (5) converges for $|x - 1| < 1$ or $0 < x < 2$, as may be confirmed by the Ratio Test.

Examples 1, 2 and 3 have meaning only if it is known that the functions are representable by means of power series. There are examples of functions for which it is possible to compute all the quantities $f^{(n)}(x)$ at a given value $a$, and yet the Taylor series about $a$ will not represent the function. (See Problem 39 at the end of this Section.)

**EXAMPLE 4**    Find the Taylor series for $\sin x$ centered at $a = \pi/3$, assuming that such an expansion is valid.

**Solution**    We proceed as in Example 1:

$$f(x) = \sin x, \qquad f\left(\frac{\pi}{3}\right) = \frac{\sqrt{3}}{2}$$

$$f'(x) = \cos x, \qquad f'\left(\frac{\pi}{3}\right) = \frac{1}{2}$$

$$f''(x) = -\sin x, \qquad f''\left(\frac{\pi}{3}\right) = -\frac{\sqrt{3}}{3}$$

$$f^{(3)}(x) = -\cos x, \qquad f^{(3)}\left(\frac{\pi}{3}\right) = -\frac{1}{2}$$

$$f^{(4)}(x) = \sin x, \qquad f^{(4)}\left(\frac{\pi}{3}\right) = \frac{\sqrt{3}}{2}$$

and therefore

$$\sin x = \frac{\sqrt{3}}{2} + \frac{1}{2}\left(x - \frac{\pi}{3}\right) - \frac{\sqrt{3}}{2(2!)}\left(x - \frac{\pi}{3}\right)^2 - \frac{1}{2(3!)}\left(x - \frac{\pi}{3}\right)^3$$

$$+ \frac{\sqrt{3}}{2(4!)}\left(x - \frac{\pi}{3}\right)^4 + \cdots.$$

It is not easy to express this series concisely. We can do it by writing

$$\sin x = \sum_{n=0}^{\infty} a_n (x - \pi/3)^n,$$

and distinguishing between the even and odd coefficients, i.e., between $a_{2n}$ and $a_{2n+1}$. For the even ones we can write:

$$a_{2n} = \frac{\sqrt{3}(-1)^n}{2(2n)!} \quad n = 0, 1, 2, 3, \ldots$$

and for the odd ones:

$$a_{2n+1} = \frac{1(-1)^n}{2(2n+1)!} \quad n = 0, 1, 2, 3, \ldots.$$

Therefore

$$\sin x = \sum_{n=0}^{\infty} \frac{\sqrt{3}}{2}\frac{(-1)^n}{(2n)!}\left(x - \frac{\pi}{3}\right)^{2n} + \sum_{n=0}^{\infty} \frac{1}{2}\frac{(-1)^n}{(2n+1)!}\left(x - \frac{\pi}{3}\right)^{2n+1} \qquad \square$$

**EXAMPLE 5**    Compute the first six terms of the Maclaurin expansion of the function

$$f(x) = \tan x,$$

assuming that such an expansion is valid.

**Solution**   We have

$$f(x) = \tan x, \qquad\qquad f(0) = 0,$$

$$f'(x) = \sec^2 x, \qquad\qquad f'(0) = 1,$$

$$f''(x) = 2 \sec^2 x \tan x, \qquad\qquad f''(0) = 0,$$

$$f^{(3)}(x) = 2 \sec^4 x + 4 \sec^2 x \tan^2 x, \qquad\qquad f^{(3)}(0) = 2,$$

$$f^{(4)}(x) = 8 \tan x \sec^2 x(2 + 3 \tan^2 x), \qquad\qquad f^{(4)}(0) = 0,$$

$$f^{(5)}(x) = 48 \tan^2 x \sec^4 x \qquad\qquad f^{(5)}(0) = 16.$$
$$+ 8 \sec^2 x(2 + 3 \tan^2 x)(\sec^2 x + 2 \tan^2 x),$$

Therefore

$$f(x) = \tan x = x + \frac{x^3}{3} + \frac{2x^5}{15} + \cdots .\qquad\qquad \square$$

*Remark.*   Example 5 shows that the general pattern for the successive derivatives may not always be readily discernible. Examples 1, 2, and 3, on the other hand, show how the general formula for the $n$th derivative may be arrived at simply.

## 7   PROBLEMS

In Problems 1 through 16, find the Taylor (Maclaurin if $a = 0$) series for each function $f$ about the given value of $a$.

**1** $f(x) = e^x, \quad a = 0$

**2** $f(x) = \cos x, \quad a = \pi/4$

**3** $f(x) = \ln(1 + x), \quad a = 0$

**4** $f(x) = \ln(1 + x), \quad a = 1$

**5** $f(x) = (1 - x)^{-2}, \quad a = 0$

**6** $f(x) = (1 - x)^{-1/2}, \quad a = 0$

**7** $f(x) = (1 + x)^{1/2}, \quad a = 0$

**8** $f(x) = e^x, \quad a = 1$

**9** $f(x) = \ln x, \quad a = 3$

**10** $f(x) = \sin x, \quad a = \pi/4$

**11** $f(x) = \cos x, \quad a = \pi/3$

**12** $f(x) = \sin x, \quad a = 2\pi/3$

**13** $f(x) = \sqrt{x}, \quad a = 4$

**14** $f(x) = \sin(x + \frac{1}{2}), \quad a = 0$

**15** $f(x) = \cos(x + \frac{1}{2}), \quad a = 0$

**16** $f(x) = x^m, \quad a = 1$

In each of Problems 17 through 37, find the first few terms of the Taylor expansion about the given value of $a$. Carry out the process to include the term $(x - a)^n$ for the given integer $n$.

**17** $f(x) = e^{-x^2}, \quad a = 0, n = 4$

**18** $f(x) = xe^x, \quad a = 0, n = 4$

**19** $f(x) = \dfrac{1}{1 + x^2}, \quad a = 0, n = 4$

**20** $f(x) = \arctan x, \quad a = 0, n = 5$

**21** $f(x) = e^{x/2}, \quad a = 1, n = 3$

**22** $f(x) = x^3 \ln x, \quad a = 1, n = 7$

**23** $f(x) = \ln(1 + 2x), \quad a = 1, n = 5$

**24** $f(x) = \cos^2 x, \quad a = \pi, n = 4$

**25** $f(x) = \sin^2 x, \quad a = \dfrac{\pi}{2}, n = 4$

**26** $f(x) = (x - 1)^{10}, \quad a = 0, n = 4$

**27** $f(x) = e^x \cos x, \quad a = 0, n = 4$

**28** $f(x) = \dfrac{1}{\sqrt{1 - x^2}}, \quad a = 0, n = 4$

**29** $f(x) = \arcsin x, \quad a = 0, n = 5$

**30** $f(x) = \tan x, \quad a = \pi/4, n = 5$

**31** $f(x) = \ln \sec x, \quad a = 0, n = 6$

**32** $f(x) = \sec x, \quad a = 0, n = 4$

**33** $f(x) = \csc x, \quad a = \pi/6, n = 4$

**34** $f(x) = \csc x, \quad a = \pi/2, n = 4$

**35** $f(x) = \sec x, \quad a = \pi/3, n = 3$

**36** $f(x) = \ln \sin x, \quad a = \pi/4, n = 4$

**37** $f(x) = \tan x, \quad a = 0, n = 7$
(*Hint:* See Example 5; express $f^{(5)}(x)$ in terms of sec $x$.)

**38** a)  Given the polynomial

$$f(x) = 3 + 2x - x^2 + 4x^3 - 2x^4, \tag{6}$$

show that $f$ may be written in the form

$$f(x) = a_0 + a_1(x - 1) + a_2(x - 1)^2$$
$$+ a_3(x - 1)^3 + a_4(x - 1)^4.$$

(*Hint:* Use the Taylor expansion (2) and (6) to get each $a_i$.)
*b)  Given the same polynomial in two forms,

$$f(x) = \sum_{k=0}^{n} a_k(x - a)^k, \qquad f(x) = \sum_{k=0}^{n} b_k(x - b)^k,$$

express each $b_k$ in terms of $a$, $b$, and the $a_k$.

*\*39** a)  Given the function (see Fig. 13)

$$F(x) = \begin{cases} e^{-1/x^2}, & x \neq 0, \\ 0, & x = 0, \end{cases}$$

use l'Hôpital's Rule to show that

$$F'(0) = 0.$$

b)  Show that $F^{(n)}(0) = 0$ for every positive integer $n$.

c)  What can be said about the Taylor series for $F$?

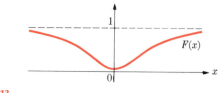

**FIGURE 13**

**40**  For any real number $n$ and any positive number $a$ show that the Maclaurin series for $(a + x)^n$ is given by

$$(a + x)^n = a^n + na^{n-1}x + \frac{n(n - 1)}{2!}a^{n-2}x^2 + \cdots$$

$$+ \frac{n(n - 1) \cdots (n - k + 1)}{k!}a^{n-k}x^k + \cdots.$$

Set $a = 1$ and $n = \frac{1}{2}$ and find the Maclaurin series for $(1 + x)^{1/2}$. Find the radius of convergence. Also, find the Maclaurin expansion and the radius of convergence for $(1 + x)^{-1}$.

*\*41**  Let $i$ be the imaginary square root of $-1$. (That is, $i^2 = -1$.) Assuming that the Maclaurin series for $e^x$ obtained in the solution of Problem 1 holds for imaginary numbers of the form $z = ix$, where $x$ is real, verify Euler's Formula:

$$e^{ix} = \cos x + i \sin x.$$

**42**  Find the Taylor expansion of $(3 + x)^{-1/2}$ about $a = 1$. (*Hint:* Write $3 + x = 4 + (x - 1)$ and use the result in Problem 34 with $a = 4$ and $x$ replaced by $x - 1$.)

---

**8**

## TAYLOR'S THEOREM WITH REMAINDER

If a function $f$ possesses only a finite number—say $n$—of derivatives, then it is clear that it is not possible to represent it by a Taylor series, since the coefficients $a_k = f^{(k)}(a)/k!$ cannot be computed beyond $a_n$. In such cases it is still possible to obtain a *finite version* of a Taylor expansion.

Suppose that $f(x)$ possesses $n$ continuous derivatives in some interval about the value $a$. Then it is always possible to write

$$f(x) = f(a) + \frac{f'(a)}{1!}(x-a) + \frac{f^{(2)}(a)}{2!}(x-a)^2 + \cdots$$

$$+ \frac{f^{(n)}(a)}{n!}(x-a)^n + R_n \qquad (1)$$

for $x$ in this interval. The right side consists of a polynomial in $x$ of degree $n$ and a **remainder** $R_n$ about which, as yet, we have no knowledge. In fact, the quantity $R_n$ is *defined* by the formula (1). For example, in the unusual case that $f$ and its $n$ derivatives all vanish at $x = a$, the "remainder" $R_n$ is just $f$ itself. The content of **Taylor's Theorem** concerns useful information about the nature of $R_n$. This theorem is not only of great theoretical value but may also be used in approximations and numerical computations.

**THEOREM 21**　　**(Taylor's Theorem with Derivative Form of Remainder)**　　*Suppose that $f$, $f'$, $f^{(2)}, \ldots, f^{(n)}, f^{(n+1)}$ are all continuous on some interval containing $a$ and $b$. Then there is a number $\xi$ between $a$ and $b$ such that*

$$f(b) = f(a) + \frac{f'(a)}{1!}(b-a) + \frac{f^{(2)}(a)}{2!}(b-a)^2 + \cdots$$

$$+ \frac{f^{(n)}(a)}{n!}(b-a)^n + \frac{f^{(n+1)}(\xi)(b-a)^{n+1}}{(n+1)!}.$$

*That is, the remainder $R_n$ is given by the formula*

$$R_n = \frac{f^{(n+1)}(\xi)(b-a)^{n+1}}{(n+1)!} \qquad (2)$$

*Remarks.*　i) We see that $R_n$ depends on both $b$ and $a$, and we write, in general, $R_n = R_n(a, b)$, a function of two variables.

ii) In the special case $n = 0$, we obtain

$$f(b) = f(a) + f'(\xi)(b-a),$$

which we recognize as the Mean Value Theorem (page 131). Thus this form of Taylor's Theorem is a direct generalization of the Mean Value Theorem.

**Proof of Theorem 21**　　The proof makes use of Rolle's Theorem (page 130). We create a function $\phi(x)$ which is zero at both $a$ and $b$ and so, by Rolle's Theorem, there must be a number $\xi$ between $a$ and $b$ where $\phi'(\xi) = 0$. The algebra is lengthy, and the reader should write out the details for the cases $n = 1, 2, 3$ in order to grasp the essence of the proof. We use the form (1) with $x = b$ and write

$$f(b) = f(a) + \frac{f'(a)(b-a)}{1!} + \frac{f^{(2)}(a)(b-a)^2}{2!} + \cdots + \frac{f^{(n)}(a)(b-a)^n}{n!} + R_n(a, b);$$

we wish to find $R_n(a, b)$. We define the function

$$\phi(x) = f(b) - f(x) - \frac{f'(x)(b-x)}{1!} - \frac{f^{(2)}(x)(b-x)^2}{2!}$$

$$- \frac{f^{(3)}(x)(b-x)^3}{3!} - \cdots - \frac{f^{(n-1)}(x)(b-x)^{n-1}}{(n-1)!}$$

$$- \frac{f^{(n)}(x)(b-x)}{n!} - R_n(a, b)\frac{(b-x)^{n+1}}{(b-a)^{n+1}}.$$

The function $\phi$ was concocted in such a way that $\phi(a) = 0$ and $\phi(b) = 0$, facts which are easily checked by straight substitution. We compute the derivative $\phi'(x)$ (using the formula for the derivative of a product wherever necessary):

$$\phi'(x) = -f'(x) + f'(x) - \frac{f^{(2)}(x)(b-x)}{1!} + \frac{2f^{(2)}(x)(b-x)}{2!}$$

$$- \frac{f^{(3)}(x)(b-x)^2}{2!} + \frac{3f^{(3)}(x)(b-x)^2}{3!} - \frac{f^{(4)}(x)(b-x)^3}{3!} + \cdots$$

$$- \frac{f^{(n+1)}(x)(b-x)^n}{n!} + \frac{R_n(a, b)(n+1)(b-x)^n}{(b-a)^{n+1}}.$$

Amazingly, all the terms cancel except the last two, and we find

$$\phi'(x) = - \frac{f^{(n+1)}(x)(b-x)^n}{n!} + R_n(a, b)(n+1)\frac{(b-x)^n}{(b-a)^{n+1}}.$$

Using Rolle's Theorem, we know there must be a value $\xi$ between $a$ and $b$ such that $\phi'(\xi) = 0$. Therefore we get

$$0 = - \frac{f^{(n+1)}(\xi)(b-\xi)^n}{n!} + R_n(a, b)(n+1)\frac{(b-\xi)^n}{(b-a)^{n+1}}$$

or, upon solving for $R_n(a, b)$, the formula (2) exactly.                      □

*Remarks.*   i) If we know that $f(x)$ has continuous derivatives of all orders and if $R_n(a, b) \to 0$ as $n \to \infty$, then we can establish the validity of the Taylor series.

ii) In any case, $R_n$ is a measure of how much $f$ differs from a certain polynomial of degree $n$. If $R_n$ is small, then the polynomial may be used for an approximation to $f$. This fact is especially useful in applications where a function which is not known precisely may describe an important quantity; while the function may be unknown it is often possible to estimate accurately its first few derivatives. If $R_n$ is known to be small, Taylor's Theorem states that a polynomial can then be used to approximate the unknown function $f$.

When we use Taylor's Theorem in the computation of functions from the approximating polynomial, errors may arise from two sources: the error $R_n$, made above by neglecting the powers of $(b-a)$ beyond the $n$th; and the "round-off error" made by expressing each term in decimal form. If we wish to compute the value of some function $f(b)$ to an accuracy of four decimal places, it is essential to be able to say for certain that $f(b)$ is between some decimal fraction with four decimals $-0.00005$ and the same decimal fraction $+0.00005$. *Time is saved by computing each term to two decimals more than are required.* Frequently $R_n$ is close to the value of the first term omitted in the formation of the approximating series, and this fact can be used as a guide in choosing the number of terms. Although we usually do not know $R_n$ exactly, we can often show that there are two numbers $m$ and $M$ with

$$m \le f^{(n+1)}(x) \le M \quad \text{for } all \ x \text{ between } a \text{ and } b.$$

Then we get for $R_n(a, b)$ the inequality

$$\frac{m(b-a)^{n+1}}{(n+1)!} \le R_n(a, b) \le \frac{M(b-a)^{n+1}}{(n+1)!}.$$

**EXAMPLE 1**    Compute $(1.1)^{1/5}$ to an accuracy of four decimal places.

**Solution**    The key to the solution, using Taylor's Theorem, is the fact that we can set

$$f(x) = (1+x)^{1/5}, \quad a = 0, b = 0.1.$$

Then

$$f(x) = (1+x)^{1/5}, \qquad\qquad f(a) = 1 \qquad\qquad = \quad 1.0000\ 00$$

$$f'(x) = \frac{1}{5}(1+x)^{-4/5}, \qquad f'(a)(b-a) = \frac{1}{5}(0.1)^1 \quad = \quad 0.0200\ 00$$

$$f''(x) = -\frac{4}{25}(1+x)^{-9/5}, \qquad \frac{f''(a)(b-a)^2}{2!} = -\frac{2}{25}(0.1)^2 = -0.0008\ 00$$

$$f'''(x) = \frac{36}{125}(1+x)^{-14/5}, \qquad \frac{f'''(a)(b-a)^3}{3!} = \frac{6}{125}(0.1)^3 \quad = \quad 0.0000\ 48.$$

For all $x$ between 0 and 1 we have

$$0 < (1+x)^{-14/5} < 1.$$

Therefore we can estimate $R_n$ for $n = 2$:

$$0 < R_2 = \frac{36}{125}(1+\xi)^{-14/5}\frac{(b-a)^3}{3!} < \frac{6}{125}(0.1)^3 = 0.0000\ 48.$$

Adding the terms in the Taylor expansion through $n = 2$, we get

$$(1.1)^{1/5} = 1.0192, \text{ approximately;}$$

in fact, a more precise statement is

$$1.0192 < (1.1)^{1/5} < 1.0192\ 48. \qquad\qquad\qquad \square$$

*Remarks.* In Example 1 we could have selected $f(x) = x^{1/5}$ with $a = 1$, $b = 1.1$. The result is the same. Hand calculators are sufficient for making the required computations.

**EXAMPLE 2**    Compute $\sqrt[3]{7}$ to an accuracy of four decimal places.

**Solution**    Set

$$f(x) = x^{1/3}, \quad a = 8, \; b = 7.$$

Then $b - a = -1$ and

| | | |
|---|---|---|
| $f(x) = x^{1/3},$ | $f(a) = 2$ | $= \;\;2.0000\,00$ |
| $f'(x) = \dfrac{1}{3}x^{-2/3},$ | $f'(a)(b-a) = -\dfrac{1}{12}$ | $= -0.0833\,33$ |
| $f''(x) = -\dfrac{2}{9}x^{-5/3},$ | $\dfrac{f''(a)(b-a)^2}{2!} = -\dfrac{1}{9\cdot 2^5}$ | $= -0.0034\,72$ |
| $f^{(3)}(x) = \dfrac{10}{27}x^{-8/3},$ | $\dfrac{f^{(3)}(a)(b-a)^3}{3!} = -\dfrac{5}{81\cdot 2^8}$ | $= -0.0002\,41$ |
| $f^{(4)}(x) = \dfrac{-80}{81}x^{-11/3},$ | $\dfrac{f^{(4)}(a)(b-a)^4}{4!} = -\dfrac{5}{243\cdot 2^{10}}$ | $= -0.0000\,20.$ |

It would appear to be sufficient to use only the terms through $(b-a)^3$. However, by computing the sum of the decimal fractions given, we obtain $1.9129\,54$. But the next term is $-0.0000\,20$ which, if included, would reduce the value to $1.9129\,34$. If we stopped with the $(b-a)^3$ term and rounded off, we would obtain $1.9130$ whereas, if we keep the next term and round off, we obtain $1.9129$. So the term in $(b-a)^4$ should be retained, and the remainder $R_4$ must be estimated. We have

$$f^{(5)}(x) = \frac{880}{243}x^{-14/3} = \frac{880x^{1/3}}{243x^5}.$$

Since we are concerned with the interval $7 < x < 8$, we see that $x^{1/3} < 2$ and $x^5 > (49)(343)$. Hence

$$0 < \frac{f^{(5)}(x)}{5!} < \frac{1760}{(243)(49)(343)(120)} < \frac{1}{270{,}000} < 0.000004.$$

Since $(b-a)^5 = -1$, we conclude that

$$-0.000004 < R_4 < 0$$

and that

$$\sqrt[3]{7} = 1.9129$$

to the required accuracy. Actually, if we merely keep an extra decimal in each term retained, we see that

$$\sqrt[3]{7} = 1.91293$$

to five decimals.     □

*Remarks.* (i) In Example 1 there is no round-off error, since each decimal fraction gave the exact value of the corresponding term. This is not true in Example 2, however. In general, the round-off error in each term may be as much as $\frac{1}{2}$ in the last decimal place retained. Round-off errors may tend to cancel each other if there is a large number of computations in a given problem. (ii) Many calculators can compute $\sqrt[3]{7}$ at the press of a button. The internal program of the calculator which performs this task must use an approximation which involves only additions, subtractions, multiplications, and divisions. Polynomials and rational functions (i.e., the quotient of two polynomials) are suited perfectly for such approximations. In many cases the simplest program (and the one used in the calculator) is given by the polynomial in Taylor's Theorem.

The remainder $R_n(a, b)$ in Taylor's Theorem may be given in many forms. The next theorem, stated without proof, gives the remainder in the form of an integral.

THEOREM 22    **(Taylor's Theorem with Integral Form of Remainder)**  *Suppose that* $f$, $f'$, $f^{(2)}, \ldots, f^{(n)}, f^{(n+1)}$ *are all continuous in some interval containing* $a$ *and* $b$. *Then* $f(b)$ *may be written in the form*

$$f(b) = f(a) + \frac{f'(a)}{1!}(b-a) + \frac{f^{(2)}(a)(b-a)^2}{2!} + \cdots + \frac{f^{(n)}(a)(b-a)^n}{n!} + R_n,$$

*where*

$$R_n = \frac{1}{n!} \int_a^b f^{(n+1)}(t)(b-t)^n \, dt.$$

EXAMPLE 3    Write $\ln(1+x)$ as a polynomial of the third degree, and estimate the remainder $R_n$ for $0 < x < \frac{1}{2}$.

Solution    We set

$$f(x) = \ln(1+x), \quad a = 0, \, b = x.$$

Then we write successive derivatives,

$$f'(x) = \frac{1}{1+x}, \qquad f'(0) = 1,$$

$$f''(x) = -\frac{1}{(1+x)^2}, \qquad f''(0) = -1,$$

$$f^{(3)}(x) = \frac{2}{(1+x)^3}, \qquad f^{(3)}(0) = 2,$$

$$f^{(4)}(x) = -\frac{6}{(1+x)^4}.$$

Therefore

$$\ln(1+x) = x - \frac{x^2}{2} + \frac{x^3}{3} + R_3,$$

with

$$R_3 = -\frac{6}{6} \int_0^x \frac{1}{(1+t)^4}(x-t)^3 \, dt.$$

A simple estimate replaces $(1+t)^4$ by its smallest value 1, and we find

$$|R_3| < \int_0^{1/2} \left( \frac{1}{2} - t \right)^3 dt = \frac{1}{64}.$$

☐

## 8  PROBLEMS

In each of Problems 1 through 8, find Taylor's Formula with Derivative Form of Remainder for the given values of $a$ and $n$, and any unknown value $b$.

1  $f(x) = \sin x$,   $a = \dfrac{\pi}{2}, n = 3$

2  $f(x) = \cos x$,   $a = \dfrac{\pi}{6}, n = 3$

3  $f(x) = \sqrt{x}$,   $a = 9, n = 2$

4  $f(x) = e^{-x}$,   $a = 0, n = 4$

5  $f(x) = \ln x$,   $a = e, n = 3$

6  $f(x) = \arctan x$,   $a = 1, n = 2$

7  $f(x) = e^{2x}$,   $a = 12, n = 3$

8  $f(x) = \dfrac{1}{x}$,   $a = -3, n = 2$

In each of Problems 9 through 28, use Taylor's Theorem with Remainder to compute the given quantities to the specified number of decimal places. Use a hand calculator in estimating the remainder.

9  $e^{-0.2}$,   5 decimals        10  $e^{-0.4}$,   4 decimals

11  $e^{0.2}$,   5 decimals        12  $\sin(0.5)$,   5 decimals

13  $\cos(0.5)$,   5 decimals      14  $\tan(0.1)$,   3 decimals

15  $\ln(1.2)$,   4 decimals       16  $\ln(0.9)$,   5 decimals

17  $e^{-1}$,   5 decimals         18  $e$,   5 decimals

19  $(1.08)^{1/4}$,   5 decimals   20  $(0.92)^{1/4}$,   5 decimals

21  $(0.91)^{1/3}$,   5 decimals   22  $(0.90)^{1/5}$,   5 decimals

23  $(30)^{1/5}$,   5 decimals     24  $(15)^{1/4}$,   5 decimals

25  $(0.8)^{1/5}$,   5 decimals    26  $(65)^{1/6}$,   5 decimals

27  $\ln(0.8)$,   5 decimals       28  $\ln(0.6)$,   3 decimals

Given that $1° = \pi/180$ radians $= 0.0174533$ radians and $5° = \pi/36$ radians $= 0.0872665$ radians, compute each of the following to the number of decimal places required without using a calculator.

29  $\sin 1°$,   6 decimals        30  $\sin 5°$,   5 decimals

31  $\cos 5°$,   5 decimals

32  Find the largest interval about $x = 0$ in which the function $f(x) = \sin x$ may be approximated to four decimal places by $x - \frac{1}{6}x^3$, the first two terms in the Maclaurin expansion.

33  Given the function $f(x) = \cos^2 x$, find an upper and lower bound for $R_n(a, b)$ where $a = 0$, $b = \pi/4$, $n = 4$.

34  Verify the identity in $x$:

$$\frac{1}{1+x} = 1 - x + x^2 - \cdots + (-1)^{n-1}x^{n-1} + \frac{(-1)^n x^n}{1+x}.$$

Integrate between 0 and $b$ to obtain the formula

$$\ln(1+b) = \sum_{k=1}^{n} (-1)^k \frac{b^k}{k} + (-1)^n \int_0^b \frac{x^n}{1+x} \, dx.$$

Show that this is Taylor's Formula with Integral Remainder. Obtain the estimate for $R_n$:

$$|R_n| \leq \frac{|b|^{n+1}}{1-|b|} \quad \text{for } |b| < 1.$$

35  Apply the method of Problem 34 to the function $f(x) = 1/(1+x^2)$ and obtain Taylor's formula for $\arctan x$.

36  Let $\Sigma_{n=0}^{\infty} u_n$ be a convergent infinite series whose terms are alternately positive and negative and also such that $|u_{n+1}| < |u_n|$ for each $n$. The Corollary to Theorem 16 shows that the error in approximating the above series by $\Sigma_{n=0}^{N} u_n$ is always less than $|u_{N+1}|$. Apply this result to approximate $\cos(\pi/6)$ by its Taylor series with remainder and thereby find $\sqrt{3}$ to three decimal places.

*37  Let $f$ be a continuous function defined on the interval $[a, b]$. Combine Theorem 21 and Theorem 22 to show that there exists a value $\xi$, $a < \xi < b$, such that

$$\int_a^b (x-t)^n f(t) \, dt = \frac{(x-a)^{n+1} f(\xi)}{n+1}.$$

---

**9**

---

## DIFFERENTIATION AND INTEGRATION OF SERIES

In Section 7 we developed the formula for the Taylor series of a function and, in so doing, we ignored the validity of the manipulations which were performed. Now we shall establish the theorems which verify the correctness of the results already obtained.

---

**THEOREM 23**    **(Term-by-Term Differentiation of Power Series)**    *If $R > 0$ and the series*

$$\sum_{n=0}^{\infty} a_n x^n \tag{1}$$

*converges for $|x| < R$, then the series obtained from (1) by term-by-term differentiation converges absolutely for $|x| < R$.*

---

**Proof**    Term-by-term differentiation of (1) yields

$$\sum_{n=1}^{\infty} n a_n x^{n-1}. \tag{2}$$

Choose any value $x$ such that $|x| < R$ and choose $x_1$ so that $|x| < |x_1| < R$. According to the Lemma of Section 6 (page 418), there is a positive number $M$ with the property that

$$|a_n x_1^n| \leq M \quad \text{for all } n.$$

We have the relation

$$|n a_n x^{n-1}| = \left| n a_n \frac{x^{n-1}}{x_1^n} \cdot x_1^n \right| \leq n \frac{M}{x_1} \left| \frac{x}{x_1} \right|^{n-1},$$

and now we can apply the Comparison Test to the series (2). The series

$$\frac{M}{x_1} \sum_{n=1}^{\infty} n \left| \frac{x}{x_1} \right|^{n-1}$$

converges by the Ratio Test, since $\rho = |x/x_1| < 1$; hence, so does the series (2). Since $x$ is any number in the interval $(-R, R)$, the interval of convergence of (2) is the same as that of (1).    □

---

**COROLLARY**    *Under the hypotheses of Theorem 23, the series (1) may be differentiated any number of times and each of the differentiated series converges for $|x| < R$.*

---

*Remarks.* i) The Corollary is obtained by induction, since each differentiated series has the same radius of convergence as the one before. (ii) The results of Theorem 23 and the Corollary are valid for a series of the form

$$\sum_{n=0}^{\infty} a_n (x - a)^n,$$

which converges for $|x - a| < R$ so long as $R > 0$. The proof is the same. (iii) The quantity $R$ may be $+\infty$, in which case the series and its derived ones converge for *all* values of $x$.

**THEOREM 24**     *If $R > 0$ and $f$ is defined by*

$$f(x) = \sum_{n=0}^{\infty} a_n x^n \quad \text{for } |x| < R,$$

*then $f$ is continuous for $|x| < R$.*

**Proof**     Let $x_0$ be any number such that $-R < x_0 < R$; we wish to show that $f$ is continuous at $x_0$. In other words we must show that

$$f(x) \to f(x_0) \quad \text{as } x \to x_0.$$

We have

$$|f(x) - f(x_0)| = \left| \sum_{n=0}^{\infty} a_n(x^n - x_0^n) \right| \le \sum_{n=0}^{\infty} |a_n| |x^n - x_0^n|.$$

We apply the Mean Value Theorem to the function $g(x) = x^n$; that is, the relation

$$g(x) - g(x_0) = g'(\xi)(x - x_0), \quad (\xi \text{ is between } x \text{ and } x_0),$$

applied to the function $g(x) = x^n$, is

$$x^n - x_0^n = n\xi_n^{n-1}(x - x_0), \quad (\xi_n \text{ between } x \text{ and } x_0).$$

The subscript $n$ has been put on $\xi$ to identify the particular exponent of the function $x^n$. We may also write

$$|x^n - x_0^n| = n|\xi_n|^{n-1}|x - x_0|.$$

Thus we find

$$|f(x) - f(x_0)| \le \sum_{n=0}^{\infty} n|a_n| |\xi_n|^{n-1}|x - x_0|.$$

So long as $x$ is in the interval of convergence there is an $x_1$ such that $|\xi_n| < x_1$ for all $n$. We deduce that

$$|f(x) - f(x_0)| \le |x - x_0| \sum_{n=0}^{\infty} n|a_n| x_1^{n-1}.$$

Now we apply Theorem 23 to conclude that the series on the right converges; call its sum $K$. Then

$$|f(x) - f(x_0)| \le |x - x_0| \cdot K.$$

As $x$ tends to $x_0$ the quantity on the right tends to zero, and so $f(x)$ tends to $f(x_0)$. That is, $f(x)$ is continuous at $x_0$. Since $x_0$ is *any* number such that $-R < x_0 < R$, the function $f$ is continuous for all $x$ in the interval $[-R, R]$.

**THEOREM 25**    **(Term-by-Term Integration of Power Series)**    *Suppose that $R > 0$ and the series*

$$f(x) = \sum_{n=0}^{\infty} a_n x^n \tag{3}$$

*converges for $|x| < R$. We define*

$$F(x) = \int_0^x f(t)\, dt.$$

*Then the formula*

$$F(x) = \sum_{n=0}^{\infty} a_n \frac{x^{n+1}}{n+1} \tag{4}$$

*holds for $|x| < R$.*

**Proof**    Let $x$ be any number such that $-R < x < R$. Choose $x_1$ so that $|x| < |x_1| < R$. Note that for any $n$

$$\int_0^x a_n t^n\, dt = \frac{a_n x^{n+1}}{n+1}.$$

Therefore

$$F(x) - \sum_{n=0}^{N} a_n \frac{x^{n+1}}{n+1} = \int_0^x \left[ f(t) - \sum_{n=0}^{N} a_n t^n \right] dt. \tag{5}$$

But now

$$f(t) - \sum_{n=0}^{N} a_n t^n = \sum_{n=0}^{\infty} a_n t^n - \sum_{n=0}^{N} a_n t^n = \sum_{n=N+1}^{\infty} a_n t^n,$$

and for all $t$ such that $-|x| < t < |x|$ and $|x| < |x_1| < R$,

$$\left| f(t) - \sum_{n=0}^{N} a_n t^n \right| \le \sum_{n=N+1}^{\infty} |a_n||x|^n. \tag{6}$$

Since the series (3) converges absolutely at $x$, the right side of (6)—being the remainder—tends to zero as $N \to \infty$. We conclude from (5) that

$$\left| F(x) - \sum_{n=0}^{N} a_n \frac{x^{n+1}}{n+1} \right| \le \int_0^x \left( \sum_{n=N+1}^{\infty} |a_n||x|^n \right) dt = \left( \sum_{n=N+1}^{\infty} |a_n||x|^n \right) \cdot x.$$

As $N$ tends to $\infty$, the right side above tends to zero, and the left side above yields (4).    □

**EXAMPLE 1**    Assuming that the function $f(x) = \sin x$ is given by the series

$$\sin x = x - \frac{x^3}{3!} + \frac{x^5}{5!} - \cdots + \frac{(-1)^n x^{2n+1}}{(2n+1)!} + \cdots,$$

find the Maclaurin series for $\cos x$.

**Solution** Applying the Ratio Test to the series

$$\sum_{n=0}^{\infty} \frac{(-1)^n x^{2n+1}}{(2n+1)!},$$

we see that it converges for all values of $x$. We define

$$F(x) = \int_0^x \sin t \, dt = -\cos x + 1.$$

Integrating the above series for $\sin x$ term by term, we obtain

$$F(x) = 1 - \cos x = \sum_{n=0}^{\infty} \frac{(-1)^n x^{2n+2}}{(2n+2)!}$$

or

$$\cos x = 1 - \frac{x^2}{2!} + \frac{x^4}{4!} - \frac{x^6}{6!} + \cdots + \frac{(-1)^n x^{2n}}{(2n)!} + \cdots = \sum_{n=0}^{\infty} \frac{(-1)^n x^{2n}}{(2n)!}. \quad \square$$

The next theorem relates the derivative of a function given by a series with the term-by-term differentiation of the series.

**THEOREM 26**  **(Term-by-Term Differentiation of Power Series)** *Suppose that $R > 0$ and*

$$f(x) = \sum_{n=0}^{\infty} a_n x^n \quad \text{for } |x| < R; \tag{7}$$

*then $f(x)$ has continuous derivatives of all orders for $|x| < R$ which are given there by series obtained by successive term-by-term differentiations of (7).*

**Proof** The fact that

$$\sum_{n=1}^{\infty} n a_n x^{n-1} \tag{8}$$

converges for $|x| < R$ was shown in Theorem 23. We must show that $g(x) = \sum_{n=1}^{\infty} n a_n x^{n-1}$ is the derivative of $f$. Theorem 24 establishes the fact that $g$ is continuous. Then, integrating the series (8) term by term we get, on the one hand,

$$a_0 + \int_0^x g(t) \, dt$$

and, on the other, the series for $f(x)$. That is,

$$f(x) = a_0 + \int_0^x g(t) \, dt.$$

The Fundamental Theorem of Calculus then asserts that

$$f'(x) = g(x),$$

which is the result we wished to establish. Now starting with $g(x) = \sum_{n=1}^{\infty} n a_n x^{n-1}$ which converges for $|x| < R$, we may repeat the argument to show that $g'(x) = \sum_{n=2}^{\infty} n(n-1) a_n x^{n-2}$ for $|x| < R$. By induction, we conclude that the series (7) differentiated $k$ times yields a convergent series for $f^{(k)}(x)$, $k = 1, 2, \ldots$. The radius of convergence is always $R$.  $\square$

*Remark.* The result of Theorem 26 holds equally well for functions $f$ given by series of the form

$$f(x) = \sum_{n=0}^{\infty} a_n (x-a)^n,$$

which converge for $|x-a| < R$ with $R > 0$. The $k$th derivative of $f$ is given by

$$f^{(k)}(x) = \sum_{n=k}^{\infty} n(n-1) \cdots (n-k+1)(x-a)^{n-k}.$$

**EXAMPLE 2**  Assuming that the expansion of $f(x) = (4+x^2)^{-1}$ is given by

$$\frac{1}{4+x^2} = \sum_{n=0}^{\infty} \frac{(-1)^n}{2^{2n+2}} x^{2n},$$

valid for $|x| < 2$, obtain an expansion for $2x/(4+x^2)^2$.

**Solution**  Differentiating term by term, we find

$$-\frac{2x}{(4+x^2)^2} = \sum_{n=1}^{\infty} \frac{(-1)^n 2n}{2^{2n+2}} x^{2n-1}.$$

We may replace $n$ by $n+1$ in the infinite series if we change the lower limit from 1 to 0. We get

$$\frac{2x}{(4+x^2)^2} = \sum_{n=0}^{\infty} \frac{(-1)^n (2n+2)}{2^{2n+4}} x^{2n+1}. \qquad \square$$

We now make use of the above theorems and the fact that the function $1/(1+x)$ may be expanded in the simple geometric series

$$\frac{1}{1+x} = \sum_{n=0}^{\infty} (-1)^n x^n, \quad |x| < 1,$$

to obtain additional series expansions.

**THEOREM 27**  *For $|x| < 1$, we have the expansion:*

$$\ln(1+x) = \sum_{n=1}^{\infty} \frac{(-1)^{n-1} x^n}{n}.$$

**Proof**  Letting $F(x) = \ln(1+x)$ and differentiating, we find

$$F'(x) = f(x) = \frac{1}{1+x} = \sum_{n=0}^{\infty} (-1)^n x^n, \quad |x| < 1.$$

Now, by Theorem 25, we may integrate term by term and get

$$F(x) = \int_0^x f(t)\,dt = \sum_{n=0}^{\infty} \frac{(-1)^n x^{n+1}}{n+1},$$

which is equivalent to the statement of the theorem. $\qquad \square$

**EXAMPLE 3**    Find the Maclaurin series for

$$f(x) = \frac{1}{(1-x)^2}.$$

**Solution**    If we define $F(x) = (1-x)^{-1}$, we see that $F'(x) = f(x)$.
Now

$$F(x) = \sum_{n=0}^{\infty} x^n, \quad |x| < 1$$

(which we obtain from the expansion for $(1+x)^{-1}$ if we replace $x$ by $-x$).
Therefore,

$$f(x) = \frac{1}{(1-x)^2} = \sum_{n=1}^{\infty} nx^{n-1} = \sum_{k=0}^{\infty} (k+1)x^k, \quad |x| < 1. \qquad \square$$

**EXAMPLE 4**    Find the Maclaurin expansion for $f(x) = (1+x^2)^{-1}$.

**Solution**    The geometric series

$$\frac{1}{1+u} = \sum_{n=0}^{\infty} (-1)^n u^n, \quad |u| < 1$$

after substitution of $x^2$ for $u$, becomes

$$\frac{1}{1+x^2} = \sum_{n=0}^{\infty} (-1)^n x^{2n},$$

valid for $x^2 < 1$. The inequality $x^2 < 1$ is equivalent to the inequality $|x| < 1$.
$\square$

For a geometric series we are able to compute the *exact* sum of the series
by using the formula

$$\sum_{n=0}^{\infty} ar^n = \frac{a}{1-r}, \quad \text{if } |r| < 1.$$

Term-by-term differentiation or integration allows us to obtain new formulas
and exact sums as the next example illustrates.

**EXAMPLE 5**    Find the exact sum of the series

$$\sum_{n=1}^{\infty} \frac{n}{2^n}.$$

**Solution**    We rewrite the series in the form

$$\sum_{n=1}^{\infty} \frac{n}{2^n} = \sum_{n=1}^{\infty} n\left(\frac{1}{2}\right)^n = \frac{1}{2} \sum_{n=1}^{\infty} n\left(\frac{1}{2}\right)^{n-1}.$$

We observe that the function $f(x) = \sum_{n=0}^{\infty} x^n$ converges for $|x| < 1$. Then we
have $f'(x) = \sum_{n=1}^{\infty} nx^{n-1}$ and therefore $\frac{1}{2} f'(\frac{1}{2})$ is our desired quantity. Since the
series $\sum_{n=0}^{\infty} x^n$ is a convergent *geometric* series, we have

$$f(x) = \frac{1}{1-x},$$

and so

$$f'(x) = \frac{1}{(1-x)^2},$$

valid for $|x| < 1$, and $\frac{1}{2}f'(\frac{1}{2}) = 2$. We conclude that

$$\sum_{n=1}^{\infty} \frac{n}{2^n} = 2.$$

In the problems below, assume that the following formulas hold. They will be proved in the next section.

$$(1-x)^{-1} = \sum_{n=0}^{\infty} x^n; \qquad\qquad e^x = \sum_{n=0}^{\infty} \frac{x^n}{n!};$$

$$\sin x = \sum_{n=0}^{\infty} \frac{(-1)^n x^{2n+1}}{(2n+1)!}; \qquad \cos x = \sum_{n=0}^{\infty} \frac{(-1)^n x^{2n}}{(2n)!};$$

$$(1+x)^m = 1 + \sum_{n=1}^{\infty} \frac{m(m-1)\cdots(m-n+1)}{n!} x^n.$$

## 9  PROBLEMS

In Problems 1 and 2, assume as known the Maclaurin series for $\sin x$ and $\cos x$.

**1** Find the Maclaurin series for

$$f(x) = \begin{cases} (\sin x)/x, & x \neq 0, \\ 1, & x = 0. \end{cases}$$

**2** Find the Maclaurin series for

$$f(x) = \begin{cases} (1 - \cos x)/x, & x \neq 0, \\ 0, & x = 0. \end{cases}$$

**3** Find the Maclaurin series for $e^x$, determine the radius of convergence, and then find the Maclaurin series for $f$ given by

$$f(x) = \begin{cases} (e^x - 1)/x, & x \neq 0, \\ 1, & x = 0. \end{cases}$$

In Problems 4 through 11, use term-by-term differentiation and integration of known series as in Examples 2, 3, and 4 to determine Taylor or Maclaurin series for the function $f$. State the radius of convergence.

**4** $f(x) = (1+x)^{-2}$     **5** $f(x) = \ln \dfrac{1+x}{1-x}$

**6** $f(x) = (1-x)^{-3}$     **7** $f(x) = (1+x)^{-3}$

**8** $f(x) = x \ln(1+x^2)$     **9** $f(x) = \ln(1-x)$

**10** $f(x) = \ln(3-2x)$     **11** $f(x) = \ln(1+x^2)$

**12** Find the Maclaurin expansion for $(1+x^2)^{-1}$ and then use term-by-term integration to get a Maclaurin expansion for arctan $x$.

**13** Find the Maclaurin expansion for $(1-x^2)^{-1/2}$, and then find one for arcsin $x$.

**14** Find the Maclaurin expansion for $\cos 2x$ and then find one for $\sin^2 x$.

**15** If $f(x) = e^{\alpha x}$, show that the Maclaurin series for $f^{(k)}(x)$ is $\alpha^k$ times the Maclaurin series for $f$.

In Problems 16 through 25, use term-by-term differentiation or integration as in Example 5 to find the exact sum of the given series.

**16** $\displaystyle\sum_{n=1}^{\infty} \frac{n}{3^{n-1}}$     **17** $\displaystyle\sum_{n=1}^{\infty} \frac{n3^n}{4^{n-1}}$

**18** $\displaystyle\sum_{n=2}^{\infty} n(n-1)3^{2-n}$     **19** $\displaystyle\sum_{n=1}^{\infty} \frac{(n^2+n)}{2^{n-1}}$

**20** $\displaystyle\sum_{n=1}^{\infty} \frac{n5^n + 3^{n-1}}{7^n}$     **21** $\displaystyle\sum_{n=0}^{\infty} \frac{1}{(n+1)2^n}$

**22** $\displaystyle\sum_{n=0}^{\infty} \frac{1}{(n^2+3n+2)3^{n+2}}$     **23** $\displaystyle\sum_{n=1}^{\infty} \frac{2^{n-1}}{(n+1)3^n}$

**24** $\displaystyle\sum_{n=0}^{\infty} \frac{(n!)2^{n-10}}{(n+1)!3^n}$     **25** $\displaystyle\sum_{n=1}^{\infty} \frac{1}{4n} 2^{-n}$

**26** Find the exact sum of the series (that is, an expression which is equal to the sum of the series for all $x$ in the interval of convergence):

$$\sum_{k=1}^{\infty} kx^{k-1} \quad \text{for } |x| < 1.$$

**27** Find the sum of the series

$$\sum_{k=1}^{\infty} \frac{x^k}{k(k+1)}.$$

**28** Find the Maclaurin expansion for $\ln(x + \sqrt{1 + x^2})$.

**29** Find the sum of the series

$$\sum_{n=0}^{\infty} \frac{(n+2)(n+1)x^n}{n!}$$

and conclude that

$$e = \frac{1}{7} \sum_{n=0}^{\infty} \frac{(n+2)(n+1)}{n!}.$$

**30** Find the first five terms of the Maclaurin expansion of

$$F(x) = \int_0^x e^{-t^2} dt.$$

Obtain an approximate value of $F(1)$. Can you approximate the error?

**\*31** Show that for every positive integer $p$, we have

$$(1-x)^{-p-1} = \sum_{n=0}^{\infty} \binom{n+p}{p} x^n, \quad |x| < 1,$$

where the symbol $\binom{n+p}{p}$ is an abbreviation of

$$\frac{(n+p)(n+p-1)\cdots(n+1)}{1 \cdot 2 \cdot 3 \cdots p}.$$

Deduce the formula $3^{p+1} = \sum_{n=0}^{\infty} \binom{n+p}{p}(\frac{2}{3})^n$.

**\*32** Let

$$f(x) = \frac{1}{2} + \frac{x}{5} + \frac{x^2}{8} + \frac{x^3}{11} + \cdots.$$

a) Show that the expansion for $x/(1-x^3)$ is precisely $(x^2 f(x^3))'$.

b) Find an expression for $f(x)$.

**33** Use the fact that $x^p = e^{p \log x}$ to express $x^p$ as a power series in $\log x$:

$$x^p = 1 + \frac{p(\log x)}{1!} + \cdots + \frac{p^n(\log x)^n}{n!} + \cdots.$$

Find the radius of convergence. For what values of $p$ is the above expansion valid?

**34** Suppose that $f(x)$ has a Maclaurin series with a positive radius of convergence: $f(x) = \sum_{n=0}^{\infty} a_n x^n$. Show that if $f$ is an even function of $x$, that is, if $f(-x) = f(x)$ for all $x$, then $a_n = 0$ for all odd integers $n$. If $f$ is odd, i.e., if $f(-x) = -f(x)$, show that $a_n = 0$ for all even $n$.

**35** Writing the identity

$$\frac{1}{1-x} = \frac{-1}{x\left(1 - \frac{1}{x}\right)},$$

we get the formal expansion

$$\frac{1}{1-x} = -\sum_{n=0}^{\infty} \frac{1}{x^{n+1}}.$$

For what values of $x$ does this series converge? Under what conditions can the series be differentiated and integrated term by term?

**36** Find an expansion for $\arctan x$ in powers of $1/x$. (*Hint:* Consider the expansion of $1/(1 + x^2)$ as in Problem 35 and then integrate.)

---

## 10

## NUMERICAL COMPUTATION WITH SERIES

In this section we present theorems that both show that certain important functions are truly represented by their Taylor series, and also give methods for making numerical computations.

**THEOREM 28**   *For any values of $a$ and $x$, we have*

$$e^x = e^a \sum_{n=0}^{\infty} \frac{(x-a)^n}{n!};$$   (1)

*i.e., the Taylor series for $e^x$ about $x = a$ converges to $e^x$ for any $a$ and $x$.*

**Proof**    For simplicity, set $a = 0$, the proof being analogous when $a \neq 0$. If we let $f(x) = e^x$, then $f^{(n)}(x) = e^x$ for all $n$. Now, using Taylor's Theorem with Derivative Form of the Remainder, we have

$$e^x = \sum_{k=0}^{n} \frac{x^k}{k!} + R_n, \qquad \text{where} \qquad R_n = \frac{e^{\xi} x^{n+1}}{(n+1)!},$$

with $\xi$ between 0 and $x$. If $x$ is positive, then $e^{\xi} < e^x$ while, if $x$ is negative, then $e^{\xi} < e^0 = 1$. In either case,

$$|R_n| \leq C \frac{|x|^{n+1}}{(n+1)!}, \tag{2}$$

where $C$ is the larger of 1 and $e^x$ but is *independent of n*. If the right side of (2) is the general term of a series then, by the Ratio Test, that series is convergent for all $x$. The general term of any convergent series must tend to zero, and so

$$C \frac{|x|^{n+1}}{(n+1)!} \to 0 \quad \text{as } n \to \infty \text{ for each } x.$$

We conclude that $R_n \to 0$ as $n \to \infty$, and so (1) is established.    □

---

**THEOREM 29**    *The following functions are given by their Maclaurin series:*

$$\sin x = \sum_{n=0}^{\infty} \frac{(-1)^n x^{2n+1}}{(2n+1)!}, \qquad \cos x = \sum_{n=0}^{\infty} \frac{(-1)^n x^{2n}}{(2n)!}.$$

---

The proofs of these results follow the same outline as the proof of Theorem 28 and are left as exercises for the reader at the end of this section.

The reader no doubt recalls the formula $(1 + x)^2 = 1 + 2x + x^2$ and is probably familiar with the binomial formula for $(1 + x)^k$ for each positive integer $k$. Infinite series allows us to provide a formula for $(1 + x)^m$, for every real number $m$, integer or not.

---

**THEOREM 30**    **(Binomial Theorem)**    *For each real number m, we have*

$$(1 + x)^m = 1 + \sum_{n=1}^{\infty} \frac{m(m-1)(m-2) \cdots (m-n+1)}{n!} x^n \quad \textit{for } |x| < 1. \tag{3}$$

---

**Proof**    To show that the series on the right converges absolutely for $|x| < 1$, we apply the Ratio Test:

$$\left| \frac{u_{n+1}}{u_n} \right| = \left| \frac{m(m-1) \cdots (m-n+1)(m-n)}{(n+1)!} \cdot \frac{n!}{m(m-1) \cdots (m-n+1)} \cdot \frac{x^{n+1}}{x_n} \right|$$

$$= \frac{|m-n|}{n+1} |x| = \frac{|1 - m/n|}{1 + 1/n} |x|.$$

The quantity on the right tends to $|x|$ as $n \to \infty$, and so the series converges for $|x| < 1$. We define

$$f(x) = 1 + \sum_{n=1}^{\infty} \frac{m(m-1) \cdots (m-n+1)}{n!} x^n, \tag{4}$$

and we still must show that $f(x) = (1 + x)^m$ if $|x| < 1$. Employing Theorem 26, we get $f'(x)$ by term-by-term differentiation of the series for $f$. We have

$$f'(x) = m + \sum_{n=2}^{\infty} \frac{m(m-1) \cdots (m-n+1)}{(n-1)!} x^{n-1}. \tag{5}$$

Multiplying both sides of (5) by $x$, we get

$$xf'(x) = \sum_{n=1}^{\infty} n \frac{m(m-1) \cdots (m-n+1)}{n!} x^n. \tag{6}$$

We add (5) and (6) to obtain

$$(1 + x)f'(x) = m \left\{ 1 + \sum_{n=1}^{\infty} \frac{m(m-1) \cdots (m-n+1)}{n!} x^n \right\} = mf(x).$$

In other words, the function $f$ satisfies the relation

$$(1 + x)f'(x) - mf(x) = 0.$$

To find $f(x)$ we use a method that comes from the theory of linear first order differential equations. We multiply this equation by $(1 + x)^{-m-1}$, getting

$$(1 + x)^{-m}f'(x) - m(1 + x)^{-m-1}f(x) = 0.$$

This last equation can be written

$$\frac{d}{dx} [(1 + x)^{-m}f(x)] = 0.$$

A function whose derivative is identically zero must be constant, so that

$$(1 + x)^{-m}f(x) = K,$$

where $K$ is some constant. Setting $x = 0$ in (4), we see that $f(0) = 1$, and this fact yields $K = 1$. We conclude that

$$f(x) = (1 + x)^m \quad \text{for } |x| < 1. \qquad \square$$

The series expansion of $(1 + x)^m$ for $|x| < 1$ given by (3) is called the **Binomial Series.**

**EXAMPLE 1**    Write the first 5 terms of the series expansion for $(1 + x)^{3/2}$.

**Solution**    We have $m = \frac{3}{2}$ in Theorem 30, and therefore

$$(1 + x)^{3/2} = 1 + \frac{\frac{3}{2}}{1!} x + \frac{(\frac{3}{2})(\frac{1}{2})}{2!} x^2 + \frac{(\frac{3}{2})(\frac{1}{2})(-\frac{1}{2})}{3!} x^3 + \frac{(\frac{3}{2})(\frac{1}{2})(-\frac{1}{2})(-\frac{3}{2})}{4!} x^4$$

$$= 1 + \frac{3}{2} x + \frac{3}{8} x^2 - \frac{3}{2^3 \cdot 3!} x^3 + \frac{3^2}{2^4 \cdot 4!} x^4 + \cdots. \qquad \square$$

**EXAMPLE 2** Write the Binomial Series for $(1 + x)^7$.

**Solution** We have

$$(1 + x)^7 = 1 + \sum_{n=1}^{\infty} \frac{7(6) \cdots (7 - n + 1)}{n!} x^n.$$

We now observe that beginning with $n = 8$ all the terms have a zero in the numerator. Therefore,

$$(1 + x)^7 = 1 + \sum_{n=1}^{7} \frac{7(6) \cdots (7 - n + 1)}{n!} x^n$$

$$= 1 + 7x + \frac{7 \cdot 6}{2!} x^2 + \frac{7 \cdot 6 \cdot 5}{3!} x^3 + \cdots + \frac{7!}{7!} x^7. \qquad \square$$

Example 2 is a special case of the following general rule:

*For m a positive integer, the Binomial Series always terminates after a finite number of terms.*

**EXAMPLE 3** Use Theorem 25 on term-by-term integration of series and a hand calculator to compute

$$\int_0^{0.5} e^{x^2} \, dx$$

to an accuracy of five decimal places.

**Solution** We have

$$e^u = \sum_{n=0}^{\infty} \frac{u^n}{n!} \quad \text{for all } u, \text{ and so} \quad e^{x^2} = \sum_{n=0}^{\infty} \frac{x^{2n}}{n!} \quad \text{for all } x.$$

By Theorem 25 we can integrate term by term to get

$$\int_0^x e^{t^2} \, dt = \sum_{n=0}^{\infty} \frac{x^{2n+1}}{n!(2n+1)}.$$

For $x = 0.5$, we now compute

$$x = 0.5 \quad = \frac{1}{2} \qquad = 0.50000\ 00,$$

$$\frac{x^3}{1! \cdot 3} = \frac{(0.5)^3}{3} = \frac{1}{24} \qquad = 0.04166\ 67^-,$$

$$\frac{x^5}{2! \cdot 5} = \frac{(0.5)^5}{10} = \frac{1}{320} \qquad = 0.00312\ 50,$$

$$\frac{x^7}{3! \cdot 7} = \frac{(0.5)^7}{42} = \frac{1}{5376} \qquad = 0.00018\ 60^+,$$

$$\frac{x^9}{4! \cdot 9} = \frac{(0.5)^9}{216} = \frac{1}{110{,}592} = 0.00000\ 90^+,$$

sum of the right column $= 0.54498\ 67.$

If we wish to stop at this point we must estimate the error made by neglecting all the remaining terms. The remainder is

$$\sum_{n=5}^{\infty} \frac{x^{2n+1}}{n!(2n+1)} \le \frac{x^{11}}{5! \cdot 11}\left(1 + \frac{x^2}{6} + \frac{x^4}{6^2} + \frac{x^6}{6^3} + \cdots\right)$$

$$\le \frac{x^{11}}{1320}\frac{1}{(1-x^2/6)} \le \frac{24}{23} \cdot \frac{1}{1320} \cdot \frac{1}{2048}$$

$$\le 0.00000\ 04.$$

Therefore

$$\int_0^{0.5} e^{x^2}\,dx = \begin{cases} 0.54499 \text{ to an accuracy of 5 decimals,} \\ 0.544987 \text{ to an accuracy of 6 decimals.} \end{cases}$$

$\square$

## 10    PROBLEMS

In each of Problems 1 through 9, write the beginning of the binomial series for the given expression to the required number of terms.

1  $(1+x)^{-3/2}$,   5 terms

2  $(1-x)^{1/2}$,   4 terms

3  $(1+x^2)^{-2/3}$,   5 terms

4  $(1-x^2)^{-1/2}$,   6 terms

5  $(1+x^3)^7$,   all terms

6  $(5+x)^{1/2}$,   4 terms

7  $(3+\sqrt{x})^{-3}$,   5 terms

8  $(1+x)^{1/2}$,   4 terms

9  $(1+x)^{-1/3}$,   4 terms

In each of Problems 10 through 12, use the Binomial Theorem and a calculator to make the appropriate estimate.

10  Estimate $7^{-1/3}$ to within 0.01

11  Estimate $\sqrt{11}$ to within 0.01

12  Estimate $\sqrt{101}$ to within 0.01. (*Hint:* Use that $\sqrt{101} = (100+1)^{1/2} = 10(1+\frac{1}{100})^{1/2}$.)

In each of Problems 13 through 22, compute the value of the definite integral to the number of decimal places specified. Estimate the remainder. Use Theorem 25 and a calculator.

13  $\int_0^1 \sin(x^2)\,dx$,   5 decimals

14  $\int_0^1 \cos(x^2)\,dx$,   5 decimals

15  $\int_0^1 e^{-x^2}\,dx$,   5 decimals

16  $\int_0^{0.5} \frac{dx}{1+x^3}$,   5 decimals

17  $\int_0^1 \frac{\sin x}{x}\,dx$,   5 decimals

18  $\int_0^1 \frac{e^x - 1}{x}\,dx$,   5 decimals

19  $\int_0^{0.5} \frac{dx}{\sqrt{1+x^3}}$,   5 decimals

20  $\int_0^{1/3} \frac{dx}{\sqrt[3]{1+x^2}}$,   5 decimals

21  $\int_0^{1/3} \frac{dx}{\sqrt[3]{1-x^2}}$,   5 decimals

22  $\int_0^{0.5} \frac{dx}{\sqrt{1-x^3}}$,   5 decimals

23  Use the series for

$$\ln\frac{1+x}{1-x}$$

to find ln 1.5 to 5 decimals of accuracy.

24  Same as Problem 23, to find ln 2.

25  Prove that

$$\sin x = \sum_{n=0}^{\infty} \frac{(-1)^n x^{2n+1}}{(2n+1)!}.$$

26  Prove that

$$\cos x = \sum_{n=0}^{\infty} \frac{(-1)^n x^{2n}}{(2n)!}.$$

27  Show, by induction, that $f^{(k)}(x) = \sin(x + \frac{1}{2}k\pi)$ if $f(x) = \sin x$.

28  Use the result of Problem 27 to show that for all $a$ and $x$ the Taylor series for $\sin x$ is given by

$$\sin x = \sum_{n=0}^{\infty} \frac{\sin(a + \frac{1}{2}n\pi)}{n!}(x-a)^n.$$

**29** Show, by induction, that $f^{(k)}(x) = \cos(x + \frac{1}{2}k\pi)$ if $f(x) = \cos x$.

**30** Use the result of Problem 29 to show that for all $a$ and $x$ the Taylor series for $\cos x$ is given by

$$\cos x = \sum_{n=0}^{\infty} \frac{\cos(a + \frac{1}{2}n\pi)}{n!}(x - a)^n.$$

**31** Use the series for $\ln[(1 + x)/(1 - x)]$ to obtain the formula

$$\ln(y + 1) = \ln y$$
$$+ 2\left[\frac{1}{2y + 1} + \frac{1}{3(2y + 1)^3} + \frac{1}{5(2y + 1)^5} + \cdots\right].$$

(*Hint:* Let $x = (2y + 1)^{-1}$.) Find the values of $y$ for which the series converges.

**32** Assuming that $\ln 2$ and $\ln 5$ are known, use the expansion in Problem 31 to estimate the error in finding $\ln 11$ when three terms of the series are used.

---

## CHAPTER 10

### REVIEW PROBLEMS

In each of Problems 1 through 9, find the limit (finite or infinite) when it exists.

**1** $\lim_{x \to 3} \dfrac{3x^2 + 6x - 45}{x^2 - 9}$

**2** $\lim_{x \to 1} \dfrac{x^2 - 1}{\sin \pi x}$

**3** $\lim_{x \to +\infty} \dfrac{\ln(1 + x^2 + 2x)}{3x^2 + 4x}$

**4** $\lim_{x \to +\infty} \dfrac{\ln(x) + x^2 + 3}{(6x + 3)^2}$

**5** $\lim_{x \to \pi/2} \dfrac{\cot x}{\ln \cos x}$

**6** $\lim_{x \to +\infty} \dfrac{\sin\left(\dfrac{1}{x^2}\right)}{\dfrac{1}{x^2}}$

**7** $\lim_{x \to 1} \dfrac{(x - 1)^2}{(x^2 - 1)^{4/3}}$

**8** $\lim_{x \to 0} \dfrac{\ln \cos x}{x^2}$

**9** $\lim_{x \to \pi/4} \dfrac{\cos x - \sin x}{\ln \tan x}$

In each of Problems 10 through 13, evaluate the limit of the sequence (or show that the sequence tends to $\infty$, $+\infty$, or $-\infty$).

**10** $\lim_{n \to \infty} \dfrac{n + 3}{n^2 + 2}$

**11** $\lim_{n \to \infty} \dfrac{1}{\sqrt[3]{n^2 + 3}}$

**12** $\lim_{n \to +\infty} \dfrac{3^n}{e^n}$

**13** $\lim_{n \to +\infty} (\sqrt{n^2 + 3n + 6} - n)$

In Problems 14 through 17, express each repeating decimal as the ratio of two integers.

**14** $0.626262\ldots$

**15** $0.272727\ldots$

**16** $0.230769230769\ldots$

**17** $0.285714285714\ldots$

In each of Problems 18 through 31, test for convergence or divergence. All summations are from $n = 1$ to $\infty$.

**18** $\sum \dfrac{1}{n^2}$

**19** $\sum \dfrac{1}{\sqrt{n^3 + 3}}$

**20** $\sum \dfrac{1}{n^2 + 10}$

**21** $\sum \left(\frac{3}{2}\right)^{-n}$

**22** $\sum \dfrac{e^n}{2^{n+1}}$

**23** $\sum \left(1 - \frac{1}{n}\right)^n$

**24** $\sum \dfrac{n^{n-1}}{n!}$

**25** $\sum \dfrac{1}{n \ln n}$

**26** $\sum \dfrac{\sqrt{n - 1}}{n}$

**27** $\sum \dfrac{n^2 + 1}{n^3 - 7}$

**28** $\sum ne^{-n^2}$

**29** $\sum \dfrac{1}{n(\ln n)^2}$

**30** $\sum \dfrac{64n^2}{\pi^n}$

**31** $\sum n^2 e^{-an} \quad (a > 0)$

In Problems 32 through 49, test for convergence or divergence. If the series is convergent, determine whether it is absolutely or conditionally convergent (all summations are from $n = 1$ to $\infty$).

**32** $\sum \dfrac{(-1)^n}{(2n)!}$

**33** $\sum \dfrac{(-1)^{n+1}n}{3^n}$

**34** $\sum \dfrac{(n - 2)(n + 3)}{2 \cdot 4 \cdot 6 \cdots (2n)}$

**35** $\sum \dfrac{1 \cdot 4 \cdot 7 \cdots (3n + 1)}{n^5}$

**36** $\sum \dfrac{(-1)^n(n + 2)}{3n - 1}$

**37** $\sum \dfrac{(-1)^{n-1} 1000n^2}{3^n}$

**38** $\sum \dfrac{\sin\left(\dfrac{\pi n}{2}\right)}{6n + 3}$

**39** $\sum \dfrac{\sin n}{n^2}$

**40** $\sum \dfrac{n \cos\left(\dfrac{\pi n}{4}\right)}{n + 1}$

**41** $\sum \dfrac{(-1)^{n+1} n\pi^n}{e^{2n} + 1}$

**42** $\sum \dfrac{(-1)^n(2n)!}{10^n}$

**43** $\sum \dfrac{(-1)^n 10^{4n}}{n!}$

**44** $\sum \dfrac{n^2}{(\ln(n+1))^n}$

**45** $\sum \left(\dfrac{4n-5}{3n+1}\right)^n$

**46** $\sum \dfrac{(\cos \pi n)n}{5^n}$

**47** $\sum \left(\dfrac{-n}{10^{10}}\right)^n$

**48** $\sum \left(\dfrac{-n}{n+1}\right)^{n^2}$

**49** $\sum \dfrac{-\cos \pi n}{2n+3}$

In Problems 50 through 59, find the values of $x$ for which the following power series converge. Include a discussion of the end-points.

**50** $\displaystyle\sum_{n=0}^{\infty} (x+1)^n$

**51** $\displaystyle\sum_{n=1}^{\infty} (3x-2)^n$

**52** $\displaystyle\sum_{n=1}^{\infty} \dfrac{n!x^n}{2n-1}$

**53** $\displaystyle\sum_{n=1}^{\infty} \dfrac{(-1)^n x^{2n}}{(2n)!}$

**54** $\displaystyle\sum_{n=1}^{\infty} \dfrac{(n+1)x^{2n}}{5^n}$

**55** $\displaystyle\sum_{n=0}^{\infty} \dfrac{(-1)^n(3n+1)x^{2n}}{2^n}$

**56** $\displaystyle\sum_{n=0}^{\infty} \dfrac{n+1}{2^{n+1}x^n}$

**57** $\displaystyle\sum_{n=1}^{\infty} \dfrac{(-1)^n}{nx^n}$

**58** $\displaystyle\sum_{n=1}^{\infty} \dfrac{n^n x^n}{n!}$

**59** $\displaystyle\sum_{n=1}^{\infty} \dfrac{n!x^n}{n^n}$

In each of Problems 60 through 65, find the first $n$ terms of the Taylor expansion about the given value of $a$.

**60** $f(x) = \ln x, \quad a=1, n=5$

**61** $f(x) = e^x, \quad a=2, n=5$

**62** $f(x) = \sqrt{x}, \quad a=2, n=3$

**63** $f(x) = \dfrac{1}{1+x}, \quad a=-4, n=3$

**64** $f(x) = \dfrac{1}{1-x}, \quad a=0, n=5$

**65** $f(x) = \dfrac{1}{\sqrt{1+x}}, \quad a=0, n=4$

In each of Problems 66 through 68, use Taylor's Theorem with Remainder to compute the given quantities to the specified number of decimal places. Use the fact that $2 < e < 4$ and a hand calculator.

**66** $e^{-0.3}, \quad$ 5 decimals      **67** $\sin(0.4), \quad$ 4 decimals

**68** $\ln(0.7), \quad$ 5 decimals

In Problems 69 through 74, find the exact sum of the given series.

**69** $\displaystyle\sum_{n=1}^{\infty} \dfrac{2^n}{3^{n+1}}$

**70** $\displaystyle\sum_{n=1}^{\infty} \dfrac{n4^n}{5^{n-1}}$

**71** $\displaystyle\sum_{n=1}^{\infty} \dfrac{4^n}{n5^{n-1}}$

**72** $\displaystyle\sum_{n=1}^{\infty} (n^2+n)2^{-n}$

**73** $\displaystyle\sum_{n=1}^{\infty} \dfrac{(n^2+n+1)2^n}{3^n}$

**74** $\displaystyle\sum_{n=2}^{\infty} (2n-2)e^{-2n}$

**75** Use infinite series to evaluate $\displaystyle\int_0^1 \dfrac{1}{\sqrt{100-x^3}}\,dx$ to four decimal places.

**76** Approximate $\int_{-1}^1 e^{-x^2/2}\,dx$ with an accuracy of 5 decimal places.

# 11

## ANALYTIC GEOMETRY IN THE PLANE

Analytic geometry shows how the techniques of algebra can be used to obtain information about geometric figures. In this chapter we study simple curves in the plane and we use the methods of both algebra and calculus to obtain important geometrical insights.

### 1

### POINTS, LINES, AND ANGLES

A first-degree equation of the form

$$Ax + By + C = 0$$

represents a straight line which we denote by $L$. Suppose that $P(x_0, y_0)$ is a point not on the line. We construct the line through $P(x_0, y_0)$ which is perpendicular to $L$, and we consider the length of the line segment extending from $P$ to the point of intersection with $L$. See Fig. 1.

**DEFINITION**  *The **distance from a point to a line** is the length of the perpendicular segment dropped from this point to the line.*

The number representing this distance is always positive. The following theorem gives a simple formula for this distance.

**THEOREM 1**  *The distance $d$ from the point $P(x_0, y_0)$ to the line $L$ whose equation is $Ax + By + C = 0$ is given by the formula*

$$d = \frac{|Ax_0 + By_0 + C|}{\sqrt{A^2 + B^2}}.$$

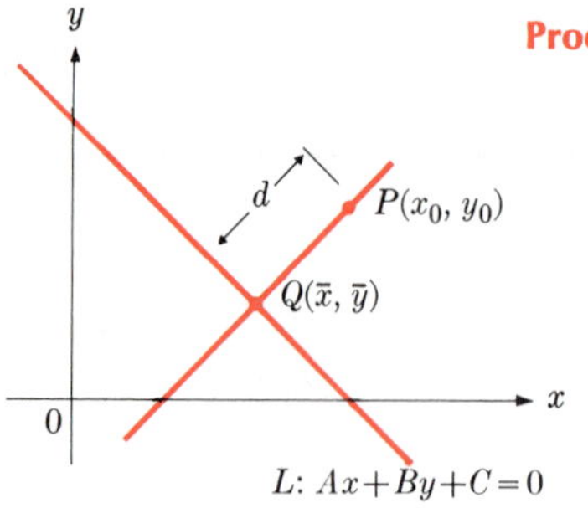

FIGURE 1

**Proof**    We shall use a proof which is easy to devise but which requires some complicated algebraic manipulation. The steps of the proof are: (1) Find the equation of the line through $P(x_0, y_0)$ which is perpendicular to $L$ (Fig. 1). (2) Find the point of intersection of this line with $L$. Call the point $Q(\bar{x}, \bar{y})$. (3) Find the distance $|PQ|$. (4) Treat vertical and horizontal lines separately. (The proof will help us review some of the facts and formulas concerning straight lines.)

**Step 1.**    The slope of the line $Ax + By + C = 0$ is $-A/B$. If the line is neither vertical nor horizontal, the slope of the line through $P$ and perpendicular to $L$ is $+B/A$, the negative reciprocal. The equation of this line, according to the point-slope formula, is

$$y - y_0 = (B/A)(x - x_0) \quad \Leftrightarrow \quad Bx - Ay + Ay_0 - Bx_0 = 0.$$

**Step 2.**    To find $Q(\bar{x}, \bar{y})$ we solve simultaneously the equations

$$Ax + By + C = 0, \qquad Bx - Ay + Ay_0 - Bx_0 = 0.$$

We multiply the first equation by $A$ and the second by $B$ and add, to get

$$\bar{x} = \frac{B^2 x_0 - ABy_0 - AC}{A^2 + B^2}.$$

Now we multiply the first equation by $B$ and the second by $A$ and subtract, to find

$$\bar{y} = \frac{-ABx_0 + A^2 y_0 - BC}{A^2 + B^2}.$$

**Step 3.**    The distance $d$, according to the formula for the distance between two points, is

$$d = \sqrt{(x_0 - \bar{x})^2 + (y_0 - \bar{y})^2}.$$

Substitution of the expressions for $\bar{x}$ and $\bar{y}$ in this formula yields

$$d^2 = \left[ x_0 - \frac{B^2 x_0 - ABy_0 - AC}{A^2 + B^2} \right]^2 + \left[ y_0 - \frac{-ABx_0 + A^2 y_0 - BC}{A^2 + B^2} \right]^2.$$

We find the least common denominator and simplify, to obtain

$$d^2 = \frac{A^2(Ax_0 + By_0 + C)^2}{(A^2 + B^2)^2} + \frac{B^2(Ax_0 + By_0 + C)^2}{(A^2 + B^2)^2}.$$

This combines even further to yield

$$d^2 = \frac{(Ax_0 + By_0 + C)^2}{A^2 + B^2},$$

and finally, taking the positive square root, we write the result:

$$d = \frac{|Ax_0 + By_0 + C|}{\sqrt{A^2 + B^2}}.$$

**Step 4.**    If the line $L$ is horizontal, its equation is of the form $y = -C/B$ and the distance from $P(x_0, y_0)$ to $L$ is the difference of the $y$ values (draw a sketch, if necessary). Therefore,

$$d = \left| y_0 - \left( -\frac{C}{B} \right) \right| = \left| \frac{By_0 + C}{B} \right|.$$

Similarly, if the line is vertical,

$$d = \left| \frac{Ax_0 + C}{A} \right|.$$

Note that the distance formula in the statement of the theorem works in *all* cases so long as $A$ and $B$ are not both zero. Furthermore if, by accident, the point $P(x_0, y_0)$ happened to be on the line $L$ the distance $d$ would be zero, since $(x_0, y_0)$ would satisfy the equation $Ax_0 + By_0 + C = 0$.  $\square$

**EXAMPLE 1**    Find the distance $d$ from the point $(2, -1)$ to the line

$$3x + 4y - 5 = 0.$$

**Solution**    We see that $A = 3$, $B = 4$, $C = -5$, $x_0 = 2$, $y_0 = -1$. Substituting in the formula, we get

$$d = \frac{|3 \cdot 2 + 4(-1) - 5|}{\sqrt{3^2 + 4^2}} = \frac{|-3|}{5} = \frac{3}{5}.$$  $\square$

---

**DEFINITION**    The **distance between parallel lines** *is the shortest distance from any point on one of the lines to the other line.*

---

**EXAMPLE 2**    Find the distance between the parallel lines

$$L_1: 2x - 3y + 7 = 0 \quad \text{and} \quad L_2: 2x - 3y - 6 = 0.$$

**Solution**    First find any point on one of the lines, say $L_1$. To do this, let $x$ be any value and solve for $y$ in the equation for $L_1$. For example, if $x = 1$ in $L_1$, then $y = 3$ and the point $(1, 3)$ is on $L_1$. The distance from $(1, 3)$ to $L_2$ is

$$d = \frac{|2(1) - 3(3) - 6|}{\sqrt{2^2 + (-3)^2}} = \frac{|-13|}{\sqrt{13}} = \frac{13}{\sqrt{13}} = \sqrt{13}.$$

The distance between the parallel lines is $\sqrt{13}$. The answer could be checked by taking any point on $L_2$ and finding its distance to $L_1$.  $\square$

An important concept in analytic geometry is that of the angle formed when two lines intersect. Recall that the inclination $\alpha$ of a line $L$ is the angle that the upward-pointing ray of $L$ makes with the positive $x$ axis. If the line is parallel to the $x$ axis, its inclination is zero. The slope $m$ of the line $L$ is $m = \tan \alpha$ ($\alpha \neq 90°$). Two nonparallel lines $L_1$ and $L_2$ make four angles at their intersection: two equal obtuse angles and two equal acute angles (unless the lines are perpendicular). The obtuse angle is the supplement of the acute angle.

The angle swept out when the line $L_1$ is rotated counterclockwise to $L_2$ about the point of intersection is called the **angle from $L_1$ to $L_2$**. Let $\phi$ be the

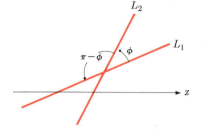

**FIGURE 2**

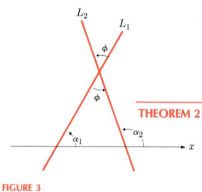

**FIGURE 3**

angle (measured in radians) from $L_1$ to $L_2$ (see Fig. 2). Then, clearly, the angle from $L_2$ to $L_1$ is $\pi - \phi$.

Let $L_1$ and $L_2$ be two intersecting lines (neither of which is vertical) with inclinations $\alpha_1$ and $\alpha_2$ and slopes $m_1 = \tan \alpha_1$ and $m_2 = \tan \alpha_2$, respectively. Let $\phi$ be the angle from $L_1$ to $L_2$ and suppose $\alpha_1 < \alpha_2$ (see Fig. 3). Then, since the exterior angle of a triangle is the sum of the remote interior angles, we have

$$\alpha_2 = \alpha_1 + \phi, \qquad \phi = \alpha_2 - \alpha_1, \qquad \tan \phi = \tan(\alpha_2 - \alpha_1).$$

Using the formula for the tangent of the difference of two angles,* we get

$$\tan \phi = \frac{\tan \alpha_2 - \tan \alpha_1}{1 + \tan \alpha_1 \tan \alpha_2}.$$

In terms of slopes, we obtain

$$\tan \phi = \frac{m_2 - m_1}{1 + m_1 m_2}.$$

We have established the following theorem.

**THEOREM 2**    *Let $L_1$ and $L_2$ be two intersecting lines with slopes $m_1$ and $m_2$, respectively. Then the tangent of the angle $\phi$ from $L_1$ to $L_2$ is*

$$\tan \phi = \frac{m_2 - m_1}{1 + m_1 m_2}. \tag{1}$$

*Remark.*   The derivation above assumes that $\alpha_2 > \alpha_1$. In case $\alpha_1 > \alpha_2$, we see that

$$\pi - \phi = \alpha_1 - \alpha_2, \qquad \tan(\pi - \phi) = \frac{m_1 - m_2}{1 + m_1 m_2}.$$

Since $\tan(\pi - \phi) = -\tan \phi$, Formula (1) holds in this case also.

**EXAMPLE 3**    Find the tangent of the angle from the line

$$L_1 = \{(x, y): 2x + 3y = 5\}$$

to the line $L_2 = \{(x, y): 4x - 3y = 2\}$.

**Solution**    Here $m_1 = -\frac{2}{3}$ and $m_2 = \frac{4}{3}$. The formula yields

$$\tan \phi = \frac{\frac{4}{3} - (-\frac{2}{3})}{1 + (-\frac{2}{3})(\frac{4}{3})} = 18. \qquad \square$$

Formula (1) fails if one of the lines is vertical. We divide both the numerator and the denominator by $m_2$ in the formula for $\tan \phi$, getting

---

*A review of all the basic formulas in trigonometry as well as the graphs of all the elementary trigonometric functions may be found in Appendix 1 at the end of the book.

$$\tan \phi = \frac{1 - (m_1/m_2)}{(1/m_2) + m_1}.$$

Consider now that $L_2$ becomes vertical; that is, $m_2 \to \infty$ as $L_2$ approaches a vertical line. Then $m_1/m_2 \to 0$ and $1/m_2 \to 0$. Therefore the formula for the tangent of the angle $\phi$ from $L_1$ to $L_2$ when $L_2$ is vertical is simply

$$\tan \phi = \frac{1}{m_1}.$$

**EXAMPLE 4**  Given the curve with equation $y = 2x^2 - 3x - 4$. Find the equations of the lines tangent to the curve at the points where $x = 1$ and $x = 2$. Find the tangent of the angle between these two tangent lines.

**Solution**  The slope at any point is found by obtaining the derivative:

$$\frac{dy}{dx} = 4x - 3.$$

At $x = 1$ the slope $m_1 = 1$; at $x = 2$ the slope $m_2 = 5$. If $x = 1$, then $y = -5$ and if $x = 2$, then $y = -2$. The first line has equation $y + 5 = 1(x - 1)$ and the second has equation $y + 2 = 5(x - 2)$. Using the formula for the tangent of the angle between two lines, we get

$$\tan \phi = \frac{5 - 1}{1 + 5 \cdot 1} = \frac{2}{3}.$$

Figure 4 shows the result. $\qquad \square$

From plane geometry we recall that the *angle bisectors* of two given intersecting lines represent the graph of all points which are equidistant from the given lines. We can find the equations of these angle bisectors by using the formula for the distance from a point to a line. The following example illustrates the technique.

**EXAMPLE 5**  Find the equations of the bisectors of the angles formed by the lines $3x + 4y - 7 = 0$ and $4x + 3y + 2 = 0$.

**Solution**  We draw a figure, as shown (Fig. 5), and let $P(x, y)$ be a point on the graph (angle bisector, in this case). Let $d_1$ be the distance from $P$ to the line $3x + 4y - 7 = 0$ and $d_2$ the distance from $P$ to the line $4x + 3y + 2 = 0$. The required condition is $d_1 = d_2$. Then, from the formula for the distance from a point to a line, we have

$$d_1 = \frac{|3x + 4y - 7|}{\sqrt{3^2 + 4^2}}$$

and

$$d_2 = \frac{|4x + 3y + 2|}{\sqrt{4^2 + 3^2}}.$$

The condition $d_1 = d_2$ becomes

$$|3x + 4y - 7| = |4x + 3y + 2|.$$

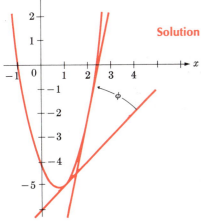

FIGURE 4

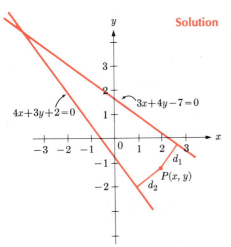

FIGURE 5

At this time we make use of our knowledge of absolute values. If $|a| = |b|$, then either $a = b$ or $a = -b$. This tells us that either

$$3x + 4y - 7 = 4x + 3y + 2$$

or

$$3x + 4y - 7 = -(4x + 3y + 2).$$

Simplifying, we obtain the equations of the two angle bisectors

$$x - y + 9 = 0$$

and

$$7x + 7y - 5 = 0.$$

Note that the angle bisectors have slopes 1 and $-1$ and are therefore perpendicular, as we know they should be.

To distinguish the bisectors from each other, we have merely to find the tangent of the angle from one of the given lines to the other, and then the tangent of the angle from one of the given lines to the angle bisectors. A comparison of the sizes indicates where the lines fall. □

## 1   PROBLEMS

In Problems 1 through 12, in each case find the distance from the given point to the given line.

**1** $(3, 2)$,   $2x + 4y - 4 = 0$      **2** $(1, 0)$,   $x - 5y + 1 = 0$

**3** $(2, -1)$,   $2x - 2y + 9 = 0$

**4** $(-3, 1)$,   $5x + y + 2 = 0$

**5** $(-1, -4)$,   $-2x + 3y - 6 = 0$

**6** $(2, -4)$,   $x - 3 = 0$

**7** $(-1, 0)$,   $y + 2 = 0$      **8** $(2, 1)$,   $x - y = 0$

**9** $(1, -2)$,   $3x + 4y = 6$      **10** $(2, 4)$,   $3x = 10 - 2y$

**11** $(2, 4)$,   $x + 2y = 10$      **12** $(4, -3)$,   $2x + y = 5$

In Problems 13 through 16, find the distance between the given parallel lines. Check your results.

**13** $3x + 4y - 7 = 0$   and   $3x + 4y + 3 = 0$

**14** $x + 2y + 4 = 0$   and   $2x + 4y - 5 = 0$

**15** $5x - 6y = 0$   and   $5x - 6y + 1 = 0$

**16** $Ax + By + 10 = 0$   and   $Ax + By + 15 = 0$

In Problems 17 through 20, the vertices of triangles are given. Find the lengths of the altitudes from the first vertex to the side joining the other two. Then find the areas of the triangles.

**17** $(3, 2), (-1, 1), (-2, 3)$      **18** $(4, -2), (-1, -3), (3, 2)$

**19** $(-2, -1), (-1, 4), (3, 2)$   **20** $(-3, 4), (1, -2), (3, 4)$

**21** Find the two points on the $x$ axis which are at a distance 2 from $3x - 4y - 6 = 0$.

**22** Find the two points on the $y$ axis which are at a distance 3 from the line $12x + 5y + 9 = 0$.

**23** Find the two points on the line $2x + 3y + 4 = 0$ which are at a distance 2 from the line $3x + 4y - 6 = 0$.

**24** Find the equations of the two lines parallel to $3x + 4y - 5 = 0$ and at a distance 2 from it.

**25** Find the equations of the two lines parallel to $x - 2y + 1 = 0$ and at a distance 3 from it.

**26** Find the distance from the midpoint of the segment joining $(3, 2)$ and $(-4, 6)$ to the line $2x - 3y + 5 = 0$.

**27** Find the point on the curve $y = x^2$ which is closest to the line $2x - y - 4 = 0$.

**\*28** Find the point on the curve $(x - 1)^2 + y^2 - 4 = 0$ which is closest to the line $4x - 3y + 12 = 0$.

**29** By using the formula for the distance from a point to a line, find the equations of the two lines which bisect the angles made by

$$L_1: x + 2y - 3 = 0 \quad \text{and} \quad L_2: 2x - y + 3 = 0.$$

**30** The lines $L_1$: $Ax + By + C_1 = 0$ and $L_2$: $Ax + By + C_2 = 0$ are parallel. Show that the distance between them is given by the formula

$$d = \frac{|C_1 - C_2|}{\sqrt{A^2 + B^2}}.$$

In Problems 31 through 37, find in each case $\tan \phi$, where $\phi$ is the angle from $L_1$ to $L_2$.

**31** $L_1$: $2x + 3y - 4 = 0$;   $L_2$: $x - 4y + 1 = 0$

**32** $L_1$: $x + 3y - 2 = 0$;   $L_2$: $3x - y + 1 = 0$

**33** $L_1$: $2x - 5y + 7 = 0$;   $L_2$: $2x - 4y - 3 = 0$

**34** $L_1$: $5x + 7y - 6 = 0$;   $L_2$: $8x + 5y - 2 = 0$

**35** $L_1$: $2x + 3y = 5$;   $L_2$: $3x + 2y = 1$

**36** $L_1$: $5x + y = 0$;   $L_2$: $3x + 2y = 5$

**37** $L_1$: $3x - 2y = 2$;   $L_2$: $y = 2$

In Problems 38 through 41, find the tangents of the angles of the triangles $ABC$ with the vertices given.

**38** $A(3, 0)$, $B(1, 2)$, $C(5, 6)$     **39** $A(-1, 2)$, $B(4, 1)$, $C(0, 5)$

**40** $A(-1, -3)$, $B(2, -1)$, $C(-3, 4)$

**41** $A(-1, 4)$, $B(5, 1)$, $C(5, 6)$

**42** Given a triangle with vertices at $A(0, 0)$, $B(5, 0)$, $C(3, 4)$; show that the three angle bisectors meet in a point.

**43** Show that the bisectors of the angles of any triangle meet in a point.

In Problems 44 through 48, find the equations of the lines through the given point $P_0$ making the given angle $\theta$ with the given line $L$.

**44** $P_0 = (2, -3)$, $\theta = \pi/4$,   $L = \{(x, y): 3x - 2y = 5\}$

**45** $P_0 = (-1, 2)$, $\tan \theta = \frac{1}{2}$,   $L = \{(x, y): 2x + y - 3 = 0\}$

**46** $P_0 = (4, 1)$, $\tan \theta = 3$,   $L = \{(x, y): 3x + 4y - 5 = 0\}$

**47** $P_0 = (-1, -2)$, $\theta = \pi/6$,   $L = \{(x, y): x + y = 1\}$

**48** $P_0 = (3, 1)$, $\theta = \pi/2$,   $L = \{(x, y): 2x - y = 3\}$

**49** Tangent lines are drawn to the curve of $y = x^2 - 2x + 4$ at the points $(1, 3)$ and $(-2, 12)$. Find the tangent of the angle at which the first tangent intersects the second. Is it necessary to find the equations of the lines?

**50** Tangent lines are drawn to the curve given by $y = x^3 - 2x^2 + 3x - 4$ at $(2, 2)$ and $(-1, -10)$. Find the tangent of the angle at which these lines intersect.

**51** Two tangent lines are drawn to the curve (a circle) $K$ given by

$$K = \{(x, y): x^2 + y^2 - 2x + 5y + 7 = 0\}$$

from the point $(-1, -2)$. Find the tangent of the angle at which these lines intersect.

---

_____2_____

## THE PARABOLA*

Straight lines correspond to equations of the first degree in $x$ and $y$. A systematic approach to the study of equations in $x$ and $y$ would proceed to equations of the second degree, third degree, and so on. Curves of the second degree are sufficiently simple that we can study them thoroughly in this course. Curves corresponding to equations of the third degree and higher will not be treated systematically, since the complications rapidly become too great. However, the study of equations of any degree—the theory of algebraic curves—is an interesting subject on its own and may be studied in advanced courses.

DEFINITIONS    _A **parabola** is the graph of all points whose distances from a fixed point equal their distances from a fixed line. The fixed point is called the **focus** and the fixed line the **directrix**._

---

*Additional material on the parabola, as well as another problem set, may be found in Appendix 2, Section 2, at the end of the book.

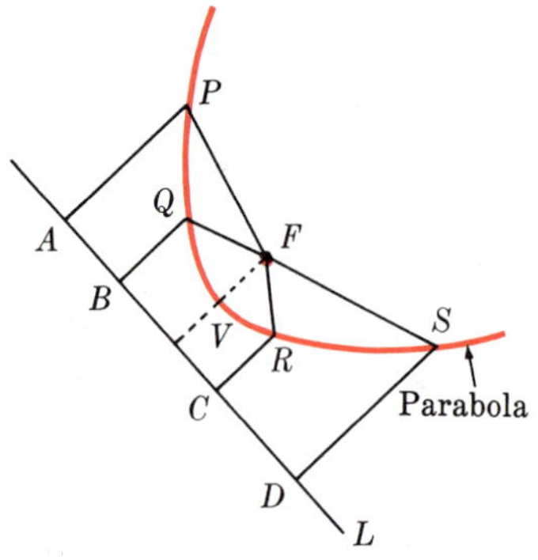

The definition of the parabola is purely geometric in the sense that it says nothing about coordinates—that is, the values of $x$ and $y$. Nevertheless we will see that parabolas can always be expressed as the graph of certain second degree equations.

Figure 6 shows a typical situation. The line $L$ is the directrix, the point $F$ is the focus, and $P$, $Q$, $R$, $S$ are points which satisfy the conditions of the graph. That is,

$$|AP| = |PF|, \quad |BQ| = |QF|, \quad |CR| = |RF|, \quad \text{and} \quad |DS| = |SF|.$$

The perpendicular to $L$ through the focus intersects the parabola at a point $V$. This point is called the **vertex** of the parabola.

To find an equation for a parabola, we set up a coordinate system and place the directrix and the focus at convenient places. Suppose that the distance from the focus $F$ to the directrix $L$ is $p$ units. We place the focus at $(p/2, 0)$ and let the line $L$ be $x = -p/2$, as shown in Fig. 7. If $P(x, y)$ is a typical point on the graph, the conditions are such that the distance $|PF| = \sqrt{[x - (p/2)]^2 + (y - 0)^2}$ is equal to the distance from $P$ to the line $L$. From an examination of the figure (or from the formula for the distance from a point to a line), we easily deduce that the distance from $P$ to the line $L$ is $|x - (-p/2)|$. Therefore we have

$$\sqrt{\left(x - \frac{p}{2}\right)^2 + y^2} = \left| x + \frac{p}{2} \right|,$$

and, since both sides are positive, we may square to obtain

$$\left(x - \frac{p}{2}\right)^2 + y^2 = \left(x + \frac{p}{2}\right)^2$$

and

$$x^2 - px + \frac{p^2}{4} + y^2 = x^2 + px + \frac{p^2}{4}.$$

FIGURE 6

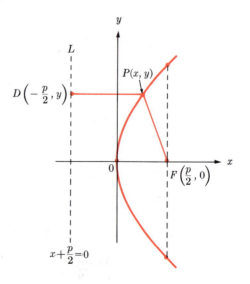

FIGURE 7

This equation yields

$$y^2 = 2px, \quad p > 0.$$

Each point on the graph satisfies this equation. Conversely, by reversing the steps, it can be shown that every point which satisfies this equation is on the graph. We call this *the equation of the parabola with focus at* $(p/2, 0)$ *and with the line* $x = -p/2$ *as directrix.* The vertex of the parabola is at the origin. It is apparent from the equation that the $x$ axis is an axis of symmetry, for if $y$ is replaced by $-y$, the equation is unchanged. The line of symmetry of a parabola is called the **axis of the parabola**.

If the focus and directrix are placed in other positions, the equation will of course be different. There are four *standard positions* for the focus-directrix combination, of which the first has just been described. With $p$ always a positive number, the second standard position is defined so that its focus is at $(-p/2, 0)$ and the line $x = p/2$ is its directrix. Using the same method as before, we solve the problem which defines the parabola. We obtain the equation

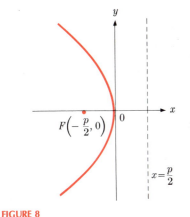

FIGURE 8

$$y^2 = -2px, \quad p > 0,$$

and a parabola in the position shown in Fig. 8. The third position has its focus at $(0, p/2)$ and the line $y = -p/2$ as the directrix. The resulting equation is

$$x^2 = 2py, \quad p > 0,$$

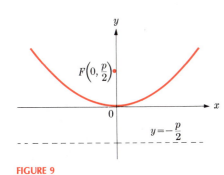

FIGURE 9

and the parabola is in the position shown in Fig. 9. In the fourth standard position the focus is at $(0, -p/2)$ and the directrix is the line $y = p/2$. The equation then becomes

$$x^2 = -2py, \quad p > 0,$$

and the parabola is in the position shown in Fig. 10. In all four positions the vertex is at the origin.

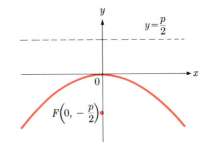

FIGURE 10

**EXAMPLE 1**  A parabola is given by

$$S = \{(x, y): y^2 = -12x\}.$$

Find the focus, directrix, and axis. Sketch the graph.

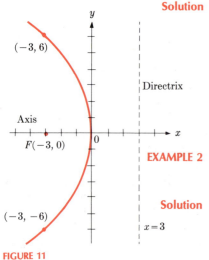

**FIGURE 11**

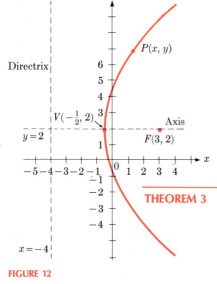

**FIGURE 12**

**Solution**  Since $2p = 12$, we have $p = 6$ and, since the equation is in the second of the standard forms, the focus is at $(-3, 0)$. The directrix is the line $x = 3$, and the $x$ axis is the axis of the parabola. The curve is sketched in Fig. 11.     □

When the parabola is not in one of the standard positions, the equation is more complicated. If the directrix is horizontal or vertical, the equation is only slightly different; however, if the directrix is parallel to neither axis, the equation is altered considerably. The following example shows how the equation of the parabola may be obtained directly from the definition.

**EXAMPLE 2**  Find the equation of the parabola with focus at $F(3, 2)$ and with the line $x = -4$ as directrix. Locate the vertex and the axis of symmetry.

**Solution**  From the definition of the parabola, $P(x, y)$ is a point on the curve if and only if $|PF| = \sqrt{(x - 3)^2 + (y - 2)^2}$ is equal to the distance from $P$ to the directrix, which is $|x + 4|$. We have

$$\sqrt{(x - 3)^2 + (y - 2)^2} = |x + 4|.$$

Since both sides are positive, we square both sides and find

$$x^2 - 6x + 9 + y^2 - 4y + 4 = x^2 + 8x + 16.$$

This yields

$$(y - 2)^2 = 14x + 7 \quad \Leftrightarrow \quad (y - 2)^2 = 14(x + \tfrac{1}{2}).$$

A sketch of the equation (Fig. 12) shows that the axis is the line $y = 2$ and the vertex is at the point $(-\tfrac{1}{2}, 2)$.     □

There are four forms for the parabola when the axis is vertical or horizontal and the vertex is at the point $(a, b)$ instead of the origin. We summarize these in the form of a theorem.

**THEOREM 3**  *If a parabola has an axis which is vertical or horizontal and has a vertex at $(a, b)$, then one of the following four equations has a graph which is the parabola (with $p > 0$):*

i) $(y - b)^2 = 2p(x - a)$,  *focus at* $(a + p/2, b)$;  *directrix*, $x = a - p/2$;
ii) $(y - b)^2 = -2p(x - a)$,  *focus at* $(a - p/2, b)$;  *directrix*, $x = a + p/2$;
iii) $(x - a)^2 = 2p(y - b)$,  *focus at* $(a, b + p/2)$;  *directrix*, $y = b - p/2$;
iv) $(x - a)^2 = -2p(y - b)$,  *focus at* $(a, b - p/2)$;  *directrix*, $y = b + p/2$.

The four formulas above may be derived directly from the definition of parabola (as in Example 2). See Problem 31 at the end of this section.

**EXAMPLE 3**  Given the parabola $x^2 = -4y - 3x + 2$, find the vertex, focus, directrix, and axis. Sketch the curve.

**Solution**  If we can express the given equation as one of the "standard forms" of Theorem 3 we will immediately obtain the focus and the directrix. We write the equation as:

$$(x^2 + 3x \qquad) = -4y + 2.$$

The process for changing $(x^2 + 3x)$ into $(x - a)^2$ is known as completing the square (see page 313). We take half the coefficient of $x$, i.e., $\frac{3}{2}$, then square it and add the result to both sides of the equation:

$$(x^2 + 3x + \tfrac{9}{4}) = -4y + 2 + \tfrac{9}{4}.$$

We obtain

$$(x + \tfrac{3}{2})^2 = -4y + 2 + \tfrac{9}{4} = -4y + \tfrac{17}{4}.$$

This yields

$$(x + \tfrac{3}{2})^2 = -4(y - \tfrac{17}{16}),$$

which is in Form (iv) in Theorem 3. We read off that the vertex is at $(-\frac{3}{2}, \frac{17}{16})$. Since $2p = 4$ we know that $p = 2$ and the focus is at $(-\frac{3}{2}, \frac{1}{16})$; the directrix is the line $y = \frac{33}{16}$, and the axis is the line $x = -\frac{3}{2}$. The curve is sketched in Fig. 13.

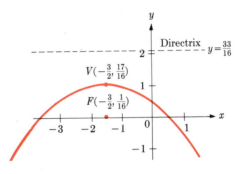

**FIGURE 13**

**EXAMPLE 4**    For every real number $k$ the equation $3x + 2y - k = 0$ is a straight line; if we let $l(k) = \{(x, y): 3x + 2y - k = 0\}$, then the collection $l(k)$ is called a **family of lines**: each value of $k$ gives a different line, and we note that all the lines are parallel. We wish to find the particular member of this family which is tangent to the parabola

$$S = \{(x, y): y = 3x^2 - 2x + 1\}.$$

**Solution**    When $l(k) \cap S$ consists of one point, the line is tangent to the parabola or is parallel to the axis of the parabola. If the line is tangent to the parabola, the slope of the line and the parabola must be the same at the point of contact. The family $l(k)$ consists of all lines of slope $-\frac{3}{2}$. The slope of the parabola at any point is

$$\frac{dy}{dx} = 6x - 2.$$

The slope of the parabola and the slope of the line are the same when

$$6x - 2 = -\tfrac{3}{2} \qquad \Leftrightarrow \qquad x = \tfrac{1}{12}.$$

Substituting $x = \frac{1}{12}$ in the equation for $S$, we get

$$y = 3(\tfrac{1}{144}) - \tfrac{2}{12} + 1 = \tfrac{123}{144}.$$

The point on the parabola at which the slope is $-\frac{3}{2}$ is $(\frac{1}{12}, \frac{123}{144})$. This point must also be on the line. Therefore

$$3(\tfrac{1}{12}) + 2(\tfrac{123}{144}) - k = 0$$

or $k = \frac{47}{24}$. The desired member of $l(k)$ is

$$l(\tfrac{47}{24}) = \{(x, y): 3x + 2y - \tfrac{47}{24} = 0\}.$$

□

**EXAMPLE 5**  Find the equation of the line tangent to the parabola

$$S = \{(x, y): y^2 + 3y - 2x + 4 = 0\},$$

at the point on the parabola at which $y = 2$.

**Solution**  When $y = 2$, by substitution in the equation for $S$, we obtain $x = 7$. The slope at any point on $S$ is given by the relation (implicit differentiation)

$$2y\frac{dy}{dx} + 3\frac{dy}{dx} - 2 = 0,$$

and at $(7, 2)$ we find

$$\frac{dy}{dx} = \frac{2}{2y + 3} = \frac{2}{7}.$$

The equation of the tangent line is $y - 2 = \frac{2}{7}(x - 7)$.

□

The parabola has many interesting geometrical properties that make it suitable for practical applications. Probably the most familiar application is the use of parabolic shapes in headlights of cars, flashlights, and so forth. If a parabola is revolved about its axis, the surface generated is called a **paraboloid**. Such surfaces are used in headlights, optical and radio telescopes, radar, etc., because of the following geometric property. If a source of light (or sound or other type of wave) is placed at the focus of a parabola, and if the parabola is a reflecting surface, then the wave will bounce back in a line parallel to the axis of the parabola (Fig. 14). This creates a parallel beam without dispersion. (In actual practice there will be some dispersion, since the source of light must occupy more than one point.) Clearly the converse is also true. If a series of incoming waves is parallel to the axis of the reflecting paraboloid, the resulting signal will be concentrated at the focus (where the receiving equipment, e.g., a radar dish, is therefore located). This reflecting property, which is vital to many applications, is established mathematically in Appendix 2.

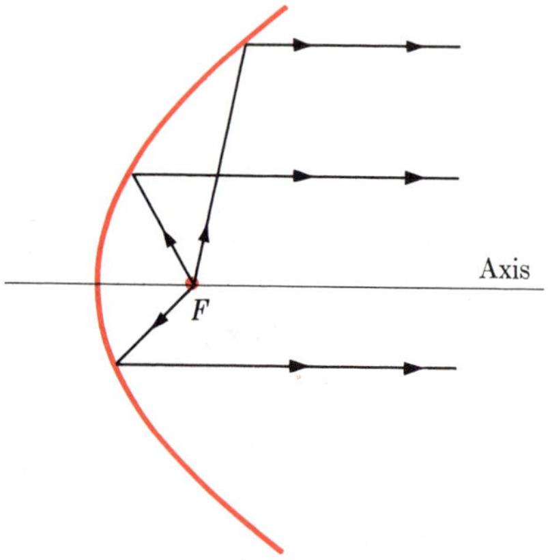

Light reflecting from a parabolic mirror

FIGURE 14

## 2  PROBLEMS

In Problems 1 through 4, find the coordinates of the foci and the equations of the directrices. Sketch the curves.

**1**  $y^2 = 8x$            **2**  $x^2 = 6y$

**3**  $y^2 = -12x$          **4**  $x^2 = -16y$

In Problems 5 through 10, find the equation of the parabola with vertex at the origin which satisfies the given additional condition.

**5**  Focus at $(-4, 0)$            **6**  Focus at $(0, 2)$

**7**   Directrix: $y = 2$      **8**   Directrix: $x = \frac{7}{3}$

**9**   Passing through $(2, 3)$ and axis along the $x$ axis

**10**   Passing through $(2, 3)$ and axis along the $y$ axis

In Problems 11 through 24, find the focus, vertex, directrix, and axis of the parabola. Sketch the curve.

**11**   $y = x^2 - 2x + 3$      **12**   $x = y^2 + 2y - 4$

**13**   $y = -4x^2 + 3x$      **14**   $x = -y^2 + 2y - 7$

**15**   $x^2 + 2y - 3x + 5 = 0$      **16**   $y^2 + 2x - 4y + 7 = 0$

**17**   $x^2 = -12y$      **18**   $y^2 = \frac{1}{3}x$

**19**   $2y^2 = -6x$      **20**   $y^2 - 4y - 2x - 4 = 0$

**21**   $y^2 + 2x + 6y + 17 = 0$      **22**   $y^2 - x + y = 0$

**23**   $x^2 + x + y = 0$      **24**   $2x^2 + 2x = 1 - y$

In Problems 25 through 30, in each case find the equation of the parabola from the definition.

**25**   Directrix $x = 0$,   focus at $(6, 0)$

**26**   Directrix $y = 0$,   focus at $(0, 5)$

**27**   Vertex $(0, 4)$,   focus at $(0, 2)$

**28**   Vertex $(-2, 0)$,   directrix $x = 1$

**29**   Focus $(-1, -2)$,   directrix $y = 2$

**30**   Focus $(2, 3)$,   directrix $x = 4$

**31**   Use the definition of a parabola to establish the formulas in Theorem 3, page 458.
     a) Formula (i)      c) Formula (iii)
     b) Formula (ii)      d) Formula (iv)

**32**   Find the member of the family

$$l(k) = \{(x, y): 4x - y - k = 0\}$$

which is tangent to the parabola $y = 2x^2 - x + 1$.

**33**   Find a member of the family

$$l(k) = \{(x, y): kx - y - 5 = 0\}$$

which is tangent to the parabola $y = 3x^2 + 2x + 4$.

**34**   Find the member of the family of parabolas $y = kx^2 + 4x - 3$ which is tangent to the line $2x - y + 1 = 0$.

In Problems 35 through 39, in each case find the equation of the line tangent to the parabola at the given point $P$ on the parabola.

**35**   $y^2 = 9x$,   $P(1, 3)$      **36**   $x^2 = 12y + 7$,   $P(1, -\frac{1}{2})$

**37**   $y^2 - 3x + 2y + 4 = 0$,   $P(\frac{4}{3}, 0)$

**38**   $(y - 2)^2 = -4(x + 3)$;   $P(-4, 4)$

**39**   $x^2 - 2x + y = 0$,   $P(-3, -15)$

**40**   The two towers of a suspension bridge are 300 meters apart and extend 80 meters above the road surface. If the cable (in the shape of a parabola) is tangent to the road at the center of the bridge, find the height of the cable above the road at 50 meters and also at 100 meters from the center of the bridge. (Assume that the road is horizontal.) See Fig. 15.

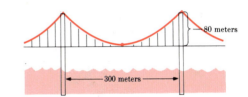

FIGURE 15

**41**   At a point $P$ on a parabola a tangent line is drawn. This line intersects the axis of the parabola at a point $A$. If $F$ denotes the focus, show that $\triangle APF$ is isosceles.

**42**   Let $P(x_1, y_1)$ be a point (not the vertex) on the parabola $y^2 = 2px$. Find the equation of the line normal to the parabola at this point. Show that this normal line intersects the $x$ axis at the point $Q(x_1 + p, 0)$.

---

## 3

## THE CIRCLE AND THE ELLIPSE*

The set of all points $(x, y)$ which are at a given distance $r$ from a fixed point $(h, k)$ represents a **circle**. We also use the statement, *the graph of all such points is a circle.* From the formula for the distance between two points we know that

$$r = \sqrt{(x - h)^2 + (y - k)^2},$$

---

*Additional material on the circle and the ellipse as well as another problem set, may be found in Appendix 2, Sections 1 and 3, at the end of the book.

or, upon squaring,

$$(x-h)^2 + (y-k)^2 = r^2.$$

More formally we have:

---

**DEFINITION**    *A **circle** is the set of all points which are at a given, fixed distance from a chosen point. The chosen point is the **center** of the circle.*

---

**THEOREM 4**    *The equation of a circle with center at $(h, k)$ and radius $r$ is*

$$(x-h)^2 + (y-k)^2 = r^2.$$

---

**EXAMPLE 1**    Find the equation of the circle with center at $(-3, 4)$ and radius 6.

**Solution**    We have $h = -3$, $k = 4$, $r = 6$. Substituting in the equation of the circle, we obtain

$$(x+3)^2 + (y-4)^2 = 36.$$

We can multiply out to get

$$x^2 + y^2 + 6x - 8y - 11 = 0$$

as another form for the answer.                                                    □

The next example starts the other way around.

**EXAMPLE 2**    Determine whether or not the equation

$$x^2 + y^2 - 4x + 7y - 8 = 0$$

represents a circle. If it does, where is the center and what is the radius?

**Solution**    The process for determining these facts consists of "completing the square." We write

$$(x^2 - 4x \qquad) + (y^2 + 7y \qquad) = 8,$$

in which we leave the appropriate spaces, as shown. We add the square of half the coefficient of $x$ and the square of half the coefficient of $y$ to both sides and so obtain

$$(x^2 - 4x + 4) + (y^2 + 7y + \tfrac{49}{4}) = 8 + 4 + \tfrac{49}{4}$$

or

$$(x - 2)^2 + (y + \tfrac{7}{2})^2 = \tfrac{97}{4}.$$

This represents a circle with center at $(2, -\tfrac{7}{2})$ and radius $\tfrac{1}{2}\sqrt{97}$.          □

A *line tangent to a circle* is defined in the usual way by the methods of calculus. We also have the elementary definition that a line perpendicular to a radius and passing through the point where the radius meets the circle is tangent to the circle at this point. We consider an example which shows that the two definitions agree.

**EXAMPLE 3**     The point $(1, 2)$ is on the circle

$$x^2 + y^2 + 2x + 3y - 13 = 0$$

which is verified by substitution of $x = 1$, $y = 2$ into the equation. We pose the problem of finding the equation of the line tangent to the circle at the point $(1, 2)$.

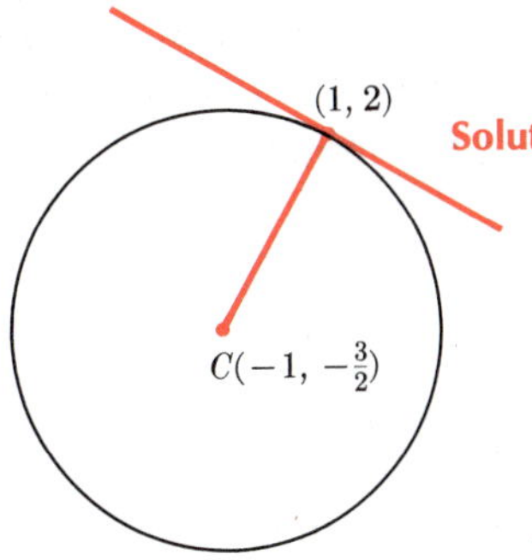

**Solution**     (See Fig. 16.) By implicit differentiation we can find the slope of the tangent line at any point. We have

$$2x + 2y\frac{dy}{dx} + 2 + 3\frac{dy}{dx} = 0$$

and

$$\frac{dy}{dx} = -\frac{2x + 2}{2y + 3}.$$

At the point $(1, 2)$ we obtain

$$\frac{dy}{dx} = -\frac{2 + 2}{4 + 3} = -\frac{4}{7},$$

and the equation of the line, according to the point-slope formula, is

$$y - 2 = -\tfrac{4}{7}(x - 1) \quad \Leftrightarrow \quad 4x + 7y - 18 = 0.$$

Any line perpendicular to this tangent line has slope $\tfrac{7}{4}$. The particular perpendicular passing through $(1, 2)$ has the equation

$$y - 2 = \tfrac{7}{4}(x - 1) \quad \Leftrightarrow \quad 7x - 4y + 1 = 0. \tag{1}$$

The center $C$ of the circle is at $(-1, -\tfrac{3}{2})$. Since

$$7(-1) - 4(-\tfrac{3}{2}) + 1 = 0,$$

the line (1) passes through the center $C$. Therefore we see that the tangent line is perpendicular to the radius at the point of contact. Thus the two definitions for a tangent to a circle agree.     □

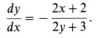

A problem that often arises in physics is that of computing the distance from a point $P$ outside a circle to a point $Q$ on the circle and such that the line through $P$ and $Q$ is tangent to the circle (see Fig. 17). Let $P(x_1, y_1)$ be a point outside the circle $K$ which has its center at $C(h, k)$ and has radius $r$. In set notation, we may write

$$K = \{(x, y): (x - h)^2 + (y - k)^2 - r^2 = 0\}.$$

A tangent from $P$ to the circle has contact at a point $Q$. Since triangle $PQC$ is a right triangle, we see that $|PQ| = \sqrt{|PC|^2 - r^2}$ and, using the distance formula for the length $|PC|$, we have

$$|PQ| = \sqrt{(x_1 - h)^2 + (y_1 - k)^2 - r^2}.$$

From the procedure for completing the square, it follows that if the circle having equation

$$x^2 + y^2 + Dx + Ey + F = 0 \tag{2}$$

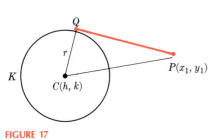

**FIGURE 16**

**FIGURE 17**

has center at $(h, k)$ and radius $r > 0$, then

$$(x - h)^2 + (y - k)^2 - r^2 \equiv x^2 + y^2 + Dx + Ey + F.$$

Hence if the circle in Fig. 17 has the equation (2), then the formula for $|PQ|$, the distance we seek, is given by

$$|PQ| = \sqrt{x_1^2 + y_1^2 + Dx_1 + Ey_1 + F}.$$

The next example illustrates the ease in using this formula.

**EXAMPLE 4**   A tangent is drawn from $P(8, 4)$ to the circle $K$ given by

$$K = \{(x, y): x^2 + y^2 + 2x + y - 3 = 0\}.$$

Find the distance from $P$ to the point of tangency $Q$.

**Solution**   Using the formula immediately above for $|PQ|$, we obtain

$$|PQ| = \sqrt{8^2 + 4^2 + 2 \cdot 8 + 4 - 3} = \sqrt{97}. \qquad \square$$

Another curve of the second degree, closely related to the circle, is the *ellipse*. The definition we provide is purely geometric.

**DEFINITIONS**   *An **ellipse** is the graph of all points the sum of whose distances from two fixed points is constant. The two fixed points are called the **foci**.*

For the definition to make sense, the constant (the sum of the distances from the foci) must be larger than the distance between the foci. Note that the definition says nothing about coordinates or equations. It will turn out that the equation of an ellipse is of the second degree in $x$ and $y$.

Suppose that the distance between the foci is $2c$ $(c > 0)$. We place the foci—label them $F_1$ and $F_2$—at convenient points, say $F_1$ at $(c, 0)$ and $F_2$ at $(-c, 0)$. Let $P(x, y)$ be a point on the ellipse, $d_1$ the distance from $P$ to $F_1$, and $d_2$ the distance from $P$ to $F_2$ (Fig. 18). The conditions state that $d_1 + d_2$ is always constant. As we said, this constant (which we shall denote by $2a$) must be larger than $2c$. Thus we have

$$d_1 + d_2 = 2a,$$

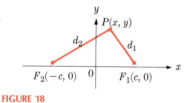

FIGURE 18

or, from the distance formula,

$$\sqrt{(x - c)^2 + (y - 0)^2} + \sqrt{(x + c)^2 + (y - 0)^2} = 2a.$$

We simplify by transferring one radical to the right and squaring both sides:

$$(x - c)^2 + y^2 = 4a^2 - 4a\sqrt{(x + c)^2 + y^2} + (x + c)^2 + y^2.$$

By multiplying out and combining, we obtain

$$\sqrt{(x + c)^2 + y^2} = a + \frac{c}{a}x.$$

We square again to find

$$(x + c)^2 + y^2 = a^2 + 2cx + \frac{c^2}{a^2}x^2,$$

and this becomes

$$\frac{x^2}{a^2} + \frac{y^2}{a^2 - c^2} = 1.$$

Since $a > c$, we can introduce a new quantity,

$$b = \sqrt{a^2 - c^2},$$

and we can write

$$\frac{x^2}{a^2} + \frac{y^2}{b^2} = 1.$$

This is **the equation of an ellipse**.

We have shown that every point on the graph satisfies this equation. Conversely (by reversing the steps*), it can be shown that each point which satisfies this equation is on the ellipse. A sketch of such an equation is shown in Fig. 19. From the equation, we see immediately that the curve is symmetric with respect to both the $x$ and $y$ axes.

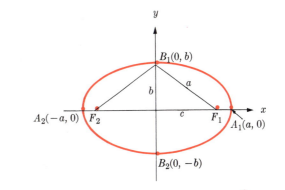

**FIGURE 19**

DEFINITIONS    *The line passing through the two foci $F_1$ and $F_2$ is called the **major axis**. The perpendicular bisector of the line segment $F_1 F_2$ is called the **minor axis**. The intersection of the major and minor axes is called the **center**.*

---

*In the process of reversing the steps, we must verify (because of the squaring operations) that

$$a + \frac{c}{a} x \geq 0 \quad \text{and} \quad 2a - \sqrt{(x + c)^2 + y^2} \geq 0.$$

However,

$$2a - \sqrt{(x + c)^2 + y^2} = 2a - \left(a + \frac{c}{a} x\right) = a - \frac{c}{a} x.$$

Since $-a \leq x \leq a$ for every point satisfying the equation, and since $c < a$, the verification follows.

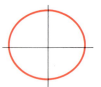

e small

FIGURE 20

e close to 1

FIGURE 21

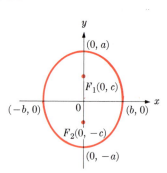

An ellipse with foci on y axis

FIGURE 22

The equation of the ellipse as we have obtained it indicates that the ellipse intersects the $x$ axis at $(a, 0)$, $(-a, 0)$. These points are called the **vertices** of the ellipse. The distance between them, $2a$, is called the *length of the major axis.* (Sometimes this segment is called simply the major axis.) The ellipse intersects the $y$ axis at $(0, b)$ and $(0, -b)$. The distance between them, $2b$, is called the *length of the minor axis.* (Sometimes this segment is called the minor axis.)

The **eccentricity** $e$ of an ellipse is defined as

$$e = c/a.$$

Note that since $c < a$, the eccentricity is always between 0 and 1. The eccentricity measures the flatness of an ellipse. If $a$ is kept fixed and $c$ is very "small," then $e$ is close to zero. But this means that the foci are close together, and the ellipse is almost a circle, as in Fig. 20. On the other hand, if $c$ is close to $a$, then $e$ is near 1 and the ellipse is quite flat, as shown in Fig. 21. The limiting position of an ellipse as $e \to 0$ is a circle of radius $a$. The limiting position as $e \to 1$ is a line segment of length $2a$.

In developing the equation of an ellipse we placed the foci at $(c, 0)$ and $(-c, 0)$. If instead we put them on the $y$ axis at $(0, c)$ and $(0, -c)$ and carry through the same argument, then the equation of the ellipse we obtain is

$$\frac{x^2}{b^2} + \frac{y^2}{a^2} = 1.$$

The $y$ axis is now the major axis, the $x$ axis is the minor axis, and the vertices are at $(0, a)$ and $(0, -a)$. The quantity $b = \sqrt{a^2 - c^2}$ and the eccentricity $e = c/a$ are defined as before. Figure 22 shows an ellipse with foci on the $y$ axis.

**EXAMPLE 5**    Given the ellipse with equation $9x^2 + 25y^2 = 225$, find the major and minor axes, the eccentricity, the coordinates of the foci and the vertices. Sketch the ellipse.

**Solution**    We put the equation in "standard form" by dividing by 225. We obtain

$$\frac{x^2}{25} + \frac{y^2}{9} = 1,$$

which tells us that $a = 5$, $b = 3$. Since $a^2 = b^2 + c^2$, we find that $c = 4$. The eccentricity $e = c/a = \frac{4}{5}$. The major axis is along the $x$ axis, the minor axis is along the $y$ axis, the vertices are at $(\pm 5, 0)$, and the foci are at $(\pm 4, 0)$. The ellipse is sketched in Fig. 23.    □

**EXAMPLE 6**    Given the ellipse with equation $16x^2 + 9y^2 = 144$, find the major and minor axes, the eccentricity, the coordinates of the foci, and the vertices. Sketch the ellipse.

**Solution**    We divide by 144 to get

$$\frac{x^2}{9} + \frac{y^2}{16} = 1.$$

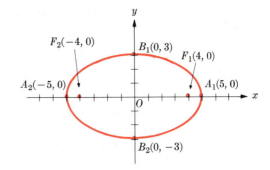

**FIGURE 23**

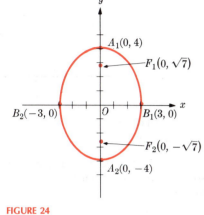

**FIGURE 24**

We note that *the number under the y term is larger*. Therefore $a = 4$, $b = 3$, $c^2 = a^2 - b^2$, and $c = \sqrt{7}$. The eccentricity $e = \sqrt{7}/4$, the major axis is along the $y$ axis, the minor axis is along the $x$ axis, the foci are at $(0, \pm\sqrt{7})$, and the vertices at $(0, \pm 4)$. The ellipse is sketched in Fig. 24. □

The foregoing examples illustrate the fact that when we have an equation of the form

$$\frac{x}{(\ \ )^2} + \frac{y^2}{(\ \ )^2} = 1,$$

the equation represents an ellipse (or a circle if the denominators are equal in size), and the *larger denominator* determines whether the foci, vertices, and major axes are along the $x$ axis or the $y$ axis.

Given that $P(x_1, y_1)$ is a point on an ellipse, it is a simple matter to find the equation of the line tangent to the ellipse at $P$. In fact, we can develop a simple formula for the equation of the line when the ellipse is given by

$$\frac{x^2}{a^2} + \frac{y^2}{b^2} = 1.$$

The slope of the ellipse is obtained by implicit differentiation:

$$\frac{2x}{a^2} + \frac{2y}{b^2}\frac{dy}{dx} = 0$$

and

$$\frac{dy}{dx} = -\frac{b^2}{a^2}\frac{x}{y}.$$

The equation of the line tangent at $P(x_1, y_1)$ is

$$y - y_1 = -\frac{b^2}{a^2}\frac{x_1}{y_1}(x - x_1) \quad \Leftrightarrow \quad a^2 y_1 y = -b^2 x_1 x + b^2 x_1^2 + a^2 y_1^2.$$

But $b^2 x_1^2 + a^2 y_1^2 = a^2 b^2$, since $(x_1, y_1)$ is on the ellipse. Therefore, substituting this relation in the preceding equation, we get

$$\frac{x_1 x}{a^2} + \frac{y_1 y}{b^2} = 1 \tag{3}$$

for the equation of the tangent line.

**EXAMPLE 7**   Find the equation of the line tangent to the ellipse

$$\frac{x^2}{9} + \frac{y^2}{16} = 1$$

at the point $(2, -\frac{4}{3}\sqrt{5})$.

**Solution**   The point $(2, -\frac{4}{3}\sqrt{5})$ is on the ellipse and, even though the major axis is along the $y$ axis, we see that the formula analogous to (3) for the tangent line holds. We obtain

$$\frac{2x}{9} + \frac{-\frac{4}{3}\sqrt{5}y}{16} = 1 \quad \Leftrightarrow \quad 8x - 3\sqrt{5}y - 36 = 0. \qquad \square$$

## 3   PROBLEMS

In Problems 1 through 14, in each case find the equation of the circle determined by the given conditions, where $C$ denotes the center and $r$ the radius.

**1** $C = (-3, 4)$, $r = 5$   **2** $C = (0, 0)$, $r = 4$

**3** $C = (2, 3)$, $r = 2$   **4** $C = (2, -1)$, $r = 3$

**5** A diameter is the segment from $(4, 2)$ to $(8, 6)$

**6** A diameter is the segment from $(-2, 3)$ to $(4, -1)$

**7** $C = (3, 4)$, tangent to the $x$ axis

**8** $C = (-2, 3)$, tangent to the $y$ axis

**9** $C = (2, 1)$, passing through $(3, 4)$

**10** $C = (-1, -2)$, passing through $(-2, 2)$

**11** $C = (2, 3)$, tangent to $3x + 4y + 2 = 0$

**12** $C = (3, -2)$, tangent to $5x - 12y = 0$

**13** Tangent to both axes, radius 6, in second quadrant

**14** Tangent to the $x$ axis, to the line $y = 2x$ and radius 5 (four solutions)

In Problems 15 through 22, in each case determine the graph of the given equation by completing the square.

**15** $x^2 + y^2 + 6x - 8y = 0$

**16** $x^2 + y^2 + 2x - 4y - 11 = 0$

**17** $x^2 + y^2 - 4x + 2y + 5 = 0$

**18** $x^2 + y^2 + 6x - 4y + 15 = 0$

**19** $x^2 + y^2 + 3x - 5y - \frac{1}{2} = 0$

**20** $x^2 + y^2 + 4x - 3y + 4 = 0$

**21** $2x^2 + 2y^2 + 3x + 5y + 2 = 0$

**22** $2x^2 + 2y^2 - 5x + 7y + 10 = 0$

In Problems 23 through 26, in each case find the equation of the line tangent to the given circle at the point $P$ on the circle.

**23** $x^2 + y^2 + 2x - 4y = 0$, $P(1, 3)$

**24** $x^2 + y^2 - 3x + 2y - 7 = 0$, $P(-1, 1)$

**25** $x^2 + y^2 + 5x - 6y - 21 = 0$, $P(2, -1)$

**26** $x^2 + y^2 + 2x - 19 = 0$, $P(-3, 4)$

In Problems 27 through 30, in each case find the equation of the line normal to the given circle at the given point $P$ on the circle. Do each problem in two ways.

**27** $x^2 + y^2 + 4y - 1 = 0$, $P(2, -3)$

**28** $x^2 + y^2 - 3x - 8y + 18 = 0$, $P(1, 4)$

**29** $x^2 + y^2 - 4x - 7y - 6 = 0$, $P(6, 1)$

**30** $x^2 + y^2 + 8x + 2y + 16 = 0$, $P(-4, -2)$

In Problems 31 through 34, in each case lines are drawn tangent to the given circle through the given point $P$, not on the circle. Find the equations of these lines.

**31** $x^2 + y^2 + 4x + 6y - 21 = 0$, $P(-4, 5)$

**32** $x^2 + y^2 - 2x + 5y + 7 = 0$, $P(-1, -2)$

**33** $x^2 + y^2 - x - 4y - 7 = 0$, $P(3, 0)$

**34** $x^2 + y^2 + 2x + 6y + 2 = 0$, $P(-3, 3)$

**35** The slope of the tangent to

$$x^2 + y^2 + 3x + y + 25 = 0$$

is

$$\frac{dy}{dx} = -\frac{2x + 3}{2y + 1}, y \neq -\frac{1}{2}.$$

What is wrong with this statement? Or is it correct?

In Problems 36 through 45, find the lengths of the major and minor axes, the coordinates of the foci and vertices, and the eccentricity of the ellipses. Sketch the curve.

**36** $16x^2 + 25y^2 = 400$   **37** $25x^2 + 169y^2 = 4225$

**38** $9x^2 + 16y^2 = 144$   **39** $25x^2 + 16y^2 = 400$

**40** $3x^2 + 2y^2 = 6$   **41** $3x^2 + 4y^2 = 12$

**42** $5x^2 + 2y^2 = 10$    **43** $5x^2 + 3y^2 = 15$
**44** $2x^2 + 3y^2 = 11$    **45** $5x^2 + 4y^2 = 17$

In Problems 46 through 56, find the equation of the ellipse satisfying the given conditions.

**46** Vertices at $(\pm 5, 0)$, foci at $(\pm 4, 0)$

**47** Vertices at $(0, \pm 5)$, foci at $(0, \pm 3)$

**48** Vertices at $(0, \pm 10)$, eccentricity 4/5

**49** Vertices at $(\pm 6, 0)$, eccentricity 2/3

**50** Foci at $(0, \pm 4)$, eccentricity 4/5

**51** Vertices at $(0, \pm 5)$, passing through $(3, 3)$

**52** Axes along the coordinate axes, passing through $(4, 3)$ and $(-1, 4)$

**53** Axes along the coordinate axes, passing through $(-5, 2)$ and $(3, 7)$

**54** Foci at $(\pm 3, 0)$, passing through $(4, 1)$

**55** Eccentricity 3/4, foci along $x$ axis, center at origin, and passing through $(6, 4)$

**56** Eccentricity 3/4, foci on the $y$ axis, center at origin, and passing through $(6, 4)$

**57** Find the equation of the graph of all points the sum of whose distances from $(6, 0)$ and $(-6, 0)$ is 20.

**58** Find the equation of the graph of all points the sum of whose distances from $(3, 0)$ and $(9, 0)$ is 12.

**59** Find the equation of the graph of all points whose distances from $(3, 0)$ are 1/3 of their distances from the line $x = 27$.

**60** Find the equation of the graph of all points whose distances from $(0, 4)$ are 2/3 of their distances from the line $y = 9$.

**61** Find the equation of the graph of all points whose distances from $(4, 0)$ are 1/2 of their distances from the line $x = 6$. Locate the major and minor axes.

**62** Find the equation of the graph of all points whose distances from $(0, 5)$ are 3/4 of their distances from the line $y = 8$. Locate the major and minor axes.

**63** Find the equation of the line tangent to the ellipse

$$\frac{x^2}{9} + \frac{y^2}{4} = 1$$

at the point $P(1, -\frac{4}{3}\sqrt{2})$.

**64** Find the equation of the line tangent to the ellipse $4x^2 + 12y^2 = 1$ at the point $P(1/4, 1/4)$.

**65** Find the equation of the line normal to the ellipse $4x^2 + 9y^2 = 36$ at the point $P(2, \frac{2}{3}\sqrt{5})$.

**66** Find the equation of the line normal to the ellipse $16x^2 + 25y^2 = 400$ at the point $P(-3, 16/5)$.

**67** Find the equation of the line normal to the ellipse $(x^2/a^2) + (y^2/b^2) = 1$ which passes through the point $P(x_1, y_1)$ on the ellipse.

**68** The tangent to an ellipse at a point $P_0$ meets the tangent at a vertex in a point $Q$. Prove that the line joining the other vertex to $P_0$ is parallel to the line joining the center to $Q$.

**69** A square is inscribed in the ellipse whose equation is $(x^2/16) + (y^2/9) = 1$. Find the coordinates of the vertices of the square; find the perimeter and area of the square.

**70** The orbit in which the Earth travels about the sun is approximately an ellipse with the sun at one focus. The major axis of the elliptical orbit is 300,000 kilometers and the eccentricity is 0.017 approximately. Find the maximum and minimum distances from the Earth to the sun.

**71** Find the dimensions of a rectangle of maximum area that can be inscribed in an ellipse of axes lengths $a$ and $b$, and if two sides of the rectangle are parallel to the major axis.

---

**4**

## THE HYPERBOLA*

A study of the hyperbola will complete our discussion of second-degree curves in the plane.

**DEFINITIONS**    *A **hyperbola** is the graph of all points the difference of whose distances from two fixed points is a positive constant. The two fixed points are called the **foci**.*

---

*Additional material on the hyperbola as well as another problem set may be found in Appendix 2, Section 4, at the end of the book.

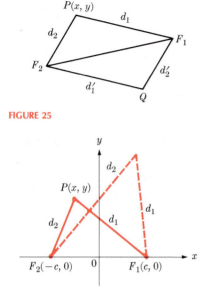

**FIGURE 25**

**FIGURE 26**

Notice that the geometric definition of a hyperbola is similar to that of an ellipse. For the hyperbola we take the *difference* of distances from two fixed points, instead of the *sum*.

A typical situation is shown in Fig. 25, where the two fixed points (foci) are labeled $F_1$ and $F_2$. For a point $P$ to be on the graph, the distance $|PF_1|$ minus the distance $|PF_2|$ must be equal to a positive constant. Also, a point, say $Q$, will be on the graph if the distance $|QF_2|$ minus $|QF_1|$ is this same constant. We remark that the definition of a hyperbola, like that of a parabola and an ellipse, is purely geometric and makes no mention of coordinate systems. We shall see that hyperbolas are second-degree curves.

Suppose that the distance between the foci is $2c$ $(c > 0)$. We place the foci at convenient points, $F_1$ at $(c, 0)$ and $F_2$ at $(-c, 0)$. Let $P(x, y)$ be a point on the hyperbola; then the conditions of the graph assert that $d_1 - d_2$ or $d_2 - d_1$ is *always a positive constant* (Fig. 26), *which we label* $2a(a > 0)$. We write the conditions of the graph as

$$d_1 - d_2 = 2a \qquad \text{or} \qquad d_2 - d_1 = 2a,$$

which may be combined as

$$d_1 - d_2 = \pm 2a.$$

Using the distance formula, we obtain

$$\sqrt{(x - c)^2 + y^2} - \sqrt{(x + c)^2 + y^2} = \pm 2a.$$

We transfer one radical to the right side and square both sides to get

$$(x - c)^2 + y^2 = (x + c)^2 + y^2 \pm 4a\sqrt{(x + c)^2 + y^2} + 4a^2.$$

Some terms may be canceled to yield

$$\pm 4a\sqrt{(x + c)^2 + y^2} = 4a^2 + 4cx.$$

We divide by $4a$ and once again square both sides, to find

$$x^2 + 2cx + c^2 + y^2 = a^2 + 2cx + \frac{c^2}{a^2}x^2 \qquad \Leftrightarrow \qquad \left(\frac{c^2}{a^2} - 1\right)x^2 - y^2 = c^2 - a^2.$$

We now divide through by $c^2 - a^2$, with the result

$$\frac{x^2}{a^2} - \frac{y^2}{c^2 - a^2} = 1.$$

We shall show that the quantity $c$ must be larger than $a$. We recall that the sum of the lengths of any two sides of a triangle must be greater than the length of the third side. When we refer to Fig. 26, we observe that

$$2c + d_2 > d_1 \qquad \text{and} \qquad 2c + d_1 > d_2.$$

Therefore

$$2c > d_1 - d_2 \qquad \text{and} \qquad 2c > d_2 - d_1;$$

combined, these inequalities give $2c > |d_1 - d_2|$. But $|d_1 - d_2| = 2a$, and so $c > a$.

We define the positive number $b$ by the relation

$$b = \sqrt{c^2 - a^2},$$

and we write **the equation of the hyperbola**:

$$\frac{x^2}{a^2} - \frac{y^2}{b^2} = 1.$$

Conversely, by showing that all the steps above are reversible, we could have demonstrated that every point which satisfies the above equation is on the graph.

We summarize the preceding discussion with a theorem.

---

**THEOREM 5**    *A hyperbola with foci at $(-c, 0)$ and $(c, 0)$ has an equation of the form:*

$$\frac{x^2}{a^2} - \frac{y^2}{b^2} = 1, \tag{1}$$

*where $a^2 + b^2 = c^2$. Moreover the graph of any equation of the form (1) is a hyperbola with foci at $(-c, 0)$ and $(c, 0)$.*

---

The equation of the hyperbola in the above form shows at once that it is symmetric with respect to both the $x$ axis and the $y$ axis.

---

**DEFINITIONS**    *The line passing through the foci $F_1$ and $F_2$ of a hyperbola is called the* **transverse axis**. *The perpendicular bisector of the segment $F_1 F_2$ is called the* **conjugate axis**. *The intersection of these axes is called the* **center**.

---

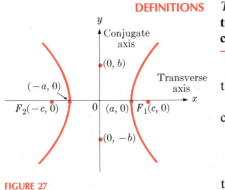

FIGURE 27

(See Fig. 27, where the foci are at $(c, 0)$ and $(-c, 0)$, the transverse axis is the $x$ axis, the conjugate axis is the $y$ axis, and the center is at the origin.)

The points of intersection of a hyperbola with the transverse axis are called its **vertices**. In Fig. 27, corresponding to the equation

$$\frac{x^2}{a^2} - \frac{y^2}{b^2} = 1,$$

these vertices occur at $(a, 0)$ and $(-a, 0)$. The length $2a$ is called the *length of the transverse axis*. Even though the points $(0, b)$ and $(0, -b)$ are not on the graph, the length $2b$ is a useful quantity. It is called the *length of the conjugate axis*.

The **eccentricity** of a hyperbola is defined as

$$e = c/a.$$

Note that since $c > a$, *the eccentricity of a hyperbola is always larger than* 1.

If the foci of the hyperbola are placed along the $y$ axis at the points $(0, c)$ and $(0, -c)$, the equation takes the form

$$\frac{y^2}{a^2} - \frac{x^2}{b^2} = 1,$$

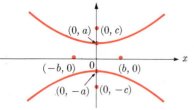

FIGURE 28

where, as before,

$$b = \sqrt{c^2 - a^2}.$$

The curve has the general appearance of the curve shown in Fig. 28.

In contrast with the equation of an ellipse, the equation for a hyperbola indicates that the relative sizes of $a$ and $b$ play no role in determining where the foci and axes are. An equation of the form

$$\frac{x^2}{(\ )^2} - \frac{y^2}{(\ )^2} = 1$$

*always* has its transverse axis (and foci) on the $x$ axis, while an equation of the form

$$\frac{y^2}{(\ )^2} - \frac{x^2}{(\ )^2} = 1$$

*always* has its transverse axis (and foci) on the $y$ axis. In the first case, the quantity under the $x^2$ term is $a^2$ and that under the $y^2$ term is $b^2$. In the second case the quantity under the $y^2$ term is $a^2$ and that under the $x^2$ term is $b^2$. In *both cases* we have $c^2 = a^2 + b^2$.

**EXAMPLE 1**    Given the hyperbola with equation $9x^2 - 16y^2 = 144$. Find the axes, the coordinates of the vertices and the foci, and the eccentricity. Sketch the curve.

**Solution**    Dividing by 144, we find that

$$\frac{x^2}{16} - \frac{y^2}{9} = 1,$$

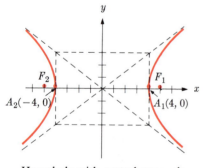

Hyperbola with central rectangle

FIGURE 29

and, therefore,

$$a = 4, \ b = 3, \ c^2 = a^2 + b^2 = 25, \ c = 5.$$

The transverse axis is along the $x$ axis; the conjugate axis is along the $y$ axis. The vertices are at $(\pm 4, 0)$. The foci are at $(\pm 5, 0)$. The eccentricity $e = c/a = 5/4$. The curve is sketched in Fig. 29, where we have constructed a rectangle by drawing parallels to the axes through the points $(a, 0)$, $(-a, 0)$, $(0, b)$, and $(0, -b)$. This is called the **central rectangle**. Note that the diagonal from the origin to one corner of this rectangle has length $c$.    □

**EXAMPLE 2**    Find the equation of the hyperbola with vertices at $(0, \pm 6)$ and eccentricity $\frac{5}{3}$. Locate the foci.

**Solution**    Since the vertices are along the $y$ axis, the equation is of the form

$$\frac{y^2}{a^2} - \frac{x^2}{b^2} = 1,$$

with $a = 6$, $e = c/a = \frac{5}{3}$, and $c = (5 \cdot 6)/3 = 10$. Therefore

$$b = \sqrt{100 - 36} = 8,$$

and the equation is

$$\frac{y^2}{36} - \frac{x^2}{64} = 1.$$

The foci are at the points $(0, \pm 10)$.

If $P(x_1, y_1)$ is a point on the hyperbola

$$\frac{x^2}{a^2} - \frac{y^2}{b^2} = 1,$$

we can get a formula for the line tangent to the hyperbola at $P$ in a way that precisely parallels the work we did to obtain a formula for the line tangent to the ellipse. The slope of the hyperbola at any point, by implicit differentiation, is

$$\frac{2x}{a^2} - \frac{2y}{b^2}\frac{dy}{dx} = 0 \qquad \Leftrightarrow \qquad \frac{dy}{dx} = \frac{b^2}{a^2}\frac{x}{y}.$$

The equation of the line tangent at $P(x_1, y_1)$ is

$$y - y_1 = \frac{b^2}{a^2}\frac{x_1}{y_1}(x - x_1) \qquad \Leftrightarrow \qquad a^2 y_1 y - a^2 y_1^2 = b^2 x_1 x - b^2 x_1^2.$$

But since $P$ is on the hyperbola, we know that $b^2 x_1^2 - a^2 y_1^2 = a^2 b^2$, and so

$$\frac{x_1 x}{a^2} - \frac{y_1 y}{b^2} = 1$$

is the equation of the tangent line.

**EXAMPLE 3**   Find the equations of the tangent line and the normal line to the hyperbola

$$\frac{x^2}{9} - \frac{y^2}{16} = 1$$

at the point $(5, -\frac{16}{3})$.

**Solution**   The point $(5, -\frac{16}{3})$ is on the hyperbola, and so the equation of the tangent line is

$$\frac{5x}{9} + \frac{\frac{16}{3}y}{16} = 1 \qquad \Leftrightarrow \qquad 5x + 3y - 9 = 0.$$

Since the tangent line has slope $-5/3$, the normal line has slope $\frac{3}{5}$, and the equation is

$$y + \frac{16}{3} = \frac{3}{5}(x - 5).$$

In studying the hyperbola

$$\frac{x^2}{a^2} - \frac{y^2}{b^2} = 1 \tag{1}$$

we shall show that the lines

$$y = \frac{b}{a}x \qquad \text{and} \qquad y = -\frac{b}{a}x$$

are of special interest. For this purpose we consider a point $\bar{P}(\bar{x}, \bar{y})$ on the hyperbola (1) and compute the distance from $\bar{P}$ to the line $y = bx/a$. According to the formula for the distance from a point to a line (see Section 1), we have for the distance $d$:

$$d = \frac{|b\bar{x} - a\bar{y}|}{\sqrt{a^2 + b^2}}.$$

Suppose we select a point $\bar{P}$ which has both of its coordinates positive, as shown in Fig. 30. Since $\bar{P}$ is a point on the hyperbola, its coordinates satisfy the equation $b^2\bar{x}^2 - a^2\bar{y}^2 = a^2b^2$, which we may write

$$(b\bar{x} - a\bar{y})(b\bar{x} + a\bar{y}) = a^2b^2 \quad \Leftrightarrow \quad b\bar{x} - a\bar{y} = a^2b^2/(b\bar{x} + a\bar{y}).$$

Substituting this last relation in the above expression for $d$, we find

$$d = \frac{a^2b^2}{\sqrt{a^2 + b^2}\,|b\bar{x} + a\bar{y}|}.$$

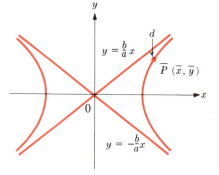

$y = \frac{b}{a}x$

$y = -\frac{b}{a}x$

$\bar{P}\,(\bar{x}, \bar{y})$

FIGURE 30

Now $a$ and $b$ are fixed positive numbers and we selected $\bar{x}$, $\bar{y}$ positive. As $P$ moves out along the hyperbola farther and farther away from the vertex, $\bar{x}$ and $\bar{y}$ increase without bound. Therefore, the number $d$ tends to zero as $\bar{x}$ and $\bar{y}$ get larger and larger. That is, the distance between the hyperbola (1) and the line $y = bx/a$ shrinks to zero as $\bar{P}$ tends to infinity. We define the line $y = bx/a$ as an **asymptote of the hyperbola** $(x^2/a^2) - (y^2/b^2) = 1$. Analogously, the line $y = -bx/a$ (see Fig. 30) is called an asymptote of the same hyperbola. We have only to recall the symmetry of the hyperbola to verify this statement.

Now we can see one of the uses of the central rectangle. *The asymptotes are the lines which contain the diagonals of the central rectangle.* In an analogous way the hyperbola $(y^2/a^2) - (x^2/b^2) = 1$ has the asymptotes

$$y = (a/b)x \quad \text{and} \quad y = -(a/b)x.$$

These lines contain the diagonals of the appropriately constructed central rectangle.

**EXAMPLE 4**    Given the hyperbola $25x^2 - 9y^2 = 225$, find the foci, vertices, eccentricity, and asymptotes. Sketch the curve.

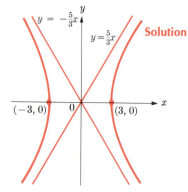

$y = -\frac{5}{3}x$

$y = \frac{5}{3}x$

$(-3, 0)$      $(3, 0)$

FIGURE 31

**Solution**    When we divide by 225, we obtain $(x^2/9) - (y^2/25) = 1$, and therefore $a = 3$, $b = 5$, $c^2 = 9 + 25$, and $c = \sqrt{34}$. The foci are at $(\pm\sqrt{34}, 0)$, vertices at $(\pm 3, 0)$; the eccentricity is $e = c/a = \sqrt{34}/3$. To find the asymptotes we merely write $y = \pm(b/a)x = \pm\frac{5}{3}x$. Since the asymptotes are a great help in sketching the curve, we draw them first (Fig. 31): Knowing the vertices and one or two additional points helps us obtain a fairly accurate graph.   □

An easy way to remember the equations of the asymptotes is to recognize that if the hyperbola is

$$\frac{x^2}{a^2} - \frac{y^2}{b^2} = 1, \quad \text{the asymptotes are} \quad \frac{x^2}{a^2} - \frac{y^2}{b^2} = 0,$$

and if the hyperbola is

$$\frac{y^2}{a^2} - \frac{x^2}{b^2} = 1, \qquad \text{the asymptotes are} \qquad \frac{y^2}{a^2} - \frac{x^2}{b^2} = 0.$$

Like the parabola, the hyperbola is useful in practical applications. A simple example is that of *range finding*. Suppose, for example, two observers at different locations at a known distance apart each hear the firing of a gun. The difference in time when they heard it multiplied by the velocity of sound gives the value $2a$ (see page 470) and hence determines a hyperbola on which the gun must be located. (See Fig. 32.)

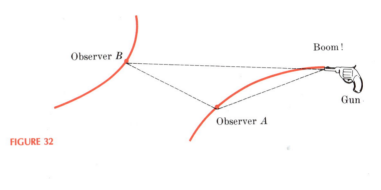

Boom!

Observer $B$

Gun

Observer $A$

**FIGURE 32**

## 4  PROBLEMS

In Problems 1 through 20, find in each case the lengths of the transverse and conjugate axes, the coordinates of the foci and the vertices, the eccentricity, and the equations for the asymptotes. Sketch the curve.

1  $\dfrac{x^2}{9} - \dfrac{y^2}{4} = 1$

2  $\dfrac{y^2}{64} - \dfrac{x^2}{25} = 1$

3  $\dfrac{y^2}{9} - \dfrac{x^2}{4} = 1$

4  $\dfrac{x^2}{64} - \dfrac{y^2}{25} = 1$

5  $16x^2 - 9y^2 = 144$

6  $16x^2 - 9y^2 + 576 = 0$

7  $25x^2 - 144y^2 + 3600 = 0$

8  $25x^2 - 144y^2 - 900 = 0$

9  $2x^2 - 3y^2 - 6 = 0$

10  $2x^2 - 3y^2 + 6 = 0$

11  $5x^2 - 2y^2 = 10$

12  $5x^2 - 4y^2 + 13 = 0$

13  $3x^2 - 2y^2 = 1$

14  $2x^2 - 3y^2 = -1$

15  $x^2 = 3y^2 - 27$

16  $121x^2 - 4y^2 = 1$

17  $y^2 = x^2 - a^2$

18  $y^2 = x^2 + a^2$

19  $11x^2 - 7y^2 = 1$

20  $2x^2 - 10y^2 = 1$

In Problems 21 through 35, find for each case the equation of the hyperbola (or hyperbolas) satisfying the given conditions.

21  Vertices at $(\pm 5, 0)$, foci at $(\pm 7, 0)$

22  Vertices at $(\pm 4, 0)$, foci at $(\pm 6, 0)$

23  Vertices at $(0, \pm 7)$, eccentricity $\frac{4}{3}$

24  Foci at $(\pm 6, 0)$, eccentricity $\frac{3}{2}$

25  Eccentricity $\sqrt{5}$, foci on $x$ axis, center at origin, passing through the point $(3, 2)$

26  Vertices at $(0, \pm 4)$, passing through $(-2, 5)$

27  Foci at $(0, \pm\sqrt{10})$, passing through $(2, 3)$

28  Axes along the coordinate axes, passing through $(4, 2)$ and $(-6, 7)$

29  Axes along the coordinate axes, passing through $(-3, 4)$ and $(5, 6)$

30  Foci at $(\pm 8, 0)$, length of conjugate axis 6

31  Asymptotes are $\dfrac{x^2}{4} - \dfrac{y^2}{16} = 0$

32  Asymptotes are $\dfrac{y^2}{4} - \dfrac{x^2}{16} = 0$

33  Asymptotes are $64x^2 = 9y^2$

34  Asymptotes are $11x^2 = 7y^2$

35  Asymptotes are $2x^2 = 10y^2$

36  Find the equation of the graph of all points such that the difference of their distances from $(4, 0)$ and $(-4, 0)$ is always equal to 2.

37  Find the equation of the graph of all points such that the difference of their distances from $(0, 7)$ and $(0, -7)$ is always equal to 3.

**38** Find the equation of the graph of all points such that the difference of their distances from $(10, 0)$ and $(2, 0)$ is always 1.

**39** Find the equation of the graph of all points such that the difference of their distances from $(2, 0)$ and $(2, 12)$ is always equal to 3.

**40** Find the equations of the tangent and normal lines to the hyperbola

$$(x^2/16) - (y^2/4) = 1 \quad \text{at the point} \quad (-5, \tfrac{3}{2}).$$

**41** Find the point of intersection of the lines tangent to the hyperbola

$$(y^2/36) - (x^2/9) = 1$$

at the points

$$(1, -2\sqrt{10}), (-4, 10).$$

**42** Given the hyperbola $(x^2/6) - (y^2/2) = 1$. Find the equations of the lines tangent to this hyperbola passing through the point $(1, -1)$.

**43** Find the equation of the graph of all points such that their distances from $(5, 0)$ are always $\tfrac{5}{3}$ times their distances from the line $x = 9/5$.

**44** A **focal radius** is the line segment from a point on a hyperbola to one of the foci. The foci of a hyperbola are at $(4, 0)$ and $(-4, 0)$. The difference in the lengths of the focal radii from any point is always $\pm 6$. Find the equation of the hyperbola.

**45** A hyperbola has its foci at $(c, 0)$ and $(-c, 0)$. The point $P(2, 4)$ is on the hyperbola and the focal radii (see Problem 44) from $P$ are perpendicular. Find the equation of the hyperbola.

**\*46** A circle $K$ touches another circle given by the equation $x^2 + y^2 + 2x = 0$ and $K$ passes through the point $(1, 0)$. Find an equation that gives all possible positions for the center of the circle $K$.

**\*47** A moving circle $K$ is tangent externally to each of the two fixed circles $(x - 2)^2 + y^2 = 1$ and $(x + 2)^2 + y^2 = 4$. Find an equation that gives all possible positions for the center of the circle $K$.

---

## 5

# TRANSLATION OF AXES

Coordinate systems were not employed in the definitions and descriptions of the principal properties of second-degree curves. However, when we want to find equations for such curves and wish to use the methods of analytic geometry, coordinate systems become essential.

When we use the method of coordinate geometry we place the axes at a position "convenient" with respect to the curve under consideration. In the examples of ellipses and hyperbolas we studied, the foci were located on one of the axes and were situated symmetrically with respect to the origin. But now suppose that we have a problem in which the curve (hyperbola, parabola, ellipse, etc.) is *not* situated so conveniently with respect to the axes. We would then like to change the coordinate system in order to have the curve at a convenient and familiar location. The process of making this change is called a **transformation of coordinates**.

The first type of transformation we consider is one of the simplest, and is called the **translation of axes**. Figure 33 shows the usual rectangular $xy$ coordinate system. We now introduce an additional coordinate system with axis $x'$ parallel to $x$ and $k$ units away, and axis $y'$ parallel to $y$ and $h$ units away. This means that the origin $O'$ of the new coordinate system has coordinates $(h, k)$ in the original system. The positive $x'$ and $y'$ directions are taken to be the same as the positive $x$ and $y$ directions. When $P$ is any point in the plane, what are its coordinates in each of the systems? Suppose that $P$ has coordinates $(x, y)$ in the original system and $(x', y')$ in the new system and *suppose that the new origin has coordinates $(h, k)$ in the original system.* An

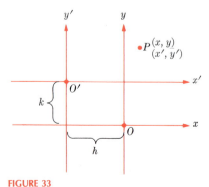

FIGURE 33

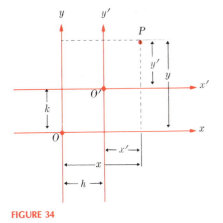

**FIGURE 34**

inspection of Fig. 34 suggests that if $x > h > 0$ and $y > k > 0$, then

$$x = x' + h \quad \text{and} \quad y = y' + k \qquad (1)$$

or equivalently

$$x' = x - h \quad \text{and} \quad y' = y - k. \qquad (2)$$

The proof of these formulas follows at once by subtraction of the appropriate directed distances shown in Fig. 35. See Problem 29 at the end of this section.

Two coordinate systems $xy$ and $x'y'$, which satisfy equations (1) or, equivalently, equations (2), are said to be related by a **translation of axes**.

The most general equation of the second degree has the form

$$Ax^2 + Bxy + Cy^2 + Dx + Ey + F = 0,$$

where $A$, $B$, $C$, $D$, $E$, and $F$ represent numbers. We assume that $A$, $B$, and $C$ are not all zero, since then the equation would be of the first degree. We recognize that the equations we have been discussing are all special cases of the above equation. For example, the circle $x^2 + y^2 - 16 = 0$ has $A = 1$, $B = 0$, $C = 1$, $D = E = 0$, $F = -16$. The ellipse $(x^2/9) + (y^2/4) = 1$ has $A = \frac{1}{9}$, $B = 0$, $C = \frac{1}{4}$, $D = E = 0$, $F = -1$. Similarly, parabolas and hyperbolas are of the above form.

An interesting case is $A = 1$, $B = 0$, $C = -1$, $D = E = F = 0$. We get $x^2 - y^2 = 0$ or $y = \pm x$, and the graph is two intersecting lines. Second-degree curves whose graphs reduce to lines or points are frequently called **degenerate**.

We now illustrate how to use translation of axes to reduce an equation of the form

$$Ax^2 + Cy^2 + Dx + Ey + F = 0$$

to an equation of the same form but with new letters $(x', y')$, and with $D$ and $E$ both equal to zero (with certain exceptions). Although presented in a new way, the equation will be easily recognized as one of the second-degree curves we have studied. *The principal tool in this process is "completing the square."*

**EXAMPLE 1**    Given the equation

$$9x^2 + 25y^2 + 18x - 100y - 116 = 0,$$

by using a translation of axes determine whether the graph of the equation is a parabola, ellipse, or hyperbola. Determine foci (or focus), vertices (or vertex), and eccentricity. Sketch the curve.

**Solution**    To complete the square in $x$ and $y$, we write the equation in the form

$$9(x^2 + 2x \quad) + 25(y^2 - 4y \quad) = 116.$$

We add 1 in the parentheses for $x$, which means adding 9 to the left side, and we add 4 in the parentheses for $y$, which means adding 100 to the left side. We obtain

$$9(x^2 + 2x + 1) + 25(y^2 - 4y + 4) = 116 + 9 + 100$$

$$\Leftrightarrow 9(x + 1)^2 + 25(y - 2)^2 = 225.$$

Now we have the clue for translation of axes. We define

$$x' = x + 1 \qquad \text{and} \qquad y' = y - 2.$$

That is, the translation is made with $h = -1$, $k = 2$. The equation becomes

$$9x'^2 + 25y'^2 = 225.$$

Dividing by 225, we find

$$\frac{x'^2}{25} + \frac{y'^2}{9} = 1,$$

which we recognize as an ellipse with $a = 5$, $b = 3$, $c^2 = a^2 - b^2 = 16$, $c = 4$, $e = \frac{4}{5}$. In the $x'y'$ system, we have: center $(0, 0)$; vertices $(\pm 5, 0)$; foci $(\pm 4, 0)$.

In the $xy$ system, we use the relations $x = x' - 1$, $y = y' + 2$ to obtain: center $(-1, 2)$; vertices $(4, 2)$, $(-6, 2)$; foci $(3, 2)$, $(-5, 2)$. The curve is sketched in Fig. 35.  □

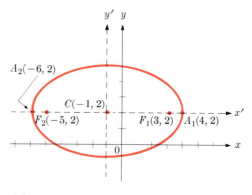

**FIGURE 35**

**EXAMPLE 2**   Discuss the properties of the graph of the equation

$$x^2 + 4x + 4y - 4 = 0.$$

**Solution**   Here we have $A = 1$, $C = 0$, $D = 4$, $E = 4$, $F = -4$. There is only one second-degree term, and we complete the square to obtain

$$(x^2 + 4x + 4) = -4y + 4 + 4 \qquad \Leftrightarrow \qquad (x + 2)^2 = -4(y - 2).$$

We read off the appropriate translation of axes. It is

$$x' = x + 2, \qquad y' = y - 2,$$

and we have

$$x'^2 = -4y',$$

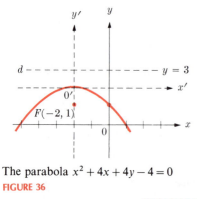

The parabola $x^2 + 4x + 4y - 4 = 0$

**FIGURE 36**

which we recognize as a parabola. In the $x'y'$ system, the vertex is at $(0, 0)$, focus at $(0, -1)$ (since $p = 2$); directrix, $y' = 1$. In the $xy$ system (since $h = -2$, $k = 2$), the vertex is at $(-2, 2)$, focus at $(-2, 1)$, and the directrix is the line $y = 3$. The curve is sketched in Fig. 36.  □

**EXAMPLE 3**   Find the equation of the ellipse with eccentricity $\frac{1}{2}$ and foci at $(4, 2)$ and $(2, 2)$.

**Solution**   The center of the ellipse is halfway between the foci and therefore is at the point $(3, 2)$. The major axis of the ellipse is along the line $y = 2$. To move the origin to the center of the ellipse by a translation of coordinates, we let

$$x' = x - 3, \qquad y' = y - 2.$$

In this system the foci are at $(\pm 1, 0)$; $e = c/a = \frac{1}{2}$ and, since $c = 1$, we have $a = 2$. Also $c^2 = a^2 - b^2$ gives us $b^2 = 3$. In the $x'y'$ system, the equation is

$$\frac{x'^2}{4} + \frac{y'^2}{3} = 1.$$

In the $xy$ system, the equation is

$$\frac{(x-3)^2}{4} + \frac{(y-2)^2}{3} = 1,$$

or

$$3(x-3)^2 + 4(y-2)^2 = 12 \qquad \Leftrightarrow \qquad 3x^2 + 4y^2 - 18x - 16y + 31 = 0.$$

$\square$

We are now in a position to discuss all the curves that an equation

$$Ax^2 + Cy^2 + Dx + Ey + F = 0$$

can possibly represent, presented here in the form of a theorem:

---

**THEOREM 6**  (a) *If A and C are both positive or both negative, then the graph is an ellipse, a circle (if A = C), a point, or nothing.* (b) *If A and C are of opposite signs, the graph is a hyperbola or two intersecting lines.* (c) *If either A or C is zero, the graph is a parabola, two parallel lines, one line, or nothing.*

---

Although we do not prove this theorem, it is within the scope of our treatment, and the interested reader should be able to provide a proof. Some of the special cases of Theorem 6 are illustrated in the following examples.

The equation $3x^2 + y^2 + 5 = 0$ has no graph, since the sum of positive quantities can never add up to zero. This exhibits the last case under (a) of Theorem 6. Under (c), the equation $x^2 - 2x = 0$ $(A = 1, C = 0, D = -2, E = 0, F = 0)$ is an example of two parallel lines, while $x^2 - 2x + 1 = 0$ $(A = 1, C = 0, D = -2, E = 0, F = 1)$ is an example of one line.

## 5  PROBLEMS

**1**  A rectangular system of coordinates is translated to a new $x'y'$ system, whose origin is at $(-3, 4)$. The points $P, Q, R,$ and $S$ have coordinates $(4, 3), (-2, 7), (-6, \frac{3}{2}),$ and $(-5, -2)$, respectively, in the original system. Find the coordinates of these points in the new system.

**2**  A translation of coordinates moves the origin to the point of intersection of the lines $2x + 3y - 4 = 0$ and $x + 4y - 1 = 0$. Find the translation of coordinates and the equations of these lines in the new system.

**3**  A translation of coordinates moves the origin to the point of intersection of the lines $3x - 7y + 1 = 0$ and $2x - 5y - 6$

$= 0$. Find the equations of these lines in the new coordinate system.

In Problems 4 through 17, translate the coordinates so as to eliminate the first-degree terms (or one first-degree term in the case of parabolas), describe the principal properties (as in Examples 1 and 2) and sketch the curves.

**4**  $y^2 + 8x - 6y + 1 = 0$

**5**  $25x^2 + 16y^2 + 50x - 64y - 311 = 0$

**6**  $9x^2 + 16y^2 - 36x - 32y - 92 = 0$

**7**  $9x^2 - 4y^2 + 18x + 16y + 29 = 0$

**8** $x^2 - 4x + 6y + 16 = 0$

**9** $4x^2 + 9y^2 + 8x - 36y + 4 = 0$

**10** $y^2 - 4x + 10y + 5 = 0$

**11** $9x^2 + 4y^2 - 18x + 8y + 4 = 0$

**12** $4x^2 - 9y^2 + 8x + 18y + 4 = 0$

**13** $3x^2 + 4y^2 - 12x + 8y + 4 = 0$

**14** $x^2 + 6x - 8y + 1 = 0$

**15** $3x^2 - 2y^2 + 6x - 8y - 17 = 0$

**16** $2x^2 + 3y^2 - 8x - 6y + 11 = 0$

**17** $4x^2 - 3y^2 + 8x + 12y - 8 = 0$

In Problems 18 through 23, find the equation of the graph indicated; sketch the curve.

**18** Parabola: vertex at $(2, 1)$; directrix $y = 3$

**19** Ellipse: vertices at $(2, -3)$ and $(2, 5)$; foci at $(2, -2)$ and $(2, 4)$

**20** Ellipse: foci at $(-2, 3)$ and $(4, 3)$; length of major axis is 10

**21** Hyperbola: vertices at $(-3, 1)$ and $(5, 1)$; passing through $(-5, 3)$

**22** Hyperbola: vertices at $(2, -5)$ and $(2, 7)$; foci at $(2, -6)$ and $(2, 8)$

**23** Parabola: axis $y = -1$; directrix $x = 2$; focus $(5, -1)$

In Problems 24 through 28, find the equation of the given graph. Identify the curve.

**24** The graph of all points whose distances from $(4, 3)$ equal their distances from the line $x = 6$

**25** The graph of all points whose distances from $(2, 1)$ are equal to one-half their distances from the line $y = -2$

**26** The graph of all points whose distances from $(-1, 2)$ are equal to twice their distances from the line $x = -4$

**27** The graph of all points whose distances from $(2, -3)$ are equal to twice their distances from $(-4, 3)$

**28** The graph of all points whose distances from $(1, 2)$ are equal to their distances from the line $3x - 4y - 5 = 0$

**29** By referring to Fig. 34, establish Formulas (1) on page 477. Use directed distances and verify that the result is the same if $P$ in Fig. 34 is located in the second, third, or fourth quadrants.

**30** Prove Theorem 6.

---

## 6

## ROTATION OF AXES. THE GENERAL EQUATION OF THE SECOND DEGREE

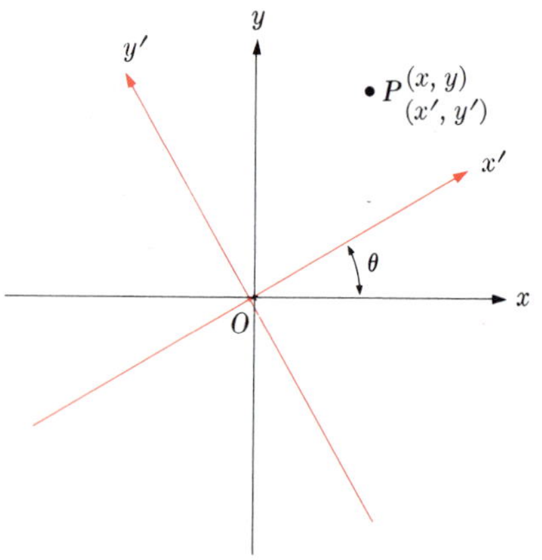

FIGURE 37

Suppose we make a transformation of coordinates from an $xy$ system to an $x'y'$ system in the following way. The origin is kept fixed, and the $x'$ and $y'$ axes are obtained by rotating the $x$ and $y$ axes counterclockwise an amount $\theta$, as shown in Fig. 37. Every point $P$ will have coordinates $(x, y)$ with respect to the original system and coordinates $(x', y')$ with respect to the new system. We now find the relationship between $(x, y)$ and $(x', y')$. Draw the lines $OP$, $AP$, and $BP$, as shown in Fig. 38, and note that $x = \overline{OA}$, $y = \overline{AP}$, $x' = \overline{OB}$, $y' = \overline{BP}$. We have

$$x = \overline{OA} = \overline{OP} \cos(\theta + \alpha), \qquad y = \overline{AP} = \overline{OP} \sin(\theta + \alpha).$$

Recalling from trigonometry* the formula for the sine and cosine of the sum of two angles, we obtain

$$\overline{OA} = x = \overline{OP} \cos \theta \cos \alpha - \overline{OP} \sin \theta \sin \alpha,$$

$$\overline{AP} = y = \overline{OP} \sin \theta \cos \alpha + \overline{OP} \cos \theta \sin \alpha.$$

---

*A review of all the basic formulas in trigonometry may be found in Appendix 1 at the end of the book.

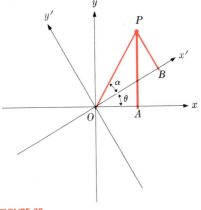

**FIGURE 38**

But $\overline{OP} \cos \alpha = \overline{OB} = x'$ and $\overline{OP} \sin \alpha = \overline{BP} = y'$. Therefore

$$x = x' \cos \theta - y' \sin \theta, \qquad y = x' \sin \theta + y' \cos \theta.$$

When two coordinate systems, $xy$ and $x'y'$, satisfy these equations, we say that they are related by a **rotation of axes**; more specifically, we can say that the $x'$ and $y'$ axes are obtained by rotating the $x$ and $y$ axes through the angle $\theta$.

To express $x'$ and $y'$ in terms of $x$ and $y$, we merely solve the above equations for $x'$ and $y'$. Multiplying the first equation by $\cos \theta$, the second by $\sin \theta$, and adding, we find

$$x \cos \theta + y \sin \theta = x'(\cos^2 \theta + \sin^2 \theta) \qquad \text{or} \qquad x' = x \cos \theta + y \sin \theta.$$

Similarly, solving for $y'$, we get the following equations.

**EQUATIONS FOR ROTATION OF AXES**

$$x' = x \cos \theta + y \sin \theta$$

$$y' = -x \sin \theta + y \cos \theta.$$

**EXAMPLE 1**   A rectangular coordinate system is rotated $\pi/3$ radians. Find the coordinates of the point $P(3, -1)$ in the new system. What is the equation of the line $2x - 3y + 1 = 0$ in the rotated system?

**Solution**   We have $\sin \pi/3 = \frac{1}{2}\sqrt{3}$, $\cos \pi/3 = \frac{1}{2}$. The equations relating the $xy$ system and the $x'y'$ system are

$$x' = \tfrac{1}{2}x + \tfrac{1}{2}\sqrt{3}y, \qquad y' = -\tfrac{1}{2}\sqrt{3}x + \tfrac{1}{2}y.$$

The coordinates of $P$ are $x' = \frac{3}{2} - \frac{1}{2}\sqrt{3}$, $y' = -\frac{3}{2}\sqrt{3} - \frac{1}{2}$. The relationships giving the $xy$ system in terms of the $x'y'$ system are

$$x = \tfrac{1}{2}x' - \tfrac{1}{2}\sqrt{3}y', \qquad y = \tfrac{1}{2}\sqrt{3}x' + \tfrac{1}{2}y';$$

and so the equation of the line in the new system is

$$2(\tfrac{1}{2}x' - \tfrac{1}{2}\sqrt{3}y') - 3(\tfrac{1}{2}\sqrt{3}x' + \tfrac{1}{2}y') + 1 = 0,$$

$$\Leftrightarrow (1 - \tfrac{3}{2}\sqrt{3})x' - (\tfrac{3}{2} + \sqrt{3})y' + 1 = 0. \qquad \square$$

We recall that the most general equation of the second degree has the form

$$Ax^2 + Bxy + Cy^2 + Dx + Ey + F = 0 \quad (A, B, C \text{ not all zero}).$$

In Section 5 we showed (when $B = 0$) how the axes could be translated so that in the new system $D = E = 0$ (except in the case of parabolas, when we could make only one of them zero).

Now we shall show that *it is always possible to rotate the coordinates in such a way that in the new system there is no $x'y'$ term.* To do this we take the equations

$$x = x' \cos \theta - y' \sin \theta, \qquad y = x' \sin \theta + y' \cos \theta,$$

and we substitute in the general equation of the second degree. Then we have

$$Ax^2 = A(x' \cos\theta - y' \sin\theta)^2,$$

$$Bxy = B(x' \cos\theta - y' \sin\theta)(x' \sin\theta + y' \cos\theta),$$

$$Cy^2 = C(x' \sin\theta + y' \cos\theta)^2,$$

$$Dx = D(x' \cos\theta - y' \sin\theta),$$

$$Ey = E(x' \sin\theta + y' \cos\theta),$$

$$F = F.$$

We add these equations and obtain (after multiplying out the right side)

$$Ax^2 + Bxy + Cy^2 + Dx + Ey + F$$
$$= A'x'^2 + B'x'y' + C'y'^2 + D'x' + E'y' + F' = 0,$$

where $A'$, $B'$, $C'$, $D'$, $E'$, and $F'$ are the abbreviations of

$$A' = A \cos^2\theta + B \sin\theta \cos\theta + C \sin^2\theta,$$

$$B' = 2(C - A) \sin\theta \cos\theta + B(\cos^2\theta - \sin^2\theta),$$

$$C' = A \sin^2\theta - B \sin\theta \cos\theta + C \cos^2\theta,$$

$$D' = D \cos\theta + E \sin\theta,$$

$$E' = -D \sin\theta + E \cos\theta,$$

$$F' = F.$$

Our purpose is to select $\theta$ so that the $x'y'$ term is missing, in order that $B'$ will be zero. Let us set $B'$ equal to zero and see what happens:

$$2(C - A) \sin\theta \cos\theta + B(\cos^2\theta - \sin^2\theta) = 0.$$

We recall from trigonometry the double-angle formulas

$$\sin 2\theta = 2 \sin\theta \cos\theta, \qquad \cos 2\theta = \cos^2\theta - \sin^2\theta,$$

and we see that $B' = 0$ becomes

$$(C - A) \sin 2\theta + B \cos 2\theta = 0$$

or

$$\cot 2\theta = \frac{A - C}{B}.$$

In other words, if we select $\theta$ so that $\cot 2\theta = (A - C)/B$ we will obtain $B' = 0$. There is always an angle $2\theta$ between $0$ and $\pi$ which solves this equation.

The next example illustrates the process.

**EXAMPLE 2**     Given the equation $8x^2 - 4xy + 5y^2 = 36$. Choose new axes by rotation so as to eliminate the $x'y'$ term; sketch the curve and locate the principal quantities.

**Solution**   We have $A = 8$, $B = -4$, $C = 5$, $D = E = 0$, $F = -36$. We select $\cot 2\theta = (A - C)/B = (8 - 5)/(-4) = -\frac{3}{4}$. This means that $2\theta$ is in the second quadrant and $\cos 2\theta = -\frac{3}{5}$.

We remember from trigonometry the half-angle formulas

$$\sin \theta = \sqrt{(1 - \cos 2\theta)/2}, \qquad \cos \theta = \sqrt{(1 + \cos 2\theta)/2},$$

and we use these to get

$$\sin \theta = \sqrt{\tfrac{4}{5}}, \qquad \cos \theta = \sqrt{\tfrac{1}{5}},$$

which gives the rotation

$$x = \frac{1}{\sqrt{5}}x' - \frac{2}{\sqrt{5}}y' = \frac{x' - 2y'}{\sqrt{5}}, \qquad y = \frac{2}{\sqrt{5}}x' + \frac{1}{\sqrt{5}}y' = \frac{2x' + y'}{\sqrt{5}}.$$

Substituting in the equation $8x^2 - 4xy + 5y^2 = 36$, we obtain

$$\tfrac{8}{5}(x' - 2y')^2 - \tfrac{4}{5}(x' - 2y')(2x' + y') + \tfrac{5}{5}(2x' + y')^2 = 36;$$

when we multiply out, we find that

$$4x'^2 + 9y'^2 = 36 \qquad \Leftrightarrow \qquad \frac{x'^2}{9} + \frac{y'^2}{4} = 1.$$

The graph is an ellipse, which is sketched in Fig. 39. The $x'y'$ coordinates of the vertices are at $(\pm 3, 0)$, and the foci are at $(\pm\sqrt{5}, 0)$. The eccentricity is $\sqrt{5}/3$. In the original system, the vertices are at

$$(3/\sqrt{5}, 6/\sqrt{5}), (-3/\sqrt{5}, -6/\sqrt{5}).$$

The foci are at $(1, 2)$, $(-1, -2)$.    ☐

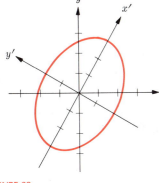

**FIGURE 39**

In Example 2 the coefficients $D$ and $E$ are zero. When a problem arises with $B$, $D$, and $E$ all different from zero, we may eliminate them by performing in succession a rotation (eliminating $B$) and a translation (eliminating the $D$ and $E$ terms). The technique of performing several transformations in succession occurs frequently in mathematical problems.

**EXAMPLE 3**   Given the equation

$$4x^2 - 12xy + 9y^2 - 52x + 26y + 81 = 0,$$

reduce it to standard form by eliminating $B$, $D$, and $E$. Identify the curve and locate the principal quantities.

**Solution**   We first rotate and choose $\theta$ so that

$$\cot 2\theta = \frac{A - C}{B} = \frac{4 - 9}{-12} = \frac{5}{12}.$$

Then $2\theta$ is in the first quadrant and $\cos 2\theta = \frac{5}{13}$. From the half-angle formulas (as in Example 2),

$$\cos \theta = \sqrt{\tfrac{9}{13}}, \qquad \sin \theta = \sqrt{\tfrac{4}{13}}.$$

The desired rotation of coordinates is

$$x = \frac{1}{\sqrt{13}}(3x' - 2y'), \qquad y = \frac{1}{\sqrt{13}}(2x' + 3y').$$

Substituting in the given equation, we obtain

$$\tfrac{4}{13}(9x'^2 - 12x'y' + 4y'^2) - \tfrac{12}{13}(6x'^2 + 5x'y' - 6y'^2)$$
$$+ \tfrac{9}{13}(4x'^2 + 12x'y' + 9y'^2) - 4\sqrt{13}(3x' - 2y')$$
$$+ 2\sqrt{13}(2x' + 3y') + 81 = 0,$$

and, after simplification,

$$13y'^2 - 8\sqrt{13}x' + 14\sqrt{13}y' + 81 = 0.$$

Note that in the process of eliminating the $x'y'$ term we also eliminated the $x'^2$ term. We now realize that the curve must be a parabola. To translate the coordinates properly, we complete the square in $y$. This yields

$$13\left( y'^2 + \frac{14}{13}\sqrt{13}y' \qquad\right) = 8\sqrt{13}x' - 81$$
$$\Leftrightarrow 13\left( y' + \frac{7}{\sqrt{13}}\right)^2 = 8\sqrt{13}\left( x' - \frac{4}{\sqrt{13}}\right).$$

The translation of axes,

$$x'' = x' - \frac{4}{\sqrt{13}}, \qquad y'' = y' + \frac{7}{\sqrt{13}},$$

leads to the equation

$$y''^2 = \frac{8}{\sqrt{13}}x''.$$

In the $x''y''$ system, $p = 4/\sqrt{13}$, the vertex is at the origin, the focus is at $(2/\sqrt{13}, 0)$, the directrix is the line $x'' = -2/\sqrt{13}$, and the $x''$ axis is the axis of the parabola. In the $x'y'$ system, the focus is at $(6/\sqrt{13}, -7/\sqrt{13})$, and the directrix is the line $x' = 2/\sqrt{13}$. As for the original $xy$ system, we find that the focus is at $(\tfrac{32}{13}, -\tfrac{9}{13})$. The directrix is the line $3x + 2y = 2$. The parabola and all sets of axes are sketched in Fig. 40; the points $(3, 2)$ and $(5, 1)$ in the sketch are merely aids in plotting the axes. $\qquad\square$

In a rotation of the coordinates, the general equation of the second degree,

$$Ax^2 + Bxy + Cy^2 + Dx + Ey + F = 0,$$

goes into the equation

$$A'x^2 + B'xy + C'y^2 + D'x + E'y + F = 0,$$

with

$$A' = A\cos^2\theta + B\sin\theta\cos\theta + C\sin^2\theta,$$
$$B' = 2(C - A)\sin\theta\cos\theta + B(\cos^2\theta - \sin^2\theta),$$
$$C' = A\sin^2\theta - B\sin\theta\cos\theta + C\cos^2\theta.$$

The quantity $A' + C'$, when calculated, is

$$A' + C' = A(\cos^2\theta + \sin^2\theta) + C(\sin^2\theta + \cos^2\theta) = A + C.$$

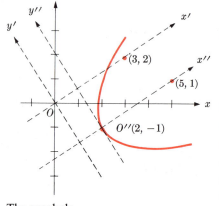

The parabola
$4x^2 - 12xy + 9y^2 - 52x + 26y + 81 = 0$,
with rotation and translation of
the coordinate axes

FIGURE 40

In other words, even though $A$ changes to $A'$ and $C$ changes to $C'$ when a rotation through *any* angle is made, the quantity $A + C$ does not change at all. We say that $A + C$ is **invariant** under a rotation of coordinates.

If we compute the expression $B'^2 - 4A'C'$ (a tedious computation) and use some trigonometry, we find that

$$B'^2 - 4A'C' = B^2 - 4AC. \tag{1}$$

In other words, $B^2 - 4AC$ is also an *invariant* under rotation of axes. The quantity $B^2 - 4AC$ can be used to determine which type of curve the general equation (1) describes as Theorem 7 below states. Thus it is important enough to merit a name: The quantity $B^2 - 4AC$ is called the **discriminant**.

It can readily be checked that $A + C$ and $B^2 - 4AC$ are invariant under translation of axes, and therefore we can formulate the following theorem for *general equations of the second degree*.

---

**THEOREM 7**    *For the general equation of the second degree, $Ax^2 + Bxy + Cy^2 + Dx + Ey + F = 0$, there are three possibilities depending on the discriminant:*
  i) *if $B^2 - 4AC < 0$, the curve is an ellipse, a circle, a point, or there is no curve;*
  ii) *if $B^2 - 4AC > 0$, the curve is a hyperbola or two intersecting straight lines;*
  iii) *if $B^2 - 4AC = 0$, the curve is a parabola, two parallel lines, one line, or there is no curve.*

---

The circle, ellipse, parabola, and hyperbola are often called **conic sections**, for all of them can be obtained as sections cut from a right circular cone by planes. The cone is thought of as extending indefinitely on both sides of its vertex; the part of the cone on one side of the vertex is called a **nappe**.

If the plane intersects only one nappe, as in Fig. 41(a), the curve of the intersection is an ellipse. (A circle is a special case of the ellipse, and occurs when the plane is perpendicular to the axis of the cone.) If the plane is parallel to one of the generators of the cone, the intersection is a parabola, as shown in Fig. 41(b). If the plane intersects both nappes, the curve is a hyperbola, one branch coming from each nappe, as in Fig. 41(c).

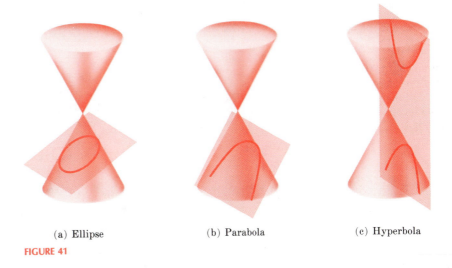

(a) Ellipse          (b) Parabola          (c) Hyperbola

FIGURE 41

Certain degenerate cases also occur; the graph is two intersecting lines when the plane intersects both nappes and also passes through the vertex. If the plane contains one of the generators, the graph of the intersection is a single straight line. Finally, the graph is a single point if the plane contains the vertex and does not intersect either nappe of the cone. The degenerate graph of two parallel lines cannot be obtained as a plane section of a cone.

# 6 PROBLEMS

In Problems 1 through 17, in each case change from an $xy$ system to an $x'y'$ system such that the $x'y'$ term is missing. For the ellipses and hyperbolas, find the coordinates of the vertices. For each parabola find the $xy$ coordinates of the focus and the $xy$ equation of the directrix. If the graph consists of lines, find their $xy$ equations.

1 $x^2 + 4xy + 4y^2 = 9$        2 $7x^2 - 4xy + 4y^2 = 240$

3 $2x^2 + 3xy - 2y^2 = 25$      4 $7x^2 - 6xy - y^2 = 0$

5 $x^2 + 2xy + y^2 - 8x + 8y = 0$

6 $x^2 + 4xy + y^2 + 32 = 0$

7 $8x^2 + 12xy + 13y^2 = 884$

8 $x^2 - 4xy + 4y^2 - 40x - 20y = 0$

9 $x^2 - 2xy = 10$            10 $xy = 4$

11 $2xy - 3y^2 = 5$          12 $xy = -3$

13 $9x^2 - 24xy + 16y^2 = 18x + 101y - 19$

14 $x^2 - 4xy - 2y^2 = 6$      15 $x^2 + 8xy + 7y^2 = 1$

16 $3x^2 + 2\sqrt{3}xy + y^2 = 12x - 12\sqrt{3}y + 24$

17 $16x^2 - 24xy + 9y^2 - 30x - 40y = 0$

In Problems 18 through 31, in each case rotate the axes to an $x'y'$ system such that the $x'y'$ term is missing. Then translate the axes so that the first-degree terms are absent. Sketch the graphs and identify the principal quantities.

18 $3x^2 + 10xy + 3y^2 - 2x - 14y - 5 = 0$

19 $4x^2 - 8xy - 2y^2 + 20x - 4y + 15 = 0$

20 $4x^2 + 4xy + y^2 - 24x + 38y - 139 = 0$

21 $16x^2 - 24xy + 9y^2 + 56x - 42y + 49 = 0$

22 $12x^2 + 24xy + 19y^2 - 12x - 40y + 31 = 0$

23 $4x^2 - 4\sqrt{5}xy + 5y^2 + 8x - 4\sqrt{5}y - 21 = 0$

24 $31x^2 + 10\sqrt{3}xy + 21y^2 - (124 - 40\sqrt{3})x + (168 - 20\sqrt{3})y + 316 - 80\sqrt{3} = 0$

25 $x^2 + 2xy + y^2 - 4x - 4y + 4 = 0$

26 $x^2 - \sqrt{3}xy + 2\sqrt{3}x - 3y - 3 = 0$

27 $3xy - 4y^2 + x - 2y + 1 = 0$

28 $11x^2 - 24xy + 4y^2 + 6x + 8y = -15$

29 $17x^2 + 18xy - 7y^2 = 80$

30 $x^2 - 4xy - 2y^2 = 6$       31 $5x^2 + 12xy = 4$

32 Prove that a second-degree equation with an $xy$ term in it can never represent a circle.

33 Prove that a second-degree equation with $D$ and $E$ absent (i.e., no $x$ and $y$ terms) cannot be a parabola.

34 Prove that $B^2 - 4AC$, the discriminant, is invariant under rotation and translation of axes.

35 Given the transformation of coordinates

$$x' = ax + by, \qquad y' = cx + dy,$$

with $a$, $b$, $c$, and $d$ numbers such that $ad - bc$ is positive. If the general equation of the second degree undergoes such a transformation, what can be said about $B^2 - 4AC$ (the discriminant)?

36 Give an example of an equation of the second degree such that the graph degenerates to (a) two intersecting straight lines; (b) two parallel lines; (c) one line; (d) a point.

37 Prove Theorem 7.

## CHAPTER 11

### REVIEW PROBLEMS

In Problems 1 through 6, find in each case the distance from the given point to the given line.

**1** $(1, 2)$　$3x + 4y - 6 = 0$　　　**2** $(-1, 3)$　$2x = 4y - 3$

**3** $(1, 1)$　$7x - 6y - 1 = 0$　　　**4** $(-1, 2)$　$8x = 7y - 22$

**5** $(2, 1)$　$x + y = 0$　　　**6** $(2, 1)$　$x - 3 = 0$

In Problems 7 through 12, find in each case $\tan \phi$, where $\phi$ is the angle from $L_1$ to $L_2$.

**7** $L_1: 2x + 3y - 2 = 0$;　$L_2: 2x - 4y + 1 = 0$

**8** $L_1: x + 7y - 1 = 0$;　$L_2: x - 7y - 1 = 0$

**9** $L_1: 3x - 4y + 6 = 0$;　$L_2: 3x - 2y - 2 = 0$

**10** $L_1: 3x - y = 0$;　$L_2: 2x + 4y = 5$

**11** $L_1: x + y + 1 = 0$;　$L_2: x - y + 1 = 0$

**12** $L_1: 3x + 3y + 5 = 0$;　$L_2: 3x - 3y + 4 = 0$

In Problems 13 through 38, determine whether the equation represents a parabola, a circle, an ellipse, or a hyperbola. In each case, sketch the graph and identify the principal quantities (*parabola:* focus, vertex, directrix, and axis; *circle:* center and radius; *ellipse:* lengths of the major and minor axes, foci, vertices, and eccentricity; *hyperbola:* lengths of the transverse and conjugate axes, foci, vertices, eccentricity, and asymptotes).

**13** $x^2 + 8y = 0$　　　　　**14** $y^2 = -20x$

**15** $y^2 - 12x + 24 = 0$　　　**16** $y^2 + 8x + 16 = 0$

**17** $x^2 + y^2 + 3x - 4y = 0$

**18** $2x + x^2 + y^2 + 6y - 4 = 0$

**19** $x^2 + y^2 - 24 = 0$　　　**20** $x^2 + 2y^2 = 6$

**21** $4x^2 + 9y^2 = 1$　　　　**22** $\dfrac{x^2}{9} + \dfrac{y^2}{16} - 25 = 0$

**23** $44 - 3x^2 - 7y^2 = 0$　　　**24** $x^2 - y^2 = 1$

**25** $-3x^2 + y^2 - 6x = 0$

**26** $4x^2 - 8x - y^2 + 6y - 1 = 0$

**27** $\dfrac{(x-1)^2}{16} - \dfrac{(y-3)^2}{9} = 1$

**28** $4x^2 - 8x + y^2 + 4y - 8 = 0$

**29** $x^2 + y^2 - 4x + 3y - 9 = 0$

**30** $y^2 = 12(y - x)$　　　　**31** $y^2 - x^2 = 1$

**32** $3x - x^2 - y^2 - 7y + 10 = 0$

**33** $16(x - 2)^2 + 25(y - 3)^2 = 400$

**34** $8x^2 + 7x + 3y + 8y^2 = 0$

**35** $2x^2 + 2x + y - 2 = 0$

**36** $9(x - 2)^2 - 16(y - 3)^2 = 144$

**37** $4x^2 + y^2 - 6y - 5 = 0$　　**38** $4x^2 + 6x - y + 2 = 0$

In Problems 39 through 50, identify whether the equation represents a point, an ellipse, a circle, a hyperbola, a parabola, two lines, one line, or no curve at all.

**39** $xy = 1$　　　　　　　**40** $3x^2 + 6xy + 3y^2 + 1 = 0$

**41** $x^2 - 2x + 3y^2 - 6y + 1 = 0$

**42** $2x^2 - y - 2x - 26 = 0$

**43** $x^2 - y^2 - 4x + 14y - 45 = 0$

**44** $41x^2 - 24xy + 34y^2 - 25 = 0$

**45** $5x^2 - 8xy + 5y^2 = 9$

**46** $11x^2 + 10\sqrt{3}xy + y^2 = 4$

**47** $16x^2 - 24xy + 9y^2 - 60x - 80y + 100 = 0$

**48** $x^2 + 4xy + 4y^2 + 6\sqrt{5}x - 18\sqrt{5}y + 45 = 0$

**49** $40x^2 - 36xy + 25y^2 - 8\sqrt{13}x - 12\sqrt{13}y = 0$

**50** $64x^2 - 240xy + 225y^2 + 1020x - 544y = 0$

# 12

## PARAMETRIC EQUATIONS. ARC LENGTH. POLAR COORDINATES

We have been considering curves in the plane which are the graphs of functions. By introducing a quantity called a parameter, it is possible to represent curves which are more complicated that those we have studied thus far. In this chapter we show how the methods of calculus combined with parametric equations expand enormously our ability to calculate geometrical quantities such as arc length, area, and volume. In addition, we show how a transformation to polar coordinates enables us to treat in a simple manner types of curves which are inordinately complicated in rectangular coordinates.

---

1

### PARAMETRIC EQUATIONS

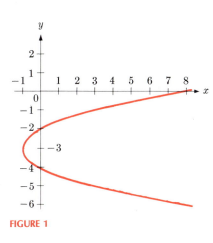

**FIGURE 1**

We consider the two equations

$$x = t^2 + 2t, \qquad y = t - 2,$$

involving the quantities $x$, $y$, and $t$. Each value of $t$ determines a value of $x$ and one of $y$ and therefore a point $(x, y)$ in the plane. The totality of these points obtained by letting $t$ take on all possible values is a relation in $R^2$. The graph of this relation may be exhibited by the following method. We set up a table of values for $t$, $x$, and $y$ by letting $t$ take on various values and then computing $x$ and $y$ from the given equations. The points $(x, y)$ are plotted in the usual way on a rectangular coordinate system. The table is shown below and the graph is sketched in Fig. 1. The $t$ scale is completely separate and does not appear in the graph. The quantity $t$ is called a **parameter** and the above equations are called the **parametric equations** of the graph shown in Fig. 1.

| $t$ | 0 | 1 | 2 | 3 | 4 | $-1$ | $-2$ | $-3$ | $-4$ |
|---|---|---|---|---|---|---|---|---|---|
| $x$ | 0 | 3 | 8 | 15 | 24 | $-1$ | 0 | 3 | 8 |
| $y$ | $-2$ | $-1$ | 0 | 1 | 2 | $-3$ | $-4$ | $-5$ | $-6$ |

In general, if $f$ and $g$ are functions with the same domain $S$ in $R^1$, and if we write

$$x = f(t), \qquad y = g(t),$$

then we say that the two equations form a set of **parametric equations**. The **graph** of the parametric equations is the set of points in the $xy$ plane which we get when $t$ takes on all possible values in the domain $S$.

Let us now consider the particularly simple case in which both $f$ and $g$ are *linear* functions. That is, we suppose that

$$f(t) = x_0 + at, \qquad g(t) = y_0 + bt,$$

where $x_0$, $y_0$, $a$, and $b$ are numbers. We establish the following result.

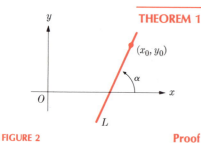

FIGURE 2

**THEOREM 1**

*The set of parametric equations*

$$x = x_0 + at, \qquad y = y_0 + bt, \tag{1}$$

*so long as $a$ and $b$ are not both zero, has as its graph a straight line in the $xy$ plane. This line (denoted by $L$ in Fig. 2) passes through the point $(x_0, y_0)$. If $a \neq 0$, then $L$ has slope $b/a$. If $a = 0$, the line is vertical.*

**Proof** Suppose that $a \neq 0$. Solving $x = x_0 + at$ for $t$ and substituting in the equation $y = y_0 + bt$, we find that each point on the graph of (1) satisfies the equation

$$y - y_0 = \frac{b}{a}(x - x_0), \tag{1'}$$

which we recognize as the equation of a straight line through the point $(x_0, y_0)$ with slope $b/a$. Moreover, if $(x, y)$ is any point on the line $(1')$ and we define $t = (x - x_0)/a$, then $x$ and $y$ satisfy Equations (1).

If $a = 0$, we see that $x = x_0$ always and, since $b \neq 0$, $y$ may have any value. Thus $L$ must be a vertical line through $(x_0, y_0)$, and the theorem is proved. □

*Remark.* The converse of Theorem 1 is not difficult to establish. That is, *if a line $L$ is given by an equation of the form $y - y_0 = m(x - x_0)$, or if the line is vertical, then it is possible to represent $L$ parametrically by Equations (1).* (See Problem 29 at the end of this section.)

If $k$ is any number different from zero, the equations

$$x = x_0 + kat, \qquad y = y_0 + kbt \tag{2}$$

represent the same line as Equations (1). To see this we observe that if $a \neq 0$, then the slope of the line determined by (2) is $kb/ka = b/a$ and the line goes through the point $(x_0, y_0)$. Also, if $a = 0$, the line must be vertical and pass through $(x_0, y_0)$. Denoting the inclination of $L$ by $\alpha$ (see Fig. 2), we may select

$k$ so that the parametric equations of $L$ have the form

$$x = x_0 + (\cos \alpha)t, \qquad y = y_0 + (\sin \alpha)t. \qquad (3)$$

Once $a$ and $b$ are given, we verify easily that the selections

$$k = 1/\sqrt{a^2 + b^2} \qquad \text{for } b > 0,$$
$$k = -1/\sqrt{a^2 + b^2} \qquad \text{for } b < 0,$$
$$k = 1/a \qquad \text{for } b = 0,$$

transform the parametric equations (2) into the form (3). Thus we see (by the Remark after Theorem 1) that *every straight line in the plane may be represented parametrically by Equations (3).* We call them the **parametric equations of a line in standard form.**

**EXAMPLE 1**    Find parametric equations of the line $L$ through the points $(2, -1)$ and $(4, 7)$. Also determine the parametric equations in standard form.

**Solution**    We may select $(2, -1)$ as $(x_0, y_0)$. The slope of $L$ is

$$\frac{7 - (-1)}{4 - 2} = 4.$$

We may choose $a$ and $b$ arbitrarily so long as $b/a = 4$. Selecting $b = 4$ and $a = 1$, we substitute in Equations (1) to get the parametric equations

$$x = 2 + t, \qquad y = -1 + 4t.$$

Since $b > 0$, we choose $k = 1/\sqrt{1 + 16} = 1/\sqrt{17}$ and the standard form of the parametric equations is

$$x = 2 + \frac{1}{\sqrt{17}}t, \qquad y = -1 + \frac{4}{\sqrt{17}}t.$$

We note that since $\cos \alpha = 1/\sqrt{17}$, $\sin \alpha = 4/\sqrt{17}$, the inclination $\alpha$ of $L$ is an acute angle whose value is $\arccos (1/\sqrt{17})$.    □

When we are given a pair of parametric equations it is sometimes possible to *eliminate the parameter* by solving for $t$ in one of the equations and then substituting in the other. In the parametric equations

$$x = t^2 + 2t, \qquad y = t - 2,$$

which we considered at the beginning of the section, we may solve the second equation for $t$, getting $t = y + 2$, and then substitute in the first equation. We obtain

$$x = (y + 2)^2 + 2(y + 2) = y^2 + 6y + 8,$$

and we recognize the equation $x = y^2 + 6y + 8$ as that of a parabola, as indeed Fig. 1 shows. We have shown that any point on the graph of the parametric equations satisfies this equation. Moreover, if $(x, y)$ is any point satisfying this equation, we may define $t = y + 2$ and find that $x$ and $y$ satisfy the parametric equations. It can happen, however that the graph of the parametric equations is not the same as the graph of the equation in $x$ and $y$

obtained by "eliminating the parameter $t$." For example, the graph of the parametric equations

$$x = \cos t, \qquad y = \cos^2 t \qquad (4)$$

is just part of the graph of the parabola $y = x^2$ obtained by eliminating $t$. In fact, since $-1 \le \cos t \le 1$ for all $t$, the graph of (4) is that part of the parabola $y = x^2$ for which $-1 \le x \le 1$. See Problem 30 at the end of this section. (Also, see Fig. 3.)

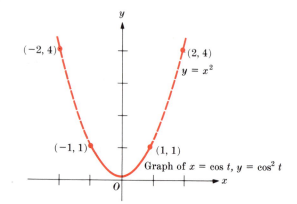

FIGURE 3          Graphs of $x = \cos t$, $y = \cos^2 t$, and $y = x^2$

The equations

$$x = t^7 + 3t^2 - 1,$$

$$y = e^t + 2t^2 - 3\sqrt{t},$$

are a pair of parametric equations. Because of the square-root sign, it is clear that the domain for $t$ is restricted to nonnegative real numbers. It is not possible to eliminate the parameter in any simple way, since solving either equation for $t$ would be extremely difficult.

**EXAMPLE 2**   Plot the graph of the parametric equations

$$x = t^2 + 2t - 1, \qquad y = t^2 + t - 2.$$

Eliminate the parameter, if possible.

**Solution**   We construct the following table:

| $t$ | $-4$ | $-3$ | $-2$ | $-1$ | $0$ | $1$ | $2$ | $3$ |
|-----|------|------|------|------|-----|-----|-----|-----|
| $x$ | 7 | 2 | $-1$ | $-2$ | $-1$ | 2 | 7 | 14 |
| $y$ | 10 | 4 | 0 | $-2$ | $-2$ | 0 | 4 | 10 |

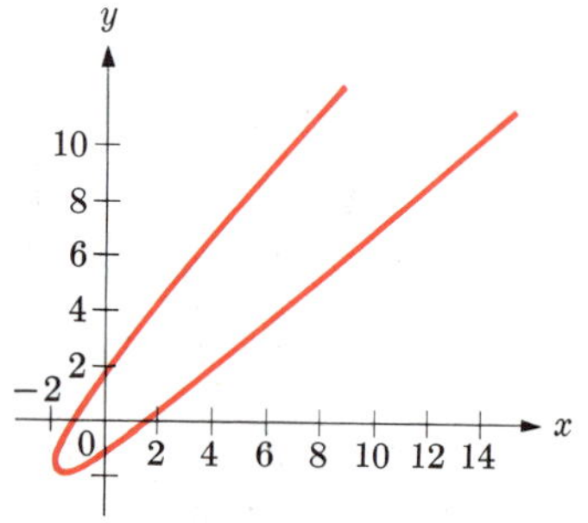

By plotting these points, we obtain the curve shown in Fig. 4. To eliminate the parameter, we first subtract the equations, obtaining

$$x - y = t + 1 \quad \Leftrightarrow \quad t = x - y - 1.$$

Substituting for $t$ in the second of the equations, we find that

$$y = (x - y - 1)^2 + (x - y - 1) - 2,$$

FIGURE 4

and, simplifying,

$$x^2 - 2xy + y^2 - x - 2 = 0.$$

**EXAMPLE 3**    A projectile moves approximately according to the law

$$x = (v_0 \cos \alpha)t, \qquad y = (v_0 \sin \alpha)t - 5t^2,$$

where $v_0$ and $\alpha$ are constants and $t$ is the time, in seconds, after the projectile is fired. The equations give the rectangular coordinates $x$, $y$ (in meters) of the center of the projectile in the vertical plane of motion with the muzzle of the gun at the origin of the coordinate system, the $x$ axis horizontal, and the $y$ axis vertical; $v_0$ is the *muzzle velocity*, i.e., the velocity of the projectile at the instant it leaves the gun; $\alpha$ is the angle of inclination of the projectile as it leaves the gun (Fig. 5). If $v_0 = 100$ m/sec, $\cos \alpha = \frac{3}{5}$, and $\sin \alpha = \frac{4}{5}$, what are the coordinates of the center of the projectile at times $t = 1, 2, 3, 4$, and 5? Find the time $T$ when the projectile hits the ground, and find the distance $R$ from the muzzle of the gun to the place where the projectile strikes the ground.

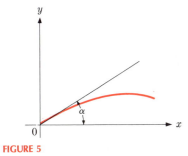

FIGURE 5

**Solution**    We have $v_0 \cos \alpha = 60$, $v_0 \sin \alpha = 80$, and so

$$x = 60t, \qquad y = 80t - 5t^2.$$

The projectile hits the ground when $y = 0$. Therefore $0 = 80t - 5t^2$ and solving gives $t = 0, 16$. Consequently, $T = 16$. To find $R$, we insert $t = 16$ in $x = 60t$, obtaining $R = 960$ m. We construct the table:

| $t$ | 0 | 1 | 2 | 3 | 4 | 5 | 6 | 7 | 8 | 16 |
|---|---|---|---|---|---|---|---|---|---|---|
| $x$ | 0 | 60 | 120 | 180 | 240 | 300 | 360 | 420 | 480 | 960 |
| $y$ | 0 | 75 | 140 | 195 | 240 | 275 | 300 | 315 | 320 | 0 |

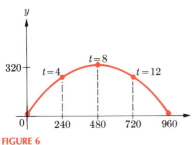

FIGURE 6

Plotting the points yields the curve shown in Fig. 6; $t$ is restricted to the interval $0 \le t \le 16$.

A circular hoop starts rolling along a stretch of level ground. A point on the rim of the hoop has a mark on it. We wish to find the path traced out by the marked point. Figure 7 shows the hoop in a number of different positions as it

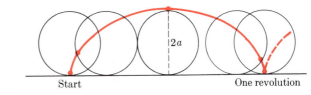

Start                              One revolution

FIGURE 7

rolls along. The curve traced out by the marked point, which we denote by $P$, may be expressed in terms of parametric equations. Let $a$ be the radius of the hoop, and suppose that when the hoop begins rolling the point $P$ is on the ground at the point labeled $O$ in Fig. 8. Figure 8 shows the position of the point $P$ after the hoop has turned through an angle $\theta$. Since the hoop is assumed to roll without slipping, we have, from the diagram,

$$|OM| = \text{arc } \overset{\frown}{MP} = a\theta, \quad \theta \text{ in radians.}$$

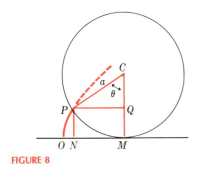

FIGURE 8

From $\triangle CPQ$ we read off

$$|PQ| = a \sin \theta, \quad 0 \le \theta \le \frac{\pi}{2},$$

$$|QC| = a \cos \theta, \quad 0 \le \theta \le \frac{\pi}{2}.$$

Denoting the coordinates of $P$ by $(x, y)$, we see from Fig. 8 that

$$x = |ON| = |OM| - |NM| = \text{arc } \widehat{MP} - |PQ| = a\theta - a \sin \theta,$$

$$y = |NP| = |MC| - |QC| = a - a \cos \theta.$$

Even though these equations were derived for $\theta$ between 0 and $\pi/2$, it can be shown that for all values of $\theta$ the equations

$$x = a(\theta - \sin \theta), \qquad y = a(1 - \cos \theta) \tag{5}$$

represent the path of the marked point on the rim. This curve is called a **cycloid**.

While the letter $t$ is commonly used as a parameter, any letter can serve as well. In the cycloid given by (5), we use the angle $\theta$ as the parameter instead of $t$.

The cycloid is a particularly good example of a curve which is obtained without too much difficulty by use of parametric equations, while any attempt to find the relation between $x$ and $y$ without resorting to a parameter would lead to an almost insurmountable problem.

## 1  PROBLEMS

In Problems 1 through 6, in each case find the set of parametric equations in standard form for the line $L$ determined by the given conditions.

1  $L$ passes through the points $(3, -7)$ and $(-4, 6)$.

2  $L$ passes through the points $(-3, 1)$ and $(-3, 8)$.

3  $L$ passes through $(4, -2)$ and is parallel to the line with equation $2x + 5y = 1$.

4  $L$ passes through $(-1, -3)$ and is perpendicular to the line with equation $x + 3y = 2$.

5  $L$ has slope $\frac{2}{3}$ and passes through the intersection of the lines $2x - 3y - 3 = 0$, $x + 2y = 5$.

6  $L$ passes through $(2, -6)$ and is perpendicular to the line $y = 7$.

In Problems 7 through 28, plot each curve as in Example 2; in each case eliminate the parameter and get an equation relating $x$ and $y$ (which is satisfied by every point on the graph).

7  $x = 2t, y = -5t$

8  $x = t - 1, y = t^2$

9  $x = t, y = 1/t$

10  $x = 2 + (s^2/2),$
     $y = -1 + \frac{1}{8}s^3$

11  $x = -1 + \cos \theta$
     $y = 2 + 2 \sin \theta$

12  $x = 3 \cos \theta,$
     $y = 2 \sin \theta$

13  $x = 2 \cos^3 \theta, y = 2 \sin^3 \theta$

14  $x = 3 \sec t, y = 2 \tan t$

15  $x = t^2 + 2t + 3,$
     $y = t^2 + t - 1$

16  $x = t^2 + t + 1,$
     $y = \frac{1}{2}t^2 + t - 1$

17  $x = \dfrac{20t}{4 + t^2},$

     $y = \dfrac{5(4 - t^2)}{4 + t^2}$

18  $x = \dfrac{3(2 - t)^2(2 + t)}{6t^2 + 8},$

     $y = \dfrac{3(2 - t)(2 + t)^2}{6t^2 + 8}$

19  $x = e^t, y = e^{-t}$

20  $x = \ln s, y = e^{2s}$

21  $x = e^t + e^{-t}, y = e^t - e^{-t}$

22  $x = t - 2, y = 2t + 3$

23  $x = t^2 + 4, y = t^2 - 4$

24  $x = 2t^2 - 1, y = 3t + 4$

25  $x = \sqrt{t}, y = 2t + 6$

26  $x = \cos 2t, y = \sin t$

27  $x = t^2, y = 2 \ln t$

28  $x = t, y = \sqrt{t^2 - 4}$ ($|t| \ge 2$)

**29** Prove that every line in the plane has a parametric representation in the form

$$x = x_0 + at, \qquad y = y_0 + bt.$$

(*Hint:* Consider two cases: one in which the line is vertical and one in which the line is not vertical.)

**30** Plot the graph of each of the following pairs of parametric equations.

  i) $x = \cos t$,         $y = \cos^2 t$
 ii) $x = \cos^2 t$,      $y = \cos^4 t$
iii) $x = t^2$,           $y = t^4$
iv) $x = -\sqrt{-t}$,   $y = -t$

Draw the graph of $y = x^2$ and compare each of the above graphs with this graph. Also find two more sets of parametric equations which yield a part of the parabola $y = x^2$ different from (i) through (iv) above.

**31** Assuming that the equations of Example 3 hold, and given that $v_0 = 500$ m/sec, $\cos \alpha = \frac{4}{5}$, $\sin \alpha = \frac{3}{5}$, find $x$ and $y$ for $t = 2, 4, 6, 8,$ and $10$. Also find $T$ and $R$ and plot the trajectory.

**32** Assuming that the equations of Example 3 hold, find $T$ and $R$ in terms of $v_0$ and $\alpha$. Find $y$ in terms of $x$ (i.e., eliminate the parameter).

**33** A wheel of radius $a$ rolls without slipping along a level stretch of ground. A point on one of the spokes of the wheel is marked. Find the path traced by this point, given that it is at a distance $b$ from the center. (Of course, $b < a$.)

**34** Find the graph of a point $P$ on a circle of radius $a$ which rolls without slipping on the inside of a circle of radius $4a$. Choose the center of the large circle as origin, choose the positive $x$ axis through a point where $P$ touches the large circle (M), and choose the parameter $\theta$ as the angle $MOC$, where $C$ is the center of the small circle (Fig. 9). (*Hint:* Note that arc $\overset{\frown}{BHP} = $ arc $\overset{\frown}{MB}$.)

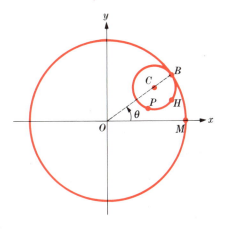

FIGURE 9

**35** A string is wound about a circle of radius $a$. The path traced by the end of the string as it is unwound is called the **involute of the circle.** Refer to Fig. 10 and show that the equations of the involute are

$$x = a \cos \theta + a\theta \sin \theta$$

$$y = a \sin \theta - a\theta \cos \theta.$$

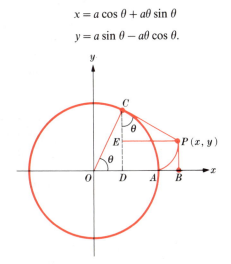

FIGURE 10

**36** Plot the graph of the equations

$$x = a \sin \theta, \qquad y = a \tan \theta (1 + \sin \theta).$$

Eliminate the parameter and show that the line $x = a$ is an asymptote. The curve is called a **strophoid.**

**37** A circle of radius $a$ is drawn tangent to the $x$ axis as shown in Fig. 11. The line $OA$ intersects the circle at point $B$. Then the projection of $AB$ on the vertical line through $A$ is the segment $AP$. The graph of the point $P$ as $B$ moves around the circle is called the **Witch of Agnesi.** Show that the curve is given by the parametric equations

$$x = 2a \cot \theta, \qquad y = 2a \sin^2 \theta.$$

Eliminate the parameter and show that the $x$ axis is an asymptote.

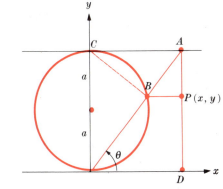

FIGURE 11

## 2

## DERIVATIVES AND PARAMETRIC EQUATIONS. ARC LENGTH

In previous chapters we learned how to use the first and second derivatives of functions as aids in drawing graphs. Such aids can also be used with parametric equations, even when the parameter cannot be eliminated. Suppose we have the parametric equations

$$x = f(t), \qquad y = g(t),$$

which represent a graph in the $xy$ plane. Usually this graph is a curve in the plane, but only rarely will the graph represent a function. No matter how complicated the shape of the curve may be, if it is "smooth," then a tangent line may be drawn at each point (see Fig. 12). The slope of the curve can be found if we can compute $dy/dx$. When $x$ and $y$ are given parametrically, this derivative may be found by the Chain Rule. We have

$$\frac{dy}{dx} = \frac{dy}{dt} \cdot \frac{dt}{dx} = \frac{dy/dt}{dx/dt}.$$

This derivative will be given in terms of $t$. If $t$ cannot be expressed in terms of $x$ or $y$—as is frequently the case—the process of getting the second derivative requires some explanation. The idea is to use the Chain Rule again. We write

$$\frac{d^2y}{dx^2} = \frac{d}{dx}\left(\frac{dy}{dx}\right) = \frac{d}{dt}\left(\frac{dy}{dx}\right)\frac{dt}{dx} = \frac{d}{dt}\left(\frac{dy}{dx}\right) \div \frac{dx}{dt}.$$

Since $dy/dx$ is given in terms of $t$, finding

$$\frac{d}{dt}\left(\frac{dy}{dx}\right)$$

is a routine matter. Furthermore, we calculated $dx/dt$ previously, when we obtained $dy/dx$. We illustrate the technique of differentiating parametric equations with three examples.

**EXAMPLE 1**   Find $dy/dx$, $d^2y/dx^2$, and $d^3y/dx^3$, given that $x = t^2 + 3t - 2$, $y = 2 - t - t^2$.

**Solution**   We have

$$\frac{dx}{dt} = 2t + 3, \qquad \frac{dy}{dt} = -1 - 2t.$$

Therefore

$$\frac{dy}{dx} = \frac{dy/dt}{dx/dt} = \frac{-(2t+1)}{2t+3}.$$

The second derivative is given by

$$\frac{d^2y}{dx^2} = \frac{\dfrac{d}{dt}\left(\dfrac{dy}{dx}\right)}{dx/dt},$$

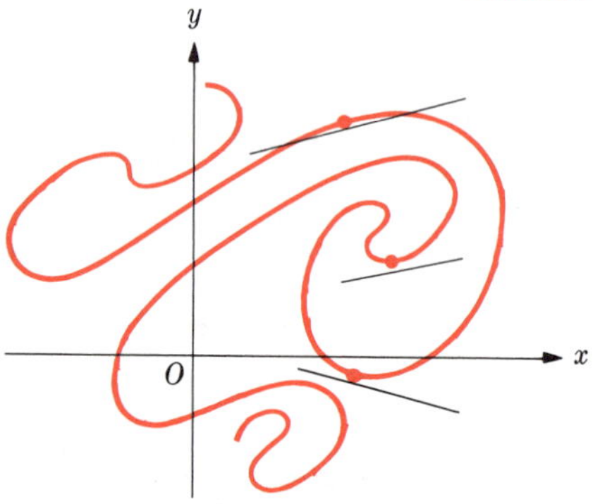

FIGURE 12

and since

$$\frac{d}{dt}\left(\frac{dy}{dx}\right) = \frac{d}{dt}\left(-\frac{2t+1}{2t+3}\right) = -\frac{(2t+3)(2) - (2t+1)(2)}{(2t+3)^2} = \frac{-4}{(2t+3)^2},$$

we can combine this with our previous calculation that $\dfrac{dx}{dt} = 2t + 3$ to get

$$\frac{d^2y}{dx^2} = -\frac{4/(2t+3)^2}{2t+3} = \frac{-4}{(2t+3)^3}.$$

We have not derived a formula for the third derivative, but the process is analogous. We write

$$\frac{d^3y}{dx^3} = \frac{d}{dx}\left(\frac{d^2y}{dx^2}\right) = \frac{d}{dt}\left(\frac{d^2y}{dx^2}\right) \cdot \frac{dt}{dx} = \frac{\dfrac{d}{dt}\left(\dfrac{d^2y}{dx^2}\right)}{dx/dt}.$$

For the numerator, we obtain

$$\frac{d}{dt}\left(\frac{d^2y}{dx^2}\right) = \frac{d}{dt}\left(\frac{-4}{(2t+3)^3}\right) = \frac{24}{(2t+3)^4},$$

and so

$$\frac{d^3y}{dx^3} = \frac{24/(2t+3)^4}{2t+3} = \frac{24}{(2t+3)^5}.$$ □

**EXAMPLE 2**   Given that $x = x(t) = t^2 - t$, $y = y(t) = t^3 - 3t$. Find $dy/dx$ and $d^2y/dx^2$. Plot the graph.

**Solution**   We compute

$$\frac{dx}{dt} = 2t - 1, \qquad \frac{dy}{dt} = 3(t^2 - 1);$$

$$\frac{dy}{dx} = \frac{3(t^2 - 1)}{2t - 1}, \qquad \frac{d^2y}{dx^2} = \frac{\dfrac{d}{dt}\left(\dfrac{dy}{dx}\right)}{dx/dt} = 6\frac{t^2 - t + 1}{(2t - 1)^3}.$$

We construct the table of values:

| $t$ | $-3$ | $-2$ | $-\sqrt{3}$ | $-1$ | $0$ | $\frac{1}{2}$ | $1$ | $\sqrt{3}$ | $2$ | $3$ |
|---|---|---|---|---|---|---|---|---|---|---|
| $x$ | $12$ | $6$ | $3+\sqrt{3}$ | $2$ | $0$ | $-\frac{1}{4}$ | $0$ | $3-\sqrt{3}$ | $2$ | $6$ |
| $x'(t)$ | $-$ | $-$ | $-$ | $-$ | $-$ | $0$ | $+$ | $+$ | $+$ | $+$ |
| $y$ | $-18$ | $-2$ | $0$ | $2$ | $0$ | $-\frac{11}{8}$ | $-2$ | $0$ | $2$ | $18$ |
| $y'(t)$ | $+$ | $+$ | $+$ | $0$ | $-$ | $-$ | $0$ | $+$ | $+$ | $+$ |

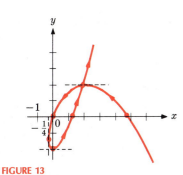

FIGURE 13

Plotting the values and then connecting them with a smooth curve, we obtain a sketch of the graph, as in Fig. 13. □

**EXAMPLE 3**  Find the equation of the line tangent to the curve

$$x = t^2 - 2, \qquad y = t^3 - 2t + 1,$$

at the point where $t = 2$.

**Solution**  When $t = 2$, we have $x = 2$, $y = 5$. Further, $dx/dt = 2t$, $dy/dt = 3t^2 - 2$, and $dy/dx = (3t^2 - 2)/2t$. The slope of the tangent line is

$$\frac{3(2)^2 - 2}{2(2)} = \frac{5}{2}.$$

The equation of the desired line is

$$y - 5 = \tfrac{5}{2}(x - 2) \qquad \Leftrightarrow \qquad 5x - 2y = 0. \qquad \square$$

Suppose we are given an arc $C$ in the parametric form

$$x = x(t), \qquad y = y(t), \quad a \leq t \leq b.$$

In Chapter 6, Section 4 we studied how to compute the length of a curve, provided the curve is the graph of a function; that is, provided the equation of the curve can be written in the form $y = f(x)$. The formula we obtained for arc length is

$$s = \int_a^b \sqrt{1 + [f'(x)]^2} \; dx \;\text{(see page 239)}.$$

Sometimes a curve given by parametric equations can have the parameter eliminated and the above formula used, but as we know this is not always possible. For this reason we now derive directly an expression for arc length for curves in parametric form.

Let the endpoints of a curve $C$ be $A = (x(a), y(a))$ and $B = (x(b), y(b))$. We divide $[a, b]$ into subintervals $a = t_0 < t_1 < \cdots < t_n = b$; let $P_i$ be the point $P_i = (x(t_i), y(t_i))$. (See Fig. 14.) The length of the line segment connecting

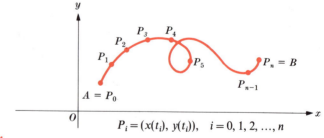

$$P_i = (x(t_i), y(t_i)), \quad i = 0, 1, 2, \ldots, n$$

**FIGURE 14**

$P_{i-1} P_i$ is given by

$$|P_{i-1} P_i| = \sqrt{(x(t_i) - x(t_{i-1}))^2 + (y(t_i) - y(t_{i-1}))^2},$$

and the sum of the lengths of the inscribed polygon to the curve $C$ is given by the formula

$$\sum_{i=1}^{n} |P_{i-1} P_i| = \sum_{i=1}^{n} \sqrt{[x(t_i) - x(t_{i-1})]^2 + [y(t_i) - y(t_{i-1})]^2}.$$

Assuming that $x(t)$ and $y(t)$ have continuous first derivatives on the interval $[a, b]$, we may apply the Mean Value Theorem to both functions, getting

$$x(t_i) - x(t_{i-1}) = x'(\xi_i)(t_i - t_{i-1}), \quad t_{i-1} < \xi_i < t_i,$$

$$y(t_i) - y(t_{i-1}) = y'(\eta_i)(t_i - t_{i-1}), \quad t_{i-1} < \eta_i < t_i.$$

We find that

$$\sum_{i=1}^{n} |P_{i-1}P_i| = \sum_{i=1}^{n} \sqrt{[x'(\xi_i)]^2 + [y'(\eta_i)]^2}(t_i - t_{i-1}).$$

Proceeding to the limit as the norm of the subdivision tends to zero and taking into account the definition of integral,* we obtain the formula for arc length $s$:

$$s = \int_a^b \sqrt{[x'(t)]^2 + [y'(t)]^2}\, dt.$$

From the Fundamental Theorem of Calculus, which states that differentiation and integration are inverse processes, the above integral formula for $s$ yields

$$s'(t) = \frac{ds}{dt} = \sqrt{\left(\frac{dx}{dt}\right)^2 + \left(\frac{dy}{dt}\right)^2}.\dagger$$

Or we may express this formula quite simply by using differentials:

$$ds^2 = dx^2 + dy^2.$$

**EXAMPLE 4**    Given that $x = t^3 + 1$, $y = 2t^{9/2} - 4$. Find the length of the arc from the point where $t = 1$ to the point where $t = 3$.

**Solution**    We have $x'(t) = 3t^2$, $y'(t) = 9t^{7/2}$. Therefore

$$s = \int_1^3 \sqrt{9t^4 + 81t^7}\, dt = 9 \int_1^3 \sqrt{\tfrac{1}{9} + t^3}\, t^2\, dt.$$

Letting $u = \frac{1}{9} + t^3$, $du = 3t^2\, dt$, and integrating, we get

$$s = [2(\tfrac{1}{9} + t^3)^{3/2}]_1^3 = \tfrac{2}{27}(244\sqrt{244} - 10\sqrt{10}). \qquad \square$$

**EXAMPLE 5**    Find the length of one arch of a cycloid, given that

$$x = a(\theta - \sin \theta), \qquad y = a(1 - \cos \theta), \quad 0 \le \theta \le 2\pi.$$

---

*The argument here is not quite precise. To apply the definition of integral directly, it is necessary that $\xi_i$ and $\eta_i$ be the same. However, the fact that the result is valid when $\xi_i \ne \eta_i$ is not difficult to establish.

†Here $s(T)$ is thought of as that function of $T$ which gives the length of the part of the arc for $t$ between $a$ and $T$.

**Solution**   We have $dx/d\theta = a(1 - \cos\theta)$, $dy/d\theta = a\sin\theta$, and therefore

$$s = \int_0^{2\pi} \sqrt{a^2(1 - \cos\theta)^2 + a^2\sin^2\theta}\; d\theta = a\int_0^{2\pi} \sqrt{2 - 2\cos\theta}\; d\theta.$$

Making use of the formula $\sin(\theta/2) = \sqrt{(1 - \cos\theta)/2}$, we obtain

$$s = 2a\int_0^{2\pi} \sin\frac{\theta}{2}\, d\theta = \left[-4a\cos\frac{\theta}{2}\right]_0^{2\pi} = 8a. \qquad \square$$

*Remark.*   If the arc is in the form $y = f(x)$ or $x = g(y)$ we obtain, respectively,

$$\frac{ds}{dx} = \sqrt{1 + (dy/dx)^2} \qquad \text{and} \qquad \frac{ds}{dy} = \sqrt{1 + (dx/dy)^2}.$$

Suppose we have a curve given in parametric form:

$$x = x(t), \qquad y = y(t), \quad a \le t \le b,$$

where $x$ has the special form $x(t) = t$. Then the parameter is $x$ itself, and we can substitute $x$ for $t$ in the second term to get $y(x)$ instead of $y(t)$. In this case the formulas agree:

$$s = \int_a^b \sqrt{x'(t)^2 + y'(t)^2}\; dt = \int_a^b \sqrt{(t')^2 + y'(t)^2}\; dt$$

$$= \int_a^b \sqrt{1 + y'(t)^2}\; dt$$

$$= \int_a^b \sqrt{1 + y'(x)^2}\; dx.$$

In Chapter 6, Section 4, we studied the area of a surface of revolution. We obtained a surface of revolution by starting with an arc $C$, described by

$$C : y = f(x), \quad a \le x \le b.$$

If the arc $C$ is situated entirely above the $x$ axis we can revolve it about the $x$ axis. The volume it sweeps out has a surface area, which we saw was given by

$$A(S) = 2\pi \int_a^b f(x)\sqrt{1 + f'(x)^2}\; dx. \tag{1}$$

If the curve $C$ were revolved about the $y$ axis, the corresponding surface area formula is

$$A(S) = 2\pi \int_a^b x\sqrt{1 + f'(x)^2}\; dx.$$

The term $\sqrt{1 + f'(x)^2}\, dx$ is readily recognized as arc length, $ds$. If we describe the curve parametrically we get an analogous formula.

THEOREM 2    *If an arc*

$$C : x = x(t), \qquad y = y(t), \quad a \le t \le b,$$

*is in the upper half-plane, then the area $A(S)$ of the surface generated by revolving C about the x axis is given by the formula*

$$A(S) = 2\pi \int_a^b y(t) \sqrt{(dx/dt)^2 + (dy/dt)^2} \, dt. \tag{2}$$

*If the arc C is located to the right of the y axis and is revolved about the y axis, the area $A(S)$ of the surface generated is given by the formula*

$$A(S) = 2\pi \int_a^b x(t) \sqrt{(dx/dt)^2 + (dy/dt)^2} \, dt.$$

The above formulas can be written in a simpler, more useful form. For revolution about the $x$ axis, we write

$$A(S) = 2\pi \int_a^b y \, ds$$

while for revolution about the $y$ axis, we write

$$A(S) = 2\pi \int_a^b x \, ds.$$

EXAMPLE 6    Find the area of the surface generated by revolving about the $x$ axis the arc $x = t^3$, $y = \frac{3}{2}t^2$, $1 \le t \le 3$.

Solution    We use the formula

$$A = 2\pi \int_1^3 y \, ds,$$

with $ds = \sqrt{(dx/dt)^2 + (dy/dt)^2} \, dt$ and $y = \frac{3}{2}t^2$. Therefore

$$A = 3\pi \int_1^3 t^2 \sqrt{9t^4 + 9t^2} \, dt = 9\pi \int_1^3 t^3 \sqrt{1 + t^2} \, dt.$$

We make the substitution $u = 1 + t^2$, $du = 2t \, dt$, and obtain

$$A = \frac{9\pi}{2} \int_2^{10} (u - 1)\sqrt{u} \, du = \frac{9}{2}\pi[\frac{2}{5}u^{5/2} - \frac{2}{3}u^{3/2}]_2^{10} = 3\pi(50\sqrt{10} - \frac{2}{5}\sqrt{2}). \quad \square$$

## 2  PROBLEMS

In Problems 1 through 10, find $dy/dx$ and $d^2y/dx^2$. Give results in terms of the parameter.

1  $x = 4t + 6$, $y = 2 - 5t$

2  $x = 2t^2 - 3$, $y = t^2 + 4t - 1$

3  $x = 2 - t^2$, $y = t^3 + 4t^2 + 2t$

4  $x = 3 \cos t$, $y = 5 \sin t$

5  $x = e^{-2t}$, $y = 1 + 3t^2$

6  $x = 2t + 1$, $y = 2e^{3t}$

7  $x = 4 \cos t$, $y = 2 \sin^2 t$

8  $x = 2 \sin 2\theta$, $y = 2 \sin \theta$

9  $x = a \cos^3 \theta$, $y = a \sin^3 \theta$

10  $x = a\theta - a \sin \theta$, $y = a - a \cos \theta$ (cycloid)

11  Find $dy/dx$, $d^2y/dx^2$, $d^3y/dx^3$, $d^4y/dx^4$, given that $x = t^3 + t$, $y = \frac{3}{2}t^4 + t^2$.

12  Same as Problem 11, given that $x = e^{2t}$, $y = e^t + e^{-t}$.

In Problems 13 through 25, find in each case the equations of the lines tangent and normal to the specified curve at the point corresponding to the given value of the parameter.

13  $x = t^2 + 1$, $y = t^3 + 2t$, $t = -2$

14  $x = e^{2t}$, $y = 2 + t^2$, $t = 1$

15  $x = 4 \cos t$, $y = 2 \sin^2 t$, $t = \pi/3$

16  $x = 5 \cos t$, $y = 4 \sin t$, $t = \pi/3$

17  $x = t^2 - 1$, $y = 2e^t$, $t = -1$

18  $x = 2 \sin 2\theta$, $y = 2 \sin \theta$, $\theta = \pi/4$

19  $x = 2 \cos^3 \theta$, $y = 2 \sin^3 \theta$, $\theta = \pi/4$

20  $x = 3\theta - 3 \sin \theta$, $y = 3 - 3 \cos \theta$, $\theta = \pi/2$

21  $x = t^3$, $y = t^2$, $t = 1$

22  $x = e^t$, $y = e^{-2t}$, $t = 3$

23  $x = \sqrt{t}$, $y = 4t + 7$, $t = \pi$

24  $x = t^2 - 2t + 1$, $y = t^4 - 4t^2 + 4$, $t = 2$

25  $x = 3 - 4 \sin t$, $y = 4 + 3 \cos t$, $t = 17\pi$

In Problems 26 through 31, find in each case the intervals of values of the parameter when $x$ and $y$ are increasing and decreasing. Plot the curve. Solve for $y$ in terms of $x$.

26  $x = 2 \tan \phi$, $y = \sec \phi$, $-\pi/2 < \phi < 3\pi/2$

27  $x = 2 \tan \theta$, $y = 4 \cos^2 \theta$, $-\pi/2 < \theta < \pi/2$

28  $x = t^2 + 2t$, $y = t^2 + t$, $-\infty < t < \infty$

29  $x = \cos \theta$, $y = \sin 2\theta$, $0 \le \theta \le 2\pi$

30  $x = e^{2t} + 1$, $y = 1 - e^{-t}$, $-\infty < t < \infty$

31  $x = t^2(t - 2)$, $y = t(t - 2)^2$, $-\infty < t < \infty$. (Do not try to solve for $y$ in terms of $x$.)

In Problems 32 through 37, in each case subdivide the given interval into the number of equal subintervals indicated by the integer $n$. Compute the quantity

$$\sum_{i=1}^{n} \sqrt{[x'(\xi_i)]^2 + [y'(\xi_i)]^2}(t_i - t_{i-1})$$

by taking the value $\xi_i$ at the midpoint of the $i$th interval $[t_{i-1}, t_i]$.

32  $x = t$, $y = x^2/2$, $0 \le t \le 2$, $n = 4$

33  $x = t^2 + t$, $y = t + 1$, $0 \le t \le 1$, $n = 5$

34  $x = \ln t$, $y = \dfrac{1}{t}$, $1 \le t \le 2$, $n = 4$

35  $x = t^3 + 1$, $y = t + 2$, $0 \le t \le 1$, $n = 5$

36  $x = 2\sqrt{1 + t}$, $y = 2(1 + t)^{-1/2}$, $0 \le t \le 1$, $n = 5$

37  $x = \dfrac{t}{t + 1}$, $y = \dfrac{2}{\sqrt{t + 1}}$, $0 \le t \le 2$, $n = 4$

In Problems 38 through 43, in each case set up the integral for arc length, but do not attempt to evaluate the integral.

38  $x = \frac{1}{6}t^3$, $y = \dfrac{1}{2t}$, $1 \le t \le 3$

39  $x = t^3 + 2t - 1$, $y = t^2 - t + 5$, $1 \le t \le 5$

40  $x = 5 \cos t$, $y = 4 \sin t$, $0 \le t \le \dfrac{\pi}{2}$

41  $x = t \cos t$, $y = t \sin t$, $0 \le t \le \pi$

42  $x = e^t$, $y = e^{-t}$, $1 \le t \le 2$

43  $x = \ln t$, $y = \cos t$, $2 \le t \le \pi$

In Problems 44 through 49, find the arc length of the curve.

44  $x = 6 \cos t$, $y = 6 \sin t$, $\dfrac{\pi}{3} \le t \le \dfrac{\pi}{2}$

45  $x = 4t + 3$, $y = 3t - 2$, $0 \le t \le 2$

46  $x = \frac{1}{3}t^3$, $y = \frac{1}{2}t^2$, $0 \le t \le 1$

47  $x = 3t^2 - 7$, $y = 2t^2 + 3$, $0 \le t \le 2$

48  $x = e^{-t} \cos t$, $y = e^{-t} \sin t$, $0 \le t \le \dfrac{\pi}{2}$

49  $x = e^{-3t}$, $y = e^{-2t}$, $0 \le t < \infty$

50  Given that

$$x = 1 + f(t), \qquad y = \frac{1 - f(t)}{1 + f(t)}, \quad \text{find } \frac{dy}{dx}$$

in terms of $f$ and $f'$.

51  Given that $x = a \cos g(t)$, $y = b \sin g(t)$. Prove that

$$xy^2 \frac{d^2y}{dx^2} = b^2 \frac{dy}{dx}.$$

**52** A projectile moves so that its parametric equations are

$$x = (v_0 \cos \alpha)t, \qquad y = (v_0 \sin \alpha)t - 5t^2$$

where $v_0$ is the initial speed and $\alpha$ is the angle the direction of motion makes with the horizontal at $t = 0$.

a) Eliminate the parameter $t$ and show that the path of the projectile is parabolic.

b) The speed $s$ at any time $t$ is given by

$$s = \left[ \left( \frac{dx}{dt} \right)^2 + \left( \frac{dy}{dt} \right)^2 \right]^{1/2}.$$

Find $s$ when $t = 1$ and $t = 2$.

c) Show that the projectile hits the ground when $t = \frac{1}{5}v_0 \sin \alpha$. Find how far the projectile has travelled horizontally at that time.

**53** A particle moves so that its parametric equations are given by

$$x = \frac{4t}{t^2 + 4}, \qquad y = \frac{t^2 - 4}{t^2 + 4}.$$

a) Show that the point moves on the circle of radius 1 with center at the origin.

b) The speed $s$ is given by $\quad s = \left[ \left( \frac{dx}{dt} \right)^2 + \left( \frac{dy}{dt} \right)^2 \right]^{1/2}.$

Find the speed in terms of $t$.

c) The magnitude of the acceleration, denoted $|\mathbf{a}|$ is given by

$$\left[ \left( \frac{d^2x}{dt^2} \right)^2 + \left( \frac{d^2y}{dt^2} \right)^2 \right]^{1/2}.$$

Find $|\mathbf{a}|$ when $t = 1$ and $t = 2$.

In Problems 54 through 57, find the area of the surface generated by revolving the given arc $C$ about the given axis.

**54** $C = \{(x, y): x = t - \sin t, y = 1 - \cos t, 0 \le t \le 2\pi\}$; $x$ axis

**55** $C = \{(x, y): x = 3 + 2 \cos t, y = 2 \sin t, 0 \le t \le 2\pi\}$; $y$ axis

**56** $C = \{(x, y): x = t^2, y = \frac{1}{6}t^6 + \frac{1}{2t^2}, 1 \le t \le 2\}$; $y$ axis

**57** $C = \{(x, y): x = t^3, y = e^{-t^3}, 0 \le t \le 2\}$; $x$ axis

---

## 3

# CURVATURE

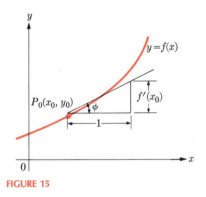

FIGURE 15

Our aim in this section is to obtain a measure of the rapidity with which curves change direction. Suppose that a curve is given by the equation $y = f(x)$ and that $f$ has a continuous second derivative. At a particular point $P_0(x_0, y_0)$ the tangent to the curve makes an angle with the positive $x$ direction which we call $\phi$ (Fig. 15). From the definition of derivative we know that $\tan \phi = f'(x_0)$ or that at a point $P(x, y)$,

$$\phi(x) = \arctan f'(x).$$

The way $\phi$ changes as we move along the curve is a measure of the sharpness of the curve. Note that for a straight line $\phi$ doesn't change at all, and it changes very little even after traversing a long section of the arc of a gradually changing curve.

**DEFINITION**    *The **curvature** $\kappa$ of an arc\* given in the form $y = f(x)$ is the rate of change of the angle $\phi$ with respect to the arc length $s$. That is,*

$$\kappa = \frac{d\phi}{ds}.$$

The following theorem tells us how to compute the curvature when we are given the function $f$.

---

\*The Greek letter kappa ($\kappa$) is customarily used to denote curvature.

**THEOREM 3**     *For an arc of the form $y = f(x)$, for which $f''(x)$ exists, the curvature $\kappa$ is given by*

$$\kappa(x) = \frac{f''(x)}{\{1 + [f'(x)]^2\}^{3/2}}. \tag{1}$$

**Proof**     From the Chain Rule, we know that

$$\kappa = \frac{d\phi}{ds} = \frac{d\phi}{dx} \cdot \frac{dx}{ds}.$$

We first find $d\phi/dx$:

$$\frac{d\phi}{dx} = \frac{d}{dx}(\arctan f'(x)) = \frac{1}{1 + [f'(x)]^2} \cdot f''(x).$$

To obtain $dx/ds$, we recall that since differentiation and integration are inverse processes, the formula

$$s = \int \sqrt{1 + [f'(x)]^2}\, dx$$

may be differentiated to give

$$ds = \sqrt{1 + [f'(x)]^2}\, dx,$$

or

$$\frac{dx}{ds} = \frac{1}{\sqrt{1 + [f'(x)]^2}}.$$

Therefore we get

$$\kappa = \frac{d\phi}{dx} \cdot \frac{dx}{ds} = \frac{f''(x)}{1 + [f'(x)]^2} \cdot \frac{1}{\sqrt{1 + [f'(x)]^2}} = \frac{f''(x)}{\{1 + [f'(x)]^2\}^{3/2}},$$

and Theorem 3 is established.      □

If a curve is given in the form $x = g(y)$, the angle $\phi$ which the tangent line forms with the positive $x$ direction may still be defined. It can be shown that for an arc of the form $x = g(y)$ the curvature is given by the formula

$$\kappa(y) = -\frac{g''(y)}{\{1 + [g'(y)]^2\}^{3/2}}, \tag{2}$$

Similar arguments can be used to establish a formula for the curvature of an arc given in parametric form. We state it as a theorem.

**THEOREM 4**     *If an arc $C$ is given in parametric form by $x = x(t)$, $y = y(t)$, and if $x''$ and $y''$ exist, then the curvature $\kappa$ at a point $P = (x(t), y(t))$ is given by*

$$\kappa(t) = \frac{x'(t)y''(t) - x''(t)y'(t)}{(x'^2 + y'^2)^{3/2}}. \tag{3}$$

It can be shown that the magnitude of the curvature $\kappa$ of a curve $C$ is an intrinsic property of the curve and does not depend on the coordinate system. However, the sign of $\kappa$ is determined by the parameter we select to describe $C$. For example, if $C$ is given in the form $y = f(x)$, then the curvature has the same sign as $f''(x)$, as can be seen from Formula (1) on page 504. This fact reflects the method we used to define arc length $s$. When defining $s$, we stated that $s$ is positive as we traverse $C$ from smaller values of $x$ to larger values of $x$. This definition of arc length amounts to the selection of a **direction** or an **orientation** of every arc $C$. It is clear that each arc has two possible orientations, according as we go along it in one direction or the opposite direction. However, once we choose an orientation, the direction of positive arc length is decided and, along with it, the sign of the curvature. For curves described by parametric equations $x = x(t)$, $y = y(t)$, the orientation of the curve and therefore the sign of $\kappa$ is determined by the direction along $C$ which corresponds to an increase in the parameter $t$.

If $\phi$ is increasing so that the curve is "turning to the left" as the parameter increases, then $\kappa$ is positive; $\kappa$ is negative if $\phi$ is decreasing. This is equivalent to saying that the concave side of the curve is on the left if $\kappa > 0$ and is on the right if $\kappa < 0$ (Fig. 16).

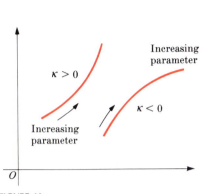

FIGURE 16

**EXAMPLE 1**   Find the curvature of the circle $x = a \cos t$, $y = a \sin t$.

**Solution**   Differentiation yields

$$x' = -a \sin t, \ x'' = -a \cos t, \ y' = a \cos t, \ y'' = -a \sin t.$$

Therefore

$$\kappa = \frac{a^2 \sin^2 t + a^2 \cos^2 t}{(a^2 \sin^2 t + a^2 \cos^2 t)^{3/2}} = \frac{a^2}{a^3} = \frac{1}{a}. \qquad \square$$

*We conclude that **a circle of radius** $a$ **has constant curvature equal to** $1/a$.*

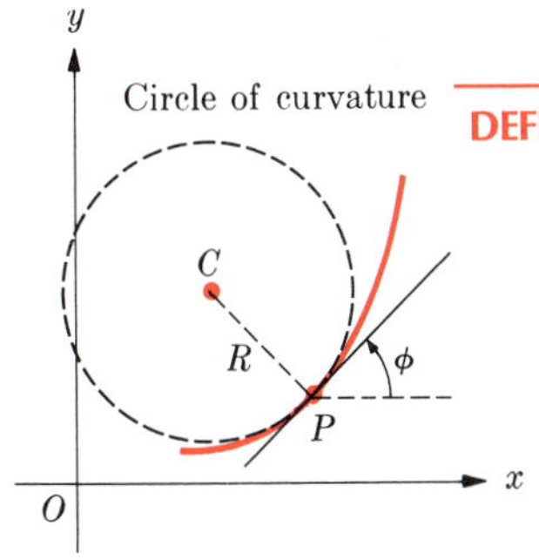

**DEFINITIONS**   *The **radius of curvature** $R$ **of an arc at a point** is defined as the reciprocal of the absolute value of the curvature at that point; that is,*

$$R = \frac{1}{|\kappa|}.$$

*The **circle of curvature of an arc at a point** $P$ is that circle passing through $P$ which has radius equal to $R$, the radius of curvature, and whose center $C$ lies on the concave side of the curve along the normal through $P$ (see Fig. 17).*

FIGURE 17

**EXAMPLE 2**   Find the curvature $\kappa$ and the radius of curvature $R$ for the parabola $y = x^2$. Find the center of the circle of curvature at the point $(1, 1)$.

**Solution**   We have $dy/dx = 2x$, $d^2y/dx^2 = 2$, and therefore

$$\kappa = \frac{2}{(1 + 4x^2)^{3/2}}; \qquad R = \tfrac{1}{2}(1 + 4x^2)^{3/2}.$$

At $(1, 1)$ the slope is 2 and the equation of the normal line is

$$y - 1 = -\tfrac{1}{2}(x - 1) \quad \Leftrightarrow \quad x + 2y - 3 = 0.$$

The radius of curvature $R$ is $\tfrac{5}{2}\sqrt{5}$. (See Fig. 18). To find the center $C$ of the circle of curvature we first write the equation of the circle with center at $(1, 1)$ and radius $\tfrac{5}{2}\sqrt{5}$. It is

$$(x - 1)^2 + (y - 1)^2 = \tfrac{125}{4}.$$

Solving this equation simultaneously with $x + 2y - 3 = 0$, the equation of the normal line, we obtain for the coordinates $x_c$, $y_c$ of $C$, the values

$$x_c = -4, \qquad y_c = \tfrac{7}{2}. \qquad \qquad \square$$

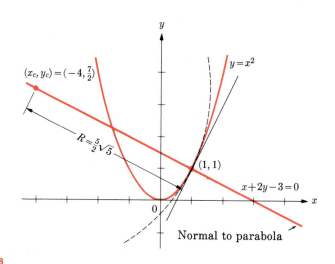

**FIGURE 18**

EXAMPLE 3    Find the curvature and radius of curvature of the cycloid

$$x = a(\theta - \sin \theta), \qquad y = a(1 - \cos \theta), \quad 0 < \theta < 2\pi.$$

Solution    We have

$$x'(\theta) = a(1 - \cos \theta), \qquad x''(\theta) = a \sin \theta,$$
$$y'(\theta) = a \sin \theta, \qquad \qquad y''(\theta) = a \cos \theta.$$

Therefore, using Formula (3) on page 504, we find

$$\kappa(\theta) = \frac{a(1 - \cos \theta)a \cos \theta - a \sin \theta\, a \sin \theta}{[a^2(1 - \cos \theta)^2 + a^2 \sin^2 \theta]^{3/2}}$$

$$= \frac{1}{a} \frac{\cos \theta - 1}{[2(1 - \cos \theta)]^{3/2}} = -\frac{1}{a2\sqrt{2}\sqrt{1 - \cos \theta}} = -\frac{1}{4a \sin (\theta/2)}.$$

$$R(\theta) = 4a \sin \frac{\theta}{2}, \quad 0 < \theta < 2\pi. \qquad \qquad \square$$

## 3    PROBLEMS

In Problems 1 through 12, find in each case the curvature $\kappa(x)$ and the radius of curvature $R(x)$.

**1** $y = 2x^2 + 1$

**2** $y = e^{-2x}$

**3** $y = 1 + 3\sqrt{x}$

**4** $y = x + \dfrac{2}{x}, \quad x > 0$

**5** $y = \sin x$

**6** $y = \frac{1}{2}\cos 3x$

**7** $y = ae^{bx}, \quad a, b$ constants

**8** $y = \tan x$

**9** $y = \arcsin x$

**10** $y = (a^{2/3} - x^{2/3})^{3/2}$

**11** $y = \dfrac{b}{a}\sqrt{a^2 - x^2}, \quad |x| < a$

**12** $y = \dfrac{b}{a}\sqrt{x^2 - a^2}, \quad |x| > a$

In Problems 13 through 16, in each case find $\kappa(y)$ and $R(y)$.

**13** $x = \cot y$

**14** $x = \ln y$

**15** $x = -\sqrt{a^2 - y^2}$

**16** $x = \frac{1}{3}y^3$

In Problems 17 through 22, in each case find $\kappa$ and $R$ in terms of the parameter.

**17** $x = \frac{1}{2}t^2, \quad y = \frac{1}{3}t^3$

**18** $x = e^t, \quad y = t^2$

**19** $x = e^t \sin t, \quad y = e^t \cos t$

**20** $x = e^{-2t}\cos t, \quad y = e^{-2t}\sin t$

**21** $x = a\cos^3\theta, \quad y = a\sin^3\theta$

**22** $x = a\cos\theta, \quad y = b\sin\theta$

In Problems 23 through 32, find the curvature $\kappa$ at the indicated point $P$.

**23** $y = 4 - 2x^3; \ P(1, 2)$

**24** $y = x^4; \ P(1, 1)$

**25** $y = \dfrac{1}{\sqrt{2\pi}} e^{-x^2/2}; \ P\left(0, \dfrac{1}{\sqrt{2\pi}}\right)$

**26** $y = \dfrac{1}{\pi(1 + x^2)}; \ P\left(0, \dfrac{1}{\pi}\right)$

**27** $y = e^{-x}; \ P\left(1, \dfrac{1}{e}\right)$

**28** $x = t - 1, \ y = \sqrt{t}; \ P(3, 2)$

**29** $x = 3t - 4, \ y = t^2 + 3t + 2; \ P(2, 12)$

**30** $x = t - \sin t, \ y = 1 - \cos t; \ P\left(\dfrac{\pi}{2} - 1, \, 1\right)$

**31** $x = \cos t, \ y = \ln t; \ P(\cos 1, 0)$

**32** $x = e^t, \ y = e^{-t}; \ P(1, 1)$

In Problems 33 through 36, find $\kappa(x)$ in each case, and locate the values of $x$ for which $\kappa$ has relative maxima and minima.

**33** $y = \frac{1}{2}x^2$

**34** $y = \ln x$

**35** $y = 1/x$

**36** $y = \ln(\sin x)$

**37** Prove that for any arc the coordinates of the center of the circle of curvature $C$ are given by

$$x_c = x - \frac{\sin \phi}{\kappa}, \qquad y_c = y + \frac{\cos \phi}{\kappa}.$$

In Problems 38 through 41, in each case find $\kappa$ and $R$, and locate the center of the circle of curvature at the point indicated.

**38** $y = x^2; \ (-1, 1)$

**39** $y = \cos x; \ (\pi/3, \frac{1}{2})$

**40** $y = x^3 - 2x^2 + 3x - 5; \ (1, -1)$

**41** $y = e^x; \ (0, 1)$

In Problems 42 and 43, find in terms of $t$ the coordinates of the center of the circle of curvature.

**42** $x = a\cos t, \quad y = b\sin t$

**43** $x = a\sec t, \quad y = b\tan t$

**44** Establish Formula (2) on page 504 for the curvature $\kappa(y)$ of an arc given by $x = g(y)$.

**45** Establish Formula (3) on page 504 for the curvature $\kappa(t)$ of an arc given parametrically by $x = x(t), \ y = y(t)$.

**46** Find the point on the graph of $y = 3\ln(2x + 1)$ where the curvature has its minimum value. Show that there is no point where the curvature is a maximum.

**47** Given the cycloid

$$x = a(\theta - \sin\theta), \qquad y = a(1 - \cos\theta),$$

show that the curvature is a maximum at the point on the curve where the tangent to the curve is horizontal. Explain why there is no point on the curve where the curvature is a minimum.

---

### 4

## POLAR COORDINATES. GRAPHS IN POLAR COORDINATES

A coordinate system in the plane allows us to associate an ordered pair of numbers with each point in the plane. Until now we have considered rectangular coordinate systems exclusively. However, there are many problems for which other systems of coordinates are more advantageous than are

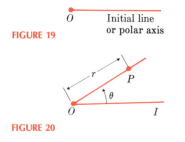

FIGURE 19

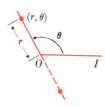

FIGURE 20

rectangular coordinates. With this in mind we now describe the system known as **polar coordinates**.

We begin by selecting a point in the plane which we call the **pole** or **origin** and label it $O$. From this point we draw a half-line starting at the pole and extending indefinitely in one direction. This line is usually drawn horizontally and to the right of the pole, as shown in Fig. 19. It is called the **initial line** or **polar axis**.

Let $P$ be any point in the plane. Its position will be determined by its distance from the pole and by the angle that the line $OP$ makes with the initial line. We measure angles $\theta$ from the initial line as in trigonometry—positive in a counterclockwise direction and negative in a clockwise direction. The distance $r$ from the origin to the point $P$ will be taken as positive. The coordinates of $P$ (Fig. 20) in the polar coordinate system are $(r, \theta)$. There is a sharp distinction between rectangular and polar coordinates, in that a point $P$ may be represented in just one way by a pair of rectangular coordinates, but it may be represented in many ways by polar coordinates. For example, the point $Q$ with polar coordinates $(2, \pi/6)$ also has polar coordinates

$$\left(2, 2\pi + \frac{\pi}{6}\right), \left(2, 4\pi + \frac{\pi}{6}\right), \left(2, 6\pi + \frac{\pi}{6}\right), \left(2, -2\pi + \frac{\pi}{6}\right), \left(2, -4\pi + \frac{\pi}{6}\right), \text{ etc.}$$

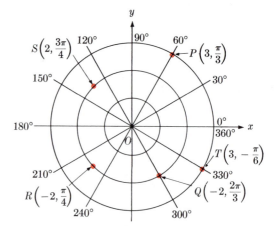

FIGURE 21

In other words, there are infinitely many representations of the same point. Furthermore, it is convenient to allow $r$, the distance from the origin, to take on negative values. We establish the convention that a pair of coordinates such as $(-3, \theta)$ is simply another representation of the point with coordinates $(3, \theta + \pi)$. Figure 21 shows the relationship of the points $(r, \theta)$ and $(-r, \theta)$.

**EXAMPLE 1**   Plot the points whose polar coordinates are

$$P\left(3, \frac{\pi}{3}\right), Q\left(-2, \frac{2\pi}{3}\right), R\left(-2, \frac{\pi}{4}\right), S\left(2, \frac{3\pi}{4}\right), T\left(3, -\frac{\pi}{6}\right).$$

**Solution**   The points are plotted in Fig. 22.                                    □

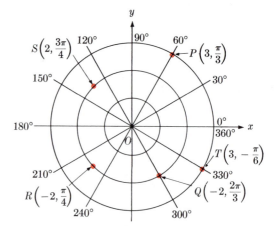

FIGURE 22

It is important to know the connection between rectangular and polar coordinate systems. To find this relationship, let us consider a plane with one system superimposed on the other in such a way that the origin of the rectangular system is at the pole and the positive $x$ axis coincides with the initial line (Fig. 23). The relationship between the rectangular coordinates $(x, y)$ and the polar coordinates $(r, \theta)$ of a point $P$ is given by the equations

$$x = r \cos \theta, \qquad y = r \sin \theta.$$

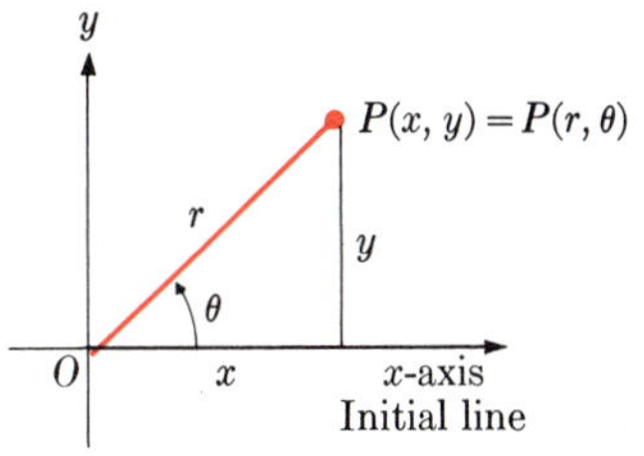

**FIGURE 23**

When we are given $r$ and $\theta$, these equations tell us how to find $x$ and $y$. We also have the formulas

$$r = \pm \sqrt{x^2 + y^2}, \qquad \tan \theta = \frac{y}{x},$$

which give us $r$ and $\theta$ when the rectangular coordinates are known.

**EXAMPLE 2**    The rectangular coordinates of a point are $(\sqrt{3}, -1)$. Find a set of polar coordinates for this point.

**Solution**    We have

$$r = \sqrt{3 + 1} = 2 \qquad \text{and} \qquad \tan \theta = -1 \sqrt{3}.$$

Since the point is in the fourth quadrant, we select for $\theta$ the value $-\pi/6$ (or $11 \pi/6$). The answer is $(2, -\pi/6)$. □

Equations can be described in polar coordinates as well as in rectangular coordinates. For example, $r$ and $\theta$ can be connected by some equation such as $r = 3 \cos 2\theta$ or $r^2 = 4 \sin 3\theta$.

---

**DEFINITION**    *The **graph of an equation in polar coordinates** is the set of all points $P$ such that each point has at least one pair of polar coordinates $(r, \theta)$ which satisfies the given equation.*

---

To plot the graph of such an equation we must find all ordered pairs $(r, \theta)$ which satisfy the given equation and then plot the points obtained. We can obtain a good approximation of the graph in polar coordinates, as we can in rectangular coordinates, by making a sufficiently complete table of values, plotting the points, and connecting them by a smooth curve.

**EXAMPLE 3**   Sketch the graph of the equation $r = 3 \cos \theta$.

**Solution**   We construct the table:

| $\theta$ | 0 | $\pi/6$ | $\pi/3$ | $\pi/2$ | $2\pi/3$ | $5\pi/6$ |
|---|---|---|---|---|---|---|
| $\theta°$ | 0 | 30 | 60 | 90 | 120 | 150 |
| $r$ | 3 | $\frac{3}{2}\sqrt{3}$ | $\frac{3}{2}$ | 0 | $-\frac{3}{2}$ | $-\frac{3}{2}\sqrt{3}$ |
| $r$ approx. | 3.00 | 2.60 | 1.50 | 0 | $-1.50$ | $-2.60$ |

| $\theta$ | $\pi$ | $7\pi/6$ | $4\pi/3$ | $3\pi/2$ | $5\pi/3$ | $11\pi/6$ |
|---|---|---|---|---|---|---|
| $\theta°$ | 180 | 210 | 240 | 270 | 300 | 330 |
| $r$ | $-3$ | $-\frac{3}{2}\sqrt{3}$ | $-\frac{3}{2}$ | 0 | $\frac{3}{2}$ | $\frac{3}{2}\sqrt{3}$ |
| $r$ approx. | $-3.00$ | $-2.60$ | $-1.50$ | 0 | 1.50 | 2.60 |

The graph is shown in Fig. 24. It can be shown to be a circle with center at $(\frac{3}{2}, 0)$ and radius $\frac{3}{2}$. It is worth noting that even though there are 12 entries in the table, only 6 points are plotted. The curve is symmetric with respect to the initial line, and we could have saved effort if this fact had been taken into account. Since $\cos(-\theta) = \cos \theta$ for all values of $\theta$, we could have obtained the points on the graph for values of $\theta$ between 0 and $-\pi$ without extra computation. □

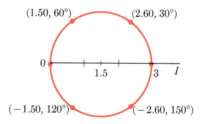

**FIGURE 24**

We shall set forth the rules of symmetry. These rules, which are useful as aids in graphing, are described in terms of symmetries with respect to $x$ and $y$ axes in rectangular coordinates. That is, we suppose that the positive $x$ axis coincides with the initial line of the polar coordinate system.

**Rule I.**   *If the substitution of $(r, -\theta)$ for $(r, \theta)$ yields the same equation, the graph is symmetric with respect to the x axis.*

**Rule II.**   *If the substitution of $(r, \pi - \theta)$ for $(r, \theta)$ yields the same equation, the graph is symmetric with respect to the y axis.*

**Rule III.**   *If the substitution of $(-r, \theta)$ or of $(r, \pi + \theta)$ for $(r, \theta)$ yields the same equation, the graph is symmetric with respect to the pole.*

These rules can be summarized in the form of a table as follows:

**Table 1**

| Substitution | Symmetry |
|---|---|
| $-\theta$ for $\theta$ | $x$-axis (line $\theta = 0$) |
| $-r$ for $r$ | pole (the origin) |
| $\pi + \theta$ for $\theta$ | pole (the origin) |
| $\pi - \theta$ for $\theta$ | $y$-axis $\left( \text{line } \theta = \pm \dfrac{\pi}{2} \right)$ |

It is true that if any two of the three **symmetries** hold, the remaining one holds automatically. However, *it is possible for a graph to have certain symmetry properties which the rules above will fail to exhibit.*

**EXAMPLE 4**     Discuss for symmetry and plot the graph of $r = 3 + 2 \cos \theta$.

**Solution**     The graph is symmetric with respect to the $x$ axis, since $\cos(-\theta) = \cos \theta$ and Rule I applies. We construct the table:

| $\theta$ | 0 | $\pm \pi/6$ | $\pm \pi/3$ | $\pm \pi/2$ | $\pm 2\pi/3$ | $\pm 5\pi/6$ | $\pm \pi$ |
|---|---|---|---|---|---|---|---|
| $\theta°$ | 0 | $\pm 30°$ | $\pm 60°$ | $\pm 90°$ | $\pm 120°$ | $\pm 150°$ | $\pm 180°$ |
| $r$ | 5 | $3 + \sqrt{3}$ | 4 | 3 | 2 | $3 - \sqrt{3}$ | 1 |
| $r$ approx. | 5.00 | 4.73 | 4.00 | 3.00 | 2.00 | 1.27 | 1.00 |

We use values of $\theta$ from $-\pi$ to $+\pi$ only, since $\cos(\theta + 2\pi) = \cos \theta$ and no new points could possibly be obtained with higher values of $\theta$. The graph, called a **limaçon**, is shown in Fig. 25.  □

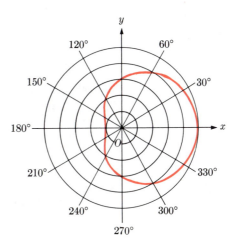

**FIGURE 25**

EXAMPLE 5    Test for symmetry and plot the graph of the equation

$$r = 2 \cos 2\theta.$$

Solution    If we replace $\theta$ by $-\theta$ the equation is unchanged and, therefore, by Rule I, we have symmetry with respect to the $x$ axis. When we replace $\theta$ by $\pi - \theta$ we get

$$\cos 2(\pi - \theta) = \cos 2\pi \cos 2\theta + \sin 2\pi \sin 2\theta = \cos 2\theta$$

and, by Rule II, the graph is symmetric with respect to the $y$ axis. Consequently, the graph is symmetric with respect to the pole. We construct the table:

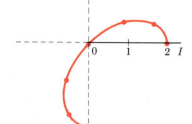

| $\theta$ | 0 | $\pi/12$ | $\pi/6$ | $\pi/4$ | $\pi/3$ | $5\pi/12$ | $\pi/2$ |
|---|---|---|---|---|---|---|---|
| $2\theta$ | 0 | $\pi/6$ | $\pi/3$ | $\pi/2$ | $2\pi/3$ | $5\pi/6$ | $\pi$ |
| $2\theta°$ | 0° | 30° | 60° | 90° | 120° | 150° | 180° |
| $r$ | 2 | $\sqrt{3}$ | 1 | 0 | $-1$ | $-\sqrt{3}$ | $-2$ |
| $r$ approx. | 2.00 | 1.73 | 1.00 | 0 | $-1.00$ | $-1.73$ | $-2.00$ |

FIGURE 26

We plot the points as shown in Fig. 26. Now, making use of the symmetries, we can easily complete the graph (Fig. 27). This curve is called a **four-leaved rose**.                                          ☐

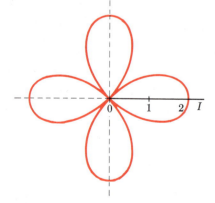

FIGURE 27

Equations of the form

$$r = a \sin n\theta, \qquad r = a \cos n\theta,$$

where $n$ is a positive integer, have graphs which are called **rose** or **petal** curves. The number of petals is equal to $n$ if $n$ is an odd integer and is equal to $2n$ if $n$ is an even integer. If $n = 1$, there is one petal and it is circular.

The graphs of equations of the form

$$r = a \pm b \cos \theta \qquad \text{or} \qquad r = a \pm b \sin \theta$$

are called **limaçons**. In the special cases in which $a = b$, the graphs are called **cardioids** (see Problems 20, 21, and 22 below). The graphs of equations of the form

$$r^2 = a^2 \cos 2\theta \qquad \text{or} \qquad r^2 = a^2 \sin 2\theta$$

are called **lemniscates**; these graphs have the appearance of "figure eights" (see Problems 29 and 30 below). Polar coordinates are particularly well suited to the study of certain curves called **spirals**. The so-called **spiral of Archimedes** has an equation of the form

$$r = k\theta$$

and its graph is drawn in Fig. 28 (the dashed portion arises from negative values of $\theta$). The **logarithmic spiral** has an equation of the form

$$\log_b r = \log_b a + k\theta$$

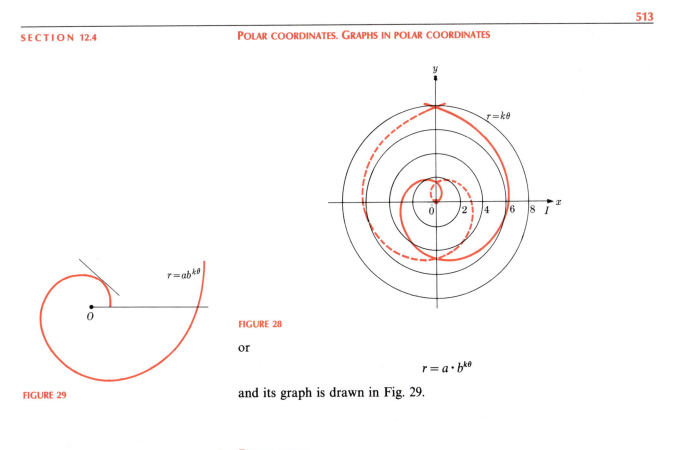

FIGURE 28

or

$$r = a \cdot b^{k\theta}$$

FIGURE 29

and its graph is drawn in Fig. 29.

## 4    PROBLEMS

In Problems 1 through 4, the polar coordinates of points are given. Find the rectangular coordinates of the same points and plot them on graphs.

**1** $(4, \pi/6)$, $(3, 3\pi/4)$, $(2, \pi)$, $(1, 0)$, $(-2, \pi)$

**2** $(2, \pi/4)$, $(1, \pi/3)$, $(3, \pi/2)$, $(4, 3\pi/2)$, $(-1, 7\pi/6)$

**3** $(-1, 0)$, $(2, -\pi/6)$, $(4, -\pi/3)$, $(-3, 3\pi/4)$, $(0, \pi/2)$

**4** $(2, -\pi/2)$, $(-1, -3\pi/2)$, $(2, 4\pi/3)$, $(-1, -\pi/4)$, $(0, -\pi)$

In Problems 5 through 8, the rectangular coordinates of points are given. Find a pair of polar coordinates for each of the points and plot them on graphs.

**5** $(3, 3)$, $(0, 4)$, $(-1, \sqrt{3})$, $(0, -1)$, $(2, 0)$

**6** $(-2, -2)$, $(-4, 0)$, $(\sqrt{3}, 1)$, $(-\sqrt{3}, -1)$, $(0, -2)$

**7** $(-2, 2)$, $(3, -3)$, $(-\sqrt{3}, 1)$, $(2\sqrt{3}, 2)$, $(2, 2\sqrt{3})$

**8** $(4, 0)$, $(0, 0)$, $(6, 6)$, $(\sqrt{6}, \sqrt{2})$, $(\frac{3}{2}, -\frac{3}{2})$

**9** Describe the graph of all points which, in polar coordinates, satisfy the condition $r = 5$; do the same for the condition $\theta = \pi/3$ and for the condition $\theta = -5\pi/6$. What can be said about the angle of intersection of the curve $r = $ const, with $\theta = $ const?

**10** Find the distance between the points with polar coordinates $(3, \pi/4)$, $(2, \pi/3)$.

**11** Find the distance between the points with polar coordinates $(1, \pi/2)$, $(4, 5\pi/6)$.

**12** Find a formula for the distance between the points $(r_1, \theta_1)$ and $(r_2, \theta_2)$.

**13** Find the polar coordinates of the midpoint of the line segment connecting the points $P$ and $Q$ whose polar coordinates are $P(3, \pi/4)$, $Q(6, \pi/6)$.

**14** Find a formula for the coordinates $Q(\bar{r}, \bar{\theta})$ of the midpoint of the line segment connecting the points $P_1(r_1, \theta_1)$ and $P_2(r_2, \theta_2)$.

In Problems 15 through 50, in each case discuss for symmetry and plot the graph of the equation.

**15** $r = 4 \cos \theta$

**16** $r = -4 \cos \theta$

**17** $r = -3 \sin \theta$

**18** $r = 2 \cos (\theta + \pi/6)$

**19** $r = 3 \sin \left(\theta + \dfrac{\pi}{4}\right)$

**20** $r = 3(1 - \cos \theta)$

**21** $r = 2(1 + \cos \theta)$

**22** $r = 2(1 + \sin \theta)$

**23** $r = 4 - 2 \sin \theta$

**24** $r \cos \theta = 2$

**25** $r = 2 + 4 \cos \theta$

**26** $r = 2 - 4 \sin \theta$

**27** $r \sin \theta = 1$

**28** $r^2 = 2 \cos \theta$

**29** $r^2 = \cos 2\theta$

**30** $r^2 = 4 \sin 2\theta$

**31** $r = 5 \cos 2\theta$

**32** $r = 5 \sin 2\theta$

**33** $r = 5 \cos 3\theta$

**34** $r = 5 \sin 3\theta$

**35** $r = 4 \sin^2 \frac{1}{2}\theta$

**36** $r = 2 \cos \frac{1}{2}\theta$

**37** $r = \tan \theta$

**38** $r = 5 \cos 4\theta$

**39** $r = 5 \sin 4\theta$

**40** $r = \cot \theta$

**41** $r = 2 \csc \theta$

**42** $r(2 - \cos \theta) = 4$

**43** $r(1 - \cos \theta) = 2$

**44** $r(1 + 2 \cos \theta) = 4$

**45** $r(1 + 2 \sin \theta) = 2$

**46** $r(1 + \cos \theta) = -1$

**47** $r = 2 + \theta$

**48** $r = 3 - 2\theta$

**49** $r = 2e^\theta$

**50** $r = 3e^{-2\theta}$

**51** Prove that if the symmetries described by Rules I and III hold, then the symmetry described by Rule II must hold. Give an example in which the symmetry of Rule II holds but the rule itself does not.

In each of Problems 52 through 59, sketch the curves and find the coordinates of the point (or points) of intersection, if any.

**52** $r = 3 \cos \theta, r = 2 \sin \theta$        **53** $r = \sin \theta, r = \sin 2\theta$

**54** $r = 2 \cos \theta, r \cos \theta = 1$        **55** $r = \cos \theta, r^2 = \cos 2\theta$

**56** $r = \tan \theta, r = 2 \sin \theta$        **57** $r = 2 + \cos \theta, r = 6 \cos \theta$

**58** $r = \sin 2\theta, r = \sin 4\theta$        **59** $r = 1 + \cos \theta, r = 1 + \sin \theta.$

**60** Sketch the graphs of the curves $r = \cos^n \theta$ for $n = 1, 2, 3, 4$. Find the limiting curve as $n \to \infty$.

---

**5**

---

# EQUATIONS AND CONIC SECTIONS IN POLAR COORDINATES

The curves discussed in Section 4 find their natural setting in polar coordinates. However, the equations of some curves may be simpler in appearance, or their properties may be more transparent in one coordinate system than in another. For this reason it is useful to be able to transform an equation given in one coordinate system into the corresponding equation in another system.

Suppose that we are given the equation of a curve in the form $y = f(x)$ in a rectangular system. If we simply make the substitution

$$x = r \cos \theta, \qquad y = r \sin \theta,$$

we have the same equation in a polar coordinate system. We can also go the other way, so that if we are given $r = g(\theta)$ in polar coordinates, the substitution

$$r^2 = x^2 + y^2, \qquad \theta = \arctan \frac{y}{x}, \quad \left(\text{i.e., } \tan \theta = \frac{y}{x}\right)$$

transforms the relation into one in rectangular coordinates. In this connection, it is frequently useful to make the substitutions

$$\sin \theta = \frac{y}{\sqrt{x^2 + y^2}}, \qquad \cos \theta = \frac{x}{\sqrt{x^2 + y^2}}, \qquad \tan \theta = \frac{y}{x},$$

rather than use the formula for $\theta$ itself.

**EXAMPLE 1**   Find the graph of the equation in polar coordinates which corresponds to the graph of the equation

$$x^2 + y^2 - 3x = 0$$

in rectangular coordinates.

**Solution**    Substituting $x = r \cos \theta$, $y = r \sin \theta$, we obtain

$$r^2 - 3r \cos \theta = 0 \qquad \text{or} \qquad r(r - 3 \cos \theta) = 0.$$

Therefore the graph is $r = 0$ or $r - 3 \cos \theta = 0$. Because the pole is part of the graph of $r - 3 \cos \theta = 0$ (since $r = 0$ when $\theta = \pi/2$), the result is

$$r = 3 \cos \theta.$$

We recognize this graph as the circle of Example 3, Section 4. [In rectangular coordinates, the equation is $(x - \frac{3}{2})^2 + y^2 = \frac{9}{4}$.]    □

**EXAMPLE 2**    Given the polar coordinate equation

$$r = \frac{1}{1 - \cos \theta},$$

find the corresponding equation in rectangular coordinates.

**Solution**    We write

$$r - r \cos \theta = 1 \qquad \text{or} \qquad r = 1 + r \cos \theta.$$

Substituting for $r$ and $\cos \theta$, we get

$$\pm \sqrt{x^2 + y^2} = 1 + x.$$

Squaring both sides (a dangerous operation, since this may introduce extraneous solutions), we obtain

$$x^2 + y^2 = 1 + 2x + x^2 \qquad \text{or} \qquad y^2 = 2x + 1.$$

When we squared both sides we introduced the extraneous solution

$$r = -(1 + r \cos \theta).$$

The graph of this equation is the same as the graph of the equation

$$r = \frac{-1}{1 + \cos \theta}$$

as is seen by solving for $r$. But this last equation can be put in the form

$$(-r) = \frac{1}{1 - \cos(\theta + \pi)},$$

which is obtained from the original by replacing $(r, \theta)$ by $(-r, \theta + \pi)$. Since we know that $(r, \theta)$ and $(-r, \theta + \pi)$ are polar coordinates of the same point, the extraneous graph coincides with the original.

Therefore the correct result is the parabola

$$y^2 = 2x + 1.$$

The graph is drawn in Fig. 30.    □

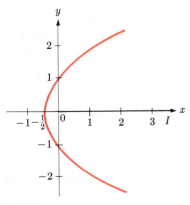

**FIGURE 30**

The straight line was discussed in detail in Chapters 1 and 11. We found that every straight line has an equation (in rectangular coordinates) of the form

$$Ax + By + C = 0.$$

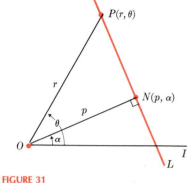

**FIGURE 31**

To obtain the equation in polar coordinates we can simply substitute $x = r \cos \theta$, $y = r \sin \theta$. However, it is also possible to find the desired equation directly. In Fig. 31, we consider a line $L$ (not passing through the pole), and we draw a perpendicular from the pole to $L$. Call the point of intersection $N$ and suppose that the polar coordinates of $N$ are $(p, \alpha)$. The quantities $p$ and $\alpha$ are sufficient to determine the line $L$ completely. To see this let $P(r, \theta)$ be any point on $L$, and note, according to Fig. 31, that $\angle PON = \theta - \alpha$. Therefore

$$r \cos(\theta - \alpha) = p.$$

When $\alpha$ and $p$ are given, this is *the equation of a straight line*. The reader may verify that the same equation results regardless of the positions of $L$ and $N$. The above equation can also be written as

$$r \cos \theta \cos \alpha + r \sin \theta \sin \alpha = p,$$

and changing to rectangular coordinates yields

$$x \cos \alpha + y \sin \alpha = p,$$

which is known as the **normal form of the equation of a straight line**.

**THEOREM 5**    *Any linear equation may be put in normal form:*

$$x \cos \alpha + y \sin \alpha = p.$$

*This equation describes a straight line, and the quantity $p$ is the distance to the line from the origin.*

Fig. 31 illustrates Theorem 5. It is simple to put a linear equation in normal form. When the coefficients of $x$ and $y$ are the cosine and sine of an angle, respectively, the constant term gives the distance from the origin. For example, the linear equation

$$3x + 4y = 7$$

may be written (upon division by $\sqrt{3^2 + 4^2} = 5$) as

$$\tfrac{3}{5}x + \tfrac{4}{5}y = \tfrac{7}{5}.$$

This equation is now in normal form, and the distance from the origin to the line is $\tfrac{7}{5}$ units. (Note that this procedure is the same as the one using the formula for the distance from a point to a line, as given in Chapter 11, Section 1.)

A line through the pole has $p = 0$, and its polar coordinate equation is

$$\cos (\theta - \alpha) = 0.$$

This is equivalent to the statement

$$\theta = \text{const},$$

a condition which we already know represents straight lines passing through the pole.

We shall employ a direct method for *obtaining the equation of a circle in*

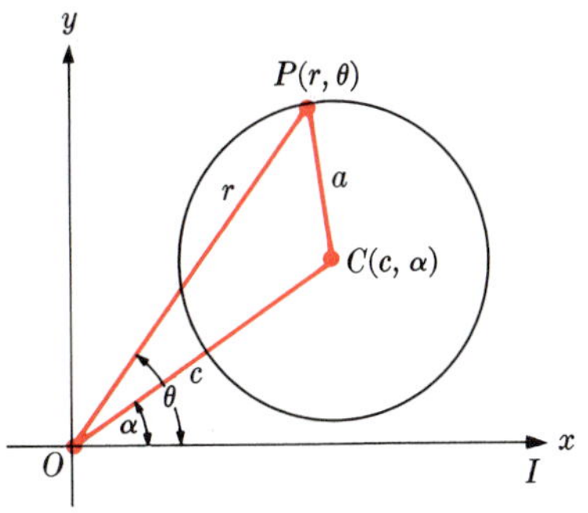

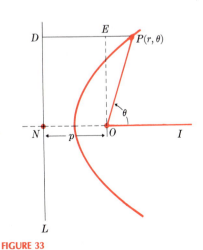

FIGURE 32

polar coordinates. Let the circle have radius $a$ and center at the point $C$ with coordinates $(c, \alpha)$ as shown in Fig. 32. If $P(r, \theta)$ is any point on the circle, then in $\triangle POC$, side $|OP| = r$, side $|OC| = c$, side $|PC| = a$, and $\angle POC = \theta - \alpha$. We use the law of cosines in triangle $POC$, obtaining

$$r^2 - 2cr \cos (\theta - \alpha) + c^2 = a^2.$$

If $c$, $\alpha$, and $a$ are given, that is, if the center and radius of the circle are known, then the equation of the circle is completely determined. It is instructive to draw a figure similar to Fig. 32 but with $a > c$ and to repeat the proof with $P(r, \theta)$ in various positions.

**EXAMPLE 3**    A circle has center at $(5, \pi/3)$ and radius 2. Find its equation in polar coordinates.

**Solution**    We have $c = 5$, $\alpha = \pi/3$, $a = 2$. Therefore

$$r^2 - 10r \cos \left( \theta - \frac{\pi}{3} \right) + 25 = 4 \quad \text{or} \quad r^2 - 10r \cos \left( \theta - \frac{\pi}{3} \right) + 21 = 0.$$

□

If a circle passes through the origin, so that $c = \pm a$, the above equation for a circle becomes particularly simple. The graph is represented by

$$r = 2c \cos (\theta - \alpha).$$

An illustration is given in Example 3, Section 4, (page 510), where we have $c = \frac{3}{2}$ and $\alpha = 0$.

We now take up the study of parabolas, ellipses, and hyperbolas, which we discussed previously in Chapter 11. Polar coordinates are particularly well suited to the discussion of these conics, since the equations of all three types have the same form in this coordinate system. To derive the equations of conics in polar coordinates, we shall solve a problem which differs from the ones we employed in rectangular coordinates. We shall show that

> *the graph of a point which moves so that the ratio of its distance from a fixed point to its distance from a fixed line remains constant is*
>
> *a parabola*    if the ratio is 1,
> *an ellipse*    if the ratio is between 0 and 1,
> *a hyperbola*    if the ratio is larger than 1.

To derive the equation of the above graph, we place the fixed point at the pole and let the fixed line $L$ be a line perpendicular to the initial line and $p$ units to the left of the pole (Fig. 33). If $P(r, \theta)$ is a point on the graph (and $r > 0$), then the conditions say that $|OP|/|PD| = $ a constant which we call the **eccentricity**, $e$. (The eccentricity $e$ should not be confused with the number $e = 2.71828 \ldots$, which is the base of the natural logarithm.) Also, we write (since $|OP| = r$),

$$r = e|PD|,$$

and we see from Fig. 33 that

$$|PD| = |PE| + |ED| = r \cos \theta + p.$$

FIGURE 33

Therefore we have

$$r = er \cos \theta + ep$$

or

$$r = \frac{ep}{1 - e \cos \theta}. \tag{1}$$

Equation (1) represents an ellipse if $0 < e < 1$, a parabola if $e = 1$, and a hyperbola if $e > 1$.

We summarize this discussion with the following theorem.

---

**THEOREM 6**    *A polar equation of the form*

$$r = \frac{ep}{1 - e \cos \theta} \tag{1}$$

*describes a conic section. It is an ellipse if $0 < e < 1$, a parabola if $e = 1$, and a hyperbola if $e > 1$.*

---

We can prove Theorem 6 most easily by transforming Equation (1) to rectangular coordinates and then employing the information on conics in Chapter 11. In the above definition of conic, the fixed point is a **focus** of the conic, and we call the fixed line a **directrix**.

In Theorem 6 we have taken the focus at 0 (the origin) and the directrix is perpendicular to the $x$ axis (the initial line); indeed, the directrix has the equation $x = -p$ in rectangular coordinates. We can obtain similar equations to describe other conics. For example, if we keep the focus at 0 but move the directrix to $x = +p$, we obtain the equation

$$r = \frac{ep}{1 + e \cos \theta}, \tag{2}$$

whereas if we take the directrix parallel to the $x$ axis (the lines $y = p$ or $y = -p$) we obtain:

$$r = \frac{ep}{1 - e \sin \theta} \quad (y = -p) \tag{3}$$

$$r = \frac{ep}{1 + e \sin \theta} \quad (y = +p). \tag{4}$$

Furthermore, it can be shown that the equation

$$r = \frac{ep}{1 - e \cos (\theta - \alpha)}$$

is a conic of eccentricity $e$ with focus at the pole and directrix the line

$$r \cos (\theta - \alpha) = -p.$$

**EXAMPLE 4**  Find the equation of the conic with focus (fixed point) at the origin, with directrix (fixed line) the line $r \cos \theta = -3$, and with eccentricity $\frac{1}{2}$.

**Solution**  The equation $r \cos \theta = -3$ tells us that $p = 3$ in Equation (1) above. Also, we have $e = \frac{1}{2}$, and so

$$r = \frac{\frac{1}{2}(3)}{1 - \frac{1}{2} \cos \theta} = \frac{3}{2 - \cos \theta}.$$

The curve is an ellipse. (See Fig. 34.)

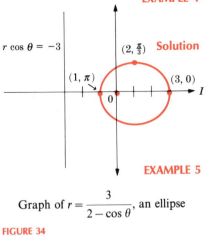

$r \cos \theta = -3$

$(2, \frac{\pi}{3})$

$(1, \pi)$

$(3, 0)$

Graph of $r = \dfrac{3}{2 - \cos \theta}$, an ellipse

**FIGURE 34**

**EXAMPLE 5**  Given the conic equation

$$r = \frac{7}{4 - 5 \cos \theta},$$

find the eccentricity and locate the directrix. Identify the conic.

**Solution**  In order to get the equation in the *exact* form of Equation (1), we divide numerator and denominator by 4, getting

$$r = \frac{\frac{7}{4}}{1 - \frac{5}{4} \cos \theta}.$$

Then we read off the eccentricity $e = \frac{5}{4}$ and $ep = \frac{7}{4}$. This gives $p = \frac{7}{5}$, and the directrix is the line $r \cos \theta = -\frac{7}{5}$. The curve is a hyperbola. (See Fig. 35.)

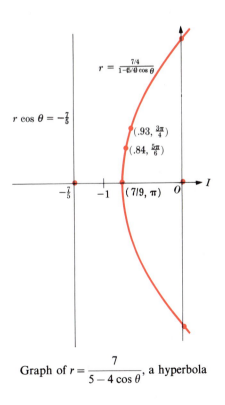

$r = \frac{7/4}{1 - 5/4 \cos \theta}$

$r \cos \theta = -\frac{7}{5}$

$(.93, \frac{3\pi}{4})$

$(.84, \frac{5\pi}{6})$

$-\frac{7}{5}$     $-1$     $(7/9, \pi)$     $0$

Graph of $r = \dfrac{7}{5 - 4 \cos \theta}$, a hyperbola

**FIGURE 35**

**EXAMPLE 6**   Given the conic with equations

$$r = \frac{3}{1 + 2\cos\theta},$$

find the eccentricity, locate the directrix, and sketch the graph.

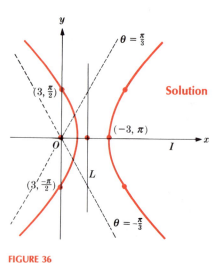

FIGURE 36

**Solution**   This equation is in the form (2), so the eccentricity is 2. Since $ep = 2p = 3$, we have $p = \frac{3}{2}$. Also since $e = 2$, the equation represents a hyperbola. The directrix is the line $x = +p$, in this case $x = \frac{3}{2}$. Making a table of values we have:

| $\theta$ | 0 | $\dfrac{-\pi}{2}$ | $\dfrac{\pi}{2}$ | $\dfrac{2\pi}{3}$ |
|---|---|---|---|---|
| $r$ | 1 | 3 | 3 | $\infty$ |

From the table it is clear that $\theta = \dfrac{2\pi}{3}$ $\left(\text{and } \theta = \dfrac{-2\pi}{3}\right)$ are asymptotes. Using symmetry we sketch a graph of the hyperbola as shown in Fig. 36. $\qquad\square$

## 5   PROBLEMS

In Problems 1 through 12, an equation is given in rectangular coordinates. In each case find an equation in polar coordinates which describes the same graph.

**1**  $x = -2$

**2**  $y = 3$

**3**  $x - y = 0$

**4**  $2x + 3y = 0$

**5**  $2x + y\sqrt{2} = 4$

**6**  $3x + 3y = 7$

**7**  $xy = 4$

**8**  $y^2 = 2x$

**9**  $x^2 + y^2 + 2x - 4y = 0$

**10**  $x^2 + y^2 - 4x = 0$

**11**  $x^2 + y^2 + 2x + 6y = 0$

**12**  $x^2 + y^2 - 4x + 4y + 4 = 0$

In Problems 13 through 32, in each case find a polynomial equation in rectangular coordinates whose graph contains that of the equation given in polar coordinates. Discuss possible extraneous solutions.

**13**  $r = 7$

**14**  $\theta = \pi/3$

**15**  $r = 3\cos\theta$

**16**  $r = 4\sin\theta$

**17**  $r\cos\theta = 5$

**18**  $r\sin\theta = -2$

**19**  $r\cos(\theta - \pi/4) = 2$

**20**  $r\cos(\theta + \pi/6) = 1$

**21**  $r^2\cos 2\theta = 4$

**22**  $r^2\sin 2\theta = 2$

**23**  $r^2 = \sin 2\theta$

**24**  $r^2 = \cos 2\theta$

**25**  $r = 2\sec\theta\tan\theta$

**26**  $r(2 - \cos\theta) = 2$

**27**  $r(1 - 2\cos\theta) = 2$

**28**  $r = a\cos 3\theta$

**29**  $r = a\sin 3\theta$

**30**  $r = 1 - \cos\theta$

**31**  $r = 1 - 2\sin\theta$

**32**  $r = a\cos 2\theta$

In Problems 33 through 38, in each case the line $L$ passes through the point $N$ and is perpendicular to the line going through $N$ and the pole $O$. Find the equation of $L$ in polar coordinates when the coordinates of $N$ are as given.

**33**  $N(2, \pi/3)$

**34**  $N(3, 5\pi/6)$

**35**  $N(-1, \pi/2)$

**36**  $N(2, 0)$

**37**  $N(-3, -\pi)$

**38**  $N(-3, -\pi/2)$

In Problems 39 through 45, in each case find in polar coordinates the equation of the circle satisfying the given conditions.

**39**  Center $(3, \pi/6)$, radius 2

**40**  Center $(-2, 5\pi/6)$, radius 3

**41**  Center $(4, 0)$, radius 4

**42**  Center $(5, \pi/2)$, radius 5

**43**  Center $(4, \pi/3)$, passing through $(7, \pi/4)$

**44**  Center $(5, \pi/4)$ passing through $(6, 0)$

**45**  Center on the line $\theta = \pi/3$ and passing through $(5, \pi/2)$ and $(0, 0)$

In Problems 46 through 51 find, in polar coordinates, the equation of the conic with focus at the pole and having the given eccentricity and directrix.

**46** $e = 1$, directrix $r \cos \theta = -4$

**47** $e = 2$, directrix $r \cos \theta = -5$

**48** $e = \frac{1}{2}$, directrix $r \cos \theta = 2$

**49** $e = \frac{1}{2}$, directrix $r \cos (\theta - \pi/2) = 1$

**50** $e = 1$, directrix $r \cos (\theta - \pi/4) = 2\sqrt{2}$

**51** $e = 2$, directrix $r \cos (\theta + \pi/3) = 3$

**52** Given the line with equation $r \cos (\theta - \pi/3) = 2$, find the distance from the point $(2, \pi/2)$ to this line.

**53** Given the line with equation $r \cos (\theta - \alpha) = p$, find a formula for the distance $d$ from a point $(r_1, \theta_1)$ to this line.

In Problems 54 through 65, conics are given. Find the eccentricity and directrix of each and sketch the curve.

**54** $r = \dfrac{6}{1 - \cos \theta}$

**55** $r = \dfrac{9}{2 - \cos \theta}$

**56** $r = \dfrac{4}{2 - 3 \cos (\theta - \pi/4)}$

**57** $r = \dfrac{2}{1 - \sin \theta} \left[ Note: \sin \theta = \cos \left( \theta - \dfrac{\pi}{2} \right). \right]$

**58** $r = \dfrac{4}{2 - 4 \sin \theta}$

**59** $r = \dfrac{5}{1 - 2 \cos (\theta - \pi/3)}$

**60** $r = \dfrac{2}{2 + \cos \theta}$

**61** $r = \dfrac{4}{6 + \sin \theta}$

**62** $r = \dfrac{4 \sec \theta}{2 \sec \theta - 1}$

**63** $r = \dfrac{3 \sec \theta}{4 \sec \theta + 1}$

**64** $r = \dfrac{2 \csc \theta}{\csc \theta + 2}$

**65** $r = \dfrac{3 \csc \theta}{6 \csc \theta - 1}$

**66** Transform the equation

$$r = \frac{ep}{1 - e \cos \theta}$$

into rectangular coordinates. Make use of the results on conics in Chapter 11 to show that the above equation is an ellipse if $0 < e < 1$, a hyperbola if $e > 1$, and a parabola if $e = 1$.

**67** Consider the line $L: r \sin \theta = p$, where $p$ is a constant. Find, in polar coordinates, the equation of the graph of all points such that the ratio of their distances from the pole to their distances from $L$ is a constant ($= e$). Use the method in Problem 66 to identify the conic for various values of $e$.

**68** Transform the equation

$$r = \frac{ep}{1 - e \cos (\theta - \alpha)}$$

into rectangular coordinates ($p, e, \alpha$ are constants). Use the results of Chapter 11, Section 6 to identify the conic for various values of $e$.

**69** If $a^2 + b^2 > 0$ and $c \neq 0$, show that the graph of

$$r = \frac{c}{a \cos \theta + b \sin \theta}$$

is always a straight line.

**70** Given the hyperbola $r = 3/(1 - 3 \cos \theta)$, show that the asymptotes have inclinations

$$\theta = \arccos (1/3), \qquad \theta = \arccos (-1/3).$$

**71** Given the equation $r = ep/(1 - e \cos \theta)$, $e > 1$, show that the asymptotes have inclinations

$$\theta = \arccos (1/e), \qquad \theta = \arccos (-1/e).$$

**72** Show that the graphs of the equations $r = a \sec^2(\theta/2)$ and $r = a \csc^2(\theta/2)$ are parabolas.

**73** The equation $r^2 = \cos 4\theta$ corresponds to a polynomial equation in $x$ and $y$. What is the degree of this polynomial?

**74** The equation $r^2 = \cos 2n\theta$, with $n$ a positive integer, corresponds to a polynomial equation in $x$ and $y$. What is the degree of this polynomial?

---

**6**

# DERIVATIVES IN POLAR COORDINATES

Consider a curve given in polar coordinates by an equation of the form $r = f(\theta)$. We wish to investigate the meaning of the derivative. Before doing so, however, we shall examine the graph of one of the simplest equations in polar coordinates. The equation

$$r = \theta$$

represents a **spiral of Archimedes**, which we discussed on page 512. We saw there that the curve is an ever-increasing spiral around the origin. The dashed portion in Fig. 37 corresponds to negative values of $\theta$.

We know that for a function given in rectangular coordinates, the derivative gives the slope of the tangent line. If we plunge ahead blindly and take derivatives in polar coordinates with the intention of obtaining the slope, trouble arises at once. In the example of the spiral of Archimedes we have $dr/d\theta = 1$, and the derivative is clearly not the slope. We need more information in order to learn about slopes in polar coordinates.

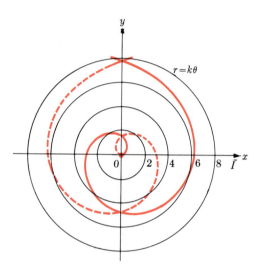

**FIGURE 37**

If $r = f(\theta)$, then the equations in rectangular coordinates,

$$x = r \cos \theta, \qquad y = r \sin \theta,$$

when we substitute $f(\theta)$ for $r$, may be considered as the *parametric equations of a curve with $\theta$ as parameter*. We then have

$$x = f(\theta) \cos \theta, \qquad y = f(\theta) \sin \theta.$$

Differentiating, we find that

$$\frac{dx}{d\theta} = f'(\theta) \cos \theta - f(\theta) \sin \theta, \qquad \frac{dy}{d\theta} = f'(\theta) \sin \theta + f(\theta) \cos \theta.$$

The slope is

$$\frac{dy}{dx} = \frac{dy/d\theta}{dx/d\theta}.$$

Suppose that the curve $r = f(\theta)$ has the appearance of the one in Fig. 38. At a point $P$ the tangent line is drawn, and we recognize the slope of this line to be $\tan \phi$. The slope is not a particularly convenient quantity in polar coordinates, but the angle $\psi$ between the tangent line and the line from $P$ passing through the pole turns out to be convenient. As seen in Fig. 38, $\psi$ and $\phi$ bear the simple relationship

$$\psi = \phi - \theta,$$

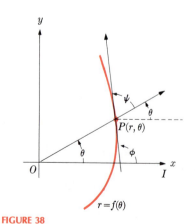

**FIGURE 38**

and so

$$\tan \psi = \tan (\phi - \theta) = \frac{\tan \phi - \tan \theta}{1 + \tan \phi \tan \theta}.$$

We know that

$$\tan \phi = \frac{dy}{dx} = \frac{dy/d\theta}{dx/d\theta} = \frac{f'(\theta) \sin \theta + f(\theta) \cos \theta}{f'(\theta) \cos \theta - f(\theta) \sin \theta}. \tag{1}$$

When we recall that $r$ and $\theta$ are connected by the relation $r = f(\theta)$, we have established the following theorem.

---

**THEOREM 7**   *Given an equation $r = f(\theta)$ in polar coordinates, if $f$ is differentiable then the slope $\tan \phi$ of the tangent line to the graph of $r = f(\theta)$ at $(r, \theta)$ is given by*

$$\tan \phi = \frac{\dfrac{dr}{d\theta} \sin \theta + r \cos \theta}{\dfrac{dr}{d\theta} \cos \theta - r \sin \theta}. \tag{2}$$

---

It is a good exercise on algebraic manipulation to substitute Formula (1) for $\tan \phi$ into the formula for $\tan \psi$ and obtain the simple relation

$$\tan \psi = \frac{f(\theta)}{f'(\theta)}.$$

We can also write

$$\cot \psi = \frac{f'(\theta)}{f(\theta)} = \frac{1}{r} \frac{dr}{d\theta}, \quad r \neq 0.$$

The significance of the derivative in polar coordinates is now becoming clear. The derivative $f'(\theta)$ at a point $P$ is related to the angle that the tangent line forms with the line through the point $P$ and the pole, according to the above formula for $\cot \psi$.

**EXAMPLE 1**   Given the circle $r = 4 \cos \theta$, find the angle between the tangent line and the line from the pole through the point of tangency. Evaluate this angle at $(2, \pi/3)$.

**Solution**   We have

$$\cot \psi = \frac{1}{r} \frac{dr}{d\theta} = \frac{1}{4 \cos \theta} (-4 \sin \theta) = -\tan \theta,$$

and so

$$\psi = \theta + \frac{\pi}{2}.$$

For $\theta = \pi/3$, we obtain $\psi = 5\pi/6$.   □

**EXAMPLE 2**    Given the curve $r = 3e^{2\theta}$. Find $\cot \psi$ at any point, and sketch the curve.

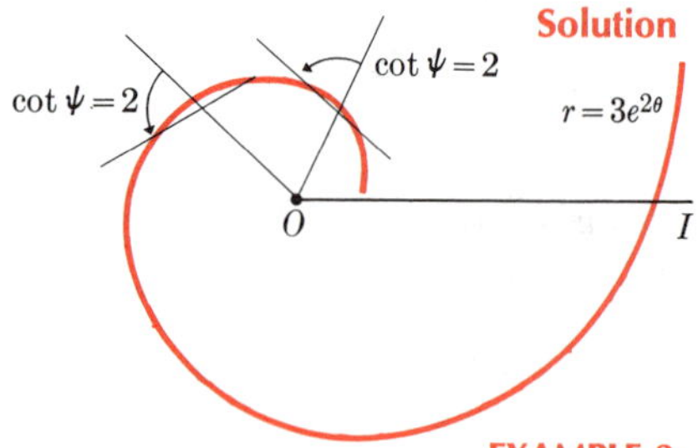

**Solution**    We have

$$\cot \psi = \frac{1}{r}\frac{dr}{d\theta} = \frac{1}{3e^{2\theta}} \cdot 3 \cdot 2e^{2\theta} = 2.$$

In other words, the angle which the tangent makes with the line from the pole through the point of tangency is always the same. The curve, called a **logarithmic spiral**, is shown in Fig. 39. (See the discussion on page 512.) □

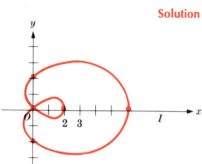

**FIGURE 39**

**EXAMPLE 3**    Given the equation $r = 2 + 4 \cos \theta$, find the slope of the tangent line at $\theta = \dfrac{\pi}{6}$.

**Solution**    The graph is sketched in Fig. 40. Since $\tan \phi$ is the slope of the tangent line, we can use Formula (2) of Theorem 7. We have

$$\tan \phi = \frac{(-4 \sin \theta) \sin \theta + (2 + 4 \cos \theta) \cos \theta}{(-4 \sin \theta) \cos \theta - (2 + 4 \cos \theta) \sin \theta}$$

$$= \frac{2(\cos^2 \theta - \sin^2 \theta) + \cos \theta}{-4 \sin \theta \cos \theta - \sin \theta}.$$

Using trigonometric identities, we simplify the above expression to

$$\tan \phi = \frac{2 \cos 2\theta + \cos \theta}{-2 \sin 2\theta - \sin \theta}.$$

Sketch of graph of $r = 2 + 4 \cos \theta$

**FIGURE 40**

Evaluating at $\theta = \dfrac{\pi}{6}$, we have

$$\tan \phi = \frac{2 \cos\left(\dfrac{\pi}{3}\right) + \cos\left(\dfrac{\pi}{6}\right)}{-2 \sin\left(\dfrac{\pi}{3}\right) - \sin\left(\dfrac{\pi}{6}\right)} = -\frac{2 + \sqrt{3}}{1 + 2\sqrt{3}}. \qquad \square$$

To obtain a formula for **arc length** in polar coordinates, we start with the formula

$$ds^2 = dx^2 + dy^2,$$

which we discussed in Section 2. If $r = f(\theta)$ has a continuous first derivative, we then write, as before,

$$\frac{dx}{d\theta} = f'(\theta) \cos \theta - f(\theta) \sin \theta, \qquad \frac{dy}{d\theta} = f'(\theta) \sin \theta + f(\theta) \cos \theta.$$

Substitution of these two expressions in the formula

$$\frac{ds}{d\theta} = \sqrt{(dx/d\theta)^2 + (dy/d\theta)^2}$$

yields (after some algebraic manipulation)

$$\frac{ds}{d\theta} = \sqrt{[f'(\theta)]^2 + [f(\theta)]^2},$$

and, upon integration,

$$s = \int_{\theta_0}^{\theta_1} \sqrt{(dr/d\theta)^2 + r^2} \; d\theta.$$

**THEOREM 8**    *Let* $r = f(\theta)$, $\alpha \le \theta \le \beta$, *describe an arc in polar coordinates. If f has a continuous first derivative, then the arc length s is given by*

$$s = \int_{\alpha}^{\beta} \sqrt{\left(\frac{dr}{d\theta}\right)^2 + r^2} \; d\theta$$

We provide an example showing how this formula may be used to compute arc length when the equation of the arc is given in polar coordinates.

**EXAMPLE 4**    Find the arc length of the curve $r = 3e^{2\theta}$ from $\theta = 0$ to $\theta = \pi/6$.

**Solution**    We have $dr/d\theta = 6e^{2\theta}$, and so

$$s = \int_0^{\pi/6} \sqrt{36e^{4\theta} + 9e^{4\theta}} \; d\theta = 3\sqrt{5} \int_0^{\pi/6} e^{2\theta} \; d\theta = \left[\tfrac{3}{2}\sqrt{5}e^{2\theta}\right]_0^{\pi/6}.$$

Therefore

$$s = \tfrac{3}{2}\sqrt{5}\left[e^{\pi/3} - 1\right] = 6.2 \text{ approx.} \qquad \square$$

## 6   PROBLEMS

In Problems 1 through 14, find $ds/d\theta$ and $\cot \psi$ for each equation.

**1**   $r = 2a \sin \theta$

**2**   $r = 3 \cos \theta + 4 \sin \theta$

**3**   $r = 3(1 - \cos \theta)$

**4**   $r = 2(1 + \sin \theta)$

**5**   $r = -5\theta$

**6**   $r = 3\theta^2$

**7**   $r = 3 + 2 \cos \theta$

**8**   $r = 1 - 2 \cos \theta$

**9**   $r = e^{5\theta}$

**10**   $r = e^{-4\theta}$

**11**   $r(2 - \sin \theta) = 2$

**12**   $r(1 + \cos \theta) = 3$

**13**   $r(1 + 2 \cos \theta) = 2$

**14**   $r(5 - 6 \sin \theta) = -1$

In Problems 15 through 19, find the slope of the tangent line to the graph of the equation at the point corresponding to the indicated value of $\theta$.

**15**   $r = 3 \cos \theta$, $\theta = \dfrac{\pi}{3}$

**16**   $r = 4(1 - \sin \theta)$, $\theta = \dfrac{-\pi}{3}$

**17**   $r = 8 \cos 3\theta$, $\theta = \dfrac{-3\pi}{4}$

**18**   $r^2 = 4 \cos 2\theta$, $\theta = \dfrac{\pi}{6}$

**19**   $r = 2^\theta$, $\theta = \pi$

Problems 20 through 23, find the length of arc as indicated.

**20**   $r = 3\theta^2$ from $\theta = 1$ to $\theta = 2$

**21**   $r = 2e^{3\theta}$ from $\theta = 0$ to $\theta = 3$

**22**   $r = 3 \cos \theta$ from $\theta = 0$ to $\theta = \pi/4$

**23**   $r = 3(1 + \cos \theta)$ from $\theta = 0$ to $\theta = \pi/2$

**24**   Suppose that $r = f(\theta)$ and $x = f(\theta) \cos \theta$, $y = f(\theta) \sin \theta$. Find an expression for $d^2y/dx^2$ in terms of $\theta$.

$$\left[ \text{Hint: } \frac{d^2y}{dx^2} = \frac{\dfrac{d}{d\theta}\left(\dfrac{dy}{dx}\right)}{dx/d\theta}. \right]$$

**25**   Using the result of Problem 24, show that the curvature $\kappa$ is given by the formula

$$\kappa = \frac{r^2 + 2(dr/d\theta)^2 - r(d^2r/d\theta^2)}{[r^2 + (dr/d\theta)^2]^{3/2}}.$$

**26** Using the formula in Problem 25, find $\kappa$, given that $r = 2a \sin \theta$.

**27** Using the formula of Problem 25, find $\kappa$, given that $r = 1 - \cos \theta$.

**28** Draw the curve of $r = 1 - \cos \theta$ and construct the tangent line at the point where $\theta = \pi/3$. Find the equation of the tangent line at this point in rectangular coordinates.

**29** Draw the curve of $r = 2 + \sqrt{2} \cos \theta$ and construct the

tangent line at the point where $\theta = 3\pi/4$. Find the equation of this tangent line in rectangular coordinates.

**30** Given that $r = f(\theta)$. Find, in polar coordinates, the formula for the equation of the line tangent to the curve at the point $(r_1, \theta_1)$ on the curve.

**31** Given that $r = f(\theta)$. Find, in polar coordinates, the formula for the line normal to the curve at the point $(r_1, \theta_1)$ on the curve.

---

**7**

---

## AREA IN POLAR COORDINATES

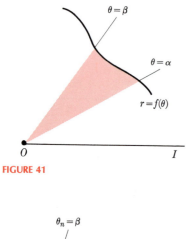

FIGURE 41

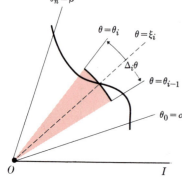

FIGURE 42

Two basic concepts are needed for the development of area in rectangular coordinate systems. They are (1) the idea of the limit of a sum, and (2) the formula for the area of a rectangle: length times width.

To find areas enclosed by curves given in polar coordinates, again two concepts are necessary. They are, first, the idea of the limit of a sum (as in rectangular coordinates) and second, the formula for the area of a sector of a circle. We recall that *the area of a sector of a circle of radius a with angle opening θ* (*measured in radians*) *is*

$$\tfrac{1}{2}\theta a^2.$$

Suppose that $r = f(\theta)$ is a continuous, positive function defined for values of $\theta$ between $\theta = \alpha$ and $\theta = \beta$, with $0 \leq \alpha < \beta \leq 2\pi$. We construct the lines $\theta = \alpha$ and $\theta = \beta$, and we pose the problem of determining the area of the region bounded by these straight lines and the curve with equation $r = f(\theta)$. (See Fig. 41). We subdivide the $\theta$ scale into $n$ parts between $\alpha$ and $\beta$ by introducing the values $\alpha = \theta_0 < \theta_1 < \theta_2 < \cdots < \theta_{n-1} < \theta_n = \beta$. In this way we obtain $n$ subintervals; in each subinterval we select a value of $\theta$ which we call $\xi_i$. We then compute the area of the circular sector of radius $f(\xi_i)$ and angle opening $\Delta_i\theta = \theta_i - \theta_{i-1}$, as shown in Fig. 42. According to the formula for the area of a sector of a circle given above, the area is

$$\tfrac{1}{2}[f(\xi_i)]^2\Delta_i\theta.$$

We add these areas for $i$ between 1 and $n$, getting

$$\tfrac{1}{2}[f(\xi_1)]^2\Delta_1\theta + \tfrac{1}{2}[f(\xi_2)]^2\Delta_2\theta + \tfrac{1}{2}[f(\xi_3)]^2\Delta_3\theta + \cdots + \tfrac{1}{2}[f(\xi_n)]^2\Delta_n\theta,$$

or, concisely,

$$\tfrac{1}{2}\sum_{i=1}^{n}[f(\xi_i)]^2\Delta_i\theta.$$

It can be shown (although we shall not do so) that as the number of subdivision points increases without bound and as the norm of the subdivision (length of the largest $\Delta_i\theta$) tends to zero, the area A, bounded by the lines $\theta = \alpha$, $\theta = \beta$, and by the curve $r = f(\theta)$, is given by

$$A = \tfrac{1}{2}\int_{\alpha}^{\beta} [f(\theta)]^2\, d\theta.$$

We have just obtained the desired formula for area in polar coordinates, and we summarize the above discussion in the following theorem.

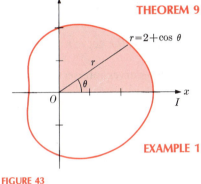

**FIGURE 43**

**THEOREM 9**    *Let f be a continuous, positive function with domain containing* $[\alpha, \beta]$, *where* $0 \le \alpha < \beta \le 2\pi$. *Then the area A of the region bounded by the lines* $\theta = \alpha$, $\theta = \beta$ *and the curve* $r = f(\theta)$ *is given by*

$$A = \tfrac{1}{2}\int_{\alpha}^{\beta} [f(\theta)]^2\, d\theta.$$

**EXAMPLE 1**    Find the area bounded by the curve

$$r = 2 + \cos\theta$$

and the lines $\theta = 0$, $\theta = \pi/2$ (see Fig. 43).

**Solution**    We have

$$A = \tfrac{1}{2}\int_{0}^{\pi/2} [2 + \cos\theta]^2\, d\theta = \tfrac{1}{2}\int_{0}^{\pi/2} [4 + 4\cos\theta + \cos^2\theta]\, d\theta$$

$$= [2\theta + 2\sin\theta]_0^{\pi/2} + \tfrac{1}{2}\int_{0}^{\pi/2} \cos^2\theta\, d\theta = \pi + 2 + \tfrac{1}{2}\int_{0}^{\pi/2} \cos^2\theta\, d\theta.$$

To perform the remaining integration we use the half-angle formula

$$\cos^2\theta = (1 + \cos 2\theta)/2,$$

and obtain

$$\tfrac{1}{2}\int_{0}^{\pi/2} \cos^2\theta\, d\theta = \tfrac{1}{4}\int_{0}^{\pi/2} (1 + \cos 2\theta)\, d\theta = [\tfrac{1}{4}(\theta + \tfrac{1}{2}\sin 2\theta)]_0^{\pi/2}.$$

Therefore

$$A = \pi + 2 + \frac{\pi}{8} = \frac{9\pi}{8} + 2. \qquad \square$$

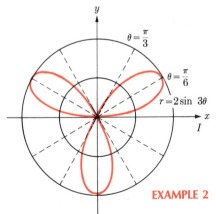

**FIGURE 44**

**EXAMPLE 2**    Find the entire area enclosed by the curve $r = 2\sin 3\theta$.

**Solution**    We first draw the curve, as shown in Fig. 44. In fact, in all area problems in polar coordinates (as well as in rectangular coordinates), a sketch should be made. The curve is a **rose curve** with three congruent petals. *The main difficulty in this problem is determining the limits of integration.* To find the area of the loop in the first quadrant, we observe that $r = 0$ when $\theta = 0$, and that the value of $r$ increases steadily to a maximum when $\theta = \pi/6$ and $r = 2$; then $r$ decreases steadily until $\theta = \pi/3$ when $r = 0$. In other words, the petal is described completely as $\theta$ goes from 0 to $\pi/3$. The total area $A$ is three times the area of this loop. Therefore we have

$$\tfrac{1}{3}A = \tfrac{1}{2}\int_0^{\pi/3} (2\sin 3\theta)^2\, d\theta \qquad \text{or} \qquad A = 6\int_0^{\pi/3} \frac{1-\cos 6\theta}{2}\, d\theta,$$

where we have used the half-angle formula, $\sin^2\phi = (1 - \cos 2\phi)/2$. Integration yields

$$A = 6\left[\frac{\theta}{2} - \frac{1}{12}\sin 6\theta\right]_0^{\pi/3} = \pi. \qquad \square$$

**EXAMPLE 3**  Find the area inside the circle $r = 5\cos\theta$ and outside the curve $r = 2 + \cos\theta$.

**Solution**  We make a sketch, as shown in Fig. 45. The two curves intersect when

$$5\cos\theta = 2 + \cos\theta \Leftrightarrow \cos\theta = \tfrac{1}{2},$$

or

$$\theta = \frac{\pi}{3},\ -\frac{\pi}{3}.$$

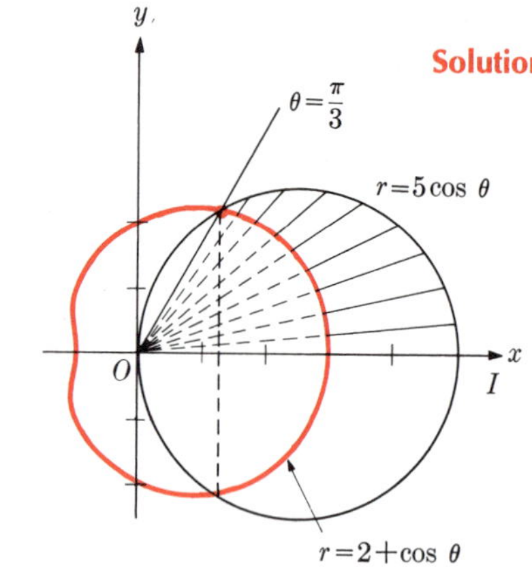

The limits of integration are $-\pi/3$ and $+\pi/3$. The area $A$ inside $5\cos\theta$ and outside $2 + \cos\theta$ is given by

$$A = \tfrac{1}{2}\int_{-\pi/3}^{+\pi/3} (5\cos\theta)^2\, d\theta - \tfrac{1}{2}\int_{-\pi/3}^{+\pi/3} (2 + \cos\theta)^2\, d\theta.$$

We can shorten the work in two ways: (1) We observe that both curves are symmetric with respect to the $x$ axis, and (2) the integrals may be combined, since the limits of integration are the same. In this way we obtain

$$A = \int_0^{\pi/3} [25\cos^2\theta - (4 + 4\cos\theta + \cos^2\theta)]\, d\theta$$

$$= \int_0^{\pi/3} (8 + 12\cos 2\theta - 4\cos\theta)\, d\theta$$

$$= [8\theta + 6\sin 2\theta - 4\sin\theta]_0^{\pi/3} = \frac{8\pi}{3} + \sqrt{3}. \qquad \square$$

FIGURE 45

## 7  PROBLEMS

In Problems 1 through 9, in each case find the area enclosed by the curve and the two lines.

1  $r = 2\theta,\ \theta = 0,\ \theta = \pi$

2  $r = 2\tan\theta,\ \theta = 0,\ \theta = \pi/6$

3  $r = e^\theta,\ \theta = 0,\ \theta = 3\pi/2$

4  $r = 2\cos 2\theta,\ \theta = -\pi/4,\ \theta = \pi/4$

5  $r = \theta^2,\ \theta = 0,\ \theta = \pi$

6  $r = -3\theta^3,\ \theta = 0,\ \theta = \pi/2$

7  $r = 1/\theta,\ \theta = \pi/4,\ \theta = \pi/2$

8  $r = 2\sin\theta,\ \theta = \dfrac{\pi}{6},\ \theta = \dfrac{\pi}{3}$

9  $r = \dfrac{4}{(1 - \cos\theta)},\ \theta = 0,\ \theta = \pi$

In Problems 10 through 21, sketch each curve and find the entire area bounded by it.

**10** $r = 3 + 2 \cos \theta$

**11** $r = 2 \cos 3\theta$

**12** $r^2 = \sin \theta$

**13** $r^2 = 2 \sin 2\theta$

**14** $r = 2 \sin 2\theta$

**15** $r = 2(1 + \cos \theta)$

**16** $r = \cos n\theta$, $n$ a positive integer

**17** $r = \sin n\theta$, $n$ a positive integer

**18** $r = 4 + 2 \sin \theta$

**19** $r = 1 + \cos \theta$

**20** $r = 3 - 3 \sin \theta$

**21** $r^2 = 4 \cos 2\theta$

In Problems 22 through 31, find the indicated area.

**22** Inside the circle $r = 4 \cos \theta$ and outside $r = 2$

**23** Inside $r = 2 + 2 \cos \theta$ and outside $r = 1$

**24** Inside $r^2 = 8 \cos 2\theta$ and outside $r = 2$

**25** Inside $r = \sqrt{2 \cos \theta}$ and outside $r = 2(1 - \cos \theta)$

**26** Inside the curves $r = 3 \cos \theta$ and $r = 2 - \cos \theta$

**27** Inside the curves $r = 2 \sin \theta$ and $r^2 = 2 \cos 2\theta$

**28** Inside the small loop of the curve $r = 1 + 2 \cos \theta$

**29** One loop of $r^2 = 4 \cos 2\theta$

**30** One loop of $r = 3 \cos 5\theta$

**31** Inside the curves $r = 2(1 + \sin \theta)$, $r = 1$

**\*32** A chord of a circle makes an angle $\alpha$ with the tangents at its ends. Find the area of the segment cut off.

**\*33** Find the area bounded by the line $y = x$, the curve $(x^2 + a^2)y^2 = 4a^2 x^2$, in the region $\{(x, y): x > 0, y > 0\}$ (i.e., the first quadrant).

# CHAPTER 12

## REVIEW PROBLEMS

In each of the Problems 1 through 13, plot the curve given by the parametric equations. In each case eliminate the parameter and get an equation relating $x$ and $y$.

**1** $x = \cos t$, $y = \sin t$

**2** $x = 4t^2 - 5$, $y = 2t + 3$

**3** $x = e^t$, $y = e^{-2t}$

**4** $x = \frac{1}{2}(e^t + e^{-t})$, $y = \frac{1}{2}(e^t - e^{-t})$

**5** $x = t^3$, $y = t^2$

**6** $x = \cos t$, $y = \cos t + \sin t$

**7** $x = 1 + 2^{-t}$, $y = 2^t$

**8** $x = \cos t$, $y = \cos t - \sin t$

**9** $x = 2 - t$, $y = t + t^3$

**10** $x = t^2 - 1$, $y = t^3 - 5t$

**11** $x = 1 - t$, $y = 1 + t$

**12** $x = \dfrac{1}{1 - t}$, $y = \dfrac{1}{1 - t^2}$

**13** $x = \dfrac{1}{1 + t}$, $y = \dfrac{2}{1 + t^2}$

In Problems 14 through 21, find the first and second derivatives of $y$ with respect to $x$. Give the results in terms of the parameter.

**14** $x = 1 + t^2$, $y = 4t - 3$

**15** $x = t^3 + 7$, $y = 6t^2 - 1$

**16** $x = 3(t - 1)^2$, $y = 8t^2 + 7$

**17** $x = \dfrac{1}{t^2}$, $y = t^2 - 4t + 2$

**18** $x = (t - 3)^{3/2}$, $y = 2t^2 - 1$

**19** $x = \frac{1}{2}(e^t - e^{-t})$, $y = \frac{1}{2}(e^t + e^{-t})$

**20** $x = 3 \cos 2\theta$, $y = 4 \cos \theta$

**21** $x = \tan \theta$, $y = \sec 2\theta$

In Problems 22 through 27, find the equations of the tangent line and the normal line to the specified curve at the point corresponding to the given value of the parameter.

**22** $x = 2 - t$, $y = 1 + 2t^2$, $t = 1$

**23** $x = e^t$, $y = 2t^2 - 3$, $t = 1$

**24** $x = 3 \cos t$, $y = 4 \sin t$, $t = \pi/6$

**25** $x = t \cos t$, $y = \sin t$, $t = \pi/2$

**26** $x = t^2$, $y = t^3 + t$, $t = 1.423$

**27** $x = \ln t$, $y = 3 + \sin t$, $t = 2$

In Problems 28 through 35, find the length of the indicated arc.

**28** $x = e^{-3t}$, $y = e^{-2t}$, $(t > 0)$

**29** $x = e^{-t}$, $y = e^{-t} - e^{-2t}$, $(t > 0)$

**30** $x = t^2$, $y = 4t^3$, $(0 \le t \le 1)$

**31** The cardioid $r = 2(1 + \cos \theta)$

**32** The cardioid $r = 7(1 - \sin \theta)$

**33** The curve $r = 2 \cos^2 \theta$

**34** The logarithmic spiral $r = 2e^{3\theta}$, $(0 \leq \theta \leq \pi)$

**35** The spiral of Archimedes $r = \pi\theta$ from $\theta = 0$ to $\theta = \pi$.

**\*36** For the cycloid $x = 7(\theta - \sin\theta)$, $y = 7(1 - \cos\theta)$, find the length of one arch.

In Problems 37 through 47, find the curvature and the radius of curvature at the indicated point $P$. If no point is given, find the curvature and radius of curvature at all points of the curve.

**37** $y = x - \frac{1}{8}x^2$; $P(1, \frac{7}{8})$

**38** $y = 2 + 2x - x^2$; $P(\frac{1}{2}, \frac{11}{4})$

**39** $y = x(x + 2)^2$; $P(0, 0)$

**40** $y = \cos x$; $P(\pi, -1)$

**41** $y = 6 \sec \dfrac{x}{6}$ at $x = \dfrac{3}{2}\pi$

**42** $y = \ln \tan\left(\dfrac{x}{2}\right)$ at $x = \dfrac{\pi}{4}$

**43** $x = 3t + 1$, $y = t^2$

**44** $x = 1 - 3t$, $y = t^2 + 2$

**45** $x = 2 \sin^3\theta$, $y = 2 \cos^3\theta$

**46** $x = \tan\theta$, $y = \cot\theta$

**47** $x = \cos^4\theta$, $y = \sin^4\theta$

In each of Problems 48 through 55, discuss for symmetry and plot the graph of the equation.

**48** $r = 2\cos\theta$          **49** $r = 3\sin 2\theta$

**50** $r = 2\cos^2\theta$        **51** $r = 2(4 + \cos\theta)$

**52** $r = 2\sin\theta - 1$      **53** $r = \frac{3}{2}(4\cos^2\theta - 3)$

**54** $r = \sec^2\theta$         **55** $r = 2\csc^2\theta$

In Problems 56 through 59, show that the two equations represent the same curve.

**56** $r = \sec\theta + 1$; $r = \sec\theta - 1$   (conchoid)

**57** $r = \sec\theta - \tan\theta$; $r = \sec\theta + \tan\theta$   (strophoid)

**58** $r = 2\cos\theta - 1$; $r = 2\cos\theta + 1$   (limaçon)

**59** $r = \sqrt{2}\,(\sec\theta - \cos\theta)$; $r = \sqrt{2}\,\sin\theta\tan\theta$   (cissoid)

In Problems 60 through 66, conics are given in polar coordinates. Find the eccentricity and directrix of each, and sketch the curve.

**60** $r = \dfrac{3}{1 - \cos\theta}$          **61** $r = \dfrac{8}{1 - 2\cos\theta}$

**62** $r = \dfrac{2}{3 + \sin\theta}$          **63** $r = \dfrac{6}{1 - 6\sin\theta}$

**64** $r(3 + 2\sin\theta) = 6$       **65** $r(2 + 3\sin\theta) = 6$

**66** $r(5 + \cos\theta) = 20$

**67** Find the area of the inner loop of the curve $r = 1 + 2\cos\theta$.

**68** Find the area between the inner and outer ovals of the curve $r^2 = 4(1 + \sin\theta)$.

**69** Find the area between the ovals of the curve $r^2 = 4(2 - \cos\theta)$.

**\*70** Use polar coordinates to find the area cut off from the parabola $y^2 = 4x$ by a chord through the vertex making an angle $\alpha$ with the axis of the parabola.

**71** Use polar coordinates to find the area of the loop of the folium $x^3 + y^3 = 3xy$.

In Problems 72 and 73, find the area of the surface generated by revolving the given arc $C$ about the given axis.

**72** $C = \{(x, y): x = t^2, y = \sqrt{2 - t^2}, 0 \leq t \leq 4\}$; $x$ axis

**73** $C = \{(x, y): x = t^4, y = \frac{1}{4}t^8 - 2\ln t, 1 \leq t \leq 2\}$; $y$ axis

# 13

# ANALYTIC GEOMETRY IN THREE DIMENSIONS

Functions of one variable are represented by graphs in the plane. We have seen how the geometric interpretation helped in understanding the calculus of functions of one variable. Functions of two variables are represented by surfaces in three dimensions. In preparation for the calculus of functions of two variables, we study the basic elements of analytic geometry in three dimensions, including lines, planes, spheres, and cylinders. The geometry of equations of the second degree, known as quadric surfaces, is provided in Appendix 2, Section 5.

## 1

## THE NUMBER SPACE $R^3$. COORDINATES. THE DISTANCE FORMULA

In the study of analytic geometry of the plane we are careful to distinguish the geometric plane from the number plane $R^2$. As we saw in Chapter 1, page 16, the number plane consists of the collection of ordered pairs of real numbers. In the study of three-dimensional geometry it is useful to introduce the set of all ordered *triples* of real numbers. Calling this set the **three-dimensional number space**, we denote it by $R^3$. Each individual number triple is a **point** in $R^3$. The three elements in each number triple are called its **coordinates**. We now show how three-dimensional number space may be represented on a geometric or Euclidean three-dimensional space.

In three-dimensional space, consider three mutually perpendicular lines which intersect in a point $O$. We designate these lines the **coordinate axes** and, starting from $O$, set up identical number scales on each of them. If the positive

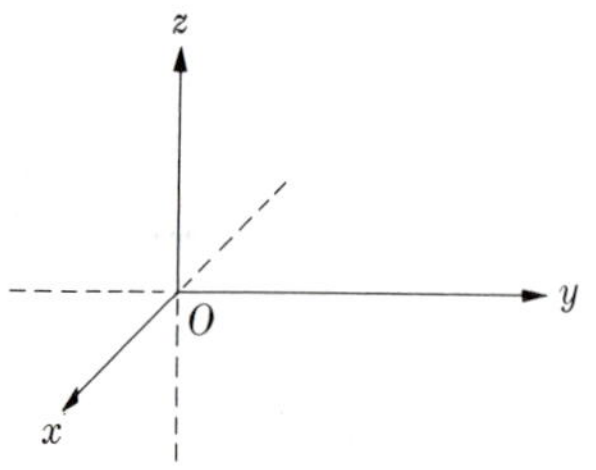

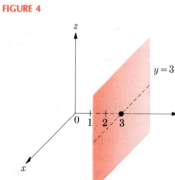

FIGURE 1

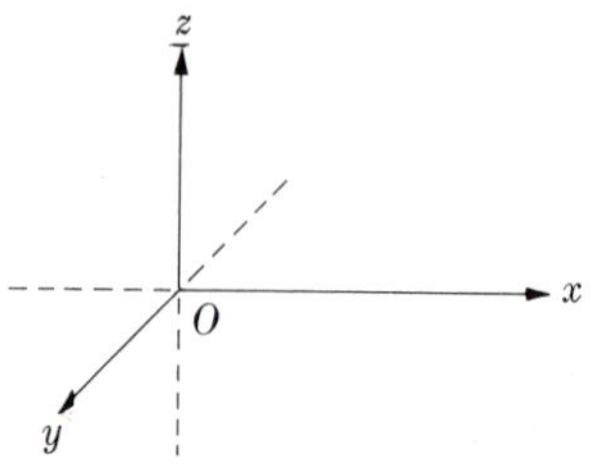

FIGURE 2

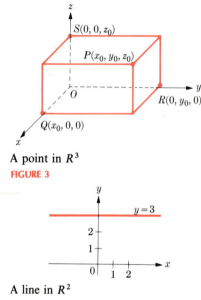

A point in $R^3$

FIGURE 3

A line in $R^2$

FIGURE 4

A plane in $R^3$

FIGURE 5

directions of the $x$, $y$, and $z$ axes are labeled $x$, $y$, and $z$, as shown in Fig. 1, we say the axes form a **right-handed system**. Figure 2 illustrates the axes in a **left-handed system**. We shall use a right-handed coordinate system throughout.

*Any two intersecting lines in space determine a plane.* A plane containing two of the coordinate axes is called a **coordinate plane**. Clearly, there are three such planes.

To each point $P$ in three-dimensional space we can assign a point in $R^3$ in the following way. Through $P$ construct three planes, each parallel to one of the coordinate planes as shown in Fig. 3. We label the intersections of the planes through $P$ with the coordinate axes $Q$, $R$, and $S$, as shown. Then, if $Q$ is $x_0$ units from the origin $O$, $R$ is $y_0$ units from $O$, and $S$ is $z_0$ units from $O$, we assign to $P$ the number triple $(x_0, y_0, z_0)$ and say that the point $P$ has **rectangular coordinates** $(x_0, y_0, z_0)$. To each point in space there corresponds exactly one ordered number triple and, conversely, to each ordered number triple there is associated exactly one point in three-dimensional space. We have just described a **rectangular** coordinate system. In Section 6 we shall discuss other coordinate systems.

In studying analytic geometry in the plane, we saw that an equation such as

$$y = 3$$

represents all points lying on a line parallel to the $x$ axis and three units above it (Fig. 4). In set notation, we write

$$\{(x, y): y = 3\}$$

to denote all the points on this line. The equation

$$y = 3$$

*in the context of three-dimensional geometry represents something entirely different.* The graph of points satisfying this equation is a plane parallel to the $xz$ plane (the $xz$ plane is the coordinate plane determined by the $x$ and the $z$ axes) and three units from it (Fig. 5). In set notation, we represent this plane by writing

$$\{(x, y, z): y = 3\}.$$

We see that set notation, by use of the symbols $(x, y)$ or $(x, y, z)$, indicates clearly when we are dealing with two- or three-dimensional geometry. The equation $y = 3$ by itself is ambiguous unless we know in advance the dimension of the geometry.

In three dimensions the plane represented by $y = 3$ is perpendicular to the $y$ axis and passes through the point $(0, 3, 0)$. Since there is exactly one plane which is perpendicular to a given line and which passes through a given point, we see that the graph of the equation $y = 3$ consists of one and only one such plane. Conversely, from the very definition of a rectangular coordinate system every point with $y$ coordinate 3 must lie in this plane. Equations such as $x = a$ or $y = b$ or $z = c$ always represent planes parallel to the coordinate planes.

We recall from Euclidean solid geometry that *any two nonparallel planes intersect in a straight line.* Therefore, the graph of all points which simultaneously satisfy the equations

$$x = a \qquad \text{and} \qquad y = b$$

is a line parallel to (or coincident with) the $z$ axis. Conversely, any such line is the graph of a pair of equations of the above form. Since the plane $x = a$ is parallel to the $z$ axis, and the plane $y = b$ is parallel to the $z$ axis, the line of intersection must be parallel to the $z$ axis also. (Corresponding statements hold with the axes interchanged.)

**THEOREM 1**  *The distance $d$ between the points $P_1(x_1, y_1, z_1)$ and $P_2(x_2, y_2, z_2)$ is*

$$d = \sqrt{(x_2 - x_1)^2 + (y_2 - y_1)^2 + (z_2 - z_1)^2}.$$

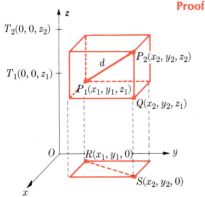

FIGURE 6

**Proof**  We make the construction shown in Fig. 6. By the Pythagorean theorem we have

$$d^2 = |P_1 Q|^2 + |Q P_2|^2.$$

Noting that $|P_1 Q| = |RS|$, we use the formula for distance in the $xy$ plane to get

$$|P_1 Q|^2 = |RS|^2 = (x_2 - x_1)^2 + (y_2 - y_1)^2.$$

Furthermore, since $P_2$ and $Q$ are on a line parallel to the $z$ axis, we see that

$$|Q P_2|^2 = |T_1 T_2|^2 = (z_2 - z_1)^2.$$

Therefore

$$d^2 = (x_2 - x_1)^2 + (y_2 - y_1)^2 + (z_2 - z_1)^2. \qquad \square$$

The midpoint $P$ of the line segment connecting the point $P_1(x_1, y_1, z_1)$ and $P_2(x_2, y_2, z_2)$ has coordinates $P(\bar{x}, \bar{y}, \bar{z})$ given by the following formula.

**THE MIDPOINT FORMULA**

$$\bar{x} = \frac{x_1 + x_2}{2}, \qquad \bar{y} = \frac{y_1 + y_2}{2}, \qquad \bar{z} = \frac{z_1 + z_2}{2}.$$

If $P_1$ and $P_2$ lie in the $xy$ plane—that is, if $z_1 = 0$ and $z_2 = 0$—then so does the midpoint $P$, and the formula is the one we learned in plane analytic geometry. The above formula for $\bar{x}$ is proved as in Chapter 1, Section 7, by passing planes through $P_1$, $P$, and $P_2$ perpendicular to the $x$ axis. The formulas for $\bar{y}$ and $\bar{z}$ are established by analogy.

**EXAMPLE 1**  Find the coordinates of the point $Q$ which divides the line segment from $P_1(1, 4, -2)$ to $P_2(-3, 6, 7)$ in the proportion 3 to 1.

**Solution**  The midpoint $P$ of the segment $P_1 P_2$ has coordinates $P(-1, 5, \frac{5}{2})$. When we find the midpoint of $PP_2$ we get $Q(-2, \frac{11}{2}, \frac{19}{4})$. $\square$

**EXAMPLE 2**  One endpoint of a segment $P_1 P_2$ has coordinates $P_1(-1, 2, 5)$. The midpoint $P$ is known to lie in the $xz$ plane, while the other endpoint is known to lie on the intersection of the planes $x = 5$ and $z = 8$. Find the coordinates of $P$ and $P_2$.

**Solution** For $P(\bar{x}, \bar{y}, \bar{z})$ we note that $\bar{y} = 0$, since $P$ is in the $xz$ plane. Similarly, for $P_2(x_2, y_2, z_2)$ we have $x_2 = 5$ and $z_2 = 8$. From the Midpoint Formula we get

$$\bar{x} = \frac{-1 + 5}{2}, \qquad 0 = \bar{y} = \frac{2 + y_2}{2}, \qquad \bar{z} = \frac{5 + 8}{2}.$$

Therefore the points have coordinates $P(2, 0, \frac{13}{2})$, $P_2(5, -2, 8)$.  □

# 1   PROBLEMS

In Problems 1 through 5, find the lengths of the sides of triangle $ABC$ and state whether the triangle is a right triangle, an isosceles triangle, or both.

**1** $A(2, 1, 3), B(3, -1, -2), C(0, 2, -1)$

**2** $A(4, 3, 1), B(2, 1, 2), C(0, 2, 4)$

**3** $A(3, -1, -1), B(1, 2, 1), C(6, -1, 2)$

**4** $A(1, 2, -3), B(4, 3, -1), C(3, 1, 2)$

**5** $A(0, 0, 0), B(4, 1, 2), C(-5, -5, -1)$

In Problems 6 and 7, find the midpoint of the segment joining the given points $A, B$.

**6** $A(4, -2, 6), B(-2, 8, 1)$

**7** $A(-2, 3, 5), B(-6, 0, 4)$

In Problems 8 and 9, in each case find the coordinates of the three points which divide the given segment $AB$ into four equal parts.

**8** $A(3, 4, -1), B(7, -2, 5)$

**9** $A(1, -6, 0), B(6, 12, 7)$

In Problems 10 through 13, find the lengths of the medians of the given triangles $ABC$.

**10** $A(2, 1, 3), B(3, -1, -2), C(0, 2, -1)$

**11** $A(4, 3, 1), B(2, 1, 2), C(0, 2, 4)$

**12** $A(3, -1, -1), B(1, 2, 1), C(6, -1, 2)$

**13** $A(1, 2, -3), B(4, 3, -1), C(3, 1, 2)$

**14** One endpoint of a line segment is at $P(4, 6, -3)$ and the midpoint is at $Q(2, 1, 6)$. Find the other endpoint.

In Problems 15 through 19, an endpoint $P$ of a line segment, and a midpoint $Q$ of the same line segment are given. Find the other endpoint.

**15** $P(1, -3, 1), Q(1, \frac{1}{2}, -\frac{1}{2})$

**16** $P(0, 0, 0), Q(0, 1, 0)$     **17** $P(0, 1, 0), Q(0, 0, 0)$

**18** $P(1, 2, 3), Q(2, 3, 1)$     **19** $P(-1, 0, 2), Q(4, 2, -1)$

**20** One endpoint of a line segment is at $P_1(-2, 1, 6)$ and the midpoint $Q$ lies in the plane $y = 3$. The other endpoint, $P_2$, lies on the intersection of the planes $x = 4$ and $z = -6$. Find the coordinates of $P_2$ and $Q$.

**21** One endpoint of a line segment is at $P_1(1, 2, 3)$ and the midpoint $Q$ lies in the plane $y = 0$. The other endpoint, $P_2$, lies on the intersection of the planes $x = 1$ and $z = 3$. Find the coordinates of $P_2$ and $Q$.

In Problems 22 through 25, determine whether or not the three given points lie on a line.

**22** $A(1, -1, 2), B(-1, -4, 3), C(3, 2, 1)$

**23** $A(2, 3, 1), B(4, 6, 5), C(-2, -2, -7)$

**24** $A(1, -1, 2), B(3, 3, 4), C(-2, -6, -1)$

**25** $A(-4, 5, -6), B(-1, 2, -1), C(3, -3, 6)$

**26** Describe the set of points in space given by $\{(x, y, z): x = 2, y = -4\}$

**27** Describe the set of points in space given by $\{(x, y, z): -2 \le y \le 5\}$.

**28** Describe the set of points in space given by $\{(x, y, z): x \ge 0, y \ge 0, z \ge 0\}$.

**29** Describe the set of points in space given by $\{(x, y, z): x^2 + y^2 + z^2 < 1\}$.

**30** Derive the formula for determining the midpoint of a line segment.

**31** The formula for the coordinates of a point $Q(x_0, y_0, z_0)$ which divides the line segment from $P_1(x_1, y_1, z_1)$ to $P_2(x_2, y_2, z_2)$ in the ratio $p$ to $q$ is

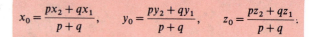

$$x_0 = \frac{px_2 + qx_1}{p + q}, \qquad y_0 = \frac{py_2 + qy_1}{p + q}, \qquad z_0 = \frac{pz_2 + qz_1}{p + q}.$$

Derive this formula.

In Problems 32 through 34, use the formula in Problem 31.

**32** Find a point $R$ along the line segment $PQ$ two thirds of the distance from $P$ to $Q$ if $P = (1, 2, 3)$ and $Q = (2, 3, 4)$.

**33** Find a point $R$ along the line segment $PQ$ that is $1/\pi$ of the distance from $P$ to $Q$ if $P = (0, \sqrt{2}, 3)$ and $Q = (-1, \sqrt{2}, -7)$.

**34** Find a point $R$ that is 53% of the distance from $Q$ to $P$ along the line segment $PQ$ if $P = (-1, 2, -3)$ and $Q = (1, 0, 17)$.

**35** Find the equation of the graph of all points equidistant from the points $(2, -1, 3)$ and $(3, 1, -1)$. Can you describe the graph?

**36** Find the equation of the graph of all points equidistant from the points $(5, 1, 0)$ and $(2, -1, 4)$. Can you describe the graph?

**37** Find the equation of the graph of all points such that the sum of the distances from $(1, 0, 0)$ and $(-1, 0, 0)$ is always equal to 4. Describe the graph.

**38** The points $A(0, 0, 0)$, $B(1, 0, 0)$, $C(\frac{1}{2}, \frac{1}{2}, 1/\sqrt{2})$, $D(0, 1, 0)$ are the vertices of a four-sided figure. Show that $|AB| = |BC| = |CD| = |DA| = 1$. Prove that the figure is not a rhombus.

**39** Prove that the diagonals joining opposite vertices of a rectangular parallelepiped (there are four of them which are interior to the parallelepiped) bisect each other.

---

**2**

## DIRECTION COSINES AND NUMBERS

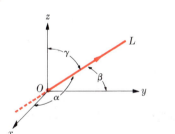

In three-dimensional space we consider a line passing through the origin $O$ and place an arrow on it so that one of the two possible directions on the line is distinguished (Fig. 7). We call such a line a **directed line**. If no arrow is placed, then $L$ is called an **undirected line**. We use the symbol $\vec{L}$ to indicate a directed line while the letter $L$ without the arrow over it indicates an undirected line. We denote by $\alpha$, $\beta$, and $\gamma$ the angles made by the directed line $\vec{L}$ and the positive directions of the $x$, $y$, and $z$ axes, respectively. We define these angles to be the **direction angles** of the directed line $\vec{L}$. The *undirected line* $L$ will have two possible sets of direction angles according to the ordering chosen. The two sets are

$$\alpha, \beta, \gamma \qquad \text{and} \qquad \pi - \alpha, \quad \pi - \beta, \quad \pi - \gamma.$$

The term "line" without further specification shall mean undirected line.

**DEFINITION**     *If $\alpha$, $\beta$, $\gamma$ are direction angles of a directed line $\vec{L}$, then $\cos \alpha$, $\cos \beta$, $\cos \gamma$ are called the **direction cosines** of $\vec{L}$.*

Since $\cos(\pi - \theta) = -\cos\theta$, we see that if $\lambda$, $\mu$, $v$ are direction cosines of a directed line $\vec{L}$, then $\lambda$, $\mu$, $v$ and $-\lambda$, $-\mu$, $-v$ are the two sets of direction cosines of the *undirected* line $L$.

**THEOREM 2**     *Let $\cos \alpha$, $\cos \beta$, and $\cos \gamma$ be the direction cosines of any line $L$. Then*

$$\cos^2 \alpha + \cos^2 \beta + \cos^2 \gamma = 1.$$

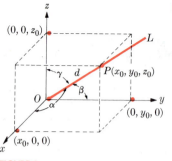

**FIGURE 8**

**Proof**  Let $P(x_0, y_0, z_0)$ be a point on a line $L$ which goes through the origin. Then the distance $d$ of $P$ from the origin is

$$d = \sqrt{x_0^2 + y_0^2 + z_0^2},$$

and (see Fig. 8) we have

$$\cos \alpha = \frac{x_0}{d}, \qquad \cos \beta = \frac{y_0}{d}, \qquad \cos \gamma = \frac{z_0}{d}.$$

Squaring and adding, we get the desired result.  □

To define the direction cosines of any line $L$ in space, we simply consider the line $L'$ parallel to $L$ which passes through the origin, and assert that *by definition $L$* **has the same direction cosines as** *$L'$. Thus all parallel lines in space have the same direction cosines.*

**DEFINITION**  *Two sets of number triples, $a, b, c$ and $a', b', c'$, neither all zero, are said to be* **proportional** *if there is a number $k$ such that*

$$a' = ka, \qquad b' = kb, \qquad c' = kc.$$

*Remark.*  The number $k$ may be positive or negative but not zero, since by hypothesis neither of the number triples is $0, 0, 0$. If none of the numbers $a, b,$ and $c$ is zero, we may write the proportionality relations as

$$\frac{a'}{a} = k, \qquad \frac{b'}{b} = k, \qquad \frac{c'}{c} = k$$

or, more simply,

$$\frac{a'}{a} = \frac{b'}{b} = \frac{c'}{c}.$$

**DEFINITION**  *Suppose that a line $L$ has direction cosines $\lambda, \mu, \nu$. Then a set of numbers $a, b, c$ is called a* **set of direction numbers** *for $L$ if $a, b, c$ and $\lambda, \mu, \nu$ are proportional.*

A line $L$ has unlimited sets of direction numbers.

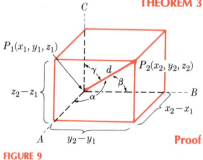

**FIGURE 9**

**THEOREM 3**  *If $P_1(x_1, y_1, z_1)$ and $P_2(x_2, y_2, z_2)$ are two points on a line $L$, and $d$ is the distance from $P_1$ to $P_2$, then*

$$\lambda = \frac{x_2 - x_1}{d}, \qquad \mu = \frac{y_2 - y_1}{d}, \qquad \nu = \frac{z_2 - z_1}{d}$$

*is a set of direction cosines of L.*

**Proof**  In Fig. 9 the angles $\alpha, \beta,$ and $\gamma$ are equal to the direction angles, since the lines $P_1A, P_1B, P_1C$ are parallel to the coordinate axes. We read off from the figure that

$$\cos \alpha = \frac{x_2 - x_1}{d}, \qquad \cos \beta = \frac{y_2 - y_1}{d}, \qquad \cos \gamma = \frac{z_2 - z_1}{d},$$

which is the desired result. □

**COROLLARY 1**  If $P_1(x_1, y_1, z_1)$ and $P_2(x_2, y_2, z_2)$ are two distinct points on a line L, then

$$x_2 - x_1, \qquad y_2 - y_1, \qquad z_2 - z_1$$

constitute a set of direction numbers for L.

**Proof**  Multiplying $\lambda, \mu, v$ of Theorem 3 by the constant $d$, we obtain the result of Corollary 1. □

**EXAMPLE 1**  Find direction numbers and direction cosines for the line L passing through the points $P_1(1, 5, 2)$ and $P_2(3, 7, -4)$.

**Solution**  From Corollary 1, it is clear that $2, 2, -6$ form a set of direction numbers. We compute

$$d = |P_1 P_2| = \sqrt{4 + 4 + 36} = \sqrt{44} = 2\sqrt{11},$$

and so

$$\frac{1}{\sqrt{11}}, \quad \frac{1}{\sqrt{11}}, \quad -\frac{3}{\sqrt{11}}$$

form a set of direction cosines. Since L is undirected, it has two such sets, the other being $-1/\sqrt{11}, -1/\sqrt{11}, 3/\sqrt{11}$. □

**EXAMPLE 2**  Do the three points $P_1(3, -1, 4)$, $P_2(1, 6, 8)$, and $P_3(9, -22, -8)$ lie on the same straight line?

**Solution**  A set of direction numbers for the line $L_1$ through $P_1$ and $P_2$ is $-2, 7, 4$. A set of direction numbers for the line $L_2$ through $P_2$ and $P_3$ is $8, -28, -16$. Since the second set is proportional to the first (with $k = -4$), we conclude that $L_1$ and $L_2$ have the same direction cosines. Therefore the two lines are certainly parallel. However, they have the point $P_2$ in common and so must coincide. The answer is yes. □

From Theorem 3 and the statements in Example 2, we easily obtain the next result.

**COROLLARY 2**  A line $L_1$ is parallel to a line $L_2$ if and only if a set of direction numbers of $L_1$ is proportional to a set of direction numbers of $L_2$.

The angle between two intersecting lines in space is defined in the same way as the angle between two lines in the plane. It may happen that two lines $L_1$ and $L_2$ in space are neither parallel nor intersecting. Such lines are said to be **skew** to each other. Nevertheless, the angle between $L_1$ and $L_2$ can still be defined. Denote by $L_1'$ and $L_2'$ the lines passing through the origin and parallel to $L_1$ and $L_2$, *respectively. The **angle between** $L_1$ and $L_2$ is defined to be the angle between the intersecting lines $L_1'$ and $L_2'$.*

**THEOREM 4**    *If $L_1$ and $L_2$ have direction cosines $\lambda_1, \mu_1, \nu_1$ and $\lambda_2, \mu_2, \nu_2$, respectively, and if $\theta$ is the angle between $L_1$ and $L_2$, then*

$$\cos \theta = \lambda_1 \lambda_2 + \mu_1 \mu_2 + \nu_1 \nu_2.$$

**Proof**    From the way the angle between two lines is defined we may consider $L_1$ and $L_2$ as lines passing through the origin. Let $P_1(x_1, y_1, z_1)$ be a point on $L_1$ and $P_2(x_2, y_2, z_2)$ a point on $L_2$, neither point being $O$ (see Fig. 10). Denote by $d_1$ the distance of $P_1$ from $O$, by $d_2$ the distance of $P_2$ from $O$; let $d = |P_1 P_2|$. We apply the Law of Cosines to triangle $OP_1P_2$, getting

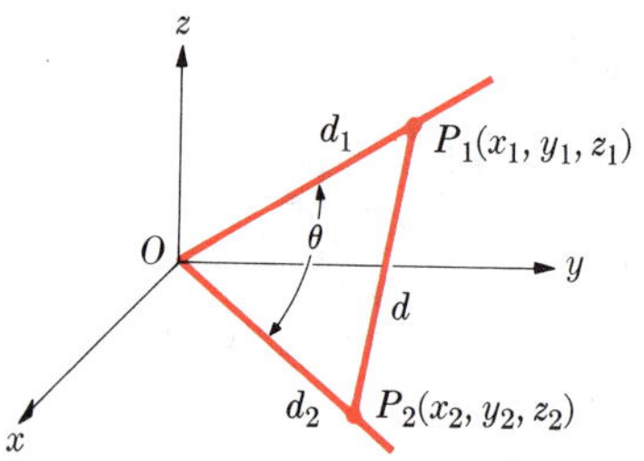

FIGURE 10

$$d^2 = d_1^2 + d_2^2 - 2d_1 d_2 \cos \theta \qquad \text{or} \qquad \cos \theta = \frac{d_1^2 + d_2^2 - d^2}{2d_1 d_2}$$

and

$$\cos \theta = \frac{x_1^2 + y_1^2 + z_1^2 + x_2^2 + y_2^2 + z_2^2 - (x_2 - x_1)^2 - (y_2 - y_1)^2 - (z_2 - z_1)^2}{2d_1 d_2}.$$

After simplification we find

$$\cos \theta = \frac{x_1 x_2 + y_1 y_2 + z_1 z_2}{d_1 d_2} = \frac{x_1}{d_1} \cdot \frac{x_2}{d_2} + \frac{y_1}{d_1} \cdot \frac{y_2}{d_2} + \frac{z_1}{d_1} \cdot \frac{z_2}{d_2}$$

$$= \lambda_1 \lambda_2 + \mu_1 \mu_2 + \nu_1 \nu_2. \qquad \square$$

**COROLLARY**    *Two lines $L_1$ and $L_2$ with direction numbers $a_1, b_1, c_1$ and $a_2, b_2, c_2$, respectively, are perpendicular if and only if*

$$a_1 a_2 + b_1 b_2 + c_1 c_2 = 0.$$

**EXAMPLE 3**    Find the cosine of the angle between the line $L_1$, passing through the points $P_1(1, 4, 2)$ and $P_2(3, -1, 3)$, and the line $L_2$, passing through the points $Q_1(3, 1, 2)$ and $Q_2(2, 1, 3)$.

**Solution**    A set of direction numbers for $L_1$ is $2, -5, 1$. A set for $L_2$ is $-1, 0, 1$. Therefore direction cosines for the two lines are

$$L_1 : \frac{2}{\sqrt{30}}, \frac{-5}{\sqrt{30}}, \frac{1}{\sqrt{30}}; \qquad L_2 : \frac{-1}{\sqrt{2}}, 0, \frac{1}{\sqrt{2}}.$$

We obtain

$$\cos \theta = -\frac{1}{\sqrt{15}} + 0 + \frac{1}{2\sqrt{15}} = -\frac{1}{2\sqrt{15}}. \qquad \square$$

We observe that two lines always have two possible supplementary angles of intersection. If $\cos \theta$ is negative, we have obtained the obtuse angle and, if it is positive, the acute angle of intersection.

## 2  PROBLEMS

In Problems 1 through 4, find a set of direction numbers and a set of direction cosines for the line passing through the given points.

**1**  $A(2, 1, 4), B(3, 5, -2)$  **2**  $A(4, 2, -3), B(1, 0, 5)$

**3**  $A(-2, 1, -4), B(0, -5, -7)$  **4**  $A(6, 7, -2), B(8, -5, 1)$

In each of Problems 5 through 8, a point $P_1$ and a set of direction numbers are given. Find the coordinates of another point on the line $L$ determined by $P_1$ and the given direction numbers.

**5**  $P_1(1, 4, -2)$,  direction numbers $2, 1, 4$

**6**  $P_1(3, 5, -1)$,  direction numbers $2, 0, 4$

**7**  $P_1(0, 4, -3)$,  direction numbers $0, 0, 5$

**8**  $P_1(1, 2, 0)$,  direction numbers $4, 0, 0$

In each of Problems 9 through 12, determine whether or not the three given points lie on a line.

**9**  $A(3, 1, 0), B(2, 2, 2), C(0, 4, 6)$

**10**  $A(2, -1, 1), B(4, 1, -3), C(7, 4, -9)$

**11**  $A(4, 2, -1), B(2, 1, 1), C(0, 0, 2)$

**12**  $A(5, 8, 6), B(-2, -3, 1), C(4, 2, 8)$

In each of Problems 13 through 17, determine whether or not the line through the points $P_1, P_2$ is parallel to the line through the points $Q_1, Q_2$.

**13**  $P_1(4, 8, 0), P_2(1, 2, 3); Q_1(0, 5, 0), Q_2(-3, -1, 3)$

**14**  $P_1(2, 1, 1), P_2(3, 2, -1); Q_1(0, 1, 4), Q_2(2, 3, 0)$

**15**  $P_1(3, 1, 4), P_2(-3, 2, 5); Q_1(4, 6, 1), Q_2(0, 5, 8)$

**16**  $P_1(0, 0, 0), P_2(1, 2, 3); Q_1(-2, 3, 4), Q_2(\pi, 1, 2)$

**17**  $P_1(1, \sqrt{2}, 3), P_2(2, \sqrt{8}, 6); Q_1(7, \sqrt{8}, 0), Q_2(8, \sqrt{18}, 3)$

In each of Problems 18 through 22, determine whether or not the line through the points $P_1, P_2$ is perpendicular to the line through the points $Q_1, Q_2$.

**18**  $P_1(2, 1, 3), P_2(4, 0, 5); Q_1(3, 1, 2), Q_2(2, 1, 6)$

**19**  $P_1(2, -1, 0), P_2(3, 1, 2); Q_1(2, 1, 4), Q_2(4, 0, 4)$

**20**  $P_1(0, -4, 2), P_2(5, -1, 0); Q_1(3, 0, 2), Q_2(2, 1, 1)$

**21**  $P_1(1, \sqrt{2}, 3), P_2(2, \sqrt{8}, 6); Q_1(2, \sqrt{18}, 7), Q_2(4, \sqrt{8}, 7)$

**22**  $P_1(0, 0, 0), P_2(1, 2, 4); Q_1(3, 1, -2), Q_2(7, 0, 0)$

In each of Problems 23 through 27, draw the indicated line.

**23**  Through the origin, with direction numbers 1, 2, 3

**24**  Through the origin, with direction numbers 2, 5, 4

**25**  Through $(1, 2, 3)$ with direction numbers 4, 2, 1

**26**  Through $(0, 1, 0)$ with direction numbers 1, 1, 1

**27**  Through $(4, 2, 2)$ with direction numbers $-1, -1, 1$

In each of Problems 28 through 32, find $\cos \theta$ where $\theta$ is the angle between the line $L_1$ passing through $P_1, P_2$ and $L_2$ passing through $Q_1, Q_2$.

**28**  $P_1(2, 1, 4), P_2(-1, 4, 1); Q_1(0, 5, 1), Q_2(3, -1, -2)$

**29**  $P_1(4, 0, 5), P_2(-1, -3, -2); Q_1(2, 1, 4), Q_2(2, -5, 1)$

**30**  $P_1(0, 0, 5), P_2(4, -2, 0); Q_1(0, 0, 6), Q_2(3, -2, 1)$

**31**  $P_1(0, 0, 0), P_2(0, 0, 1); Q_1(1, 2, 3), Q_2(2, 2, 4)$

**32**  $P_1(1, 2, 3), P_2(-2, 3, 4); Q_1(-1, 0, 1), Q_2(-2, 1, 2)$

**33**  **A regular tetrahedron** is a 4-sided figure each side of which is an equilateral triangle. Find 4 points in space which are the vertices of a regular tetrahedron with each edge of length 2 units.

**34**  **A regular pyramid** is a 5-sided figure with a square base and sides consisting of 4 congruent isosceles triangles. If the base has a side of length 4 and if the height of the pyramid is 6 units, find the area of each of the triangular faces.

**35**  The points $P_1(1, 2, 3), P_2(2, 1, 2), P_3(3, 0, 1), P_4(5, 2, 7)$ are the vertices of a plane quadrilateral. Find the coordinates of the midpoints of the sides. What kind of quadrilateral do these four midpoints form?

**36**  Prove that the four interior diagonals of a parallelepiped bisect each other.

**37**  Given the set $S = \{(x, y, z): x^2 + y^2 + z^2 = 1\}$. Find the points of $S \cap L$ where $L$ is the line passing through the origin and having direction numbers 2, 1, 3.

**38**  Let $P, Q, R, S$ be the vertices of any quadrilateral in three-space. Prove that the lines joining the midpoints of the sides form a parallelogram.

**39**  The vertices of a tetrahedron are at $(3, 0, 0)$, $(6, 0, 0)$, $(0, 9, 0)$, and $(6, 12, 15)$. Show that the three lines joining the midpoints of opposite sides bisect each other. How general is this result?

_____3

## EQUATIONS OF LINES AND PLANES IN $R^3$

In plane analytic geometry a single equation of the first degree,

$$Ax + By + C = 0,$$

is the equation of a line (so long as $A$ and $B$ are not both zero). *In the geometry of three dimensions such an equation represents a plane.* Although we shall postpone a systematic study of planes until later in this section, we assert now that in three-dimensional geometry it is not possible to represent a line by a single first-degree equation.

A line in space is determined by two points. If $P_0(x_0, y_0, z_0)$ and $P_1(x_1, y_1, z_1)$ are given points, we seek an analytic method of representing the line $L$ determined by these points. The result is obtained by solving a geometric problem. A point $P(x, y, z)$, different from $P_0$, is on $L$ if and only if the direction numbers determined by $P$ and $P_0$ are proportional to those determined by $P_1$ and $P_0$ (Fig. 11). Calling the proportionality constant $t$, we see that the conditions are

$$x - x_0 = t(x_1 - x_0), \; y - y_0 = t(y_1 - y_0), \; z - z_0 = t(z_1 - z_0).$$

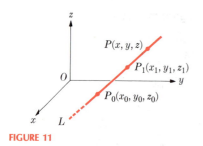

**FIGURE 11**

Thus we obtain the following theorem.

**THEOREM 5**    *The two-point form of the parametric equations of a line is*

$$
\begin{aligned}
x &= x_0 + (x_1 - x_0)t, \\
y &= y_0 + (y_1 - y_0)t, \\
z &= z_0 + (z_1 - z_0)t.
\end{aligned}
\tag{1}
$$

**DEFINITION**    *Given two points $P_0(x_0, y_0, z_0)$ and $P_1(x_1, y_1, z_1)$, the point which is $t$ **of the way from** $P_0$ to $P_1$ is the point $P(x, y, z)$ whose coordinates are given by equations (1).*

**EXAMPLE 1**    Find the parametric equations of the line through the points $A(3, 2, -1)$ and $B(4, 4, 6)$. Locate three additional points on the line.

**Solution**    Substituting in (1) we obtain

$$x = 3 + t, \qquad y = 2 + 2t, \qquad z = -1 + 7t.$$

To get an additional point on the line we let $t = 2$ and obtain $P_1(5, 6, 13)$; $t = -1$ yields $P_2(2, 0, -8)$ and $t = 3$ gives $P_3(6, 8, 20)$.    □

**THEOREM 6**     *The parametric equations of a line L through the point $P_0(x_0, y_0, z_0)$ with direction numbers a, b, c are given by*

$$x = x_0 + at, \qquad y = y_0 + bt, \qquad z = z_0 + ct. \tag{2}$$

**Proof**     The point $P_1(x_0 + a, y_0 + b, z_0 + c)$ must be on $L$, since the direction numbers formed by $P_0$ and $P_1$ are just $a, b, c$. Using the two-point form (1) for the equations of a line through $P_0$ and $P_1$, we get (2) precisely.     □

If $\cos \alpha$, $\cos \beta$, $\cos \gamma$ are the direction cosines of a line $L$ passing through the point $(x_0, y_0, z_0)$, the equations of the line are

$$x = x_0 + t \cos \alpha, \qquad y = y_0 + t \cos \beta, \qquad z = z_0 + t \cos \gamma.$$

**EXAMPLE 2**     Find the parametric equations of the line $L$ through the point $A(3, -2, 5)$ with direction numbers $4, 0, -2$. What is the relation of $L$ to the coordinate planes?

**Solution**     Substituting in (2), we obtain

$$x = 3 + 4t, \qquad y = -2, \qquad z = 5 - 2t.$$

Since all points on the line must satisfy all three of the above equations, $L$ must lie in the plane $y = -2$. This plane is parallel to the $xz$ plane. Therefore $L$ is parallel to the $xz$ plane.     □

If none of the direction numbers is zero, the parameter $t$ may be eliminated from the system of equations (2). We may write

$$\frac{x - x_0}{a} = \frac{y - y_0}{b} = \frac{z - z_0}{c} \tag{3}$$

for the equations of a line. For any value of $t$ in (2) the ratios in (3) are equal. Conversely, if the ratios in (3) are all equal we may set the common value equal to $t$ and (2) is satisfied.

If one of the direction numbers is zero, the form (3) may still be used *if the zero in the denominator is interpreted properly*. The equations

$$\frac{x - x_0}{a} = \frac{y - y_0}{b} = \frac{z - z_0}{0}, \qquad a \neq 0, b \neq 0$$

*are understood to stand for the equations*

$$\frac{x - x_0}{a} = \frac{y - y_0}{b} \qquad \text{and} \qquad z = z_0.$$

The two-point form for the equations of a line also may be written *symmetrically*. The equations

$$\frac{x - x_0}{x_1 - x_0} = \frac{y - y_0}{y_1 - y_0} = \frac{z - z_0}{z_1 - z_0}$$

are called the **symmetric form for the equations of a line.**

**EXAMPLE 3**   Find the point of intersection of the line

$$L = \{(x, y, z): x = 3 - t, y = 2 + 3t, z = -1 + 2t\}$$

with the plane $S = \{(x, y, z): z = 5\}$.

**Solution**   Denoting by $P(x_0, y_0, z_0)$ the point of intersection of $L$ and $S$, we see that $z_0$ must have the value 5. From the equation $z = -1 + 2t$, we conclude that at the point of intersection, $5 = -1 + 2t$ or $t = 3$. Then $x_0 = 3 - 3 = 0$ and $y_0 = 2 + 3(3) = 11$. Hence $P = L \cap S$ has coordinates $(0, 11, 5)$.   □

**EXAMPLE 4**   Find the equations of the line through the point $P(2, -1, 3)$ and parallel to the line through the points $Q(1, 4, -6)$ and $R(-2, -1, 5)$.

**Solution**   The line through $Q$ and $R$ has direction numbers: $-2 - 1 = -3$, $-1 - 4 = -5$, $5 - (-6) = 11$. Therefore using Formula (2) the parallel line through $P$ is given by the equations

$$x = 2 - 3t, \qquad y = -1 - 5t, \qquad z = 3 + 11t.$$   □

Any three points not on a straight line determine a plane. While this characterization of a plane is quite simple, it is not convenient for beginning the study of planes. Instead we use the fact that *there is exactly one plane which passes through a given point and is perpendicular to a given line.*

Let $P_0(x_0, y_0, z_0)$ be a given point, and suppose that a given line $L$ goes through the point $P_1(x_1, y_1, z_1)$ and has direction numbers $A, B, C$.

**THEOREM 7**   *The equation of the plane passing through $P_0$ and perpendicular to $L$ is*

$$A(x - x_0) + B(y - y_0) + C(z - z_0) = 0.$$

**Proof**   We establish the result by using the above characterization of a plane. Let $P(x, y, z)$ be a point on the plane (Fig. 12). From three-dimensional Euclidean geometry we recall that if a line $L_1$ through $P_0$ and $P$ is perpendicular to $L$, then $P$ must be in the desired plane. A set of direction numbers for the line $L_1$ is

$$x - x_0, \quad y - y_0, \quad z - z_0.$$

Since $L$ has direction numbers $A, B, C$, we recall from the Corollary to Theorem 4 that the two lines $L$ and $L_1$ are perpendicular if and only if their direction numbers satisfy the relation

$$A(x - x_0) + B(y - y_0) + C(z - z_0) = 0.$$

This is the equation we seek.   □

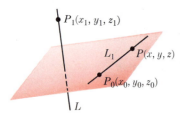

**FIGURE 12**

*Remark.*   Note that only the direction of $L$—and not the coordinates of $P_1$—enters the above equation. We obtain the same plane and the same equation if any line parallel to $L$ is used in its stead.

**EXAMPLE 5**   Find the equation of the plane through the point $P_0(5, 2, -3)$ which is perpendicular to the line through the points $P_1(5, 4, 3)$ and $P_2(-6, 1, 7)$.

**Solution**   The line through the points $P_1$ and $P_2$ has direction numbers $-11, -3, 4$. The equation of the plane is

$$-11(x - 5) - 3(y - 2) + 4(z + 3) = 0$$

$$\Leftrightarrow \qquad 11x + 3y - 4z - 73 = 0. \qquad \square$$

All lines perpendicular to the same plane are parallel and therefore have proportional direction numbers.

**DEFINITION**   *A set of* **attitude numbers** *of a plane is any set of direction numbers of a line perpendicular to the plane.*

In Example 5 above, $11, 3, -4$ form a set of attitude numbers of the plane.

**EXAMPLE 6**   What are sets of attitude numbers for planes parallel to the coordinate planes?

**Solution**   A plane parallel to the $yz$ plane has an equation of the form $x - c = 0$, where $c$ is a constant. A set of attitude numbers for this plane is $1, 0, 0$. A plane parallel to the $xz$ plane has attitude numbers $0, 1, 0$, and any plane parallel to the $xy$ plane has attitude numbers $0, 0, 1$. $\qquad \square$

Since lines perpendicular to the same or parallel planes are themselves parallel, we get at once the next theorem.

**THEOREM 8**   *Two planes are parallel if and only if their attitude numbers are proportional.*

**THEOREM 9**   *If $A$, $B$, and $C$ are not all zero, the graph of an equation of the form*

$$Ax + By + Cz + D = 0 \qquad (1)$$

*is a plane.*

**Proof**   Suppose that $C \neq 0$, for example. Then the point $P_0(0, 0, -D/C)$ is on the graph, as its coordinates satisfy the above equation. Therefore we may write

$$A(x - 0) + B(y - 0) + C\left(z + \frac{D}{C}\right) = 0,$$

and the graph is the plane passing through $P_0$ perpendicular to any line with direction numbers $A, B, C$. $\qquad \square$

An equation of the plane through three points not on a line can be found by assuming that the plane has an equation of the form (1), substituting in turn the coordinates of the three points, and solving simultaneously the three resulting equations. The fact that there are four constants, $A$, $B$, $C$, $D$, and only three equations is illusory, since we may divide through by one of them (say $D$) and obtain three equations in the unknowns $A/D$, $B/D$, $C/D$. This is equivalent to setting $D$ (or one of the other constants) equal to some convenient value. An example illustrates the procedure.

**EXAMPLE 7**    Find an equation of the plane passing through the points $(2, 1, 3)$, $(1, 3, 2)$, $(-1, 2, 4)$.

**Solution**    Since the three points lie in the plane, each of them satisfies Equation (1). We have

$$(2, 1, 3): \quad 2A + B + 3C + D = 0,$$

$$(1, 3, 2): \quad A + 3B + 2C + D = 0,$$

$$(-1, 2, 4): \quad -A + 2B + 4C + D = 0.$$

Solving for $A, B, C$ in terms of $D$, we obtain

$$A = -\tfrac{3}{25}D, \qquad B = -\tfrac{4}{25}D, \qquad C = -\tfrac{5}{25}D.$$

We can now choose any value for $D$. It is convenient to take $D = -25$. We get the equation

$$3x + 4y + 5z - 25 = 0.  \qquad \square$$

## 3   PROBLEMS

In each of Problems 1 through 4, find the equations of the line going through the given points.

**1** $A(1, 3, 2)$, $B(2, -1, 4)$      **2** $A(1, 0, 5)$, $B(-2, 0, 1)$

**3** $A(4, -2, 0)$, $B(3, 2, -1)$      **4** $A(5, 5, -2)$, $B(6, 4, -4)$

In each of Problems 5 through 9, find the equations of the line passing through the given point with the given direction numbers.

**5** $P_1(1, 0, -1)$,   direction numbers $2, 1, -3$

**6** $P_1(-2, 1, 3)$,   direction numbers $3, -1, -2$

**7** $P_1(4, 0, 0)$,   direction numbers $2, -1, -3$

**8** $P_1(1, 2, 0)$,   direction numbers $0, 1, 3$

**9** $P_1(3, -1, -2)$,   direction numbers $2, 0, 0$

In each of Problems 10 through 14, decide whether or not $L_1$ and $L_2$ are perpendicular.

**10** $L_1: \dfrac{x-2}{2} = \dfrac{y+1}{-3} = \dfrac{z-1}{4}$;   $L_2: \dfrac{x-2}{-3} = \dfrac{y+1}{2} = \dfrac{z-1}{3}$

**11** $L_1: \dfrac{x}{1} = \dfrac{y+1}{2} = \dfrac{z+1}{3}$;   $L_2: \dfrac{x-3}{0} = \dfrac{y+1}{-3} = \dfrac{z+4}{2}$

**12** $L_1: \dfrac{x+2}{-1} = \dfrac{y-2}{2} = \dfrac{z+3}{3}$;   $L_2: \dfrac{x+2}{1} = \dfrac{y-2}{2} = \dfrac{z+3}{-1}$

**13** $L_1: \dfrac{x+5}{4} = \dfrac{y-1}{3} = \dfrac{z+8}{5}$;   $L_2: \dfrac{x-4}{3} = \dfrac{y+7}{2} = \dfrac{z+4}{1}$

**14** $L_1: \dfrac{x+1}{0} = \dfrac{y-2}{1} = \dfrac{z+8}{0}$;   $L_2: \dfrac{x-3}{1} = \dfrac{y+2}{0} = \dfrac{z-1}{0}$

**15** Find the equations of the medians of the triangle with vertices at $A(4, 0, 2)$, $B(3, 1, 4)$, $C(2, 5, 0)$.

**16** Find the equations of any line through $P_1(2, 1, 4)$ and perpendicular to any line having direction numbers $4, 1, 3$.

**17** Find the equations of any line through $P_1(2, -1, 5)$ and perpendicular to any line having direction numbers $2, -3, 1$.

**18** Find the points of intersection of the line

$$L = \{(x, y, z): x = 3 + 2t,\ y = 7 + 8t,\ z = -2 + t\}$$

with each of the coordinate planes.

**19** Find the points of intersection of the line

$$L = \left\{(x, y, z): \dfrac{x+1}{-2} = \dfrac{y+1}{3} = \dfrac{z-1}{7}\right\}$$

with each of the coordinate planes.

**20** Show that the following lines are coincident:

$$\dfrac{x-1}{2} = \dfrac{y+1}{-3} = \dfrac{z}{4}; \quad \dfrac{x-5}{2} = \dfrac{y+7}{-3} = \dfrac{z-8}{4}.$$

**21** Find the equations of the line through $(3, 1, 5)$ which is parallel to the line

$$L = \{(x, y, z): x = 4 - t,\ y = 2 + 3t,\ z = -4 + t.\}$$

**22** Find the equations of the line through $(3, 1, -2)$ which is perpendicular and intersects the line

$$\dfrac{x+1}{1} = \dfrac{y+2}{1} = \dfrac{z+1}{1}.$$

(*Hint:* Let $(x_0, y_0, z_0)$ be the point of intersection and determine its coordinates.)

**23** A triangle has vertices at $A(2, 1, 6)$, $B(-3, 2, 4)$, and $C(5, 8, 7)$. Perpendiculars are drawn from these vertices to the $xz$ plane. Locate the points $A'$, $B'$, and $C'$ which are the intersections of the perpendiculars through $A$, $B$, $C$ and the $xz$ plane. Find the equations of the sides of the triangle $A'B'C'$.

The point $(\bar{x}, \bar{y}, \bar{z})$ which is $h$ of the way from $(x_1, y_1, z_1)$ to $(x_2, y_2, z_2)$ has its coordinates given by

$$\bar{x} = x_1 + h(x_2 - x_1),$$
$$\bar{y} = y_1 + h(y_2 - y_1), \qquad (4)$$
$$\bar{z} = z_1 + h(z_2 - z_1).$$

Equations (4) are known as the **point of division formula**.

**24** Use Equations (4) to find the point $\frac{1}{3}$ of the way from $P_1(2, -1, 3)$ to $P_2(6, 2, -5)$.

**25** Use Equations (4) to find the point $\sqrt{2}/\pi$ of the way from $P_1(0, e, 3)$ to $P_2(\pi, 4, -\sqrt{2})$.

**26** Show that the medians of any triangle in three-space intersect at a point which divides each median in the proportion 2:1.

In each of Problems 27 through 30, find the equation of the plane which passes through the given point $P_0$ and has the given attitude numbers.

**27** $P_0(1, 4, 2)$; 3, 1, $-4$     **28** $P_0(2, 1, -5)$; 3, 0, 2

**29** $P_0(4, -2, -5)$; 0, 3, $-2$

**30** $P_0(-1, -2, -3)$; 4, 0, 0

In each of Problems 31 through 34, find the equation of the plane which passes through the three points.

**31** $(1, -2, 1), (2, 0, 3), (0, 1, -1)$

**32** $(2, 2, 1), (-1, 2, 3), (3, -5, -2)$

**33** $(3, -1, 2), (1, 2, -1), (2, 3, 1)$

**34** $(-1, 3, 1), (2, 1, 2), (4, 2, -1)$

In each of Problems 35 through 38, find the equation of the plane passing through $P_1$ and perpendicular to the line $L_1$.

**35** $P_1(2, -1, 3)$;   $L_1: x = -1 + 2t, y = 1 + 3t, z = -4t$

**36** $P_1(1, 2, -3)$;   $L_1: x = t, y = -2 - 2t, z = 1 + 3t$

**37** $P_1(2, -1, -2)$;   $L_1: x = 2 + 3t, y = 0, z = -1 - 2t$

**38** $P_1(-1, 2, -3)$;   $L_1: x = -1 + 5t, y = 1 + 2t, z = -1 + 3t$

In each of Problems 39 through 42, find the equations of the line through $P_1$ and perpendicular to the given plane $M_1$.

**39** $P_1(-2, 3, 1)$;   $M_1: 2x + 3y + z - 3 = 0$

**40** $P_1(1, -2, -3)$;   $M_1: 3x - y - 2z + 4 = 0$

**41** $P_1(-1, 0, -2)$;   $M_1: x + 2z + 3 = 0$

**42** $P_1(2, -1, -3)$;   $M_1: x = 4$

In each of Problems 43 through 46, find an equation of the plane through $P_1$ and parallel to the plane $\Phi$.

**43** $P_1(1, -2, -1)$;   $\Phi: 3x + 2y - z + 4 = 0$

**44** $P_1(-1, 3, 2)$;   $\Phi: 2x + y - 3z + 5 = 0$

**45** $P_1(2, -1, 3)$;   $\Phi: x - 2y - 3z + 6 = 0$

**46** $P_1(3, 0, 2)$;   $\Phi: x + 2y + 1 = 0$

In each of Problems 47 through 50, find the equations of the line through $P_1$ parallel to the given line $L$.

**47** $P_1(2, -1, 3)$;   $L: \dfrac{x-1}{3} = \dfrac{y+2}{-2} = \dfrac{z-2}{4}$

**48** $P_1(0, 0, 1)$;   $L: \dfrac{x+2}{1} = \dfrac{y-1}{3} = \dfrac{z+1}{-2}$

**49** $P_1(1, -2, 0)$;   $L: \dfrac{x-2}{2} = \dfrac{y+2}{-1} = \dfrac{z-3}{4}$

**50** $P_1(3, 1, 1)$;   $L: \dfrac{x-3}{2} = \dfrac{y-1}{3} = \dfrac{z-2}{1}$

In each of Problems 51 through 55, find the equation of the plane containing $L_1$ and $L_2$.

**51** $L_1: \dfrac{x+1}{2} = \dfrac{y-2}{3} = \dfrac{z-1}{1}$;

    $L_2: \dfrac{x+1}{1} = \dfrac{y-2}{-1} = \dfrac{z-1}{2}$

**52** $L_1: \dfrac{x-1}{3} = \dfrac{y+2}{2} = \dfrac{z-2}{2}$;

    $L_2: \dfrac{x-1}{1} = \dfrac{y+2}{1} = \dfrac{z-2}{0}$

**53** $L_1: \dfrac{x+2}{1} = \dfrac{y}{0} = \dfrac{z+1}{2}$;   $L_2: \dfrac{x+2}{2} = \dfrac{y}{3} = \dfrac{z+1}{1}$

**54** $L_1: \dfrac{x}{2} = \dfrac{y-1}{3} = \dfrac{z+2}{-1}$;

    $L_2: \dfrac{x-2}{2} = \dfrac{y+1}{3} = \dfrac{z}{-1}$    $(L_1 \| L_2)$

**55** $L_1: \dfrac{x-2}{2} = \dfrac{y+1}{-1} = \dfrac{z}{3}$;

    $L_2: \dfrac{x+1}{2} = \dfrac{y}{-1} = \dfrac{z+2}{3}$    $(L_1 \| L_2)$

In Problems 56 through 59, find the equation of the plane through $P_1$ and the given line $L$.

**56** $P_1(3, -1, 2)$;   $L = \left\{ (x, y, z): \dfrac{x-2}{2} = \dfrac{y+1}{3} = \dfrac{z}{-2} \right\}$

**57** $P_1(1, -2, 3)$;

    $L = \{ (x, y, z): x = -1 + t, y = 2 + 2t, z = 2 - 2t \}$

**58** $P_1(1, -1, 1)$;   $L = \left\{ (x, y, z): \dfrac{x-1}{3} = \dfrac{y+1}{2} = \dfrac{z-2}{4} \right\}$

**59** $P_1(0, 0, 0)$;

    $L = \{ (x, y, z): x = 1 + t, y = 1 - t, z = 2 - 3t \}$

**60** Show that the plane $2x - 3y + z - 2 = 0$ is parallel to the line

$$\frac{x-2}{1} = \frac{y+2}{1} = \frac{z+1}{1}.$$

**61** Show that the plane $5x - 3y - z - 6 = 0$ contains the line

$$x = 1 + 2t, \qquad y = -1 + 3t, \qquad z = 2 + t.$$

**62** A plane has attitude numbers $A, B, C$, and a line has direction numbers $a, b, c$. What condition must be satisfied in order that the plane and line be parallel?

**63** Show that the three planes

$$7x - 2y - 2z - 5 = 0,$$
$$3x + 2y - 3z - 10 = 0,$$
$$7x + 2y - 5z - 16 = 0$$

all contain a common line. Find the coordinates of two points on this line.

**\*64** Find a condition that three planes

$$A_1 x + B_1 y + C_1 z + D_1 = 0,$$
$$A_2 x + B_2 y + C_2 z + D_2 = 0,$$
$$A_3 x + B_3 y + C_3 z + D_3 = 0$$

either have a line in common or have no point in common.

---

**4**

---

## ANGLES. DISTANCE FROM A POINT TO A PLANE

The angle between two lines was defined in Section 2. We recall that if line $L_1$ has direction cosines $\lambda_1, \mu_1, \nu_1$ and line $L_2$ has direction cosines $\lambda_2, \mu_2, \nu_2$, then

$$\cos \theta = \lambda_1 \lambda_2 + \mu_1 \mu_2 + \nu_1 \nu_2,$$

where $\theta$ is the angle between $L_1$ and $L_2$.

**DEFINITION**  *Let $\Phi_1$ and $\Phi_2$ be two planes, and let $L_1$ and $L_2$ be two lines which are perpendicular to $\Phi_1$ and $\Phi_2$, respectively. Then the **angle between $\Phi_1$ and $\Phi_2$** is, by definition, the angle between $L_1$ and $L_2$. (See Fig. 13.) Furthermore, we make the convention that we always select the acute angle between these lines as the angle between $\Phi_1$ and $\Phi_2$.*

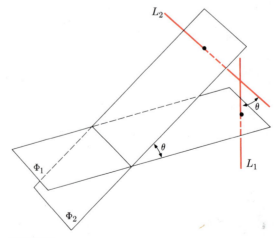

FIGURE 13

**THEOREM 10** *The angle $\theta$ between the planes $A_1 x + B_1 y + C_1 z + D_1 = 0$ and $A_2 x + B_2 y + C_2 z + D_2 = 0$ is given by*

$$\cos\theta = \frac{|A_1 A_2 + B_1 B_2 + C_1 C_2|}{\sqrt{A_1^2 + B_1^2 + C_1^2}\sqrt{A_2^2 + B_2^2 + C_2^2}}.$$

**Proof** From the definition of attitude numbers for a plane, we know that they are direction numbers of any line perpendicular to the plane. Converting to direction cosines, we get the above formula. □

**COROLLARY** *Two planes with attitude numbers $A_1$, $B_1$, $C_1$ and $A_2$, $B_2$, $C_2$ are perpendicular if and only if*

$$A_1 A_2 + B_1 B_2 + C_1 C_2 = 0.$$

**EXAMPLE 1** Find $\cos\theta$ where $\theta$ is the angle between the planes $3x - 2y + z = 4$ and $x + 4y - 3z - 2 = 0$.

**Solution** Substituting in the formula of Theorem 10, we have

$$\cos\theta = \frac{|3 - 8 - 3|}{\sqrt{9 + 4 + 1}\sqrt{1 + 16 + 9}} = \frac{4}{\sqrt{91}}. \qquad \square$$

Two nonparallel planes intersect in a line. Every point on the line satisfies the equations of both planes and, conversely, every point which satisfies the equations of both planes must be on the line. *Therefore we may characterize any line in space by finding two planes which contain it.* Since every line has an unlimited number of planes which pass through it and since *any* two of them are sufficient to determine the line uniquely, we see that there is an unlimited number of ways of writing the equations of a line. The next example shows how to transform one representation into another.

**EXAMPLE 2** The two planes

$$2x + 3y - 4z - 6 = 0 \qquad \text{and} \qquad 3x - y + 2z + 4 = 0$$

intersect in a line. (That is, the points which satisfy *both* equations constitute the line.) Find a set of parametric equations of the line of intersection.

**Solution** We solve simultaneously the above equations for $x$ and $y$ in terms of $z$, getting

$$x = -\tfrac{2}{11}z - \tfrac{6}{11}, \qquad y = \tfrac{16}{11}z + \tfrac{26}{11}.$$

Then we express $z$ in terms of $x$ and express $z$ in terms of $y$, to get the symmetric form for the equations of the line:

$$\frac{x + \tfrac{6}{11}}{-\tfrac{2}{11}} = \frac{y - \tfrac{26}{11}}{\tfrac{16}{11}} = \frac{z}{1}.$$

Setting $z = t$, we can also write

$$x = -\tfrac{6}{11} - \tfrac{2}{11}t, \qquad y = \tfrac{26}{11} + \tfrac{16}{11}t, \quad z = t,$$

which are the desired parametric equations. □

Three planes may be parallel, may pass through a common line, may have no common points, or may have a unique point of intersection. If they have a unique point of intersection, the intersection point may be found by solving simultaneously the three equations of the planes. If they have no common point, an attempt to solve simultaneously will fail. A further examination will show whether or not two or more of the planes are parallel.

**EXAMPLE 3**   Determine whether or not the planes $\Phi_1: 3x - y + z - 2 = 0$; $\Phi_2: x + 2y - z + 1 = 0$; $\Phi_3: 2x + 2y + z - 4 = 0$ intersect. If so, find the point of intersection.

**Solution**   Eliminating $z$ between $\Phi_1$ and $\Phi_2$, we have

$$4x + y - 1 = 0. \tag{1}$$

Eliminating $z$ between $\Phi_2$ and $\Phi_3$, we find

$$3x + 4y - 3 = 0. \tag{2}$$

We solve equations (1) and (2) simultaneously to get

$$x = \tfrac{1}{13}, \qquad y = \tfrac{9}{13}.$$

Substituting in the equation for $\Phi_1$, we obtain $z = \tfrac{32}{13}$. Therefore the single point of intersection of the three planes is $(\tfrac{1}{13}, \tfrac{9}{13}, \tfrac{32}{13})$.   □

**EXAMPLE 4**   Find the point of intersection of the plane

$$3x - y + 2z - 3 = 0$$

and the line

$$\frac{x+1}{3} = \frac{y+1}{2} = \frac{z-1}{-2}.$$

**Solution**   We write the equations of the line in parametric form:

$$x = -1 + 3t, \qquad y = -1 + 2t, \qquad z = 1 - 2t.$$

The point of intersection is given by a value of $t$; call it $t_0$. This point must satisfy the equation of the plane. We have

$$3(-1 + 3t_0) - (-1 + 2t_0) + 2(1 - 2t_0) - 3 = 0 \qquad \Leftrightarrow \qquad t_0 = 1.$$

The desired point is $(2, 1, -1)$.   □

We now derive an important formula which tells us how to find the perpendicular distance from a point in space to a plane.

**THEOREM 11**   *The distance $d$ from the point $P_1(x_1, y_1, z_1)$ to the plane*

$$Ax + By + Cz + D = 0$$

*is given by*

$$d = \frac{|Ax_1 + By_1 + Cz_1 + D|}{\sqrt{A^2 + B^2 + C^2}}.$$

**Proof**    We write the equations of the line $L$ through $P_1$ which is perpendicular to the plane. They are

$$L: x = x_1 + At, \qquad y = y_1 + Bt, \qquad z = z_1 + Ct.$$

Denote by $(x_0, y_0, z_0)$ the point of intersection of $L$ and the plane. Then

$$d^2 = (x_1 - x_0)^2 + (y_1 - y_0)^2 + (z_1 - z_0)^2. \tag{3}$$

Also $(x_0, y_0, z_0)$ is on both the line and the plane. Therefore, for some value $t_0$ we have

$$\begin{aligned}
x_0 &= x_1 + At_0, \\
y_0 &= y_1 + Bt_0, \\
z_0 &= z_1 + Ct_0,
\end{aligned} \tag{4}$$

and

$$Ax_0 + By_0 + Cz_0 + D = 0 = A(x_1 + At_0) + B(y_1 + Bt_0) + C(z_1 + Ct_0) + D.$$

Solving for $t_0$, we find

$$t_0 = \frac{-(Ax_1 + By_1 + Cz_1 + D)}{A^2 + B^2 + C^2}.$$

Thus, from (3) and (4), we write

$$d = \sqrt{A^2 + B^2 + C^2}\, |t_0|,$$

and now, inserting the relation for $t_0$ in the expression for $d$, we obtain the desired formula.    ☐

**EXAMPLE 5**    Find the distance from the point $(2, -1, 5)$ to the plane

$$3x + 2y - 2z - 7 = 0.$$

**Solution**

$$d = \frac{|6 - 2 - 10 - 7|}{\sqrt{9 + 4 + 4}} = \frac{13}{\sqrt{17}}. \qquad ☐$$

## 4   PROBLEMS

In each of Problems 1 through 4, find $\cos \theta$ where $\theta$ is the angle between the given planes.

**1**   $2x - y + 2z - 3 = 0, \; 3x + 2y - 6z - 11 = 0$

**2**   $x + 2y - 3z + 6 = 0, \; x + y + z - 4 = 0$

**3**   $2x - y + 3z - 5 = 0, \; 3x - 2y + 2z - 7 = 0$

**4**   $x + 4z - 2 = 0, \; y + 2z - 6 = 0$

In each of Problems 5 through 8, find the equations in parametric form of the line of intersection of the given planes.

**5**   $3x + 2y - z + 5 = 0, \; 2x + y + 2z - 3 = 0$

**6**   $x + 2y + 2z - 4 = 0, \; 2x + y - 3z + 5 = 0$

**7**   $x + 2y - z + 4 = 0, \; 2x + 4y + 3z - 7 = 0$

**8**   $2x + 3y - 4z + 7 = 0, \; 3x - 2y + 3z - 6 = 0$

In each of Problems 9 through 12, find the point of intersection of the given line and the given plane.

**9**   $3x - y + 2z - 5 = 0, \quad \dfrac{x-1}{2} = \dfrac{y+1}{3} = \dfrac{z-1}{-2}$

**10**   $2x + 3y - 4z + 15 = 0, \quad \dfrac{x+3}{2} = \dfrac{y-1}{-2} = \dfrac{z+4}{3}$

**11**   $x + 2z + 3 = 0, \quad \dfrac{x+1}{1} = \dfrac{y}{0} = \dfrac{z+2}{2}$

**12**   $2x + 3y + z - 3 = 0, \quad \dfrac{x+2}{2} = \dfrac{y-3}{3} = \dfrac{z-1}{1}$

In each of Problems 13 through 16, find the distance from the given point to the given plane.

**13** $(2, 1, -1)$, $x - 2y + 2z + 5 = 0$

**14** $(3, -1, 2)$, $3x + 2y - 6z - 9 = 0$

**15** $(-1, 3, 2)$, $2x - 3y + 4z - 5 = 0$

**16** $(0, 4, -3)$, $3y + 2z - 7 = 0$

**17** Find the equation of the plane through the line

$$\frac{x+1}{3} = \frac{y-1}{2} = \frac{z-2}{4}$$

which is perpendicular to the plane

$$2x + y - 3z + 4 = 0.$$

**18** Find the equation of the plane through the line

$$\frac{x-2}{2} = \frac{y-2}{3} = \frac{z-1}{-2}$$

which is parallel to the line

$$\frac{x+1}{3} = \frac{y-1}{2} = \frac{z+1}{1}.$$

**19** Find the equation of the plane through the line

$$\frac{x+2}{3} = \frac{y}{-2} = \frac{z+1}{2}$$

which is parallel to the line

$$\frac{x-1}{2} = \frac{y+1}{3} = \frac{z-1}{4}.$$

**20** Find the equations of any line through the point $(1, 4, 2)$ which is parallel to the plane

$$2x + y + z - 4 = 0.$$

**21** Find the equation of the plane through $(3, 2, -1)$ and $(1, -1, 2)$ which is parallel to the line

$$\frac{x-1}{3} = \frac{y+1}{2} = \frac{z}{-2}.$$

**22** Find the equation of the plane through $P(5, -3, 7)$ which is parallel to (a) the $xy$-plane; (b) the $yz$-plane; (c) the $xz$-plane.

**23** Find the equation of the plane through $P(1, 1, 1)$ and parallel to the plane $x - 2y + 3z = 4$.

In each of Problems 24 through 28, find all the points of intersection of the three given planes. If the three planes pass through a line, find the equations of the line in parametric form.

**24** $2x + y - 2z - 1 = 0$, $3x + 2y + z - 10 = 0$,
$x + 2y - 3z + 2 = 0$

**25** $x + 2y + 3z - 4 = 0$, $2x - 3y + z - 2 = 0$,
$3x + 2y - 2z - 5 = 0$

**26** $3x - y + 2z - 4 = 0$, $x + 2y - z - 3 = 0$,
$3x - 8y + 7z + 1 = 0$

**27** $2x + y - 2z - 3 = 0$, $x - y + z + 1 = 0$, $x + 5y - 7z - 3 = 0$

**28** $x + 2y + 3z - 5 = 0$, $2x - y - 2z - 2 = 0$,
$x - 8y - 13z + 11 = 0$

In each of Problems 29 through 33, find the equations in parametric form of the line through the given point $P_1$ which intersects and is perpendicular to the given line $L$.

**29** $P_1(3, -1, 2)$;   $L: \dfrac{x-1}{2} = \dfrac{y+1}{-1} = \dfrac{z}{3}$

**30** $P_1(-1, 2, 3)$;   $L: \dfrac{x}{2} = \dfrac{y-2}{0} = \dfrac{z+3}{-3}$

**31** $P_1(0, 2, 4)$;   $L: \dfrac{x-1}{3} = \dfrac{y-2}{1} = \dfrac{z-3}{4}$

**32** $P_1(0, 0, 0)$;   $L: \dfrac{x-1}{2} = \dfrac{y-2}{-1} = \dfrac{z-1}{3}$

**33** $P_1(0, 7, -1)$;   $L: \dfrac{x}{2} = \dfrac{y-3}{0} = \dfrac{z-1}{4}$

**34** If $A_1x + B_1y + C_1z + D_1 = 0$ and $A_2x + B_2y + C_2z + D_2 = 0$ are two intersecting planes, what is the graph of all points which satisfy

$$A_1x + B_1y + C_1z + D_1 + k(A_2x + B_2y + C_2z + D_2) = 0,$$

where $k$ is a constant?

**35** Find the equation of the plane passing through the point $(2, 1, -3)$ and the intersection of the planes $3x + y - z - 2 = 0$, $2x + y + 4z - 1 = 0$. (*Hint:* See Problem 34.)

**36** Given a regular tetrahedron with each edge $a$ units in length. (See Problem 33 at the end of Section 2.) Find the distance from a vertex to the opposite face.

**37** Given a regular pyramid with each edge of the base 2 units in length and each lateral edge 4 units in length. (See Problem 34 at the end of Section 2.) Find the distance from the top of the pyramid to the base. Also find the distance from one of the vertices of the base to a face opposite to that vertex.

**38** A slice is made in a cube by cutting through a diagonal of one face and proceeding through one of the vertices of the opposite face as shown in Fig. 14. Find the angle between the planar slice and the first face which is cut.

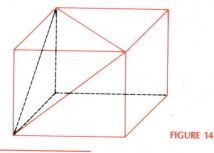

FIGURE 14

_____5_____

## THE SPHERE.* CYLINDERS

A **sphere** is the graph of all points at a given distance from a fixed point. The fixed point is called the **center** and the fixed distance is called the **radius**.

If the center is at the point $(h, k, l)$, the radius is $r$, and $(x, y, z)$ is any point on the sphere, then, from the formula for the distance between two points, we obtain the relation

$$(x - h)^2 + (y - k)^2 + (z - l)^2 = r^2.$$

---

**THEOREM 12**    *An equation of a sphere of radius $r$ and center $C(h, k, l)$ is*

$$(x - h)^2 + (y - k)^2 + (z - l)^2 = r^2. \tag{1}$$

---

Equation (1) is the **equation of a sphere**. If it is multiplied out and the terms collected we have an equivalent form which we state as a theorem:

---

**THEOREM 13**    *An equation of the form*

$$x^2 + y^2 + z^2 + Dx + Ey + Fz + G = 0 \tag{2}$$

*is an equation of a sphere if $D^2 + E^2 + F^2 > 4G$.*

---

**Proof**    From Theorem 12, the equation of a sphere is $(x - h)^2 + (y - k)^2 + (z - l)^2 = r^2$. This is equivalent to

$$x^2 - 2hx + h^2 + y^2 - 2ky + k^2 + z^2 - 2lz + l^2 = r^2.$$

We set $D = -2h$, $E = -2k$, $F = -2l$, and we have

$$x^2 + y^2 + z^2 + Dx + Ey + Fz + (h^2 + k^2 + l^2) - r^2 = 0.$$

We must set $G = h^2 + k^2 + l^2 - r^2$. We then have $G = \frac{1}{4}(D^2 + E^2 + F^2) - r^2$, or $4G - (D^2 + E^2 + F^2) = -4r^2 < 0$, which is the condition of the theorem. □

**EXAMPLE 1**    Find the center and the radius of the sphere with equation

$$x^2 + y^2 + z^2 + 4x - 6y + 9z - 6 = 0.$$

**Solution**    We complete the square by first writing

$$x^2 + 4x \quad + y^2 - 6y \quad + z^2 + 9z \quad = 6;$$

---

*Additional material on second degree equations in three variables will be found in Appendix 2, Section 5.

then, adding the appropriate quantities to both sides, we have

$$(x^2 + 4x + 4) + (y^2 - 6y + 9) + (z^2 + 9z + \tfrac{81}{4}) = 6 + 4 + 9 + \tfrac{81}{4}$$

$$\Leftrightarrow \quad (x+2)^2 + (y-3)^2 + (z+\tfrac{9}{2})^2 = \tfrac{157}{4}.$$

The center is at $(-2, 3, -\tfrac{9}{2})$ and the radius is $\tfrac{1}{2}\sqrt{157}$.

The next example shows that 4 points on a sphere are sufficient to determine its equation.

**EXAMPLE 2**  Find the equation of the sphere which passes through $(2, 1, 3)$, $(3, 2, 1)$, $(1, -2, -3)$, $(-1, 1, 2)$.

**Solution**  Substituting these points in the form (2) above for the equation of a sphere, we obtain

$$(2, 1, 3): \quad 2D + \phantom{2}E + 3F + G = -14,$$
$$(3, 2, 1): \quad 3D + 2E + \phantom{3}F + G = -14,$$
$$(1, -2, -3): \quad D - 2E - 3F + G = -14,$$
$$(-1, 1, 2): \quad -D + \phantom{2}E + 2F + G = -6.$$

Solving these by elimination (first $G$, then $D$, then $F$) we obtain, successively,

$$D + E - 2F = 0, \qquad D + 3E + 6F = 0, \qquad 3D + F = -8,$$

and

$$2E + 8F = 0, \qquad 3E - 7F = 8; \qquad \text{and so} \qquad -38E = -64.$$

Therefore

$$E = \tfrac{32}{19}, \qquad F = -\tfrac{8}{19}, \qquad D = -\tfrac{48}{19}, \qquad G = -\tfrac{178}{19}.$$

The desired equation is

$$x^2 + y^2 + z^2 - \tfrac{48}{19}x + \tfrac{32}{19}y - \tfrac{8}{19}z - \tfrac{178}{19} = 0.$$

**DEFINITION**  *A **cylindrical surface** is a surface which consists of a collection of parallel lines. Each of the parallel lines is called a **generator** of the **cylinder** or cylindrical surface.*

The customary right circular cylinder of elementary geometry is clearly a special case of the type of cylinder we are considering. Figure 15 shows some examples of cylindrical surfaces. Note that a plane is a cylindrical surface.

FIGURE 15

> **THEOREM 14** *An equation of the form*
>
> $$f(x, y) = 0$$
>
> *is a cylindrical surface with generators all parallel to the z axis. The surface intersects the xy plane in the curve*
>
> $$f(x, y) = 0, \qquad z = 0.$$
>
> *A similar result holds with axes interchanged.*

**Proof** Suppose that $x_0, y_0$ satisfies $f(x_0, y_0) = 0$. Then any point $(x_0, y_0, z)$ for $-\infty < z < \infty$ satisfies the same equation, since $z$ is absent. Therefore the line parallel to the $z$ axis through $(x_0, y_0, 0)$ is a generator. □

**EXAMPLE 3** Describe and sketch the graph of the equation $x^2 + y^2 = 9$.

**Solution** The graph is a right circular cylinder with generators parallel to the $z$ axis (Theorem 14). It is sketched in Fig. 16. □

**EXAMPLE 4** Describe and sketch the graph of the equation $y^2 = 4z$.

**Solution** According to Theorem 14 the graph is a cylindrical surface with generators parallel to the $x$ axis. The intersection with the $yz$ plane is a parabola. The graph, called a **parabolic cylinder**, is sketched in Fig. 17. □

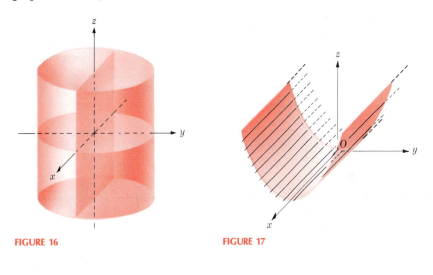

FIGURE 16          FIGURE 17

## 5 PROBLEMS

In each of Problems 1 through 4, find the equation of the sphere with center at $C$ and given radius $r$.

**1** $C(1, 4, -2), r = 3$

**2** $C(2, 0, -3), r = 5$

**3** $C(0, 1, 4), r = 6$

**4** $C(-2, 1, -3), r = 1$

In each of Problems 5 through 13, determine the graph of the equation. If it is a sphere, find its center and radius.

**5** $x^2 + y^2 + z^2 + 2x - 4z + 1 = 0$

**6** $x^2 + y^2 + z^2 - 4x + 2y + 6z - 2 = 0$

**7** $x^2 + y^2 + z^2 - 2x + 4y - 2z + 7 = 0$

**8** $x^2 + y^2 + z^2 + 4x - 2y + 4z + 9 = 0$

**9** $x^2 + y^2 + z^2 - 6x + 4y + 2z + 10 = 0$

**10** $x^2 + y^2 + z^2 + 4x - 2y + 2z + 2 = 0$

**11** $x^2 + y^2 + z^2 - 8x + 8z + 16 = 0$

**12** $x^2 + y^2 + z^2 + 4y = 0$     **13** $x^2 + y^2 + z^2 - 2z = 0$

**14** Find the equation of the graph of all points which are twice as far from $A(3, -1, 2)$ as from $B(0, 2, -1)$.

**15** Find the equation of the graph of all points which are three times as far from $A(2, 1, -3)$ as from $B(-2, -3, 5)$.

**16** Find the equation of the graph of all points whose distances from the point $(0, 0, 4)$ are equal to their perpendicular distances from the $xy$ plane.

In each of Problems 17 through 28, describe and sketch the graph of the given equation.

**17** $x = 3$                    **18** $x^2 + y^2 = 16$

**19** $2x + y = 3$              **20** $x + 2z = 4$

**21** $x^2 = 4z$                **22** $z = 2 - y^2$

**23** $4x^2 + y^2 = 16$        **24** $4x^2 - y^2 = 16$

**25** $y^2 + x^2 = 9$          **26** $z^2 = 2 - 2x$

**27** $x^2 + y^2 - 2x = 0$     **28** $z^2 = y^2 + 4$

In each of Problems 29 through 32, describe the curve of intersection, if any, of the given surface $S$ and the given plane $\Phi$. That is, describe the set $S \cap \Phi$.

**29** $S: x^2 + y^2 + z^2 = 25;$   $\Phi: z = 3$

**30** $S: 4x = y^2 + z^2;$   $\Phi: x = 4$

**31** $S: x^2 + 2y^2 + 3z^2 = 12;$   $\Phi: y = 4$

**32** $S: x^2 + y^2 = z^2;$   $\Phi: 2x + z = 4$

**33** Verify that the graph of the equation $(x - 2)^2 + (y - 1)^2 = 0$ is a straight line. Show that every straight line parallel to one of the coordinate axes can be represented by a *single* equation of the *second* degree.

The following concept is used in Problems 34 through 37. If

$$S_1 = \{(x, y, z): x^2 + y^2 + z^2 + A_1 x + B_1 y + C_1 z + D_1 = 0\}$$

and

$$S_2 = \{(x, y, z): x^2 + y^2 + z^2 + A_2 x + B_2 y + C_2 z + D_2 = 0\}$$

are two spheres, then the **radical plane** is obtained by subtraction of the equations for $S_1$ and $S_2$. It is

$$\Phi = \{(x, y, z): (A_1 - A_2)x + (B_1 - B_2)y + (C_1 - C_2)z + (D_1 - D_2) = 0\}.$$

**34** Show that the radical plane is perpendicular to the line joining the centers of the spheres $S_1$ and $S_2$.

**\*35** Show that the three radical planes of three spheres intersect in a common line or are parallel.

**36** Show that if two spheres intersect in a circle $C$, then their radical plane contains $C$.

**37** Show that if two spheres are tangent to each other, then their radical plane is tangent to each of the spheres.

---

**6**

## CYLINDRICAL AND SPHERICAL COORDINATES

In plane analytic geometry we employed a rectangular coordinate system for certain types of problems and a polar coordinate system for others. We saw that there are circumstances in which one system is more convenient than the other. A similar situation prevails in three-dimensional geometry, and we now take up systems of coordinates other than the rectangular one which we have studied exclusively so far. One such system, known as **cylindrical coordinates**, is described in the following way. A point $P$ in space with rectangular coordinates $(x, y, z)$ may also be located by replacing the $x$ and $y$ values with the corresponding polar coordinates $r, \theta$ and by allowing the $z$ value to remain unchanged. In other words, to each ordered number triple of the form $(r, \theta, z)$, there is associated a point in space. The transformation from cylindrical to rectangular coordinates is given by the equations

$$x = r \cos \theta, \qquad y = r \sin \theta, \qquad z = z.$$

The transformation from rectangular to cylindrical coordinates is given by

$$r^2 = x^2 + y^2, \qquad \tan \theta = y/x, \qquad z = z.$$

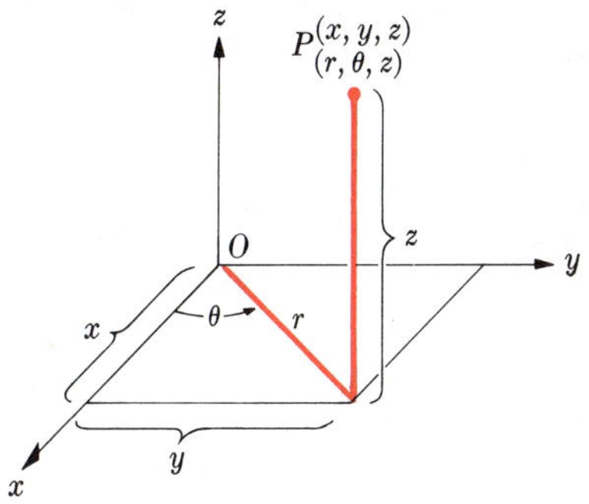

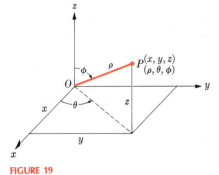

FIGURE 18

If the coordinates of a point are given in one system, the above equations show how to get the coordinates in the other. Figure 18 exhibits the relation between the two systems. It is always assumed that the origins of the systems coincide and that $\theta = 0$ corresponds to the $xz$ plane. We see that the graph $\theta = $ const consists of all points in a plane containing the $z$ axis. The graph of $r = $ const consists of all points on a right circular cylinder with the $z$ axis as its central axis. (The term "cylindrical coordinates" comes from this fact.) The graph of $z = $ const consists of all points in a plane parallel to the $xy$ plane.

**EXAMPLE 1**   Find cylindrical coordinates of the points whose rectangular coordinates are $P(3, 3, 5)$, $Q(2, 0, -1)$, $R(0, 4, 4)$, $S(0, 0, 5)$, $T(2, 2\sqrt{3}, 1)$.

**Solution**   For the point $P$ we have $r = \sqrt{9 + 9} = 3\sqrt{2}$, $\tan \theta = 1$, $\theta = \pi/4$, $z = 5$. Therefore one set of coordinates is $(3\sqrt{2}, \pi/4, 5)$. For $Q$ we have $r = 2$, $\theta = 0$, $z = -1$. The coordinates are $(2, 0, -1)$. For $R$ we get $r = 4$, $\theta = \pi/2$, $z = 4$. The result is $(4, \pi/2, 4)$. For $S$ we see at once that the coordinates are $(0, \theta, 5)$ for any $\theta$. For $T$ we get $r = \sqrt{4 + 12} = 4$, $\tan \theta = \sqrt{3}$, $\theta = \pi/3$. The answer is $(4, \pi/3, 1)$.   $\square$

*Remark.*   Just as polar coordinates do not give a one-to-one correspondence between ordered number pairs and points in the plane, so cylindrical coordinates do not give a one-to-one correspondence between ordered number triples and points in space.

A **spherical coordinate system** is defined in the following way. A point $P$ with rectangular coordinates $(x, y, z)$ has spherical coordinates $(\rho, \theta, \phi)$ where $\rho$ is the distance of the point $P$ from the origin, $\theta$ is the same quantity as in cylindrical coordinates, and $\phi$ is the angle that the directed line $\overrightarrow{OP}$ makes with the positive $z$ direction. Figure 19 exhibits the relation between rectangular and spherical coordinates. The transformation from spherical to rectangular coordinates is given by the equations

$$x = \rho \sin \phi \cos \theta, \qquad y = \rho \sin \phi \sin \theta, \qquad z = \rho \cos \phi.$$

FIGURE 19

The transformation from rectangular to spherical coordinates is given by

$$\rho^2 = x^2 + y^2 + z^2,$$

$$\tan \theta = \frac{y}{x},$$

$$\cos \phi = \frac{z}{\sqrt{x^2 + y^2 + z^2}}.$$

We note that the graph of $\rho = $ const is a sphere with center at the origin (from which the term "spherical coordinates" is derived). The graph of $\theta = $ const is a plane through the $z$ axis, as in cylindrical coordinates. The graph of $\phi = $ const is a **cone** with vertex at the origin and angle opening $2\phi$ if $0 < \phi < \pi/2$. (See Fig. 20.) The lower portion of the cone in Fig. 20 is or is not included according as negative values of $\rho$ are or are not allowed.

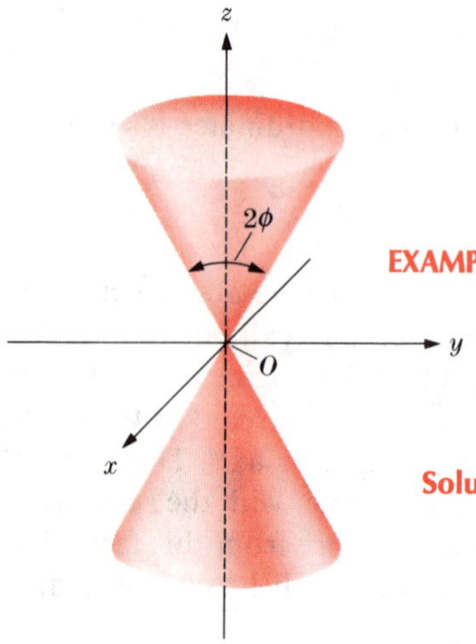

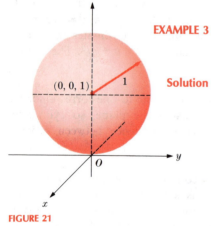

Graph of $\phi = $ const in spherical coordinates

FIGURE 20

**EXAMPLE 2**    Find an equation in spherical coordinates of the sphere

$$x^2 + y^2 + z^2 - 2z = 0.$$

Sketch the graph.

**Solution**    Since $\rho^2 = x^2 + y^2 + z^2$ and $z = \rho \cos \phi$, we have

$$\rho^2 - 2\rho \cos \phi = 0 \qquad \Leftrightarrow \qquad \rho(\rho - 2 \cos \phi) = 0.$$

The graph of this equation is the graph of $\rho = 0$ and $\rho - 2 \cos \phi = 0$. The graph of $\rho = 0$ is on the graph of $\rho - 2 \cos \phi = 0$ (with $\phi = \pi/2$). Plotting the surface

$$\rho = 2 \cos \phi,$$

we get the surface shown in Fig. 21.    □

**EXAMPLE 3**    Find an equation in rectangular coordinates for $\rho = 2 \sec \phi$, and describe the graph of this equation.

**Solution**    Multiplying both sides of the equation by $\cos \phi$ yields the equation $\rho \cos \phi = 2$. Since $z = \rho \cos \phi$ in spherical coordinates, the equation $\rho = 2 \sec \phi$ is equivalent to the plane $z = 2$ in rectangular coordinates. (See Fig. 22.)    □

FIGURE 21

If $\rho$ is a constant, say $c$, then the quantities $(\theta, \phi)$ form a coordinate system on the surface of the sphere $\rho = c$. Latitude and longitude on the surface of the Earth (which is considered a sphere) also form a coordinate system. If we restrict $\theta$ so that $-\pi < \theta \leq \pi$, then $\theta$ is called the **longitude** of the point in spherical coordinates. If $\phi$ is restricted so that $0 \leq \phi \leq \pi$, then $\phi$ is called the **colatitude** of the point. That is, $\phi$ is $(\pi/2) - $ latitude, where latitude is taken in the ordinary sense—i.e., positive north of the equator and negative south of it.

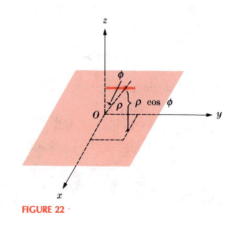

FIGURE 22

## 6    PROBLEMS

**1** Find a set of cylindrical coordinates for each of the points whose rectangular coordinates are
a) $(3, 3, 7)$,
b) $(4, 8, 2)$,
c) $(-2, 3, 1)$.

**2** Find the rectangular coordinates of the points whose cylindrical coordinates are
a) $(2, \pi/3, 1)$,
b) $(3, -\pi/4, 2)$,
c) $(7, 2\pi/3, -4)$.

**3** Find a set of spherical coordinates for each of the points whose rectangular coordinates are
a) $(2, 2, 2)$,
b) $(2, -2, -2)$,
c) $(-1, \sqrt{3}, 2)$.

**4** Find the rectangular coordinates of the points whose spherical coordinates are
a) $(4, \pi/6, \pi/4)$,
b) $(6, 2\pi/3, \pi/3)$,
c) $(8, \pi/3, 2\pi/3)$.

**5** Find a set of cylindrical coordinates for each of the points whose spherical coordinates are
a) $(4, \pi/3, \pi/2)$,
b) $(2, 2\pi/3, 5\pi/6)$,
c) $(7, \pi/2, \pi/6)$.

**6** Find a set of spherical coordinates for each of the points whose cylindrical coordinates are
a) $(2, \pi/4, 1)$,
b) $(3, \pi/2, 2)$,
c) $(1, 5\pi/6, -2)$.

In each of Problems 7 through 16, find an equation in cylindrical coordinates of the graph whose $(x, y, z)$ equation is given. Sketch.

**7** $x^2 + y^2 + z^2 = 9$        **8** $x^2 + y^2 + 2z^2 = 8$
**9** $x^2 + y^2 = 4z$            **10** $x^2 + y^2 - 2x = 0$

**11** $x^2 + y^2 = z^2$          **12** $x^2 + y^2 + 2z^2 + 2z = 0$
**13** $x^2 - y^2 = 4$            **14** $xy + z^2 = 5$
**15** $x^2 + y^2 - 4y = 0$
**16** $x^2 + y^2 + z^2 - 2x + 3y - 4z = 0$

In each of Problems 17 through 22, find an equation in spherical coordinates of the graph whose $(x, y, z)$ equation is given. Sketch.

**17** $x^2 + y^2 + z^2 - 4z = 0$   **18** $x^2 + y^2 + z^2 + 2z = 0$
**19** $x^2 + y^2 = z^2$          **20** $x^2 + y^2 = 4$
**21** $x^2 + y^2 = 4z + 4$   (Solve for $\rho$ in terms of $\phi$.)
**22** $x^2 + y^2 - z^2 + z - y = 0$

In each of Problems 23 through 38, describe the graphs either directly or after transforming the equation to rectangular coordinates.

**23** $\rho = 3$                 **24** $r = 4$

**25** $\theta = \dfrac{\pi}{3}$   **26** $\phi = \dfrac{\pi}{4}$

**27** $\rho \sin \phi = 6$        **28** $\rho = 2 \sin \phi$
**29** $\rho \sin \phi = \tan \phi$  **30** $\rho^2 - 3\rho + 2 = 0$
**31** $z = 1 - r^2$              **32** $\rho \cos \phi = 2$
**33** $\rho^2 \sin 2\phi \sin \theta = 4$  **34** $\rho \cos \phi = \tan \theta$
**35** $r = 2 \sin \theta$         **36** $r \cos \theta = z$
**37** $r^2 = z$                  **38** $r^2 + z^2 = 2z$

**39** Theorem 12 on page 553 describes a surface when the equation of the surface has one of the rectangular coordinates absent. Describe, in the form of a theorem, the nature of a surface with equation $f(r, z) = 0$ where $r, \theta, z$ are cylindrical coordinates. Do the same when the equation is of the form $f(\theta, z) = 0$.

**40** Same as Problem 39 for spherical coordinates when the equation is of the form $f(\rho, \phi) = 0$.

---

## CHAPTER 13

## REVIEW PROBLEMS

In Problems 1 through 3, decide whether or not the three given points lie on a line.

**1** $A(0, 0, 0)$,  $B(1, 1, 2)$,  $C(-3, 2, 1)$

**2** $A(1, 2, -1)$,  $B(3, 5, -1)$,  $C(5, 8, -5)$

**3** $A(-1, 3, 0)$,  $B(-2, 5, 1)$,  $C(-3, 7, 2)$

**4** One endpoint of a line segment is at $P_1(-1, 3, 0)$ and the midpoint $Q$ lies in the plane $y = 4$. The other endpoint, $P_2$, lies on the intersection of the planes $x = 2$ and $z = 4$. Find the coordinates of $P_2$ and $Q$.

In Problems 5 through 8, find the cosine of the angle between the given lines.

**5** $\dfrac{x+1}{1} = \dfrac{y}{2} = \dfrac{z-7}{2}$; $\dfrac{x-6}{-3} = \dfrac{y-2}{4} = \dfrac{z}{5}$

**6** $\dfrac{x}{2} = \dfrac{y}{3} = \dfrac{z+6}{4}$; $\dfrac{x-3}{1} = \dfrac{y+4}{-2} = \dfrac{z}{1}$

**7** $\dfrac{x-4}{1} = \dfrac{y+2}{2} = \dfrac{z-4}{3}$, $\dfrac{x}{3} = \dfrac{y+8}{2} = \dfrac{z}{-1}$

**8** $\dfrac{x+5}{3} = \dfrac{y-1}{2} = \dfrac{z-4}{-2}$; $\dfrac{x}{2} = \dfrac{y}{1} = \dfrac{z-5}{4}$

**9** Find equations of the line through $(3, 1, 5)$ intersecting the line

$$\frac{x+2}{4} = \frac{y+3}{2} = \frac{z+2}{-1}$$

at right angles.

**\*10** Find the shortest distance between the skew lines

$$\frac{x}{1} = \frac{y+3}{1} = \frac{z-2}{1} \quad \text{and} \quad \frac{x}{1} = \frac{y+1}{2} = \frac{z}{-1}.$$

(*Hint:* Construct a plane through one line that is parallel to the other line.)

**11** Find the intersection of the perpendicular line from the origin to the plane $x + 2y + z = 12$.

**12** Find the intersection of the perpendicular line from the point $(2, 4, 3)$ to the plane $x + 2y + 3z = 5$.

In Problems 13 through 18, find the equations of the line (in any form) as instructed:

**13** Passing through $A(1, 2, 3)$, $B(\sqrt{2}, -2, 0)$

**14** Passing through $A(0, 3, -2)$, $B(1, 6, 4)$

**15** Passing through $A(1, -2, 3)$, $B(2, -4, 6)$

**16** Passing through $A(-2, 0, 2)$, $B(-1, 0, -3)$

**17** Passing through $P(1, 0, 2)$ with direction numbers $2, -1, 3$

**18** Passing through $P(2, -1, 3)$ with direction numbers $0, 2, 3$

**19** Find the equation of the plane passing through $(3, 2, -1)$ and perpendicular to the line $\dfrac{x-2}{2} = \dfrac{y-1}{1} = \dfrac{z+3}{4}$.

**20** Find the equation of the plane passing through $(2, -1, 4)$ and perpendicular to the line $\dfrac{x-1}{5} = -\dfrac{y}{3} = \dfrac{z+2}{1}$.

**\*21** Find the shortest distance between the skew lines $x = y + 3 = z - 2$ and $x = \dfrac{y+1}{2} = -z$.

**22** Find the shortest distance between the two parallel lines $x = y - 1 = -z + 4$ and $x = y = -z$.

**23** The lines $x = 1 + 2t$, $y = 3t$, $z = -1 - t$, and $x = 4t$, $y = -2 + 6t$, $z = 3 - 2t$ determine a plane. Find an equation for this plane.

In Problems 24 through 29, find the center and radius of the given sphere.

**24** $x^2 + y^2 + z^2 - 4x - 6y + 2z - 11 = 0$

**25** $x^2 + y^2 + z^2 + 8x - 2y + 1 = 0$

**26** $x^2 + y^2 + z^2 = 6y - 8z$     **27** $x^2 + y^2 + z^2 = 12x$

**28** $2x^2 + 2y^2 + 2z^2 + 14x - 6y = 1$

**29** $4x^2 + 4y^2 + 4z^2 - x + 2y + 4z - 8 = 0$

**30** Find the equation of a sphere touching the plane $x + y + 2z = 1$ at $(0, -1, 1)$ and passing through $(-2, 0, 0)$.

**\*31** Find the equation of the sphere which has its center on the line $x = y = z$ and which passes through the points $(5, 3, 0)$ and $(-1, 4, 1)$.

In Problems 32 through 35, describe and sketch the graph of the given equation.

**32** $x^2 + y^2 = 4x$        **33** $4y = 1 - x^2$

**34** $y^2 - z^2 = 16$        **35** $y^2 = 4 - 3x$

In Problems 36 through 50, describe the graphs either directly or after transforming the equation to rectangular coordinates.

**36** $r \sin \theta = 2$        **37** $r \cos \theta = 2$

**38** $r = 2 \cos \theta$        **39** $r^2 \sin 2\theta = z$

**40** $r^2 = 4(1 - z^2)$      **41** $r^2 + z^2 = 16z$

**42** $z = 3 \cot \theta$        **43** $z = 3(1 - \tan \theta)$

**44** $\phi^2 - \pi\phi + (3\pi^2/16) = 0$     **45** $\phi = 0$

**46** $\rho = 2 \sin \phi$        **47** $\rho \sin \phi = 4 \sin \theta$

**48** $\rho^2 \cos 2\phi = 16$      **49** $\rho^2 \sin 2\phi = 3$

**50** $\rho \sin \phi = 3 \tan \theta$

**51** Show that an equation in $\rho$ and $\phi$, but not containing $\theta$, represents a surface of revolution about the $z$ axis.

**52** If an equation involves only $\rho$ and $\theta$, but not $\phi$, show that sections of the graph made by planes through the $z$ axis are circles.

In Problems 53 through 63, find equations in cylindrical or in spherical coordinates (whichever is appropriate) of the graph whose equation is given in rectangular coordinates.

**53** $x^2 + y^2 + z^2 = 1$       **54** $x^2 + y^2 = 9$

**55** $x^2 + y^2 = 9z$        **56** $3x = x^2 + y^2$

**57** $y = xz$             **58** $x^2 - y^2 - z^2 = 4$

**59** $x^2 - 4z^2 + y^2 = 0$      **60** $y = x$

**61** $x^2 + y^2 + z^2 = -4z$     **62** $x^2 + y^2 = 8z + 8$

**63** $z = x^2 - y^2$

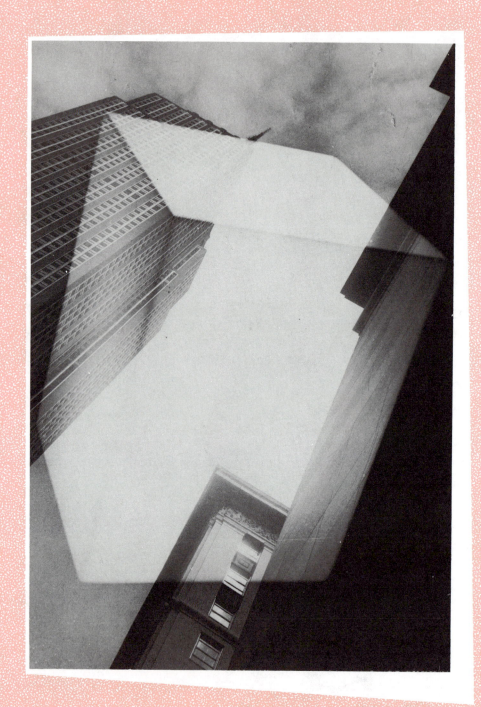

# 14

## VECTORS IN THE PLANE AND IN THREE-SPACE

In this chapter we develop the elementary operations of vectors including the scalar and vector product of two vectors. We also show how vector functions can be differentiated and we use these processes to describe the motion of objects in the plane and in three-space.

---

### 1

#### VECTORS IN THE PLANE

Let $A$ and $B$ be two points in a plane. We recall that the length of the line segment joining $A$ and $B$ is denoted $|AB|$.

**DEFINITIONS** **The directed segment from $A$ to $B$ is defined as the line segment $AB$ which is ordered so that $A$ precedes $B$. We use the symbol $\overrightarrow{AB}$ to denote such a directed segment. We call $A$ its base and $B$ its head.**

If we select the opposite ordering of the line segment $AB$, then we get the directed segment $\overrightarrow{BA}$. In this case $B$ precedes $A$ and we call $B$ the base and $A$ the head. To distinguish geometrically the base and head of a directed line segment, we usually draw an arrow at the head, as shown in Fig. 1.

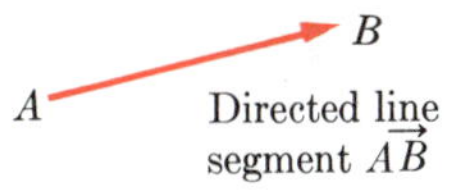

Directed line
segment $\overrightarrow{AB}$

FIGURE 1

**DEFINITIONS**

*The **magnitude** of a directed line segment $\vec{AB}$ is its length $|AB|$. Two directed line segments $\vec{AB}$ and $\vec{CD}$ are said to **have the same magnitude and direction** if and only if either one of the following two conditions holds:*

*i) $\vec{AB}$ and $\vec{CD}$ are both on the same line L, their magnitudes are equal, and the heads B and D are pointing in the same direction, as shown in Fig. 2.*
*ii) The points A, C, D, and B are the vertices of a parallelogram, as shown in Fig. 3.*

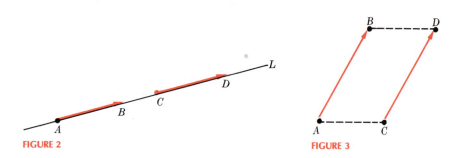

**FIGURE 2**                                            **FIGURE 3**

Figure 4 shows several line segments having the same magnitude and direction. Whenever two directed line segments $\vec{AB}$ and $\vec{CD}$ have the same magnitude and direction, we say that they are **equivalent** and write

$$\vec{AB} \approx \vec{CD}.$$

**FIGURE 4**

We now prove a theorem which expresses this equivalence relationship in terms of coordinates.

**THEOREM 1**

*Suppose that A, B, C, and D are points in a plane. Denote the coordinates of A, B, C, and D by $(x_A, y_A)$, $(x_B, y_B)$, $(x_C, y_C)$, $(x_D, y_D)$, respectively. (See Fig. 5.)*

*i) If the coordinates above satisfy the equations*

$$x_B - x_A = x_D - x_C \qquad and \qquad y_B - y_A = y_D - y_C, \qquad (1)$$

*then*

$$\vec{AB} \approx \vec{CD}.$$

*ii) Conversely, if $\vec{AB}$ is equivalent to $\vec{CD}$, then the coordinates of the four points satisfy Equations (1).*

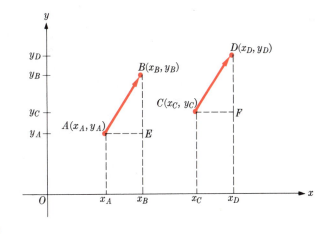

FIGURE 5

**Proof** First, we suppose that Equations (1) hold, and we wish to show that $\vec{AB} \approx \vec{CD}$. If $x_B - x_A$ and $y_B - y_A$ are positive, as shown in Fig. 5, then both $\vec{AB}$ and $\vec{CD}$ are directed segments which are pointing upward and to the right. It is a simple result from plane geometry that $\triangle ABE$ is congruent to $\triangle CDF$; hence $|AB| = |CD|$ and the line through $AB$ is parallel to (or coincides with) the line through $CD$. Therefore $A, B, D, C$ are the vertices of a parallelogram or the line through $AB$ coincides with the line through $CD$. In either case, $\vec{AB}$ is equivalent to $\vec{CD}$. The reader can easily draw the same conclusion if either $x_B - x_A$ or $y_B - y_A$ is negative or if both quantities are negative. See Problem 48 at the end of the section.

To prove the converse, we assume that $\vec{AB} \approx \vec{CD}$. That is, we suppose that $|AB| = |CD|$ and that the line through $AB$ is parallel to (or coincides with) the line through $CD$. If the directed line segments appear as in Fig. 5, we conclude from plane geometry that $\triangle ABE$ is congruent to $\triangle CDF$. Consequently $x_B - x_A = x_D - x_C$ and $y_B - y_A = y_D - y_C$. The reader may verify that the result holds generally if $\vec{AB}$ and $\vec{CD}$ are pointing in directions other than upward and to the right. See Problem 49 at the end of the section. ☐

If we are given a directed line segment $\vec{AB}$, we see at once that there is an unlimited number of equivalent ones. In fact, if $C$ is any given point in the plane, we can use Equations (1) of Theorem 1 to find the coordinates of the unique point $D$ such that $\vec{CD} \approx \vec{AB}$. Another name for a directed line segment is a **vector**. Since directed line segments with the same direction and magnitude are equivalent, *it is convenient to denote any two such equivalent line segments as the same vector.*

We are free to choose any point in $R^2$ as the base of a vector. Often it is convenient to select the origin $(0, 0)$ as the base. Vectors as we defined them are more properly called **free vectors** and a precise definition involves an abstract concept known as an **equivalence class**. We omit the technical details required for the definition of an equivalence class. Since a vector is a directed line segment, its **length** is just the length of the line segment. Two vectors are **orthogonal** (or **perpendicular**) if the lines on which they lie are perpendicular. A **unit vector** is a vector of length one. The **zero vector**, denoted by **0**, is a line segment of zero length: that is, a point. We make the convention that the zero vector is orthogonal to every vector.

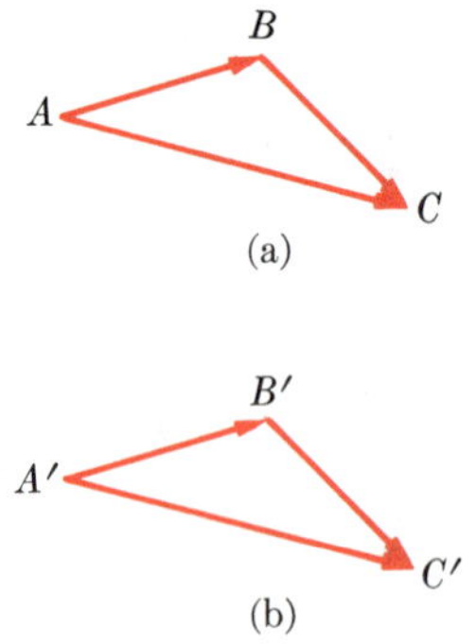

**FIGURE 6**

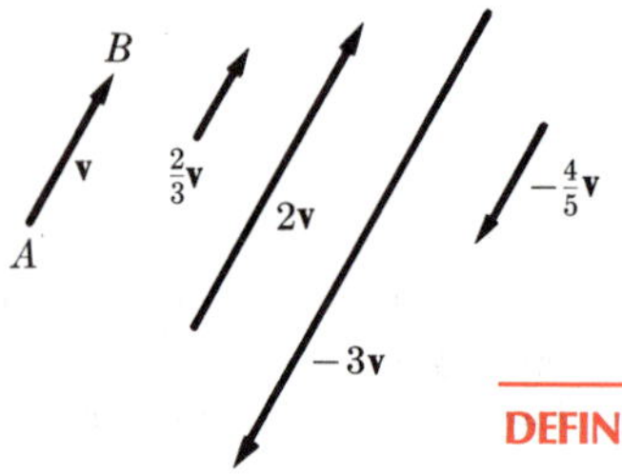

**FIGURE 7**

Vectors may be added to yield other vectors. Suppose **u** and **v** are vectors. To add **u** and **v**, first choose a base $A$ for **u** and consider it as a directed line segment $\vec{AB}$, as shown in Fig. 6(a). Next choose $B$ as the base for **v** and consider it as a directed line segment $\vec{BC}$. Then draw the directed line segment $\vec{AC}$. The sum **w** of **u** and **v** is the vector, or directed line segment, $\vec{AC}$. We write

$$\mathbf{u} + \mathbf{v} = \mathbf{w}.$$

It is important to note that we could have started with any base for **u**, say $A'$ in Fig. 6(b). Then we would have **v** with base at $B'$. The directed line segments $\vec{A'C'}$ and $\vec{AC}$ represent the same vector, as is easily seen from Theorem 1.

Vectors may be multiplied by numbers to yield new vectors. When we do so we refer to the number multiplying the vector as a **scalar**. If **v** is a vector and $c$ is a positive real number, then $c$ is a positive scalar, and $c\mathbf{v}$ is a vector having the same direction as **v** but with magnitude $c$ times as long as **v**. If $c$ is a negative number then $c\mathbf{v}$ has the opposite direction to **v** and its magnitude $|c|$ times as long as that of **v**. If $c$ is zero, we get the vector **0**. Figure 7 shows various multiples of a vector **v**. We write $-\mathbf{v}$ for the vector $(-1)\mathbf{v}$.

**DEFINITIONS**   *Suppose we are given a rectangular coordinate system in the plane. We call I the point with coordinates $(1, 0)$ and J the point with coordinates $(0, 1)$ as shown in Fig. 8. The **unit vector i** is defined as the vector which has the same direction and magnitude as $\vec{OI}$. The **unit vector j** is defined as the vector which has the same direction and magnitude as $\vec{OJ}$.*

**THEOREM 2**   Suppose a vector **w** *has the same direction and magnitude as a particular directed line segment $\vec{AB}$. Denote the coordinates of $A$ and $B$ by $(x_A, y_A)$ and $(x_B, y_B)$, respectively. Then **w** may be expressed in the form*

$$\mathbf{w} = (x_B - x_A)\mathbf{i} + (y_B - y_A)\mathbf{j}.$$

**Proof**   From Equations (1) in Theorem 1, we know that **w** is equivalent to $\vec{OP}$ where $P$ has coordinates $(x_B - x_A, y_B - y_A)$. (See Fig. 8.) Let $Q(x_B - x_A, 0)$ and $R(0, y_B - y_A)$ be the points on the coordinate axes as shown in Fig. 8. It is clear geometrically that $\vec{OQ}$ is $(x_B - x_A)$ times as long as $\vec{OI}$ and that $\vec{OR}$ is $(y_B - y_A)$ times as long as $\vec{OJ}$. We denote by **u** the vector $\vec{OQ}$ and by **v** the

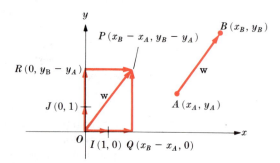

**FIGURE 8**

vector $\vec{OR}$. Using the rule for addition of vectors, we obtain

$$\mathbf{w} = \mathbf{u} + \mathbf{v},$$

since $\vec{QP}$ as well as $\vec{OR}$ is equivalent to $\mathbf{v}$. Since $\mathbf{u} = (x_B - x_A)\mathbf{i}$ and $\mathbf{v} = (y_B - y_A)\mathbf{j}$, the result of the theorem is established.  □

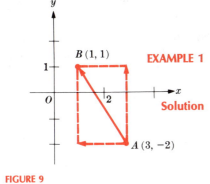

**FIGURE 9**

**EXAMPLE 1**    A vector $\mathbf{v}$ is equivalent to $\vec{AB}$. Given that $A$ has coordinates $(3, -2)$ and $B$ has coordinates $(1, 1)$, express $\mathbf{v}$ in terms of $\mathbf{i}$ and $\mathbf{j}$. Draw a figure.

**Solution**    Using the formula in Theorem 2, we have (see Fig. 9)

$$\mathbf{v} = (1 - 3)\mathbf{i} + (1 + 2)\mathbf{j} = -2\mathbf{i} + 3\mathbf{j}.$$  □

**Notation.** *The length of a vector $\mathbf{v}$ will be denoted by $|\mathbf{v}|$.*

**THEOREM 3**    *If $\mathbf{v} = a\mathbf{i} + b\mathbf{j}$, then*

$$|\mathbf{v}| = \sqrt{a^2 + b^2}.$$

*Therefore $\mathbf{v} = \mathbf{0}$ if and only if $a = b = 0$.*

**Proof**    By means of Theorem 2, we know that $\mathbf{v}$ is equivalent to the directed segment $\vec{OP}$ where $P$ has coordinates $(a, b)$. (See Fig. 10.) Then $|\vec{OP}| = \sqrt{a^2 + b^2}$, since, by definition, the length of a vector is the length of any of its directed line segments; the result follows.  □

The next theorem is useful for problems concerned with the addition of vectors and the multiplication of vectors by numbers.

**THEOREM 4**    *If $\mathbf{v} = a\mathbf{i} + b\mathbf{j}$ and $\mathbf{w} = c\mathbf{i} + d\mathbf{j}$, then*

$$\mathbf{v} + \mathbf{w} = (a + c)\mathbf{i} + (b + d)\mathbf{j}.$$

*Further, if $h$ is any number, then*

$$h\mathbf{v} = (ha)\mathbf{i} + (hb)\mathbf{j}.$$

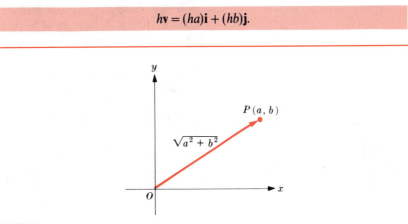

**FIGURE 10**

**Proof**   Let $P$, $Q$, and $S$ have coordinates as shown in Fig. 11. Then $\vec{OP}$ and $\vec{OQ}$ are equivalent to **v** and **w**, respectively. Since $\vec{PS} \approx \vec{OQ}$, we use the rule for addition of vectors to find that $\vec{OS}$ is equivalent to **v** + **w**. The point $R$ (see Fig. 12) is $h$ of the way from $O$ to $P$. Hence $\vec{OR}$ is equivalent to $h\mathbf{v}$.   □

We conclude from the theorems above that the addition of vectors and their multiplication by numbers satisfy the following laws which we collect together as a theorem.

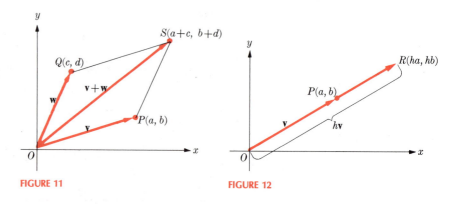

FIGURE 11                                            FIGURE 12

**THEOREM 5**   *Let **u**, **v**, **w** be vectors, and let $c$, $d$ be scalars. Then the following relations hold:*

$$\mathbf{u} + (\mathbf{v} + \mathbf{w}) = (\mathbf{u} + \mathbf{v}) + \mathbf{w} \quad\left.\begin{array}{l}\\\\\end{array}\right\} \textit{Associative laws}$$
$$c(d\mathbf{v}) = (cd)\mathbf{v}$$

$$\mathbf{u} + \mathbf{v} = \mathbf{v} + \mathbf{u} \qquad\qquad \textit{Commutative law}$$

$$(c + d)\mathbf{v} = c\mathbf{v} + d\mathbf{v} \quad\left.\begin{array}{l}\\\\\end{array}\right\} \textit{Distributive laws}$$
$$c(\mathbf{u} + \mathbf{v}) = c\mathbf{u} + c\mathbf{v}$$

$$1 \cdot \mathbf{u} = \mathbf{u}, \qquad 0 \cdot \mathbf{u} = \mathbf{0}, \qquad (-1)\mathbf{u} = -\mathbf{u}$$

*where $-\mathbf{u}$ denotes that vector such that $\mathbf{u} + (-\mathbf{u}) = \mathbf{0}$.*

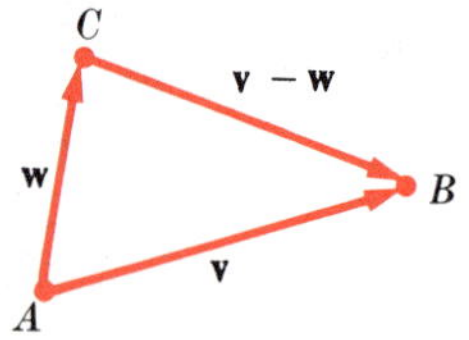

FIGURE 13

It is important to note that multiplication and division of vectors in the ordinary sense is not (and will not be) defined. However, subtraction is defined. If $\vec{AB}$ is equivalent to **v** and $\vec{AC}$ is equivalent to **w**, then $\vec{CB}$ is equivalent to **v** − **w** (see Fig. 13).

**EXAMPLE 2**   Given the vectors $\mathbf{u} = 2\mathbf{i} - 3\mathbf{j}$, $\mathbf{v} = -4\mathbf{i} + \mathbf{j}$. Express the vector $2\mathbf{u} - 3\mathbf{v}$ in terms of **i** and **j**.

**Solution**   $2\mathbf{u} = 4\mathbf{i} - 6\mathbf{j}$ and $-3\mathbf{v} = 12\mathbf{i} - 3\mathbf{j}$. Adding these vectors, we get $2\mathbf{u} - 3\mathbf{v} = 16\mathbf{i} - 9\mathbf{j}$.   □

**DEFINITION**   *Let **v** be any vector except **0**. The unit vector **u** in the direction of **v** is defined by*

$$\mathbf{u} = \left(\frac{1}{|\mathbf{v}|}\right)\mathbf{v}.$$

**EXAMPLE 3**    Given the vector $\mathbf{v} = -2\mathbf{i} + 3\mathbf{j}$, find a unit vector in the direction of $\mathbf{v}$.

**Solution**    We have $|\mathbf{v}| = \sqrt{4+9} = \sqrt{13}$. The desired vector $\mathbf{u}$ is

$$\mathbf{u} = \frac{1}{\sqrt{13}}\mathbf{v} = -\frac{2}{\sqrt{13}}\mathbf{i} + \frac{3}{\sqrt{13}}\mathbf{j}. \qquad \square$$

**EXAMPLE 4**    Given the vector $\mathbf{v} = 2\mathbf{i} - 4\mathbf{j}$. If the base of $\mathbf{v}$ is taken to be $(3, -5)$, find the head of $\mathbf{v}$.

**Solution**    Denote $A = (3, -5)$ and let $\mathbf{v}$ be equivalent to $\vec{AB}$. Denote the coordinates of $B$ by $(x_B, y_B)$. Then we have (by Theorem 2)

$$x_B - 3 = 2 \quad \text{and} \quad y_B + 5 = -4.$$

Therefore $x_B = 5$, $y_B = -9$. $\qquad \square$

## 1    PROBLEMS

In Problems 1 through 5, express $\mathbf{v}$ in terms of $\mathbf{i}$ and $\mathbf{j}$, given that the endpoints $A$ and $B$ of $\mathbf{v}$ have the given coordinates. Draw a figure.

**1**  $A(3, -2)$, $B(1, 5)$

**2**  $A(-4, 1)$, $B(2, -1)$

**3**  $A(5, -6)$, $B(0, -2)$

**4**  $A(4, 4)$, $B(8, -3)$

**5**  $A(7, -2)$, $B(-5, -6)$

In Problems 6 through 9, in each case find a unit vector $\mathbf{u}$ in the direction of $\mathbf{v}$. Express $\mathbf{u}$ in terms of $\mathbf{i}$ and $\mathbf{j}$.

**6**  $\mathbf{v} = 4\mathbf{i} + 3\mathbf{j}$

**7**  $\mathbf{v} = -5\mathbf{i} - 12\mathbf{j}$

**8**  $\mathbf{v} = 2\mathbf{i} - 2\sqrt{3}\mathbf{j}$

**9**  $\mathbf{v} = -2\mathbf{i} + 5\mathbf{j}$

In Problems 10 through 15, find the vector $\mathbf{v}$ with base at $A$ and head at $B$ from the information given. Draw a figure.

**10**  $\mathbf{v} = 7\mathbf{i} - 3\mathbf{j}$, $A(2, -1)$

**11**  $\mathbf{v} = -2\mathbf{i} + 4\mathbf{j}$, $A(6, 2)$

**12**  $\mathbf{v} = 3\mathbf{i} + 2\mathbf{j}$, $B(-2, 1)$

**13**  $\mathbf{v} = -4\mathbf{i} - 2\mathbf{j}$, $B(0, 5)$

**14**  $\mathbf{v} = 3\mathbf{i} + 2\mathbf{j}$, midpoint of segment $AB$ has coordinates $(3, 1)$

**15**  $\mathbf{v} = -2\mathbf{i} + 3\mathbf{j}$, midpoint of segment $AB$ has coordinates $(-4, 2)$

**16**  Find the vector $\mathbf{v}$ of unit length making an angle of $30°$ with the positive $x$ direction. Express $\mathbf{v}$ in terms of $\mathbf{i}$ and $\mathbf{j}$.

**17**  Find the vector $\mathbf{v}$ (in terms of $\mathbf{i}$ and $\mathbf{j}$) which has length $2\sqrt{2}$ and makes an angle of $45°$ with the positive $y$ axis (two solutions).

**18**  Given that $\mathbf{u} = 3\mathbf{i} - 2\mathbf{j}$, $\mathbf{v} = 4\mathbf{i} + 3\mathbf{j}$. Find $\mathbf{u} + \mathbf{v}$ in terms of $\mathbf{i}$ and $\mathbf{j}$. Draw a figure.

**19**  Given that $\mathbf{u} = -2\mathbf{i} + 3\mathbf{j}$, $\mathbf{v} = \mathbf{i} - 2\mathbf{j}$. Find $\mathbf{u} + \mathbf{v}$ in terms of $\mathbf{i}$ and $\mathbf{j}$. Draw a figure.

**20**  Given that $\mathbf{u} = -3\mathbf{i} - 2\mathbf{j}$, $\mathbf{v} = 2\mathbf{i} + \mathbf{j}$. Find $3\mathbf{u} - 2\mathbf{v}$ in terms of $\mathbf{i}$ and $\mathbf{j}$. Draw a figure.

In Problems 21 through 26, the vectors $\mathbf{v}$ and $\mathbf{w}$ have their bases at the origin, $(0, 0)$, and hence may be fully described by the coordinates of their heads which we designate as $B$ and $B'$, respectively. In each problem find (a) $\mathbf{v} + \mathbf{w}$, (b) $\mathbf{v} - \mathbf{w}$, and (c) $3\mathbf{v} - 4\mathbf{w}$.

**21**  $B = (2, -3)$, $B' = (1, 2)$

**22**  $B = (0, 3)$, $B' = (-3, 2)$

**23**  $B = (0, 0)$, $B' = (1, -1)$

**24**  $B = (3, 2\sqrt{2})$  $B' = (3, 2\sqrt{2})$

**25**  $B = (6, 2)$, $B' = (-1, 2)$

**26**  $B = (1, 1)$, $B' = (0, 2)$

In Problems 27 through 35, find the length of the vector $\mathbf{v}$.

**27**  $\mathbf{v} = 3\mathbf{i} + 4\mathbf{j}$

**28**  $\mathbf{v} = -2(3\mathbf{i} + 4\mathbf{j})$

**29**  $\mathbf{v} = 2\mathbf{i} - 6\mathbf{j}$

**30**  $\mathbf{v} = -(3\mathbf{i} - 4\mathbf{j})$

**31**  $\mathbf{v} = \pi(2\mathbf{i} - 4\mathbf{j})$

**32**  $\mathbf{v} = -17\mathbf{j}$

**33**  $\mathbf{v} = 10\mathbf{i} - 10\mathbf{j}$

**34**  $\mathbf{v} = 0\mathbf{i} + 0\mathbf{j}$

**35**  $\mathbf{v} = \mathbf{i}$

In Problems 36 through 43, find the length of $\mathbf{v}$ if there is enough information; otherwise indicate there is insufficient information.

**36**  $\mathbf{v}$ is the directed line segment $\vec{AB}$, and $|\vec{AB}| = 3$.

**37**  $\mathbf{v} = -3\mathbf{u}$ and $|\mathbf{u}| = 2$

**38**  $\mathbf{v} = 2\mathbf{i} + 3\mathbf{j} - \mathbf{u}$, and $|\mathbf{u}| = 2$

**39**  $\mathbf{v} = \mathbf{u} + \mathbf{w}$ and $|\mathbf{u}| = 1$ and $|\mathbf{w}| = 2$

**40**  $\mathbf{v} = \mathbf{u} - 2\mathbf{w}$ and $|\mathbf{u}| = 3$ and $|\mathbf{w}| = 0$

**41**  $\mathbf{v} = \mathbf{u} + 2\mathbf{w}$, $|\mathbf{u}| = \sqrt{2}$, $|\mathbf{w}| = 3$, and $\mathbf{u}$ and $\mathbf{w}$ are orthogonal.

**42**  $\mathbf{v} = \mathbf{u} - 2\mathbf{w}$, $|\mathbf{u}| = \sqrt{2}$, $|\mathbf{w}| = 3$, and $\mathbf{u}$ and $\mathbf{w}$ are not orthogonal.

**43**  $\mathbf{v} = 3\mathbf{u} + 2\mathbf{w}$, $|\mathbf{u}| = 1$, $|\mathbf{w}| = 1$, and $\mathbf{u}$ and $\mathbf{v}$ are othogonal.

**44** Suppose $\vec{AB}$ and $\vec{DE}$ are both equivalent to a vector **u**, and that $\vec{AC}$ and $\vec{DF}$ are also both equivalent to another vector **v**. Show that there is a third vector **w** such that $\vec{AC}$ and $\vec{DF}$ are both equivalent to **w**. Draw a picture.

**45** Show that the vectors $\mathbf{v} = 2\mathbf{i} + 4\mathbf{j}$ and $\mathbf{w} = 10\mathbf{i} - 5\mathbf{j}$ are orthogonal.

**46** Show that the vectors $\mathbf{v} = -3\mathbf{i} + \sqrt{2}\mathbf{j}$ and $\mathbf{w} = 4\sqrt{2}\mathbf{i} + 12\mathbf{j}$ are orthogonal.

**47** Let **u** and **v** be two nonzero vectors. Show that they are orthogonal if and only if the equation

$$|\mathbf{u} + \mathbf{v}|^2 = |\mathbf{u}|^2 + |\mathbf{v}|^2$$

holds.

**\*48** Vectors in the plane may be divided into categories as follows: pointing upward to the right, upward to the left, downward to the right, downward to the left; also those pointing vertically and those horizontally. Theorem 1, part (i) was established for vectors pointing upward to the right (Fig. 5). Write a proof of Theorem 1(i) for three of the remaining five possibilities. Draw a figure and devise a general proof.

**\*49** The same as Problem 48 for part (ii) of Theorem 1.

**50** Write a proof of Theorem 5.

---

**2**

## OPERATIONS WITH PLANE VECTORS. THE SCALAR PRODUCT

Two vectors **v** and **w** are said to be **parallel** or **proportional** when each is a scalar multiple of the other (and neither is zero).

We are already familiar with angles between intersecting lines (Fig. 14(a)) and with angles between directed line segments having the same base (Fig. 14(b)).

We now want to define the angle between two vectors, **v** and **w**. The problem is that **v** and **w** may not have the same base (Fig. 14(c)). We solve this

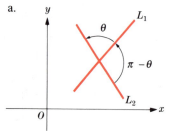

a.

Angles between intersecting lines

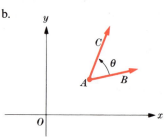

b.

Angle between directed line segments with the same base

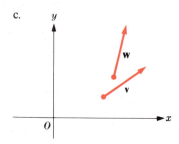

c.

Two vectors without common base

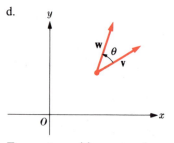

d.

Two vectors with common base and angle $\theta$ between them

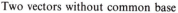

**FIGURE 14**

problem by recalling that we are free to choose any point as a base for a vector so long as we preserve the direction and magnitude. We therefore choose as base for **w** the point that serves as base for **v** (Fig. 14(d)). We then define the angle between two vectors **v** and **w** to be the angle between the two directed line segments with common base having the same direction and magnitude as **v** and **w**.

Suppose that **i** and **j** are the usual unit vectors pointing in the direction of the $x$ and $y$ axes, respectively. Then we have the following theorem.

**THEOREM 6**  *If $\theta$ is the angle between the vectors*

$$\mathbf{v} = a\mathbf{i} + b\mathbf{j} \qquad and \qquad \mathbf{w} = c\mathbf{i} + d\mathbf{j},$$

*then*

$$\cos \theta = \frac{ac + bd}{|\mathbf{v}|\,|\mathbf{w}|}.$$

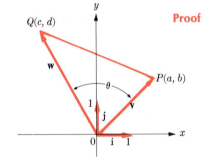

FIGURE 15

**Proof**  We draw **v** and **w** with base at the origin of the coordinate system, as shown in Fig. 15. Using Theorem 2 of Section 1, we see that the coordinates of $P$ are $(a, b)$ and those of $Q$ are $(c, d)$. The length of **v** is $|OP|$ and the length of **w** is $|OQ|$. We apply the law of cosines to $\triangle OPQ$, obtaining

$$\cos \theta = \frac{|OP|^2 + |OQ|^2 - |QP|^2}{2|OP|\,|OQ|}.$$

Therefore

$$\cos \theta = \frac{a^2 + b^2 + c^2 + d^2 - (a-c)^2 - (b-d)^2}{2|OP|\,|OQ|} = \frac{ac + bd}{|\mathbf{v}|\,|\mathbf{w}|}. \qquad \square$$

**EXAMPLE 1**  Given the vectors $\mathbf{v} = 2\mathbf{i} - 3\mathbf{j}$ and $\mathbf{w} = \mathbf{i} - 4\mathbf{j}$, compute the cosine of the angle between **v** and **w**.

**Solution**  We have

$$|\mathbf{v}| = \sqrt{4 + 9} = \sqrt{13}, \qquad |\mathbf{w}| = \sqrt{1 + 16} = \sqrt{17}.$$

Therefore

$$\cos \theta = \frac{2 \cdot 1 + (-3)(-4)}{\sqrt{17}\sqrt{13}} = \frac{14}{\sqrt{221}}. \qquad \square$$

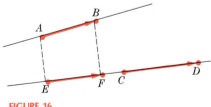

FIGURE 16

Suppose that we have two directed line segments $\vec{AB}$ and $\vec{CD}$, as shown in Fig. 16. The **projection of $\vec{AB}$ in the direction $\vec{CD}$** is defined as the directed length $\overline{EF}$ of the line segment $EF$ obtained by dropping perpendiculars from $A$ and $B$ to the line containing $\vec{CD}$. We can find the projection of a vector **v** along a vector **w** by taking **v** and finding its projection (call it $\overline{EF}$) onto a line in the direction of **w**.

If $\theta$ is the angle between two vectors **v** and **w**, the quantity

$$|\mathbf{v}| \cos \theta$$

is positive if $\theta$ is an acute angle and the projection of **v** on **w** is positive. If $\theta$ is an obtuse angle, then $|\mathbf{v}| \cos \theta$ is negative, and the projection of $\mathbf{v}$ on $\mathbf{w}$ is negative.

**DEFINITION**   *The quantity $|\mathbf{v}| \cos \theta$ is called the* **projection of v on w.** *We write* $\text{Proj}_{\mathbf{w}}\mathbf{v}$ *for this quantity. If $\mathbf{v} = a\mathbf{i} + b\mathbf{j}$ and $\mathbf{w} = c\mathbf{i} + d\mathbf{j}$, then*

$$\text{Proj}_{\mathbf{w}}\mathbf{v} = |\mathbf{v}| \cos \theta = \frac{ac + bd}{|\mathbf{w}|}.$$

**EXAMPLE 2**   Find the projection of $\mathbf{v} = 3\mathbf{i} - 2\mathbf{j}$ on $\mathbf{w} = -2\mathbf{i} - 4\mathbf{j}$.

**Solution**   We obtain the desired projection by using the formula

$$\text{Proj}_{\mathbf{w}}\mathbf{v} = \frac{-6 + 8}{\sqrt{20}} = \frac{1}{\sqrt{5}}.$$

**DEFINITION**   *The* **scalar product** *of two nonzero vectors $\mathbf{v}$ and $\mathbf{w}$, written $\mathbf{v} \cdot \mathbf{w}$, is defined by the formula*

$$\mathbf{v} \cdot \mathbf{w} = |\mathbf{v}||\mathbf{w}| \cos \theta,$$

*where $\theta$ is the angle between $\mathbf{v}$ and $\mathbf{w}$. If one of the vectors is $\mathbf{0}$, the scalar product is defined to be 0.*

The terms **dot product** and **inner product** are also used to designate scalar product.

**THEOREM 7**   *Let $\mathbf{u}$, $\mathbf{v}$, $\mathbf{w}$ be any vectors. The following hold:*
   i) $\mathbf{v} \cdot \mathbf{w} = \mathbf{w} \cdot \mathbf{v}$
   ii) $\mathbf{v} \cdot \mathbf{v} = |\mathbf{v}|^2$
   iii) *if $\mathbf{v}$ and $\mathbf{w}$ are orthogonal, then $\mathbf{v} \cdot \mathbf{w} = 0$;*
   iv) *if $\mathbf{v}$ and $\mathbf{w}$ are parallel, then $\mathbf{v} \cdot \mathbf{w} = \pm |\mathbf{v}||\mathbf{w}|$, and conversely;*
   v) *if $\mathbf{v} = a\mathbf{i} + b\mathbf{j}$ and $\mathbf{w} = c\mathbf{i} + d\mathbf{j}$, then $\mathbf{v} \cdot \mathbf{w} = ac + bd$;*
   vi) $\mathbf{u} \cdot (\mathbf{v} + \mathbf{w}) = \mathbf{u} \cdot \mathbf{v} + \mathbf{u} \cdot \mathbf{w}$ *(the distributive law).*

Parts (i) through (iv) of Theorem 7 follow easily from the definition of the scalar product. Part (v) follows from the definition and Theorem 6. The proof of part (vi) is a little more difficult. We leave it as an exercise for the reader (Problem 25 at the end of this section.)

**EXAMPLE 3**   Given the vectors $\mathbf{u} = 3\mathbf{i} + 2\mathbf{j}$ and $\mathbf{v} = 2\mathbf{i} + a\mathbf{j}$. Determine the scalar $a$ so that $\mathbf{u}$ and $\mathbf{v}$ are orthogonal. Determine $a$ so that $\mathbf{u}$ and $\mathbf{v}$ are parallel. For what value of $a$ will $\mathbf{u}$ and $\mathbf{v}$ make an angle of $\pi/4$?

**Solution**   If $\mathbf{u}$ and $\mathbf{v}$ are orthogonal, we have

$$3 \cdot 2 + 2 \cdot a = 0 \qquad \text{and} \qquad a = -3.$$

For **u** and **v** to be parallel, we must have

$$\mathbf{u} \cdot \mathbf{v} = \pm |\mathbf{u}||\mathbf{v}|$$

or

$$6 + 2a = \pm \sqrt{13} \cdot \sqrt{4 + a^2}.$$

Solving, we obtain $a = \frac{4}{3}$.

Employing the formula for $\cos \theta$ in Theorem 6, we find

$$\tfrac{1}{2}\sqrt{2} = \cos\frac{\pi}{4} = \frac{6 + 2a}{\sqrt{13} \cdot \sqrt{4 + a^2}},$$

so that **u** and **v** make an angle of $\pi/4$ when

$$a = 10, \ -\tfrac{2}{5}. \qquad \square$$

Because they are geometric quantities which are independent of the coordinate system, vectors are well suited for establishing certain types of theorems in plane geometry. We give two examples to exhibit the technique.

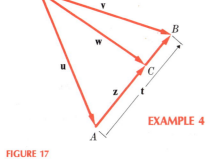

FIGURE 17

**EXAMPLE 4**  Let $\vec{OA}$ and $\vec{OB}$ be vectors, and let $C$ be the point on the line $\vec{AB}$ which is $\frac{2}{3}$ of the way from $A$ to $B$ (Fig. 17). Let $\mathbf{u} = \vec{OA}$, $\mathbf{v} = \vec{OB}$, and $\mathbf{w} = \vec{OC}$. Express **w** in terms of **u** and **v**.

**Solution**  Let **z** be the vector $\vec{AC}$, and **t** the vector $\vec{AB}$. We have

$$\mathbf{w} = \mathbf{u} + \mathbf{z} = \mathbf{u} + \tfrac{2}{3}\mathbf{t}.$$

Also we know that $\mathbf{t} = \mathbf{v} - \mathbf{u}$, and so

$$\mathbf{w} = \mathbf{u} + \tfrac{2}{3}(\mathbf{v} - \mathbf{u}) = \tfrac{1}{3}\mathbf{u} + \tfrac{2}{3}\mathbf{v}. \qquad \square$$

It is often convenient to think of a directed line segment as a vector, as Example 4 shows. For example, we cannot "subtract" directed line segments unless the base of one is at the head of the other, whereas any two vectors can be subtracted. If $\vec{AB}$ denotes a directed line segment, we introduce a convenient symbol to indicate we are treating $\vec{AB}$ as a vector with base at $A$ and head at $B$.

> **Notation.** *The symbol* $\mathbf{v}[\vec{AB}]$ *denotes the vector having the same magnitude and direction as* $\vec{AB}$.

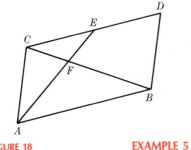

FIGURE 18

**EXAMPLE 5**  Let $ABDC$ be a parallelogram, as shown in Fig. 18. Suppose that $E$ is the midpoint of $CD$ and $F$ is $\frac{2}{3}$ of the way from $A$ to $E$ on $AE$. Show that $F$ is $\frac{2}{3}$ of the way from $B$ to $C$.

**Solution**  Let $\vec{AB}$, $\vec{AC}$, $\vec{AF}$, $\vec{AE}$, $\vec{BF}$, $\vec{BC}$, and $\vec{CE}$ all be directed line segments. Then $\mathbf{v}[\vec{AB}], \mathbf{v}[\vec{AC}], \mathbf{v}[\vec{AF}]$, etc., are the vectors which correspond to the directed line segments shown in brackets. By hypothesis, we have

$$\mathbf{v}[\vec{AB}] = \mathbf{v}[\vec{CD}] \qquad \text{and} \qquad \mathbf{v}[\vec{CE}] = \tfrac{1}{2}\mathbf{v}[\vec{CD}].$$

The rule for addition of vectors gives us

$$\mathbf{v}[\vec{AE}] = \mathbf{v}[\vec{AC}] + \mathbf{v}[\vec{CE}] = \mathbf{v}[\vec{AC}] + \tfrac{1}{2}\mathbf{v}[\vec{AB}].$$

Also, since $\mathbf{v}[\vec{AF}] = \tfrac{2}{3}\mathbf{v}[\vec{AE}]$, we obtain

$$\mathbf{v}[\vec{AF}] = \tfrac{2}{3}\mathbf{v}[\vec{AC}] + \tfrac{1}{3}\mathbf{v}[\vec{AB}].$$

The rule for subtraction of vectors yields

$$\mathbf{v}[\vec{BF}] = \mathbf{v}[\vec{AF}] - \mathbf{v}[\vec{AB}] = \tfrac{2}{3}\mathbf{v}[\vec{AC}] - \tfrac{2}{3}\mathbf{v}[\vec{AB}] = \tfrac{2}{3}(\mathbf{v}[\vec{AC}] - \mathbf{v}[\vec{AB}]).$$

Since $\mathbf{v}[\vec{BC}] = \mathbf{v}[\vec{AC}] - \mathbf{v}[\vec{AB}]$, we conclude that

$$\mathbf{v}[\vec{BF}] = \tfrac{2}{3}\mathbf{v}[\vec{BC}],$$

which is the desired result.    □

**EXAMPLE 6**    Let $A = (0, 0)$, $B = (1, 2)$, and $C = (-2, 1)$. Show that $\vec{AB}$ and $\vec{AC}$ are perpendicular.

**Solution**    Let $\mathbf{v} = \mathbf{v}[\vec{AB}]$ and $\mathbf{w} = \mathbf{v}[\vec{AC}]$. It is enough to show that $\mathbf{v}$ and $\mathbf{w}$ are orthogonal (i.e., perpendicular). Since $\vec{AB}$ has its base at $(0, 0)$ and its head at $(1, 2)$, we deduce $\mathbf{v} = \mathbf{i} + 2\mathbf{j}$ and similarly $\mathbf{w} = -2\mathbf{i} + \mathbf{j}$. By part (v) of Theorem 6 $\mathbf{v} \cdot \mathbf{w} = (1)(-2) + (2)(1) = -2 + 2 = 0$. By definition, $\mathbf{v} \cdot \mathbf{w} = |\mathbf{v}||\mathbf{w}| \cos \theta$ and since $|\mathbf{v}||\mathbf{w}| \neq 0$, we see that $\cos \theta = 0$, where $\theta$ is the angle between $\mathbf{v}$ and $\mathbf{w}$. Therefore

$$\theta = \frac{\pi}{2} \quad \text{or} \quad \frac{-\pi}{2},$$

and so $\mathbf{v}$ and $\mathbf{w}$ are orthogonal.    □

## 2    PROBLEMS

In Problems 1 through 6, given that $\theta$ is the angle between $\mathbf{v}$ and $\mathbf{w}$, find $|\mathbf{v}|$, $|\mathbf{w}|$, $\cos \theta$, and the projection of $\mathbf{v}$ on $\mathbf{w}$.

1   $\mathbf{v} = 4\mathbf{i} - 3\mathbf{j}$, $\mathbf{w} = -4\mathbf{i} - 3\mathbf{j}$      2   $\mathbf{v} = 2\mathbf{i} + \mathbf{j}$, $\mathbf{w} = \mathbf{i} - \mathbf{j}$

3   $\mathbf{v} = 3\mathbf{i} - 4\mathbf{j}$, $\mathbf{w} = 5\mathbf{i} + 12\mathbf{j}$      4   $\mathbf{v} = 4\mathbf{i} + \mathbf{j}$, $\mathbf{w} = 6\mathbf{i} - 8\mathbf{j}$

5   $\mathbf{v} = 3\mathbf{i} + 2\mathbf{j}$, $\mathbf{w} = 2\mathbf{i} - 3\mathbf{j}$      6   $\mathbf{v} = 6\mathbf{i} + 5\mathbf{j}$, $\mathbf{w} = 2\mathbf{i} + 5\mathbf{j}$

In Problems 7 through 12, find the projection of $\mathbf{v}[\vec{AB}]$ onto $\mathbf{v}[\vec{CD}]$. Draw a figure in each case.

7   $A(1, 0)$, $B(2, 3)$, $C(1, 1)$, $D(-1, 1)$

8   $A(0, 0)$, $B(1, 4)$, $C(0, 0)$, $D(2, 7)$

9   $A(3, 1)$, $B(5, 2)$, $C(-2, -1)$, $D(-1, 3)$

10   $A(2, 4)$, $B(4, 7)$, $C(6, -1)$, $D(2, 2)$

11   $A(2, -1)$, $B(1, 3)$, $C(5, 2)$, $D(9, 3)$

12   $A(1, 6)$, $B(2, 5)$, $C(5, 2)$, $D(9, 3)$

In Problems 13 through 17, find $\cos \theta$ and $\cos \alpha$, given that $\theta = \angle ABC$ and $\alpha = \angle BAC$. Use vector methods and draw figures.

13   $A(-1, 1)$, $B(3, -1)$, $C(3, 4)$

14   $A(2, 1)$, $B(-1, 2)$, $C(1, 3)$

15   $A(3, 4)$, $B(5, 1)$, $C(4, 1)$

16   $A(4, 1)$, $B(1, -1)$, $C(3, 3)$

17   $A(0, 0)$, $B(3, -5)$, $C(6, -10)$

In Problems 18 through 24, determine the number $a$ (if possible) such that the given condition for $\mathbf{v}$ and $\mathbf{w}$ is satisfied.

18   $\mathbf{v} = 2\mathbf{i} + a\mathbf{j}$, $\mathbf{w} = \mathbf{i} + 3\mathbf{j}$, $\mathbf{v}$ and $\mathbf{w}$ orthogonal

19   $\mathbf{v} = \mathbf{i} - 3\mathbf{j}$, $\mathbf{w} = 2a\mathbf{i} + \mathbf{j}$, $\mathbf{v}$ and $\mathbf{w}$ orthogonal

20   $\mathbf{v} = 3\mathbf{i} - 4\mathbf{j}$, $\mathbf{w} = 2\mathbf{i} + a\mathbf{j}$, $\mathbf{v}$ and $\mathbf{w}$ parallel

**21** $v = ai + 2j$, $w = 2i - aj$, $v$ and $w$ parallel

**22** $v = ai$, $w = 2i - 3j$, $v$ and $w$ parallel

**23** $v = 5i + 12j$, $w = i + aj$, $v$ and $w$ make an angle of $\pi/3$

**24** $v = 4i - 3j$, $w = 2i + aj$, $v$ and $w$ make an angle of $\pi/6$

**25** Prove the distributive law for the scalar product (part (vi) of Theorem 6).

In Problems 26 through 30, let $v = v[\vec{AB}]$ and $w = v[\vec{CD}]$. Determine when $v$ and $w$ are orthogonal.

**26** $A = (0, 0)$, $B = (2, 3)$; $C = (3, -2)$, $D = (0, 0)$

**27** $A = (1, 2)$, $B = (3, 4)$; $C = (-7, 3)$, $D = (6, -10)$

**28** $A = (2, 1)$, $B = (6, 5)$; $C = (1, -2)$, $D = (5, 2)$

**29** $A = (0, -3)$, $B = (7, 2)$; $C = (3, 2)$, $D = (3, 2)$

**30** $A = (0, -3)$, $B = (0, -3)$; $C = (4, 2)$, $D = (1, 3)$

**31** Let $i$ and $j$ be the usual unit vectors of one coordinate system, and let $i_1$ and $j_1$ be the unit orthogonal vectors corresponding to another rectangular system of coordinates. Given that

$$v = ai + bj, \quad w = ci + dj, \quad v = a_1 i_1 + b_1 j_1, \quad w = c_1 i_1 + d_1 j_1,$$

show that

$$ac + bd = a_1 c_1 + b_1 d_1.$$

In Problems 32 through 35, the quantity $|AB|$ denotes (as is customary) the length of the line segment $AB$, the quantity $|AC|$, the length of $AC$, etc.

**32** Given $\triangle ABC$, in which $\angle A = 120°$, $|AB| = 4$, and $|AC| = 7$. Find $|BC|$ and the projections of $\vec{AB}$ and $\vec{AC}$ on $\vec{BC}$. Draw a figure.

**33** Given $\triangle ABC$, with $\angle A = 45°$, $|AB| = 8$, $|AC| = 6\sqrt{2}$. Find $|BC|$ and the projections of $\vec{AB}$ and $\vec{AC}$ on $\vec{BC}$. Draw a figure.

**34** Given $\triangle ABC$, with $|AB| = 10$, $|AC| = 9$, $|BC| = 7$. Find the projections of $\vec{AC}$ and $\vec{BC}$ on $\vec{AB}$. Draw a figure.

**35** Given $\triangle ABC$, with $|AB| = 5$, $|AC| = 7$, $|BC| = 9$. Find the projections of $\vec{AB}$ and $\vec{AC}$ on $\vec{CB}$. Draw a figure.

**36** Given the line segments $AB$ and $AC$, with $D$ on $AB$ $\frac{2}{3}$ of the way from $A$ to $B$. Let $E$ be the midpoint of $AC$. Express $v[\vec{DE}]$ in terms of $v[\vec{AB}]$ and $v[\vec{AC}]$. Draw a figure.

**37** Suppose that $v[\vec{AD}] = \frac{1}{4}v[\vec{AB}]$ and $v[\vec{BE}] = \frac{1}{2}v[\vec{BC}]$. Find $v[\vec{DE}]$ in terms of $v[\vec{AB}]$ and $v[\vec{BC}]$. Draw a figure.

**38** Given $\square ABDC$, a parallelogram, with $E$ $\frac{2}{3}$ of the way from $B$ to $D$, and $F$ as the midpoint of segment $CD$. Find $v[\vec{EF}]$ in terms of $v[\vec{AB}]$ and $v[\vec{AC}]$. Draw a figure.

**39** Given parallelogram $ABDC$, with $E$ $\frac{1}{4}$ of the way from $B$ to $C$, and $F$ $\frac{1}{4}$ of the way from $A$ to $D$. Find $v[\vec{EF}]$ in terms of $v[\vec{AB}]$ and $v[\vec{AC}]$. Draw a figure.

**40** Given parallelogram $ABDC$, with $E$ $\frac{1}{3}$ of the way from $B$ to $D$, and $F$ $\frac{1}{4}$ of the way from $B$ to $C$. Show that $F$ is $\frac{3}{4}$ of the way from $A$ to $E$. Draw a figure.

**41** Suppose that on the sides of $\triangle ABC$, $v[\vec{BD}] = \frac{2}{3}v[\vec{BC}]$, $v[\vec{CE}] = \frac{2}{3}v[\vec{CA}]$, and $v[\vec{AF}] = \frac{2}{3}v[\vec{AB}]$. Draw a figure and show that

$$v[\vec{AD}] + v[\vec{BE}] + v[\vec{CF}] = 0.$$

**42** Show that the conclusion of Problem 41 holds when the fraction $\frac{2}{3}$ is replaced by any real number $h$.

**43** Let $a = v[\vec{OA}]$, $b = v[\vec{OB}]$, and $c = v[\vec{OC}]$. Show that the medians of $\triangle ABC$ meet at a point $P$, and express $v[\vec{OP}]$ in terms of $a$, $b$ and $c$. Draw a figure.

---

## 3

# VECTORS IN THREE DIMENSIONS

The development of vectors in three-dimensional space is a direct extension of the theory of vectors in the plane as given in Sections 1 and 2. This section, which is completely analogous to part of Section 1, may be read quickly. The essential distinction between two- and three-dimensional vectors appears in Sections 4, 5, and 6.

A **directed line segment** $\vec{AB}$ is defined as before, except that now the **base** $A$ and the **head** $B$ may be situated anywhere in three-space. The **magnitude** of a directed line segment is its length. Two directed line segments $\vec{AB}$ and $\vec{CD}$

are said to **have the same magnitude and direction** if and only if either one of the following two conditions holds:

i) $\vec{AB}$ and $\vec{CD}$ are both on the same directed line $\vec{L}$ and their directed lengths are equal; or

ii) the points $A$, $C$, $D$, and $B$ are the vertices of a parallelogram as shown in Fig. 19.

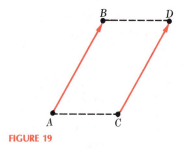

FIGURE 19

We note that the above definition is the same as that given in Section 1.

Whenever two directed line segments $\vec{AB}$ and $\vec{CD}$ have the same magnitude and direction, we say they are **equivalent** and write

$$\vec{AB} \approx \vec{CD}.$$

We shall next state a theorem which is a direct extension of Theorem 1 in Section 1.

**THEOREM 8**   *Suppose that A, B, C, and D are points in space. Denote the coordinates of A, B, C, and D by $(x_A, y_A, z_A)$, $(x_B, y_B, z_B)$, and so forth. (i) If the coordinates satisfy the equations*

$$x_B - x_A = x_D - x_C, \qquad y_B - y_A = y_D - y_C, \qquad z_B - z_A = z_D - z_C, \qquad (1)$$

*then $\vec{AB} \approx \vec{CD}$. (ii) Conversely, if $\vec{AB} \approx \vec{CD}$, the coordinates satisfy the equations in (1).*

**Sketch of proof**   (i) We assume that the equations in (1) hold. Then also,

$$x_C - x_A = x_D - x_B, \; y_C - y_A = y_D - y_B, \; z_C - z_A = z_D - z_B. \qquad (2)$$

As in the proof of Theorem 1 in Section 1, we can conclude that the lines $AB$ and $CD$ are parallel because their direction numbers are equal. Similarly, $AC$ and $BD$ are parallel. Therefore $\vec{AB} \approx \vec{CD}$ or all the points are on one directed line $\vec{L}$. The proof that $\vec{AB} \approx \vec{CD}$ in this latter case is given in Appendix 5. (ii) To prove the converse, assume that $\vec{AB} \approx \vec{CD}$. Then either $ACDB$ is a parallelogram or all the points are on a directed line $\vec{L}$. When $ACDB$ is a parallelogram there is a unique point $E$ such that

$$x_E - x_C = x_B - x_A, \; y_E - y_C = y_B - y_A, \; z_E - z_C = z_B - z_A.$$

As in the proof of Theorem 1 in Section 1, we conclude from part (i) that $ACEB$ is a parallelogram and $D = E$; hence Equations (1) hold. The proof that Equations (1) hold when the points are on a directed line $\vec{L}$ is given in Appendix 5. □

If we are given a directed line segment $\vec{AB}$, it is clear that there is an unlimited number of equivalent ones. In fact, if $C$ is any given point in three-space, we can use Equations (1) of Theorem 8 to find the coordinates of the unique point $D$ such that $\vec{CD} \approx \vec{AB}$.

As in the case of directed line segments in $R^2$, we use the term **vector** as another name for a directed line segment. Moreover, since directed line segments with the same direction and magnitude are equivalent, *it is convenient to denote any two such equivalent line segments as the same vector.* We are therefore free to choose any point in $R^3$ as the base of a vector. Often it is convenient to choose the origin $(0, 0, 0)$ as the base.

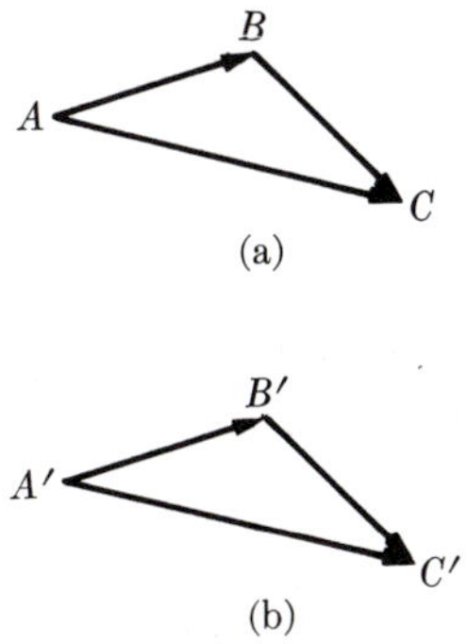

Since a vector is a directed line segment, its **length** is the length of the line segment. Two vectors are **orthogonal** (or **perpendicular**) if, when the base for each is taken to be the origin (0, 0, 0), the resulting directed line segments are perpendicular. A **unit vector** is a vector of length one, and the **zero vector**, denoted by **0**, is a line segment of zero length; that is, a point. By convention, the zero vector is orthogonal to every other vector.

As in Section 1, we can define the sum of two vectors. Given **u** and **v**, choose a base for **u** and consider it as a directed line segment $\vec{AB}$, as shown in Fig. 20(a). Next choose B as the base for **v** and consider it as the directed line segment $\vec{BC}$. Then **u** + **v** is the vector $\vec{AC}$ as shown in Fig. 20(a). If $\vec{A'B'}$ and $\vec{B'C'}$ are other directed line segments equivalent to **u** and **v**, respectively, it follows from Theorem 8 that $\vec{A'C'} \approx \vec{AC}$. Therefore $\vec{A'C'}$ is equivalent to **u** + **v** (Fig. 20(b)). In other words, the rule for forming the sum of two vectors does not depend on the location of the bases we select when making the calculation.

Vectors may be multiplied by numbers, which are referred to as **scalars**. Given a vector **u** and a scalar $c$, then $c\mathbf{u}$ is a vector in the same direction as **u** if $c$ is positive, and with magnitude $c|\mathbf{u}|$. If $c$ is negative, then the direction of $c\mathbf{u}$ is the opposite to **u**, and the magnitude is $|c||\mathbf{u}|$.

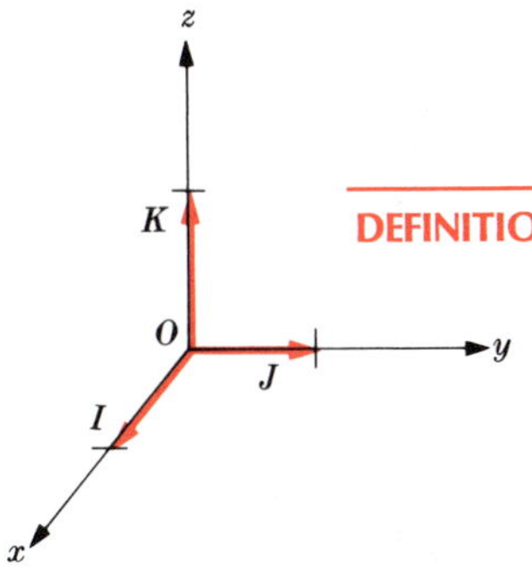

**FIGURE 20**

**DEFINITIONS**    *Suppose that a rectangular coordinate system is given. Figure 21 shows such a system with the points I(1, 0, 0), J(0, 1, 0), and K(0, 0, 1) identified. The **unit vector i** is defined as the vector having the same direction and magnitude as $\vec{OI}$. The **unit vector j** is defined as the vector which has the same direction and magnitude as $\vec{OJ}$. The **unit vector k** is defined as the vector which has the same direction and magnitude as $\vec{OK}$.*

**FIGURE 21**

We now establish a direct extension of Theorem 2 in Section 1.

**THEOREM 9**    *Suppose a vector **w** has the same direction and magnitude as a particular directed line segment $\vec{AB}$. Denote the coordinates of A and B by $(x_A, y_A, z_A)$ and $(x_B, y_B, z_B)$, respectively. Then **w** may be expressed in the form*

$$\mathbf{w} = (x_B - x_A)\mathbf{i} + (y_B - y_A)\mathbf{j} + (z_B - z_A)\mathbf{k}.$$

**Proof**    From equations (1) of Theorem 8 we know that **w** is equivalent to $\vec{OP}$ where P has coordinates $(x_B - x_A, y_B - y_A, z_B - z_A)$. Let

$$Q(x_B - x_A, 0, 0), \quad R(0, y_B - y_A, 0),$$
$$S(0, 0, z_B - z_A), \quad T(x_B - x_A, y_B - y_A, 0)$$

be as shown in Fig. 22. Then Q is $x_B - x_A$ of the way from O to I(1, 0, 0), and similarly for R and S with regard to J and K. Therefore,

$$\mathbf{v}[\vec{OQ}] = (x_B - x_A)\mathbf{i},$$
$$\mathbf{v}[\vec{QT}] = \mathbf{v}[\vec{OR}] = (y_B - y_A)\mathbf{j},$$
$$\mathbf{v}[\vec{TP}] = \mathbf{v}[\vec{OS}] = (z_B - z_A)\mathbf{k}.$$

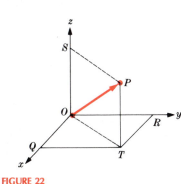

**FIGURE 22**

Using the rule for addition of vectors, we find

$$\mathbf{v}[\vec{OP}] = \mathbf{v}[\vec{OQ}] + \mathbf{v}[\vec{QT}] + \mathbf{v}[\vec{TP}],$$

and the proof is complete.  ☐

**EXAMPLE 1**   A vector $\mathbf{v}$ is equivalent to $\vec{AB}$. If $A$ and $B$ have coordinates $(3, -2, 4)$ and $(2, 1, 5)$, respectively, express $\mathbf{v}$ in terms of $\mathbf{i}$, $\mathbf{j}$, and $\mathbf{k}$.

**Solution**   From Theorem 9, we obtain

$$\mathbf{v}(\vec{AB}) = (2-3)\mathbf{i} + (1+2)\mathbf{j} + (5-4)\mathbf{k} = -\mathbf{i} + 3\mathbf{j} + \mathbf{k}.$$  ☐

The next two theorems are direct extensions of Theorems 3 and 4 in Section 1. The proofs are left to the reader.

The **length** of a vector $\mathbf{v}$ is denoted by $|\mathbf{v}|$.

**THEOREM 10**   *If* $\mathbf{v} = a\mathbf{i} + b\mathbf{j} + c\mathbf{k}$, *then*

$$|\mathbf{v}| = \sqrt{a^2 + b^2 + c^2}.$$

**THEOREM 11**   *If* $\mathbf{v} = a_1\mathbf{i} + b_1\mathbf{j} + c_1\mathbf{k}$, $\mathbf{w} = a_2\mathbf{i} + b_2\mathbf{j} + c_2\mathbf{k}$, *then*

$$\mathbf{v} + \mathbf{w} = (a_1 + a_2)\mathbf{i} + (b_1 + b_2)\mathbf{j} + (c_1 + c_2)\mathbf{k}.$$

*If* $h$ *is any scalar, then*

$$h\mathbf{v} = ha_1\mathbf{i} + hb_1\mathbf{j} + hc_1\mathbf{k}.$$

In complete analogy with vectors in the plane, we conclude from the theorems above that the addition of vectors and their multiplication by numbers satisfy the following laws, which we present as a theorem.

**THEOREM 12**   *Let* $\mathbf{u}, \mathbf{v}, \mathbf{w}$ *be vectors, and let* $c$, $d$ *be scalars. Then the following relations hold:*

$$\left.\begin{array}{c} \mathbf{u} + (\mathbf{v} + \mathbf{w}) = (\mathbf{u} + \mathbf{v}) + \mathbf{w} \\ c(d\mathbf{v}) = (cd)\mathbf{v} \end{array}\right\} \text{Associative laws}$$

$$\mathbf{u} + \mathbf{v} = \mathbf{v} + \mathbf{u} \qquad \text{Commutative law}$$

$$\left.\begin{array}{c} (c + d)\mathbf{v} = c\mathbf{v} + d\mathbf{v} \\ c(\mathbf{u} + \mathbf{v}) = c\mathbf{u} + c\mathbf{v} \end{array}\right\} \text{Distributive laws}$$

$$1 \cdot \mathbf{u} = \mathbf{u}, \qquad 0 \cdot \mathbf{u} = \mathbf{0}, \qquad (-1)\mathbf{u} = -\mathbf{u}.$$

DEFINITION   *Let* **v** *be any vector except* **0**. *The* **unit vector u in the direction of v** *is defined by*

$$u = \frac{1}{|v|} v.$$

EXAMPLE 2   Given the vectors $u = 3i - 2j + 4k$ and $v = 6i - 4j - 2k$, express the vector $3u - 2v$ in terms of **i**, **j**, and **k**.

Solution   $3u = 9i - 6j + 12k$ and $-2v = -12i + 8j + 4k$. Adding these vectors, we get $3u - 2v = -3i + 2j + 16k$.   ☐

EXAMPLE 3   Given the vector $v = 2i - 3j + k$, find a unit vector in the direction of **v**.

Solution   We have $|v| = \sqrt{4 + 9 + 1} = \sqrt{14}$. The desired vector **u** is

$$u = \frac{1}{\sqrt{14}} v = \frac{2}{\sqrt{14}} i - \frac{3}{\sqrt{14}} j + \frac{1}{\sqrt{14}} k.$$   ☐

EXAMPLE 4   Find the head $B$ of the vector $v = 2i + 4j - 3k$ if the base $A$ has rectangular coordinates $(2, 1, -5)$.

Solution   Denote the coordinates of $B$ by $x_B$, $y_B$, $z_B$. Then we have

$$x_B - 2 = 2, \qquad y_B - 1 = 4, \qquad z_B + 5 = -3.$$

Therefore, $x_B = 4$, $y_B = 5$, $z_B = -8$.   ☐

## 3   PROBLEMS

In Problems 1 through 10, express **v** in terms of **i**, **j**, and **k**, given that **v** is equivalent to the directed line segment $\vec{PQ}$, where the rectangular coordinates of $P$ and $Q$ are given. Also find another directed line segment equivalent to **v**.

1   $P(2, 0, 3)$, $Q(1, 4, -3)$

2   $P(1, 1, 0)$, $Q(-1, 2, 0)$

3   $P(-4, -2, 1)$, $Q(1, -3, 4)$

4   $P(3, 2, 1)$, $Q(3, 3, 3)$

5   $P(2, 0, 0)$, $Q(0, 0, -3)$

6   $P(4, -5, -1)$, $Q(-2, 1, -3)$

7   $P(0, 1, 0)$, $Q(0, 3, 0)$

8   $P(1, 0, 1)$, $Q(0, 2, 0)$

9   $P(1, 1, 1)$, $Q(2, 2, 3)$

10   $P(1, -1, 1)$, $Q(4, 1, 5)$

In Problems 11 through 18, in each case find a unit vector **u** in the direction of **v**. Express **u** in terms of **i**, **j**, and **k**.

11   $v = 3i + 2j - 4k$

12   $v = i - j + k$

13   $v = 2i - 4j - k$

14   $v = -2i + 3j + 5k$

15   $v = i - 3j$

16   $v = 2i - k$

17   $v = 2i + 2j + 2k$

18   $v = i - 3j - \sqrt{2}k$

In Problems 19 through 25, find the directed line segment $\vec{AB}$ equivalent to the vector **v** from the information given.

19   $v = 2i + j - 3k$, $A(1, 2, -1)$

20   $v = -i + 3j - 2k$, $A(2, 0, 4)$

21   $v = 3i + 2j - 4k$, $B(2, 0, -4)$

22   $v = -2i + 4j + k$, $B(0, 0, -5)$

23   $v = i - 2j + 2k$; the midpoint of the segment $AB$ has coordinates $(2, -1, 4)$.

**24**  $\mathbf{v} = 3\mathbf{i} + 4\mathbf{k}$; the midpoint of the segment $AB$ has coordinates $(1, 2, -5)$.

**25**  $\mathbf{v} = -\mathbf{i} + \mathbf{j} - 2\mathbf{k}$; the point $\frac{3}{4}$ of the distance from $A$ to $B$ has coordinates $(1, 0, 2)$.

**26**  Find a vector $\mathbf{u}$ in the direction of $\mathbf{v} = -\mathbf{i} + \mathbf{j} - \mathbf{k}$ and having half the length of $\mathbf{v}$.

**27**  Given $\mathbf{u} = \mathbf{i} + 2\mathbf{j} - 4\mathbf{k}$, $\mathbf{v} = 3\mathbf{i} - 7\mathbf{j} + 5\mathbf{k}$, find $\mathbf{u} + \mathbf{v}$ in terms of $\mathbf{i}$, $\mathbf{j}$, and $\mathbf{k}$. Sketch a figure.

**28**  Given that $\mathbf{u} = -3\mathbf{i} + 7\mathbf{j} - 4\mathbf{k}$, $\mathbf{v} = 2\mathbf{i} + \mathbf{j} - 6\mathbf{k}$, find $3\mathbf{u} - 7\mathbf{v}$ in terms of $\mathbf{i}$, $\mathbf{j}$, and $\mathbf{k}$.

**29**  Let $a$ and $b$ be any real numbers. Show that the vector $\mathbf{k}$ is orthogonal to $a\mathbf{i} + b\mathbf{j}$.

**30**  Show that the vector $\mathbf{u} = \mathbf{i} + 2\mathbf{j} - 3\mathbf{k}$ is orthogonal to $a\mathbf{v} + b\mathbf{w}$ where $\mathbf{v} = 2\mathbf{i} + 2\mathbf{j} + 2\mathbf{k}$, $\mathbf{w} = -\mathbf{i} + 2\mathbf{j} + \mathbf{k}$ and $a$, $b$ are any real numbers. Interpret this statement geometrically.

**31**  Discuss the relationship between the direction numbers of a line and the representation of a vector $\mathbf{v}$ in terms of the vectors $\mathbf{i}$, $\mathbf{j}$, and $\mathbf{k}$.

**32**  Suppose two directed line segments of $\mathbf{u}$ and $\mathbf{v}$ determine a plane. Discuss the relationship between the attitude numbers of this plane and the representations of $\mathbf{u}$ and $\mathbf{v}$ in terms of $\mathbf{i}$, $\mathbf{j}$, and $\mathbf{k}$.

---

**4**

# THE SCALAR (INNER OR DOT) PRODUCT

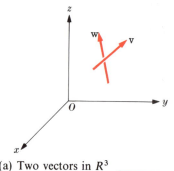

(a) Two vectors in $R^3$

Two vectors are said to be **parallel** or **proportional** when each is a scalar multiple of the other (and neither is zero). As in the case for $R^2$, by **the angle between two vectors v and w** (neither $= 0$), we mean the measure of the angle between two directed line segments having a common base and equivalent to $\mathbf{v}$ and $\mathbf{w}$, respectively. (Fig. 23). Two parallel vectors make an angle of $0$ or $\pi$, depending on whether they are pointing in the same or in the opposite direction.

**THEOREM 13**   *If $\theta$ is the angle between the vectors*

$$\mathbf{v} = a_1\mathbf{i} + a_2\mathbf{j} + a_3\mathbf{k} \qquad and \qquad \mathbf{w} = b_1\mathbf{i} + b_2\mathbf{j} + b_3\mathbf{k},$$

*then*

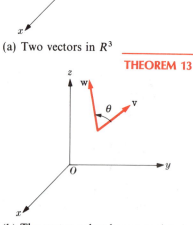

$$\cos\theta = \frac{a_1 b_1 + a_2 b_2 + a_3 b_3}{|\mathbf{v}| \cdot |\mathbf{w}|}.$$

(b) The vector $\mathbf{w}$ has been translated to have a base common with $\mathbf{v}$

**FIGURE 23**

The proof is a straightforward extension of the proof of the analogous theorem in the plane (Theorem 6 of Section 2) and will therefore be omitted.

**EXAMPLE 1**   Given the vectors $\mathbf{v} = 2\mathbf{i} + \mathbf{j} - 3\mathbf{k}$ and $\mathbf{w} = -\mathbf{i} + 4\mathbf{j} - 2\mathbf{k}$, find the cosine of the angle between $\mathbf{v}$ and $\mathbf{w}$.

**Solution**   We have

$$|\mathbf{v}| = \sqrt{4 + 1 + 9} = \sqrt{14}, \qquad |\mathbf{w}| = \sqrt{1 + 16 + 4} = \sqrt{21}.$$

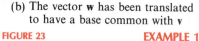

Therefore

$$\cos \theta = \frac{-2 + 4 + 6}{\sqrt{14} \cdot \sqrt{21}} = \frac{8}{7\sqrt{6}}.$$

□

**DEFINITIONS**    *Given the vectors* **u** *and* **v**, *the* **scalar (inner** *or* **dot) product u · v** *is defined by the formula*

$$\mathbf{u} \cdot \mathbf{v} = |\mathbf{u}||\mathbf{v}| \cos \theta,$$

*where* $\theta$ *is the angle between the vectors. If either* **u** *or* **v** *is* **0**, *we define* **u · v** = 0.

If we make the convention that the vector **0** is orthogonal to every vector, we observe that *two vectors* **u** *and* **v** *are orthogonal if and only if* **u · v** = 0. Note that this definition is identical to the one for plane vectors.

**THEOREM 14**    *The scalar product satisfies the laws*

     (i)   **u · v = v · u**;

     (ii)   $\mathbf{u} \cdot \mathbf{u} = |\mathbf{u}|^2.$

     (iii)   *If* $\mathbf{u} = a_1 \mathbf{i} + b_1 \mathbf{j} + c_1 \mathbf{k}$     *and*     $\mathbf{v} = a_2 \mathbf{i} + b_2 \mathbf{j} + c_2 \mathbf{k},$   *then*

$$\mathbf{u} \cdot \mathbf{v} = a_1 a_2 + b_1 b_2 + c_1 c_2.$$

**Proof**    Parts (i) and (ii) are direct consequences of the definition; part (iii) follows from Theorem 13 since

$$\mathbf{u} \cdot \mathbf{v} = |\mathbf{u}||\mathbf{v}| \cos \theta = |\mathbf{u}||\mathbf{v}| \frac{a_1 a_2 + b_1 b_2 + c_1 c_2}{|\mathbf{u}||\mathbf{v}|}.$$

□

**COROLLARY**    (i) *If c and d are any scalars and if* **u**, **v**, **w** *are any vectors, then*

$$\mathbf{u} \cdot (c\mathbf{v} + d\mathbf{w}) = c(\mathbf{u} \cdot \mathbf{v}) + d(\mathbf{u} \cdot \mathbf{w}).$$

     (ii) *We have*

$$\mathbf{i} \cdot \mathbf{i} = \mathbf{j} \cdot \mathbf{j} = \mathbf{k} \cdot \mathbf{k} = 1 \qquad \mathbf{i} \cdot \mathbf{j} = \mathbf{i} \cdot \mathbf{k} = \mathbf{j} \cdot \mathbf{k} = 0.$$

**EXAMPLE 2**    Find the scalar product of the vectors

$$\mathbf{u} = 3\mathbf{i} + 2\mathbf{j} - 4\mathbf{k} \qquad \text{and} \qquad \mathbf{v} = -2\mathbf{i} + \mathbf{j} + 5\mathbf{k}.$$

**Solution**    $\mathbf{u} \cdot \mathbf{v} = 3(-2) + 2 \cdot 1 + (-4)(5) = -24.$

□

**EXAMPLE 3**    Express $|3\mathbf{u} + 5\mathbf{v}|^2$ in terms of $|\mathbf{u}|^2$, $|\mathbf{v}|^2$, and **u · v**.

**Solution**    $|3\mathbf{u} + 5\mathbf{v}|^2 = (3\mathbf{u} + 5\mathbf{v}) \cdot (3\mathbf{u} + 5\mathbf{v})$

$$= 9(\mathbf{u} \cdot \mathbf{u}) + 15(\mathbf{u} \cdot \mathbf{v}) + 15(\mathbf{v} \cdot \mathbf{u}) + 25(\mathbf{v} \cdot \mathbf{v})$$

$$= 9|\mathbf{u}|^2 + 30(\mathbf{u} \cdot \mathbf{v}) + 25|\mathbf{v}|^2.$$

□

**DEFINITION**   *Let* **v** *and* **w** *be two vectors which make an angle* $\theta$. *We denote by* $|\mathbf{v}| \cos \theta$ *the* **projection of v on w.** *We also call this quantity the* **component of v along w.** *As before we denote this quantity by* $\text{Proj}_{\mathbf{w}}\mathbf{v}$.

From the formula for $\cos \theta$, we may write

$$\text{Proj}_{\mathbf{w}}\mathbf{v} = |\mathbf{v}| \cos \theta = |\mathbf{v}| \frac{\mathbf{v} \cdot \mathbf{w}}{|\mathbf{v}||\mathbf{w}|} = \frac{\mathbf{v} \cdot \mathbf{w}}{|\mathbf{w}|}.$$

**EXAMPLE 4**   Find the projection of $\mathbf{v} = -\mathbf{i} + 2\mathbf{j} + 3\mathbf{k}$ on $\mathbf{w} = 2\mathbf{i} - \mathbf{j} - 4\mathbf{k}$.

**Solution**   We have $\mathbf{v} \cdot \mathbf{w} = (-1)(2) + (2)(-1) + (3)(-4) = -16$; $|\mathbf{w}| = \sqrt{21}$. Therefore, the projection of $\mathbf{v}$ on $\mathbf{w}$ is $|\mathbf{v}| \cos \theta = -16\sqrt{21}$.  □

An application* of scalar product to mechanics occurs in the calculation of work done by a *constant* force **F** when its point of application moves along a segment from $A$ to $B$. The **work done** in this case is defined as the product of the distance from $A$ to $B$ and the projection of **F** on **v** ($\vec{AB}$). We have

$$\text{Proj}_{\mathbf{v}} \mathbf{F} = \text{Projection of } \mathbf{F} \text{ on } \mathbf{v} = \frac{\mathbf{F} \cdot \mathbf{v}}{|\mathbf{v}|}.$$

Since the distance from $A$ to $B$ is exactly $|\mathbf{v}|$, we conclude that

$$\text{Work done by } \mathbf{F} = \mathbf{F} \cdot \mathbf{v}.$$

**EXAMPLE 5**   Find the work done by the force

$$\mathbf{F} = 5\mathbf{i} - 3\mathbf{j} + 2\mathbf{k}$$

as its point of application moves from the point $A(2, 1, 3)$ to $B(4, -1, 5)$.

**Solution**   We have

$$\mathbf{v}[\vec{AB}] = (4 - 2)\mathbf{i} + (-1 - 1)\mathbf{j} + (5 - 3)\mathbf{k} = 2\mathbf{i} - 2\mathbf{j} + 2\mathbf{k}.$$

Therefore, work done $= 5 \cdot 2 + 3 \cdot 2 + 2 \cdot 2 = 20$.  □

**THEOREM 15**   *If* **u** *and* **v** *are not* **0**, *there is a unique scalar* $k$ *such that* $\mathbf{v} - k\mathbf{u}$ *is orthogonal to* **u**. *In fact,* $k$ *can be found from the formula*

$$k = \frac{\mathbf{u} \cdot \mathbf{v}}{|\mathbf{u}|^2}.$$

**Proof**   $(\mathbf{v} - k\mathbf{u})$ is orthogonal to **u** if and only if $\mathbf{u} \cdot (\mathbf{v} - k\mathbf{u}) = 0$. But

$$\mathbf{u} \cdot (\mathbf{v} - k\mathbf{u}) = \mathbf{u} \cdot \mathbf{v} - k|\mathbf{u}|^2 = 0.$$

*This paragraph and Problems 11–14 at the end of this section may be omitted by readers who have not studied the previous applications of calculus to problems on work (Section 7, Chapter 9).

Therefore, selection of $k = \mathbf{u} \cdot \mathbf{v}/|\mathbf{u}|^2$ yields the result. $\square$

Figure 24 shows geometrically how $k$ is to be selected. We drop a perpendicular from the head of $\mathbf{v}$ (point $B$) to the line containing $\mathbf{u}$ (point $D$). The directed segment $\vec{AD}$ gives the proper multiple of $\mathbf{u}$ ($\vec{AC}$), and the directed segment $\vec{DB}$ is the orthogonal vector.

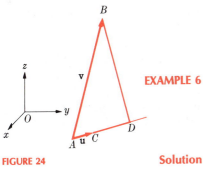

**FIGURE 24**

**EXAMPLE 6**   Find a linear combination of

$$\mathbf{u} = 2\mathbf{i} + 3\mathbf{j} - \mathbf{k} \quad \text{and} \quad \mathbf{v} = \mathbf{i} + 2\mathbf{j} + \mathbf{k}$$

which is orthogonal to $\mathbf{u}$.

**Solution**   We select $k = \dfrac{\mathbf{u} \cdot \mathbf{v}}{|\mathbf{u}|^2} = (2 + 6 - 1)/14 = \frac{1}{2}$, and the desired vector is

$$\mathbf{v} - \tfrac{1}{2}\mathbf{u} = \tfrac{1}{2}\mathbf{j} + \tfrac{3}{2}\mathbf{k}. \qquad \square$$

## 4   PROBLEMS

In each of Problems 1 through 5, find $\cos \theta$ where $\theta$ is the angle between $\mathbf{v}$ and $\mathbf{w}$.

**1**  $\mathbf{v} = \mathbf{i} + 3\mathbf{j} - 2\mathbf{k}, \ \mathbf{w} = 2\mathbf{i} + 4\mathbf{j} - \mathbf{k}$

**2**  $\mathbf{v} = -\mathbf{i} + 2\mathbf{j} + 3\mathbf{k}, \ \mathbf{w} = 3\mathbf{i} - 2\mathbf{j} - 2\mathbf{k}$

**3**  $\mathbf{v} = 4\mathbf{i} - 3\mathbf{j} + 5\mathbf{k}, \ \mathbf{w} = 2\mathbf{i} + \mathbf{j} + \mathbf{k}$

**4**  $\mathbf{v} = 2\mathbf{i} + 3\mathbf{k}, \ \mathbf{w} = \mathbf{j} + 4\mathbf{k}$

**5**  $\mathbf{v} = -2\mathbf{i} - 3\mathbf{j} - 4\mathbf{k}, \ \mathbf{w} = 2\mathbf{i} - 3\mathbf{j} + 4\mathbf{k}$

In each of Problems 6 through 10, find the projection of the vector $\mathbf{v}$ on $\mathbf{u}$.

**6**  $\mathbf{u} = 2\mathbf{i} - 6\mathbf{j} + 3\mathbf{k}, \ \mathbf{v} = \mathbf{i} + 2\mathbf{j} - 2\mathbf{k}$

**7**  $\mathbf{u} = 6\mathbf{i} + 2\mathbf{j} - 3\mathbf{k}, \ \mathbf{v} = -\mathbf{i} + 8\mathbf{j} + 4\mathbf{k}$

**8**  $\mathbf{u} = 12\mathbf{i} + 3\mathbf{j} + 4\mathbf{k}, \ \mathbf{v} = 4\mathbf{i} + 8\mathbf{j} + \mathbf{k}$

**9**  $\mathbf{u} = 3\mathbf{i} + 5\mathbf{j} - 4\mathbf{k}, \ \mathbf{v} = 4\mathbf{i} - 3\mathbf{j} + 5\mathbf{k}$

**10**  $\mathbf{u} = 2\mathbf{i} - 5\mathbf{j} + 3\mathbf{k}, \ \mathbf{v} = -\mathbf{i} + 2\mathbf{j} + 7\mathbf{k}$

In each of Problems 11 through 14, find the work done by the force $\mathbf{F}$ when its point of application moves from $A$ to $B$.

**11**  $\mathbf{F} = -32\mathbf{k}, \ A(-1, 1, 2), B(3, 2, -1)$

**12**  $\mathbf{F} = 5\mathbf{i} - 2\mathbf{j} + 3\mathbf{k}, \ A(1, -2, 2), \ B(3, 1, -1)$

**13**  $\mathbf{F} = -2\mathbf{i} + 3\mathbf{j} + 4\mathbf{k}, \ A(2, -1, -2), B(-1, 2, 3)$

**14**  $\mathbf{F} = 3\mathbf{i} - 2\mathbf{j} - 3\mathbf{k}, \ A(-1, 2, 3), B(2, 1, -1)$

In each of Problems 15 through 17, find a unit vector in the direction of $\mathbf{u}$.

**15**  $\mathbf{u} = 2\mathbf{i} - 6\mathbf{j} + 3\mathbf{k}$      **16**  $\mathbf{u} = -\mathbf{i} + 2\mathbf{k}$

**17**  $\mathbf{u} = 3\mathbf{i} - 2\mathbf{j} + 7\mathbf{k}$

In each of Problems 18 through 21, find the value of $k$ so that $\mathbf{v} - k\mathbf{u}$ is orthogonal to $\mathbf{u}$. Also, find the value $h$ so that $\mathbf{u} - h\mathbf{v}$ is orthogonal to $\mathbf{v}$.

**18**  $\mathbf{u} = 2\mathbf{i} - \mathbf{j} + 2\mathbf{k}, \ \mathbf{v} = 3\mathbf{i} + \mathbf{j} + 2\mathbf{k}$

**19**  $\mathbf{u} = 2\mathbf{i} - 3\mathbf{j} + 6\mathbf{k}, \ \mathbf{v} = 7\mathbf{i} + 14\mathbf{k}$

**20**  $\mathbf{u} = 3\mathbf{i} + 4\mathbf{j} - 5\mathbf{k}, \ \mathbf{v} = 9\mathbf{i} + 12\mathbf{j} - 5\mathbf{k}$

**21**  $\mathbf{u} = \mathbf{i} + 3\mathbf{j} - 2\mathbf{k}, \ \mathbf{v} = 6\mathbf{i} + 10\mathbf{j} - 3\mathbf{k}$

**22**  Write a detailed proof of Theorem 13.

**23**  Show that if $\mathbf{u}$ and $\mathbf{v}$ are any vectors ($\neq \mathbf{0}$), then $\mathbf{u}$ and $\mathbf{v}$ make equal angles with $\mathbf{w}$ if

$$\mathbf{w} = \left( \frac{|\mathbf{v}|}{|\mathbf{u}| + |\mathbf{v}|} \right) \mathbf{u} + \left( \frac{|\mathbf{u}|}{|\mathbf{u}| + |\mathbf{v}|} \right) \mathbf{v}.$$

**24**  Show that if $\mathbf{u}$ and $\mathbf{v}$ are any vectors, the vectors $|\mathbf{v}|\mathbf{u} + |\mathbf{u}|\mathbf{v}$ and $|\mathbf{v}|\mathbf{u} - |\mathbf{u}|\mathbf{v}$ are orthogonal.

In each of Problems 25 through 27, determine the relation between $g$ and $h$ so that $g\mathbf{u} + h\mathbf{v}$ is orthogonal to $\mathbf{w}$.

**25**  $\mathbf{u} = 3\mathbf{i} - 2\mathbf{j} + \mathbf{k}, \ \mathbf{v} = \mathbf{i} + 2\mathbf{j} - 3\mathbf{k}, \ \mathbf{w} = -\mathbf{i} + \mathbf{j} + 2\mathbf{k}$

**26**  $\mathbf{u} = 2\mathbf{i} + \mathbf{j} - 2\mathbf{k}, \ \mathbf{v} = \mathbf{i} - \mathbf{j} + \mathbf{k}, \ \mathbf{w} = -\mathbf{i} + 2\mathbf{j} + 3\mathbf{k}$

**27**  $\mathbf{u} = \mathbf{i} + 2\mathbf{j} - 3\mathbf{k}, \ \mathbf{v} = 3\mathbf{i} + \mathbf{j} - \mathbf{k}, \ \mathbf{w} = 4\mathbf{i} - \mathbf{j} + 2\mathbf{k}$

In each of Problems 28 through 30, determine scalars $g$ and $h$ so that $\mathbf{w} - g\mathbf{u} - h\mathbf{v}$ is orthogonal to both $\mathbf{u}$ and $\mathbf{v}$.

**28**  $\mathbf{u} = 2\mathbf{i} - \mathbf{j} + \mathbf{k}, \ \mathbf{v} = \mathbf{i} + \mathbf{j} + 2\mathbf{k}, \ \mathbf{w} = 2\mathbf{i} - \mathbf{j} + 4\mathbf{k}$

**29**  $\mathbf{u} = \mathbf{i} + \mathbf{j} - 2\mathbf{k}, \ \mathbf{v} = -\mathbf{i} + 2\mathbf{j} + 3\mathbf{k}, \ \mathbf{w} = 5\mathbf{i} + 8\mathbf{k}$

**30**  $\mathbf{u} = 3\mathbf{i} - 2\mathbf{j}, \ \mathbf{v} = 2\mathbf{i} - \mathbf{k}, \ \mathbf{w} = 4\mathbf{i} - 2\mathbf{k}$

In Problems 31 through 34, establish the relationships given. Do not introduce the unit vectors **i**, **j**, and **k**.

**31** $(\mathbf{u} + \mathbf{v}) \cdot (\mathbf{u} - \mathbf{v}) = \mathbf{u} \cdot \mathbf{u} - \mathbf{v} \cdot \mathbf{v}$

**32** $|\mathbf{u} + \mathbf{v}|^2 = |\mathbf{u}|^2 + 2\mathbf{u} \cdot \mathbf{v} + |\mathbf{v}|^2$

**33** $|\mathbf{u} + \mathbf{v}|^2 + |\mathbf{u} - \mathbf{v}|^2 = 2(|\mathbf{u}|^2 + |\mathbf{v}|^2)$

**34** $|\mathbf{u} \cdot \mathbf{v}| = \frac{1}{4}(|\mathbf{u} + \mathbf{v}|^2 - |\mathbf{u} - \mathbf{v}|^2)$

**35** Show that if **w** is orthogonal both to **u** and to **v**, then **w** is orthogonal to $\alpha\mathbf{u} + \beta\mathbf{v}$ for all scalars $\alpha$ and $\beta$.

**36** If **u** and **v** are nonzero vectors, under what conditions is it true that $|\mathbf{u} + \mathbf{v}| = |\mathbf{u}| + |\mathbf{v}|$?

**37** Suppose that $\vec{AB}$, $\vec{AC}$, and $\vec{AD}$ are directed line segments of **u**, **v**, and $\mathbf{u} + \mathbf{v}$, respectively, with $|\mathbf{u}| = |\mathbf{v}|$. Show that $\vec{AD}$ bisects the angle between $\vec{AB}$ and $\vec{AC}$.

**38** Let $P$ be a vertex of a cube. Draw a diagonal of the cube from $P$ and a diagonal of one of the faces from $P$. Use vectors to find the cosine of the angle between these two diagonals.

**39** Use vectors to find the cosine of the angle between two faces of a regular tetrahedron. (See Problem 33, Chapter 13, Section 2.)

---

**5**

# THE VECTOR OR CROSS PRODUCT

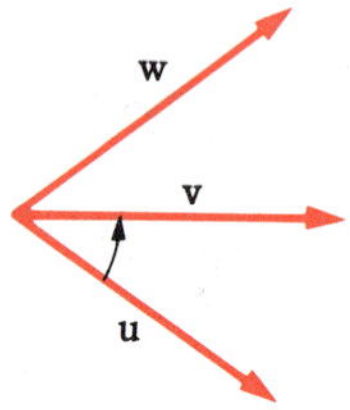

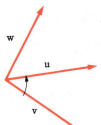

FIGURE 25

We saw in Section 4 that the scalar product of two vectors **u** and **v** associates an ordinary number, i.e., a scalar, with each pair of vectors. The vector or cross product, on the other hand, associates a *vector* with each ordered pair of vectors. However, before defining the cross product, we shall discuss the notion of "right-handed" and "left-handed" triples of vectors.

An **ordered triple** {**u**, **v**, **w**} of three independent vectors is said to be **right-handed** if the vectors are situated as in Fig. 25. If the ordered triple is situated as in Fig. 26, the vectors are said to form a **left-handed triple**. The notion of left-handed and right-handed triple is not defined if the three vectors all lie in one plane.

FIGURE 26

**DEFINITION**     *Two sets of ordered triples of vectors are said to be **similarly oriented** if and only if both sets are right-handed or both are left-handed. Otherwise they are **oppositely oriented**.*

If we replace **w** in Figure 26 with $-\mathbf{w}$, a moment's reflection shows that the triple $(\mathbf{u}, \mathbf{v}, -\mathbf{w})$ has changed from a left-handed to a right-handed triple (see Fig. 27). This type of reasoning leads to the following result, which we state without proof.

FIGURE 27

**THEOREM 16**     *If {**u**, **v**, **w**} is a right-handed triple, then (i) {**v**, **u**, $-\mathbf{w}$} is a right-handed triple, and (ii) {$c_1\mathbf{u}$, $c_2\mathbf{v}$, $c_3\mathbf{w}$} is a right-handed triple provided that $c_1 c_2 c_3 > 0$.*

DEFINITION    *Given the vectors* **u** *and* **v**, *the* **vector** *or* **cross product** **u** × **v** *is defined as follows:*

i) *if either* **u** *or* **v** *is* **0**, *then*

$$\mathbf{u} \times \mathbf{v} = \mathbf{0};$$

ii) *if* **u** *is proportional to* **v**, *then*

$$\mathbf{u} \times \mathbf{v} = \mathbf{0};$$

iii) *otherwise,*

$$\mathbf{u} \times \mathbf{v} = \mathbf{w},$$

*where* **w** *has the three properties: (a) it is orthogonal to both* **u** *and* **v**; *(b) it has magnitude* $|\mathbf{w}| = |\mathbf{u}||\mathbf{v}| \sin\theta$, *where* $\theta$ *is the angle between* **u** *and* **v**, *and (c) it is directed so that* {**u**, **v**, **w**} *is a right-handed triple.*

*Remark.* We shall always assume that any coordinate triple {**i**, **j**, **k**} is right-handed. (We have assumed this up to now without pointing out this fact specifically.)

The proofs of the next two theorems are given in Appendix 5.

THEOREM 17    *Suppose that* **u** *and* **v** *are any vectors, that* {**i**, **j**, **k**} *is a right-handed triple, and that t is any number. Then*

i) $\mathbf{v} \times \mathbf{u} = -(\mathbf{u} \times \mathbf{v})$,
ii) $(t\mathbf{u}) \times \mathbf{v} = t(\mathbf{u} \times \mathbf{v}) = \mathbf{u} \times (t\mathbf{v})$,
iii) $\mathbf{i} \times \mathbf{j} = -\mathbf{j} \times \mathbf{i} = \mathbf{k}$,
$\quad \mathbf{j} \times \mathbf{k} = -\mathbf{k} \times \mathbf{j} = \mathbf{i}$,
$\quad \mathbf{k} \times \mathbf{i} = -\mathbf{i} \times \mathbf{k} = \mathbf{j}$,
iv) $\mathbf{i} \times \mathbf{i} = \mathbf{j} \times \mathbf{j} = \mathbf{k} \times \mathbf{k} = \mathbf{0}$.

THEOREM 18    *If* **u**, **v**, **w** *are any vectors, then*

i) $\mathbf{u} \times (\mathbf{v} + \mathbf{w}) = (\mathbf{u} \times \mathbf{v}) + (\mathbf{u} \times \mathbf{w})$ *and*
ii) $(\mathbf{v} + \mathbf{w}) \times \mathbf{u} = (\mathbf{v} \times \mathbf{u}) + (\mathbf{w} \times \mathbf{u})$.

With the aid of Theorems 17 and 18, the next theorem, an extremely useful one, is easily established.

THEOREM 19    *If*

$$\mathbf{u} = a_1\mathbf{i} + a_2\mathbf{j} + a_3\mathbf{k} \qquad and \qquad \mathbf{v} = b_1\mathbf{i} + b_2\mathbf{j} + b_3\mathbf{k},$$

*then*

$$\mathbf{u} \times \mathbf{v} = (a_2 b_3 - a_3 b_2)\mathbf{i} + (a_3 b_1 - a_1 b_3)\mathbf{j} + (a_1 b_2 - a_2 b_1)\mathbf{k}. \tag{1}$$

**Proof**   By using the laws in Theorems 17 and 18 we obtain (being careful to keep the order of the factors)

$$\mathbf{u} \times \mathbf{v} = a_1 b_1 (\mathbf{i} \times \mathbf{i}) + a_1 b_2 (\mathbf{i} \times \mathbf{j}) + a_1 b_3 (\mathbf{i} \times \mathbf{k})$$
$$+ a_2 b_1 (\mathbf{j} \times \mathbf{i}) + a_2 b_2 (\mathbf{j} \times \mathbf{j}) + a_2 b_3 (\mathbf{j} \times \mathbf{k})$$
$$+ a_3 b_1 (\mathbf{k} \times \mathbf{i}) + a_3 b_2 (\mathbf{k} \times \mathbf{j}) + a_3 b_3 (\mathbf{k} \times \mathbf{k}).$$

The result follows from Theorem 17, parts (iii) and (iv) by collecting terms.

$\square$

It is convenient to use **determinants** when working with vector products. A review of determinants is given in Appendix 4. For the reader's convenience, however, we recall here the definition of $2 \times 2$ and $3 \times 3$ determinants. A **determinant of the second order** is denoted by

$$\begin{vmatrix} a & b \\ c & d \end{vmatrix}$$

and is defined by

$$\begin{vmatrix} a & b \\ c & d \end{vmatrix} = ad - bc.$$

A **determinant of the third order** is denoted

$$\begin{vmatrix} a & b & c \\ d & e & f \\ g & h & j \end{vmatrix}$$

and is defined in terms of the second-order determinants by the formula:

$$\begin{vmatrix} a & b & c \\ d & e & f \\ g & h & j \end{vmatrix} = \begin{vmatrix} e & f \\ h & j \end{vmatrix} a - \begin{vmatrix} d & f \\ g & j \end{vmatrix} b + \begin{vmatrix} d & e \\ g & h \end{vmatrix} c.$$

This formula is known as an **expansion of a determinant in terms of the first row**. (In Appendix 4 we describe an expansion of a determinant in terms of the last column, an equivalent definition.)

**EXAMPLE 1**   Calculate the determinant $\begin{vmatrix} 2 & -3 \\ 1 & 4 \end{vmatrix}$.

**Solution**   $$\begin{vmatrix} 2 & -3 \\ 1 & 4 \end{vmatrix} = (2)(4) - (-3)(1) = 8 - (-3) = 11. \qquad \square$$

**EXAMPLE 2**   Calculate the determinant

$$\begin{vmatrix} 2 & -1 & 3 \\ -2 & 4 & 0 \\ 1 & 3 & 6 \end{vmatrix}.$$

Solution

$$\begin{vmatrix} 2 & -1 & 3 \\ -2 & 4 & 0 \\ 1 & 3 & 6 \end{vmatrix} = \begin{vmatrix} 4 & 0 \\ 3 & 6 \end{vmatrix}(2) - \begin{vmatrix} -2 & 0 \\ 1 & 6 \end{vmatrix}(-1) + \begin{vmatrix} -2 & 4 \\ 1 & 3 \end{vmatrix}(3)$$

$$= (4 \cdot 6 - 0 \cdot 3) \cdot 2 - ((-2)6 - 0 \cdot 1)(-1) + ((-2) \cdot 3 - 4 \cdot 1) \cdot 3$$

$$= 24 \cdot 2 - (-12)(-1) + (-10)(3)$$

$$= 48 - 12 - 30 = 6. \qquad \square$$

Determinants can be used as an aid to calculate the cross product, since it is helpful in remembering Formula (1) of Theorem 19. Indeed, Formula (1) can be written in *symbolic form* as follows:

$$\mathbf{u} \times \mathbf{v} = \begin{vmatrix} \mathbf{i} & \mathbf{j} & \mathbf{k} \\ a_1 & a_2 & a_3 \\ b_1 & b_2 & b_3 \end{vmatrix},$$

where it is understood that this "determinant" is to be expanded formally according to its first row. The reader may easily verify that when the above expression is expanded, it is equal to Formula (1) for the cross product of **u** and **v**.

EXAMPLE 3    Find $\mathbf{u} \times \mathbf{v}$ if $\mathbf{u} = 2\mathbf{i} - 3\mathbf{j} + \mathbf{k}$, $\mathbf{v} = \mathbf{i} + \mathbf{j} - 2\mathbf{k}$.

Solution    Carrying out the formal expansion, we obtain

$$\begin{vmatrix} \mathbf{i} & \mathbf{j} & \mathbf{k} \\ 2 & -3 & 1 \\ 1 & 1 & -2 \end{vmatrix} = \begin{vmatrix} -3 & 1 \\ 1 & -2 \end{vmatrix}\mathbf{i} - \begin{vmatrix} 2 & 1 \\ 1 & -2 \end{vmatrix}\mathbf{j} + \begin{vmatrix} 2 & -3 \\ 1 & 1 \end{vmatrix}\mathbf{k} = 5\mathbf{i} + 5\mathbf{j} + 5\mathbf{k}. \quad \square$$

*Remarks.* In mechanics the cross product is used for the computation of the vector moment of a force **F** applied at a point $B$, about a point $A$. There are also applications of cross product to problems in electricity and magnetism. However, we shall confine our attention to applications in geometry.

THEOREM 20    *The area of a parallelogram with adjacent sides AB and AC is given by\**

$$|\mathbf{v}[\vec{AB}] \times \mathbf{v}[\vec{AC}]|.$$

*The area of $\triangle ABC$ is then $\frac{1}{2}|\mathbf{v}[\vec{AB}] \times \mathbf{v}[\vec{AC}]|$.*

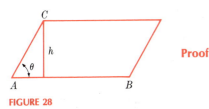

FIGURE 28

Proof    From Fig. 28, we see that the area of the parallelogram is

$$|AB|h = |AB||AC| \sin \theta.$$

The result then follows from the definition of cross product.    $\square$

---

\*The notation $\mathbf{v}[\vec{AB}]$ to denote a vector corresponding to the directed line segment $\vec{AB}$ was introduced in Section 2 (page 573).

**EXAMPLE 4**    Find the area of $\triangle ABC$ with $A(-2, 1, 3)$, $B(1, -1, 1)$, $C(3, -2, 4)$.

**Solution**    We have $\mathbf{v}[\vec{AB}] = 3\mathbf{i} - 2\mathbf{j} - 2\mathbf{k}$, $\mathbf{v}[\vec{AC}] = 5\mathbf{i} - 3\mathbf{j} + \mathbf{k}$. From Theorem 20 we obtain

$$\mathbf{v}[\vec{AB}] \times \mathbf{v}[\vec{AC}] = -8\mathbf{i} - 13\mathbf{j} + \mathbf{k}$$

and

$$\tfrac{1}{2}|-8\mathbf{i} - 13\mathbf{j} + \mathbf{k}| = \tfrac{1}{2}\sqrt{64 + 169 + 1} = \tfrac{3}{2}\sqrt{26}. \qquad \square$$

The vector product may be used to find the equation of a plane through three points. The next example illustrates the technique.

**EXAMPLE 5**    Find the equation of the plane through the points $A(-1, 1, 2)$, $B(1, -2, 1)$, $C(2, 2, 4)$.

**Solution**    A vector normal to the plane will be perpendicular to both the vectors

$$\mathbf{v}[\vec{AB}] = 2\mathbf{i} - 3\mathbf{j} - \mathbf{k} \qquad \text{and} \qquad \mathbf{v}[\vec{AC}] = 3\mathbf{i} + \mathbf{j} + 2\mathbf{k}.$$

One such vector is the cross product

$$\mathbf{v}[\vec{AB}] \times \mathbf{v}[\vec{AC}] = -5\mathbf{i} - 7\mathbf{j} + 11\mathbf{k}.$$

Therefore the numbers $-5$, $-7$, $11$ form a set of *attitude numbers* (see Chapter 13, Section 3) of the desired plane. Using $A(-1, 1, 2)$ as a point on the plane, we get for the equation

$$-5(x + 1) - 7(y - 1) + 11(z - 2) = 0$$

or

$$5x + 7y - 11z + 20 = 0. \qquad \square$$

**EXAMPLE 6**    Find the perpendicular distance between the skew lines

$$L_1: \frac{x+2}{2} = \frac{y-1}{3} = \frac{z+1}{-1}, \qquad L_2: \frac{x-1}{-1} = \frac{y+1}{2} = \frac{z-2}{4}.$$

**Solution**    The vector

$$\mathbf{v}_1 = 2\mathbf{i} + 3\mathbf{j} - \mathbf{k}$$

is a vector along $L_1$. The vector

$$\mathbf{v}_2 = -\mathbf{i} + 2\mathbf{j} + 4\mathbf{k}$$

is a vector along $L_2$. A vector perpendicular to both $\mathbf{v}_1$ and $\mathbf{v}_2$ (i.e., to both $L_1$ and $L_2$) is

$$\mathbf{v}_1 \times \mathbf{v}_2 = 14\mathbf{i} - 7\mathbf{j} + 7\mathbf{k}.$$

Call this common perpendicular $\mathbf{w}$. The desired length may be obtained as a *projection*. Select any point on $L_1$ (call it $P_1$) and any point on $L_2$ (call it $P_2$). Then the desired length is the projection of the vector $\mathbf{v}[\vec{P_1 P_2}]$ on $\mathbf{w}$. To get this, we select $P_1(-2, 1, -1)$ on $L_1$ and $P_2(1, -1, -2)$ on $L_2$; and so

$$\mathbf{v}[\vec{P_1 P_2}] = 3\mathbf{i} - 2\mathbf{j} + 3\mathbf{k}.$$

Therefore,

$$\text{Projection of } \mathbf{v}[\overrightarrow{P_1 P_2}] \text{ on } \mathbf{w} = \frac{\mathbf{v}[\overrightarrow{P_1 P_2}] \cdot \mathbf{w}}{|\mathbf{w}|}$$

$$= \frac{3 \cdot 14 + (-2)(-7) + 3(7)}{7\sqrt{6}} = \frac{11}{\sqrt{6}}. \qquad \square$$

## 5   PROBLEMS

In Problems 1 through 12, calculate the indicated determinants.

**1.** $\begin{vmatrix} 1 & 2 \\ 3 & 4 \end{vmatrix}$

**2** $\begin{vmatrix} 3 & 4 \\ 1 & 2 \end{vmatrix}$

**3** $\begin{vmatrix} 1 & 3 \\ 2 & 4 \end{vmatrix}$

**4** $\begin{vmatrix} 0 & -1 \\ 2 & 3 \end{vmatrix}$

**5** $\begin{vmatrix} 3 & 4 \\ 6 & 8 \end{vmatrix}$

**6** $\begin{vmatrix} 7 & 2 \\ 1 & 3 \end{vmatrix}$

**7** $\begin{vmatrix} 1 & 3 & 4 \\ 2 & 6 & 8 \\ 2 & -2 & 1 \end{vmatrix}$

**8** $\begin{vmatrix} 1 & 3 & 4 \\ -2 & 1 & 0 \\ 3 & 2 & 7 \end{vmatrix}$

**9** $\begin{vmatrix} 2 & -1 & 3 \\ -2 & 5 & 1 \\ 1 & 2 & -4 \end{vmatrix}$

**10** $\begin{vmatrix} 2 & -2 & 1 \\ -1 & 5 & 2 \\ 3 & 1 & -4 \end{vmatrix}$

**11** $\begin{vmatrix} 1 & 0 & 1 \\ 3 & 7 & 2 \\ \pi & 0 & \sqrt{2} \end{vmatrix}$

**12** $\begin{vmatrix} 1 & 3 & 2 \\ 4 & 6 & 3 \\ 2 & 3 & -1 \end{vmatrix}$

In each of Problems 13 through 23, find the cross product $\mathbf{u} \times \mathbf{v}$.

**13** $\mathbf{u} = \mathbf{i} + 3\mathbf{j} - \mathbf{k}, \mathbf{v} = 2\mathbf{i} - \mathbf{j} + \mathbf{k}$

**14** $\mathbf{u} = 4\mathbf{i} - 2\mathbf{j} + 3\mathbf{k}, \mathbf{v} = -\mathbf{i} - 2\mathbf{j} - \mathbf{k}$

**15** $\mathbf{u} = -\mathbf{i} + 2\mathbf{j}, \mathbf{v} = \mathbf{i} + 3\mathbf{j} - 2\mathbf{k}$

**16** $\mathbf{u} = 3\mathbf{j} + 2\mathbf{k}, \mathbf{v} = 2\mathbf{i} - 3\mathbf{j}$

**17** $\mathbf{u} = -2\mathbf{i} + 4\mathbf{j} + 5\mathbf{k}, \mathbf{v} = 4\mathbf{i} + 5\mathbf{k}$

**18** $\mathbf{u} = 2\mathbf{i} - 3\mathbf{j} + \mathbf{k}, \mathbf{v} = 4\mathbf{k}$

**19** $\mathbf{u} = \mathbf{i} - 2\mathbf{j} + 3\mathbf{k}, \mathbf{v} = 2\mathbf{i} + \mathbf{j} - 4\mathbf{k}$

**20** $\mathbf{u} = \mathbf{i} + 2\mathbf{j} + \mathbf{k}, \mathbf{v} = 2\mathbf{i} + 4\mathbf{j} + 2\mathbf{k}$

**21** $\mathbf{u} = \mathbf{i} + 2\mathbf{j}, \mathbf{v} = \mathbf{j} + 2\mathbf{k}$

**22** $\mathbf{u} = 3\mathbf{i}, \mathbf{v} = 2\mathbf{k}$

**23** $\mathbf{u} = 7\mathbf{i} - 2\mathbf{j} + \mathbf{k}, \mathbf{v} = 5\mathbf{j}$

In Problems 24 through 35, find in each case the area of $\triangle ABC$ and the equation of the plane through $A$, $B$, and $C$. Use vector methods.

**24** $A(1, -2, 3), B(3, 1, 2), C(2, 3, -1)$

**25** $A(3, 2, -2), B(4, 1, 2), C(1, 2, 3)$

**26** $A(2, -1, 1), B(3, 2, -1), C(-1, 3, 2)$

**27** $A(1, -2, 3), B(2, -1, 1), C(4, 2, -1)$

**28** $A(-2, 3, 1), B(4, 2, -2), C(2, 0, 1)$

**29** $A(0, 0, 0), B(0, 1, 0), C(0, 0, 1)$

**30** $A(-1, 0, 1), B(-2, 0, 2), C(-3, 0, 3)$

**31** $A(1, 0, 4), B(3, 2, -1), C(6, 4, -2)$

**32** $A(0, 1, 0), B(0, 0, 1), C(7, 2, -1)$

**33** $A(2, -1, 1), B(3, -1, 1), C(1, 2, 3)$

**34** $A(1, -1, 2), B(0, 3, -1), C(3, -4, 1)$

**35** $A(4, 0, 0), B(0, 5, 0), C(0, 0, 2)$

In Problems 36 through 40, find in each case the perpendicular distance between the given lines.

**36** $\dfrac{x+1}{2} = \dfrac{y-3}{-3} = \dfrac{z+2}{4}; \dfrac{x-2}{3} = \dfrac{y+1}{2} = \dfrac{z-1}{5}$

**37** $\dfrac{x-1}{3} = \dfrac{y+1}{2} = \dfrac{z-1}{5}; \dfrac{x+2}{4} = \dfrac{y-1}{3} = \dfrac{z+1}{-2}$

**38** $\dfrac{x+1}{2} = \dfrac{y-1}{-4} = \dfrac{z+2}{3}; \dfrac{x}{3} = \dfrac{y}{5} = \dfrac{z-2}{-2}$

**39** $\dfrac{x+1}{2} = \dfrac{y}{3} = \dfrac{z-2}{4}; \dfrac{x}{4} = \dfrac{y+3}{6} = \dfrac{z-1}{3}$

**40** $\dfrac{x-2}{-2} = \dfrac{y+2}{1} = \dfrac{z+1}{2}; \dfrac{x+3}{3} = \dfrac{y-2}{-1} = \dfrac{z-2}{4}$

In Problems 41 through 45, use vector methods to find, in each case, the equations in symmetric form of the line through the given point $P$ and parallel to the two given planes.

**41** $P(-1, 3, 2), 3x - 2y + 4z + 2 = 0, 2x + y - z = 0$

**42** $P(2, 3, -1), x + 2y + 2z - 4 = 0, 2x + y - 3z + 5 = 0$

**43** $P(1, -2, 3)$, $3x + y - 2z + 3 = 0$, $2x + 3y + z - 6 = 0$

**44** $P(-1, 0, -2)$, $2x + 3y - z + 4 = 0$, $3x - 2y + 2z - 5 = 0$

**45** $P(3, 0, 1)$, $x + 2y = 0$, $3y - z = 0$

In Problems 46 through 50, find in each case equations in symmetric form of the line of intersection of the given planes. Use the method of vector products.

**46** $2(x - 1) + 3(y + 1) - 4(z - 2) = 0$
$3(x - 1) - 4(y + 1) + 2(z - 2) = 0$

**47** $3(x + 2) - 2(y - 1) + 2(z + 1) = 0$
$(x + 2) + 2(y - 1) - 3(z + 1) = 0$

**48** $(x - 2) + (y + 1) + 3(z + 2) = 0$
$7(x - 2) + 2(y + 1) - 2(z + 2) = 0$

**49** $(x + 1) + 2(y - 2) + 3(z - 1) = 0$
$2(x + 1) + 4(y - 2) + 5(z - 1) = 0$

**50** $2(x + 1) + 3(y - 1) - (z + 2) = 0$
$3(x + 1) + 2(y - 1) - 4(z + 2) = 0$

In each of Problems 51 through 55, find an equation of the plane through the given point or points and parallel to the given line or lines.

**51** $(1, 3, 2)$; $\dfrac{x + 1}{2} = \dfrac{y - 2}{-1} = \dfrac{z + 3}{3}$; $\dfrac{x - 2}{1} = \dfrac{y + 1}{-2} = \dfrac{z + 2}{2}$

**52** $(2, -1, -3)$; $\dfrac{x - 1}{3} = \dfrac{y + 2}{2} = \dfrac{z}{-4}$; $\dfrac{x}{2} = \dfrac{y - 1}{-3} = \dfrac{z - 2}{2}$

**53** $(2, 1, -2)$; $(1, -1, 3)$; $\dfrac{x + 1}{3} = \dfrac{y - 1}{2} = \dfrac{z - 2}{2}$

**54** $(1, -2, 3)$; $(-1, 2, -1)$; $\dfrac{x - 2}{2} = \dfrac{y + 1}{3} = \dfrac{z - 1}{4}$

**55** $(0, 1, 2)$; $(2, 0, 1)$; $\dfrac{x - 1}{3} = \dfrac{y + 1}{0} = \dfrac{z + 1}{1}$

In Problems 56 through 58, find in each case the equation of the plane through the line $L_1$ which also satisfies the additional condition.

**56** $L_1$: $\dfrac{x - 1}{2} = \dfrac{y + 1}{3} = \dfrac{z - 2}{1}$;  through $(2, 1, 1)$

**57** $L_1$: $\dfrac{x - 2}{2} = \dfrac{y - 2}{3} = \dfrac{z - 1}{-2}$;

parallel to $\dfrac{x + 1}{3} = \dfrac{y - 1}{2} = \dfrac{z + 1}{1}$

**58** $L_1$: $\dfrac{x + 1}{1} = \dfrac{y - 1}{2} = \dfrac{z - 2}{-2}$;

perpendicular to $2x + 3y - z + 4 = 0$

In Problems 59 and 60, find the equation of the plane through the given points and perpendicular to the given planes.

**59** $(1, 2, -1)$; $2x - 3y + 5z - 1 = 0$; $3x + 2y + 4z + 6 = 0$

**60** $(-1, 3, 2)$; $(1, 6, 1)$; $3x - y + 4z - 7 = 0$

In Problems 61 and 62, find equations in symmetric form of the line through the given point $P$, which is perpendicular to and intersects the given line. Use the cross product.

**61** $P(3, 3, -1)$; $\dfrac{x}{-1} = \dfrac{y - 3}{1} = \dfrac{z + 1}{1}$

**62** $P(3, -2, 0)$; $\dfrac{x - 4}{3} = \dfrac{y - 4}{-4} = \dfrac{z - 5}{-1}$

**63** Suppose that $\mathbf{a}$ is a nonzero vector. If $\mathbf{a} \cdot \mathbf{x} = \mathbf{a} \cdot \mathbf{y}$ and $\mathbf{a} \times \mathbf{x} = \mathbf{a} \times \mathbf{y}$, is it true that $\mathbf{x} = \mathbf{y}$? Justify your answer.

**64** Given that $\mathbf{u} + \mathbf{v} + \mathbf{w} - \mathbf{r} = \mathbf{0}$ and $\mathbf{u} - \mathbf{v} + \mathbf{w} + 2\mathbf{r} = \mathbf{0}$. Prove that $\mathbf{r} \times \mathbf{v} = \mathbf{0}$, that $\mathbf{v} \times \mathbf{w} = \frac{3}{2}\mathbf{r} \times \mathbf{w}$, and that

$$2\mathbf{u} \times \mathbf{w} = \mathbf{w} \times \mathbf{r} = \mathbf{r} \times \mathbf{u}.$$

___

**6**

# PRODUCTS OF THREE VECTORS

Since two types of multiplication, the scalar product and the cross product, may be performed on vectors, we can combine three vectors in several ways. For example, we can form the product

$$(\mathbf{u} \times \mathbf{v}) \cdot \mathbf{w}$$

and the product

$$(\mathbf{u} \times \mathbf{v}) \times \mathbf{w}.$$

Also, we can consider the combinations

$$\mathbf{u} \cdot (\mathbf{v} \times \mathbf{w}) \qquad \text{and} \qquad \mathbf{u} \times (\mathbf{v} \times \mathbf{w}).$$

The next theorem gives a simple rule for computing $(\mathbf{u} \times \mathbf{v}) \cdot \mathbf{w}$ and also an elegant geometric interpretation of the quantity $|(\mathbf{u} \times \mathbf{v}) \cdot \mathbf{w}|$.

**THEOREM 21**  *Suppose that $\mathbf{u}_1, \mathbf{u}_2, \mathbf{u}_3$, are vectors and that the points $A, B, C, D$ are chosen so that*

$$\mathbf{v}[\overrightarrow{AB}] = \mathbf{u}_1, \qquad \mathbf{v}[\overrightarrow{AC}] = \mathbf{u}_2, \qquad \mathbf{v}[\overrightarrow{AD}] = \mathbf{u}_3.$$

*Then*

*i) the quantity $|(\mathbf{u}_1 \times \mathbf{u}_2) \cdot \mathbf{u}_3|$ is the volume of the parallelepiped with one vertex at $A$ and adjacent vertices at $B$, $C$, and $D$. (See Fig. 29.) This volume is zero if and only if the four points $A, B, C, D$ lie in a plane;*
*ii) if $\{\mathbf{i}, \mathbf{j}, \mathbf{k}\}$ is a right-handed coordinate triple and if*

$$\mathbf{u}_1 = a_1 \mathbf{i} + b_1 \mathbf{j} + c_1 \mathbf{k}, \qquad \mathbf{u}_2 = a_2 \mathbf{i} + b_2 \mathbf{j} + c_2 \mathbf{k}, \qquad \mathbf{u}_3 = a_3 \mathbf{i} + b_3 \mathbf{j} + c_3 \mathbf{k},$$

*then*

$$(\mathbf{u}_1 \times \mathbf{u}_2) \cdot \mathbf{u}_3 = \begin{vmatrix} a_1 & b_1 & c_1 \\ a_2 & b_2 & c_2 \\ a_3 & b_3 & c_3 \end{vmatrix};$$

*iii) $(\mathbf{u}_1 \times \mathbf{u}_2) \cdot \mathbf{u}_3 = \mathbf{u}_1 \cdot (\mathbf{u}_2 \times \mathbf{u}_3).$*

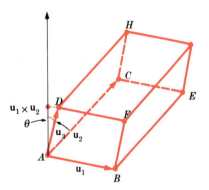

FIGURE 29

**Proof**  To prove (i), note that $|\mathbf{u}_1 \times \mathbf{u}_2|$ is the area of the parallelogram $ABEC$ and that

$$|(\mathbf{u}_1 \times \mathbf{u}_2) \cdot \mathbf{u}_3| = |\mathbf{u}_1 \times \mathbf{u}_2||\mathbf{u}_3||\cos \theta|,$$

where $\theta$ is the angle between the two vectors $\mathbf{u}_3$ and $\mathbf{u}_1 \times \mathbf{u}_2$. The quantity $|\mathbf{u}_3||\cos \theta|$ is the length of the projection of $\mathbf{u}_3$ on the normal to the plane of

*ABEC*. Clearly, $(\mathbf{u}_1 \times \mathbf{u}_2) \cdot \mathbf{u}_3 = 0 \Leftrightarrow \mathbf{u}_1 \times \mathbf{u}_2 = \mathbf{0}$ or $\mathbf{u}_3 = \mathbf{0}$ or $\cos\theta = 0$. If $\cos\theta = 0$, then $\mathbf{u}_3$ is parallel to the plane of $\mathbf{u}_1$ and $\mathbf{u}_2$ and all four points lie in a plane. The proof of parts (ii) and (iii) follow from Theorems 14 and 19 and are left to the reader. □

**THEOREM 22**

*If $\mathbf{u}$, $\mathbf{v}$, and $\mathbf{w}$ are any vectors, then*

i) $(\mathbf{u} \times \mathbf{v}) \times \mathbf{w} = (\mathbf{u} \cdot \mathbf{w})\mathbf{v} - (\mathbf{v} \cdot \mathbf{w})\mathbf{u}$,
ii) $\mathbf{u} \times (\mathbf{v} \times \mathbf{w}) = (\mathbf{u} \cdot \mathbf{w})\mathbf{v} - (\mathbf{u} \cdot \mathbf{v})\mathbf{w}$.

**Proof**   If $\mathbf{u}$ and $\mathbf{v}$ are proportional or if $\mathbf{w}$ is orthogonal to both $\mathbf{u}$ and $\mathbf{v}$, then both sides of (i) are zero. Otherwise, we see that $(\mathbf{u} \times \mathbf{v}) \times \mathbf{w}$ is orthogonal to the perpendicular to the plane determined by $\mathbf{u}$ and $\mathbf{v}$. Hence $(\mathbf{u} \times \mathbf{v}) \times \mathbf{w}$ is in the plane of $\mathbf{u}$ and $\mathbf{v}$. We choose a right-handed coordinate triple $\{\mathbf{i}, \mathbf{j}, \mathbf{k}\}$ so that $\mathbf{i}$ is in the direction of $\mathbf{u}$ and $\mathbf{j}$ is in the plane of $\mathbf{u}$ and $\mathbf{v}$. Then there are numbers $a_1$, $a_2$, $a_3$, $b_2$, $b_3$, $c_3$, so that

$$\mathbf{u} = a_1\mathbf{i}, \qquad \mathbf{v} = a_2\mathbf{i} + b_2\mathbf{j}, \qquad \mathbf{w} = a_3\mathbf{i} + b_3\mathbf{j} + c_3\mathbf{k}.$$

The reader may now compute both sides of (i) to see that they are equal. The proof of (ii) is left to the reader. □

**EXAMPLE 1**   Given $A(3, -1, 2)$, $B(1, 2, -2)$, $C(2, 1, -2)$, and $D(-1, 3, 2)$, find the volume of the parallelepiped having $AB$, $AC$, and $AD$ as edges.

**Solution**   We have

$$\mathbf{u}_1 = \mathbf{v}[\overrightarrow{AB}] = -2\mathbf{i} + 3\mathbf{j} - 4\mathbf{k},$$

$$\mathbf{u}_2 = \mathbf{v}[\overrightarrow{AC}] = -\mathbf{i} + 2\mathbf{j} - 4\mathbf{k},$$

$$\mathbf{u}_3 = \mathbf{v}[\overrightarrow{AD}] = -4\mathbf{i} + 4\mathbf{j}.$$

We compute $\mathbf{u}_2 \times \mathbf{u}_3 = 16\mathbf{i} + 16\mathbf{j} + 4\mathbf{k}$. Therefore

$$|\mathbf{u}_1 \cdot (\mathbf{u}_2 \times \mathbf{u}_3)| = |-32 + 48 - 16| = 0.$$

Hence the four points are in a plane. The volume is zero. □

**EXAMPLE 2**   Find the equations of the line through the point $P_0\,(3, -2, 1)$ perpendicular to the line $L$ (and intersecting it) given by

$$L = \left\{(x, y, z): \frac{x-2}{2} = \frac{y+1}{-2} = \frac{z}{1}\right\}.$$

**Solution**   The point $P_1(2, -1, 0)$ is on $L$, and the direction numbers of $L$ show that

$$\mathbf{u} = 2\mathbf{i} - 2\mathbf{j} + \mathbf{k}$$

is a vector in the direction of $L$. Also, we denote by $\mathbf{v}$ the vector

$$\mathbf{v} = \mathbf{v}(\overrightarrow{P_0 P_1}) = -\mathbf{i} + \mathbf{j} - \mathbf{k}.$$

The plane containing $L$ and $P_0$ has a normal perpendicular to $\mathbf{u}$ and $\mathbf{v}$. Hence this normal is proportional to $\mathbf{u} \times \mathbf{v}$. The desired line is in this plane and perpendicular to $L$. Therefore it has a direction $\mathbf{w}$ perpendicular to $\mathbf{u}$ and $\mathbf{u} \times \mathbf{v}$. Thus for some number $c$, we have

$$c\mathbf{w} = \mathbf{u} \times (\mathbf{u} \times \mathbf{v}) = (\mathbf{u} \cdot \mathbf{v})\mathbf{u} - (\mathbf{u} \cdot \mathbf{u})\mathbf{v}$$

$$= -5(2\mathbf{i} - 2\mathbf{j} + \mathbf{k}) - 9(-\mathbf{i} + \mathbf{j} - \mathbf{k}) = -\mathbf{i} + \mathbf{j} + 4\mathbf{k}.$$

Consequently, the desired line has equations

$$\frac{x-3}{-1} = \frac{y+2}{1} = \frac{z-1}{4}. \qquad \square$$

## 6  PROBLEMS

In Problems 1 through 4, find the volume of the parallelepiped having edges $AB$, $AC$, and $AD$, or else show that $A$, $B$, $C$, and $D$ lie on a plane or on a line. If they lie on a plane, find its equation; if they lie on a line, find its equations.

1  $A = (2, -1, 3)$, $B = (-1, 2, 2)$, $C = (1, 0, 1)$, $D = (4, 1, -1)$

2  $A = (3, 1, -2)$, $B = (1, 2, 1)$, $C = (2, -1, 3)$, $D = (4, 3, -7)$

3  $A = (1, 2, -3)$, $B = (3, 1, -2)$, $C = (-1, 3, 1)$, $D = (-3, 4, 3)$

4  $A = (-1, -2, 2)$, $B = (2, -1, 1)$, $C = (0, 1, 3)$, $D = (3, 2, -1)$

5  Prove Theorem 21, parts (ii) and (iii).

6  Complete the proof of Theorem 22.

In Problems 7 through 10, compute $(\mathbf{u} \times \mathbf{v}) \times \mathbf{w}$ directly and by using Theorem 22.

7  $\mathbf{u} = 2\mathbf{i} + 3\mathbf{j} - \mathbf{k}$, $\mathbf{v} = \mathbf{i} - 2\mathbf{j} + \mathbf{k}$, $\mathbf{w} = -\mathbf{i} + \mathbf{j} + 2\mathbf{k}$

8  $\mathbf{u} = 3\mathbf{i} - 2\mathbf{j} + \mathbf{k}$, $\mathbf{v} = \mathbf{i} + \mathbf{j} + 2\mathbf{k}$, $\mathbf{w} = 2\mathbf{i} - \mathbf{j} + 3\mathbf{k}$

9  $\mathbf{u} = \mathbf{i} + 2\mathbf{j} - 3\mathbf{k}$, $\mathbf{v} = -\mathbf{i} + \mathbf{j} - 2\mathbf{k}$, $\mathbf{w} = 3\mathbf{i} - \mathbf{j} + \mathbf{k}$

10  $\mathbf{u} = 2\mathbf{i} - \mathbf{j} + 3\mathbf{k}$, $\mathbf{v} = \mathbf{i} + 2\mathbf{j} + \mathbf{k}$, $\mathbf{w} = 3\mathbf{i} - 2\mathbf{j} - \mathbf{k}$

11  Show that every vector $\mathbf{v}$ satisfies the identity

$$\mathbf{i} \times (\mathbf{v} \times \mathbf{i}) + \mathbf{j} \times (\mathbf{v} \times \mathbf{j}) + \mathbf{k} \times (\mathbf{v} \times \mathbf{k}) = 2\mathbf{v}.$$

In Problems 12 through 14 find, in each case, the equations of the line through the given point and perpendicular to the given line and intersecting it.

12  $(1, 3, -2)$, $\dfrac{x-2}{3} = \dfrac{y+1}{-2} = \dfrac{z-1}{4}$

13  $(2, -1, 3)$, $\dfrac{x+1}{2} = \dfrac{y-2}{3} = \dfrac{z+1}{-5}$

14  $(-1, 2, 4)$, $\dfrac{x-1}{4} = \dfrac{y+2}{-3} = \dfrac{z-1}{2}$

*15  Let $\mathbf{t}$, $\mathbf{u}$, $\mathbf{v}$ be vectors. Show that the volume of the tetrahedron with sides formed by $\mathbf{t}$, $\mathbf{u}$, $\mathbf{v}$ is $\frac{1}{6}|(\mathbf{u} \times \mathbf{v}) \cdot \mathbf{t}|$ (see Fig. 30). (*Hint*: The volume of a tetrahedron is $\frac{1}{3}$ (area of base) (height))

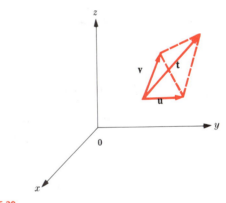

FIGURE 30

In Problems 16 through 20, use the formula of Problem 15 to compute the volume of the tetrahedron with the given vertices.

16  $P(0, 0, 0)$, $Q(1, 0, 0)$, $R(2, 3, 4)$, $T(-1, 2, 3)$

17  $P(4, 2, 1)$, $Q(4, 2, 3)$, $R(2, 4, 1)$, $T(-3, 0, -2)$

18  $P(1, 2, 1)$, $Q(2, 4, 3)$, $R(1, 3, 2)$, $T(0, 4, 0)$

19  $P(-2, 0, 1)$, $Q(1, 1, 1)$, $R(3, -2, 1)$, $T(2, 2, 3)$

20  $P(4, 1, 2)$, $Q(2, 1, 0)$, $R(0, 0, 2)$, $T(-1, 2, -3)$

In Problems 21 and 22, express $(\mathbf{t} \times \mathbf{u}) \times (\mathbf{v} \times \mathbf{w})$ in terms of $\mathbf{v}$ and $\mathbf{w}$.

21  $\mathbf{t} = \mathbf{i} + \mathbf{j} - 2\mathbf{k}$, $\mathbf{u} = 3\mathbf{i} - \mathbf{j} + 2\mathbf{k}$, $\mathbf{v} = 2\mathbf{i} + 2\mathbf{j} - \mathbf{k}$, $\mathbf{w} = -\mathbf{i} + \mathbf{j} + 2\mathbf{k}$

22  $\mathbf{t} = 2\mathbf{i} - \mathbf{j} + \mathbf{k}$, $\mathbf{u} = \mathbf{i} + 2\mathbf{j} - 3\mathbf{k}$, $\mathbf{v} = 3\mathbf{i} + \mathbf{j} + 2\mathbf{k}$, $\mathbf{w} = -\mathbf{i} + 2\mathbf{j} - 2\mathbf{k}$

23 Derive a formula expressing $(\mathbf{t} \times \mathbf{u}) \times (\mathbf{v} \times \mathbf{w})$ in terms of $\mathbf{v}$ and $\mathbf{w}$.

24 Given $\mathbf{a} = \mathbf{i} - \mathbf{j} + \mathbf{k}$, $\mathbf{b} = 2\mathbf{i} + 3\mathbf{j} + \mathbf{k}$, $p = 1$. Solve the equations $\mathbf{a} \cdot \mathbf{v} = p$, $\mathbf{a} \times \mathbf{v} = \mathbf{b}$ for $\mathbf{v}$.

25 Given that $\mathbf{a} \cdot \mathbf{b} = 0$, $\mathbf{a} \neq 0$, $\mathbf{b} \neq 0$, find a formula for the solution $\mathbf{v}$ of the equations

$$\mathbf{a} \cdot \mathbf{v} = p, \qquad \mathbf{a} \times \mathbf{v} = \mathbf{b}.$$

(*Hint:* Note that $\mathbf{a}$, $\mathbf{b}$, and $\mathbf{a} \times \mathbf{b}$ are mutually orthogonal.)

26 Show that for any vectors $\mathbf{u}$, $\mathbf{v}$, $\mathbf{w}$, we have
i) $(\mathbf{u} \pm \mathbf{v}) \cdot [(\mathbf{u} + \mathbf{v}) \times (\mathbf{u} - \mathbf{v})] = 0$
ii) $(\mathbf{u} + \mathbf{v}) \cdot [(\mathbf{u} \times \mathbf{w}) \times (\mathbf{u} + \mathbf{v})] = 0$.

27 If $\mathbf{u}$, $\mathbf{v}$, $\mathbf{w}$ are any vectors show that

$$[\mathbf{u} \times (\mathbf{v} \times \mathbf{w})] + [\mathbf{v} \times (\mathbf{w} \times \mathbf{u})] + [\mathbf{w} \times (\mathbf{u} \times \mathbf{v})] = 0.$$

*28 Show that the points $A$, $B$, $C$, $D$ are all in the same plane if $\mathbf{v}[\vec{AB}] \cdot (\mathbf{v}[\vec{AC}] \times \mathbf{v}[\vec{AD}]) = 0$.

*29 Prove the identity $|\mathbf{u} \times \mathbf{v}|^2 = |\mathbf{u}|^2 |\mathbf{v}|^2 - (\mathbf{u} \cdot \mathbf{v})^2$ (*Hint:* Use the identity $\sin^2 \theta + \cos^2 \theta = 1$.)

*30 Let $\mathscr{F}$ be a collection of objects with $\mathbf{A}$, $\mathbf{B}$, $\mathbf{C}$ members of $\mathscr{F}$. Let $\oplus$ be an operation between members of $\mathscr{F}$ satisfying the relation

$$(\mathbf{A} \oplus \mathbf{B}) \oplus \mathbf{C} = \alpha \mathbf{B} - \beta \mathbf{A}.$$

State general conditions on the numbers $\alpha$ and $\beta$ such that the formula

$$[(\mathbf{A} \oplus \mathbf{B}) \oplus \mathbf{C}] + [(\mathbf{B} \oplus \mathbf{C}) \oplus \mathbf{A}] + [(\mathbf{C} \oplus \mathbf{A}) \oplus \mathbf{B}] = 0$$

should hold.

_____ 7

## LINEAR DEPENDENCE AND INDEPENDENCE (OPTIONAL)*

Two vectors $\mathbf{u}$ and $\mathbf{v}$, neither zero, are said to be **proportional** if and only if there is a number $c$ such that $\mathbf{u} = c\mathbf{v}$; that is, each vector is a scalar multiple of the other. If $\mathbf{v}_1, \mathbf{v}_2, \ldots, \mathbf{v}_k$ are any vectors and $c_1, c_2, \ldots, c_k$ are numbers, we call an expression of the form

$$c_1 \mathbf{v}_1 + c_2 \mathbf{v}_2 + \cdots + c_k \mathbf{v}_k$$

a **linear combination** of the vectors $\mathbf{v}_1, \mathbf{v}_2, \ldots, \mathbf{v}_k$. If two vectors $\mathbf{u}$ and $\mathbf{v}$ are proportional, the definition shows that a linear combination of them is the zero vector. In fact, $\mathbf{u} - c\mathbf{v} = 0$. A set of vectors $\{\mathbf{v}_1, \mathbf{v}_2, \ldots, \mathbf{v}_k\}$ is **linearly dependent** if and only if there is a set of constants $\{c_1, c_2, \ldots, c_k\}$, *not all zero*, such that

$$c_1 \mathbf{v}_1 + c_2 \mathbf{v}_2 + \cdots + c_k \mathbf{v}_k = 0. \tag{1}$$

If no such set of constants exists, then the set $\{\mathbf{v}_1, \mathbf{v}_2, \ldots, \mathbf{v}_k\}$ is said to be **linearly independent**.

It is clear that any two proportional vectors are linearly dependent. As another example, the vectors

$$\mathbf{v}_1 = 2\mathbf{i} + 3\mathbf{j} - \mathbf{k}, \qquad \mathbf{v}_2 = -2\mathbf{i} - \mathbf{j} + \mathbf{k}, \qquad \mathbf{v}_3 = 2\mathbf{i} + 7\mathbf{j} - \mathbf{k}$$

_____

*For an understanding of Section 7, it is assumed that the reader is acquainted with determinants of the second and third order. For those unfamiliar with the subject a brief discussion is provided in Appendix 4.

form a linearly dependent set since the selection $c_1 = 3, c_2 = 2, c_3 = -1$ shows that

$$c_1\mathbf{v}_1 + c_2\mathbf{v}_2 + c_3\mathbf{v}_3 = 3(2\mathbf{i} + 3\mathbf{j} - \mathbf{k}) + 2(-2\mathbf{i} - \mathbf{j} + \mathbf{k}) - (2\mathbf{i} + 7\mathbf{j} - \mathbf{k}) = \mathbf{0}.$$

A set $\{\mathbf{v}_1, \mathbf{v}_2, \ldots, \mathbf{v}_k\}$ is linearly dependent if and only if one member of the set can be expressed as a linear combination of the remaining members. To see this, we observe that in Equation (1) one of the terms on the left side, say $\mathbf{v}_i$, must have a nonzero coefficient and so may be transferred to the right side. Dividing by the coefficient $-c_i$, we express this particular $\mathbf{v}_i$ as a linear combination of the remaining $\mathbf{v}$'s. If some $\mathbf{v}_i$ is expressible in terms of the others, it follows by transposing $\mathbf{v}_i$ that $\mathbf{v}_1, \mathbf{v}_2, \ldots, \mathbf{v}_k$ are linearly dependent.

The following statement, a direct consequence of the definition of linear dependence, is often useful in proofs of theorems. If $\{\mathbf{v}_1, \mathbf{v}_2, \ldots, \mathbf{v}_k\}$ is a linearly independent set and if

$$c_1\mathbf{v}_1 + c_2\mathbf{v}_2 + \cdots + c_k\mathbf{v}_k = \mathbf{0},$$

then it follows that $c_1 = c_2 = \cdots = c_k = 0$.

*The set $\{\mathbf{i}, \mathbf{j}, \mathbf{k}\}$ is linearly independent.* To show this observe that the equation

$$c_1\mathbf{i} + c_2\mathbf{j} + c_3\mathbf{k} = \mathbf{0} \tag{2}$$

holds if and only if $|c_1\mathbf{i} + c_2\mathbf{j} + c_3\mathbf{k}| = 0$. But

$$|c_1\mathbf{i} + c_2\mathbf{j} + c_3\mathbf{k}| = \sqrt{c_1^2 + c_2^2 + c_3^2},$$

and this last expression is zero if and only if $c_1 = c_2 = c_3 = 0$. Thus no nonzero constants satisfying (2) exist and $\{\mathbf{i}, \mathbf{j}, \mathbf{k}\}$ is a linearly independent set.

The proof of the next theorem employs the tools on determinants given in Appendix 4. The proof of Theorem 23 is given in Appendix 5.

**THEOREM 23**    *Let*

$$\mathbf{u} = a_{11}\mathbf{i} + a_{12}\mathbf{j} + a_{13}\mathbf{k},$$

$$\mathbf{v} = a_{21}\mathbf{i} + a_{22}\mathbf{j} + a_{23}\mathbf{k},$$

$$\mathbf{w} = a_{31}\mathbf{i} + a_{32}\mathbf{j} + a_{33}\mathbf{k},$$

*and denote by D the determinant*

$$D = \begin{vmatrix} a_{11} & a_{12} & a_{13} \\ a_{21} & a_{22} & a_{23} \\ a_{31} & a_{32} & a_{33} \end{vmatrix}.$$

*Then the set $\{\mathbf{u}, \mathbf{v}, \mathbf{w}\}$ is linearly independent if and only if $D \neq 0$.*

**EXAMPLE 1**    Determine whether or not the vectors

$$\mathbf{u} = 2\mathbf{i} - \mathbf{j} + \mathbf{k}, \qquad \mathbf{v} = \mathbf{i} + 2\mathbf{j} + \mathbf{k}, \qquad \mathbf{w} = -\mathbf{i} + \mathbf{j} + 3\mathbf{k}$$

form a linearly independent set.

**Solution**    Expanding $D$ by its first row, we have

$$D = \begin{vmatrix} 2 & -1 & 1 \\ 1 & 2 & 1 \\ -1 & 1 & 3 \end{vmatrix} = 2\begin{vmatrix} 2 & 1 \\ 1 & 3 \end{vmatrix} + \begin{vmatrix} 1 & 1 \\ -1 & 3 \end{vmatrix} + \begin{vmatrix} 1 & 2 \\ -1 & 1 \end{vmatrix}.$$

Therefore $D = 2(5) + 4 + 3 = 17 \neq 0$. The set is linearly independent. ☐

---

**THEOREM 24**    *If $\{\mathbf{u}, \mathbf{v}, \mathbf{w}\}$ is a linearly independent set and $\mathbf{r}$ is any vector, then there are constants $A_1$, $A_2$, and $A_3$ such that*

$$\mathbf{r} = A_1\mathbf{u} + A_2\mathbf{v} + A_3\mathbf{w}. \tag{3}$$

---

**Proof**    We know by Theorem 9, page 577, that *every* vector can be expressed as a linear combination of $\mathbf{i}$, $\mathbf{j}$, and $\mathbf{k}$. Therefore

$$\mathbf{u} = a_{11}\mathbf{i} + a_{12}\mathbf{j} + a_{13}\mathbf{k},$$

$$\mathbf{v} = a_{21}\mathbf{i} + a_{22}\mathbf{j} + a_{23}\mathbf{k},$$

$$\mathbf{w} = a_{31}\mathbf{i} + a_{32}\mathbf{j} + a_{33}\mathbf{k},$$

$$\mathbf{r} = b_1\mathbf{i} + b_2\mathbf{j} + b_3\mathbf{k}.$$

When we insert all these expressions in (3) and collect all terms on one side, we get a linear combination of $\mathbf{i}$, $\mathbf{j}$, and $\mathbf{k}$ equal to zero. Since $\{\mathbf{i}, \mathbf{j}, \mathbf{k}\}$ is a linearly independent set, the coefficients of $\mathbf{i}$, $\mathbf{j}$, and $\mathbf{k}$ are equal to zero separately. Computing these coefficients, we get the equations

$$a_{11}A_1 + a_{21}A_2 + a_{31}A_3 = b_1,$$

$$a_{12}A_1 + a_{22}A_2 + a_{32}A_3 = b_2, \tag{4}$$

$$a_{13}A_1 + a_{23}A_2 + a_{33}A_3 = b_3.$$

We have here three equations in the three unknowns $A_1$, $A_2$, $A_3$. The determinant $D'$ of the coefficients in (4) differs from the determinant $D$ of Theorem 23 in that the rows and columns are interchanged. Since $\{\mathbf{u}, \mathbf{v}, \mathbf{w}\}$ is an independent set, we know that $D \neq 0$; also Theorem 4 of Appendix 4 proves that $D = D'$, and so $D' \neq 0$. We now use Cramer's rule (Theorem 11, Appendix 4) to solve for $A_1$, $A_2$, $A_3$. ☐

Note that the proof of Theorem 24 gives the method for finding $A_1$, $A_2$, $A_3$. We work an example.

**EXAMPLE 2**    Given the vectors

$$\mathbf{u} = 2\mathbf{i} + 3\mathbf{j} + \mathbf{k}, \qquad \mathbf{v} = -\mathbf{i} + \mathbf{j} + 2\mathbf{k},$$

$$\mathbf{w} = 3\mathbf{i} - \mathbf{j} + 3\mathbf{k}, \qquad \mathbf{r} = \mathbf{i} + 2\mathbf{j} - 6\mathbf{k},$$

show that $\mathbf{u}$, $\mathbf{v}$, and $\mathbf{w}$ are linearly independent and express $\mathbf{r}$ as a linear combination of $\mathbf{u}$, $\mathbf{v}$, and $\mathbf{w}$.

**Solution**   Expanding the determinant $D$ for **u**, **v**, and **w** by its first row, we obtain

$$D = \begin{vmatrix} 2 & 3 & 1 \\ -1 & 1 & 2 \\ 3 & -1 & 3 \end{vmatrix} = 2 \begin{vmatrix} 1 & 2 \\ -1 & 3 \end{vmatrix} - 3 \begin{vmatrix} -1 & 2 \\ 3 & 3 \end{vmatrix} + \begin{vmatrix} -1 & 1 \\ 3 & -1 \end{vmatrix}$$

$$= 2(5) - 3(-9) + (-2) = 35.$$

Hence $D \neq 0$ and so $\{$**u**, **v**, **w**$\}$ is a linearly independent set. Using equations (4), we obtain the equations

$$2A_1 - A_2 + 3A_3 = 1,$$

$$3A_1 + A_2 - A_3 = 2,$$

$$A_1 + 2A_2 + 3A_3 = -6.$$

Solving these, we find that $A_1 = 1$, $A_2 = -2$, $A_3 = -1$. Finally,

$$\mathbf{r} = \mathbf{u} - 2\mathbf{v} - \mathbf{w}. \qquad \square$$

## 7   PROBLEMS

In Problems 1 through 10, state whether or not the given vectors are linearly independent.

**1**  $\mathbf{u} = 2\mathbf{i} + \mathbf{j} - \mathbf{k}$, $\mathbf{v} = \mathbf{i} - 2\mathbf{j} + 5\mathbf{k}$, $\mathbf{w} = 2\mathbf{i} - 7\mathbf{j} + \mathbf{k}$

**2**  $\mathbf{u} = \mathbf{i} + 2\mathbf{j} + 3\mathbf{k}$, $\mathbf{v} = 2\mathbf{i} + \mathbf{j} + 4\mathbf{k}$, $\mathbf{w} = 3\mathbf{j} + 2\mathbf{k}$

**3**  $\mathbf{u} = 2\mathbf{i} + 3\mathbf{j}$, $\mathbf{v} = \mathbf{i} - 4\mathbf{j}$, $\mathbf{w} = \mathbf{i} + 2\mathbf{j}$

**4**  $\mathbf{u} = -\mathbf{i} + 2\mathbf{j}$, $\mathbf{v} = \mathbf{i} + \mathbf{j} + \mathbf{k}$, $\mathbf{w} = -2\mathbf{j} + 6\mathbf{k}$

**5**  $\mathbf{u} = \mathbf{i} + \mathbf{j}$, $\mathbf{v} = 2\mathbf{i} - 6\mathbf{j} + 3\mathbf{k}$, $\mathbf{w} = -\mathbf{i} + \mathbf{j}$, $\mathbf{r} = 4\mathbf{k}$

**6**  $\mathbf{u} = \mathbf{i} + \mathbf{j}$, $\mathbf{v} = \mathbf{i} - \mathbf{j}$

**7**  $\mathbf{u} = \mathbf{i} + \mathbf{j} + \mathbf{k}$, $\mathbf{v} = \mathbf{i} - \mathbf{j} + \mathbf{k}$, $\mathbf{w} = \mathbf{k}$

**8**  $\mathbf{u} = \mathbf{i} + \mathbf{j} + \mathbf{k}$, $\mathbf{v} = 2\mathbf{i} + 2\mathbf{j} + 2\mathbf{k}$, $\mathbf{w} = \mathbf{i} - \mathbf{j}$

**9**  $\mathbf{u} = \mathbf{i}$, $\mathbf{v} = \mathbf{j}$, $\mathbf{w} = \mathbf{i} + \mathbf{j} - \mathbf{k}$

**10**  $\mathbf{u} = \mathbf{i}$, $\mathbf{v} = \mathbf{j}$, $\mathbf{w} = \mathbf{k}$, $\mathbf{r} = \mathbf{i} + \mathbf{j} - \mathbf{k}$

In Problems 11 through 16, show that **u**, **v**, and **w** are linearly independent and express **r** in terms of **u**, **v**, and **w**.

**11**  $\mathbf{u} = 2\mathbf{i} - \mathbf{j} + \mathbf{k}$, $\mathbf{v} = -\mathbf{i} + \mathbf{j} - 2\mathbf{k}$,
     $\mathbf{w} = 2\mathbf{i} - \mathbf{j} + 2\mathbf{k}$, $\mathbf{r} = 3\mathbf{i} - \mathbf{j} + 2\mathbf{k}$

**12**  $\mathbf{u} = \mathbf{i} - \mathbf{j} + \mathbf{k}$, $\mathbf{v} = -\mathbf{i} + 2\mathbf{j} - \mathbf{k}$,
     $\mathbf{w} = 2\mathbf{i} - \mathbf{j} + \mathbf{k}$, $\mathbf{r} = 2\mathbf{i} + 3\mathbf{j} + 4\mathbf{k}$

**13**  $\mathbf{u} = 3\mathbf{i} + \mathbf{j} - 2\mathbf{k}$, $\mathbf{v} = 2\mathbf{i} - \mathbf{k}$,
     $\mathbf{w} = -\mathbf{i} + 2\mathbf{j} + \mathbf{k}$, $\mathbf{r} = \mathbf{i} + 2\mathbf{j} - 3\mathbf{k}$

**14**  $\mathbf{u} = 2\mathbf{i} - \mathbf{j} + \mathbf{k}$, $\mathbf{v} = \mathbf{i} + \mathbf{j}$,
     $\mathbf{w} = -\mathbf{i} + \mathbf{j} + 2\mathbf{k}$, $\mathbf{r} = 2\mathbf{i} - \mathbf{j} - 2\mathbf{k}$

**15**  $\mathbf{u} = \mathbf{i} - 2\mathbf{j} - 3\mathbf{k}$, $\mathbf{v} = 2\mathbf{i} - \mathbf{j} - 2\mathbf{k}$,
     $\mathbf{w} = -\mathbf{i} + \mathbf{j} + \mathbf{k}$, $\mathbf{r} = 2\mathbf{i} + 3\mathbf{j} + 4\mathbf{k}$

**16**  $\mathbf{u} = 2\mathbf{i} - 3\mathbf{k}$, $\mathbf{v} = \mathbf{i} + 4\mathbf{j} - \mathbf{k}$,
     $\mathbf{w} = -2\mathbf{i} + 5\mathbf{j} + 3\mathbf{k}$, $\mathbf{r} = -\mathbf{i} + 20\mathbf{j} + 3\mathbf{k}$

**17**  Show that if **u** and **v** are linearly dependent, then $\mathbf{u} \times \mathbf{v} = \mathbf{0}$.

**18**  Show that if **u** and **v** are linearly independent, then $\{$**u**, **v**, $\mathbf{u} \times \mathbf{v}\}$ are linearly independent.

In each of Problems 19 through 24, find a vector **w** such that $\{$**u**, **v**, **w**$\}$ is a linearly independent set for the given vectors **u** and **v**.

**19**  $\mathbf{u} = \mathbf{i}$, $\mathbf{v} = \mathbf{j}$

**20**  $\mathbf{u} = \mathbf{i} + \mathbf{j}$, $\mathbf{v} = \mathbf{i} - \mathbf{j}$

**21**  $\mathbf{u} = \mathbf{i} + \mathbf{j} + \mathbf{k}$, $\mathbf{v} = \mathbf{i} + \mathbf{j} - \mathbf{k}$

**22**  $\mathbf{u} = 2\mathbf{i} - 3\mathbf{j} + \mathbf{k}$, $\mathbf{v} = \mathbf{i} - 3\mathbf{j} - \mathbf{k}$

**23**  $\mathbf{u} = \mathbf{i} + \mathbf{j} + 2\mathbf{k}$, $\mathbf{v} = 2\mathbf{i} + \mathbf{j} + \mathbf{k}$

**24**  $\mathbf{u} = \mathbf{i}$, $\mathbf{v} = \mathbf{i} - \mathbf{k}$

**25**  Show that in three dimensions any set of four vectors must be linearly dependent.

**26**  Show that if $\vec{OA}$, $\vec{OB}$, and $\vec{OC}$ are directed line segments of **u**, **v**, and **w**, respectively, and if $\{$**u**, **v**, **w**$\}$ is a linearly dependent set, then the three directed line segments lie in one plane.

**27**  Find the equations of the line passing through the point $P(2, 1, -3)$ and parallel to $\mathbf{v} = 3\mathbf{i} - 2\mathbf{j} + 7\mathbf{k}$.

**28**  Find the equations of the line through the point $A(1, -4, 0)$ and perpendicular to any plane determined by $\mathbf{v} = \mathbf{i} + 2\mathbf{j} - \mathbf{k}$ and $\mathbf{w} = 3\mathbf{i} - \mathbf{j} + \mathbf{k}$.

_____ 8 _____

## VECTOR FUNCTIONS IN THE PLANE AND THEIR DERIVATIVES

We shall extend the definition of function to the case where the range may be a vector. In this section we restrict ourselves to vectors in the plane. The more general situation in which the range consists of vectors in three-space is discussed in Section 11.

We consider the collection of all vectors in the plane and we denote this set by $V^2$. That is, any vector in the plane is a member of $V^2$.

**DEFINITION**    *A function $\mathbf{f}$ with domain contained in $R^1$ and range contained in $V^2$ is called a **vector function**. That is, for some set of numbers $S$ in $R^1$, and each $t \in S$ there is exactly one $\mathbf{v} \in V^2$ such that $\mathbf{f}(t) = \mathbf{v}$.*

We see that in terms of mappings, a vector function is a mapping from a set in $R^1$ into a set in $V^2$.

Many of the properties of ordinary functions extend easily to vector functions when suitably interpreted. We shall use boldface letters such as $\mathbf{f}$, $\mathbf{g}$, $\mathbf{v}$, $\mathbf{F}$, $\mathbf{G}$ to represent vector functions. If the dependence on the independent variable is to be indicated, we shall write $\mathbf{f}(t)$, $\mathbf{g}(s)$, $\mathbf{F}(x)$, and so forth, for the function values.

**DEFINITION**    *A vector function $\mathbf{f}$ is continuous at $t = a$ if $\mathbf{f}(a)$ is defined and if for each $\varepsilon > 0$ there is a $\delta > 0$ such that*

$$|\mathbf{f}(t) - \mathbf{f}(a)| < \varepsilon \quad \text{for all } t \text{ such that} \quad 0 < |t - a| < \delta.$$

We note that the form of the definition of continuity for vector functions is identical with that for ordinary functions. (see page 83). We must realize, however, that $\mathbf{f}(t) - \mathbf{f}(a)$ is a *vector*, and that the symbol $|\mathbf{f}(t) - \mathbf{f}(a)|$ stands for the length of a vector, whereas in the case of ordinary functions, $|f(t) - f(a)|$ is the absolute value of a number. In words, the continuity of a vector function asserts that as $t \to a$ the vector $\mathbf{f}(t)$ approaches the vector $\mathbf{f}(a)$, in *both* length and direction. When $\mathbf{f}$ is continuous at $a$, we also write

$$\lim_{t \to a} \mathbf{f}(t) = \mathbf{f}(a).$$

If $\mathbf{i}$ and $\mathbf{j}$ are the customary unit vectors associated with a rectangular coordinate system in the plane, a vector function $\mathbf{f}$ can be written in the form

$$\mathbf{f}(t) = f_1(t)\mathbf{i} + f_2(t)\mathbf{j}$$

where $f_1$ and $f_2$ are functions in the ordinary sense. Statements about vector functions $\mathbf{f}$ may always be interpreted as statements about a pair of functions $(f_1, f_2)$.

**THEOREM 25**    *A function $\mathbf{f}$ is continuous at $t = a$ if and only if $f_1$ and $f_2$ are continuous at $t = a$.*

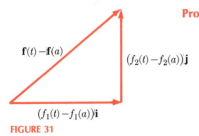

**FIGURE 31**

**Proof**  We have $\mathbf{f}(a) = f_1(a)\mathbf{i} + f_2(a)\mathbf{j}$ and

$$|\mathbf{f}(t) - \mathbf{f}(a)| = |(f_1(t) - f_1(a))\mathbf{i} + (f_2(t) - f_2(a))\mathbf{j}|.$$

Also, (see Fig. 31)

$$|\mathbf{f}(t) - \mathbf{f}(a)| = \sqrt{|f_1(t) - f_1(a)|^2 + |f_2(t) - f_2(a)|^2}.$$

If

$$\lim_{t \to a} f_1(t) = f_1(a) \qquad \text{and} \qquad \lim_{t \to a} f_2(t) = f_2(a),$$

it follows that

$$\lim_{t \to a} \mathbf{f}(t) = \mathbf{f}(a).$$

On the other hand, the inequalities

$$|f_1(t) - f_1(a)| \le |\mathbf{f}(t) - \mathbf{f}(a)|, \qquad |f_2(t) - f_2(a)| \le |\mathbf{f}(t) - \mathbf{f}(a)|$$

show that if

$$\lim_{t \to a} \mathbf{f}(t) = \mathbf{f}(a),$$

then *both*

$$\lim_{t \to a} f_1(t) = f_1(a) \qquad \text{and} \qquad \lim_{t \to a} f_2(t) = f_2(a). \qquad \square$$

**DEFINITION**

*If $\mathbf{f}$ is a vector function, we define the* **derivative $\mathbf{f}'$** *as*

$$\mathbf{f}'(t) = \lim_{h \to 0} \frac{\mathbf{f}(t + h) - \mathbf{f}(t)}{h},$$

*whenever the limit exists.*

While the above definition is exactly analogous to the definition of the derivative of a real-valued function, it is not practical to use the definition to compute the derivative of any specific vector function. Instead we note that if $\mathbf{f}$ is given in terms of functions $f_1$ and $f_2$ by

$$\mathbf{f}(t) = f_1(t)\mathbf{i} + f_2(t)\mathbf{j},$$

then the derivative may be computed by the simple formula

$$\mathbf{f}'(t) = f_1'(t)\mathbf{i} + f_2'(t)\mathbf{j}. \tag{1}$$

The quantities $f_1'(t)$ and $f_2'(t)$ are derivatives in the ordinary sense. Formula (1) follows directly from the definition of derivative and the theorem on the limit of a sum.

**EXAMPLE 1**  Find the derivative $\mathbf{f}'(t)$ if $\mathbf{f}(t) = (t^2 + 2t - 1)\mathbf{i} + (3t^3 - 2)\mathbf{j}$.

**Solution**  $\mathbf{f}'(t) = (2t + 2)\mathbf{i} + 9t^2\mathbf{j}.$ $\qquad \square$

**EXAMPLE 2**  Given that $\mathbf{f}(t) = (\sin t)\mathbf{i} + (3 - 2 \cos t)\mathbf{j}$, find $\mathbf{f}''(t)$.

**Solution** $\mathbf{f}'(t) = \cos t\mathbf{i} + 2 \sin t\mathbf{j}$. Hence $\mathbf{f}''(t) = -\sin t\mathbf{i} + 2 \cos t\mathbf{j}$. □

**EXAMPLE 3** Given $\mathbf{f}(t) = (3t - 2)\mathbf{i} + (2t^2 + 1)\mathbf{j}$. Find the value of

$$\mathbf{f}'(t) \cdot \mathbf{f}''(t).$$

**Solution** $\mathbf{f}'(t) = 3\mathbf{i} + 4t\mathbf{j}$, $\mathbf{f}''(t) = 4\mathbf{j}$, and

$$\mathbf{f}'(t) \cdot \mathbf{f}''(t) = (3\mathbf{i} + 4t\mathbf{j}) \cdot (4\mathbf{j}) = 16t.$$ □

The derivative of a vector function has a simple geometric interpretation in terms of directed line segments. Draw the particular directed line segment equivalent to $\mathbf{f}(t)$ which has its base at the origin of the coordinate system. Then the head of this line segment will trace out a curve $C$ as $t$ takes on all possible values in its domain (Fig. 32). The directed line segment $\overrightarrow{OP}$ represents $\mathbf{f}(t)$. Let $\overrightarrow{OQ}$ represent $\mathbf{f}(t + h)$. Then $\mathbf{f}(t + h) - \mathbf{f}(t)$ is the directed line segment $\overrightarrow{PQ}$. Multiplying $\mathbf{f}(t + h) - \mathbf{f}(t)$ by $1/h$ gives a vector in the direction of $\overrightarrow{PQ}$, but $1/h$ times as long. As $h$ tends to zero, the quantity

$$\frac{\mathbf{f}(t + h) - \mathbf{f}(t)}{h}$$

tends to a vector tangent to the curve $C$ at the point $P$. We conclude that $\mathbf{f}'(t)$ *is the vector tangent to the curve* $\mathbf{f}(t)$.

We can use these facts to find the equation of a line tangent to a given curve at a point. We illustrate this with an example.

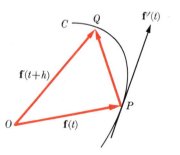

**FIGURE 32**

**EXAMPLE 4** Let $C$ be a curve described by the parametric equations

$$x(t) = t^2 + 6, \qquad y(t) = t^3 - 3t$$

where $t$ is in $R^1$. Find parametric equations for the tangent line to $C$ at the point on the curve corresponding to $t = 3$.

**Solution** A point $P(t)$ on $C$ is given by $P(t) = (t^2 + 6, t^3 - 3t)$, and $P(3) = (15, 18)$. The vector $\mathbf{f}(t) = (t^2 + 6)\mathbf{i} + (t^3 - 3t)\mathbf{j}$ has derivative $\mathbf{f}'(t) = (2t)\mathbf{i} + (3t^2 - 3)\mathbf{j}$, and the tangent vector at $t = 3$ is $\mathbf{f}'(3) = 6\mathbf{i} + 24\mathbf{j}$. Using the parametric equations for a line passing through $(15, 18)$ in the direction $6\mathbf{i} + 24\mathbf{j}$ yields

$$x(t) = 15 + 6t$$

$$y(t) = 18 + 24t$$

as the equations for the tangent line to $C$ at the point $P(15, 18)$.    □

## 8   PROBLEMS

In Problems 1 through 6, calculate $\mathbf{f}'(t)$ and $\mathbf{f}''(t)$.

**1**   $\mathbf{f}(t) = (t^2 + 1)\mathbf{i} + (t^3 - 3t)\mathbf{j}$

**2**   $\mathbf{f}(t) = \left(\dfrac{t-3}{t+1}\right)\mathbf{i} - \dfrac{t^2+1}{t^2+t+1}\mathbf{j}$   **3**   $\mathbf{f}(t) = (\tan 3t)\mathbf{i} + (\cos \pi t)\mathbf{j}$

**4**   $\mathbf{f}(t) = \dfrac{1}{t}\mathbf{i} - \dfrac{1}{t^2+1}\mathbf{j}$      **5**   $\mathbf{f}(t) = e^{2t}\mathbf{i} + e^{-2t}\mathbf{j}$

**6**   $\mathbf{f}(t) = (e^t + e^{-t})\mathbf{i} + (e^t - e^{-t})\mathbf{j}$

**7**   Find $\mathbf{f}(t) \cdot \mathbf{f}'(t)$ if $\mathbf{f}(t) = \dfrac{t^2+1}{t^2+2}\mathbf{i} + 2t\mathbf{j}$.

**8**   Find $\dfrac{d}{dt}(\mathbf{f}(t) \cdot \mathbf{f}'(t))$ if $\mathbf{f}(t) = 2t\mathbf{i} + \dfrac{1}{t+1}\mathbf{j}$.

**9**   Find $\dfrac{d}{dt}|\mathbf{f}(t)|$ if $\mathbf{f}(t) = \sin 2t\mathbf{i} + \cos 3t\mathbf{j}$.

**10**   Find $\dfrac{d}{dt}|\mathbf{f}(t)|$ if $\mathbf{f}(t) = (t^2 + 1)\mathbf{i} + (3 + 2t^2)\mathbf{j}$.

**11**   Find $\dfrac{d}{dt}|\mathbf{f}(t)|$ if $\mathbf{f}(t) = \cos \dfrac{1}{t}\mathbf{i} + \sin \dfrac{1}{t}\mathbf{j}$.

**12**   Find $\dfrac{d}{dt}(\mathbf{f}'(t) \cdot \mathbf{f}''(t))$ if $\mathbf{f}(t) = (3t + 1)\mathbf{i} + (2t^2 - t^3)\mathbf{j}$.

**13**   Find $\dfrac{d}{dt}(\mathbf{f}'(t) \cdot \mathbf{f}''(t))$ if $\mathbf{f}(t) = (\log t)\mathbf{i} + \dfrac{2}{t}\mathbf{j}$.

**14**   Find $\dfrac{d}{dt}(\mathbf{f}''(t) \cdot \mathbf{f}'''(t))$ if $\mathbf{f}(t) = e^{3t}\mathbf{i} + e^{-3t}\mathbf{j}$.

In each of Problems 15 through 20, find parametric equations for the line tangent to $C$ at the point $P$.

**15**   $x = 2t^3 - 4, \ y = t + 3; \ P(-2, 4)$

**16**   $x = \cos t, \ y = t \sin t; \ P\left(\frac{1}{2}, \frac{\pi\sqrt{3}}{6}\right)$

**17**   $x = te^t, \ y = e^t; \ P(0, 1)$

**18**   $x = 1 + \sin t, \ y = \cos t; \ P(1, 1)$

**19**   $x = t^2 + t \cos t, \ y = e^t - t^2; \ P(0, 1)$

**20**   $x = t^3 + t^2 + t, \ y = (t^2 + 1)^2; \ P(-1, 4)$

In each of Problems 21 through 25, find parametric equations for the line tangent to $C$ at the point $P$. The curve $C$ is described in vector notation.

**21**   $\mathbf{f}(t) = t^2\mathbf{i} + 2t\mathbf{j}; \ P(1, 2)$

**22**   $\mathbf{f}(t) = (t^3 - 1)\mathbf{i} - 4t\mathbf{j}; \ P(0, -4)$

**23**   $\mathbf{f}(t) = e^t(\mathbf{i} - \mathbf{j}); \ P(1, -1)$

**24**   $\mathbf{f}(t) = (\sin t)\mathbf{i} + (\tan t)\mathbf{j}; \ P\left(\dfrac{\sqrt{2}}{2}, 1\right)$

**25**   $\mathbf{f}(t) = \sqrt{t^2 - 1}\mathbf{i} + t^3\mathbf{j}; \ P(\sqrt{3}, 8)$

**26**   Given that $\mathbf{f}(t) = (2t + 1)\mathbf{i} + 3t\mathbf{j}, \mathbf{g}(t) = 4t\mathbf{j}$. Find $d\theta/dt$, where $\theta = \theta(t)$ is the angle between $\mathbf{f}$ and $\mathbf{g}$.

**27**   Write out a proof establishing Formula (1) on page 599.

**28**   Prove the formula

$$\dfrac{d}{dt}(F(t)\mathbf{f}(t)) = F(t)\mathbf{f}'(t) + F'(t)\mathbf{f}(t).$$

**29**   Show that the following Chain Rule holds:

$$\dfrac{d}{dt}\{\mathbf{f}[g(t)]\} = \mathbf{f}'[g(t)]g'(t).$$

**30**   Given the vector $\mathbf{f}(t) = (2t + 1)\mathbf{i} + 2t\mathbf{j}$. Describe the curve traced out by the head of the directed line segment which has its base at the origin.

**31**   Given the vector $\mathbf{g}(t) = \cos t\mathbf{i} + \sin t\mathbf{j}$. Describe the curve traced out by the head of the directed line segment which has its base at the origin.

**32**   Prove that if $\mathbf{f}(t) = \sin 2t\mathbf{i} + \cos 2t\mathbf{j}$, then $\mathbf{f}(t) \cdot \mathbf{f}'(t) = 0$. What is the geometric interpretation of this result?

**33**   Prove that if

$$\mathbf{f}(t) = \dfrac{\mathbf{g}(t)}{h(t)}, \quad \text{then} \quad \mathbf{f}'(t) = \dfrac{h(t)\mathbf{g}'(t) - \mathbf{g}(t)h'(t)}{h^2(t)}$$

**34**   Show that if

$$\mathbf{f}(t) = f_1(t)\mathbf{i} + f_2(t)\mathbf{j}, \quad \text{then} \quad \dfrac{d}{dt}(\mathbf{f}(t) \cdot \mathbf{f}(t)) = 2\mathbf{f}(t) \cdot \mathbf{f}'(t).$$

**35**   Show that

$$\dfrac{d}{dt}(\mathbf{f}(t) \cdot \mathbf{g}(t)) = \mathbf{f}(t) \cdot \mathbf{g}'(t) + \mathbf{f}'(t) \cdot \mathbf{g}(t).$$

(*Hint:* Write $\mathbf{f}(t) = f_1(t)\mathbf{i} + f_2(t)\mathbf{j}$ and $\mathbf{g}(t) = g_1(t)\mathbf{i} + g_2(t)\mathbf{j}$.)

---

### 9

## APPLICATION OF VECTORS TO MOTION IN THE PLANE

The vector function

$$\mathbf{f}(t) = x(t)\mathbf{i} + y(t)\mathbf{j}$$

is equivalent to the pair of parametric equations

$$x = x(t), \qquad y = y(t),$$

since the head of the directed line segment with base at the origin, which represents $\mathbf{f}$, traces out a curve which is identical to the curve given in parametric form by these equations.

We recall (Chapter 12, Section 2, page 499) that the arc length $s(t)$ of a curve in parametric form satisfies the relation

$$\frac{ds}{dt} = \sqrt{(dx/dt)^2 + (dy/dt)^2},$$

and therefore

$$\frac{ds}{dt} = |\mathbf{f}'(t)|.$$

In our earlier study of the motion of a particle (Chapter 2, Section 5), we were concerned primarily with motion along a straight line. Now we shall take up the motion of a particle along a curve $C$ in the plane given by the parametric equations

$$C: x = x(t), \qquad y = y(t).$$

Letting $t$ denote the time, we define the **velocity vector v** at the time $t$ as

$$\mathbf{v}(t) \equiv \frac{d\mathbf{f}}{dt} \equiv \mathbf{f}'(t) = x'(t)\mathbf{i} + y'(t)\mathbf{j}.$$

According to the geometric interpretation of the derivative of a vector function given in the preceding section, the velocity vector is always tangent to the path describing the motion. We define the **speed** of the particle to be *the magnitude of the velocity vector*. The speed is

$$|\mathbf{v}(t)| = |\mathbf{f}'(t)| = \sqrt{(dx/dt)^2 + (dy/dt)^2},$$

which tells us that the speed is identical to the quantity $ds/dt$; in other words, the speed measures the rate of change in arc length $s$ with respect to time $t$.

The **acceleration vector a**$(t)$ is defined as the derivative of the velocity vector, or

$$\mathbf{a}(t) = \mathbf{v}'(t) = \mathbf{f}''(t).$$

**EXAMPLE 1**    Suppose that a particle $P$ moves according to the law

$$\mathbf{f}(t) = (3 \cos 2t)\mathbf{i} + (3 \sin 2t)\mathbf{j}.$$

Find $\mathbf{v}(t)$, $\mathbf{a}(t)$, $s'(t)$, $s''(t)$, $|\mathbf{a}(t)|$, and $\mathbf{v}(t) \cdot \mathbf{a}(t)$.

**Solution**    We have

$$\mathbf{v}(t) = \mathbf{f}'(t) = (-6 \sin 2t)\mathbf{i} + (6 \cos 2t)\mathbf{j},$$

$$\mathbf{a}(t) = (-12 \cos 2t)\mathbf{i} - (12 \sin 2t)\mathbf{j},$$

$$s'(t) = |\mathbf{f}'(t)| = [(-6 \sin 2t)^2 + (6 \cos 2t)^2]^{1/2} = 6,$$

$$s''(t) = 0,$$

$$|\mathbf{a}(t)| = [(-12 \cos 2t)^2 + (12 \sin 2t)^2]^{1/2} = 12,$$

$$\mathbf{v}(t) \cdot \mathbf{a}(t) = 0. \qquad \square$$

*Remark.*    We note that $P$ is moving around a circle with center at $O$ and radius 3, with a constant speed but a changing velocity vector! Since

$$\mathbf{v}(t) \cdot \mathbf{a}(t) = 0,$$

and since $\mathbf{v}(t)$ is always tangent to the circle, we conclude that the acceleration vector is always pointing toward the center of the circle.

**EXAMPLE 2**    Suppose that a particle $P$ moves according to the law

$$x(t) = t \cos t, \qquad y(t) = t \sin t,$$

or, equivalently,

$$\mathbf{f}(t) = (t \cos t)\mathbf{i} + (t \sin t)\mathbf{j}.$$

Find $\mathbf{v}(t)$, $\mathbf{a}(t)$, $s'(t)$, $s''(t)$, and $|\mathbf{a}(t)|$.

**Solution**

$$\mathbf{v}(t) = (-t \sin t + \cos t)\mathbf{i} + (t \cos t + \sin t)\mathbf{j},$$

$$\mathbf{a}(t) = (-t \cos t - 2 \sin t)\mathbf{i} + (-t \sin t + 2 \cos t)\mathbf{j}.$$

Therefore

$$s'(t) = |\mathbf{v}(t)| = \sqrt{1 + t^2},$$

$$s''(t) = \tfrac{1}{2}(1 + t^2)^{-1/2}(2t) = \frac{t}{\sqrt{1 + t^2}},$$

$$|\mathbf{a}(t)| = \sqrt{4 + t^2}. \qquad \square$$

# 9    PROBLEMS

In Problems 1 through 16, assume that a particle $P$ moves according to the given law, $t$ denoting the time. Compute $\mathbf{v}(t)$, $|\mathbf{v}(t)|$, $\mathbf{a}(t)$, $|\mathbf{a}(t)|$, $s'(t)$, and $s''(t)$.

1    $\mathbf{f}(t) = t^2\mathbf{i} - 3t\mathbf{j}$

2    $\mathbf{f}(t) = \tfrac{1}{2}t^2\mathbf{i} + \tfrac{1}{3}t^3\mathbf{j}$

3    $\mathbf{f}(t) = 3t\mathbf{i} + (1 - t^{-1})\mathbf{j}$

4    $\mathbf{f}(t) = (t - \tfrac{1}{2}t^2)\mathbf{i} + 2t^{1/2}\mathbf{j}$

5    $\mathbf{f}(t) = (\log \sec t)\mathbf{i} + t\mathbf{j}$

6    $\mathbf{f}(t) = (3 \cos t)\mathbf{i} + (2 \sin t)\mathbf{j}$

7    $\mathbf{f}(t) = (t^2 + 3)\mathbf{i} + (-t^3 + 3t + 1)\mathbf{j}$

8    $\mathbf{f}(t) = 2t\mathbf{i} + \sin t\mathbf{j}$

9    $\mathbf{f}(t) = te^t\mathbf{i} + (t - e^t)\mathbf{j}$

**10** $\mathbf{f}(t) = (t^2 + 1)^2\mathbf{i} + (t^2 + 2)\mathbf{j}$

**11** $x = 2e^t,\ y = 3e^{-t}$

**12** $x = e^{-t}\cos t,\ y = e^{-t}\sin t$

**13** $x = t,\ y = t^2$

**14** $x = \sqrt{t},\ y = t^3 + 2$

**15** $x = t^2 + 2,\ y = -t^3 + 3t^2$

**16** $x = t^2 + 1,\ y = \sqrt{t} + 3$

In Problems 17 through 25, assume that a particle $P$ moves according to the given law. Find $\mathbf{v}(t)$, $\mathbf{a}(t)$, $s'(t)$, and $s''(t)$ at the given time $t$.

**17** $x = t^2 - t - 1,\ y = t^2 - 2t,\ t = 1$

**18** $\mathbf{f}(t) = (5\cos t)\mathbf{i} + (4\sin t)\mathbf{j},\ t = 2\,\pi/3$

**19** $x = 3\sec t,\ y = 2\tan t,\ t = \pi/6$

**20** $\mathbf{f}(t) = 4(\pi t - \sin \pi t)\mathbf{i} + 4(1 - \cos \pi t)\mathbf{j},\ t = \frac{3}{4}$

**21** $x = \log(1 + t),\ y = 3/t,\ t = 2$

**22** $x = t^2 + 2,\ y = t^3 + t^2 + 2,\ t = 1$

**23** $x = e^t,\ y = te^t,\ t = 0$

**24** $x = \sqrt{t},\ y = t^2 + 1,\ t = 4$

**25** $x = \log t,\ y = e^{-t},\ t = 1$

**26** Suppose that $P$ moves according to the law

$$\mathbf{f}(t) = (R\cos wt)\mathbf{i} + (R\sin wt)\mathbf{j},$$

where $R > 0$ and $w$ are constants. Find $\mathbf{v}(t)$, $\mathbf{a}(t)$, $s'(t)$, and $s''(t)$. Show that

$$|\mathbf{a}(t)| = \frac{1}{R}|\mathbf{v}(t)|^2.$$

**27** Let $\mathbf{T}(t)$ be a vector one unit long and parallel to the *velocity vector*. Show that

$$\mathbf{T}(t) = \frac{\mathbf{v}(t)}{ds/dt} = \frac{x'(t)}{s'(t)}\mathbf{i} + \frac{y'(t)}{s'(t)}\mathbf{j}.$$

**28** Using the formula of Problem 27, compute $\mathbf{T}(t)$ if the law of motion is

$$\mathbf{f}(t) = (3t - 1)\mathbf{i} + (t^2 + 2)\mathbf{j}.$$

**29** Using the formula of Problem 27, compute $\mathbf{T}(t)$ if the law of motion is

$$\mathbf{f}(t) = (4e^{2t})\mathbf{i} + (3e^{-2t})\mathbf{j}.$$

**30** A formula for $\mathbf{T}(t)$ is given in Problem 27. Compute $\mathbf{T}'(t)$ and show that $\mathbf{T}(t) \cdot \mathbf{T}'(t) = 0$. (*Hint:* Use the fact that $x'^2 + y'^2 = s'^2$.)

---

## 10

# NEWTON'S LAWS OF MOTION. DIFFERENTIAL EQUATIONS

An equation which contains derivatives is called a **differential equation.** In the preceding section we defined velocity and acceleration in terms of derivatives. We say that any equation involving either of these quantities is a differential equation. In this section we shall present some applications of integration which are related to physical concepts and the differential equations connecting them. We begin by stating Newton's first two laws (axioms) of motion.

**FIRST LAW**   *A body at rest remains at rest and a body in motion moves in a straight line with unchanging velocity, unless some external force acts on it.*

**SECOND LAW**   *The rate of change of the momentum of a body is proportional to the resultant external force that acts on the body.*

We shall study the behavior of objects which move along curved paths, with the restriction that *the motion lies in a plane.* The velocity will be a vector quantity which we denote by **v**. If $m$ is the mass, the **momentum vector** is defined as the vector $m\mathbf{v}$. If the mass is constant during the motion—the only type of case we shall consider—Newton's second law becomes

**SECOND LAW**

$$m\frac{d\mathbf{v}}{dt} = m\mathbf{a} = k\mathbf{F},$$

where **a** is the acceleration vector, **F** is the force vector, and $k$ is a proportionality constant that depends on the units used. A unit of force, called a **newton**, is the force exerted by gravity on a body with mass $1/g$ kilograms, where the number $g$ is the acceleration due to gravity. The customary value used for $g$ is 9.80 meters/sec². For motion near the surface of the earth the constant $k$ in Newton's Second Law is the number $g$.

If a rectangular coordinate system $(x, y)$ is introduced, the vector equation $m\mathbf{a} = k\mathbf{F}$ becomes

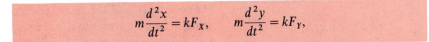

$$m\frac{d^2x}{dt^2} = kF_X, \qquad m\frac{d^2y}{dt^2} = kF_Y,$$

where $d^2x/dt^2$, $d^2y/dt^2$ are the components of **a** along the $x$ and $y$ axes, and $F_X$ and $F_Y$ are the components of **F** along these axes.

**EXAMPLE 1**   A ball is shot vertically upward with a speed of 100 m/sec from a point 40 meters above level ground. Express its height above the ground as a function of time; neglect air resistance.

**Solution**   According to Newton's law, the motion will take place in a vertical line. Let $m$ be the mass of the ball. The only force acting on it is that of gravity, which acts directly downward, so that

$$F_Y = -m; \qquad F_X = 0.$$

Then we have

$$m\frac{d^2y}{dt^2} = -9.8m,$$

with the auxiliary conditions that

$$y = 40 \text{ when } t = 0; \qquad \frac{dy}{dt} = 100 \text{ when } t = 0.$$

Integrating the differential equation $d^2y/dt^2 = -9.8$, we obtain

$$\frac{dy}{dt} = -9.8t + C.$$

Inserting the auxiliary condition for the initial velocity, we find that

$$100 = -9.8(0) + C \qquad \text{or} \qquad C = 100.$$

The differential equation $dy/dt = -9.8t + 100$ may be integrated once more to give

$$y = -4.9t^2 + 100t + k.$$

The auxiliary condition for the initial position of the ball is now used to determine that $k = 40$. We finally obtain

$$y = -4.9t^2 + 100t + 40.$$

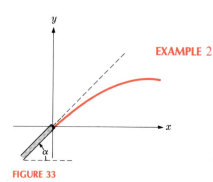

FIGURE 33

**EXAMPLE 2**  A projectile is fired from a gun with a velocity of $v_0$ meters/sec. The barrel of the gun is inclined at an angle $\alpha$ from the horizontal, as in Fig. 33. Assuming that there is no air resistance and that the motion of the projectile is in the vertical plane through the barrel of the gun, show that the equations of motion are

$$x = (v_0 \cos \alpha)t, \qquad y = (v_0 \sin \alpha)t - \tfrac{1}{2}gt^2,$$

where we are using rectangular coordinates with the origin chosen to be at the muzzle of the gun, $y$ axis vertical and $x$ axis horizontal.

**Solution**  The components of the velocity vector in the $x$ and $y$ directions at time $t = 0$ (**initial velocity vector**) are $v_0 \cos \alpha$ and $v_0 \sin \alpha$, respectively. By hypothesis, the only force acting is that of gravity. Therefore $F_X = 0$, $F_Y = -mg$. The differential equations governing the motion are

$$\frac{d^2x}{dt^2} = 0, \qquad \frac{d^2y}{dt^2} = -g.$$

The auxiliary conditions are

$$x = 0 \quad \text{and} \quad y = 0 \quad \text{when } t = 0;$$

also,

$$\frac{dx}{dt} = v_0 \cos \alpha \quad \text{and} \quad \frac{dy}{dt} = v_0 \sin \alpha \quad \text{when } t = 0.$$

Integrating the differential equations, we find that

$$\frac{dx}{dt} = C_1, \qquad \frac{dy}{dt} = -gt + C_2.$$

The auxiliary conditions yield $C_1 = v_0 \cos \alpha$, $C_2 = v_0 \sin \alpha$. Integrating once again, we obtain

$$x = (v_0 \cos \alpha)t + C_3, \qquad y = -\tfrac{1}{2}gt^2 + (v_0 \sin \alpha)t + C_4.$$

Since $x = y = 0$ when $t = 0$, the constants $C_3$ and $C_4$ are both zero.

A large class of problems in physics, chemistry, biology, economics, etc., involves a differential equation of the form

$$\frac{dy}{dt} = ky,$$

in which $k$ is a constant and $y$ is a quantity which is a *positive* function of the time $t$. The above differential equation expresses the fact that *the rate at which y changes is proportional to y itself*. To solve this equation, we write

$$\frac{dy}{y} = k \ dt$$

and integrate to obtain

$$\ln y = kt + C.$$

Since the natural logarithm and the exponential functions are inverses, we can use the definition of natural logarithm to get

$$y = e^{kt + C}.$$

From the law of exponents, $e^{kt+C} = e^{kt} \cdot e^C$ and, writing $A$ for the (positive) constant $e^C$, we obtain the more convenient form

$$y = Ae^{kt}.$$

The law is completely determined once the constants $A$ and $k$ are known. If $k$ is positive, the law is one of **exponential growth**. If $k$ is negative, the law is that of **exponential decay**. Applications of this law are illustrated in the following three examples.

**EXAMPLE 3**     In a favorable environment the number of bacteria increases at a rate proportional to the number present. If 1,000,000 bacteria are present at a certain time and 2,000,000 are present an hour later, find the number present four hours later.

**Solution**     Let $y =$ number of bacteria present at time $t$. Then the growth law states that

$$\frac{dy}{dt} = ky;$$

integrating this equation, we find that

$$y = Ae^{kt}.$$

Letting $t = 0$ correspond to the time when 1,000,000 bacteria are present, we see that $1{,}000{,}000 = Ae^{k \cdot 0}$, and therefore $A = 1{,}000{,}000$. Further, at $t = 1$, $y = 2{,}000{,}000$, and so

$$2{,}000{,}000 = 1{,}000{,}000 e^{k \cdot 1} \qquad \text{or} \qquad 2 = e^k.$$

Taking logarithms, we find that $\ln 2 = k$. The equation for $y$ is

$$y = 1{,}000{,}000 e^{t \ln 2}.$$

Since

$$t \ln 2 = \ln(2^t) \qquad \text{and} \qquad e^{\ln(2^t)} = 2^t,$$

the equation for $y$ may be written

$$y = 1{,}000{,}000 \cdot 2^t.$$

Letting $t = 4$, we find that $y = 16{,}000{,}000$.                    □

**EXAMPLE 4**  A radioactive substance decays at a rate proportional to the amount present. If one gram of a radioactive substance reduces to $\frac{1}{4}$ gram in four hours, find how long it will be until $\frac{1}{10}$ gram remains.

**Solution**  Letting $y$ = amount of substance remaining at time $t$, and using our knowledge of the differential equation $dy/dt = ky$, we write

$$y = Ae^{kt}.$$

The conditions $y = 1$ when $t = 0$ and $y = \frac{1}{4}$ when $t = 4$ yield

$$A = 1 \qquad \text{and} \qquad \tfrac{1}{4} = e^{4k}.$$

Therefore $k = \frac{1}{4} \ln \frac{1}{4} = -\frac{1}{4} \ln 4$, and we find $y = e^{-(1/4)t \ln 4}$. Since $-\frac{1}{4}t \ln 4 = \ln(4^{-t/4})$ and $e^{\ln 4^{-t/4}} = 4^{-t/4}$, we obtain $y$ in the form

$$y = 4^{-t/4}.$$

We wish to find $t$ when $y = 0.1$. Upon taking logarithms, we find that $\ln(0.1) = -(t \ln 4)/4$, and that

$$t = -\frac{4 \ln 0.1}{\ln 4} = \frac{4 \ln 10}{\ln 4} = 6.65 \text{ hours, approximately.} \qquad \square$$

**EXAMPLE 5**  A tank initially contains 600 liters of brine in which 450 grams of salt are dissolved. Pure water is run into the tank at the rate of 15 liters/min, and the mixture, kept uniform by stirring, is withdrawn at the same rate. How many grams of salt remain after 20 min?

**Solution**  Let $y$ be the number of grams of salt remaining after $t$ minutes. At that instant there are $y/600$ grams of salt per liter in the mixture. Therefore at that instant $y$ is decreasing at the rate of $15(y/600)$ gm/min. The corresponding differential equation is

$$\frac{dy}{dt} = -\frac{15y}{600} = -\frac{1}{40}y.$$

Integrating this equation, we have

$$y = Ae^{-t/40}.$$

Since $y = 450$ when $t = 0$, the resulting expression for $y$ at time $t$ is $y = 450e^{-t/40}$, and when $t = 20$, $y = 450(e^{-1/2}) = 273$ grams, approximately.  $\square$

## 10  PROBLEMS

**1**  A ball is shot upward with a speed of 150 m/sec from a point 80 meters above level ground. Express its height above the ground as a function of time. What is the highest point it reaches? Neglect air resistance.

**2**  A ball is dropped from a balloon which is stationary at an altitude of 1000 meters. How long does it take for the ball to reach the ground? Neglect air resistance.

In Problems 3 through 7, assume that a particle moves along the $x$ axis with the given value of its acceleration $a_x$ (which may be positive or negative). Find $x$ and $dx/dt = v$ in terms of $t$, given the stated auxiliary conditions.

**3**   $a_x = 2$; $v = 12$ and $x = 1$ when $t = 0$

**4**   $a_x = -t$; $v = 12$ and $x = -2$ when $t = 0$

**5**   $a_x = 2v$; $v = 5$ and $x = 1$ when $t = 0$

**6**   $a_x = -kv$; $v = 10$ and $x = 0$ when $t = 0$

**7**   $a_x = -kv^2$; $v = v_0$ and $x = x_0$ when $t = 0$

**8**   An automobile is traveling in a straight line at a speed of $v_0$ m/sec. Suddenly the driver applies the brakes and the car stops in $T$ sec after traveling $S$ meters. Assuming that the brakes produce a constant negative acceleration $-k$, find formulas for $S$ and $T$ in terms of $v_0$ and $k$.

**9**   A projectile is fired at an angle of $60°$ with the horizontal and with an initial velocity of 1500 m/sec. Determine how far the projectile travels and the length of time it takes to strike the ground. Neglect air resistance.

**10**   An airplane flying horizontally over level ground at an altitude of 3000 meters drops a projectile. If the plane is traveling 1000 km/hr find the equation of the trajectory in terms of the time $t$. What is the horizontal distance traveled by the projectile? Neglect air resistance.

**11**   An airplane is climbing at an angle of $30°$ with the horizontal at a speed of 750 km/hr. At the instant a projectile is dropped, the plane is 6000 meters above level ground. How long does it take for the projectile to strike the ground? Neglect air resistance.

**12**   A baseball is dropped from a window and hits the ground exactly 4 seconds later. How high is the window? Neglect air resistance.

**13**   A crystalline chemical present in a solution is such that crystals adhere to it at a rate proportional to the amount present. If there are 2 gm initially and 5 gm one hour later, find the amount of crystalline material present at any time $t$. What is the approximate amount after three hours?

**14**   The number of bacteria in a certain culture grows at a rate which is exactly equal to $\frac{1}{2}$ the number present. If there are 10,000 bacteria initially, find the number at any time $t$.

**15**   The *half-life* of a radioactive substance (see Example 4) is the time it takes for the original amount of material to reduce to one-half that amount. The half-life of radium is 1690 yrs. Find the approximate time it would take for 1 gm of radium to reduce to 0.1 gm.

**16**   Rework Problem 15, given that the radioactive substance has a half-life of 3.5 min.

**17**   If the half-life of a radioactive substance is 1 week, how long does it take to arrive at the point where only 1% of the original amount remains? (See Problem 15.)

**18**   Rework Example 5, given that the tank initially contains 800 liters of brine in which 400 grams of salt are dissolved, and pure water is run in at the rate of 20 liters/min, the solution being drawn off at the same rate.

**19**   Rework Example 5, given that the tank initially contains the mixture of the preceding problem, but a brine of 0.2 grams of salt per liter is run in at 20 liters/min, the solution being drawn off at the same rate.

**20**   Air is pumped into a balloon at the rate of $100\sqrt{t}$ cubic centimeters per minute at $t$ minutes after beginning the inflation. The balloon breaks at 4050 cc. How long does it take to break the balloon?

**21**   Assume that the rate at which a body cools is proportional to the difference between its temperature and that of the surrounding air. A body originally at $40°C$ cools to $30°C$ in 10 min in air at $20°C$. Find an expression for the temperature of the body at any time $t$.

**22**   A body cools at a rate (in minutes) which is exactly equal to $\frac{1}{3}$ the difference between its temperature and the temperature of the surrounding air. If the body is originally at $50°C$ and the air is constantly at $20°C$, find the temperature of the body after 20 min. How long will it take (approximately) for the body to reach a temperature of $21°C$?

**23**   A department store is having a big sale. The manager approximates the rate of sales by $At - Bt^2$ dollars per hour, $t$ hours after the store opens. If the store is open 12 hours, calculate the revenue, in terms of $A$ and $B$.

**24**   Carbon 14, a radioactive substance, has a half-life of approximately 5600 years. It is useful in geology since Carbon 14 in a living body begins to decay only after the death of its host. Therefore, in archaeological finds, an examination of the decayed Carbon 14 yields information about when death occurred. Find how long ago an animal died if its present Carbon 14 concentration is one third that of a similar animal living now.

**25**   The growth of tissue in a body is a linear function of the amount of tissue present. That is, the amount $y$ satisfies a differential equation of the form

$$\frac{dy}{dt} = k_1 y + k_2.$$

If $t$ is measured in days, $y$ in grams, then find the amount of tissue after ten days if there are three grams initially and $k_1 = 1.2$, $k_2 = 3$.

**26**   An example of exponential growth occurs when a bank offers to compound interest "continuously." (An offer to compound it "daily" is practically the same.) Find what interest rate is required if a sum of money compounded continuously is to double in twelve years (see page 267).

### 11

## VECTOR FUNCTIONS IN SPACE. DERIVATIVES. TANGENTS. ARC LENGTH

The collection of all vectors in three-space is denoted by $V^3$. We now extend the definition of vector function given in Section 8 (page 598) to the case where the range is an element of $V^3$.

**DEFINITION**    *A function* **f** *with domain contained in* $R^1$ *and range contained in* $V^3$ *is called a* **vector function in three-space,** *or simply a* **vector function.** *That is, for some set of numbers S in* $R^1$ *and each* $t \in S$ *there is exactly one* $\mathbf{v} \in V^3$ *such that* $\mathbf{f}(t) = \mathbf{v}$.

The definition of continuity for a vector function in three-space is identical with that given in Section 8 for vector functions in the plane. Part (b) of the next theorem is proved exactly as is Theorem 25 in Section 8. The proofs of the remaining parts are left to the reader.

**THEOREM 26**    *Suppose that*

$$\mathbf{f}(t) = f_1(t)\mathbf{i} + f_2(t)\mathbf{j} + f_3(t)\mathbf{k}$$

*is a vector function and that the vector* $\mathbf{c} = c_1\mathbf{i} + c_2\mathbf{j} + c_3\mathbf{k}$ *is a constant. Then*

a) $\mathbf{f}(t) \to \mathbf{c}$ *as* $t \to a$ *if and only if*

$$f_1(t) \to c_1 \qquad and \qquad f_2(t) \to c_2 \qquad and \qquad f_3(t) \to c_3.$$

b) **f** *is continuous at a if and only if* $f_1, f_2,$ *and* $f_3$ *are.*
c) $\mathbf{f}'(t)$ *exists if and only if* $f_1'(t), f_2'(t),$ *and* $f_3'(t)$ *do.*
d) *We have the formula*

$$\mathbf{f}'(t) = f_1'(t)\mathbf{i} + f_2'(t)\mathbf{j} + f_3'(t)\mathbf{k}.$$

e) *If* $\mathbf{v}(t) = a\mathbf{w}(t),$ *then* $\mathbf{v}'(t) = a\mathbf{w}'(t)$ *where a is a constant.*
f) *If* $\mathbf{v}(t) = c(t)\mathbf{w}(t),$ *then* $\mathbf{v}'(t) = c(t)\mathbf{w}'(t) + c'(t)\mathbf{w}(t).$
g) *If* $\mathbf{v}(t) = \mathbf{w}(t)/c(t),$ *then*

$$\mathbf{v}'(t) = \frac{c(t)\mathbf{w}'(t) - c'(t)\mathbf{w}(t)}{[c(t)]^2}.$$

**EXAMPLE 1**    Given $\mathbf{f}(t) = 3t^2\mathbf{i} - 2t^3\mathbf{j} + (t^2 + 3)\mathbf{k}$, find $\mathbf{f}'(t), \mathbf{f}''(t), \mathbf{f}'''(t)$.

**Solution**
$$\mathbf{f}'(t) = 6t\mathbf{i} - 6t^2\mathbf{j} + 2t\mathbf{k},$$
$$\mathbf{f}''(t) = 6\mathbf{i} - 12t\mathbf{j} + 2\mathbf{k},$$
$$\mathbf{f}'''(t) = -12\mathbf{j}. \qquad \square$$

The next theorem shows how to differentiate functions involving scalar and vector products. The proofs are left to the reader.

**THEOREM 27**    *If $\mathbf{u}(t)$ and $\mathbf{v}(t)$ are differentiable, then the derivative of $f(t) = \mathbf{u}(t) \cdot \mathbf{v}(t)$ is given by the formula*

$$f'(t) = \mathbf{u}(t) \cdot \mathbf{v}'(t) + \mathbf{u}'(t) \cdot \mathbf{v}(t). \tag{1}$$

*The derivative of $\mathbf{w}(t) = \mathbf{u}(t) \times \mathbf{v}(t)$ is given by the formula*

$$\mathbf{w}'(t) = \mathbf{u}(t) \times \mathbf{v}'(t) + \mathbf{u}'(t) \times \mathbf{v}(t). \tag{2}$$

The proofs of Formulas (1) and (2) follow from the corresponding differentiation formulas for ordinary functions. Note that in Formula (2) it is essential to retain the order of the factors in each vector product.

**EXAMPLE 2**    Find the derivative $f'(t)$ and $\mathbf{w}'(t)$ of

$$f(t) = \mathbf{u}(t) \cdot \mathbf{v}(t) \qquad \text{and} \qquad \mathbf{w}(t) = \mathbf{u}(t) \times \mathbf{v}(t)$$

when

$$\mathbf{u}(t) = (t+3)\mathbf{i} + t^2\mathbf{j} + (t^3 - 1)\mathbf{k}$$

and

$$\mathbf{v}(t) = 2t\mathbf{i} + (t^4 - 1)\mathbf{j} + (2t + 3)\mathbf{k}.$$

**Solution**    By Formula (1) we have

$$f'(t) = [(t+3)\mathbf{i} + t^2\mathbf{j} + (t^3 - 1)\mathbf{k}] \cdot (2\mathbf{i} + 4t^3\mathbf{j} + 2\mathbf{k})$$
$$+ (\mathbf{i} + 2t\mathbf{j} + 3t^2\mathbf{k}) \cdot [2t\mathbf{i} + (t^4 - 1)\mathbf{j} + (2t + 3)\mathbf{k}]$$
$$= (2t + 6) + 4t^5 + 2t^3 - 2 + 2t + 2t^5 - 2t + 6t^3 + 9t^2$$
$$= 6t^5 + 8t^3 + 9t^2 + 2t + 4.$$

According to (2), we have

$$\mathbf{w}'(t) = [(t+3)\mathbf{i} + t^2\mathbf{j} + (t^3 - 1)\mathbf{k}] \times (2\mathbf{i} + 4t^3\mathbf{j} + 2\mathbf{k})$$
$$+ (\mathbf{i} + 2t\mathbf{j} + 3t^2\mathbf{k}) \times [2t\mathbf{i} + (t^4 - 1)\mathbf{j} + (2t + 3)\mathbf{k}].$$

Computing the two cross products on the right, we get

$$\mathbf{w}'(t) = (-7t^6 + 4t^3 + 9t^2 + 6t)\mathbf{i} + (8t^3 - 4t - 11)\mathbf{j}$$
$$+ (5t^4 + 12t^3 - 6t^2 - 1)\mathbf{k}. \qquad \square$$

Consider a rectangular coordinate system and a directed line segment from the origin $O$ to a point $P$ in space. As in two dimensions, we denote the vector $\mathbf{v}[\overrightarrow{OP}]$ by $\mathbf{r}$. We define an **arc** $C$ in space in a way completely analogous to that in which an arc in the plane was defined. (See page 238 and 498.) The vector equation

$$\mathbf{r}(t) = x(t)\mathbf{i} + y(t)\mathbf{j} + z(t)\mathbf{k}$$

is considered to be equivalent to the parametric equations

$$x = x(t), \qquad y = y(t), \qquad z = z(t),$$

which represent an arc $C$ in space.

The definition of **length** of an arc $C$ in three-space is identical with its definition in the plane. If we denote the length of an arc by $l(C)$ we get, in a way that is similar to the two-dimensional analysis, the formulas

$$l(C) = \int_a^b \sqrt{[x'(t)]^2 + [y'(t)]^2 + [z'(t)]^2}\, dt \qquad (3)$$

and

$$s'(t) = \sqrt{[x'(t)]^2 + [y'(t)]^2 + [z'(t)]^2}, \qquad (4)$$

where $s(t)$ is an arc length function. We could also write the above Formulas (3) and (4) in the vector form

$$l(C) = \int_a^b |\mathbf{r}'(t)|\, dt, \qquad s'(t) = |\mathbf{r}'(t)|.$$

**DEFINITIONS** *If $\mathbf{r}'(t) \neq \mathbf{0}$, we define the vector $\mathbf{T}(t) = \mathbf{r}'(t)/|\mathbf{r}'(t)|$ as the* **unit tangent vector** *to the path corresponding to the value $t$. The line through the point $P_0$ corresponding to $\mathbf{r}(t_0)$ and parallel to $\mathbf{T}(t_0)$ is called the* **tangent line to the arc** *at $t_0$; the line directed in the same way as $\mathbf{T}(t_0)$ is called the* **directed tangent line** *at $t_0$ (Fig. 34).*

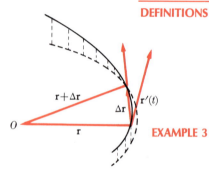

$\mathbf{r} + \Delta\mathbf{r}$

$\Delta\mathbf{r}$  $\mathbf{r}'(t)$

$0$  $\mathbf{r}$

**FIGURE 34**

**EXAMPLE 3**  The graph of the equations

$$x = a \cos t, \qquad y = a \sin t, \qquad z = bt \qquad (5)$$

is called a **helix**.

i) Find $s'(t)$.

ii) Find the length of that part of the helix for which $0 \leq t \leq 2\pi$.

iii) Show that the unit tangent vector makes a constant angle with the $z$ axis.

**Solution**  (See Fig. 35 for the graph.)

i) $x'(t) = -a \sin t$, $y'(t) = a \cos t$, $z'(t) = b$. Therefore,

$$s'(t) = \sqrt{a^2 + b^2}.$$

ii) $l(C) = 2\pi\sqrt{a^2 + b^2}.$

iii) $\mathbf{T}(t) = \dfrac{-y\mathbf{i} + x\mathbf{j} + b\mathbf{k}}{\sqrt{a^2 + b^2}}.$

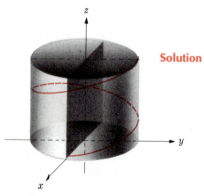

**FIGURE 35**

Letting $\phi$ be the angle between $\mathbf{T}$ and $\mathbf{k}$, we get

$$\cos \phi = b/\sqrt{a^2 + b^2}. \qquad \square$$

*Remark.*  We note that the helix (5) winds around the cylinder $x^2 + y^2 = a^2$.

**DEFINITIONS**   *If t denotes time in the parametric equations of an arc* $\mathbf{r}(t)$, *then* $\mathbf{r}'(t)$ *is the* **velocity vector** $\mathbf{v}(t)$, *and* $\mathbf{r}''(t) = \mathbf{v}'(t)$ *is called the* **acceleration vector**. *The quantity* $s'(t)$ *(a scalar) is called the* **speed** *of a particle moving according to the law*

$$\mathbf{r}(t) = x(t)\mathbf{i} + y(t)\mathbf{j} + z(t)\mathbf{k}.$$

## 11  PROBLEMS

In each of Problems 1 through 11, find the derivatives $\mathbf{f}'(t)$ and $\mathbf{f}''(t)$.

**1** $\mathbf{f}(t) = t^2\mathbf{i} + (t^2 + 1)\mathbf{j} + (2 - 3t)\mathbf{k}$

**2** $\mathbf{f}(t) = (2t^3 + t - 1)\mathbf{i} + (t^{-1} + 1)\mathbf{j} + (t^2 + t^{-2})\mathbf{k}$

**3** $\mathbf{f}(t) = (\cos 2t)\mathbf{i} + (\sin 2t)\mathbf{j} + 2t\mathbf{k}$

**4** $\mathbf{f}(t) = e^{2t}\mathbf{i} + t^2\mathbf{j} + e^{-2t}\mathbf{k}$

**5** $\mathbf{f}(t) = \left(\dfrac{t^2}{t^2 + 1}\right)\mathbf{i} + \dfrac{1}{t^2 + 1}\mathbf{j} + (t^2 + 1)\mathbf{k}$

**6** $\mathbf{f}(t) = (\log 2t)\mathbf{i} + e^{3t}\mathbf{j} + (t \log t)\mathbf{k}$

**7** $\mathbf{f}(t) = (\cos t)\mathbf{i} + (\tan t)\mathbf{j} + (\sin t)\mathbf{k}$

**8** $\mathbf{f}(t) = t^3\mathbf{i} + t^{-3}\mathbf{j} + 3\mathbf{k}$

**9** $\mathbf{f}(t) = (t^2 + 6)\mathbf{i} + (\cos t)\mathbf{j} + (te^{-t})\mathbf{k}$

**10** $\mathbf{f}(t) = (t^3 + t)^2\mathbf{i} + (\cos^{101} t)\mathbf{j} + 3t\mathbf{k}$

**11** $\mathbf{f}(t) = [(t^2 + 1)(t^2 - 2)]\mathbf{i} + (t^3 + 2t^{-3})\mathbf{j} + [\ln(t^2 + 1)]\mathbf{k}$

In each of Problems 12 through 17, find the domain of $\mathbf{f}$ and determine where $\mathbf{f}$ is continuous.

**12** $\mathbf{f}(t) = t^3\mathbf{i} + \tan t\mathbf{j} + 3\mathbf{k}$

**13** $\mathbf{f}(t) = t^{1/3}\mathbf{i} + t^{1/2}\mathbf{j} + \cos t\mathbf{k}$

**14** $\mathbf{f}(t) = e^{-t}\mathbf{i} + (2 + t)^{-1}\mathbf{j} + \left(\dfrac{t^2 - 9}{t - 3}\right)\mathbf{k}$

**\*15** $\mathbf{f}(t) = \dfrac{\sin t}{t}\mathbf{i} + e^t\mathbf{j} + t \sin t\mathbf{k}$

**16** $\mathbf{f}(t) = \ln(1 - t)\mathbf{i} + t^2\mathbf{j} + \sqrt[3]{t}\mathbf{k}$

**17** $\mathbf{f}(t) = (e^t + e^{-t})\mathbf{i} + \left(\dfrac{1}{t}\right)\mathbf{j} + (\cos t)\mathbf{k}$

In Problems 18 through 23, find in each case either $f'(t)$ or $\mathbf{f}'(t)$, whichever is appropriate.

**18** $f(t) = \mathbf{u}(t) \cdot \mathbf{v}(t)$, where

$$\mathbf{u}(t) = 3t\mathbf{i} + 2t^2\mathbf{j} + \frac{1}{t}\mathbf{k}, \quad \mathbf{v}(t) = t^2\mathbf{i} + \frac{1}{t}\mathbf{j} + t^3\mathbf{k}.$$

**19** $f(t) = \mathbf{u}(t) \cdot \mathbf{v}(t)$, where

$$\mathbf{u}(t) = (t^2 + 2)\mathbf{i} + t\mathbf{j} + 3\mathbf{k}, \quad \mathbf{v}(t) = e^t\mathbf{i} + 3\mathbf{j} + t^2\mathbf{k}.$$

**20** $\mathbf{f}(t) = \mathbf{u}(t) \times \mathbf{v}(t)$, where

$$\mathbf{u}(t) = (\cos t)\mathbf{i} + (\sin t)\mathbf{j} + t\mathbf{k}, \quad \mathbf{v}(t) = (\sin t)\mathbf{i} + (\cos t)\mathbf{j} + t^2\mathbf{k}.$$

**21** $\mathbf{f}(t) = \mathbf{u}(t) \times \mathbf{v}(t)$, where

$$\mathbf{u}(t) = t\mathbf{i} + (t^2 + 2)\mathbf{j} + t^{72}\mathbf{k}, \quad \mathbf{v}(t) = (t^2 + 2)\mathbf{i} + 4\mathbf{j} + t\mathbf{k}.$$

**22** $f(t) = \mathbf{u}(t) \cdot [\mathbf{v}(t) \times \mathbf{w}(t)]$, where

$$\mathbf{u}(t) = t\mathbf{i} + (t + 1)\mathbf{k}, \quad \mathbf{v}(t) = t^2\mathbf{j}, \quad \mathbf{w}(t) = \frac{1}{t^2}\mathbf{k}.$$

**23** $\mathbf{f}(t) = \mathbf{u}(t) \cdot [\mathbf{v}(t) \times \mathbf{w}(t)]$, where

$$\mathbf{u}(t) = (\cos t)\mathbf{i} + (\sin t)\mathbf{k}, \quad \mathbf{v}(t) = e^t\mathbf{j}, \quad \mathbf{w}(t) = (\tan t)\mathbf{k}.$$

In Problems 24 through 31, find $l(C)$.

**24** $C: x = t, y = t^2/\sqrt{2}, z = t^3/3; 0 \le t \le 2$

**25** $C: x = t, y = 3t^2/2, z = 3t^3/2; 0 \le t \le 2$

**26** $C: x = t, y = \ln(\sec t + \tan t), z = \ln \sec t; 0 \le t \le \pi/4$

**27** $C: x = t \cos t, y = t \sin t, z = t; 0 \le t \le \pi/2$

**28** $C: x = 5t, y = 4t^2, z = 3t^2; 0 \le t \le 2$

**29** $C: x = t^2, y = t \sin t, z = t \cos t; 0 \le t \le 1$

**30** $C: x = e^t \cos t, y = e^t, z = e^t \sin t; 0 \le t \le \pi$

**31** $C: x = 7t, y = \sqrt{2} \sin 6t, z = \sqrt{2} \sin 6t; 0 \le t \le 2\pi$

In Problems 32 through 34, the parameter $t$ is the time in seconds. Taking $s$ to be the length in meters, in each case find the velocity, speed, and acceleration of a particle moving according to the given law.

**32** $\mathbf{r}(t) = t^2\mathbf{i} + 2t\mathbf{j} + (t^3 - 1)\mathbf{k}$

**33** $\mathbf{r}(t) = (t \sin t)\mathbf{i} + (t \cos t)\mathbf{j} + t\mathbf{k}$

**34** $\mathbf{r}(t) = e^{3t}\mathbf{i} + e^{-3t}\mathbf{j} + te^{3t}\mathbf{k}$

**35** a) Given $\mathbf{f}(t) = (3 \cos t)\mathbf{i} + (4 \cos t)\mathbf{j} + (5 \sin t)\mathbf{k}$, show that

$$\mathbf{f}(t) \cdot \mathbf{f}'(t) = 0$$

for all $t$.

b) Let $\mathbf{f}(t)$ be a vector function such that $|\mathbf{f}(t)| = 1$ for all $t$. Show that $\mathbf{f}(t)$ and $\mathbf{f}'(t)$ are orthogonal for all $t$.

**36** Suppose that $\mathbf{f}(t) = \alpha(t)\mathbf{i} + \beta(t)\mathbf{j} + \gamma(t)\mathbf{k}$ has the properties:

$\mathbf{f}(t) = \mathbf{f}'(t)$ for all $t$ and $\alpha(0) = \beta(0) = \gamma(0) = 2$. Find $\mathbf{f}(t)$.

**37** Show that every differentiable vector function $\mathbf{v}(t)$ satisfies the identity

$$2\mathbf{v}'(t) = \mathbf{i} \times (\mathbf{v}' \times \mathbf{i}) + \mathbf{j} \times (\mathbf{v}' \times \mathbf{j}) + \mathbf{k} \times (\mathbf{v}' \times \mathbf{k}).$$

**38** Write out a complete proof of Theorem 26.

**39** Write out a complete proof of Theorem 27.

---

**12**

## TANGENTIAL AND NORMAL COMPONENTS. THE MOVING TRIHEDRAL (OPTIONAL)

We consider the graph of the equation

$$\mathbf{r} = \mathbf{r}(t) \qquad \Leftrightarrow \qquad \mathbf{r}(t) = x(t)\mathbf{i} + y(t)\mathbf{j} + z(t)\mathbf{k}.$$

In the last section, we defined the *unit tangent vector* $\mathbf{T}(t)$ by the relation

$$\mathbf{T}(t) = \frac{\mathbf{r}'(t)}{|\mathbf{r}'(t)|},$$

which we can also write

$$\mathbf{T}(t) = \frac{\mathbf{r}'(t)}{s'(t)} = \frac{d\mathbf{r}}{ds}.$$

Since $\mathbf{T} \cdot \mathbf{T} = 1$ for all $t$, we differentiate this relation with respect to $t$ to obtain

$$\mathbf{T}(t) \cdot \mathbf{T}'(t) + \mathbf{T}'(t) \cdot \mathbf{T}(t) = 2\mathbf{T}(t) \cdot \mathbf{T}'(t) = 0.$$

Therefore, the vector $\mathbf{T}'(t)$ is orthogonal to $\mathbf{T}(t)$. We define the vector $\kappa(t)$ by the relation

$$\kappa(t) = \frac{d\mathbf{T}}{ds} = \frac{\mathbf{T}'(t)}{s'(t)}.$$

Taking the scalar product of $\kappa$ and $\mathbf{T}$, we see that

$$\kappa(t) \cdot \mathbf{T}(t) = \frac{1}{s'(t)} \mathbf{T}'(t) \cdot \mathbf{T}(t) = 0,$$

and so $\kappa$ is orthogonal to $\mathbf{T}$. The vector $\kappa$ is the vector rate of change of direction of the curve with respect to arc length. Bearing in mind the definition of curvature of a plane curve (see Chapter 12, Section 3), we make the following definitions.

**DEFINITIONS**    *The vector* $\kappa(t)$, *defined above, is called the* **curvature vector*** *of the curve* $\mathbf{r} = \mathbf{r}(t)$. *The magnitude* $|\kappa(t)| = \kappa(t)$ *is called the* **curvature.** *If* $\kappa(t) \neq \mathbf{0}$, *we define the* **principal normal vector** $\mathbf{N}(t)$ *and the* **binormal vector** $\mathbf{B}(t)$ *by the relations*

$$\mathbf{N}(t) = \frac{\kappa(t)}{|\kappa(t)|}, \qquad \mathbf{B}(t) = \mathbf{T}(t) \times \mathbf{N}(t).$$

*The* **center of curvature** $C(t)$ *is defined by the equation*

$$\mathbf{v}[\vec{OC}(t)] = \mathbf{r}(t) + \frac{1}{\kappa(t)} \mathbf{N}(t).$$

*The* **radius of curvature** *is* $R(t) = 1/\kappa(t)$. *The* **osculating plane** *corresponding to the value* $t$ *is the plane which contains the tangent line to the path and the center of curvature at* $t$; *it is defined only for those values of* $t$ *for which* $\kappa(t) \neq 0$.

**EXAMPLE 1**    Given the curve

$$x = t, \qquad y = t^2, \qquad z = 1 + t^2,$$

find $\mathbf{T}(t)$ and $\kappa(t)$.

**Solution**    We may write

$$\mathbf{r}(t) = t\mathbf{i} + t^2\mathbf{j} + (1 + t^2)\mathbf{k}.$$

Therefore

$$\mathbf{r}'(t) = \mathbf{i} + 2t\mathbf{j} + 2t\mathbf{k} \qquad \text{and} \qquad |\mathbf{r}'(t)| = \sqrt{1 + 8t^2} = s'(t).$$

Since $\mathbf{T} = \mathbf{r}'(t)/|\mathbf{r}'(t)|$, we have

$$\mathbf{T}(t) = \frac{1}{\sqrt{1 + 8t^2}} (\mathbf{i} + 2t\mathbf{j} + 2t\mathbf{k}).$$

We differentiate to get

$$\mathbf{T}'(t) = \frac{-8t}{(1 + 8t^2)^{3/2}} (\mathbf{i} + 2t\mathbf{j} + 2t\mathbf{k}) + \frac{1}{\sqrt{1 + 8t^2}} (2\mathbf{j} + 2\mathbf{k}).$$

Hence

$$\kappa(t) = \frac{\mathbf{T}'(t)}{s'(t)} = \frac{-8t}{(1 + 8t^2)^2} (\mathbf{i} + 2t\mathbf{j} + 2t\mathbf{k}) + \frac{1}{1 + 8t^2} (2\mathbf{j} + 2\mathbf{k})$$

$$= \frac{-8t}{(1 + 8t^2)^2} \mathbf{i} + \frac{2}{(1 + 8t^2)^2} \mathbf{j} + \frac{2}{(1 + 8t^2)^2} \mathbf{k}. \qquad \square$$

It is clear from the definitions of $\mathbf{T}$ and $\mathbf{N}$ that they are unit vectors which are orthogonal. From the definition of cross product, we see at once that $\mathbf{B}$ is

---

*The Greek letter kappa is customarily used to denote curvature.

a unit vector orthogonal to both **T** and **N**. The triple $\mathbf{T}(t), \mathbf{N}(t), \mathbf{B}(t)$ forms a mutually orthogonal triple of unit vectors at each point, called the **trihedral** at the point. See Fig. 36 where $\mathbf{T}(t), \mathbf{N}(t)$, and $\mathbf{B}(t)$ are drawn with their bases at the point $P$ on the curve corresponding to the value $t$ of the parameter. (It is assumed that $\kappa(t) \neq 0$.) We may also write the formulas

$$\frac{d\mathbf{T}}{ds} = \kappa\mathbf{N} \quad \text{and} \quad \mathbf{T}'(t) = \kappa(t)s'(t)\mathbf{N}(t).$$

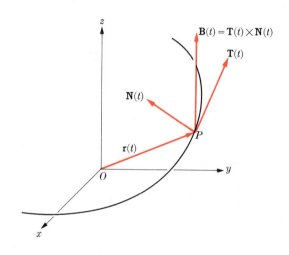

**FIGURE 36**

The equation of the osculating plane at $t = t_0$ is given by

$$b_1(x - x_0) + b_2(y - y_0) + b_3(z - z_0) = 0,$$

where $(x_0, y_0, z_0)$ is the point on the curve corresponding to

$$t = t_0 \quad \text{and} \quad \mathbf{B}(t_0) = b_1\mathbf{i} + b_2\mathbf{j} + b_3\mathbf{k}.$$

**EXAMPLE 2**   Given the helix

$$x = 4 \cos t, \qquad y = 4 \sin t, \qquad z = 2t$$

or, equivalently,

$$\mathbf{r}(t) = (4 \cos t)\mathbf{i} + (4 \sin t)\mathbf{j} + (2t)\mathbf{k},$$

find **T**, **N**, and **B** for $t = 2\pi/3$. Also find the curvature $\kappa$, the equation of the osculating plane, and the equations of the tangent line.

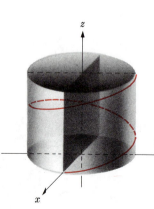

**FIGURE 37**

**Solution**   The curve is shown in Fig. 37. We have

$$x' = -4 \sin t, \qquad y' = 4 \cos t, \qquad z' = 2, \qquad s' = 2\sqrt{5},$$

$$\mathbf{T}(t) = \frac{1}{2\sqrt{5}}(-y\mathbf{i} + x\mathbf{j} + 2\mathbf{k}), \qquad \mathbf{T}'(t) = \frac{1}{2\sqrt{5}}(-x\mathbf{i} - y\mathbf{j}).$$

We compute

$$\kappa(t) = \frac{|\mathbf{T}'(t)|}{s'(t)} = \frac{1}{5}$$

and, using the formula $\mathbf{T}'(t) = \kappa(t)s'(t)\mathbf{N}(t)$, we get

$$\mathbf{N}(t) = -(\cos t)\mathbf{i} - (\sin t)\mathbf{j}.$$

For $t = 2\pi/3$, $x_0 = -2$, $y_0 = 2\sqrt{3}$, $z_0 = 4\pi/3$, we obtain

$$\mathbf{T} = \frac{1}{\sqrt{5}}(-\sqrt{3}\mathbf{i} - \mathbf{j} + \mathbf{k}), \qquad \mathbf{N} = \frac{1}{2}\mathbf{i} - \frac{\sqrt{3}}{2}\mathbf{j},$$

$$\mathbf{B} = \frac{1}{2\sqrt{5}}(\sqrt{3}\mathbf{i} + \mathbf{j} + 4\mathbf{k}), \qquad \kappa\left(\frac{2\pi}{3}\right) = \frac{1}{5}.$$

The tangent line is

$$\frac{x+2}{\sqrt{3}} = \frac{y - 2\sqrt{3}}{1} = \frac{z - 4\pi/3}{-1},$$

and, since $\mathbf{B}$ is perpendicular to the osculating plane, the equation of the osculating plane is

$$\sqrt{3}(x+2) + (y - 2\sqrt{3}) + 4(z - 4\pi/3) = 0. \qquad \square$$

---

**DEFINITIONS**    *If a particle moves according to the law $\mathbf{r} = \mathbf{r}(t)$, the **tangential** and **normal** components of any vector are its components along $\mathbf{T}$ and $\mathbf{N}$, respectively.*

---

**THEOREM 28**    *If a particle moves according to the law $\mathbf{r} = \mathbf{r}(t)$, with $t$ the time and $\mathbf{T}(t)$, $\mathbf{N}(t)$, $R(t)$ defined as on pages 614–615, then the acceleration vector $\mathbf{a}(t) = \mathbf{v}'(t)$ satisfies the equation*

$$\mathbf{a}(t) = s''(t)\mathbf{T}(t) + \frac{|\mathbf{v}(t)|^2}{R(t)}\mathbf{N}(t). \tag{1}$$

---

**Proof**    Since $\mathbf{v}(t) = \mathbf{r}'(t)$, we have, from the definition of $\mathbf{T}(t)$, that

$$\mathbf{v}(t) = s'(t)\mathbf{T}(t).$$

We differentiate this equation and obtain

$$\mathbf{a}(t) = s''(t)\mathbf{T}(t) + s'(t)\mathbf{T}'(t).$$

The relation $\mathbf{T}'(t) = \kappa(t)s'(t)\mathbf{N}(t)$ yields

$$\mathbf{a}(t) = s''(t)\mathbf{T}(t) + \kappa(t)[s'(t)]^2\mathbf{N}(t).$$

We observe that $|\mathbf{v}(t)| = s'(t)$ and $\kappa(t) = 1/R(t)$, and the proof is complete.

$\square$

*Remark.* Since **T**, **N**, and **B** are mutually orthogonal vectors, Formula (1) shows that $s''(t)$ and $|\mathbf{v}(t)|^2/R(t)$ are the tangential and normal components, respectively, of $\mathbf{a}(t)$.

**EXAMPLE 3**  A particle moves according to the law

$$x = t, \qquad y = t^2, \qquad z = t^3.$$

Find (a) its vector velocity, (b) its acceleration, (c) its speed. Also find (d) the unit vector **T**, and (e) the unit vector **N**; find (f) the normal and (g) the tangential components of the acceleration vector. Evaluate all these quantities at $t = 1$.

**Solution**  Since $\mathbf{r}(t) = t\mathbf{i} + t^2\mathbf{j} + t^3\mathbf{k}$, we have

$$\mathbf{v}(t) = \mathbf{r}'(t) = \mathbf{i} + 2t\mathbf{j} + 3t^2\mathbf{k}, \qquad \mathbf{a}(t) = \mathbf{v}'(t) = 2\mathbf{j} + 6t\mathbf{k}.$$

Also,

$$s'(t) = (1 + 4t^2 + 9t^4)^{1/2}; \qquad s''(t) = \frac{4t + 18t^3}{(1 + 4t^2 + 9t^4)^{1/2}}.$$

The path of the particle is sketched in Fig. 38. For $t = 1$,

a)                          $$\mathbf{v} = \mathbf{i} + 2\mathbf{j} + 3\mathbf{k},$$

b)                          $$\mathbf{a} = 2\mathbf{j} + 6\mathbf{k}.$$

c)  The speed is $s'(1) = \sqrt{14}$. We obtain at $t = 1$

$$s''(1) = \frac{22}{\sqrt{14}};$$

d)                  $$\mathbf{T} = \frac{1}{\sqrt{14}}(\mathbf{i} + 2\mathbf{j} + 3\mathbf{k});$$

$$\mathbf{a} - s''\mathbf{T} = \frac{(s')^2}{R}\mathbf{N} = 2\mathbf{j} + 6\mathbf{k} - \frac{11}{7}(\mathbf{i} + 2\mathbf{j} + 3\mathbf{k})$$

$$= \frac{1}{7}(-11\mathbf{i} - 8\mathbf{j} + 9\mathbf{k}).$$

Therefore

$$|\mathbf{a} - s''\mathbf{T}| = \frac{1}{7}\sqrt{266} = \frac{14}{R};$$

e)                  $$\mathbf{N} = \frac{1}{\sqrt{266}}(-11\mathbf{i} - 8\mathbf{j} + 9\mathbf{k});$$

$$R = \frac{98}{\sqrt{266}}.$$

At $t = 1$, the tangential component is $s''(1)$, and so for (g) we find $22/\sqrt{14}$. To find the normal component (f), we compute $|\mathbf{v}(1)|^2/R(1)$ and get the value $\sqrt{266}/7$.  □

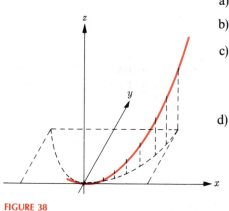

**FIGURE 38**

## 12　PROBLEMS

In each of Problems 1 through 10, find the unit tangent vector **T**(t).

**1** $\mathbf{r}(t) = t^3\mathbf{i} + (1-t)\mathbf{j} + (2t+1)\mathbf{k}$

**2** $\mathbf{r}(t) = (t^2+2)\mathbf{i} + (t^2+2)^2\mathbf{j} + t(t^2+2)^3\mathbf{k}$

**3** $\mathbf{r}(t) = 3t\mathbf{i} + \mathbf{j} + e^t\mathbf{k}$

**4** $\mathbf{r}(t) = t\sin t\mathbf{i} + t\cos t\mathbf{j} + 3\mathbf{k}$

**5** $\mathbf{r}(t) = t\mathbf{i} + 2t\mathbf{j} + 3t\mathbf{k}$

**6** $\mathbf{r}(t) = (\sin t)\mathbf{i} + (\cos t)\mathbf{j} + 3t^4\mathbf{k}$

**7** $x(t) = e^{-2t}, y(t) = e^{2t}, z(t) = 1 + t^2$

**8** $\mathbf{r}(t) = \dfrac{t}{1+t}\mathbf{i} + \dfrac{t^2}{1+t}\mathbf{j} + \dfrac{1-t}{1+t}\mathbf{k}$

**9** $x(t) = e^t\sin t, y(t) = e^{2t}\cos t, z(t) = e^{-t}$

**10** $\mathbf{r}(t) = \ln(1+t)\mathbf{i} + \dfrac{t}{1+t^2}\mathbf{j} - 2t^3\mathbf{k}$

In each of Problems 11 through 16, for the particular value of t given, find the vectors **T**, **N**, and **B**; the curvature $\kappa$; the equations of the tangent line; and the equation of the osculating plane to the given curve.

**11** $x = 1 + t, y = 3 - t, z = 2t + 4, t = 3$

**12** $\mathbf{r}(t) = t\mathbf{i} + t^2\mathbf{j} + \frac{1}{3}t^3\mathbf{k}, t = 0$

**13** $x = e^t\cos t, y = e^t\sin t, z = e^t, t = 0$

**14** $x = \frac{1}{3}t^3, y = 2t, z = 2/t, t = 2$

**15** $x = t, y = 3t, z = 4t, t = 1$

**16** $x = (e^{t/2} + e^{-t/2}), y = (e^{t/2} - e^{-t/2}), z = 2t, t = 0$

**17** Given the path $x = e^t\cos t, y = e^t\sin t, z = e^t$, show that the path lies on the upper half of the cone $z^2 = x^2 + y^2$, and that the tangent vector **T**(t) cuts the generators of the cone at a constant angle and makes a constant angle with the z axis for all t. Also, find **N** and $\kappa$ in terms of t.

In Problems 18 through 21, a particle moves according to the law given. In each case, find its vector velocity and acceleration, its speed, the radius of curvature of its path, the unit vectors **T** and **N**, and the tangential and normal components of acceleration at the given time.

**18** $x = t, y = \frac{3}{2}t^2, z = \frac{3}{2}t^3, t = 2$

**19** $x = t, y = \ln(\sec t + \tan t), z = \ln\sec t, t = \pi/3$

**20** $x = t^3/3, y = 2t, z = 2/t, t = 1$

**21** $x = t\cos t, y = t\sin t, z = t, t = 0$

## CHAPTER 14

### REVIEW PROBLEMS

In each of Problems 1 through 5, express the vector **v** in terms of **i** and **j**, given that the endpoints A and B of **v** have the given coordinates.

**1** $A(2, -1), B(0, 3)$　　　　**2** $A(1, 3), B(0, 0)$

**3** $A(6, 2), B(3, -1)$　　　　**4** $A(-2, 0), B(0, 5)$

**5** $A(1, 2), B(-7, -3)$

In Problems 6 through 9, let $\mathbf{u} = \pi\mathbf{i}, \mathbf{v} = 2\mathbf{i} + 3\mathbf{j}, \mathbf{w} = \mathbf{i} + e\mathbf{j}$.

**6** Find unit vectors in the directions of each of **u**, **v** and **w**.

**7** Find (a) $\mathbf{u} + \mathbf{v}$, (b) $\mathbf{u} - \mathbf{v}$, (c) $2\mathbf{u} - 3\mathbf{v}$.

**8** Find the lengths of the following vectors:
(a) **u**, (b) $\mathbf{u} + \mathbf{v}$, (c) $3(\mathbf{v} - \mathbf{w})$.

**9** Show that **u** is orthogonal to $\mathbf{v} - 2\mathbf{w}$.

In Problems 10 through 13, find the projection of $\mathbf{v}[\vec{AB}]$ onto $\mathbf{v}[\vec{CD}]$.

**10** $A(0, 0), B(1, 1), C(1, 2), D(2, 1)$

**11** $A(2, 3), B(4, 5), C(-3, -1), D(-1, 3)$

**12** $A(3, -2), B(1, 3), C(-1, 6), D(1, 1)$

**13** $A(1, 2), B(4, 5), C(3, 0), D(-4, -7)$

In Problems 14 through 21, find the indicated vectors or scalars if $\mathbf{u} = \mathbf{i} + 3\mathbf{j}$ and $\mathbf{v} = 4\mathbf{i} - 2\mathbf{j}$.

**14** $2\mathbf{u} + 3\mathbf{v}$　　　　　　**15** $|\mathbf{u}| + |\mathbf{v}|$

**16** $|\mathbf{u}| - |\mathbf{v}|$　　　　　　**17** $|\mathbf{u} - \mathbf{v}|$

**18** $2\mathbf{u} \cdot (\mathbf{v} - 4\mathbf{i})$

**19** Show that there are non-zero scalars $\alpha$ and $\beta$ such that $\alpha\mathbf{u}$ is orthogonal to $\beta\mathbf{v}$.

**20** Find the angle between $\mathbf{u}$ and $\mathbf{i}$.

**21** Find the cosine of the angle between $\mathbf{u}$ and $\mathbf{v}$.

In Problems 22 through 25, let $\mathbf{u} = 2\mathbf{i} - \mathbf{j} - 3\mathbf{k}$ and $\mathbf{v} = \mathbf{i} + 5\mathbf{j} - 2\mathbf{k}$.

**22** Find $3\mathbf{u} - 2\mathbf{v}$.　　　　**23** Find $|\mathbf{u} + \mathbf{v}|$.

**24** Find $\mathbf{u} \cdot (\mathbf{v} - \mathbf{u})$.

**25** Find unit vectors in the direction of (a) $\mathbf{u}$, and (b) $\mathbf{v}$. Express the unit vector in terms of $\mathbf{i}$, $\mathbf{j}$ and $\mathbf{k}$.

In Problems 26 through 29, find the directed line segment $\overrightarrow{AB}$ equivalent to $\mathbf{u}$ from the information given.

**26** $\mathbf{u} = 2\mathbf{i} - \mathbf{j} - 3\mathbf{k}$, $A(1, 2, -5)$

**27** $\mathbf{u} = \mathbf{i} + 5\mathbf{j} - 2\mathbf{k}$, $A(-2, 1, 3)$

**28** $\mathbf{u} = 2\mathbf{i} + \mathbf{j} + 3\mathbf{k}$, $B(4, 3, 7)$

**29** $\mathbf{u} = \mathbf{i} - 5\mathbf{j} - 2\mathbf{k}$, $B(-1, 7, 2)$

In Problems 30 through 47, let $\mathbf{u} = 3\mathbf{i} - 2\mathbf{j} + \mathbf{k}$, $\mathbf{v} = -2\mathbf{i} + 3\mathbf{j} + 2\mathbf{k}$, $\mathbf{w} = 6\mathbf{i} - \mathbf{j} + \mathbf{k}$, $\mathbf{r} = \mathbf{i} - \mathbf{j} - 2\mathbf{k}$. Find the vectors or scalars indicated.

**30** $\mathbf{u} \cdot \mathbf{v}$　　　　　　　　**31** $\mathbf{v} \cdot \mathbf{w}$

**32** $\mathbf{u} \times \mathbf{v}$　　　　　　　　**33** $\mathbf{v} \times \mathbf{w}$

**34** $\mathbf{v} \cdot (\mathbf{u} \times \mathbf{w})$　　　　　**35** $\mathbf{w} \times (\mathbf{r} \times \mathbf{u})$

**36** The angle between $\mathbf{u}$ and $\mathbf{v}$.

**37** The angle between $\mathbf{v}$ and $\mathbf{w}$.

**38** $\cos \theta$, where $\theta$ is the angle between $\mathbf{u}$ and $\mathbf{w}$.

**39** $\cos \theta$, where $\theta$ is the angle between $\mathbf{r}$ and $\mathbf{w}$.

**40** The projection of $\mathbf{v}$ on $\mathbf{u}$.

**41** The projection of $\mathbf{w}$ on $\mathbf{v}$.

**42** The projection of $\mathbf{r}$ on $\mathbf{u}$.

**43** Find a unit vector in the direction of (a) $\mathbf{u}$, (b) $\mathbf{v}$.

**44** Find a unit vector in the direction of (a) $\mathbf{w}$, (b) $\mathbf{r}$.

**45** Find scalars $\alpha$ and $\beta$ such that $\alpha\mathbf{u} + \beta\mathbf{v}$ is orthogonal to $\mathbf{w}$.

**46** Find scalars $\alpha$ and $\beta$ such that $\alpha\mathbf{u} + \beta\mathbf{v}$ is orthogonal to $\mathbf{r}$.

**47** Find two unit vectors orthogonal to both $\mathbf{u}$ and $\mathbf{v}$.

In Problems 48 and 49, find in each case the area of $\triangle ABC$ and the equation of the plane through $A$, $B$ and $C$. (Use vector methods.)

**48** $A(1, -2, 2)$, $B(3, 1, 1)$, $C(2, 3, -2)$

**49** $A(6, 0, 1)$, $B(5, 1, 2)$, $C(3, 2, -2)$

**50** Find the angle between the two lines

$$L_1: \frac{x-3}{2} = \frac{y+1}{-4} = \frac{z-5}{8}$$

$$L_2: \frac{x+1}{4} = \frac{2-y}{3} = \frac{2z+1}{11}$$

**51** Find the equations in symmetric form of the line through $(3, 1, 5)$ intersecting at a right angle the line

$$\frac{x+2}{4} = \frac{y+3}{2} = \frac{z+2}{-1}.$$

**52** The two equations $4x + 4y + 3z = 14$ and $x - 2y + 3z = 2$ represent a line. Find the equations of this line in symmetric form.

In Problems 53 and 54, find in each case equations in symmetric form of the line of intersection of the given planes. Use the method of vector products.

**53**　$2x + 2y + 3z - 8 = 0$
　　　$x + 4y + 6z - 10 = 0$

**54**　$3(x + 2) - 4(y - 4) - z = 0$
　　　$-2(x + 2) + 3(y - 4) + z = 0$

In Problems 55 through 58, calculate $\mathbf{f}'(t)$ and $\mathbf{f}''(t)$.

**55** $\mathbf{f}(t) = (t^3 + t + 3)\mathbf{i} + (t^2 - 2)\mathbf{j}$

**56** $\mathbf{f}(t) = (t^2 + 3)^3\mathbf{i} + (t^2 + 3t)\mathbf{j}$

**57** $\mathbf{f}(t) = \cos t\,\mathbf{i} + t \sin t\,\mathbf{j}$

**58** $\mathbf{f}(t) = \ln t\,\mathbf{i} + e^{-t}\mathbf{j}$

In Problems 59 through 61, find parametric equations for the line tangent to the curve $C$ at the point $P$.

**59** $x = e^t \sin t$, $y = e^t \cos t$, $z = e^t$, $P(0, 1, 1)$

**60** $x = 3t$, $y = t^3$, $z = t^4$, $P(3, 1, 1)$

**61** $x = 3t^2 + 1$, $y = 4t - 2$, $z = e^{t-1}$, $P(4, 2, 1)$

In Problems 62 through 65, assume that a particle $P$ moves according to the given law, $t$ denoting time. Compute $\mathbf{v}(t)$, $|\mathbf{v}(t)|$, $\mathbf{a}(t)$, $|\mathbf{a}(t)|$, $s'(t)$, $s''(t)$.

**62** $\mathbf{f}(t) = (t - \sin t)\mathbf{i} + (1 - \cos t)\mathbf{j}$

**63** $\mathbf{f}(t) = 3t\mathbf{i} + t^3\mathbf{j} + t^4\mathbf{k}$

**64** $\mathbf{f}(t) = e^{2t}\mathbf{j} + e^{-t}\mathbf{k}$

**65** $\mathbf{f}(t) = \cos t\,\mathbf{i} + 2 \sin t\,\mathbf{j}$

**66** The number of bacteria in a certain culture grows at a rate which is exactly equal to 1/5 the number present. If there are 1000 bacteria initially, find the number at any time $t$.

**67** If the half-life of a radioactive substance is 14,000 years, how long does it take to arrive at the point where only 7% of the original amount remains? (See Problem 14 of Section 10.)

**68** A scientist modeling a certain phenomenon decides that the rate of change is proportional to a quadratic function of the amount present. Describe a differential equation that would serve to describe this model.

In Problems 69 through 71, an arc $C$ in three-space is described. Find $l(C)$, the length of the arc.

**69**  $C: x = 2,\ y = t,\ z = \ln \cos t;\quad 0 \le t \le 2\pi$

**70**  $C: x = 3t,\ y = t^3,\ z = 4t;\quad 0 \le t \le 1$

**71**  $C: x = 4(t - \sin t),\ y = 4(1 - \cos t),\ z = 3;\ 0 \le t \le 2\pi/3$

**72**  A particle moves according to the law: $x = e^t,\ y = e^{-t},\ z = t$.

For the time $t = 1$, find the particle's

a) vector velocity,
b) acceleration,
c) speed,
d) radius of curvature of its path,
e) unit vectors $\mathbf{T}$ and $\mathbf{N}$ (the tangential and normal components of acceleration at time $t = 1$).

# 15

## PARTIAL DERIVATIVES. APPLICATIONS

Until now we have discussed functions in which there is one independent and one or more dependent variables: the simplest case. Most functions which arise in the applications of calculus to problems in physics, engineering, economics and, in fact, all other subjects, involve many independent variables. In this chapter we begin the study of the calculus of functions of two or more variables.

### 1

### LIMITS AND CONTINUITY. PARTIAL DERIVATIVES

A function $f$ may be considered as a mapping which takes each element of the domain $D$ into an element of the range $S$. We write $f: D \to S$. If the domain $D$ consists of ordered pairs of numbers, then we have a function on $R^2$, or a function of two variables. We employ the notation

$$z = f(x, y)$$

to indicate a typical function on $R^2$ when the elements of the range are real numbers (elements of $R^1$).

A definition of a function of two variables was given on page 156 and we repeat it here.

**DEFINITIONS** *Let $A$ be a set of ordered pairs $(x, y)$ with $x, y$ real numbers, and let $B$ be a set of real numbers. A* **function** *$f$ from a set $A$ to a set $B$ is a correspondence that assigns to each element $(x, y)$ of $A$ a unique value $f(x, y) = z$, an element of $B$. The set $A$ is the* **domain** *of $f$ and the set $B$ is the* **range** *of $f$.*

*Remark.* The definition of a *function of three variables* is precisely the same as that for a function of two variables except that the elements of $A$ are ordered triples rather than ordered pairs. A *function of n variables*, where $n$ is any positive integer, is defined by considering the elements of $A$ to be ordered $n$ tuples of real numbers.

A function of three variables is frequently denoted by

$$w = G(x, y, z)$$

where $(x, y, z)$ is an element of $R^3$ and $w$ is an element of $R^1$. We can consider functions of four, five, or of any number of variables. If the exact number of variables is $n$, we write

$$y = f(x_1, x_2, \ldots, x_n)$$

in which $(x_1, x_2, \ldots, x_n)$ is an element of $R^n$ and $y$ is an element of $R^1$. The same letters $f, F, g, G, \phi$, and so forth, are used for functions of any number of variables.

If $f$ is a function of two variables, we can write $z = f(x, y)$, which suggests that a function of two variables can be graphed in three dimensions in a way analogous to the graphs of functions of one variable in the plane. The graph of $z = f(x, y)$ is a two-dimensional surface, but it is usually not easy to graph. Fairly simple functions may give rise to complicated surfaces, as Figs. 1 through 4 illustrate.

An aid both to visualizing and to drawing the graph of a function of two variables is the identification of the *level curves* of a surface.

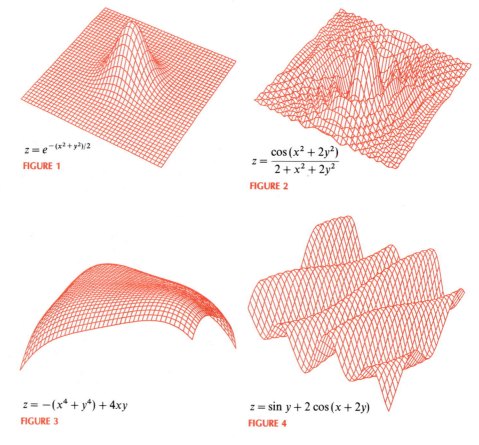

$z = e^{-(x^2 + y^2)/2}$

**FIGURE 1**

$z = \dfrac{\cos(x^2 + 2y^2)}{2 + x^2 + 2y^2}$

**FIGURE 2**

$z = -(x^4 + y^4) + 4xy$

**FIGURE 3**

$z = \sin y + 2\cos(x + 2y)$

**FIGURE 4**

| DEFINITION | Let $z = f(x, y)$ be a function of two variables, and let $c$ be a number in the range of $f$. Then the curve $f(x, y) = c$ is called a **level curve** for $f$. |

Starting with $z = f(x, y)$, we see that the level curve is the curve of intersection of the plane $z = c$ with the graph of the function.

Level curves are used in topographical maps to show curves of constant elevation. (See Fig. 5.)

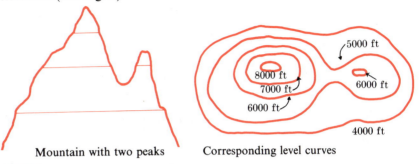

Mountain with two peaks          Corresponding level curves

**FIGURE 5**

**EXAMPLE 1**  Use level curves to construct the graph of the function $z = 5 - 2x^2 - 3y^2$.

**Solution**  We choose simple values of $c$ to get equations for the level curves. Notice that $5 - 2x^2 - 3y^2 = 5 - (2x^2 + 3y^2)$ is always less than 5, so we must choose values of $c \le 5$. For a particular $c$, we have $c = 5 - (2x^2 + 3y^2)$, or

$$2x^2 + 3y^2 = 5 - c, \quad \text{or}$$

$$\frac{x^2}{5 - \dfrac{c}{2}} + \frac{y^2}{5 - \dfrac{c}{3}} = 1,$$

which is the equation of an ellipse with vertices at

$$\left(-\frac{(5-c)}{2}, 0\right), \quad \left(\frac{5-c}{2}, 0\right), \quad \left(0, \frac{-(5-c)}{3}\right), \quad \left(0, \frac{5-c}{3}\right).$$

A sketch of the graph of $z = 5 - 2x^2 - 3y^2$ is shown in Fig. 6.  □

In Figs. 7(a) and 7(b) we indicate the level curves and give a computer-drawn sketch of the surface $z = 2 - (x^{1/3} + y^3)$.

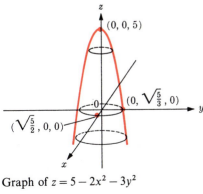

Graph of $z = 5 - 2x^2 - 3y^2$

**FIGURE 6**

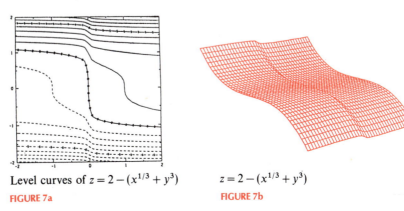

Level curves of $z = 2 - (x^{1/3} + y^3)$           $z = 2 - (x^{1/3} + y^3)$

**FIGURE 7a**                                          **FIGURE 7b**

The study of limits and continuity for functions of two variables is somewhat more complicated than that for functions of one variable. While the definitions of limit and continuity are similar in appearance to the definitions we learned in the one-variable case, there are many examples of functions of two variables which appear at first glance to be continuous, but after a close examination turn out to have discontinuities. (See Problems 47–53 at the end of the section.)

We now define a limit of a function of two variables. If $f$ is such a function, we write

$$z = f(x, y)$$

and we wish to examine the behavior of $f$ as the pair $(x, y)$ tends to the pair $(a, b)$; equivalently, we consider $x$ tending to $a$ and $y$ tending to $b$.

**DEFINITION**

*We say that $f(x, y)$ **tends to the number** $L$ **as** $(x, y)$ **tends to** $(a, b)$, and we write*

$$f(x, y) \to L \qquad as \qquad (x, y) \to (a, b)$$

*if and only if for each $\varepsilon > 0$ there is a $\delta > 0$ such that*

$$|f(x, y) - L| < \varepsilon$$

*whenever*

$$|x - a| < \delta \qquad and \qquad |y - b| < \delta \qquad and \qquad (x, y) \neq (a, b).$$

A geometric interpretation of this definition is exhibited in Fig. 8. The definition asserts that whenever $(x, y)$ are in the shaded square, as shown, then the function values, which we represent by $z$, must lie in the rectangular box of height $2\varepsilon$ between the values $L - \varepsilon$ and $L + \varepsilon$. This interpretation is an extension of the one given in Chapter 2 for functions of one variable.

We do not have to use rectangular boxes in the definition of limit. We could, for example, use disks. In this case we change the definition by requiring $|f(x, y) - L| < \varepsilon$ whenever $0 < \sqrt{(x - a)^2 + (y - b)^2} < \delta$. A geometric interpretation of this "disk" definition is shown in Fig. 9.

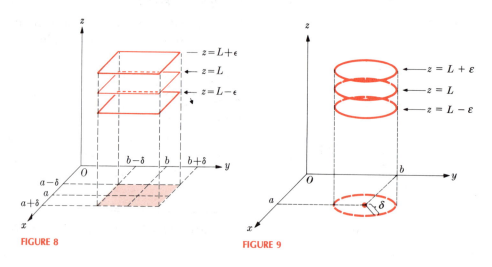

FIGURE 8                                                          FIGURE 9

*Remark.* It is not necessary that $(a, b)$ be in the domain of $f$. That is, a limit may exist with $f(a, b)$ being undefined.

DEFINITION

*We say that $f$ is **continuous at** $(a, b)$ if and only if*

i) $f(a, b)$ *is defined, and*
ii) $f(x, y) \to f(a, b)$ *as* $(x, y) \to (a, b)$.

Definitions of limits and continuity for functions of three, four, and more variables are completely similar.

Functions of two or more variables do not have ordinary derivatives of the type we studied for functions of one variable. If $f$ is a function of two variables, say $x$ and $y$, then for each *fixed* value of $y$, $f$ is a function of a single variable $x$. The derivative with respect to $x$ (keeping $y$ fixed) is then called the partial derivative with respect to $x$. For $x$ fixed and $y$ varying, we also obtain a partial derivative, one with respect to $y$.

DEFINITIONS

*We define **the partial derivatives of a function $f$ on $R^2$** by*

$$f_x(x, y) = \lim_{h \to 0} \frac{f(x + h, y) - f(x, y)}{h},$$

$$f_y(x, y) = \lim_{h \to 0} \frac{f(x, y + h) - f(x, y)}{h},$$

*where $y$ is kept fixed in the first limit and $x$ is kept fixed in the second. We use the symbols $f_x$ and $f_y$ for these partial derivatives.*

*If $F$ is a function on $R^3$, we define the **partial derivatives $F_x$, $F_y$, and $F_z$** by*

$$F_x(x, y, z) = \lim_{h \to 0} \frac{F(x + h, y, z) - F(x, y, z)}{h}, \quad (y, z \text{ fixed})$$

$$F_y(x, y, z) = \lim_{h \to 0} \frac{F(x, y + h, z) - F(x, y, z)}{h}, \quad (x, z \text{ fixed})$$

$$F_z(x, y, z) = \lim_{h \to 0} \frac{F(x, y, z + h) - F(x, y, z)}{h}, \quad (x, y \text{ fixed}).$$

In other words, *to find the partial derivative $F_x$ of a function $F(x, y, z)$ of three variables, regard $y$ and $z$ as constants and find the **usual derivative** with respect to $x$; the derivatives $F_y$ and $F_z$ are found correspondingly. The same procedure applies for functions of any number of variables.*

EXAMPLE 2    Given $f(x, y) = x^3 + 7x^2 y + 8y^3 + 3x - 2y + 7$, find $f_x$ and $f_y$ and evaluate them at $(2, -1)$.

Solution    Keeping $y$ fixed and differentiating with respect to $x$, we find that

$$f_x = 3x^2 + 14xy + 3.$$

Similarly, keeping $x$ fixed, we get

$$f_y = 7x^2 + 24y^2 - 2.$$

Using these results we have

$$f_x(2, -1) = 3(2)^2 + 14(2)(-1) + 3 = -13$$

and

$$f_y(2, -1) = 7(2)^2 + 24(-1)^2 - 2 = 50. \qquad \square$$

**EXAMPLE 3**   Given $f(x, y, z) = x^3 + y^3 + z^3 + 3xyz$, find $f_x(x, y, z)$, $f_y(x, y, z)$, and $f_z(x, y, z)$.

**Solution**   We have

$$f_x(x, y, z) = 3x^2 + 3yz, \; f_y(x, y, z) = 3y^2 + 3xz, \; f_z(x, y, z) = 3z^2 + 3xy.$$

$\square$

*Remarks on notation.* The notation used here has the disadvantage of implying that the first variable must be the letter $x$ and the second must be the letter $y$. A better notation, but one we shall not employ, is $f_{,1}$ instead of $f_x$ and $f_{,2}$ instead of $f_y$. The subscripts 1 and 2 indicate the variable being differentiated without prescribing the letter being used. For example, if $z = f(y, u)$ is a function of two variables, the symbol $f_{,1}$ means the derivative with respect to the first variable, while the symbol $f_x$ is ambiguous.

    Another common symbol is

$$\frac{\partial f}{\partial x} \quad \text{(read: partial of } f \text{ with respect to } x\text{)}.$$

This notation has the double disadvantage of using the letter $x$ and of giving the impression (incorrectly) that the partial derivative is a fraction with $\partial f$ and $\partial x$ having independent meanings (which they do not). If we write $z = f(x, y)$, still another symbol for partial derivative is the expression

$$\frac{\partial z}{\partial x}.$$

    Because of the multiplicity of symbols for partial derivatives used in texts on mathematics and various related branches of technology, it is important that the reader be familiar with all of them. We shall use

$$f_x, \frac{\partial f}{\partial x}, \frac{\partial z}{\partial x}, \qquad \text{and} \qquad f_y, \frac{\partial f}{\partial y}, \frac{\partial z}{\partial y}$$

interchangeably.

**EXAMPLE 4**   Given $f(x, y) = e^{xy} \cos x \sin y$, find $f_x(x, y)$ and $f_y(x, y)$.

**Solution**   We have

$$f_x(x, y) = e^{xy} \sin y(-\sin x) + \cos x \sin y \, e^{xy} \cdot y$$

$$= e^{xy} \sin y(y \cos x - \sin x),$$

and

$$f_y(x, y) = e^{xy} \cos x(x \sin y + \cos y). \qquad \square$$

**EXAMPLE 5**    Given that $z = x \arctan(y/x)$, find $\partial z/\partial x$ and $\partial z/\partial y$.

**Solution**    We have

$$\frac{\partial z}{\partial x} = x \cdot \frac{-(y/x^2)}{1 + (y^2/x^2)} + \arctan(y/x) = \arctan(y/x) - \frac{xy}{x^2 + y^2},$$

and

$$\frac{\partial z}{\partial y} = x \frac{1/x}{1 + (y^2/x^2)} = \frac{x^2}{x^2 + y^2}. \qquad \square$$

# 1    PROBLEMS

In Problems 1 and 2, draw level curves and then sketch the graph of $z = f(x, y)$ for the given $f$.

**1**   $f(x, y) = 3x + 2y + 6$      **2**   $f(x, y) = \sqrt{4 - (x^2 + y^2)}$

In Problems 3 through 13, draw in the $xy$ plane several level curves of the given function $z = f(x, y)$.

**3**   $f(x, y) = xy$             **4**   $f(x, y) = x - y$

**5**   $f(x, y) = \dfrac{x}{x + y}$        **6**   $f(x, y) = \dfrac{y}{x^2}$

**7**   $f(x, y) = e^{xy}$          **8**   $f(x, y) = x^2 - y^2$

**9**   $f(x, y) = \dfrac{x^2}{x^2 + y^2}$     **10**   $f(x, y) = x^2 y^2$

**11**   $f(x, y) = e^x - y$       **12**   $f(x, y) = y - \cos x$

**13**   $f(x, y) = x - \sin y$

In each of Problems 14 through 27, find $f_x$ and $f_y$.

**14**   $f(x, y) = 2x^2 - 3xy + 4x$    **15**   $f(x, y) = x^2 y^4 + 2xy - 6$

**16**   $f(x, y) = \sqrt{x^2 + 1} + y^3$    **17**   $f(x, y) = (x^2 - y^3)^2$

**18**   $f(x, y) = e^x \ln xy$

**19**   $f(x, y) = x^3 + y^3 + 3x^2 y - 3xy^2 + 7$

**20**   $f(x, y) = \sqrt{x^2 + y^2}$

**21**   $f(x, y) = \dfrac{xy}{x^2 + y^2}$

**22**   $f(x, y) = \ln(x^2 + y^2)$

**23**   $f(x, y) = \ln \sqrt{x^2 + 3y^2}$

**24**   $f(x, y) = \arctan(y/x)$

**25**   $f(x, y) = \arcsin \dfrac{x}{1 + y}$

**26**   $f(x, y) = xye^{x^2 + y^2}$

**27**   $f(x, y) = \cos(xe^y)$

In each of Problems 28 through 33, find both first derivatives at the values indicated.

**28**   $f(x, y) = x \arcsin(x - y)$, $x = 1$, $y = 2$

**29**   $f(u, v) = e^{uv} \sec\left(\dfrac{u}{v}\right)$, $u = v = 3$

**30**   $f(x, z) = e^{\sin x} \tan xz$, $x = \dfrac{\pi}{4}$, $z = 1$

**31**   $f(t, u) = \dfrac{\cos 2tu}{t^2 + u^2}$, $t = 0$, $u = 1$

**32**   $f(y, x) = x^{xy}$, $x = y = 2$

**33**   $f(s, t) = s^t + t^s$, $s = t = 3$

In each of Problems 34 through 40, find $f_x(x, y, z)$, $f_y(x, y, z)$ and $f_z(x, y, z)$.

**34**   $f(x, y, z) = x^2 y - 2x^2 z + 3xyz - y^2 z + 2xz^2$

**35**   $f(x, y, z) = \dfrac{xyz}{x^2 + y^2 + z^2}$

**36**   $f(x, y, z) = e^{xyz} \sin xy \cos 2xz$

**37**   $f(x, y, z) = x^2 + z^2 + y^3 + 2x - 3y + 4z$

**38**   $f(x, y, z) = 2 + \tan x + y \sin z$

**39**   $f(x, y, z) = x^2 + y^2 + z^2 - \cos(xyz)$

**40**   $f(x, y, z) = (x^2 + xyz + z)^2$

In Problems 41 through 44, find in each case the indicated partial derivatives.

**41**   $w = \ln\left(\dfrac{xy}{x^2 + y^2}\right)$;    $\dfrac{\partial w}{\partial x}, \dfrac{\partial w}{\partial y}$

**42**  $w = (r^2 + s^2 + t^2)\cos(rst);\quad \dfrac{\partial w}{\partial r}, \dfrac{\partial w}{\partial t}$

**43**  $w = e^{\sin(y/x)};\quad \dfrac{\partial w}{\partial y}, \dfrac{\partial w}{\partial x}$

**44**  $w = (\sec tu)\arcsin tv;\quad \dfrac{\partial w}{\partial t}, \dfrac{\partial w}{\partial u}, \dfrac{\partial w}{\partial v}$

**45**  If $z = \ln(y/x)$, show that

$$x\frac{\partial z}{\partial x} + y\frac{\partial z}{\partial y} = 0.$$

**46**  Let $\quad P(x_1, x_2, \ldots, x_n) = a_1 x_1^k + a_2 x_2^k + \cdots + a_n x_n^k, \quad$ where $a_1, a_2, \ldots, a_n$ are numbers. Show that

$$x_1 P_{x_1} + x_2 P_{x_2} + \cdots + x_n P_{x_n} = kP.$$

**47**  Let $Q(x, y, z) = \Sigma_{k=1}^n a_k x^{2n-2k} y^{k+1} z^{k-1}$, where the $a_k$ are numbers and $n$ is an integer greater than 1. Show that

$$xQ_z + yQ_y + zQ_z = 2nQ.$$

**48**  i) Given $f(x, y) = xy$. Show that $f$ is continuous at $(0, 0)$ by finding a value of $\delta$ corresponding to each given $\varepsilon$ such that $|xy - 0| < \varepsilon$ for all $(x, y)$ for which $|x - 0| < \delta$ and $|y - 0| < \delta$.

ii) For the same function $f$, show that it is continuous at $(a, b)$ by finding a value of $\delta$ corresponding to each given $\varepsilon$ such that $|xy - ab| < \varepsilon$ for all $(x, y)$ for which $|x - a| < \delta$ and $|y - b| < \delta$. (*Hint:* Write $xy - ab = xy - ay + ay - ab$ and

$$|xy - ab| \le |xy - ay| + |ay - ab| \le |x - a| \cdot |y| + |a| \cdot |y - b|.)$$

**49**  Consider the function

$$g(x, y) = \begin{cases} \dfrac{xy}{x^2 + y^2} & \text{for } (x, y) \neq (0, 0), \\ 0 & \text{for } x = y = 0. \end{cases}$$

i) If $x$ tends to zero and if we select $y = x^2$, then $y$ also tends

to zero. Show that under these conditions

$$g(x, y) \to 0 \quad \text{as} \quad x \to 0, y \to 0.$$

ii) If $x$ tends to zero and we select $y = x$, show that

$$g(x, y) \to \tfrac{1}{2} \quad \text{as} \quad x \text{ and } y \text{ tend to } 0.$$

Conclude that $g$ is not continuous at $(0, 0)$.

**50**  Use the method of Problem 49 to show that

$$h(x, y) = \begin{cases} \dfrac{x^2 - y^2}{x^2 + y^2} & \text{for } (x, y) \neq (0, 0), \\ 0 & \text{for } x = y = 0 \end{cases}$$

is not continuous at $(0, 0)$.

**51**  Show that the function

$$f(x, y) = \begin{cases} \dfrac{x^2 y^2}{x^2 + y^2} & \text{for } (x, y) \neq (0, 0), \\ 0 & \text{for } x = y = 0 \end{cases}$$

is continuous at $(0, 0)$. (*Hint:* $x^2 y^2 / (x^2 + y^2) \le x^2$.)

**52**  Show that the function $f(x, y) = \sqrt{|x|}\sqrt{|y|}$ is continuous at $(0, 0)$, but that $f_x$ and $f_y$ are not continuous at $(0, 0)$.

**53**  Let $f(x, y) = \sin(xy)$; $g(x, y) = \sin(x/y)$ for $y \neq 0$, $g(x, 0) = 0$. Is $f$ continuous at $(0, 0)$? Is $g$ continuous at $(0, 0)$?

**54**  Given the function

$$g(x, y) = \begin{cases} 0 & \text{for } x = y = 0, \\ \dfrac{xy}{(x^2 + y^2)} & \text{otherwise.} \end{cases}$$

Show that $g_x$ and $g_y$ exist at $x = y = 0$. Use the result of Problem 49 to conclude that a function may have a partial derivative at a point and yet not be continuous there. (This contrasts with Theorem 13 in Chapter 2 which shows that for functions of one variable differentiability at a point implies continuity at that point.)

---

## 2

## IMPLICIT DIFFERENTIATION

An equation involving $x$, $y$, and $z$ establishes a relation among the variables. If we can solve for $z$ in terms of $x$ and $y$, then we may have one or more functions determined by the relation. For example, the equation

$$2x^2 + y^2 + z^2 - 16 = 0 \tag{1}$$

may be solved for $z$ to give

$$z = \pm\sqrt{16 - 2x^2 - y^2}. \tag{2}$$

If one or more functions are determined by a relation, it is possible to compute partial derivatives implicitly in a way that is completely similar to the methods used for ordinary derivatives. (See Chapter 3, page 111.) Being able to compute partial derivatives implicitly is important because we may not be able to solve for $z$ explicitly.

For example, in Equation (1) above, considering $z$ as a function of $x$ and $y$, we can compute $\partial z/\partial x$ directly from (1) without resorting to (2). We keep $y$ fixed and, in (1), differentiate implicitly with respect to $x$, getting

$$4x + 2z\frac{\partial z}{\partial x} = 0 \quad \text{and} \quad \frac{\partial z}{\partial x} = -\frac{2x}{z}.$$

Further examples exhibit the method.

**EXAMPLE 1**    Suppose that $x$, $y$, and $z$ are variables and that $z$ is a function of $x$ and $y$ which satisfies

$$x^3 + y^3 + z^3 + 3xyz = 5.$$

Find $\partial z/\partial x$ and $\partial z/\partial y$.

**Solution**    Holding $y$ constant and differentiating $z$ with respect to $x$ implicitly, we obtain

$$3x^2 + 3z^2\frac{\partial z}{\partial x} + 3xy\frac{\partial z}{\partial x} + 3yz = 0.$$

Therefore

$$\frac{\partial z}{\partial x} = -\frac{x^2 + yz}{xy + z^2}.$$

Holding $x$ constant and differentiating with respect to $y$, we get

$$3y^2 + 3z^2\frac{\partial z}{\partial y} + 3xy\frac{\partial z}{\partial y} + 3xz = 0$$

and

$$\frac{\partial z}{\partial y} = -\frac{y^2 + xz}{xy + z^2}. \qquad \square$$

The same technique works with equations relating four or more variables, as the next example shows.

**EXAMPLE 2**    If $r$, $s$, $t$, and $w$ are variables and if $w$ is a function of $r$, $s$, and $t$ which satisfies

$$e^{rt} - 2se^w + wt - 3w^2r = 5,$$

find $\partial w/\partial r$ and $\partial w/\partial s$.

**Solution**    To find $\partial w/\partial r$, we keep $s$ and $t$ fixed and differentiate implicitly with respect to $r$. The result is

$$te^{rt} - 2se^w\frac{\partial w}{\partial r} + t\frac{\partial w}{\partial r} - 3w^2 - 6rw\frac{\partial w}{\partial r} = 0$$

or

$$\frac{\partial w}{\partial r} = \frac{3w^2 - te^{rt}}{t - 2se^w - 6rw}.$$

Keeping $r$ and $t$ fixed, we obtain

$$-2e^w - 2se^w \frac{\partial w}{\partial s} + t \frac{\partial w}{\partial s} - 6rw \frac{\partial w}{\partial s} = 0$$

and

$$\frac{\partial w}{\partial s} = \frac{2e^w}{t - 2se^w - 6rw}.$$

## 2    PROBLEMS

In each of Problems 1 through 18, assume that $w$ is a function of all other variables. Find the partial derivatives as indicated in each case.

1  $3x^2 + 2y^2 + 6w^2 - x + y - 12 = 0;$   $\dfrac{\partial w}{\partial x}, \dfrac{\partial w}{\partial y}$

2  $x^2 + y^2 + w^2 + 3xy - 2xw + 3yw = 36;$   $\dfrac{\partial w}{\partial x}, \dfrac{\partial w}{\partial y}$

3  $x^2 - 2xy + 2xw + 3y^2 + w^2 = 21;$   $\dfrac{\partial w}{\partial x}, \dfrac{\partial w}{\partial y}$

4  $x^2y - x^2w - 2xy^2 - yw^2 + w^3 = 7;$   $\dfrac{\partial w}{\partial x}, \dfrac{\partial w}{\partial y}$

5  $w - (r^2 + s^2) \cos rw = 0;$   $\dfrac{\partial w}{\partial r}, \dfrac{\partial w}{\partial s}$

6  $w - e^{w \sin (y/x)} = 1;$   $\dfrac{\partial w}{\partial x}, \dfrac{\partial w}{\partial y}$

7  $e^{xyw} \sin xy \cos 2xw - 4 = 0;$   $\dfrac{\partial w}{\partial x}, \dfrac{\partial w}{\partial y}$

8  $w^2 - 3xw - \ln\left(\dfrac{xy}{x^2 + y^2}\right) = 0;$   $\dfrac{\partial w}{\partial x}, \dfrac{\partial w}{\partial y}$

9  $xyz + x^2z + xzw - yzw + yz^2 - w^3 = 3;$   $\dfrac{\partial w}{\partial x}, \dfrac{\partial w}{\partial y}, \dfrac{\partial w}{\partial z}$

10  $r^2 + 3s^2 - 2t^2 + 6tw - 8w^2 + 12sw^3 = 4;$   $\dfrac{\partial w}{\partial r}, \dfrac{\partial w}{\partial s}, \dfrac{\partial w}{\partial t}$

11  $we^{xw} - ye^{yw} + e^{xy} = 1;$   $\dfrac{\partial w}{\partial x}, \dfrac{\partial w}{\partial y}$

12  $x^3 + 3x^2y + 2z^2t - 4zt^3 + 7xw - 8yw^2 + w^4 = 5;$   $\dfrac{\partial w}{\partial x}, \dfrac{\partial w}{\partial y}, \dfrac{\partial w}{\partial z}, \dfrac{\partial w}{\partial t}$

13  $xw^4 + y \cos w = 0;$   $\dfrac{\partial w}{\partial x}$

14  $e^{xw} + zwe^{-w} = 3;$   $\dfrac{\partial w}{\partial x}$

15  $yx - \sqrt{xw} + 4w = 12;$   $\dfrac{\partial w}{\partial x}$

16  $xyw + e^w = z;$   $\dfrac{\partial w}{\partial x}, \dfrac{\partial w}{\partial y}$

17  $(x + yw)^2 = e^z + 2;$   $\dfrac{\partial w}{\partial z}, \dfrac{\partial w}{\partial x}$

18  $(xyw) + (x - zw)^2 = 23;$   $\dfrac{\partial w}{\partial x}, \dfrac{\partial w}{\partial y}, \dfrac{\partial w}{\partial z}$

19  Given that $x^2 + y^2 - z^2 = 4$. Show that

$$\frac{\partial z}{\partial x} \cdot \frac{\partial x}{\partial y} \cdot \frac{\partial y}{\partial z} = -1.$$

*20  Suppose that the relation $f(x, y, z) = 0$ may be solved so that $z$ is a function of $x$ and $y$. Then we may compute $\partial z/\partial x$. If, also, we may compute $\partial x/\partial y$ and $\partial y/\partial z$, show that

$$\frac{\partial z}{\partial x} \cdot \frac{\partial x}{\partial y} \cdot \frac{\partial y}{\partial z} = -1.$$

21  The transformation of rectangular coordinates to spherical coordinates is given by

$$x = \rho \sin \phi \cos \theta, \qquad y = \rho \sin \phi \sin \theta, \qquad z = \rho \cos \phi.$$

Find

$$\frac{\partial x}{\partial \rho}, \frac{\partial x}{\partial \phi}, \frac{\partial x}{\partial \theta}, \frac{\partial y}{\partial \rho}, \frac{\partial y}{\partial \phi}, \frac{\partial y}{\partial \theta}, \frac{\partial z}{\partial \rho}, \frac{\partial z}{\partial \phi}, \frac{\partial z}{\partial \theta}.$$

**22** In Problem 21 assume that $\rho, \phi, \theta$ are functions of $x, y, z$ and find in terms of $x$, $y$, and $z$:

(a) $\dfrac{\partial \rho}{\partial x}, \dfrac{\partial \rho}{\partial y}, \dfrac{\partial \rho}{\partial z}$,   (b) $\dfrac{\partial \phi}{\partial x}, \dfrac{\partial \phi}{\partial y}, \dfrac{\partial \phi}{\partial z}$,   (c) $\dfrac{\partial \theta}{\partial x}, \dfrac{\partial \theta}{\partial y}, \dfrac{\partial \theta}{\partial z}$.

**23** An *affine transformation* in $R^3$ is one which changes a rectangular $(x, y, z)$ coordinate system into a $(u, v, w)$

coordinate system by the equations

$$u = a_{11}x + a_{12}y + a_{13}z$$

$$v = a_{21}x + a_{22}y + a_{23}z$$

$$w = a_{31}x + a_{32}y + a_{33}z$$

where all the $a_{ij}$ are numbers. Find $\partial v/\partial y$ and $\partial y/\partial v$. How are these quantities related?

---

## 3

## THE CHAIN RULE

The Chain Rule is one of the most effective devices for calculating ordinary derivatives. (See Chapter 3, page 105.) In this section we show how to extend the Chain Rule for the computation of partial derivatives. The basis of the Chain Rule in the case of functions of one variable is the Fundamental Lemma on Differentiation, which we now recall.

---

**THEOREM 1**   *If $F$ has a derivative at a value $a$ so that $F'(a)$ exists, then*

$$F(a + h) - F(a) = [F'(a) + G(h)]h,$$

*where $G(h)$ tends to zero as $h$ tends to zero and $G(0) = 0$.*

The proof of this theorem is on page 104.
      The above theorem has a natural generalization for functions of two variables.

---

**THEOREM 2**   **(Fundamental Lemma on Differentiation)**   *Suppose that $f$ is a continuous function of two variables (say $x$ and $y$) and that $f_x$ and $f_y$ are continuous at $(x_0, y_0)$. Then there are two functions, $G_1(h, k)$ and $G_2(h, k)$ continuous at $(0, 0)$ with $G_1(0, 0) = G_2(0, 0) = 0$, such that*

$$f(x_0 + h, y_0 + k) - f(x_0, y_0) = f_x(x_0, y_0)h + f_y(x_0, y_0)k$$
$$+ G_1(h, k)h + G_2(h, k)k. \tag{1}$$

---

**Proof**   The proof depends on writing the left side of (1) in a more complicated way:

$$f(x_0 + h, y_0 + k) - f(x_0, y_0) = [f(x_0 + h, y_0 + k) - f(x_0 + h, y_0)]$$
$$+ [f(x_0 + h, y_0) - f(x_0, y_0)]. \tag{2}$$

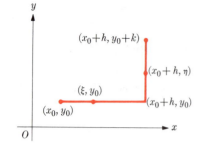

**FIGURE 10**

Figure 10 shows the points at which $f$ is evaluated in (2) ($h$ and $k$ are taken to be positive in the figure). We apply the Mean Value Theorem to each of the quantities in brackets in (2) above. The result for the first quantity is

$$\frac{f(x_0 + h, y_0 + k) - f(x_0 + h, y_0)}{(y_0 + k) - y_0} = f_y(x_0 + h, \eta), \tag{3}$$

and for the second quantity,

$$\frac{f(x_0 + h, y_0) - f(x_0, y_0)}{(x_0 + h) - x_0} = f_x(\xi, y_0), \tag{4}$$

where $\eta$ is between $y_0$ and $y_0 + k$ and $\xi$ is between $x_0$ and $x_0 + h$. Typical locations for $\xi$ and $\eta$ are shown in Fig. 10. Substituting (3) and (4) into the right side of (2), we find

$$f(x_0 + h, y_0 + k) - f(x_0, y_0) = f_x(\xi, y_0)h + f_y(x_0 + h, \eta)k. \tag{5}$$

The quantities $G_1$ and $G_2$ are defined by*

$$G_1 = f_x(\xi, y_0) - f_x(x_0, y_0),$$

$$G_2 = f_y(x_0 + h, \eta) - f_y(x_0, y_0).$$

Multiplying the expression for $G_1$ by $h$ and that for $G_2$ by $k$ and inserting the result in the right side of (5), we obtain the statement of the theorem. Since

$$\xi \to x_0 \quad \text{as } h \to 0 \qquad \text{and} \qquad \eta \to y_0 \quad \text{as } k \to 0,$$

it follows (since $f_x$ and $f_y$ are continuous) that $G_1$ and $G_2$ tend to zero as $h$ and $k$ tend to zero. Thus $G_1$ and $G_2$ are continuous at $(0, 0)$ if we define them as being equal to zero there. ☐

---

**THEOREM 3**   **(Chain Rule)**   *Suppose that $z = f(x, y)$ is continuous and that $\partial f/\partial x$, $\partial f/\partial y$ are continuous. Assume that $x = x(r, s)$ and $y = y(r, s)$ are functions of $r$ and $s$ such that $\partial x/\partial r$, $\partial x/\partial s$, $\partial y/\partial r$, $\partial y/\partial s$ all exist. Then $z$ is a function of $r$ and $s$ and the following formulas hold:*

$$\left.\begin{aligned}
\frac{\partial z}{\partial r} &= \left(\frac{\partial f}{\partial x}\right)\left(\frac{\partial x}{\partial r}\right) + \left(\frac{\partial f}{\partial y}\right)\left(\frac{\partial y}{\partial r}\right) \\
\frac{\partial z}{\partial s} &= \left(\frac{\partial f}{\partial x}\right)\left(\frac{\partial x}{\partial s}\right) + \left(\frac{\partial f}{\partial y}\right)\left(\frac{\partial y}{\partial s}\right)
\end{aligned}\right\} \tag{6}$$

---

**Proof**   The first formula will be established; the second is proved similarly. We use the $\Delta$ notation. A change $\Delta r$ in $r$ induces a change $\Delta x$ in $x$ and a change $\Delta y$ in $y$. That is,

$$\Delta x = x(r + \Delta r, s) - x(r, s),$$

$$\Delta y = y(r + \Delta r, s) - y(r, s).$$

The function $z$ has the partial derivative

---

*Of course, if $k = 0$, we define $G_2 = f_y(x_0 + h, y_0) - f_y(x_0, y_0)$, etc.

$$\frac{\partial z}{\partial r} = \lim_{\Delta r \to 0} \frac{\Delta z}{\Delta r},$$

where $\Delta z$, the change in $z$ due to the change $\Delta r$ in $r$, is given by

$$\Delta z = f(x + \Delta x, y + \Delta y) - f(x, y).$$

The latter expression is also denoted by $\Delta f$. The Fundamental Lemma on Differentiation with $h = \Delta x$ and $k = \Delta y$ reads

$$\Delta f = \frac{\partial f}{\partial x} \Delta x + \frac{\partial f}{\partial y} \Delta y + G_1 \Delta x + G_2 \Delta y,$$

where we have changed notation by using $\partial f / \partial x$ in place of $f_x(x, y)$ and $\partial f / \partial y$ for $f_y(x, y)$. Dividing by $\Delta r$ in the above equation for $\Delta f$, we get

$$\frac{\Delta z}{\Delta r} \equiv \frac{\Delta f}{\Delta r} = \frac{\partial f}{\partial x} \frac{\Delta x}{\Delta r} + \frac{\partial f}{\partial y} \frac{\Delta y}{\Delta r} + G_1 \frac{\Delta x}{\Delta r} + G_2 \frac{\Delta y}{\Delta r}.$$

Letting $\Delta r$ tend to zero and remembering that $G_1 \to 0$, $G_2 \to 0$, we obtain the first formula in (6), as desired.  □

*Remarks.*  i) The formulas (6) may be written in various notations. Two common expressions are

$$\left.\begin{aligned}
\frac{\partial z}{\partial r} &= \left(\frac{\partial z}{\partial x}\right)\left(\frac{\partial x}{\partial r}\right) + \left(\frac{\partial z}{\partial y}\right)\left(\frac{\partial y}{\partial r}\right) \\
\frac{\partial z}{\partial s} &= \left(\frac{\partial z}{\partial x}\right)\left(\frac{\partial x}{\partial s}\right) + \left(\frac{\partial z}{\partial y}\right)\left(\frac{\partial y}{\partial s}\right)
\end{aligned}\right\} \tag{7}$$

and

$$\left.\begin{aligned}
f_r &= f_x x_r + f_y y_r \\
f_s &= f_x x_s + f_y y_s
\end{aligned}\right\}. \tag{8}$$

ii)  For functions of *one* variable, the Chain Rule is easily remembered as the rule which allows us to think of derivatives as fractions. The formula

$$\frac{dy}{dx} = \frac{dy}{du} \cdot \frac{du}{dx}$$

is an example. The symbol $du$ has a meaning of its own. To attempt to draw such an analogy with the Chain Rule for Partial Derivatives leads to disaster. Formulas (6) are the ones we usually employ in the applications. The parentheses around the individual terms are used to indicate the inseparable nature of each item. Actually, the form (8) for the same formulas avoids the danger of erroneously treating partial derivatives as fractions.

**EXAMPLE 1**   Suppose that $z = x^3 + y^3$, $x = 2r + s$, $y = 3r - 2s$. Find $\partial z / \partial r$ and $\partial z / \partial s$.

**Solution**   We employ the Chain Rule and obtain

$$\frac{\partial z}{\partial x} = 3x^2, \qquad \frac{\partial z}{\partial y} = 3y^2,$$

$$\frac{\partial x}{\partial r} = 2, \quad \frac{\partial x}{\partial s} = 1, \quad \frac{\partial y}{\partial r} = 3, \quad \frac{\partial y}{\partial s} = -2.$$

Therefore

$$\frac{\partial z}{\partial r} = (3x^2)(2) + (3y^2)(3) = 6x^2 + 9y^2 = 6(2r + s)^2 + 9(3r - 2s)^2,$$

$$\frac{\partial z}{\partial s} = (3x^2)(1) + (3y^2)(-2) = 3x^2 - 6y^2 = 3(2r + s)^2 - 6(3r - 2s)^2. \quad \square$$

In Theorem 3 (the Chain Rule), the variables $r$ and $s$ are the **independent variables**; we denote the variables $x$ and $y$ **intermediate variables**. The formulas we derived extend easily to any number of independent variables and any number of intermediate variables. For example,

if   $w = f(x, y, z)$       and if       $x = x(r, s), \quad y = y(r, s), \quad z = z(r, s),$

then

$$\frac{\partial w}{\partial r} = \left(\frac{\partial f}{\partial x}\right)\left(\frac{\partial x}{\partial r}\right) + \left(\frac{\partial f}{\partial y}\right)\left(\frac{\partial y}{\partial r}\right) + \left(\frac{\partial f}{\partial z}\right)\left(\frac{\partial z}{\partial r}\right),$$

and there is a similar formula for $\partial w/\partial s$. The case of four intermediate variables and one independent variable—that is,

$$w = f(x, y, u, v), \quad x = x(t), \quad y = y(t), \quad u = u(t), \quad v = v(t)$$

—leads to the formula

$$\frac{dw}{dt} = \frac{\partial f}{\partial x}\frac{dx}{dt} + \frac{\partial f}{\partial y}\frac{dy}{dt} + \frac{\partial f}{\partial u}\frac{du}{dt} + \frac{\partial f}{\partial v}\frac{dv}{dt}.$$

The ordinary $d$ (rather than the round $\partial$) is used for derivatives with respect to $t$, since $w, x, y, u,$ and $v$ are all functions of the *single* variable $t$.

> *Remark.*   As an aid in remembering the Chain Rule, we note that *there are as many terms in the formula as there are intermediate variables.*

**EXAMPLE 2**   If $z = f(x, y) = 2x^2 + xy - y^2 + 2x - 3y + 5$, $x = 2s - t$, $y = s + t$, find $\partial z/\partial t$.

**Solution**   We use the Chain Rule:

$$\frac{\partial f}{\partial x} = 4x + y + 2, \quad \frac{\partial f}{\partial y} = x - 2y - 3, \quad \frac{\partial x}{\partial t} = -1, \quad \frac{\partial y}{\partial t} = 1.$$

Therefore

$$\frac{\partial z}{\partial t} = (4x + y + 2)(-1) + (x - 2y - 3)(1) = -3x - 3y - 5$$

$$= -3(2s - t) - 3(s + t) - 5$$

$$= -9s - 5. \quad \square$$

**EXAMPLE 3**   Given   $w = f(x, y, z) = x^2 + 3y^2 - 2z^2 + 4x - y + 3z - 1, \quad x = t^2 - 2t + 1,$ $y = 3t - 2$, $z = t^2 + 4t - 3$, find $dw/dt$ when $t = 2$.

**Solution**    Employing the Chain Rule, we find

$$\frac{\partial f}{\partial x} = 2x + 4, \quad \frac{\partial f}{\partial y} = 6y - 1, \quad \frac{\partial f}{\partial z} = -4z + 3,$$

$$\frac{dx}{dt} = 2t - 2, \quad \frac{dy}{dt} = 3, \quad \frac{dz}{dt} = 2t + 4.$$

Therefore

$$\frac{dw}{dt} = (2x + 4)(2t - 2) + (6y - 1)(3) + (-4z + 3)(2t + 4).$$

When $t = 2$, we have $x = 1$, $y = 4$, $z = 9$, and so

$$\frac{dw}{dt} = (6)(2) + (23)(3) + (-33)(8) = -183. \qquad \square$$

## 3  PROBLEMS

In each of Problems 1 through 18, use the Chain Rule to obtain the indicated partial derivatives.

1  $z = f(x, y) = x^2 + y^2$; $x = s - 2t$, $y = 2s + t$;  $\dfrac{\partial z}{\partial s}, \dfrac{\partial z}{\partial t}$

2  $z = f(x, y) = x^2 - xy - y^2$; $x = s + t$, $y = -s + t$;  $\dfrac{\partial z}{\partial s}, \dfrac{\partial z}{\partial t}$

3  $z = f(x, y) = x^2 + y^2$; $x = s^2 - t^2$, $y = 2st$;  $\dfrac{\partial z}{\partial s}, \dfrac{\partial z}{\partial t}$

4  $z = f(x, y) = \dfrac{x}{x^2 + y^2}$; $x = s \cos t$, $y = s \sin t$;  $\dfrac{\partial z}{\partial s}, \dfrac{\partial z}{\partial t}$

5  $z = f(x, y) = \dfrac{x}{\sqrt{x^2 + y^2}}$; $x = 2s - t$, $y = s + 2t$;  $\dfrac{\partial z}{\partial s}, \dfrac{\partial z}{\partial t}$

6  $z = f(x, y) = x \cos y$; $x = s^2 t^2$, $y = st$;  $\dfrac{\partial z}{\partial s}, \dfrac{\partial z}{\partial t}$

7  $z = f(x, y) = xy + y^3$; $x = s \sin t$, $y = t \sin s$;  $\dfrac{\partial z}{\partial s}, \dfrac{\partial z}{\partial t}$

8  $z = f(x, y) = x^2 + 3xy + y^2$; $x = s \ln t$, $y = t + 2s$;  $\dfrac{\partial z}{\partial s}, \dfrac{\partial z}{\partial t}$

9  $z = f(x, y) = ye^{-x} - xe^{-y}$; $x = u + v$, $y = (u + 2v)^2$;  $\dfrac{\partial z}{\partial u}, \dfrac{\partial z}{\partial v}$

10  $z = f(x, y) = e^{\sin(x/y)}$; $x = 3u + v$, $y = 4uv$;  $\dfrac{\partial z}{\partial u}, \dfrac{\partial z}{\partial v}$

11  $z = f(x, y) = x^2 \cos(xy)$; $x = uv^2$, $y = u^2(v + 3u)^{102}$;  $\dfrac{\partial z}{\partial u}, \dfrac{\partial z}{\partial v}$

12  $z = f(x, y) = e^x \cos y$; $x = s^2 - t^2$, $y = 2st$;  $\dfrac{\partial z}{\partial s}, \dfrac{\partial z}{\partial t}$

13  $w = f(x, y, z) = x^2 + y^2 + z^2 + 3xy - 2xz + 4$; $x = 3s + t$, $y = 2s - t$, $z = s + 2t$;  $\partial w/\partial s$, $\partial w/\partial t$

14  $w = f(x, y, z) = x^3 + 2y^3 + z^3$; $x = s^2 - t^2$, $y = s^2 + t^2$, $z = 2st$;  $\partial w/\partial s$, $\partial w/\partial t$

15  $w = f(x, y, z) = x^2 - y^2 + 2z^2$; $x = r^2 + 1$, $y = r^2 - 2r + 1$, $z = r^2 - 2$;  $dw/dr$

16  $z = f(x, y) = \dfrac{x - y}{1 + x^2 + y^2}$; $x = r + 3s - t$, $y = r - 2s + 3t$;  $\partial z/\partial r$, $\partial z/\partial s$, $\partial z/\partial t$

17  $w = f(u) = u^3 + 2u^2 - 3u + 1$; $u = r^2 - s^2 + t^2$;  $\dfrac{\partial w}{\partial r}, \dfrac{\partial w}{\partial s}, \dfrac{\partial w}{\partial t}$

18  $w = f(x, y, u, v) = x^2 + y^2 - u^2 - v^2 + 3x - 2y + u - v$; $x = 2r + s - t$, $y = r - 2s + t$, $u = 3r - 2s + t$, $v = r - s - t$;  $\dfrac{\partial w}{\partial r}, \dfrac{\partial w}{\partial s}, \dfrac{\partial w}{\partial t}$

In Problems 19 through 28, use the Chain Rule to find the indicated derivatives at the values given.

19  $z = x^2 - y^2$; $x = r \cos \theta$, $y = r \sin \theta$;  $\dfrac{\partial z}{\partial r}, \dfrac{\partial z}{\partial \theta}$  where

$$r = \sqrt{2}, \theta = \frac{\pi}{4}$$

**20** $w = x^2 + y^2 - z^2$; $x = 1 - t^2$, $y = 2t + 3$, $z = t^2 + t$;
$\dfrac{dw}{dt}$ where $t = -1$

**21** $w = xy + yz + zx$; $x = t \cos t$, $y = t \sin t$, $z = t$;
$\dfrac{dw}{dt}$ where $t = \dfrac{\pi}{4}$

**22** $z = u^3 + 2u - 3$; $u = s^2 + t^2 - 4$; $\dfrac{\partial z}{\partial s}, \dfrac{\partial z}{\partial t}$ where
$s = 1$ and $t = 2$

**23** $z = \dfrac{xy}{x^2 + y^2}$; $x = r \cos \theta$, $y = r \sin \theta$; $\dfrac{\partial z}{\partial r}, \dfrac{\partial z}{\partial \theta}$ where
$r = 3$, $\theta = \dfrac{\pi}{6}$

**24** $f = e^{xyz} \cos xyz$; $x = r^2 + t^2$, $y = t^2 + r$, $z = r + t$;
$\dfrac{\partial f}{\partial x}, \dfrac{\partial f}{\partial y}, \dfrac{\partial f}{\partial z}$, where $r = 2, t = -2$

**25** $w = x^3 + y^3 + z^3 - u^2 - v^2$; $x = r^2 + s^2 + t^2$, $y = r^2 + s^2 - t^2$,
$z = r^2 - s^2 - t^2$, $u = r^2 + t^2$, $v = r^2 - s^2$;
$\dfrac{\partial w}{\partial s}, \dfrac{\partial w}{\partial t}$ where $r = 1, s = 0, t = -1$

**26** $z = x(3x + y)^{101}$; $x = 2^{uv}$, $y = (3u + v)^2$; $\dfrac{\partial z}{\partial u}, \dfrac{\partial z}{\partial v}$, where
$u = 1, v = 2$

**27** $w = r^3 - s \tan v$; $r = \sin^2 t$, $s = \cos 2tv$, $v = 4t$; $\dfrac{dw}{dt}$,
where $t = 0$

**28** $w = x^4 - y^4 - z^4$; $x = 5r + 3s - 2t + u - v$,
$y = 2r - 4s + t - u^2 + v^2$, $z = s^3 - 2t^2 + 3v^2$;
$\dfrac{\partial w}{\partial s}, \dfrac{\partial w}{\partial v}$ where $r = 1, s = -1, t = 0, u = 3, v = -2$

**29** Suppose that $w = f(x, y, z)$ is continuous and that $\partial f/\partial x$, $\partial f/\partial y$, $\partial f/\partial z$ are continuous. Let $x = x(r, s)$, $y = y(r, s)$, $z = z(r, s)$ be functions all of whose first partial derivatives exist. Derive the Chain Rule formula for $\partial w/\partial s$.

**30** State and prove the Fundamental Lemma on Differentiation (Theorem 2) for functions $f$ of three variables (say $x$, $y$, and $z$).

**31** Let $f = f(x, y, z)$ be a function with continuous partial derivatives and define $g(t) = f(tx, ty, tz)$. Show that

$$g'(t) = xf_x + yf_y + zf_z.$$

---

**4**

## APPLICATIONS OF THE CHAIN RULE

The Chain Rule may be employed profitably in many types of applications. These are best illustrated with examples, and we begin with two problems in related rates, a topic discussed in Chapter 4, page 169.

**EXAMPLE 1**    At a certain instant the altitude of a right circular cone is 30 cm and is increasing at the rate of 2 cm/sec. At the same instant, the radius of the base is 20 cm and is increasing at the rate of 1 cm/sec. At what rate is the volume increasing at that instant? (See Fig. 11.)

**Solution**    The volume $V$ is given by $V = \frac{1}{3}\pi r^2 h$, with $r$ and $h$ functions of the time $t$. We can apply the Chain Rule to obtain

$$\frac{dV}{dt} = \frac{\partial V}{\partial r}\frac{dr}{dt} + \frac{\partial V}{\partial h}\frac{dh}{dt}$$

$$= \frac{2}{3}\pi rh\frac{dr}{dt} + \frac{1}{3}\pi r^2\frac{dh}{dt}.$$

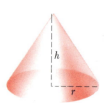

**FIGURE 11**

At the given instant,

$$\frac{dV}{dt} = \frac{2}{3}\pi(20)(30)(1) + \frac{1}{3}\pi(20)^2(2) = \frac{2000\pi}{3} \text{ cm}^3/\text{sec.} \qquad \square$$

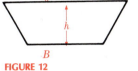

**FIGURE 12**

**EXAMPLE 2**    The base $B$ of a trapezoid increases in length at the rate of 2 cm/sec and the base $b$ decreases in length at the rate of 1 cm/sec. If the altitude $h$ is increasing at the rate of 3 cm/sec, how rapidly is the area $A$ changing when $B = 30$ cm, $b = 50$ cm, and $h = 10$ cm? (See Fig. 12.)

**Solution**    The area $A$ is given by $A = \frac{1}{2}(B + b)h$, with $B$, $b$, and $h$ functions of time. We apply the Chain Rule to get

$$\frac{dA}{dt} = \frac{\partial A}{\partial B}\frac{dB}{dt} + \frac{\partial A}{\partial b}\frac{db}{dt} + \frac{\partial A}{\partial h}\frac{dh}{dt} = \frac{1}{2}h\frac{dB}{dt} + \frac{1}{2}h\frac{db}{dt} + \frac{1}{2}(B+b)\frac{dh}{dt}$$

$$= (5)(2) + (5)(-1) + (40)(3)$$

$$= 125 \text{ cm}^2/\text{sec.}$$

Note that since $b$ is decreasing, $db/dt$ is negative.      $\square$

The next example shows that many applications require a clear understanding of the symbolism in partial differentiation.

**EXAMPLE 3**    Suppose that $z = f(x + at)$ and $a$ is constant. Show that

$$\frac{\partial z}{\partial t} = a\frac{\partial z}{\partial x}.$$

**Solution**    We observe that $f$ is a function of *one argument* (in which, however, two variables occur in a particular combination). We let $u = x + at$ and, if we now write

$$z = f(u), \qquad u = x + at,$$

we recognize the applicability of the Chain Rule. Therefore

$$\frac{\partial z}{\partial x} = \frac{dz}{du}\frac{\partial u}{\partial x} = f'(u) \cdot 1,$$

$$\frac{\partial z}{\partial t} = \frac{dz}{du}\frac{\partial u}{\partial t} = f'(u) \cdot a.$$

We conclude that

$$\frac{\partial z}{\partial t} = a\frac{\partial z}{\partial x}. \qquad \square$$

**EXAMPLE 4**    An airplane is traveling directly east at 300 km/hr and is climbing at the rate of 600 meters/min. At a certain instant, the airplane is 12,000 meters above ground and 5 km directly west of an observer on the ground. How fast is the distance changing between the airplane and the observer at this instant?

**Solution**    Referring to Fig. 13, with the observer at $O$ and the airplane at $A$, we see that $x$, $y$, and $s$ are functions of the time $t$. The distance $s$ between the airplane and the observer is given by $s = (x^2 + y^2)^{1/2}$, and we wish to find $ds/dt$. Using the Chain Rule, we get

$$\frac{ds}{dt} = \frac{\partial s}{\partial x}\frac{dx}{dt} + \frac{\partial s}{\partial y}\frac{dy}{dt} = \frac{x}{\sqrt{x^2+y^2}}\frac{dx}{dt} + \frac{y}{\sqrt{x^2+y^2}}\frac{dy}{dt}.$$

From the given data we see that, at the instant in question,

$$y = 12{,}000, \quad x = 5{,}000, \quad \frac{dx}{dt} = -83.3 \text{ m/sec.} \quad \text{and} \quad \frac{dy}{dt} = +10 \text{ m/sec.}$$

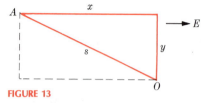

FIGURE 13

Therefore

$$\frac{ds}{dt} = \frac{5{,}000}{13{,}000}(-83.3) + \frac{12{,}000}{13{,}000} \tag{10}$$

$$= -22.8 \text{ m/sec.}$$

The negative sign indicates that the airplane is approaching the observer.

## 4   PROBLEMS

**1** Find the rate at which the lateral area of the cone in Example 1 is increasing at the given instant.

**2** At a certain instant a right circular cylinder has radius of base 10 cm and altitude 15 cm. At this instant the radius is decreasing at the rate of 5 cm/sec and the altitude is increasing at the rate of 4 cm/sec. How rapidly is the volume changing at this moment?

**3** A gas obeys the law $pv = kT (k = \text{const})$. At a certain instant while the gas is being compressed, $v = 15 \text{ m}^3$, $p = 25 \text{ kg/cm}^2$, $v$ is decreasing at the rate of 3 m³/min, and $p$ is increasing at the rate of $6\frac{2}{3}$ kg/cm²/min. Find $dT/dt$. (Answer in terms of $k$.)

**4** a) In Problem 2, find how rapidly the lateral surface area of the cylinder is changing at the same instant. (b) What would the result be for the area $A$ consisting of the top and bottom of the cylinder as well as the lateral surface?

**5** At a certain instant of time, the angle $A$ of a triangle $ABC$ is 60° and increasing at the rate of 5°/sec, the side $AB$ is 10 cm and increasing at the rate of 1 cm/sec, and side $AC$ is 16 cm and decreasing at the rate of $\frac{1}{2}$ cm/sec. Find the rate of change of side $BC$.

**6** A point moves along the surface $z = x^2 + 2y^2 - 3x + y$ in such a way that $dx/dt = 3$ and $dy/dt = 2$. Find how $z$ changes with time when $x = 1$, $y = 4$.

**7** Water is leaking out of a conical tank at the rate of 0.5 m³/min. The tank is also stretching in such a way that, while it remains conical, the distance across the top at the water surface is increasing at the rate of 0.2 m/min. How fast is the height $h$ of water changing at the instant when $h = 10$ and the volume of water is 75 cubic meters?

**8** A rectangular bin is changing in size in such a way that its length is increasing at the rate of 3 cm/sec, its width is decreasing at the rate of 2 cm/sec, and its height is increasing at the rate of 1 cm/sec. (a) How fast is the volume changing at the instant when the length is 15, the width is 10, and the height is 8? (b) How fast is the total surface area changing at the same instant?

**9** a) Given $z = f(y/x)$. Find $\partial z/\partial x$ and $\partial z/\partial y$ in terms of $f'(y/x)$ and $x$ and $y$. (*Hint:* Let $u = y/x$.) (b) Show that $x(\partial z/\partial x) + y(\partial z/\partial y) = 0$.

**10** a) Given that $w = f(y - x - t, z - y + t)$. By letting $u = y - x - t$, $v = z - y + t$, find $\partial w/\partial x$, $\partial w/\partial y$, $\partial w/\partial z$, $\partial w/\partial t$ in terms of $f_u$ and $f_v$. (b) Show that

$$\frac{\partial w}{\partial x} + 2\frac{\partial w}{\partial y} + \frac{\partial w}{\partial z} + \frac{\partial w}{\partial t} = 0.$$

**11** Suppose that $z = f(x, y)$ and $x = r \cos \theta$, $y = r \sin \theta$.
(a) Express $\partial z/\partial r$ and $\partial z/\partial \theta$ in terms of $\partial z/\partial x$ and $\partial z/\partial y$.
(b) Show that

$$\left(\frac{\partial z}{\partial r}\right)^2 + \frac{1}{r^2}\left(\frac{\partial z}{\partial \theta}\right)^2 = \left(\frac{\partial z}{\partial x}\right)^2 + \left(\frac{\partial z}{\partial y}\right)^2.$$

**12** Suppose that $z = f(x, y)$, $x = e^s \cos t$, $y = e^s \sin t$. Show that

$$\left(\frac{\partial z}{\partial s}\right)^2 + \left(\frac{\partial z}{\partial t}\right)^2 = e^{2s}\left[\left(\frac{\partial z}{\partial x}\right)^2 + \left(\frac{\partial z}{\partial y}\right)^2\right].$$

**13** Suppose that $u = f(x + at, y + bt)$, with $a$ and $b$ constants. Show that

$$\frac{\partial u}{\partial t} = a\frac{\partial u}{\partial x} + b\frac{\partial u}{\partial y}.$$

**14** Given that

$$f(x, y) = \frac{x + y}{x^2 - xy + y^2},$$

show that

$$xf_x + yf_y = -f.$$

**15** Given that $f(x, y) = x^2 - y^2 + xy \ln (y/x)$, show that $xf_x + yf_y = 2f$.

**16** If in Example 4 a second observer is situated at a point $O'$, 12 kilometers west of the one at $O$, find the rate of change of the distance between $A$ and $O'$ at the same instant.

**17** If in Example 2 one of the acute angles is held constant at $\pi/3$ radians, find the rate of change of the perimeter of the trapezoid at the instant in question.

**18** A function of two variables satisfies the relation $f(tx, ty) = t^n f(x, y)$, for some number $n$ and all $t$. Show that for such functions $xf_x(x, y) + yf_y(x, y) = nf(x, y)$.

**19** Suppose $f = f(u)$ is a differentiable function of one variable. Let $z = f(xy)$ and show that

$$x\frac{\partial z}{\partial x} - y\frac{\partial z}{\partial y} = 0.$$

**20** Suppose $u = f(x, y)$ and $v = g(x, y)$ satisfy the relations $u_x = v_y$ and $u_y = -v_x$ (these are known as the *Cauchy-Riemann equations*). If $x = r \cos \theta$, $y = r \sin \theta$, show that

$$\frac{\partial u}{\partial r} = \frac{1}{r}\frac{\partial v}{\partial \theta} \quad \text{and} \quad \frac{\partial v}{\partial r} = \frac{-1}{r}\frac{\partial u}{\partial \theta}.$$

**21** A rectangular bin (without a top) is changing in size in such a way that its total surface area always maintains the value of 100 sq. centimeters. The length and width are each increasing at the rate of 1 cm/sec. Find how fast the height and the volume are changing when the length and width are each 2 cm.

**22** The number $A$ of bacteria at time $t$ (in hours) of a certain type follows a growth law given by

$$\frac{\partial A}{\partial t} = x\frac{\partial A}{\partial x} + 2y\frac{\partial A}{\partial y},$$

where $x$ and $y$ are two types of liquid (in cc) in which the culture is grown. Show that $A(x, y, t) = c(x^2 + y)e^{2t}$, where $c$ is a constant, is a possible law of growth. If there are 4 cc of $x$-type liquid and 10 cc of $y$-type and if there are 1000 bacteria at time $t = 0$, find the number of bacteria after four hours.

---

# 5

# DIRECTIONAL DERIVATIVES. GRADIENT

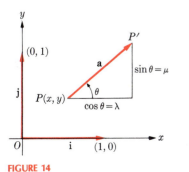

**FIGURE 14**

The partial derivative of a function with respect to $x$ may be considered as the derivative in the $x$ direction; the partial derivative with respect to $y$ is the derivative in the $y$ direction. We now show how we may define the derivative in *any direction*. To see this, we consider a function $f(x, y)$ and a point $P(x, y)$ in the $xy$ plane. A particular direction is singled out by specifying the angle $\theta$ which a line through $P$ makes with the positive $x$ axis (Fig. 14). We may also prescribe the direction by drawing the directed line segment $\overrightarrow{PP'}$ of *unit* length, as shown, and defining the vector $\mathbf{a}$ by the relation

$$\mathbf{a} = \lambda\mathbf{i} + \mu\mathbf{j},$$

with $\lambda = \cos\theta$, $\mu = \sin\theta$, and $\mathbf{i}$ and $\mathbf{j}$ the customary unit vectors. Note that $\mathbf{a}$ is a vector of unit length. The vector $\mathbf{a}$ determines the same direction as the angle $\theta$.

**DEFINITION**

*Let $f$ be a function of two variables. The **directional derivative $D_{\mathbf{a}}f$ of $f$ in the direction of $\mathbf{a}$** is defined by*

$$D_{\mathbf{a}}f(x, y) = \lim_{h \to 0} \frac{f(x + \lambda h, y + \mu h) - f(x, y)}{h}$$

*whenever the limit exists.*

*Remark.* We note that, when $\theta = 0$, then $\lambda = 1$, $\mu = 0$, and the direction is the positive $x$ direction. The directional derivative is exactly $\partial f / \partial x$. Similarly, if $\theta = \pi/2$ we have $\lambda = 0$, $\mu = 1$, and the directional derivative is $\partial f / \partial y$.

The working formula for directional derivatives is established in the next theorem.

**THEOREM 4**

*If $f(x, y)$ and its partial derivatives are continuous and*

$$\mathbf{a} = (\cos\theta)\mathbf{i} + (\sin\theta)\mathbf{j},$$

*then*

$$D_{\mathbf{a}}f(x, y) = f_x(x, y)\cos\theta + f_y(x, y)\sin\theta.$$

**Proof**   The proof uses the following device. We define the function $g(s)$ by

$$g(s) = f(x + s\cos\theta, y + s\sin\theta),$$

in which we keep $x$, $y$, and $\theta$ fixed and allow $s$ to vary. The Chain Rule now yields

$$g'(s) = f_x(x + s\cos\theta, y + s\sin\theta)\cos\theta + f_y(x + s\cos\theta, y + s\sin\theta)\sin\theta,$$

and setting $s = 0$ we get

$$g'(0) = f_x(x, y)\cos\theta + f_y(x, y)\sin\theta.$$

By definition we know that

$$g'(0) = \lim_{s \to 0} \frac{g(s) - g(0)}{s} = \lim_{s \to 0} \frac{f(x + s\cos\theta, y + s\sin\theta) - f(x, y)}{s}.$$

and hence $g'(0)$ is precisely $D_{\mathbf{a}}f(x, y)$. The result is established.   ☐

**EXAMPLE 1**   Given $f(x, y) = x^2 + 2y^2 - 3x + 2y$, find the directional derivative of $f$ in the direction $\theta = \pi/6$. What is the value of this derivative at the point $(2, -1)$?

**Solution**   We compute

$$\frac{\partial f}{\partial x} = 2x - 3, \qquad \frac{\partial f}{\partial y} = 4y + 2.$$

Therefore, since $\cos (\pi/6) = \sqrt{3}/2$, $\sin (\pi/6) = 1/2$, we find

$$D_{\mathbf{a}} f = (2x - 3) \cdot \tfrac{1}{2}\sqrt{3} + (4y + 2) \cdot \tfrac{1}{2}.$$

In particular, when $x = 2$ and $y = -1$, we obtain

$$D_{\mathbf{a}} f = \tfrac{1}{2}\sqrt{3} - 1. \qquad \Box$$

An alternate notation for directional derivative for functions of two variables is the symbol

$$d_{\theta} f(x, y),$$

in which $\theta$ is the angle the direction makes with the positive $x$ axis. For a fixed value of $x$ and $y$, the directional derivative is a function of $\theta$. It is an ordinary problem in maxima and minima to find the value of $\theta$ which makes the directional derivative at a given point the largest or the smallest. The next example shows the method.

EXAMPLE 2    Given $f(x, y) = x^2 - xy - y^2$, find $d_{\theta} f(x, y)$ at the point $(2, -3)$. For what value of $\theta$ does $d_{\theta}(2, -3)$ take on its maximum value?

Solution    We have

$$\frac{\partial f}{\partial x} = 2x - y, \qquad \frac{\partial f}{\partial y} = -x - 2y.$$

Therefore for $x = 2$, $y = -3$,

$$d_{\theta} f(2, -3) = 7 \cos \theta + 4 \sin \theta.$$

To find the maximum of this function of $\theta$ we differentiate the function

$$k(\theta) = 7 \cos \theta + 4 \sin \theta$$

and set the derivative equal to zero. We get

$$k'(\theta) = -7 \sin \theta + 4 \cos \theta = 0 \qquad \text{or} \qquad \tan \theta = \tfrac{4}{7}.$$

The result is $\cos \theta = \pm 7/\sqrt{65}$, $\sin \theta = \pm 4/\sqrt{65}$, and

$$\tan \theta = \tfrac{4}{7}, \qquad \theta = \begin{cases} 0.5191 \text{ radians, approximately,} \\ 3.6607 \text{ radians, approximately.} \end{cases}$$

It is clear by substitution that the first choice for $\theta$ makes $k(\theta)$ a maximum, while the second makes it a minimum. $\qquad \Box$

The definition of directional derivative for functions of two variables has a natural extension to functions of three variables. In three dimensions, a direction is determined by a set of direction cosines $\lambda$, $\mu$, $v$ or, equivalently, by a vector

$$\mathbf{a} = \lambda \mathbf{i} + \mu \mathbf{j} + v \mathbf{k}.$$

We recall that $\lambda^2 + \mu^2 + v^2 = 1$, and so **a** is a *unit* vector.

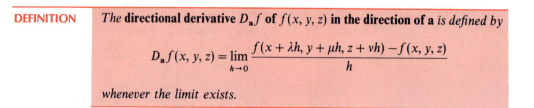

**DEFINITION**    *The* **directional derivative** $D_\mathbf{a} f$ *of* $f(x, y, z)$ **in the direction of a** *is defined by*

$$D_\mathbf{a} f(x, y, z) = \lim_{h \to 0} \frac{f(x + \lambda h, y + \mu h, z + vh) - f(x, y, z)}{h}$$

*whenever the limit exists.*

The proof of the next theorem is entirely analogous to (and employs the same device as) the proof of Theorem 4.

**THEOREM 5**    *If* $f(x, y, z)$ *and its partial derivatives are continuous and*

$$\mathbf{a} = \lambda\mathbf{i} + \mu\mathbf{j} + v\mathbf{k}$$

*is a unit vector, then*

$$D_\mathbf{a} f(x, y, z) = \lambda f_x(x, y, z) + \mu f_y(x, y, z) + v f_z(x, y, z).$$

**EXAMPLE 3**    Find the directional derivative of

$$f(x, y, z) = x^2 + y^2 + z^2 - 3xy + 2xz - yz$$

at the point $(1, 2, -1)$.

**Solution**    We have

$$\frac{\partial f}{\partial x} = 2x - 3y + 2z; \qquad \frac{\partial f}{\partial y} = 2y - 3x - z; \qquad \frac{\partial f}{\partial z} = 2z + 2x - y.$$

Denoting the direction by $\mathbf{a} = \lambda\mathbf{i} + \mu\mathbf{j} + v\mathbf{k}$, we get

$$D_\mathbf{a} f(1, 2, -1) = -6\lambda + 2\mu - 2v. \qquad \square$$

**EXAMPLE 4**    Given the function

$$f(x, y, z) = xe^{yz} + ye^{xz} + ze^{xy},$$

find the directional derivative at $P(1, 0, 2)$ in the direction going from $P$ to $P'(5, 3, 3)$.

**Solution**    A set of direction numbers for the line through $P$ and $P'$ is $4, 3, 1$. The corresponding direction cosines are $4/\sqrt{26}, 3/\sqrt{26}, 1/\sqrt{26}$, which we denote by $\lambda, \mu, v$. The direction **a** is given by

$$\mathbf{a} = \frac{4}{\sqrt{26}}\mathbf{i} + \frac{3}{\sqrt{26}}\mathbf{j} + \frac{1}{\sqrt{26}}\mathbf{k}.$$

We find that

$$\frac{\partial f(x, y, z)}{\partial x} = e^{yz} + yze^{xz} + zye^{xy} \qquad \text{and} \qquad \frac{\partial f(1, 0, 2)}{\partial x} = 1,$$

$$\frac{\partial f(x, y, z)}{\partial y} = xze^{yz} + e^{xz} + xze^{xy} \qquad \text{and} \qquad \frac{\partial f(1, 0, 2)}{\partial y} = 4 + e^2,$$

$$\frac{\partial f(x, y, z)}{\partial z} = xye^{yz} + xye^{xz} + e^{xy} \qquad \text{and} \qquad \frac{\partial f(1, 0, 2)}{\partial z} = 1.$$

Therefore

$$D_{\mathbf{a}} f(1, 0, 2) = \frac{4}{\sqrt{26}} + (4 + e^2)\frac{3}{\sqrt{26}} + \frac{1}{\sqrt{26}} = \frac{17 + 3e^2}{\sqrt{26}}. \qquad \square$$

As the next definition shows, the *gradient* of a function is a vector containing the partial derivatives of the function.

---

**DEFINITIONS**

i) *If $f(x, y)$ has partial derivatives, the **gradient vector** is defined by*

$$\textbf{grad } f(x, y) = f_x(x, y)\mathbf{i} + f_y(x, y)\mathbf{j}.$$

ii) *If $g(x, y, z)$ has partial derivatives, the **gradient vector** is defined by*

$$\textbf{grad } g(x, y, z) = g_x(x, y, z)\mathbf{i} + g_y(x, y, z)\mathbf{j} + g_z(x, y, z)\mathbf{k}.$$

---

The symbol $\nabla$, an inverted delta, is called "del" and is a common one used to denote the gradient. We will frequently write $\nabla f$ for **grad** $f$.

If $\mathbf{b}$ and $\mathbf{c}$ are two vectors, the scalar product $\mathbf{b} \cdot \mathbf{c}$ of $\mathbf{b} = b_1\mathbf{i} + b_2\mathbf{j} + b_3\mathbf{k}$ and $\mathbf{c} = c_1\mathbf{i} + c_2\mathbf{j} + c_3\mathbf{k}$ is given by

$$\mathbf{b} \cdot \mathbf{c} = b_1 c_1 + b_2 c_2 + b_3 c_3.$$

For two-dimensional vectors the result is the same, with $b_3 = c_3 = 0$.

We recognize that if $\mathbf{a}$ is a *unit* vector so that $\mathbf{a} = \lambda\mathbf{i} + \mu\mathbf{j} + \nu\mathbf{k}$, then

$$D_{\mathbf{a}} f = \lambda f_x + \mu f_y + \nu f_z = \mathbf{a} \cdot \nabla f.$$

Looked at another way, the scalar product of $\mathbf{a}$ and $\nabla f$ is given by

$$\mathbf{a} \cdot \nabla f = |\mathbf{a}||\nabla f| \cos \phi = D_{\mathbf{a}} f,$$

where $\phi$ is the angle between the vector $\mathbf{a}$ and $\nabla f$.

From the above formula we can conclude that $D_{\mathbf{a}} f$ is a *maximum when $\phi$ is zero*—i.e., when $\mathbf{a}$ is in the direction of **grad** $f$.

**EXAMPLE 5**    Given the function

$$f(x, y, z) = x^3 + 2y^3 + z^3 - 4xyz,$$

find the maximum value of $D_{\mathbf{a}} f$ at the point $P = (-1, 1, 2)$.

**Solution**    We have

$$\frac{\partial f}{\partial x} = 3x^2 - 4yz; \qquad \frac{\partial f}{\partial y} = 6y^2 - 4xz; \qquad \frac{\partial f}{\partial z} = 3z^2 - 4xy.$$

Therefore

$$\nabla f(-1, 1, 2) = -5\mathbf{i} + 14\mathbf{j} + 16\mathbf{k},$$

and a unit vector **a** in the direction of $\nabla f$ is

$$\mathbf{a} = -\frac{5}{3\sqrt{53}}\mathbf{i} + \frac{14}{3\sqrt{53}}\mathbf{j} + \frac{16}{3\sqrt{53}}\mathbf{k}.$$

The maximum value of $D_{\mathbf{a}}f$ is given by

$$D_{\mathbf{a}}f = -5\left(\frac{-5}{3\sqrt{53}}\right) + 14\left(\frac{14}{3\sqrt{53}}\right) + 16\left(\frac{16}{3\sqrt{53}}\right) = 3\sqrt{53}. \qquad \Box$$

## 5   PROBLEMS

In Problems 1 through 10, find in each case $d_\theta f(x, y)$ at the given point.

**1** $f(x, y) = x^2 + y^2$; (3, 4)

**2** $f(x, y) = \sqrt{x^2 + y^2}$; $P(-2, 3)$

**3** $f(x, y) = 6x - 3y$; $P(1, 1)$

**4** $f(x, y) = x^2 \ln(x - y)$; $P(7, 6)$

**5** $f(x, y) = e^{-3x} \tan y$; $P(0, \pi/4)$

**6** $f(x, y) = x^3 + y^3 - 3x^2y - 3xy^2$; $(1, -2)$

**7** $f(x, y) = \arctan(y/x)$; (4, 3)

**8** $f(x, y) = \sin(xy)$; $(2, \pi/4)$

**9** $f(x, y) = e^x \cos y$; $(0, \pi/3)$

**10** $f(x, y) = (\sin x)^{xy}$; $(\pi/2, 0)$

In each of Problems 11 through 17, find the value of $d_\theta f(x, y)$ at the given point. Also, find the value of $\theta$ which makes $d_\theta f$ a maximum at this point. Express your answer in terms of $\sin \theta$ and $\cos \theta$.

**11** $f(x, y) = x^2 + y^2 - 2x + 3y$; $(2, -1)$

**12** $f(x, y) = \arctan(x/y)$; (3, 4)

**13** $f(x, y) = e^x \sin y$; $(0, \pi/6)$

**14** $f(x, y) = (\sin y)^{xy}$; $(0, \pi/2)$

**15** $f(x, y) = 17x - 3y$; $P(0, \sqrt{2})$

**16** $f(x, y) = 3x + 3y$; $P(0, 0)$

**17** $f(x, y) = \dfrac{e^x - e^{-y}}{2}$; $P(0, \ln 2)$

In each of Problems 18 through 23, find $D_{\mathbf{a}}f$ at the given point $P$.

**18** $f(x, y, z) = x^2 + xy - xz + y^2 - z^2$; $P(2, 1, -2)$

**19** $f(x, y, z) = x^2y + xze^y - xye^z$; $P(-2, 3, 0)$

**20** $f(x, y, z) = \cos xy + \sin xz$; $P(0, 2, -1)$

**21** $f(x, y, z) = \ln(x + y + z) - xyz$; $P(-1, 2, 1)$

**22** $f(x, y, z) = z^2e^{xy}$; $P(0, 0, 1)$

**23** $f(x, y, z) = \sqrt{xy} \cos z$; $P(9, 16, \pi/4)$

In Problems 24 through 32, in each case find $D_{\mathbf{a}}f$ at the given point $P$ when **a** is the given vector.

**24** $f(x, y, z) = x^2 + 2xy - y^2 + xz + z^2$; $P(2, 1, 1)$; $\mathbf{a} = \frac{1}{3}\mathbf{i} - \frac{2}{3}\mathbf{j} + \frac{2}{3}\mathbf{k}$

**25** $f(x, y, z) = x^2y + xye^z - 2xze^y$; $P(1, 2, 0)$; $\mathbf{a} = \frac{2}{7}\mathbf{i} - \frac{3}{7}\mathbf{j} + \frac{6}{7}\mathbf{k}$

**26** $f(x, y, z) = \sin xz + \cos xy$; $P(0, -1, 2)$; $\mathbf{a} = \frac{1}{\sqrt{6}}\mathbf{i} - \frac{1}{\sqrt{6}}\mathbf{j} + \frac{2}{\sqrt{6}}\mathbf{k}$

**27** $f(x, y, z) = \tan xyz + \sin xy - \cos xz$; $P(0, 1, 1)$; $\mathbf{a} = \frac{1}{\sqrt{26}}\mathbf{i} + \frac{3}{\sqrt{26}}\mathbf{j} + \frac{4}{\sqrt{26}}\mathbf{k}$

**28** $f(x, y) = x \sin^2 y$; $P(3, \pi/4)$;   $\mathbf{a} = \frac{4}{\sqrt{17}}\mathbf{i} + \frac{1}{\sqrt{17}}\mathbf{j}$

**29** $f(x, y) = \arctan(y/x)$; $P(2, -2)$;   $\mathbf{a} = \frac{3}{\sqrt{13}}\mathbf{i} - \frac{2}{\sqrt{13}}\mathbf{j}$

**30** $f(x, y) = x^2 \ln(xy)$; $P(5, \frac{1}{5})$;  $\mathbf{a} = \dfrac{1}{\sqrt{10}}\mathbf{i} - \dfrac{3}{\sqrt{10}}\mathbf{j}$

**31** $f(x, y, z) = \sqrt{xy} \sin z$; $P(1, 4, \pi/4)$;

$$\mathbf{a} = \dfrac{1}{\sqrt{11}}\mathbf{i} + \dfrac{1}{\sqrt{11}}\mathbf{j} - \dfrac{3}{\sqrt{11}}\mathbf{k}$$

**32** $f(x, y, z) = (x + y)^2(y + z)$; $P(1, 3, 2)$;  $\mathbf{a} = -\dfrac{2}{3}\mathbf{i} + \dfrac{1}{3}\mathbf{j} + \dfrac{2}{3}\mathbf{k}$

In each of Problems 33 through 36, find the vector $\nabla f$ at the given point.

**33** $f(x, y) = x^3 - 2x^2y + xy^2 - y^3$; $P(3, -2)$

**34** $f(x, y) = \ln(x^2 + y^2 + 1) + e^{2xy}$; $P(0, -2)$

**35** $f(x, y, z) = \sin xy + \sin xz + \sin yz$; $P(1, 2, -1)$

**36** $f(x, y, z) = xze^{xy} + yze^{xz} + xye^{yz}$; $P(-1, 2, 1)$

In each of Problems 37 through 40, find $D_{\mathbf{a}} f$ at the given point $P$ where $\mathbf{a}$ is a unit vector in the direction $\overrightarrow{PP'}$. Also, find at $P$ the value of $D_{\bar{\mathbf{a}}} f$ where $\bar{\mathbf{a}}$ is a unit vector such that $D_{\bar{\mathbf{a}}} f$ is a maximum.

**37** $f(x, y, z) = x^2 + 3xy + y^2 + z^2$; $P(1, 0, 2)$; $P'(-1, 3, 4)$

**38** $f(x, y, z) = e^x \cos y + e^y \sin z$; $P(2, 1, 0)$; $P'(-1, 2, 2)$

**39** $f(x, y, z) = \ln(x^2 + y^2) + e^z$; $P(0, 1, 0)$; $P'(-4, 2, 3)$

**40** $f(x, y, z) = x \cos y + y \cos z + z \cos x$; $P(2, 1, 0)$; $P'(1, 4, 2)$

**41** Prove Theorem 5.

**42** Given $g(x, y, z) = 2x^2 + y^2 - z^2 + 2xy - 3x + 2y + z$, find the points, if any, such that $|\nabla g| = 0$.

**43** The temperature at any point of a rectangular plate in the $xy$ plane is given by the formula $T = 50(x^2 - y^2)$ (in degrees Celsius). Find $d_\theta T(4, 3)$, and find $\tan \theta$ when $d_\theta T(4, 3) = 0$. Find also the slope of the curve $T = \text{const}$ which passes through that point.

**44** The temperature at any point $(x, y)$ on a rectangular metal plate in the $xy$ plane is given by the formula

$$T(x, y) = \frac{xy}{1 + x^2 + y^2}$$

in degrees Celsius.

a) Find the rate of change of temperature at $(1, 1)$ in the direction of $\mathbf{a} = \mathbf{i} - 2\mathbf{j}$.

b) Determine in which direction the temperature is changing most rapidly at the point $(1, 1)$ on the plate.

**45** Given $g(x, y, z) = x^3 - 2y^2 + z^2 - 2xz + 4y^2z$, find the points, if any, such that $|\nabla g| = 0$.

**46** Suppose that $f(x, y, z)$ and $g(x, y, z)$ are given functions and that $a$ and $b$ are constants. Prove the formulas:

$$\nabla(af + bg) = a\nabla f + b\nabla g,$$

$$\nabla(fg) = f\nabla g + g\nabla f.$$

**47** Suppose that the domain of $F = F(u)$ contains the range of $g = g(x, y, z)$. Prove that

$$\nabla(F(g)) = F'(u)\nabla g.$$

**48** Suppose that $\mathbf{F}(x, y, z) = f_1(x, y, z)\mathbf{i} + f_2(x, y, z)\mathbf{j} + f_3(x, y, z)\mathbf{k}$ is a vector function. We define the vector $\nabla$ by the symbolic formula

$$\nabla = \frac{\partial}{\partial x}\mathbf{i} + \frac{\partial}{\partial y}\mathbf{j} + \frac{\partial}{\partial z}\mathbf{k}$$

and the cross product $\nabla \times \mathbf{F}$ is given by

$$\nabla \times \mathbf{F} = \left(\frac{\partial f_3}{\partial y} - \frac{\partial f_2}{\partial z}\right)\mathbf{i} + \left(\frac{\partial f_1}{\partial z} - \frac{\partial f_3}{\partial x}\right)\mathbf{j} + \left(\frac{\partial f_2}{\partial x} - \frac{\partial f_1}{\partial y}\right)\mathbf{k}.$$

If $g(x, y, z)$ is a scalar function show that

$$\nabla \times (g\mathbf{F}) = g(\nabla \times \mathbf{F}) + (\nabla g) \times \mathbf{F}.$$

**49** Refer to Problem 48 for definitions and show that

$$\nabla \cdot (\mathbf{F} \times \mathbf{G}) = \mathbf{G} \cdot \nabla \times \mathbf{F} - \mathbf{F} \cdot \nabla \times \mathbf{G}.$$

**50** Suppose that $f : R^3 \to R^1$ has the property that $\nabla f \equiv 0$. Show that $f$ is a constant.

**\*51** Let $\mathbf{F} = x^3\mathbf{i} + y^3\mathbf{j} + z^3\mathbf{k}$. Show that $\nabla \times \mathbf{F} \equiv \mathbf{0}$; conclude that $\nabla \times \mathbf{F} = \mathbf{0}$ does not imply $\mathbf{F} = \text{const}$. (See Problem 50.) (*Hint:* See the proof of Theorem 2.)

---

**6**

## GEOMETRIC INTERPRETATION OF PARTIAL DERIVATIVES. TANGENT PLANES

From the geometric point of view, a function of one variable represents a curve in the plane. The derivative at a point on the curve is the slope of the line tangent to the curve at this point. A function of two variables $z = f(x, y)$ represents a surface in three-dimensional space. If $(x_0, y_0)$ are the coordinates of a point in the $xy$ plane, then $P(x_0, y_0, z_0)$, with $z_0 = f(x_0, y_0)$, is a point on the surface. Consider the vertical plane $y = y_0$, as shown in Fig. 15. This plane cuts the surface $z = f(x, y)$ in a curve $C_1$ which contains the point $P$. From the definition of partial derivative we see that

$$f_x(x_0, y_0)$$

is the slope of the line tangent to the curve $C_1$ at the point $P$. This line is labeled $L_1$ in Fig. 15 and, of course, is in the plane $y = y_0$. In a completely analogous manner we construct a plane $x = x_0$ intersecting the surface with equation $z = f(x, y)$ in a curve $C_2$. The partial derivative

$$f_y(x_0, y_0)$$

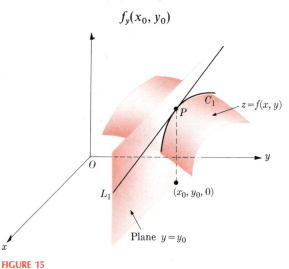

FIGURE 15

is the slope of the line tangent to $C_2$ at the point $P$. The line is denoted $L_2$ in Fig. 16, and its slope is the tangent of the angle $\beta$, as shown.

The directional derivative $d_\theta f(x_0, y_0)$ has a similar interpretation. We construct the vertical plane through $(x_0, y_0, 0)$ which makes an angle $\theta$ with the positive $x$ direction. Such a plane is shown in Fig. 17. The curve $C_3$ is the intersection of this plane with the surface, and the line $L_3$ is the line tangent to $C_3$ at $P$. Then $d_\theta f(x_0, y_0)$ is the slope of $L_3$.

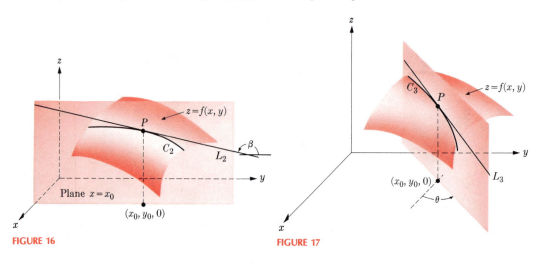

FIGURE 16

FIGURE 17

According to Theorem 2 (The Fundamental Lemma on Differentiation), we may write Formula (1) in the theorem in the form

$$f(x, y) - f(x_0, y_0) = f_x(x_0, y_0)(x - x_0) + f_y(x_0, y_0)(y - y_0)$$
$$+ G_1(x - x_0) + G_2(y - y_0),$$

in which we take

$$x - x_0 = h \quad \text{and} \quad y - y_0 = k.$$

Since $G_1$ and $G_2$ tend to zero as $(x, y) \to (x_0, y_0)$, the next definition has an intuitive geometric meaning.

**DEFINITION**   *The plane whose equation is*

$$m_1(x - x_0) + m_2(y - y_0) - (z - z_0) = 0$$

*where*

$$z_0 = f(x_0, y_0), \qquad m_1 = f_x(x_0, y_0), \qquad m_2 = f_y(x_0, y_0)$$

*is called the* **tangent plane** *to the surface* $z = f(x, y)$ *at* $(x_0, y_0)$.

*Remarks.*   i) According to the geometric interpretation of partial derivative which we gave, it is easy to verify that the tangent plane contains the lines $L_1$ and $L_2$, tangents to $C_1$ and $C_2$, respectively. (See Fig. 18.) Here $L_1$ is the line in Fig. 15 and $L_2$ is the line in Fig. 16.

ii)   From the definition of tangent plane we observe at once that

$$f_x(x_0, y_0), \quad f_y(x_0, y_0), \quad -1$$

is a set of attitude numbers for the plane. (Attitude numbers were defined on page 545.) The equation of the tangent plane is conveniently expressed in vector notation. Define

$$\mathbf{n} = m_1 \mathbf{i} + m_2 \mathbf{j} - \mathbf{k} \quad \text{where} \quad m_1 = f_x(x_0, y_0), m_2 = f_y(x_0, y_0).$$

Also, let $\mathbf{v} = x\mathbf{i} + y\mathbf{j} + z\mathbf{k}$ and $\mathbf{v}_0 = x_0\mathbf{i} + y_0\mathbf{j} + z_0\mathbf{k}$. Then the definition of the tangent plane takes the form

$$\mathbf{n} \cdot (\mathbf{v} - \mathbf{v}_0) = 0.$$

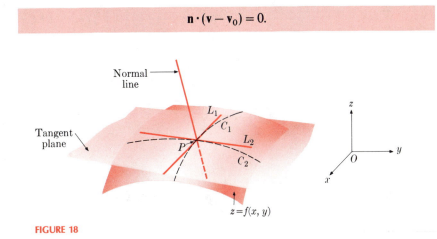

FIGURE 18

Geometrically, the plane is traced by the heads of the directed line segments having base at the origin which form the vector **v**. The geometric importance of the vector **n** is given in the next definition.

**DEFINITION**     *The line with equations*

$$\frac{x - x_0}{m_1} = \frac{y - y_0}{m_2} = \frac{z - z_0}{-1},$$

*with* $z_0 = f(x_0, y_0)$, $m_1 = f_x(x_0, y_0)$, $m_2 = f_y(x_0, y_0)$, *is called the* **normal line** *to the surface at the point* $P(x_0, y_0, z_0)$. *Clearly, the normal line is perpendicular to the tangent plane.* (See Fig. 18.)

*Remark.*   In addition to $f_x(x_0, y_0)$ we shall use the symbols

$$\frac{\partial f(x_0, y_0)}{\partial x}, \qquad \text{and} \qquad \frac{\partial z}{\partial x}\bigg|_{\substack{x = x_0 \\ y = y_0}}$$

for $m_1$, and analogous notations for $m_2$.

The normal line is also expressible in parametric form by the equations

$$x = x_0 + m_1 t$$

$$y = y_0 + m_2 t$$

$$z = z_0 - t$$

or, in terms of vectors, by the single vector equation

$$\mathbf{v} = \mathbf{v}_0 + \mathbf{a}t.$$

**EXAMPLE 1**     Find the equation of the tangent plane and the equations of the normal line to the surface

$$z = x^2 + xy - y^2$$

at the point where $x = 2$, $y = -1$. Express the results in vector form.

**Solution**     We have $z_0 = f(2, -1) = 1$; $f_x(x, y) = 2x + y$, $f_y(x, y) = x - 2y$. Therefore $m_1 = 3$, $m_2 = 4$, and the desired equation for the tangent plane is

$$z - 1 = 3(x - 2) + 4(y + 1).$$

In vector form the tangent plane is

$\mathbf{n} \cdot (\mathbf{v} - \mathbf{v}_0) = 0$   with   $\mathbf{v} = x\mathbf{i} + y\mathbf{j} + z\mathbf{k}$, $\mathbf{n} = 3\mathbf{i} + 4\mathbf{j} - \mathbf{k}$, and $\mathbf{v}_0 = 2\mathbf{i} - \mathbf{j} + \mathbf{k}$.

The equations of the normal line are

$$\frac{x - 2}{3} = \frac{y + 1}{4} = \frac{z - 1}{-1}.$$

The vector equation of the line is

$$\mathbf{v} = \mathbf{v}_0 + \mathbf{n}t \quad \text{with} \quad \mathbf{v}, \mathbf{n}, \text{and } \mathbf{v}_0 \text{ as above.} \qquad \square$$

If the equation of the surface is given in implicit form it is possible to use the methods of Section 2 to find $\partial z/\partial x$ and $\partial z/\partial y$ at the desired point. The next example shows how we obtain the equations of the tangent plane and normal line under such circumstances.

**EXAMPLE 2**    Find the equation of the tangent plane and the equations of the normal line at $(3, -1, 2)$ to the graph of

$$xy + yz + xz - 1 = 0.$$

**Solution**    Holding $y$ constant and differentiating with respect to $x$, we get

$$y + y\frac{\partial z}{\partial x} + z + x\frac{\partial z}{\partial x} = 0 \quad \text{and} \quad \frac{\partial z}{\partial x} = -\frac{y+z}{y+x} = -\frac{1}{2} = m_1.$$

Similarly, holding $x$ constant, we obtain

$$x + y\frac{\partial z}{\partial y} + z + x\frac{\partial z}{\partial y} = 0 \quad \text{and} \quad \frac{\partial z}{\partial y} = -\frac{x+z}{x+y} = -\frac{5}{2} = m_2.$$

The equation of the tangent plane is

$$z - 2 = -\tfrac{1}{2}(x - 3) - \tfrac{5}{2}(y + 1) \quad \Leftrightarrow \quad x + 5y + 2z - 2 = 0.$$

The normal line has equations

$$\frac{x-3}{-1/2} = \frac{y+1}{-5/2} = \frac{z-2}{-1} \quad \Leftrightarrow \quad \frac{x-3}{1} = \frac{y+1}{5} = \frac{z-2}{2}. \qquad \square$$

The methods of the calculus of functions of several variables enable us to establish purely geometric facts, as the next example shows.

**EXAMPLE 3**    Show that any line normal to the sphere

$$x^2 + y^2 + z^2 = a^2$$

always passes through the center of the sphere.

**Solution**    Since the sphere has center at $(0, 0, 0)$, we must show that the equations of every normal line are satisfied for $x = y = z = 0$. Let $(x_0, y_0, z_0)$ be a point on the sphere. Differentiating implicitly, we find

$$\frac{\partial z}{\partial x} = -\frac{x}{z}, \qquad \frac{\partial z}{\partial y} = -\frac{y}{z}.$$

The normal line to the sphere at $(x_0, y_0, z_0)$ has equations

$$\frac{x - x_0}{-x_0/z_0} = \frac{y - y_0}{-y_0/z_0} = \frac{z - z_0}{-1}. \tag{1}$$

Letting $x = y = z = 0$, we get an identity for (1); hence the line passes through the origin.    $\square$

Let the equations

$$z = f(x, y) \quad \text{and} \quad z = g(x, y)$$

represent surfaces which intersect in a curve $C$. The **tangent line** to the intersection at a point $P$ on $C$ is by definition the line of intersection of the tangent planes to $f$ and $g$ at $P$ (Fig. 19).

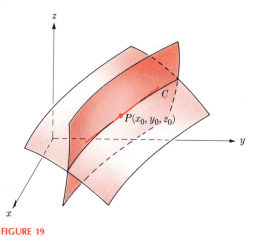

**FIGURE 19**

We can use vector algebra in the following way to find the equations of this tangent line. If $P$ has coordinates $(x_0, y_0, z_0)$, then

$$\mathbf{u} = f_x(x_0, y_0)\mathbf{i} + f_y(x_0, y_0)\mathbf{j} + (-1)\mathbf{k}$$

is the vector normal to the plane tangent to $f$ at $P$. Similarly,

$$\mathbf{v} = g_x(x_0, y_0)\mathbf{i} + g_y(x_0, y_0)\mathbf{j} + (-1)\mathbf{k}$$

is the vector normal to the plane tangent to $g$ at $P$. The line of intersection of these tangent planes is perpendicular to both $\mathbf{u}$ and $\mathbf{v}$. We recall from the study of vectors that, if $\mathbf{u}$ and $\mathbf{v}$ are nonparallel vectors, the vector $\mathbf{u} \times \mathbf{v}$ is perpendicular to both $\mathbf{u}$ and $\mathbf{v}$. Defining the vector $\mathbf{w} = a\mathbf{i} + b\mathbf{j} + c\mathbf{k}$ by the relation

$$\mathbf{w} = \mathbf{u} \times \mathbf{v},$$

we see that the equations of the line of intersection of the two tangent planes are

$$\frac{x - x_0}{a} = \frac{y - y_0}{b} = \frac{z - z_0}{c}.$$

The next example shows how the method works.

**EXAMPLE 4**   Find the equations of the line tangent to the intersection of the surfaces

$$z = f(x, y) = x^2 + 2y^2, \qquad z = g(x, y) = 2x^2 - 3y^2 + 1$$

at the point $(2, 1, 6)$.

**Solution**   We have

$$f_x = 2x, \quad f_y = 4y; \qquad g_x = 4x, \quad g_y = -6y.$$

Therefore

$$\mathbf{u} = 4\mathbf{i} + 4\mathbf{j} - \mathbf{k}; \qquad \mathbf{v} = 8\mathbf{i} - 6\mathbf{j} - \mathbf{k};$$

$$\mathbf{w} \equiv \mathbf{u} \times \mathbf{v} = -10\mathbf{i} - 4\mathbf{j} - 56\mathbf{k}.$$

The desired equations are

$$\frac{x-2}{10} = \frac{y-1}{4} = \frac{z-6}{56}.$$

## 6   PROBLEMS

In Problems 1 through 20, find in each case the equation of the tangent plane and the equations of the normal line to the given surface at the given point. Also express the results in vector form.

1  $z = x^2 + 2y^2; (2, -1, 6)$

2  $z = 3x^2 - y^2 - 2; (-1, 2, -3)$

3  $z = xy; (2, -1, -2)$     4  $z = x^2 y^2; (-2, 2, 16)$

5  $z = e^x \sin y; (1, \pi/2, e)$

6  $z = e^{2x} \cos 3y; (1, \pi/3, -e^2)$

7  $z = \ln \sqrt{x^2 + y^2}; (-3, 4, \ln 5)$

8  $x^2 + 2y^2 + 3z^2 = 6; (1, 1, -1)$

9  $x^2 + 2y^2 - 3z^2 = 3; (2, 1, -1)$

10  $x^2 + 3y^2 - z^2 = 0; (2, -2, 4)$

11  $x^2 + z^2 = 25; (4, -2, -3)$

12  $xy + yz + xz = 1; (2, 3, -1)$

13  $x^{1/2} + y^{1/2} + z^{1/2} = 6; (4, 1, 9)$

14  $y^{1/2} + z^{1/2} = 7; (3, 16, 9)$

15  $x = \ln\left(\dfrac{y}{2z}\right); (0, 4, 2)$

16  $z = \ln(xy); (e, e, 2)$

17  $xyz - 3xz^3 + y^3 = 9; (1, 2, -1)$

18  $z = \sqrt{2} e^{-x} \cos y; (0, \pi/4, 1)$

19  $z = x^2 - y^2; (3, 3, 0)$

20  $x^2 - 3y^2 - z^2 + 3 = 0; (2, -1, -2)$

21  Show that the equation of the plane tangent at $(x_1, y_1, z_1)$ to the ellipsoid

$$\frac{x^2}{A^2} + \frac{y^2}{B^2} + \frac{z^2}{C^2} = 1 \qquad \text{is} \qquad \frac{x_1 x}{A^2} + \frac{y_1 y}{B^2} + \frac{z_1 z}{C^2} = 1.$$

22  Show that the equation of the plane tangent at $(x_1, y_1, z_1)$ to the hyperboloid

$$\frac{x^2}{A^2} - \frac{y^2}{B^2} + \frac{z^2}{C^2} = 1 \qquad \text{is} \qquad \frac{xx_1}{A^2} - \frac{yy_1}{B^2} + \frac{zz_1}{C^2} = 1.$$

*23  Let  $Ax^2 + By^2 + Cz^2 = D$  denote a quadric surface ($A, B, C, D$ can be positive or negative). Show that the equation of the plane tangent at $(x_1, y_1, z_1)$ is

$$Axx_1 + Byy_1 + Czz_1 = D.$$

24  Show that every plane tangent to the cone

$$x^2 + y^2 = z^2$$

passes through the origin.

25  Show that every line normal to the cone

$$z^2 = 3x^2 + 3y^2$$

intersects the $z$ axis.

26  Show that the sum of the squares of the intercepts on the coordinate axes of any plane tangent to the surface

$$x^{2/3} + y^{2/3} + z^{2/3} = a^{2/3}$$

is constant.

In Problems 27 through 31, find the equations of the line tangent to the intersection of the two surfaces at the given point. Write the equations of the line in vector form.

27  $z = x^2 + y^2, z = 2x + 4y + 20; (4, -2, 20)$

28  $z = \sqrt{x^2 + y^2}, z = 2x - 3y - 13; (3, -4, 5)$

29  $z = x^2, z = 25 - y^2; (4, -3, 16)$

30  $z = \sqrt{25 - 9x^2}, z = e^{xy} + 3; (1, 0, 4)$

31  $x^2 + y^2 + z^2 = 9, x + y + z = 5; (1, 2, 2)$

32  Show that the surfaces $x^2 - y^2 + 4z^2 = 1$ and $x^2 - y^2 + 2(z+1)^2 = 5$ are tangent at the point $(1, 2, 1)$.

33  Show that the surfaces $x^2 + y^2 + z^2 = 8$ and $xy = 4$ are tangent to each other at the point $(2, 2, 0)$.

34  Find the point or points on the surface

$$z = x^2 - 2y^2 + 3y - 6$$

where the tangent plane is parallel to the plane $2x + 3y + z = 5$.

**\*35** Consider the hyperboloid $x^2 + y^2 - z^2 = 1$. Show that at each point parts of this surface lie on both sides of the tangent plane.

**36** Given the surface

$$x^\alpha + y^\alpha + z^\alpha = a^\alpha$$

where $\alpha \, (\neq 1)$ and $a$ are positive constants. Let $\bar{x}, \bar{y}, \bar{z}$ be the intercepts of any tangent plane to the surface with the coordinate axes. Show that

$$\bar{x}^{\alpha/(1-\alpha)} + \bar{y}^{\alpha/(1-\alpha)} + \bar{z}^{\alpha/(1-\alpha)} = \text{const.}$$

Find the value of the constant.

**37** Find the point or points (if any) on the surface

$$z = x^2 + 2xy - y^2 + 3x - 2y - 4$$

where the tangent plane is parallel to the $xy$-plane.

**38** Find the point or points (if any) on the surface

$$x^2 + 2xy - y^2 + 3z^2 - 2x + 2y - 6z - 2 = 0$$

where the tangent plane is parallel to the $yz$-plane.

---

## 7

## THE TOTAL DIFFERENTIAL. APPROXIMATION

The differential of a function of one variable is a function of two variables selected in a special way. We recall that if $y = f(x)$, then the quantity $df$, called the differential of $f$, is defined by the relation

$$df = f'(x)h,$$

where $h$ and $x$ are independent variables. (See Chapter 4, page 156.)

Let $f$ be a function of several variables; the next definition is the appropriate one for generalizing the notion of differential to such functions.

**DEFINITIONS**    *The **total differential** of $f(x, y)$ is the function $df$ of four variables $x, y, h, k$ given by the formula*

$$df(x, y, h, k) = f_x(x, y)h + f_y(x, y)k.$$

*If $F$ is a function of three variables—say $x$, $y$, and $z$—we define the **total differential** as the function of six variables $x, y, z, h, k, l$ given by*

$$dF(x, y, z, h, k, l) = F_x(x, y, z)h + F_y(x, y, z)k + F_z(x, y, z)l.$$

A quantity related to the total differential is the *difference* of the values of a function at two nearby points. As is customary, we use $\Delta$ notation and, for functions of two variables, we define the quantity $\Delta f$ by the relation

$$\Delta f \equiv \Delta f(x, y, h, k) = f(x + h, y + k) - f(x, y).$$

Here $\Delta F$ is a function of four variables, as is $df$. If $F$ is a function of $x$, $y$, and $z$, then $\Delta F$ is a function of six variables defined by

$$\Delta F \equiv \Delta F(x, y, z, h, k, l) = F(x + h, y + k, z + l) - F(x, y, z).$$

EXAMPLE 1　Given the function

$$f(x, y) = x^2 + xy - 2y^2 - 3x + 2y + 4,$$

find $df(a, b, h, k)$ and $\Delta f(a, b, h, k)$ with $a = 3$, $b = 1$.

Solution　We have

$$f(3, 1) = 7; \quad f_x(x, y) = 2x + y - 3; \quad f_y(x, y) = x - 4y + 2;$$

$$f_x(3, 1) = 4; \quad f_y(3, 1) = 1.$$

Also

$$f(3 + h, 1 + k) = (3 + h)^2 + (3 + h)(1 + k) - 2(1 + k)^2 - 3(3 + h)$$

$$+ 2(1 + k) + 4 = h^2 + hk - 2k^2 + 4h + k + 7.$$

Therefore

$$df(3, 1, h, k) = 4h + k; \quad \Delta f(3, 1, h, k) = 4h + k + h^2 + hk - 2k^2. \quad \square$$

　　　The close relationship between $df$ and $\Delta f$ is exhibited in Theorem 2, page 633. Equation (1) of that theorem may be written in the form

$$\Delta f(x_0, y_0, h, k) = df(x_0, y_0, h, k) + G_1(h, k)h + G_2(h, k)k.$$

The conclusion of the theorem implies that

$$\frac{\Delta f - df}{|h| + |k|} \to 0 \quad \text{as } h, k \to 0,$$

since both $G_1$ and $G_2$ tend to zero with $h$ and $k$. In many problems $\Delta f$ is difficult to calculate, while $df$ is easy. If $h$ and $k$ are both "small," we can use $df$ as an approximation to $\Delta f$. The next example shows the technique.

EXAMPLE 2　Find, approximately, the value of $\sqrt{(5.98)^2 + (8.01)^2}$.

Solution　We consider the function

$$z = f(x, y) = \sqrt{x^2 + y^2},$$

and we wish to find $f(5.98, 8.01)$. We see at once that $f(6, 8) = 10$; hence we may write

$$f(5.98, 8.01) = f(6, 8) + \Delta f,$$

where $\Delta f$ is defined as above with $x_0 = 6$, $y_0 = 8$, $h = -0.02$, $k = 0.01$. The approximation consists of replacing $\Delta f$ by $df$. We have

$$f_x = \frac{x}{\sqrt{x^2 + y^2}}, \quad f_y = \frac{y}{\sqrt{x^2 + y^2}},$$

and so

$$df(6, 8, -0.02, 0.01) = \tfrac{6}{10}(-0.02) + \tfrac{8}{10}(0.01) = -0.004.$$

We conclude that $\sqrt{(5.98)^2 + (8.01)^2} = 10 - 0.004 = 9.996$, approximately. The true value is $9.9960242 +$. $\quad \square$

*Remark.* While the result in Example 2 may be obtained easily with a hand calculator, there are complex situations where replacing $\Delta f$ by $df$ results in an enormous reduction in computation with little loss in accuracy.

As in the case of functions of one variable, the symbolism for the total differential may be used as an aid in differentiation. We let $z = f(x, y)$ and employ the symbols

$$dz \text{ for } df, \qquad dx \text{ for } h, \qquad \text{and} \qquad dy \text{ for } k.$$

As in the case of one variable, there is a certain ambiguity, since $dz$ has a precise definition as the total differential, while $dx$ and $dy$ are used as independent variables. The next theorem shows how the Chain Rule comes to our rescue and removes all difficulties when $dx$ and $dy$ are in turn functions of other variables (i.e., $dx$ and $dy$ are what we call intermediate variables).

**THEOREM 6**  *Suppose that $z = f(x, y)$, and $x$ and $y$ are functions of some other variables. Then*\*

$$dz = \frac{\partial z}{\partial x} dx + \frac{\partial z}{\partial y} dy.$$

*The result for $w = F(x, y, z)$ is similar. That is, the formula*

$$dw = \frac{\partial w}{\partial x} dx + \frac{\partial w}{\partial y} dy + \frac{\partial w}{\partial z} dz$$

*holds when $x$, $y$, and $z$ are either independent or intermediate variables.*

**Proof**  We establish the result for $z = f(x, y)$ with $x$ and $y$ functions of two variables, say $r$ and $s$. The proof in all other cases is analogous. We write

$$x = x(r, s), \qquad y = y(r, s),$$

and then

$$z = f(x, y) = f[x(r, s), y(r, s)] \equiv g(r, s).$$

The definition of total differential yields

$$dz = g_r(r, s)h + g_s(r, s)k$$

$$dx = x_r(r, s)h + x_s(r, s)k$$

$$dy = y_r(r, s)h + y_s(r, s)k.$$

According to the Chain Rule, we have

$$g_r(r, s) = \frac{\partial f}{\partial x} \frac{\partial x}{\partial r} + \frac{\partial f}{\partial y} \frac{\partial y}{\partial r} = f_x(x, y)x_r(r, s) + f_y(x, y)y_r(r, s);$$

$$g_s(r, s) = \frac{\partial f}{\partial x} \frac{\partial x}{\partial s} + \frac{\partial f}{\partial y} \frac{\partial y}{\partial s} = f_x(x, y)x_s(r, s) + f_y(x, y)y_s(r, s).$$

---

\*Of course, $\partial z/\partial x = f_x(x, y)$; $\partial w/\partial x = F_x(x, y, z)$, etc.

Substituting the above expressions for $g_r$ and $g_s$ into that for $dz$, we obtain

$$dz = \frac{\partial f}{\partial x}[x_r(r, s)h + x_s(r, s)k] + \frac{\partial f}{\partial y}[y_r(r, s)h + y_s(r, s)k]$$

or

$$dz = \frac{\partial f}{\partial x}\,dx + \frac{\partial f}{\partial y}\,dy.$$

We recognize this last formula as the statement of the theorem.  □

**EXAMPLE 3**  Given

$$z = e^x \cos y + e^y \sin x, \qquad x = r^2 - t^2, \qquad y = 2rt,$$

find $dz(r, t, h, k)$ in two ways and verify that the results coincide.

**Solution**  We have, by one method,

$$dz = \frac{\partial z}{\partial x}\,dx + \frac{\partial z}{\partial y}\,dy = (e^x \cos y + e^y \cos x)\,dx + (-e^x \sin y + e^y \sin x)\,dy;$$

$$dx = 2rh - 2tk;$$

$$dy = 2th + 2rk.$$

Therefore

$$dz = (e^x \cos y + e^y \cos x)(2rh - 2tk) + (-e^x \sin y + e^y \sin x)(2th + 2rk)$$

$$= 2[(e^x \cos y + e^y \cos x)r + (e^y \sin x - e^x \sin y)t]h$$

$$+ 2[(-e^y \cos x - e^x \cos y)t + (e^y \sin x - e^x \sin y)r]k. \qquad (1)$$

On the other hand, the second method gives

$$dz = \frac{\partial z}{\partial r}h + \frac{\partial z}{\partial t}k, \qquad (2)$$

and using the Chain Rule, we find

$$\frac{\partial z}{\partial r} = \frac{\partial z}{\partial x}\frac{\partial x}{\partial r} + \frac{\partial z}{\partial y}\frac{\partial y}{\partial r}, \qquad \frac{\partial z}{\partial t} = \frac{\partial z}{\partial x}\frac{\partial x}{\partial t} + \frac{\partial z}{\partial y}\frac{\partial y}{\partial t}.$$

We compute the various quantities in the two formulas above and find that

$$\frac{\partial z}{\partial r} = (e^x \cos y + e^y \cos x)(2r) + (e^y \sin x - e^x \sin y)(2t),$$

$$\frac{\partial z}{\partial t} = (e^x \cos y + e^y \cos x)(-2t) + (e^y \sin x - e^x \sin y)(2r).$$

Substituting these expressions in (2), we get (1) precisely. Finally, setting $x = r^2 - t^2$, $y = 2rt$ in (1), we obtain $dz$ in terms of $r$, $t$ $h$, and $k$.  □

## 7  PROBLEMS

In each of Problems 1 through 6, find $df(x, y, h, k)$ and $\Delta f(x, y, h, k)$ for the given values of $x$, $y$, $h$, and $k$.

1  $f(x, y) = x^2 - xy + 2y^2$; $x = 2$, $y = -1$, $h = -0.01$, $k = 0.02$

2  $f(x, y) = 2x^2 + 3xy - y^2$; $x = 1$, $y = 2$, $h = 0.02$, $k = -0.01$

3  $f(x, y) = \sin xy + \cos(x + y)$; $x = \pi/6$, $y = 0$, $h = 2\pi$, $k = 3\pi$

4  $f(x, y) = e^{xy} \sin(x + y)$; $x = \pi/4$, $y = 0$, $h = -\pi/2$, $k = 4\pi$

5  $f(x, y) = x^3 - 3xy + y^3$; $x = -2$, $y = 1$, $h = -0.03$, $k = -0.02$

6  $f(x, y) = x^2 y - 2xy^2 + 3x$; $x = 1$, $y = 1$, $h = 0.02$, $k = 0.01$

In each of Problems 7 through 10, find $df(x, y, z, h, k, l)$ and $\Delta f(x, y, z, h, k, l)$ for the given values of $x$, $y$, $z$, $h$, $k$, and $l$.

7  $f(x, y, z) = x^2 - 2y^2 + z^2 - xz$; $(x, y, z) = (2, -1, 3)$; $(h, k, l) = (0.01, -0.02, 0.03)$

8  $f(x, y, z) = xy - xz + yz + 2x - 3y + 1$; $(x, y, z) = (2, 0, -3)$; $(h, k, l) = (0.1, -0.2, 0.1)$

9  $f(x, y, z) = x^2 y - xyz + z^3$; $(x, y, z) = (1, 2, -1)$; $(h, k, l) = (-0.02, 0.01, 0.02)$

10  $f(x, y, z) = \sin(x + y) - \cos(x - z) + \sin(y + 2z)$; $(x, y, z) = (\pi/3, \pi/6, 0)$; $(h, k, l) = (\pi/4, \pi/2, 2\pi)$

In each of Problems 11 through 19, find the total differential.

11  $z = y^5 - 5xy^4$

12  $w = x^2 y - z^2 + x^4$

13  $z = \sqrt{x^2 + y^2}$

14  $w = \sqrt{x^2 + y^2 + z^2}$

15  $\theta = \arctan(y/x)$

16  $\theta = \arcsin(y/r)$

17  $w = \ln\sqrt{x^2 + y^2 + z^2}$

18  $w = \ln(xyz)$

19  $w = x^2 e^{xy} + (1/y^2) + z^3$

20  We define the **approximate percentage error** of a function $f$ by the relation

$$\text{Approximate percentage error} = 100\left(\frac{df}{f}\right).$$

Find the approximate percentage error if $f(x, y, z) = 3x^3 y^7 z^4$.

21  Find the approximate percentage error (see Problem 20) if $f = cx^m y^n z^p$, $c$, $m$, $n$, $p$ constants.

22  A box has square ends, 11.98 cm on each side, and has a length of 30.03 cm. Find its approximate volume, using differentials.

23  Use differentials to find the approximate value of

$$\sqrt{(5.02)^2 + (11.97)^2}.$$

24  The legs of a right triangle are measured and found to be 6.0 and 8.0 cm, with a possible error of 0.1 cm. Find approximately the maximum possible value of the error in computing the hypotenuse. What is the maximum approximate percentage error? (See Problem 20.)

25  Find in degrees the maximum possible approximate error in the computed value of the smaller acute angle in the triangle of Problem 24.

26  The diameter and height of a right circular cylinder are found by measurement to be 8.0 and 12.5 cm, respectively, with possible errors of 0.05 cm in each measurement. Find the maximum possible approximate error in the computed volume.

27  A right circular cone is measured, and the radius of the base is 12.0 cm with the height 16.0 cm. If the possible error in each measurement is 0.06, find the maximum possible error in the computed volume. What is the maximum possible approximate error in the lateral surface area?

28  By measurement, a triangle is found to have two sides of length 50 cm and 70 cm; the angle between them is 30°. If there are possible errors of $\frac{1}{2}\%$ in the measurements of the sides and $\frac{1}{2}$ degree in that of the angle, find the maximum approximate percentage error in the measurement of the area. (See Problem 20.)

29  A sprinter runs 100 meters in about 11 seconds. If the time may be in error by as much as $\frac{1}{20}$ sec., and the distance by as much as 10 centimeters, find the greatest possible error in the velocity. (Assume runner has constant velocity.)

30  Use differentials to find the approximate value of

$$\sqrt{(3.02)^2 + (1.99)^2 + (5.97)^2}.$$

31  Use differentials to find the approximate value of

$$[(3.01)^2 + (3.98)^2 + (6.02)^2 + 5(1.97)^2]^{-1/2}.$$

In each of Problems 32 through 36, find $dz$ in two ways (as in Example 3) in terms of the independent variables.

32  $z = x^2 + xy - y^2$; $x = r^2 + 2s^2$, $y = rs + 2$

33  $z = 2x^2 + 3xy + y^2$; $x = t^3 + 2t - 1$, $y = t^2 + t - 3$

34  $z = x^3 + y^3 - x^2 y$; $x = r + 2s - t$, $y = r - 3s + 2t$

35  $z = u^2 + 2v^2 - x^2 + 3y^2$; $u = r^2 - s^2$, $v = r^2 + s^2$, $x = 2rs$, $y = 2r/s$

**36**  $z = u^3 + v^3 + w^3$; $u = r^2 + s^2 + t^2$, $v = r^2 - s^2 + t^2$,
$w = r^2 + s^2 - t^2$

**37**  Using the formulas in the proof of Theorem 2, find explicit expressions for the functions $G_1$ and $G_2$ if

$$z = x^2 + 2y^2 + 6xy.$$

Conclude that $(\Delta z - dz)/(|h| + |k|)$ tends to zero as $h$, $k \to 0$.

**38**  Same as Problem 37 for $z = 3x^3 + 2y^3 + 2xy$.

**\*39**  Let $z = f(x_1, x_2, \ldots, x_n)$ be a function of $n$ variables. Define the **total differential** $df$ by the formula

$$df = f_{x_1} h_1 + f_{x_2} h_2 + \cdots + f_{x_n} h_n$$

where $h_1$, $h_2$, ..., $h_n$ are independent variables. Assume each $x_i$ is a function of the $m$ variables $y_1, y_2, \ldots, y_m$. State and prove the analog of Theorem 6.

**40**  Use the formula in Problem 39 to find approximately the value of

$$[(3.01)^2 + (2.97)^2 + (5.02)^2 + (3.99)^2$$
$$+ (7.01)^2 + (6.02)^2]^{1/2}.$$

**41**  The period $T$ of a simple pendulum is given by $T = 2\pi(l/g)^{1/2}$ where $l$ is the length and $g$ is the gravitational constant. If $l$ is measured to be 20 cm with an error of 0.2, if $g$ is 980 with an error of 7, and if $\pi$ is computed as 3.14 with an error of 0.002, use differentials to find an approximate value of $T$.

---

## 8

## APPLICATIONS OF THE TOTAL DIFFERENTIAL

Once we clearly understand the concept of function we are able to use the notation of the total differential to obtain a number of useful differentiation formulas.

In the study of implicit differentiation in Chapter 3, p. 111ff, we learned how to calculate the derivative of a function when it is given in the form $f(x, y) = 0$. Now we show how partial derivatives can be used for the same purpose. We employ the fact that the *total differential of a constant is zero*. If $x$ and $y$ are related by an equation such as

$$f(x, y) = 0$$

and it turns out that $y$ is a function of $x$, we can compute $dy/dx$ by using the fact that since $f = 0$, then $df = 0$. The next theorem provides the main result.

**THEOREM 7**  *Suppose that $f(x, y)$, $f_x(x, y)$ and $f_y(x, y)$ are continuous functions and that $f(x, y) = 0$; we assume that $y$ is a function of $x$. If at a point $(x_0, y_0)$ we have $f_y(x_0, y_0) \neq 0$, then*

$$\frac{dy}{dx} = -\frac{\dfrac{\partial f}{\partial x}}{\dfrac{\partial f}{\partial y}}. \tag{1}$$

*for all $(x, y)$ near $(x_0, y_0)$.*

**Proof**   Since $f(x, y)$ is a constant, we know $df = 0$. On the other hand, from Theorem 6, we find

$$df = \frac{\partial f}{\partial x} dx + \frac{\partial f}{\partial y} dy.$$

Therefore,

$$\frac{\partial f}{\partial x} dx + \frac{\partial f}{\partial y} dy = 0,$$

and since

$$\frac{\partial f}{\partial y} \neq 0,$$

we get

$$\frac{dy}{dx} = \frac{-\dfrac{\partial f}{\partial x}}{\dfrac{\partial f}{\partial y}}. \qquad \square$$

*Remark.*   The hypotheses in Theorem 7 that $f_x$ and $f_y$ are continuous, and that $f_y \neq 0$, actually imply that $y$ is a differentiable function of $x$ near $(x_0, y_0)$. This fact is usually proved in more advanced courses.

**EXAMPLE 1**   Use the methods of partial differentiation to compute $dy/dx$ if

$$x^4 + 3x^2 y^2 - y^4 + 2x - 3y = 5.$$

**Solution**   Setting

$$f(x, y) = x^4 + 3x^2 y^2 - y^4 + 2x - 3y - 5 = 0,$$

we find

$$f_x = 4x^3 + 6xy^2 + 2, \qquad f_y = 6x^2 y - 4y^3 - 3.$$

Therefore, using (1) above,

$$\frac{dy}{dx} = -\frac{4x^3 + 6xy^2 + 2}{6x^2 y - 4y^3 - 3}.$$

Of course, the same result is obtained by the customary process of implicit differentiation as described in Chapter 3, Section 3.   $\square$

An analogous procedure can be used to compute partial derivatives implicitly when we are given a function of three variables. (See Section 2 of this chapter.) The proof of Theorem 8 below is similar to that of Theorem 7. If $F(x, y, z) = 0$, then

$$dF = F_x dx + F_y dy + F_z dz = 0$$

or

$$dz = \left( \frac{-F_x}{F_z} \right) dx + \left( \frac{-F_y}{F_z} \right) dy.$$

From Theorem 6 we know that

$$dz = \frac{\partial z}{\partial x} dx + \frac{\partial z}{\partial y} dy.$$

Comparing these two formulas for $dz$, we get the following result (although we omit the details).

---

**THEOREM 8**   *Suppose that $F(x, y, z) = 0$, that $F_x, F_y, F_z$ are continuous, and that $z$ is a function of $x$ and $y$ such that the partial derivatives of $z$ exist. If $(x_0, y_0, z_0)$ is a point where $F_z(x_0, y_0, z_0) \neq 0$, then for all $(x, y, z)$ near this point*

$$\frac{\partial z}{\partial x} = -\frac{\dfrac{\partial F}{\partial x}}{\dfrac{\partial F}{\partial z}}, \quad \text{and} \quad \frac{\partial z}{\partial y} = -\frac{\dfrac{\partial F}{\partial y}}{\dfrac{\partial F}{\partial z}}. \tag{2}$$

---

We exhibit the utility of Formulas (2) in the next example.

**EXAMPLE 2**   Use Formulas (2) to find $\partial z / \partial x$ and $\partial z / \partial y$ if

$$e^{xy} \cos z + e^{-xz} \sin y + e^{yz} \cos x = 0.$$

**Solution**   We set

$$F(x, y, z) = e^{xy} \cos z + e^{-xz} \sin y + e^{yz} \cos x,$$

and compute

$$F_x = y e^{xy} \cos z - z e^{-xz} \sin y - e^{yz} \sin x,$$

$$F_y = x e^{xy} \cos z + e^{-xz} \cos y + z e^{yz} \cos x,$$

$$F_z = -e^{xy} \sin z - x e^{-xz} \sin y + y e^{yz} \cos x.$$

Therefore

$$\frac{\partial z}{\partial x} = -\frac{y e^{xy} \cos z - z e^{-xz} \sin y - e^{yz} \sin x}{-e^{xy} \sin z - x e^{-xz} \sin y + y e^{yz} \cos x}.$$

$$\frac{\partial z}{\partial y} = -\frac{x e^{xy} \cos z + e^{-xz} \cos y + z e^{yz} \cos x}{-e^{xy} \sin z - x e^{-xz} \sin y + y e^{yz} \cos x}. \qquad \square$$

*Remarks.*   (i) Note that we also could have found the derivatives by the implicit methods described in Section 2. (ii) Formulas similar to (2) may be established for a single relation with any number of variables. For example, if we are given $G(x, y, u, v, w) = 0$ and we assume $w$ is a function of the remaining variables with $G_w \neq 0$, then

$$\frac{\partial w}{\partial x} = -\frac{G_x}{G_w}, \quad \frac{\partial w}{\partial y} = -\frac{G_y}{G_w}, \quad \frac{\partial w}{\partial u} = -\frac{G_u}{G_w}, \quad \frac{\partial w}{\partial v} = -\frac{G_v}{G_w}.$$

A more complicated application of differentials is exhibited in the derivation of the next set of formulas. Suppose, for example, that $x, y, u, v$ are related by *two* equations, so that

$$F(x, y, u, v) = 0 \quad \text{and} \quad G(x, y, u, v) = 0.$$

If we could solve one of them for, say $u$, and substitute in the other, we would get a single equation for $x$, $y$, $v$. Then, solving this equation for $v$, we would find that $v$ is a function of $x$ and $y$. Similarly, we might find $u$ as a function of $x$ and $y$. Of course, all this work is purely fictitious, since we have no intention of carrying out such a process. In fact, it may be impossible. The main point is that we know that under appropriate circumstances the process is *theoretically feasible*. (This fact is usually proved in an advanced course in analysis.) Therefore, whenever $u = u(x, y)$ and $v = v(x, y)$, it makes sense to write the symbols

$$\frac{\partial u}{\partial x}, \frac{\partial u}{\partial y}, \frac{\partial v}{\partial x}, \frac{\partial v}{\partial y}. \tag{3}$$

Furthermore, the selection of $u$ and $v$ in terms of $x$ and $y$ is arbitrary. We could equally well attempt to solve for $v$ and $x$ in terms of $u$ and $y$ or for any two of the variables in terms of the remaining two variables.

The problem we pose is that of determining the quantities in (3) without actually finding the functions $u(x, y)$ and $v(x, y)$. We use the total differential. Since $F = 0$ and $G = 0$, so are $dF$ and $dG$. We have

$$dF = F_x dx + F_y dy + F_u du + F_v dv = 0,$$

$$dG = G_x dx + G_y dy + G_u du + G_v dv = 0.$$

We write these two equations,

$$F_u du + F_v dv = -F_x dx - F_y dy, \qquad G_u du + G_v dv = -G_x dx - G_y dy,$$

and consider $du$ and $dv$ as **unknowns** with everything else known. Solving two linear equations in two unknowns is easy. We obtain

$$du = \frac{G_x F_v - G_v F_x}{F_u G_v - F_v G_u} dx + \frac{G_y F_v - G_v F_y}{F_u G_v - F_v G_u} dy, \tag{4}$$

$$dv = \frac{G_u F_x - G_x F_u}{F_u G_v - F_v G_u} dx + \frac{G_u F_y - G_y F_u}{F_u G_v - F_v G_u} dy. \tag{5}$$

(We suppose, of course, that $F_u G_v - F_v G_u \neq 0$.) On the other hand, we know that if $u = u(x, y)$, $v = v(x, y)$, then from Theorem 6

$$du = \frac{\partial u}{\partial x} dx + \frac{\partial u}{\partial y} dy, \tag{6}$$

$$dv = \frac{\partial v}{\partial x} dx + \frac{\partial v}{\partial y} dy. \tag{7}$$

Therefore, comparing (4) with (6) and (5) with (7), we find

$$\frac{\partial u}{\partial x} = \frac{G_x F_v - G_v F_x}{F_u G_v - F_v G_u}, \tag{8}$$

and similar formulas for $\partial u / \partial y$, $\partial v / \partial x$, and $\partial v / \partial y$. If $F$ and $G$ are specific functions, the right side of (8) is computable.

We summarize the preceding discussion in the form of a theorem.

**THEOREM 9**    *Suppose that* $F(x, y, u, v) = 0$, $G(x, y, u, v) = 0$, *and that all the first partial derivatives of both F and G are continuous. Suppose further that u and v are each functions of x and y such that their first partial derivatives exist. If* $F_u G_v - F_v G_u \neq 0$ *then*

$$\frac{\partial u}{\partial x} = \frac{G_x F_v - G_v F_x}{F_u G_v - F_v G_u}, \qquad \frac{\partial u}{\partial y} = \frac{G_y F_v - G_v F_y}{F_u G_v - F_v G_u}$$

$$\frac{\partial v}{\partial x} = \frac{G_u F_x - G_x F_u}{F_u G_v - F_v G_u}, \qquad \frac{\partial v}{\partial y} = \frac{G_u F_y - G_y F_u}{F_u G_v - F_v G_u}$$

(9)

**EXAMPLE 3**    Given the relations for $x, y, u, v$:

$$u^2 - uv - v^2 + x^2 + y^2 - xy = 0,$$

$$uv - x^2 + y^2 = 0,$$

and assuming that $u = u(x, y)$, $v = v(x, y)$, find $\partial u / \partial x$, $\partial u / \partial y$, $\partial v / \partial x$, and $\partial v / \partial y$.

**Solution**    Denoting the first equation by $F = 0$ and the second by $G = 0$, we could find $F_x, F_y, \ldots, G_u, G_v$ and then substitute for the coefficients in (4) and (5) to obtain the result or, equivalently, use Formulas (9). Instead we make use of the fact that we can treat differentials both as independent variables and as total differentials, with no fear of difficulty (because of the Chain Rule). Taking such differentials in each of the equations given, we get

$$2u\,du - u\,dv - v\,du - 2v\,dv + 2x\,dx + 2y\,dy - x\,dy - y\,dx = 0,$$

$$v\,du + u\,dv - 2x\,dx + 2y\,dy = 0.$$

Solving the two equations simultaneously for $du$ and $dv$ in terms of $dx$ and $dy$, we obtain

$$du = \frac{uy + 4xv}{2(u^2 + v^2)}\,dx + \frac{ux - 4y(u + v)}{2(u^2 + v^2)}\,dy,$$

$$dv = \frac{4xu - yv}{2(u^2 + v^2)}\,dx + \frac{4y(v - u) - xv}{2(u^2 + v^2)}\,dy.$$

From these equations and equations (6) and (7) we read off the results. For example,

$$\frac{\partial u}{\partial y} = \frac{ux - 4y(u + v)}{2(u^2 + v^2)},$$

and there are corresponding expressions for $\partial u / \partial x$, $\partial v / \partial x$, $\partial v / \partial y$.    □

# 8  PROBLEMS

In each of Problems 1 through 10, find the derivative $dy/dx$ by the methods of partial differentiation.

1  $x^2 + 3xy - 4y^2 + 2x - 6y + 7 = 0$

2  $x^3 + 3x^2y - 4xy^2 + y^3 - x^2 + 2y - 1 = 0$

3  $\ln(1 + x^2 + y^2) + e^{xy} = 5$

4  $x^4 - 3x^2y^2 + y^4 - x^2y + 2xy^2 = 3$

5  $e^{xy} + \sin xy + \frac{1}{2} = 0$

6  $xe^y + ye^x + \sin(x + y) - 2 = 0$

7  $\arctan(y/x) + (x^2 + y^2)^{3/2} = 2$

8  $\sqrt{x} + \sqrt{y} = \sqrt{2}$        9  $(2y)^{1/2} + (3y)^{1/3} = x$

10  $\ln(xy + 3) - y = 0$

In each of Problems 11 through 19, assume that $w$ is a function of the remaining variables. Find the partial derivatives as indicated by the method of Example 2.

11  $x^2 + y^2 + w^2 - 3xyw - 4 = 0$; $\dfrac{\partial w}{\partial y}$

12  $x^3 + 3x^2w - y^2w + 2yw^2 - 3w + 2x = 8$; $\dfrac{\partial w}{\partial x}$

13  $e^{xy} + e^{yw} - e^{xw} + xyw = 4$; $\dfrac{\partial w}{\partial y}$

14  $\sin(xyw) + x^2 + y^2 + w^2 = 3$; $\dfrac{\partial w}{\partial x}$

15  $(w^2 - y^2)(w^2 + x^2)(x^2 - y^2) = 1$; $\dfrac{\partial w}{\partial y}$

16  $\ln(xyw + 4) + 2(x + y) = 0$; $\dfrac{\partial w}{\partial y}$

17  $2xw^3 - 2yw^2 + x^3y^2 + 4w = 6$; $\dfrac{\partial w}{\partial x}$

18  $xe^{yw} - 2ye^{xw} + we^{xy} = 14$; $\dfrac{\partial w}{\partial y}$

19  $w + \cos(xy) - \sin(wx) = 6$; $\dfrac{\partial w}{\partial y}$

In Problems 20 and 21, use the methods of this section to find the partial derivatives as indicated.

20  $x^2 + y^2 - z^2 - w^2 + 3xy - 2xz + 4xw - 3zw + 2x = 3y$; $\dfrac{\partial w}{\partial y}$

21  $x^2y^2z^2w^2 + x^2z^2w^4 - y^4w^4 + x^6w^2 - 2y^3w^3 = 8$; $\dfrac{\partial w}{\partial z}$

If $F(x, y, z) = 0$ and $G(x, y, z) = 0$, then we may consider $z$ and $y$

as functions of the single variable $x$; that is, $z = z(x)$, $y = y(x)$. Using differentials, we obtain

$$F_x\,dx + F_y\,dy + F_z\,dz = 0, \qquad G_x\,dx + G_y\,dy + G_z\,dz = 0,$$

and so by elimination we can get the ordinary derivatives $dz/dx$ and $dy/dx$. Use this method in Problems 22 through 26 to obtain these derivatives.

22  $z = x^2 + y^2$, $y^2 = 4x + 2z$

23  $x^2 - y^2 + z^2 = 7$, $2x + 3y + 4z = 15$

24  $2x^2 + 3y^2 + 4z^2 = 12$, $x = yz$

25  $xyz = 5$, $x^2 + y^2 - z^2 = 16$

26  $z = \cos(xy)$, $x\cos(xy) + z^2 = 5\sqrt{2}$

In each of Problems 27 through 31, find $\partial u/\partial x$, $\partial u/\partial y$, $\partial v/\partial x$, and $\partial v/\partial y$ by the method of differentials.

27  $x = u^2 - v^2$, $y = 2uv$    28  $x = u + v$, $y = uv$

29  $u + v - x^2 = 0$, $u^2 - v^2 - y = 0$

30  $u^3 + xv^2 - xy = 0$, $u^2y + v^3 + x^2 - y^2 = 0$

31  $u^2 + v^2 + x^2 - y^2 = 4$, $u^2 - v^2 - x^2 - y^2 = 1$

32  Given that $F(x, y, z, u) = 0$, show that if each of the partial derivatives in the expressions below actually exists, then the two formulas are valid:

$$\frac{\partial x}{\partial v} \cdot \frac{\partial y}{\partial z} \cdot \frac{\partial z}{\partial x} = -1; \qquad \frac{\partial x}{\partial y} \cdot \frac{\partial y}{\partial z} \cdot \frac{\partial z}{\partial u} \cdot \frac{\partial u}{\partial x} = 1.$$

33  Given that $x = f(u, v)$, $y = g(u, v)$, find $\partial u/\partial x$, $\partial u/\partial y$, $\partial v/\partial x$, $\partial v/\partial y$ in terms of $u$ and $v$ and the derivatives of $f$ and $g$.

34  Given that $F(u, v, x, y, z) = 0$ and $G(u, v, x, y, z) = 0$, assume that $u = u(x, y, z)$ and $v = v(x, y, z)$ and find formulas for $\partial u/\partial x$, $\partial v/\partial x$, ..., $\partial v/\partial z$ in terms of the derivatives of $F$ and $G$.

35  In the theory of electricity, if $n$ resistances $R_1, R_2, ..., R_n$ are connected in parallel, then the total resistance $R$ is given by

$$\frac{1}{R} = \sum_{i=1}^{n} \frac{1}{R_i}.$$

Prove that for $i = 1, 2, ..., n$,

$$\frac{\partial R}{\partial R_i} = \left(\frac{R}{R_i}\right)^2.$$

36  Given $F(x_1, x_2, ..., x_n, u_1, u_2) = 0$ and $G(x_1, x_2, ..., x_n, u_1, u_2) = 0$. Find formulas for

$$\frac{\partial u_1}{\partial x_i}, \frac{\partial u_2}{\partial x_i}, \quad i = 1, 2, ..., n,$$

assuming that $u_1$ and $u_2$ are functions of $x_1, x_2, ..., x_n$.

**37** Given the equations $F_i(x_1, x_2, \ldots, x_n, u_1, u_2, \ldots, u_k) = 0$ where $i = 1, 2, \ldots, k$. Assuming that $u_i = u_i(x_1, x_2, \ldots, x_n)$ where $i = 1, 2, \ldots, k$, find formulas for

$$\frac{\partial u_i}{\partial x_j}, \quad i = 1, 2, \ldots, k, \quad j = 1, 2, \ldots, n.$$

**38** Given $F(x, y, z, u, v, w) = 0$, $G(x, y, z, u, v, w) = 0$, and $H(x, y, z, u, v, w) = 0$. Use the total differentials $dF$, $dG$, and $dH$ to derive formulas for

$$\frac{\partial u}{\partial x}, \frac{\partial u}{\partial y}, \ldots, \frac{\partial w}{\partial z}$$

similar to Formula (8) in the case of two functions.

*Answer:* Let

$$D = \begin{vmatrix} F_u & F_v & F_w \\ G_u & G_v & G_w \\ H_u & H_v & H_w \end{vmatrix} \quad \text{and} \quad D_1 = \begin{vmatrix} F_x & F_v & F_w \\ G_x & G_v & G_w \\ H_x & H_v & H_w \end{vmatrix}.$$

Then $\partial u / \partial x = -(D_1/D)$ if the determinant $D$ is not zero. Similar formulas hold for the remaining partial derivatives.

**39** Given

$$x^2 - y^2 + z^2 + 2u - v^2 + w^2 - 3 = 0,$$

$$x^2 - z^2 + 2u^2 - w^3 - 1 = 0,$$

$$y^2 + 2z^2 - u^2 + 2v - 3w^3 - 4 = 0.$$

Use the results in Problem 38 to find $\partial v / \partial y$.

---
**9**
---

## SECOND AND HIGHER DERIVATIVES

If $f$ is a function of two variables—say $x$ and $y$—then $f_x$ and $f_y$ are also functions of the same two variables. When we differentiate $f_x$ and $f_y$, we obtain second partial derivatives.

**DEFINITIONS**  **The second partial derivatives of $f$ are defined by the formulas:**

$$f_{xx}(x, y) = \lim_{h \to 0} \frac{f_x(x + h, y) - f_x(x, y)}{h};$$

$$f_{xy}(x, y) = \lim_{k \to 0} \frac{f_x(x, y + k) - f_x(x, y)}{k};$$

$$f_{yx}(x, y) = \lim_{h \to 0} \frac{f_y(x + h, y) - f_y(x, y)}{h};$$

$$f_{yy}(x, y) = \lim_{k \to 0} \frac{f_y(x, y + k) - f_y(x, y)}{k}.$$

Note that if $f$ is a function of two variables, there are four second partial derivatives.

There is a multiplicity of notations for partial derivatives which at times may lead to confusion. For example, if we write $z = f(x, y)$, then the following four symbols all have the same meaning:

$$\frac{\partial^2 z}{\partial x^2}; \frac{\partial^2 f}{\partial x^2}; f_{xx}; z_{xx}.$$

For other partial derivatives we have the variety of expressions:

$$f_{xy} = \frac{\partial}{\partial y}\left(\frac{\partial z}{\partial x}\right) = \frac{\partial^2 z}{\partial y \partial x} = \frac{\partial^2 f}{\partial y \partial x} = z_{xy},$$

$$f_{yx} = \frac{\partial}{\partial x}\left(\frac{\partial z}{\partial y}\right) = \frac{\partial^2 z}{\partial x \partial y} = \frac{\partial^2 f}{\partial x \partial y} = z_{yx},$$

$$\frac{\partial}{\partial x}\left(\frac{\partial^2 z}{\partial y \partial x}\right) = \frac{\partial^3 z}{\partial x \partial y \partial x} = \frac{\partial^3 f}{\partial x \partial y \partial x} = f_{xyx} = z_{xyx},$$

and so forth. Note that, in the subscript notation, symbols such as $f_{xyy}$ or $z_{xyy}$ mean that the order of partial differentiation is taken from left to right—that is, first with respect to $x$ and then twice with respect to $y$. On the other hand, the symbol

$$\frac{\partial^3 z}{\partial x \partial y \partial y}$$

asserts that we first take two derivatives with respect to $y$ and then one with respect to $x$. *The denominator symbol and the subscript symbol are the reverse of each other.*

**EXAMPLE 1**   Given $z = x^3 + 3x^2 y - 2x^2 y^2 - y^4 + 3xy$, find

$$\frac{\partial z}{\partial x}, \frac{\partial z}{\partial y}, \frac{\partial^2 z}{\partial x^2}, \frac{\partial^2 z}{\partial x \partial y}, \frac{\partial^2 z}{\partial y \partial x}, \frac{\partial^2 z}{\partial y^2}.$$

**Solution**   We have

$$\frac{\partial z}{\partial x} = 3x^2 + 6xy - 4xy^2 + 3y; \qquad \frac{\partial z}{\partial y} = 3x^2 - 4x^2 y - 4y^3 + 3x;$$

$$\frac{\partial^2 z}{\partial x^2} = 6x + 6y - 4y^2; \qquad \frac{\partial^2 z}{\partial y^2} = -4x^2 - 12y^2;$$

$$\frac{\partial^2 z}{\partial y \partial x} = 6x - 8xy + 3; \qquad \frac{\partial^2 z}{\partial x \partial y} = 6x - 8xy + 3. \qquad \square$$

In the example above, it is not accidental that $\partial^2 z/\partial y \partial x = \partial^2 z/\partial x \partial y$, as the next theorem shows.

**THEOREM 10**   *Assume that $f(x, y)$, $f_x$, $f_y$, $f_{xy}$, and $f_{yx}$ are all continuous at $(x_0, y_0)$. Then*

$$f_{xy}(x_0, y_0) = f_{yx}(x_0, y_0).$$

*(The order of partial differentiation may be reversed without affecting the result.)*

**Proof**   The result is obtained by use of a quantity we call the **double difference**, denoted by $\Delta_2 f$, and defined by the formula

$$\Delta_2 f = [f(x_0 + h, y_0 + h) - f(x_0 + h, y_0)] - [f(x_0, y_0 + h) - f(x_0, y_0)].$$

$$(1)$$

We shall show that, as $h$ tends to zero, the quantity $\Delta_2 f / h^2$ tends to $f_{xy}(x_0, y_0)$. On the other hand, we shall also show that the same quantity tends to $f_{yx}(x_0, y_0)$. The principal tool is the repeated application of the Mean Value Theorem. (See page 131.) We may write $\Delta_2 f$ in a more transparent way by defining

$$\phi(s) = f(x_0 + s, y_0 + h) - f(x_0 + s, y_0), \tag{2}$$

$$\psi(t) = f(x_0 + h, y_0 + t) - f(x_0, y_0 + t). \tag{3}$$

(The quantities $x_0$, $y_0$, $h$ are considered fixed in the definition of $\phi$ and $\psi$.) Then straight substitution in (1) shows that

$$\Delta_2 f = \phi(h) - \phi(0) \tag{4}$$

and

$$\Delta_2 f = \psi(h) - \psi(0). \tag{5}$$

We apply the Mean Value Theorem in (4) and (5), getting two expressions for $\Delta_2 f$. They are (assuming $h$ is positive)

$$\Delta_2 f = \phi'(s_1) \cdot h \quad \text{with } 0 < s_1 < h,$$

$$\Delta_2 f = \psi'(t_1) \cdot h \quad \text{with } 0 < t_1 < h.$$

The derivatives $\phi'(s_1)$ and $\psi'(t_1)$ are easily computed from (2) and (3). They are

$$\phi'(s_1) = f_x(x_0 + s_1, y_0 + h) - f_x(x_0 + s_1, y_0),$$

$$\psi'(t_1) = f_y(x_0 + h, y_0 + t_1) - f_y(x_0, y_0 + t_1),$$

and the two expressions for $\Delta_2 f$ yield

$$\frac{1}{h} \Delta_2 f = [f_x(x_0 + s_1, y_0 + h) - f_x(x_0 + s_1, y_0)], \tag{6}$$

$$\frac{1}{h} \Delta_2 f = [f_y(x_0 + h, y_0 + t_1) - f_y(x_0, y_0 + t_1)]. \tag{7}$$

In (6) the Mean Value Theorem may be applied to the expression on the right with respect to $y_0 + h$ and $y_0$. We get

$$\frac{1}{h} \Delta_2 f = f_{xy}(x_0 + s_1, y_0 + t_2)h \quad \text{with } 0 < t_2 < h. \tag{8}$$

Similarly, the Mean Value Theorem may be applied in (7) to the expression on the right with respect to $x_0 + h$ and $x_0$. The result is

$$\frac{1}{h} \Delta_2 f = f_{yx}(x_0 + s_2, y_0 + t_1)h \quad \text{with } 0 < s_2 < h. \tag{9}$$

Dividing by $h$ in (8) and (9), we find that

$$\frac{1}{h^2} \Delta_2 f = f_{xy}(x_0 + s_1, y_0 + t_2) = f_{yx}(x_0 + s_2, y_0 + t_1).$$

Letting $h$ tend to zero and noticing that $s_1$, $s_2$, $t_1$, and $t_2$ all tend to zero with $h$, we obtain the result. The argument is similar if $h$ is negative. $\quad\square$

COROLLARY 1    *Suppose that f is a function of any number of variables and s and t are any two of them. Then with the hypotheses on the continuity of the partial derivatives as in Theorem 10, we have*

$$f_{st} = f_{ts}.$$

For example, if the function is $f(x, y, s, t, u, v)$, then

$$f_{xt} = f_{tx}, \quad f_{yu} = f_{uy}, \quad f_{yv} = f_{vy}, \quad \text{etc.}$$

The proof of the corollary is identical to the proof of the theorem.

COROLLARY 2    *For derivatives of the third, fourth, or any order, it does not matter in what order the differentiations with respect to the various variables are performed. For instance, assuming that all fourth-order partial derivatives are continuous, we have*

$$\frac{\partial^4 z}{\partial x \partial x \partial y \partial y} = \frac{\partial^4 z}{\partial x \partial y \partial x \partial y} = \frac{\partial^4 z}{\partial x \partial y \partial y \partial x}$$

$$= \frac{\partial^4 z}{\partial y \partial x \partial x \partial y} = \frac{\partial^4 z}{\partial y \partial x \partial y \partial x} = \frac{\partial^4 z}{\partial y \partial y \partial x \partial x}.$$

*Remarks.*   (i) It is true that there are functions for which $f_{xy}$ is not equal to $f_{yx}$. Of course, the hypotheses of Theorem 10 are violated for such functions. (ii) All the functions we have considered thus far and all the functions we shall consider from now on will always satisfy the hypotheses of Theorem 10. Therefore the order of differentiation will be reversible throughout. (See, however, Problem 44 at the end of this section.)

EXAMPLE 2    Given $u = e^x \cos y + e^y \sin z$, find all first partial derivatives and verify that

$$\frac{\partial^2 u}{\partial x \partial y} = \frac{\partial^2 u}{\partial y \partial x}, \qquad \frac{\partial^2 u}{\partial x \partial z} = \frac{\partial^2 u}{\partial z \partial x}, \qquad \frac{\partial^2 u}{\partial y \partial z} = \frac{\partial^2 u}{\partial z \partial y}.$$

Solution    We have

$$\frac{\partial u}{\partial x} = e^x \cos y; \qquad \frac{\partial u}{\partial y} = -e^x \sin y + e^y \sin z; \qquad \frac{\partial u}{\partial z} = e^y \cos z.$$

Therefore

$$\frac{\partial^2 u}{\partial y \partial x} = -e^x \sin y = \frac{\partial^2 u}{\partial x \partial y};$$

$$\frac{\partial^2 u}{\partial z \partial x} = \quad 0 \quad = \frac{\partial^2 u}{\partial x \partial z};$$

$$\frac{\partial^2 u}{\partial z \partial y} = e^y \cos z \quad = \frac{\partial^2 u}{\partial y \partial z}.$$

□

**EXAMPLE 3**    Suppose that $u = F(x, y, z)$ and $z = f(x, y)$. Obtain a formula for $\partial^2 u / \partial x^2$ in terms of the derivatives of $F$ (that is, $F_x$, $F_y$, $F_z$, $F_{xx}$, etc.) and the derivatives of $f$ (or, equivalently, $z$). That is, in the expression for $F$ we consider $x$, $y$, $z$ *intermediate variables*, while in the expression for $f$ we consider $x$ and $y$ *independent variables*.

**Solution**    We apply the Chain Rule to $F$ to obtain $\partial u / \partial x$ with $x$ and $y$ as independent variables. We get

$$\frac{\partial u}{\partial x} = F_x \frac{\partial x}{\partial x} + F_y \frac{\partial y}{\partial x} + F_z \frac{\partial z}{\partial x}.$$

Since $x$ and $y$ are independent, $\partial y / \partial x = 0$; also, $\partial x / \partial x = 1$. Therefore

$$\frac{\partial u}{\partial x} = F_x + F_z \frac{\partial z}{\partial x}.$$

In order to differentiate a second time, we must recognize that $F_x$ and $F_z$ are again functions of the three intermediate variables. We find that

$$\frac{\partial^2 u}{\partial x^2} = F_{xx} \frac{\partial x}{\partial x} + F_{xy} \frac{\partial y}{\partial x} + F_{xz} \frac{\partial z}{\partial x}$$

$$+ \frac{\partial z}{\partial x}\left( F_{zx} \frac{\partial x}{\partial x} + F_{zy} \frac{\partial y}{\partial x} + F_{zz} \frac{\partial z}{\partial x} \right) + F_z \frac{\partial^2 z}{\partial x^2}.$$

The result is

$$\frac{\partial^2 u}{\partial x^2} = F_{xx} + 2 F_{xz} \frac{\partial z}{\partial x} + F_{zz} \left( \frac{\partial z}{\partial x} \right)^2 + F_z \frac{\partial^2 z}{\partial x^2}. \qquad \square$$

# 9   PROBLEMS

In each of Problems 1 through 10, verify that $f_{xy} = f_{yx}$.

**1** $f(x, y) = x^3 + 7x^2 y - 2xy^2 + y^2$

**2** $f(x, y) = x^4 + 4x^2 y^2 - 2y^4 + 4x^2 y - 3xy^2$

**3** $f(x, y) = x^5 + 2y^4 - 3x^3 y^2 + x^2 y^3 - xy^3 + 2x$

**4** $f(x, y, z) = x^2 + 2xyz + y^2 + z^2 + 3xz - 2xy$

**5** $f(x, y) = e^{xy} \sin x + e^{xy} \cos y$

**6** $f(x, y) = \arctan\left( \dfrac{xy}{x + y} \right)$

**7** $f(x, y) = e^{x/y} \cos(xy) + e^x \sin y$

**8** $f(x, y, z) = \ln \dfrac{1 + x}{1 + y} - e^{xyz}$

**9** $f(x, y, z) = (x^2 + y^2 - z^2)^{1/2}$

**10** $f(x, y, z, t) = x^3 - 2y^3 + z^3 + 2t^3 - 3xyzt$

In each of Problems 11 through 19, verify that $u_{xy} = u_{yx}$ and $u_{xz} = u_{zx}$.

**11** $u = \ln \sqrt{x^2 + y^2 + z^2}$      **12** $u = \ln(x + \sqrt{y^2 + z^2})$

**13** $u = x^3 + y^3 + z^3 - 3xyz$      **14** $u = e^{xy} + e^{2xz} - e^{3yz}$

**15** $u = e^{xy} / \sqrt{x^2 + z^2}$      **16** $u = xyz + x^2 - 3$

**17** $u = \dfrac{x + yz - y}{x^2 + y^2 + 2}$      **18** $u = \sin(xyz)$

**19** $u = x \cos y + y \tan z + z \sin x$

**20** Given that $u = 1/\sqrt{x^2 + y^2 + z^2}$, verify that

$$\frac{\partial^2 u}{\partial x^2} + \frac{\partial^2 u}{\partial y^2} + \frac{\partial^2 u}{\partial z^2} = 0.$$

**21** Given that $u = xe^x \cos y$, verify that

$$\frac{\partial^4 u}{\partial x^4} + 2\frac{\partial^4 u}{\partial x^2 \partial y^2} + \frac{\partial^4 u}{\partial y^4} = 0.$$

In each of Problems 22 through 25, $r$ and $s$ are independent variables. Find $\partial^2 z/\partial r^2$ (a) by the Chain Rule and (b) by finding $z$ in terms of $r$ and $s$ first.

**22** $z = x^2 - xy - y^2$, $x = r + s$, $y = s - r$

**23** $z = x^2 - y^2$, $x = r \cos s$, $y = r \sin s$

**24** $z = x^3 - y^3$, $x = 2r - s$, $y = s + 2r$

**25** $z = x^2 - 2xy - y^2$, $x = r^2 - s^2$, $y = 2rs$

In each of Problems 26 through 31, $r$ and $s$ are independent variables. Find $\partial^2 z/\partial r^2$ by any method.

**26** $z = xy$, $x = r^2$, $y = rs$

**27** $z = (x + y)^2(x - y)^{4/3}$, $x = \cos r$, $y = \sin(rs)$

**28** $z = x \ln y$, $x = 3r + 2s$, $y = r - s$

**29** $z = x^2 \sin y$, $x = r$, $y = s^2$

**30** $z = \ln(x + y)$, $x = e^{-2r}$, $y = e^{2s}$

**31** $z = \dfrac{e^x + e^{-x}}{2y}$, $x = r + s$, $y = e^s$

**32** Given that $u = F(x, y, z)$ and $z = f(x, y)$, find $\partial^2 u/\partial y \partial x$ with all variables as in Example 3.

**33** Given that $u = F(x, y, z)$ and $z = f(x, y)$, find $\partial^2 u/\partial y^2$ with all variables as in Example 3.

**34** If $u = F(x, y)$, $y = f(x)$, find $d^2u/dx^2$.

**35** Given that $u = f(x + 2y) + g(x - 2y)$, show that

$$u_{xx} - \tfrac{1}{4}u_{yy} = 0.$$

**\*36** Let $f$ and $g$ be twice differentiable functions of one variable. Use $f$ and $g$ to construct a solution to the partial differential equation $u_{xx} = a^2 u_{yy}$, where $a$ is a constant. (*Hint:* Refer to Problem 35.)

**37** Given that $u = F(x, y)$, $x = r \cos \theta$, $y = r \sin \theta$, find $\partial^2 u/\partial r^2$, $r$ and $\theta$ being independent variables.

**38** Given $u = F(x, y)$, $x = f(r, s)$, $y = g(r, s)$, find $\partial^2 u/\partial r \partial s$, $r$ and $s$ being independent variables.

**39** If $F(x, y) = 0$, find $d^2y/dx^2$ in terms of partial derivatives of $F$.

**40** Given that $u = F(x, y)$, $x = e^s \cos t$, $y = e^s \sin t$, use the Chain Rule to show that

$$\frac{\partial^2 u}{\partial s^2} + \frac{\partial^2 u}{\partial t^2} = e^{2s}\left(\frac{\partial^2 u}{\partial x^2} + \frac{\partial^2 u}{\partial y^2}\right),$$

where $s$ and $t$ are independent variables and $x$ and $y$ are intermediate variables.

**41** Given $V = F(x, y)$, $x = \tfrac{1}{2}r(e^s + e^{-s})$, $y = \tfrac{1}{2}r(e^s - e^{-s})$, show that

$$V_{xx} - V_{yy} = V_{rr} + \frac{1}{r}V_r - \frac{1}{r^2}V_{ss}.$$

**\*42** If $u = f(x - ut)$, show that $u_t + uu_x = 0$.

**43** Assume that $f(x, y)$ has all continuous partial derivatives to the $n$th order. Use mathematical induction to prove that every $n$th partial derivative is independent of the order in which it is taken.

**\*44** Consider the function

$$f(x, y) = \begin{cases} xy\dfrac{x^2 - y^2}{x^2 + y^2}, & x^2 + y^2 > 0, \\ 0, & x = y = 0. \end{cases}$$

a) Show that $f_x(0, 0) = 0$ and $f_y(0, 0) = 0$.

b) Show, by appealing to the definition of derivative, that

$$f_{xy}(0, 0) = \lim_{k \to 0}\frac{f_x(0, k) - f_x(0, 0)}{k} = -1,$$

$$f_{yx}(0, 0) = \lim_{h \to 0}\frac{f_y(h, 0) - f_y(0, 0)}{h} = 1.$$

c) Show that the result of (b) follows because $f_{xy}$ and $f_{yx}$ are not continuous at $x = y = 0$.

**45** a) Show that $f(x_1, x_2, \dots, x_n) = [x_1^2 + x_2^2 + \cdots + x_n^2]^{(2-n)/2}$ for $n > 2$ is a solution of the equation

$$\frac{\partial^2 f}{\partial x_1^2} + \cdots + \frac{\partial^2 f}{\partial x_n^2} = 0$$

except when $x_1 = x_2 = \cdots = x_n = 0$.
  b) Show that $g(x_1, x_2, \dots, x_n) = [x_1^2 + \cdots + x_{n-1}^2 - x_n^2]^{(2-n)/2}$ for $n > 2$ is a solution of the equation

$$\frac{\partial^2 f}{\partial x_1^2} + \cdots + \frac{\partial^2 f}{\partial x_{n-1}^2} - \frac{\partial^2 f}{\partial x_n^2} = 0$$

except when $(x_1, \dots, x_n)$ is on the cone $x_1^2 + x_2^2 + \cdots + x_{n-1}^2 = x_n^2$.

**46** Suppose that $u = F(x, y)$, $x = f(s, t)$, $y = g(s, t)$. Show that

$$\frac{\partial^2 u}{\partial s \partial t} = F_{xx}\frac{\partial x}{\partial s}\frac{\partial x}{\partial t} + F_{xy}\left(\frac{\partial x}{\partial s}\frac{\partial y}{\partial t} + \frac{\partial x}{\partial t}\frac{\partial y}{\partial s}\right)$$

$$+ F_{yy}\frac{\partial y}{\partial s}\frac{\partial y}{\partial t} + F_x\frac{\partial^2 x}{\partial s \partial t} + F_y\frac{\partial^2 y}{\partial s \partial t}.$$

Also find an expression for $\partial^2 u/\partial s^2$.

<u>**10**</u>

## TAYLOR'S THEOREM WITH REMAINDER (OPTIONAL)

Taylor's theorem for functions of one variable was established in Chapter 10 on page 427. There we found that if $F(x)$ has $n + 1$ derivatives in an interval containing a value $x_0$, then we can obtain the expansion

$$F(x) = F(x_0) + F'(x_0)(x - x_0) + \cdots + \frac{F^{(n)}(x_0)(x - x_0)^n}{n!} + R_n, \qquad (1)$$

where the remainder $R_n$ is given by the formula

$$R_n = \frac{F^{(n+1)}(\xi)(x - x_0)^{n+1}}{(n+1)!},$$

with $\xi$ some number between $x_0$ and $x$.

Taylor's theorem in several variables is a generalization of the expansion (1). We carry out the procedure for a function of two variables $f(x, y)$, the process for functions of more variables being completely analogous. Consider the function

$$\phi(t) = f(x + \lambda t, y + \mu t),$$

in which the quantities $x$, $y$, $\lambda$, and $\mu$ are temporarily kept constant. Then $\phi$ is a function of the single variable $t$, and we may compute its derivative. Using the Chain Rule, we obtain

$$\phi'(t) = f_x(x + \lambda t, y + \mu t)\lambda + f_y(x + \lambda t, y + \mu t)\mu.$$

We will simplify matters by omitting the arguments in $f$. We write

$$\phi'(t) = f_x\lambda + f_y\mu.$$

It is important to compute second, third, fourth, etc., derivatives of $\phi$. We do so by applying the Chain Rule repeatedly. The results are

$$\phi^{(2)}(t) = \lambda^2 f_{xx} + 2\lambda\mu f_{xy} + \mu^2 f_{yy},$$

$$\phi^{(3)}(t) = \lambda^3 f_{xxx} + 3\lambda^2\mu f_{xxy} + 3\lambda\mu^2 f_{xyy} + \mu^3 f_{yyy},$$

$$\phi^{(4)}(t) = \lambda^4 f_{xxxx} + 4\lambda^3\mu f_{xxxy} + 6\lambda^2\mu^2 f_{xxyy} + 4\lambda\mu^3 f_{xyyy} + \mu^4 f_{yyyy}.$$

Examining the pattern in each of the above derivatives, we see that the coefficients in $\phi^{(2)}$ are formed by the **symbolic** expression

$$\left(\lambda\frac{\partial}{\partial x} + \mu\frac{\partial}{\partial y}\right)^2 f,$$

provided that the exponent applied to a partial derivative is interpreted as *repeated differentiation instead of multiplication*. Using this new symbolism we can easily write any derivative of $\phi$. The $k$th derivative is

$$\phi^{(k)}(t) = \left(\lambda\frac{\partial}{\partial x} + \mu\frac{\partial}{\partial y}\right)^k f. \qquad (2)$$

For instance, with $k = 7$ we obtain

$$\phi^{(7)}(t) = \lambda^7 f_{xxxxxxx} + 7\lambda^6 \mu f_{xxxxxxy} + \frac{7 \cdot 6}{1 \cdot 2} \lambda^5 \mu^2 f_{xxxxxyy} + \cdots + \mu^7 f_{yyyyyyy}.$$

Of course all derivatives are evaluated at $(x + \lambda t, y + \mu t)$. The validity of (2) may be established by mathematical induction.

Before stating Taylor's theorem for functions of two variables, we introduce more symbols. The quantity

$$\sum_{1 \leq r+s \leq p} (\quad)$$

means that the sum of the terms in parentheses is taken over all possible nonnegative integers $r$ and $s$ which add up to a number between 1 and $p$. Neither $r$ nor $s$ is allowed to be negative. For example, if $p = 3$ the combinations are

$$(r = 0, s = 1), \quad (r = 0, s = 2), \quad (r = 0, s = 3),$$
$$(r = 1, s = 0), \quad (r = 1, s = 1), \quad (r = 1, s = 2),$$
$$(r = 2, s = 0), \quad (r = 2, s = 1),$$
$$(r = 3, s = 0).$$

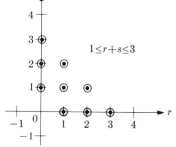

$1 \leq r+s \leq 3$

FIGURE 20

The terms in the sum are shown schematically in Fig. 20 as the circled lattice points.

The symbol

$$\sum_{r+s=p} (\quad)$$

means that the sum is taken over all possible nonnegative integers $r$ and $s$ which add up to $p$ exactly. For instance, if $p = 3$, then the combinations are

$$(r = 0, s = 3), \quad (r = 1, s = 2), \quad (r = 2, s = 1), \quad (r = 3, s = 0).$$

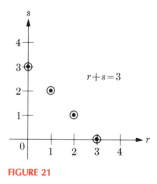

$r+s = 3$

FIGURE 21

The terms in this sum are indicated schematically in Fig. 21.

THEOREM 11       **(Taylor's Theorem)**   *Suppose that $f$ is a function of two variables and that $f$ and all of its partial derivatives of order up to $p + 1$ are continuous in a neighborhood of the point $(a, b)$. Then we have the expansion*

$$f(x, y) = f(a, b) + \sum_{1 \leq r+s \leq p} \frac{\partial^{r+s} f(a, b)}{\partial x^r \partial y^s} \frac{(x - a)^r}{r!} \cdot \frac{(y - b)^s}{s!} + R_p, \qquad (3)$$

*where the remainder $R_p$ is given by the formula*

$$R_p = \sum_{r+s=p+1} \frac{\partial^{r+s} f(\xi, \eta)}{\partial x^r \partial y^s} \frac{(x - a)^r}{r!} \frac{(y - b)^s}{s!},$$

*with the value $(\xi, \eta)$ situated on the line segment joining the points $(a, b)$ and $(x, y)$. (See Fig. 22.)*

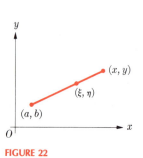

FIGURE 22

**Proof**   We let $d = \sqrt{(x-a)^2 + (y-b)^2}$ and define

$$\lambda = \frac{x-a}{d}, \qquad \mu = \frac{y-b}{d}.$$

The function

$$\phi(t) = f(a + \lambda t, b + \mu t), \quad 0 \le t \le d$$

may be differentiated according to the rules described at the beginning of the section. Taylor's theorem for $\phi(t)$ (a function of *one* variable) taken about $t = 0$ and evaluated at $d$ yields

$$\phi(d) = \phi(0) + \phi'(0)d + \phi^{(2)}(0)\frac{d^2}{2!} + \cdots + \frac{\phi^{(p)}(0)d^p}{p!} + \frac{\phi^{(p+1)}(\tau)d^{p+1}}{(p+1)!}.$$

$$(4)$$

The $k$th derivative of $\phi$ evaluated at 0 is given symbolically by

$$\phi^{(k)}(0) = \left(\lambda\frac{\partial}{\partial x} + \mu\frac{\partial}{\partial y}\right)^k f(a, b).$$

We now recall the binomial formula:

$$(A + B)^k = A^k + \frac{k}{1}A^{k-1}B + \frac{k(k-1)}{1 \cdot 2}A^{k-2}B^2 + \cdots + B^k$$

$$= \sum_{q=0}^{k} \frac{k!}{q!(k-q)!}A^{k-q}B^q.$$

Applying the binomial formula for the symbolic expression for $\phi^{(k)}(0)$, we obtain

$$\phi^{(k)}(0) = \sum_{q=0}^{k} \frac{k!}{q!(k-q)!} \frac{\partial^k f(a, b)}{\partial x^{k-q}\partial y^q} \lambda^{k-q}\mu^q.$$

$$(5)$$

Noting that $\phi(d) = f(a + \lambda d, b + \mu d) = f(x, y)$ and that $\phi(0) = f(a, b)$, we find, upon substitution of (5) into (4):

$$f(x, y) = f(a, b) + \frac{\partial f(a, b)}{\partial x}(x - a) + \frac{\partial f(a, b)}{\partial y}(y - b)$$

$$+ \frac{\partial^2 f(a, b)}{\partial x^2}\frac{(x-a)^2}{2!} + \frac{\partial^2 f(a, b)}{\partial x\partial y}(x - a)(y - b)$$

$$+ \frac{\partial^2 f(a, b)}{\partial y^2}\frac{(y-b)^2}{2!} + \cdots,$$

which is precisely the formula in the statement of the theorem. The remainder term shows that if $0 < \tau < d$, then

$$\xi = a + \frac{(x-a)}{d}\tau, \qquad \eta = b + \frac{(y-b)}{d}\tau,$$

which places $(\xi, \eta)$ on the line segment joining $(a, b)$ and $(x, y)$.   □

*Remarks.*    (i) **Taylor's formula** (as (3) is usually called) is also often written in the form

$$f(x, y) = f(a, b) + \sum_{q=1}^{p} \frac{1}{q!} \left[ \sum_{r=0}^{q} \frac{q!}{(q-r)!r!} \frac{\partial^q f(a, b)}{\partial x^{q-r} \partial y^r} (x - a)^{q-r} (y - b)^r \right] + R_p.$$

ii) For some functions $f$, if we let $p$ tend to infinity we may find that $R_p \to 0$. We thereby obtain a representation of $f$ as an infinite series in $x$ and $y$. Such a series is called a **double series**, and we say that $f$ is **expanded about the point** $(a, b)$. For functions of three variables we obtain a triple sum, the Taylor formula in this case being

$$f(x, y, z) = f(a, b, c)$$

$$+ \sum_{1 \le r+s+t \le p} \frac{\partial^{r+s+t} f(a, b, c)}{\partial x^r \partial y^s \partial z^t} \frac{(x-a)^r}{r!} \frac{(y-b)^s}{s!} \frac{(z-c)^t}{t!} + R_p,$$

with

$$R_p = \sum_{r+s+t=p+1} \frac{\partial^{p+1} f(\xi, \eta, \zeta)}{\partial x^r \partial y^s \partial z^t} \frac{(x-a)^r}{r!} \frac{(y-b)^s}{s!} \frac{(z-c)^t}{t!},$$

where $(\xi, \eta, \zeta)$ is on the line segment joining $(a, b, c)$ and $(x, y, z)$.

**EXAMPLE 1**    Expand $x^2 y$ about the point $(1, -2)$ up to and including the terms of the second degree. Find $R_2$.

**Solution**    Setting $f(x, y) = x^2 y$, we obtain

$$f_x = 2xy, f_y = x^2, f_{xx} = 2y, f_{xy} = 2x, f_{yy} = 0,$$

$$f_{xxx} = 0, f_{xxy} = 2, f_{xyy} = f_{yyy} = 0.$$

Noting that $f(1, -2) = -2$, we find

$$x^2 y = -2 - 4(x - 1) + (y + 2)$$

$$+ \frac{1}{2!} [-4(x - 1)^2 + 4(x - 1)(y + 2)] + R_2,$$

with

$$R_2 = \frac{1}{2!} 2(x - 1)^2 (y + 2) = (x - 1)^2 (y + 2). \qquad \square$$

**EXAMPLE 2**    Given $f(x, y, z) = e^x \cos y + e^y \cos z + e^z \cos x$. Define

$$\phi(t) = f(x + \lambda t, y + \mu t, z + \nu t).$$

Find $\phi'(0)$ and $\phi^{(2)}(0)$ in terms of $x, y, z, \lambda, \mu, \nu$.

**Solution**    We have

$$\phi'(0) = f_x(x, y, z)\lambda + f_y(x, y, z)\mu + f_z(x, y, z)\nu.$$

Computing the derivatives, we obtain

$$\phi'(0) = (e^x \cos y - e^z \sin x)\lambda + (e^y \cos z - e^x \sin y)\mu$$
$$+ (e^z \cos x - e^y \sin z)v.$$

The formula for $\phi^{(2)}(0)$ is

$$\phi^{(2)}(0) = \left(\lambda \frac{\partial}{\partial x} + \mu \frac{\partial}{\partial y} + v \frac{\partial}{\partial z}\right)^2 f$$

$$= \lambda^2(e^x \cos y - e^z \cos x) + \mu^2(e^y \cos z - e^x \cos y)$$
$$+ v^2(e^z \cos x - e^y \cos z) + 2\lambda\mu(-e^x \sin y)$$
$$+ 2\lambda v(-e^z \sin x) + 2\mu v(-e^y \sin z). \qquad \square$$

## 10    PROBLEMS

**1** Expand $x^3 + xy^2$ about the point $(2, 1)$.

**2** Expand $x^4 + x^2y^2 - y^4$ about the point $(1, 1)$ up to terms of the second degree. Find the form of $R_2$.

**3** Find the expansion of $\sin(x + y)$ about $(0, 0)$ up to and including the terms of the third degree in $(x, y)$. Compare the result with that obtained by writing $\sin u \approx u - \frac{1}{6}u^3$ and setting $u = x + y$.

**4** Find the expansion of $\cos(x + y)$ about $(0, 0)$ up to and including terms of the fourth degree in $(x, y)$. Compare the result with that obtained by writing $\cos u \approx 1 - \frac{1}{2}u^2 + \frac{1}{24}u^4$ and setting $u = x + y$.

**5** Find the expansion of $e^{x+y}$ about $(0, 0)$ up to and including the terms of the third degree in $(x, y)$. Compare the result with that obtained by setting $e^u \approx 1 + u + \frac{1}{2}u^2 + \frac{1}{6}u^3$, and then setting $u = x + y$. Next compare the result with that obtained by multiplying the series for $e^x$ by that for $e^y$ and keeping terms up to and including the third degree.

**6** Find the expansion of $\sin x \sin y$ about $(0, 0)$ up to and including the terms of the fourth degree in $(x, y)$. Compare the result with that obtained by multiplying the series for $\sin x$ and $\sin y$.

**7** Do the same as Problem 6 for $\cos x \cos y$.

**8** Expand $e^x \arctan y$ about $(1, 1)$ up to and including the terms of the second degree in $(x - 1)$ and $(y - 1)$.

**9** Expand $x^2 + 2xy + yz + z^2$ about $(1, 1, 0)$.

**10** Expand $x^3 + x^2y - yz^2 + z^3$ about $(1, 0, 1)$ up to and including the terms of the second degree in $(x - 1)$, $y$, and $(z - 1)$.

**11** If $f(x, y) = x^2 + 4xy + y^2 - 6x$ and $\phi(t) = f(x + \lambda t, y + \mu t)$, find $\phi^{(2)}(0)$ when $x = -1$ and $y = 2$. Is $\phi^{(2)}(0) > 0$

when $(x, y) = (-1, 2)$ if $\lambda$ and $\mu$ are related so that

$$\lambda^2 + \mu^2 = 1?$$

**12** If $f(x, y) = x^3 + 3xy^2 - 3x^2 - 3y^2 + 4$ and $\phi(t) = f(x + \lambda t, y + \mu t)$ find $\phi^{(2)}(0)$ for $(x, y) = (2, 0)$. Show that $\phi^{(2)}(0) > 0$ for all $\lambda$ and $\mu$ such that $\lambda^2 + \mu^2 = 1$.

**\*13** a) Write the appropriate expansion formula using binomial coefficients for

$$(A + B + C)^k,$$

with $k$ a positive integer. (b) If $f(x, y, z)$ and $\phi(t) = f(x + \lambda t, y + \mu t, z + vt)$ are sufficiently differentiable, show the relationship between the symbolic expression

$$\left(\lambda \frac{\partial}{\partial x} + \mu \frac{\partial}{\partial y} + v \frac{\partial}{\partial z}\right)^k f(x + \lambda t, y + \mu t, z + vt)$$

and

$$\phi^{(k)}(t).$$

**14** Write Taylor's formula for a function $f(x, y, u, v)$ of four variables expanded about the point $a, b, c, d$. How many second-derivative terms are there? Third-derivative terms?

**\*15** Use mathematical induction to establish Formula (2) on page 671.

**16** Referring to the symbol $\Sigma_{1 \le r+s+t \le p}$, find the number of terms when $p = 3$; when $p = 4$. Exhibit the corresponding lattices with three-dimensional diagrams.

**17** Write Taylor's formula for a function $f(x_1, x_2, \ldots, x_k)$ of $k$ variables. Give an expression for $R_p$, the remainder.

**18** Write a proof of Taylor's Theorem (Theorem 11) for a function $f(x, y, z)$ of three variables.

*19 a) Given the positive integer $p$, how many solutions are there of the equation

$$r_1 + r_2 + \cdots + r_k = p,$$

where $r_1, r_2, \ldots, r_k$ are positive integers? (b) Same problem with $r_1, r_2, \ldots, r_k$ nonnegative integers. (c) How many different partial derivatives of order $p$ are there for a function $f(x_1, x_2, \ldots, x_k)$?

20 Expand the function $f(x, y) = \sin x \cos y$ about the values $x = \pi/6$ and $y = \pi/3$ to terms of order 2 and write an approximate formula for $\sin 29° \cos 61°$. Estimate the remainder term. Evaluate the third derivatives at the point $(\pi/6, \pi/3)$.

21 Let $f(x, y, z)$ be a polynomial in $(x, y, z)$. That is,

$$f(x, y, z) = \sum_{i+j+k=0}^{n} a_{ijk} x^i y^j z^k, \quad i, j, k \geq 0.$$

Show that if all partial derivatives of $f$ up to and including order $n$ vanish at some point $(x_0, y_0, z_0)$, then $f \equiv 0$.

---

## 11

## MAXIMA AND MINIMA

One of the principal applications of differentiation of functions of one variable occurs in the study of maxima and minima. In Chapter 4 we derived various tests using first and second derivatives which enable us to determine relative maxima and minima of functions of a single variable. These tests are useful for graphing functions, for solving problems involving related rates, and for attacking a variety of geometrical and physical problems. (See Chapter 4, Sections 3, 4, 6, and 7.)

The study of maxima and minima for functions of two, three, or more variables has its basis in the following theorem, which we state without proof.

---

**THEOREM 12**    **(Extreme Value Theorem)**  *Let R be a region in the xy plane with the boundary curve of R considered as part of R also (Fig. 23). If f is a function of two variables defined and continuous on R, then there is (at least) one point in R where f takes on a maximum value and there is (at least) one point in R where f takes on a minimum value.*

---

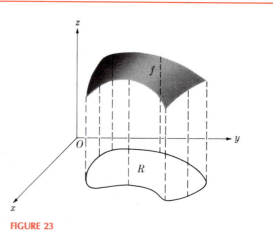

FIGURE 23

*Remarks.* (i) Theorem 12 is a straightforward generalization of Theorem 1 given on page 125 (Extreme Value Theorem). (ii) Analogous theorems may be stated for functions of three, four, or more variables. (iii) The maximum and minimum may occur on the boundary of $R$. Thus, as in the case of one variable where the interval must contain its *endpoints*, the region $R$ must contain its *boundary* in order to guarantee the validity of the result.

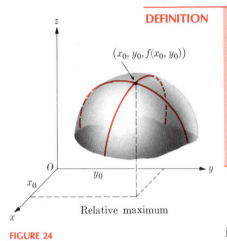

**FIGURE 24**

**DEFINITION**

*A function $f(x, y)$ is said to have a **relative maximum** at $(x_0, y_0)$ if there is some region containing $(x_0, y_0)$ in its interior such that*

$$f(x, y) \le f(x_0, y_0)$$

*for all $(x, y)$ in this region. (See Fig. 24). More precisely, there must be some positive number $\delta$ (which may be "small") such that the above inequality holds for all $(x, y)$ in the square*

$$|x - x_0| < \delta, \qquad |y - y_0| < \delta.$$

A similar definition holds for **relative minimum** when the inequality

$$f(x, y) \ge f(x_0, y_0)$$

is satisfied in a square about $(x_0, y_0)$. (See Fig. 25.) The above definitions are easily extended to functions of three, four, or more variables.

**THEOREM 13**

*Suppose that $f(x, y)$ is defined in a region $R$ containing $(x_0, y_0)$ in its interior. Suppose that $f_x(x_0, y_0)$ and $f_y(x_0, y_0)$ are defined and that*

$$f(x, y) \le f(x_0, y_0)$$

*for all $(x, y)$ in $R$; that is, $f(x_0, y_0)$ is a relative maximum. Then*

$$f_x(x_0, y_0) = f_y(x_0, y_0) = 0.$$

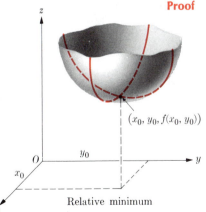

**FIGURE 25**

**Proof**  We show that $f_x(x_0, y_0) = 0$, the proof for $f_y$ being analogous. By definition,

$$f_x(x_0, y_0) = \lim_{h \to 0} \frac{f(x_0 + h, y_0) - f(x_0, y_0)}{h}.$$

By hypothesis,

$$f(x_0 + h, y_0) - f(x_0, y_0) \le 0$$

for all $h$ sufficiently small so that $(x_0 + h, y_0)$ is in $R$. If $h$ is positive, then

$$\frac{f(x_0 + h, y_0) - f(x_0, y_0)}{h} \le 0,$$

and as $h \to 0$ we conclude that $f_x$ must be nonpositive. On the other hand, if $h$ is negative, then

$$\frac{f(x_0 + h, y_0) - f(x_0, y_0)}{h} \ge 0,$$

since division of both sides of an inequality by a negative number reverses its direction. Letting $h \to 0$, we conclude that $f_x$ is nonnegative. A quantity which is both nonnegative and nonpositive must be zero. ☐

| | |
|---|---|
| **COROLLARY** | *The conclusion of Theorem 13 holds at a relative minimum.* |

| | |
|---|---|
| **DEFINITION** | *A value $(x_0, y_0)$ at which both $f_x$ and $f_y$ are zero is called a **critical point** of $f$.* |

*Discussion.* The conditions that $f_x$ and $f_y$ vanish at a point are *necessary* conditions for a relative maximum or a relative minimum. It is easy to find a function for which $f_x$ and $f_y$ vanish at a point, with the function having neither a relative maximum nor a relative minimum at that point. A critical point at which $f$ is neither a maximum nor a minimum may be a **"saddle point."** A simple example of a function which has such a point is given by

$$f(x, y) = x^2 - y^2.$$

We see that $f_x = 2x$, $f_y = -2y$, and $(0, 0)$ is a critical point. However, as Fig. 26 shows, the function is "saddle-shaped" in a neighborhood of $(0, 0)$.

Level curves (see Section 1 of this chapter) can be an aid in deciding if a critical point is indeed a saddle point. Figures 27 and 28 illustrate this fact.

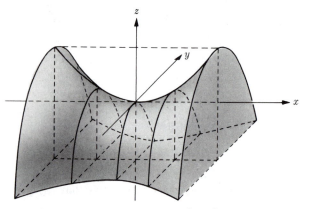

Drawing of graph of $z = f(x, y) = x^2 - y^2$

**FIGURE 26**

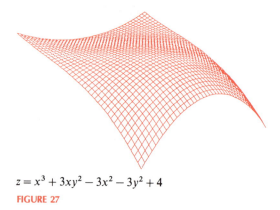

$z = x^3 + 3xy^2 - 3x^2 - 3y^2 + 4$

**FIGURE 27**

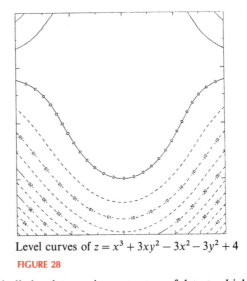

Level curves of $z = x^3 + 3xy^2 - 3x^2 - 3y^2 + 4$

**FIGURE 28**

While we shall develop an important, useful test which, under certain conditions, guarantees that a function has a maximum or minimum at a critical point, it is sometimes possible to make this decision from the nature of the problem itself. We exhibit such an example.

**EXAMPLE 1**    In three-dimensional space find the point on the plane

$$S = \{(x, y, z): 2x + 3y - z = 1\}$$

which is closest to the origin.

**Solution**    The function $d = \sqrt{x^2 + y^2 + z^2}$ represents a distance function which has a specific value at each point on the plane $S$. The minimum of the function

$$f(x, y, z) = x^2 + y^2 + z^2$$

occurs at the same point as the minimum of $d$, and $f$ is simpler to handle. We substitute for $z$ from the equation of the plane, and so we must minimize

$$f(x, y) = x^2 + y^2 + (1 - 2x - 3y)^2$$
$$= 5x^2 + 10y^2 + 12xy - 4x - 6y + 1.$$

A critical point must be a solution of the equations

$$f_x = 10x + 12y - 4 = 0, \qquad f_y = 20y + 12x - 6 = 0.$$

Solving these equations simultaneously, we find

$$x = \tfrac{1}{7}, \qquad y = \tfrac{3}{14}.$$

From the geometric character of the problem we know that $(\tfrac{1}{7}, \tfrac{3}{14})$ corresponds to a minimum. The point on $S$ corresponding to $x = \tfrac{1}{7}$, $y = \tfrac{3}{14}$ is found by substitution in the equation for $S$. The answer is $(\tfrac{1}{7}, \tfrac{3}{14}, -\tfrac{1}{14})$.    □

The basic criterion for finding maxima and minima for functions of two variables is the **Second Derivative Test**, which we now establish. The Proof of Theorem 14 uses Taylor's theorem for two variables which is established in Section 10. Since Section 10 is optional the proof here is for those readers who have studied Taylor's theorem for two variables.

**THEOREM 14**    **(Second Derivative Test)**    *Suppose that $f$ and its partial derivatives up to and including those of the third order are continuous near the point $(a, b)$, and suppose that*

$$f_x(a, b) = f_y(a, b) = 0;$$

*that is, $(a, b)$ is a critical point. Then we have*

i) *a local minimum if*

$$f_{xx}(a, b) f_{yy}(a, b) - f_{xy}^2(a, b) > 0 \qquad \text{and} \qquad f_{xx}(a, b) > 0;$$

ii) *a local maximum if*

$$f_{xx}(a, b) f_{yy}(a, b) - f_{xy}^2(a, b) > 0 \qquad \text{and} \qquad f_{xx}(a, b) < 0;$$

iii) *a saddle point if*

$$f_{xx}(a, b) f_{yy}(a, b) - f_{xy}^2(a, b) < 0;$$

iv) *no information if*

$$f_{xx}(a, b) f_{yy}(a, b) - f_{xy}^2(a, b) = 0.$$

**Proof (Optional)**    For convenience, we define

$$A = f_{xx}(a, b), \qquad B = f_{xy}(a, b), \qquad C = f_{yy}(a, b).$$

Recalling the Taylor expansion (Theorem 11, page 672) of $f(x, y)$ about the point $(a, b)$, we find

$$f(x, y) = f(a, b) + \tfrac{1}{2}[A(x - a)^2 + 2B(x - a)(y - b) + C(y - b)^2] + R_2. \quad (1)$$

The first derivative terms are absent because $(a, b)$ is a critical point. The term $R_2$ is given by

$$R_2 = \frac{1}{6}\left[ \frac{\partial^3 f(\xi, \eta)}{\partial x^3}(x - a)^3 + 3 \frac{\partial^3 f(\xi, \eta)}{\partial x^2 \partial y}(x - a)^2(y - b) \right.$$

$$\left. + 3 \frac{\partial^3 f(\xi, \eta)}{\partial x \partial y^2}(x - a)(y - b)^2 + \frac{\partial^3 f(\xi, \eta)}{\partial y^3}(y - b)^3 \right].$$

We define $r = \sqrt{(x - a)^2 + (y - b)^2}$ and the quantities $\lambda, \mu$ by the relations

$$\lambda = \frac{x - a}{r}, \qquad \mu = \frac{y - b}{r}.$$

Note that for any $x, y, a, b$ the relation $\lambda^2 + \mu^2 = 1$ prevails. It is not difficult to verify that the Taylor expansion (1) now becomes

$$f(x, y) - f(a, b) = \tfrac{1}{2} r^2 (A\lambda^2 + 2B\lambda\mu + C\mu^2 + r\rho), \qquad (2)$$

where

$$\rho = \frac{1}{3}\left( \frac{\partial^3 f}{\partial x^3} \lambda^3 + 3 \frac{\partial^3 f}{\partial x^2 \partial y} \lambda^2 \mu + 3 \frac{\partial^3 f}{\partial x \partial y^2} \lambda \mu^2 + \frac{\partial^3 f}{\partial y^3} \mu^3 \right)_{\substack{x = \xi \\ y = \eta}}.$$

The quantity $\rho$ is bounded since, by hypothesis, $f$ has continuous third derivatives. The behavior of $f(x, y) - f(a, b)$ is determined completely by the size of $r\rho$ and the size of the quadratic expression

$$A\lambda^2 + 2B\lambda\mu + C\mu^2, \tag{3}$$

with $\lambda^2 + \mu^2 = 1$. If

$$B^2 - AC < 0 \qquad \text{and} \qquad A > 0,$$

then there are no real roots to (3) and it has a positive minimum value. (Call it $m$.) Now, selecting $r$ so small that $r\rho$ is negligible compared with $m$, we deduce that the right side of (2) is always positive if $(x, y)$ is sufficiently close to $(a, b)$. Hence

$$f(x, y) - f(a, b) > 0$$

and $f$ is a minimum at $(a, b)$. We have just established part (i) of the theorem. By the same argument, if

$$B^2 - AC < 0 \qquad \text{and} \qquad A < 0,$$

then (3) is always negative and

$$f(x, y) - f(a, b) < 0$$

for $(x, y)$ near $(a, b)$. Thus the statement of (ii) follows. Part (iii) results when $B^2 - AC > 0$, in which case (3) (and therefore (2)) is sometimes positive and sometimes negative. Then $f$ can have neither a maximum nor a minimum at $(a, b)$ and the surface $z = f(x, y)$ can be shown to be saddle-shaped near $(a, b)$. Part (iv) is provided for completeness.     □

**EXAMPLE 2**    Test for relative maxima and minima the function $f$ defined by

$$f(x, y) = x^3 + 3xy^2 - 3x^2 - 3y^2 + 4.$$

**Solution**    We have

$$f_x = 3x^2 + 3y^2 - 6x$$

and

$$f_y = 6xy - 6y.$$

We set these equations equal to zero and solve simultaneously. Writing

$$x^2 + y^2 - 2x = 0, \qquad y(x - 1) = 0,$$

we see that the second equation vanishes only when $y = 0$ or $x = 1$. If $y = 0$, the first equation gives $x = 0$ or $2$; if $x = 1$, the first equation gives $y = \pm 1$. The critical points are

$$(0, 0), \quad (2, 0), \quad (1, 1), \quad (1, -1).$$

To apply the Second Derivative Test, we compute

$$A = f_{xx} = 6x - 6, \qquad B = f_{xy} = 6y, \qquad C = f_{yy} = 6x - 6.$$

At $(0, 0)$:     $AC - B^2 > 0$    and $A < 0$, a maximum.

At $(2, 0)$:     $AC - B^2 > 0$    and $A > 0$, a minimum.

At $(1, 1)$:     $AC - B^2 < 0$,   saddle point.

At $(1, -1)$:    $AC - B^2 < 0$,   saddle point.     □

**EXAMPLE 3**   Test at $(0, 0)$ for relative maxima and minima the functions $f, g,$ and $h$ defined by

$$f(x, y) = x^6 + y^8$$

$$g(x, y) = -(x^8 + y^6)$$

$$h(x, y) = 8x^2 - 6xy^2 + y^4$$

**Solution**   We have $f_x = 6x^5$, $f_y = 8y^7$, and therefore $(0, 0)$ is the only critical point for $f$. Note that $f_{xx} = 30x^4$, $f_{yy} = 56y^6$ and $f_{xy} = 0$. Therefore at $(0, 0)$ the Second Derivative Test gives us no information. However $f(0, 0) = 0$, and since $f(x, y) = x^6 + y^8$ is always strictly positive when $(x, y) \neq (0, 0)$, we conclude that $(0, 0)$ is a local as well as an absolute minimum for $f$. Since $g_x = -8x^7$ and $g_y = -6y^5$, $(0, 0)$ is also the only critical point of $g$. Analogously, the Second Derivative Test gives no information for $g$, but it is clear that $(0, 0)$ is an absolute maximum.

We now turn our attention to $h$. Since $h_x = 16x - 6y^2$, $h_y = -12xy + 4y^3$, we see that $(0, 0)$ is also the only critical point for $h$. However

$$h_{xx} = 16$$

$$h_{yy} = -12x + 12y^2$$

$$h_{xy} = -12y$$

and again the Second Derivative Test gives us no information at $(x, y) = (0, 0)$. In this case it is still not clear what is happening at the critical point $(0, 0)$. We note that $h(x, y) = 8x^2 - 6xy^2 + y^4 = (y^2 - 2x)(y^2 - 4x)$, and that $h(0, 0) = 0$. If we draw the level curve $h(x, y) = 0$, we are graphing the two parabolas $y^2 = 2x$ and $y^2 = 4x$. (See Fig. 29.) We then indicate on the graph by $+$ and $-$ which values of $(x, y)$ give rise to positive values of $h$, and which give rise to negative values of $h$, respectively. It is then simple to observe that any circle drawn around $(0, 0)$, however small, contains both positive and negative values of $h$. Therefore $(0, 0)$ must be a saddle point.    □

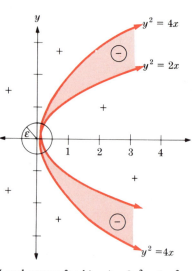

Level curves for $h(x, y) = 8x^2 - 6xy^2 + y^4 = c$ with $c = 0$. Regions in the $(x, y)$ plane where $z = h(x, y)$ is positive or negative are indicated

**FIGURE 29**

**EXAMPLE 4**   Examine $z = f(x, y) = \cos x + \cos y$ on the square

$$-\frac{\pi}{2} < x < \frac{\pi}{2}, \qquad -\frac{\pi}{2} < y < \frac{\pi}{2},$$

for relative maxima and minima.

**Solution**   We have

$$f_x = -\sin x \qquad f_y = -\sin y$$

$$f_{xx} = -\cos x \qquad f_{yy} = -\cos y$$

$$f_{xy} = 0.$$

For $(x, y)$ to be a critical point we must have $-\sin x = -\sin y = 0$. Since we are restricted to the square

$$-\frac{\pi}{2} < x < \frac{\pi}{2}, \qquad -\frac{\pi}{2} < y < \frac{\pi}{2},$$

we find that only $(x, y) = (0, 0)$ is a critical point. The Second Derivative Test gives

$$(-1)(-1) - (0)^2 = 1 > 0$$

and $f_{xx}(0, 0) = -\cos(0) = -1$. Therefore the point $(0, 0)$ is a local maximum. Actually, it is an absolute maximum since $z(0, 0) = \cos(0) + \cos(0) = 2$, which is as large as $z$ can be. Figure 30 shows a graph of part of $z = \cos x + \cos y$. □

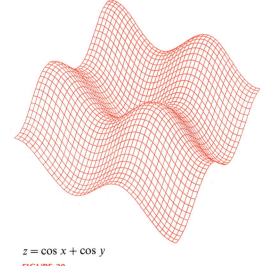

$z = \cos x + \cos y$

**FIGURE 30**

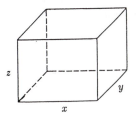

**FIGURE 31(a)**

**EXAMPLE 5**  Find the dimensions of the rectangular box, open at the top, which has maximum volume if the surface area is 12.

**Solution**  Let $V$ be the volume of the box; let $(x, y)$ be the horizontal directions and $z$ the height. See Fig. 31(a). Then $V = xyz$, and the surface area is given by

$$xy + 2xz + 2yz = 12. \tag{4}$$

Solving this equation for $z$ and substituting its value in the expression for $V$, we get

$$V = \frac{xy(12 - xy)}{2(x + y)} = \frac{12xy - x^2y^2}{2(x + y)}.$$

The domains for $x, y, z$ are restricted by the inequalities

$$x > 0, \qquad y > 0, \qquad xy < 12.$$

In other words, $x$ and $y$ must lie in the shaded region shown in Fig. 31(b). To find the critical points, we compute

$$V_x = \frac{y^2(12 - x^2 - 2xy)}{2(x + y)^2}, \qquad V_y = \frac{x^2(12 - y^2 - 2xy)}{2(x + y)^2}$$

and set these expressions equal to zero. We obtain (excluding $x = y = 0$)

$$x^2 + 2xy = 12, \qquad y^2 + 2xy = 12.$$

**FIGURE 31(b)**

Subtracting, we find that $x = \pm y$. If $x = y$, then the positive solution is $x = y = 2$. We reject $x = -y$, since both quantities must be positive. From formula (4) for surface area we conclude that $z = 1$ when $x = y = 2$. From geometrical considerations, we see that these are the dimensions which give a maximum volume. □

*Remarks.* (i) The determination of maxima and minima hinges on our ability to solve the two simultaneous equations in two unknowns resulting when we set $f_x = 0$ and $f_y = 0$. In Example 1 these equations are linear and so quite easy to solve. In Examples 2 and 5, however, the equations are nonlinear, and there are no routine methods for solving nonlinear simultaneous equations. Elementary courses in algebra usually avoid such topics, and the reader is left to his or her own devices. The only general rule we can state is: Try to solve one of the equations for one of the unknowns in terms of the other. Substitute this value in the second equation and try to find all solutions of the second equation. Otherwise use trickery and guesswork. In actual practice, systems of nonlinear equations may be solved by a variety of numerical techniques. Computers and even sophisticated hand calculators are particularly valuable in such problems. (ii) Definitions of critical point, relative maximum and minimum, etc., for functions of three, four, and more variables are simple extensions of the two-variable case. If $f(x, y, z)$ has first derivatives, then a point where

$$f_x = 0, \qquad f_y = 0, \qquad f_z = 0$$

is a **critical point**. We obtain such points by solving simultaneously three equations in three unknowns. To obtain the critical points for functions of $n$ variables, we set all $n$ first derivatives equal to zero and solve simultaneously the $n$ equations in $n$ unknowns. (iii) Extensions of the Second Derivative Test (Theorem 13) for functions of three or more variables are given in advanced courses.

## 11   PROBLEMS

In each of Problems 1 through 22, test the functions $f$ for relative maxima and minima.

1  $f(x, y) = x^2 + 2y^2 - 4x + 4y - 3$

2  $f(x, y) = x^2 - y^2 + 2x - 4y - 2$

3  $f(x, y) = x^2 + 2xy + 3y^2 + 2x + 10y + 9$

4  $f(x, y) = x^2 - 3xy + y^2 + 13x - 12y + 13$

5  $f(x, y) = y^3 + x^2 - 6xy + 3x + 6y - 7$

6  $f(x, y) = x^3 + y^2 + 2xy + 4x - 3y - 5$

7  $f(x, y) = 3x^2 y + x^2 - 6x - 3y - 2$

8  $f(x, y) = xy + 4/x + 2/y$

9  $f(x, y) = \sin x + \sin y + \sin(x + y)$

10  $f(x, y) = x^3 - 6xy + y^3$

11  $f(x, y) = 8^{2/3} - x^{2/3} - y^{2/3}$

12  $f(x, y) = e^x \cos y$

13  $f(x, y) = e^{-x} \sin^2 y$

14  $f(x, y) = x \sin y$

15  $f(x, y) = (x + y)(xy + 1)$

16  $f(x, y) = xy + 1/x + 8/y$

17  $f(x, y) = xy + 1/x + 1/y$

18  $f(x, y) = (x - y)(xy - 1)$

19  $f(x, y) = x + y \sin x$

20  $f(x, y) = y^4 - 4xy^2 + 3x^2$

21  $f(x, y) = -9y^2 + 14x^2 + y^4$

22  $f(x, y) = 2y^4 - 11xy^2 + 5x^2$

In each of Problems 23 through 27, test the function $g$ for relative maxima and minima within the indicated region.

**23** $g(x, y) = \sin x + \sin y; \quad 0 < x < \pi, 0 < y < \pi$

**24** $g(x, y) = \sin x + \sin y + \sin(x + y);$
$0 < x < \pi, 0 < y < \pi$

**25** $g(x, y) = \sin x + \sin y + \sin(x + y);$
$0 < x < 2\pi, 0 < y < 2\pi$

**26** $g(x, y) = \cos x - \sin y; \quad 0 < x < \pi/2, 0 < y < \pi/2$

**27** $g(x, y) = (y - \pi/2) \cos x + \sin y;$
$\pi/4 < x < 3\pi/4, \pi/4 < y < 3\pi/4$

In each of Problems 28 through 31, find the critical points.

**28** $f(x, y, z) = x^2 + 2y^2 + z^2 - 6x + 3y - 2z - 5$

**29** $f(x, y, z) = x^2 + y^2 - 2z^2 + 3x + y - z - 2$

**30** $f(x, y, z) = x^2 + y^2 + z^2 + 2xy - 3xz + 2yz - x$
$+ 3y - 2z - 5$

**31** $f(x, y, z, t) = x^2 + y^2 + z^2 - t^2 - 2xy + 4xz + 3xt - 2yt$
$+ 4x - 5y - 3$

**32** In three-dimensional space find the minimum distance from the origin to the plane

$$S = \{(x, y, z): 3x + 4y + 2z = 6\}.$$

**33** In the plane find the minimum distance from the point $(-1, -3)$ to the line

$$l = \{(x, y): x + 3y = 7\}.$$

**34** In three-dimensional space find the minimum distance from the point $(-1, 3, 2)$ to the plane

$$S = \{(x, y, z): x + 3y - 2z = 8\}.$$

**35** In three-dimensional space find the minimum distance from the origin to the cone

$$C = \{(x, y, z): z^2 = (x - 1)^2 + (y - 2)^2\}.$$

**36** For a package to go by parcel post, the sum of the length and girth (perimeter of cross section) must not exceed 100 cm. Find the dimensions of the package of largest volume which can be sent; assume the package has the shape of a rectangular box.

**\*37** Show that $\sqrt[3]{xyz} \le \frac{1}{3}(x + y + z)$ in the region $0 \le x, 0 \le y, 0 \le z$. (*Hint:* Find the minimum of $x + y + z$ for $xyz = \alpha$, where $\alpha$ is a constant, by setting $z = \alpha/xy$ and finding this minimum for $x > 0, y > 0$.)

**38** Find the dimensions of the rectangular parallelepiped of maximum volume with edges parallel to the axes which can be inscribed in the ellipsoid

$$E = \left\{(x, y, z): \frac{x^2}{9} + \frac{y^2}{4} + \frac{z^2}{16} = 1\right\}.$$

**39** Find the shape of the closed rectangular box of largest volume with a surface area of 16 sq cm.

**40** The base of an open rectangular box costs half as much per square cm as the sides. Find the dimensions of the box of largest volume which can be made for $D$ dollars.

**\*41** The cross section of a trough is an isosceles trapezoid (see Fig. 32). If the trough is made by bending up the sides of a strip of metal 18 cm wide, what should the dimensions be in order for the area of the cross section to be a maximum? Choose $h$ and $l$ as independent variables.

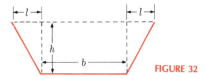

FIGURE 32

**\*42** A pentagon is composed of a rectangle surmounted by an isosceles triangle (see Fig. 33). If the pentagon has a given perimeter $P$, find the dimensions for maximum area. Choose variables as indicated in Fig. 33.

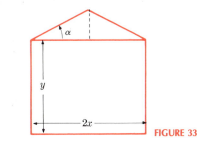

FIGURE 33

**43** A curve $C$ in three-space is given by

$$C = \{(x, y, z): x^2 - xy + y^2 - z^2 = 1 \quad \text{and} \quad x^2 + y^2 = 1\}.$$

Find the point or points on $C$ closest to the origin.

**44** a) Find a function $f(x, y)$ with $f_{xx}f_{yy} - f_{xy}^2 = 0$ at a point $(a, b)$ and such that $f$ has a maximum at $(a, b)$.
b) Same as part (a) except that $f$ has a minimum at $(a, b)$.
c) Same as part (a) except that $f$ has neither a maximum nor a minimum at $(a, b)$.

**\*45** State and prove a Second Derivative Test such as Theorem 14 for functions $f(x, y, z)$ of three variables.

**46** Let $f(x_1, x_2, \ldots, x_n)$ be a function with the property that $f_{x_i} = 0$ at a point $P(a_1, a_2, \ldots, a_n)$ for $i = 1, 2, \ldots, n$. Show that if at the point $P$ we have $f_{x_i x_j} = 0, i \ne j; i, j = 1, 2, \ldots, n$ and $f_{x_i x_i} < 0, i = 1, 2, \ldots, n$, then $f$ has a relative maximum at $P$.

**47** Show that of all triangles having a given perimeter $L$, the equilateral triangle has the largest area. (*Hint:* Use two sides and the included angle as variables.)

**48** Suppose that a rectangular parallelepiped is inscribed in a sphere of radius $R$. That is, all eight vertices lie on the sphere. Show that the inscribed parallelepiped of largest volume is a cube.

## 12

### MAXIMA AND MINIMA; LAGRANGE MULTIPLIERS (OPTIONAL)

In Example 2 of Section 11 (page 681), we solved the problem of finding the relative maxima and minima of the function

$$f(x, y) = x^3 + 3xy^2 - 3x^2 - 3y^2 + 4. \tag{1}$$

In Example 1 of the same section (page 679), we solved the problem of finding the minimum of the function

$$f(x, y, z) = x^2 + y^2 + z^2, \tag{2}$$

*subject to the condition* that $(x, y, z)$ is on the plane

$$S = \{(x, y, z): 2x + 3y - z - 1 = 0\}. \tag{3}$$

The problem of finding the critical points of (1) is quite different from that of finding those of (2) because, in the latter case, the additional condition (3) is attached. This distinction leads to the following definitions.

**DEFINITIONS**    *The problem of finding maxima and minima of a function of several variables (such as* (1) *above) without added conditions is called a problem in* **free maxima and minima**. *When a condition such as* (3) *is imposed on a function such as* (2) *above, the problem of determining the maximum and minimum of that function is called a problem in* **constrained maxima and minima**. *The added condition is called a* **side condition**.

Problems in maxima and minima may have one or more side conditions. When side conditions occur, they are crucial. For example, the minimum of the function $f$ given by (2) without a side condition is obviously zero.

While the problem of minimizing (2) with the side condition (3) has already been solved, we shall do it again by a new and important method. This method, due to Lagrange, changes a problem in constrained maxima and minima to a problem in free maxima and minima.

We first introduce a new variable, traditionally denoted by $\lambda$ and called the **Lagrange multiplier**, and form the function

$$F(x, y, z, \lambda) = (x^2 + y^2 + z^2) + \lambda(2x + 3y - z - 1).$$

The problem of finding the critical points of (2) with side condition (3) can be shown to be equivalent (under rather general circumstances) to that of finding the critical points of $F$ considered as a function of the *four* variables $x, y, z, \lambda$. (See Theorem 15, page 690.) We proceed by computing $F_x, F_y, F_z,$ and $F_\lambda$ and setting each of these expressions equal to zero. We obtain

$$F_x = 2x + 2\lambda = 0,$$

$$F_y = 2y + 3\lambda = 0,$$

$$F_z = 2z - \lambda = 0,$$

$$F_\lambda = 2x + 3y - z - 1 = 0.$$

Note that the equation $F_\lambda = 0$ is precisely the side condition (3). That is, any solution to the problem will automatically satisfy the side condition. We solve these equations simultaneously by writing

$$x = -\lambda, \qquad y = -\tfrac{3}{2}\lambda, \qquad z = \tfrac{1}{2}\lambda,$$

$$2(-\lambda) + 3(-\tfrac{3}{2}\lambda) - (\tfrac{1}{2}\lambda) - 1 = 0,$$

and we get $\lambda = -\tfrac{1}{7}$, $x = \tfrac{1}{7}$, $y = \tfrac{3}{14}$, $z = -\tfrac{1}{14}$. The solution satisfies $F_\lambda = 0$ and so is on the plane (2). The point we seek is $(\tfrac{1}{7}, \tfrac{3}{14}, -\tfrac{1}{14})$.

The general method, known as the **method of Lagrange multipliers**, may be stated as follows: In order to find the critical points of a function

$$f(x, y, z)$$

subject to the side condition

$$\phi(x, y, z) = 0,$$

form the function

$$F(x, y, z, \lambda) = f(x, y, z) + \lambda \phi(x, y, z)$$

and find the critical points of $F$ considered as a function of the four variables $x, y, z, \lambda$.

The method is quite general in that several "multipliers" may be introduced if there are several side conditions. To find the critical points of

$$f(x, y, z),$$

subject to the conditions

$$\phi_1(x, y, z) = 0 \qquad \text{and} \qquad \phi_2(x, y, z) = 0, \tag{4}$$

form the function

$$F(x, y, z, \lambda_1, \lambda_2) = f(x, y, z) + \lambda_1 \phi_1(x, y, z) + \lambda_2 \phi_2(x, y, z)$$

and find the critical points of $F$ as a function of the five variables $x, y, z, \lambda_1$, and $\lambda_2$. Here there are two **Lagrange multipliers**, $\lambda_1$ and $\lambda_2$.

We exhibit the method by working several examples.

**EXAMPLE 1**    Find the minimum of the function

$$f(x, y) = x^2 + 2y^2 + 2xy + 2x + 3y,$$

subject to the condition that $x$ and $y$ satisfy the equation

$$x^2 - y = 1.$$

**Solution**    We form the function

$$F(x, y, \lambda) = (x^2 + 2y^2 + 2xy + 2x + 3y) + \lambda(x^2 - y - 1).$$

Then

$$F_x = 2x + 2y + 2 + 2x\lambda = 0,$$

$$F_y = 4y + 2x + 3 - \lambda = 0,$$

$$F_\lambda = x^2 - y - 1 = 0.$$

Substituting $y = x^2 - 1$ in the first two equations, we get

$$x + x^2 - 1 + 1 + \lambda x = 0, \qquad 4x^2 - 4 + 2x + 3 = \lambda.$$

Solving these two equations in two unknowns, we obtain

$$x = 0, \qquad y = -1, \qquad \lambda = -1,$$

or

$$x = -\tfrac{3}{4}, \quad y = -\tfrac{7}{16}, \qquad \lambda = -\tfrac{1}{4}.$$

Evaluating $f$ at these points, we find that a lower value occurs when $x = -\tfrac{3}{4}$, $y = -\tfrac{7}{16}$. From geometrical considerations we conclude that $f$ is a minimum at this value. □

*Remarks.* We could have solved this problem as a simple maximum and minimum problem by substituting $y = x^2 - 1$ in the equation for $f$ and finding the critical points of the resulting function of the single variable $x$. However, in some problems the side condition may be so complicated that we cannot easily solve for one of the variables in terms of the others, although it may be possible to do so theoretically. It is in such cases that the power of the method of Lagrange multipliers becomes apparent. The system of equations obtained by setting the first derivatives equal to zero may be solvable even though the side condition alone may not be. The next example illustrates this point.

**EXAMPLE 2**   Find the critical values of

$$f(x, y) = x^2 + y^2 \tag{5}$$

subject to the condition that

$$x^3 + y^3 - 6xy = 0. \tag{6}$$

**Solution**   We form the function

$$F(x, y, \lambda) = x^2 + y^2 + \lambda(x^3 + y^3 - 6xy)$$

and obtain the derivatives

$$F_x = 2x + 3x^2\lambda - 6y\lambda = 0,$$

$$F_y = 2y + 3y^2\lambda - 6x\lambda = 0,$$

$$F_\lambda = x^3 + y^3 - 6xy = 0.$$

Solving simultaneously, we find from the first two equations that

$$\lambda = \frac{-2x}{3x^2 - 6y}, \qquad \lambda = \frac{-2y}{3y^2 - 6x}, \qquad \text{and hence} \qquad x(3y^2 - 6x) = y(3x^2 - 6y).$$

This last equation and $F_\lambda = 0$ can be written

$$x^2y - xy^2 + 2x^2 - 2y^2 = 0, \qquad x^3 + y^3 - 6xy = 0.$$

These equations can be solved simultaneously by a trick. Factoring the first equation, we see that

$$(x - y)(2x + 2y + xy) = 0$$

and $x = y$ is a solution. When $x = y$, the second equation yields

$$2x^3 - 6x^2 = 0, \qquad x = 0, 3.$$

We discard the *complex* solutions obtained by setting $2x + 2y + xy = 0$. The values $x = 0$, $y = 0$ clearly yield a minimum, while from geometric considerations (Fig. 34), the point $x = 3$, $y = 3$ corresponds to a relative maximum. There is no true maximum of $f$, since $x^2 + y^2$ (the square of the distance from the origin to the curve) grows without bound if either $x$ or $y$ does. □

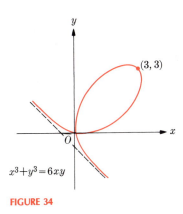

$x^3 + y^3 = 6xy$

**FIGURE 34**

Observe that it is not easy to solve Equation (6) for either $x$ or $y$ and substitute in (5) to get a function of one variable. Therefore the methods of one-dimensional calculus are not readily usable in this problem.

The next example illustrates the technique of Lagrange multipliers when there are two side conditions.

**EXAMPLE 3**    Find the minimum of the function

$$f(x, y, z, t) = x^2 + 2y^2 + z^2 + t^2,$$

subject to the conditions

$$x + 3y - z + t = 2, \tag{7}$$

$$2x - y + z + 2t = 4. \tag{8}$$

**Solution**    We form the function

$$F(x, y, z, t, \lambda_1, \lambda_2) = (x^2 + 2y^2 + z^2 + t^2) + \lambda_1(x + 3y - z + t - 2)$$
$$+ \lambda_2(2x - y + z + 2t - 4).$$

We have

$$F_x = 2x + \lambda_1 + 2\lambda_2 = 0, \qquad F_t = 2t + \lambda_1 + 2\lambda_2 = 0,$$

$$F_y = 4y + 3\lambda_1 - \lambda_2 = 0, \qquad F_{\lambda_1} = x + 3y - z + t - 2 = 0,$$

$$F_z = 2z - \lambda_1 + \lambda_2 = 0, \qquad F_{\lambda_2} = 2x - y + z + 2t - 4 = 0.$$

Solving these six linear equations in six unknowns is tedious but routine. We obtain

$$x = \tfrac{67}{69}, \qquad y = \tfrac{6}{69}, \qquad z = \tfrac{14}{69}, \qquad t = \tfrac{67}{69}.$$

The corresponding values of $\lambda_1$ and $\lambda_2$ are:

$$\lambda_1 = -\tfrac{26}{69}, \qquad \lambda_2 = -\tfrac{54}{69}. \qquad \square$$

The validity of the method of Lagrange multipliers hinges on the ability to solve an equation for a side condition such as

$$\phi(x, y, z) = 0 \tag{9}$$

for one of the unknowns in terms of the other two. Theorems which state when such a process can be performed (theoretically, that is, not actually) are called *implicit function theorems* and are studied in advanced courses. Here we are able to establish the validity of the method of Lagrange multipliers subject to assumptions concerning implicit function theorems.

**THEOREM 15**      *Suppose we are given $f(x, y, z)$ subject to the side condition $\phi(x, y, z) = 0$. Suppose also that all the first partial derivatives of $f$ and $\phi$ are continuous. Assume that $(x_0, y_0, z_0)$ is a critical point of $f$ subject to $\phi = 0$ and that $\phi_z(x_0, y_0, z_0) \neq 0$. We suppose that $\phi(x, y, z) = 0$ yields $z$ as a function of $x$ and $y$ near the point $(x_0, y_0, z_0)$. Then there exists a $\lambda_0$ such that $(x_0, y_0, z_0, \lambda_0)$ is a critical point of the function $F(x, y, z, \lambda) = f(x, y, z) + \lambda\phi(x, y, z)$.*

Theorem 15 says that if we can find the critical points of $F$, we have also found the critical points of $f$ subject to $\phi = 0$. Usually, solving a problem of free maxima and minima is easier than solving a problem in constrained maxima and minima.

**Proof of Theorem 15**      Since $\phi(x, y, z) = 0$ implies that $z$ is a function of $x$ and $y$, we can write $z = g(x, y)$. Next, we set

$$H(x, y) = f[x, y, g(x, y)],$$

and we note that $H$ has a critical point at $(x_0, y_0)$. Therefore

$$H_x = f_x + f_z g_x = 0, \qquad H_y = f_y + f_z g_y = 0. \tag{10}$$

But, by differentiating (9) implicitly, we obtain

$$\frac{\partial z}{\partial x} = g_x = -\frac{\phi_x}{\phi_z}, \qquad \frac{\partial z}{\partial y} = g_y = -\frac{\phi_y}{\phi_z}, \tag{11}$$

(since by hypothesis $\phi_z \neq 0$ near $(x_0, y_0, z_0)$). Substituting (11) into (10), we find

$$f_x - \frac{f_z}{\phi_z}\phi_x = 0 \qquad \text{and} \qquad f_y - \frac{f_z}{\phi_z}\phi_y = 0.$$

We add to these equations the obvious identity

$$f_z - \frac{f_z}{\phi_z}\phi_z = 0,$$

and then we set $\lambda_0 = -f_z(x_0, y_0, z_0)/\phi_z(x_0, y_0, z_0)$. In this way we obtain the equations

$$f_x + \lambda_0\phi_x = 0, \qquad f_y + \lambda_0\phi_y = 0, \qquad f_z + \lambda_0\phi_z = 0, \qquad \phi = 0,$$

which are just the equations satisfied at a critical point of $F = f + \lambda\phi$.    □

Similar theorems, with similar but more complicated proofs, are valid when there are several side conditions and therefore several Lagrange multipliers.

## 12  PROBLEMS

Solve the following problems by the method of Lagrange multipliers.

**1** Find the minimum of $f(x, y, z) = x^2 + y^2 + z^2$ subject to the condition that $x + 3y - 2z = 4$.

**2** Find the minimum of $f(x, y, z) = 3x^2 + 2y^2 + 4z^2$ subject to the condition that $2x + 4y - 6z + 5 = 0$.

**3** Find the minimum of $f(x, y, z) = x^2 + y^2 + z^2$ subject to the condition that $ax + by + cz = d$.

**4** Find the minimum of $f(x, y, z) = ax^2 + by^2 + cz^2$ subject to the condition that $dx + ey + gz + h = 0$ ($a, b, c$ positive).

**5** Find the minimum of $f(x, y, z) = x^2 + y^2 + z^2$ if $(x, y, z)$ is on the line of intersection of the planes

$$x + 2y + z - 1 = 0, \qquad 2x - y - 3z - 4 = 0.$$

**6** Find the minimum of $f(x, y, z) = 2x^2 + y^2 + 3z^2$ if $(x, y, z)$ is on the line of intersection of the planes

$$2x + y - 3z = 4, \qquad x - y + 2z = 6.$$

**7** Find the points on the curve $x^2 + 2xy + 2y^2 = 100$ which are closest to the origin.

In each of Problems 8 through 18, find the local constrained maxima and minima of the function $f$, as indicated.

**8** $f(x, y) = y^2 - 4xy + 4x^2$; $x^2 + y^2 = 1$;  min and max

**9** $f(x, y) = xy$; $x^2 + y^2 = 1$;  min and max

**10** $f(x, y) = x^2 + (y - 2)^2$; $x^2 - y^2 = 1$;  min

**11** $f(x, y, z) = xyz$; $x^3 + y^3 + z^3 = 1$;  $x \geq 0, y \geq 0, z \geq 0$; max

**12** $f(x, y, z) = x^3 y^3 z^3$; $x + y + z = 4$;  min and max

**13** $f(x, y) = x^2 + y^2$; $xy = 1$;  min

**14** $f(x, y) = xy^2$; $9x^2 + 4y^2 = 36$;  max

**15** $f(x, y, z) = x + 2y + 3z$; $x^2 + y^2 + z^2 = 17$;  max

**16** $f(x, y, z) = x^2 + y^2 + z^2$; $x - y + z = 1$;  min

**17** $f(x, y, z) = x^2 + y^2 + z^2$; $x - y = 1$; $y^2 - z^2 = 1$;  min

**18** $f(x, y, z) = x^4 + y^4 + z^4$; $x + y + z = 1$;  min

**19** Find the dimensions of the rectangular box, open at the top, which has maximum volume if the surface area is 12. (Compare with Example 5, page 683.)

**20** A tent is made in the form of a cylinder surmounted by a cone (Fig. 35). If the cylinder has radius 5 and the total

surface area is 100, find the height $H$ of the cylinder and the height $h$ of the cone which make the volume a maximum.

**21** A container is made of a right circular cylinder with radius 5 and with a conical cap at each end. If the volume is given, find the height $H$ of the cylinder and the height $h$ of each of the conical caps which together make the total surface area as small as possible.

**22** Find the minimum of the function

$$f(x, y, z, t) = x^2 + y^2 + z^2 + t^2$$

subject to the condition $3x + 2y - 4z + t = 2$.

**23** Find the minimum of the function

$$f(x, y, z, t) = x^2 + y^2 + z^2 + t^2$$

subject to the conditions

$$x + y - z + 2t = 2, \qquad 2x - y + z + 3t = 3.$$

**24** Find the minimum and maximum of the function

$$f(x, y, z, t) = xyzt$$

subject to the conditions $x - z = 2, y^2 + t = 4$.

**25** Find the minimum of the function

$$f(x, y, z, t) = 2x^2 + y^2 + z^2 + 2t^2$$

subject to the conditions

$$x + y + z + 2t = 1, \qquad 2x + y - z + 4t = 2,$$

$$x - y + z - t = 4.$$

**26** Find the points on the curve $x^4 + y^4 + 3xy = 2$ which are closest to the origin; find those which are farthest from the origin.

**27** Find three critical points of the function $x^4 + y^4 + z^4 + 3xyz$ subject to the condition that $(x, y, z)$ is on the plane $x + y + z = 3$. Can you identify these points?

**28** Work Exercise 36 of Section 11 by the method of Lagrange multipliers.

**29** Find the dimensions of the rectangular parallelepiped of maximum volume with edges parallel to the axes which can be inscribed in the ellipsoid

$$E = \left\{ (x, y, z) : \frac{x^2}{a^2} + \frac{y^2}{b^2} + \frac{z^2}{c^2} = 1 \right\}.$$

**30** If the base of an open rectangular box costs three times as much per square cm as the sides, find the dimensions of the box of largest volume which can be made for $D$ dollars.

**31** Find and identify the critical points of the function

$$f(x, y, z) = 2x^2 + y^2 + z^2$$

subject to the condition that $(x, y, z)$ is on the surface $x^2 yz = 1$.

**FIGURE 35**

**32** Find the critical points of the function $f(x, y, z) = x^a y^b z^c$ if

$$x + y + z = A,$$

where $a$, $b$, $c$, $A$ are given positive numbers.

**33** Let $b_1, b_2, ..., b_k$ be $k$ positive numbers. Find the maximum of

$$f(x_1, x_2, ..., x_k) = b_1 x_1 + b_2 x_2 + \cdots + b_k x_k,$$

subject to the condition

$$x_1^2 + x_2^2 + \cdots + x_k^2 = 1.$$

**34** Find the maximum of the function

$$f(x_1, x_2, ..., x_k) = x_1^2 \cdot x_2^2 \cdots x_k^2$$

subject to the condition

$$x_1^2 + x_2^2 + \cdots + x_k^2 = 1.$$

**\*35** If $a_1, a_2, ..., a_k$ are positive numbers, prove that

$$(a_1 \cdot a_2 \cdots a_k)^{1/k} \leq \frac{a_1 + a_2 + \cdots + a_k}{k}. \qquad (*)$$

(*Hint:* Define $x_1, x_2, ..., x_k$ by the relations $x_i^2 = \alpha a_i$, $i = 1, 2, ..., k$ where $\alpha = 1/\sum_{i=1}^{k} a_i$. Then (*) is equivalent to the inequality

$$(x_1^2 \cdot x_2^2 \cdots x_k^2)^{1/k} \leq 1/k,$$

with the side condition

$$\sum_{i=1}^{k} x_i^2 = 1.$$

Now use Lagrange multipliers—see Problem 34.)

**36** State and prove a theorem on the validity of the Lagrange multiplier method for obtaining the critical points of the function

$$f(x_1, x_2, ..., x_k)$$

subject to the side condition $\phi(x_1, x_2, ..., x_k) = 0$.

**37** Given the quadratic function

$$f(x_1, x_2, ..., x_n) = \sum_{i,j=1}^{n} a_{ij} x_i x_j$$

where $a_{ij}$, $i, j = 1, 2, ..., n$ are numbers such that $a_{ij} = a_{ji}$. Write the $n$ equations which must be satisfied by a maximum of $f$ on the unit sphere in $R^n$: $x_1^2 + x_2^2 + \cdots + x_n^2 = 1$. The $n$ values of $\lambda$ obtained when these equations are solved are called **eigenvalues**. If $n = 2$ and $f(x_1, x_2) = 2x_1^2 + 3x_1 x_2 + 4x_2^2$, find the two eigenvalues.

**38** A manufacturer makes three types of automobile tires labeled A, B, and C. Let $x$, $y$, and $z$ be the number of tires made each day of types A, B, and C, respectively. The profit on each tire of type A is \$2, on type B it is \$3, and on type C it is \$5. The total number of tires which can be produced each day is subject to the constraint $2x^2 + y^2 + 3z^2 = 500$. Find how many tires of each type should be produced in order to maximize profit.

---

## 13

## EXACT DIFFERENTIALS

In Section 7 we saw that the total differential of a function $f(x, y)$ is given by

$$df = \frac{\partial f}{\partial x} dx + \frac{\partial f}{\partial y} dy. \qquad (1)$$

The quantity $df$ is a function of four variables, since $\partial f / \partial x$ and $\partial f / \partial y$ are functions of $x$ and $y$ and $dx$ and $dy$ are additional independent variables. It turns out that expressions of the form

$$P(x, y) \, dx + Q(x, y) \, dy$$

occur frequently in problems in engineering, physics, and chemistry. It is natural to ask when such an expression is the total differential of a function $f$. For example, if we are given

$$(3x^2 + 2y) \, dx + (2x - 3y^2) \, dy,$$

we may guess (correctly) that the function $f(x, y) = x^3 + 2xy - y^3$ has the above expression as its total differential, $df$. On the other hand, if we are given

$$(2x^2 - 3y)\,dx + (2x - y^3)\,dy, \tag{2}$$

then it can be shown that *there is no function $f$ whose total differential is the expression (2).*

---

**DEFINITION**

*If there is a function $f(x, y)$ such that*

$$df = P(x, y)\,dx + Q(x, y)\,dy$$

*for all $(x, y)$ in some region and for all values of $dx$ and $dy$, we say that*

$$P(x, y)\,dx + Q(x, y)\,dy$$

*is an* **exact differential***. If there is a function $F(x, y, z)$ such that*

$$dF = P(x, y, z)\,dx + Q(x, y, z)\,dy + R(x, y, z)\,dz$$

*for all $(x, y, z)$ in some region and for all values of $dx$, $dy$, and $dz$, we say that $P\,dx + Q\,dy + R\,dz$ is an* **exact differential***. For functions with any number of variables the extension is immediate.*

---

The next theorem gives a precise criterion for determining when a differential expression is an exact differential.

---

**THEOREM 16**

*Suppose that $P(x, y)$, $Q(x,y)$, $\partial P/\partial y$, $\partial Q/\partial x$ are continuous in a rectangle $S$. Then the expression*

$$P(x, y)\,dx + Q(x, y)\,dy \tag{3}$$

*is an exact differential for $(x, y)$ in the region $S$ if and only if*

$$\frac{\partial P}{\partial y} = \frac{\partial Q}{\partial x} \qquad \text{for all} \qquad (x, y) \text{ in } S. \tag{4}$$

---

**Proof**    The theorem has two parts: We must show (a), that if (3) is an exact differential, then (4) holds; and (b), that if (4) holds, then the expression (3) is an exact differential.

To establish (a) we start with the assumption that there is a function $f$ such that

$$df = P\,dx + Q\,dy,$$

and so $\partial f/\partial x = P(x, y)$ and $\partial f/\partial y = Q(x, y)$. We differentiate and obtain

$$\frac{\partial^2 f}{\partial y \partial x} = \frac{\partial P}{\partial y} \qquad \text{and} \qquad \frac{\partial^2 f}{\partial x \partial y} = \frac{\partial Q}{\partial x}.$$

Now Theorem 10, page 666, which states that the order of differentiation is immaterial, may be invoked to conclude that (4) holds.

To prove (b) we assume that (4) holds, and we must construct a function $f$ such that $df$ is equal to the differential expression (3). That is, we must find a function $f$ such that

$$\frac{\partial f}{\partial x} = P(x, y) \qquad \text{and} \qquad \frac{\partial f}{\partial y} = Q(x, y). \tag{5}$$

Let $(a, b)$ be a point of $S$; suppose we try to solve these two partial differential equations for the function $f$. We integrate the first with respect to $x$, getting

$$f(x, y) = C(y) + \int_a^x P(\xi, y) \, d\xi$$

where, instead of a "constant" of integration, we get a function of the remaining variable. Letting $x = a$, we find that $f(a, y) = C(y)$, and we can write

$$f(x, y) = f(a, y) + \int_a^x P(\xi, y) \, d\xi. \tag{6}$$

Setting $x = a$ in the second equation of (5) and integrating with respect to $y$, we obtain

$$f(a, y) = C_1 + \int_b^y Q(a, \eta) \, d\eta;$$

letting $y = b$, we see that $C_1 = f(a, b)$ and we conclude that

$$f(a, y) = f(a, b) + \int_b^y Q(a, \eta) \, d\eta.$$

Substitution of this expression for $f(a, y)$ into Equation (6) yields (Fig. 36)

$$f(x, y) = f(a, b) + \int_b^y Q(a, \eta) \, d\eta + \int_a^x P(\xi, y) \, d\xi. \tag{7}$$

We may repeat the entire process by integrating with respect to $y$ first and with respect to $x$ second. The three equations are

$$f(x, y) = f(x, b) + \int_b^y Q(x, \eta) \, d\eta,$$

$$f(x, b) = f(a, b) + \int_a^x P(\xi, b) \, d\xi,$$

and

$$f(x, y) = f(a, b) + \int_a^x P(\xi, b) \, d\xi + \int_b^y Q(x, \eta) \, d\eta. \tag{8}$$

The two expressions for $f$ given by (7) and (8) will be identical if and only if (after subtraction) the equation

$$\int_a^x [P(\xi, y) - P(\xi, b)] \, d\xi = \int_b^y [Q(x, \eta) - Q(a, \eta)] \, d\eta \tag{9}$$

holds. To establish (9), we start with the observation that

$$P(\xi, y) - P(\xi, b) = \int_b^y \frac{\partial P(\xi, \eta)}{\partial y} \, d\eta = \int_b^y \frac{\partial Q(\xi, \eta)}{\partial x} \, d\eta,$$

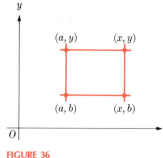

$y$

$(a, y)$   $(x, y)$

$(a, b)$   $(x, b)$

$O$           $x$

FIGURE 36

where, for the first time, we have used the hypothesis that $\partial P/\partial y = \partial Q/\partial x$. Therefore, upon integration,

$$\int_a^x [P(\xi, y) - P(\xi, b)]\, d\xi = \int_a^x \left[ \int_b^y \frac{\partial Q(\xi, \eta)}{\partial x}\, d\eta \right] d\xi.$$

It will be shown in Chapter 16, Section 3, that the order of integration in the term on the right may be interchanged, so that

$$\int_a^x [P(\xi, y) - P(\xi, b)]\, d\xi = \int_b^y \left[ \int_a^x \frac{\partial Q(\xi, \eta)}{\partial x}\, d\xi \right] d\eta$$

$$= \int_b^y [Q(x, \eta) - Q(a, \eta)]\, d\eta.$$

But this equality is (9) precisely; the theorem is established when we observe that as a result of (7) or (8), the relations

$$\frac{\partial f}{\partial x} = P \qquad \text{and} \qquad \frac{\partial f}{\partial y} = Q$$

hold.                                                                    □

The proof of Theorem 16 contains in it the method for finding the function $f$ when it exists. Examples illustrate the technique.

**EXAMPLE 1**   Show that

$$(3x^2 + 6y)\, dx + (3y^2 + 6x)\, dy$$

is an exact differential, and find the function $f$ of which it is the total differential.

**Solution**   Setting $P = 3x^2 + 6y$, $Q = 3y^2 + 6x$, we obtain

$$Q_x = 6, \qquad P_y = 6,$$

so that $P\, dx + Q\, dy$ is an exact differential. We write (as in the proof of the theorem)

$$f_x = 3x^2 + 6y$$

and integrate to get

$$f = x^3 + 6xy + C(y).$$

We differentiate with respect to $y$. We find

$$f_y = 6x + C'(y),$$

and this expression must be equal to $Q$. Therefore

$$6x + C'(y) = 3y^2 + 6x \qquad \text{or} \qquad C'(y) = 3y^2, \qquad C(y) = y^3 + C_1.$$

Thus

$$f(x, y) = x^3 + 6xy + y^3 + C_1.$$                                    □

A constant of integration will always appear in the integration of exact differentials.

The next theorem is an extension of Theorem 16 to functions of three variables.

**THEOREM 17**

*Suppose that $P(x, y, z)$, $Q(x, y, z)$, $R(x, y, z)$ are continuous on some rectangular parallelepiped S. Then*

$$P(x, y, z)\, dx + Q(x, y, z)\, dy + R(x, y, z)\, dz$$

*is an exact differential on S if and only if*

$$\frac{\partial P}{\partial y} = \frac{\partial Q}{\partial x}, \qquad \frac{\partial R}{\partial x} = \frac{\partial P}{\partial z}, \qquad \frac{\partial Q}{\partial z} = \frac{\partial R}{\partial y}.$$

*It is assumed that all the above partial derivatives are continuous functions of $(x, y, z)$ on S.*

The proof of this theorem follows the lines (and uses the proof) of Theorem 16.

The next example shows how to integrate an exact differential in three variables.

**EXAMPLE 2**   Determine whether

$$(3x^2 - 4xy + z^2 + yz - 2)\, dx + (xz - 6y^2 - 2x^2)\, dy$$
$$+ (9z^2 + 2xz + xy + 6z)\, dz$$

is an exact differential and, if so, find the function $f$ of which it is the total differential.

**Solution**   Setting $P$, $Q$, $R$ equal to the coefficients of $dx$, $dy$, and $dz$, respectively, we obtain

$$P_y = -4x + z = Q_x,$$
$$P_z = 2z + y = R_x,$$
$$Q_z = x = R_y.$$

Therefore $P\, dx + Q\, dy + R\, dz$ is an exact differential, and we proceed to find $f$. Writing $f_x = P$, we integrate to get

$$f(x, y, z) = x^3 - 2x^2y + xz^2 + xyz - 2x + C(y, z).$$

We differentiate with respect to $y$:

$$f_y(x, y, z) = -2x^2 + xz + C_y(y, z) = Q = xz - 6y^2 - 2x^2.$$

Hence

$$C_y(y, z) = -6y^2$$

and, upon integration with respect to $y$,

$$C(y, z) = -2y^3 + C_1(z).$$

We may write

$$f(x, y, z) = x^3 - 2x^2y + xz^2 + xyz - 2x - 2y^3 + C_1(z),$$

and we wish to find $C_1(z)$. We differentiate $f$ with respect to $z$:

$$f_z = 2xz + xy + C_1'(z) = R = 9z^2 + 2xz + xy + 6z.$$

We obtain

$$C_1'(z) = 9z^2 + 6z \qquad \text{and} \qquad C_1(z) = 3z^3 + 3z^2 + C_2.$$

Therefore

$$f(x, y, z) = x^3 - 2y^3 + 3z^3 - 2x^2 y + xz^2 + xyz + 3z^2 - 2x + C_2. \qquad \square$$

## 13   PROBLEMS

In each of Problems 1 through 23, determine which of the differentials are exact. In case a differential is exact, find the functions of which it is the total differential.

**1** $(x^3 + 3x^2 y)\, dx + (x^3 + y^3)\, dy$

**2** $(2x + 3y)\, dx + (3x + 2y)\, dy$

**3** $(2y - 1/x)\, dx + (2x + 1/y)\, dy$

**4** $(x^2 + 2xy)\, dx + (y^3 - x^2)\, dy$

**5** $x^2 \sin y\, dx + x^2 \cos y\, dy$

**6** $\dfrac{x^2 + y^2}{2y^2}\, dx - \dfrac{x^3}{3y^3}\, dy$

**7** $2xe^{x^2} \sin y\, dx + e^{x^2} \cos y\, dy$

**8** $(ye^{xy} + 3x^2)\, dx + (xe^{xy} - \cos y)\, dy$

**9** $\dfrac{x\, dy - y\, dx}{x^2 + y^2},\ x > 0$

**10** $(2x \ln y)\, dx + \dfrac{x^2}{y}\, dy,\ y > 0$

**11** $(x + \cos x \tan y)\, dx + (y + \tan x \cos y)\, dy$

**12** $\dfrac{1}{y} e^{2x/y}\, dx - \dfrac{1}{y^3} e^{2x/y}(y + 2x)\, dy$

**13** $(3x^2 \ln y - x^3)\, dx + \dfrac{3x^2}{y}\, dy$

**14** $\dfrac{x\, dx}{\sqrt{x^2 + y^2}} + \left( \dfrac{y}{\sqrt{x^2 + y^2}} - 2 \right) dy$

**15** $(\sin(xy) + xy \cos(xy) - 3y^2)\, dx + (x^2 \cos(xy) - 6xy)\, dy$

**16** $\left( \dfrac{1}{y} e^{xy} + xe^{xy} \right) dx + \left( \dfrac{x^2}{y} e^{xy} - \dfrac{x}{y^2} e^{xy} \right) dy$

**17** $(y^3 - 1/x)\, dx + (3xy^2 - 1/y)\, dy$

**18** $(y^2 - 4y^3 e^x)\, dx + (2xy - 12y^2)\, dy$

**19** $y^x \ln y\, dx + xy^{x-1}\, dy$

**20** $(2x - y + 3z)\, dx + (3y + 2z - x)\, dy + (2x + 3y - z)\, dz$

**21** $(2xy + z^2)\, dx + (2yz + x^2)\, dy + (2xz + y^2)\, dz$

**22** $(e^x \sin y \cos z)\, dx + (e^x \cos y \cos z)\, dy - (e^x \sin y \sin z)\, dz$

**23** $\left( \dfrac{1}{y^2} - \dfrac{y}{x^2 z} - \dfrac{z}{x^2 y} \right) dx + \left( \dfrac{1}{xz} - \dfrac{x}{y^2 z} - \dfrac{z}{xy^2} \right) dy$
$$+ \left( \dfrac{1}{xy} - \dfrac{x}{yz^2} - \dfrac{y}{xz^2} \right) dz$$

**\*24** a) Given the differential expression

$$P\, dx + Q\, dy + R\, dz + S\, dt,$$

where $P, Q, R, S$ are functions of $x, y, z, t$, state a theorem which is a plausible generalization of Theorem 17 in order to decide when the above expression is exact. (b) Use the result of part (a) to show that the following expression is exact. Find the function $f$ of which it is the total differential.

$$(3x^2 + 2z + 3)\, dx + (2y - t - 2)\, dy + (3z^2 + 2x)\, dz$$
$$+ (4 - 3t^2 - y)\, dt.$$

**\*25** Prove Theorem 17.

**\*26** If $\omega = P(x, y, z)\, dx + Q(x, y, z)\, dy + R(x, y, z)\, dz$, define

$$d\omega = \left( \dfrac{\partial P}{\partial y} - \dfrac{\partial Q}{\partial x} \right) dx\, dy + \left( \dfrac{\partial P}{\partial z} - \dfrac{\partial R}{\partial x} \right) dx\, dz$$
$$+ \left( \dfrac{\partial Q}{\partial z} - \dfrac{\partial R}{\partial y} \right) dy\, dz.$$

Verify that if $\omega$ is an exact differential so that $\omega = df$, then Theorem 17 states that $d(df) = 0$. Now by analogy state a formula for $d\omega$ if $\omega = \Sigma_{i=1}^{k} P_i(x_1, x_2, \ldots, x_k)\, dx_i$. Under what conditions is $d\omega = 0$?

**27** Let $f(x), g(y)$ be arbitrary integrable functions. Show that

$$\left[ yf(x) + \int_a^y g(\eta)\, d\eta \right] dx + \left[ \int_b^x f(\xi)\, d\xi + xg(y) \right] dy$$

is an exact differential. If $f(x), g(y), h(z)$ are integrable, generalize the above result to functions of three variables.

It may happen that the expression $P(x, y)\, dx + Q(x, y)\, dy$ is not exact but that a function $I(x, y)$ can be found so that $I(x, y)[P(x, y)\, dx + Q(x, y)\, dy]$ is an exact differential. The function $I$ is called an **integrating factor**. In each of Problems 28 through 30, show that the function $I$ is an integrating factor. Then find the function $f$ which yields the total differential.

**28** $(xy + x^2 + 1)\, dx + x^2\, dy,\ I = 1/x$

**29** $x\, dy - y\, dx - 2x^2 \log y\, dy,\ I = 1/x^2$

**30** $(2 - xy)y\, dx + (2 + xy)x\, dy,\ I = 1/x^2 y^2$

If the left side of the differential equation of the form
$$P(x, y)\, dx + Q(x, y)\, dy = 0$$
is an exact differential and if $f(x, y)$ is the function which yields the total differential, then $f(x, y) = c$ where $c$ is any constant is the **general solution** of the differential equation. In each of Problems 31 through 33, solve the differential equation by first finding an integrating factor and then determining the function $f$.

**31** $ye^{-x/y}\, dx - (xe^{-x/y} + y^3)\, dy = 0$

**32** $x\, dx + y\, dy + (x^2 + y^2)(y\, dx - x\, dy) = 0$

**33** $x\, dy - (y + x^3 e^{2x})\, dx = 0$

---

# CHAPTER 15

## REVIEW PROBLEMS

In each of Problems 1 through 10, draw four level curves of the surface $z = f(x, y)$.

**1** $f(x, y) = x^2 + y^2$

**2** $f(x, y) = y/x$

**3** $f(x, y) = x^2 + 2y$

**4** $f(x, y) = -x^2 + y^2$

**5** $f(x, y) = xe^y$

**6** $f(x, y) = y \csc x$

**7** $f(x, y) = \sqrt{\dfrac{x + y}{x - y}}$

**8** $f(x, y) = xy^2 + 2$

**9** $f(x, y) = x + \sin y$

**10** $f(x, y) = y + \cos x$

In each of Problems 11 through 17, calculate the partial derivatives indicated.

**11** $f(x, y) = x^2 + 3xy + y^4;\quad f_x, f_y, f_{xx}, f_{xy}$

**12** $f(x, y) = (x - 2y)^2(x + 3y);\quad f_x, f_y$

**13** $f(x, y) = y \arctan(x + 2y);\quad f_x, f_y$

**14** $f(x, y) = \sin x + \sin y + \sin(x + y);\quad f_x, f_y, f_{xy}$

**15** $f(x, y) = e^{\sin(y/x^2)};\quad f_x, f_y$

**16** $f(x, y, z) = \dfrac{xy}{x^2 + y^2 + z^2};\quad \dfrac{\partial f}{\partial x}, \dfrac{\partial f}{\partial y}, \dfrac{\partial f}{\partial z}$

**17** $f(x, y, z) = \cos(\sqrt{x^2 + y^2 + z^4});\quad \dfrac{\partial f}{\partial z}, \dfrac{\partial^2 f}{\partial x^2}, \dfrac{\partial^2 f}{\partial x \partial y}$

**18** Show that the function $g(x, y)$ defined by
$$g(x, y) = \begin{cases} \dfrac{x^4}{x - y} & \text{for } x \neq y \\ 0 & \text{for } x = y \end{cases}$$

is not continuous on the line $y = x$.

**19** Show that the function $g(x, y, z)$ defined by

$$g(x, y, z) = \begin{cases} \dfrac{x^4 - y}{2x + 3y + 4z} & \text{for } 2x + 3y + 4z \neq 0 \\ 0 & \text{for } 2x + 3y + 4z = 0 \end{cases}$$

is not continuous at any point on the plane $2x + 3y + 4z = 0$.

In each of Problems 20 through 25, find the indicated partial derivatives.

**20** $z \equiv f(x, y) = e^{xy};\quad x = 2st + t,\ y = t/s;\quad \dfrac{\partial z}{\partial s}, \dfrac{\partial z}{\partial t}$

**21** $z \equiv f(x, y) = x^2 + 3xy;\quad x = st,\ y = \cos t;\quad \dfrac{\partial z}{\partial s}, \dfrac{\partial z}{\partial t}$

**22** $z \equiv f(x, y) = 2 \cos x - 4 \cos xy;\quad x = 1/s,\ y = 4s;\quad \dfrac{dz}{ds}$

**23** $z \equiv f(x, y) = \arctan(x^2 + y^2);\quad x = s^2 t,\ y = 2st + 3t;\quad \dfrac{\partial z}{\partial s}, \dfrac{\partial z}{\partial t}$

**24** $z \equiv f(x, y) = (x - 2y)^{101}(x + 3y^2)^{52};\quad x = \cos(st),\ y = 4st;\quad \dfrac{\partial z}{\partial s}, \dfrac{\partial z}{\partial t}$

**25** $z \equiv f(x, y) = xe^y;\quad x = \ln(st),\ y = st;\quad \dfrac{\partial z}{\partial s}, \dfrac{\partial z}{\partial t}$

**26** At a certain instant the upper radius of the frustum of a cone is 10 cm, the lower radius is 12 cm, and the height is 18 cm. At what rate is the volume $V$ changing if the upper radius decreases at the rate of 1 cm per minute, the lower radius increases at the rate of 2 cm per minute, and the height decreases at the rate of 3 cm per minute? (See Fig. 37.) The formula for the volume $V$ is

$$V = \tfrac{1}{3}\pi h(u^2 + uv + v^2)$$

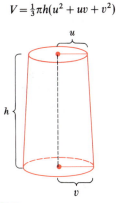

**FIGURE 37**

**27** Two straight roads intersect at right angles. Car $C$ approaches on the first road at 80 km/hour and truck $T$ approaches on the second road at 60 km/hour. At what rate is the distance between the car and the truck changing when $C$ is 1 km from the intersection and $T$ is 0.8 km from the intersection?

In each of Problems 28 through 32, find the directional derivative $D_{\mathbf{a}}f$ at the given point $P$ when $\mathbf{a}$ is the given vector.

**28** $f(x, y) = x + y;$   $P(1, 1);$   $\mathbf{a} = \mathbf{i} + \mathbf{j}$

**29** $f(x, y) = \cos(xy);$   $P(0, \pi);$   $\mathbf{a} = \mathbf{i} - \mathbf{j}$

**30** $f(x, y) = e^{-(x^2 + y^2)};$   $P(0, 0);$   $\mathbf{a} = \mathbf{i}$

**31** $f(x, y, z) = (x + y)^3(x - z)^2(y + z);$   $P(2, 1, 1);$
   $\mathbf{a} = \mathbf{i} - \mathbf{j} + \mathbf{k}$

**32** $f(x, y, z) = 2^{xyz};$   $P(0, 0, 0);$   $\mathbf{a} = \mathbf{j}$

In each of Problems 33 through 37, find the equation of the tangent plane and the equations of the normal line to the given surface at the given point.

**33** $z = x^2 + xy;$   $(1, 0, 1)$

**34** $z = e^{-x}\cos y;$   $(1, \pi, -e^{-1})$

**35** $z = \ln\sqrt{x^4 + y^4};$   $(-1, 2, \tfrac{1}{2}\ln 17)$

**36** $z = \cos x + \cos y;$   $(\pi/3, \pi/3, 1)$

**37** $z = x^2 y^3;$   $(-2, 2, 32)$

In each of Problems 38 through 42, find the total differential.

**38** $z = xy - y^4 + (x^3 + 7)^{20}$   **39** $z = \sqrt{x^4 + y^6}$

**40** $w = \ln(\sqrt{x^2 + y^4 + z^2})$

**41** $w = \ln((x + y)\cos z)$   **42** $w = ze^{\cos(xy)}$

**43** Use differentials to find an approximate value of $\sqrt{(2.98)^2 + (4.01)^2}$.

**44** Suppose $f(x, y) = yx^{1/4} + x\sqrt{y}$. Use differentials to estimate the change in $f$ from $(16, 16)$ to $(17, 18)$.

**45** The specific gravity of a solid is given by the formula $s = \alpha(\alpha - w)^{-1}$ where $\alpha$ is the weight in air and $w$ is the weight in water. Estimate $s$ if $\alpha$ is measured to be 4 kg (within a tolerance of 100 grams), and $w$ is measured to be 2.2 kg (within a tolerance of 200 grams).

In each of Problems 46 through 56, find the indicated derivative or partial derivative.

**46** $x^4 + 4x^2 y^2 - y^3 + 3x - y = 6;$   $\dfrac{dy}{dx}$

**47** $\ln(1 + x^2 + y^4) + e^{-xy^2} = 17;$   $\dfrac{dy}{dx}$

**48** $\cos x + \cos y = 0;$   $\dfrac{dy}{dx}$

**49** $e^{-xy} - \sin x + \cos y + 3 = 0;$   $\dfrac{dy}{dx}$

**50** $xyw + w^2 - y^2 = 0;$   $\dfrac{\partial w}{\partial y}$

**51** $we^{x^2 + y^2} - xyw = 0;$   $\dfrac{\partial w}{\partial y}$

**52** $w - \cos(xy) + \sin(xyw) = \tfrac{1}{2};$   $\dfrac{\partial w}{\partial y}, \dfrac{\partial w}{\partial x}$

**53** $x^2 + y^2 + z^2 - w^2 + xy - w - x + y - z = 0;$   $\dfrac{\partial w}{\partial y}$

**54** $x^2 y^2 z^3 w^4 - 2y^3 w^3 = 6;$   $\dfrac{\partial w}{\partial y}$

**55** $\cos(xyzw) + (xy - zw)^2 = 4;$   $\dfrac{\partial w}{\partial x}$

**56** $xye^{-(w+z)} = 3;$   $\dfrac{\partial w}{\partial y}, \dfrac{\partial w}{\partial z}$

In each of Problems 57 through 63, find the indicated partial derivatives.

**57** $f(x, y) = x^2 + 7xy + y^3;$   $f_{xy}, f_{yx}$

**58** $f(x, y) = 7x^3 + \cos(xy) + ye^x;$   $f_{xy}, f_{yx}$

**59** $u = \ln(x^2 + y^2 + z^3) - xyz;$   $u_{xx}, u_{xy}, u_{yx}, u_{zx}$

**60** $z = \dfrac{x + y}{xy};$   $x = r^2, y = rs^2, \dfrac{\partial^2 z}{\partial r^2}$

**61** $z = (x - y)^2(3x + 4y)^2;$   $x = \cos r, y = \sin(rs), \dfrac{\partial^2 z}{\partial r^2}$

**62** $z = \ln\left(\dfrac{xy}{\sqrt{x^2 + y^2}}\right);$   $x = 2st + r, y = s^2(r + 3), \dfrac{\partial^2 z}{\partial r^2}$

63 $z = ye^{-(x^2+y^2)}$; $\quad x = 2rs$, $y = (r+s)^2$, $\dfrac{\partial^2 z}{\partial r^2}$

64 Suppose $f(x, y) = \ln(\cos x) - \ln(\cos y)$. Show that $f$ satisfies the following partial differential equation:

$$0 = \left(1 + \left(\frac{\partial f}{\partial y}\right)^2\right)\frac{\partial^2 f}{\partial x^2} - 2\frac{\partial f}{\partial x}\frac{\partial f}{\partial y}\frac{\partial^2 f}{\partial x \partial y}$$
$$+ \left(1 + \left(\frac{\partial f}{\partial x}\right)^2\right)\frac{\partial^2 f}{\partial y^2}$$

(The minimal surface differential equation.)

In each of Problems 65 through 71, test the functions $f$ for relative maxima and minima.

65 $f(x, y) = 2x - x^2 - y^2$  66 $f(x, y) = x^3 - 3xy + y^3$

67 $f(x, y) = xy - x^3 y - xy^3$  68 $f(x, y) = e^{-x}\cos y$

69 $f(x, y) = (y/x) - (x/y)$  70 $f(x, y) = (x - y)(xy - 1)$

71 $f(x, y) = \ln(3 + x^2 + y^2) - xy$

72 A triangular metal plate occupies the region $R$ shown in Fig. 38. The temperature at each point of the metal plate is given by

$$T = x^2 + xy + y^2 - 2x + y.$$

Find the hottest and coldest points of the plate.

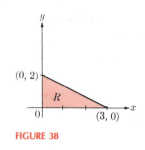

FIGURE 38

In each of Problems 73 through 80, determine which of the differentials are exact. If a differential is exact, find the function of which it is the total differential.

73 $(y - y\sin xy)\,dx + (x - x\sin xy)\,dy$

74 $(y^3 + 2(x + 2y))\,dx + (3xy^2 + 4(x + 2y))\,dy$

75 $(3x^2 + 2y^2)\,dx + (2x + 2y)\,dy$

76 $(1 + xy)e^{xy}\,dx + x^2 e^{xy}\,dy$

77 $(\sin x + \ln y)\,dx + (x/y)\,dy$

78 $(x^2\arcsin y)\,dx + \left(\dfrac{x^3}{3\sqrt{1 - y^2}} - \ln y\right)dy$

79 $(\cos x - y\sin x)\,dx + \cos x\,dy$

80 $ye^x(1 + x)\,dx + (xe^x - e^{-y})\,dy$

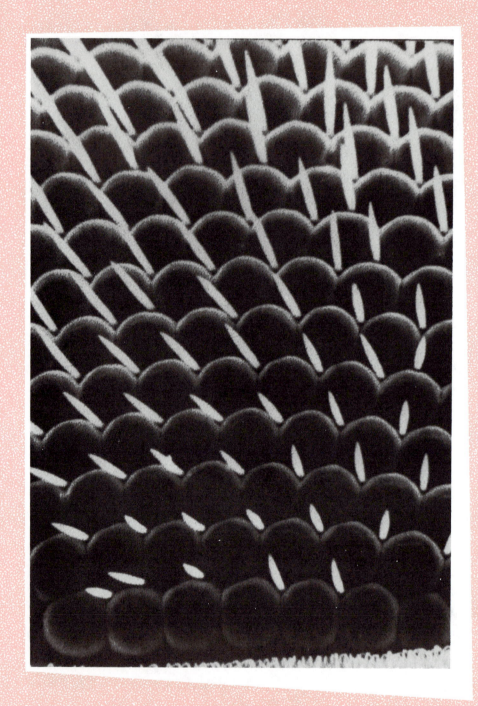

# 16

# MULTIPLE INTEGRATION

In this chapter we define the integral for functions of two
and three variables. We show that the computation of such
integrals can be reduced to successive integrals of functions
of one variable. These facts enable us to calculate areas,
volumes, and other geometric quantities for various regions
in the plane and solids in three-space.

## 1

### DEFINITION OF THE DOUBLE INTEGRAL

Let $F$ be a region of area $A$ situated in the $xy$ plane. We shall always assume
that a region includes its boundary curve. Such regions are called **closed
regions** in analogy with closed intervals on the real line—that is, ones which
include their endpoints. We subdivide the $xy$ plane into rectangles by drawing
lines parallel to the coordinate axes. These lines may or may not be equally
spaced (Fig. 1). Starting in some convenient place (such as the upper left

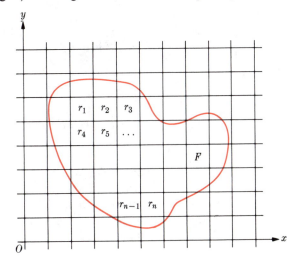

**FIGURE 1**

corner of $F$), we systematically number all the rectangles *lying entirely within* $F$. Suppose there are $n$ such and we label them $r_1, r_2, \ldots, r_n$. We use the symbols $A(r_1), A(r_2), \ldots, A(r_n)$ for the areas of these rectangles. The collection of $n$ rectangles $\{r_1, r_2, \ldots, r_n\}$ is called a **subdivision** $\Delta$ of $F$. The **norm of the subdivision**, denoted as usual by $\|\Delta\|$, is the length of the diagonal of the largest rectangle in the subdivision $\Delta$.

Suppose that $f(x, y)$ is a function defined for all $(x, y)$ in the region $F$. The definition of the *double integral of $f$ over the region $F$* is similar to the definition of the integral for functions of one variable. (See page 187.) Select arbitrarily a point in each of the rectangles of the subdivision $\Delta$, denoting the coordinates of the point in the rectangle $r_i$ by $(\xi_i, \eta_i)$. (See Fig. 2.) Now form the sum

$$f(\xi_1, \eta_1)A(r_1) + f(\xi_2, \eta_2)A(r_2) + \cdots + f(\xi_n, \eta_n)A(r_n)$$

or, more compactly,

$$\sum_{i=1}^{n} f(\xi_i, \eta_i)A(r_i). \tag{1}$$

This sum is an approximation to the double integral we shall define. Sums such as (1) may be formed for subdivisions with any positive norm and with the $i$th point $(\xi_i, \eta_i)$ chosen in any way whatsoever in the rectangle $r_i$.

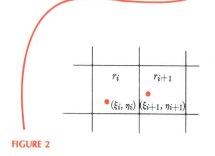

**FIGURE 2**

**DEFINITION**

*We say that **a number $L$ is the limit of sums of type** (1) and write*

$$\lim_{\|\Delta\| \to 0} \sum_{i=1}^{n} f(\xi_i, \eta_i)A(r_i) = L$$

*if the number $L$ has the property: for each $\varepsilon > 0$ there is a $\delta > 0$ such that*

$$\left| \sum_{i=1}^{n} f(\xi_i, \eta_i)A(r_i) - L \right| < \varepsilon$$

*for every subdivision $\Delta$ with $\|\Delta\| < \delta$ and for all possible choices of the points $(\xi_i, \eta_i)$ in the rectangles $r_i$.*

It can be shown that if the number $L$ exists, then it must be unique.

**DEFINITION**

*If $f$ is defined on a region $F$ and the number $L$ defined above exists, we say that $f$ **is integrable over** $F$ and write*

$$L = \iint_F f(x, y)\, dA.$$

*We also call the expression above the **double integral of $f$ over** $F$.*

The double integral has a geometric interpretation in terms of the volume of a solid. We recall the methods of finding volumes of solids of revolution developed on page 228. Now we shall discuss the notion of volume in somewhat more detail. The definition of volume depends on (i), the definition of the volume of a cube—namely, length times width times height, and (ii), a limiting process.

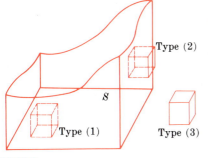

Type (2)

$S$

Type (1)

Type (3)

**FIGURE 3**

Let $S$ be a solid in three-space. We divide all of space into cubes by constructing planes parallel to the coordinate planes at a distance apart of $1/2^n$ units, with $n$ some positive integer. In such a network, the cubes are of three kinds: type (1), those cubes entirely within $S$; type (2), those cubes partly in $S$ and partly outside $S$; and type (3), those cubes entirely outside $S$ (Fig. 3). We define

$$V_n^-(S) = \left(\frac{1}{2^n}\right)^3 \text{ times the number of cubes of type (1),}$$

$$V_n^+(S) = V_n^-(S) + \left(\frac{1}{2^n}\right)^3 \text{ times the number of cubes of type (2).}$$

Intuitively we expect that, however the volume of $S$ is defined, the number $V_n^-(S)$ would be smaller than the volume, while the number $V_n^+(S)$ would be larger. It can be shown that, as $n$ increases, $V_n^-(S)$ gets larger or at least does not decrease, while $V_n^+(S)$ gets smaller or at least is nonincreasing. Clearly,

$$V_n^-(S) \le V_n^+(S),$$

for all $n$. Since bounded increasing sequences and bounded decreasing sequences tend to limits, the following definitions are appropriate.

**DEFINITIONS**   *The **inner volume** of a solid $S$ denoted $V^-(S)$, is $\lim_{n \to \infty} V_n^-(S)$. The **outer volume**, denoted $V^+(S)$, is $\lim_{n \to \infty} V_n^+(S)$. A set of points $S$ in three-space has a volume whenever $V^-(S) = V^+(S)$. This common value is denoted by $V(S)$ and is called the volume of $S$.*

*Remark.*   It is not difficult to construct point sets for which $V^-(S) < V^+(S)$. For example, take $S$ to be all points $(x, y, z)$ such that $x$, $y$, and $z$ are rational and $0 \le x \le 1, 0 \le y \le 1, 0 \le z \le 1$. The reader can verify that $V_n^-(S) = 0$ for every $n$, while $V_n^+(S) = 1$ for every $n$. However, throughout this text we shall discuss only regions for which $V^-(S) = V^+(S)$, i.e., those which have volume.

If $S_1$ and $S_2$ are two solids with no points in common, it can be shown, as expected, that $V(S_1 \cup S_2) = V(S_1) + V(S_2)$. Also, the subdivision of all of space into cubes is not vital. Rectangular parallelepipeds would do equally well, with the formula for the volume of a rectangular parallelepiped taken as length times width times height.

The volume of a solid is intimately connected with the double integral in the same way that the area of a region is connected with the single integral. We now exhibit this connection.

Suppose that $f(x, y)$ is a positive function defined for $(x, y)$ in some region $F$ (Fig. 4). An item in the sum (1) approximating the double integral is

$$f(\xi_i, \eta_i) A(r_i),$$

which we recognize as the volume of the rectangular column of height $f(\xi_i, \eta_i)$ and area of base $A(r_i)$ (Fig. 4). The sum of such columns is an approximation to the volume of the cylindrical solid bounded by the surface $z = f(x, y)$, the plane figure $F$, and lines parallel to the $z$ axis through the boundary of $F$

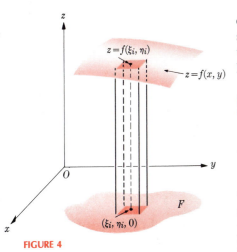

$z$

$z = f(\xi_i, \eta_i)$

$z = f(x, y)$

$O$

$y$

$F$

$(\xi_i, \eta_i, 0)$

$x$

**FIGURE 4**

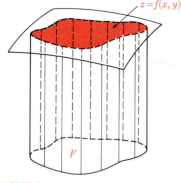

FIGURE 5

(Fig. 5). It can be shown that, with appropriate hypotheses on the function $f$, the double integral

$$\iint_F f(x, y) \, dA$$

measures the "volume under the surface" in the same way that a single integral of a positive function $f$

$$\int_a^b f(x) \, dx$$

measures the area under the curve.

The precise result is given in the next theorem which we state without proof.

**THEOREM 1**  *If $f(x, y)$ is continuous for $(x, y)$ in a closed region $F$, then $f$ is integrable over $F$. Furthermore, if $f(x, y) > 0$ for $(x, y)$ in $F$, then*

$$V(S) = \iint_F f(x, y) \, dA,$$

*where $V(S)$ is the volume of the solid defined by*

$$S = \{(x, y, z) : (x, y) \text{ in } F \text{ and } 0 \le z \le f(x, y)\}.$$

Methods for the evaluation of double integrals are discussed in Section 3.

**EXAMPLE**  Given $f(x, y) = 1 + xy$ and the region $F$ bounded by the lines $y = 0$, $y = x$, and $x = 1$ (Fig. 6), let $\Delta$ be the subdivision formed by the lines $x = 0, 0.2, 0.5, 0.8, 1$ and $y = 0, 0.2, 0.5, 0.7, 1$. Find the value of the sum

$$\sum_{i=1}^n f(\xi_i, \eta_i) A(r_i)$$

if the points $(\xi_i, \eta_i)$ are selected at the centers of the rectangles. Note that this sum approximates the double integral

$$\iint_F f(x, y) \, dA.$$

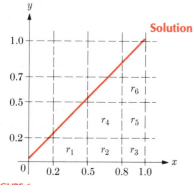

FIGURE 6

**Solution**  Referring to Fig. 6, we see that there are 6 rectangles in the subdivision which we label $r_1, r_2, \ldots, r_6$, as shown. We compute:

$$A(r_1) = 0.06, \qquad f(0.35, 0.1) = 1.035$$
$$A(r_2) = 0.06, \qquad f(0.65, 0.1) = 1.065$$
$$A(r_3) = 0.04, \qquad f(0.9, 0.1) = 1.090$$
$$A(r_4) = 0.09, \qquad f(0.65, 0.35) = 1.2275$$
$$A(r_5) = 0.06, \qquad f(0.9, 0.35) = 1.315$$
$$A(r_6) = 0.04, \qquad f(0.9, 0.6) = 1.540$$

Multiplying and adding, we find that

$$\sum_{i=1}^{n} f(\xi_i, \eta_i) A(r_i) = 0.420575 \quad \text{(Answer)}.$$ □

## 1   PROBLEMS

In each of Problems 1 through 12, calculate $\sum_{i=1}^{n} f(\xi_i, \eta_i) A(r_i)$, the sum for the subdivision $\Delta$ of the region $F$ formed by the given lines and with the points $(\xi_i, \eta_i)$ selected as directed in each case.

**1**   $f(x, y) = x^2 + 2y^2$; $F$ is the rectangle $0 \le x \le 1$, $0 \le y \le 1$. The subdivision $\Delta$ is: $x = 0, 0.4, 0.8, 1$; $y = 0, 0.3, 0.7, 1$. For each $i$ the point $(\xi_i, \eta_i)$ is taken at the center of the rectangle $r_i$.

**2**   Same as Problem 1, with $(\xi_i, \eta_i)$ taken at the point of $r_i$ which is closest to the origin.

**3**   $f(x, y) = 1 + x^2 - y^2$; $F$ is the triangular region formed by the lines $y = 0$, $y = x$, $x = 2$. The subdivision $\Delta$ is: $x = 0, 0.5, 1, 1.6, 2$; $y = 0, 0.6, 1, 1.5, 2$. For each $i$, the point $(\xi_i, \eta_i)$ is taken at the center of the rectangle $r_i$.

**4**   Same as Problem 3, with $(\xi_i, \eta_i)$ taken at the point of $r_i$ which is closest to the origin.

**5**   Same as Problem 3, with $(\xi_i, \eta_i)$ selected on the lower edge of $r_i$, midway between the vertical subdivision lines.

**6**   Let $F = \{(x, y): 0 \le x \le 6, 0 \le y \le \sqrt{36 - x^2}\}$. The subdivision $\Delta$ is determined by vertical lines with $x$ intercepts 0, 2, 3, 4, 6 and by horizontal lines with $y$ intercepts 0, 2, 4, 5, 6.

Let $f(x, y) = 40 - xy$. For each $i$, the point $(\xi_i, \eta_i)$ is taken at the lower right corner of the rectangle $r_i$.

**7**   Same as Problem 6, with $(\xi_i, \eta_i)$ taken at the upper left corner of the rectangle $r_i$.

**8**   $f(x, y) = x^2 - 2xy + 3x - 2y$; $F$ is the trapezoid bounded by the lines $x = 0$, $x = 2$, $y = 0$, $y = x + 1$. The subdivision $\Delta$ is: $x = 0, 0.4, 1, 1.5, 2$; $y = 0, 0.6, 1, 1.4, 1.8, 2, 3$. For each $i$ the point $(\xi_i, \eta_i)$ is taken at the center of the rectangle $r_i$.

**9**   Same as Problem 8, with $(\xi_i, \eta_i)$ taken at the point of $r_i$ farthest from the origin.

**10**   Same as Problem 8, with $(\xi_i, \eta_i)$ taken at the point of $r_i$ closest to the origin.

**11**   Let

$$f(x, y) = \frac{x - y}{1 + x + y};$$

$F$ is the region bounded by the line $y = 0$ and the curve $y = 2x - x^2$. The subdivision $\Delta$ is: $x = 0, 0.5, 1.0, 1.5, 2$; $y = 0, 0.2, 0.4, 0.6, 0.8, 1$. For each $i$, the point $(\xi_i, \eta_i)$ is taken at the center of the rectangle $r_i$.

**12**   Same as Problem 11 with the point $(\xi_i, \eta_i)$ taken at the point of $r_i$ closest to the origin.

## 2

## PROPERTIES OF THE DOUBLE INTEGRAL

In analogy with the properties of the definite integral of functions of one variable, we state several basic properties of the double integral. The simplest properties are given in the two following theorems.

**THEOREM 2**   *If $c$ is any number and $f$ is integrable over a closed region $F$, then $cf$ is integrable and*

$$\iint_F cf(x, y) \, dA = c \iint_F f(x, y) \, dA.$$

**THEOREM 3**    *If f and g are integrable over a closed region F, then*

$$\iint_F [f(x, y) + g(x, y)]\, dA = \iint_F f(x, y)\, dA + \iint_F g(x, y)\, dA.$$

The result holds for the sum of any finite number of integrable functions. The proofs of Theorems 2 and 3 follow directly from the definition.

**THEOREM 4**    *Suppose that f is integrable over a closed region F and*

$$m \leq f(x, y) \leq M \text{ for all } (x, y) \text{ in } F.$$

*Then, if A(F) denotes the area of F, we have*

$$mA(F) \leq \iint_F f(x, y)\, dA \leq MA(F).$$

The proof of Theorem 4 follows exactly the same pattern as does the proof in the one-variable case. (See page 197, Theorem 7.)

**THEOREM 5**    *If f and g are integrable over F and f(x, y) ≤ g(x, y) for all (x, y) in F, then*

$$\iint_F f(x, y)\, dA \leq \iint_F g(x, y)\, dA.$$

The proof is established by the same argument used in the one-variable case (page 198, Theorem 8).

**THEOREM 6**    *If the closed region F is decomposed into (nonoverlapping) regions $F_1$ and $F_2$ and if f is continuous over F, then*

$$\iint_F f(x, y)\, dA = \iint_{F_1} f(x, y)\, dA + \iint_{F_2} f(x, y)\, dA.$$

The proof depends on the definition of double integral and on the basic theorems on limits.

## 2    PROBLEMS

In Problems 1 through 7, use Theorem 4 to find in each case estimates for the largest and smallest values the given double integrals can possibly have.

1  $\iint_F xy\, dA$ where $F$ is the region bounded by the lines $x = 0$, $y = 0$, $x = 2$, $y = x + 3$.

2  $\iint_F (x^2 + y^2)\, dA$ where $F$ is the region bounded by the lines $x = -2$, $x = 3$, $y = x + 2$, $y = -2$.

3  $\iint_F (1 + 2x^2 + y^2)\, dA$ where $F$ is the region bounded by the lines $x = -3$, $x = 3$, $y = 4$, $y = -4$.

4  $\iint_F y^4\, dA$ where $F$ is the region bounded by the line $y = 0$ and the curve $y = 2x - x^2$.

5  $\iint_F (x - y)\, dA$ where $F$ is the region enclosed in the circle $x^2 + y^2 = 9$.

6  $\iint_F [1/(1 + x^2 + y^2)]\, dA$ where $F$ is the region enclosed in the ellipse $4x^2 + 9y^2 = 36$.

7  $\iint_F \sqrt{1 + x^2 + y^2}\, dA$ where $F$ is the region bounded by the curves $y = 3x - x^2$ and $y = x^2 - 3x$.

8  Write a proof of Theorem 3.

9  Write a proof of Theorem 4.

10  Write a proof of Theorem 5.

11  Write a proof of Theorem 6.

12  Let $F_1$, $F_2$, ..., $F_n$ be $n$ nonoverlapping regions. State and prove a generalization of Theorem 6.

13  Let $f(x, y)$ be continuous on a closed region $F$.

a)  Show that $\left| \iint_F f(x, y)\, dA \right| \leq \iint_F |f(x, y)|\, dA$.

b)  If $g(x, y)$ is also continuous on $F$, show that

$$\left| \iint_F [f(x, y) + g(x, y)]\, dA \right| \leq \iint_F [|f(x, y)| + |g(x, y)|]\, dA.$$

14  Use the definition of double integral to show that the volume of a right circular cylinder with height $h$ and radius of base $r$ is $\pi r^2 h$.

---

### 3

## EVALUATION OF DOUBLE INTEGRALS. ITERATED INTEGRALS

The definition of the double integral is almost useless as a tool for evaluation in any particular case. Of course, it may happen that the function $f(x, y)$ and the region $F$ are particularly simple, so that the limit of the sum

$$\sum_{i=1}^{n} f(\xi_i, \eta_i) A(r_i)$$

can be found directly. However, such limits cannot generally be found. As in the case of ordinary integrals and line integrals, it is important to develop simple and routine methods for determining the value of a given double integral. In this section we show how the evaluation of a double integral may be performed by successive evaluations of single integrals. In other words, we reduce the problem to one we have already studied extensively.

Let $F$ be the rectangle with sides $x = a$, $x = b$, $y = c$, $y = d$, as shown in Fig. 7. Suppose that $f(x, y)$ is *continuous* for $(x, y)$ in $F$. We form the ordinary integral with respect to $x$.

$$\int_a^b f(x, y)\, dx,$$

in which we keep $y$ fixed when performing the integration. Of course, the value of the above integral will depend on the value of $y$ used, and so, using $A$

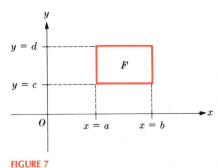

FIGURE 7

for this value, we may write

$$A(y) = \int_a^b f(x, y) \, dx.$$

The function $A(y)$ is defined for $c \leq y \leq d$ and, in fact, it can be shown that if $f(x, y)$ is continuous on $F$, then $A(y)$ is continuous on $[c, d]$. The integral of $A(y)$ may be computed, and we write

$$\int_c^d A(y) \, dy = \int_c^d \left[ \int_a^b f(x, y) \, dx \right] dy. \tag{1}$$

We could start the other way around by fixing $x$ and forming the integral

$$B(x) = \int_c^d f(x, y) \, dy.$$

Then

$$\int_a^b B(x) \, dx = \int_a^b \left[ \int_c^d f(x, y) \, dy \right] dx. \tag{2}$$

Note that the integrals are computed *successively*; in (1) we first integrate with respect to $x$ (keeping $y$ constant) and then with respect to $y$; in (2) we first integrate with respect to $y$ (keeping $x$ constant) and then with respect to $x$.

**DEFINITION**   *The integrals*

$$\int_c^d \left[ \int_a^b f(x, y) \, dx \right] dy, \qquad \int_a^b \left[ \int_c^d f(x, y) \, dy \right] dx$$

*are called the* **iterated integrals of** $f$. *The terms* **repeated** *integrals and* **successive** *integrals are also used.*

**Notation.**   The brackets in iterated integrals are unwieldy, and we will write

$$\int_c^d \int_a^b f(x, y) \, dx \, dy \qquad \text{to mean} \qquad \int_c^d \left[ \int_a^b f(x, y) \, dx \right] dy,$$

$$\int_a^b \int_c^d f(x, y) \, dy \, dx \qquad \text{to mean} \qquad \int_a^b \left[ \int_c^d f(x, y) \, dy \right] dx.$$

Iterated integrals are computed in the usual way, as the next example shows.

**EXAMPLE 1**   Evaluate

$$\int_1^4 \int_{-2}^3 (x^2 - 2xy^2 + y^3) \, dx \, dy.$$

**Solution**   Keeping $y$ fixed, we have

$$\int_{-2}^3 (x^2 - 2xy^2 + y^3) \, dx = \left[ \tfrac{1}{3}x^3 - x^2 y^2 + y^3 x \right]_{-2}^3$$

$$= 9 - 9y^2 + 3y^3 - \left( -\tfrac{8}{3} - 4y^2 - 2y^3 \right)$$

$$= \tfrac{35}{3} - 5y^2 + 5y^3.$$

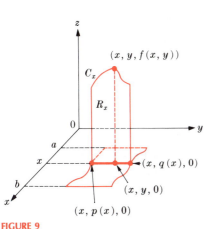

**FIGURE 8**

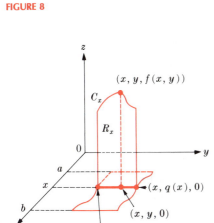

**FIGURE 9**

Therefore

$$\int_1^4 \int_{-2}^3 (x^2 - 2xy^2 + y^3) \, dx \, dy = \int_1^4 (\tfrac{35}{3} - 5y^2 + 5y^3) \, dy$$

$$= [\tfrac{35}{3}y - \tfrac{5}{3}y^3 + \tfrac{5}{4}y^4]_1^4 = \tfrac{995}{4}. \qquad \square$$

Iterated integrals may be defined over regions $F$ which have curved boundaries. This situation is more complicated than the one just discussed. Consider a region $F$ such as that shown in Fig. 8, in which the boundary consists of the lines $x = a$, $x = b$ and the graphs of the functions $p(x)$ and $q(x)$ with $p(x) \le q(x)$ for $a \le x \le b$. We may define

$$\int_a^b \int_{p(x)}^{q(x)} f(x, y) \, dy \, dx, \qquad (3)$$

in which we first integrate (for fixed $x$) from the lower curve to the upper curve, i.e., along a typical line as shown in Fig. 8; then we integrate with respect to $x$ over all such typical segments from $a$ to $b$.

When we perform the first integration in (3),

$$\int_{p(x)}^{q(x)} f(x, y) \, dy,$$

we are actually finding the area of a region $R_x$, as is represented schematically in Fig. 9. We then "continuously add up" the region $R_x$ as $x$ ranges from $a$ to $b$ to get the volume. The curve $C_x$ that forms the top boundary of $R_x$ is simply the function $g(y) \equiv f(x, y)$, with $x$ fixed.

In Fig. 8 two of the borders of $F$ are irregular; that is, $p(x)$, $q(x)$ are continuous functions of $x$. We may also allow the other two borders to be irregular: that is, the "sides" of $F$ may be continuous functions of $y$. For example, iterated integrals may be defined over a region $F$ such as the one shown in Fig. 10. Integrating first with respect to $y$, we have

$$\int_a^b \int_{p(x)}^{q(x)} f(x, y) \, dy \, dx.$$

On the other hand, the integral taken first with respect to $x$ requires that we represent $F$ as shown in Fig. 11. Then we have

$$\int_c^d \int_{r(y)}^{s(y)} f(x, y) \, dx \, dy.$$

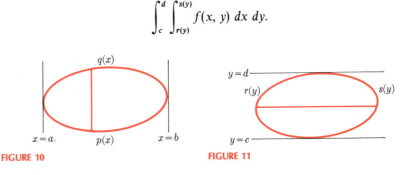

**FIGURE 10**          **FIGURE 11**

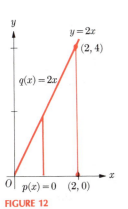

**FIGURE 12**

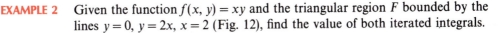

**EXAMPLE 2**  Given the function $f(x, y) = xy$ and the triangular region $F$ bounded by the lines $y = 0$, $y = 2x$, $x = 2$ (Fig. 12), find the value of both iterated integrals.

**Solution**  Referring to Fig. 12, we see that for

$$\int_a^b \int_{p(x)}^{q(x)} xy \, dy \, dx,$$

we have $p(x) = 0$, $q(x) = 2x$, $a = 0$, $b = 2$. Therefore

$$\int_0^2 \int_0^{2x} xy \, dy \, dx = \int_0^2 [\tfrac{1}{2}xy^2]_0^{2x} \, dx$$

$$= \int_0^2 2x^3 \, dx = [\tfrac{1}{2}x^4]_0^2 = 8.$$

Integrating with respect to $x$ first (Fig. 13), we have

$$\int_c^d \int_{r(y)}^{s(y)} xy \, dx \, dy \qquad \text{with} \qquad r(y) = \tfrac{1}{2}y,\ s(y) = 2,\ c = 0,\ d = 4.$$

Therefore

$$\int_0^4 \int_{y/2}^2 xy \, dx \, dy = \int_0^4 [\tfrac{1}{2}x^2 y]_{y/2}^2 \, dy$$

$$= \int_0^4 (2y - \tfrac{1}{8}y^3) \, dy = [y^2 - \tfrac{1}{32}y^4]_0^4 = 8. \qquad \square$$

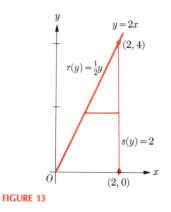

**FIGURE 13**

It is not accidental that the two integrals in Example 2 have the same value. The next theorem describes the general situation. It is a special case of a result known as *Fubini's Theorem*.

---

**THEOREM 7**  *Suppose that F is a closed region given by*

$$F = \{(x, y): a \le x \le b, \quad p(x) \le y \le q(x)\},$$

*where p and q are continuous and $p(x) \le q(x)$ for $a \le x \le b$. Suppose that $f(x, y)$ is continuous on F. Then*

$$\iint_F f(x, y) \, dA = \int_a^b \int_{p(x)}^{q(x)} f(x, y) \, dy \, dx.$$

*The corresponding result holds if the closed region F has the representation*

$$F = \{(x, y): c \le y \le d,\ r(y) \le x \le s(y)\}$$

*where $r(y) \le s(y)$ for $c \le y \le d$. In such a case,*

$$\iint_F f(x, y) \, dA = \int_c^d \int_{r(y)}^{s(y)} f(x, y) \, dx \, dy.$$

*In other words, both iterated integrals, when computable, are equal to the double integral and therefore equal to each other.*

---

**Partial Proof**  We shall discuss the first result, the second being similar. Suppose first that $f(x, y)$ is positive. A plane $x = \text{const}$ intersects the surface $z = f(x, y)$ in a curve (Fig. 14). The area under this curve in the $x = \text{const}$ plane is shown as a

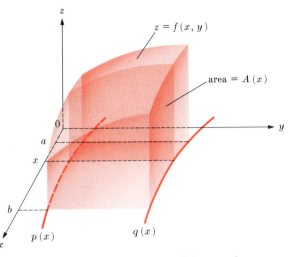

Volume by cross sections parallel to $yz$ plane

**FIGURE 14**

shaded region. Denoting the area of this region by $A(x)$, we have the formula

$$A(x) = \int_{p(x)}^{q(x)} f(x, y) \, dy.$$

It can be shown that $A(x)$ is continuous. Furthermore, it can also be shown that if $A(x)$ is integrated between $x = a$ and $x = b$, the volume $V$ under the surface $f(x, y)$ is swept out. We recall that a similar argument was used in obtaining volumes of revolution by the disk method. (See page 228.) The double integral yields the volume under the surface, and so we write

$$V = \iint_F f(x, y) \, dA.$$

On the other hand, we obtain the volume by integrating $A(x)$; that is,

$$V = \int_a^b A(x) \, dx = \int_a^b \int_{p(x)}^{q(x)} f(x, y) \, dy \, dx.$$

If $f(x, y)$ is not positive but is bounded from below by the plane $z = c$, then subtraction of the volume of the cylinder of height $c$ and cross-section $F$ leads to the same result.                                                    □

*Remarks.* We have considered two ways of expressing a region $F$ in the $xy$ plane. They are

$$F = \{(x, y): a \le x \le b, \quad p(x) \le y \le q(x)\} \tag{4}$$

and

$$F = \{(x, y): c \le y \le d, \quad r(y) \le x \le s(y)\}. \tag{5}$$

It frequently happens that a region $F$ is expressible more simply in one of the above forms than in the other. In doubtful cases, a sketch of $F$ may show which is simpler and, therefore, which of the iterated integrals is evaluated more easily.

A region $F$ may not be expressible in either the form (4) or the form (5).

In such cases, $F$ may sometimes be subdivided into a number of regions, each having one of the two forms. The integrations are then performed for each subregion and the results added. Figure 15 gives examples of how the subdivision process might take place.

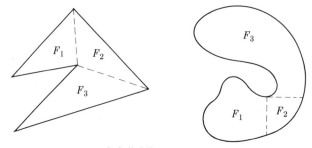

Subdividing a region

**FIGURE 15**

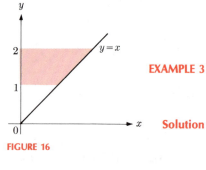

**FIGURE 16**

**EXAMPLE 3**   Evaluate $\iint\limits_F x^2 y^2 \, dA$ where $F$ is the figure bounded by the lines $y = 1$, $y = 2$, $x = 0$, and $x = y$ (Fig. 16).

**Solution**   The closed region $F$ is the set

$$F = \{(x, y): 1 \le y \le 2, 0 \le x \le y\}.$$

We use Theorem 7 and evaluate the iterated integral, to find

$$\iint\limits_F x^2 y^2 \, dA = \int_1^2 \int_0^y x^2 y^2 \, dx \, dy = \int_1^2 [\tfrac{1}{3} x^3 y^2]_0^y \, dy$$

$$= \tfrac{1}{3} \int_1^2 y^5 \, dy = \tfrac{7}{2}. \qquad \square$$

Note that in the above example the iterated integral in the other order is more difficult, since the curves $p(x)$, $q(x)$ are

$$p(x) = \begin{cases} 1 \text{ for } 0 \le x \le 1 \\ x \text{ for } 1 \le x \le 2 \end{cases}, \qquad q(x) = 2, \, 0 \le x \le 2.$$

The evaluation would have to take place in two parts, so that

$$\iint\limits_F x^2 y^2 \, dA = \int_0^1 \int_1^2 x^2 y^2 \, dy \, dx + \int_1^2 \int_x^2 x^2 y^2 \, dy \, dx.$$

**EXAMPLE 4**   Evaluate

$$\int_0^2 \int_0^{x^2/2} \frac{x}{\sqrt{1 + x^2 + y^2}} \, dy \, dx.$$

**Solution**   Carrying out the integration first with respect to $y$ is possible but difficult and leads to a complicated integral for $x$. Therefore we shall try to express the integral as an iterated integral in the opposite order and use Theorem 7. We

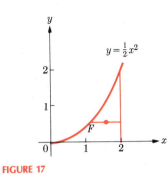

FIGURE 17

construct the region $F$ as shown in Fig. 17. The region is expressed by

$$F = \{(x, y): 0 \le x \le 2 \quad \text{and} \quad 0 \le y \le \tfrac{1}{2}x^2\}.$$

However, it is also expressed by

$$F = \{(x, y): 0 \le y \le 2 \quad \text{and} \quad \sqrt{2y} \le x \le 2\}.$$

Therefore, integrating with respect to $x$ first, we have

$$\int_0^2 \int_0^{x^2/2} \frac{x}{\sqrt{1 + x^2 + y^2}} \, dy \, dx$$

$$= \iint_F \frac{x}{\sqrt{1 + x^2 + y^2}} \, dA = \int_0^2 \int_{\sqrt{2y}}^2 \frac{x}{\sqrt{1 + x^2 + y^2}} \, dx \, dy$$

$$= \int_0^2 \left[ \sqrt{1 + x^2 + y^2} \right]_{\sqrt{2y}}^2 dy = \int_0^2 \left[ \sqrt{5 + y^2} - (1 + y) \right] dy$$

$$= [\tfrac{5}{2} \ln (y + \sqrt{y^2 + 5}) + \tfrac{1}{2} y \sqrt{y^2 + 5} - y - \tfrac{1}{2} y^2]_0^2$$

$$= \tfrac{5}{2} \ln 5 + 3 - 4 - \tfrac{5}{2} \ln \sqrt{5} = -1 + \tfrac{5}{4} \ln 5. \qquad \square$$

The next example shows how the volume of a solid may be found by iterated integration.

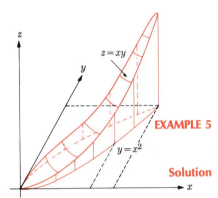

FIGURE 18

**EXAMPLE 5**   Let $S$ be the solid bounded by the surface $z = xy$, the cylinders $y = x^2$ and $y^2 = x$, and the plane $z = 0$. Find the volume $V(S)$.

**Solution**   The solid $S$ is shown in Fig. 18. It consists of all points "under" the surface $z = xy$, bounded by the cylinders, and "above" the $xy$ plane. The region $F$ in the $xy$ plane is bounded by the curves $y = x^2$, $y^2 = x$ and is shown in Fig. 19. Therefore

$$V(S) = \iint_F xy \, dA = \int_0^1 \int_{x^2}^{\sqrt{x}} xy \, dy \, dx = \int_0^1 \left[ \frac{xy^2}{2} \right]_{x^2}^{\sqrt{x}} dx$$

$$= \tfrac{1}{2} \int_0^1 (x^2 - x^5) \, dx = \tfrac{1}{12}. \qquad \square$$

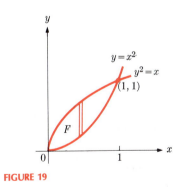

FIGURE 19

If a solid $S$ is bounded by two surfaces of the form $z = f(x, y)$ and $z = g(x, y)$ with $f(x, y) \le g(x, y)$, then the volume between the surfaces may be found as a double integral, and that integral in turn may be evaluated by iterated integrals. The closed region $F$ over which the integration is performed is found by the projection onto the $xy$ plane of the curve of intersection of the two surfaces. To find this projection we set

$$f(x, y) = g(x, y)$$

and trace this curve in the $xy$ plane. The next example shows the method.

**EXAMPLE 6**   Find the volume bounded by the surfaces

$$z = x^2$$

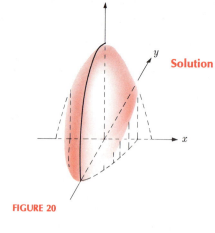

**FIGURE 20**

and

$$z = 4 - x^2 - y^2.$$

**Solution**   A portion of the solid $S$ (the part corresponding to $y \leq 0$) is shown in Fig. 20. We set

$$x^2 = 4 - x^2 - y^2$$

and find that the closed region $F$ in the $xy$ plane is the elliptical disk (Fig. 21)

$$F = \left\{ (x, y): \frac{x^2}{2} + \frac{y^2}{4} \leq 1 \right\}.$$

Note that the surface $z = 4 - x^2 - y^2 \equiv g(x, y)$ is above the surface $z = x^2 \equiv f(x, y)$ for $(x, y)$ inside the above ellipse. Therefore

$$V(S) = \iint_F (4 - y^2 - x^2 - x^2) \, dA$$

$$= \int_{-2}^2 \int_{-\sqrt{4-y^2}/\sqrt{2}}^{+\sqrt{4-y^2}/\sqrt{2}} (4 - y^2 - 2x^2) \, dx \, dy.$$

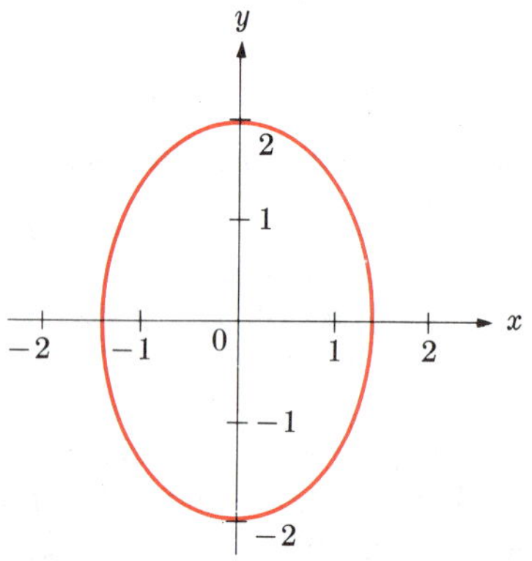

We now use the fact that $(4 - y^2)^{3/2}$ is an even function of $y$. Hence

$$V(S) = \int_{-2}^2 \frac{2\sqrt{2}}{3} (4 - y^2)^{3/2} \, dy = \frac{4\sqrt{2}}{3} \int_0^2 (4 - y^2)^{3/2} \, dy$$

$$= \frac{64\sqrt{2}}{3} \int_0^{\pi/2} \cos^4 \theta \, d\theta = \frac{16\sqrt{2}}{3} \int_0^{\pi/2} (1 + 2 \cos 2\theta + \cos^2 2\theta) \, d\theta$$

$$= \frac{8\pi\sqrt{2}}{3} + \left[ \frac{16\sqrt{2}}{3} \sin 2\theta \right]_0^{\pi/2} + \frac{8\sqrt{2}}{3} \int_0^{\pi/2} (1 + \cos 4\theta) \, d\theta = 4\pi\sqrt{2}.$$

□

**FIGURE 21**

# 3   PROBLEMS

In Problems 1 through 18, evaluate the iterated integrals as indicated. Sketch the region $F$ in the $xy$ plane over which the integration is taken. Describe $F$ using set notation.

**1** $\displaystyle\int_1^4 \int_2^5 (x^2 - y^2 + xy - 3) \, dx \, dy$

**2** $\displaystyle\int_0^2 \int_{-3}^2 (x^3 + 2x^2 y - y^3 + xy) \, dy \, dx$

**3** $\displaystyle\int_0^1 \int_0^1 e^{2x + 3y} \, dx \, dy$   **4** $\displaystyle\int_2^3 \int_{-1}^1 e^x e^{-y} \, dy \, dx$

**5** $\displaystyle\int_1^4 \int_{\sqrt{x}}^{x^2} (x^2 + 2xy - 3y^2) \, dy \, dx$

**6** $\displaystyle\int_0^1 \int_{x^3}^{x^2} (x^2 - xy) \, dy \, dx$   **7** $\displaystyle\int_2^3 \int_{1+y}^{\sqrt{y}} (x^2 y + xy^2) \, dx \, dy$

**8** $\displaystyle\int_{-2}^2 \int_{-\sqrt{4-x^2}}^{+\sqrt{4-x^2}} y \, dy \, dx$   **9** $\displaystyle\int_{-3}^3 \int_{-\sqrt{18-2y^2}}^{+\sqrt{18-2y^2}} x \, dx \, dy$

**10** $\displaystyle\int_{-3}^3 \int_{x^2}^{18-x^2} xy^3 \, dy \, dx$   **11** $\displaystyle\int_0^2 \int_{x^2}^{2x^2} x \cos y \, dy \, dx$

**12** $\displaystyle\int_1^2 \int_{x^3}^{4x^3} \frac{1}{y} \, dy \, dx$   **13** $\displaystyle\int_1^2 \int_{x^3}^x e^{y/x} \, dy \, dx$

**14** $\displaystyle\int_0^1 \int_x^1 e^{y^2} \, dx \, dy$

**15** $\displaystyle\int_0^3 \int_{-\pi/2}^0 (x \cos y - y \cos x) \, dy \, dx$

**16** $\displaystyle\int_1^{e^2} \int_1^y \ln x \, dx \, dy$

**17** $\displaystyle\int_{\pi/6}^{\pi/4}\int_{\tan x}^{\sec x}(y+\sin x)\,dy\,dx$   **18** $\displaystyle\int_{\pi/6}^{\pi/4}\int_{0}^{\sin x}e^{y}\cos x\,dy\,dx$

In Problems 19 through 25, evaluate the double integrals as indicated. Sketch the region $F$.

**19** $\displaystyle\iint_{F}(x^{2}+y^{2})\,dA;\ F=\{(x,y):y^{2}\le x\le 4,\,0\le y\le 2\}$

**20** $\displaystyle\iint_{F}x\cos y\,dA;\quad F=\{(x,y):0\le x\le\sqrt{\pi/2},\,0\le y\le x^{2}\}$

**21** $\displaystyle\iint_{F}\frac{x}{x^{2}+y^{2}}\,dA;\quad F=\{(x,y):1\le x\le\sqrt{3},\,0\le y\le x\}$

**22** $\displaystyle\iint_{F}\ln y\,dA;\quad F=\{(x,y):2\le x\le 3,\,1\le y\le x-1\}$

**23** $\displaystyle\iint_{F}\frac{x}{\sqrt{1-y^{2}}}\,dA;\quad F=\{(x,y):0\le x\le\tfrac{1}{2},\,x\le y\le\tfrac{1}{2}\}$

**24** $\displaystyle\iint_{F}\frac{x}{\sqrt{x^{2}+y^{2}}}\,dA;\quad F=\{(x,y):1\le x\le 2,\,1\le y\le x\}$

**25** $\displaystyle\iint_{F}\frac{1}{y^{2}}e^{x/\sqrt{y}}\,dA;\quad F=\{(x,y):1\le x\le\sqrt{2},\,x^{2}\le y\le 2\}$

In each of Problems 26 through 37, (a) sketch the domain $F$ over which the integration is performed; (b) write the equivalent iterated integral in the reverse order; (c) evaluate the integral obtained in (b). Describe $F$ using set notation (in both orders).

**26** $\displaystyle\int_{1}^{2}\int_{1}^{x}\frac{x^{2}}{y^{2}}\,dy\,dx$   **27** $\displaystyle\int_{-2}^{2}\int_{-\sqrt{4-x^{2}}}^{+\sqrt{4-x^{2}}}xy\,dy\,dx$

**28** $\displaystyle\int_{0}^{a}\int_{0}^{\sqrt{a^{2}-x^{2}}}(a^{2}-y^{2})^{3/2}\,dy\,dx$

**29** $\displaystyle\int_{0}^{1}\int_{y}^{1}\sqrt{1+x^{2}}\,dx\,dy$   **30** $\displaystyle\int_{0}^{1}\int_{\sqrt{x}}^{1}\sqrt{1+y^{3}}\,dy\,dx$

**31** $\displaystyle\int_{0}^{1}\int_{x}^{1}x\sin(y^{3})\,dy\,dx$   **32** $\displaystyle\int_{0}^{1}\int_{y}^{\sqrt{y}}\frac{\sin x}{x}\,dx\,dy$

**33** $\displaystyle\int_{0}^{1}\int_{y}^{1}\frac{y}{\sqrt{x^{2}+y^{2}}}\,dx\,dy$   **34** $\displaystyle\int_{-1}^{0}\int_{-x}^{1}\frac{x^{2}}{1+y^{4}}\,dy\,dx$

**35** $\displaystyle\int_{0}^{1}\int_{x}^{1}\frac{1}{y}\sin y\cos\left(\frac{\pi x}{2y}\right)dy\,dx$

**36** $\displaystyle\int_{0}^{2}\int_{y/2}^{1}\frac{1}{(1+x^{2})^{3}}\,dx\,dy$

**37** $\displaystyle\int_{0}^{1}\int_{x}^{1}x(x^{2}+y^{2})^{3/2}\,dy\,dx$

In each of Problems 38 through 47, find the volume $V(S)$ of the solid $S$ described. Sketch the domain $F$ of integration and describe it using set notation.

**38** $S$ is bounded by the surfaces $z=0$, $z=x$, and $y^{2}=2-x$.

**39** $S$ is bounded by the planes $z=0$, $y=0$, $y=x$, $x+y=2$, and $x+y+z=3$.

**40** $S$ is bounded by the surfaces $x=0$, $z=0$, $y^{2}=4-x$, and $z=y+2$.

**41** $S$ is bounded by the surfaces $x^{2}+z^{2}=4$, $y=0$, and $x+y+z=3$.

**42** $S$ is bounded by the surfaces $y^{2}=z$, $y=z^{3}$, $z=x$, and $y^{2}=2-x$.

**43** $S$ is bounded by the coordinate planes and the surface $x^{1/2}+y^{1/2}+z^{1/2}=a^{1/2}$.

**44** $S$ is bounded by the surfaces $y=x^{2}$ and $z^{2}=4-y$.

**45** $S$ is bounded by the surfaces $y^{2}=x$, $x+y=2$, $x+z=0$, and $z=x+1$.

**46** $S$ is bounded by the surfaces $x^{2}=y+z$, $y=0$, $z=0$, and $x=2$.

**47** $S$ is bounded by the surfaces $y^{2}+z^{2}=2x$ and $y=x-\tfrac{3}{2}$.

**\*48** Let $F=\{(x,y):0\le x\le 1,\,0\le y\le 1\}$. We define

$$f(x,y)=\begin{cases}1 & \text{if }y\text{ is rational,}\\ 0 & \text{otherwise.}\end{cases}$$

Show that $\displaystyle\iint_{F}f(x,y)\,dA$ does not exist.

**\*49** Let $F=\{(x,y):0\le x\le 1,\,0\le y\le 1\}$. We define

$$f(x,y)=\begin{cases}0 & \text{if }x=1/n,\ n=1,2,3,\dots,\\ 1 & \text{otherwise.}\end{cases}$$

Show that $\displaystyle\iint_{F}f(x,y)\,dA$ exists and has the value 1.

**50** Let $F=\{(x,y):a\le x\le b,\,c\le y\le d\}$ and suppose $f(x,y)=g(x)\cdot h(y)$. Show that

$$\iint_{F}f(x,y)\,dA=\left(\int_{a}^{b}g(x)\,dx\right)\left(\int_{c}^{d}h(y)\,dy\right).$$

**\*51** Suppose that $f(x,y)$ is continuous on a closed, bounded region $F$. Assume that

$$\iint_{F}f(x,y)\phi(x,y)\,dA=0$$

for *all* functions $\phi(x,y)$ which are continuous on $F$. Show that $f(x,y)\equiv 0$ on $F$. (*Hint:* Assume there is a point in $F$ where $f$ is positive, choose $\phi$ carefully, and then use Theorem 4 with $m>0$ thereby reaching a contradiction.)

**52** Find the volume of the ellipsoid

$$F=\left\{(x,y,z):\frac{x^{2}}{a^{2}}+\frac{y^{2}}{b^{2}}+\frac{z^{2}}{c^{2}}-1=0\right\}.$$

4

## AREA, DENSITY, AND MASS

The double integral of a nonnegative function $z = f(x, y)$ taken over a region $F$ may be interpreted as a volume. The value of such an integral is the volume of the cylinder having generators parallel to the $z$ axis and situated between the surface $z = f(x, y)$ and the region $F$ in the $xy$ plane.

If we select for the surface $f$ the particularly simple function $z = 1$, then the volume $V$ is given by the formula

$$V = \iint_F 1 \; dA.$$

On the other hand, the volume of a right cylinder of cross section $F$ and height 1 is

$$V = A(F) \cdot 1.$$

(See Fig. 22) Therefore

$$A(F) = \iint_F dA.$$

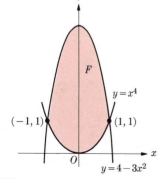

FIGURE 22

*The double integral of the function 1 taken over $F$ is precisely the area of $F$.*

By Theorem 7, we conclude that the iterated integral of the function 1 also yields the area of $F$.

**EXAMPLE 1**   Use iterated integration to find the area of the region $F$ given by

$$F = \{(x, y): -1 \le x \le 1, \; x^4 \le y \le 4 - 3x^2\}.$$

**Solution**   The region $F$ is shown in Fig. 23. One of the iterated integrals for the area is

$$A(F) = \int_{-1}^{1} \int_{x^4}^{4-3x^2} dy \; dx,$$

and its evaluation gives

$$A(F) = \int_{-1}^{1} [y]_{x^4}^{4-3x^2} dx = \int_{-1}^{1} (4 - 3x^2 - x^4) \; dx$$

$$= [4x - x^3 - \tfrac{1}{5}x^5]_{-1}^{1} = \tfrac{28}{5}.$$

Note that the iterated integral in the other direction is more difficult to evaluate. ☐

If a flat object is made of an extremely thin uniform material, then the mass of the object is just a multiple of the area of the plane region on which the object rests. (The multiple depends on the units used.) If a thin object resting on the $xy$ plane is made of a nonuniform material, then the mass of the object may be expressed in terms of the density $\rho(x, y)$ of the material at

$F$

$y = x^4$

$(-1, 1)$     $(1, 1)$

$x$

$O$

$y = 4 - 3x^2$

FIGURE 23

any point. We assume that the material is uniform in the $z$ direction. Letting $F$ denote the region occupied by the object, we decompose $F$ into rectangles $r_1, r_2, \ldots, r_n$ in the usual way. Then an approximation to the mass of the $i$th rectangle is given by

$$\rho(\xi_i, \eta_i) A(r_i),$$

where $A(r_i)$ is the area of $r_i$ and $(\xi_i, \eta_i)$ is a point in $r_i$. The total mass of $F$ is approximated by

$$\sum_{i=1}^{n} \rho(\xi_i, \eta_i) A(r_i),$$

and when we proceed to the limit in the customary manner, the mass $M(F)$ is

$$M(F) = \iint_F \rho(x, y) \, dA.$$

We summarize this discussion with a definition.

**DEFINITION**   *If an object occupying a region $F$ in the $xy$ plane has a density at each point $(x, y)$ given by $\rho(x, y)$, then the **mass** $M(F)$ is defined to be*

$$M(F) = \iint_F \rho(x, y) \, dA.$$

In other words, the double integral is a useful device for finding the mass of a thin flat object with variable density.

**EXAMPLE 2**   A thin object occupies the region

$$F = \{(x, y): y^2 \leq x \leq 4 - y^2, \ -\sqrt{2} \leq y \leq \sqrt{2}\}.$$

The density is given by $\rho(x, y) = 1 + 2x + y$. Find the total mass.

**Solution**   We have

$$M(F) = \iint_F (1 + 2x + y) \, dA.$$

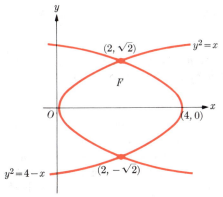

FIGURE 24

Sketching the region $F$ (Fig. 24), we obtain for $M(F)$ the iterated integral

$$M(F) = \int_{-\sqrt{2}}^{\sqrt{2}} \int_{y^2}^{4-y^2} (1 + 2x + y) \, dx \, dy$$

$$= \int_{-\sqrt{2}}^{\sqrt{2}} [x + x^2 + xy]_{y^2}^{4-y^2} \, dy$$

$$= \int_{-\sqrt{2}}^{\sqrt{2}} (20 + 4y - 10y^2 - 2y^3) \, dy$$

$$= [20y + 2y^2 - \tfrac{10}{3}y^3 - \tfrac{1}{2}y^4]_{-\sqrt{2}}^{\sqrt{2}} = \tfrac{80}{3}\sqrt{2}. \qquad \square$$

## 4　PROBLEMS

In each of Problems 1 through 8, use iterated integration to find the area of the given region $F$. Subdivide $F$ and do each part separately whenever necessary. Sketch the region $F$.

**1** $F = \{(x, y): 0 \le x \le 1, x^3 \le y \le \sqrt{x}\}$.

**2** $F = \{(x, y): \frac{1}{4}y^2 \le x \le y, 1 \le y \le 4\}$.

**3** $F$ is determined by the inequalities
$$xy \le 4, \quad y \le x, \quad 27y \ge 4x^2.$$

**4** $F$ consists of all $(x, y)$ which satisfy the inequalities
$$y^2 \le x, \quad y^2 \le 6 - x, \quad y \le x - 2.$$

**5** $F$ is the region lying between the curves $\sqrt{x} + \sqrt{y} = 4$ and $x + y = 4$.

**6** $F$ is the bounded region determined by the curves $y = x$ and $x = 4y - y^2$.

**7** $F$ is the bounded region determined by the curves $x + y = 5$, $xy = 6$.

**8** $F$ consists of all $(x, y)$ which satisfy the inequalities
$$x^2 + y^2 \le 9, \quad y \le x + 3, \quad x + y \le 0.$$

In each of Problems 9 through 22, find the mass of the given region $F$. Draw a sketch of $F$.

**9** $F = \{(x, y): x^2 + y^2 \le 64\}; \quad \rho = x^2 + y^2$.

**10** $F = \{(x, y): 0 \le x \le 1, x^2 \le y \le \sqrt{x}\}; \quad \rho = 3y$.

**11** $F = \{(x, y): -1 \le x \le 2, x^2 \le y \le x + 2\}; \quad \rho = x^2y$.

**12** $F = \{(x, y): 0 \le x \le 1, x^3 \le y \le \sqrt{x}\}; \quad \rho = 2x$.

**13** $F = \{(x, y): 1 \le x \le 4, \frac{4}{x} \le y \le 5 - x\}; \quad \rho = 4y$.

**14** $F = \{(x, y): y^2 \le x \le y + 2, -1 \le y \le 2\}; \quad \rho = x^2y^2$.

**15** $F$ is the interior of the triangle with vertices at $(0, 0)$, $(a, 0)$, $(b, c)$, $a > b > 0$, $c > 0$; $\quad \rho = 2x$.

**16** $F = \{(x, y): -a \le x \le a, 0 \le y \le \sqrt{a^2 - x^2}\}; \quad \rho = 3y$.

**17** $F$ is the interior of the rectangle with vertices at $(0, 0)$, $(a, 0)$, $(a, b)$, $(0, b)$; $\quad \rho = 3x/(1 + x^2y^2)$.

**18** $F = \{(x, y): -1 \le x \le 1, 0 \le y \le 4\}; \quad \rho = e^{x+y}$.

**19** $F = \{(x, y): 0 \le x \le 1, 0 \le y \le x\}; \quad \rho = y + \sin(\pi x^2)$.

**20** $F = \{(x, y): 0 \le y \le 1, y^2 \le x \le y\}; \rho = \sqrt{xy}$.

**21** $F = \{(x, y): 0 \le y \le 1, 0 \le x \le y^2\}; \quad \rho = ye^x$.

**22** $F$ is the interior of the triangle formed by the lines $2x = y$, $x = 1$, and the $x$-axis; $\quad \rho = e^{-x^2/2}$.

**23** Suppose the density $\rho(x, y)$ of a region $F$ satisfies the inequalities $m_1 \le \rho(x, y) \le m_2$. Show that $M(F)$, the mass of $F$, is between the limits $m_1 A \le M(F) \le m_2 A$, where $A$ is the area of $F$.

**\*24** Suppose that $\rho(x, y)$, the density, is continuous on a region $F$ which has positive area. Show that if $\iint\limits_{F} \rho(x, y)\, dA = 0$, then $\rho(x, y) \equiv 0$ on $F$.

**\*25** Suppose that the density $\rho(x, y)$ of a region $F$ in the $xy$ plane is of the form $\rho(x, y) = \rho_1(x)\rho_2(y)$. Let $M(F)$ be the total mass of $F$. Show that
$$M(F) \le \tfrac{1}{2}(M_1(F) + M_2(F))$$
where $M_1$ is the mass of $F$ with density $\rho_1^2$ and $M_2$ is the mass of $F$ with density $\rho_2^2$.

---

## 5

## EVALUATION OF DOUBLE INTEGRALS BY POLAR COORDINATES

The polar coordinates $(r, \theta)$ of a point in the plane are related to the rectangular coordinates $(x, y)$ of the same point by the equations

$$x = r \cos \theta, \qquad y = r \sin \theta, \quad r \ge 0. \tag{1}$$

We recall that certain problems in finding areas by integration are solved more easily in polar coordinates than in rectangular coordinates. (See page 526). The same situation prevails in problems involving double integration.

Instead of considering (1) as a means of representing a point in two different coordinate systems, we interpret the equations as a mapping between the $xy$ plane and the $r\theta$ plane. We draw the $r\theta$ plane as shown in Fig. 25, treating $r = 0$ and $\theta = 0$ as perpendicular straight lines. A rectangle $G$

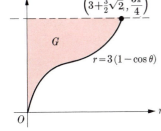

FIGURE 25

in the $r\theta$ plane bounded by the lines $r = r_1$, $r = r_2$, and $\theta = \theta_1$, $\theta = \theta_2$ ($\theta$ in radians) with $2\pi > \theta_2 > \theta_1 \geq 0$, $r_2 > r_1 > 0$ has an image $F$ in the $xy$ plane bounded by two circular arcs and two rays. For the area of $F$, denoted $A_{x,y}$, we have

$$A_{x,y}(F) = \tfrac{1}{2}(r_2^2 - r_1^2)(\theta_2 - \theta_1).$$

This area may be written as an iterated integral. A calculation shows that

$$A_{x,y}(F) = \int_{\theta_1}^{\theta_2}\left[\int_{r_1}^{r_2} r\, dr\right] d\theta.$$

Because double integrals and iterated integrals are equivalent for evaluation purposes, we can also write

$$A_{x,y}(F) = \iint_G r\, dA_{r,\theta}, \tag{2}$$

where $dA_{r,\theta}$ is an element of area in the $r\theta$ plane, $r$ and $\theta$ being treated as rectangular coordinates. That is, $dA_{r,\theta} = dr\, d\theta$.

More generally, it can be shown that if $G$ is *any region* in the $r\theta$ plane and $F$ is its image under the transformation (1), then the area of $F$ may be found by formula (2). Thus, areas of regions may be determined by expressing the double integral in polar coordinates as in (2) and then evaluating the double integral by iterated integrals in the usual way.

**EXAMPLE 1**   A region $F$ above the $x$ axis is bounded on the left by the line $y = -x$, and on the right by the curve

$$C = \{(x, y): x^2 + y^2 = 3\sqrt{x^2 + y^2} - 3x\},$$

as shown in Fig. 26. Find its area.

**Solution**   We employ polar coordinates to describe the region. The curve $C$ is the cardioid $r = 3(1 - \cos\theta)$, and the line $y = -x$ is the ray $\theta = 3\pi/4$. The region $F$ in Fig. 26 is the image under the mapping (1) of the set $G$ in the $(r, \theta)$ plane (Fig. 27), given by

$$G = \{(r, \theta): 0 \leq r \leq 3(1 - \cos\theta),\ 0 \leq \theta \leq 3\pi/4\}.$$

Therefore, for the area $A(F)$ we obtain

$$A(F) = \iint_F dA_{x,y} = \iint_G r\, dA_{r,\theta} = \int_0^{3\pi/4}\int_0^{3(1-\cos\theta)} r\, dr\, d\theta$$

$$= \int_0^{3\pi/4} \tfrac{1}{2}\left[r^2\right]_0^{3(1-\cos\theta)} d\theta = \frac{9}{2}\int_0^{3\pi/4}(1 - \cos\theta)^2\, d\theta.$$

To perform the integration we multiply out and find that

$$A(F) = \tfrac{9}{2}\int_0^{3\pi/4}(1 - 2\cos\theta + \cos^2\theta)\, d\theta$$

$$= \tfrac{9}{2}[\theta - 2\sin\theta + \tfrac{1}{2}\theta + \tfrac{1}{4}\sin 2\theta]_0^{3\pi/4}$$

$$= \tfrac{9}{8}(\tfrac{9}{2}\pi - 4\sqrt{2} - 1).$$

FIGURE 26

FIGURE 27

The transformation of regions from the $xy$ plane to the $r\theta$ plane is useful because general double integrals as well as areas may be evaluated by means of polar coordinates. The theoretical basis for the method is the Fundamental Lemma on Integration which we state without proof.

**THEOREM 8**    **(Fundamental Lemma on Integration)**    *Assume that $f$ and $g$ are continuous on some region $F$. Then for each $\varepsilon > 0$ there is a $\delta > 0$ such that*

$$\left| \sum_{i=1}^{n} f_i g_i A(F_i) - \int\int_F f(x, y) g(x, y)\, dA \right| < \varepsilon$$

*for every subdivision $F_1, F_2, \ldots, F_n$ of $F$ with norm less than $\delta$ and any numbers $f_1, f_2, \ldots, f_n, g_1, g_2, \ldots, g_n$ where each $f_i$ and each $g_i$ is between the minimum and maximum values of $f$ and $g$, respectively, on $F_i$.*

The fundamental lemma is the basis for the next theorem, the proof of which we sketch.

**THEOREM 9**    *Suppose $F$ and $G$ are regions related according to the mapping $x = r \cos \theta$, $y = r \sin \theta$, and $f(x, y)$ is continuous on $F$. Then the function $g(r, \theta) = f(r \cos \theta, r \sin \theta)$ is defined and continuous on $G$ and*

$$\int\int_F f(x, y)\, dA_{x,y} = \int\int_G g(r, \theta) r\, dA_{r,\theta}.$$

*Alternatively, we may write*

$$\int\int_F f(x, y)\, dx\, dy = \int\int_G f(r \cos \theta, r \sin \theta) r\, dr\, d\theta \tag{3}$$

The form (3) is the one most often used in actual computations of integrals in polar coordinates.

**Sketch of proof**    Consider a subdivision of $G$ into "figures" $G_1, \ldots, G_n$. (See the discussion of volume in Section 1.) Let $(r_i, \theta_i)$ be any point in $G_i$, and let $(\xi_i, \eta_i)$ and $F_i$ be the respective images of $(r_i, \theta_i)$ and $G_i$. Then $(F_1, \ldots, F_n)$ is a subdivision of $F$. From the expression for area in the $xy$ plane as an integral, we obtain

$$A_{x,y}(F_i) = \int\int_{G_i} r\, dA_{r,\theta}.$$

Using Theorem 4 concerning bounds for integrals, we find

$$\int\int_{G_i} r\, dA_{r,\theta} = \bar{r}_i A_{r,\theta}(G_i),$$

where $\bar{r}_i$ is between the minimum and maximum of $r$ on $G_i$. Thus

$$\sum_{i=1}^{n} f(\xi_i, \eta_i) A_{x,y}(F_i) = \sum_{i=1}^{n} g(r_i, \theta_i) \cdot \bar{r}_i \cdot A_{r,\theta}(G_i).$$

The theorem follows by letting the norms of the subdivisions tend to zero, using the Fundamental Lemma to evaluate the limit of the sum on the right.

□

In terms of iterated integrals, the result in Theorem 9 yields the useful formula

$$\iint_F f(x, y) \, dA = \iint_G g(r, \theta) \, r \, dr \, d\theta,$$

which is equivalent to (3). Of course, the integral on the right is equal to the iterated integral in the reverse order. In any specific case the order of integration will usually depend on the determination of the limits of integration. One way may be easier than the other. We give an example

EXAMPLE 2    Use polar coordinates to evaluate

$$\iint_F \sqrt{x^2 + y^2} \, dA_{x,y},$$

where $F$ is the region inside the circle $x^2 + y^2 = 2x$.

Solution    The region $F$ (see Fig. 28) is the image of the set $G$ (Fig. 29) given by

$$G = \left\{ (r, \theta): -\frac{\pi}{2} \le \theta \le \frac{\pi}{2}, \, 0 \le r \le 2 \cos \theta \right\}.$$

Therefore, using Formula (3) we obtain

$$\iint_F \sqrt{x^2 + y^2} \, dA_{x,y} = \iint_G r \cdot r \, dA_{r,\theta} = \int_{-\pi/2}^{\pi/2} \int_0^{2 \cos \theta} r^2 \, dr \, d\theta$$

$$= \int_{-\pi/2}^{\pi/2} \tfrac{8}{3} \cos^3 \theta \, d\theta$$

$$= \tfrac{16}{3} \int_0^{\pi/2} (1 - \sin^2 \theta) \cos \theta \, d\theta = \tfrac{32}{9}.$$

□

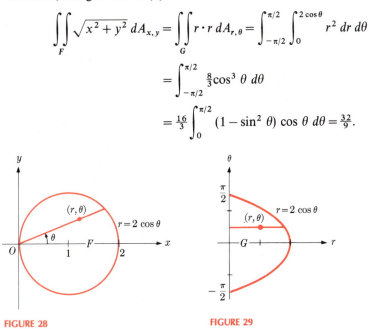

FIGURE 28                                      FIGURE 29

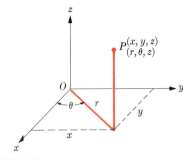

FIGURE 30

Although the construction of the region $G$ in the $r\theta$ plane is helpful in understanding the transformation (1), *it is not necessary for determining the limits in the iterated integrals.* The limits of integration in polar coordinates may be found by using rectangular and polar coordinates in the same plane and using a sketch of the region $F$ to read off the limits for $r$ and $\theta$.

Double integrals are useful for finding volumes bounded by surfaces. Cylindrical coordinates $(r, \theta, z)$ are a natural extension to three-space of polar coordinates in the plane. The $z$ direction is selected as in rectangular coordinates, as shown in Fig. 30. If a closed surface in space is expressed in cylindrical coordinates, we may find the volume enclosed by this surface by evaluating a double integral in polar coordinates. Here we have $z = f(r, \theta)$ giving the "height" of the object above the point $(r, \theta)$, as shown in Fig. 31. The next example illustrates how we can use cylindrical coordinates to find the volume of an object.

**EXAMPLE 3**   A region $S$ is bounded by the surfaces $x^2 + y^2 - 2x = 0$, $4z = x^2 + y^2$, $z^2 = x^2 + y^2$. Use cylindrical coordinates to find the volume $V(S)$.

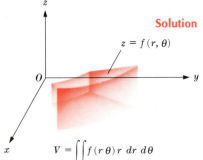

**Solution**   In cylindrical coordinates, the paraboloid $4z = x^2 + y^2$ has equation $4z = r^2$; the cylinder $x^2 + y^2 - 2x = 0$ has equation $r = 2\cos\theta$; and the cone $z^2 = x^2 + y^2$ has equation $z^2 = r^2$. The region is shown in Fig. 32, and we note that the projection of $S$ on the $xy$ plane is precisely the plane region $F$ of Example 2. (See Fig. 28.) We obtain

$$V(S) = \iint_F \left[ \sqrt{x^2 + y^2} - \frac{x^2 + y^2}{4} \right] dA_{x,y}$$

$$= \iint_G \left( r - \frac{1}{4}r^2 \right) r\, dA_{r,\theta}$$

$$= \int_{-\pi/2}^{\pi/2} \int_0^{2\cos\theta} \left( r^2 - \frac{1}{4}r^3 \right) dr\, d\theta$$

$$= \int_{-\pi/2}^{\pi/2} \left( \frac{8}{3}\cos^3\theta - \cos^4\theta \right) d\theta$$

$$= \frac{32}{9} - \frac{1}{2}\int_0^{\pi/2} \left( 1 + 2\cos 2\theta + \frac{1 + \cos 4\theta}{2} \right) d\theta = \frac{32}{9} - \frac{3\pi}{8}. \quad \square$$

$V = \iint_F f(r\,\theta) r\, dr\, d\theta$

FIGURE 31

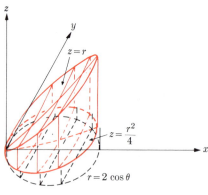

FIGURE 32

Polar coordinates are also useful for calculating definite integrals of certain functions of one variable that might otherwise seem impossible. One of the most important examples is the integral $\int_{-\infty}^{\infty} e^{-x^2/2}\, dx$, which is of fundamental importance to probability and statistics. (See Section 10 of Chapter 9.)

**EXAMPLE 4**   Use polar coordinates to evaluate the integral $\int_{-\infty}^{\infty} e^{-x^2/2}\, dx$.

**Solution**   We begin by observing that since the function $f(x) = e^{-x^2/2}$ is symmetric with respect to the $y$ axis, we have

$$\int_{-\infty}^{\infty} e^{-x^2/2}\, dx = 2\int_0^{\infty} e^{-x^2/2}\, dx \text{ (see Fig. 33.)}$$

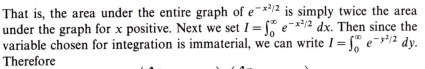
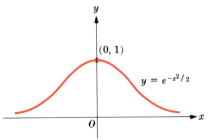

The graph of $e^{-x^2/2}$ is symmetric about the $y$ axis

**FIGURE 33**

That is, the area under the entire graph of $e^{-x^2/2}$ is simply twice the area under the graph for $x$ positive. Next we set $I = \int_0^\infty e^{-x^2/2}\, dx$. Then since the variable chosen for integration is immaterial, we can write $I = \int_0^\infty e^{-y^2/2}\, dy$. Therefore

$$I^2 = \left( \int_0^\infty e^{-x^2/2}\, dx \right)\left( \int_0^\infty e^{-y^2/2}\, dy \right) \qquad (4)$$

and it can be shown that (4) is the same as the double integral

$$I^2 = \int_0^\infty \int_0^\infty e^{-x^2/2}\, e^{-y^2/2}\, dx\, dy$$

$$= \int_0^\infty \int_0^\infty e^{-(x^2+y^2)/2}\, dx\, dy$$

$$= \iint_F e^{-(x^2+y^2)/2}\, dA_{x,y}$$

where the region $F$ is the entire first quadrant in the $xy$ plane. Changing to polar coordinates we see that $e^{-(x^2+y^2)/2}$ becomes $e^{-r^2/2}$ and $dA_{x,y}$ becomes $r\, dr\, d\theta$. The region $F$ transforms into the region

$$G = \{(r, \theta): 0 < r < \infty, 0 < \theta < \pi/2\}.$$

Therefore

$$I^2 = \iint_G e^{-r^2/2}\, dA_{r,\theta}$$

$$= \int_0^{\pi/2} \int_0^\infty e^{-r^2/2}\, r\, dr\, d\theta$$

$$= \int_0^{\pi/2} [e^{-r^2/2}]_0^\infty\, d\theta$$

$$= \int_0^{\pi/2} 1\, d\theta = \frac{\pi}{2}.$$

We take the square root, and since $I$ must be positive, we conclude that $I = \sqrt{\pi/2}$. Therefore

$$\int_{-\infty}^\infty e^{-x^2/2}\, dx = 2I = \sqrt{2\pi}.$$ □

## 5   PROBLEMS

In each of Problems 1 through 12, evaluate the given integral by using polar coordinates.

**1** $\displaystyle\int_0^2 \int_0^{\sqrt{4-y^2}} \sqrt{x^2+y^2}\, dx\, dy$

**2** $\displaystyle\int_{-2}^2 \int_{-\sqrt{4-x^2}}^{\sqrt{4-x^2}} e^{-(x^2+y^2)}\, dy\, dx$

**3** $\displaystyle\int_{-\sqrt{\pi}}^{\sqrt{\pi}} \int_{-\sqrt{\pi-y^2}}^{\sqrt{\pi-y^2}} \sin(x^2+y^2)\, dx\, dy$

**4** $\displaystyle\int_0^4 \int_{-\sqrt{4x-x^2}}^{\sqrt{4x-x^2}} \sqrt{x^2+y^2}\, dy\, dx$

**5** $\displaystyle\int_{-2}^2 \int_{2-\sqrt{4-x^2}}^{2+\sqrt{4-x^2}} \sqrt{16-x^2-y^2}\, dy\, dx$

**6** $\displaystyle\int_0^2 \int_0^x (x^2+y^2)\, dy\, dx$

**7** $\displaystyle\int_0^1 \int_y^{\sqrt{y}} (x^2+y^2)^{-1/2}\, dx\, dy$

**8** $\displaystyle\int_0^1 \int_0^{\sqrt{1-x^2}} e^{x^2+y^2} \, dy \, dx$

**9** $\displaystyle\int_0^2 \int_0^{\sqrt{4-x^2}} x^2\sqrt{x^2+y^2} \, dy \, dx$

**10** $\displaystyle\int_0^2 \int_0^x \frac{x^2}{\sqrt{x^2+y^2}} \, dy \, dx$

**11** $\displaystyle\int_0^1 \int_x^1 x \sin(y^3) \, dy \, dx$

**12** $\displaystyle\int_0^1 \int_0^x (1+x^2+y^2)^{-3/2} \, dy \, dx$

***13** Integrate $g(x, y) = \sin(\sqrt{x^2+y^2})$ over the unit disk: $\{(x, y): 0 \le x^2+y^2 \le 1\}$.

In each of Problems 14 through 20, use polar coordinates to find the area of the region given.

**14** The region inside the circle $x^2+y^2-8y=0$ and outside the circle $x^2+y^2=9$.

**15** The region $F = \{(x, y): \frac{1}{4}y^2 \le x \le 2y, \, 0 \le y \le 8\}$.

**16** The region interior to the curve $(x^2+y^2)^3 = 16x^2$.

**17** The region inside the circle $r = 4\cos\theta$ but outside the circle $r = 2$.

**18** The region inside the circle $r = 3\cos\theta$ but outside the circle $r = \cos\theta$.

**19** The region inside the cardioid $r = 1+\cos\theta$ but outside the circle $r = 1$.

**20** The region inside the circle $r = 1$ but outside the parabola $r(1+\cos\theta) = 1$.

In each of Problems 21 through 36, find the volume of $S$.

**21** $S$ is the set bounded by the surfaces $z = 0$, $2z = x^2+y^2$, and $x^2+y^2 = 4$.

**22** $S$ is the set bounded by the cone $z^2 = x^2+y^2$ and the cylinder $x^2+y^2 = 4$.

**23** $S$ is the set cut from a sphere of radius 4 by a cylinder of radius 2 whose axis is a diameter of the sphere.

**24** $S$ is the set above the cone $z^2 = x^2+y^2$ and inside the sphere

$$x^2+y^2+z^2 = a^2.$$

**25** $S$ is the set bounded by the cone $z^2 = x^2+y^2$ and the cylinder

$$x^2+y^2-2y = 0.$$

**26** $S$ is the set bounded by the sphere $x^2+y^2+z^2 = 4$ and the cylinder

$$x^2+y^2 = 2x.$$

**27** $S$ is the set bounded by the cone $z^2 = x^2+y^2$ and the

paraboloid

$$3z = x^2+y^2.$$

**28** $S$ is bounded by the surfaces $z = 0$, $2z = x^2+y^2$, and $2y = x^2+y^2$.

**29** $S$ is bounded by the cylinder $x^2+y^2 = 4$ and the hyperboloid

$$x^2+y^2-z^2 = 1.$$

**30** $S$ is bounded by the cone $z^2 = x^2+y^2$ and the cylinder $r = 1+\cos\theta$.

**31** $S$ is bounded by the surfaces $z = x$ and $2z = x^2+y^2$.

**32** $S$ is bounded by the surfaces $z = 0$, $z = x^2+y^2$, and $r = 2(1+\cos\theta)$.

**33** $S$ is inside the sphere $x^2+y^2+z^2 = a^2$ and inside the cylinder erected on one loop of the curve $z = 0$, $r = a\cos 2\theta$.

**34** $S$ is inside the sphere $x^2+y^2+z^2 = 4$ and inside the cylinder erected on one loop of the curve $z = 0$, $r^2 = 4\cos 2\theta$.

**35** $S$ is bounded by the surfaces $z^2 = x^2+y^2$, $y = 0$, $y = x$, and $x = a$.

***36** $S$ is bounded by the surfaces $z^2 = x^2+y^2$ and $x-2z+2 = 0$.

**37** A wedge is cut from a spherical ball of radius $C$ by two planes which intersect on a diameter of the ball. If the angle between the planes is $\pi/3$, find the volume of the wedge. What is the volume if the angle is $\phi$?

**38** A torus is generated by revolving a circular disk of radius $a$ about an axis outside the disk. If the distance of the axis from the center of the disk is $b$, use polar coordinates to find the volume of the torus.

**39** a) Suppose that $g(r, \theta) = g(r, -\theta)$ for all $r, \theta$. Let $F$ be a region which is symmetric with respect to the $x$ axis. Let $F_0$ be the portion of $F$ above the $x$ axis. Show that

$$\iint_F g(r, \theta) r \, dr \, d\theta = 2 \iint_{F_0} g(r, \theta) r \, dr \, d\theta.$$

b) If $g(r, \theta) = -g(r, -\theta)$, show that

$$\iint_F g(r, \theta) r \, dr \, d\theta = 0.$$

In each of Problems 40 through 42, use polar coordinates to evaluate the given integral.

**40** $\displaystyle\int_0^\infty e^{-4x^2} \, dx$

**41** $\displaystyle\int_0^\infty \sqrt{x}\, e^{-3x} \, dx$ ( *Hint:* Let $x = u^2$).

**42** $\displaystyle\int_0^1 \left\{ \ln\left(\frac{1}{x}\right) \right\}^{1/2} dx.$ ( *Hint:* Let $x = e^{-u}$).

_____ **6**

## MOMENT OF INERTIA AND CENTER OF MASS (OPTIONAL)

Consider the idealized situation in which an object of mass $m$ occupies a single point. Let $L$ be a line which we designate as an axis.

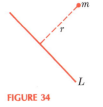

FIGURE 34

**DEFINITION**    *The* **moment of inertia of a particle of mass** $m$ **about the axis** $L$ *is* $mr^2$, *where* $r$ *is the perpendicular distance of the object from the axis (Fig. 34). If a system of particles* $m_1, m_2, \ldots, m_n$ *is at perpendicular distances, respectively, of* $r_1, r_2, \ldots, r_n$ *from the axis* $L$, *then the* **moment of inertia of the system,** $I$, *is given by*

$$I = m_1 r_1^2 + m_2 r_2^2 + \cdots + m_n r_n^2 = \sum_{i=1}^{n} m_i r_i^2.$$

Let $F$ be an object made of thin material occupying a region in the $xy$ plane (Fig. 35). We wish to define the moment of inertia of $F$ about an axis $L$. The axis $L$ may be any line in three-dimensional space. We proceed as in the definition of integration. First we make a subdivision of the plane into rectangles or squares. We designate the rectangles either wholly or partly in $F$ by $F_1, F_2, \ldots, F_n$. Since the object $F$ may be of irregular shape and of variable density, the mass of the subregions may not be calculable exactly. We select a point in each subregion $F_i$ and denote its coordinates $(\xi_i, \eta_i)$. We assume that the entire mass of $F_i$, denoted $m(F_i)$, is concentrated at the point $(\xi_i, \eta_i)$. Letting $r_i$ be the perpendicular distance of $(\xi_i, \eta_i)$ from the line $L$, we form the sum

$$\sum_{i=1}^{n} m(F_i) r_i^2 \,.$$

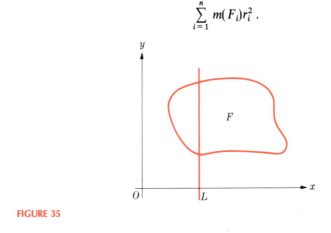

FIGURE 35

**DEFINITION**    *If the above sums tend to a limit (called* $I$) *as the norms of the subdivisions tend to zero, and if this limit is independent of the manner in which the* $(\xi_i, \eta_i)$ *are selected within the* $F_i$, *then we say that* $I$ *is the* **moment of inertia of the mass distribution about the axis** $L$.

The above definition of moment of inertia leads in a natural way to the next theorem.

**THEOREM 10**  *Given a mass distribution occupying a region F in the xy plane and having a continuous density $\rho(x, y)$. Then the moment of inertia about the y axis (denoted by $I_1$) is given by*

$$I_1 = \iint\limits_F x^2 \rho(x, y) \, dA.$$

*Similarly, the moments of inertia about the x axis and the z axis are, respectively,*

$$I_2 = \iint\limits_F y^2 \rho(x, y) \, dA, \quad I_3 = \iint\limits_F (x^2 + y^2)\rho(x, y) \, dA.$$

The proof depends on the Fundamental Lemma on Integration. (See Problem 29 at the end of this section.)

**COROLLARY**  *The moments of inertia of F about the lines*

$$L_1 = \{(x, y, z): x = a, z = 0\}, \qquad L_2 = \{(x, y, z): y = b, z = 0\},$$

$$L_3 = \{(x, y, z): x = a, y = b\}$$

*are, respectively,*

$$I_1^a = \iint\limits_F (x - a)^2 \rho(x, y) \, dA,$$

$$I_2^b = \iint\limits_F (y - b)^2 \rho(x, y) \, dA,$$

$$I_3^{a,b} = \iint\limits_F [(x - a)^2 + (y - b)^2]\rho(x, y) \, dA.$$

**EXAMPLE 1**  Find the moment of inertia about the x axis of the homogeneous plate (i.e., $\rho = \text{constant}$) bounded by the line $y = 0$ and the curve $y = 4 - x^2$ (Fig. 36).

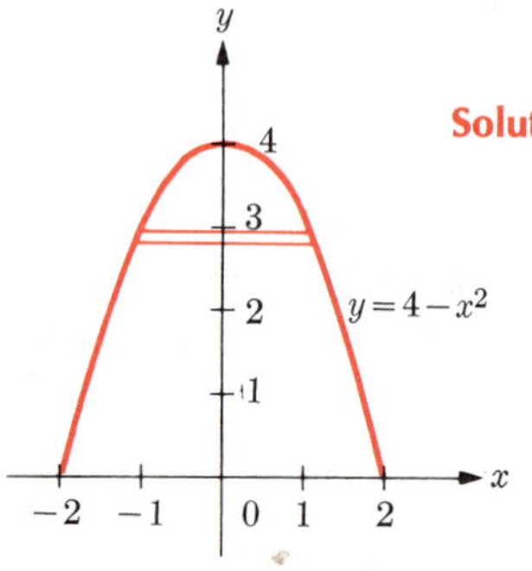

**Solution**  According to Theorem 10, we have

$$I_2 = \iint\limits_F y^2 \rho \, dA = \rho \iint\limits_F y^2 \, dy \, dx$$

$$= \rho \int_{-2}^{2} \int_{0}^{4-x^2} y^2 \, dy \, dx = \frac{\rho}{3} \int_{-2}^{2} (4 - x^2)^3 \, dx$$

$$= \frac{\rho}{3} \int_{-2}^{2} (64 - 48x^2 + 12x^4 - x^6) \, dx = \frac{4096\rho}{105}.$$

**FIGURE 36**

**EXAMPLE 2**    Find the moment of inertia about the $z$ axis of the homogeneous triangular plate bounded by the lines $y = 0$, $y = x$, and $x = 4$.

**Solution 1**   We have

$$I_3 = \iint_F (x^2 + y^2)\rho \, dA = \rho \int_0^4 \int_0^x (x^2 - y^2) \, dy \, dx$$

$$= \rho \int_0^4 \left[ x^2 y + \frac{1}{3} y^3 \right]_0^x dx = \frac{4\rho}{3} \int_0^4 x^3 \, dx = \frac{256\rho}{3}. \qquad \square$$

**Solution 2**   We may introduce polar coordinates as shown in Fig. 37. Then

$$I_3 = \iint_F (x^2 + y^2)\rho \, dA_{x,y} = \rho \iint_G r^2 \cdot r \, dr \, d\theta = \rho \int_0^{\pi/4} \int_0^{4\sec\theta} r^3 \, dr \, d\theta$$

$$= 64\rho \int_0^{\pi/4} \sec^4\theta \, d\theta = 64\rho \left[ \tan\theta + \frac{1}{3}\tan^3\theta \right]_0^{\pi/4} = \frac{256\rho}{3}. \qquad \square$$

The moment of inertia about the $z$ axis of two-dimensional objects in the $xy$ plane is called the **polar moment of inertia**. Since the combination $x^2 + y^2 = r^2$ is always present in calculating polar moments, a change to polar coordinates is frequently advantageous.

A quantity known as the radius of gyration is intimately connected with the notion of moment of inertia. It is defined in terms of the total mass $m$ of an object and the moment of inertia $I$ about a specific axis.

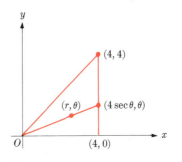

**FIGURE 37**

**DEFINITION**   *The **radius of gyration of an object about an axis** $L$ is the number $R$ such that*

$$R^2 = \frac{I}{m},$$

*where $I$ is the moment of inertia about $L$, and $m$ is the total mass of the object.*

If we imagine the total mass of a body as concentrated at one point which is at distance $R$ from the axis $L$, then the moment of inertia of this idealized "point mass" will be the same as the moment of inertia of the original body.

**EXAMPLE 3**    Find the radius of gyration for the problem in Example 1. Do the same for Example 2.

**Solution**  The mass $m$ of the homogeneous plate in Example 1 is its area multiplied by $\rho$. We have

$$m = \rho \int_{-2}^{2} \int_{0}^{4-x^2} dy\, dx = \rho \int_{-2}^{2} (4 - x^2)\, dx = \rho \left[ 4x - \frac{x^3}{3} \right]_{-2}^{2} = \frac{32\rho}{3}.$$

Therefore

$$R = \sqrt{\frac{I}{m}} = \left( \frac{4096\rho}{105} \cdot \frac{3}{32\rho} \right)^{1/2} = 8\sqrt{\frac{2}{35}}.$$

The mass of the triangular plate in Example 2 is $8\rho$, and so

$$R = \sqrt{\frac{I}{m}} = \left( \frac{256\rho}{3} \cdot \frac{1}{8\rho} \right)^{1/2} = 4\sqrt{\frac{2}{3}}. \qquad \Box$$

The center of mass of an object was defined on page 329 ff. To calculate the center of mass we make use of the *moment of a mass m* with respect to one of the coordinate axes. We recall that if particles of masses $m_1, m_2, \ldots, m_n$ are situated at the points $(x_1, y_1), (x_2, y_2), \ldots, (x_n, y_n)$ respectively, then the **algebraic moment** (sometimes called **first moment** or simply **moment**) of this system about the $y$ axis is

$$m_1 x_1 + m_2 x_2 + \cdots + m_n x_n \equiv \sum_{i=1}^{n} m_i x_i.$$

Its algebraic moment about the $x$ axis is

$$\sum_{i=1}^{n} m_i y_i.$$

More generally, the algebraic moments about the line $x = a$ and about the line $y = b$ are, respectively,

$$\sum_{i=1}^{n} m_i(x_i - a) \qquad \text{and} \qquad \sum_{i=1}^{n} m_i(y_i - b).$$

We now define the moment of a thin object occupying a region $F$ in the $xy$ plane.

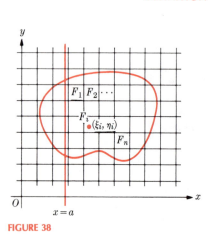

FIGURE 38

**DEFINITION**  *Assume that a thin mass occupies a region $F$ in the $xy$ plane. Let $F_1, F_2, \ldots, F_n$ be a subdivision of $F$ as shown in Fig. 38. Choose a point $(\xi_i, \eta_i)$ in each $F_i$ and replace the mass in $F_i$ by a particle of mass $m(F_i)$ located at $(\xi_i, \eta_i)$. The $n$ idealized masses have moment about the line $x = a$ of*

$$\sum_{i=1}^{n} (\xi_i - a)m(F_i).$$

*If the above sums tend to a limit $M_1$ as the norms of the subdivisions tend to zero and for any choices of the points $(\xi_i, \eta_i)$ in $F_i$, then the limit $M_1$ is defined as the* **moment of the mass distribution about the line** $x = a$. *An analogous definition for the limit $M_2$ of sums of the form*

$$\sum_{i=1}^{n} (\eta_i - b)m(F_i)$$

*yields the first moment about the line $y = b$.*

The definition of first moment and the Fundamental Lemma on Integration yield the next theorem.

**THEOREM 11**    *If a distribution of mass over a region $F$ in the $xy$ plane has a continuous density $\rho(x, y)$, then the moments $M_1$ and $M_2$ of $F$ about the lines $x = a$ and $y = b$ are given by the formulas*

$$M_1 = \iint_F (x - a)\rho(x, y)\, dA, \qquad M_2 = \iint_F (y - b)\rho(x, y)\, dA.$$

**COROLLARY**    *Given a distribution of mass over a region $F$ in the $xy$ plane as in Theorem 11, then there are unique values of $a$ and $b$ (denoted $\bar{x}$ and $\bar{y}$, respectively) such that $M_1 = M_2 = 0$. In fact, the values of $\bar{x}$ and $\bar{y}$ are given by*

$$\bar{x} = \frac{\displaystyle\iint_F x\rho(x, y)\, dA}{m(F)}, \qquad \bar{y} = \frac{\displaystyle\iint_F y\rho(x, y)\, dA}{m(F)},$$

*where*

$$m(F) = \iint_F \rho(x, y)\, dA.$$

**Proof**    If we set $M_1 = 0$, we get

$$0 = \iint_F (x - a)\rho(x, y)\, dA = \iint_F x\rho(x, y)\, dA - a\iint_F \rho(x, y)\, dA.$$

Since $m(F) = \iint_F \rho(x, y)\, dA$, we find for the value of $a$:

$$a = \frac{\displaystyle\iint_F x\rho(x, y)\, dA}{m(F)} = \bar{x}.$$

The value $\bar{y}$ is found similarly.    □

**DEFINITION**    *The point $(\bar{x}, \bar{y})$ is called the **center of mass** of the distribution over $F$.*

**EXAMPLE 4**    Find the center of mass of the region

$$F = \{(x, y): 0 \le x \le 1,\ x^3 \le y \le \sqrt{x}\}$$

if the density of $F$ is given by $\rho = 3x$.

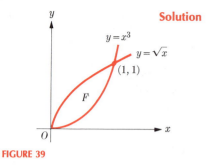

FIGURE 39

**Solution** (See Fig. 39.) For the first moments, we have

$$M_1 = \iint_F x\rho \, dA = 3 \int_0^1 \int_{x^3}^{\sqrt{x}} x^2 \, dy \, dx$$

$$= 3 \int_0^1 x^2 \Big[ y \Big]_{x^3}^{\sqrt{x}} \, dx = 3 \int_0^1 (x^{5/2} - x^5) \, dx = \tfrac{5}{14},$$

$$M_2 = \iint_F y\rho \, dA = 3 \int_0^1 \int_{x^3}^{\sqrt{x}} yx \, dy \, dx = \tfrac{3}{2} \int_0^1 x \Big[ y^2 \Big]_{x^3}^{\sqrt{x}} \, dx$$

$$= \tfrac{3}{2} \int_0^1 (x^2 - x^7) \, dx = \tfrac{5}{16}.$$

$$m(F) = \iint_F \rho \, dA = 3 \int_0^1 \int_{x^3}^{\sqrt{x}} x \, dy \, dx = 3 \int_0^1 x \Big[ y \Big]_{x^3}^{\sqrt{x}} \, dx$$

$$= 3 \int_0^1 (x^{3/2} - x^4) \, dx = \tfrac{3}{5}.$$

Therefore

$$\bar{x} = \frac{M_1}{m(F)} = \frac{5}{14} \cdot \frac{5}{3} = \frac{25}{42}; \qquad \bar{y} = \frac{M_2}{m(F)} = \frac{5}{16} \cdot \frac{5}{3} = \frac{25}{48}. \qquad \square$$

**EXAMPLE 5** Find the center of mass of a plate in the form of a circular sector of radius $a$ and central angle $2\alpha$ if its thickness is proportional to its distance from the center of the circle from which the sector is taken.

**Solution** Select the sector so that the vertex is at the origin and the $x$ axis bisects the region (Fig. 40). Then the density is given by $\rho = kr$, where $k$ is a proportionality constant. By symmetry we have $\bar{y} = 0$. Using polar coordinates, we obtain

$$M_1 = \iint_F xkr \, dA_{x,y} = k \int_{-\alpha}^{\alpha} \int_0^a r^2 \cos\theta \, r \, dr \, d\theta$$

$$= \frac{ka^4}{4} \int_{-\alpha}^{\alpha} \cos\theta \, d\theta = \frac{1}{2} ka^4 \sin\alpha,$$

$$m(F) = k \iint_F r \, dA_{x,y} = k \int_{-\alpha}^{\alpha} \int_0^a r^2 \, dr \, d\theta = \tfrac{2}{3} ka^3 \alpha.$$

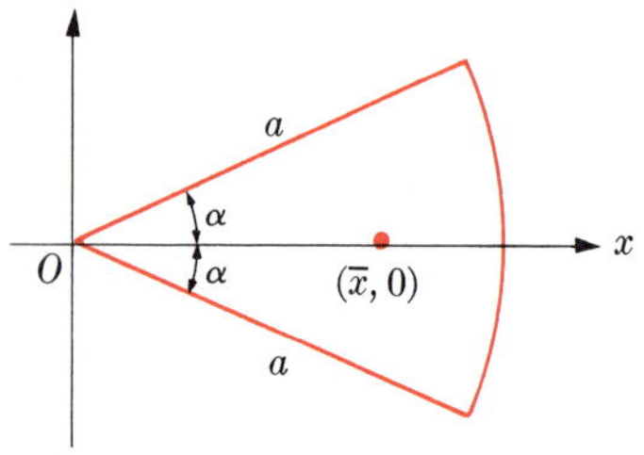

FIGURE 40

Therefore

$$\bar{x} = \frac{M_1}{m(F)} = \frac{3a \sin \alpha}{4\alpha}.$$

□

## 6 PROBLEMS

In each of Problems 1 through 7, find the moment of inertia and the radius of gyration about the given axis of the plate $F$ whose density is given

**1**  $F$ is the square with vertices $(0, 0)$, $(a, 0)$, $(a, a)$, $(0, a)$, $\rho = $ constant;  $y$ axis.

**2**  $F$ is the triangle with vertices $(0, 0)$, $(a, 0)$, $(b, c)$, with $a > 0$, $c > 0$, $\rho = $ constant;  $x$ axis.

**3**  $F = \{(x, y): 0 \leq x \leq 1, x^2 \leq y \leq \sqrt{x}\}$, $\rho = $ constant;  $y$ axis.

**4**  $F = \{(x, y): 1 \leq x \leq 4, \frac{4}{x} \leq y \leq 5 - x\}$, $\rho = ky$;  $x$ axis.

**5**  $F = \{(x, y):  -a \leq x \leq a,  0 \leq y \leq \sqrt{a^2 - x^2}\}$,  $\rho = ky$;  $x$ axis.

**6**  $F = \{(x, y):  1 \leq x \leq 4,  \frac{4}{x} \leq y \leq 5 - x\}$,  $\rho = $ constant;  $x$ axis.

**7**  $F = \{(x, y): 0 \leq x \leq \pi, 0 \leq y \leq \sin x\}$, $\rho = $ constant;  $y$ axis.

In each of Problems 8 through 17, find the moment of inertia about the given axis of the plate $F$ whose density is given.

**8**  $F = \{(x, y): x^2 + y^2 \leq a^2\}$, $\rho = k\sqrt{x^2 + y^2}$;  $z$ axis.

**9**  $F = \{(x, y):  -1 \leq x \leq 2,  x^2 \leq y \leq x + 2\}$,  $\rho = $ constant;  $x$ axis.

**10**  $F$ is the interior of the square with vertices $(0, 0)$, $(a, 0)$, $(a, a)$, $(0, a)$, $\rho = $ constant;  $z$ axis.

**11**  $F = \{(x, y): 0 \leq x \leq 1, x^2 \leq y \leq \sqrt{x}\}$, $\rho = ky$;  $y$ axis.

**12**  $F = \{(x, y):  -1 \leq x \leq 2,  x^2 \leq y \leq x + 2\}$,  $\rho = $ constant; axis is line $y = 4$.

**13**  $F = \{(x, y): x^2 + y^2 \leq a^2\}$, $\rho = k\sqrt{x^2 + y^2}$;  $x$ axis.

**14**  $F$ is bounded by the closed curve $r = 2a \cos \theta$, $\rho = kr$;  $z$ axis.

**15**  $F$ is bounded by one loop of $r^2 = a^2 \cos 2\theta$, $\rho = $ constant;  $z$ axis.

**16**  $F$ is bounded by one loop of $r^2 = a^2 \cos 2\theta$, $\rho = $ constant;  $x$ axis.

**17**  $F$ is the region in the first quadrant inside the circle $r = 1$, and bounded by $r = 1$, $\theta = r$, and $\theta = \pi/2$;  $\rho = $ constant;  $z$ axis.

In each of Problems 18 through 28, find the center of mass of the plate $F$ described.

**18**  $F = \{(x, y): 1 \leq x \leq 4, \frac{4}{x} \leq y \leq 5 - x\}$, $\rho = ky$.

**19**  $F = \{(x, y): y^2 \leq x \leq y + 2, -1 \leq y \leq 2\}$, $\rho = kx$.

**20**  $F$ is the interior of the triangle with vertices at $(0, 0)$, $(a, 0)$, $(b, c)$, with $0 < b < a$, $0 < c$, $\rho = kx$.

**21**  $F = \{(x, y): 0 \leq x \leq 1, x^2 \leq y \leq \sqrt{x}\}$, $\rho = ky$.

**22**  $F = \{(x, y): -1 \leq x \leq 2, x^2 \leq y \leq x + 2\}$, $\rho = $ constant.

**23**  $F$ is the square with vertices at $(0, 0)$, $(a, 0)$, $(a, a)$, $(0, a)$, $\rho = k(x^2 + y^2)$.

**24**  $F$ is the triangle with vertices at $(0, 0)$, $(1, 0)$, $(1, 1)$, $\rho = kr^2$.

**25**  $F$ is bounded by the cardioid $r = 2(1 + \cos \theta)$, $\rho = $ constant.

**26**  $F$ is bounded by one loop of the curve $r = 2 \cos 2\theta$, $\rho = $ constant.

**27**  $F$ is bounded by $3x^2 + 4y^2 = 48$ and $(x - 2)^2 + y^2 = 1$, $\rho = $ constant.

**28**  $F$ is bounded by one loop of the curve $r^2 = a^2 \cos 2\theta$, $\rho = $ constant.

**29**  The **Mean Value Theorem** for double integrals states that *if $f$ is integrable over a region $F$ of area $A(F)$ and if $m \leq f(x, y) \leq M$ for all $(x, y)$ on $F$, then there is a number $\bar{f}$ between $m$ and $M$ such that*

$$\iint\limits_{F} f(x, y)\, dA = \bar{f} A(F).$$

Use the Fundamental Lemma on Integration and the Mean Value Theorem to establish Theorem 10. Use the idea of the proof of Theorem 9.

**30**  Show that if a mass distribution $F$ lies between the lines $x = a$ and $x = b$ and has center of mass $(\bar{x}, \bar{y})$, then $a \leq \bar{x} \leq b$. Similarly, if $F$ lies between the lines $y = c$ and $y = d$, then $c \leq \bar{y} \leq d$.

**31** Let $F_1, F_2, \ldots, F_n$ be regions no two of which have any points in common, and let $(\bar{x}_1, \bar{y}_1), (\bar{x}_2, \bar{y}_2), \ldots, (\bar{x}_n, \bar{y}_n)$ be their respective centers of mass. Denote the mass of $F_i$, by $m_i$. If $F$ is the region consisting of all the points in every $F_i$, show that the center of mass $(\bar{x}, \bar{y})$ of $F$ is given by

$$\bar{x} = \frac{m_1 \bar{x}_1 + m_2 \bar{x}_2 + \cdots + m_n \bar{x}_n}{m_1 + m_2 + \cdots + m_n},$$

$$\bar{y} = \frac{m_1 \bar{y}_1 + m_2 \bar{y}_2 + \cdots + m_n \bar{y}_n}{m_1 + m_2 + \cdots + m_n}.$$

**32** Show that if $F$ is symmetric with respect to the $x$ axis and $\rho(x, -y) = \rho(x, y)$ for all $(x, y)$ on $F$, then $\bar{y} = 0$. A similar result holds for symmetry with respect to the $y$ axis.

**33** Find the moment of inertia about the $z$ axis of a ring of uniform density in the $xy$ plane bounded by the circles $x^2 + y^2 = r_1^2$ and $x^2 + y^2 = r_2^2$ with $r_1 < r_2$. Find the radius of the solid disk with the same density having the same moment of inertia about the $z$ axis.

**34** Let $F$ be a region in the $xy$ plane with mass $m(F)$. Show that in the notation of this section

$$I_1^a = I_1 - 2aM_1 + a^2 m(F)$$

where $M_1$ is the first moment of $F$ about the $y$ axis. Also, show that

$$I_3^{(a,b)} = I_3 - 2aM_1 - 2bM_2 + (a^2 + b^2)m(F)$$

where $M_2$ is the first moment of $F$ with respect to the $x$ axis.

---

## 7

## SURFACE AREA

To define surface area we employ a procedure similar to that used for defining area in the plane. First, we define surface area in the simplest case, and second, we employ a limiting process for the definition of surface area of a general curved surface.

Suppose two planes $\Gamma_1$ and $\Gamma_2$ intersect at an angle $\phi$ (Fig. 41). From each point of $G_1$, a region in the plane $\Gamma_1$, we drop a perpendicular to the plane $\Gamma_2$. The set of points of intersection of these perpendiculars with $\Gamma_2$ forms a region which we denote $G_2$. The set $G_2$ is called the **projection of $G_1$ on $\Gamma_2$**. We shall now determine the relationship between the area $A(G_1)$ of $G_1$ and the area $A(G_2)$ of $G_2$. If $G_1$ is a rectangle—the simplest possible case—the problem may be solved by elementary geometry. For convenience, select the rectangle in $\Gamma_1$ so that one side is parallel to the line of intersection of the two planes (Fig. 42). Let the lengths of the sides of the rectangle be $a$ and $b$, as shown. The projection of the rectangle in $\Gamma_1$ onto $\Gamma_2$ is a rectangle, as the reader may easily verify. The lengths of the sides of the rectangle in $\Gamma_2$ are $a$ and $b \cos \phi$. The area $A_1 = ab$ of the rectangle in $\Gamma_1$ and the area $A_2 = ab \cos \phi$ of the rectangle in $\Gamma_2$ satisfy the relation

$$A_2 = A_1 \cos \phi. \tag{1}$$

Equation (1) is the basis of the next useful result.

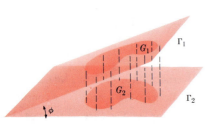

**FIGURE 41**

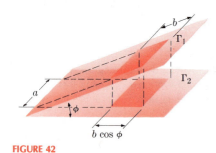

**FIGURE 42**

---

**LEMMA**    *Let $G_1$ be a region in a plane $\Gamma_1$ and let $G_2$ be the projection of $G_1$ onto a plane $\Gamma_2$. Then*

$$A(G_2) = A(G_1) \cos \phi, \tag{2}$$

*where $\phi$ is the angle between the planes $\Gamma_1$ and $\Gamma_2$.*

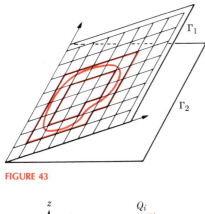

FIGURE 43

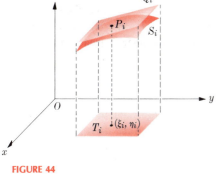

FIGURE 44

This lemma is proved by subdividing the plane $\Gamma_1$ into a network of rectangles and observing that these rectangles project onto rectangles in $\Gamma_2$ with areas related by Equation (1). Since the areas of $G_1$ and $G_2$ are obtained as limits of sums of the areas of rectangles, Formula (2) holds in the limit (Fig. 43). We observe that if the planes are parallel, then $\phi$ is zero, the region and its projection are congruent, and the areas are equal. If the planes are perpendicular, then $\phi = \pi/2$, the projection of $G_1$ degenerates into a line segment, and $A(G_2)$ is zero. Thus Formula (2) is valid for all angles $\phi$ such that $0 \le \phi \le \pi/2$.

Suppose we have a surface $S$ represented by an equation

$$z = f(x, y)$$

for $(x, y)$ on some region $F$ in the $xy$ plane. We shall consider only functions $f$ which have continuous first partial derivatives for all $(x, y)$ on $F$.

To define the area of the surface $S$ we begin by subdividing the $xy$ plane into a rectangular mesh. Suppose $T_i$, a rectangle of the subdivision, is completely contained in $F$. Select a point $(\xi_i, \eta_i)$ in $T_i$. This selection may be made in any manner whatsoever. The point $P_i(\xi_i, \eta_i, \zeta_i)$, with $\zeta_i = f(\xi_i, \eta_i)$, is on the surface $S$. Construct the plane tangent to the surface $S$ at $P_i$ (Fig. 44). Planes parallel to the $z$ axis and through the edges of $T_i$ cut out a portion (denoted $S_i$) of the surface, and they cut out a quadrilateral, denoted $Q_i$, from the tangent plane. The projection of $Q_i$ on the $xy$ plane is $T_i$. If the definition of surface area is to satisfy our intuition, then the area of $S_i$ must be close to the area of $Q_i$ whenever the subdivision in the $xy$ plane is sufficiently fine. However, $Q_i$ is a plane region, and its area can be found exactly. We recall from Chapter 15 (page 649) that we can determine the equation of a plane tangent to a surface $z = f(x, y)$ at a given point on the surface. Such a determination is possible because the quantities

$$f_x(\xi_i, \eta_i), f_y(\xi_i, \eta_i), -1$$

form a set of attitude numbers for the tangent plane at the point $(\xi_i, \eta_i, \zeta_i)$ where $\zeta_i = f(\xi_i, \eta_i)$.

On page 549 we showed that the formula for the angle between two planes is

$$\cos \phi = \frac{|a_1 b_1 + a_2 b_2 + a_3 b_3|}{\sqrt{a_1^2 + a_2^2 + a_3^2}\sqrt{b_1^2 + b_2^2 + b_3^2}},$$

where $a_1, a_2, a_3$ and $b_1, b_2, b_3$ are sets of attitude numbers of the two planes. We now find the cosine of the angle between the plane tangent to the surface and the $xy$ plane. Letting $\phi$ denote the angle between the tangent plane and the $xy$ plane and recalling that the $xy$ plane has attitude numbers $0, 0, -1$, we get

$$\cos \phi = \frac{|0 \cdot f_x + 0 \cdot f_y + 1 \cdot 1|}{\sqrt{1 + f_x^2 + f_y^2}} = (1 + f_x^2 + f_y^2)^{-1/2}.$$

According to the above lemma, we have

$$A(T_i) = A(Q_i) \cos \phi$$

or

$$A(Q_i) = A(T_i)\sqrt{1 + f_x^2(\xi_i, \eta_i) + f_y^2(\xi_i, \eta_i)}.$$

We add all expressions of the above type for rectangles $T_i$ which are in $F$. We obtain the sum

$$\sum_{i=1}^{n} A(Q_i) = \sum_{i=1}^{n} A(T_i)\sqrt{1 + f_x^2(\xi_i, \eta_i) + f_y^2(\xi_i, \eta_i)}, \tag{3}$$

and we expect that this sum is a good approximation to the (as yet undefined) surface area if the norm of the rectangular subdivision in the $xy$ plane is sufficiently small.

**DEFINITION**    *If the limit of the sums (3) exists as the norms of the subdivisions tend to zero and for arbitrary selections of the values $(\xi_i, \eta_i)$ in $T_i$, then we say that the surface $z = f(x, y)$ has* **surface area**. *The* **value of the surface area** $A(S)$ *of $S$ is the limit of the sum (3).*

**THEOREM 12**    *Suppose $z = f(x, y)$ is defined on a region $F$ in the $(x, y)$ plane. If the first derivatives $f_x$ and $f_y$ are continuous on $F$, then the surface area $A(S)$ of $z = f(x, y)$ exists and it is given by:*

$$A(S) = \iint_F \sqrt{1 + [f_x(x, y)]^2 + [f_y(x, y)]^2}\, dA.$$

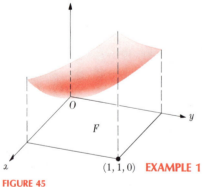

FIGURE 45

Theorem 12 is equivalent to stating that the sums in (3) tend to

$$\iint_F \sqrt{1 + [f_x(x, y)]^2 + [f_y(x, y)]^2}\, dA$$

when $f_x$ and $f_y$ are continuous. It is an immediate consequence of Theorem 1 and the fact that $\sqrt{1 + f_x^2 + f_y^2}$ is continuous if $f_x$ and $f_y$ are. The integration formula of the theorem may be used to calculate surface area, as the next examples show.

**EXAMPLE 1**    Find the area of the surface $z = \frac{2}{3}(x^{3/2} + y^{3/2})$ situated above the square $F = \{(x, y): 0 \le x \le 1, 0 \le y \le 1\}$ (Fig. 45).

**Solution**    Setting $z = f(x, y)$, we have $f_x = x^{1/2}, f_y = y^{1/2}$, and

$$A(S) = \iint_F (1 + x + y)^{1/2}\, dA$$

$$= \int_0^1 \int_0^1 (1 + x + y)^{1/2}\, dy\, dx.$$

Therefore

$$A(S) = \int_0^1 \tfrac{2}{3}[(1+x+y)^{3/2}]_0^1 \, dx = \tfrac{2}{3}\int_0^1 (2+x)^{3/2} - (1+x)^{3/2}] \, dx$$

$$= \tfrac{4}{15}[(2+x)^{5/2} - (1+x)^{5/2}]_0^1 = \tfrac{4}{15}(1+9\sqrt{3} - 8\sqrt{2}). \qquad \square$$

**EXAMPLE 2**   Find the area of the part of the cylinder $z = \tfrac{1}{2}x^2$ cut out by the planes $y = 0$, $y = x$, $x = 2$.

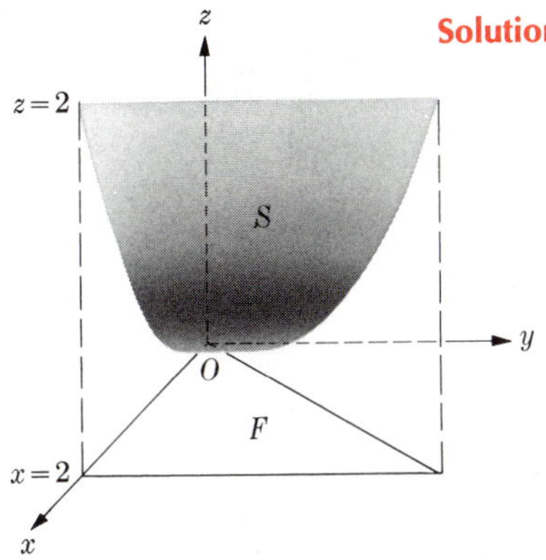

**Solution**   See Figs. 46 and 47, which show the surface $S$ and the projection $F$. We have $\partial z/\partial x = x$, $\partial z/\partial y = 0$. Therefore

$$A(S) = \iint_F \sqrt{1+x^2} \, dA = \int_0^2 \int_0^x \sqrt{1+x^2} \, dy \, dx$$

$$= \int_0^2 x\sqrt{1+x^2} \, dx = \tfrac{1}{3}[(1+x^2)^{3/2}]_0^2$$

$$= \tfrac{1}{3}(5\sqrt{5} - 1). \qquad \square$$

The next example shows that it is sometimes useful to use polar coordinates for the evaluation of the double integral.

**EXAMPLE 3**   Find the surface area of the part of the sphere $x^2 + y^2 + z^2 = a^2$ cut out by the vertical cylinder erected on one loop of the curve whose equation in polar coordinates is $r = a \cos 2\theta$.

**Solution**   (See Fig. 48.) The surface consists of two parts, one above and one below the $xy$ plane, symmetrically placed. The area of the upper half will be found. We have

$$z = \sqrt{a^2 - x^2 - y^2}, \qquad \frac{\partial z}{\partial x} = \frac{-x}{\sqrt{a^2 - x^2 - y^2}}, \qquad \frac{\partial z}{\partial y} = \frac{-y}{\sqrt{a^2 - x^2 - y^2}}.$$

Therefore (Fig. 48), we obtain

$$A(S) = \iint_F \frac{a}{\sqrt{a^2 - x^2 - y^2}} \, dA_{x,y}$$

$$= \int_{-\pi/4}^{\pi/4} \int_0^{a\cos 2\theta} \frac{ar}{\sqrt{a^2 - r^2}} \, dr \, d\theta.$$

This integral is an improper integral, but it can be shown to be convergent. Taking this fact for granted, we get

$$A(S) = 2a \int_0^{\pi/4} [-\sqrt{a^2 - r^2}]_0^{a\cos 2\theta} \, d\theta$$

$$= 2a^2 \int_0^{\pi/4} (1 - \sin 2\theta) \, d\theta = \tfrac{1}{2}a^2(\pi - 2).$$

The total surface area is $a^2(\pi - 2)$. $\qquad \square$

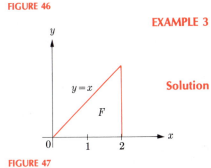

**FIGURE 46**

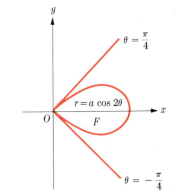

**FIGURE 47**

**FIGURE 48**

If the given surface is of the form $y = f(x, z)$ or $x = f(y, z)$, we get similar formulas for the surface area. The three basic formulas are

$$A(S) = \iint_F \sqrt{1 + \left(\frac{\partial z}{\partial x}\right)^2 + \left(\frac{\partial z}{\partial y}\right)^2}\ dA_{x,y} \quad \text{if } z = f(x, y),$$

$$A(S) = \iint_F \sqrt{1 + \left(\frac{\partial y}{\partial x}\right)^2 + \left(\frac{\partial y}{\partial z}\right)^2}\ dA_{x,z} \quad \text{if } y = f(x, z),$$

$$A(S) = \iint_F \sqrt{1 + \left(\frac{\partial x}{\partial y}\right)^2 + \left(\frac{\partial x}{\partial z}\right)^2}\ dA_{y,z} \quad \text{if } x = f(y, z).$$

## 7   PROBLEMS

In each of Problems 1 through 20, find the area of the surface described.

**1**  The portion of the surface $z = \frac{2}{3}(x^{3/2} + y^{3/2})$ situated above the triangle

$$F = \{(x, y): 0 \le x \le y,\ 0 \le y \le 1\}.$$

**2**  The portion of the plane $x/a + y/b + z/c = 1$ in the first octant ($a > 0,\ b > 0,\ c > 0$).

**3**  The part of the cylinder $x^2 + z^2 = a^2$ inside the cylinder $x^2 + y^2 = a^2$.

**4**  The part of the cylinder $x^2 + z^2 = a^2$ above the square $|x| \le \frac{1}{2}a,\ |y| \le \frac{1}{2}a$.

**5**  The part of the cone $z^2 = x^2 + y^2$ inside the cylinder $x^2 + y^2 = 2x$.

**6**  The part of the cone $z^2 = x^2 + y^2$ above the figure bounded by one loop of the curve $r^2 = 4 \cos 2\theta$.

**7**  The part of the cone $x^2 = y^2 + z^2$ between the cylinder $y^2 = z$ and the plane $y = z - 2$.

**8**  The part of the cone $y^2 = x^2 + z^2$ cut off by the plane $2y = (x + 2)\sqrt{2}$.

**9**  The part of the cone $x^2 = y^2 + z^2$ inside the sphere $x^2 + y^2 + z^2 = 2z$.

**10**  The part of the surface $z = xy$ inside the cylinder $x^2 + y^2 = a^2$.

**11**  The part of the surface $4z = x^2 - y^2$ above the region bounded by the curve $r^2 = 4 \cos \theta$.

**12**  The part of the surface of a sphere of radius $2a$ inside a cylinder of radius $a$ if the center of the sphere is on the surface of the cylinder.

**13**  The part of the surface of a sphere of radius $a$, center at the origin, inside the cylinder erected on one loop of the curve $r = a \cos 3\theta$.

**14**  The part of the sphere $x^2 + y^2 + z^2 = 4z$ inside the paraboloid $x^2 + y^2 = z$.

**15**  The part of the cylinder $y^2 + z^2 = 2z$ cut off by the cone $x^2 = y^2 + z^2$.

**16**  The part of the cylinder $x^2 + y^2 = 2ax$ inside the sphere $x^2 + y^2 + z^2 = 4a^2$.

**17**  The part of the cylinder $y^2 + z^2 = 4a^2$ above the $xy$ plane and bounded by the planes $y = 0$, $x = a$, and $y = x$.

**18**  The part of the paraboloid $y^2 + z^2 = 4ax$ cut off by the cylinder $y^2 = ax$ and the plane $x = 3a$; outside the cylinder $y^2 = ax$.

**19**  The part of the parabolic cylinder $z = y^2$ which lies over the triangle in the $xy$ plane having vertices $(0, 0)$, $(1, 0)$, $(1, 1)$.

**20**  The part of the surface $9(z - y)^2 = 4x^3$ which lies over the triangle in the $xy$ plane having vertices $(0, 0)$, $(0, 2)$, $(2, 2)$.

**21**  a) Use elementary geometry (and trigonometry) to establish equation (2) for an arbitrary triangle.
b) Use the result of (a) to establish equation (2) for an arbitrary polygon.

**22**  If $r = \sqrt{x^2 + y^2}$, $\theta = \arctan(y/x)$, and $z = f(x, y)$, establish the formula for surface area in polar coordinates:

$$A(S) = \iint_F \sqrt{1 + f_r^2 + \frac{1}{r^2} f_\theta^2}\ r\ dr\ d\theta.$$

Use the polar coordinates formula (Problem 22) to find the surface area in Problems 23 through 25.

**23** The area of the surface of the paraboloid $z = x^2 + y^2$ which is inside the cylinder $x^2 + y^2 = 4$.

**24** The portion of the cone $x^2 + y^2 = z^2$ inside the cylinder $(x^2 + y^2)^2 = 2xy$.

**25** The portion of the cone $x^2 + y^2 = z^2$ inside the cylinder $x^2 + y^2 = 1$.

---

## 8

## VOLUMES OF SOLIDS OF REVOLUTION (OPTIONAL)

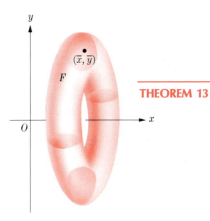

**FIGURE 49**

We previously developed methods for finding the volume of certain solids of revolution. These techniques were applicable whenever the resulting process reduced to a single integration. (See page 228 ff.) Now that areas may be determined in a more general way by double integration, we can calculate the volume of a greater variety of solids of revolution. The basic tool is the Theorem of Pappus which we now state. (See also page 345.)

**THEOREM 13**    **(Theorem of Pappus)**    *If a plane figure F lies on one side of a line L in its plane, the volume of the set S generated by revolving F around L is equal to the product of A(F), the area of F, and the length of the path described by the center of mass of F; in other words, if F is in the xy plane and L is the x axis (see Fig. 49), then*

$$V(S) = 2\pi\bar{y}A(F) = \iint_F 2\pi y \, dA. \tag{1}$$

The second equality in (1) above follows from the definition of $\bar{y}$. In case $F$ is a rectangle of the form $a \le x \le b, c \le y \le d$, where $c \ge 0$, then $S$ is just a circular ring of altitude $b - a$, inner radius of base $c$ (if $c > 0$) and outer radius $d$. Thus

$$V(S) = \pi(d^2 - c^2)(b - a) = 2\pi\left(\frac{c+d}{2}\right)A(F) = 2\pi\bar{y}A(F) = \iint_F 2\pi y \, dA.$$

The theorem is proved in general by subdividing $F$ and noting that

$$\sum_{i=1}^{n} V(S_i) = \sum_{i=1}^{n} \iint_{F_i} 2\pi y \, dA = \iint_{F_n^*} 2\pi y \, dA,$$

where $F_n^*$ is the union of the $F_i$; we then pass to the limit.

**EXAMPLE 1**    The region $F = \{(x, y): 0 \le x \le 1, x^3 \le y \le \sqrt{x}\}$ is revolved about the $x$ axis. Find the volume generated.

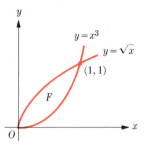

**FIGURE 50**

**Solution**    See Fig. 50. We have

$$V(S) = 2\pi \iint_F y \, dA = 2\pi \int_0^1 \int_{x^3}^{\sqrt{x}} y \, dy \, dx$$

$$= \pi \int_0^1 (x - x^6) \, dx = \tfrac{5}{14}\pi. \qquad \square$$

The Theorem of Pappus in this more general form is especially useful whenever the transformation to polar coordinates is appropriate. The next example illustrates this point.

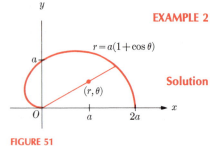

**FIGURE 51**

**EXAMPLE 2**    Find the volume of the set $S$ generated by revolving around the $x$ axis the upper half $F$ of the area bounded by the cardioid $r = a(1 + \cos \theta)$. (See Fig. 51.)

**Solution**    We have, since $y$ transforms to $r \sin \theta$ in polar coordinates,

$$V(S) = \iint_F 2\pi y \, dA = \iint_F 2\pi r \sin \theta \, dA_{x,y}$$

$$= 2\pi \int_0^\pi \int_0^{a(1+\cos\theta)} r^2 \sin \theta \, dr \, d\theta$$

$$= \tfrac{2}{3}\pi a^3 \int_0^\pi (1 + \cos \theta)^3 \sin \theta \, d\theta$$

$$= \frac{\pi a^3}{6} [-(1 + \cos \theta)^4]_0^\pi = \frac{8\pi a^3}{3}. \qquad \square$$

The formula analogous to (1) for revolving a region $F$ in the $xy$ plane about the $y$ axis is

$$V(S) = 2\pi \bar{x} A(F) = 2\pi \iint_F x \, dA.$$

## 8    PROBLEMS

In each of Problems 1 through 18, find the volume of the set obtained by revolving the region described about the axis indicated. Sketch the region. Express in set notation each of the plane regions being revolved.

**1**    The upper half of the ellipse $x^2/a^2 + y^2/b^2 = 1$; the $x$ axis.

**2**    The region in the first quadrant inside $3x^2 + 4y^2 = 48$ and outside $(x - 2)^2 + y^2 = 1$; the $x$ axis.

**3**    The region satisfying the inequalities $xy \le 4$, $y \le x$, $27y \ge 4x^2$; the $x$ axis.

**4**    The upper half of the figure bounded by the right-hand loop of the curve $r^2 = a^2 \cos \theta$; the $x$ axis.

**5**    The upper half of the right-hand loop of the curve $r = a \cos^2 \theta$; the $x$ axis.

**6**    The region inside $x^2 + y^2 = 64$ and outside $x^2 + y^2 = 8x$; the line $x = 8$.

**7**    The region bounded by the curve $y = x^3$ and the lines $y = 0$, $x = 1$; the $x$ axis.

**8** The region bounded by the curve $y = e^x$ and the lines $y = 0$, $x = 1$; the $x$ axis.

**9** The region bounded by the curve $y = \ln x$ and the lines $y = 0$, $x = e$; the $x$ axis.

**10** Same region as Problem 9; the $y$ axis.

**11** The region in the first quadrant between the parabolas $x^2 = 4y$, $x^2 = 8y - 4$; the $x$ axis.

**12** Same region as Problem 11; the $y$ axis.

**13** The loop of $r^3 = \sin 2\theta$ in the first quadrant; the $y$ axis.

**14** The upper half of the area outside the circle $r = 4$ and inside the limaçon $r = 3 + 2 \cos \theta$; the $x$ axis.

**15** The upper half of the area to the right of the line $x = \frac{3}{2}$ and inside the cardioid $r = 2(1 + \cos \theta)$; the $x$ axis.

**16** The upper half of the area to the right of the parabola $r = 9/(1 + \cos \theta)$ and inside the cardioid $r = 4(1 + \cos \theta)$; the $x$ axis.

**17** The right-hand horizontal loop of the curve $r = 2 \cos 2\theta$; the $y$ axis. Find the center of mass of that loop.

**18** The upper half of the right-hand loop of the curve $r^2 = a^2 \cos 2\theta$; the $x$ axis.

**19** A triangle $T$ with base of length $a$ on the $x$ axis and vertex at a point $(b, c)$ in the first quadrant of the $xy$ plane is revolved about the $x$ axis. Use the formulas for the area of a triangle and the volume of a right circular cone to find the center of mass of $T$.

**20** A semicircle with diameter $d$ on the $x$ axis is revolved about the $x$ axis. Use the formulas for the area of a semicircle and the volume of a sphere to find the center of mass of the semicircle.

**21** a) A triangle in the first quadrant of the $xy$ plane with base parallel to the $x$ axis is revolved about the $x$ axis. Find the volume generated. (*Hint*: See Problem 19.)
b) A trapezoid in the first quadrant of the $xy$ plane with bases parallel to the $x$ axis is revolved about the $x$ axis. Use the results in (a) to find the volume generated and the center of mass of the trapezoid.

---

## 9

## THE TRIPLE INTEGRAL

The definition of the triple integral parallels that of the double integral. In the simplest case, we consider a rectangular box $R$ bounded by the six planes $x = a_0$, $x = a_1$, $y = b_0$, $y = b_1$, $z = c_0$, $z = c_1$ (Fig. 52). Let $f(x, y, z)$ be a function of three variables defined for $(x, y, z)$ in $R$. We subdivide the entire three-dimensional space into rectangular boxes by constructing planes

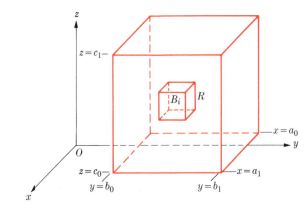

**FIGURE 52**

parallel to the coordinate planes. Let $B_1$, $B_2$, ..., $B_n$ be those boxes of the subdivision which contain points of R. Denote by $V(B_i)$ the volume of the $i$th box, $B_i$. We now select a point $P_i(\xi_i, \eta_i, \zeta_i)$ in $B_i$; this selection may be made in any manner whatsoever. The sum

$$\sum_{i=1}^{n} f(\xi_i, \eta_i, \zeta_i) V(B_i)$$

is an approximation to the triple integral. The **norm of the subdivision** is the length of the longest diagonal of the boxes $B_1$, $B_2$, ..., $B_n$. If the above sums tend to a limit as the norms of the subdivisions tend to zero and for any choices of the points $P_i$, we call this limit the **triple integral of $f$ over R**. The expression

$$\iiint_R f(x, y, z) \, dV$$

is used to represent this limit.

Just as the double integral is equal to a twice-iterated integral, so the triple integral has the same value as a threefold-iterated integral. In the case of the rectangular box R, we obtain

$$\iiint_R f(x, y, z) \, dV = \int_{a_0}^{a_1} \left\{ \int_{b_0}^{b_1} \left[ \int_{c_0}^{c_1} f(x, y, z) \, dz \right] dy \right\} dx.$$

Suppose a region $S$ is bounded by the planes $x = a_0$, $x = a_1$, $y = b_0$, $y = b_1$, and by the surfaces $z = r(x, y)$, $z = s(x, y)$, as shown in Fig. 53. The triple integral may be defined in the same way as for a rectangular box R, and once again it is equal to the iterated integral. We have

$$\iiint_S f(x, y, z) \, dV = \int_{a_0}^{a_1} \left\{ \int_{b_0}^{b_1} \left[ \int_{r(x, y)}^{s(x, y)} f(x, y, z) \, dz \right] dy \right\} dx.$$

We state without proof the following theorem, which applies in the general case.

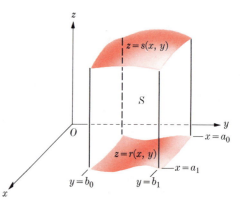

**FIGURE 53**

**THEOREM 14**    *Suppose that S is a region defined by the inequalities*

$$S = \{(x, y, z): a \leq x \leq b, p(x) \leq y \leq q(x), r(x, y) \leq z \leq s(x, y)\},$$

*where the functions p, q, r, and s are continuous. If f is a continuous function on S, then*

$$\iiint\limits_{S} f(x, y, z) \, dV = \int_a^b \left\{ \int_{p(x)}^{q(x)} \left[ \int_{r(x, y)}^{s(x, y)} f(x, y, z) \, dz \right] dy \right\} dx.$$

The iterated integrations are performed in turn by holding all variables constant except the one being integrated. Brackets and braces in multiple integrals will be omitted unless there is danger of confusion.

**EXAMPLE 1**    Evaluate the iterated integral

$$\int_0^3 \int_0^{6-2z} \int_0^{4-(2/3)y-(4/3)z} yz \, dx \, dy \, dz.$$

**Solution**    We have

$$\int_0^3 \int_0^{6-2z} \int_0^{4-(2/3)y-(4/3)z} yz \, dx \, dy \, dz$$

$$= \int_0^3 \int_0^{6-2z} \left[ xyz \right]_0^{4-(2/3)y-(4/3)z} dy \, dz$$

$$= \int_0^3 \int_0^{6-2z} yz(4 - \tfrac{2}{3}y - \tfrac{4}{3}z) \, dy \, dz$$

$$= \int_0^3 [2zy^2 - \tfrac{2}{9}zy^3 - \tfrac{2}{3}y^2z^2]_0^{6-2z} \, dz$$

$$= \int_0^3 [(2z - \tfrac{2}{3}z^2)(6 - 2z)^2 - \tfrac{2}{9}z(6 - 2z)^3] \, dz$$

$$= \tfrac{1}{9} \int_0^3 z(6 - 2z)^3 \, dz.$$

The integration may be performed by the substitution $u = 6 - 2z$. The result is 54/5.    □

The determination of the limits of integration is the principal difficulty in reducing a triple integral to an iterated integral. The reader who works a large number of problems will develop good powers of visualization of three-dimensional figures. There is no simple mechanical technique for determining the limits of integration in the wide variety of problems we encounter. The next examples illustrate the process.

**EXAMPLE 2**    Evaluate

$$\iiint\limits_{S} x \, dV,$$

where $S$ is the region bounded by the surfaces $y = x^2$, $y = x + 2$, $4z = x^2 + y^2$, and $z = x + 3$.

**Solution**  To transform the triple integral into an iterated integral, we must determine the limits of integration. The region $S$ is sketched in Fig. 54.

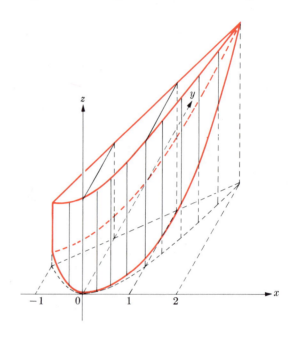

**FIGURE 54**

The projection of $S$ on the $xy$ plane is the region $F$ bounded by the curves $y = x^2$ and $y = x + 2$, as shown in Fig. 55. From this projection, the region rises with vertical walls, bounded from below by the paraboloid $z = \frac{1}{4}(x^2 + y^2)$ and above by the plane $z = x + 3$. Since $F$ is described by the inequalities

$$F = \{(x, y): -1 \le x \le 2, \ x^2 \le y \le x + 2\},$$

we have

$$S = \{(x, y, z): -1 \le x \le 2, \ x^2 \le y \le x + 2, \ \tfrac{1}{4}(x^2 + y^2) \le z \le x + 3\}.$$

Therefore

$$\iiint_S x \, dV = \int_{-1}^{2} \int_{x^2}^{x+2} \int_{(x^2+y^2)/4}^{x+3} x \, dz \, dy \, dx$$

$$= \int_{-1}^{2} \int_{x^2}^{x+2} [x^2 + 3x - \tfrac{1}{4}(x^3 + xy^2)] \, dy \, dx$$

$$= \int_{-1}^{2} \left\{ \left( 3x + x^2 - \frac{1}{4}x^3 \right)(2 + x - x^2) - \frac{x}{12}[(2 + x)^3 - x^6] \right\} dx$$

$$= \frac{837}{160}. \qquad \square$$

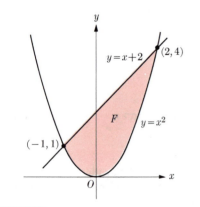

$y = x + 2$ $(2, 4)$

$F$ $y = x^2$

$(-1, 1)$

$O$

**FIGURE 55**

In the case of double integrals there are two possible orders of integration, one of them often being easier to calculate than the other. In the case of triple integrals there are six possible orders of integration. It becomes a matter of practice and trial and error to find which order is the most convenient.

The limits of integration may sometimes be found by projecting the region on one of the coordinate planes and then finding the equations of the "bottom" and "top" surfaces. This method was used in Example 2. *If part of the boundary is a cylinder perpendicular to one of the coordinate planes, that fact can be used to help determine the limits of integration.*

**EXAMPLE 3**   Express the integral

$$I = \iiint_S f(x, y, z) \, dV$$

as an iterated integral in six different ways if $S$ is the region bounded by the surfaces

$$z = 0, \qquad z = x, \qquad \text{and} \qquad y^2 = 4 - 2x.$$

**Solution**   The region $S$ is shown in Fig. 56. The projection of $S$ on the $xy$ plane is the two-dimensional region $F_{xy}$ bounded by $x = 0$ and $y^2 = 4 - 2x$, as shown in Fig. 57. Therefore the integral may be written

$$I = \int_0^2 \int_{-\sqrt{4-2x}}^{+\sqrt{4-2x}} \int_0^x f(x, y, z) \, dz \, dy \, dx$$

$$= \int_{-2}^2 \int_0^{2-(1/2)y^2} \int_0^x f(x, y, z) \, dz \, dx \, dy.$$

The projection of $S$ on the $xz$ plane is the triangular region bounded by the lines $z = 0$, $z = x$, and $x = 2$, as shown in Fig. 58. The iterated integral in this case becomes

$$I = \int_0^2 \int_0^x \int_{-\sqrt{4-2x}}^{+\sqrt{4-2x}} f(x, y, z) \, dy \, dz \, dx$$

$$= \int_0^2 \int_z^2 \int_{-\sqrt{4-2x}}^{+\sqrt{4-2x}} f(x, y, z) \, dy \, dx \, dz.$$

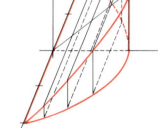

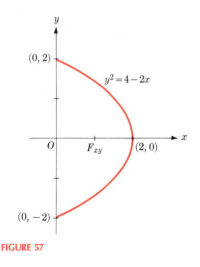

**FIGURE 56**

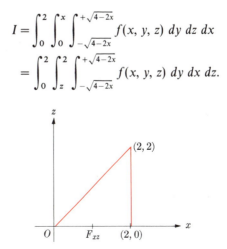

**FIGURE 57**

**FIGURE 58**

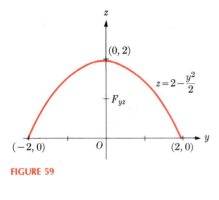

**FIGURE 59**

The projection of $S$ on the $yz$ plane is the plane region bounded by $z = 0$ and $z = 2 - \frac{1}{2}y^2$ (Fig. 59). Then $I$ takes the form

$$I = \int_{-2}^{2} \int_{0}^{2-(1/2)y^2} \int_{z}^{2-(1/2)y^2} f(x, y, z) \, dx \, dz \, dy$$

$$= \int_{0}^{2} \int_{-\sqrt{4-2z}}^{+\sqrt{4-2z}} \int_{z}^{2-(1/2)y^2} f(x, y, z) \, dx \, dy \, dz. \qquad \square$$

## 9    PROBLEMS

In each of Problems 1 through 11, find the value of the iterated integral. Express each region of integration in set notation.

**1** $\displaystyle\int_{0}^{1} \int_{0}^{x} \int_{0}^{x-y} x \, dz \, dy \, dx$

**2** $\displaystyle\int_{-1}^{1} \int_{0}^{1-y^2} \int_{-\sqrt{x}}^{\sqrt{x}} 2y^2\sqrt{x} \, dz \, dx \, dy$

**3** $\displaystyle\int_{0}^{1} \int_{y^2}^{\sqrt{y}} \int_{0}^{y+z} xy \, dx \, dz \, dy$

**4** $\displaystyle\int_{0}^{4} \int_{0}^{\sqrt{16-x^2}} \int_{0}^{\sqrt{16-x^2-y^2}} (x + y + z) \, dz \, dy \, dx$

**5** $\displaystyle\int_{0}^{2} \int_{0}^{\sqrt{4-z^2}} \int_{0}^{2-z} z \, dx \, dy \, dz$

**6** $\displaystyle\int_{0}^{1} \int_{0}^{x} \int_{0}^{y} \frac{1 + \sqrt[3]{z}}{\sqrt{z}} \, dz \, dy \, dx$

**7** $\displaystyle\int_{0}^{1} \int_{0}^{1-x} \int_{0}^{1-x-y} 3 \, dz \, dy \, dx$

**8** $\displaystyle\int_{0}^{1} \int_{0}^{1-x} \int_{0}^{1-x-y} xy \, dz \, dy \, dx$

**9** $\displaystyle\int_{0}^{r} \int_{0}^{\sqrt{r^2-x^2}} \int_{0}^{4} z \, dz \, dy \, dx$

**10** $\displaystyle\int_{-r}^{r} \int_{-\sqrt{r^2-x^2}}^{\sqrt{r^2-x^2}} \int_{0}^{4} z \, dz \, dy \, dx$ (*Hint:* Use symmetry to reduce this integral to the one of Problem 9.)

**11** $\displaystyle\int_{-2}^{2} \int_{-(1/2)\sqrt{4-x^2}}^{\sqrt{4-x^2}} \int_{x^2+3y^2}^{4-y^2} dz \, dy \, dx$

In Problems 12 through 22, evaluate

$$\iiint_{S} f(x, y, z) \, dV$$

where $S$ is bounded by the given surfaces and $f$ is the given function. In each case, express $S$ in set notation.

**12** $z = 0$, $y = 0$, $y = x$, $x + y = 2$, $x + y + z = 3$; $f(x, y, z) = x$

**13** $x = 0$, $x = \sqrt{a^2 - y^2 - z^2}$; $f(x, y, z) = x$

**14** $z = 0$, $x^2 + z = 1$, $y^2 + z = 1$; $f(x, y, z) = z^2$

**15** $x^2 + z^2 = a^2$, $y^2 + z^2 = a^2$; $f(x, y, z) = x^2 + y^2$

**16** $x = 0$, $y = 0$, $z = 0$, $(x/a) + (y/b) + (z/c) = 1$, $(a, b, c > 0)$; $f(x, y, z) = z$

**17** $y = z^2$, $y^2 = z$, $x = 0$, $x = y - z^2$; $f(x, y, z) = y + z^2$

**18** $x = 0$, $y = 0$, $z = 0$, $x^{1/2} + y^{1/2} + z^{1/2} = a^{1/2}$; $f(x, y, z) = z$

**19** $x = 0$, $y = 0$, $z = 0$, $y^2 = 4 - z$, $x = y + 2$; $f(x, y, z) = x^2$; $y \geq 0$

**20** $z = x^2 + y^2$, $z = 27 - 2x^2 - 2y^2$; $f(x, y, z) = 1$

**21** $z^2 = 4ax$, $x^2 + y^2 = 2ax$; $f(x, y, z) = 1$

**\*22** $y^2 + z^2 = 4ax$, $y^2 = ax$, $x = 3a$; $f(x, y, z) = x^2$, $y^2 \leq ax$

In Problems 23 through 30, express each iterated integral as a triple integral by describing the set $S$ over which the integration is performed. Sketch the set $S$ and then express the iterated integral in two orders differing from the original. Do not evaluate the integrals.

**23** $\displaystyle\int_{0}^{1} \int_{0}^{x} \int_{0}^{x-y} x \, dz \, dy \, dz$

**24** $\displaystyle\int_{-1}^{1} \int_{0}^{1-y^2} \int_{-\sqrt{x}}^{\sqrt{x}} 2y^2\sqrt{x} \, dz \, dx \, dy$

**25** $\displaystyle\int_0^1 \int_{y^2}^{\sqrt{y}} \int_0^{y+z} xy\, dx\, dz\, dy$

**26** $\displaystyle\int_{-2}^2 \int_0^{4-y^2} \int_0^{y+2} (y^2 + z^2)\, dz\, dx\, dy$

**27** $\displaystyle\int_{-2}^2 \int_0^{4-y^2} \int_0^{y+2} xyz\, dz\, dx\, dy$

**28** $\displaystyle\int_0^1 \int_0^1 \int_0^{x^2+y^2} ye^{xz}\, dz\, dy\, dx$

**29** $\displaystyle\int_0^1 \int_0^{1-x} \int_0^{x+y} (x-y)z\, dz\, dy\, dx$

**30** $\displaystyle\int_0^1 \int_{x^2}^{\sqrt{x}} \int_0^{y-x^2} f(x, y, z)\, dz\, dy\, dx$

**31** Express the integral of $f(x, y, z)$ over the region $S$ bounded by the surface $z = \sqrt{16 - x^2 - y^2}$ and the plane $z = 2$ in 6 ways.

**32** Express the integral

$$\int_0^2 \int_0^z \int_0^x (x^2 + y^2 + z^2)\, dy\, dx\, dz$$

in 5 additional ways.

**33** Suppose $S$ is the rectangular box $a \le x \le b$, $c \le y \le d$, $e \le z \le f$. Show that when the integrand has the special form $f(x, y, z) = g(x)h(y)j(z)$, then

$$\iiint_S g(x)h(y)j(z)\, dx\, dy\, dz$$

$$= \left[\int_a^b g(x)\, dx\right]\left[\int_c^d h(y)\, dy\right]\left[\int_e^f j(z)\, dz\right].$$

In Problems 34 through 37, use the result of Problem 33 to evaluate the integrals given.

**34** $\iiint_S xz^2 \sin y\, dx\, dy\, dz$; $S = \{x, y, z:\ 0 \le x \le 3,\ 1 \le y \le 5, -3 \le z \le 0\}$.

**35** $\iiint_S (x^2yz + yz)\, dx\, dy\, dz$; $S = \{x, y, z:\ 0 \le x \le 4,\ 0 \le y \le 6, 0 \le z \le 10\}$.

**36** $\iiint_S e^{2x+3y-4z}\, dx\, dy\, dz$; $\quad S = \{x,\quad y,\quad z:\quad -4 \le x \le 4, -3 \le y \le 3,\ -2 \le z \le 2\}$.

**37** $\iiint_S \ln(x^{zy})\, dx\, dy\, dz$; $S = \{x,\quad y,\quad z:\ 1 \le x \le 2,\ 1 \le y \le 2, 1 \le z \le 2\}$.

**38** Let $S$ be the solid tetrahedron with vertices at $(0, 0, 0)$, $(1, 0, 0)$, $(0, 1, 0)$, and $(0, 0, 1)$. Find the value of $\iiint_S f(x, y, z)\, dV$ where $f(x, y, z) = (x^2 + 2xz + y^2)$.

**39** Let $S$ be the solid pyramid with vertices $(0, 0, 0)$, $(1, 0, 0)$, $(1, 1, 0)$, $(0, 1, 0)$, and $(0, 0, 1)$. Find the value of $\iiint_S f(x, y, z)\, dV$ where $f(x, y, z) = xyz + 2yz$.

---

**10**

---

## MASS OF A SOLID. TRIPLE INTEGRALS IN CYLINDRICAL AND SPHERICAL COORDINATES

From the definition of triple integral we see that if $f(x, y, z) \equiv 1$, **then the triple integral taken over a region $S$ is precisely the volume $V(S)$.** In general, if a solid object occupies a region $S$, and if the density at any point is given by $\delta(x, y, z)$, then the total mass, $m(S)$, is given by the triple integral

$$m(S) = \iiint_S \delta(x, y, z)\, dV.$$

**Notation.** For the remainder of this chapter the symbol $\delta$ will be used for density. The quantity $\rho$, which we previously used for density, will denote one of the variables in spherical coordinates.

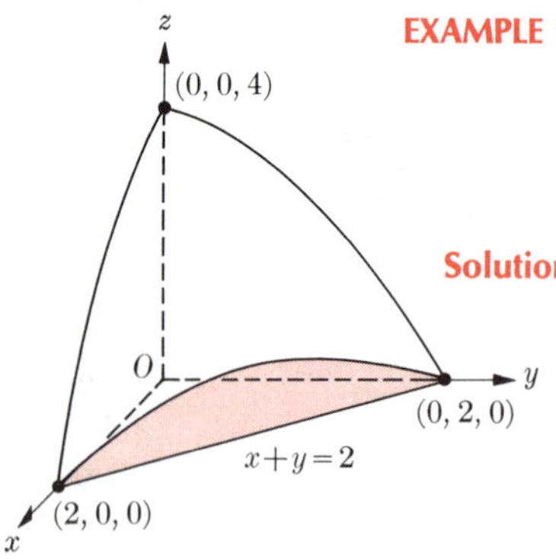

**FIGURE 60**

**EXAMPLE 1**   The solid $S$ in the first octant is bounded by the surfaces

$$z = 4 - x^2 - y^2, \qquad z = 0, \qquad x + y = 2, \qquad x = 0, \qquad y = 0.$$

The density is given by $\delta(x, y, z) = 2z$. Find the total mass.

**Solution**   We have (Fig. 60)

$$m(S) = \iiint_S 2z \, dV = 2 \int_0^2 \int_0^{2-x} \int_0^{4-x^2-y^2} z \, dz \, dy \, dx$$

$$= \int_0^2 \int_0^{2-x} (4 - x^2 - y^2)^2 \, dy \, dx$$

$$= \int_0^2 \int_0^{2-x} (16 + x^4 + y^4 - 8x^2 - 8y^2 + 2x^2 y^2) \, dy \, dx$$

$$= \int_0^2 \left[ (4 - x^2)^2 y - \tfrac{2}{3}(4 - x^2) y^3 + \tfrac{1}{5} y^5 \right]_0^{2-x} dx$$

$$= \int_0^2 \left[ 2(4 - x^2)^2 - x(4 - x^2)^2 - \tfrac{2}{3}(4 - x^2)(2 - x)^3 + \tfrac{1}{5}(2 - x)^5 \right] dx.$$

The above integral, a polynomial in $x$, can be evaluated. *Answer:*   704/45.

□

We found that certain double integrals are easy to evaluate if a polar coordinate system is used. Similarly, there are triple integrals which, although difficult to evaluate in rectangular coordinates, are simple integrations when transformed into other systems. The most useful transformations are those to cylindrical and spherical coordinates. (See Chapter 13, Section 6.)

Cylindrical coordinates consist of polar coordinates in the plane and a $z$ coordinate as in a rectangular system. The transformation from rectangular to cylindrical coordinates is

$$x = r \cos \theta, \qquad y = r \sin \theta, \qquad z = z. \tag{1}$$

A region $S$ in $(x, y, z)$ space corresponds to a region $U$ in $(r, \theta, z)$ space. The volume of $S$, $V(S)$, may be found in terms of a triple integral in $(r, \theta, z)$ space by the formula (Fig. 61)

$$V(S) = \iiint_U r \, dV_{r\theta z}.$$

This formula is a natural extension of the formula relating area in rectangular and in polar coordinates. (See page 526 ff.)

More generally, if $f(x, y, z)$ is a continuous function, and if we define

$$g(r, \theta, z) = f(r \cos \theta, r \sin \theta, z),$$

then we have the following relationship between triple integrals:

$$\iiint_S f(x, y, z) \, dV_{xyz} = \iiint_U g(r, \theta, z) \, r \, dV_{r\theta z}.$$

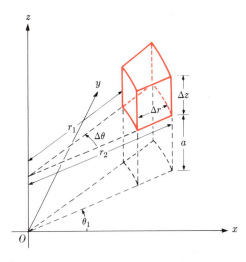

FIGURE 61

A triple integral in cylindrical coordinates may be evaluated by iterated integrals. We obtain

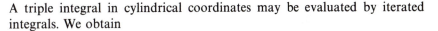

$$\iiint g(r, \theta, z)r \, dV_{r\theta z} = \iiint g(r, \theta, z)r \, dr \, d\theta \, dz,$$

and, as before, there are five other orders of integration possible. Once again the major problem is the determination of the limits of integration. For this purpose it is helpful to superimpose cylindrical coordinates on a rectangular system, sketch the surface, and read off the limits of integration. The next example shows the method.

**EXAMPLE 2**    Find the mass of the solid bounded by the cylinder $x^2 + y^2 = ax$ and the cone $z^2 = x^2 + y^2$ if the density is $\delta = k\sqrt{x^2 + y^2}$ (Fig. 62).

**Solution**    We change to cylindrical coordinates. The cylinder is $r = a \cos \theta$ and the cone is $z^2 = r^2$. The density is $kr$. The region $S$ corresponds to the region $U$ given by

$$U = \left\{ (r, \theta, z): 0 \leq r \leq a \cos \theta, \ -\frac{\pi}{2} \leq \theta \leq \frac{\pi}{2}, \ -r \leq z \leq r \right\}.$$

Therefore

$$m(S) = \iiint_S k\sqrt{x^2 + y^2} \, dV_{xyz} = k \iiint_U r \cdot r \cdot dV_{r\theta z}$$

$$= k \int_{-\pi/2}^{\pi/2} \int_0^{a \cos \theta} \int_{-r}^r r^2 \, dz \, dr \, d\theta$$

$$= 2k \int_{-\pi/2}^{\pi/2} \int_0^{a \cos \theta} r^3 \, dr \, d\theta = \frac{1}{2} ka^4 \int_{-\pi/2}^{\pi/2} \cos^4 \theta \, d\theta$$

$$= \frac{1}{8} ka^4 \int_{-\pi/2}^{\pi/2} \left( 1 + 2 \cos 2\theta + \frac{1 + \cos 4\theta}{2} \right) d\theta = \frac{3k\pi a^4}{16}. \qquad \square$$

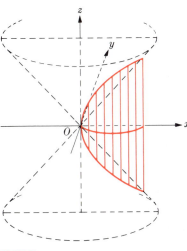

FIGURE 62

The transformation from rectangular to **spherical coordinates** is given by the equations

$$x = \rho \cos \theta \sin \phi, \qquad y = \rho \sin \theta \sin \phi, \qquad z = \rho \cos \phi. \qquad (2)$$

The region $U$ given by

$$U = \{(\rho, \theta, \phi): \rho_1 \le \rho \le \rho_2, \theta_1 \le \theta \le \theta_2, \phi_1 \le \phi \le \phi_2\}$$

corresponds to a rectangular box in $(\rho, \theta, \phi)$ space. We wish to find the volume of the solid $S$ in $(x, y, z)$ space which corresponds to $U$ under the transformation (2). Referring to Fig. 63, we see that $S$ is a region such as $ABCDA'B'C'D'$ between the spheres $\rho = \rho_1$ and $\rho = \rho_2$, between the planes $\theta = \theta_1$ and $\theta = \theta_2$, and between the cones $\phi = \phi_1(OADD'A')$ and $\phi = \phi_2(OBCC'B')$. The region $S$ is obtained by sweeping the plane region $F(ABCD$ shown in Fig. 64) through an angle $\Delta\theta = \theta_2 - \theta_1$. The Theorem of Pappus (page 739) for determining volumes of solids of revolution applies also for areas swept through any angle about an axis. We obtain

$$V(S) = \Delta\theta \iint_F x \, dA_{zx}.$$

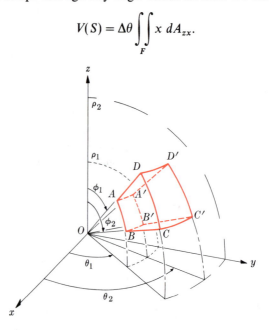

FIGURE 63

Since $\rho, \phi$ are polar coordinates in the $zx$ plane (Fig. 64), we have

$$x = \rho \sin \phi \qquad \text{and} \qquad dA_{zx} = \rho dA_{\rho\phi} = \rho \, d\rho \, d\phi.$$

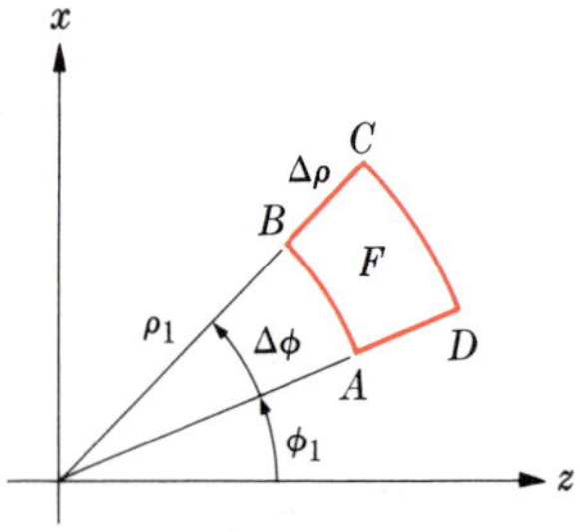

Therefore

$$V(S) = \Delta\theta \int_{\phi_1}^{\phi_2} \int_{\rho_1}^{\rho_2} \rho^2 \sin \phi \, d\rho \, d\phi$$

$$= \int_{\theta_1}^{\theta_2} \int_{\phi_1}^{\phi_2} \int_{\rho_1}^{\rho_2} \rho^2 \sin \phi \, d\rho \, d\phi \, d\theta$$

$$= \iiint_U \rho^2 \sin \phi \, dV_{\rho\theta\phi}.$$

FIGURE 64

More generally, if $f(x, y, z)$ is continuous on a region $S$ and if

$$g(\rho, \theta, \phi) = f(\rho \cos \theta \sin \phi, \rho \sin \theta \sin \phi, \rho \cos \phi),$$

then the triple integral of $f$ may be transformed according to the formula

$$\iiint_S f(x, y, z)\, dV_{xyz} = \iiint_U g(\rho, \theta, \phi)\rho^2 \sin \phi\, dV_{\rho\theta\phi}$$

$$= \iiint_U g(\rho, \theta, \phi)\rho^2 \sin \phi\, d\rho\, d\theta\, d\phi.$$

The triple integral is evaluated by iterated integrations. The next example illustrates the process.

**EXAMPLE 3**    Find the volume above the cone $z^2 = x^2 + y^2$ and inside the sphere $x^2 + y^2 + z^2 = 2az$ (Fig. 65).

**Solution**    In spherical coordinates the cone and sphere have the equations

$$\phi = \frac{\pi}{4} \qquad \text{and} \qquad \rho = 2a \cos \phi,$$

respectively. Therefore

$$V(S) = \iiint_S dV_{xyz} = \iiint_U \rho^2 \sin \phi\, dV_{\rho\theta\phi}$$

$$= \int_0^{\pi/4} \int_0^{2a\cos\phi} \int_0^{2\pi} \rho^2 \sin \phi\, d\theta\, d\rho\, d\phi$$

$$= 2\pi \int_0^{\pi/4} \int_0^{2a\cos\phi} \rho^2 \sin \phi\, d\rho\, d\phi = \frac{16a^3\pi}{3} \int_0^{\pi/4} \cos^3 \phi \sin \phi\, d\phi$$

$$= \frac{4a^3\pi}{3} [-\cos^4 \phi]_0^{\pi/4} = \pi a^3. \qquad \square$$

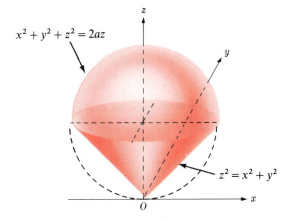

$x^2 + y^2 + z^2 = 2az$

$z^2 = x^2 + y^2$

FIGURE 65

## 10   PROBLEMS

In each of Problems 1 through 16, find the mass of the solid having the given density $\delta$ and bounded by the surfaces whose equations are given.

**1** $z^2 = x^2 + y^2$, $x^2 + y^2 + z^2 = a^2$, above the cone, $\delta = $ const.

**2** The rectangular parallelepiped bounded by $x = -a$, $x = a$, $y = -b$, $y = b$, $z = -c$, $z = c$, $\delta = k(x^2 + y^2 + z^2)$.

**3** $x^2 + y^2 + z^2 = a^2$, $x^2 + y^2 + z^2 = b^2$, $a < b$, $\delta = k\sqrt{x^2 + y^2 + z^2}$.

**4** The rectangular parallelepiped bounded by $x = 0$, $x = 2a$, $y = 0$, $y = 2b$, $z = 0$, $z = 2c$, $\delta = k(x^2 + y^2 + z^2)$.

**5** $x^2 + y^2 = a^2$, $x^2 + y^2 + z^2 = 4a^2$; $\delta = kz^2$; outside the cylinder.

**6** The tetrahedron bounded by the coordinate planes and $x + y + z = 1$; $\delta = kxyz$.

**7** $x^2 + y^2 = 2ax$, $x^2 + y^2 + z^2 = 4a^2$; $\delta = k(x^2 + y^2)$.

**8** $z^2 = 25(x^2 + y^2)$, $z = x^2 + y^2 + 4$; $\delta = $ const; above the paraboloid.

**9** $z^2 = x^2 + y^2$, $x^2 + y^2 + z^2 = 2az$; above the cone; $\delta = kz$.

**10** Interior of $x^2 + y^2 + z^2 = a^2$; $\delta = k(x^2 + y^2 + z^2)^n$, $n$ a positive number.

**11** $x^2 + y^2 = az$, $x^2 + y^2 + z^2 = 2az$; above the paraboloid; $\delta = $ const.

**12** $2z = x^2 + y^2$, $z = 2x$; $\delta = k\sqrt{x^2 + y^2}$.

**13** $x^2 + y^2 + z^2 = a^2$, $r^2 = a^2 \cos 2\theta$ (cylindrical coordinates); $\delta = $ const.

**14** $x^2 + y^2 + z^2 = 4az$, $z = 3a$, above the plane; $\delta = k\sqrt{x^2 + y^2 + z^2}$.

**15** $z^2 = x^2 + y^2$, $(x^2 + y^2)^2 = a^2(x^2 - y^2)$; $\delta = k\sqrt{x^2 + y^2}$.

**\*16** $z^2 = x^2 + y^2$, $x^2 + y^2 + z^2 = 2ax$; $\delta = $ const; above the cone.

Evaluate the integrals in Problems 17 through 21 by using cylindrical coordinates.

**17** $\iiint_S \sqrt{x^2 + y^2} \; dx \; dy \; dz$; $S$ is bounded by $z = \sqrt{x^2 + y^2}$, $z = 0$, and $x^2 + y^2 = 1$.

**18** $\iiint_S z^2 \; dx \; dy \; dz$; $S$ is the hemisphere bounded by $z = \sqrt{1 - x^2 - y^2}$ and $z = 0$.

**19** $\iiint_S 1 \; dx \; dy \; dz$; $S$ is in the first octant, inside the cylinder $x^2 + y^2 = ay$ and the paraboloid $x^2 + y^2 + az = a^2$, where $a$ is a constant. (See Fig. 66.)

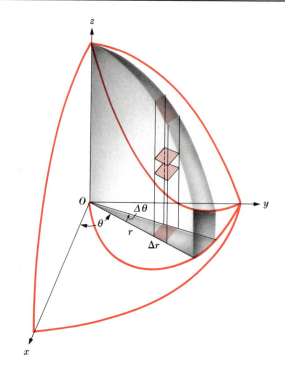

**FIGURE 66**

**20** $\iiint_S z\sqrt{x^2 + y^2} \; dx \; dy \; dz$; $S$ is bounded by $z = 4$, $z = x^2 + y^2$.

**21** $\iiint_S 1 \; dx \; dy \; dz$; $S$ bounded by $z = e^{x^2 + y^2}$, $z = 0$, and $x^2 + y^2 = 16$.

Evaluate the integrals in Problems 22 through 25 by using spherical coordinates.

**22** $\iiint_S 1 \; dx \; dy \; dz$ where $S$ is the intersection of the regions $x^2 + y^2 + z^2 \le 2$ and $x^2 + y^2 + z^2 \le 2z$.

**23** $\iiint_S (x^2 + y^2 + z^2)^3 \; dx \; dy \; dz$; $S$ is determined by $0 \le x^2 + y^2 \le \frac{1}{2}$, $(x^2 + y^2)^{1/2} \le z \le (1 - x^2 - y^2)^{1/2}$.

**24** $\iiint_S x^2 \; dx \; dy \; dz$; $S$ is determined by $\{0 \le x^2 + y^2 \le 1, \sqrt{x^2 + y^2} \le z \le 1\}$. (A truncated solid cone with vertex at $(0, 0, 0)$).

**25** $\iiint_S \sin([x^2 + y^2 + z^2])^{3/2} \; dx \; dy \; dz$; $S$ is the spherical shell between the spheres $x^2 + y^2 + z^2 = 1$ and $x^2 + y^2 + z^2 = 9$.

_____ **11** _____

## MOMENT OF INERTIA. CENTER OF MASS (OPTIONAL)

The definition of moment of inertia of a solid body is similar to the definition of moment of mass of a plane region (page 329). The following definition is basic for this section.

**DEFINITION** *Suppose that a solid body occupies a region S and let L be any line in three-space. We make a subdivision of space into rectangular boxes and let $S_1, S_2, \ldots,$ $S_n$ be those boxes which contain points of S. For each i, select any point $P_i(\xi_i, \eta_i,$ $\zeta_i)$ in $S_i$. If the sums*

$$\sum_{i=1}^{n} r_i^2 m(S_i) \qquad (r_i = \text{distance of } P_i \text{ from } L)$$

*tend to a limit I as the norms of the subdivisions tend to zero, and for any choices of the $P_i$, then I is called the* **moment of inertia of the solid S about L.**

It can be shown that if a solid S has continuous density $\delta(x, y, z)$, then the moments of inertia $I_x$, $I_y$, and $I_z$ about the $x$, $y$, and $z$ axes, respectively, are given by the triple integrals

$$I_x = \iiint_S (y^2 + z^2)\delta(x, y, z)\, dV, \qquad I_y = \iiint_S (x^2 + z^2)\delta(x, y, z)\, dV,$$

$$I_z = \iiint_S (x^2 + y^2)\delta(x, y, z)\, dV.$$

**DEFINITION** If a solid S has a density $\delta(x, y, z)$ and a mass $m(S)$, the point $(\bar{x}, \bar{y}, \bar{z})$, defined by the formulas

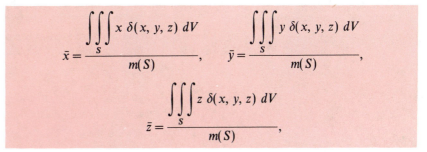

$$\bar{x} = \frac{\iiint_S x\, \delta(x, y, z)\, dV}{m(S)}, \qquad \bar{y} = \frac{\iiint_S y\, \delta(x, y, z)\, dV}{m(S)},$$

$$\bar{z} = \frac{\iiint_S z\, \delta(x, y, z)\, dV}{m(S)},$$

is called the **center of mass of S.**

In determining the center of mass, it is helpful to take into account all available symmetries. The following rules are useful:
a) *If S is symmetric in the xy plane and $\delta(x, y, -z) = \delta(x, y, z)$, then $\bar{z} = 0$. A similar result holds for other coordinate planes.*

b) *If S is symmetric in the x axis and* $\delta(x, -y, -z) = \delta(x, y, z)$, *then* $\bar{y} = \bar{z} = 0$.
*A similar result holds for the other axes.*

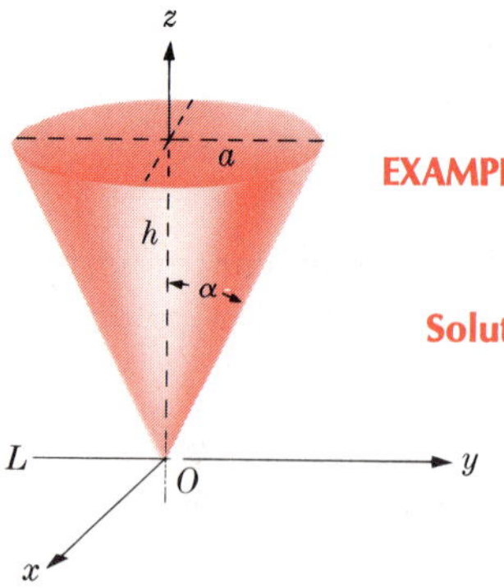

**EXAMPLE 1** Find the moment of inertia of a homogeneous solid cone of base radius $a$ and altitude $h$ about a line through the vertex and perpendicular to the axis.

**Solution** We take the vertex at the origin, the $z$ axis as the axis of the cone, and the $x$ axis as the line $L$ about which the moment of inertia is to be computed (Fig. 67).

Let $\alpha = \arctan (a/h)$ be half the angle opening of the cone. In spherical coordinates we get (see Fig. 68)

$$I = \iiint\limits_{S} (y^2 + z^2) \, dV_{xyz}$$

$$= \int_0^\alpha \int_0^{h\sec\phi} \int_0^{2\pi} \rho^2(\sin^2\phi \sin^2\theta + \cos^2\phi)\rho^2 \sin\phi \, d\theta \, d\rho \, d\phi.$$

Since

$$\int_0^{2\pi} \sin^2\theta \, d\theta = \pi,$$

we obtain

$$I = \pi \int_0^\alpha \int_0^{h\sec\phi} \rho^4(1 + \cos^2\phi) \sin\phi \, d\rho \, d\phi$$

$$= \frac{\pi h^5}{5} \int_0^\alpha [(\cos\phi)^{-5} + (\cos\phi)^{-3}] \sin\phi \, d\phi$$

$$= \frac{\pi h^5}{5}\left[\frac{\sec^4\alpha - 1}{4} + \frac{\sec^2\alpha - 1}{2}\right] = \frac{\pi h a^2}{20}(4h^2 + a^2),$$

since $\tan\alpha = a/h$. ☐

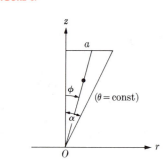

FIGURE 67

FIGURE 68

**EXAMPLE 2** Find the center of mass of a solid hemisphere of radius $a$ which has density proportional to the distance from the center.

**Solution** We select the hemisphere so that the plane section is in the $xy$ plane and the $z$ axis is an axis of symmetry (Fig. 69). Then $\bar{x} = \bar{y} = 0$. Changing to spherical coordinates, we have

$$m(S) = \int_0^{2\pi} \int_0^{\pi/2} \int_0^a k\rho \cdot \rho^2 \sin\phi \, d\rho \, d\phi \, d\theta = \tfrac{1}{2}\pi k a^4$$

and from this we obtain

$$\bar{z} = \frac{\displaystyle\int_0^{2\pi} \int_0^{\pi/2} \int_0^a \rho\cos\phi \cdot k\rho \cdot \rho^2 \sin\phi \, dp \, d\phi \, d\theta}{\tfrac{1}{2}\pi k a^4}$$

$$= \frac{2}{\pi a^4}\cdot\frac{1}{5}a^5\cdot 2\pi \int_0^{\pi/2} \cos\phi \sin\phi \, d\phi = \frac{4}{5}a\left[\frac{\sin^2\phi}{2}\right]_0^{\pi/2} = \frac{2a}{5}.$$

The center of mass is at $(0, 0, 2a/5)$. ☐

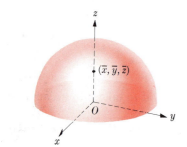

FIGURE 69

## 11   PROBLEMS

In each of Problems 1 through 15, find the moment of inertia about the given axis of the solid having the specified density $\delta$ and bounded by the surfaces as described.

1  A cube of side $a$; $\delta =$ const; about an edge.

2  A cube of side $a$; $\delta =$ const; about a line parallel to an edge, at distance 2 from it, and in a plane of one of the faces.

3  Bounded by $x = 0$, $y = 0$, $z = 0$, $x + z = a$, $y = z$; $\delta = kx$; about the $x$ axis.

4  $x^2 + y^2 = a^2$; $x^2 + z^2 = a^2$; $\delta =$ const; about the $z$ axis.

5  $z = x$, $y^2 = 4 - 2z$, $x = 0$; $\delta =$ const; about the $z$ axis.

6  $x = 0$, $y = 0$, $z^2 = 1 - x - y$; $\delta =$ const; about the $z$ axis.

7  $z^2 = y^2(1 - x^2)$; $y = 1$; $\delta =$ const; about the $x$ axis.

8  $x^2 + y^2 = a^2$, $x^2 + y^2 = b^2$, $z = 0$, $z = h$; $\delta = k\sqrt{x^2 + y^2}$; about the $x$ axis $(a < b)$.

9  $x^2 + y^2 + z^2 = a^2$, $x^2 + y^2 + z^2 = b^2$; $\delta = k\sqrt{x^2 + y^2 + z^2}$; about the $z$ axis $(a < b)$.

10  $\rho = 4$, $\rho = 5$, $z = 1$, $z = 3$ $(1 \le z \le 3)$; $\delta =$ const; about the $z$ axis.

11  $z^2 = x^2 + y^2$, $(x^2 + y^2)^2 = a^2(x^2 - y^2)$; $\delta = k\sqrt{x^2 + y^2}$; about the $z$ axis.

12  $z = 0$, $z = \sqrt{a^2 - x^2 - y^2}$; $\delta = kz$; about the $x$ axis.

13  $x^2/a^2 + y^2/b^2 + z^2/c^2 = 1$; $\delta =$ const; about the $z$ axis. (Divide the integral into two parts.)

14  $(r - b)^2 + z^2 = a^2 (0 < a < b)$; $\delta = kz$; about the $z$ axis. Assume $z \ge 0$.

15  $\rho = b$, $\rho = c$, $r = a$ $(a < b < c)$; outside $r = a$, between $\rho = b$ and $\rho = c$; $\delta =$ const; about the $z$ axis.

In each of Problems 16 through 32, find the center of mass of the solid having the given density and bounded by the surfaces as described. Describe each solid in set notation.

16  $x = 0$, $y = 0$, $z = 0$, $x/a + y/b + z/c = 1$; $\delta =$ const.

17  $x = 0$, $y = 0$, $z = 0$, $x + z = a$, $y = z$; $\delta = kx$.

18  $x^2 + y^2 = a^2$, $x^2 + z^2 = a^2$, $\delta =$ const; the portion where $x \ge 0$.

19  $z = x$, $z = -x$, $y^2 = 4 - 2x$; $\delta =$ const.

20  $z^2 = y^2(1 - x^2)$, $y = 1$; $\delta =$ const.

21  $z = 0$, $x^2 + z = 1$, $y^2 + z = 1$; $\delta =$ const.

22  $x = 0$, $y = 0$, $z = 0$, $x^{1/2} + y^{1/2} + z^{1/2} = a^{1/2}$; $\delta =$ const.

23  $y^2 + z^2 = 4ax$, $y^2 = ax$, $x = 3a$; $\delta =$ const. (Inside $y^2 = ax$.)

24  $z^2 = 4ax$, $x^2 + y^2 = 2ax$; $\delta =$ const.

25  $z^2 = x^2 + y^2$, $x^2 + y^2 + z^2 = a^2$, above the cone; $\delta =$ const.

26  $z^2 = x^2 + y^2$, $x^2 + y^2 = 2ax$; $\delta = k(x^2 + y^2)$.

27  $z^2 = x^2 + y^2$, $x^2 + y^2 + z^2 = 2az$, above the cone; $\delta = kz$.

28  $x^2 + y^2 = az$, $x^2 + y^2 + z^2 = 2az$, above the paraboloid; $\delta =$ const.

29  $x^2 + y^2 + z^2 = 4az$, $z = 3a$, above the plane; $\delta = k\sqrt{x^2 + y^2 + z^2}$.

30  $\rho = 4$, $\rho = 5$, $z = 1$, $z = 3$ $(1 \le z \le 3)$; $\delta =$ const.

*31  $z^2 = x^2 + y^2$, $(x^2 + y^2)^2 = a^2(x^2 - y^2)$; $\delta = k\sqrt{x^2 + y^2}$; the part for which $x \ge 0$.

*32  $z^2 = x^2 + y^2$, $x^2 + y^2 + z^2 = 2ax$, above the cone; $\delta =$ const.

---

## CHAPTER 16

## REVIEW PROBLEMS

In Problems 1 through 11, evaluate the iterated integrals. Change order of integration, if necessary.

1  $\displaystyle\int_0^1 \int_2^3 (x^2 + xy - y^3)\, dx\, dy$

2  $\displaystyle\int_0^2 \int_{-1}^1 (\sin(xy) + xy)\, dy\, dx$

3  $\displaystyle\int_{-1}^2 \int_{x-1}^{x^2} (x^2 - 3y)\, dy\, dx$

4  $\displaystyle\int_1^2 \int_1^{\ln y} \frac{1}{y}\, dx\, dy$

5  $\displaystyle\int_0^1 \int_{x^2}^{x^{1/4}} (x^{1/2} - y^2)\, dy\, dx$

6  $\displaystyle\int_0^1 \int_y^1 \cos(\tfrac{1}{2}\pi x^2)\, dx\, dy$

7  $\displaystyle\int_0^1 \int_{\sqrt{x}}^1 \sin\left(\frac{y^3 + 1}{2}\right)\, dy\, dx$

8  $\displaystyle\int_3^4 \int_0^{\ln x} e^{-y}\, dy\, dx$

9  $\displaystyle\int_0^1 \int_x^1 x^2 e^{y^4}\, dy\, dx$

10  $\displaystyle\int_0^1 \int_0^{\arccos y} e^{\sin x}\, dx\, dy$

11  $\displaystyle\int_0^1 \int_0^2 \int_0^3 xy(z + x^2)\, dz\, dx\, dy$

In Problems 12 through 17, evaluate the multiple integrals as indicated.

12  $\displaystyle\iint_F (x + 3)^2 y\, dx\, dy$; $F = \{(x, y): 0 \le x \le 2,\ -1 \le y \le 2\}$

**13** $\iint\limits_{F} e^{(x-y)} \, dx \, dy$; $F = \{(x, y): 1 \leq x \leq 2, 0 \leq y \leq 2\}$

**14** $\iint\limits_{F} \cos(\tfrac{1}{2}\pi x) \cos(\tfrac{1}{2}\pi y) \, dx \, dy$;

$F = \{(x, y): -1 \leq x \leq 1, 0 \leq y \leq 3\}$

**15** $\iint\limits_{F} \sin(x + y) \, dx \, dy$; $F = \left\{(x, y): 0 \leq x \leq \dfrac{\pi}{2}, 0 \leq y \leq \dfrac{\pi}{2}\right\}$

**16** $\iint\limits_{F} (2x^2 + 3y^2) \, dx \, dy$; $F = \{(x, y): 0 \leq x^2 + y^2 \leq 1\}$

**17** $\iint\limits_{F} ye^x \, dx \, dy$; $F = \{(x, y): 0 \leq x \leq y^2, 0 \leq y \leq 1\}$

In each of Problems 18 through 22, find the area of the bounded region $F$ determined by the given curves.

**18** $xy = 2$, $y = 1$, $y = 3x + 1$

**19** $x^2 = 4y$, $2y - x - 4 = 0$

**20** $y^2 = 5 + x$, $y^2 = 5 - x$

**21** $y = x^2$, $y^2 = e^x$

**22** $y - \tfrac{1}{2} = x^2$, $x^2 + y^2 = 1$

In each of Problems 23 through 27, evaluate the given integral by using polar coordinates.

**23** $\iint\limits_{F} \cos(x^2 + y^2) \, dx \, dy$; $F$ is the unit disk of radius 1 centered at the origin.

**24** $\displaystyle\int_{-2}^{0} \int_{-\sqrt{4-x^2}}^{0} \sqrt{x^2 + y^2} \, dy \, dx$

**25** $\iint\limits_{F} e^{-(x^2 + y^2)} \, dx \, dy$; $F$ as in Problem 23.

**26** $\displaystyle\int_{0}^{1} \int_{2y}^{2} y\sqrt{x^2 + y^2} \, dx \, dy$

**27** $\displaystyle\int_{0}^{1} \int_{x}^{1} x \sin(y^3) \, dy \, dx$

In each of Problems 28 through 32, use polar coordinates to find the area, or the volume, as indicated, of the region given.

**28** The region inside the circle $r = \tfrac{3}{2}$ but to the right of the line $4r \cos \theta = 3$. (Area)

**29** The region $\left\{(r, \theta): 0 \leq r \leq 2\sqrt{\cos \theta}, 0 \leq \theta \leq \dfrac{\pi}{2}\right\}$. (Area)

**30** The region $\left\{(r, \theta): 1 \leq r \leq 1 + \cos \theta, 0 \leq \theta \leq \dfrac{\pi}{4}\right\}$. (Area)

**31** The region bounded below by the $xy$ plane and above by the paraboloid $z = 1 - (x^2 + y^2)$. (Volume)

**32** The region $\left\{(x, y, z): \dfrac{x^2}{4} + \dfrac{y^2}{4} + \dfrac{z^2}{3} \leq 1\right\}$. (Volume)

In each of Problems 33 through 38, find the moment of inertia and the radius of gyration about the given axis of the plate $F$ whose density is given.

**33** Let $F$ be the triangular region of Fig. 70 with density $\rho = 3xy$; $y$ axis.

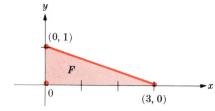

**FIGURE 70**

**34** Let $F$ be the half-disk of Fig. 71 with density $\rho = 2(x^2 + y^2)$; $x$ axis.

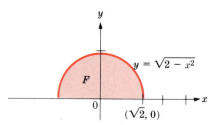

**FIGURE 71**

**35** Let $F$ be the region inside $r = 2 + \sin \theta$ and outside $r = 1$. The density $\rho = \rho(r, \theta)$ is inversely proportional to the distance from the pole, with proportionality constant $k$; polar moment of inertia.

**36** Let $F$ be the region between two concentric circles of radii $a$ and $b$ and center at the origin. The density is constant. Find the polar moment of inertia.

**37** Let $F$ be bounded by the curves $y = \sin^2 x$ and $y = -\sin^2 x$, $-\pi \leq x \leq \pi$. The density $\rho$ is constant and equal to one; polar moment of inertia.

**38** Let $F$ be bounded by the curves $y = e^x$, $y = 0$, $0 \leq x \leq 14$; $\rho(x, y) = xy$; $x$ axis.

In each of Problems 39 through 43, find the center of mass of the plate $F$ described.

**39** $F$ is the bounded region described by $y = x^2$ and $x + y = 2$.

**40** $F$ is the region bounded by $x - 2y + 8 = 0$; $x + 3y + 5 = 0$; $x = -2$; and $x = 4$.

**41** $F$ is the region bounded by $y = \sin x$, $y = \cos x$, $0 \leq x \leq \dfrac{\pi}{4}$.

**42** $F$ is bounded by $y = \ln x$; $y = 0$; $x = 1$; and $x = 2$.

**43** $F$ is bounded by the curves $\sqrt{x} + \sqrt{y} = 1$; $x = 0$; and $y = 0$.

In each of Problems 44 through 46, find the area of the surface described.

**44** The part of the plane $x + y + z - 4 = 0$ cut by the cylinder $x^2 + y^2 = 16$.

**45** That part of the surface $z = \frac{1}{2}(x^2 + y^2)$ below the plane $z = 1$.

**46** That part of the cone $x^2 + y^2 = z^2$ lying between the two planes $z = 0$ and $x - 3 = -2z$.

**47** Find the volume of the solid generated by revolving the triangle with vertices $(2, 2), (5, 2), (3, 3)$ about: (a) the $x$ axis; (b) the $y$ axis; (c) the line $y = x$.

**48** Let $F$ be the region in Fig. 72, with center of mass $(\bar{x}, \bar{y}) = (2, 1)$, and area $\frac{2}{3}$. Find the volume of the solid generated by revolving $F$ about: (a) the $x$ axis; (b) the $y$ axis; and (c) the line $y = -x + \frac{4}{7}$.

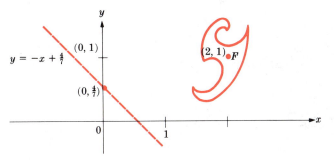

$$y = -x + \tfrac{4}{7}$$

**FIGURE 72**

**\*49** Let $f$ be a continuous function of one variable, and let $a$ be a constant. Show that the following holds:

$$\int_0^a \int_0^y \int_0^z f(x)\, dx\, dz\, dy = \frac{1}{2} \int_0^a (a - x)^2 f(x)\, dx.$$

**50** Show that the following formula is valid:

$$\int_0^1 \int_0^y e^{n(a-x)} f(x)\, dx\, dy = \int_0^1 (1 - x) e^{n(a-x)} f(x)\, dx,$$

for constants $a > 0$, $n$, and a continuous function $f$.

In each of Problems 51 through 61, find the value of the iterated integral.

**51** $\displaystyle\int_0^1 \int_0^{1-x} \int_0^{x+y} (x + y - z)\, dz\, dy\, dx$

**52** $\displaystyle\int_0^1 \int_0^{3y} \int_0^x (x + 4z)\, dz\, dx\, dy$

**53** $\displaystyle\int_3^0 \int_0^{x^2} \int_0^x (x + z)\, dy\, dz\, dx$

**54** $\displaystyle\int_0^4 \int_{\sqrt{x}}^1 \int_{y^2}^x zx^2 y^3\, dz\, dy\, dx$

**55** $\displaystyle\int_0^1 \int_0^{\sqrt{1-x^2}} \int_0^{\sqrt{4-(x^2+y^2)}} dz\, dy\, dx$

**56** $\displaystyle\int_0^{\pi/2} \int_0^{\pi/4} \int_0^{2\cos\phi} \rho^2 \sin\phi\, d\rho\, d\phi\, d\theta$

**57** $\displaystyle\int_{-1}^1 \int_0^{\sqrt{1-x^2}} \int_{\sqrt{x^2+y^2}}^1 z^3\, dz\, dy\, dx$

**58** $\displaystyle\int_1^2 \int_0^{\pi/4} \int_4^6 \ln(x^{z\sin y})\, dz\, dy\, dx$

**59** $\displaystyle\iiint_S e^{(x^2+y^2+z^2)^{3/2}}\, dx\, dy\, dz$, where $S$ is the solid bounded below by the nappe of the cone $z = \sqrt{x^2 + y^2}$ and above by the sphere $x^2 + y^2 + z^2 = 1$.

**\*60** $\displaystyle\iiint_S [(x - 2)^2 + (y - 3)^2 + (z - 2)^2]^{-1/2}\, dx\, dy\, dz$, where $S$ is the unit sphere centered at the origin.

**61** $\displaystyle\iiint_S dx\, dy\, dz$, where $S$ is the solid between two concentric spheres of radii $a$ and $b$, $0 \le a < b$, centered at the origin.

**62** Compute the mass of the solid described in Problem 61 if the density $\gamma$ at each point is equal to the square of the distance from this point to the center of the two concentric spheres.

In each of Problems 63 through 67, find the center of mass of the solid having the given density and bounded by the surfaces as described.

**63** The solid is the tetrahedron with vertices $(0, 0, 0), (2, 0, 0), (0, 3, 0)$, and $(0, 0, 1)$. The density $\delta = yz$.

**64** The solid is as in Problem 63; $\delta = xyz$.

**65** The solid is as in Problem 47; $\delta = y$.

**66** The solid is bounded above by $x^2 + z = 4$, below by $x + z = 2$, and on the sides by $y = 0$ and $y = 3$; $\delta = 1 + y$.

**67** The solid is bounded above by $z = y$; below by the $xy$ plane; and on the sides by $x = 0, x = 1, y = 0$, and $y = 1$; $\delta$ at a given point is proportional to the square of the distance from the point to the origin.

**\*68** Let $S$ be a sphere of radius $a$ centered at the origin. Show that

$$\iiint_S \frac{|xyz|}{\sqrt{x^2 + y^2 + z^2}}\, dx\, dy\, dz = \frac{a^5}{5}.$$

# 17

# VECTOR FIELD THEORY

Line integrals are an extension of the ordinary integral to curves in the plane and in three-space. We establish two theorems, Green's Theorem and Stokes' Theorem, which exhibit an important relationship between double integrals and line integrals. We also establish the Divergence Theorem which connects a triple integral over a region in three-space with an integral taken over the boundary of the region.

## 1

### DEFINITION OF A LINE INTEGRAL

Let $C$ be an arc in the plane extending from the point $A(a, b)$ to the point $B(c, d)$, as shown in Fig. 1. Suppose that $f(x, y)$ is a continuous function defined in a region which contains the arc $C$ in its interior. We make a decomposition of the arc $C$ by introducing $n - 1$ points between $A$ and $B$ along $C$. We label these points $P_1, P_2, \ldots, P_{n-2}, P_{n-1}$, and set $A = P_0, B = P_n$. Denote the coordinates of the point $P_i$ by $(x_i, y_i)$, $i = 0, 1, 2, \ldots, n$. (See Fig. 2.) Between each two successive points of the subdivision we select a point on the curve. Call these points $Q_1, Q_2, \ldots, Q_n$, and denote the coordinates of $Q_i$ by $(\xi_i, \eta_i)$, $i = 1, 2, \ldots, n$. This selection may be made in any manner whatsoever so long as $Q_i$ is on the part of $C$ between $P_{i-1}$ and $P_i$ (Fig. 3). We form the sum

$$f(\xi_1, \eta_1)(x_1 - x_0) + f(\xi_2, \eta_2)(x_2 - x_1) + \cdots + f(\xi_n, \eta_n)(x_n - x_{n-1}),$$

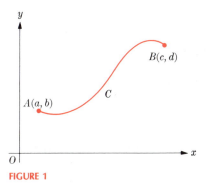

**FIGURE 1**

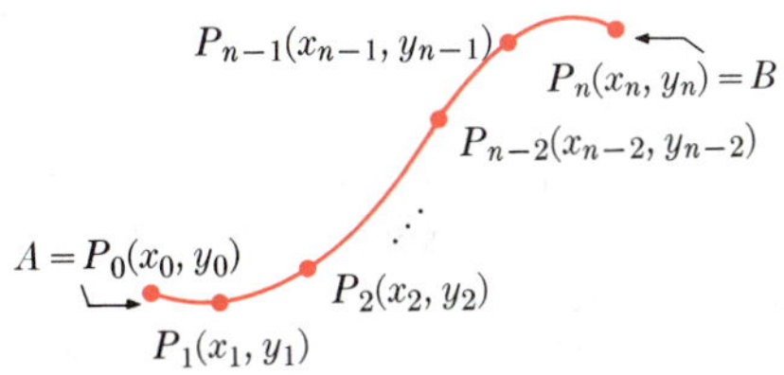

**FIGURE 2**

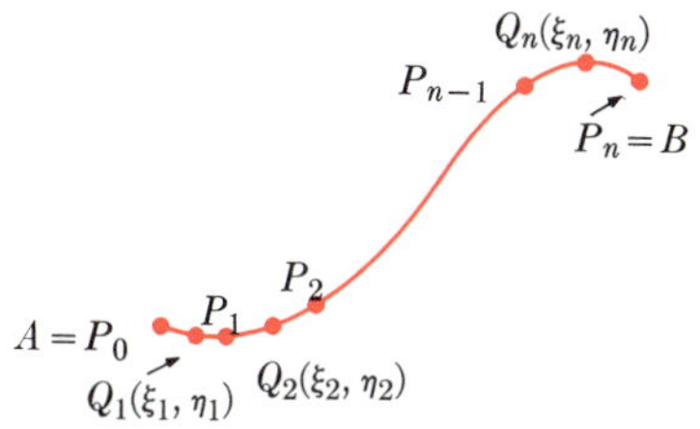

**FIGURE 3**

or, written more compactly,

$$\sum_{i=1}^{n} f(\xi_i, \eta_i)(x_i - x_{i-1}) \equiv \sum_{i=1}^{n} f(\xi_i, \eta_i)\,\Delta_i x. \tag{1}$$

As in the case of subdivisions of an interval along the $x$ axis (see page 188), we define the **norm of the subdivision** $P_0, P_1, P_2, \ldots, P_n$ of the curve $C$ to be the maximum distance between any two successive points of the subdivision. We denote the norm by $\|\Delta\|$.

**DEFINITION**

*Suppose there is a number $L$ with the following property: for each $\varepsilon > 0$ there is a $\delta > 0$ such that*

$$\left| \sum_{i=1}^{n} f(\xi_i, \eta_i)(x_i - x_{i-1}) - L \right| < \varepsilon$$

*for every subdivision with $\|\Delta\| < \delta$ and for any choices of the $(\xi_i, \eta_i)$ as described above. Then we say that **the line integral of $f$ with respect to $x$ along the curve $C$ exists and its value is $L$**. There are various symbols for this line integral such as*

$$\int_C f(x, y)\, dx \qquad \text{and} \qquad (C) \int_A^B f(x, y)\, dx. \tag{2}$$

*Note that the value of the integral will depend, in general, not only on $f$ and the points $A$ and $B$ but also on the particular arc $C$ selected. The number $L$ is unique if it exists at all.*

The expression (1) is one of several types of sums which are commonly formed in line integrations. We also introduce the sum

$$\sum_{i=1}^{n} f(\xi_i, \eta_i)(y_i - y_{i-1})$$

in which the points $(\xi_i, \eta_i)$ are selected as before. The limit, if it exists (as $\|\Delta\| \to 0$), is the line integral

$$\int_C f(x, y)\, dy \tag{3}$$

and will generally have a value different from (2).

In Section 3 we shall give an application of line integration to a physical problem.

If the arc $C$ happens to be a segment of the $x$ axis, then the line integral $\int_C f(x, y)\, dx$ reduces to an ordinary integral. To see this observe that in the approximating sums all the $\eta_i = 0$. Therefore

$$\int_C f(x, y)\, dx = \int_a^b f(x, 0)\, dx.$$

On the other hand, when $C$ is a segment of the $x$ axis, the integral $\int_C f(x, y)\, dy$ always vanishes, since in each approximating sum $y_i - y_{i-1} = 0$ for every $i$.

Simple properties of line integrals, analogous to those for ordinary integrals, may be derived directly from the definition. For example, if the arc $C$ is traversed in the opposite direction, the line integral changes sign. We let $-C$ denote the arc $C$ traversed in the opposite direction. We can write:

$$\int_C f(x, y)\, dx = -\int_{-C} f(x, y)\, dx.$$

If $C_1$ is an arc extending from $A_1$ to $A_2$ and $C_2$ is an arc extending from $A_2$ to $A_3$, then

$$\int_{C_1} f(x, y)\, dx + \int_{C_2} f(x, y)\, dx = \int_{C_1 \cup C_2} f(x, y)\, dx, \tag{4}$$

where the symbol $C_1 \cup C_2$ means the arc comprised of the two arcs from $A_1$ to $A_2$ and from $A_2$ to $A_3$. As in the case of ordinary integrals, line integrals satisfy the additive property:

$$\int_C [f(x, y) + g(x, y)]\, dx = \int_C f(x, y)\, dx + \int_C g(x, y)\, dx.$$

Statements similar to those above hold for integrals of the type $\int_C f(x, y)\, dy$.

There is one more type of line integral we can define. If the arc $C$ and the function $f$ are as before and if $s$ denotes arc length along $C$ measured from the point $A$ to the point $B$, we can define **the line integral with respect to the arc length** $s$. We use the symbol

$$\int_C f(x, y)\, ds$$

for this line integral. If $C$ is given in the form $y = g(x)$, we can use the relation $ds = [1 + (g'(x))^2]^{1/2}\, dx$ to define:

$$\int_C f(x, y)\, ds = \int_C f[x, g(x)]\sqrt{1 + (g'(x))^2}\, dx,$$

in which the right side has already been defined. If the curve $C$ is in the form $x = h(y)$, we may write

$$\int_C f(x, y)\, ds = \int_C f[h(y), y]\sqrt{1 + (h'(y))^2}\, dy.$$

If $C$ is in neither the form $y = g(x)$ nor the form $x = h(y)$, it may be broken up into a sum of arcs, each one of which does have the appropriate functional behavior. Then the integrals over each piece may be calculated and the results added.

For ordinary integrals we stated a simple theorem to the effect that if a function $f$ is continuous on an interval $[a, b]$, then it is integrable there. (See page 192.) It can be shown that if $f(x, y)$ is continuous and if the arc $C$ is rectifiable (i.e., has finite length), then the line integrals exist. We shall consider throughout only functions and arcs which are sufficiently smooth so that the line integrals always exist. It is worth remarking that if $C$ consists of a collection of smooth arcs joined together (Fig. 4), then because of (4) the line integral along $C$ exists as the sum of the line integrals taken along each of the subarcs.

Line integrals in three dimensions may be defined similarly to the way they are defined in the plane. An arc $C$ joining the points $A$ and $B$ in three-space may be given parametrically by three equations,

$$x = x(t), \qquad y = y(t), \qquad z = z(t), \quad t_0 \le t \le t_1.$$

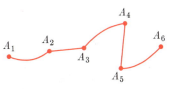

**FIGURE 4**

In some cases they may be given nonparametrically by two equations,

$$y = g_1(x), \qquad z = g_2(x).$$

If $f(x, y, z)$ is a function defined along $C$, then a subdivision of the arc $C$ leads to a sum of the form

$$\sum_{i=1}^{n} f(\xi_i, \eta_i, \zeta_i)(x_i - x_{i-1})$$

which, in turn, is an approximation to the line integral

$$\int_C f(x, y, z)\, dx.$$

Line integrals such as $\int_C f(x, y, z)\, dy$, $\int_C f(x, y, z)\, dz$, and $\int_C f(x, y, z)\, ds$ are defined similarly.

---
**2**
---

## CALCULATION OF LINE INTEGRALS

In the study of integration of functions of one variable, we saw that the definition of integral turned out to be fairly worthless as a tool for computing the value of any specific integral. The methods we employ for performing integration most often use certain properties of integrals, special formulas for antiderivatives, and the techniques described in Chapter 8.

The situation with line integrals is similar. In the last section we defined various types of line integrals, and now we shall exhibit methods for calculating the value of these integrals when the curve $C$ and the function $f$ are given specifically. It is an important fact that *all such integrals may be reduced to ordinary integrations of the type we have already studied.* Once the reduction is made, the problem becomes routine and all the formulas we learned for evaluation of integrals may be used.

The next theorem, stated without proof, establishes the rule for reducing a line integration to an ordinary integration of a function of a single variable.

**THEOREM 1**    *Let $C$ be a (directed) arc of finite length given in the form*

$$C = \{(x, y): x = x(t),\, y = y(t),\quad t_0 \le t \le t_1\}, \tag{1}$$

*so that the point $A(a, b)$ corresponds to $t_0$, and $B(c, d)$ corresponds to $t_1$. Suppose $f(x, y)$ is a continuous function along $C$, and $x'(t)$, $y'(t)$ are continuous. Then*

$$\int_C f(x, y)\, dx = \int_{t_0}^{t_1} f[x(t), y(t)]\, x'(t)\, dt,$$

$$\int_C f(x, y)\, dy = \int_{t_0}^{t_1} f[x(t), y(t)]\, y'(t)\, dt,$$

$$\int_C f(x, y)\, ds = \int_{t_0}^{t_1} f[x(t), y(t)]\sqrt{(x'(t))^2 + (y'(t))^2}\, dt.$$

A similar theorem is valid for line integrals in three-space.

**COROLLARY**  *If the arc C in Theorem 1 is in the form* $y = g(x)$, *then*

$$\int_C f(x, y)\, dy = \int_b^d f[h(y), y]\, dy.$$

For, if $y = g(x)$, then $x$ may be used as a parameter in place of $t$ in (1) and the Corollary is a restatement of Theorem 1. A similar statement holds if $C$ is given by an equation of the type $x = h(y)$. In this case the formula is

$$\int_C f(x, y)\, dy = \int_b^d f[h(y), y]\, dy.$$

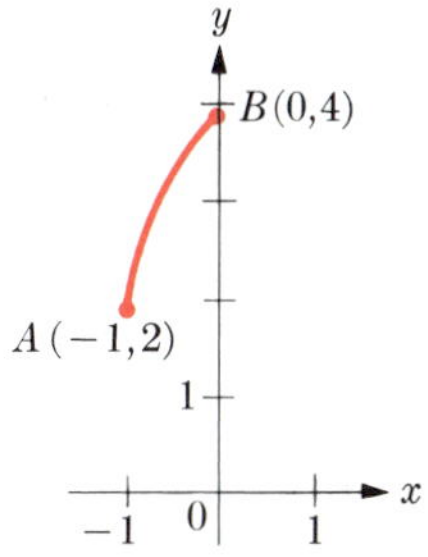

**EXAMPLE 1**  Evaluate the integrals

$$\int_C (x^2 - y^2)\, dx - \int_C 2xy\, dy$$

where $C$ is the arc (Fig. 5):

$$C = \{(x, y): x = t^2 - 1,\ y = t^2 + t + 2,\ 0 \le t \le 1\}.$$

*y*

*B*(0,4)

*A*(−1,2)

1

−1   0   1   *x*

**FIGURE 5**

**Solution**  According to Theorem 1, we compute

$$x'(t) = 2t, \qquad y'(t) = 2t + 1$$

and make the appropriate substitutions. We get

$$\int_C (x^2 - y^2)\, dx = \int_0^1 [(t^2 - 1)^2 - (t^2 + t + 2)^2] \cdot 2t\, dt,$$

$$- \int_C 2xy\, dy = -2 \int_0^1 (t^2 - 1)(t^2 + t + 2)(2t + 1)\, dt.$$

Multiplying out the integrands, we find

$$\int_C (x^2 - y^2)\, dx = 2 \int_0^1 (-2t^3 - 7t^2 - 4t - 3)t\, dt,$$

$$-2 \int_C xy\, dy = -2 \int_0^1 (t^4 + t^3 + t^2 - t - 2)(2t + 1)\, dt.$$

The integration is now routine, and the final result is

$$\int_C [(x^2 - y^2)\, dx - 2xy\, dy] = -2 \int_0^1 (2t^5 + 5t^4 + 10t^3 + 3t^2 - 2t - 2)\, dt$$

$$= -\tfrac{11}{3}. \qquad \square$$

**EXAMPLE 2**  Evaluate the integral

$$\int_C (x^2 - 3xy + y^3)\, dx,$$

where $C$ is the arc

$$C = \{(x, y): y = 2x^2,\ 0 \le x \le 2\}.$$

**Solution**    We have

$$\int_C (x^2 - 3xy + y^3)\,dx = \int_0^2 [x^2 - 3x(2x^2) + (2x^2)^3]\,dx$$

$$= \left[\frac{x^3}{3} - \frac{3}{2}x^4 + \frac{8}{7}x^7\right]_0^2 = \frac{2624}{21}. \qquad \square$$

**EXAMPLE 3**    Evaluate

$$\int_C y\,ds$$

where $C$ is the arc

$$C = \{(x, y): y = \sqrt{x}, 0 \le x \le 6\}.$$

**Solution**    We have

$$ds = \sqrt{1 + \left(\frac{dy}{dx}\right)^2}\,dx = \frac{1}{2}\sqrt{\frac{1 + 4x}{x}}\,dx,$$

and therefore

$$\int_C y\,ds = \frac{1}{2}\int_0^6 \sqrt{x}\,\sqrt{\frac{1 + 4x}{x}}\,dx = \frac{1}{8}\int_0^6 \sqrt{1 + 4x}\,d(1 + 4x)$$

$$= \left[\frac{1}{8}\cdot\frac{2}{3}(1 + 4x)^{3/2}\right]_0^6 = \frac{31}{3}. \qquad \square$$

The next example shows how to evaluate integrals when the arc $C$ consists of several parts.

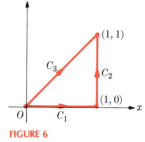

FIGURE 6

**EXAMPLE 4**    Evaluate

$$\int_C [(x + 2y)\,dx + (x^2 - y^2)\,dy],$$

where $C$ is the line segment $C_1$ from $(0, 0)$ to $(1, 0)$ followed by the line segment $C_2$ from $(1, 0)$ to $(1, 1)$ (Fig. 6).

**Solution**    Along $C_1$ we have $x = x$, $y = 0$, $0 \le x \le 1$, so $dy = 0$, and

$$\int_{C_1} [(x + 2y)\,dx + (x^2 - y^2)\,dy] = \int_0^1 x\,dx = \tfrac{1}{2}.$$

Along $C_2$ we have $x = 1$, $y = y$, and so $dx = 0$. We obtain

$$\int_{C_2} [(x + 2y)\,dx + (x^2 - y^2)\,dy] = \int_0^1 (1 - y^2)\,dy = \tfrac{2}{3}.$$

Therefore

$$\int_C [(x + 2y)\,dx + (x^2 - y^2)\,dy] = \tfrac{1}{2} + \tfrac{2}{3} = \tfrac{7}{6}. \qquad \square$$

The next example illustrates the method for evaluation of line integrals in three-space.

**EXAMPLE 5** Evaluate the integral

$$\int_C [(x^2 + y^2 - z^2)\, dx + yz\, dy + (x - y)\, dz]$$

where $C$ is the arc

$$C = \{(x, y, z): x = t^2 + 2, y = 2t - 1, z = 2t^2 - t, 0 \le t \le 1\}. \qquad (2)$$

**Solution** We substitute for $x$, $y$, $z$ from (2) and insert the values $dx = 2t\, dt$, $dy = 2\, dt$, $dz = (4t - 1)\, dt$, to obtain

$$\int_0^1 \{[(t^2 + 2)^2 + (2t - 1)^2 - (2t^2 - t)^2]2t\, dt + (2t + 1)(2t^2 - t)2\, dt$$

$$+ (t^2 - 2t + 3)(4t - 1)\, dt\}.$$

Upon multiplying out all the terms and performing the resulting routine integration we get the value 263/30. ☐

**EXAMPLE 6** Evaluate the integral

$$\int_C (xy + z)\, ds$$

where $C$ is the helix

$$C = \{(x, y, z): x = a \cos t, y = a \sin t, z = bt, 0 \le t \le 2\pi\}.$$

**Solution** Using Theorem 1 in three dimensions, we have

$$\int_C (xy + z)\, ds = \int_0^{2\pi} (a^2 \cos t \sin t + bt)\sqrt{a^2(-\sin t)^2 + a^2 \cos^2 t + b^2}\, dt$$

$$= \int_0^{2\pi} (a^2 \cos t \sin t + bt)\sqrt{a^2 + b^2}\, dt$$

$$= \sqrt{a^2 + b^2}\left[\frac{a^2 \sin^2 t}{2} + \frac{bt^2}{2}\right]_0^{2\pi}$$

$$= 2\pi^2 b\sqrt{a^2 + b^2}. \qquad ☐$$

## 2 PROBLEMS

In each of Problems 1 through 10, evaluate $\int_C (P\, dx + Q\, dy)$ and draw a sketch of the arc $C$.

**1** $\int_C [(x + y)\, dx + (x - y)\, dy]$ where $C$ is the line segment from $(0, 0)$ to $(2, 1)$.

**2** $\int_C [(x + y)\, dx + (x - y)\, dy]$ where $C$ consists of the line segment from $(0, 0)$ to $(2, 0)$ followed by that from $(2, 0)$ to $(2, 1)$.

**3** $\int_C [(x^2 - 2y)\, dx + (2x + y^2)\, dy]$ where $C$ is the arc of

$y^2 = 4x - 1$ going from $(\frac{1}{2}, -1)$ to $(\frac{5}{4}, 2)$.

**4** $\int_C [(x^2 - 2y)\, dx + (2x + y^2)\, dy]$ where $C$ is the line segment going from $(\frac{1}{2}, -1)$ to $(\frac{5}{4}, 2)$.

**5** $\int_C [y\, dx + (x^2 + y^2)\, dy]$ where $C$ is the arc of the circle $y = \sqrt{4 - x^2}$ from $(-2, 0)$ to $(0, 2)$.

**6** $\int_C [y\, dx + (x^2 + y^2)\, dy]$ where $C$ consists of the line segment from $(-2, 0)$ to $(0, 0)$ followed by that from $(0, 0)$ to $(0, 2)$.

7   $\int_C \left( \dfrac{x^2}{\sqrt{x^2 - y^2}}\, dx + \dfrac{2y}{4x^2 + y^2}\, dy \right)$

where $C$ is the arc $y = \frac{1}{2}x^2$ from $(0, 0)$ to $(2, 2)$.

8   $\int_C \left( \dfrac{x^2}{\sqrt{x^2 - y^2}}\, dx + \dfrac{2y}{4x^2 + y^2}\, dy \right)$

where $C$ consists of the line segment from $(0, 0)$ to $(2, 0)$, followed by the line segment from $(2, 0)$ to $(2, 2)$.

9   $\int_C \left( \dfrac{-y}{x\sqrt{x^2 - y^2}}\, dx + \dfrac{1}{\sqrt{x^2 - y^2}}\, dy \right)$

where $C$ is the arc of $x^2 - y^2 = 9$ from $(3, 0)$ to $(5, 4)$.

10   Same integral as in Problem 9, where $C$ consists of the line segment from $(3, 0)$ to $(5, 0)$, followed by the line segment from $(5, 0)$ to $(5, 4)$.

11   Calculate $\int_C \sqrt{x + (3y)^{5/3}}\, ds$ where $C$ is the arc $y = \frac{1}{3}x^3$ going from $(0, 0)$ to $(3, 9)$.

12   Calculate $\int_C \sqrt{x + 3y}\, ds$ where $C$ is the straight line segment going from $(0, 0)$ to $(3, 9)$.

13   Calculate $\int_C y^2 \sin^3 x \sqrt{1 + \cos^2 x}\, ds$ where $C$ is the arc $y = \sin x$ going from $(0, 0)$ to $(\pi/2, 1)$.

14   Calculate $\int_C (2x^2 + 3y^2 - xy)\, ds$ where $C$ is the arc

$$\left. \begin{array}{l} x = 3\cos t \\ y = 3\sin t \end{array} \right\}, \quad 0 \leq t \leq \frac{\pi}{4}.$$

15   Calculate $\int_C x^2\, ds$ where $C$ is the arc $x = 2y^{3/2}$ going from $(2, 1)$ to $(16, 4)$.

16   Calculate $\int_C xyz\, ds$ where $C$ is the line segment from $(0, 0, 0)$ to $(2, 3, -1)$.

17   Calculate $\int_C e^{2x+y}\, ds$ where $C$ is the line segment from $(0, 0)$ to $(6, -5)$.

18   Calculate $\int_C \sqrt{y + z}\, ds$ where $C$ is the line segment from $(0, 0, 0)$ to $(2, 1, 3)$.

*19   Calculate $\int_C xy^2\, ds$ where $C$ is the arc given by $x = t^3$, $y = t^2$ where $-1 \leq t \leq 1$.

20   Calculate $\int_C xy^2\, ds$ where $C$ is the same arc as in Problem 19, but traversed in the opposite direction.

21   Calculate $\int_C (xy/z)\, ds$ where $C$ is the arc given by $x = t$, $y = t^2$, $z = \frac{2}{3}t^3$, $2 \geq t \geq 1$ (that is, $t = 2$ is the starting point of $C$, and $t = 1$ is the termination point).

*22   A spring has the shape of a helix: $x = a\cos t$, $y = a\sin t$, $z = bt$. One complete coil of the spring therefore corresponds to $0 \leq t \leq 2\pi$. Calculate the mass $m$ of one coil of the spring if the density at $(x, y, z)$ is $\rho = x^2 + y^2 + z^2$.

*23   Let $d_z(x, y, z)$ denote the perpendicular distance of the spring of Problem 22 from the $z$ axis. The *moment of inertia* $I_z$ of the spring about the $z$ axis is defined to be $\int_C d_z^2(x, y, z)\rho(x, y, z)\, ds$, where $\rho$ denotes the density of the spring, and $C$ is the arc describing the spring as given in

Problem 22. Calculate the moment of inertia $I_z$ of the spring if $d_z^2(x, y, z) = x^2 + y^2 - a^2$.

*24   Calculate the moments of inertia $I_x$ and $I_y$ of the spring described in Problem 22 (see also Problem 23).

25   Calculate $\int_C [(x^2 + y^2)\, dx + (x^2 - y^2)\, dy]$ where $C$ is the arc

$$C = \{(x, y): x = t^2 + 3,\, y = t - 1,\, 1 \leq t \leq 2\}.$$

26   Calculate $\int_C [\sin x\, dy + \cos y\, dx]$ where $C$ is the arc

$$C = \{(x, y): x = t^2 + 3,\, y = 2t^2 - 1,\, 0 \leq t \leq 2\}.$$

27   Calculate $\int_C [(x - y)\, dx + (y - z)\, dy + (z - x)\, dz]$ where $C$ is the line segment extending from $(1, -1, 2)$ to $(2, 3, 1)$.

28   Calculate $\int_C [(x^2 - y^2)\, dx + 2xz\, dy + (xy - yz)\, dz]$ where $C$ is the line segment

$$C = \{(x, y, z): x = 2t - 1,\, y = t + 1,\, z = t - 2,\, 0 \leq t \leq 3\}.$$

29   Calculate $\int_C [(x - y + z)\, dx + (y + z - x)\, dy + (z + x - y)\, dz]$ where $C$ consists of straight line segments connecting the points $(1, -1, 2)$, $(2, -1, 2)$, $(2, 3, 2)$, and $(2, 3, 1)$, in that order.

30   Calculate

$$\int_C \frac{x\, dx + y\, dy + z\, dz}{x^2 + y^2 + z^2}$$

where $C$ is the arc $x = 2t$, $y = 2t + 1$, $z = t^2 + t$, joining the points $(0, 1, 0)$ and $(2, 3, 2)$.

31   Same as Problem 30, where $C$ is the straight line segment joining $(0, 1, 0)$ and $(2, 3, 2)$.

32   Evaluate

$$\int_C \frac{y\, dx + x\, dy}{\sqrt{x^2 + y^2}}$$

where $C$ is the *closed curve*

$$C = \{(x, y): x = \cos t,\, y = \sin t,\, -\pi \leq t \leq \pi\}.$$

33   Evaluate

$$\int_C \frac{-y\, dx + x\, dy}{\sqrt{x^2 + y^2}}$$

where $C$ is the same curve as in Problem 32.

34   Write out a proof of the formula

$$\int_C (f + g)\, dx = \int_C f\, dx + \int_C g\, dx.$$

35   Suppose $|f(x, y)| \leq M$ for all points $(x, y)$ on an arc $C$. If $C$ is of length $L$, establish the inequality

$$\left| \int_C f(x, y)\, ds \right| \leq ML.$$

**36** Suppose a closed curve $C$ consists of the segment $L = \{(x, y): a \le x \le b, y = 0\}$ and the arc $K = \{(x, y): y = f(x), a \le x \le b\}$ where the endpoints of $K$ are at $(a, 0)$ and $(b, 0)$ and where $f(x) \ge 0$. Show that the area enclosed by $C$ is given by $\int_C x\, dy$ where the integral is taken counterclockwise. Extend the result to a general closed curve $C$.

**37** Let $f(x, y)$ have continuous first derivatives in a region containing a smooth arc $C$ with endpoints $A(x_0, y_0)$, $B(x_1, y_1)$. Show that

$$\int_C [f_x\, dx + f_y\, dy] = f(x_1, y_1) - f(x_0, y_0).$$

(*Hint:* Assume $C$ is given parametrically by $C = \{(x, y): x = x(t), y = y(t), t_0 \le t \le t_1\}$. Then define $\phi(t) = f[x(t), y(t)]$ and use the Chain Rule.)

**38** Suppose that $\mathbf{f}$ is a vector function in the plane. That is, $\mathbf{f}(x, y) = P(x, y)\mathbf{i} + Q(x, y)\mathbf{j}$. Define the vector differential $d\mathbf{v} = (dx)\mathbf{i} + (dy)\mathbf{j}$. Then it is natural to define

$$\int_C \mathbf{f} \cdot d\mathbf{v} = \int_C [P\, dx + Q\, dy].$$

If $g(x, y)$ is a scalar function which has two derivatives, show that

$$\int_C \nabla g \cdot d\mathbf{v} = 0$$

if $C$ is a smooth closed curve. (*Hint:* Recall the theorem on exact differentials of Chapter 15, Section 13, or use the result in Problem 37.)

---

**3**

---

## PATH-INDEPENDENT LINE INTEGRALS

In general, the value of a line integral depends on the integrand, on the two endpoints, and on the arc connecting these endpoints. However, there are special circumstances when the value of a line integral depends solely on the integrand and endpoints but *not* on the arc on which the integration is performed. When such conditions prevail, we say that the integral is **independent of the path**. The next theorem establishes the connection between path-independent integrals and exact differentials. (See Section 13 of Chapter 15.)

We recall here, for the reader's convenience, that if there is a function $f(x, y)$ such that $df = P(x, y)\, dx + Q(x, y)\, dy$ for all $(x, y)$ in some region and for all values of $dx$ and $dy$, then the expression $P(x, y)\, dx + Q(x, y)\, dy$ is an **exact differential**. We also saw in Chapter 15 (Theorem 16, page 693) that if

$$P, \quad Q, \quad \frac{\partial P}{\partial y}, \quad \text{and} \quad \frac{\partial Q}{\partial x}$$

are continuous in a rectangle $S$, then the expression $P(x, y)\, dx + Q(x, y)\, dy$ is an exact differential in $S$ if and only if

$$\frac{\partial P}{\partial y} = \frac{\partial Q}{\partial x}$$

for all $(x, y)$ in $S$.

**THEOREM 2**

*Suppose that $P(x, y)\,dx + Q(x, y)\,dy$ is an exact differential. That is, there is a function $f(x, y)$ with*

$$df = P\,dx + Q\,dy.$$

*Let C be an arc given parametrically by*

$$x = x(t), \qquad y = y(t), \quad t_0 \leq t \leq t_1$$

*where $x'(t)$, $y'(t)$ are continuous. Then*

$$\int_C (P\,dx + Q\,dy) = f[x(t_1), y(t_1)] - f[x(t_0), y(t_0)].$$

*Thus the integral depends only on the endpoints and not on the arc C joining them.*

**Proof**    We define the function $F(t)$ by

$$F(t) = f[x(t), y(t)], \quad t_0 \leq t \leq t_1.$$

We use the Chain Rule to calculate the derivative:

$$F'(t) = (f_x)x'(t) + (f_y)y'(t)$$

so that

$$F'(t) = P[x(t), y(t)]x'(t) + Q[x(t), y(t)]y'(t). \tag{1}$$

Integrating both sides of (1) with respect to $t$ and employing Theorem 1 we find that

$$F(t_1) - F(t_0) = \int_C (P\,dx + Q\,dy).$$

The result follows when we note that

$$F(t_1) = f[x(t_1), y(t_1)] \qquad \text{and} \qquad F(t_0) = f[x(t_0), y(t_0)]. \qquad \square$$

**COROLLARY 1**

*If $P\,dx + Q\,dy + R\,dz$ is an exact differential, then*

$$\int_C (P\,dx + Q\,dy + R\,dz) = f[x(t_1), y(t_1), z(t_1)] - f[x(t_0), y(t_0), z(t_0)]$$

*where $df = P\,dx + Q\,dy + R\,dz$ and the curve C is given by*

$$C = \{(x, y, z): x = x(t), y = y(t), z = z(t), t_0 \leq t \leq t_1\}.$$

Theorem 2 and Corollary 1 may be expressed in vector form. Define the vector $d\mathbf{r} = (dx)\mathbf{i} + (dy)\mathbf{j}$. Then $df = \nabla f \cdot d\mathbf{r}$ where $\nabla f = f_x\mathbf{i} + f_y\mathbf{j} = P\mathbf{i} + Q\mathbf{j}$. If the endpoints of the arc $C$ are denoted by $A$ and $B$, then Theorem 2 takes the form of Corollary 2 below. If the gradient of $f$, $\nabla f$, is thought of as the "derivative" of $f$, then Corollary 2 is analogous to the Fundamental Theorem of Calculus.

**COROLLARY 2**     **(The Fundamental Theorem for Line Integrals)**     *Let* $C: \mathbf{r} = \mathbf{r}(t) = x(t)\mathbf{i} + y(t)\mathbf{j}$, $t_0 \leq t \leq t_1$, *where* $x'(t)$, $y'(t)$ *are continuous. Denote the endpoints of* $C$ *by* $A$ *and* $B$. *Let* $f$ *be defined and continuous on a region containing* $C$. *If* $\nabla f$ *is also defined and continuous, then*

$$\int_C \nabla f \cdot d\mathbf{r} = f(B) - f(A). \tag{2}$$

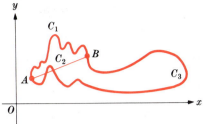

**FIGURE 7**

Three different smooth curves connecting $A$ and $B$. The line integral is the same along all three curves if the integrand is an exact differential.

Similarly, if $d\mathbf{r} = (dx)\mathbf{i} + (dy)\mathbf{j} + (dz)\mathbf{k}$ and $\nabla f = P\mathbf{i} + Q\mathbf{j} + R\mathbf{k}$, then Corollary 1 is the assertion (2) above where now $C$ is an arc in $R^3$.

The last sentence in Theorem 2 states that the line integral does not depend on the actual arc chosen, but only on the endpoints of the arc. (See Fig. 7.) This property of not depending on the actual arc traversed is known as **path independence** of the line integral. We can use path independence to calculate line integrals. Examples 1 and 2 illustrate two methods.

**EXAMPLE 1**     Show that the integrand of

$$\int_C [(2x + 3y)\, dx + (3x - 2y)\, dy]$$

is an exact differential and find the value of the integral over any arc $C$ going from the point $(1, 3)$ to the point $(-2, 5)$.

**Solution**     Setting $P = 2x + 3y$, $Q = 3x - 2y$, we have

$$\frac{\partial P}{\partial y} = 3 = \frac{\partial Q}{\partial x}.$$

By Theorem 16 of Chapter 15, the integrand is an exact differential. Using the methods of Section 13 of Chapter 15 for integrating exact differentials, we find that

$$f(x, y) = x^2 + 3xy - y^2 + C_1.$$

Therefore

$$\int_C [(2x + 3y)\, dx + (3x - 2y)\, dy] = f(-2, 5) - f(1, 3) = -52. \qquad \square$$

Notice that in the evaluation process the constant $C_1$ disappears.

**EXAMPLE 2**     Show that the integrand of

$$\int_C [(3x^2 + 6xy)\, dx + (3x^2 - 3y^2)\, dy]$$

is an exact differential, and find the value of the integral over any arc $C$ going from the point $(1, 1)$ to the point $(2, 3)$.

**Solution**     Setting $P = 3x^2 + 6xy$, $Q = 3x^2 - 3y^2$, we have

$$\frac{\partial P}{\partial y} = 6x = \frac{\partial Q}{\partial x}.$$

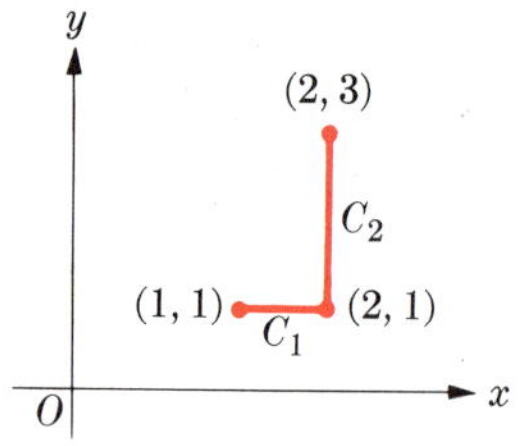

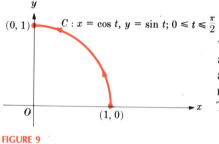

**FIGURE 8**

Instead of finding the function $f$ with the property that $df = P\,dx + Q\,dy$, we may pick *any* simple path joining $(1, 1)$ and $(2, 3)$ and evaluate the integral along that path. We select the horizontal path $C_1$ from $(1, 1)$ to $(2, 1)$, followed by the vertical path $C_2$ from $(2, 1)$ to $(2, 3)$, as shown in Fig. 8. The result is

$$\int_{C_1} (3x^2 + 6x)\,dx + \int_{C_2} (12 - 3y^2)\,dy = [x^3 + 3x^2]_1^2 + [12y - y^3]_1^3 = 14.$$

When working with vectors, Corollary 2 is useful in evaluating line integrals, as the next example illustrates.

**EXAMPLE 3**   Calculate the line integral $\int_C \mathbf{g} \cdot d\mathbf{r}$, where $\mathbf{g}(x, y) = (e^y - 8xy)\mathbf{i} + (xe^y - 4x^2)\mathbf{j}$, and $\mathbf{r}(t) = \cos t\mathbf{i} + \sin t\mathbf{j}$, $0 \le t \le \pi/2$.

**Solution**   We first check to see if $\mathbf{g}$ is a gradient. We note that $\mathbf{g}(x, y)$ has the form $P(x, y)\mathbf{i} + Q(x, y)\mathbf{j}$, with $P(x, y) = e^y - 8xy$, and $Q(x, y) = xe^y - 4x^2$. Since

$$\frac{\partial P}{\partial y} = e^y - 8x = \frac{\partial Q}{\partial x},$$

which is continuous everywhere, we conclude that $\mathbf{g}$ is an exact differential and that there exists a function $f$ such that $\nabla f = \mathbf{g}$. Moreover, antidifferentiation shows that $f(x, y) = xe^y - 4x^2y + k$. The endpoints of $C$ are $\mathbf{r}(0)$ and $\mathbf{r}(\pi/2)$ which in rectangular coordinates are $(1, 0)$ and $(0, 1)$. (See Fig. 9.) Therefore by Corollary 2 we have

$$\int_C \mathbf{g} \cdot d\mathbf{r} = \int_C \nabla f \cdot d\mathbf{r} = f(1, 0) - f(0, 1) = 1.$$

**FIGURE 9**

An application of line integrals* occurs in the determination of the work done by a force acting on a particle in motion along a path. We recall that we considered the work done in the case of the motion of objects along a straight line. (See page 353.) The basic formula depends on the elementary idea that if the force $F$ is constant and if the distance the particle moves (along a straight line) is $d$, then the work $W$ is given by the relation

$$W = Fd.$$

For motion in the plane, *force* is a vector quantity, since it has both *magnitude* and, at the point of application, a *direction* along which it acts. We employ the representation

$$\mathbf{F} = P\mathbf{i} + Q\mathbf{j}$$

for a force vector in the plane. Suppose a particle is constrained to move in the direction of a straight line $l$. If the motion is caused by a force $\mathbf{F}$, then the only portion of $\mathbf{F}$ which has any effect on the motion is the component of $\mathbf{F}$ in

---

*The remainder of this section as well as Problems 22 through 28 at the end of the section is an optional topic and may be omitted without loss of continuity.

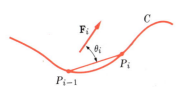

Projection of **F** on *l*

**FIGURE 10**

the direction of *l*. We recall that this quantity is called the **projection** of **F** on *l* (Fig. 10), and is given by

$$|\mathbf{F}| \cos \theta,$$

where $\theta$ is the angle between the direction of **F** and the direction of *l*.

If the particle moves along *l* a distance *d* and if the force **F** is constant then, according to the elementary principle, the work done is

$$W = |\mathbf{F}| \cos \theta \cdot d.$$

We may represent the quantity above in another way. Let **r** be a vector of length *d* in the direction of *l*. Then we may write $|\mathbf{r}| = d$ and

$$W = |\mathbf{F}||\mathbf{r}| \cos \theta.$$

We recognize this quantity as the inner, or scalar, product of the vectors **F** and **r**. That is,

$$W = \mathbf{F} \cdot \mathbf{r}, \tag{3}$$

where **r** is a vector of the appropriate length in the direction of motion. Note that work is a scalar quantity.

Using (3) as the basic formula for motion in the plane, we proceed to *define* work when the force is variable and the motion is along a curved path. We write

$$\mathbf{F} = P(x, y)\mathbf{i} + Q(x, y)\mathbf{j}$$

for a force which may vary from point to point. We let *C* be an arc and suppose that a particle is constrained so that it must move along *C* (as a bead on a wire); we suppose the motion is caused by a force **F**. We let the arc *C* be given by the equations

$$x = x(t), \qquad y = y(t), \quad a \le t \le b,$$

and make a subdivision $a = t_0 < t_1 < \cdots < t_{n-1} < t_n = b$ of the interval $a \le t \le b$. We thus get a subdivision $P_0, P_1, \ldots, P_{n-1}, P_n$ of the arc *C*. We now replace each subarc by a straight line segment and *assume* that the force is approximately constant along each such subarc. Denoting the *i*th subarc by $P_{i-1}P_i$ and the force along this subarc by $\mathbf{F}_i$, we obtain, for the work done along this arc, the approximate quantity (Fig. 11),

**FIGURE 11**

$$W_i = |\mathbf{F}_i| \cos \theta_i \sqrt{[x(t_i) - x(t_{i-1})]^2 + [y(t_i) - y(t_{i-1})]^2}.$$

If we introduce the vector

$$(\Delta_i x)\mathbf{i} + (\Delta_i y)\mathbf{j}$$

and write $\mathbf{F}_i = P_i \mathbf{i} + Q_i \mathbf{j}$ then, taking (3) into account, the work is

$$W_i = P_i \Delta_i x + Q_i \Delta_i y.$$

The total work done is the sum of the amounts of work done on the individual subarcs. (See page 353, Principle 1.) With $\tau_i$ denoting any point in the interval $[t_{i-1}, t_i]$, we have, approximately,

$$W = \sum_{i=1}^{n} \{P[x(\tau_i), y(\tau_i)]\Delta_i x + Q[x(\tau_i), y(\tau_i)]\Delta_i y\}.$$

Proceeding to the limit, we *define* the **total work** by the formula

$$\text{Total Work} = W = \int_C (P\,dx + Q\,dy).$$

For motion in three dimensions with the force given by a vector

$$\mathbf{F} = P(x, y, z)\mathbf{i} + Q(x, y, z)\mathbf{j} + R(x, y, z)\mathbf{k},$$

work is defined by the formula

$$\text{Total Work} = W = \int_C (P\,dx + Q\,dy + R\,dz).$$

**EXAMPLE 4**   A particle moves along the curve $y = x^2$ from the point $(1, 1)$ to the point $(3, 9)$. If the motion is caused by the force $\mathbf{F} = (x^2 - y^2)\mathbf{i} + x^2 y\mathbf{j}$ applied to the particle, find the total work done.

**Solution**   Using the formula for work, we have

$$W = \int_C [(x^2 - y^2)\,dx + x^2 y\,dy].$$

Employing the normal methods for calculating such integrals, we find

$$W = \int_1^3 [(x^2 - x^4)\,dx + x^2(x^2)(2x)\,dx]$$

$$= [\tfrac{1}{3}x^6 - \tfrac{1}{5}x^5 + \tfrac{1}{3}x^3]_1^3 = \tfrac{3044}{15}. \qquad \square$$

## 3   PROBLEMS

In each of Problems 1 through 11, show that the integrand is an exact differential and evaluate the integral.

1   $\int_C [(x^2 + 2y)\,dx + (2y + 2x)\,dy]$ where $C$ is any arc from $(2, 1)$ to $(4, 2)$.

2   $\int_C [(3x^2 + 4xy - 2y^2)\,dx + (2x^2 - 4xy - 3y^2)\,dy]$ where $C$ is any arc from $(1, 1)$ to $(3, 2)$.

3   $\int_C (e^x \cos y\,dx - e^x \sin y\,dy)$ where $C$ is any arc from $(1, 0)$ to $(0, 1)$.

4   $\int_C \left[ \left( \dfrac{2xy^2}{1 + x^2} + 3 \right) dx + (2y \ln(1 + x^2) - 2)\,dy \right]$

where $C$ is any arc from $(0, 2)$ to $(5, 1)$.

5   $\int_C \left[ \dfrac{y^2}{(x^2 + y^2)^{3/2}}\,dx - \dfrac{xy}{(x^2 + y^2)^{3/2}}\,dy \right]$

where $C$ is any arc from $(4, 3)$ to $(-3, 4)$ which does not pass through the origin.

6   $\int_C \left[ \dfrac{x}{\sqrt{1 + x^2 + y^2}}\,dx + \dfrac{y}{\sqrt{1 + x^2 + y^2}}\,dy \right]$

where $C$ is any arc from $(-2, -2)$ to $(4, 1)$.

7   $\int_C \{[ye^{xy}(\cos xy - \sin xy) + \cos x]\,dx$
$\qquad\qquad + [xe^{xy}(\cos xy - \sin xy) + \sin y]\,dy\}$

where $C$ is any arc from $(0, 0)$ to $(3, -2)$.

8   $\int_C [(2x - 2y + z + 2)\,dx + (2y - 2x - 1)\,dy + (-2z + x)\,dz]$

where $C$ is any arc from $(1, 0, 2)$ to $(3, -1, 4)$.

9   $\int_C [(2x + y - z)\,dx + (-2y + x + 2z + 3)\,dy$
$\qquad\qquad + (4z - x + 2y - 2)\,dz]$

where $C$ is any arc from $(0, 2, -1)$ to $(1, -2, 4)$.

10   $\int_C [(3x^2 - 3yz + 2xz)\,dx + (3y^2 - 3xz + z^2)\,dy$
$\qquad\qquad + (3z^2 - 3xy + x^2 + 2yz)\,dz]$

where $C$ is any arc from $(-1, 2, 3)$ to $(3, 2, -1)$.

**11** $\int_C [(yze^{xyz} \cos x - e^{xyz} \sin x + y \cos xy + z \sin xz) \, dx$
$+ (xze^{xyz} \cos x + x \cos xy) \, dy$
$+ (xye^{xyz} \cos x + x \sin xz) \, dz]$
where $C$ is any arc from $(0, 0, 0)$ to $(-1, -2, -3)$.

In each of Problems 12 through 17, calculate the line integral $\int_C \mathbf{g} \cdot d\mathbf{r}$, for the given vector function $\mathbf{g}$ and the arc described by $\mathbf{r}$.

**12** $\mathbf{g} = x^2\mathbf{i} - 3y^2\mathbf{j}$; $\mathbf{r}(t) = 2 \cos t\mathbf{i} + 3 \sin t\mathbf{j}$, $0 \le t \le 2\pi$.

**13** $\mathbf{g} = (2y + \ln y)\mathbf{i} + (2x + x/y)\mathbf{j}$; $\mathbf{r}(t) = 2t\mathbf{i} + 3t^2\mathbf{j}$; $1 \le t \le 2$.

**14** $\mathbf{g} = (e^{x+y} + \sin(\pi y))\mathbf{i} + (e^x e^y + \pi x \cos(\pi y))\mathbf{j}$; $\mathbf{r}(t)$
$= 2t\mathbf{i} + \frac{1}{3}t^2\mathbf{j}$, $0 \le t \le 9$.

**15** $\mathbf{g} = (2xy - y^2)\mathbf{i} + (x^2 - 2xy)\mathbf{j}$; $\mathbf{r}(t) = \cos t\mathbf{i}$
$+ \sin t\mathbf{j}$, $0 \le t \le \pi$.

**16** $\mathbf{g} = 3x(x^2 + y^4)^{1/2}\mathbf{i} + 6y^3(x^2 + y^4)^{1/2}\mathbf{j}$; $\mathbf{r}(t)$ describes the arc $y = (1 - x^2)^{1/2}$, from $(1, 0)$ to $(-1, 0)$.

**17** $\mathbf{g} = (y \cos(xy) + 2xy^3)\mathbf{i} + (x \cos(xy) + 3x^2y^2)\mathbf{j}$; $\mathbf{r}(t)$ is any differentiable arc (of finite length) from $(0, 0)$ to $(2, 3\pi)$.

**18** Find the value of the integral in Problem 5 where $C$ is a circle to which the origin is exterior. Show that the result is independent of the size of the circle selected.

**19** Let $\alpha_i$, $\beta_i$, $i = 1, 2, \ldots, k$ be constants, and suppose that $P_i \, dx + Q_i \, dy$, $i = 1, 2, \ldots, k$ are exact differentials. Show that

$$\sum_{i=1}^{k} [(\alpha_i P_i) \, dx + (\beta_i Q_i) \, dy]$$

is exact if $\alpha_i = \beta_i$, $i = 1, 2, \ldots, k$. Is the condition $\alpha_i = \beta_i$ necessary?

**20** Suppose that $P_i \, dx + Q_i \, dy$, $i = 1, 2$ are exact differentials. Now show that $P_1 P_2 \, dx + Q_1 Q_2 \, dy$ is exact if

$$(P_2 - Q_2)\frac{\partial P_1}{\partial y} = (Q_1 - P_1)\frac{\partial Q_2}{\partial x}.$$

Is this condition necessary? Extend the result to the product of two exact differentials in three variables.

**\*21** Let $P_1(x_1, x_2, x_3, x_4)$, $P_2(x_1, x_2, x_3, x_4)$, $P_3(x_1, x_2, x_3, x_4)$, $P_4(x_1, x_2, x_3, x_4)$ be four functions of four variables such that

$$\frac{\partial P_i}{\partial x_j} = \frac{\partial P_j}{\partial x_i}, \quad i, j = 1, 2, 3, 4.$$

State and prove a theorem similar to Theorem 2. Note that it is first necessary to define a line integral in $R^4$.

Problems 22 through 28 involve the concept of work, a topic in this section which is optional.

**22** A particle is moving along the path

$$x = t + 1, \qquad y = 2t^2 + t + 2$$

from the point $(1, 2)$ to the point $(2, 5)$, subject to the force

$$\mathbf{F} = (x^2 + y)\mathbf{i} + 2xy\mathbf{j}.$$

Find the total work done.

**23** A particle is moving in the $xy$ plane along a straight line from the point $A(a, b)$ to the point $B(c, d)$, subject to the force

$$\mathbf{F} = \frac{-x}{x^2 + y^2}\mathbf{i} - \frac{y}{x^2 + y^2}\mathbf{j}.$$

Find the work done. Show that the work done is the same if a different path between $A$ and $B$ is selected. (The path should not go through the origin.)

**24** A particle moves in the $xy$ plane along the straight line connecting $A(a, b)$ and $B(c, d)$, subject to the force

$$\mathbf{F} = \frac{-x}{(x^2 + y^2)^{3/2}}\mathbf{i} + \frac{-y}{(x^2 + y^2)^{3/2}}\mathbf{j}.$$

Find the work done. Show that the work done is the same if a different path is selected joining the points $A$ and $B$ but not going through the origin.

**25** A particle moves along the straight line (in three space) joining the points $A(a, b, c)$ and $B(d, e, f)$, subject to the force

$$\mathbf{F} = \frac{-x}{(x^2 + y^2 + z^2)^{3/2}}\mathbf{i} - \frac{y}{(x^2 + y^2 + z^2)^{3/2}}\mathbf{j}$$

$$- \frac{z}{(x^2 + y^2 + z^2)^{3/2}}\mathbf{k}.$$

Find the work done. Show that the work done is the same if a different path is selected joining the points $A$ and $B$ but not going through the origin.

**26** Same as Problem 25, with

$$\mathbf{F} = \frac{x}{(x^2 + y^2 + z^2)^2}\mathbf{i} + \frac{y}{(x^2 + y^2 + z^2)^2}\mathbf{j}$$

$$+ \frac{z}{(x^2 + y^2 + z^2)^2}\mathbf{k}.$$

**27** Find the work done on an object that traverses the helix $\mathbf{r}(t) = \cos t\mathbf{i} + \sin t\mathbf{j} + t\mathbf{k}$, $0 \le t \le 2\pi$, if it is subject to a force

$$\mathbf{F} = \frac{k}{|\mathbf{r}|^2}\mathbf{r}.$$

**28** Find the work done on an object that traverses the circular arc $\mathbf{r}(t) = \cos t\mathbf{i} + \sin t\mathbf{j}$, $0 \le t \le 3\pi/2$, subject to a force $\mathbf{F} = 6x(x^2 + y)^2\mathbf{i} + (4y^3 + 3(x^2 + y)^2)\mathbf{j}$.

4

## GREEN'S THEOREM

The Fundamental Theorem of Calculus states that differentiation and integration are inverse processes. An appropriate extension of this theorem to double integrals of functions of two variables is known as Green's Theorem. Suppose that $P$ and $Q$ are smooth (i.e., differentiable, with continuous first derivatives) functions defined in some region $R$ of the plane. **A simple closed curve** is a curve that can be obtained as the union of two arcs which have only their endpoints in common. Thus a circle is the union of two half circles. Of course, any two points on a simple closed curve divide it into two arcs in this way. It is intuitively clear that a simple closed curve in the plane divides the plane into two regions, constituting the "interior" and the "exterior" of the curve. This fact, which is surprisingly hard to prove, is not used in the proofs of any theorems. A **smooth simple closed curve** is one which has a parametric representation $x = x(t)$, $y = y(t)$, $a \le t \le b$, in which $x$, $y$, $x'$, and $y'$ are continuous and $[x'(t)]^2 + [y'(t)]^2 > 0$ and $x(b) = x(a)$, $x'(b) = x'(a)$, $y(b) = y(a)$, $y'(b) = y'(a)$. If $\Gamma$ is a smooth simple closed curve which, together with its interior $G$, is in $R$, then the basic formula associated with Green's Theorem is

$$\iint\limits_G \left( \frac{\partial Q}{\partial x} - \frac{\partial P}{\partial y} \right) dA = \oint_\Gamma (P\,dx + Q\,dy). \tag{1}$$

*The symbol on the right represents the line integral taken in the counter-clockwise sense*, so that $\Gamma$ is traversed with the interior of $G$ always on the left.

We first establish Green's Theorem for regions which have a special shape. Then we indicate how the result for these special regions may be extended to yield the theorem for general domains.

**THEOREM 3**　*Suppose that $G$ is a region bounded by the straight lines $x = a$, $x = b$, $y = c$, and by an arc (situated above the line $y = c$) with equation*

$$y = f(x), \quad a \le x \le b.$$

*Assume that $f$ is smooth ($f$ has a continuous derivative). If $P(x, y)$ and $Q(x, y)$ have continuous derivatives in a region which contains $G$ and its boundary then*

$$\iint\limits_G \left( \frac{\partial Q}{\partial x} - \frac{\partial P}{\partial y} \right) dA = \oint_{\partial G} (P\,dx + Q\,dy), \tag{2}$$

*where the symbol $\partial G$ denotes the boundary of $G$ and the line integral is traversed in a counterclockwise sense.*

**Proof**　We establish the result by proving separately each of the formulas

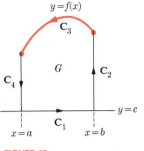

**FIGURE 12**

$$-\iint_G \frac{\partial P}{\partial y} \, dA = \oint_{\partial G} P \, dx, \tag{2a}$$

$$\iint_G \frac{\partial Q}{\partial x} \, dA = \oint_{\partial G} Q \, dy. \tag{2b}$$

Figure 12 shows a typical region $G$ with the arcs directed as shown by the arrows. To prove (2a), we change the double integral to an iterated integral and then employ the Fundamental Theorem of Calculus. We get

$$-\iint_G \frac{\partial P}{\partial y} \, dA = -\int_a^b \int_c^{f(x)} \frac{\partial P}{\partial y} \, dy \, dx$$

$$= -\int_a^b \{P[x, f(x)] - P(x, c)\} \, dx. \tag{3}$$

The right side of (3) may be written as the line integrals

$$\int_{C_3} P(x, y) \, dx + \int_{C_1} P(x, y) \, dx.$$

Since $x$ is constant along $C_2$ and $C_4$, we may write

$$\int_{C_2} P \, dx = \int_{C_4} P \, dx = 0,$$

and so

$$-\iint_G \frac{\partial P}{\partial y} \, dA = \oint_{C_1 \cup C_2 \cup C_3 \cup C_4} P \, dx = \oint_{\partial G} P \, dx.$$

The proof of (2b) is more difficult.

To prove (2b), we define

$$U(x, y) = \int_c^y Q(x, \eta) \, d\eta. \tag{4}$$

We next wish to calculate $U_x(x, y)$. It is shown in more advanced courses that since both $Q$ and $Q_x$ are continuous in the region containing $G$ and its boundary, we can "differentiate under the integral sign" to obtain

$$U_x(x, y) = \int_c^y Q_x(x, \eta) \, d\eta. \tag{5}$$

The Fundamental Theorem of Calculus applied in (4) and (5) yield $U_y(x, y) = Q(x, y)$ and $U_{xy} = Q_x$. Therefore $U_{yx} = Q_x = U_{xy}$. This allows us to apply Theorem 2 to get $\oint_{\partial G} (U_x \, dx + U_y \, dy) = 0$. Therefore

$$\oint_{\partial G} U_x \, dx = -\oint_{\partial G} U_y \, dy = -\oint_{\partial G} Q \, dy. \tag{6}$$

Now we use (2a) with $P = U_x$ to find

$$\iint_G Q_x \, dA = \iint_G U_{xy} \, dA = -\oint_{\partial G} U_x \, dx.$$

Taking (6) into account, we conclude that

$$\iint_G Q_x \, dA = \oint_{\partial G} Q \, dy. \qquad \square$$

By means of a simple change of coordinates or other minor adjustment, we see easily that the above lemma holds for all regions of the type shown in Fig. 13. The next important step is the observation that the result of

FIGURE 13    Shapes of regions to which Theorem 3 applies

Theorem 3 (i.e., Equation (2)) may be established for any region which can be divided up into a finite number of regions, each of the type considered in the theorem. How this may be done is suggested in Fig. 14, which shows a region with a smooth simple closed curve as boundary divided by straight-line segments into a number of regions of special type. It is clear that the double integral over the whole region is the sum of the double integral over the parts. In adding the line integrals we observe that the integrals over the interior segments must cancel, since each segment is directed in opposite senses when considered as part of the boundary of two adjacent special regions. Unfortunately, it is not true that every region for which Formula (2) holds can be divided into special regions in the manner described above.

We shall not attempt to examine the most general type of domain for which Formula (2) is valid, but shall note some examples of regions to which we may apply Theorem 3. The shaded domain shown in Fig. 15 is bounded by four smooth simple closed curves. It may be subdivided into regions to which Theorem 3 applies. *It is important to notice in such a case that the line integral as given in (2) must be traversed so that the region G always remains on the left—counterclockwise for the outer boundary curve and clockwise for the three inner boundary curves.*

A simple closed curve is **piecewise smooth** if it is made up of a finite number of smooth arcs; that is, arcs that have a parametric representation of the form $C: x = x(t)$, $y = y(t)$, $a \leq t \leq b$, where $x$, $x'$, $y$, $y'$ are all continuous. Moreover these arcs are joined at points called corners. A **corner** is the juncture of two smooth arcs which have limiting tangent lines *making a positive angle* (Fig. 16). For a piecewise smooth boundary with corners, it is possible to use Theorem 3, thus establishing Formula (2) for any region which has a boundary consisting of a piecewise smooth simple closed curve. Since polygons have piecewise smooth boundaries, they are included in the collection of regions for which (2) is valid. We now state Green's Theorem in a form which is sufficiently general for most applications.

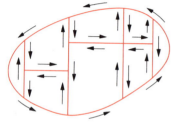

Region subdivided into subregions for which Theorem 3 applies

FIGURE 14

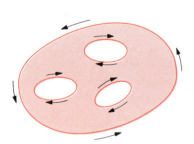

FIGURE 15

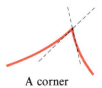

A corner

FIGURE 16

**THEOREM 4**    **(Green's Theorem in the Plane)**   *Suppose that G is a region with a boundary consisting of a finite number of piecewise smooth simple closed curves, no two of which intersect. Suppose that P, Q are continuously differentiable functions defined in a region which contains G and $\partial G$. Then*

$$\iint_G \left( \frac{\partial Q}{\partial x} - \frac{\partial P}{\partial y} \right) dA = \oint_{\partial G} (P\, dx + Q\, dy), \tag{7}$$

*where the integral on the right is defined to be the sum of the integrals over the boundary curves, each of which is directed so that G is on the left.*

The proof of Green's Theorem for all possible regions which may be subdivided into regions of special type has been described above in a discursive manner. Green's Theorem may be stated in vector form. To do this (and in preparation for Stokes' Theorem), we first define the **curl** of a vector function **v**. Suppose $\mathbf{V}(x, y, z) = v_1(x, y, z)\mathbf{i} + v_2(x, y, z)\mathbf{j} + v_3(x, y, z)\mathbf{k}$ where $v_1, v_2, v_3$ and all their first partial derivatives are continuous. We **define** a new vector function $\nabla \times \mathbf{v}$ by the *symbolic* determinant, in which we expand the determinant in minors of the first row.

**DEFINITION**

$$\nabla \times \mathbf{v} = \begin{vmatrix} \mathbf{i} & \mathbf{j} & \mathbf{k} \\ \dfrac{\partial}{\partial x} & \dfrac{\partial}{\partial y} & \dfrac{\partial}{\partial z} \\ v_1 & v_2 & v_3 \end{vmatrix}$$

$$= \left( \frac{\partial v_3}{\partial y} - \frac{\partial v_2}{\partial z} \right)\mathbf{i} + \left( \frac{\partial v_1}{\partial z} - \frac{\partial v_3}{\partial x} \right)\mathbf{j} + \left( \frac{\partial v_2}{\partial x} - \frac{\partial v_1}{\partial y} \right)\mathbf{k}.$$

**DEFINITION**   *Suppose **v** is a vector function $\mathbf{v} = v_1\mathbf{i} + v_2\mathbf{j} + v_3\mathbf{k}$ such that $v_1, v_2, v_3$ and all of their first partial derivatives are continuous. Then the **curl** of the vector function **v** is defined to be*

$$\operatorname{curl} \mathbf{v} = \nabla \times \mathbf{v}.$$

To relate the curl to Green's Theorem we write

$$\mathbf{v} = P\mathbf{i} + Q\mathbf{j} \equiv P\mathbf{i} + Q\mathbf{j} + 0\mathbf{k}$$

and denote by $\mathbf{r}(t) = x\mathbf{i} + y\mathbf{j}$ the vector from the origin $O$ of a rectangular coordinate system in the plane to the boundary $\partial G$ of $G$. In this case

$$\operatorname{curl} \mathbf{v} = \left( \frac{\partial Q}{\partial x} - \frac{\partial P}{\partial y} \right)\mathbf{k}.$$

We can now restate Theorem 4 in vector form:

**THEOREM 4′**    **(Green's Theorem in vector form)**    *With the same hypotheses as in Theorem 4, and letting* $\mathbf{v} = P\mathbf{i} + Q\mathbf{j} + O\mathbf{k}$, *we have*

$$\iint_G \operatorname{curl} \mathbf{v} \cdot \mathbf{k}\, dA = \oint_{\partial G} \mathbf{v} \cdot d\mathbf{r}.$$

The curl of a vector function can be shown to be independent of the coordinate system chosen. Therefore the vector formulation has the advantage of exhibiting the invariance of the result under a change of coordinates. Furthermore, the nature of the extension to three-space will be apparent from the vector formulation.

We now illustrate Green's Theorem in the plane with several examples.

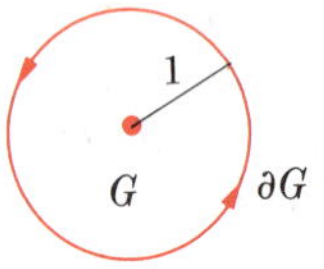

The unit disk traversed counterclockwise

**FIGURE 17**

**EXAMPLE 1**    Verify Green's Theorem when $P(x, y) = 2y$, $Q(x, y) = 3x$ and $G$ is the unit disk,

$$G = \{(x, y): x^2 + y^2 \leq 1\}.$$

(See Fig. 17.)

**Solution**    We have

$$\frac{\partial Q}{\partial x} - \frac{\partial P}{\partial y} = 3 - 2 = 1.$$

Therefore

$$\iint_G \left( \frac{\partial Q}{\partial x} - \frac{\partial P}{\partial y} \right) dA = \iint_G dA = \pi \cdot 1^2 = \pi.$$

The boundary $\partial G$ is given by $x = \cos\theta$, $y = \sin\theta$, $-\pi \leq \theta \leq \pi$. Hence

$$\oint_{\partial G} (P\, dx + Q\, dy) = \int_{-\pi}^{\pi} \{2(\sin\theta)\, d(\cos\theta) + 3(\cos\theta)\, d(\sin\theta)\}$$

$$= \int_{-\pi}^{\pi} (-2\sin^2\theta + 3\cos^2\theta)\, d\theta = -2\pi + 3\pi = \pi. \quad \square$$

**EXAMPLE 2**    If $G$ is the unit disk as in Example 1, use Green's Theorem to evaluate

$$\oint_{\partial G} [(x^2 - y^3)\, dx + (y^2 + x^3)\, dy].$$

**Solution**    Here, $P = x^2 - y^3$, $Q = y^2 + x^3$. Therefore

$$\oint_{\partial G} [(x^2 - y^3)\, dx + (y^2 + x^3)\, dy] = \iint_G 3(x^2 + y^2)\, dA$$

$$= 3 \int_0^{2\pi} \int_0^1 r^2 \cdot r\, dr\, d\theta = \frac{3\pi}{2}. \quad \square$$

**EXAMPLE 3**  Let $G$ be the region outside the unit circle which is bounded on the left by the parabola $y^2 = 2(x + 2)$ and on the right by the line $x = 2$. (See Fig. 18.) Use Green's Theorem to evaluate

$$\int_{C_1} \left( \frac{-y}{x^2 + y^2} dx + \frac{x}{x^2 + y^2} dy \right),$$

where $C_1$ is the outer boundary of $G$ oriented as shown in Fig. 18.

**Solution**  We write $P = -y/(x^2 + y^2)$, $Q = x/(x^2 + y^2)$ and observe that $P$ and $Q$ have singularities at the origin. Denoting the boundary of the unit disk oriented *clockwise* by $C_2$ and noting that $(\partial Q/\partial x) - (\partial P/\partial y) = 0$, we use Green's Theorem to write

$$\int_{C_1 \cup C_2} \left( -\frac{y}{x^2 + y^2} dx + \frac{x}{x^2 + y^2} dy \right) = 0.$$

Therefore

$$\int_{C_1} (P \, dx + Q \, dy) = -\int_{C_2} (P \, dx + Q \, dy) = \int_{-C_2} (P \, dx + Q \, dy),$$

where $-C_2$ is the unit circle oriented *counterclockwise*. Since $x = \cos \theta$, $y = \sin \theta$, $-\pi \leq \theta \leq \pi$ on the unit circle $-C_2$, we obtain $P = -\sin \theta$, $Q = \cos \theta$, $dx = -\sin \theta \, d\theta$, $dy = \cos \theta \, d\theta$ and

$$\int_{C_1} (P \, dx + Q \, dy) = \int_{-\pi}^{\pi} (\sin^2 \theta + \cos^2 \theta) \, d\theta = 2\pi. \qquad \square$$

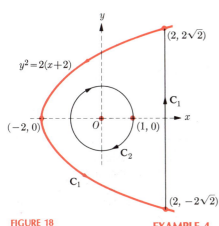

$y^2 = 2(x+2)$

$(2, 2\sqrt{2})$

$(-2, 0)$   $O$   $(1, 0)$   $C_1$

$C_2$

$C_1$

$(2, -2\sqrt{2})$

**FIGURE 18**

**EXAMPLE 4**  Let $\mathbf{v} = -\frac{1}{2}y\mathbf{i} + \frac{1}{2}x\mathbf{j}$ be defined in a region $G$ with area $A$. Show that

$$A = \oint_{\partial G} \mathbf{v} \cdot d\mathbf{r}.$$

**Solution**  We apply Green's Theorem and obtain

$$\oint_{\partial G} \mathbf{v} \cdot d\mathbf{r} = \iint_G \text{curl } \mathbf{v} \cdot \mathbf{k} \, dA = \iint_G \left( \tfrac{1}{2} + \tfrac{1}{2} \right) dA = A,$$

in which we have used the scalar interpretation of curl $\mathbf{v}$ by writing

$$\mathbf{v} = -\tfrac{1}{2}y\mathbf{i} + \tfrac{1}{2}x\mathbf{j} + 0\mathbf{k}. \qquad \square$$

## 4   PROBLEMS

In each of Problems 1 through 8, verify Green's Theorem.

**1** $P(x, y) = -y$, $Q(x, y) = x$; $G: 0 \leq x \leq 1, 0 \leq y \leq 1$

**2** $P = 0$, $Q = x$; $G$ is the region outside the unit circle, bounded below by the parabola $y = x^2 - 2$ and bounded above by the line $y = 2$.

**3** $P = xy$, $Q = -2xy$; $G = \{(x, y): 1 \leq x \leq 2, 0 \leq y \leq 3\}$

**4** $P = e^x \sin y$, $Q = e^x \cos y$; $G = \{(x, y): 0 \leq x \leq 1, 0 \leq y \leq \pi/2\}$

**5** $P = \frac{2}{3}xy^3 - x^2y$, $Q = x^2y^2$; $G$ is the triangle with vertices at $(0, 0)$, $(1, 0)$, and $(1, 1)$.

**6** $P = 0$, $Q = x$; $G$ is the region inside the circle $x^2 + y^2 = 4$ and outside the circles $(x - 1)^2 + y^2 = \frac{1}{4}$, $(x + 1)^2 + y^2 = \frac{1}{4}$.

**7** $\mathbf{v} = (x^2 + y^2)^{-1}(-y\mathbf{i} + x\mathbf{j})$; $G$ is the region between the circles $x^2 + y^2 = 1$ and $x^2 + y^2 = 4$.

**8** $P = 4x - 2y$, $Q = 2x + 6y$; $G$ is the interior of the ellipse $x = 2 \cos \theta$, $y = \sin \theta$, $-\pi \leq \theta \leq \pi$.

In each of Problems 9 through 18, use Green's Theorem to evaluate the given line integrals in a counterclockwise direction.

**9** $\oint_C (xy^2\,dx + x^2y\,dy)$ where $C$ is the circle $x^2 + y^2 = 4$.

**10** $\oint_C [(4xy + y^2)\,dx - (xy + 3x^2)\,dy]$, where $C$ is the circle $(x+2)^2 + (y-3)^2 = 9$.

**11** $\oint_C [(2 + 20xy + 2y^2)\,dx + (12xy + 10x^2)\,dy]$, where $C$ is the square with vertices $(0,0), (3,0), (3,3), (0,3)$.

**12** $\oint_C [5x\,dx + 3y\,dy]$, where $C$ is the circle $(x-1)^2 + (y+1)^2 = 1$.

**13** $\oint_C [y^2\,dx + x^2\,dy]$, where $C$ is the triangle with vertices $(1,1), (4,1), (2,2)$.

**14** $\oint_C [(x^2 + y)\,dx + xy^3\,dy]$, where $C$ is the closed curve determined by $y^2 = x$ and $y = -x$ from $(0,0)$ to $(1,-1)$.

**15** $\oint_C [(x-3y)^2\,dx + (xy)\,dy]$, where $C$ is the closed curve determined by $y = x^3$ and $y = 4x$ from $(0,0)$ to $(2,8)$.

**16** $\oint_C [(x+y)\,dx + (x^2 - y^3)\,dy]$, where $C$ is the boundary of the annulus (ring-shaped region) between $x^2 + y^2 = 1$ and $x^2 + y^2 = 4$.

**17** Find $\oint_C [(x+y)\,dx + (x^2 - y^3)\,dy]$, where $C$ is the curve of Problem 16.

**18** $\oint_C [(e^x \sin y)\,dx + (e^x \cos y)\,dy]$ where $C$ is the ellipse $3x^2 + 8y^2 = 24$.

**19** Let $G$ be a region in the $xy$ plane such that $\partial G$ consists of a finite number of piecewise smooth simple closed curves. Use Green's Theorem to show that the area of $G$, $A(G)$, is given by the formula

$$A(G) = \oint_{\partial G} x\,dy. \qquad (8)$$

In each of Problems 20 through 28, compute the area, $A(G)$, of $G$, using Formula (8) of Problem 19.

**20** $G$ is the triangle with vertices at $(1,1), (4,1)$, and $(4,9)$.

**21** $G$ is the triangle with vertices at $(2,1), (3,4)$, and $(1,5)$.

**22** $G$ is the region given by $G = \{(x,y): 0 \le x \le y^2, 1 \le y \le 3\}$.

**23** $G$ is the region bounded by the line $y = x + 2$ and the parabola $y = x^2$.

**24** $G$ is the region in the first quadrant bounded by the lines $4y = x$ and $y = 4x$ and the hyperbola $xy = 4$.

**25** $G$ is the region interior to the ellipse

$$\frac{x^2}{16} + \frac{y^2}{9} = 1.$$

**26** $G$ is bounded by $x = 2\cos^3 t$, $y = 2\sin^3 t$; $0 \le t \le 2\pi$.

**27** $G$ is the region bounded by one arch of the cycloid $x = 3(t - \sin t)$, $y = 3(1 - \cos t)$, $0 \le t \le 2\pi$ and the $x$ axis.

**28** $G$ is the region bounded by the graphs of the equations $y = x^3$, $y^2 = x$.

In each of Problems 29 through 34, compute $\oint_{\partial G} \mathbf{v} \cdot d\mathbf{r}$, using Green's Theorem.

**29** $\mathbf{v} = (\frac{4}{5}xy^5 + 2y - e^x)\mathbf{i} + (2xy^4 - 4\sin y)\mathbf{j}$;
   $G: 1 \le x \le 2, 1 \le y \le 3$

**30** $\mathbf{v} = (2xe^y - x^2y - \frac{1}{3}y^3)\mathbf{i} + (x^2 e^y + \sin y)\mathbf{j}$; $G: x^2 + y^2 \le 1$

**31** $\mathbf{v} = 2xy^2\mathbf{i} + 3x^2 y\mathbf{j}$; $G$ is the interior of the ellipse

$$\frac{x^2}{a^2} + \frac{y^2}{b^2} = 1.$$

**32** $\mathbf{v} = 2\arctan(y/x)\mathbf{i} + \log(x^2 + y^2)\mathbf{j}$;
   $G = \{(x,y): 1 \le x \le 2, -1 \le y \le 1\}$

**33** $\mathbf{v} = -3x^2 y\mathbf{i} + 3xy^2\mathbf{j}$;
   $G = \{(x,y): -a \le x \le a, 0 \le y \le \sqrt{a^2 - x^2}\}$

**34** $\mathbf{v} = f(x)\mathbf{i} + g(y)\mathbf{j}$; $G$ any region with a finite, smooth boundary.

**\*35** Evaluate

$$\oint_C \frac{x\,dy - y\,dx}{x^2 + y^2},$$

where $C$ consists of the arc of the parabola $y = x^2 - 1$, $-1 \le x \le 2$, followed by the straight segment from $(2,3)$ to $(-1,0)$. Do this by applying Green's Theorem to the region $G$ interior to $C$ and exterior to a small circle of radius $\rho$ centered at the origin.

**\*36** Evaluate

$$\oint_C \frac{x\,dx + y\,dy}{x^2 + y^2},$$

where $C$ is the path described in Problem 35.

**37** Suppose that $f(x,y)$ satisfies the Laplace equation ($f_{xx} + f_{yy} = 0$) in a region $G$. Show that

$$\oint_{\partial G^*} (f_y\,dx - f_x\,dy) = 0,$$

where $G^*$ is any region interior to $G$.

**38** If $(\bar{x}, \bar{y})$ is the location of the center of gravity of a plane region $G$ of uniform density, show that

$$\bar{x} = \frac{\oint_{\partial G} x^2\,dy}{2\oint_{\partial G} x\,dy}, \qquad \bar{y} = \frac{\oint_{\partial G} y^2\,dx}{2\oint_{\partial G} y\,dx}.$$

**\*39** If $f_{xx} + f_{yy} = 0$ in a region $R$ and $v$ is any smooth function, use the identity $(vf_x)_x = vf_{xx} + v_x f_x$ and a similar one for the derivative with respect to $y$, to show that

$$-\oint_{\partial G} v(f_y\,dx - f_x\,dy) = \iint_G (v_x f_x + v_y f_y)\,dA,$$

where $G$ is any region interior to $R$.

### 5

## SURFACES. SURFACE INTEGRALS (OPTIONAL)*

So far we have interpreted the term "surface in three-space" as the graph in a given rectangular coordinate system of an equation of the form $z = f(x, y)$. The definition and computation of areas of general surfaces were taken up in Section 7 of Chapter 16. Now we are interested in studying surfaces which have a structure more complicated than those we have considered previously.

In order to simplify the study of surfaces, we decompose a given surface into a large number of small pieces and examine each piece separately. It may happen that a particular surface has an unusual or bizarre appearance and yet each small piece has a fairly simple structure. In fact, for most surfaces we shall investigate, the "local" behavior will be much like that of a piece of a sphere, a hyperboloid, a cylinder, or some other smooth surface.

Some surfaces have boundaries and others do not. For example, a hemisphere has a boundary consisting of its equatorial rim. An entire sphere, an ellipsoid, and the surface of a cube are examples of surfaces *without* boundary.

We now describe the "small pieces" into which we divide a surface. More precisely, a **smooth surface element** is the graph of a system of equations of the form

$$x = x(u, v), \qquad y = y(u, v), \qquad z = z(u, v), \quad (u, v) \in G \cup \partial G, \qquad (1)$$

in which $x, y, z$ are continuously differentiable functions on a domain $D$ in $R^2$ containing $G$ and its boundary $\partial G$. It is convenient to use vector notation, and we write the above system in the form

$$\mathbf{v}[\overrightarrow{OQ}] = \mathbf{r}(u, v), \qquad (u, v) \in G \cup \partial G, \quad O \text{ is the origin of coordinates} \qquad (2)$$

where $\mathbf{r}(u, v) = x(u, v)\mathbf{i} + y(u, v)\mathbf{j} + z(u, v)\mathbf{k}$. We assume that $G$ is a domain in $R^2$ whose boundary consists of a finite number of piecewise smooth simple closed curves. See Fig. 19, in which $\partial G$ consists of one piecewise smooth curve.

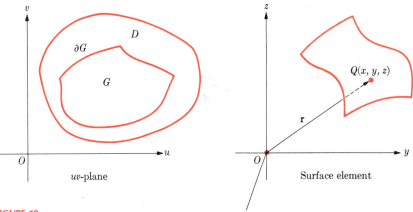

FIGURE 19

---

*Except for three definitions: (a) that of piecewise smooth surface; (b) that of area as defined in Eq. (6); and (c) that of surface integral as defined in eq. (9), this section may be omitted without loss of continuity.

We shall suppose that

$$\mathbf{r}_u \times \mathbf{r}_v \neq \mathbf{0} \quad \text{for } (u, v) \in D, \tag{3}$$

and that $\mathbf{r}(u_1, v_1) \neq \mathbf{r}(u_2, v_2)$ whenever $(u_1, v_1) \neq (u_2, v_2)$. In other words, Equations (1) (or (2)) define a one-to-one, continuously differentiable* transformation from $G \cup \partial G$ onto the points of the surface element. We observe that since

$$\mathbf{r}(u, v) = x(u, v)\mathbf{i} + y(u, v)\mathbf{j} + z(u, v)\mathbf{k},$$

then

$$\mathbf{r}_u \times \mathbf{r}_v = (y_u z_v - y_v z_u)\mathbf{i} + (z_u x_v - z_v x_u)\mathbf{j} + (x_u y_v - x_v y_u)\mathbf{k}.$$

We also say that Equations (1) or (2) form a **parametric representation** of the smooth surface element. Such a parametric representation is not unique. This does not cause problems, however, because it can be shown that any two such parametric representations are simply related.

Suppose that $S$ is a smooth surface element given by

$$\mathbf{r}(u, v) = x(u, v)\mathbf{i} + y(u, v)\mathbf{j} + z(u, v)\mathbf{k}, \quad (u, v) \in G \cup \partial G.$$

Then the **boundary** of $S$ is the image of $\partial G$ in the above parametric representation. The boundary of a smooth surface element consists of a finite number of piecewise smooth, simple closed curves, no two of which intersect.

**DEFINITION**

*A **piecewise smooth surface** $S$ is the union $S_1 \cup S_2 \cup \cdots \cup S_n$ of a finite number of smooth surface elements $S_1, S_2, \ldots, S_n$ satisfying the following conditions:*

i) *no two of the $S_i$ have common interior points;*
ii) *the intersection of the boundaries of two elements, $\partial S_i \cap \partial S_j$, is either empty, or a single point, or a piecewise smooth arc (see Fig. 20);*
iii) *the boundaries of any three elements have at most one point in common;*
iv) *any two points of $S$ can be joined by an arc in $S$.*

A piecewise smooth surface $S$ may be decomposed into a finite union of smooth surface elements in many ways. A smooth arc which is part of the boundary of two of the $S_i$ is called an **edge** of the decomposition. A corner of one of the boundaries of an $S_i$ is called a **vertex**.

A piecewise smooth surface $S$ is said to be a **smooth surface** if for every point $P$ not on $\partial S$ a decomposition of $S$ into piecewise smooth surface elements can be found such that $P$ is an interior point of one of the surface elements. It can be shown that *if $F$ is a smooth scalar field on some domain and $F(P)$ and $\nabla F(P)$ are never simultaneously zero, then the graph of the equation $F(P) = 0$ is a smooth surface provided that the graph is a bounded set which contains its boundary.* Thus by choosing, for example,

$$F(P) = x^2 + y^2 + z^2 - a^2 \quad \text{or} \quad F(P) = \frac{x^2}{a^2} + \frac{y^2}{b^2} + \frac{z^2}{c^2} - 1,$$

we see that spheres and ellipsoids are smooth surfaces. Any polyhedron is a piecewise smooth surface.

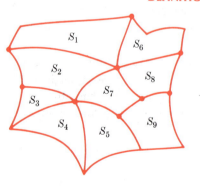

**FIGURE 20**

A piecewise smooth surface decomposed into smooth surface elements whose boundaries are points or piecewise smooth arcs

---

*"Continuously differentiable" is a convenient way of saying, "differentiable, with continuous first partial derivatives."

**EXAMPLE 1**　Suppose that a smooth scalar field is defined in all of $R^3$ by the formula

$$F(P) = x^2 + 4y^2 + 9z^2 - 44.$$

Show that the set $S: F(P) = 0$ is a smooth surface.

**Solution**　We compute the gradient:

$$\nabla F = 2x\mathbf{i} + 8y\mathbf{j} + 18z\mathbf{k}.$$

Then $\nabla F$ is zero only at $(0, 0, 0)$. Since $S$ is not void (the point $(\sqrt{44}, 0, 0)$ is on it) and since $(0, 0, 0)$ is not a point of $S$ we see that $F$ and $\nabla F$ are never both zero on $S$. Hence the surface is smooth.　　□

If a surface $S$ is given by

$$S = \{(x, y, z): z = f(x, y), \quad (x, y) \in G \cup \partial G\},$$

we know that if $f$ has continuous first derivatives, the area of $S$ may be computed by the method described in Chapter 16, Section 7. We begin the study of area for more general surfaces by defining the area of a smooth surface element.

Let $\sigma$ be a smooth surface element given by

$$\sigma: \mathbf{v}(\overrightarrow{OQ}) = \mathbf{r}(u, v), \quad \text{where } (u, v) \in G \cup \partial G, \tag{4}$$

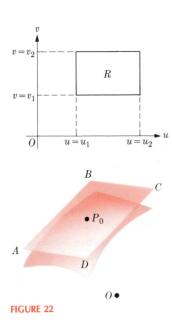

FIGURE 21

with $\mathbf{r}$ and $G$ satisfying the conditions stated in the definition of a surface element. If $v$ is held constant, equal to $v_0$, say, then the graph of (4) is a smooth curve on the surface element $\sigma$. Therefore the vector $\mathbf{r}_u(u, v_0)$ is a vector tangent to the curve on the surface. Similarly, the vector $\mathbf{r}_v(u_0, v)$ is tangent to the smooth curve on the surface element $\sigma$ obtained when $u$ is set equal to the constant $u_0$ (Fig. 21). The vectors $\mathbf{r}_u(u_0, v_0)$ and $\mathbf{r}_v(u_0, v_0)$ lie in the plane tangent to the surface at the point $P_0$ corresponding to $\mathbf{r}(u_0, v_0)$. Since the cross product of two vectors is orthogonal to each of them, it follows that *the vector $\mathbf{r}_u(u_0, v_0) \times \mathbf{r}_v(u_0, v_0)$ is a vector normal to the surface of* $P_0$. That is, the vector $\mathbf{r}_u(u_0, v_0) \times \mathbf{r}_v(u_0, v_0)$ is perpendicular to the plane tangent to the surface at $P_0$.

For convenience, we write $\mathbf{a} = \mathbf{r}_u(u_0, v_0)$, $\mathbf{b} = \mathbf{r}_v(u_0, v_0)$, and we consider the **tangent linear transformation** defined by

$$\mathbf{v}(\overrightarrow{OP}) = \mathbf{r}(u_0, v_0) + (u - u_0)\mathbf{a} + (v - v_0)\mathbf{b}. \tag{5}$$

We notice that the image of the rectangle

$$R = \{(u, v): u_1 \leq u \leq u_2, \quad v_1 \leq v \leq v_2\}$$

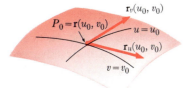

under (5) is a parallelogram $ABCD$ in the plane tangent to the surface at the point $P_0 = \mathbf{r}(u_0, v_0)$ (Fig. 22). The points $A$, $B$, $C$, and $D$ are determined by

$$\mathbf{v}(\overrightarrow{OA}) = \mathbf{r}(u_0, v_0) + (u_1 - u_0)\mathbf{a} + (v_1 - v_0)\mathbf{b},$$

$$\mathbf{v}(\overrightarrow{OB}) = \mathbf{r}(u_0, v_0) + (u_2 - u_0)\mathbf{a} + (v_1 - v_0)\mathbf{b},$$

$$\mathbf{v}(\overrightarrow{OC}) = \mathbf{r}(u_0, v_0) + (u_1 - u_0)\mathbf{a} + (v_2 - v_0)\mathbf{b},$$

$$\mathbf{v}(\overrightarrow{OD}) = \mathbf{r}(u_0, v_0) + (u_2 - u_0)\mathbf{a} + (v_2 - v_0)\mathbf{b}.$$

Using vector subtraction, we find that

$$\mathbf{v}(\overrightarrow{AB}) = (u_2 - u_1)\mathbf{a}, \qquad \mathbf{v}(\overrightarrow{AD}) = (v_2 - v_1)\mathbf{b}.$$

FIGURE 22

Denoting the area of $R$ by $A(R)$, we use the definition of cross product to obtain

$$\text{Area } ABCD = |\mathbf{v}(\overrightarrow{AB}) \times \mathbf{v}(\overrightarrow{AD})| = |(u_2 - u_1)(v_2 - v_1)| \cdot |\mathbf{a} \times \mathbf{b}|$$
$$= A(R)|\mathbf{r}_u(u_0, v_0) \times \mathbf{r}_v(u_0, v_0)|.$$

In defining the area of a surface element, we would expect that for a very small piece near $P_0$, the area of $ABCD$ would be a good approximation to the (as yet undefined) area of the portion of the surface which is the image of $R$. We use this fact to define the area of a surface element $\sigma$. We subdivide the region $G$ in the $uv$ plane into a number of subregions $G_1, G_2, \ldots, G_k$ and define the norm $\|\Delta\|$ as the maximum diameter of any of the subregions. We define

$$\text{Area of } \sigma = \lim_{\|\Delta\| \to 0} \sum_{i=1}^{n} A(G_i)|\mathbf{r}_u(u_i, v_i) \times \mathbf{r}_v(u_i, v_i)|,$$

where $(u_i, v_i)$ is any point in $G_i$ and where the limit has the usual interpretation as given in the definition of integral. Therefore the **definition of area of** $\sigma$ is

$$A(\sigma) = \iint\limits_{G} |\mathbf{r}_u(u, v) \times \mathbf{r}_v(u, v)| \, dA_{uv}. \tag{6}$$

It would appear that this definition of area depends on the particular parameterization of the surface element $\sigma$. It is shown in more advanced courses that the area is unchanged under a different parametric representation.

If the parametric representation is the simple one $z = f(x, y)$, we may set $x = u$, $y = v$, $z = f(u, v)$; then $\mathbf{r}(u, v) = u\mathbf{i} + v\mathbf{j} + f(u, v)\mathbf{k}$, and $|\mathbf{r}_u \times \mathbf{r}_v| = \sqrt{1 + (f_u)^2 + (f_v)^2}$. Then (6) becomes

$$A(\sigma) = \iint\limits_{G} \sqrt{1 + (f_u)^2 + (f_v)^2} \, dA_{uv}, \tag{7}$$

which is the formula we established in Section 7, Chapter 16.

Unfortunately, not every surface can be covered by a single smooth surface element, and so (7) cannot be used exclusively for the computation of surface area. In fact, it can be shown that even a simple surface such as a sphere cannot be part of a single smooth surface element. (See, however, Example 2 below.)

We define figures and their areas on more general piecewise continuous surfaces as follows: We say that a set $F$ *on a smooth surface element* is a **figure** if and only if it is the image under (1) of a region $E$ in $G \cup \partial G$ which has an area, and we define its area by the formula (6). We say that a set $F$ is a **figure** $\Leftrightarrow F = F_1 \cup \cdots \cup F_k$ where each $F_i$ is a figure contained in some smooth surface element $\sigma_i$ and no two have common interior points; in this case we define

$$A(F) = A(F_1) + \cdots + A(F_k).$$

We shall not prove the fact that two different decompositions of this sort yield the same result for $A(F)$.

If $F$ is a closed figure (i.e., one which contains its boundary) on a piecewise smooth surface and $f$ is a continuous scalar field on $F$, we define

$$\iint_F f\, dS = \sum_{i=1}^{k} \iint_{F_i} f\, dS, \tag{8}$$

where $F = \{F_1, \ldots, F_k\}$ is a subdivision of $F$ as above; the terms in the right side of (8) are given by

$$\iint_{F_i} f\, dS = \iint_{E_i} f[\mathbf{r}_i(u, v)] \cdot |\mathbf{r}_{iu} \times \mathbf{r}_{iv}|\, du\, dv, \tag{9}$$

where $F_i$ is the image of $E_i$ under the transformation

$$\mathbf{v}(\overrightarrow{OQ}) = \mathbf{r}_i(u, v), \quad (u, v) \in G_i \cup \partial G_i.$$

If a surface element $\sigma$ has the representation $z = \phi(x, y)$ in a given coordinate system, then (9) becomes

$$\iint_F f(Q)\, dS = \iint_G f[x, y, \phi(x, y)]\sqrt{1 + \phi_x^2 + \phi_y^2}\, dA_{xy}, \tag{10}$$

with corresponding formulas in the cases $x = \phi(y, z)$ or $y = \phi(x, z)$. The evaluation of the integral on the right in (10) follows the usual rules for evaluation of double integrals which we studied earlier. Some examples illustrate the technique.

EXAMPLE 2  Evaluate $\iint_F z^2\, dS$, where $F$ is the part of the lateral surface of the cylinder $x^2 + y^2 = 4$ between the planes $z = 0$ and $z = x + 3$.

Solution  (See Fig. 23.) If we transform to cylindrical coordinates $r, \theta, z$ then $F$ lies on the surface $r = 2$. We may choose $\theta$ and $z$ as parametric coordinates on $F$ and write

$$F = \{(x, y, z): x = 2\cos\theta,\ y = 2\sin\theta,\ z = z,\ (\theta, z)\ \text{in}\ G\},$$

$$G = \{(\theta, z): -\pi \le \theta \le \pi,\ 0 \le z \le 3 + 2\cos\theta\}.$$

See Fig. 24. The element of surface area $dS$ is given by

$$dS = |\mathbf{r}_\theta \times \mathbf{r}_z|\, dA_{\theta z},$$

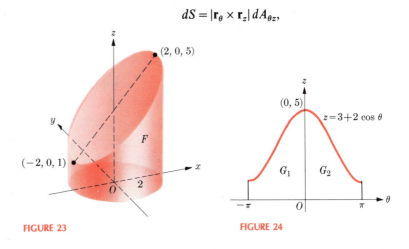

FIGURE 23                    FIGURE 24

and since

$$\mathbf{r}_\theta \times \mathbf{r}_z = (2 \cos \theta)\mathbf{i} + (2 \sin \theta)\mathbf{j} + 0 \cdot \mathbf{k},$$

we have

$$dS = 2 \, dA_{\theta z}.$$

Then

$$\iint_F z^2 \, dS = 2 \iint_G z^2 \, dA_{\theta z} = 2 \int_{-\pi}^{\pi} \int_0^{3 + 2 \cos \theta} z^2 \, dz \, d\theta$$

$$= \tfrac{2}{3} \int_{-\pi}^{\pi} (3 + 2 \cos \theta)^3 \, d\theta \tag{11}$$

$$= \tfrac{2}{3} \int_{-\pi}^{\pi} (27 + 54 \cos \theta + 36 \cos^2 \theta + 8 \cos^3 \theta) \, d\theta$$

$$= \tfrac{2}{3}(54\pi + 36\pi) = 60\pi. \qquad \square$$

*Remark.* Strictly speaking, the whole of $F$ is not a smooth surface element, since the transformation from $G$ to $F$ shows that the points $(-\pi, z)$ and $(\pi, z)$ of $G$ are carried into the same points on $F$. For a smooth surface element, the condition that the transformation be one-to-one is therefore violated. However, if we subdivide $G$ into $G_1$ and $G_2$ as shown in Fig. 24, the images are smooth surface elements. The evaluation as given in (11) is unaffected.

Surface integrals can be used to make approximate computations of various physical quantities. The center of mass and the moment of inertia of a thin curvilinear plate are sometimes computable in the form of surface integrals. The potential due to a distribution of an electric charge over a surface may be expressed in the form of an integral. (See Problems 18 through 21 at the end of this section.) With each surface $F$ we may associate a mass (assuming it to be made of a thin material), and this mass may be given by a density function $\delta(x, y, z)$. The density will be assumed continuous but not necessarily constant. The total mass $M(F)$ of a surface $F$ is given by

$$M(F) = \iint_F \delta(x, y, z) \, dS.$$

The formula for the moment of inertia of $F$ about the $z$ axis becomes

$$I_z = \iint_F \delta(x, y, z)(x^2 + y^2) \, dS,$$

with corresponding formulas for $I_x$ and $I_y$. The formula for the center of mass is analogous to those which we studied earlier. (See Chapter 16, Section 6, and also Example 3 below.)

EXAMPLE 3    Find the moment of inertia about the $x$ axis of the part of the surface of the unit sphere $x^2 + y^2 + z^2 = 1$ which is above the cone $z^2 = x^2 + y^2$. Assume $\delta = \text{const}.$

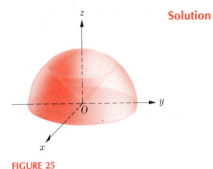

**FIGURE 25**

**Solution**    (See Fig. 25.) In spherical coordinates $(\rho, \phi, \theta)$, the portion $F$ of the surface is given by $\rho = 1$. Then $(\phi, \theta)$ are parametric coordinates and

$$F = \{(x, y, z): x = \sin \phi \cos \theta, y = \sin \phi \sin \theta, z = \cos \phi, (\phi, \theta) \in G\};$$

$$G = \{(\phi, \theta): 0 \le \phi \le \pi/4, 0 \le \theta \le 2\pi\}.$$

We compute $\mathbf{r}_\phi \times \mathbf{r}_\theta$, getting

$$\mathbf{r}_\phi \times \mathbf{r}_\theta = (\sin^2 \phi \cos \theta)\mathbf{i} + (\sin^2 \phi \sin \theta)\mathbf{j} + (\sin \phi \cos \phi)\mathbf{k}.$$

Then

$$|\mathbf{r}_\phi \times \mathbf{r}_\theta|^2 = (\sin^2 \phi \cos \theta)^2 + (\sin^2 \phi \sin \theta)^2 + (\sin \phi \cos \phi)^2$$

$$= \sin^4 \phi + \sin^2 \phi \cos^2 \phi = \sin^2 \phi,$$

and we obtain

$$dS = |\mathbf{r}_\phi \times \mathbf{r}_\theta| \, dA_{\phi\theta} = |\sin \phi| \, dA_{\phi\theta}.$$

Since $0 \le \phi \le \pi/4$, we can replace $|\sin \phi|$ by $\sin \phi$, and therefore

$$I_x = \iint\limits_G (y^2 + z^2) \delta \sin \phi \, dA_{\phi\theta}$$

$$= \delta \int_0^{\pi/4} \int_0^{2\pi} (\sin^2 \phi \sin^2 \theta + \cos^2 \phi) \sin \phi \, d\theta \, d\phi$$

$$= \pi\delta \int_0^{\pi/4} (\sin^2 \phi + 2 \cos^2 \phi) \sin \phi \, d\phi$$

$$= \pi\delta \int_0^{\pi/4} (1 + \cos^2 \phi) \sin \phi \, d\phi = \pi\delta \left[ -\cos \phi - \frac{1}{3} \cos^3 \phi \right]_0^{\pi/4}$$

$$= \frac{\pi\delta}{12}(16 - 7\sqrt{2}). \qquad \square$$

*Remark.*    The parametric representation of $F$ by spherical coordinates does not fulfill the conditions required for such representations; in particular it violates Condition (3) that $\mathbf{r}_u \times \mathbf{r}_v \neq \mathbf{0}$ for all $(u, v) \in D$, since $|\mathbf{r}_\phi \times \mathbf{r}_\theta| = \sin \phi$ vanishes for $\phi = 0$. However, the same surface, with a small hole cut out around the $z$ axis, is of the required type. A limiting process in which the size of the hole tends to zero yields the above result for the moment of inertia $I_x$.

**EXAMPLE 4**    Let $R$ be the region in three-space bounded by the cylinder $x^2 + y^2 = 1$ and the planes $z = 0$, $z = x + 2$. Evaluate $\iint_F x \, dS$, where $F$ is the surface made up of the entire boundary of $R$ (Fig. 26). If $F$ is made of thin material of uniform density $\delta$, find its mass. Also, find $\bar{x}$, the $x$-coordinate of the center of mass.

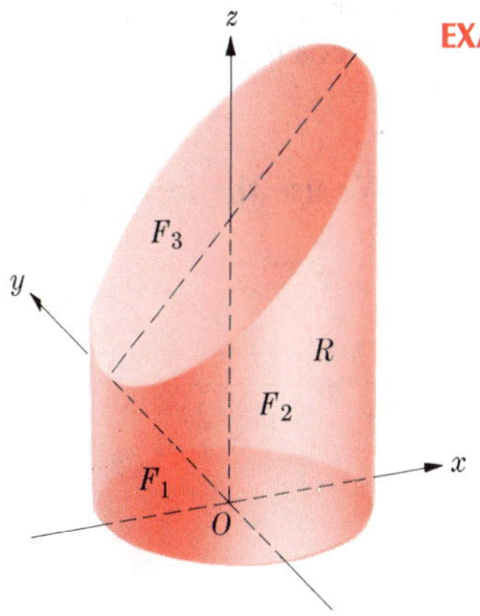

**Solution**    As shown in Fig. 26, the surface is composed of three parts: $F_1$, the circular base in the $xy$ plane; $F_2$, the lateral surface of the cylinder; and $F_3$, the part of the plane $z = x + 2$ inside the cylinder. We have

$$\iint\limits_{F_1} x \, dS = \iint\limits_{G_1} x \, dA_{xy} = 0,$$

where

$$G_1 = \{(x, y, z): x^2 + y^2 \le 1, z = 0\},$$

**FIGURE 26**

and the integrand is an odd function on this disk. On $F_3$, we see that $z = f(x, y) = x + 2$, so that $dS = \sqrt{2}\, dA_{xy}$ and

$$\iint\limits_{F_3} x\, dS = \sqrt{2} \iint\limits_{G_1} x\, dA_{xy} = 0.$$

On $F_2$, we choose coordinates $(\theta, z)$ as in Example 2 and obtain $x = \cos\theta$, $y = \sin\theta$, $z = z$, $dS = dA_{\theta z}$. We write

$$\iint\limits_{F_2} x\, dS = \iint\limits_{G_2} \cos\theta\, dA_{\theta z} = \int_{-\pi}^{\pi} \int_{0}^{2 + \cos\theta} \cos\theta\, dz\, d\theta,$$

where

$$G_2 = \{(\theta, z): -\pi \le \theta \le \pi, 0 \le z \le 2 + \cos\theta\}.$$

Therefore

$$\iint\limits_{F_2} x\, dS = \int_{-\pi}^{\pi} (2\cos\theta + \cos^2\theta)\, d\theta = \pi.$$

We conclude that

$$\iint\limits_{F} x\, dS = \iint\limits_{F_1 \cup F_2 \cup F_3} x\, dS = \pi.$$

We observe that the surface $F$ is a piecewise smooth surface without boundary. There are no vertices on $F$, but there are two smooth edges, namely the intersections of $x^2 + y^2 = 1$ with the planes $z = 0$ and $z = x + 2$. To compute the mass $M(F)$, we have

$$M(F) = \iint\limits_{F} \delta\, dS = \delta A(F) = \delta[A(F_1) + A(F_2) + A(F_3)].$$

Clearly, $A(F_1) = \pi$, $A(F_3) = \pi\sqrt{2}$. To find $A(F_2)$, we write

$$A(F_2) = \iint\limits_{G_2} dA_{\theta z} = \int_{-\pi}^{\pi} \int_{0}^{2 + \cos\theta} dz\, d\theta = \int_{-\pi}^{\pi} (2 + \cos\theta)\, d\theta = 4\pi.$$

Therefore $M(F) = \delta[\pi + 4\pi + \pi\sqrt{2}] = \pi(5 + \sqrt{2})\,\delta$. We use the formula

$$\bar{x} = \frac{\iint\limits_{F} \delta x\, dA}{M(F)} \qquad \text{to get} \qquad \bar{x} = \frac{\delta\pi}{\delta\pi(5 + \sqrt{2})} = \frac{1}{5 + \sqrt{2}}. \qquad \square$$

## 5   PROBLEMS

In each of Problems 1 through 9, evaluate

$$\iint_F f(x, y, z)\, dS.$$

**1** $f(x, y, z) = x$, $F$ is the part of the plane $x + y + z = 1$ in the first octant.

**2** $f(x, y, z) = x^2$, $F$ is the part of the plane $z = x$ inside the cylinder $x^2 + y^2 = 1$.

**3** $f(x, y, z) = x^2$, $F$ is the part of the cone $z^2 = x^2 + y^2$ between the planes $z = 1$ and $z = 2$.

**4** $f(x, y, z) = x^2$, $F$ is the part of the cylinder $z = x^2/2$ cut out by the planes $y = 0$, $x = 2$, and $y = x$.

**5** $f(x, y, z) = xz$, $F$ is the part of the cylinder $x^2 + y^2 = 1$ between the planes $z = 0$ and $z = x + 2$.

**6** $f(x, y, z) = x$, $F$ is the part of the cylinder $x^2 + y^2 = 2x$ between the lower and upper nappes of the cone $z^2 = x^2 + y^2$.

**7** $f(x, y, z) = 1$, $F$ is the part of the vertical cylinder erected on the spiral $r = \theta$, $0 \le \theta \le \pi/2$ (polar coordinates in the $xy$ plane), bounded below by the $xy$ plane and above by the cone $z^2 = x^2 + y^2$.

**8** $f(x, y, z) = x^2 + y^2 - 2z^2$, $F$ is the surface of the sphere $x^2 + y^2 + z^2 = a^2$.

**9** $f(x, y, z) = x^2$, $F$ is the total boundary of the region $R$ in three-space bounded by the cone $z^2 = x^2 + y^2$ and the planes $z = 1$, $z = 2$. (See Problem 3.)

In each of Problems 10 through 14, find the moment of inertia of $F$ about the indicated axis, assuming that $\delta = $ const.

**10** The surface $F$ of Problem 3; $x$ axis.

**11** The surface $F$ of Problem 6; $x$ axis.

**12** The surface $F$ of Problem 7; $z$ axis.

**13** The total surface of the tetrahedron $L$ bounded by the coordinate planes and the plane $x + y + z = 1$, having vertices $B(1, 0, 0)$, $E(0, 1, 0)$, $R(0, 0, 1)$, and $S(0, 0, 0)$; $y$ axis.

**14** The torus $(r - b)^2 + z^2 = a^2$, $0 < a < b$; $z$ axis. Note that

$$r^2 = x^2 + y^2$$

and that we may introduce parameters

$$x = (b + a \cos \phi) \cos \theta,$$
$$y = (b + a \cos \phi) \sin \theta,$$
$$z = a \sin \phi,$$

with the torus swept out by

$$0 \le \phi \le 2\pi, \qquad 0 \le \theta \le 2\pi.$$

(See Fig. 27.)

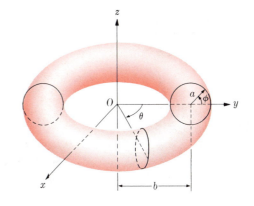

**FIGURE 27**

In each of Problems 15 through 17, find the center of mass, assuming $\delta = $ const.

**15** $F$ is the surface of Example 2.

**16** $F$ is the surface of Problem 13.

**17** $F$ is the part of the sphere $x^2 + y^2 + z^2 = 4a^2$ inside the cylinder $x^2 + y^2 = 2ax$.

The electrostatic potential $V(Q)$ at a point $Q$ due to a distribution of charge (with charge density $\delta$) on a surface $F$ is given by

$$V(Q) = \iint_F \frac{\delta(P)\, dS}{|PQ|},$$

where $|PQ|$ is the distance from $Q$ in space not on $F$ to a point $P$ on the surface $F$. In Problems 18 through 21, find $V(Q)$ at the point given, assuming $\delta$ constant.

**18** $Q = (0, 0, 0)$, $F$ is the part of the cylinder $x^2 + y^2 = 1$ between the planes $z = 0$ and $z = 1$.

**19** $Q = (0, 0, c)$; $F$ is the surface of the sphere $x^2 + y^2 + z^2 = a^2$. Do two cases: (i), $c > a > 0$ and (ii), $a > c > 0$.

**20** $Q = (0, 0, c)$; $F$ is the upper half of the sphere $x^2 + y^2 + z^2 = a^2$; $0 < c < a$.

**21** $Q = (0, 0, 0)$; $F$ is the surface of Problem 3.

_____ 6 _____

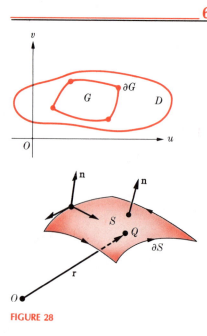

FIGURE 28

## ORIENTABLE SURFACES. STOKES' THEOREM

Suppose $S$ is a smooth surface element represented by the equation

$$\mathbf{v}[\overrightarrow{OQ}] = \mathbf{r}(u, v) \quad \text{with } (u, v) \text{ in } G \cup \partial G \tag{1}$$

where $G$ is a region in the $(u, v)$ plane with piecewise smooth boundary $\partial G$. According to the definition of smooth surface element, we know that $\mathbf{r}_u \times \mathbf{r}_v \neq \mathbf{0}$ for $(u, v)$ in a domain $D$ containing $G$ and its boundary. Therefore we can define a **unit normal function to the surface** $S$ by the formula (Fig. 28)

$$\mathbf{n} = \frac{\mathbf{r}_u \times \mathbf{r}_v}{|\mathbf{r}_u \times \mathbf{r}_v|}, \quad (u, v) \text{ in } G. \tag{2}$$

Whenever $S$ is a smooth surface element, the vector $\mathbf{n}$ is a continuous function of $u$ and $v$. Such a definition of $\mathbf{n}$ as given in (2) depends on the parametric representation $\mathbf{r}$ chosen. It can be shown, however, that if

$$\mathbf{v}[\overrightarrow{OQ}] = \mathbf{r}_1(s, t), \quad (s, t) \in G_1 \cup \partial G_1,$$

is another parametric representation of the same smooth surface element $S$, and if $\mathbf{n}_1$ is the unit normal obtained from $\mathbf{r}_1$, then either $\mathbf{n}_1(P) = \mathbf{n}(P)$, or $\mathbf{n}_1(P) = -\mathbf{n}(P)$, for all $P$ on $S$.

**DEFINITIONS**    *A smooth surface $S$ is **orientable** if and only if there exists a continuous unit normal function defined over the whole of $S$. Such a normal function is called an **orientation** of $S$.*

Since any unit vector normal to $S$ at a point $P$ is either $\mathbf{n}(P)$ or $-\mathbf{n}(P)$, we see that *each smooth orientable surface possesses exactly two orientations each of which is the negative of the other.* We call a pair $(S, \mathbf{n})$, where $\mathbf{n}$ is an orientation of $S$, an **oriented surface** and denote it by $\vec{S}$.

It is easy to see that smooth surfaces such as spheres, ellipsoids, tori (see Problem 14, Section 5), and so forth, are all orientable. However, there are smooth surfaces for which there is no way to choose a continuous normal function over the whole surface. One such surface is the **Möbius strip** drawn in Fig. 29. The reader can make a model of such a surface from a long, narrow, rectangular strip of paper by giving one end a half-twist and then gluing the ends together. A Möbius strip can be represented parametrically on a

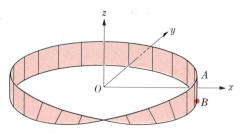

FIGURE 29

rectangle

$$R = \{(\theta, s): 0 \le \theta \le 2\pi, \ -h \le s \le h\}, \quad (0 < h < a)$$

(see Fig. 30) by $\mathbf{r}(\theta, s) = x\mathbf{i} + y\mathbf{j} + z\mathbf{k}$ with

$$x = \left(a + s \sin \frac{\theta}{2}\right) \cos \theta, \qquad y = \left(a + s \sin \frac{\theta}{2}\right) \sin \theta, \qquad z = s \cos \frac{\theta}{2}. \tag{3}$$

FIGURE 30

To obtain the unit normal to the surface, we compute $\mathbf{r}_\theta \times \mathbf{r}_s$ to get

$$\mathbf{r}_\theta \times \mathbf{r}_s = f_1(\theta, s)\mathbf{i} + f_2(\theta, s)\mathbf{j} + f_3(\theta, s)\mathbf{k},$$

where

$$f_1(\theta, s) = \left(a + s \sin \frac{\theta}{2}\right) \cos \frac{\theta}{2} \cos \theta + \frac{s}{2} \sin \theta,$$

$$f_2(\theta, s) = \left(a + s \sin \frac{\theta}{2}\right) \cos \frac{\theta}{2} \sin \theta - \frac{s}{2} \cos \theta,$$

$$f_3(\theta, s) = -\left(a + s \sin \frac{\theta}{2}\right) \sin \frac{\theta}{2},$$

and

$$|\mathbf{r}_\theta \times \mathbf{r}_s|^2 = \left(a + s \sin \frac{\theta}{2}\right)^2 + \frac{1}{4}s^2.$$

Therefore $\mathbf{r}_\theta \times \mathbf{r}_s$ is never $\mathbf{0}$, so that if we define $\mathbf{n}(\theta, s)$ by (2), we see that $\mathbf{n}(\theta, s)$ is continuous. However, $\mathbf{n}(0, 0) = \mathbf{i}$, $\mathbf{n}(2\pi, 0) = -\mathbf{i}$, and the points $(0, 0)$, $(2\pi, 0)$ in the parametric plane correspond to the same point on the surface. The transformation (3) is one-to-one, except that the points of $R$ given by $(0, s)$ and $(2\pi, -s)$, $-h \le s \le h$, are always carried into the same point on the surface (Fig. 30). The Möbius strip, often called a "one-sided surface," is not a smooth surface element (defined in Section 4). If a pencil line is drawn down the center of the strip (corresponding to $s = 0$) then, after one complete trip around, the line is on the "opposite side." Two circuits are needed to "close up" the curve made by the pencil line. The Möbius strip is not an orientable surface.

A smooth surface element $S$ which, according to (1) is the image of a plane region $G$, has a boundary (denoted $\partial S$) which is the image of the boundary $\partial G$ of $G$. The closed curve $\partial G$, which is piecewise smooth, can be made into a directed curve by traversing it in a given direction. We say that $\partial G$ is **positively directed** when we travel along it so that the interior of $G$ is

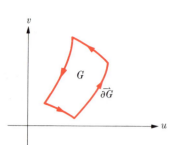

**FIGURE 31**

always on the left. We write $\overrightarrow{\partial G}$ for a positively directed closed curve. In Fig. 31 the curve $\partial G$ is shown positively directed. The closed curve $\partial S$, the image of $\partial G$, becomes a directed curve in space, with the direction the one induced by $\overrightarrow{\partial G}$. We say the curve $\overrightarrow{\partial S}$ is **positively directed** when its direction corresponds to the positively directed curve $\overrightarrow{\partial G}$. Geometrically the curve $\overrightarrow{\partial S}$ is directed so that when one walks along $\overrightarrow{\partial S}$ in an upright position with his head in the direction of the positive normal **n** to the surface, then the surface is on his left.

In terms of right- and left-handed coordinate systems, if **t** is tangent to $\overrightarrow{\partial S}$ pointing in the positive direction, if **n** is perpendicular to **t** and in the direction of the positive normal, and if **b** is perpendicular to the vectors **t** and **n** and pointing toward the surface, then the triple **t**, **b**, and **n** is a right-handed triple (Fig. 32). If $\partial G$ and $\partial S$ contain several closed curves, this argument holds for each curve.

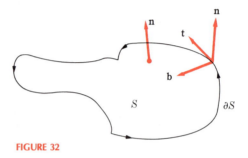

**FIGURE 32**

We wish to extend the notion of orientable surface to *piecewise smooth surfaces*. Such surfaces have edges, and so a continuous unit normal vector field cannot be defined over the entirety of such a surface. However, a piecewise smooth surface $F$ can be subdivided into a finite number of smooth surface elements $F_1, F_2, \ldots, F_k$. Each such surface element may be oriented and each boundary $\partial F_i$ may be correspondingly positively directed. Suppose that $\gamma_{ij}$ is a smooth arc which is the common boundary of two surface elements $F_i$ and $F_j$. If the positive direction of $\gamma_{ij}$ as part of $\overrightarrow{\partial F_i}$ is the opposite of the positive direction of $\gamma_{ij}$ as part of $\overrightarrow{\partial F_j}$ for all arcs $\gamma_{ij}$, we say that the surface $F$ is an **orientable, piecewise smooth surface**. In this case, the collection of orientations of the $F_i$ form an orientation of $F$ and the normal function defined at each interior point of $F_i$ is called the **positive normal**.

Figure 33 shows a piecewise smooth, orientable surface and its decomposition into smooth surface elements. The boundary $\overrightarrow{\partial F}$ is a positively directed, closed, piecewise smooth curve. Those boundary arcs of $F_1, F_2, \ldots, F_k$ which are traversed only once comprise $\overrightarrow{\partial F}$.

**FIGURE 33**

A Möbius strip is not orientable even if it is treated as a piecewise smooth surface. On the other hand, the surface of a cube, which is piecewise smooth, is an orientable surface. We select the "outward" pointing normal on each face and traverse the boundary of any face in a counterclockwise direction as we view it from outside the cube. We easily verify that all edges are traversed twice, once in each direction.

In Section 4 we gave a vector formulation of Green's Theorem in the plane: For a vector function **v** of the form $\mathbf{v} = P\mathbf{i} + Q\mathbf{j} + O\mathbf{k}$, Green's Theorem takes the form

$$\iint_G \text{curl } \mathbf{v} \cdot \mathbf{k} \, dA = \oint_{\partial G} \mathbf{v} \cdot d\mathbf{r}.$$

It is this form of Green's Theorem that generalizes to surfaces in $R^3$. The region $G$ can be thought of as a surface in the $xy$ plane; the vector $\mathbf{k}$ is then its "outward pointing normal." The analogous theorem for surfaces in $R^3$ is known as **Stokes' Theorem** and yields the formula

$$\iint_{\vec{F}} \text{curl } \mathbf{v} \cdot \mathbf{n} \, dS = \oint_{\partial \vec{F}} \mathbf{v} \cdot d\mathbf{r}$$

where the oriented surface $\vec{F}$ replaces $G$, and a surface integral "$dS$" replaces the double integral "$dA$".

To be more concrete, let us begin by letting $\vec{F}$ be a smooth oriented surface element and $(x, y, z)$ a fixed rectangular coordinate system. We represent $\vec{F}$ parametrically:

$$\vec{F}: \mathbf{r}(u, v) = x(u, v)\mathbf{i} + y(u, v)\mathbf{j} + z(u, v)\mathbf{k}.$$

Then the positive unit normal $\mathbf{n}(u, v)$ is given by

$$\mathbf{n}(u, v) = \frac{\mathbf{r}_u \times \mathbf{r}_v}{|\mathbf{r}_u \times \mathbf{r}_v|}.$$

If $\mathbf{v}$ is a continuous vector field defined on $\vec{F}$ with coordinate functions

$$\mathbf{v}(x, y, z) = v_1(x, y, z)\mathbf{i} + v_2(x, y, z)\mathbf{j} + v_3(x, y, z)\mathbf{k},$$

we can compute $\mathbf{v} \cdot \mathbf{n}$, and since it is a continuous function on $\vec{F}$, we may define the integral

$$\iint_{\vec{F}_i} \mathbf{v} \cdot \mathbf{n}_i dS_i.$$

The surface element $dS$ can be computed in terms of the parameters $(u, v)$ according to the formula

$$dS = |\mathbf{r}_u \times \mathbf{r}_v| \, dA_{uv}.$$

When $\vec{F}$ is a piecewise smooth orientable surface, we can represent it as the union of a number of smooth surface elements $\vec{F}_1, \vec{F}_2, \ldots, \vec{F}_k$. Then we define

$$\iint_{\vec{F}} \mathbf{v} \cdot \mathbf{n} \, dS = \sum_{i=1}^{k} \iint_{\vec{F}_i} \mathbf{v} \cdot \mathbf{n}_i \, dS_i.$$

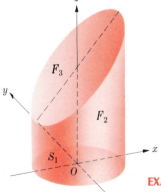

FIGURE 34

**EXAMPLE 1**  Let $R$ be the region bounded by the cylinder $x^2 + y^2 = 1$ and the planes $z = 0$ and $z = x + 2$ (Fig. 34). Let $\vec{F}$ be the entire boundary of $R$, positively directed. Find the value of $\iint_{\vec{F}} \mathbf{v} \cdot \mathbf{n} \, dS$ where $\mathbf{n}$ is the outward directed unit normal on $\vec{F}$ and

$$\mathbf{v} = 2x\mathbf{i} - 3y\mathbf{j} + z\mathbf{k}.$$

**Solution**  The surface $\vec{F}$ is piecewise smooth and we label the smooth portions $\vec{F}_1$, $\vec{F}_2$, and $\vec{F}_3$, as shown in Fig. 34. On $\vec{F}_1$ we have

$$\mathbf{n} = -\mathbf{k}, \qquad \mathbf{v} \cdot \mathbf{n} \, dS = -z \, dA_{xy} = 0,$$

as $\vec{F}_1$ is in the plane $z = 0$. Therefore we have $\iint_{\vec{F}} \mathbf{v} \cdot \mathbf{n} \, dS = 0$. On $\vec{F}_3$, we have $z = x + 2$ and

$$\mathbf{n} = \frac{1}{\sqrt{2}}(-\mathbf{i} + \mathbf{k}),$$

$$\mathbf{v} \cdot \mathbf{n} = \frac{1}{\sqrt{2}}(-2x + z), \qquad dS = \sqrt{2} \, dA_{xy},$$

$$\mathbf{v} \cdot \mathbf{n} \, dS = (-2x + z) \, dA_{xy}.$$

We obtain $\iint_{\vec{F}_3} \mathbf{v} \cdot \mathbf{n} \, dS = \iint_{\vec{F}_1} (-x + 2) \, dA_{xy}$. We see that $\iint_{\vec{F}_1} x \, dA_{xy} = 0$, since the integrand is an odd function. Also, $2 \iint_{\vec{F}_1} dA_{xy} = 2\pi$.

On $\vec{F}_2$, we select cylindrical coordinates

$$F_2: x = \cos\theta, \qquad y = \sin\theta, \qquad z = z, \quad (\theta, z) \text{ in } H_2, \text{ where}$$

$$H_2 = \{(\theta, z): -\pi \le \theta \le \pi, 0 \le z \le 2 + \cos\theta\}.$$

We find

$$\mathbf{n} = (\cos\theta)\mathbf{i} + (\sin\theta)\mathbf{j}, \qquad \mathbf{v} \cdot \mathbf{n} = 2\cos^2\theta - 3\sin^2\theta.$$

Therefore

$$\iint_{\vec{F}_2} \mathbf{v} \cdot \mathbf{n} \, dS = \iint_{H_2} (2\cos^2\theta - 3\sin^2\theta) \, dA_{\theta z}$$

$$= \int_{-\pi}^{\pi} \int_0^{2 + \cos\theta} (2 - 5\sin^2\theta) \, dz \, d\theta$$

$$= \int_{-\pi}^{\pi} (4 - 10\sin^2\theta + 2\cos\theta - 5\sin^2\theta\cos\theta) \, d\theta$$

$$= 8\pi - 10\pi + 0 + 0 = -2\pi.$$

Hence

$$\iint_{\vec{F}} \mathbf{v} \cdot \mathbf{n} \, dS = \iint_{\vec{F}_1} \mathbf{v} \cdot \mathbf{n} \, dS + \iint_{\vec{F}_2} \mathbf{v} \cdot \mathbf{n} \, dS + \iint_{\vec{F}_3} \mathbf{v} \cdot \mathbf{n} \, dS = 0. \qquad \square$$

Suppose that $\vec{F}$ is an oriented, piecewise smooth surface with the boundary $\partial\vec{F}$ positively directed and made up of a finite number of smooth arcs $\vec{C}_1, \vec{C}_2, \ldots, \vec{C}_k$. If $\mathbf{v}$ is a continuous vector field defined on and near $\partial\vec{F}$, we define

$$\int_{\partial\vec{F}} \mathbf{v} \cdot d\mathbf{r} = \sum_{i=1}^{k} \int_{\vec{C}_i} \mathbf{v} \cdot d\mathbf{r}.$$

With the aid of this definition we can state Stokes' Theorem.

THEOREM 5 **(Stokes' Theorem)** *Suppose that $\vec{F}$ is a bounded, closed, orientable, piecewise smooth surface and that* $\mathbf{v}$ *is a smooth vector field defined in a region containing* $\vec{F}$. *Then*

$$\iint\limits_{\vec{F}} (\text{curl } \mathbf{v}) \cdot \mathbf{n} \, dS = \int_{\partial \vec{F}} \mathbf{v} \cdot d\mathbf{r}. \qquad (4)$$

COROLLARY *Suppose that* $\vec{F}$ *is a bounded, closed, orientable, piecewise smooth surface without boundary and that* $\mathbf{v}$ *is a smooth vector field defined in a region containing* $\vec{F}$. *Then*

$$\iint\limits_{\vec{F}} (\text{curl } \mathbf{v}) \cdot \mathbf{n} \, dS = 0. \qquad (5)$$

Stokes' Theorem in this form can be viewed as an extension of Green's Theorem from plane surfaces to surfaces in three dimensions. Since the boundary of $F$ has an orientation, we do not need the symbol $\circlearrowleft$ to indicate the counterclockwise direction for the line integral.

We require the next lemma to prove Stokes' Theorem.

LEMMA 1 *Suppose that a rectangular coordinate system in three-space is given with orthogonal unit vectors* $\mathbf{i}, \mathbf{j}, \mathbf{k}$. *Suppose that* $\vec{S}$ *is a smooth oriented surface element with a parametric representation*

$$\mathbf{r}(u, v) = x(u, v)\mathbf{i} + y(u, v)\mathbf{j} + z(u, v)\mathbf{k}, \quad (u, v) \in G \cup \partial G. \qquad (6)$$

*Let* $\mathbf{v}$ *be a continuous vector field defined in a region E containing S, and suppose* $\mathbf{v}$ *is given by*

$$\mathbf{v}(x, y, z) = v_1(x, y, z)\mathbf{i} + v_2(x, y, z)\mathbf{j} + v_3(x, y, z)\mathbf{k}, \quad (x, y, z) \in E. \qquad (7)$$

*Then*

$$\int_{\partial \vec{S}} \mathbf{v} \cdot d\mathbf{r} = \int_{\partial \vec{G}} [(v_1 x_u + v_2 y_u + v_3 z_u) \, du + (v_1 x_v + v_2 y_v + v_3 z_v) \, dv]. \qquad (8)$$

Proof We establish the result for the case that $\partial \vec{G}$ consists of a single piecewise smooth simple closed curve; the extension to several such curves is clear. Let

$$u = u(s), \qquad v = v(s), \quad a \le s \le b$$

be a parametric representation of the closed curve $\partial \vec{G}$. Then

$$x = x[u(s), v(s)], \qquad y = y[u(s), v(s)], \qquad z = z[u(s), v(s)], \quad a \le s \le b$$

is a parametric representation of $\partial \vec{S}$. Using the Chain Rule, we find

$$d\mathbf{r} = dx\mathbf{i} + dy\mathbf{j} + dz\mathbf{k}$$

$$= (x_u u_s + x_v v_s) \, ds \, \mathbf{i} + (y_u u_s + y_v v_s) \, ds \, \mathbf{j} + (z_u u_s + z_v v_s) \, ds \, \mathbf{k}.$$

Therefore

$$\int_{\overline{\partial S}} \mathbf{v} \cdot d\mathbf{r} = \int_a^b \left[ (v_1 x_u + v_2 y_u + v_3 z_u) u_s + (v_1 x_v + v_2 y_v + v_3 z_v) v_s \right] ds. \tag{9}$$

Making the insertions

$$du = u_s\, ds, \qquad dv = v_s\, ds$$

in the right side of (8), we see that (8) and (9) are identical.  ☐

We now establish Stokes' Theorem (Theorem 5).

**Proof of Stokes' Theorem**    We prove this theorem for the case that $\vec{S}$ is a single smooth oriented surface element. Let a rectangular coordinate system with corresponding vectors $\mathbf{i}, \mathbf{j}$, and $\mathbf{k}$ be given, and let $\mathbf{v}$ be given by (7) and let (6) be a parametric representation of $\vec{S}$. Finally we assume that $x, y, z, x_u, y_u, z_u, x_v, y_v$, and $z_v$ are all smooth on $D$. Then (8) holds on account of Lemma 1. Applying Green's Theorem to the integral on the right side of (8), we obtain

$$\int_{\overline{\partial S}} \mathbf{v} \cdot d\mathbf{r} = \iint_G \left[ \frac{\partial}{\partial u}(v_1 x_v + v_2 y_v + v_3 z_v) - \frac{\partial}{\partial v}(v_1 x_u + v_2 y_u + v_3 z_u) \right] dA_{uv}$$

$$= \iint_G \left[ x_v(v_{1x} x_u + v_{1y} y_u + v_{1z} z_u) + y_v(v_{2x} x_u + v_{2y} y_u + v_{2z} z_u) \right. \tag{10}$$

$$+ z_v(v_{3x} x_u + v_{3y} y_u + v_{3z} z_u) - x_u(v_{1x} x_v + v_{1y} y_v + v_{1z} z_v)$$

$$\left. - y_u(v_{2x} x_v + v_{2y} y_v + v_{2z} z_v) - z_u(v_{3x} x_v + v_{3y} y_v + v_{3z} z_v) \right] dA_{uv}.$$

It is easy to verify that all the terms in the right side of (10) involving the derivatives $x_{uv}, y_{uv}$, and $z_{uv}$ cancel each other. Collecting terms in (10), we see that the integrand on the right side equals $\mathbf{v} \cdot (\mathbf{r}_u \times \mathbf{r}_v)\, dA_{uv}$. However

$$\mathbf{v} \cdot (\mathbf{r}_u \times \mathbf{r}_v)\, dA_{uv} = \mathbf{v} \cdot \frac{(\mathbf{r}_u \times \mathbf{r}_v)}{|\mathbf{r}_u \times \mathbf{r}_v|} |\mathbf{r}_u \times \mathbf{r}_v|\, dA_{uv}$$

$$= \mathbf{v} \cdot \mathbf{n}\, dS,$$

and thus the right side of (10) is the left side of (4).

If $x(u, v), y(u, v), z(u, v)$ and their derivatives are piecewise smooth on $D$, then it can be shown that these functions and their derivatives can be approximated arbitrarily closely by smooth functions. Thus Stokes' Theorem can be shown to be valid for piecewise smooth functions. The details are usually given in more advanced treatments of Stokes' Theorem.

Finally, if $\vec{S}$ is any piecewise smooth oriented surface, we may express $\vec{S} = \vec{S}_1 \cup \cdots \cup \vec{S}_k$, where each $\vec{S}_i$ is an oriented smooth surface element, oriented in such a way that any piecewise smooth arc on the boundary of $\vec{S}_i$ and $\vec{S}_j$ is directed oppositely as part of $\overline{\partial S}_i$ from the way it is directed as a part of $\overline{\partial S}_j$. Formula (4) follows for each $i$ from the proof above. Adding these results, we obtain

$$\iint_{\vec{S}} (\text{curl } \mathbf{v}) \cdot \mathbf{n}\, dS = \sum_{i=1}^k \int_{\overline{\partial S}_i} \mathbf{v} \cdot d\mathbf{r}. \tag{11}$$

In the sum on the right the integrals over arcs which are on the boundaries of two $\overline{S}_i$ cancel, leaving only the integrals over those arcs only on the boundary of one $\overline{S}_i$. These latter arcs comprise $\overline{\partial S}$.    □

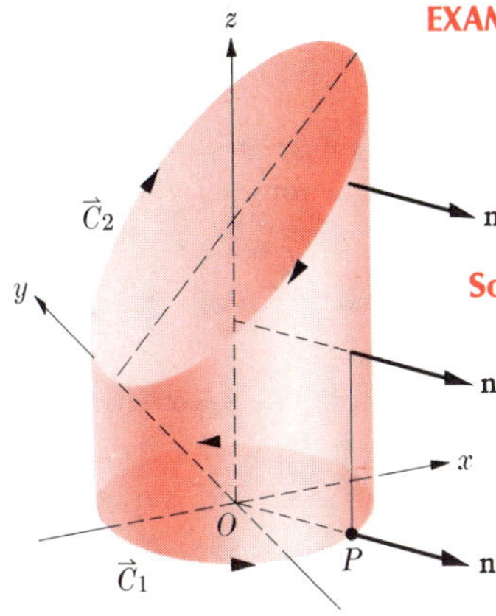

**EXAMPLE 2**    Verify Stokes' Theorem, given that

$$\mathbf{v} = y\mathbf{i} + z\mathbf{j} + x\mathbf{k}$$

and that $\overline{S}$ is the part of the surface of the cylinder $x^2 + y^2 = 1$ between the planes $z = 0$ and $z = x + 2$, oriented with $\mathbf{n}$ pointing outward (Fig. 35).

**Solution**    An examination of Fig. 35 shows that the curves $\overline{C}_1$ and $\overline{C}_2$ must be directed as exhibited. We choose cylindrical coordinates and write

$$\overline{S} = \{(x, y, z): x = \cos\theta, y = \sin\theta, z = z, (\theta, z) \text{ on } F\};$$

$$F = \{(\theta, z): -\pi \le \theta \le \pi, 0 \le z \le 2 + \cos\theta\}.$$

We think of $S$ as made up of two smooth surface elements corresponding to $F_1$ and $F_2$, as shown in Fig. 36. We must make this subdivision because the representation of $\overline{S}$ by $F$ is not one-to-one. The orientation of the boundary of $\overline{S}$ is determined by the orientations of the boundaries of $F_1$ and $F_2$. We note that when we form the boundary integral, those parts taken over vertical segments cancel. Now

$$\text{curl } \mathbf{v} = -\mathbf{i} - \mathbf{j} - \mathbf{k}, \qquad \mathbf{n} = (\cos\theta)\mathbf{i} + (\sin\theta)\mathbf{j}, \qquad \text{and} \qquad dS = dA_{\theta z}.$$

We obtain

$$\iint_{\overline{S}} (\text{curl } \mathbf{v} \cdot \mathbf{n}) \, dS = \int_{-\pi}^{\pi} \int_0^{2+\cos\theta} (-\cos\theta - \sin\theta) \, dz \, d\theta$$

$$= -\int_{-\pi}^{\pi} [(2\cos\theta + \cos^2\theta) + (2 + \cos\theta)\sin\theta] \, d\theta \qquad (12)$$

$$= -\pi.$$

For the boundary integral, we have

$$\int_{\overline{\partial S}_i} \mathbf{v} \cdot d\mathbf{r} = \int_{\overline{C}_1} \mathbf{v} \cdot d\mathbf{r} - \int_{-\overline{C}_2} \mathbf{v} \cdot d\mathbf{r},$$

in which the integrals on the right are taken in a counterclockwise direction. We find:

On $\overline{C}_1$:    $\mathbf{v} = (\sin\theta)\mathbf{i} + (\cos\theta)\mathbf{k}, \qquad d\mathbf{r} = (-\sin\theta\mathbf{i} + \cos\theta\mathbf{j}) \, d\theta.$

On $-\overline{C}_2$:    $\mathbf{v} = (\sin\theta)\mathbf{i} + (2 + \cos\theta)\mathbf{j} + (\cos\theta)\mathbf{k},$

$$d\mathbf{r} = (-\sin\theta\mathbf{i} + \cos\theta\mathbf{j} - \sin\theta\mathbf{k}) \, d\theta.$$

Taking the scalar products, we get

$$\int_{\overline{C}_1} \mathbf{v} \cdot d\mathbf{r} = \int_{-\pi}^{\pi} (-\sin^2\theta) \, d\theta = -\pi, \qquad (13)$$

$$\int_{\overline{C}_2} \mathbf{v} \cdot d\mathbf{r} = \int_{-\pi}^{\pi} (-\sin^2\theta + 2\cos\theta + \cos^2\theta - \sin\theta\cos\theta) \, d\theta = 0. \qquad (14)$$

A comparison of (12) with (13) and (14) verifies Stokes' Theorem.    □

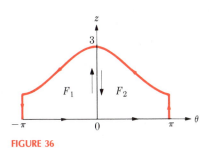

**FIGURE 35**

**FIGURE 36**

## 6  PROBLEMS

In each of Problems 1 through 7, compute $\iint_{\vec{S}} \mathbf{v} \cdot \mathbf{n} \, dS$.

1  $\mathbf{v} = (x+1)\mathbf{i} - (2y+1)\mathbf{j} + z\mathbf{k}$; $\vec{S}$ is the triangle with vertices $(1, 0, 0), (0, 1, 0)$, and $(0, 0, 1)$, with $\mathbf{n}$ pointing away from the origin.

2  $\mathbf{v} = x\mathbf{i} + y\mathbf{j} + z\mathbf{k}$; $\vec{S}$ is the part of the paraboloid $2z = x^2 + y^2$ inside the cylinder $x^2 + y^2 = 2x$ with $\mathbf{n} \cdot \mathbf{k} > 0$ ($\mathbf{n}$ pointing upward).

3  $\mathbf{v} = x^2\mathbf{i} + y^2\mathbf{j} + z^2\mathbf{k}$; $\vec{S}$ is the part of the cone

$$z^2 = x^2 + y^2 \qquad \text{for which } 1 \le z \le 2,$$

with $\mathbf{n} \cdot \mathbf{k} > 0$.

4  $\mathbf{v} = xy\mathbf{i} + xz\mathbf{j} + yz\mathbf{k}$; $\vec{S}$ is the part of the cylinder $y^2 = 2 - x$ cut out by the cylinders $y^2 = z$ and $y = z^3$.

5  $\mathbf{v} = y^2\mathbf{i} + z\mathbf{j} - x\mathbf{k}$; $\vec{S}$ is the part of the cylinder $y^2 = 1 - x$ between the planes $z = 0$ and $z = x$; $x \ge 0$, with $\mathbf{n} \cdot \mathbf{i} > 0$.

6  $\mathbf{v} = 2x\mathbf{i} - y\mathbf{j} + 3z\mathbf{k}$; $\vec{S}$ is the part of the cylinder $z^2 = x$ to the left of the cylinder $y^2 = 1 - x$ and $\mathbf{n} \cdot \mathbf{i} > 0$, $\mathbf{n}$ pointing to the right.

*7  $\mathbf{v} = x\mathbf{i} + y\mathbf{j} - 2z\mathbf{k}$; $\vec{S}$ is the part of the cylinder $x^2 + y^2 = 2x$ between the two nappes of the cone $z^2 = x^2 + y^2$, $\mathbf{n}$ pointing outward.

In each of Problems 8 through 13, verify Stokes' Theorem.

8  $\mathbf{v} = z\mathbf{i} + x\mathbf{j} + y\mathbf{k}$; $\vec{S}$ is the part of the paraboloid $z = 1 - x^2 - y^2$ for which $z \ge 0$ and $\mathbf{n} \cdot \mathbf{k} > 0$.

9  $\mathbf{v} = y^2\mathbf{i} + xy\mathbf{j} - 2xz\mathbf{k}$; $\vec{S}$ is the hemisphere $x^2 + y^2 + z^2 = a^2$, $z \ge 0$ with $\mathbf{n} \cdot \mathbf{k} > 0$.

10  $\mathbf{v} = -yz\mathbf{i}$; $\vec{S}$ is the part of the sphere $x^2 + y^2 + z^2 = 4$ outside the cylinder $x^2 + y^2 = 1$, $\mathbf{n}$ pointing outward.

11  $\mathbf{v} = -z\mathbf{j} + y\mathbf{k}$; $\vec{S}$ is the part of the vertical cylinder $r = \theta$ (cylindrical coordinates), $0 \le \theta \le \pi/2$, bounded below by the $(x, y)$ plane and above by the cone $z^2 = x^2 + y^2$, $\mathbf{n} \cdot \mathbf{i} > 0$ for $\theta > 0$.

12  $\mathbf{v} = y\mathbf{i} + z\mathbf{j} + x\mathbf{k}$; $\vec{S}$ is the part of the surface $z^2 = 4 - x$ to the right of the cylinder $y^2 = x$, $\mathbf{n} \cdot \mathbf{i} > 0$, $\mathbf{n}$ pointing to the right.

13  $\mathbf{v} = z\mathbf{i} - x\mathbf{k}$; $\vec{S}$ is the part of the cylinder $r = 2 + \cos \theta$ above the $(x, y)$ plane and below the cone $z^2 = x^2 + y^2$, $\mathbf{n}$ pointing outward.

In each of Problems 14 through 16, compute $\int_{\partial \vec{S}} \mathbf{v} \cdot d\mathbf{r}$, using Stokes' Theorem.

14  $\mathbf{v} = r^{-3}\mathbf{r}$, $\mathbf{r} = x\mathbf{i} + y\mathbf{j} + z\mathbf{k}$, $r = |\mathbf{r}|$; $\vec{S}$ is the surface of Example 2.

15  $\mathbf{v} = (e^x \sin y)\mathbf{i} + (e^x \cos y - z)\mathbf{j} + y\mathbf{k}$; $\vec{S}$ is the surface of Problem 3.

16  $\mathbf{v} = (x^2 + z)\mathbf{i} + (y^2 + x)\mathbf{j} + (z^2 + y)\mathbf{k}$; $\vec{S}$ is the part of the sphere $x^2 + y^2 + z^2 = 1$ above the cone $z^2 = x^2 + y^2$; $\mathbf{n} \cdot \mathbf{k} > 0$.

17  Show that if $\vec{S}$ is given by $z = f(x, y)$ for $x^2 + y^2 \le 1$, where $f$ is smooth and if $\mathbf{v} = (1 - x^2 - y^2)\mathbf{w}(x, y, z)$, where $\mathbf{w}$ is any smooth vector field defined on an open set containing $S \cup \partial S$, then $\iint_{\vec{S}} (\text{curl } \mathbf{v} \cdot \mathbf{n}) \, dS = 0$.

18  Suppose that $\mathbf{v} = r^{-3}(y\mathbf{i} + z\mathbf{j} + x\mathbf{k})$, where $r = |\mathbf{r}| = |x\mathbf{i} + y\mathbf{j} + z\mathbf{k}|$ and $\vec{S}$ is the unit sphere with $\mathbf{n}$ directed outward. Show by direct calculation that

$$\iint_{\vec{S}} (\text{curl } \mathbf{v}) \cdot \mathbf{n} \, dS = 0.$$

## 7

## THE DIVERGENCE THEOREM

Stokes' Theorem, which relates an integral over a surface in space to a line integral over the boundary of the surface, is a generalization of Green's Theorem. Another type of generalization, known as the Divergence Theorem, establishes a connection between an integral over a three-dimensional domain and an integral over the surface which forms the boundary of the domain. It can be shown that the Divergence Theorem and the theorems of Green and Stokes are all special cases of a general formula which connects an integral over a set of points in some $n$-dimensional space with another integral over the boundary of that set of points.

**DEFINITION**   *Let* **w** *be a vector function of the form* $\mathbf{w} = f(x, y, z)\mathbf{i} + g(x, y, z)\mathbf{j} + h(x, y, z)\mathbf{k}$ *with* $f, g, h$ *and their first partial derivatives all continuous. The* **divergence of w**, *denoted* div **w**, *is the scalar function given by*

$$\operatorname{div} \mathbf{w} = \frac{\partial f}{\partial x} + \frac{\partial g}{\partial y} + \frac{\partial h}{\partial z}.$$

div **w** *is also symbolically written* $\nabla \cdot \mathbf{w}$.

From the definition it would appear that the divergence of a vector function is only defined for a rectangular coordinate system. It can be shown, however, that there is an equivalent coordinate-free definition of divergence, as is implied by the notation $\nabla \cdot \mathbf{w}$. Theorem 6 gives some simple properties of the divergence operator. The proof is left to the reader.

**THEOREM 6**   *If* **u** *and* **v** *are vector fields and* $f$ *is a scalar field, all continuously differentiable,\** *then*

a) $\operatorname{div}(\mathbf{u} + \mathbf{v}) = \operatorname{div} \mathbf{u} + \operatorname{div} \mathbf{v}$,
b) $\operatorname{div}(f\mathbf{u}) = f \operatorname{div} \mathbf{u} + \nabla f \cdot \mathbf{u}$.

**EXAMPLE 1**   Given the vector field

$$\mathbf{w} = \frac{x + z}{x^2 + y^2 + z^2}\mathbf{i} + \frac{y - x}{x^2 + y^2 + z^2}\mathbf{j} + \frac{z - y}{x^2 + y^2 + z^2}\mathbf{k},$$

find div **w**. Evaluate div **w** at $P_0(1, 0, -2)$.

**Solution**   We have

$$\frac{\partial}{\partial x}\left(\frac{x + z}{x^2 + y^2 + z^2}\right) = \frac{(x^2 + y^2 + z^2) - (x + z)(2x)}{(x^2 + y^2 + z^2)^2} = \frac{y^2 + z^2 - x^2 - 2xz}{(x^2 + y^2 + z^2)^2},$$

$$\frac{\partial}{\partial y}\left(\frac{y - x}{x^2 + y^2 + z^2}\right) = \frac{x^2 + z^2 - y^2 + 2xy}{(x^2 + y^2 + z^2)^2},$$

$$\frac{\partial}{\partial z}\left(\frac{z - y}{x^2 + y^2 + z^2}\right) = \frac{x^2 + y^2 - z^2 + 2yz}{(x^2 + y^2 + z^2)^2}.$$

Therefore

$$\operatorname{div} \mathbf{w} = \frac{x^2 + y^2 + z^2 + 2(xy - xz + yz)}{(x^2 + y^2 + z^2)^2}.$$

At $P_0$ we obtain div **w** $= 2/5$.   □

**EXAMPLE 2**   Suppose $R$ is the radius of the Earth, $O$ is its center, and $g$ is the acceleration due to gravity at the surface of the Earth. If $P$ is a point in space near the surface, we denote by **r** the directed line segment $\overline{OP}$. The length $|\mathbf{r}|$ of the vector **r** we denote simply by $r$. From classical physics it is known that the

---

\*Recall that a field is continuously differentiable if the field and all first partial derivatives are continuous.

vector field $\mathbf{v}(P)$ due to gravity (called the *gravitational field of the Earth*) is given (approximately) by the equation

$$\mathbf{v}(P) = \frac{-gR^2}{r^3}\mathbf{r}.$$

Show that for $r > R$, we have div $\mathbf{v}(P) = 0$.

**Solution**    A computation yields

$$\text{div } \mathbf{v} = \text{div}\left(\frac{-gR^2}{r^3}\mathbf{r}\right),$$

and using Formula (b) of Theorem 6, we find

$$\text{div } \mathbf{v} = \frac{-gR^2}{r^3}\text{div }\mathbf{r} + \nabla\left(\frac{-gR^2}{r^3}\right)\cdot\mathbf{r}$$

$$= -gR^2\left[\frac{1}{r^3}\text{div }\mathbf{r} + \frac{d}{dr}\left(\frac{1}{r^3}\right)\nabla r\cdot\mathbf{r}\right]$$

$$= -gR^2(r^{-3}\text{div }\mathbf{r} - 3r^{-4}\nabla r\cdot\mathbf{r}).$$

Introducing a rectangular $(x, y, z)$ system with origin at $O$, we write

$$\mathbf{r} = x\mathbf{i} + y\mathbf{j} + z\mathbf{k},$$

from which we obtain div $\mathbf{r} = 3$. Also, the length $r$ of $\mathbf{r}$ is given by

$$r = \sqrt{x^2 + y^2 + z^2},$$

and so

$$\frac{\partial r}{\partial x} = \frac{x}{r}, \qquad \frac{\partial r}{\partial y} = \frac{y}{r}, \qquad \frac{\partial r}{\partial z} = \frac{z}{r}.$$

Hence

$$\nabla r = \frac{1}{r}(x\mathbf{i} + y\mathbf{j} + z\mathbf{k}) = \frac{\mathbf{r}}{r}$$

and

$$\nabla r\cdot\mathbf{r} = \frac{\mathbf{r}\cdot\mathbf{r}}{r} = r.$$

We conclude that

$$\text{div }\mathbf{v} = -gR^2(3r^{-3} - 3r^{-4}\cdot r) = 0. \qquad \square$$

We next state the Divergence Theorem. We provide a general version of the theorem, one which is proved in more advanced courses. However, in Theorem 8 we give a proof of the Divergence Theorem when $G$ and $\mathbf{v}$ are both of special form. By decomposing a region into those of the type considered in Theorem 8, we can obtain a form of the Divergence Theorem which is applicable widely.

**THEOREM 7**    **(The Divergence Theorem)** *Suppose that a bounded domain G in three-space is bounded by one or more disjoint piecewise smooth, orientable surfaces without boundary and suppose that* **v** *is a smooth vector function defined on an open set containing G and ∂G. Then*

$$\iiint_G \operatorname{div} \mathbf{v}\, dV = \iint_{\partial G} \mathbf{v} \cdot \mathbf{n}\, dS, \tag{1}$$

*where the boundary ∂G is oriented by taking* **n** *as the exterior normal.*

In Lemma 2 and Theorem 8 below we prove important special cases of the Divergence Theorem.

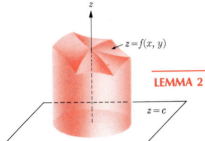

FIGURE 37

**LEMMA 2**    *Suppose that* **v** *and G are such that there exists a rectangular coordinate system in which* $\mathbf{v}(x, y, z) = R(x, y, z)\mathbf{k}$ *and G is of the form*

$$G = \{(x, y, z): (x, y) \in D, c < z < f(x, y)\},$$

*where f is piecewise smooth (as indicated in Fig. 37).* \* *If R and $R_z$ are continuous on a set containing G and ∂G, then (1) holds.*

**Proof**    We note that $\operatorname{div} \mathbf{v} = R_z$ and, therefore, that

$$\iiint_G \operatorname{div} \mathbf{v}\, dV = \iiint_G R_z\, dV_{xyz} = \iint_D \int_c^{f(x,y)} R_z\, dz\, dA_{xy}.$$

Upon performing the integration with respect to $z$, we find

$$\iiint_G \operatorname{div} \mathbf{v}\, dV = \iint_D \{R[x, y, f(x, y)] - R(x, y, c)\}\, dA_{xy}. \tag{2}$$

Let $\partial G = S_1 \cup S_2 \cup S_3$, where $S_1$ is the domain $D$ in the plane $z = c$, $S_2$ is the lateral surface of the cylinder, and $S_3$ is the top surface:

$$S_3 = \{(x, y, z): (x, y) \in D, z = f(x, y)\}.$$

We have

$$\text{On } S_2: \quad \mathbf{n} \cdot \mathbf{k} = 0, \qquad \iint_{S_2} \mathbf{v} \cdot \mathbf{n}\, dS = 0. \tag{3}$$

$$\text{On } S_1: \quad \mathbf{n} = -\mathbf{k}, \qquad \iint_{S_1} \mathbf{v} \cdot \mathbf{n}\, dS = -\iint_D R(x, y, c)\, dA_{xy}. \tag{4}$$

$$\text{On } S_3: \quad \mathbf{n} = (1 + f_x^2 + f_y^2)^{-1/2}(-f_x\mathbf{i} - f_y\mathbf{j} + \mathbf{k}),$$

$$dS = (1 + f_x^2 + f_y^2)^{1/2}\, dA_{xy}; \tag{5}$$

$$\iint_{S_3} \mathbf{v} \cdot \mathbf{n}\, dS = \iint_D R[x, y, f(x, y)]\, dA_{xy}.$$

A comparison of (2) with (3), (4), and (5) yields the result.    □

---

\* That is, the graph $S$ of the equation $z = f(x, y), (x, y) \in D \cup \partial D$ is a piecewise smooth surface and $z = f(x, y), (x, y) \in D_i \cup \partial D_i$ is a parametric representation of each smooth part $S_i$.

**THEOREM 8**   *Suppose that the hypotheses of Lemma 2 hold, except that $\mathbf{v}$ has the form*

$$\mathbf{v} = P(x, y, z)\mathbf{i} + Q(x, y, z)\mathbf{j}$$

*with $P, Q$ smooth on a region containing $G$ and $\partial G$. Then (1) holds.*

**Proof**   We define the functions $U, V$ by the formulas

$$U(x, y, z) = \int_c^z Q(x, y, t)\, dt, \qquad V(x, y, z) = -\int_c^z P(x, y, t)\, dt.$$

Also, we set

$$\mathbf{w} = U\mathbf{i} + V\mathbf{j} \qquad \text{and} \qquad -R = V_x - U_y, \qquad \mathbf{u} = R\mathbf{k}.$$

Then $\mathbf{w}$ is smooth and $R, R_z$ are continuous, so that

$$\operatorname{curl}\mathbf{w} = -V_z\mathbf{i} + U_z\mathbf{j} + (V_x - U_y)\mathbf{k} = \mathbf{v} - \mathbf{u},$$

$$\operatorname{div}\mathbf{v} = P_x + Q_y = R_z = \operatorname{div}\mathbf{u}. \tag{6}$$

We apply the Corollary to Stokes' Theorem and obtain

$$\iint_{\partial G} (\operatorname{curl}\mathbf{w}) \cdot \mathbf{n}\, dS = 0 = \iint_{\partial G} (\mathbf{v} - \mathbf{u}) \cdot \mathbf{n}\, dS.$$

Therefore

$$\iint_{\partial G} \mathbf{v} \cdot \mathbf{n}\, dS = \iint_{\partial G} \mathbf{u} \cdot \mathbf{n}\, dS = \iiint_G \operatorname{div}\mathbf{u}\, dV; \tag{7}$$

the last equality holds because Lemma 2 may be applied to $\mathbf{u} = R\mathbf{k}$. Taking (6) and (7) into account, we get

$$\iint_{\partial G} \mathbf{v} \cdot \mathbf{n}\, dS = \iiint_G \operatorname{div}\mathbf{v}\, dV. \qquad \square$$

The Divergence Theorem holds not only for domains which are parallel to the $z$ axis, but also for cylindrical domains which are parallel to the $x$ or $y$ axes. By addition, the result is valid for smooth vector fields $\mathbf{v}$ defined over regions $G$ which are the sum of cylindrical domains of the kind just described. These regions can be quite general.

In two dimensions the Divergence Theorem is a direct consequence of Green's Theorem. To see this we choose the usual coordinate system in the $xy$ plane and suppose that $\mathbf{v} = P\mathbf{i} + Q\mathbf{j}$. We define $\mathbf{u} = Q\mathbf{i} - P\mathbf{j}$. As Fig. 38 shows, the exterior normal $\mathbf{n}$ to a region $G$ makes an angle of $-\pi/2$ with the tangent vector $\mathbf{T}$, directed so that the interior of $G$ is always on the left as we proceed around the boundary. If $\phi$ is the angle the vector $\mathbf{T}$ makes with the positive $x$-direction, we may write

$$\mathbf{T} = (\cos\phi)\mathbf{i} + (\sin\phi)\mathbf{j}, \qquad \mathbf{n} = (\sin\phi)\mathbf{i} - (\cos\phi)\mathbf{j},$$

$$\mathbf{v} \cdot \mathbf{T} = \mathbf{u} \cdot \mathbf{n}, \qquad \operatorname{div}\mathbf{u} = Q_x - P_y.$$

The Divergence Theorem (Equation (1)) applied to $\mathbf{u}$ is, when translated in terms of $\mathbf{v}$ (or $P, Q$), a restatement of Green's Theorem.

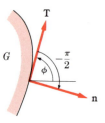

**FIGURE 38**

**EXAMPLE 3**   Verify the Divergence Theorem, given that $G$ is the domain between the concentric spheres $S_1$ and $S_2$ of radius 1 and 2, respectively, and center at $O$; the vector $\mathbf{v}$ is $\mathbf{v} = \mathbf{r}/r^3$, with $\mathbf{r} = \mathbf{v}(\overrightarrow{OP})$, $P$ a point in the domain, and $r = |\mathbf{r}|$.

**Solution**   (See Fig. 39.) Let $\vec{S}_1$ and $\vec{S}_2$ be the spheres, both oriented with $\mathbf{n}$ pointing outward from $O$. Then

$$\iint_{\partial G} \mathbf{v} \cdot \mathbf{n} \, dS = \iint_{\vec{S}_2} \mathbf{v} \cdot \mathbf{n} \, dS - \iint_{\vec{S}_1} \mathbf{v} \cdot \mathbf{n} \, dS,$$

since the normal on $\vec{S}_1$, as part of $\partial \vec{G}$, points toward $O$ when the Divergence Theorem is used. According to Example 2 we easily find that

$$\operatorname{div} \mathbf{v} = 0$$

and hence

$$\iiint_G \operatorname{div} \mathbf{v} \, dV = 0.$$

Now, on both $\vec{S}_1$ and $\vec{S}_2$, we have $\mathbf{n} = r^{-1}\mathbf{r}$, and so

$$\iint_{\vec{S}_2} \mathbf{v} \cdot \mathbf{n} \, dS - \iint_{\vec{S}_1} \mathbf{v} \cdot \mathbf{n} \, dS = \tfrac{1}{4}A(S_2) - 1 \cdot A(S_1) = 0. \qquad \square$$

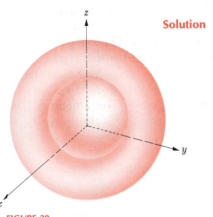

**FIGURE 39**

**EXAMPLE 4**   Given that $G$ is the domain inside the cylinder

$$x^2 + y^2 = 1$$

and between the planes $z = 0$ and $z = x + 2$, and given that

$$\mathbf{v} = (x^2 + ye^z)\mathbf{i} + (y^2 + ze^x)\mathbf{j} + (z^2 + xe^y)\mathbf{k},$$

use the Divergence Theorem to evaluate $\iint_{\partial G} \mathbf{v} \cdot \mathbf{n} \, dS$.

**Solution**   We have $\operatorname{div} \mathbf{v} = 2x + 2y + 2z$. Therefore, letting $F$ denote the unit disk, we obtain (Fig. 34, p. 793)

$$\iint_{\partial G} \mathbf{v} \cdot \mathbf{n} \, dS = \iiint_G 2(x + y + z) \, dV = 2 \iint_F \left[(x + y)z + \tfrac{1}{2}z^2\right]_0^{x+2} \, dA_{xy}$$

$$= \iint_F \left[2y(x + 2) + 2x^2 + 4x + (x^2 + 4x + 4)\right] dA_{xy}$$

$$= \iint_F (3x^2 + 4) \, dA_{xy} = \int_0^1 \int_0^{2\pi} \left[(3r^3 \cos^2 \theta) + 4r\right] dr \, d\theta$$

$$= \pi \int_0^1 (3r^3 + 8r) \, dr = \tfrac{19}{4}\pi. \qquad \square$$

The Divergence Theorem has an important physical interpretation in connection with problems in fluid flow. We suppose that a fluid (liquid or gas) is flowing through a region in space. In general, the density $\rho$ and the

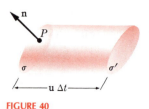

**FIGURE 40**

velocity vector **u** will depend not only on the point $P$ in space but also on the time $t$. We select a point $P$ in space, a value of $t$, and a small plane surface $\sigma$ through $P$ (Fig. 40). For a moment we suppose that $\rho$ and **u** are constant. Then, after a short time $\Delta t$, all the particles on $\sigma$ at time $t$ would be in the shaded region $\sigma'$ shown in Fig. 40. The particles sweep out a small cylindrical region as they travel from $\sigma$ to $\sigma'$. The total mass of fluid flowing across $\sigma$ in time $\Delta t$ is just enough to fill up the oblique cylinder between $\sigma$ and $\sigma'$. We denote this cylindrical region by $G$, its boundary by $\partial G$, and the outward normal on the boundary by **n**. If $\mathbf{u} \cdot \mathbf{n} > 0$ on $\sigma$, then the fluid flow is *out of* $G$ across $\sigma$, while if $\mathbf{u} \cdot \mathbf{n} < 0$ on $\sigma$, the flow is *into* $G$ across $\sigma$. The amount of fluid leaving (or entering, if the value is negative) $G$ across $\sigma$ is then given by

$$(\rho \mathbf{u})\, \Delta t \cdot \mathbf{n} A(\sigma).$$

The net *rate* of flow of mass out of $G$ across $\sigma$ (per unit time) is just

$$\rho \mathbf{u} \cdot \mathbf{n} A(\sigma).$$

If $\rho$ and **u** are continuous on $\partial G$, a piecewise smooth boundary, and if we divide up $\partial G$ into small, smooth surface elements such as $\sigma$ and add the results, we obtain (in the limit)

$$\iint_{\partial G} (\rho \mathbf{u}) \cdot \mathbf{n}\, dS$$

for the rate of flow of the total amount of mass out of $G$. We denote by $M(G, t)$ the total mass in the region $G$ at time $t$. Using the definition of density, we may write

$$M(G, t) = \iiint_{G} \rho(P, t)\, dV_p.$$

If we now assume that we can differentiate "through the integral" (a procedure justified in more advanced courses, and known as Leibniz' Rule), we find

$$\frac{d}{dt} M(G, t) = \iiint_{G} \rho_t(P, t)\, dV_p.$$

Since $(d/dt)M$ is the amount of mass flowing into $G$, we obtain

$$\iiint_{G} \rho_t(P, t)\, dV_P = -\iint_{\partial G} (\rho \mathbf{u}) \cdot \mathbf{n}\, dS.$$

If $\rho$ and **u** are smooth, we may apply the Divergence Theorem to the boundary integral and get the equation

$$\iiint_{G} [\rho_t(P, t) + \operatorname{div}(\rho \mathbf{u})]\, dV = 0.$$

Since $G$ is arbitrary, we can divide the above equation by $V(G)$, the volume of $G$, and let $G$ shrink to a point $P$. The result is the **equation of continuity**

$$\rho_t + \operatorname{div}(\rho \mathbf{u}) = 0.$$

If the fluid is an incompressible liquid, so that $\rho = \text{const}$, the equation of continuity becomes

$$\text{div } \mathbf{u} = 0.$$

A vector field $\mathbf{u}$ with the property that div $\mathbf{u} = 0$ is called **divergence free**. In electricity and magnetism divergence free vector fields occur, and they are called **solenoidal**.

## 7  PROBLEMS

In each of Problems 1 through 10, find div $\mathbf{v}$.

1  $\mathbf{v} = (a_{11}x + a_{12}y + a_{13}z)\mathbf{i} + (a_{21}x + a_{22}y + a_{23}z)\mathbf{j}$
   $+ (a_{31}x + a_{32}y + a_{33}z)\mathbf{k}$

2  $\mathbf{v} = (x^2 - y^2)\mathbf{i} + (x^2 - z^2)\mathbf{j} + (y^2 - z^2)\mathbf{k}$

3  $\mathbf{v} = (x^2 + 1)\mathbf{i} + (y^2 - 1)\mathbf{j} + z^2\mathbf{k}$

4  $\mathbf{v} = 4xz\mathbf{i} - 2yz\mathbf{j} + (2x^2 - y^2 - z^2)\mathbf{k}$

5  $\mathbf{v} = e^{xz}(\cos yz\mathbf{i} + \sin yz\mathbf{j} - \mathbf{k})$

6  $\mathbf{v} = y\log(1 + x)\mathbf{i} + z\log(1 + y)\mathbf{j} + x\log(1 + z)\mathbf{k}$

7  $\mathbf{v} = \nabla u, u = x^3 - 3xy^2$

8  $\mathbf{v} = \nabla u, u = a_{11}x^2 + a_{22}y^2 + a_{33}z^2 + 2a_{12}xy$
   $+ 2a_{13}xz + 2a_{23}yz$

9  $\mathbf{v} = \nabla u, u = e^x\cos y + e^y\cos z + e^z\cos x$

10  $\mathbf{v} = 2x\mathbf{i} + 3y\mathbf{j} + (\nabla^2 u)\mathbf{k}$,  where $u = x^3 + y^3 + z^3$

In each of Problems 11 through 20, verify the Divergence Theorem by computing separately each side of Eq. (1).

11  $\mathbf{v} = xy\mathbf{i} + yz\mathbf{j} + zx\mathbf{k}$; $G$ is bounded by the coordinate planes and the plane $x + y + z = 1$.

12  $\mathbf{v} = x^2\mathbf{i} - y^2\mathbf{j} + z^2\mathbf{k}$; $G$ is bounded by: $x^2 + y^2 = 4$, $z = 0$, $z = 2$.

13  $\mathbf{v} = 2x\mathbf{i} + 3y\mathbf{j} - 4z\mathbf{k}$; $G$ is the ball $x^2 + y^2 + z^2 \le 4$.

14  $\mathbf{v} = x^2\mathbf{i} + y^2\mathbf{j} + z^2\mathbf{k}$; $G$ is bounded by: $y^2 = 2 - x$, $z = 0$, $z = x$.

15  $\mathbf{v} = x\mathbf{i} + y\mathbf{j} + z\mathbf{k}$; $G$ is the domain outside $x^2 + y^2 = 1$ and inside

$$x^2 + y^2 + z^2 = 4.$$

16  $\mathbf{v} = x\mathbf{i} - 2y\mathbf{j} + 3z\mathbf{k}$; $G$ is bounded by $y^2 = x$ and $z^2 = 4 - x$.

17  $\mathbf{v} = r^{-3}(z\mathbf{i} + x\mathbf{j} + y\mathbf{k})$; $G$ is the domain outside $x^2 + y^2 + z^2 = 1$ and inside $x^2 + y^2 + z^2 = 4$.

18  $\mathbf{v} = 3x\mathbf{i} - 2y\mathbf{j} + z\mathbf{k}$; $G$ is bounded by $x^2 + z^2 = 4$, $y = 0$,

$$x + y + z = 3.$$

19  $\mathbf{v} = 2x\mathbf{i} + y\mathbf{j} + z\mathbf{k}$; $G$ is bounded by $z = x^2 + y^2$ and $z = 2x$.

20  $\mathbf{v} = x\mathbf{i} + y\mathbf{j} + z\mathbf{k}$; $G$ is bounded by $x^2 + y^2 = 4$ and $x^2 + y^2 - z^2 = 1$.

In each of Problems 21 through 25, evaluate $\iint_{\partial G} \mathbf{v} \cdot \mathbf{n}\, dS$, using the Divergence Theorem.

21  $\mathbf{v} = ye^z\mathbf{i} + (y - ze^x)\mathbf{j} + (xe^y - z)\mathbf{k}$; $G$ is the interior of the torus $(r - b)^2 + z^2 \le a^2$, $0 < a < b$;  $r, z$  cylindrical coordinates.

22  $\mathbf{v} = x^3\mathbf{i} + y^3\mathbf{j} + z^3\mathbf{k}$; $G$ is the ball $x^2 + y^2 + z^2 \le 1$.

23  $\mathbf{v} = (x^4 + \cos z)\mathbf{i} + (x^3y + \sin z)\mathbf{j} + e^{xy}\mathbf{k}$; $G$ is the region bounded by the cylinder $z = 4 - x^2$, the plane $y + z = 5$, and the $xy$ and $xz$ planes (Fig. 41).

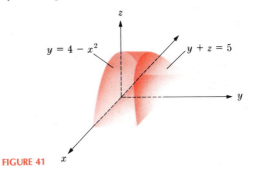

$y = 4 - x^2$

$y + z = 5$

**FIGURE 41**

24  $\mathbf{v} = x^5\mathbf{i} + y^5\mathbf{j} + z^5\mathbf{k}$; $G$ is the region bounded by the cylinder $x^2 + y^2 = 9$ and the planes $z = 0$ and $z = 5$.

25  $\mathbf{v} = x^3\mathbf{i} + y^3\mathbf{j} + z\mathbf{k}$; $G$ is bounded by: $x^2 + y^2 = 1$, $z = 0$, and $z = x + 2$.

26  Suppose that $G$ is a region in three-space with a boundary $\partial G$ for which the Divergence Theorem is applicable. Establish the following formula for integration by parts if $u$ and $v$ are smooth on a region $D$ containing $G$ and $\partial G$:

$$\iiint_G u\,\text{div}\,\mathbf{v}\, dV = \iint_{\partial G} u\mathbf{v} \cdot \mathbf{n}\, dS - \iiint_G \nabla u \cdot \mathbf{v}\, dV.$$

27  Suppose that $D$ and $G$ are as in Problem 26 and that $u$, $\nabla u$, and $v$ are smooth in $D$. Let $\partial/\partial n$ denote the directional derivative on $\partial G$ in the direction of $\mathbf{n}$. Show that

$$\iiint_G v\nabla^2 u\, dV = \iint_{\partial G} v\frac{\partial u}{\partial n}\, dS - \iiint_G \nabla v \cdot \nabla u\, dV.$$

**28** Evaluate $\iint_S \vec{v} \cdot \mathbf{n} \, dS$, where $S$ is the torus

$$S = \{(r, \theta, z): (r - 3)^2 + z^2 = 1\}$$

$((r, \theta, z)$ cylindrical coordinates);

$\mathbf{n}$ is the outward normal on $S$; $\mathbf{R} = (x - 3)\mathbf{i} + y\mathbf{j} + z\mathbf{k}$; $R = |\mathbf{R}|$, $\mathbf{v} = R^{-3}\mathbf{R}$. [*Hint*: Use the Divergence Theorem, with $G$ as the part of the interior of the torus which is outside a small sphere of radius $\rho$ and center at $(3, 0, 0)$.]

---

## CHAPTER 17

### REVIEW PROBLEMS

In each of Problems 1 through 15, evaluate the line integrals as indicated.

**1** $\int_C [xy^2 \, dx + (x + y) \, dy]$, where $C$ consists of the two line segments from $(3, 1)$ to $(2, 1)$ and then from $(2, 1)$ to $(2, 4)$.

**2** $\int_C [x^2 y \, dx + yx^2 \, dy]$, where $C$ is part of the parabola $y = x^2 + 3$ from $(0, 3)$ to $(2, 7)$.

**3** $\int_C (x - y)^2 \, dx$, where $C$ is the arc described by $x = 2t - 1$, $y = 5t^2 - 3t$; $1 \leq t \leq 2$.

**4** $\int_C [yz \, dx + xz \, dy + xy \, dz]$, where $C$ is described by $x = t^2$, $y = t$, $z = t^3$; $0 \leq t \leq 3$.

**5** $\int_C [xy \, dx + (x^2 + xy^3) \, dy]$, where $C$ is the arc described by $y = x^3$ from $(0, 0)$ to $(1, 1)$.

**6** $\int_C z \, ds$, where $C$ is described by $x = \cos t$, $y = \sin t$, $z = t$; $0 \leq t \leq \pi$.

**7** $\int_C (x^2 + y^2 + z^2) \, ds$, where $C$ is as in Problem 6.

**8** $\int_C (x + y) \, ds$, where $C$ is the triangle with vertices $(0, 0)$, $(1, 0)$, and $(0, 1)$, traversed in a counterclockwise direction.

**9** $\int_C x \, dx$, where $C$ is the semicircle $y = \sqrt{1 - x^2}$, traversed *clockwise*.

**10** $\int_C x \, ds$, where $C$ is as in Problem 9.

**11** $\int_C [(3x^2 - 6xy) \, dx + (4y^3 - 3x^2) \, dy]$, where $C$ is any smooth arc from $(0, 0)$ to $(1, 2)$.

**12** $\int_C [2xy \cos(x^2 y) \, dx + x^2 \cos(x^2 y) \, dy]$, where $C$ is any smooth arc from $(0, 0)$ to $(-1, \pi/2)$.

**13** $\int_C e^{2x + 3y} (2 \, dx + 3 \, dy)$, where $C$ is any smooth arc from $(0, 0)$ to $(-1, 1)$.

**14** $\int_C [f_x \, dx + f_y \, dy]$, where $f(x, y)$ has continuous partial derivatives, and $C$ is a smooth arc as described in Fig. 42,

starting and ending at the point $(1, 2)$.

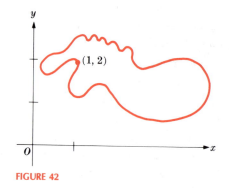

**FIGURE 42**

**15** $\int_C (x\mathbf{i} + y\mathbf{j}) \cdot d\mathbf{r}$, where $\mathbf{r}$ describes the arc $y = 1 + x^3$, from $(0, 1)$ to $(2, 9)$.

**16** Find an arc described by $\mathbf{r}$, beginning and ending at the same point, such that

$$\int_C (x\mathbf{i} - y\mathbf{j}) \cdot d\mathbf{r} \neq 0.$$

**17** For which of the vector functions $\mathbf{v}$ in (a) through (e) do there exist closed arcs $\mathbf{r}$ (starting and ending at the same point) such that $\int_C \mathbf{v} \cdot d\mathbf{r} \neq 0$? If there exists such arcs $\mathbf{r}$, exhibit one.

a) $\mathbf{v} = y\mathbf{i} + (xy - x)\mathbf{j}$
b) $\mathbf{v} = (2xy - 3yx^2)\mathbf{i} + (x^2 - x^3)\mathbf{j}$
c) $\mathbf{v} = y\mathbf{i} + x\mathbf{j} + x\mathbf{k}$
d) $\mathbf{v} = xy\mathbf{i} + (x^2 + 1)\mathbf{j} + z^2\mathbf{k}$
e) $\mathbf{v} = 2xy^3 z^2 \mathbf{i} + 3x^2 y^2 z^2 \mathbf{j} + 2x^2 y^3 z\mathbf{k}$

In each of Problems 18 through 23, use Green's Theorem to evaluate the given line integrals. Take all integrals counterclockwise.

**18** $\oint_C [(y + 3x)\, dx + (2y - x)\, dy]$, where $C$ is the ellipse $4x^2 + y^2 = 4$.

**19** $\oint_C [(5 - xy - y^2)\, dx + (x^2 - 2xy)\, dy]$, where $C$ is the rectangle with vertices $(0, 0)$, $(1, 0)$, $(1, 3)$, $(0, 3)$.

**20** $\oint_C [y\, dx + x\, dy]$, where $C$ is the circle of radius 3 and center at the origin.

**21** $\oint_C [(e^x + y^2)\, dx + (e^y + x^2)\, dy]$, where $C$ is the boundary of the region bounded by the curves $y = x$ and $y = x^2$.

**22** $\oint_C \left[ \ln(1 + y)\, dx - \dfrac{xy}{1 + y}\, dy \right]$,

where $C$ is the triangle with vertices $(0, 0)$, $(2, 0)$, and $(0, 4)$.

**23** $\oint_C [\cos x \sin y\, dx + \sin x \cos y\, dy]$, where $C$ is the triangle with vertices $(0, 0)$, $(2, 2)$, and $(0, 2)$.

**24** Show that the area of a region $G$ with a smooth boundary, $\partial G$, is given by the line integral $\frac{1}{2} \oint_{\partial G} [-y\, dx + x\, dy]$.

**25** Find $\oint_{\partial G} (e^{x^2} \mathbf{i} + \sin y \mathbf{j}) \cdot d\mathbf{r}$, where $G$ is the disk of radius 3, centered at $(1, 2)$.

**26** Find $\oint_{\partial G} \mathbf{v} \cdot d\mathbf{r}$, where $G$ is the unit disk centered at the origin, if $\mathbf{v}$ is given by:
a) $\mathbf{v} = -yx^2 \mathbf{i} + xy^2 \mathbf{j}$
b) $\mathbf{v} = (e^x - y^3)\mathbf{i} + (\cos y + x^3)\mathbf{j}$
c) $\mathbf{v} = -y\mathbf{i} + x\mathbf{j}$

**27** Find the area of the region enclosed by the curve $x = 2\cos^3 t$, $y = 2\sin^3 t$, $0 \leq t \leq 2\pi$.

In each of Problems 28 through 35, evaluate the surface integral

$$\iiint_F f(x, y, z)\, dS$$

for the given function $f$ and region $F$.

**28** $f(x, y, z) = 1$; and $F$ is the surface bounded by $z = \frac{1}{2}y^2$, $0 \leq x \leq 1$, $0 \leq y \leq 1$.

**29** $f(x, y, z) = 2y$; $F$ is as in Problem 28.

**30** $f(x, y, z) = \sqrt{2z}$; $F$ is as in Problem 28.

**31** $f(x, y, z) = \sqrt{1 + y^2}$; $F$ is as in Problem 28.

**32** $f(x, y, z) = 1$, $F$ is the surface given by $z = xy$, $0 \leq x^2 + y^2 \leq 1$.

**33** $f(x, y, z) = 4\sqrt{x^2 + y^2}$; $F$ is as in Problem 32.

**34** $f(x, y, z) = z$; $F$ is the top hemisphere of the sphere of radius $a$, centered at the origin.

**35** $f(x, y, z) = x^4 - y^4 + y^2z^2 - z^2x^2 + 1$; $F$ is the part of the surface of the upper nappe of the cone $x^2 + y^2 = z^2$ (i.e., $z \geq 0$), cut out by the cylinder $x^2 + y^2 = 2x$.

In each of Problems 36 through 40, compute $\iint_{\vec{F}} \mathbf{v} \cdot \mathbf{n}\, dS$, for the given vector function $\mathbf{v}$ and the given oriented surface $\vec{F}$.

**36** $\mathbf{v} = z^2 \mathbf{i} - 2x\mathbf{j} + y^3 \mathbf{k}$; $F$ is the top half of the unit sphere $x^2 + y^2 + z^2 = 1$, and $\mathbf{n}$ points away from the origin.

**37** $\mathbf{v} = 3x\mathbf{i} + (x + \frac{2}{3}x^3 + 2xy^2)\mathbf{j} + z\mathbf{k}$; $F$ is the top half of the ellipsoid $2x^2 + 2y^2 + z^2 = 1$; $\mathbf{n}$ points away from the origin.

**38** $\mathbf{v} = y^2 \mathbf{i} + xy\mathbf{j} + xz\mathbf{k}$; $F$ is as in Problem 36.

**39** $\mathbf{v} = (y - z)\mathbf{i} + yz\mathbf{j} - xz\mathbf{k}$; $F$ is the five faces of the cube $0 \leq x \leq 2$, $0 \leq y \leq 2$, $0 \leq z \leq 2$, not in the $xy$ plane; $\mathbf{n}$ is the outward pointing normal.

**40** $\mathbf{v} = y^2 \mathbf{i} + z^2 \mathbf{j} + x^2 \mathbf{k}$; $F$ is that part of the plane $x + y + z = 1$ that is in the first octant (i.e., $x > 0$, $y > 0$, and $z > 0$); $\mathbf{n}$ points away from the origin.

**41** If $\mathbf{v} = 2y\mathbf{i} + e^z \mathbf{j} - \arctan x\mathbf{k}$, use Stokes' Theorem to evaluate $\iint_F \text{curl } \mathbf{v} \cdot \mathbf{n}\, dS$, where $F$ is bounded by the paraboloid $z = 4 - x^2 - y^2$ and the $xy$ plane.

**42** Let $\mathbf{v} = y^3 \mathbf{i} + (xy + 3xy^2)\mathbf{j} + z^4 \mathbf{k}$, and use Stokes' Theorem to help to find $\oint_C \mathbf{v} \cdot d\mathbf{r}$, where $C$ is the curve created by the intersection of the upper half of the ellipsoid $x^2 + y^2 + 2z^2 = 2$ with the cylinder $x^2 + y^2 - y = 0$.

In each of Problems 43 through 48, find div $\mathbf{v}$.

**43** $\mathbf{v} = 2x\mathbf{i} + 3y\mathbf{j}$

**44** $\mathbf{v} = \dfrac{x}{x^2 + y^2}\mathbf{i} + \dfrac{y}{x^2 + y^2}\mathbf{j}$

**45** $\mathbf{v} = e^{x+y}\mathbf{i} + e^{x-y}\mathbf{j}$

**46** $\mathbf{v} = (9x - 2y^2)\mathbf{i} + (x - 6y)\mathbf{j}$

**47** $\mathbf{v} = \nabla u$, $u = x^4 - 2xy^2$

**48** $\mathbf{v} = \nabla u$, $u = e^{x^2 + 2xy - y^2}$

In each of Problems 49 through 53, use the Divergence Theorem to help in the evaluation of $\iint_{\partial G} \mathbf{v} \cdot \mathbf{n}\, dS$.

**49** $\mathbf{v} = e^{y^2}\mathbf{i} - xy\mathbf{j} + x \arcsin y\mathbf{k}$; $\partial G$ is the surface bounded by the coordinate planes and the plane $x + y + z = 1$.

**50** $\mathbf{v} = (x^2 + \cos yz)\mathbf{i} + (y - xe^{-z^2})\mathbf{j} + z^2 \mathbf{k}$; $\partial G$ is the surface of the region bounded by the graphs of $x^2 + y^2 = 9$, $x + z = 3$, and $z = 0$.

**51** $\mathbf{v} = x^2 \mathbf{i} - xz\mathbf{j} + z^2 \mathbf{k}$; $\partial G$ is the surface of the unit cube: $0 \leq x \leq 1$, $0 \leq y \leq 1$, $0 \leq z \leq 1$.

**52** $\mathbf{v} = x^2 \mathbf{i} - 2y\mathbf{j} + 3z^2 \mathbf{k}$; $\partial G$ is the surface of the truncated right circular cylinder: $0 \leq x^2 + y^2 \leq 9$; $0 \leq z \leq 1$.

**53** $\mathbf{v} = 3xy\mathbf{i} + y^2 \mathbf{j} - 3yz\mathbf{k}$; $\partial G$ is the surface of the ball of radius $a$, centered at the origin: $x^2 + y^2 + z^2 \leq a^2$.

# 18

# DIFFERENTIAL EQUATIONS

A differential equation is an equation which involves derivatives. Such equations occur when one applies mathematical methods to science and engineering. In this chapter we give an introduction to methods for solving elementary differential equations, and we point out several instances of particular equations which arise in applications.

## 1

### INTRODUCTION AND EXAMPLES

The equation

$$\frac{dy}{dx} = 2x \tag{1}$$

is an example of a differential equation. The function $y = x^2 + c$ (where $c$ is any constant) is a solution of this equation. In general, any equation which contains derivatives is called a **differential equation**. Other examples of differential equations are

$$2x\frac{d^2y}{dx^2} + \left(\frac{dy}{dx}\right)^2 = \frac{1}{y} + 2x, \tag{2}$$

$$\frac{d^4y}{dx^4} + x\left(\frac{dy}{dx}\right)^2 - y^3 = \tan x, \tag{3}$$

$$x^2\frac{\partial z}{\partial x} + 2\frac{\partial z}{\partial y} - 3xy + 4z \sec y = 0. \tag{4}$$

Equations (2) and (3) contain ordinary derivatives; from the way they are written, we see that $x$ is the independent variable and that $y$ is a function of $x$. Equation (4) involves partial derivatives of the variable $z$ in terms of the independent variables $x$ and $y$.

**DEFINITIONS**   *An* **ordinary differential equation** *is an equation which involves one unknown function, say y, and one or more derivatives of y taken with respect to an independent variable x. A* **partial differential equation** *is one which contains partial derivatives.*

Equations (1), (2), and (3) are ordinary differential equations.

We shall restrict our attention entirely to the problem of finding solutions of ordinary differential equations. (The systematic study of partial differential equations is usually taken up in more advanced courses). When we use the term "differential equation" we shall always mean ordinary differential equation.

The **order** of a differential equation is the order of the highest derivative which appears. Equation (2) is of the second order, and equation (3) is of the fourth order.

A differential equation in the form

$$c_0(x)\frac{d^n y}{dx^n} + c_1(x)\frac{d^{n-1}y}{dx^{n-1}} + \cdots + c_{n-1}(x)\frac{dy}{dx} + c_n(x)y = f(x) \qquad (5)$$

is said to be a **linear differential equation**. The functions $c_0(x), c_1(x), \ldots, c_n(x)$, $f(x)$ may be any functions of $x$, and their character does not affect the linearity of the equation. Equation (5) is of the $n$th order. The equations

$$x^2\frac{d^2 y}{dx^2} + 2\frac{dy}{dx} - x^3 y = \tan x, \qquad \sqrt{x}\frac{dy}{dx} + 8y = 2,$$

$$\frac{d^3 y}{dx^3} + 2x^4\frac{dy}{dx} + 8y = 6x$$

are examples of linear differential equations. On the other hand, the equations

$$\frac{d^2 y}{dx^2} + y\frac{dy}{dx} = 1, \qquad \left(\frac{dy}{dx}\right)^2 + 2y = 8x,$$

$$\left(\frac{dy}{dx}\right)\left(\frac{d^2 y}{dx^2}\right) + 8y = x^2, \qquad \frac{dy}{dx} + y^2 = 2$$

are all examples of nonlinear differential equations.

**To solve a differential equation** means to find all functions $y$ such that, for each $x$ (in some set), the numerical values of $y$ and its derivatives satisfy the given equation. In the following sections, we shall often be content to find relations which define the desired solutions $y$ implicitly. Some particularly simple differential equations may be solved by inspection. For example, the equation

$$\frac{dy}{dx} = f(x) \qquad (6)$$

has the solution

$$y = \int f(x)\,dx + C. \qquad (7)$$

**EXAMPLE 1**   Solve the differential equation

$$\frac{dy}{dx} = x + \sin x \tag{8}$$

**Solution**   Equation (8) is a linear differential equation of the first order. We solve the equation by integrating both sides of (8) to obtain:

$$y = \frac{x^2}{2} - \cos x + C.$$

The constant of integration $C$ cannot be determined without supplementary information.   □

While it is desirable to evaluate the indefinite integral of $f$ in (7) and so obtain an explicit answer, we call $\int f(x)\, dx + C$ a solution even when the actual integration cannot be performed. When the integration is not carried out we say that the solution has been obtained except for a **quadrature**.

**EXAMPLE 2**   Solve the differential equation

$$\frac{dy}{dx} = e^{-x^2}. \tag{9}$$

**Solution**   Equation (9) is a first order linear differential equation. We integrate both sides of (9) to obtain

$$y = \int_a^x e^{-u^2}\, du + C.$$

The integral $\int_a^x e^{-u^2}\, du$ has the variable $x$ as its upper limit of integration and is a function of $x$. By the Fundamental Theorem of Calculus its derivative is $e^{-x^2}$. The lower limit of integration, $a$, is arbitrary. However, if we were to choose a different value for the lower limit, we would change the value of the solution, $y$. We can compensate for changing the value of the lower limit by changing the constant of integration, $C$. Finally, it can be shown that the function $e^{-u^2}$ has no explicit antiderivative, and hence we must be content to write the solution in the form

$$y(x) = \int_a^x e^{-u^2}\, du + C,$$

which we say *is* the solution, except for a quadrature.   □

The equation

$$\frac{d^2y}{dx^2} = f(x) \tag{10}$$

may be integrated to give

$$\frac{dy}{dx} = F(x) + C,$$

where $F(x) = \int f(x)\, dx$. Integrating once more, we find

$$y = G(x) + Cx + D,$$

in which $G$ is an antiderivative of $F$, and $C$ and $D$ are constants.

**EXAMPLE 3**   Solve the differential equation

$$\frac{d^2y}{dx^2} = e^{-x^2} + \sin x \tag{11}$$

**Solution**   We integrate both sides of (11) to obtain

$$\frac{dy}{dx} = \int_a^x e^{-u^2}\, du - \cos x + C, \tag{12}$$

and integrating both sides of (12) we have

$$y = \int_a^x \int_a^u e^{-v^2}\, dv\, du - \sin x + Cx + D,$$

which is a solution, except for a quadrature. The constants $C$ and $D$ cannot be determined without additional information.   □

Another example of an equation which we can solve easily is the linear first-order equation

$$\frac{dy}{dx} = ay + b, \quad a \neq 0 \tag{13}$$

with $a$ and $b$ constant. We write

$$\frac{dy}{y + (b/a)} = a\, dx, \quad y + \frac{b}{a} > 0,$$

and obtain, upon integration,

$$\ln\left(y + \frac{b}{a}\right) = ax + C.$$

Therefore

$$y + \frac{b}{a} = e^{ax+C} = De^{ax}, \quad \text{where*} \ D = e^C.$$

We note that the solutions of Equations (9) and (13), which are both of the first order, contain one constant of integration, while the solution of (10), a second-order equation, contains two arbitrary constants of integration.

The motion of one or several particles is described by equations which involve both the velocity and the acceleration of the individual particles. Since velocity and acceleration are defined in terms of derivatives, these equations of motion are differential equations. Newton's law states that the force acting on a particle is proportional to the acceleration; therefore it is

---

*The interested reader can show (recalling that $d \ln|u| = u^{-1}\, du$) that $y = (-b/a) + De^{ax}$ is a solution of (13) even if $D$ is negative or 0, and that all the solutions are obtained in this way.

easy to see how differential equations arise in mechanics and many branches of science and engineering. The subject of differential equations has been studied extensively by mathematicians, physicists, and engineers for several centuries. More recently, differential equations have come to play a crucial role in many subjects including chemistry, biology, economics, ecology and psychology. The material presented in the next sections is a brief introduction to the subject.

## 1   PROBLEMS

In Problems 1 through 7, show by differentiating the given function $y$ that it is a solution of the given differential equation. Determine whether or not the differential equation is linear, and give its order.

**1**   $y = \sin x + 3;\ \dfrac{dy}{dx} = \cos x$

**2**   $y = \sin x + 3x - 2;\ \dfrac{dy}{dx} = \cos x + 3$

**3**   $y = 5e^x + 7e^{2x};\ \dfrac{d^2y}{dx^2} - 3\dfrac{dy}{dx} + 2y = 0$

**4**   $y = 4e^{-2x};\ \dfrac{dy}{dx} = -2y$

**5**   $y = x^3;\ x^3\dfrac{d^3y}{dx^3} + x^2\dfrac{d^2y}{dx^2} - 3x\dfrac{dy}{dx} - 3y = 0$

**6**   $y = 0;\ \dfrac{dy}{dx} + e^x y = (\sin x)y^2.$

**7**   $y = \left(3\sqrt{\dfrac{x}{2}} + C\right)^{2/3};\ \sqrt{2xy}\,\dfrac{dy}{dx} = 1.$

In each of Problems 8 through 22, solve the given differential equation, except for a quadrature, if necessary.

**8**   $\dfrac{dy}{dx} = \dfrac{1}{\sqrt{1 - x^2}}$

**9**   $\dfrac{dy}{dx} = 2y$

**10**   $\dfrac{dy}{dx} = \cos x + x$

**11**   $\dfrac{dy}{dx} = \sec x \tan x + 2$

**12**   $\dfrac{dy}{dx} = \tan x + \sin^2 x + 6$

**13**   $\dfrac{dy}{dx} = e^x \sin 2x + 3$

**14**   $\dfrac{dy}{dx} = e^{x^2}$

**15**   $\dfrac{dy}{dx} = x^2(1 + x)^5$

**16**   $\dfrac{dy}{dx} = (\sin^2 x)/x^2$

**17**   $\dfrac{d^2y}{dx^2} = \sin x$

**18**   $\dfrac{d^2y}{dx^2} = x^3 + x^2 + \cos x$

**19**   $\dfrac{d^2y}{dx^2} = e^{x^2} + 3x$

**20**   $\dfrac{d^2y}{dx^2} = (2x - e^{-x})^2$

**21**   $\dfrac{dy}{dx} = 2y + 3$

**22**   $\dfrac{dy}{dx} = 4y - 3$

## 2

## SEPARABLE DIFFERENTIAL EQUATIONS

We consider a few types of equations which can be solved by special devices. From the point of view of the theory of differential equations, those equations which are amenable to these tricks are not particularly far-reaching. Some of these *methods*, however, are useful in more advanced topics in differential equations. Moreover, in studying these special methods, the reader will understand more clearly the reasons for the various classifications of differential equations.

If a first-order differential equation is of the form

$$\frac{dy}{dx} = f(x)g(y), \quad g(y) \neq 0, \tag{1}$$

we say that the **variables in the equation are separable**. That is, we may write

$$\frac{dy}{g(y)} = f(x)\, dx,$$

and the solution, except for the quadratures involved, is

$$\int \frac{dy}{g(y)} = \int f(x)\, dx + C,$$

**EXAMPLE 1**    Solve

$$\frac{dy}{dx} = \frac{x \cos y}{e^x \sin y}, \qquad y \neq n\pi,\left(n + \frac{1}{2}\right)\pi,\, n = 0,\, \pm 1,\, \pm 2,\, \dots .$$

**Solution**    We can separate variables by writing

$$\frac{\sin y\, dy}{\cos y} = xe^{-x}\, dx,$$

from which we conclude that

$$-\ln \cos y = -(x + 1)e^{-x} + C.$$

The solution, $y$, is given in implicit form. Using the properties of the logarithm and cosine functions, we find that

$$y = \text{arc} \cos\{e^{(x+1)e^{-x}} - C\}. \qquad \square$$

Occasionally differential equations of the first order are given in terms of differentials. Thus an expression such as

$$P(x, y)\, dx + Q(x, y)\, dy = 0$$

is equivalent to the differential equation (supposing $Q \neq 0$)

$$\frac{dy}{dx} = -\frac{P(x, y)}{Q(x, y)}.$$

No distinction will be made between equations with derivatives and those with differentials.

**EXAMPLE 2**    Solve: $(1 + x^2)(1 + y^2)\, dx - xy\, dy = 0$.

**Solution**    We have

$$\frac{(1 + x^2)\, dx}{x} = \frac{y\, dy}{1 + y^2},$$

and the variables are separated. The solution is

$$\ln x + \tfrac{1}{2}x^2 = \tfrac{1}{2} \ln (1 + y^2) + C. \tag{2}$$

$\square$

*Remarks.*    Equation (2) yields the solution in implicit form. We shall frequently perform the necessary algebra to obtain an explicit result. Writing

$$\ln(1 + y^2) = 2 \ln x + x^2 - 2C,$$

we find

$$1 + y^2 = e^{2 \ln x + x^2 - 2C}$$
$$= e^{2 \ln x} \cdot e^{x^2} \cdot e^{-2C}$$
$$= Dx^2 e^{x^2},$$

where we have written $D = e^{-2C}$. Hence the explicit solution to Example 2 is

$$y = \pm \sqrt{Dx^2 e^{x^2} - 1}.$$

Since at present we restrict ourselves to real solutions, the domain of the solution above consists of those $x$ which satisfy the inequality $x^2 e^{x^2} \geq 1/D$.

Equations which fall into the "variables separable" class are of an extremely special character. It is interesting to compare equations of type (1) with an equation of the form

$$\frac{dy}{dx} = f(x) + g(y). \tag{3}$$

The variables are not separable in this case and, in fact, there is no known way to reduce every such equation to a simple problem in quadratures.

　　Sometimes a differential equation can be reduced to an equation in which the variables are separable. Theorem 1 below provides an example of such an equation. Before stating Theorem 1, however, we introduce the idea of a homogeneous function.

---

**DEFINITION**　　*A function $f(x, y)$ is said to be **homogeneous of degree** $n$ if and only if*

$$f(tx, ty) = t^n f(x, y) \qquad \text{for all } t > 0 \text{ and all } (x, y) \neq (0, 0).$$

---

For example, the function

$$f(x, y) = x^3 + 2x^2 y - 3xy^2 + y^3$$

is homogeneous of degree 3, since

$$f(tx, ty) = (tx)^3 + 2(tx)^2(ty) - 3(tx)(ty)^2 + (ty)^3$$
$$= t^3(x^3 + 2x^2 y - 3xy^2 + y^3) = t^3 f(x, y).$$

The function

$$f(x, y) = \sqrt{x^2 + y^2 - 8xy}$$

is homogeneous of degree 1, and the function

$$f(x, y) = \frac{x^2 + 2y^2}{8xy + 4y^2}$$

is homogeneous of degree zero. The function $f(x, y) = 1 + x^2 + y^2$ is not homogeneous.

**DEFINITION**    *A function f of the k variables $x_1, x_2, \ldots, x_k$ is* **homogeneous of degree** *n if and only if*

$$f(tx_1, tx_2, \ldots, tx_k) = t^n f(x_1, x_2, \ldots, x_k) \qquad \textit{for all } t > 0 \textit{ and all } (x_1, x_2, \ldots, x_k)$$
$$\neq (0, 0, \ldots, 0).$$

Note that $n$ is not restricted to positive values. The function

$$f = (x_1^2 + x_2^2 + x_3^2)^{-1/3}$$

is homogeneous of degree $-\frac{2}{3}$.

**THEOREM 1**    *A differential equation of the form*

$$P(x, y)\, dx + Q(x, y)\, dy = 0, \tag{4}$$

*in which P and Q are homogeneous of the* **same degree** *is reducible to the variables separable case.*

**Proof**    We assume that $P$ and $Q$ are homogeneous of degree $n$ and set

$$y(x) = xv(x),$$

where $v$ is a new unknown function. Since

$$dy = x\, dv + v\, dx,$$

Equation (4) becomes

$$P(x, vx)\, dx + Q(x, vx)(x\, dv + v\, dx) = 0$$

or

$$x^n P(1, v)\, dx + x^n Q(1, v)(x\, dv + v\, dx) = 0 \quad \text{if } x > 0.$$

Therefore

$$[P(1, v) + vQ(1, v)]\, dx + xQ(1, v)\, dv = 0 \quad \text{if } x > 0$$

and

$$[P(-1, -v) + vQ(-1, -v)]\, dx + xQ(-1, -v)\, dv = 0 \quad \text{if } x < 0.$$

In either case, we have

$$\frac{dx}{x} = -\frac{Q(\pm 1, \pm v)}{P(\pm 1, \pm v) + vQ(\pm 1, \pm v)}\, dv,$$

and the variables are separated.                                    ☐

**EXAMPLE 3**    Solve: $(x^2 - xy + y^2)\, dx + x^2\, dy = 0$.

**Solution**    $P = x^2 - xy + y^2$ and $Q = x^2$ are homogeneous of degree 2. Setting $y = xv$, we obtain

$$x^2(1 - v + v^2)\, dx + x^2(x\, dv + v\, dx) = 0$$

or

$$\frac{dv}{1+v^2} = -\frac{dx}{x}.$$

Integrating, we find

$$\arctan v = -\ln(x) + C \quad \text{and} \quad v = \tan(C - \ln|x|).$$

Finally, setting $v = y/x$, we get

$$y = x\tan(C - \ln|x|). \qquad \square$$

## 2  PROBLEMS

In each of Problems 1 through 15, solve the given differential equation.

1  $x^2\,dy - \cos^2 y\,dx = 0$

2  $(1-x)y^2\,dx + x\,dy = 0$

3  $y^2\,dx + 2y^2\,dy = dy$

4  $\dfrac{dy}{dx} + 2y = 3$

5  $e^y\,dx + x^2(2+e^y)\,dy = 0$

6  $\dfrac{dy}{dx} = \dfrac{y}{x}$

7  $\dfrac{dy}{dx} = xy$

8  $\sqrt{a^2-x^2}\,dy - y\sqrt{a^2-y^2}\,dx = 0$

9  $e^{2x+y}\,dx - 2e^{x-y}\,dy = 0$

10  $\dfrac{dy}{dx} = xe^{x+y}$

11  $(xy^2+x)\,dx + (x^2y+y)\,dy = 0$

12  $(xy^2+2y^2)\,dx + (xy^2-x)\,dy = 0$

13  $2y\,dy + 4x^3\sqrt{4-y^4}\,dx = 0$

14  $(1+x^2)\dfrac{dy}{dx} + 2xy = x$

15  $\dfrac{\ln y}{\ln x}dy - \dfrac{x^4}{y^2}dx = 0$

In each of Problems 16 through 25, decide whether or not the given function is homogeneous. If so, find the degree.

16  $f(x, y) = x^4 + 8x^3y - 2x^2y^2 + y^4$

17  $f(x, y) = 8x^2$

18  $f(x, y) = y^3 + 1$

19  $f(x, y, z) = x + y + z$

20  $f(x, y) = \dfrac{(x^2+y^2-xy)^{1/2}}{x^3+y^3+x^2y}$

21  $f(x, y) = x^2 + 2xy + x(x+y)$

22  $f(x, y, z) = xyz + 4$

23  $f(x, y) = \dfrac{x^2(x+y)}{x^4+x^3y+x^2(x^2+y^2)}$

24  $f(x, y) = \ln\dfrac{y}{x} + \arctan\dfrac{x}{y}$

25  $f(x, y) = x\ln y - y\ln x$

*26  Let $f(x, y)$ be homogeneous of degree $n$. Show that $xf_x + yf_y = nf(x, y)$ (**Euler's Formula** for homogeneous functions).

In each of Problems 27 through 40, solve the differential equation.

27  $(x+2y)\,dx + (-2x+y)\,dy = 0$

28  $(2x-y)\,dx + (x-2y)\,dy = 0$

29  $(2x+y)\,dx - y\,dy = 0$

30  $(2x^2-y^2)\,dx - xy\,dy = 0$

31  $\dfrac{dy}{dx} = \dfrac{3x^2+6xy-y^2}{5x^2+2xy+y^2}$

32  $\dfrac{dy}{dx} = \dfrac{x^2+y^2}{2xy}$

33  $x\dfrac{dy}{dx} = y - \sqrt{x^2-y^2}$

34  $\dfrac{dy}{dx} = \dfrac{y}{x} + \sin\dfrac{y}{x}$

35  $(x\sqrt{x^2+y^2}-y^2)\,dx + xy\,dy = 0$

36  $[x+(x-y)e^{y/x}]\,dx + xe^{y/x}\,dy = 0$

37  $\left(\dfrac{1}{x} - \dfrac{y}{x^2}e^{y/x}\right)dx + \left(\dfrac{1}{x}e^{y/x} - \dfrac{1}{y}\right)dy = 0$

38  $\left(\dfrac{1}{x-y} + \dfrac{y}{x^2+y^2}\right)dx + \left(\dfrac{1}{y-x} - \dfrac{x}{x^2+y^2}\right)dy = 0$

39  $\left(y - x\cot\dfrac{x}{y}\right)dy + y\cot\dfrac{x}{y}\,dx = 0$

**40** $\left( y \sin \dfrac{y}{x} + x \cos \dfrac{y}{x} \right) dx - x \sin \left( \dfrac{y}{x} \right) dy = 0$

**41** If $f(x)$ and $g(y)$ are each homogeneous of degree zero, show that the equation

$$\frac{dy}{dx} = f(x) + g(y)$$

may be solved explicitly.

**\*42** Given the equation

$$(a_1 x + b_1 y + c_1)\, dx + (a_2 x + b_2 y + c_2)\, dy = 0,$$

with $a_1, b_1, c_1, b_2, c_2$ constant, show how $h$ and $k$ may be selected so that the change in variables $x' = x + h$ and $y' = y + k$ reduces the equation to one of homogeneous type. Express $h$ and $k$ in terms of the $a_i, b_i,$ and $c_i, i = 1, 2$. Are any conditions necessary?

---

## 3

## EXACT DIFFERENTIALS. INTEGRATING FACTORS

Suppose that the differential equation
$$P(x, y)\, dx + Q(x, y)\, dy = 0 \tag{1}$$
satisfies the condition

$$\frac{\partial P}{\partial y} = \frac{\partial Q}{\partial x}.$$

Then the left side of (1) is an *exact differential* and there is a function $f(x, y)$ such that

$$df = P\, dx + Q\, dy.$$

Methods for finding the function $f$ were discussed in Chapter 15, Section 13. Once the function $f$ is obtained, we see that any differentiable function $y$ which satisfies the equation

$$f(x, y) = c, \tag{2}$$

where $c$ is any constant, is a solution of (1); this follows immediately from the Chain Rule. In general, the relation (2) will only give $y$ as a function of $x$ in implicit form.

We can rewrite Equation (1) in the form $P(x, y) + Q(x, y)\dfrac{dy}{dx} = 0$. We summarize these results as a theorem.

---

**THEOREM 2**    *Let $P(x, y)$ and $Q(x, y)$ have continuous first partial derivatives. If*

$$\frac{\partial P}{\partial y} = \frac{\partial Q}{\partial x},$$

*then there exists a function $f(x, y)$ such that for any constant $C$, the equation $f(x, y) = C$ defines implicitly a function $y = y(x)$ which is a solution of*

$$P(x, y) + Q(x, y)\frac{dy}{dx} = 0.$$

**EXAMPLE 1**   Solve

$$y \cos x\, dx + (\sin x - \sin y)\, dy = 0.$$

**Solution**   Here $P(x, y) = y \cos x$ and $Q(x, y) = \sin x - \sin y$. Then

$$\frac{\partial P}{\partial y} = \cos x = \frac{\partial Q}{\partial x},$$

and the differential equation is exact. Solving the exact differential (using the method of Chapter 15, Section 13) yields

$$f(x, y) = \int y \cos x\, dx + C(y)$$

or

$$f(x, y) = y \sin x + C(y). \tag{3}$$

Then $f_y(x, y) = \sin x + C'(y) = Q(x, y)$; that is,

$$\sin x + C'(y) = \sin x - \sin y.$$

We see that $C'(y) = -\sin y$, which implies that $C(y) = \cos y + C_1$.
  Finally, substituting $C(y)$ into (3) we conclude that if $x$ and $y$ satisfy

$$y \sin x + \cos y = C, \tag{4}$$

then the solution $y$ in terms of $x$ is expressed implicitly by (4). □

**EXAMPLE 2**   Solve

$$(3x^2 + 4xy - y^2 - 2)\, dx + (2x^2 - 2xy + 3y^2 + 3)\, dy = 0.$$

**Solution**   We have

$$\frac{\partial}{\partial y}(3x^2 + 4xy - y^2 - 2) = 4x - 2y = \frac{\partial}{\partial x}(2x^2 - 2xy + 3y^2 + 3).$$

Then

$$f(x, y) = \int [3x^2 + 4xy - y^2 - 2]\, dx + C(y)$$

$$= x^3 + 2x^2 y - xy^2 - 2x + C(y)$$

and

$$f_y(x, y) = 2x^2 - 2xy + C'(y) = 2x^2 - 2xy + 3y^2 + 3.$$

Therefore

$$C'(y) = 3y^2 + 3, \qquad C(y) = y^3 + 3y + C_1.$$

We conclude that if $y$ is any differentiable function satisfying the equation

$$f(x, y) \equiv x^3 + 2x^2 y - xy^2 - 2x + y^3 + 3y = C \tag{5}$$

then $y$ satisfies the differential equation. The solution of $y$ in terms of $x$ is expressed implicity by (5). □

It sometimes happens that an expression

$$P(x, y)\, dx + Q(x, y)\, dy$$

is not an exact differential, but that a function $I(x, y)$ can be found so that

$$I(x, y)(P\, dx + Q\, dy) \qquad\qquad (6)$$

is an exact differential. If $I \neq 0$, then the solutions of $P\, dx + Q\, dy = 0$ and $I(P\, dx + Q\, dy) = 0$ are the same.

**DEFINITION**   *If a nonvanishing function $I(x, y)$ makes the expression (6) an exact differential, then I is called an* **integrating factor***.*

There are no general rules for finding integrating factors, and usually none can be found. However, there are a few simple cases where known expressions can be employed as integrating factors. We illustrate with two examples, and remark that the problems at the end of this section (Problems 16–25) were concocted so that integrating factors for them can be found.

**EXAMPLE 3**   Solve $x^2\, dy + (xy + x^2 + 1)dx = 0$.

**Solution**   We observe that the equation is not an exact differential. However, we note that

$$x^2\, dy + xy\, dx = x(x\, dy + y\, dx) = x\, d(xy).$$

Therefore, if we divide the original equation by $x$, we obtain

$$x\, dy + y\, dx + \left( x + \frac{1}{x} \right) dx = 0.$$

We may write this equation

$$d(xy) + \left( x + \frac{1}{x} \right) dx = 0,$$

and we can integrate to get the solution

$$xy + \frac{1}{2}x^2 + \ln |x| = C \qquad \text{or} \qquad y = -\frac{1}{2}x - \frac{1}{x}\ln |x| + \frac{C}{x}.$$

The integrating factor we employed is $I(x, y) = 1/x$.   $\square$

In Example 3 the key to the solution was the observation that the combination $x\, dy + y\, dx$ is $d(xy)$. There are a few other simple differentials which, when we recognize their form in an equation, may lead to an integrating factor. Some of these differentials are:

$$\frac{x\, dy - y\, dx}{x^2} = d\left( \frac{y}{x} \right), \qquad \frac{x\, dy - y\, dx}{x^2 + y^2} = d\left( \arctan \frac{y}{x} \right)$$

$$x\, dx + y\, dy = \tfrac{1}{2}\, d(x^2 + y^2).$$

**EXAMPLE 4**   Solve

$$x\, dy - y\, dx = 2x^2 \ln y\, dy.$$

**Solution**   We observe that

$$\frac{x\,dy - y\,dx}{x^2} = d\left(\frac{y}{x}\right),$$

and so we try $I(x, y) = 1/x^2$ as an integrating factor. The result is

$$\frac{x\,dy - y\,dx}{x^2} - 2\ln y\,dy = 0, \tag{7}$$

which is exact. To obtain the solution, we may employ the methods of exact differentials or proceed in the following way. Equation (7) may be written

$$d\left(\frac{y}{x}\right) = 2\ln y\,dy$$

and, upon integration of both sides, we find

$$\frac{y}{x} - 2(y \ln y - y) = C. \qquad \square$$

## 3   PROBLEMS

In each of Problems 1 through 15, use the method of exact differentials to solve the given differential equations.

**1**  $(x + 2y - 2)\,dx + (2x - y + 3)\,dy = 0$

**2**  $(x + y - 4)\,dx + (x + y + 2)\,dy = 0$

**3**  $(y - 3x^2 + 2)\,dx + (x - y^2 + 2y)\,dy = 0$

**4**  $(3x^2 - 4xy + y^2 - 3y)\,dx + (2xy - 2x^2 + 6y^2 - 3x)\,dy = 0$

**5**  $(2x^3 + 6xy^2 - 2y^3 + 4x + 3y)\,dx + (6x^2y - 6xy^2 - 4y^3 + 3x - 2y)\,dy = 0$

**6**  $(2x \sin y + e^x \cos y)\,dx + (x^2 \cos y - e^x \sin y)\,dy = 0$

**7**  $\arcsin y\,dx + \dfrac{x + 2\sqrt{1 - y^2}\,\cos y}{\sqrt{1 - y^2}}\,dy = 0$

**8**  $(2x + ye^{xy})\,dx + (\cos y + xe^{xy})\,dy = 0$

**9**  $\left(\dfrac{1}{x - y} + \dfrac{x}{x^2 + y^2}\right)dx + \left(\dfrac{1}{y - x} + \dfrac{y}{x^2 + y^2}\right)dy = 0$

**10**  $e^x(x^2 + y^2 + 2x)\,dx + 2ye^x\,dy = 0$

**11**  $(x \ln y + y \ln x + y)\,dx + \left(\dfrac{x^2}{2y} + x \ln x\right)dy = 0$

**12**  $\arctan(y^2)\,dx + \dfrac{2xy}{1 + y^4}\,dy = 0$

**13**  $\left(\sin y + \ln(xy) + \dfrac{1}{x}\right)dx + \left(x \cos y + \dfrac{x}{y}\right)dy = 0$

**14**  $(2x \sin y - y \sin x)\,dx + (x^2 \cos y + \cos x)\,dy = 0$

**15**  $\left(\dfrac{y}{1 + x^2y^2}\right)dx + \left(\dfrac{x}{1 + x^2y^2} + 3y^2\right)dy = 0$

In each of Problems 16 through 25, use the method of integrating factors to solve the given differential equations.

**16**  $x\,dy - y\,dx = (x^2 + y^2)\,dx$

**17**  $x\,dy - y\,dx = (x^2y + x^2y^3)\,dy$

**18**  $x\,dy - y\,dx = (x^2 + y^2)\,dy$

**19**  $x\,dy - (y + x^3e^{2x})\,dx = 0$

**20**  $ye^{-x/y}\,dx - (xe^{-x/y} + y^3)\,dy = 0$

**21**  $x \cos^2 y\,dx + 2 \csc x\,dy = 0$

**22**  $dx + (x \tan y - 2 \sec y)\,dy = 0$

**23**  $x\,dx + y\,dy + (x^2 + y^2)(y\,dx - x\,dy) = 0$

**24**  $(2 - xy)y\,dx + (2 + xy)x\,dy = 0$

**25**  $\sqrt{x^2 - y^2} - y)\,dx + x\,dy = 0$.
   (*Hint:* Compute $d \arcsin(y/|x|)$ for $x > 0$ and for $x < 0$.)

**26**  Show that $I_1(x, y) = y$ and $I_2(x, y) = 2xy^3$ are both integrating factors for the equation $y\,dx + 2x\,dy = 0$.

**\*27**  Suppose $I(x, y)$ is an integrating factor for the equation $P(x, y)\,dx + Q(x, y)\,dy = 0$. Show that

$$\frac{\partial P}{\partial y} - \frac{\partial Q}{\partial x} = Q\frac{\partial}{\partial x}\log|I| - P\frac{\partial}{\partial y}\log|I|.$$

_____**4**

## FIRST ORDER LINEAR DIFFERENTIAL EQUATIONS

The most general linear equation of the first order has the form

$$c_0(x)\frac{dy}{dx} + c_1(x)y = f(x). \tag{1}$$

Dividing this equation by $c_0(x)$, we obtain

$$\frac{dy}{dx} + P(x)y = Q(x), \tag{2}$$

which is equivalent to (1) so long as $c_0(x) \neq 0$, with $P(x) = c_1(x)/c_0(x)$ and $Q(x) = f(x)/c_0(x)$. We will assume that $c_0(x)$ is never 0 and consider only equations of the form (2).

**DEFINITION**     A **first order linear differential equation** _is an equation that can be written in the form_

$$\frac{dy}{dx} + P(x)y = Q(x). \tag{2}$$

It turns out that every equation of the form (2) has an integrating factor $I$ which is a function of $x$ alone. To see this, we multiply (2) by $I$, getting

$$I\frac{dy}{dx} + IPy = Q(x)I. \tag{3}$$

If we write

$$I\frac{dy}{dx} + \frac{dI}{dx}y \equiv \frac{d}{dx}(Iy) = Q(x)I, \tag{4}$$

then (3) and (4) will be identical if and only if

$$\frac{dI}{dx} = IP(x). \tag{5}$$

We can solve Equation (5) by separating variables. We find

$$\frac{dI}{I} = P(x)\,dx, \qquad \ln I = \int P(x)\,dx.$$

Therefore the function

$$I = e^{\int P\,dx} \tag{6}$$

puts (3) into the form (4). However, (4) is easily integrated. We see at once that

$$\int \frac{d}{dx}(Iy) = \int QI\,dx,$$

and so

$$e^{\int P\,dx}y = \int Qe^{\int P\,dx}\,dx + C.$$

The explicit solution of Equation (2) is

$$y = e^{-\int P\,dx}\int^{x} Q(\xi)e^{\int^{\xi} P\,dx}\,d\xi + Ce^{-\int P\,dx}.$$

It is much easier to remember the form of the integrating factor (6) and work through the development of the solution than it is to try to remember the formula for the explicit answer.

---

**THEOREM 3**    *The first order linear differential equation*

$$\frac{dy}{dx} + P(x)y = Q(x),$$

*with P and Q continuous functions, can be solved by using the integrating factor*

$$I = e^{\int P(x)dx}.$$

---

An example shows the method.

**EXAMPLE**    Solve

$$\frac{dy}{dx} + 2xy = 2x^3.$$

**Solution**    We have $P = 2x$ and $\int P\,dx = x^2$, and we take $I = e^{x^2}$ as an integrating factor. We write

$$e^{x^2}\frac{dy}{dx} + 2xe^{x^2}y = 2x^3e^{x^2},$$

and so

$$\frac{d}{dx}(e^{x^2}y) = 2x^3e^{x^2}.$$

Therefore

$$e^{x^2}y = \int 2x^3e^{x^2}\,dx + C = (x^2 - 1)e^{x^2} + C$$

and

$$y = (x^2 - 1) + Ce^{-x^2}. \qquad \square$$

## 4   PROBLEMS

In each of Problems 1 through 19, use the method of integrating factors to solve the following differential equations.

**1** $\dfrac{dy}{dx} - \dfrac{y}{x} = x^n, \quad n \neq 0$     **2** $\dfrac{dy}{dx} - \dfrac{y}{x} = 1$

**3** $\dfrac{dy}{dx} - y = (3x^2 + 4x - 3)e^x$    **4** $\dfrac{dy}{dx} - 2y = e^x$

**5** $x\dfrac{dy}{dx} + y = e^{2x} - \sin x$     **6** $x\dfrac{dy}{dx} + 2y = e^x$

**7** $\dfrac{dy}{dx} - \dfrac{xy}{x^2 - 1} = x$     **8** $\dfrac{dy}{dx} + y \sin x = xe^{\cos x}$

**9** $\dfrac{dy}{dx} + \dfrac{3y}{2x - 3} = \dfrac{x}{\sqrt{2x - 3}}$

**10** $\dfrac{dy}{dx} - ry = \cos sx, \quad r, s \text{ constant}$

**11** $\dfrac{dy}{dx} + (\cot x)y = \csc x$     **12** $x\dfrac{dy}{dx} - 3y = x^6$

**13** $\dfrac{dy}{dx} + (\tan x)y = \sin x$     **14** $\dfrac{dy}{dx} + (\cot x)y = 4x^2 \csc x$

**15** $(2y - x^2 + 1)\,dx + (x^2 - 1)\,dy = 0$

**16** $(3xy + 2e^{x^3})x\,dx - dy = 0$

**17** $[y \cos 2x + 2(\sin 2x)^{3/2}]\,dx + \sin 2x\,dy = 0$

**18** $(x^2 + 1)\dfrac{dy}{dx} - 2xy = 6x^2 + 2x$

**19** $(a^2 - x^2)\dfrac{dy}{dx} + 2ay = (a^2 - x^2)^2$

---

## 5

## APPLICATIONS

Differential equations may be applied in such disciplines as mechanics, physics, biology, chemistry, economics, statistics, and engineering as well as in various branches of mathematics. In this section we restrict our attention to some applications to geometric problems, to population problems in ecology theory, and to problems of physics. In the applications to be considered we must first determine the differential equation which describes the situation and then find a solution which takes into account the various conditions prescribed in the particular application. Since each problem seems unlike all others, at least in appearance, it is difficult to give general rules of procedure which would cover large classes of applications. All we can do is study a number of illustrative examples and then solve some typical exercises.

**EXAMPLE 1**   Find a differential equation which describes the growth rate of a population of animals under the assumptions: (a) the population has no enemies and (b) there is an absolute limit $L$ to the number of animals that the environment can sustain. Solve the differential equation.

**Solution**   Let $P(t)$ denote the population at time $t$. Then $dP/dt$ is the rate of growth of $P$. We make additional assumptions that the rate of growth of $P$ is proportional to both $P$ itself and to $L - P(t)$ which tells how many additional animals the environment can accommodate. Thus we get the equation

$$\frac{dP}{dt} = kP(t)(L - P(t)),$$

APPLICATIONS

where $k$ is the proportionality constant. We can separate variables and get

$$\frac{dP}{P(L-P)} = k\ dt.$$

The left side can be integrated by the method of partial fractions. We find

$$\frac{1}{L}\ln\left|\frac{P}{L-P}\right| = kt + C_1$$

or

$$\ln\left|\frac{P}{L-P}\right| = kLt + C_2, \quad (C_2 = LC_1).$$

Hence, solving for $P$, we obtain

$$P = \frac{LC_3 e^{kLt}}{1 + C_3 e^{kLt}}, \quad C_3 = e^{C_2},$$

where $C_3$ is an arbitrary positive constant. □

**EXAMPLE 2**  Find the equation of the curve such that its slope at any point $P(x, y)$ is equal to the difference of the squares of the distances of $P(x, y)$ from the points $(1, 0)$ and $(4, 0)$.

**Solution**  See Fig. 1 in which the distances from $P$ to $(1, 0)$ and $(4, 0)$ are labeled $d_1$ and $d_2$, respectively. The conditions of the problem state that

$$\frac{dy}{dx} = d_1^2 - d_2^2,$$

which is a differential equation. Therefore

$$\frac{dy}{dx} = (x-1)^2 + y^2 - (x-4)^2 - y^2$$

$$= 6x - 15.$$

Integrating, we find

$$y = 3x^2 - 15x + C.$$

The solution consists of a family of parabolas. □

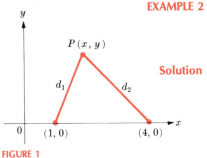

**FIGURE 1**

**EXAMPLE 3**  Find the equation in polar coordinates of the curve such that the tangent of the angle between the radius vector and the tangent to the curve is equal to the square of the radius vector.

**Solution**  Let $P(r, \theta)$ be a typical point on the curve and denote by $\psi$ the angle between the radius vector and the tangent. (See Fig. 2.) The condition of the problem states that

$$\tan \psi = r^2.$$

In the study of curves in polar coordinates we learned (page 523) that

$$\cot \psi = \pm \frac{1}{r}\frac{dr}{d\theta},$$

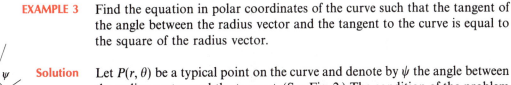

**FIGURE 2**

and so the condition of the problem yields the differential equation

$$r\frac{d\theta}{dr} = \pm r^2 \qquad \text{or} \qquad r\frac{dr}{d\theta} = \pm 1.$$

The solution is the family of curves

$$r^2 = \pm 2\theta + C. \qquad\qquad \square$$

Suppose we are given a relation

$$\phi(x, y, c) = 0, \qquad\qquad (1)$$

in which $c$ is an arbitrary constant. In general, Equation (1) will represent a family of curves, one for each value of the constant $c$. Suppose we wish to find a first-order differential equation which is satisfied by every curve of the family. If (1) can be solved for $c$, giving

$$f(x, y) = c,$$

then such a differential equation is obtained by differentiating with respect to $x$ or $y$. We write

$$f_x dx + f_y dy = 0,$$

which is the desired equation. If it is not convenient to solve for $c$ first, we may differentiate (1) with respect to $x$, getting

$$\phi_x(x, y, c) + \phi_y(x, y, c)\frac{dy}{dx} = 0. \qquad\qquad (2)$$

Elimination of $c$ between (1) and (2) yields the differential equation. In the applications it is essential that the constant $c$ appearing in the family of curves be absent from the differential equation.

**EXAMPLE 4**   The equation

$$\frac{x^2}{c^2 + 4} + \frac{y^2}{c^2} = 1 \qquad\qquad (3)$$

represents a family of confocal ellipses (i.e., all ellipses have the same foci). Find a first-order differential equation satisfied by all such ellipses.

**Solution**   Multiplying by $c^2(c^2 + 4)$, we obtain

$$c^2 x^2 + (c^2 + 4)y^2 = c^2(c^2 + 4).$$

Differentiating with respect to $x$, we find

$$c^2 x + (c^2 + 4)y\frac{dy}{dx} = 0 \qquad \text{or} \qquad c^2 = -\frac{4y(dy/dx)}{x + y(dy/dx)}.$$

Substituting in (4), we get (after some algebraic work)

$$\frac{x^2[x + y(dy/dx)]}{4x} - \frac{y^2[x + y(dy/dx)]}{4y(dy/dx)} = 1.$$

A simplification yields

$$xy\left(\frac{dy}{dx}\right)^2 + (x^2 - y^2 - 4)\frac{dy}{dx} - xy = 0.$$

APPLICATIONS

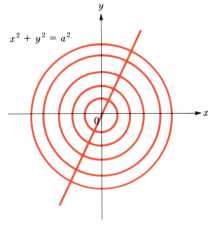

$x^2 + y^2 = a^2$

**FIGURE 3**

We could have obtained the same result by first solving (3) for $c^2$ and then differentiating with respect to $x$.  □

Suppose that $\phi(x, y, c) = 0$ represents a family of curves. An **orthogonal trajectory** to this family is a curve which intersects every member of the family at right angles. For example, the family of curves $x^2 + y^2 = a^2$ has as an orthogonal trajectory any line through the origin (Fig. 3). We can use the methods of differential equations to find the orthogonal trajectories of a given family. The general procedure depends on the known fact that two curves which intersect at right angles have slopes at the point of intersection which are the negative reciprocals of each other. Therefore, to find the orthogonal trajectories of a family $\phi(x, y, c) = 0$, we first find the differential equation which the family satisfies (eliminating $c$ in the process), and we then replace $dy/dx$ by $-dx/dy$ in this equation. The solution of the new differential equation gives the orthogonal trajectories. An example illustrates the process.

**EXAMPLE 5**  Find the orthogonal trajectories of the family of parabolas $y^2 = cx$. Sketch these trajectories.

**Solution**  We have

$$2y\frac{dy}{dx} = c \qquad \text{or} \qquad y = 2x\frac{dy}{dx}$$

as the differential equation of the parabolas. For the orthogonal trajectories, we write

$$y = -2x\frac{dx}{dy} \qquad \text{or} \qquad \frac{dy}{dx} = -\frac{2x}{y} \qquad \text{or} \qquad 2x\,dx + y\,dy = 0.$$

The solution is (writing $c^2$ for the constant of integration)

$$2x^2 + y^2 = 2c^2 \qquad \text{or} \qquad \frac{x^2}{c^2} + \frac{y^2}{2c^2} = 1.$$

The orthogonal trajectories are confocal ellipses, i.e., a family of ellipses all having the same foci. These ellipses and the family of parabolas are sketched in Fig. 4.  □

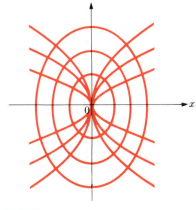

**FIGURE 4**

We next give an example from physics.

**EXAMPLE 6**  Find a differential equation describing the velocity of a particle directed radially outward from the Earth and acted upon only by gravity. Use this equation to determine the **escape velocity** of Earth, i.e., the velocity at which an object will leave the Earth's gravitational field.

**Solution**  Newton's law of gravitation states that the acceleration of a particle is inversely proportional to the square of the distance from the particle to the center of the Earth. Let $t$ denote time, $a$ denote acceleration, $v$ denote velocity, and $r = r(t)$ the distance of the particle from the center of Earth. Newton's law states that

$$a = \frac{dv}{dt} = \frac{k}{r^2}, \tag{4}$$

where $k$ is the constant of proportionality. Let $R$ denote the radius of the Earth. When $r = R$, then $a = -g$, where $g$ is the gravitational pull of the Earth, a known constant ($g = 980$ cm/sec²). Therefore $-g = k/R^2$ by Equation (4), allowing us to solve for $k$, to get

$$a = \frac{-gR^2}{r^2}.$$

To express the acceleration in terms of $v$ and $r$, we use the Chain Rule to find

$$a = \frac{dv}{dt} = \frac{dv}{dr}\frac{dr}{dt} = \frac{dv}{dr}v,$$

since $v = \dfrac{dr}{dt}$; now Equation (4) becomes

$$v\frac{dv}{dr} = \frac{-gR^2}{r^2}, \tag{5}$$

which is the differential equation we seek. Equation (5) can be solved by separation of variables to obtain

$$v^2 = \frac{2gR^2}{r} + C. \tag{6}$$

If we assume the initial velocity of an object on the Earth is $v_0$, we have $v(R) = v_0$ and we can solve for $C$. We get

$$C = v_0^2 - 2gR.$$

Thus we have, using (6),

$$v^2 = \frac{2gR^2}{r} + v_0^2 - 2gR. \tag{7}$$

If the particle does *not* escape the Earth's gravitational pull, then it will eventually fall back to Earth, and the velocity will reverse directions; that is, the velocity will change from positive to negative. Therefore to escape, the right side of (7) must always be positive. Since $2gR^2/r$ becomes arbitrarily close to 0 when $r$ becomes large, we can guarantee that the right side of (7) is positive only if $v_0^2 - 2gR \geq 0$. Thus a particle with initial velocity $v_0 \geq \sqrt{2gR}$ will escape from the Earth's gravitational pull. The minimum such velocity is called the **escape velocity**, $v_e$. Clearly then, $v_e = \sqrt{2gR}$. (The radius $R$ of Earth is approximately 6373 kilometers, and $g = 980$ cm/sec². Therefore $v_e = 11.18$ km/sec, approximately.) ☐

## 5  PROBLEMS

1  Find the equation of the curve such that the tangent to the curve at $P(x, y)$, the line $OP$ joining the origin to $P$, and the $x$ axis bound an isosceles triangle having base along the $x$ axis.

2  Find the equation in polar coordinates of a curve if the angle from the radius vector to the tangent is half the angle from the $x$ axis to the radius vector.

3  Find the equation of a graph which is such that the tangent to the curve at any point $P = (x, y)$ on it and the line joining $P$ to the origin make complementary angles with the positive $x$ axis.

4  Show that if the normal to a curve always passes through a fixed point the curve is a circle (or part of one).

5  Find the function $y(x)$ such that for any $X(>a)$ the area bounded by the line $x = a$, $x = X$, and $y = 0$ and the curve $y = y(x)$ is $Ks(X)$ where $s(X)$ is the length of the arc $y = y(x)$ for $a \leq x \leq X$.

6  Find the equation of a curve in which the perpendicular from the origin to the tangent is equal to the $x$ value of the point of contact.

In Problems 7 through 11, find the differential equation of the first order for each given family of curves.

7  $y = Cx + e^x$         8  $y = x + Ce^x$

9  $x^2 y + x = Cy^2$      10  $y = \sin(x + C)$

11  The family of all parabolas having focus at the origin and the $x$ axis as axis.

In each of Problems 12 through 15, find the orthogonal trajectories of the given family of curves, and sketch.

12  $x^2 + y^2 = C$        13  $2x^2 + 3y^2 = C$

14  $x^{2/3} + y^{2/3} = C^{2/3}$    15  $x^2 + y^2 = Cx$

16  A sociologist wishes to study the propagation of rumors. The sociologist assumes that the rate the rumor spreads is proportional to the product of the number who know the rumor at a given time with the number who do not yet know it, for a given, fixed, total population, $L$. Find a differential equation for $N(t)$, the number of people knowing the rumor at time $t$.

17  The differential equation $dP/dt = \alpha P + \beta, \alpha, \beta$ constants, has been proposed as a model describing population growth when there is a constant flow of new immigrants into a country. Explain why this is a reasonable model and solve the equation.

18  A bacterial population $B$ is known to have a rate of growth proportional to $B$ itself. Suppose between 1 P.M. and 3 P.M. the population triples. At what time will $B$ become 100 times what it was at 1 P.M.?

19  The rate of reproduction in a certain strain of bacteria is subject to periodic changes. It can be modeled as the rate of change being $(2 + \sin(\alpha t))B$, if $B$ is the population at time $t$ and $\alpha$ is a constant. If $B = 100$ at time $t = 0$, what is $B(t)$ for $t > 0$?

20  Suppose a population has a growth rate satisfying the logistic equation $dP/dt = kP(t)(L - P(t))$ of Example 1. What is the population when its rate of growth is a maximum?

21  The rate of the formation of ice on a lake is assumed to be inversely proportional to current thickness of the ice. Find a differential equation describing the thickness of ice formation and solve it.

22  Let $p(t)$ be the price of an item at time $t$. The consumer demand, $D$, for the item is a function of its price: $D = D(p)$. The proportion $I$ by which demand increases is given by

$$I = \frac{D(p + \Delta p) - D(p)}{D(p)},$$

and the proportion by which price increases is

$$J = \frac{\Delta p}{p}.$$

The **elasticity of demand**, $E(p)$, is defined to be the limit as $\Delta p$ tends to 0 of $-I/J$. Show that $E(p)$ satisfies

$$E(p) = \frac{-p}{D(p)} \frac{dD}{dp}.$$

Find a demand function such that the elasticity $E(p)$ is a positive constant.

23  A heavy object is dropped from rest in a resisting medium in which the resistance is directly proportional to the velocity. Determine its motion.

24  A heavy object is thrown downward into the medium of Problem 23 with a velocity $v_0$. Determine its motion.

25  A body falls from rest through the air. If air resistance is proportional to the square of the velocity, determine its motion.

26  A body is projected vertically upward with a velocity of $v_0$. If air resistance is as in Problem 25, determine its motion.

27  Determine the escape velocity of the moon if its radius is about 1740 kilometers and the gravitational pull at its surface is approximately 162 cm/sec$^2$.

28  Determine the escape velocity of the planet Jupiter if its radius is approximately 69,200 kilometers and the gravitational pull at its surface is approximately 2550 cm/sec$^2$.

29  Determine the escape velocity of the sun if its radius is approximately 695,000 kilometers and the gravitational pull at its surface is approximately 27,500 cm/sec$^2$.

30  In a chemistry experiment, two substances, $A$ and $B$, are being converted into a single compound $C$. It is assumed the following law of conversion holds: The rate of change of the amount $x$ of compound $C$ is proportional to the product of the amounts of the unconverted substances $A$ and $B$. Assume that one unit of $C$ is formed from one unit of $A$ together with one unit of $B$. If one begins with $a$ units of $A$ and $b$ units of $B$ and 0 units of $C$, find a differential equation describing the law of conversion. Solve this equation.

**31** In Problem 30 assume the proportionality constant is positive. Determine the behavior of the compound $C$ as time increases.

**32** A particle of mass $m$ moving along a straight line is acted on by a force of magnitude $km|x|$ directed toward the origin and by a resistance force of magnitude $lm|dx/dt|$ and directed opposite to the motion. Write the differential equation of the motion.

---

**6**

## SECOND ORDER LINEAR DIFFERENTIAL EQUATIONS WITH CONSTANT COEFFICIENTS

In Section 4 of this chapter we studied first order differential equations. Now we take up the study of differential equations of the second order.

**DEFINITION**   *Let $a_0, a_1, a_2, \ldots, a_n$ be defined and continuous on some interval $I$ of the x axis. A **linear differential equation of order** $n$ is an equation of the form*

$$a_0(x)\frac{d^n y}{dx^n} + a_1(x)\frac{d^{n-1}y}{dx^{n-1}} + \cdots + a_{n-1}(x)\frac{dy}{dx} + a_n(x)y = f(x). \qquad (1)$$

*If $f(x) = 0$ for all $x$ in $I$, then the equation is said to be **homogeneous.***

If $y$ and its first $n$ derivatives are continuous on $I$, then the left side of (1) is a continuous function of $x$. In this section we shall be concerned only with the case where the order $n$ is not greater than 2 and where the functions $a_0$, $a_1$, and $a_2$ are all constants.

We define differentiation operators $D^0$, $D^1$, and $D^2$ by

$$D^2 = \frac{d^2}{dx^2}$$

$$D^1 = \frac{d}{dx}$$

$$D^0 = \text{the identity.}$$

Thus if $y$ is a twice differentiable function of $x$,

$$D^2(y) = \frac{d^2 y}{dx^2}, \; D^1(y) = \frac{dy}{dx}, \text{ and } D^0(y) = y.$$

We can add these operators, and we can multiply them by functions. This allows us to write a linear differential equation in a new way, as the following example illustrates.

**EXAMPLE 1**   Write the equation $(D^2 + D^1 + 4D^0)(y) = x$ in traditional notation.

**Solution**   We have

$$D^2 y = \frac{d^2 y}{dx^2}, \qquad D^1 y = \frac{dy}{dx}, \qquad D^0 y = y.$$

The equation is

$$\frac{d^2 y}{dx^2} + \frac{dy}{dx} + 4y = x \qquad \text{or} \qquad y'' + y' + 4y = x. \qquad \square$$

The general second order linear differential equation with constant coefficients can now be written

$$(C_0 D^2 + C_1 D^1 + C_2 D^0)(y) = f(x). \tag{2}$$

---

**DEFINITION**   *The polynomial $C_0 r^2 + C_1 r + C_2$ corresponding to the differential expression $C_0 D^2 + C_1 D^1 + C_2 D_0$ is called the* **auxiliary polynomial.**

---

We wish to consider only second order equations (and not first order equations), so we assume $C_0 \neq 0$. Also, we will consider in this section only *homogeneous equations*. Dividing both sides of (2) by $C_0$ and using the fact that $f(x) = 0$, we shall solve the second order equation

$$(D^2 + a_1 D + a_2)y = 0, \tag{3}$$

in which $a_1$ and $a_2$ are real constants. The auxiliary polynomial equation is

$$r^2 + a_1 r + a_2 = 0, \tag{4}$$

and we denote the roots of this quadratic equation by $r_1$ and $r_2$. We consider three cases which depend on the nature of the roots.

**Case I.**   The roots $r_1$ and $r_2$ are real and unequal. We may write Equation (3) in the form

$$(D - r_1)(D - r_2)y = 0. \tag{5}$$

If $y$ is a solution of (5), we set $u = (D - r_2)y$, and then

$$(D - r_1)u = 0 \qquad \text{or} \qquad \frac{du}{dx} = r_1 u.$$

We separate variables and obtain

$$u = c e^{r_1 x}.$$

Then $y$ satisfies the equation

$$(D - r_2)y = c e^{r_1 x}$$

or

$$\frac{dy}{dx} - r_2 y = c e^{r_1 x}.$$

This linear equation has $e^{-r_2 x}$ as an integrating factor. Therefore

$$\frac{d}{dx}(e^{-r_2 x}y) = c e^{(r_1 - r_2)x}, \tag{6}$$

from which we conclude that

$$ye^{-r_2 x} = \frac{c}{r_1 - r_2} e^{(r_1 - r_2)x} + C_2, \quad r_1 \neq r_2.$$

The solution of (3) is

$$y = C_1 e^{r_1 x} + C_2 e^{r_2 x}, \quad r_1 \neq r_2 \tag{7}$$

in which we have set $c/(r_1 - r_2)$ equal to $C_1$.

**Case II.**   The roots are real and equal. The argument is the same as in Case I until we get to Equation (6). Then, since $r_1 = r_2$, we replace (6) by

$$\frac{d}{dx}(e^{-r_2 x} y) = c$$

and we integrate, to obtain

$$y = (C_1 x + C_2)e^{rx} \quad \text{with } r = r_1 = r_2. \tag{8}$$

There is one more case to consider: The auxiliary polynomial equation has two roots which are complex conjugates; that is, roots $r_1$ and $r_2$ where $r_1 = a + bi$, $r_2 = a - bi$, where $a$ and $b$ are real numbers and $i = \sqrt{-1}$.*

We wish to define $e^z$ for an arbitrary complex number $z$. We can assume $z$ has the form $z = a + bi$ for some real numbers $a$ and $b$. Since $e^{x+y} = e^x e^y$ for real $x$, $y$, it is reasonable to define $e^z = e^{a+bi} = e^a e^{bi}$. Therefore we need only give a meaning to $e^{bi}$, where $b$ is a real number. Using the ratio test one can show that the Maclaurin type expansion:

$$e^{ib} = 1 + ib - \frac{b^2}{2!} - \frac{ib^3}{3!} + \frac{b^4}{4!} + \frac{ib^5}{5!} + \cdots$$

$$= \sum_{n=0}^{\infty} \frac{(ib)^n}{n!}$$

converges for all real numbers $b$. To see this we rewrite the above expansion:

$$e^{ib} = \left(1 - \frac{b^2}{2!} + \frac{b^4}{4!} - \cdots\right) + i\left(b - \frac{b^3}{3!} + \frac{b^5}{5!} - \cdots\right). \tag{9}$$

Now we recognize the right side of (9) as the power series expansions of the cosine and sine functions, respectively. Hence, we obtain the following theorem, which is known as **Euler's Formula**.

**THEOREM 4**   **(Euler's Formula)** *For any real number $b$, $e^{ib} = \cos b + i \sin b$.*

**COROLLARY**   *If $x$ is a real variable and $z$ is a complex number, then*

$$\frac{d}{dx}e^{zx} = ze^{zx}.$$

---

*We assume that the reader is familiar with complex numbers.

**Proof**   Let $z = a + bi$ with $a$, $b$ real. Then

$$e^{zx} = e^{(a + bi)x} = e^{ax}e^{bix}$$

$$= e^{ax}(\cos bx + i \sin bx).$$

Therefore

$$\frac{d}{dx} e^{zx} = ae^{ax}(\cos bx + i \sin bx) + e^{ax}(-b \sin bx + ib \cos bx).$$

Since $-i^2 = 1$, we find

$$\frac{d}{dx} e^{zx} = ae^{x(a + bi)} + e^{ax}ib(+i \sin bx - i^2 \cos bx)$$

$$= ae^{xz} + e^{ax}ib(\cos bx + i \sin bx)$$

$$= ae^{xz} + ibe^{xa}e^{xbi} = ae^{xz} + ibe^{xz}$$

$$= ze^{xz}. \qquad \square$$

**Case III.**   The roots $r_1$ and $r_2$ are complex conjugates. By the Corollary to Theorem 4 the formula

$$\frac{d}{dx} e^{ax} = ae^{ax}$$

holds for complex constants $a$ as well as for real ones. Hence the analysis for Case I applies equally well when $r_1$ and $r_2$ are complex. Writing

$$r_1 = \alpha + i\beta, \qquad r_2 = \alpha - i\beta, \quad \beta \neq 0$$

where $\alpha$ and $\beta$ are real, we obtain from (7) the solution

$$y = C_1 e^{(\alpha + i\beta)x} + C_2 e^{(\alpha - i\beta)x}, \qquad (10)$$

where $C_1$ and $C_2$ are constants. In Equation (3) the coefficients $a_1$ and $a_2$ are real, and we are interested in real solutions $y(x)$. On the other hand, (10) gives the appearance of a complex-valued function. If we write (10) in the form

$$y = C_1 e^{\alpha x}(\cos \beta x + i \sin \beta x) + C_2 e^{\alpha x}(\cos \beta x - i \sin \beta x)$$

$$= D_1 e^{\alpha x} \cos \beta x + D_2 e^{\alpha x} \sin \beta x$$

or

$$y = e^{\alpha x}(D_1 \cos \beta x + D_2 \sin \beta x), \qquad (11)$$

the solution will be real whenever $D_1$ and $D_2$ are. We observe that

$$D_1 = C_1 + C_2, \qquad D_2 = i(C_1 - C_2),$$

and therefore $C_1$ and $C_2$ will be complex, in general, in order to have the constants $D_1$ and $D_2$ be real. $\qquad \square$

The preceding analysis leads to the following theorem.

**THEOREM 5**    *The function y is a solution of the equation*

$$\frac{d^2y}{dx^2} + a_1\frac{dy}{dx} + a_2 y = 0$$

*if and only if $r_1$, $r_2$ are roots of the auxiliary polynomial equation*

$$r^2 + a_1 r + a_2 = 0,$$

*and*

i) *y is given by $y = C_1 e^{r_1 x} + C_2 e^{r_2 x}$ when $r_1 \neq r_2$ and $r_1$, $r_2$ are real;*
ii) *y is given by $y = (C_1 x + C_2)e^{rx}$ when $r = r_1 = r_2$;*
iii) *y is given by $y = e^{ax}(C_1 \cos bx + C_2 \sin bx)$ when the roots $r_1$ and $r_2$ are complex with $r_1 = a + ib$, $r_2 = a - ib$. ($C_1$ and $C_2$ are real constants.)*

**EXAMPLE 1**    Solve the equation

$$\frac{d^2y}{dx^2} - 3\frac{dy}{dx} + 2y = 0.$$

**Solution**    The equation can be written in the form

$$(D^2 - 3D + 2)y = 0.$$

The auxiliary polynomial equation is

$$r^2 - 3r + 2 = (r - 1)(r - 2) = 0.$$

We have Case I, and the solution is

$$y = C_1 e^x + C_2 e^{2x}$$

where $C_1$ and $C_2$ are arbitrary constants.                    □

**EXAMPLE 2**    Solve

$$y'' - 2y' + 2y = 0.$$

**Solution**    The roots of the auxiliary polynomial are $1 \pm i$. Using (11) with $\alpha = 1$, $\beta = 1$, we obtain

$$y = e^x(D_1 \cos x + D_2 \sin x).$$                    □

In Examples 1 and 2 we solved the differential equations with arbitrary constants in the solution. To determine the value of these constants we need more information. Such information is often given at a particular value of $x$, and hence are called **initial conditions**.

**EXAMPLE 3**    Solve the differential equation

$$\frac{d^2y}{dx^2} - 3\frac{dy}{dx} - 10y = 0$$

if $y(0) = 0$ and $y'(0) = 7$.

**Solution**   The auxiliary equation is $r^2 - 3r - 10 = 0$ which factors into $(r - 5)(r + 2)$ $= 0$, with roots $r_1 = 5$, $r_2 = -2$. By Theorem 5 the solution is $y(x) = C_1 e^{5x}$ $+ C_2 e^{-2x}$. Then $y(0) = C_1 + C_2 = 0$ implies $C_1 = -C_2$. Also $y'(x) = 5C_1 e^{5x}$ $- 2C_2 e^{-2x}$ and $y'(0) = 5C_1 - 2C_2 = 5C_1 + 2C_1 = 7$. Therefore $C_1 = 1$ and $C_2 = -1$. The solution is

$$y(x) = e^{5x} - e^{-2x}.$$

□

**EXAMPLE 4**   Find the solution of the equation

$$y'' + 6y' + 9y = 0$$

which satisfies the conditions that $y = 2$ when $x = 0$ and $y = 0$ when $x = 1$.

**Solution**   We have

$$r^2 + 6r + 9 = 0 \qquad \text{and} \qquad r_1 = r_2 = -3.$$

Therefore

$$y = (C_1 x + C_2) e^{-3x}.$$

Substituting $y = 2$, $x = 0$, we get $C_2 = 2$. Then, for $y = 0$, $x = 1$, we have

$$0 = (C_1 + 2) e^{-3} \qquad \text{or} \qquad C_1 = -2.$$

The result is

$$y = (-2x + 2) e^{-3x}.$$

□

## 6  PROBLEMS

In each of Problems 1 through 25, solve the given differential equation.

**1** $y'' - 4y = 0$

**2** $y'' + y' - 2y = 0$

**3** $(D^2 - 4D + 3)(y) = 0$

**4** $(D^2 - 3D + 1)(y) = 0$

**5** $y'' + 2y' + 2y = 0$

**6** $(D^2 + 4D + 4)(y) = 0$

**7** $(D^2 - 2D + 3)(y) = 0$

**8** $y'' - 8y' + 16y = 0$

**9** $y'' + 6y' + 10y = 0$

**10** $y'' + y' + y = 0$

**11** $y'' + k^2 y = 0$

**12** $y'' - k^2 y = 0$

**13** $\dfrac{d^2 y}{dx^2} + 2\dfrac{dy}{dx} + 5y = 0$

**14** $\dfrac{d^2 y}{dx^2} - 3\dfrac{dy}{dx} + 4y = 0$

**15** $y'' - 2y' + 3y = 0$

**16** $y'' + 2y' - 3y = 0$

**17** $y'' - 2y' + y = 0$

**18** $y'' - 2y' + 2y = 0$

**19** $y'' + 2y' + y = 0$

**20** $y'' - 2y' + 5y = 0$

**21** $4y'' - 8y' + 7y = 0$

**22** $2y'' + 3y' = 0$

**23** $y'' + 2\sqrt{2}y' + 2y = 0$

**24** $6y'' - 5y' + 3y = 0$

**25** $y'' - 4y' = 2y$

In each of Problems 26 through 37 find the solution which satisfies the given initial conditions.

**26** $y'' - y' - 2y = 0$, $y(0) = 1$, $y'(0) = -2$

**27** $(D^2 + D - 6)(y) = 0$, $y(0) = -1$, $y'(0) = 3$

**28** $(D^2 + 4)(y) = 0$, $y(0) = 3$, $y'(0) = -2$

**29** $y'' - 6y' + 9y = 0$, $y(0) = -1$, $y'(0) = 2$

**30** $2y'' + 3y' = 0$, $y(0) = 1$, $y'(0) = 1$

**31** $6y'' - 7y' + 2y = 0$, $y(0) = 0$, $y'(0) = -1$

**32** $y'' - 2y' + y = 0$, $y(0) = 1$, $y'(0) = 2$

**33** $y'' + 2y' - 3y = 0$, $y(0) = 0$, $y(1) = \dfrac{2}{e}\left(\dfrac{1}{e^2} - 1\right)$

**34** $y'' + 5y = 0$, $y(0) = 4$, $y'(0) = 2$

**35** $y'' - 25y = 0$, $y(3) = -1$, $y'(3) = 0$

**36** $y'' - 4y' - y = 0$, $y(1) = 2$, $y'(1) = -1$

**\*37** $y'' + 4y' + 5y = 0$, $y(0) = 2$, $y'(0) = y''(0)$

_____ **7**

## NONHOMOGENEOUS LINEAR DIFFERENTIAL EQUATIONS

In the previous section, we saw that a linear differential equation of order two is of the form

$$a_0(x)\frac{d^2y}{dx^2} + a_1(x)\frac{dy}{dx} + a_2(x)y = f(x). \tag{1}$$

Letting $D^n$ denote the operator $D^n = d^n/dx^n$ for $n = 0, 1, 2, \ldots$, Equation (1) can be written

$$(a_0(x)D^2 + a_1(x)D^1 + a_2(x))y = f(x). \tag{2}$$

In Section 6 we considered only equations where $f(x) = 0$, which are said to be **homogeneous**. Equations of the form (1) or (2) where $f(x)$ is not identically zero are called **nonhomogeneous equations**.

In this section we consider second order linear equations *with constant coefficients* that are nonhomogeneous. That is, we shall show how to solve nonhomogeneous second-order equations of the form

$$(D^2 + a_1D^1 + a_2)y = f(x), \tag{3}$$

in which $a_1$ and $a_2$ are constants and $f(x)$ is a given function.

The first step in obtaining a solution of (3) is to find a solution of the same equation with $f = 0$. That is, we find a solution of the corresponding homogeneous equation.

**DEFINITION**    *The solution of the corresponding homogeneous equation is called the* **complementary function**.

The solution of nonhomogeneous equations is obtained by means of the following theorem.

**THEOREM 6**    *Any solution of a nonhomogeneous linear equation of the form* (1) *is the sum of a particular solution of the nonhomogeneous equation and a solution of the corresponding homogeneous equation (complementary function).*

**Proof**    Let $y_1$ be any fixed solution of (1) (that is, $y_1$ is a **particular solution**). Next let $y$ be any other solution of (1).

Since $y$ and $y_1$ each solve the equation we have

$$a_0(x)D^2y_1 + a_1(x)D^1y_1 + a_2(x)y_1 = f(x) \tag{4}$$

and

$$a_0(x)D^2y + a_1(x)D^1y + a_2(x)y = f(x). \tag{5}$$

Subtracting Equation (5) from (4), we obtain

$$a_0(x)D^2(y_1 - y) + a_1(x)D^1(y_1 - y) + a_0(x)(y_1 - y) = 0,$$

and so the function $v = y - y_1$ is a complementary function. Therefore

$$y = y_1 + v,$$

and the theorem is established.                    $\square$

We next exhibit the technique for finding particular solutions of (2) when the function $f$ has various forms.

*For notational convenience we write $L(y)$ to denote*

$$L(y) = D^2 y + a_1 D^1 y + a_2 y$$

$$= (D^2 + a_1 D^1 + a_2)y.$$

Equation (3) then takes the form

$$L(y) = f(x). \tag{6}$$

When the function $f$ has a simple form there are elementary methods for finding a particular solution. We illustrate with two cases.

**Case I.** The function $f$ has the form $ae^{bx}$, where $b$ is not a root of the auxiliary polynomial. In this case we observe that

$$L(ce^{bx}) = c(b^2 + a_1 b + a_2)e^{bx}$$

and so, if we select

$$c = \frac{a}{b^2 + a_1 b + a_2}, \tag{7}$$

then $y_1 = ce^{bx}$ is a particular solution of

$$L(y_1) = ae^{bx}.$$

Notice that if $b$ were a root of the auxiliary polynomial, the selection of $c$ as in (7) would be impossible.

**EXAMPLE 1** Solve the equation

$$y'' + y' - 6y = 2e^{-x}.$$

**Solution** The auxiliary polynomial $r^2 + r - 6 = 0$ has roots $2, -3$ and, since $b = -1$, $b$ is not a root. We select

$$c = \frac{2}{1 + 1 \cdot (-1) + (-6)} = \frac{-1}{3},$$

and a particular solution is $y_1 = -\frac{1}{3}e^{-x}$. The complementary function is

$$v = C_1 e^{2x} + C_2 e^{-3x},$$

and so the general solution is

$$y = C_1 e^{2x} + C_2 e^{-3x} - \frac{1}{3}e^{-x}.                    \square$$

**Case II.** The function $f$ is a polynomial. To obtain a particular solution $y_1$ of Equation (3), we observe that if $y_1$ is a polynomial, then $L(y_1)$ will be a polynomial of the same degree (*unless $a_2 = 0$, when the equation can be* solved anyway by the methods of Section 4). *Thus, if f is a polynomial and $a_2 \neq 0$, a particular solution can be found by setting $y_1$ equal to a general polynomial of the same degree as f with unknown coefficients.* The coefficients are then obtained by computing $L(y_1)$, setting the result equal to $f$, and equating coefficients of like powers of $x$. An example illustrates the procedure.

**EXAMPLE 2**   Solve the equation $(D^2 + D^1 - 2)y = x^2$.

**Solution**   Writing $(D + 2)(D - 1)y = x^2$, we see that the complementary function is $v = C_1 e^{-2x} + C_2 e^x$. To get a particular solution, we set $y_1$ equal to *a general polynomial of the second degree*. We choose the second degree because $f(x) = x^2$ is a polynomial of the second degree. Continuing, we set

$$y_1 = ax^2 + bx + c.$$

Then

$$y_1' = 2ax + b, \qquad y_1'' = 2a,$$

$$L(y_1) = 2a + 2ax + b - 2ax^2 - 2bx - 2c$$

$$= (-2a)x^2 + (2a - 2b)x + (2a + b - 2c).$$

We set this equal to $x^2(=f)$ and equate coefficients. The result is

$$-2a = 1, \qquad 2a - 2b = 0, \qquad 2a + b - 2c = 0,$$

or

$$a = -\tfrac{1}{2}, \qquad b = -\tfrac{1}{2}, \qquad c = -\tfrac{3}{4}.$$

The desired solution is

$$y = v + y_1 = C_1 e^{-2x} + C_2 e^x - \tfrac{1}{2}(x^2 + x + \tfrac{3}{2}). \qquad \square$$

The methods described in the two cases above come under the general heading of the **method of undetermined coefficients**. We now describe another method, which sometimes works even when $f$ does not satisfy the conditions of either Case I or Case II.

Suppose that $v$ is a solution of the homogeneous equation $L(v) = 0$ and we wish to find a solution which satisfies $L(y) = f$. That is, we first solve the homogeneous equation obtaining $v$, and we then use this solution $v$ to find a particular solution. We set

$$y = vw,$$

where $w$ is to be determined. Once we find $w$, the problem is solved. To compute $L(y)$ we write

$$y' = vw' + v'w, \qquad y'' = vw'' + 2v'w' + v''w$$

and

$$L(y) = vw'' + (2v' + a_1 v)w' + L(v)w.$$

Since we assumed that $v$ is a solution of the homogeneous equation, we have $L(v) = 0$, and therefore

$$L(y) = vw'' + (2v' + a_1 v)w'.$$

If $y$ is to be a solution, we must have

$$vw'' + (2v' + a_1v)w' = f(x).$$

Next we multiply the above equation by $v$ and get

$$\frac{d}{dx}(v^2w') + a_1(v^2w') = vf(x).$$

Now $e^{a_1x}$ is an integrating factor, and we obtain

$$\frac{d}{dx}(e^{a_1x}v^2w') = e^{a_1x}vf,$$

and

$$e^{a_1x}v^2w' = C + \int e^{a_1x}vf(x)\,dx.$$

Since $v$ is a known function, we can integrate once more to find $w$, and the problem is solved. An explicit result can be obtained only when all the integrations can be performed.

*Remark.*  The reader should not try to memorize the last formula above. It is better to proceed by setting $y = vw$ and making the necessary computations in each problem.

**EXAMPLE 3**  Solve $(D^2 + D^1 - 2)y = xe^x$.

**Solution**  The function $f(x) = xe^x$ does not fit into the form of either Case I or Case II. We begin by solving the homogeneous equation $(D^2 + D^1 - 2)y = 0$. The auxiliary polynomial factors into: $(r + 2)(r - 1) = 0$, and we see that $v = e^x$ is a solution of the homogeneous equation. We set

$$y = e^xw$$

and compute

$$y' = e^x(w' + w), \qquad y'' = e^x(w'' + 2w' + w)$$

$$L(y) = e^x(w'' + 3w') = xe^x.$$

We now solve the equation

$$w'' + 3w' = x$$

or, since $e^{3x}$ is an integrating factor,

$$\frac{d}{dx}(e^{3x}w') = xe^{3x}.$$

This result is

$$e^{3x}w' = C + \int xe^{3x}\,dx = C + \tfrac{1}{3}xe^{3x} - \tfrac{1}{9}e^{3x}$$

and

$$w' = Ce^{-3x} + \tfrac{1}{3}x - \tfrac{1}{9}.$$

Another integration yields

$$w = C_1 e^{-3x} + C_2 + \tfrac{1}{6}x^2 - \tfrac{1}{9}x$$

and

$$y = e^x w = C_1 e^{-2x} + C_2 e^x + (\tfrac{1}{6}x^2 + \tfrac{1}{9}x)e^x. \qquad \square$$

Sometimes the function $f$ on the right side of Equation (3) may be a sum of a function of the type in Case I and a polynomial. The following theorem allows us to treat such situations.

---

**THEOREM 7**    *If $f(x) = c_1 f_1(x) + c_2 f_2(x)$ and if $y_1(x)$, $y_2(x)$ are solutions of*

$$L(y_1) = f_1(x), \qquad L(y_2) = f_2(x),$$

*respectively, then $y = c_1 y_1(x) + c_2 y_2(x)$ is a solution of $L(y) = f$.*

---

**Proof**    We observe that for linear operators $L$ of the form

$$L(y) = D^2 y + a_1 D^1 y + a_2 y$$

we have

$$L(y) = L(c_1 y_1 + c_2 y_2) = c_1 L(y_1) + c_2 L(y_2) = c_1 f_1(x) + c_2 f_2(x) = f(x). \quad \square$$

Theorem 7 also holds if $f$ is the finite sum of any number of functions. If $f$ is the sum of a polynomial and several functions of the type in Case I, Theorem 7 yields a quick method for finding a solution.

We can also solve nonhomogeneous equations when the function $f$ has the form $f(x) = e^x \cos x$ or $e^x \sin x$. We are able to solve these equations by using complex numbers. Our first step is the observation that the methods of this section work equally well when $f(x)$ is a complex-valued function of the real variable $x$. Of course, in such a case the solution will be complex even when $a_1$ and $a_2$ are real. That is, if

$$f(x) = f_1(x) + if_2(x)$$

and if $y$ satisfies the equation

$$L(y) = y'' + a_1 y' + a_2 y = f, \quad a_1, a_2 \text{ real},$$

we may write $y(x) = y_1(x) + iy_2(x)$ and, therefore,

$$L(y) = L(y_1 + iy_2) = L(y_1) + iL(y_2) = f_1(x) + if_2(x).$$

*Two complex numbers are equal if and only if their real and imaginary parts are.* Hence,

$$L(y_1) = f_1(x) \qquad \text{and} \qquad L(y_2) = f_2(x).$$

The fact that complex solutions can be separated into real and imaginary parts is employed in the next example.

**EXAMPLE 4**    Solve $y'' + 4y = e^x \cos x$.

**Solution**    The complementary function is $v = C_1 \cos 2x + C_2 \sin 2x$. We could proceed by the method of undetermined coefficients, but instead we employ the following trick.

By Euler's Formula we know $e^{x+ix} = e^x \cos x + i e^x \sin x$, for real $x$. That is, $e^x \cos x$ is the real part of $e^{(1+i)x}$. The equation

$$y'' + 4y = e^{(1+i)x}$$

can be solved by the method of Case I. The solution will be complex, *but the real part of the solution will give the answer to the original problem.* We set

$$y_1 = ce^{(1+i)x}$$

and compute

$$y_1'' + 4y_1 = (4+2i)ce^{(1+i)x} = e^{(1+i)x}.$$

Therefore $c = 1/(4+2i) = (2-i)/10$, and we write

$$y_1 = \frac{2-i}{10}e^{(1+i)x} = \frac{2-i}{10}e^x(\cos x + i \sin x)$$

$$= \frac{e^x}{10}[(2 \cos x + \sin x) + i(2 \sin x - \cos x)].$$

The general solution of the given equation is

$$y = C_1 \cos 2x + C_2 \sin 2x + \frac{e^x}{10}(2 \cos x + \sin x). \qquad \Box$$

## 7   PROBLEMS

In each of Problems 1 through 23, solve the given differential equation.

**1**   $y'' - 2y' - 3y = 27x^2$

**2**   $y'' - 2y' - 3y = 2 - 4x - 3x^2$

**3**   $y'' + 2y' + 2y = 2x + 3 + e^{-x}$

**4**   $y'' + 2y' + y = x + e^x$     **5**   $y'' - 4y' + 4y = x^2 - e^{-2x}$

**6**   $(D^2 - D^1)(y) = e^x$     **7**   $(D^1 - 1)^2(y) = e^x$

**8**   $(D^2 - 1)(y) = e^{-x}$     **9**   $y'' - y = \cos x$

**10**   $(D^2 + 4)(y) = \cos x$     **11**   $(D^2 + 2D^1 + 2)(y) = \sin 2x$

**12**   $(D^2 + 2D^1 + 5)(y) = e^x \sin x$

**13**   $y'' + y = \sin x$     **14**   $y'' - 2y' + 2y = e^x \cos x$

**15**   $y'' + 2y' = e^x$     **16**   $y'' + y' - 2y = e^{2x} - 2$

**17**   $y'' + 4y = 15e^x - 8x$     **18**   $y'' + 4y = 15e^x - 8x^2$

**19**   $y'' - 4y = e^{2x} + 2$

**20**   $y'' - y = e^{-x}(2 \sin x + 4 \cos x)$

**21**   $y'' - y = 10 \sin^2 x$
    (*Hint*: Use the identity $\sin^2 x = \frac{1}{2}(1 - \cos 2x)$.)

**22**   $y'' + y = 12 \cos^2 x$ (cf. Problem 21)

**23**   $y'' + 4y = 4 \sin^2 x$ (cf. Problem 21)

In Problems 24 through 30, find in each case that solution which satisfies the initial conditions $y(0) = y'(0) = 0$.

**24**   $y'' - y = x$     **25**   $(D^2 - D^1 - 2)(y) = e^{3x}$

**26**   $(D^2 + D^1 - 6)(y) = -4e^x$     **27**   $y'' + y = x$

**28**   $(D^2 + 4)(y) = e^x$     **29**   $y'' + 2y' = x$

**30**   $y'' + y = 10e^{2x}$

In each of Problems 31 through 33, find the particular solution determined by the initial conditions given

**31**   $y'' - 4y = 2 - 8x$, $y(0) = 0$ and $y'(0) = 5$

**32**   $y'' + 4y' + 5y = 10e^{-3x}$, $y(0) = 4$ and $y'(0) = 0$

**33**   $y'' + 4y' + 5y = 10$, $y(0) = 0$ and $y'(0) = 0$

**34**   Show that the equation

$$y'' + y = x^3, \quad y(0) = 0, \quad y(\pi) = 0$$

has no solution.

**35**   Show that the equation

$$y'' + y = 2 \cos x, \quad y(0) = 0 \text{ and } y(\pi) = 0$$

does not have a unique solution. Indeed, find an infinite number of different solutions.

_____ **8**

## SERIES SOLUTIONS OF DIFFERENTIAL EQUATIONS

Linear equations and linear systems with constant coefficients can frequently be solved explicitly. For linear equations with variable coefficients and for nonlinear equations, it is generally true that explicit solutions in terms of elementary functions do not exist. In this section we describe a method for obtaining solutions of differential equations in the form of convergent infinite series. The process is particularly useful in those cases in which explicit solutions cannot be found. We describe the method by working two illustrative examples.

**EXAMPLE 1**    Express the general solution of the equation

$$L(y) \equiv y'' + xy' + y = 0$$

as a Maclaurin series.

**Solution**    We write the solution in the form

$$y = a_0 + a_1 x + a_2 x^2 + \cdots + a_n x^n + \cdots = \sum_{n=0}^{\infty} a_n x^n. \qquad (1)$$

The solution is determined when the coefficients $a_i$, $i = 0, 1, 2, \ldots$ are found and when the convergence of the series is verified. Assuming that the series (1) has a positive radius of convergence, we may differentiate term by term and obtain

$$y' = a_1 + 2a_2 x + 3a_3 x^2 + \cdots + na_n x^{n-1} + \cdots = \sum_{n=1}^{\infty} na_n x^{n-1}$$

and

$$y'' = 2a_2 + 3 \cdot 2a_3 x + 4 \cdot 3a_4 x^2 + \cdots + n(n-1)a_n x^{n-2} + \cdots$$

$$= \sum_{n=2}^{\infty} n(n-1)a_n x^{n-2}.$$

Using the series for $y$, $y'$, and $y''$, we find

$$L(y) = \sum_{n=2}^{\infty} n(n-1)a_n x^{n-2} + x \sum_{n=1}^{\infty} na_n x^{n-1} + \sum_{n=0}^{\infty} a_n x^n = 0. \qquad (2)$$

The above series may be rewritten so that corresponding powers of $x$ can be combined. In a moment we shall see the importance of this rearrangement. If we replace $n$ by $k + 2$, the first series in (2) becomes

$$\sum_{k=0}^{\infty} (k+2)(k+1)a_{k+2} x^k.$$

For the second series in (2), we replace $n$ by $k$, allow the sum to begin with $k = 0$, and move the $x$ inside the sum sign. The result is

$$\sum_{k=0}^{\infty} ka_k x^k.$$

The last series in (2) is left unaltered except that the index $n$ is replaced by $k$.

With these changes in notation, we find

$$L(y) = \sum_{k=0}^{\infty} [(k+2)(k+1)a_{k+2} + (k+1)a_k]x^k = 0. \tag{3}$$

Before continuing with this example, we state *the basic tool in determining solutions in infinite series:*

---

**THEOREM 8**    *A power series with a positive radius of convergence can vanish identically if and only if the coefficient of every power of x vanishes.*

---

**Proof**    Theorem 8 can be established by using the theory of Taylor series. As a consequence of Theorem 21 of Chapter 10 (page 427) we have for the power series $f(x) = \sum_{n=0}^{\infty} a_n x^n$ that

$$a_n = \frac{f^{(n)}(0)}{n!},$$

where $f^{(n)}(0)$ is the $n$th derivative of $f$ at 0. If $f$ vanishes, then $f^{(n)}(0) = 0$ for all $n$, hence $a_n = 0$ for all $n$. On the other hand, if $a_n = 0$ for all $n$, then $f^{(n)}(0) = 0$ for all $n$ and $f$ vanishes.    $\square$

We now continue with Example 1 by applying this fact to (3) to obtain

$$a_{k+2} = -\frac{1}{k+2} a_k, \quad k = 0, 1, 2, \ldots.$$

This formula, called a **recurrence relation**, is typical of solutions determined by infinite series. It enables us to express $a_2, a_3, \ldots, a_n, \ldots$ in terms of $a_0$ and $a_1$. In particular, selecting $k = 2, 4, 6, \ldots$, we find

$$a_2 = -\frac{a_0}{2}, \qquad a_4 = -\frac{a_2}{4} = (-1)^2 \frac{a_0}{2 \cdot 4},$$

$$a_6 = -\frac{a_4}{6} = (-1)^3 \frac{a_0}{2 \cdot 4 \cdot 6}, \ldots.$$

For $k = 1, 3, 5, \ldots$, we have

$$a_3 = -\frac{a_1}{3}, \qquad a_5 = -\frac{a_3}{5} = (-1)^2 \frac{a_1}{3 \cdot 5}, \qquad a_7 = (-1)^3 \frac{a_1}{3 \cdot 5 \cdot 7}, \ldots.$$

The general expressions for $a_{2k}$ and $a_{2k+1}$ are

$$a_{2k} = (-1)^k \frac{a_0}{2 \cdot 4 \cdot 6 \cdots 2k} = (-1)^k \frac{a_0}{2^k k!},$$

$$a_{2k+1} = (-1)^k \frac{a_1}{1 \cdot 3 \cdot 5 \cdots (2k+1)} = (-1)^k \frac{2^k k! a_0}{(2k+1)!}.$$

We conclude that the solution is given by

$$y = a_0 \left[ 1 - \frac{x^2}{2} + \cdots + \frac{(-1)^k x^{2k}}{2^k \cdot k!} + \cdots \right]$$

$$+ a_1 \left[ x - \frac{x^3}{1 \cdot 3} + \cdots + \frac{(-1)^k 2^k k! x^{2k+1}}{(2k+1)!} + \cdots \right].$$

Note that $a_0$ and $a_1$ are constants of integration and that, by the ratio test, each series is convergent for all values of $x$.     □

**EXAMPLE 2**     Solve Legendre's equation

$$L(y) \equiv (1 - x^2)y'' - 2xy' + p(p + 1)y = 0$$

by the method of power series; the number $p$ is a constant but not necessarily an integer.

**Solution**     We set $y = \sum_{n=0}^{\infty} a_n x^n$ and, as in Example 1, we find

$$y' = \sum_{n=0}^{\infty} n a_n x^{n-1}, \qquad y'' = \sum_{n=0}^{\infty} (n + 2)(n + 1)a_{n+2}x^n.$$

Also,

$$-2xy' = \sum_{n=0}^{\infty} (-2na_n x^n),$$

$$-x^2 y'' = \sum_{n=0}^{\infty} -(n + 2)(n + 1)a_{n+2}x^{n+2} = \sum_{k=0}^{\infty} -k(k - 1)a_k x^k.$$

Inserting the various series into the expression for $L(y)$ and changing the summation index to $k$ throughout, we find

$$L(y) = \sum_{k=0}^{\infty} \{(k + 2)(k + 1)a_{k+2} + [p(p + 1) - k(k + 1)]a_k\}x^k = 0.$$

Using the fact that the coefficient of $x^k$ must vanish for every $k$, we get the recurrence formula

$$a_{k+2} = \frac{(k - p)(k + p + 1)}{(k + 2)(k + 1)}a_k.$$

Evaluating $a_2, a_4, a_6, \ldots$ in terms of $a_0$, and evaluating $a_3, a_5, a_7, \ldots$ in terms of $a_1$, we obtain

$$y = a_0 \left[ 1 - \frac{p(p + 1)}{2!}x^2 + \frac{p(p - 2)(p + 1)(p + 3)}{4!}x^4 - + \cdots \right]$$

$$+ a_1 \left[ x - \frac{(p - 1)(p + 2)}{3!}x^3 + \frac{(p - 1)(p - 3)(p + 2)(p + 4)}{5!}x^5 - + \cdots \right].$$

If $p$ is an even integer the series in the first bracket is finite, the expression being a polynomial of degree $p$. If $p$ is an odd integer the series in the second bracket is a polynomial, all terms beyond the $p$th being zero. These polynomials (when $p$ is an integer) are called the **Legendre polynomials.**     □

The validity of the results in the two examples above is established in more advanced courses.

We raise the question: For what values of $x$ do the power series solutions converge? In Example 1 the ratio test shows that the series expansion for the solution converges for all values of $x$. If we solve Legendre's equation as given in Example 2, for $y''$, we obtain

$$y'' = \frac{2x}{1 - x^2}y' - \frac{p(p + 1)}{1 - x^2}y.$$

Since the Maclaurin expansions of both $2x/(1-x^2)$ and $p(p+1)/(1-x^2)$ have radius 1, it follows that the solution $y$ as given in Example 2 converges at least for $|x| < 1$.

For both linear and nonlinear equations it is frequently quite difficult or impossible to obtain a general formula for the $n$th term in the power series expansion of the solution. If we seek only the first few terms of the Taylor series of a *particular* solution, then we just need to know $y$ and its first few derivatives at $x^0$, the value about which the Taylor expansion is taken. In such cases the method exhibited in the next example is valuable. Of course, the convergence of the series must be considered separately.

**EXAMPLE 3**  Find the terms out to that of $x^4$ of the Maclaurin series for the particular solution of

$$y' = x^2 + y^2$$

which satisfies the condition $y = 1$ when $x = 0$.

**Solution**  Substituting $y = 1$, $x = 0$ in the differential equation, we find $y'(0) = 1$. We differentiate the differential equation, getting

$$y'' = 2x + 2yy'.$$

Substituting $x = 0$, $y = 1$, $y' = 1$ in this equation, we obtain $y''(0) = 2$. Repeating the process once again, we see that

$$y''' = 2 + 2yy'' + 2y'^2$$

and

$$y'''(0) = 2 + 2 \cdot 2 + 2 \cdot 1 = 8.$$

A continuation yields

$$y^{(4)} = 2yy''' + 6y'y''$$

and

$$y^{(4)}(0) = 2(8) + 6(1)(2) = 28.$$

Therefore

$$y = 1 + x + \frac{2x^2}{2!} + 8\frac{x^3}{3!} + 28\frac{x^4}{4!} + \cdots$$

$$= 1 + x + x^2 + \tfrac{4}{3}x^3 + \tfrac{7}{6}x^4 + \cdots. \qquad \square$$

## 8   PROBLEMS

In each of Problems 1 through 13, find the general solution by the method of power series. Find a formula for the general term as in Example 1.

1  $y'' - y = 0$

2  $y'' + y = 0$

3  $y'' + xy' + 2y = 0$

4  $y'' + 2xy' + 2y = 0$

5  $y'' + 2xy' + 4y = 0$

6  $y'' - xy' + y = 0$

7  $y'' - xy = 0$

8  $y' - 2xy = 2x$, $y(0) = 0$

9  $y'' - xy' - y = 0$, $y(0) = 1$, $y'(0) = 0$

10  $y'' - xy = x^4$

**11** $y'' - (x-1)y = 0$; set $y = \sum_{n=0}^{\infty} a_n(x-1)^n$. (*Hint*: To make certain formulas hold for all $n$, define $a_k = 0$ for $k < 0$.)

**12** $y'' - xy' - 2y = 1 + x$　　**13** $y'' - x^2y = 0$

**14** Obtain the recurrence relation for the coefficients in the solution of

$$(x^2 - 3x + 2)y'' + (x^2 - 2x - 1)y' + (x - 3)y = 0$$

and find $a_2, a_3, a_4, a_5$. For what values of $x$ will the series surely converge?

**15** For what values of $x$ will the series for the solutions of the equation

$$(2 - x)y'' + (x - 1)y' - y = 0$$

surely converge? Show that the recurrence relation is

$$a_{n+2} = \frac{(n+1)^2 a_{n+1} - (n-1)a_n}{2(n+1)(n+2)}.$$

Show that if the $a_n$ satisfy this relation for every $n$, then

$$a_n = \frac{(a_0 + a_1)}{2 \cdot n!} \quad \text{for all } n \geq 2.$$

What is the general solution?

In each of Problems 16 through 28, find the terms of the series solution out to and including the term in $(x - a)^n$ for the given $n$ and $a$.

**16** $y' = xy^2 + 1$; $y = 1$ for $x = 1$; $a = 1$, $n = 4$

**17** $y' = \sin(xy) + x$; $y = 1$ for $x = 0$; $a = 0$, $n = 4$

**18** $y' = \sin(x^2 + y)$; $y = \pi/2$ for $x = 0$; $a = 0$, $n = 4$

**19** $y'' + 2xy' + y = 0$; $y = 0$, $y' = 1$ for $x = 0$; $a = 0$, $n = 5$

**20** $y'' = y^2$, $y(0) = 1$, $y'(0) = 2$　$a = 0$, $n = 4$

**21** $y' = x^2 + y^2$, $y(0) = 1$, $a = 0$, $n = 4$

**22** $y'' = x^2 - y^2$; $y = 1$, $y' = 0$ for $x = 0$; $a = 0$, $n = 5$

**23** $(1 + x^2)y'' + x(y' + y) = 0$; $y = 0$, $y' = 1$ for $x = 0$; $a = 0$, $n = 4$

**24** $\dfrac{dy_1}{dx} = xy_1 + 2y_2$;

$$\frac{dy_2}{dx} = -2y_1 + xy_2; \; y_1 = 0, \; y_2 = 1 \text{ for } x = 0; \; a = 0, \; n = 3$$

(*Hint*: Find expansions for $y_1$ and $y_2$.)

**25** $\dfrac{dy_1}{dx} = y_1^2 + y_2^2$;

$$\frac{dy_2}{dx} = y_2^2 - y_1; \; y_1 = y_2 = 1 \text{ for } x = 0; \; a = 0, \; n = 2$$

**26** $y''' = xy + yy'$; $y = 0$, $y' = 1$, $y'' = 2$ for $x = 0$; $a = 0$, $n = 5$

**27** $y' = \cos y$, $y(0) = 0$, $a = 0$, $n = 4$

**28** $y'' = (y + 1)^{1/2}$, $y(0) = 0$, $y'(0) = 2$, $a = 0$, $n = 4$

---

# CHAPTER 18

## REVIEW PROBLEMS

In each of Problems 1 through 5, solve the given differential equations, except for a quadrature, if necessary.

**1** $\dfrac{dy}{dx} = 3y$　　　　　**2** $\dfrac{dy}{dx} = \sin x - x$

**3** $\dfrac{dy}{dx} = e^{-x^2}$　　　　**4** $\dfrac{d^2y}{dx^2} = \cos x$

**5** $\dfrac{dy}{dx} = \dfrac{\cos^2 x}{x^2} + 3x$

In each of Problems 6 through 29, solve the given differential equations.

**6** $(1 - x)y' = y^2$　　　　**7** $xy^3 dx + e^{x^2} dy = 0$

**8** $2y\,dx = 3x\,dy$　　　　　**9** $xy\,dx = (4 + e^{x^2})\,dy$

**10** $x \cos^2 y\,dx + \tan y\,dy = 0$

**11** $y\,dx = (e^{3x} + 1)\,dy$　　**12** $y \ln x \ln y\,dx + dy = 0$

**13** $(1 + x^2)\,dy + (1 + y^2)\,dx = 0$

**14** $\left(x \csc \dfrac{y}{x} - y\right) dx + x\,dy = 0$

**15** $x\,dx + \sin^2 \dfrac{y}{x}(y\,dx - x\,dy) = 0$

**16** $\left(x - y \arctan \dfrac{y}{x}\right) dx + \arctan \dfrac{y}{x}\,dy = 0$

**17** $(x + y)\,dx + (x - y)\,dy = 0$

**18** $(\cos 2y - 3x^2y^2)\,dx + (\cos 2y - 2x \sin 2y - 2x^3y)\,dy = 0$

**19** $(1 + y^2 + xy^2)\,dx + (x^2y + y + 2xy)\,dy = 0$

**20** $(2xy - \tan y)\,dx + (x^2 - x \sec^2 y)\,dy = 0$

**21** $(ye^{xy} - 2y^3)\,dx + (xe^{xy} - 6xy^2 - 2y)\,dy = 0$; $y(0) = 2$.

**22** $(x^4 + 2y)\,dx - x\,dy = 0$　　**23** $y' = \csc x - y \cot x$

**24** $y' = x - 2xy$

**25** $(2xy + x^2 + x^4)\,dx - (1 + x^2)\,dy = 0$

**26** $dx - (1 + 2x \tan y)\, dy = 0$

**27** $\dfrac{dy}{dx} = 1 + 3y \tan x$

**28** $(x + a)\dfrac{dy}{dx} = bx - ny$; $a$, $b$, $n$ are constants with $n \neq 0$, $n \neq -1$.

**29** $\dfrac{dy}{dx} = x^3 - 2xy$, $y(1) = 1$

**30** Determine the escape velocity of the planet Mars if its radius is approximately 3400 kilometers and the gravitational pull is approximately 990 cm/sec$^2$.

**31** The growth rate of a certain plant varies corresponding to different times of the year. Suppose it can be modelled by the differential equation

$$\frac{dM}{dt} = \frac{1}{12} M \sin\!\left(\frac{\pi t}{6}\right)$$

where $t$ is time measured in months. If there are 600 units of this plant at time $t = 0$, how many are there nine months later?

**32** An electric circuit has a capacitor of capacitance $C$ (in farads) and a resistor with a resistance $R(t)$ that varies with time (in ohms). (See Figure 5.)

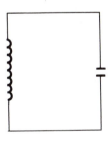

**FIGURE 5**

The charge $q(t)$ (in coulombs) on the capacitor at time $t$ (in seconds) satisfies the differential equation

$$R(t)\frac{dq}{dt} + \frac{1}{C} q = 0.$$

Find $q(t)$ in terms of $q(0)$ and $C$ under the assumption that $r(t) = (2 - \cos t)^{-1}$.

**33** Find the orthogonal trajectories of the family of curves $y = ce^{-x/4}$.

**34** Find the orthogonal trajectories of the following family of curves: $x^2 - y^2 = C$.

**35** Find the orthogonal trajectories of the following family of curves: $x^4 + y^4 = C$.

In each of Problems 36 through 52, solve the given differential equation.

**36** $y'' - y' - 2y = 0$      **37** $y'' - y' - 6y = 0$

**38** $y'' - 2y' - 3y = 0$, $y(0) = 0$, $y'(0) = -4$

**39** $4y'' - 4y' + y = 0$, $y(0) = -2$, $y'(0) = 2$

**40** $y'' - 2y' + 5y = 0$      **41** $y'' - 4y' + 7y = 0$

**42** $y'' - y = 0$, $y(0) = 1$, $y'(0) = 0$

**43** $y'' + y = 0$, $y(0) = 2$, $y'(0) = 0$

**44** $y'' + 2ay' + b^2 y = 0$, $b > a > 0$, $y(0) = 0$, $y'(0) = v_0$

**45** $y'' + y = -\cos x$

**46** $y'' + 6y' + 9y = e^x$

**47** $y'' + 4y' + 5y = 50x + 13e^{3x}$

**48** $y'' + y = 10e^{2x}$, $y(0) = 0$, $y'(0) = 0$

**49** $y'' + 2y' + y = e^{-x}$

**50** $y'' - 2y' + y = x + 2xe^x$

**51** $y'' + y' - 2y = e^x + e^{2x}$

**52** $y'' + 2y' + y = 7 + 75 \sin 2x$

# TRIGONOMETRY REVIEW

**1**

## BASIC FORMULAS

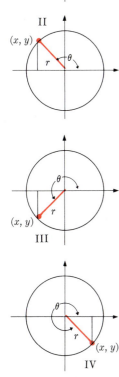

**FIGURE 1**

We assume that the reader has successfully completed a course in trigonometry. However, no matter how well he or she has learned the formulas of trigonometry, they are easy to forget unless they are used frequently. We shall therefore devote this appendix to a review of the basic notions of trigonometry and a compilation of many of the elementary formulas, for we shall need to use them often. In any case, the formulas serve as handy tools for future use, and the reader who has difficulty remembering them should relearn them systematically.

A circle of radius $r$ is drawn and a rectangular coordinate system constructed with origin at the center of this circle. An angle $\theta$ is drawn, starting from the positive direction of the $x$ axis, and is measured counterclockwise. Angles in the first, second, third, and fourth quadrants are shown in Fig. 1. The basic definitions of the trigonometric functions are

$$\sin\theta = \frac{y}{r}, \qquad \cos\theta = \frac{x}{r},$$

where $(x, y)$ are the coordinates of the point where the terminating side of the angle $\theta$ intersects the circle. These definitions depend only on the angle $\theta$ and not on the size of the circle since, by similar triangles, the ratios are independent of the size of $r$. The quantity $r$ is always positive, while $x$ and $y$ have the sign that goes with the quadrant: $x$ and $y$ are positive in the first quadrant; $x$ is negative, $y$ is positive in the second quadrant, and so on.

The remaining trigonometric functions are defined by the formulas

$$\tan\theta = \frac{\sin\theta}{\cos\theta}, \qquad \cot\theta = \frac{\cos\theta}{\sin\theta},$$

$$\sec\theta = \frac{1}{\cos\theta}, \qquad \csc\theta = \frac{1}{\sin\theta}.$$

We also can define them by the formulas

$$\tan \theta = \frac{y}{x}, \qquad \cot \theta = \frac{x}{y}, \qquad \sec \theta = \frac{r}{x}, \qquad \csc \theta = \frac{r}{y}.$$

For angles larger than 360°, the trigonometric functions repeat, and for this reason they are called **periodic** functions. This means that for all $\theta$

$$\sin (360° + \theta) = \sin \theta, \qquad \cos (360° + \theta) = \cos \theta.$$

**Negative angles** are measured by starting from the positive $x$ direction and going *clockwise*. Figure 2 shows a negative angle $-\theta$ and its corresponding positive angle $\theta$. By congruent triangles it is easy to see that for *every* $\theta$ we have

$$\sin (-\theta) = -\sin \theta \qquad \text{and} \qquad \cos (-\theta) = \cos \theta,$$

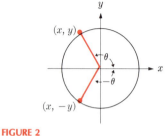

**FIGURE 2**

which says that the sine function is an **odd** function of $\theta$, while the cosine function is an **even** function of $\theta$. It is easy to show that

$$\tan (-\theta) = -\tan \theta, \qquad \cot (-\theta) = -\cot \theta,$$

$$\sec (-\theta) = \sec \theta, \qquad \csc (-\theta) = -\csc \theta.$$

No matter what quadrant the angle $\theta$ is in, the Pythagorean Theorem tells us that

$$x^2 + y^2 = r^2,$$

or, upon dividing by $r^2$,

$$\left(\frac{x}{r}\right)^2 + \left(\frac{y}{r}\right)^2 = 1.$$

This formula asserts that for every angle $\theta$,

$$\cos^2 \theta + \sin^2 \theta = 1.$$

Let $ABC$ be any triangle with sides $a, b, c$, as shown in Fig. 3. We have the **Law of Sines**:

$$\frac{\sin A}{a} = \frac{\sin B}{b} = \frac{\sin C}{c};$$

and the **Law of Cosines**:

$$a^2 = b^2 + c^2 - 2bc \cos A,$$

$$b^2 = a^2 + c^2 - 2ac \cos B,$$

$$c^2 = a^2 + b^2 - 2ab \cos C.$$

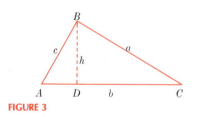

**FIGURE 3**

The area of $\triangle ABC$ is: Area $= \frac{1}{2}ab \sin C = \frac{1}{2}bc \sin A = \frac{1}{2}ac \sin B$. The proofs of these statements should be reviewed. An ambitious reader may want to rediscover these proofs for himself. As a help in establishing the Law of Sines, the reader should drop the perpendicular from $B$ to side $b$ in Fig. 3.

The most common unit for measurement of angles is the degree. The circumference of a circle is divided into 360 equal parts, and the angle from the center to two adjacent subdivision points is **one degree**. Each degree is divided into 60 **minutes** and each minute into 60 **seconds**.

BASIC FORMULAS

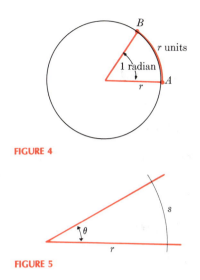

**FIGURE 4**

**FIGURE 5**

A second type of unit for measuring angles is the *radian*. A circle is drawn and an arc is measured which is equal in length to the radius, as in Fig. 4. Note that the arc $\overset{\frown}{AB}$ (*not* the chord) is $r$ units in length. *The angle subtended from the center of the circle is* 1 **radian**, by definition. This angle is the same regardless of the size of the circle and so is a perfectly good unit for measurement. To measure any angle $\theta$ in radian units (Fig. 5), we measure the arc $s$ intercepted by a circle of radius $r$. Then division by $r$ gives the radian measure of the angle:

$$\theta = \frac{s}{r} \text{ radians.}$$

We are all familiar with the definition of the number $\pi$, which is

$$\pi = \frac{\text{length of circumference of circle}}{\text{length of diameter of circle}}.$$

If $C$ is the length of the circumference and $r$ is the radius, then

$$2\pi = C/r.$$

This is the same as saying that the radian measure of the angle we know as $360°$ is exactly $2\pi$ radians. We have the basic formula

$$\pi \text{ radians} = 180 \text{ degrees.}$$

The fundamental formulas upon which many of the trigonometric identities and relations are based are the **addition formulas for the sine and cosine**. These are

$$\sin(\phi + \theta) = \sin\phi\cos\theta + \cos\phi\sin\theta;$$
$$\sin(\phi - \theta) = \sin\phi\cos\theta - \cos\phi\sin\theta;$$
$$\cos(\phi + \theta) = \cos\phi\cos\theta - \sin\phi\sin\theta;$$
$$\cos(\phi - \theta) = \cos\phi\cos\theta + \sin\phi\sin\theta.$$

Proofs of these formulas should be reviewed. See Problems 4 and 5 at the end of this section. Many of the formulas listed below are special cases of the above formulas. Before listing them, we shall indicate some special values of the three principal trigonometric functions in a way that somewhat reduces the strain on memorization. The tangent function tends to infinity as $\theta \to 90°$ and we sometimes say, speaking loosely, that "tan $90°$ is infinite."

| Function \ Angle | 0° | 30° | 45° | 60° | 90° | 120° | 135° | 150° | 180° |
|---|---|---|---|---|---|---|---|---|---|
| sin | $\frac{\sqrt{0}}{2}$ | $\frac{\sqrt{1}}{2}$ | $\frac{\sqrt{2}}{2}$ | $\frac{\sqrt{3}}{2}$ | $\frac{\sqrt{4}}{2}$ | $\frac{\sqrt{3}}{2}$ | $\frac{\sqrt{2}}{2}$ | $\frac{\sqrt{1}}{2}$ | $\frac{\sqrt{0}}{2}$ |
| cos | $\frac{\sqrt{4}}{2}$ | $\frac{\sqrt{3}}{2}$ | $\frac{\sqrt{2}}{2}$ | $\frac{\sqrt{1}}{2}$ | $\frac{\sqrt{0}}{2}$ | $-\frac{\sqrt{1}}{2}$ | $-\frac{\sqrt{2}}{2}$ | $-\frac{\sqrt{3}}{2}$ | $-\frac{\sqrt{4}}{2}$ |
| tan | $\sqrt{\frac{0}{4}}$ | $\sqrt{\frac{1}{3}}$ | $\sqrt{\frac{2}{2}}$ | $\sqrt{\frac{3}{1}}$ | $*$ | $-\sqrt{\frac{3}{1}}$ | $-\sqrt{\frac{2}{2}}$ | $-\sqrt{\frac{1}{3}}$ | $-\sqrt{\frac{0}{4}}$ |

From the relation

$$\sin^2 \theta + \cos^2 \theta = 1,$$

we obtain, when we divide by $\cos^2 \theta$, the formula

$$\tan^2 \theta + 1 = \sec^2 \theta.$$

If we divide by $\sin^2 \theta$, we have the identity

$$1 + \cot^2 \theta = \csc^2 \theta.$$

The **double-angle formulas** are obtained by letting $\theta = \phi$ in the addition formulas. We find that

$$\sin 2\theta = 2 \sin \theta \cos \theta;$$

$$\cos 2\theta = \cos^2 \theta - \sin^2 \theta = 2 \cos^2 \theta - 1 = 1 - 2 \sin^2 \theta.$$

Since $\tan (\phi + \theta) = \sin (\phi + \theta)/\cos(\phi + \theta)$, we have

$$\tan(\phi + \theta) = \frac{\tan \phi + \tan \theta}{1 - \tan \phi \tan \theta};$$

$$\tan(\phi - \theta) = \frac{\tan \phi - \tan \theta}{1 + \tan \phi \tan \theta}.$$

The double-angle formula becomes

$$\tan 2\theta = \frac{2 \tan \theta}{1 - \tan^2 \theta}.$$

The **half-angle formulas** are derived from the double-angle formulas by writing $\theta/2$ for $\theta$. We obtain

$$\sin^2 \frac{\theta}{2} = \frac{1 - \cos \theta}{2}, \qquad \cos^2 \frac{\theta}{2} = \frac{1 + \cos \theta}{2};$$

$$\tan \frac{\theta}{2} = \frac{1 - \cos \theta}{\sin \theta} = \frac{\sin \theta}{1 + \cos \theta}.$$

The following relations may be verified from the addition formulas:

$$\sin \phi + \sin \theta = 2 \sin \left( \frac{\phi + \theta}{2} \right) \cos \left( \frac{\phi - \theta}{2} \right);$$

$$\sin \phi - \sin \theta = 2 \cos \left( \frac{\phi + \theta}{2} \right) \sin \left( \frac{\phi - \theta}{2} \right);$$

$$\cos \phi + \cos \theta = 2 \cos \left( \frac{\phi + \theta}{2} \right) \cos \left( \frac{\phi - \theta}{2} \right);$$

$$\cos \phi - \cos \theta = - 2 \sin \left( \frac{\phi + \theta}{2} \right) \sin \left( \frac{\phi - \theta}{2} \right).$$

The addition formulas, with one of the angles taken as a special value, yield the following useful relations. Starting with the formula

$$\sin \left( \frac{\pi}{2} - \theta \right) = \sin \frac{\pi}{2} \cos \theta - \cos \frac{\pi}{2} \sin \theta, \qquad \left( \frac{\pi}{2} = 90° \right),$$

and noting that $\sin \dfrac{\pi}{2} = 1$, $\cos \dfrac{\pi}{2} = 0$, we obtain

$$\sin\left(\frac{\pi}{2} - \theta\right) = \cos \theta.$$

In a similar way, we have

$$\cos\left(\frac{\pi}{2} - \theta\right) = \sin \theta;$$

$$\sin\left(\frac{\pi}{2} + \theta\right) = \cos \theta, \qquad \cos\left(\frac{\pi}{2} + \theta\right) = -\sin \theta;$$

$$\cos(\pi - \theta) = -\cos \theta, \qquad \sin(\pi - \theta) = \sin \theta;$$

$$\cos(\pi + \theta) = -\cos \theta, \qquad \sin(\pi + \theta) = -\sin \theta.$$

## 1   PROBLEMS

**1** Prove the Law of Sines.

**2** Prove the Law of Cosines.

**3** Establish the formula: Area $= \frac{1}{2}ab \sin C$, for the area of a triangle.

**4** Given the diagram in Fig. 6 (circle has radius $= 1$). Apply the Law of Cosines to triangle $ABO$ to obtain a formula for $\cos(\phi - \theta)$. Then use the distance formula for the length $|AB|$ to derive

$$\cos(\phi - \theta) = \cos \phi \cos \theta + \sin \phi \sin \theta.$$

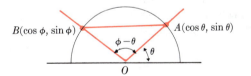

$B(\cos \phi, \sin \phi)$     $A(\cos \theta, \sin \theta)$

$\phi - \theta$   $\theta$

$O$

**FIGURE 6**

**5** Derive formulas for $\sin 3\theta$ and $\cos 3\theta$ in terms of $\sin \theta$ and $\cos \theta$.

**6** Derive formulas for $\sin 4\theta$ and $\cos 4\theta$ in terms of $\sin \theta$ and $\cos \theta$.

**7** Write out the derivation of the formulas for $\sin(\theta/2)$ and $\cos(\theta/2)$ in terms of $\cos \theta$.

**8** Derive formulas for $\sin \frac{1}{4}\theta$ and $\cos \frac{1}{4}\theta$ in terms of $\cos \theta$.

**9** Without using tables, find the values of the following:

$\sin 15°$, $\cos 15°$, $\sin 22\frac{1}{2}°$, $\cos 22\frac{1}{2}°$, $\tan 75°$, $\sec 75°$.

**10** Without using tables, find the values of the following:

$\tan 22\frac{1}{2}°$, $\sin 7\frac{1}{2}°$, $\cos 7\frac{1}{2}°$, $\sin 37\frac{1}{2}°$, $\tan 37\frac{1}{2}°$.

**11** Write formulas in terms of $\sin \theta$ and $\cos \theta$ for

$$\sin\left(\frac{3\pi}{2} - \theta\right), \cos\left(\frac{3\pi}{2} - \theta\right), \sin\left(\frac{3\pi}{2} + \theta\right), \cos\left(\frac{3\pi}{2} + \theta\right),$$

$$\sin(2\pi - \theta), \cos(2\pi - \theta).$$

**12** Since $\pi = 3.14^+$ approximately, let

$$x = 0, \frac{\pi}{6}, \frac{\pi}{4}, \frac{\pi}{3}, \frac{\pi}{2}, \frac{2\pi}{3}, \frac{3\pi}{4}, \frac{5\pi}{6}, \pi, \frac{7\pi}{6}, \text{ etc.,}$$

and plot the curve $y = \sin x$ on a rectangular coordinate system. Extend the graph from $-2\pi$ to $+4\pi$.

**13** Work Problem 12 again for $y = \cos x$.

**14** Work Problem 12 again for $y = \tan x$.

**15** Derive (from the addition formulas) the identity

$$\sin \phi + \sin \theta = 2 \sin\left(\frac{\phi + \theta}{2}\right) \cos\left(\frac{\phi - \theta}{2}\right).$$

**16** Work Problem 15 again for

$$\cos \phi + \cos \theta = 2 \cos\left(\frac{\phi + \theta}{2}\right) \cos\left(\frac{\phi - \theta}{2}\right).$$

**17** In terms of $\sin \theta$ and $\cos \theta$, find

$$\sin\left(\frac{\pi}{6} - \theta\right), \cos\left(\frac{\pi}{3} + \theta\right), \cos\left(\frac{5\pi}{6} - \theta\right).$$

**18** In terms of $\sin \theta$ and $\cos \theta$, find

$$\sin\left(\frac{\pi}{12} - \theta\right), \cos\left(\frac{\pi}{3} + \frac{\pi}{4} - \theta\right), \sin\left(\frac{\pi}{4} - \frac{\pi}{24} - \theta\right).$$

**19** Sketch, in rectangular coordinates, the graph of $y = \cot x$.

**20** Sketch, in rectangular coordinates, the graph of $y = \sec x$.

**21** Sketch, in rectangular coordinates, the graph of $y = \csc x$.

_____ **2**

## GRAPHS OF TRIGONOMETRIC FUNCTIONS

In the study of trigonometry we occasionally have use for a short table of the values of trigonometric functions in terms of radian measure. However, the most common measure used for angles is the degree, and most tables for the sine, cosine, etc., are in degrees, minutes, and perhaps seconds.

In calculus, however, the situation is reversed: The *natural unit* for the study of trigonometric and related functions is the *radian*. Measurement of angles in degrees is awkward and seldom used in calculus, as is apparent in Chapter 3. In this section we use radians to graph trigonometric functions, and in the process we shall become familiar with many of the properties of these functions.

The graph of the function $y = \sin x$ in a rectangular coordinate system is made by drawing up a table of values and sketching the curve, as shown in Fig. 7. It is necessary to make a table of values only between 0 and $2\pi$. From the relation $\sin(2\pi + x) = \sin x$, we know that the curve then repeats indefinitely to the left and right.

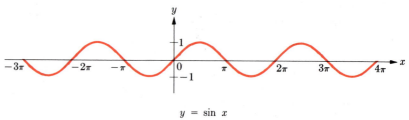

$$y = \sin x$$

**FIGURE 7**

The graph of $y = \cos x$ is found in the same way. A table of values between 0 and $2\pi$ and a knowledge of the periodic nature of the function yield the graph shown in Fig. 8. Note that the curves in Figs. 7 and 8 are identical except that the cosine function is shifted to the left by an amount $\pi/2$. This fact is not surprising if we recall the relation $\sin[(\pi/2) + x] = \cos x$.

| $x$ | 0 | $\pi/6$ | $\pi/4$ | $\pi/3$ | $\pi/2$ | $2\pi/3$ | $3\pi/4$ | $5\pi/6$ | $\pi$ |
|---|---|---|---|---|---|---|---|---|---|
| $y$ | 0 | $1/2$ | $\sqrt{2}/2$ | $\sqrt{3}/2$ | 1 | $\sqrt{3}/2$ | $\sqrt{2}/2$ | $1/2$ | 0 |

| $x$ | $7\pi/6$ | $5\pi/4$ | $4\pi/3$ | $3\pi/2$ | $5\pi/3$ | $7\pi/4$ | $11\pi/6$ | $2\pi$ |
|---|---|---|---|---|---|---|---|---|
| $y$ | $-1/2$ | $-\sqrt{2}/2$ | $-\sqrt{3}/2$ | $-1$ | $-\sqrt{3}/2$ | $-\sqrt{2}/2$ | $-1/2$ | 0 |

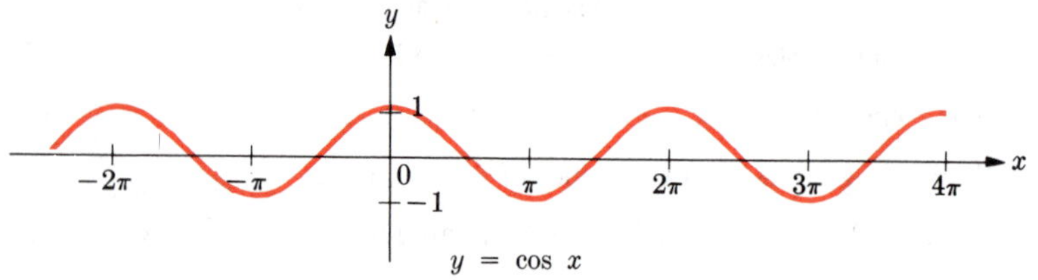

$$y = \cos x$$

**FIGURE 8**

A function $f$ is called **periodic** if there is a number $l$ such that

$$f(x+l) = f(x)$$

for all $x$ (in the domain of $f$), in which case the number $l$ is called a **period** of the function $f$. For the sine function, $2\pi$ is a period, but so are $4\pi$, $6\pi$, $8\pi$, etc., since $\sin(x+4\pi) = \sin x$, $\sin(x+6\pi) = \sin x$, and so on. For periodic functions there is always a smallest number $l$ for which $f(x+l) = f(x)$. This number is called the **least period** of $f$. The least period of $\sin x$ is $2\pi$; the same is true for $\cos x$. The function $\tan x$ has period $2\pi$, but this is not the least period. A graph of $y = \tan x$ is given in Fig. 9, which shows clearly that $\pi$ is a period of $\tan x$. This is the least period for the tangent function.

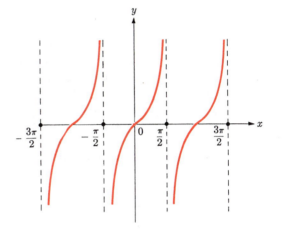

**FIGURE 9**

The function $y = \sin kx$ repeats when $kx$ changes by an amount $2\pi$. If $k$ is a constant, then $\sin kx$ repeats when $x$ changes by an amount $2\pi/k$. The least period of the function $\sin kx$ is $2\pi/k$. The same is true for $y = \cos kx$. For example, $y = \cos 5x$ has as its least period $2\pi/5$.

A periodic function $f$, which never becomes larger than some number $b$ in value and never becomes smaller than some number $c$, is said to have **amplitude** $\frac{1}{2}(b-c)$, provided that there are points at which $f$ actually takes on the values $b$ and $c$. For example, the function $y = 3\sin 2x$ is never larger than 3 or smaller than $-3$. The amplitude is 3. Note that $y = 3$ when $x = \pi/4$. The tangent function has no amplitude.

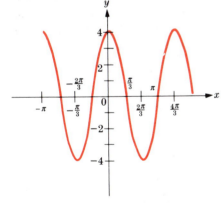

**FIGURE 10**

**EXAMPLE 1**    Find the amplitude and period of $y = 4\cos(3x/2)$. Sketch the graph.

**Solution**    The amplitude is 4; the period is $2\pi/(\frac{3}{2}) = 4\pi/3$. The graph is sketched in Fig. 10.    □

**EXAMPLE 2**    Find the amplitude and period of $y = 2.5\sin[2x + (\pi/3)]$ and sketch its graph.

**Solution**    This function is similar to $2.5\sin 2x$, except that it is shifted horizontally. We write $y = 2.5\sin 2[x + (\pi/6)]$. When $x = -\pi/6$, $y = 0$, and we see that the "zero point" is moved to the left an amount $\pi/6$. The period is $2\pi/2 = \pi$, and the amplitude is 2.5. With this information it is now easy to sketch the curve, as shown in Fig. 11. ☐

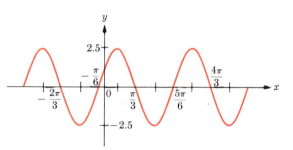

**FIGURE 11**

Combinations of trigonometric functions may be sketched quickly by a method known as **addition of ordinates**. If the sum of two functions is to be sketched, we start by sketching each function separately. Then, at a particular value of $x$, the $y$ values of each of the functions are added to give the result. Incidentally, this method is quite general and may be used for the sum or difference of any functions. An example illustrates the technique.

**EXAMPLE 3**  Sketch the graph of

$$y = 2 \sin x + \sin 2x.$$

**Solution**  The function $2 \sin x$ has period $2\pi$ and amplitude 2, while $\sin 2x$ has period $\pi$ and amplitude 1. We sketch the graph of each of these functions on the same coordinate system from 0 to $2\pi$, as shown in Fig. 12. At convenient values of $x$, the $y$ values of the two graphs are added and the results plotted either on the same graph or on a similar one, as shown in Fig. 13. ☐

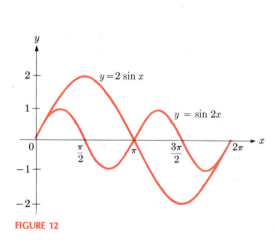

**FIGURE 12**

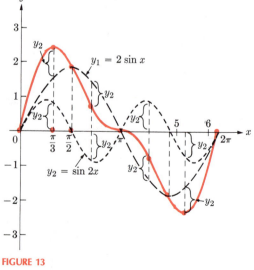

**FIGURE 13**

# 2   PROBLEMS

In Problems 1 through 16, find the amplitudes and the periods of the functions defined by the given expressions and sketch their graphs.

**1**   $2 \sin \frac{1}{2} x$

**2**   $3 \cos \frac{3}{2} x$

**3**   $4 \cos \frac{1}{2} \pi x$

**4**   $2.5 \sin \frac{1}{3} \pi x$

**5**   $3 \sin \frac{2}{3} x$

**6**   $5 \sin \frac{1}{2} \pi x$

**7**   $2 \cos \left( 2x + \dfrac{\pi}{3} \right)$

**8**   $3 \sin \left( \frac{1}{2} \pi x + \frac{1}{4} \pi \right)$

**9**   $-2 \sin 2x$

**10**   $-3 \cos \frac{2}{3} \pi x$

**11**   $4 \sin \left( 2x - \dfrac{\pi}{3} \right)$

**12**   $3 \cos \left( 2x - \dfrac{\pi}{3} \right)$

**13**   $-3 \sin \left( \frac{1}{2} x + 7 \right)$

**14**   $\frac{1}{3} \sin \left( 2x + \frac{1}{2} \pi \right)$

**15**   $1 + \sin 2x$

**16**   $-3 + 2 \cos (x - 1)$

In Problems 17 through 24, find the periods and amplitudes (if any), and sketch the graphs.

**17**   $\tan 2x$

**18**   $\cot 3x$

**19**   $\sin x + \cos x$

**20**   $\sec 2x$

**21**   $\csc \frac{1}{2} x$

**22**   $2 \tan (3x - 1)$

**23**   $2 \cos 2x - 2\sqrt{3} \sin 2x$

**24**   $-4 \cot (2x - \pi)$

In Problems 25 through 32, sketch the graphs of the curves, using the method of addition of ordinates.

**25**   $y = 2 \sin x + \cos 2x$

**26**   $y = 3 \cos x - 2 \sin 2x$

**27**   $y = 3 \sin \frac{1}{2} x - \frac{3}{2} \sin x$

**28**   $y = 2 \sin \frac{1}{2} \pi x - 3 \cos \pi x$

**29**   $y = 3 \sin 2x - 2 \cos 3x$

**30**   $y = 2 \cos 3x - 3 \sin 4x$

**31**   $y = x + \sin x$

**32**   $y = 1 - x + \cos x$

**33**   Sketch the graph of $y = 2 \sin^2 x$. Show that it has period $\pi$. How does it compare with the graph of $\cos 2x$?

# CONIC SECTIONS. QUADRIC SURFACES

## 1

### THE CIRCLE (continued from Chapter 11, Section 3)

We consider the general equation

$$x^2 + y^2 + Dx + Ey + F = 0$$

where $D$, $E$, $F$ are any numbers. We perform the process of completing the square and obtain

$$x^2 + Dx + \frac{D^2}{4} + y^2 + Ey + \frac{E^2}{4} = -F + \frac{D^2}{4} + \frac{E^2}{4},$$

or

$$\left(x + \frac{D}{2}\right)^2 + \left(y + \frac{E}{2}\right)^2 = \frac{1}{4}(D^2 + E^2 - 4F).$$

*We conclude that such an equation represents a circle with center at $(-D/2, -E/2)$ and radius $\frac{1}{2}\sqrt{D^2 + E^2 - 4F}$.* But there is a problem: If the quantity under the radical is negative, there is no circle and consequently no graph. If it is zero, the circle reduces to a point. *The equation does represent a circle of radius $\frac{1}{2}\sqrt{D^2 + E^2 - 4F}$ if $D^2 + E^2 > 4F$, and only then.*

A circle is determined if the numbers $D$, $E$, and $F$ are given. This means that in general three conditions determine a circle. We know that there is exactly one circle through three points (not all on a straight line). The following example shows how the equation is found when three points are prescribed.

**EXAMPLE 1**  Find the equation of the circle passing through the points $(-3, 4)$, $(4, 5)$, and

(1, −4). Locate the center and determine the radius.

**Solution**    If the circle has the equation

$$x^2 + y^2 + Dx + Ey + F = 0,$$

then each point on the circle must satisfy this equation. By substituting, we get

$$(-3, 4): -3D + 4E + F = -25;$$

$$(4, 5): \quad 4D + 5E + F = -41;$$

$$(1, -4): \quad D - 4E + F = -17.$$

We subtract the second equation from the first and the third from the second and eliminate $F$, to obtain

$$7D + E = -16, \qquad 3D + 9E = -24.$$

We solve these simultaneously and get $D = -2$, $E = -2$. Substituting back into any one of the original three equations gives $F = -23$. The equation of the circle is

$$x^2 + y^2 - 2x - 2y - 23 = 0.$$

The center and radius are obtained by completing the square. We write

$$(x^2 - 2x \quad) + (y^2 - 2y \quad) = 23 \qquad \text{and} \qquad (x-1)^2 + (y-1)^2 = 25.$$

The center is at (1, 1); the radius is 5. ☐

The problem of Example 1 could be solved in another way: (1) Find the perpendicular bisector of the line segment joining the first two points. (2) Do the same with the first and third (or second and third) points. (3) The point of intersection of the two perpendicular bisectors is the center of the circle. (4) The formula for the distance from the center of the circle to any of the original three points gives the radius (5). Knowing the center and radius, we write the equation of the circle. (See Fig. 1.)

The equation

$$x^2 + (y - k)^2 = 25$$

represents a **family of circles**. All circles of this family have radius 5 and, furthermore, the centers are always on the $y$ axis. Since a line is determined by two conditions while a circle is fixed by three conditions, families of circles are more extensive than families of lines, from some points of view. For example, all circles of radius 3 comprise a "larger" class than all lines of slope 7. This has meaning in the following sense: Consider any point in the plane; only one member of the above family of lines passes through this point; however, an infinite number of circles of radius 3 pass through this same point. The family of lines of slope 7 is called a *one-parameter family*. The equation $y = 7x + k$ has *one* constant $k$ which can take on any real value. The equation

$$(x - h)^2 + (y - k)^2 = 9$$

is a *two-parameter family* of circles. The numbers $h$ and $k$ may independently take on any numerical value.

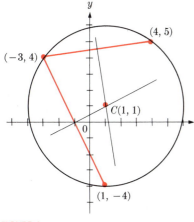

**FIGURE 1**

EXAMPLE 2    Find the equation of the family of circles with centers on the line $y = x$ and radius 7.

Solution    The center $C(h, k)$ must have $h = k$ if the center is on the line $y = x$. The required equation is

$$(x - h)^2 + (y - h)^2 = 49.$$

This is a one-parameter family of circles.    □

EXAMPLE 3    Find the equation of the family of circles with center at $(3, -2)$.

Solution    The equation is

$$(x - 3)^2 + (y + 2)^2 = r^2$$

This is a one-parameter family of concentric circles.    □

The two circles

$$x^2 + y^2 + Dx + Ey + F = 0,$$

$$x^2 + y^2 + D'x + E'y + F' = 0,$$

may or may not intersect. If we subtract one of these equations from the other, we obtain

$$(D - D')x + (E - E')y + (F - F') = 0,$$

which is the equation of a straight line unless $D' = D$ and $E' = E$. This straight line has slope $-(D - D')/(E - E')$ (if $E \neq E'$). The circles have centers at $(-D/2, -E/2)$ and $(-D'/2, -E'/2)$, respectively. The line joining the centers has slope

$$\frac{(-E'/2) + (E/2)}{(-D'/2) + (D/2)} = \frac{E - E'}{D - D'},$$

which is the negative reciprocal of $-(D - D')/(E - E')$. *The equation of the line obtained by subtraction of the equations of two circles* is called the **radical axis**, and it *is perpendicular to the line joining the centers of the circles.* Furthermore, if the circles intersect, the points of intersection satisfy *both* equations of circles. These points must therefore be on the radical axis.

EXAMPLE 4    Find the points of intersection of the circles:

$$x^2 + y^2 - 2x - 4y - 4 = 0, \qquad x^2 + y^2 - 6x - 2y - 8 = 0.$$

Solution    By subtraction we get

$$4x - 2y + 4 = 0,$$

and we solve this equation simultaneously with the equation of either circle. The substitution gives

$$5x^2 - 2x - 8 = 0, \qquad \text{or} \qquad x = \tfrac{1}{5} \pm \tfrac{1}{5}\sqrt{41}.$$

The points of intersection are

$$(\tfrac{1}{5} + \tfrac{1}{5}\sqrt{41}, \ \tfrac{12}{5} + \tfrac{2}{5}\sqrt{41}) \qquad \text{and} \qquad (\tfrac{1}{5} - \tfrac{1}{5}\sqrt{41}, \ \tfrac{12}{5} - \tfrac{2}{5}\sqrt{41}).$$    □

# 1　PROBLEMS

**1** Show that when $D = E$ the graph of $x^2 + y^2 + Dx + Ey + F = 0$ is always a circle if $F < D^2/2$. Note that if $F < 0$ the graph is always a circle, regardless of the sizes of $D$ and $E$.

In Problems 2 through 5, in each case find the equation of the circle through the given points by the method used in Example 1. Find the center and radius.

**2** $(-5, 2), (-3, 4), (1, 2)$　　**3** $(-3, 1), (-2, 2), (6, -2)$

**4** $(7, -2), (3, -4), (0, -3)$　　**5** $(1, -1), (0, 1), (-3, -3)$

In Problems 6 and 7, find the equations of the circles by the method described after Example 1.

**6** $(-2, -4), (2, -6), (4, 2)$　　**7** $(-2, -1), (0, 2), (4, 1)$

**8** Determine whether or not the points $(2, 0), (0, 4), (2, 2)$, and $(1, 1)$ lie on a circle.

**9** Find the equation of the circle or circles with center on the $y$ axis and passing through $(-2, 3)$ and $(4, 1)$.

**10** Find the equation of the circle with center on the line $x + y - 1 = 0$ and passing through the points $(0, 5)$ and $(2, 1)$.

In Problems 11 through 16, in each case write the equation of the family of circles with the given properties.

**11** center on the $y$ axis, radius 4

**12** center on the line $x + y - 6 = 0$, radius 1

**13** center on the $x$ axis, tangent to the $y$ axis

**14** center on the line $y = x$, tangent to both coordinate axes

**15** center on the line $y = 3x$, tangent to the line $x + 3y + 4 = 0$

**16** center on the line $x + 2y - 7 = 0$ and tangent to the line $2x + 3y - 1 = 0$

**17** Give a reasonable definition for the statement that two circles intersect at right angles. Do the circles $x^2 + y^2 - 6y = 0$ and $x^2 + y^2 - 6x = 0$ satisfy your definition at their intersection points?

In Problems 18 through 21, find the points of intersection of the two circles, if they intersect. If they do not intersect, draw a graph showing the location of the radical axis.

**18** $x^2 + y^2 + 6x - 9y = 0, \ x^2 + y^2 - 2x + 3y - 7 = 0$

**19** $x^2 + y^2 + x - 3y + 1 = 0, \ x^2 + y^2 - 4x + 2y - 3 = 0$

**20** $x^2 + y^2 + 2x + 3y + 1 = 0, \ x^2 + y^2 - 8y - 7 = 0$

**21** $2x^2 + 2y^2 + 4x - 2y - 1 = 0, \ 3x^2 + 3y^2 + 6x - 9y + 2 = 0$

**22** Given the circles $x^2 + y^2 + 2x - 4 = 0$ and $x^2 + y^2 - x - y - 2 = 0$. What is the geometric property of the family of circles

$$(x^2 + y^2 - x - y - 2) + k(x^2 + y^2 + 2x - 4) = 0?$$

Sketch a few members of this family.

**23** Find the graph of all points $P(x, y)$ such that the distance from $(3, 2)$ is always twice the distance from $(4, 1)$. Describe the graph.

**24** Find the graph of all points $P(x, y)$ such that the distance from $(1, -4)$ is always $\frac{2}{3}$ the distance from $(2, 1)$. Describe the graph.

**25** Given the three circles

$$x^2 + y^2 - x - y = 0,$$
$$x^2 + y^2 + 2x - y - 3 = 0,$$
$$x^2 + y^2 + x - 2y - 4 = 0,$$

show that the three radical axes obtained from each pair of circles are concurrent (i.e., the three lines meet in one point).

**26** Work Problem 25 again for the circles

$$x^2 + y^2 + Dx + Ey + F = 0,$$
$$x^2 + y^2 + D'x + E'y + F' = 0,$$
$$x^2 + y^2 + D''x + E''y + F'' = 0.$$

What conditions are necessary for the result? (*Hint*: Is there a pair of circles which do not have a radical axis?)

---

# 2

## THE PARABOLA (continued from Chapter 11, Section 2)

We consider a parabola and a point $(x_1, y_1)$ on it. We show how to derive a formula which gives the equation of the line tangent to the parabola at $(x_1, y_1)$. If, for example, the parabola has the equation

$$y^2 = 2px,$$

then the slope at any point is given (by implicit differentiation) as

$$2y\frac{dy}{dx} = 2p \qquad \text{and} \qquad \frac{dy}{dx} = \frac{p}{y}.$$

The equation of any line through $(x_1, y_1)$ is

$$y - y_1 = m(x - x_1).$$

If the slope is $p/y_1$ the equation becomes

$$y - y_1 = \frac{p}{y_1}(x - x_1).$$

This formula can be simplified. Since $(x_1, y_1)$ is on the parabola, we know that $y_1^2 = 2px_1$. Therefore the above equation may be written

$$y_1 y - y_1^2 = px - px_1$$

or

$$y_1 y - 2px_1 = px - px_1 \qquad \Leftrightarrow \qquad y_1 y = p(x + x_1).$$

**EXAMPLE 1**   Find the equation of the line tangent to $y^2 = 20x$ at the point $(3, 2\sqrt{15})$.

**Solution**   In this case $x_1 = 3$, $y_1 = 2\sqrt{15}$, $p = 10$, and the equation is

$$2\sqrt{15}\,y = 10(x + 3). \qquad \square$$

A somewhat more difficult problem is that of finding the line tangent to a parabola when the given point $(x_1, y_1)$ is *not* on the parabola. There will be either two solutions or no solution to this problem, as Fig. 2 shows. Let the parabola have equation $x^2 = 2py$, and suppose that $y - y_1 = m(x - x_1)$ is the equation of the desired line. The slope of the parabola at any point is

$$\frac{dy}{dx} = \frac{x}{p}.$$

Let $(x_0, y_0)$ be a point of tangency. Since this point is on both the parabola and the line, we have $x_0^2 = 2py_0$ and $y_0 - y_1 = (x_0/p)(x_0 - x_1)$. We have two equations in the unknowns $x_0$ and $y_0$. Once these are solved simultaneously the problem is done, since the value $x_0/p$ may be substituted in the equation of the line. The following example illustrates the method in a particular case.

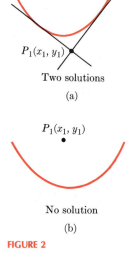

$P_1(x_1, y_1)$

Two solutions

(a)

$P_1(x_1, y_1)$

No solution

(b)

**FIGURE 2**

**EXAMPLE 2**   Find the equations of the lines which are tangent to the parabola $y^2 - y - 2x + 4 = 0$ and which pass through the point $(-5, 2)$.

**Solution**   Any line through $(-5, 2)$ has equation

$$y - 2 = m(x + 5)$$

and the slope of the curve at any point on the parabola is given by

$$2y\frac{dy}{dx} - \frac{dy}{dx} - 2 = 0 \qquad \text{or} \qquad \frac{dy}{dx} = \frac{2}{2y - 1}.$$

If $(x_0, y_0)$ is the point of tangency, then $m = 2/(2y_0 - 1)$. We have the equations

$$y_0^2 - y_0 - 2x_0 + 4 = 0 \qquad \text{and} \qquad y_0 - 2 = \frac{2}{2y_0 - 1}(x_0 + 5).$$

These are solved simultaneously by eliminating $x_0$, and the two solutions are $x_0 = 17$, $y_0 = 6$ and $x_0 = 5$, $y_0 = -2$; corresponding slopes are $m_0 = \frac{2}{11}$, $-\frac{2}{5}$. The desired lines have equations

$$y - 2 = \frac{2}{11}(x + 5) \qquad \text{and} \qquad y - 2 = -\frac{2}{5}(x + 5). \qquad \square$$

If there were no solution, the attempt to solve the quadratic equation for $y_0$ would give complex numbers for the roots. That there is no solution may also be determined in another way. As we know, a parabola such as $x^2 = 8y$ divides the plane into two regions; in one of them $x^2 > 8y$ and in the other $x^2 < 8y$. Any point $(x_1, y_1)$ such that $x_1^2 > 8y_1$ is a point such that two tangents to the parabola may be constructed (Fig. 3). If $x_1^2 < 8y_1$, no tangents can be drawn. More generally, if the equation of the parabola is

$$(x - h)^2 = 2p(y - k),$$

then two tangents may be drawn if $(x_1 - h)^2 > 2p(y_1 - k)$ and none can be drawn if $(x_1 - h)^2 < 2p(y - k)$.

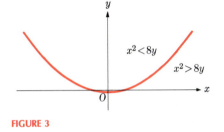

$x^2 < 8y$

$x^2 > 8y$

FIGURE 3

EXAMPLE 3    Given the parabola $-2x^2 + 3y - 2x + 1 = 0$ and the points $P(1, -2)$, $Q(2, 9)$, decide whether or not tangents to the parabola may be drawn from $P$ and $Q$.

Solution    First complete the square, getting

$$(x + \tfrac{1}{2})^2 = \tfrac{3}{2}(y + \tfrac{1}{2}).$$

Now substitute the coordinates of $P(1, -2)$ in the equation. This yields

$$(\tfrac{3}{2})^2 > -\tfrac{9}{4},$$

so that two tangents may be drawn. For the point $Q(2, 9)$ the result is

$$\tfrac{25}{4} < \tfrac{57}{4},$$

so that no tangents can be constructed.    $\square$

The parabola has many interesting geometrical properties that make it suitable for practical applications. Probably the most familiar application is the use of parabolic shapes in headlights of cars, flashlights, and so forth. If a parabola is revolved about its axis, the surface generated is called a **paraboloid**. Such surfaces are used in headlights, optical and radio telescopes, radar, etc., because of the following geometric property: If a source of light (or sound or other type of wave) is placed at the focus of a parabola, and if the parabola is a reflecting surface, then the wave will bounce back in a line parallel to the axis of the parabola (Fig. 4). This creates a parallel beam without dispersion. (In actual practice there will be some dispersion, since the source of light must occupy more than one point.) Clearly the converse is also true. If a series of incoming waves is parallel to the axis of the reflecting paraboloid, the resulting signal will be concentrated at the focus (where the receiving equipment is therefore located).

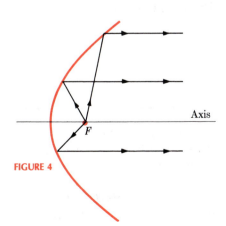

Axis

$F$

FIGURE 4

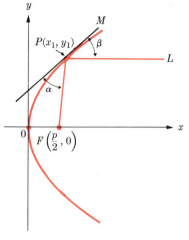

**FIGURE 5**

We shall establish the reflecting property for the parabola $y^2 = 2px$. Let $P(x_1, y_1)$ be a point (not the origin) on the parabola, and draw the line segment from $P$ to the focus (Fig. 5). Now construct a line $L$ through $P$ parallel to the $x$ axis and the line $M$ tangent to the parabola at $P$. The law for the reflecting property of the parabola states that the angle formed by $M$ and $PF$ ($\alpha$ in the figure) must be equal to the angle formed by $L$ and $M$ ($\beta$ in the figure). The slope of the line $M$ is $\tan \beta$, which is also the slope of the parabola at $(x_1, y_1)$. The result is

$$2y\frac{dy}{dx} = 2p \qquad \text{and} \qquad \tan \beta = \frac{p}{y_1}.$$

We now find $\tan \alpha$ from the formula for the angle between two lines. The slope of $PF$ is

$$\frac{y_1 - 0}{x_1 - \dfrac{p}{2}}.$$

Therefore

$$\tan \alpha = \frac{\dfrac{y_1}{x_1 - \dfrac{p}{2}} - \dfrac{p}{y_1}}{1 + \dfrac{y_1}{x_1 - \dfrac{p}{2}}\dfrac{p}{y_1}} \quad \Leftrightarrow \quad \tan \alpha = \frac{y_1^2 - px_1 + \frac{1}{2}p^2}{x_1 y_1 - \frac{1}{2}py_1 + py_1}.$$

Since $y_1^2 = 2px_1$, this simplifies to

$$\tan \alpha = \frac{p}{y_1},$$

and the result is established.

The main cables of suspension bridges are parabolic in shape, since it can be shown that if the total weight of a bridge is distributed uniformly along its length, a cable in the shape of a parabola bears the load evenly. It is a remarkable fact that no other shape will perform this task.

## 2   PROBLEMS

In Problems 1 through 3, find the equation of the line tangent to the parabola at the given point $P$ on the parabola.

**1** $y^2 = -3x$, $P(-3, -3)$

**2** $x^2 - 2y + 3x - 4 = 0$, $P(-2, -3)$

**3** $-x^2 - 2y + 2x + 4 = 0$, $P(4, -2)$

**4** Find the equation of the line tangent to the parabola $y^2 - 3y + 4x - 8 = 0$ at the point on the parabola at which $x = 1$.

In Problems 5 through 14, first decide whether lines tangent to the given parabola which pass through the given point $P$ can be constructed. When they can, determine the equations of the lines.

**5** $y^2 = 8x$, $P(-3, 1)$          **6** $y^2 = 8x$, $P(3, 1)$

**7** $y^2 = -4x$, $P(1, -5)$         **8** $x^2 = -4y$, $P(4, 1)$

**9** $x^2 = 12y$, $P(-1, -3)$

**10** $x^2 - 4x + 2y - 1 = 0$, $P(-1, 3)$

11  $2y^2 - 3y + x - 4 = 0$, $P(1, -1)$

12  $x^2 - 4x + 2y - 1 = 0$, $P(1, -8)$

13  $x + 2y - 2x^2 + 1 = 0$, $P(1, 3)$

14  $x + 2y - 2x^2 + 1 = 0$, $P(1, -3)$

15  For the parabola $x^2 = 2py$, establish the "optical property." That is, if $(x_1, y_1)$ is a point on the parabola and $M$ is the line tangent at $(x_1, y_1)$, show that the angle formed between $M$ and a line from $(x_1, y_1)$ through the focus is equal to the angle $M$ forms with a line through $(x_1, y_1)$ parallel to the axis.

16  A line through the focus of the parabola $x^2 = -12y$ intersects the parabola at the point $(5, -\frac{25}{12})$. Find the other point of intersection of this line with the parabola.

17  An arch has the shape of a parabola with vertex at the top and axis vertical. If it is 45 meters wide at the base and 18 meters high in the center, how wide is it 10 meters above the base?

18  The line through the focus of a parabola and parallel to the directrix is called the *line of the latus rectum*. The portion of this line cut off by the parabola is called the **latus rectum**. Show that for the parabola $y^2 = 2px$ the length of the latus rectum is $2p$. Is this true for any parabola where the distance from the focus to the directrix is $p$ units?

19  Find the graph of all points whose distances from $(4, 0)$ are 1 unit less than their distances from the line $x = -6$. Describe the curve.

## 3

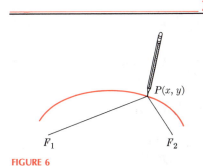

FIGURE 6

# THE ELLIPSE (continued from Chapter 11, Section 3)

A mechanical construction of an ellipse can be made directly from the facts given in the definition, as shown in Fig. 6. Select a piece of string of length $2a$. Fasten its ends with thumbtacks at the foci, insert a pencil so as to draw the string taut, and trace out an arc. The curve will be an ellipse.

The ellipse has many interesting geometrical properties. The paths of the planets about the sun and those of the man-made satellites about the earth are approximately elliptical; whispering galleries are ellipsoidal (an ellipse revolved about the major axis) in shape. If a signal (in the form of a light, sound, or other type of wave) has its source at one focus of an ellipse, *all* reflected waves will pass through the other focus.

We have seen that a parabola has a focus and a directrix. The ellipse has two foci and also two directrices. If the ellipse has foci at $(\pm c, 0)$ and vertices at $(\pm a, 0)$, the **directrices** *are the lines**

$$x = \frac{a^2}{c} = \frac{a}{e} \qquad \text{and} \qquad x = -\frac{a^2}{c} = -\frac{a}{e}.$$

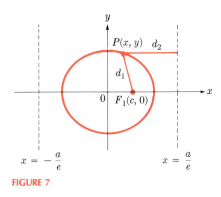

FIGURE 7

To draw an analogy with the parabola, we can show that such an ellipse is the graph of all points whose distances from a fixed point $(F_1)$ are equal to $e$ $(0 < e < 1)$ times their distances from a fixed line $(x = a/e)$. As Fig. 7 shows, the conditions of the graph are

$$d_1 = ed_2 \quad \Leftrightarrow \quad \sqrt{(x-c)^2 + y^2} = e\left|x - \frac{a}{e}\right|.$$

If we square both sides and simplify, we obtain $(x^2/a^2) + (y^2/b^2) = 1$, which we know to be the equation of an ellipse. The same situation holds true if the focus $F_2(-c, 0)$ and the directrix $x = -a/e$ are used.

*The letter $e$ is used to denote the eccentricity of an ellipse and should not be confused with the number $e = 2.71828...$, which is the base of natural logarithms.

**EXAMPLE 1**   Find the equation of the ellipse with foci at $(\pm 5, 0)$ and with the line $x = 36/5$ as one directrix.

**Solution**   We have $c = 5$, $a/e = 36/5$. Since $e = c/a$, we obtain $a^2/c = 36/5$, or $a = 6$. Then $b = \sqrt{36 - 25} = \sqrt{11}$, and the equation is

$$\frac{x^2}{36} + \frac{y^2}{11} = 1. \qquad \square$$

The above example shows that if the foci and one directrix are specified, then the equation of the ellipse is determined. The next example shows that if the axes of the ellipse are specified, then two additional points on the ellipse may be used to find its equation.

**EXAMPLE 2**   Find the equation of the ellipse with axes along the $x$ axis and $y$ axis which passes through the points $P(6, 2)$ and $Q(-4, 3)$.

**Solution**   Let the equation be

$$\frac{x^2}{A^2} + \frac{y^2}{B^2} = 1;$$

we don't know whether $A$ is larger than $B$ or $B$ is larger than $A$. Since the points $P$ and $Q$ are on the ellipse, we have

$$36\left(\frac{1}{A^2}\right) + 4\left(\frac{1}{B^2}\right) = 1, \qquad 16\left(\frac{1}{A^2}\right) + 9\left(\frac{1}{B^2}\right) = 1,$$

which gives us two equations in the unknowns $1/A^2$, $1/B^2$. Solving simultaneously, we obtain

$$\frac{1}{A^2} = \frac{1}{52}, \qquad \frac{1}{B^2} = \frac{1}{13},$$

and the resulting ellipse is

$$\frac{x^2}{52} + \frac{y^2}{13} = 1.$$

The major axis is along the $x$ axis. $\qquad \square$

The reflected-wave property of ellipses is exhibited by the following procedure. Let $P(x_1, y_1)$ be a point on the ellipse (not a vertex), as shown in Fig. 8. Construct a tangent line $M$ at this point. Then the angle made by $M$ and the line $PF_1$ ($\alpha$ in the figure) is equal to the angle made by $M$ and the line $PF_2$ ($\beta$ in the figure). This result may be established by setting up the formulas for $\tan \alpha$ and $\tan \beta$ as the angles between two lines and showing them to be equal.

An ellipse divides the plane into two regions, one inside and one outside the curve itself. Every point inside the ellipse $(x^2/a^2) + (y^2/b^2) = 1$ satisfies the inequality

$$\frac{x^2}{a^2} + \frac{y^2}{b^2} < 1.$$

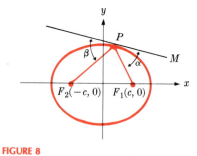

**FIGURE 8**

From any point outside the ellipse two tangents may be drawn, and clearly no tangents to the ellipse can be constructed which pass through a point inside the ellipse itself.

**EXAMPLE 3**  Given the ellipse $(x^2/25) + (y^2/9) = 1$, and the points $P(1, 2)$, $Q(0, 6)$, and $R(3, 2)$. For each point, decide whether tangents to the ellipse passing through the point can be constructed. If they can, find the equations of the tangent lines.

**Solution**  For the point $P$ we have

$$\frac{1^2}{25} + \frac{2^2}{9} < 1,$$

and no tangents exist. Similarly, for the point $R$ we have $(9/25) + (4/9) < 1$. The point $Q$ is outside the ellipse, since $(0^2/25) + (6^2/9) > 1$, and we proceed to find the tangent lines through $Q$. Any line through $Q$ has the equation

$$y - 6 = m(x - 0).$$

The slope of the ellipse at any point is, by implicit differentiation,

$$\frac{2x}{25} + \frac{2y}{9}\frac{dy}{dx} = 0 \qquad \text{or} \qquad \frac{dy}{dx} = -\frac{9}{25}\frac{x}{y}.$$

Suppose that $P_1(x_1, y_1)$ is a point where the desired line is tangent to the ellipse. We have $m = -9x_1/25y_1$ and

$$\frac{x_1^2}{25} + \frac{y_1^2}{9} = 1, \qquad y_1 - 6 = -\frac{9}{25}\frac{x_1}{y_1}x_1.$$

We solve these equations simultaneously and obtain

$$(\tfrac{5}{2}\sqrt{3}, \tfrac{3}{2}) \qquad \text{and} \qquad (-\tfrac{5}{2}\sqrt{3}, \tfrac{3}{2})$$

as the points of tangency. The equations of the lines are

$$y - 6 = \frac{9}{25} \cdot \frac{\tfrac{5}{2}\sqrt{3}}{\tfrac{3}{2}}x, \qquad y - 6 = -\frac{9}{25} \cdot \frac{-\tfrac{5}{2}\sqrt{3}}{\tfrac{3}{2}}x,$$

and, when we simplify, we find

$$3\sqrt{3}x + 5y - 30 = 0, \qquad 3\sqrt{3}x - 5y + 30 = 0. \qquad \square$$

## 3   PROBLEMS

In Problems 1 through 6, find the equation of the ellipse satisfying the given conditions.

1  Foci at $(\pm 4, 0)$, directrices $x = \pm 6$

2  Eccentricity $\frac{1}{2}$, directrices $x = \pm 8$

3  Foci at $(0, \pm 4)$, directrices $y = \pm 7$

4  Vertices at $(\pm 6, 0)$, directrices $x = \pm 9$

5  Vertices at $(0, \pm 7)$, directrices $y = \pm 12$

6  Eccentricity $\frac{2}{3}$, directrices $y = \pm 9$

7  Prove the reflected-wave property of an ellipse, as described in this section.

8  Prove that the segment of a tangent to an ellipse between the point of contact and a directrix subtends a right angle at the corresponding focus.

9  A line through a focus parallel to a directrix is called a **latus rectum**. The length of the segment of this line cut off by the ellipse is called the *length of the latus rectum*. Show that the length of the latus rectum of

$$\frac{x^2}{a^2} + \frac{y^2}{b^2} = 1$$

is $2b^2/a$.

10  Prove that the line joining a point $P_0$ of an ellipse to the center and the line through a focus and perpendicular to the tangent at $P_0$ intersects on the corresponding directrix.

11  The tangent to an ellipse at a point $P_0$ meets the tangent at a vertex in a point $Q$. Prove that the line joining the other vertex to $P_0$ is parallel to the line joining the center to $Q$.

12  Given the ellipse $(x^2/4) + (y^2/2) = 1$. Decide whether or not the points $P(1, \frac{1}{2})$, $Q(1, 4)$, and $R(0, 1)$ are outside the ellipse. Construct the tangent lines passing through the points wherever possible.

13  Given the ellipse $(x^2/1) + (y^2/9) = 1$. Rework Problem 12 for the points $P(\frac{1}{2}, 1)$, $Q(2, 3)$, and $R(5, 0)$.

---

**4**

## THE HYPERBOLA (continued from Chapter 11, Section 4)

The hyperbola, like the ellipse, has two directrices. For the hyperbola

$$\frac{x^2}{a^2} - \frac{y^2}{b^2} = 1, \qquad \text{the directrices are the lines} \qquad x = \pm \frac{a}{e},$$

and for the hyperbola

$$\frac{y^2}{a^2} - \frac{x^2}{b^2} = 1, \qquad \text{the directrices are the lines} \qquad y = \pm \frac{a}{e}.$$

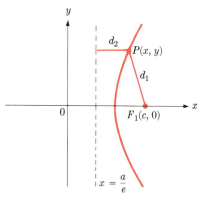

FIGURE 9

A hyperbola has the property that it is the graph of all points $P(x, y)$ whose distances from a fixed point ($F_1$) are equal to $e$ ($e > 1$) times their distances from a fixed line (say $x = a/e$). As Fig. 9 shows, the conditions of the graph are

$$d_1 = ed_2 \qquad \Leftrightarrow \qquad \sqrt{(x-c)^2 + y^2} = e \left| x - \frac{a}{e} \right|.$$

If we square both sides and simplify, we obtain $(x^2/a^2) - (y^2/b^2) = 1$. The same is true if we start with $F_2(-c, 0)$ and the line $x = -a/e$.

EXAMPLE 1   Find the equation of the hyperbola if the foci are at the points $(\pm 6, 0)$ and one directrix is the line $x = 3$.

Solution   We have $c = 6$ and $a/e = 3$. Since $e = c/a$, we may also write $a^2/c = 3$ and $a^2 = 18$, $a = \sqrt{18}$. In addition, $c^2 = a^2 + b^2$, from which we obtain $b^2 = 18$. The equation is

$$\frac{x^2}{18} - \frac{y^2}{18} = 1.$$

In studying the hyperbola

$$\frac{x^2}{a^2} - \frac{y^2}{b^2} = 1 \tag{1}$$

we shall show that the lines

$$y = \frac{b}{a}x \qquad \text{and} \qquad y = -\frac{b}{a}x$$

are of special interest. For this purpose we consider a point $\bar{P}(\bar{x}, \bar{y})$ on the hyperbola (1) and compute the distance from $\bar{P}$ to the line $y = bx/a$. According to the formula for the distance from a point to a line (see Section 1 of Chapter 11), we have for the distance $d$:

$$d = \frac{|b\bar{x} - a\bar{y}|}{\sqrt{a^2 + b^2}}.$$

Suppose we select a point $\bar{P}$ which has both of its coordinates positive, as shown in Fig. 10. Since $\bar{P}$ is a point on the hyperbola, its coordinates satisfy the equation $b^2\bar{x}^2 - a^2\bar{y}^2 = a^2b^2$, which we may write

$$(b\bar{x} - a\bar{y})(b\bar{x} + a\bar{y}) = a^2b^2 \qquad \Leftrightarrow \qquad b\bar{x} - a\bar{y} = a^2b^2/(b\bar{x} + a\bar{y}).$$

Substituting this last relation in the above expression for $d$, we find

$$d = \frac{a^2b^2}{\sqrt{a^2 + b^2}\,|b\bar{x} + a\bar{y}|}.$$

Now $a$ and $b$ are fixed positive numbers and we selected $\bar{x}$, $\bar{y}$ positive. As $P$ moves out along the hyperbola farther and farther away from the vertex, $\bar{x}$ and $\bar{y}$ increase without bound. Therefore, the number $d$ tends to zero as $\bar{x}$ and $\bar{y}$ get larger and larger. That is, the distance between the hyperbola (1) and the line $y = bx/a$ shrinks to zero as $\bar{P}$ tends to infinity. We define the line $y = bx/a$ as an **asymptote of the hyperbola** $(x^2/a^2) - (y^2/b^2) = 1$. Analogously, the line $y = -bx/a$ (see Fig. 10) is called an asymptote of the same hyperbola. We have only to recall the symmetry of the hyperbola to verify this statement.

   Now we can see one of the uses of the central rectangle. *The asymptotes are the lines which contain the diagonals of the central rectangle.* In an analogous way the hyperbola $(y^2/a^2) - (x^2/b^2) = 1$ has the asymptotes

$$y = (a/b)x \qquad \text{and} \qquad y = -(a/b)x.$$

These lines contain the diagonals of the appropriately constructed central rectangle.

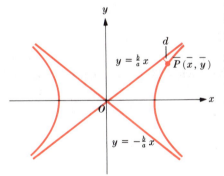

FIGURE 10

EXAMPLE 2   Given the hyperbola $25x^2 - 9y^2 = 225$, find the foci, vertices, eccentricity, directrices, and asymptotes. Sketch the curve.

Solution   When we divide by 225, we obtain $(x^2/9) - (y^2/25) = 1$, and therefore $a = 3$, $b = 5$, $c^2 = 9 + 25$, and $c = \sqrt{34}$. The foci are at $(\pm\sqrt{34}, 0)$, vertices at $(\pm 3, 0)$; the eccentricity is $e = c/a = \sqrt{34}/3$, and the directrices are the lines $x = \pm a/e = \pm 9/\sqrt{34}$. To find the asymptotes we merely write $y = \pm(b/a)x =$

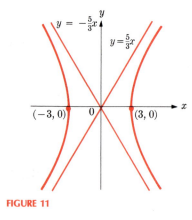

FIGURE 11

$\pm \frac{5}{3}x$. Since the asymptotes are a great help in sketching the curve, we draw them first (Fig. 11): Knowing the vertices and one or two additional points helps us obtain a fairly accurate graph. □

An easy way to remember the equations of the asymptotes is to recognize that if the hyperbola is

$$\frac{x^2}{a^2} - \frac{y^2}{b^2} = 1, \quad \text{the asymptotes are} \quad \frac{x^2}{a^2} - \frac{y^2}{b^2} = 0,$$

and if the hyperbola is

$$\frac{y^2}{a^2} - \frac{x^2}{b^2} = 1, \quad \text{the asymptotes are} \quad \frac{y^2}{a^2} - \frac{x^2}{b^2} = 0.$$

A hyperbola divides the plane into *three* regions. For example, the hyperbola

$$S = \left\{ (x, y) : \frac{x^2}{4} - \frac{y^2}{16} = 1 \right\}$$

divides the plane into the regions

$$S_1 = \left\{ (x, y) : \frac{x^2}{4} - \frac{y^2}{16} < 1 \right\},$$

$$S_2 = \left\{ (x, y) : \frac{x^2}{4} - \frac{y^2}{16} > 1 \text{ and } x > 2 \right\},$$

$$S_3 = \left\{ (x, y) : \frac{x^2}{4} - \frac{y^2}{16} > 1 \text{ and } x < -2 \right\}.$$

For any given point $P$ in the first of these regions, tangents to the hyperbola which pass through this point may be constructed, while no tangents can be constructed which pass through a point (such as $Q$) of the other two regions (Fig. 12).

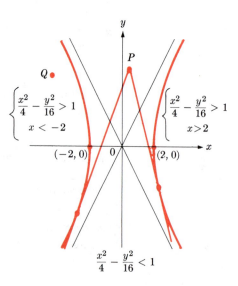

FIGURE 12

**EXAMPLE 3**   Given the hyperbola $(x^2/4) - (y^2/3) = 1$. Decide whether or not tangents to the hyperbola can be constructed which pass through any of the points $P(3, 1)$, $Q(-4, 2)$, $R(1, 3)$. If they can, find the equations of the tangent lines.

**Solution**   For the point $P$ we have $(3^2/4) - (1^2/3) > 1$, and no tangents can be constructed. Similarly, no tangents can be constructed which pass through $Q$, since $((-4)^2/4) - (2^2/3) > 1$. The point $R$ satisfies the inequality $(1^2/4) - (3^2/3) < 1$, and we proceed to find the tangent lines. Any line through $R$ has the equation

$$y - 3 = m(x - 1),$$

and the slope of the hyperbola at any point is

$$\frac{dy}{dx} = \frac{3}{4}\frac{x}{y}.$$

If $(x_0, y_0)$ is a point of tangency, then $m = \frac{3}{4}(x_0/y_0)$, and the two equations

$$\frac{x_0^2}{4} = \frac{y_0^2}{3} + 1, \qquad y_0 - 3 = \frac{3}{4}\frac{x_0}{y_0}(x_0 - 1)$$

hold. We solve these simultaneously and find

$$\left(\frac{-4 + 12\sqrt{5}}{11}, \frac{-12 + 3\sqrt{5}}{11}\right), \qquad \left(\frac{-4 - 12\sqrt{5}}{11}, \frac{-12 - 3\sqrt{5}}{11}\right).$$

The equations of the lines are

$$(1 - 3\sqrt{5})x + (-4 + \sqrt{5})y + 11 = 0,$$

$$(1 + 3\sqrt{5})x - (4 + \sqrt{5})y + 11 = 0. \qquad \square$$

## 4   PROBLEMS

In Problems 1 through 8, find the vertices, foci, eccentricity, equations of the asymptotes, and directrices. Construct the central rectangle and sketch the hyperbola.

1  $4x^2 - 9y^2 = 36$      2  $9x^2 - 4y^2 = 36$

3  $x^2 - 4y^2 = 4$       4  $9x^2 - y^2 = -9$

5  $x^2 - y^2 = 1$        6  $x^2 - y^2 = -1$

7  $25x^2 - 8y^2 = -200$   8  $2x^2 - 3y^2 = 8$

In Problems 9 through 17, find in each case the equation of the hyperbola (or hyperbolas) satisfying the given conditions.

9  Asymptotes $y = \pm 2x$, vertices $(\pm 6, 0)$

10  Asymptotes $y = \pm 2x$, vertices $(0, \pm 4)$

11  Eccentricity $\frac{4}{3}$, directrices $x = \pm 2$

12  Foci at $(\pm 6, 0)$, directrices $x = \pm 4$

13  Asymptotes $3y = \pm 4x$, foci at $(\pm 6, 0)$

14  Directrices $y = \pm 4$, asymptotes $y = \pm \frac{3}{2}x$

15  Asymptotes $2y = \pm 3x$, passing through $(5, 3)$

16  Directrices $x = \pm 1$, passing through $(-4, 6)$, center at origin

17  Directrices $y = \pm 1$, passing through $(1, \sqrt{3})$, center at origin

18  Prove that the lines $y = \pm(a/b)x$ are asymptotes of the hyperbola

$$(y^2/a^2) - (x^2/b^2) = 1.$$

**19** A line through a focus of a hyperbola and parallel to the directrix is called a **latus rectum**. The length of the segment cut off by the hyperbola is called the *length of the latus rectum*. Show that for the hyperbola

$$(x^2/a^2) - (y^2/b^2) = 1$$

the length of the latus rectum is $2b^2/a$.

**20** Find the graph of all points whose distances from the point $(0, \sqrt{41})$ are always $\sqrt{41}/5$ times their distances from the line $y = 25/\sqrt{41}$.

**21** Find the graph of all points whose distances from the point $(8, 0)$ are always 3 times their distances from the line $x = 1$.

**22** Show that an asymptote, a directrix, and the line through the corresponding focus perpendicular to the asymptote pass through a point.

**23** Given the hyperbola $(y^2/4) - (x^2/6) = 1$. Decide whether or not tangents to this hyperbola can be drawn which pass through any of the points $P(0, 8)$, $Q(1, 5)$, $R(4, 0)$, and $S(-8, -9)$. Find the equations of the lines, wherever possible.

-----

## 5

## QUADRIC SURFACES

In the plane any equation of the form

$$Ax^2 + Bxy + Cy^2 + Dx + Ey + F = 0$$

is the equation of a curve. More specifically, we have found that circles, parabolas, ellipses, and hyperbolas, i.e., all conic sections, are represented by such second-degree equations.

In three-space the most general equation of the second degree in $x$, $y$, and $z$ has the form

$$ax^2 + by^2 + cz^2 + dxy + exz + fyz + gx + hy + kz + l = 0, \tag{1}$$

where the quantities $a, b, c, \ldots, l$ are positive or negative numbers or zero. The points in space satisfying such an equation all lie on a surface. Certain special cases, such as spheres and cylinders, were discussed in Chapter 13. Any second-degree equation which does not reduce to a cylinder, plane, line, or point corresponds to a surface which we call **quadric**. Quadric surfaces are classified into six types, and it can be shown that every second-degree equation which does not degenerate into a cylinder, a plane, etc., corresponds to one of these six types. The proof of this result involves the study of translation and rotation of coordinates in three-dimensional space, a topic beyond the scope of this Appendix.

DEFINITIONS  *The* $x$, $y$, **and** $z$ **intercepts** *of a surface are, respectively, the* $x$, $y$, *and* $z$ *coordinates of the points of intersection of the surface with the respective axes. When we are given an equation of a surface, we get the* $x$ *intercept by setting* $y$ *and* $z$ *equal to zero and solving for* $x$. *We proceed analogously for the* $y$ *and* $z$ *intercepts.*

QUADRIC SURFACES

The **traces** of a surface on the coordinate planes are the curves of intersections of the surface with the coordinate planes. When we are given a surface, we obtain the trace on the $xz$ plane by first setting $y$ equal to zero and then considering the resulting equation in $x$ and $z$ as the equation of a curve in the plane, as in plane analytic geometry. A **section of a surface** by a plane is the curve of intersection of the surface with the plane.

**EXAMPLE 1**   Find the $x$, $y$, and $z$ intercepts of the surface

$$3x^2 + 2y^2 + 4z^2 = 12.$$

Describe the traces of this surface. Find the section of this surface by the plane $z = 1$ and by the plane $x = 3$.

**Solution**   We set $y = z = 0$, getting $3x^2 = 12$; the $x$ intercepts are at 2 and $-2$. Similarly, the $y$ intercepts are at $\pm\sqrt{6}$, the $z$ intercepts at $\pm\sqrt{3}$. To find the trace on the $xy$ plane, we set $z = 0$, getting $3x^2 + 2y^2 = 12$. It is convenient to use set notation, and for this trace we write

$$\left\{ (x, y, z) : \frac{x^2}{4} + \frac{y^2}{6} = 1, z = 0 \right\}.$$

We recognize this curve as an ellipse, with major semi-axis $\sqrt{6}$, minor semi-axis 2, foci at $(0, \sqrt{2}, 0), (0, -\sqrt{2}, 0)$. Similarly, the trace on the $xz$ plane is the ellipse

$$\left\{ (x, y, z) : \frac{x^2}{4} + \frac{z^2}{3} = 1, y = 0 \right\},$$

and the trace on the $yz$ plane is the ellipse

$$\left\{ (x, y, z) : \frac{y^2}{6} + \frac{z^2}{3} = 1, x = 0 \right\}.$$

The section of the surface by the plane $z = 1$ is the curve

$$\begin{cases} 3x^2 + 2y^2 + 4 = 12, \\ z = 1 \end{cases} \Leftrightarrow \begin{cases} \dfrac{x^2}{8/3} + \dfrac{y^2}{4} = 1, \\ z = 1 \end{cases}$$

which we recognize as an ellipse. The section by the plane $x = 3$ is the curve

$$\begin{cases} 27 + 2y^2 + 4z^2 = 12, \\ x = 3 \end{cases} \Leftrightarrow \begin{cases} 2y^2 + 4z^2 + 15 = 0, \\ x = 3 \end{cases}.$$

Since the sum of three positive quantities can never be zero, we conclude that the plane $x = 3$ does not intersect the surface. The section is empty.   $\square$

**DEFINITIONS**   *A surface is **symmetric with respect to the** $xy$ **plane** if and only if the point $(x, y, -z)$ lies on the surface whenever $(x, y, z)$ does; it is **symmetric with respect to the** $x$ **axis** if and only if the point $(x, -y, -z)$ is on the graph whenever $(x, y, z)$ is. Similar definitions are easily formulated for symmetry with respect to the remaining coordinate planes and axes.*

The notions of intercepts, traces, and symmetry are useful in the following description of the six types of quadric surfaces.

i) An **ellipsoid** is the graph of an equation of the form

$$\frac{x^2}{A^2} + \frac{y^2}{B^2} + \frac{z^2}{C^2} = 1.$$

The surface is sketched in Fig. 13. The $x$, $y$, $z$ intercepts are the numbers $\pm A$, $\pm B$, $\pm C$, respectively, and the traces on the $xy$, $xz$, and $yz$ planes are, respectively, the ellipses

$$\frac{x^2}{A^2} + \frac{y^2}{B^2} = 1, \qquad \frac{x^2}{A^2} + \frac{z^2}{C^2} = 1, \qquad \frac{y^2}{B^2} + \frac{z^2}{C^2} = 1.$$

Sections made by the planes $y = k$ ($k$ a constant) are the similar ellipses

$$\frac{x^2}{A^2(1 - k^2/B^2)} + \frac{z^2}{C^2(1 - k^2/B^2)} = 1, \qquad y = k, \qquad -B < k < B.$$

Several such ellipses are drawn in Fig. 13.

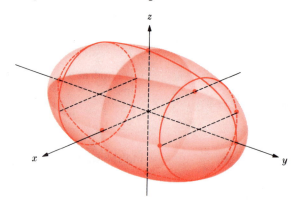

**FIGURE 13**

If $A = B = C$, we obtain a sphere, while if two of the three numbers are equal, the surface is an **ellipsoid of revolution**, also called a **spheroid**. If, for example, $A = B$ and $C > A$, the surface is called a **prolate spheroid**, exemplified by a football. On the other hand, if $A = B$ and $C < A$, we have an **oblate spheroid**. The earth is approximately the shape of an oblate spheroid, with the section at the equator being circular and the distance between the North and South poles being smaller than the diameter of the equatorial circle.

ii) An **elliptic hyperboloid of one sheet** is the graph of an equation of the form

$$\frac{x^2}{A^2} + \frac{y^2}{B^2} - \frac{z^2}{C^2} = 1.$$

The graph of such a surface is sketched in Fig. 14. The $x$ intercepts are at $\pm A$ and the $y$ intercepts at $\pm B$. As for the $z$ intercepts, we must solve the equation $-z^2/C^2 = 1$, which has no real solutions. Therefore the surface does not intersect the $z$ axis. The trace on the $xy$ plane is an ellipse, while the traces on the $yz$ and $xz$ planes are hyperbolas. The sections made by any plane $z = k$ are the ellipses

$$\frac{x^2}{A^2(1 + k^2/C^2)} + \frac{y^2}{B^2(1 + k^2/C^2)} = 1, \qquad z = k,$$

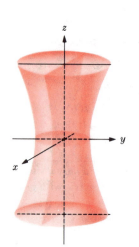

**FIGURE 14**

and the sections made by the planes $y = k$ are the hyperbolas

$$\frac{x^2}{A^2(1 - k^2/B^2)} - \frac{z^2}{C^2(1 - k^2/B^2)} = 1, \qquad y = k.$$

iii) An **elliptic hyperboloid of two sheets** is the graph of an equation of the form

$$\frac{x^2}{A^2} - \frac{y^2}{B^2} - \frac{z^2}{C^2} = 1.$$

Such a graph is sketched in Fig. 15. We observe that we must have $|x| \geq A$, for otherwise the quantity $(x^2/A^2) < 1$ and the left side of the above equation will always be less than the right side. The $x$ intercepts are at $x = \pm A$. There are no $y$ and $z$ intercepts. The traces on the $xz$ and $xy$ planes are hyperbolas; there is no trace on the $yz$ plane. The sections made by the planes $x = k$ are the ellipses

$$\frac{y^2}{B^2(k^2/A^2 - 1)} + \frac{z^2}{C^2(k^2/A^2 - 1)} = 1, \qquad x = k, \text{ if } |k| > A,$$

while the section is empty if $|k| < A$. The sections made by the planes $y = k$ and $z = k$ are hyperbolas.

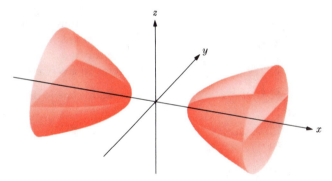

iv) An **elliptic paraboloid** is the graph of an equation of the form

$$\frac{x^2}{A^2} + \frac{y^2}{B^2} = z.$$

A typical elliptic paraboloid is sketched in Fig. 16.

All three intercepts are 0; the traces on the $yz$ and $xz$ planes are parabolas, while the trace on the $xy$ plane consists of a single point, the origin. Sections made by planes $z = k$ are ellipses if $k > 0$, empty if $k < 0$. Sections made by planes $x = k$ and $y = k$ are parabolas.

If $A = B$, we have a **paraboloid of revolution**, and the sections made by the planes $z = k$, $k > 0$ are circles. The reflecting surfaces of telescopes, automobile headlights, etc., are always paraboloids of revolution. (See Section 2.)

v) A **hyperbolic paraboloid** is the graph of an equation of the form

$$\frac{x^2}{A^2} - \frac{y^2}{B^2} = z.$$

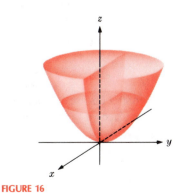

Such a graph is sketched in Fig. 17. As in the elliptic paraboloid, all intercepts are zero. The trace on the $xz$ plane is a parabola opening upward; the trace on the $yz$ plane is a parabola opening downward; and the trace on the $xy$ plane is the pair of intersecting straight lines

$$y = \pm (B/A)x.$$

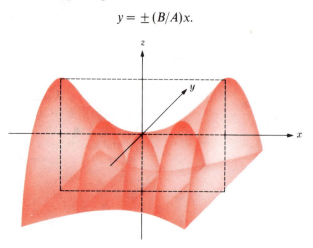

FIGURE 17

As Fig. 17 shows, the surface is "saddle-shaped"; sections made by planes $x = k$ are parabolas opening downward and those made by planes $y = k$ are parabolas opening upward. The sections made by planes $z = k$ are hyperbolas facing one way if $k < 0$ and the other way if $k > 0$. The trace on the $xy$ plane corresponds to $k = 0$ and, as we saw, consists of two intersecting lines.

vi) An **elliptic cone** is the graph of an equation of the form

$$\frac{x^2}{A^2} + \frac{y^2}{B^2} = \frac{z^2}{C^2}.$$

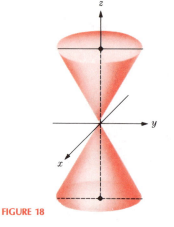

FIGURE 18

A typical cone of this type is shown in Fig. 18. Once again all intercepts are zero. The traces on the $xz$ and $yz$ planes are pairs of intersecting straight lines, while the trace on the $xy$ plane is a single point, the origin. Planes parallel to the coordinate planes yield sections which are the familiar conic sections of plane analytic geometry.

EXAMPLE 2    Name and sketch the graph of $4x^2 - 9y^2 + 8z^2 = 72$. Indicate a few sections parallel to the coordinate planes.

Solution    We divide by 72, getting

$$\frac{x^2}{18} - \frac{y^2}{8} + \frac{z^2}{9} = 1,$$

which is an elliptic hyperboloid of one sheet. The $x$ intercepts are $\pm 3\sqrt{2}$, the $z$ intercepts are $\pm 3$, and there are no $y$ intercepts. The trace on the $xy$ plane is the hyperbola

$$\frac{x^2}{18} - \frac{y^2}{8} = 1,$$

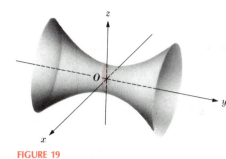

**FIGURE 19**

the trace on the $xz$ plane is the ellipse

$$\frac{x^2}{18} + \frac{z^2}{9} = 1,$$

and the trace on the $yz$ plane is the hyperbola

$$\frac{z^2}{9} - \frac{y^2}{8} = 1.$$

The surface is sketched in Fig. 19.

## 5   PROBLEMS

Name and sketch the graph of each of the following equations.

1 $\dfrac{x^2}{16} + \dfrac{y^2}{9} + \dfrac{z^2}{4} = 1$    2 $\dfrac{x^2}{9} + \dfrac{y^2}{12} + \dfrac{z^2}{9} = 1$

3 $\dfrac{x^2}{16} + \dfrac{y^2}{20} + \dfrac{z^2}{20} = 1$    4 $\dfrac{x^2}{16} + \dfrac{y^2}{9} - \dfrac{z^2}{4} = 1$

5 $\dfrac{x^2}{16} - \dfrac{y^2}{9} - \dfrac{z^2}{4} = 1$    6 $\dfrac{x^2}{16} - \dfrac{y^2}{9} + \dfrac{z^2}{4} = 1$

7 $\dfrac{y^2}{9} + \dfrac{z^2}{4} = 1 + \dfrac{x^2}{16}$    8 $-\dfrac{x^2}{16} - \dfrac{y^2}{9} + \dfrac{z^2}{4} = 1$

9 $\dfrac{x}{4} = \dfrac{y^2}{4} + \dfrac{z^2}{9}$    10 $z = \dfrac{y^2}{4} - \dfrac{x^2}{9}$

11 $y = \dfrac{x^2}{8} + \dfrac{z^2}{8}$    12 $z^2 = 4x^2 + 4y^2$

13 $y^2 = x^2 + z^2$    14 $z^2 = x^2 - y^2$

15 $2x^2 + 6y^2 - 3z^2 = 8$    16 $4x^2 - 3y^2 + 2z^2 = 0$

17 $3x = 2y^2 - 5z^2$    18 $\dfrac{y}{5} = 8z^2 - 2x^2$

19 Show that the intersection of the hyperbolic paraboloid $x^2 - y^2 = z$ with the plane $z = x + y$ consists of two intersecting straight lines. Establish the same result for the intersection of this quadric with the plane $z = ax + ay$, where $a$ is any number.

*20 Using the result of Problem 19 as a guide, show that any hyperbolic paraboloid is "composed entirely of straight lines."

# Properties of hyperbolic functions

---
1

## THE HYPERBOLIC FUNCTIONS

Exponential functions appear in many investigations in engineering, physics, chemistry, biology, and the social sciences. Certain combinations of them have been tabulated and given special names.

**DEFINITIONS** *Two functions, denoted* sinh *and* cosh, *are defined by the formulas*

$$\sinh x = \tfrac{1}{2}(e^x - e^{-x}), \qquad \cosh x = \tfrac{1}{2}(e^x + e^{-x}).$$

We read these "hyperbolic sine of $x$" and "hyperbolic cosine of $x$." We shall see that these functions satisfy many relations which are reminiscent of the sine and cosine functions.

In analogy with trigonometric functions, there are four more hyperbolic functions defined in terms of $\sinh x$ and $\cosh x$, as follows:

$$\tanh x = \frac{\sinh x}{\cosh x}, \qquad \coth x = \frac{\cosh x}{\sinh x},$$

$$\operatorname{sech} x = \frac{1}{\cosh x}, \qquad \operatorname{csch} x = \frac{1}{\sinh x}.$$

These are called "hyperbolic tangent of $x$," etc.

Sketches of the hyperbolic functions are shown in Fig. 1. It is evident that, unlike the trigonometric functions, none of the hyperbolic functions is periodic. Sinh $x$, tanh $x$, coth $x$, and csch $x$ are odd functions, while cosh $x$ and sech $x$ are even. To see this, we appeal directly to the definition, noting for example that

$$\sinh(-x) = \tfrac{1}{2}(e^{-x} - e^{-(-x)}) = -\tfrac{1}{2}(e^x - e^{-x}) = -\sinh x$$

and
$$\cosh(-x) = \tfrac{1}{2}(e^{-x} + e^{-(-x)}) = \tfrac{1}{2}(e^x + e^{-x}) = \cosh x$$

The relation
$$\cosh^2 x - \sinh^2 x = 1$$

is established by observing that

$$[\tfrac{1}{2}(e^x + e^{-x})]^2 - [\tfrac{1}{2}(e^x - e^{-x})]^2 = \tfrac{1}{4}(e^{2x} + 2 + e^{-2x}) - \tfrac{1}{4}(e^{2x} - 2 + e^{-2x}) = 1.$$

The formulas

$$\tanh^2 x + \operatorname{sech}^2 x = 1, \quad \coth^2 x - \operatorname{csch}^2 x = 1,$$

are then easily verified from the definitions of the functions and from the relation $\cosh^2 x - \sinh^2 x = 1$.

The addition formulas for the hyperbolic functions are:

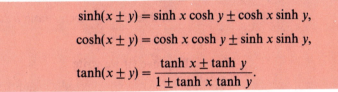

$$\sinh(x \pm y) = \sinh x \cosh y \pm \cosh x \sinh y,$$

$$\cosh(x \pm y) = \cosh x \cosh y \pm \sinh x \sinh y,$$

$$\tanh(x \pm y) = \frac{\tanh x \pm \tanh y}{1 \pm \tanh x \tanh y}.$$

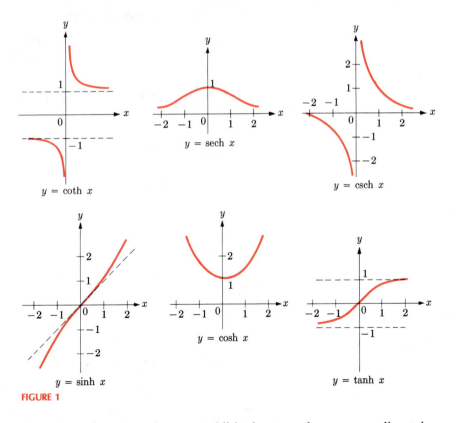

$y = \coth x$

$y = \operatorname{sech} x$

$y = \operatorname{csch} x$

$y = \sinh x$

$y = \cosh x$

$y = \tanh x$

FIGURE 1

These formulas are easier to establish than are the corresponding trig-onometric addition formulas. For example, to prove the first one we merely substitute the appropriate exponentials in the left and right sides and see if they are equal. We have

$$\sinh(x+y) = \tfrac{1}{2}(e^{x+y} - e^{-x-y}),$$

$\sinh x \cosh y + \cosh x \sinh y$

$$= \tfrac{1}{2}(e^x - e^{-x})\tfrac{1}{2}(e^y + e^{-y}) + \tfrac{1}{2}(e^x + e^{-x})\tfrac{1}{2}(e^y - e^{-y})$$

$$= \tfrac{1}{4}(e^{x+y} + e^{x-y} - e^{-x+y} - e^{-x-y} + e^{x+y} - e^{x-y} + e^{-x+y} - e^{-x-y})$$

$$= \tfrac{1}{2}(e^{x+y} - e^{-x-y}) = \sinh(x+y).$$

The remaining formulas are obtained in a similar way.

**THEOREM 1**  *The following differentiation formulas hold:*

i)  *d* sinh *u* = cosh *u du*                ii)  *d* cosh *u* = sinh *u du*

iii) *d* tanh *u* = sech² *u du*              iv) *d* coth *u* = − csch² *u du*

v) *d* sech *u* = − sech *u* tanh *u du*      vi) *d* csch *u* = − csch *u* coth *u du*.

**Proof**  To prove (i) we write

$$\sinh u = \tfrac{1}{2}(e^u - e^{-u}),$$

$$d \sinh u = \tfrac{1}{2}d(e^u - e^{-u}) = \tfrac{1}{2}(e^u + e^{-u})du = \cosh u\, du.$$

The remaining formulas, (ii) through (vi), follow in the same way, by appeal to the definition and by differentiation of the exponential function.  □

**EXAMPLE 1**  Given that $\sinh x = -\tfrac{3}{4}$, find the values of the other hyperbolic functions.

**Solution**  $\cosh^2 x = 1 + \sinh^2 x = \tfrac{25}{16}$. Since cosh *x* is always positive, we have $\cosh x = \tfrac{5}{4}$. Therefore

$$\tanh x = \frac{\sinh x}{\cosh x} = -\frac{3}{5}, \qquad \coth x = -\frac{5}{3},$$

$$\operatorname{sech} x = \frac{1}{\cosh x} = \frac{4}{5}, \qquad \operatorname{csch} x = \frac{1}{\sinh x} = -\frac{4}{3}. \qquad □$$

**EXAMPLE 2**  Given $f(x) = \sinh^3(2x^2 + 3)$, find $f'(x)$.

**Solution**  We have, by the Chain Rule,

$$df = 3 \sinh^2(2x^2 + 3)d[\sinh(2x^2 + 3)],$$

and from the formula for the derivative of the sinh function,

$$df = 3 \sinh^2(2x^2 + 3)\cosh(2x^2 + 3)d(2x^2 + 3),$$

$$f'(x) = 12x \sinh^2(2x^2 + 3)\cosh(2x^2 + 3). \qquad □$$

## 1   PROBLEMS

In Problems 1 through 17, establish the formulas.

**1**   $\sinh(x - y) = \sinh x \cosh y - \cosh x \sinh y$

**2**   $\cosh(x + y) = \cosh x \cosh y + \sinh x \sinh y$

**3**   $\cosh(x - y) = \cosh x \cosh y - \sinh x \sinh y$

**4**   $\tanh(x + y) = \dfrac{\tanh x + \tanh y}{1 + \tanh x \tanh y}$

**5**   $\tanh(x - y) = \dfrac{\tanh x - \tanh y}{1 - \tanh x \tanh y}.$

**6**   $\sinh 2x = 2 \sinh x \cosh x$

**7**   $\cosh 2x = \cosh^2 x + \sinh^2 x = 2 \cosh^2 x - 1 = 2 \sinh^2 x + 1$

**8**   $\tanh 2x = \dfrac{2 \tanh x}{1 + \tanh^2 x}$

**9**   $\sinh A + \sinh B = 2 \sinh \dfrac{A + B}{2} \cosh \dfrac{A - B}{2}$

**10**   $\cosh A + \cosh B = 2 \cosh \dfrac{A + B}{2} \cosh \dfrac{A - B}{2}$

**11**   $\cosh(x/2) = \sqrt{(1 + \cosh x)/2}$

**12**   $\sinh(x/2) = \pm \sqrt{(\cosh x - 1)/2}$

**13**   $d \cosh u = \sinh u \, du$

**14**   $d \tanh u = \operatorname{sech}^2 u \, du$

**15**   $d \coth u = -\operatorname{csch}^2 u \, du$

**16**   $d \operatorname{sech} u = -\operatorname{sech} u \tanh u \, du$

**17**   $d \operatorname{csch} u = -\operatorname{csch} u \coth u \, du$

In Problems 18 through 23, find the values of the remaining hyperbolic functions.

**18**   $\sinh x = -\frac{12}{5}$

**19**   $\tanh x = -\frac{4}{5}$

**20**   $\coth x = 2$

**21**   $\operatorname{csch} x = -2$

**22**   $\cosh x = 4,\ x > 0$

**23**   $\cosh x = 2,\ x < 0$

In Problems 24 through 29, examine the curves for maximum points, minimum points, and points of inflection. State where the curves are concave upward and where they are concave downward. Sketch the graphs.

**24**   $y = \sinh 2x$

**25**   $y = \cosh(x/2)$

**26**   $y = \tanh 3x$

**27**   $y = \coth \frac{1}{3} x$

**28**   $y = \operatorname{sech}(x + 1)$

**29**   $y = \operatorname{csch}(1 - x)$

## 2

## THE INVERSE HYPERBOLIC FUNCTIONS

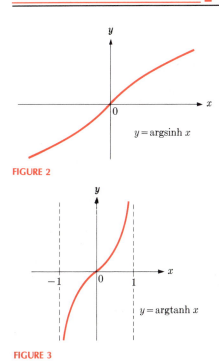

$y = \operatorname{argsinh} x$

FIGURE 2

$y = \operatorname{argtanh} x$

FIGURE 3

The function $f(x) = \sinh x$ is an increasing function for all values of $x$. This is so because the derivative of $\sinh x$ is $\cosh x$, a positive quantity for all values of $x$, and we know that a function with a positive derivative is an increasing function.

The inverse relation of a function which is increasing is a function (Inverse Function Theorem). We define

$$y = \operatorname{argsinh} x$$

to be the inverse of the sinh function. Its domain consists of all real numbers. The graph is sketched in Fig. 2.

Similarly, the functions $\tanh x$, $\coth x$, and $\operatorname{csch} x$ are all increasing or decreasing functions for all values of $x$ and so have inverse functions, which we denote by

$$\operatorname{argtanh} x, \quad \operatorname{argcoth} x, \quad \operatorname{argcsch} x.$$

The domain of $\operatorname{argtanh} x$ is the set of numbers $-1 < x < 1$, while the range consists of all real numbers. Its graph is sketched in Fig. 3.

The function cosh $x$ is an increasing function, and sech $x$ is a decreasing function, for nonnegative values of $x$. We define

$$\text{argcosh } x, \quad \text{argsech } x$$

to be the inverse of the cosh and sech functions respectively, restricted to nonnegative values. The domain of $f(x) = \text{argcosh } x$ consists of all numbers $x \geq 1$, and the range is the set of nonnegative real numbers.

The hyperbolic functions are defined in terms of exponentials. The exponential and logarithmic functions are the inverse of each other. It is therefore natural to ask what relation, if any, exists between the logarithmic function and the inverse hyperbolic functions. The direct connection between the natural logarithm and the inverse hyperbolic functions is exhibited in the next theorem.

---

**THEOREM 2**  i)  $\text{argsinh } x = \ln(x + \sqrt{x^2 + 1})$ *for all* $x$,

ii)  $\text{argtanh } x = \frac{1}{2}\ln[(1 + x)/(1 - x)], \quad -1 < x < 1,$

iii)  $\text{argcoth } x = \frac{1}{2}\ln[(x + 1)/(x - 1)], \quad |x| > 1,$

iv)  $\text{argcosh } x = \ln(x + \sqrt{x^2 - 1}), \quad x \geq 1.$

---

**Proof**  To prove (i), we write $y = \text{argsinh } x$ and, equivalently, $x = \sinh y$. We wish to find an expression for $y$ in terms of the logarithm. The definition of sinh $y$ gives us

$$x = \tfrac{1}{2}(e^y - e^{-y}) \quad \Leftrightarrow \quad e^y - \frac{1}{e^y} - 2x = 0.$$

We solve this for $e^y$ by first writing

$$(e^y)^2 - 2x(e^y) - 1 = 0$$

and then applying the quadratic formula to obtain

$$e^y = x \pm \sqrt{x^2 + 1}.$$

The plus sign must be chosen, since $e^y > 0$ always. Taking natural logarithms of both sides, we find that

$$y = \ln(x + \sqrt{x^2 + 1}),$$

which proves (i).

The proofs of (ii), (iii), and (iv) follow the same pattern: First, we express the hyperbolic function in terms of exponentials; second, we solve for the exponential by the quadratic formula; third, we make the correct choice of the plus or minus sign in the quadratic formula; and fourth, we take logarithms of both sides. We shall exhibit the steps in the proof of (iv). Let $y = \text{argcosh } x$, $x \geq 1$, and write $\cosh y = x$, $y \geq 0$: Then we have

$$e^y + e^{-y} = 2x, \quad y \geq 0,$$

and

$$(e^y)^2 - 2x(e^y) + 1 = 0, \quad y \geq 0.$$

Solving by the quadratic formula, we obtain

$$e^y = x \pm \sqrt{x^2 - 1}, \quad y \geq 0.$$

For $y \geq 0$ we must have $e^y \geq 1$. Since

$$(x + \sqrt{x^2 - 1})(x - \sqrt{x^2 - 1}) = 1,$$

it follows that for $x > 1$, $x + \sqrt{x^2 - 1} > 1$, and $x - \sqrt{x^2 - 1} < 1$. We therefore choose the plus sign. Taking logarithms, we get

$$y = \ln(x + \sqrt{x^2 - 1}), \quad x \geq 1. \qquad \square$$

**EXAMPLE 1**  Express argcosh 2, argtanh $(-\frac{1}{2})$ in terms of the logarithm function.

**Solution**  From Theorem 2, we obtain

$$\text{argcosh } 2 = \ln(2 + \sqrt{3}),$$

$$\text{argtanh}\left(-\frac{1}{2}\right) = \frac{1}{2}\ln\frac{1/2}{3/2} = -\frac{1}{2}\ln 3 = \ln\frac{1}{\sqrt{3}}. \qquad \square$$

By expressing the inverse hyperbolic functions in terms of logarithms, we have simplified the problem of computing the derivatives of these functions.

**THEOREM 3**  *The inverse hyperbolic functions have the following differentiation formulas:*

i)  $d \text{ argsinh } u = \dfrac{du}{\sqrt{u^2 + 1}},$

ii)  $d \text{ argcosh } u = \dfrac{du}{\sqrt{u^2 - 1}}, \quad u > 1,$

iii)  $d \text{ argtanh } u = \dfrac{du}{1 - u^2}, \quad -1 < u < 1,$

iv)  $d \text{ argcoth } u = \dfrac{du}{1 - u^2}, \quad |u| > 1,$

v)  $d \text{ argsech } u = \dfrac{-du}{u\sqrt{1 - u^2}}, \quad 0 < u < 1,$

vi)  $d \text{ argcsch } u = -\dfrac{du}{|u|\sqrt{u^2 + 1}}, \quad u \neq 0.$

**Proof**  To prove (i), we write

$$y = \text{argsinh } u = \ln(u + \sqrt{u^2 + 1}).$$

Therefore

$$dy = \frac{1}{u + \sqrt{u^2 + 1}} d(u + \sqrt{u^2 + 1})$$

$$= \frac{1}{u + \sqrt{u^2 + 1}}\left(1 + \frac{u}{\sqrt{u^2 + 1}}\right) du = \frac{du}{\sqrt{u^2 + 1}}.$$

The remaining formulas, (ii) through (vi), are proved in a similar manner. ☐

**EXAMPLE 2**   Given $f(x) = \text{argcosh}(x^2 + 2x + 2)$, find $f'(x)$.

**Solution**   Letting $u = x^2 + 2x + 2$, we have

$$df = \frac{du}{\sqrt{u^2 - 1}} = \frac{2(x+1)\,dx}{\sqrt{(x^2 + 2x + 2)^2 - 1}},$$

and, since $x^2 + 2x + 2 = (x+1)^2 + 1$,

$$f'(x) = \frac{2(x+1)}{\sqrt{(x+1)^4 + 2(x+1)^2}} = \frac{2}{\sqrt{(x+1)^2 + 2}}, \quad x + 1 > 0. \qquad ☐$$

**EXAMPLE 3**   Evaluate

$$\int_{-1}^{2} \frac{dx}{\sqrt{1 + x^2}}$$

and express the result in terms of the logarithm function.

**Solution**   Since every differentiation formula carries with it an integration formula, Theorem 3(i) gives us

$$\int_{-1}^{2} \frac{dx}{\sqrt{1 + x^2}} = \text{argsinh } x \Big]_{-1}^{2} = \text{argsinh } 2 - \text{argsinh}(-1)$$

$$= \ln(2 + \sqrt{5}) - \ln(-1 + \sqrt{2}) = \ln\frac{2 + \sqrt{5}}{\sqrt{2} - 1}. \qquad ☐$$

## 2   PROBLEMS

In Problems 1 through 6, express the given quantities in terms of the logarithm function.

**1**  $\text{argsinh}\left(\frac{1}{2}\right)$  **2**  $\text{argtanh}\left(\frac{3}{5}\right)$

**3**  $\text{argcosh } 2$  **4**  $\text{argcoth}(-2)$

**5**  $\text{argsech}\left(\frac{3}{5}\right)$  **6**  $\text{argcosh}(2.6)$

In Problems 7 through 12, examine the curves for maximum points, minimum points, and points of inflection. State where the curves are concave upward and where they are concave downward. Sketch the graphs.

**7**  $y = \text{argsinh } 2x$  **8**  $y = \text{argcosh}(x+1)$

**9**  $y = \text{argtanh } \frac{1}{2}x$  **10**  $y = \text{argcoth}(x-1)$

**11**  $y = \text{argsech } 4x$  **12**  $y = \text{argcsch}(2x-1)$

In Problems 13 through 20, perform the differentiations.

**13**  $f(x) = \text{argsinh } 2x$  **14**  $f(x) = x^2 \text{ argcosh } 3x$

**15**  $g(x) = x^{-1} \text{ argtanh}(x^2)$

**16**  $f(x) = e^{-2x} \text{ argsinh}(3x-2)$

**17**  $G(x) = \text{argsinh}(\tan x)$  **18**  $F(x) = (\text{argsinh}\sqrt{x})/\sqrt{x}$

**19**  $F(x) = \text{argsinh}^2(2x)$

**20**  $H(x) = xe^{-x} \text{ argcosh}(1-x)$

In Problems 21 through 26, evaluate the integrals, giving the answers in terms of the logarithm function.

**21**  $\int_{2}^{5} \frac{dx}{\sqrt{x^2 - 1}}$  **22**  $\int_{-1/2}^{3/5} \frac{dx}{1 - x^2}$

**23**  $\int_{-2}^{-1} \frac{dx}{\sqrt{x^2 + 1}}$  **24**  $\int_{-3}^{-2} \frac{dx}{1 - x^2}$

**25**  $\int_{3}^{5} \frac{dx}{\sqrt{x^2 - 4}}$  **26**  $\int_{1}^{3} \frac{dx}{16 - x^2}$

**27**  Prove that $\text{argtanh } x = \frac{1}{2}\ln[(1+x)/(1-x)]$, $-1 < x < 1$.

**28**  Prove that $\text{argcoth } x = \frac{1}{2}\ln[(x+1)/(x-1)]$, $|x| > 1$.

**29**  Sketch the graph of $y = \text{argcoth } x$; of $y = \text{argcsch } x$.

**30** Prove that $d\operatorname{argcosh} u = \dfrac{du}{\sqrt{u^2 - 1}},\quad u > 1.$

**31** Prove that $d\operatorname{argtanh} u = \dfrac{du}{1 - u^2},\quad -1 < u < 1.$

**32** Prove that $d\operatorname{argcoth} u = \dfrac{du}{1 - u^2},\quad |u| > 1.$

**33** Prove that $d\operatorname{argsech} u = -\dfrac{du}{u\sqrt{1 - u^2}},\quad 0 < u < 1.$

**34** Prove that $d\operatorname{argcsch} u = -\dfrac{du}{|u|\sqrt{u^2 + 1}},\quad u \neq 0.$

**35** Let $x = \cosh t$, $y = \sinh t$. Make a table of values for $x$ and $y$ as $t$ assumes values in the interval $-10 \le t \le 10$. Sketch the graph in the $xy$ plane. Can you identify the curve?

# THEOREMS ON DETERMINANTS

## DETERMINANTS OF SECOND AND THIRD ORDER

In this appendix we present an introduction to determinants of the second and third order. A determinant of the second order is denoted by

$$\begin{vmatrix} a & b \\ c & d \end{vmatrix},$$

in which $a$, $b$, $c$, and $d$ are numbers. A determinant of the third order is denoted by

$$\begin{vmatrix} a & b & c \\ d & e & f \\ g & h & j \end{vmatrix},$$

in which $a$, $b$, ..., $j$ are numbers.

Determinants are actually functions of the elements which appear. That is, a second-order determinant is a function of the four quantities $a$, $b$, $c$, and $d$, while a third-order determinant is a function of the nine entries $a$, $b$, $c$, ..., $j$. Using the terminology of variables, we say that a second-order determinant is a function of four variables and a third-order determinant is a function of nine variables.

For a second-order determinant, we define

$$\begin{vmatrix} a & b \\ c & d \end{vmatrix} = ad - bc.$$

For example,

$$\begin{vmatrix} 2 & 1 \\ -5 & 6 \end{vmatrix} = 12 - (-5) = 17.$$

A third-order determinant is defined in terms of second-order determinants by the formula:

$$\begin{vmatrix} a & b & c \\ d & e & f \\ g & h & j \end{vmatrix} = c\begin{vmatrix} d & e \\ g & h \end{vmatrix} - f\begin{vmatrix} a & b \\ g & h \end{vmatrix} + j\begin{vmatrix} a & b \\ d & e \end{vmatrix}.$$

This formula is known as an **expansion in terms of the last column**. For example,

$$\begin{vmatrix} 2 & 1 & -1 \\ 0 & 3 & 2 \\ 1 & 4 & 3 \end{vmatrix} = -1\begin{vmatrix} 0 & 3 \\ 1 & 4 \end{vmatrix} - 2\begin{vmatrix} 2 & 1 \\ 1 & 4 \end{vmatrix} + 3\begin{vmatrix} 2 & 1 \\ 0 & 3 \end{vmatrix}$$

$$= -(-3) - 2(8-1) + 3(6) = 7.$$

---

**THEOREM 1**   **(Cramer's Rule for $n = 2$)**   *Given the two equations*

$$\left. \begin{array}{l} ax + by = e, \\ cx + dy = f, \end{array} \right\} \tag{1}$$

*if the determinant of the coefficients*

$$\begin{vmatrix} a & b \\ c & d \end{vmatrix}$$

*is not zero, then the two simultaneous equations* (1) *have a unique solution given by*

$$x = \frac{\begin{vmatrix} e & b \\ f & d \end{vmatrix}}{\begin{vmatrix} a & b \\ c & d \end{vmatrix}}, \qquad y = \frac{\begin{vmatrix} a & e \\ c & f \end{vmatrix}}{\begin{vmatrix} a & b \\ c & d \end{vmatrix}}.$$

---

The proof is left to the reader. See also Theorem 11 which states Cramer's Rule for $n = 3$.

---

**THEOREM 2**   *Two sets of numbers $\{a_1, b_1, c_1\}$ and $\{a_2, b_2, c_2\}$, neither all zero, are proportional if and only if all three determinants*

$$\begin{vmatrix} a_1 & b_1 \\ a_2 & b_2 \end{vmatrix}, \qquad \begin{vmatrix} a_1 & c_1 \\ a_2 & c_2 \end{vmatrix}, \qquad \begin{vmatrix} b_1 & c_1 \\ b_2 & c_2 \end{vmatrix}$$

*are zero.*

**Proof**    If the sets are proportional, then there is a number $k$ such that

$$a_2 = ka_1, \qquad b_2 = kb_1, \qquad c_2 = kc_1. \tag{2}$$

It is immediate that the three determinants vanish. Now suppose all determinants vanish. We wish to prove the sets are proportional. Suppose $c_1 \neq 0$, for instance. Then we have

$$\begin{vmatrix} a_1 & c_1 \\ a_2 & c_2 \end{vmatrix} = a_1 c_2 - a_2 c_1 = 0 \qquad \Leftrightarrow \qquad a_2 = \left(\frac{c_2}{c_1}\right) a_1,$$

$$\begin{vmatrix} b_1 & c_1 \\ b_2 & c_2 \end{vmatrix} = b_1 c_2 - b_2 c_1 = 0 \qquad \Leftrightarrow \qquad b_2 = \left(\frac{c_2}{c_1}\right) b_1.$$

We define $k = c_2/c_1$ and we see that (2) holds. If $c_1 = 0$, then either $a_1 \neq 0$ or $b_1 \neq 0$ and a similar argument leads to the same conclusion. $\square$

It is clear that the value of a determinant depends on the arrangement of the elements which enter it. It is simpler to state theorems about determinants if we introduce a systematic notation for the entries. The standard method consists of a double subscript for each element, the first subscript identifying the row in which the element is situated, and the second subscript identifying the column. That is, we denote the element in the $i$th row and $j$th column by the symbol $a_{ij}$. The evaluation of a third-order determinant may now be written

$$\begin{vmatrix} a_{11} & a_{12} & a_{13} \\ a_{21} & a_{22} & a_{23} \\ a_{31} & a_{32} & a_{33} \end{vmatrix} = a_{13} \begin{vmatrix} a_{21} & a_{22} \\ a_{31} & a_{32} \end{vmatrix} - a_{23} \begin{vmatrix} a_{11} & a_{12} \\ a_{31} & a_{32} \end{vmatrix} + a_{33} \begin{vmatrix} a_{11} & a_{12} \\ a_{21} & a_{22} \end{vmatrix}.$$

Using the definition of second-order determinants, we obtain for the value of a third-order determinant:

$$\begin{vmatrix} a_{11} & a_{12} & a_{13} \\ a_{21} & a_{22} & a_{23} \\ a_{31} & a_{32} & a_{33} \end{vmatrix} = a_{11} a_{22} a_{33} + a_{12} a_{31} a_{23} + a_{13} a_{21} a_{32} \tag{3}$$
$$- a_{11} a_{32} a_{23} - a_{12} a_{21} a_{33} - a_{13} a_{22} a_{31}.$$

It is a simple matter to verify that the interchange of each $a_{ij}$ with $a_{ji}$ in the right side of (3) does not alter it in any way. Therefore we conclude the following result for third-order determinants.

**THEOREM 3**    *If a new determinant is obtained from a given one by an interchange of its rows and columns, the value of the new determinant is equal to the value of the given one.*

**DEFINITIONS**     *Given a determinant*

$$D \equiv \begin{vmatrix} a_{11} & a_{12} & a_{13} \\ a_{21} & a_{22} & a_{23} \\ a_{31} & a_{32} & a_{33} \end{vmatrix}, \tag{4}$$

*we define the quantity $D_{ij}$ as the second-order determinant obtained from D by crossing out the ith row and jth column in D. For example,*

$$D_{21} = \begin{vmatrix} a_{12} & a_{13} \\ a_{32} & a_{33} \end{vmatrix}.$$

The **cofactor** of the element $a_{ij}$ in $D$ is defined to be the number $A_{ij}$ given by

$$A_{ij} = (-1)^{i+j} D_{ij}.$$

For example, the cofactor of $a_{13}$ is

$$A_{13} = (-1)^{1+3} D_{13} = (+1) \begin{vmatrix} a_{21} & a_{22} \\ a_{31} & a_{32} \end{vmatrix} = a_{21}a_{32} - a_{31}a_{22}.$$

**THEOREM 4**     *If D denotes the determinant in (4), we have*

$$D = a_{11}A_{11} + a_{12}A_{12} + a_{13}A_{13} = a_{11}A_{11} + a_{21}A_{21} + a_{31}A_{31}, \tag{5a}$$

$$D = a_{21}A_{21} + a_{22}A_{22} + a_{23}A_{23} = a_{12}A_{12} + a_{22}A_{22} + a_{32}A_{32}, \tag{5b}$$

$$D = a_{31}A_{31} + a_{32}A_{32} + a_{33}A_{33} = a_{13}A_{13} + a_{23}A_{23} + a_{33}A_{33}. \tag{5c}$$

**Proof**     We verify the first formula on the right:

$$a_{11}A_{11} + a_{21}A_{21} + a_{31}A_{31}$$

$$= a_{11} \begin{vmatrix} a_{22} & a_{23} \\ a_{32} & a_{33} \end{vmatrix} - a_{21} \begin{vmatrix} a_{12} & a_{13} \\ a_{32} & a_{33} \end{vmatrix} + a_{31} \begin{vmatrix} a_{12} & a_{13} \\ a_{22} & a_{23} \end{vmatrix}$$

$$= a_{11}(a_{22}a_{33} - a_{32}a_{23}) - a_{21}(a_{12}a_{33} - a_{32}a_{13}) + a_{31}(a_{12}a_{23} - a_{22}a_{13})$$

$$= D.$$

Theorem 3 and a computation may be used to verify the remaining formulas in (5). The details are left to the reader.     □

**DEFINITIONS**     *The first sums in (5a), (5b), and (5c), are called the **expansions of D according to the first, second, and third rows,** respectively. The second sums in (5a), (5b), and (5c) are called the **expansions of D according to the first, second, and third columns,** respectively.*

**EXAMPLE 1**   Evaluate the following determinant by expanding it according to (i) the second column, and (ii) the third row:

$$\begin{vmatrix} 1 & -2 & 3 \\ 2 & 1 & -1 \\ -2 & -1 & 2 \end{vmatrix}.$$

**Solution**   i) Expanding according to the second column, we obtain

$$\begin{vmatrix} 1 & -2 & 3 \\ 2 & 1 & -1 \\ -2 & -1 & 2 \end{vmatrix} = -(-2)\begin{vmatrix} 2 & -1 \\ -1 & 2 \end{vmatrix} + 1\cdot\begin{vmatrix} 1 & 3 \\ -2 & 2 \end{vmatrix} - (-1)\begin{vmatrix} 1 & 3 \\ 2 & -1 \end{vmatrix}$$

$$= 2(4-2)+(2+6)+(-1-6)=5.$$

ii) Expanding according to the third row, we obtain

$$\begin{vmatrix} 1 & -2 & 3 \\ 2 & 1 & -1 \\ -2 & -1 & 2 \end{vmatrix} = (-2)\begin{vmatrix} -2 & 3 \\ 1 & -1 \end{vmatrix} - (-1)\begin{vmatrix} 1 & 3 \\ 2 & -1 \end{vmatrix} + 2\begin{vmatrix} 1 & -2 \\ 2 & 1 \end{vmatrix}$$

$$= (-2)(-1)+1\cdot(-7)+2\cdot 5 = 5. \qquad \square$$

The definitions and theorems given for determinants of the second and third order may be extended to determinants of order $n$, where $n$ is any positive integer. A determinant of order $n$ has $n$ rows and $n$ columns. The evaluation of such determinants, a process beyond the scope of this appendix, is most easily accomplished by an induction technique. We now state a number of general properties of determinants, valid for any order, which are most useful in such evaluations. The proofs depend on the expansion given in Theorem 4 which we have proved only for $n = 3$. Actually the appropriate extension of Theorem 4 is valid for determinants of any order.

**THEOREM 5**   *Let D be a determinant and suppose that each element of the kth row $a_{kj}$ is the sum of two numbers: $a_{kj} = a'_{kj} + a''_{kj}$. Then $D = D' + D''$ where $D'$ is the determinant obtained from D by inserting $a'_{kj}$ instead of $a_{kj}$ for each element of the kth row. Similarly, $D''$ is obtained by inserting $a''_{kj}$ for $a_{kj}$. The result also holds if the elements of a column are treated analogously.*

**Proof**   We expand $D$ according to the elements of the $k$th row:

$$D = a_{k1}A_{k1} + a_{k2}A_{k2} + a_{k3}A_{k3}$$

$$= (a'_{k1} + a''_{k1})A_{k1} + (a'_{k2} + a''_{k2})A_{k2} + (a'_{k3} + a''_{k3})A_{k3}$$

$$= (a'_{k1}A_{k1} + a'_{k2}A_{k2} + a'_{k3}A_{k3}) + (a''_{k1}A_{k1} + a''_{k2}A_{k2} + a''_{k3}A_{k3})$$

$$= D' + D''. \qquad \square$$

Theorem 4 shows that the same analysis works for columns.

Theorem 5 shows that, for example,

$$\begin{vmatrix} a_{11} & a_{12} & a_{13} \\ a_{21} & a_{22} & a_{23} \\ a'_{31}+a''_{31} & a'_{32}+a''_{32} & a'_{33}+a''_{33} \end{vmatrix} = \begin{vmatrix} a_{11} & a_{12} & a_{13} \\ a_{21} & a_{22} & a_{23} \\ a'_{31} & a'_{32} & a'_{33} \end{vmatrix} + \begin{vmatrix} a_{11} & a_{12} & a_{13} \\ a_{21} & a_{22} & a_{23} \\ a''_{31} & a''_{32} & a''_{33} \end{vmatrix}.$$

**THEOREM 6**   *If each element $a_{kj} = ca'_{kj}$ for k fixed and $j = 1, 2,$ and 3, then $D = cD'$ where $D'$ is obtained from D as in Theorem 5.*

The proof is similar to that of Theorem 5. In words, Theorem 6 states that if all the elements of a row are multiplied by a constant, then the determinant is multiplied by that constant.

**THEOREM 7**   *If $D'$ is obtained from D by interchanging any two rows, then $D' = -D$. The same result holds for the interchange of two columns.*

**Proof**   The result is a direct consequence of Formula (3). The details are left to the reader.  □

**THEOREM 8**   *If any two rows (or columns) of a determinant D are proportional, then $D = 0$.*

**Proof**   If two rows are identical, then interchanging them has no effect. On the other hand, Theorem 7 shows that $D' = -D$. Thus $D' = -D = D$ and $D = 0$. If two rows are proportional, one is $c$ times the other, and by Theorem 6, $D = cD''$ where $D''$ has two identical rows. Hence $D = 0$. The proof for columns is similar.  □

The next lemma and theorem are very useful in evaluating determinants, especially those of high order.

**LEMMA 1**   *If $D'$ is obtained from D by multiplying the kth row by the constant c and adding the result to the ith row, where $i \neq k$, then $D' = D$. The same result holds for columns.*

**Proof**   The elements of the $i$th row of $D'$, denoted by $a'_{ij}$, have the form

$$a'_{ij} = a_{ij} + ca_{kj}.$$

Using Theorem 5, we find

$$D' = D + cD''$$

where $D''$ is obtained by replacing the $i$th row of $D$ by the $k$th row. But then $D''$ has two identical rows (the $i$th and $k$th), so that $D'' = 0$. Hence

$$D' = D.$$  □

**THEOREM 9**    *If $D'$ is obtained from $D$ by multiplying the kth row by $c_i$ and adding the result to the ith row for all $i \neq k$ in turn, then $D' = D$. The same is true for columns.*

**Proof**    Each step in the process is one for which the lemma applies leaving the value unchanged.                                                                      □

The next example shows how to use Theorem 9 to simplify the determinant before applying the expansion theorem.

**EXAMPLE 2**    Simplify by using Theorem 9 and use the expansion theorem to evaluate the determinant

$$D = \begin{vmatrix} 2 & 1 & 3 \\ -1 & 2 & 2 \\ 2 & -3 & 1 \end{vmatrix}.$$

**Solution**    Multiplying the first row by $-2$ and adding the result to the second row, and then multiplying it by 3 and adding to the third row, we obtain

$$D = \begin{vmatrix} 1 & 1 & 3 \\ -5 & 0 & -4 \\ 8 & 0 & 10 \end{vmatrix}$$

We can expand the new determinant according to the second column, thus getting

$$D = -1 \cdot \begin{vmatrix} -5 & -4 \\ 8 & 10 \end{vmatrix} = (-1) \cdot (-18) = 18. \qquad \square$$

The following theorem includes and supplements the expansion theorem (Theorem 4).

**THEOREM 10**    *If $D$ is any third-order determinant, then*

a) $a_{i1} A_{k1} + a_{i2} A_{k2} + a_{i3} A_{k3} = \begin{cases} D & \text{if } k = i, \\ 0 & \text{if } k \neq i, \end{cases}$

b) $a_{1j} A_{1k} + a_{2j} A_{2k} + a_{3j} A_{3k} = \begin{cases} D & \text{if } k = j, \\ 0 & \text{if } k \neq j. \end{cases}$

**Proof**    The cases $k = i$ in (a) and $k = j$ in (b) restate Theorem 4. If $k \neq i$ in (a), then we see from the expansion theorem that the left side in (a) is the expansion of a determinant $D'$ obtained from $D$ by replacing the $k$th row by the $i$th row. But then $D'$ has two identical rows and so is zero. The proof of (b) is the same.

□

**THEOREM 11**    **(Cramer's Rule for $n = 3$)**    *If the determinant*

$$D = \begin{vmatrix} a_{11} & a_{12} & a_{13} \\ a_{21} & a_{22} & a_{23} \\ a_{31} & a_{32} & a_{33} \end{vmatrix}$$

*of the coefficients of the three equations*

$$\left.\begin{array}{l} a_{11}x + a_{12}y + a_{13}z = b_1, \\ a_{21}x + a_{22}y + a_{23}z = b_2, \\ a_{31}x + a_{32}y + a_{33}z = b_3, \end{array}\right\} \tag{6}$$

*is not zero, then the equations have one and only one solution given by*

$$x = \frac{D_1}{D}, \qquad y = \frac{D_2}{D}, \qquad z = \frac{D_3}{D}, \tag{7}$$

*where*

$$D_1 = \begin{vmatrix} b_1 & a_{12} & a_{13} \\ b_2 & a_{22} & a_{23} \\ b_3 & a_{32} & a_{33} \end{vmatrix}, \qquad D_2 = \begin{vmatrix} a_{11} & b_1 & a_{13} \\ a_{21} & b_2 & a_{23} \\ a_{31} & b_3 & a_{33} \end{vmatrix}, \qquad D_3 = \begin{vmatrix} a_{11} & a_{12} & b_1 \\ a_{21} & a_{22} & b_2 \\ a_{31} & a_{32} & b_3 \end{vmatrix}.$$

**Proof**    a) Suppose Equations (6) hold. If we multiply the first, second, and third equations by $A_{11}$, $A_{21}$, and $A_{31}$, respectively, and then add, we get

$$Dx + (a_{12}A_{11} + a_{22}A_{21} + a_{32}A_{31})y + (a_{13}A_{11} + a_{23}A_{21} + a_{33}A_{31})z$$
$$= (b_1 A_{11} + b_2 A_{21} + b_3 A_{31}). \tag{8}$$

By Theorem 10, the coefficients of $y$ and $z$ are zero. The right side of the above equation is the expansion of $D_1$ according to its first column. Hence $x = D_1/D$. Multiplying Equations (6) by $A_{12}, A_{22}, A_{32}$ and proceeding similarly, we get the result $y = D_2/D$. The value for $z$ is obtained in the same way.

b) Now assume that Equations (7) hold and that $D \neq 0$. Then (8) holds; similarly for equations for $y$ and $z$. Therefore, by addition,

$$D(a_{11}x + a_{12}y + a_{13}z) = b_1(a_{11}A_{11} + a_{12}A_{12} + a_{13}A_{13})$$
$$+ b_2(a_{11}A_{21} + a_{12}A_{22} + a_{13}A_{23})$$
$$+ b_3(a_{11}A_{31} + a_{12}A_{32} + a_{13}A_{33})$$
$$= Db_1.$$

The last equality is valid because of Theorem 10. In a similar way, we see that the remaining equations of (6) are satisfied.    □

**EXAMPLE 3**　Use Cramer's Rule to solve the following equations:

$$3x - 2y + 4z = 5,$$
$$x + y + 3z = 2,$$
$$-x + 2y - z = 1.$$

**Solution**　To simplify and evaluate $D$, we multiply the second row by 2 and add it to the first row and then multiply the second row by $-2$ and add it to the third row. We obtain

$$D = \begin{vmatrix} 3 & -2 & 4 \\ 1 & 1 & 3 \\ -1 & 2 & -1 \end{vmatrix} = \begin{vmatrix} 5 & 0 & 10 \\ 1 & 1 & 3 \\ -3 & 0 & -7 \end{vmatrix}.$$

Now expanding according to the second column is easy. We get

$$D = 1 \cdot \begin{vmatrix} 5 & 10 \\ -3 & -7 \end{vmatrix} = -5.$$

Proceeding similarly for $D_1$, $D_2$, and $D_3$, we find

$$D_1 = \begin{vmatrix} 5 & -2 & 4 \\ 2 & 1 & 3 \\ 1 & 2 & -1 \end{vmatrix} = \begin{vmatrix} 9 & 0 & 10 \\ 2 & 1 & 3 \\ -3 & 0 & -7 \end{vmatrix} = 1 \cdot \begin{vmatrix} 9 & 10 \\ -3 & -7 \end{vmatrix} = -33.$$

$$D_2 = \begin{vmatrix} 3 & 5 & 4 \\ 1 & 2 & 3 \\ -1 & 1 & -1 \end{vmatrix} = \begin{vmatrix} 8 & 5 & 9 \\ 3 & 2 & 5 \\ 0 & 1 & 0 \end{vmatrix} = -1 \cdot \begin{vmatrix} 8 & 9 \\ 3 & 5 \end{vmatrix} = -13,$$

$$D_3 = \begin{vmatrix} 3 & -2 & 5 \\ 1 & 1 & 2 \\ -1 & 2 & 1 \end{vmatrix} = \begin{vmatrix} 5 & -2 & 9 \\ 0 & 1 & 0 \\ -3 & 2 & -3 \end{vmatrix} = 1 \cdot \begin{vmatrix} 5 & 9 \\ -3 & -3 \end{vmatrix} = 12.$$

Therefore $x = \frac{33}{5}$, $y = \frac{13}{5}$, $z = -\frac{12}{5}$.　□

**COROLLARY**　*If there are numbers x, y, and z, not all zero, which satisfy Equations (6) with* $b_1 = b_2 = b_3 = 0$, *then* $D = 0$.

# 1　PROBLEMS

**1**　Prove Theorem 1.

**2**　Complete the proof of Theorem 4.

In Problems 3 through 5, evaluate the given determinant by expanding it according to (a) the second row, (b) the third column.

**3**　$\begin{vmatrix} 3 & 2 & -1 \\ -1 & 0 & 1 \\ 2 & 1 & -2 \end{vmatrix}$　**4**　$\begin{vmatrix} 1 & 0 & 3 \\ 2 & -1 & -2 \\ 1 & 3 & 2 \end{vmatrix}$　**5**　$\begin{vmatrix} 2 & 3 & 1 \\ 1 & 2 & -2 \\ -2 & 1 & 3 \end{vmatrix}$

In Problems 6 through 8, simplify the determinant, using Theorem 9, and then evaluate it by the expansion theorem.

6
$$\begin{vmatrix} 2 & -1 & 3 \\ 3 & -1 & 2 \\ -1 & 2 & 3 \end{vmatrix}$$
7
$$\begin{vmatrix} 1 & -1 & -2 \\ 2 & 3 & -2 \\ -1 & -3 & 2 \end{vmatrix}$$
8
$$\begin{vmatrix} 2 & 1 & -2 \\ -1 & 3 & 2 \\ -3 & 1 & -2 \end{vmatrix}$$

In Problems 9 through 11, solve by Cramer's Rule and check.

9
$$\begin{aligned} 2x - y + 3z &= 1 \\ 3x + y - z &= 2 \\ x + 2y + 3z &= -6 \end{aligned}$$

10
$$\begin{aligned} 2x - y + z &= -3 \\ x + 3y - 2z &= 0 \\ x - y + z &= -2 \end{aligned}$$

11
$$\begin{aligned} x + 3y - 2z &= 4 \\ -2x + y + 3z &= 2 \\ 2x + 4y - z &= -1 \end{aligned}$$

# PROOFS OF THEOREMS 8, 17, 18, AND 23 OF CHAPTER 14

In this appendix we give statements and proofs of several of
the more difficult theorems on vectors in three dimensions.
The proofs make use of the material on determinants given
in Appendix 4.

**THEOREM 8**   *Suppose that A, B, C, and D are points in space and that a rectangular
coordinate system is introduced in space. Denote the coordinates of A, B, C, and
D by $(x_A, y_A, z_A)$, $(x_B, y_B, z_B)$, and so forth. (i) If the coordinates satisfy the
equations*

$$x_B - x_A = x_D - x_C, \qquad y_B - y_A = y_D - y_C, \qquad z_B - z_A = z_D - z_C, \qquad (1)$$

*then $\vec{AB} \approx \vec{CD}$. (ii) Conversely, if $\vec{AB} \approx \vec{CD}$, the coordinates satisfy the
equations in (1).*

**Proof**   i) We assume that the equations in (1) hold. Then also,

$$x_C - x_A = x_D - x_B, \qquad y_C - y_A = y_D - y_B, \qquad z_C - z_A = z_D - z_B. \qquad (2)$$

From (1) and Corollary 2 on p. 539, it follows that either $AB \parallel CD$ or $A, B, C,$
and $D$ are on a line. From Equations (2), we conclude that either $AC \parallel BD$ or
$A, B, C,$ and $D$ are on a line. Thus, either $ACDB$ is a parallelogram or $A, B, C,$
and $D$ are on a line $\vec{L}$, which we may assume is directed.

   If $ACDB$ is a parallelogram then $\vec{AB} \approx \vec{CD}$ by definition. If $A, B, C,$ and $D$
are on a line $L$, let $\vec{L}$ have the parametric equations

$$x = x_0 + t \cos \alpha, \qquad y = y_0 + t \cos \beta, \qquad z = z_0 + t \cos \gamma \qquad (3)$$

and let $A, B, C,$ and $D$ have $t$ coordinates $t_A, t_B, t_C,$ and $t_D$, respectively. Thus

$x_A = x_0 + t_A \cos \alpha$, $x_B = x_0 + t_B \cos \alpha$, etc. Subtracting, we get

$$
\left.
\begin{aligned}
x_B - x_A &= (t_B - t_A) \cos \alpha, & x_D - x_C &= (t_D - t_C) \cos \alpha, \\
y_B - y_A &= (t_B - t_A) \cos \beta, & y_D - y_C &= (t_D - t_C) \cos \beta, \\
z_B - z_A &= (t_B - t_A) \cos \gamma, & z_D - z_C &= (t_D - t_C) \cos \gamma.
\end{aligned}
\right\} \quad (4)
$$

From the equations in (1) and (4), we conclude that

$$
\left.
\begin{aligned}
(t_B - t_A) \cos \alpha &= (t_D - t_C) \cos \alpha, \\
(t_B - t_A) \cos \beta &= (t_D - t_C) \cos \beta, \\
(t_B - t_A) \cos \gamma &= (t_D - t_C) \cos \gamma.
\end{aligned}
\right\} \quad (5)
$$

Since $\cos \alpha$, $\cos \beta$, and $\cos \gamma$ are never simultaneously zero (as $\cos^2 \alpha + \cos^2 \beta + \cos^2 \gamma = 1$), it follows from (5) that $t_B - t_A = t_D - t_C$, so that $\overline{AB} = \overline{CD}$ and hence $\overrightarrow{AB} \approx \overrightarrow{CD}$ in this case also.

ii) To prove the converse, we assume that $\overrightarrow{AB} \approx \overrightarrow{CD}$. Then either $ACDB$ is a parallelogram or $A$, $B$, $C$, and $D$ are on a directed line $\vec{L}$. Let us first assume the former; we wish to show that the equations in (1) hold. Suppose they do not. It is clear that there are unique numbers $x_E$, $y_E$, $z_E$, coordinates of a point $E \neq D$ such that

$$
x_E - x_C = x_B - x_A, \qquad y_E - y_C = y_B - y_A,
$$

and

$$
z_E - z_C = z_B - z_A.
$$

Then, by part (i), we know that $ACEB$ is a parallelogram (since $C$ is not on line $AB$ because $ACDB$ is a parallelogram). But then $D$ and $E$ must coincide, thus contradicting the fact above that $D \neq E$. Accordingly, the equations in (1) must hold.

To consider the other case, let $\vec{L}$ have the parametric equations (3). If we use our previous notation, we conclude that the equations in (4) hold. But, since $\overrightarrow{AB} \approx \overrightarrow{CD}$, we know by definition that $\overline{AB} = \overline{CD}$, i.e., that $t_B - t_A = t_D - t_C$. But then the equations in (1) follow from those in (4), and the proof is complete. □

---

**THEOREM 17**   *Suppose that **u** and **v** are any vectors, that $\{\mathbf{i}, \mathbf{j}, \mathbf{k}\}$ is a right-handed coordinate triple, and that $t$ is any number. Then*

i) $\mathbf{v} \times \mathbf{u} = -\mathbf{u} \times \mathbf{v}$,

ii) $(t\mathbf{u}) \times \mathbf{v} = t(\mathbf{u} \times \mathbf{v}) = \mathbf{u} \times (t\mathbf{v})$,

iii) $\mathbf{i} \times \mathbf{j} = -\mathbf{j} \times \mathbf{i} = \mathbf{k}$,

  $\mathbf{j} \times \mathbf{k} = -\mathbf{k} \times \mathbf{j} = \mathbf{i}$,

  $\mathbf{k} \times \mathbf{i} = -\mathbf{i} \times \mathbf{k} = \mathbf{j}$,

iv) $\mathbf{i} \times \mathbf{i} = \mathbf{j} \times \mathbf{j} = \mathbf{k} \times \mathbf{k} = \mathbf{0}$.

**Proofs**   i)  By definition, $|\mathbf{v} \times \mathbf{u}| = |\mathbf{u} \times \mathbf{v}|$ and $\mathbf{v} \times \mathbf{u}$ and $\mathbf{u} \times \mathbf{v}$ are both orthogonal to both $\mathbf{u}$ and $\mathbf{v}$ (or are both zero if $\mathbf{u}$ and $\mathbf{v}$ are proportional). Thus $\mathbf{v} \times \mathbf{u} = \pm \mathbf{u} \times \mathbf{v}$. If we let $\mathbf{w} = \mathbf{u} \times \mathbf{v}$, then $\{\mathbf{u}, \mathbf{v}, \mathbf{w}\}$ and $\{\mathbf{v}, \mathbf{u}, -\mathbf{w}\}$ are right-handed (see Theorem 16, Chapter 14), so $\mathbf{v} \times \mathbf{u}$ must equal $-\mathbf{w}$.

ii)  If $t = 0$ or $\mathbf{u}$ and $\mathbf{v}$ are proportional, (ii) certainly holds. Otherwise, let us set $\mathbf{w} = \mathbf{u} \times \mathbf{v}$. Then (ii) follows since all the terms in (ii) have the same magnitude, all are orthogonal to both $\mathbf{u}$ and $\mathbf{v}$, and $\{t\mathbf{u}, \mathbf{v}, t\mathbf{w}\}$ and $\{\mathbf{u}, t\mathbf{v}, t\mathbf{w}\}$ are right-handed by Theorem 16, Chapter 14.

iv)  This follows, since we must have $\mathbf{i} \times \mathbf{i} = -\mathbf{i} \times \mathbf{i} = \mathbf{0}$, etc.

iii)  To prove (iii), we note that, since $\{\mathbf{i}, \mathbf{j}, \mathbf{k}\}$ is right-handed, $\theta = \pi/2$, $|\mathbf{i}| = |\mathbf{j}| = |\mathbf{k}| = 1$, and $\mathbf{k}$ is orthogonal to both $\mathbf{i}$ and $\mathbf{j}$, it follows that $\mathbf{i} \times \mathbf{j} = \mathbf{k}$. That $\mathbf{j} \times \mathbf{i} = -\mathbf{k}$ follows from this and from (i). Since $\{\mathbf{i}, \mathbf{j}, \mathbf{k}\}$ is a coordinate triple, it follows as above that $\mathbf{j} \times \mathbf{k} = \pm \mathbf{i}$. Setting

$$\mathbf{u}_1 = \mathbf{i}, \, \mathbf{v}_1 = \mathbf{j}, \, \mathbf{w}_1 = \mathbf{k}, \, \mathbf{u}_2 = \mathbf{j}, \, \mathbf{v}_2 = \mathbf{k}, \, \mathbf{w}_2 = \mathbf{i},$$

we see that

$$\mathbf{u}_2 = 0 \cdot \mathbf{u}_1 + 1 \cdot \mathbf{v}_1 + 0 \cdot \mathbf{w}_1,$$

$$\mathbf{v}_2 = 0 \cdot \mathbf{u}_1 + 0 \cdot \mathbf{v}_1 + 1 \cdot \mathbf{w}_1,$$

$$\mathbf{w}_2 = 1 \cdot \mathbf{u}_1 + 0 \cdot \mathbf{v}_1 + 0 \cdot \mathbf{w}_1$$

**Since** $\{\mathbf{i}, \mathbf{j}, \mathbf{k}\}$ was given as right-handed, it follows from the discussion in Chapter 14, Section 5, that $\{\mathbf{u}_2, \mathbf{v}_2, \mathbf{w}_2\}$ is right-handed since

$$D = \begin{vmatrix} 0 & 1 & 0 \\ 0 & 0 & 1 \\ 1 & 0 & 0 \end{vmatrix} = +1.$$

The proof that $\mathbf{k} \times \mathbf{i} = \mathbf{j}$ is similar.   □

---

**THEOREM 18**   **(Distributive law)**   *If* $\mathbf{u}$, $\mathbf{v}$, *and* $\mathbf{w}$ *are any vectors,*

i)  $\mathbf{u} \times (\mathbf{v} + \mathbf{w}) = (\mathbf{u} \times \mathbf{v}) + (\mathbf{u} \times \mathbf{w})$ *and*

ii)  $(\mathbf{v} + \mathbf{w}) \times \mathbf{u} = (\mathbf{v} \times \mathbf{u}) + (\mathbf{w} \times \mathbf{u}).$

---

**Proof**   Part (ii) follows from part (i) and part (i) of Theorem 16, for

$$(\mathbf{v} + \mathbf{w}) \times \mathbf{u} = -[\mathbf{u} \times (\mathbf{v} + \mathbf{w})] = -[(\mathbf{u} \times \mathbf{v}) + (\mathbf{u} \times \mathbf{w})]$$

$$= [-(\mathbf{u} \times \mathbf{v})] + [-(\mathbf{u} \times \mathbf{w})] = (\mathbf{v} \times \mathbf{u}) + (\mathbf{w} \times \mathbf{u}).$$

It is clear that (i) holds if $\mathbf{u} = \mathbf{0}$. Otherwise, let $\{\mathbf{i}', \mathbf{j}', \mathbf{k}'\}$ be a right-handed coordinate triple such that $\mathbf{u} = |\mathbf{u}|\mathbf{i}'$ (i.e., $\mathbf{i}'$ is the unit vector in the direction of $\mathbf{u}$). Suppose that

$$\mathbf{v} = a_1\mathbf{i}' + b_1\mathbf{j}' + c_1\mathbf{k}', \qquad \mathbf{u} \times \mathbf{v} = \mathbf{V} = A_1\mathbf{i}' + B_1\mathbf{j}' + C_1\mathbf{k}'.$$

We first find $A_1$, $B_1$, $C_1$ in terms of $a_1$, $b_1$, and $c_1$.

Since $\mathbf{V}$ is orthogonal to both $\mathbf{u}$ and $\mathbf{v}$, we must have

$$\mathbf{V} \cdot \mathbf{u} = A_1|\mathbf{u}| = 0, \qquad \mathbf{V} \cdot \mathbf{v} = A_1 a_1 + B_1 b_1 + C_1 c_1 = 0.$$

Thus

$$A_1 = 0 \quad \text{and} \quad b_1 B_1 + c_1 C_1 = 0 \quad \text{so that} \quad B_1 = -kc_1, \quad C_1 = kb_1$$

for some $k$. Moreover,

$$|\mathbf{V}| = |\mathbf{u}| \cdot \sqrt{a_1^2 + b_1^2 + c_1^2} \sin \theta$$

and

$$\mathbf{u} \cdot \mathbf{v} = |\mathbf{u}| \cdot \sqrt{a_1^2 + b_1^2 + c_1^2} \cos \theta = |\mathbf{u}|a_1.$$

Since $0 \le \theta \le \pi$ and $\cos \theta = a_1/|\mathbf{v}|$, it follows that

$$\sin \theta = \sqrt{b_1^2 + c_1^2}/|\mathbf{v}|.$$

Thus

$$|\mathbf{V}| = |k| \cdot \sqrt{b_1^2 + c_1^2} = |\mathbf{u}| \cdot \sqrt{b_1^2 + c_1^2} \quad \text{so } k = \pm|\mathbf{u}|.$$

Finally $\{\mathbf{u}, \mathbf{v}, \mathbf{V}\}$ must be right-handed, so that

$$\begin{vmatrix} |\mathbf{u}| & 0 & 0 \\ a_1 & b_1 & c_1 \\ 0 & -kc_1 & kb_1 \end{vmatrix} = k|\mathbf{u}| \cdot (b_1^2 + c_1^2) > 0$$

(unless $\mathbf{v} = \mathbf{0}$ or $\mathbf{v}$ is proportional to $\mathbf{u}$). Hence $k = +1$ *and*

$$\mathbf{V} = |\mathbf{u}| \cdot (-c_1 \mathbf{j}' + b_1 \mathbf{k}'). \tag{12}$$

The result in (12) evidently holds also if $\mathbf{v} = \mathbf{0}$ or is proportional to $\mathbf{u}$.

In like manner, if we let

$$\mathbf{w} = a_2 \mathbf{i}' + b_2 \mathbf{j}' + c_2 \mathbf{k}', \qquad \mathbf{W} = \mathbf{u} \times \mathbf{w}, \qquad \mathbf{X} = \mathbf{u} \times (\mathbf{v} + \mathbf{w}),$$

we see that

$$\mathbf{W} = |\mathbf{u}| \cdot (-c_2 \mathbf{j}' + b_2 \mathbf{k}'), \qquad \mathbf{X} = |\mathbf{u}| \cdot [-(c_1 + c_2)\mathbf{j}' + (b_1 + b_2)\mathbf{k}']$$

from which (i) follows.     □

We saw in Chapter 14, Section 7, that the vectors $\mathbf{i}, \mathbf{j}$, and $\mathbf{k}$ corresponding to any given coordinate system in space are linearly independent. Consequently Theorem 23 is a special case of the following Theorem 23′:

---

**THEOREM 23′**    *Suppose that $\mathbf{u}_1, \mathbf{v}_1$, and $\mathbf{w}_1$ are linearly independent and suppose that*

$$\begin{aligned} \mathbf{u}_2 &= a_{11}\mathbf{u}_1 + a_{12}\mathbf{v}_1 + a_{13}\mathbf{w}_1, \\ \mathbf{v}_2 &= a_{21}\mathbf{u}_1 + a_{22}\mathbf{v}_1 + a_{23}\mathbf{w}_1, \qquad D = \begin{vmatrix} a_{11} & a_{12} & a_{13} \\ a_{21} & a_{22} & a_{23} \\ a_{31} & a_{32} & a_{33} \end{vmatrix}. \\ \mathbf{w}_2 &= a_{31}\mathbf{u}_1 + a_{32}\mathbf{v}_1 + a_{33}\mathbf{w}_1, \end{aligned} \tag{6}$$

*Then the set $\{\mathbf{u}_2, \mathbf{v}_2, \mathbf{w}_2\}$ is linearly dependent*    $\Leftrightarrow$    $D = 0$.

---

**Proof**    a) Suppose the set is linearly dependent. Then there are constants $c_1, c_2$, and $c_3$, not all zero, such that

$$c_1 \mathbf{u}_2 + c_2 \mathbf{v}_2 + c_3 \mathbf{w}_2 = \mathbf{0}. \tag{7}$$

If we substitute (6) into (7), we obtain

$$(c_1 a_{11} + c_2 a_{21} + c_3 a_{31})\mathbf{u}_1 + (c_1 a_{12} + c_2 a_{22} + c_3 a_{32})\mathbf{v}_1$$

$$+ (c_1 a_{13} + c_2 a_{23} + c_3 a_{33})\mathbf{w}_1 = \mathbf{0}. \quad (8)$$

Since $\mathbf{u}_1$, $\mathbf{v}_1$, and $\mathbf{w}_1$ are linearly independent, their coefficients must all vanish. That is, we must have

$$a_{11} c_1 + a_{21} c_2 + a_{31} c_3 = 0,$$

$$a_{12} c_1 + a_{22} c_2 + a_{32} c_3 = 0, \quad (9)$$

$$a_{13} c_1 + a_{23} c_2 + a_{33} c_3 = 0.$$

But if (9) holds with $c_1$, $c_2$, and $c_3$ not all zero, the determinant $D'$ of the coefficients must vanish according to the Corollary to Cramer's Rule (Theorem 11, Appendix 4). But $D'$ is obtained from $D$ by interchanging rows and columns. Accordingly, $D = D' = 0$.

b) Now suppose $D = 0$. If all the cofactors $A_{ij}$ are zero, any two rows of $D$ and hence any two of the vectors $\mathbf{u}_2$, $\mathbf{v}_2$, and $\mathbf{w}_2$ are proportional and the set is linearly dependent. Otherwise, some $A_{pq} \neq 0$. By interchanging the order of the vectors, if necessary, we may assume that $p = 3$. From the expansion theorem (Theorem 10, Appendix 4), we conclude that

$$a_{11} A_{1q} + a_{21} A_{2q} + a_{31} A_{3q} = 0,$$

$$a_{12} A_{1q} + a_{22} A_{2q} + a_{32} A_{3q} = 0, \quad (10)$$

$$a_{13} A_{1q} + a_{23} A_{2q} + a_{33} A_{3q} = 0.$$

Since $A_{3q} \neq 0$, we can solve equations (10) for the $a_{3j}$, obtaining

$$a_{31} = k a_{11} + l a_{21}, \qquad a_{32} = k a_{12} + l a_{22}, \qquad a_{33} = k a_{13} + l a_{23}, \Big\}$$
$$k = -A_{1q}/A_{3q}, \qquad l = -A_{2q}/A_{3q}. \qquad (11)$$

In this case it follows from (11) and (6) that $\mathbf{w}_2 = k\mathbf{u}_2 + l\mathbf{v}_2$, and the vectors $\mathbf{u}_2$, $\mathbf{v}_2$, and $\mathbf{w}_2$ are linearly dependent. $\qquad \square$

# INTRODUCTION TO THE USE OF A TABLE OF INTEGRALS

We recall some of the methods of integration with which the reader should be familiar. These devices, together with the integrals listed below, enable the reader to perform expeditiously any integration that is required in order to work the problems in this text.

## 1

### SUBSTITUTION IN A TABLE OF INTEGRALS

**EXAMPLE** Letting $u = \tan x$, $du = \sec^2 x \, dx$, we find that $\int e^{\tan x} \sec^2 x \, dx = \int e^u \, du = e^u + C = e^{\tan x} + C$.

## 2

### CERTAIN TRIGONOMETRIC AND HYPERBOLIC INTEGRALS

We illustrate with trigonometric integrals; the corresponding hyperbolic forms are treated similarly.

a) $\int \sin^m u \cos^n u \, du$.

i) $n$ **an odd positive integer**, $m$ **arbitrary**. Factor out cos $u$ $du$ and express the remaining cosines in terms of sines.

**EXAMPLE 1**

$$\int \sin^4 2x \cos^3 2x \, dx = \frac{1}{2} \int \sin^4 2x (1 - \sin^2 2x) \cdot (2 \cos 2x \, dx)$$

$$= \frac{1}{2} \int (\sin^4 2x - \sin^6 2x) d(\sin 2x)$$

$$= \frac{1}{10} \sin^5 2x - \frac{1}{14} \sin^7 2x + C. \qquad \square$$

ii) $m$ **an odd positive integer**, $n$ **arbitrary**. Factor out sin $u$ $du$ and express the remaining sines in terms of cosines.

iii) $m$ **and** $n$ **both even integers** $\geq 0$. Reduce the degree of the expression by the substitutions

$$\sin^2 u = \frac{1 - \cos 2u}{2}, \qquad \cos^2 u = \frac{1 + \cos 2u}{2}.$$

**EXAMPLE 2**

$$\int \sin^4 u \, du = \frac{1}{4} \int (1 - \cos 2u)^2 \, du$$

$$= \frac{1}{4} \int (1 - 2 \cos 2u) \, du + \frac{1}{8} \int (1 + \cos 4u) \, du$$

$$= \frac{3u}{8} - \frac{\sin 2u}{4} + \frac{\sin 4u}{32} + C. \qquad \square$$

b) $\int \tan^m u \sec^n u \, du$.

i) $n$ **an even positive integer**, $m$ **arbitrary**. Factor out $\sec^2 u \, du$ and express the remaining secants in terms of tan $u$.

**EXAMPLE 3**

$$\int \frac{\sec^4 u \, du}{\sqrt{\tan u}} = \int (\tan u)^{-1/2} (1 + \tan^2 u) \cdot (\sec^2 u \, du)$$

$$= \int [(\tan u)^{-1/2} + (\tan u)^{3/2}] d(\tan u)$$

$$= 2(\tan u)^{1/2} + \frac{2}{5} (\tan u)^{5/2} + C. \qquad \square$$

ii) $m$ **an odd positive integer**, $n$ **arbitrary**. Factor out sec $u$ tan $u$ $du$ and express the remaining tangents in terms of the secants.

**EXAMPLE 4**

$$\int \frac{\tan^3 u \, du}{\sqrt[3]{\sec u}} = \int (\sec u)^{-4/3} \tan^2 u \cdot (\sec u \tan u \, du)$$

$$= \int [(\sec u)^{2/3} - (\sec u)^{-4/3}] d(\sec u)$$

$$= \frac{3}{5} (\sec u)^{5/3} + 3(\sec u)^{-1/3} + C. \qquad \square$$

c) $\int \cot^m u \csc^n u \, du$. These are treated like those in (b).

_____ 3 _____

## TRIGONOMETRIC AND HYPERBOLIC SUBSTITUTIONS

a) If $\sqrt{a^2 - u^2}$ occurs (or $a^2 - u^2$ occurs in the denominator), set $u = a \sin \theta$, $\theta = \arcsin (u/a)$, $du = a \cos \theta \, d\theta$, $\sqrt{a^2 - u^2} = a \cos \theta$.

**EXAMPLE 1**

$$\int \sqrt{a^2 - u^2} \, du = a^2 \int \cos^2 \theta \, d\theta = \frac{a^2}{2} \int (1 + \cos 2\theta) \, d\theta$$

$$= \frac{a^2}{2} (\theta + \sin \theta \cos \theta) + C$$

$$= \frac{a^2}{2} \arcsin \frac{u}{a} + u \sqrt{a^2 - u^2} + C. \qquad \square$$

b) If $\sqrt{a^2 + u^2}$ occurs, set $u = a \tan \theta$, etc.

**EXAMPLE 2**

$$\int u^3 \sqrt{a^2 + u^2} \, du = a^5 \int \tan^3 \theta \sec^3 \theta \, d\theta. \qquad \square$$

The last integral is of type (2)(b)(ii) above.

c) If $\sqrt{u^2 - a^2}$ occurs, set $u = a \sec \theta$, etc.

**EXAMPLE 3**

$$\int \frac{\sqrt{u^2 - a^2}}{u} \, du = \int \frac{a \tan \theta}{a \sec \theta} \cdot a \sec \theta \tan \theta \, d\theta$$

$$= \int a \tan^2 \theta \, d\theta = a \int (\sec^2 \theta - 1) \, d\theta$$

$$= a (\tan \theta - \theta) + C$$

$$= \sqrt{u^2 - a^2} - a \operatorname{arcsec} (u/a) + C. \qquad \square$$

A hyperbolic substitution is sometimes more effective:

**EXAMPLE 4**   If we let $u = a \sinh v$, then

$$\int \sqrt{a^2 + u^2} \, du = \int a^2 \cosh^2 v \, dv = \frac{a^2}{2} \int (1 + \cosh 2v) \, dv$$

$$= \frac{a^2}{2} (v + \sinh v \cosh v) + C$$

$$= \frac{a^2}{2} \operatorname{argsinh} \left( \frac{u}{a} \right) + u \sqrt{a^2 + u^2} + C. \qquad \square$$

_____ 4

## INTEGRALS INVOLVING QUADRATIC FUNCTIONS

Complete the square in the quadratic function and introduce a simple change of variable to reduce the quadratic to one of the forms $a^2 - u^2$, $a^2 + u^2$, or $u^2 - a^2$.

**EXAMPLE**

$$\int \frac{(2x-3)\,dx}{x^2+2x+2} = \int \frac{(2x-3)\,dx}{(x+1)^2+1}.$$

Let $u = x + 1$. Then $x = u - 1$, $dx = du$, and

$$\int \frac{(2x-3)\,dx}{(x+1)^2+1} = \int \frac{(2u-5)\,du}{u^2+1} = \ln(u^2+1) - 5 \arctan u + C$$

$$= \ln(x^2+2x+2) - 5 \arctan(x+1) + C. \qquad \square$$

_____ 5

## INTEGRATION BY PARTS

$$\int u\,dv = uv - \int v\,du.$$

**EXAMPLE**   If we let $u = x$ and $v = e^x$ in the integral $\int xe^x\,dx$, then the formula for integration by parts gives

$$\int xe^x\,dx = \int u\,dv = uv - \int v\,du = xe^x - \int e^x\,dx = xe^x - e^x + C. \qquad \square$$

_____ 6

## INTEGRATION OF RATIONAL FUNCTIONS (QUOTIENTS OF POLYNOMIALS)

If the degree of the numerator $\geq$ that of the denominator, divide out, thus expressing the given function as a polynomial plus a "proper fraction." Each proper fraction can be expressed as a sum of simpler "proper partial fractions":

$$\frac{P(x)}{Q(x)} = \frac{P_1(x)}{Q_1(x)} + \cdots + \frac{P_n(x)}{Q_n(x)},$$

in which no two $Q_i$ have common factors and each $Q_i$ is of the form $(x - a)^k$ or $(ax^2 + bx + c)^k$; of course $Q = Q_1 \cdot Q_2 \cdots Q_n$. Each of these fractions can be expressed uniquely in terms of still simpler fractions as follows:

$$\frac{P_i(x)}{(x - a)^k} = \frac{A_1}{x - a} + \frac{A_2}{(x - a)^2} + \cdots + \frac{A_k}{(x - a)^k},$$

$$\frac{P_i(x)}{(ax^2 + bx + c)^k} = \frac{A_1 x + B_1}{ax^2 + bx + c} + \frac{A_2 x + B_2}{(ax^2 + bx + c)^2} + \cdots$$

$$+ \frac{A_k x + B_k}{(ax^2 + bx + c)^k}.$$

Each of these simplest fractions can be integrated by methods already described. The constants are obtained by multiplying up the denominators and either equating coefficients of like powers of $x$ in the resulting polynomials or by substituting a sufficient number of values of $x$ to determine the coefficients.

**EXAMPLE 1**    Integrate

$$\int \frac{x^2 + 2x + 3}{x(x - 1)(x + 1)} \, dx.$$

**Solution**    According to the results above, there exist constants $A$, $B$, and $C$ such that

$$\frac{x^2 + 2x + 3}{x(x - 1)(x + 1)} = \frac{A}{x} + \frac{B}{x - 1} + \frac{C}{x + 1}.$$

Multiplying up, we see that we must have

$$A(x - 1)(x + 1) + Bx(x + 1) + Cx(x - 1) \equiv x^2 + 2x + 3.$$

The constants are most easily found by substituting $x = 0, 1,$ and $-1$ in turn in this identity, yielding $A = -3$, $B = 3$, $C = 1$.    □

**EXAMPLE 2**    Integrate

$$\int \frac{3x^2 + x - 2}{(x - 1)(x^2 + 1)}.$$

**Solution**    There are constants $A$, $B$, and $C$ such that

$$\frac{3x^2 + x - 2}{(x - 1)(x^2 + 1)} = \frac{A}{x - 1} + \frac{Bx + C}{x^2 + 1},$$

or

$$A(x^2 + 1) + Bx(x - 1) + C(x - 1) \equiv 3x^2 + x - 2.$$

Setting $x = 1, 0,$ and $-1$ in turn, we find that $A = 1$, $C = 3$, $B = 2$.    □

**EXAMPLE 3**    Show how to break up the fraction

$$\frac{2x^6 - 3x^5 + x^4 - 4x^3 + 2x^2 - x + 1}{(x - 2)^3(x^2 + 2x + 2)^2} \equiv \frac{P(x)}{Q(x)}$$

into simplest partial fractions. Do not determine the constants.

**Solution**

$$\frac{P(x)}{Q(x)} = \frac{P_1(x)}{(x-2)^3} + \frac{P_2(x)}{(x^2+2x+2)^2}$$

$$= \frac{A_1}{x-2} + \frac{A_2}{(x-2)^2} + \frac{A_3}{(x-2)^3} + \frac{A_4 x + A_5}{x^2+2x+2}$$

$$+ \frac{A_6 x + A_7}{(x^2+2x+2)^2}.$$

&#9633;

---

**7**

## THREE RATIONALIZING SUBSTITUTIONS

a) If the integrand contains a single irrational expression of the form $(ax+b)^{p/q}$, the substitution $z = (ax+b)^{1/q}$ will convert the integral into that of a rational function of $z$.

**EXAMPLE 1**    If we let $z = (x+1)^{1/3}$ so that $x = z^3 - 1$ and $dx = 3z^2\, dz$, then

$$\int \frac{\sqrt[3]{x+1}}{x}\, dx = \int \frac{z}{z^3-1} \cdot 3z^2\, dz = \int 3\, dz + \int \frac{3\, dz}{(z-1)(z^2+z+1)}.$$

&#9633;

b) If a single irrational expression of the form $\sqrt{a^2 - x^2}$, $\sqrt{a^2 + x^2}$, or $\sqrt{x^2 - a^2}$ occurs with an odd power of $x$ outside, the substitution $z = \sqrt{a^2 - x^2}$ (or etc.) reduces the given integral to that of a rational function.

**EXAMPLE 2**    If we let $z = \sqrt{a^2 - x^2}$, then

$$x^2 = a^2 - z^2, \qquad x\, dx = -z\, dz,$$

$$\int \frac{\sqrt{a^2 - x^2}}{x^3}\, dx = -\int \frac{z^2\, dz}{(z^2 - a^2)^2}.$$

c) In case a given integrand is a rational function of trigonometric functions, the substitution

$$t = \tan(\theta/2), \qquad \theta = 2\arctan t, \qquad d\theta = \frac{2\, dt}{1+t^2},$$

$$\cos\theta = \frac{1-t^2}{1+t^2}, \qquad \sin\theta = \frac{2t}{1+t^2},$$

reduces the integral to one of a rational function of $t$.

&#9633;

**EXAMPLE 3**  Making these substitutions, we obtain the following simplification of an integral.

$$\int \frac{d\theta}{5 - 4 \cos \theta} = \int \frac{(2 \, dt)/(1 + t^2)}{5 - 4[(1 - t^2)/(1 + t^2)]} = \int \frac{2 \, dt}{1 + 9t^2}. \qquad \square$$

## A SHORT TABLE OF INTEGRALS

· The constant of integration is omitted.

### Elementary formulas

1. $\int u^n \, du = \dfrac{u^{n+1}}{n+1}, \quad n \neq -1$

2. $\int \dfrac{du}{u} = \ln |u|$

3. $\int e^u \, du = e^u$

4. $\int a^u \, du = \dfrac{a^u}{\ln a}, \quad a > 0, \, a \neq 1$

5. $\int \sin u \, du = -\cos u$

6. $\int \cos u \, du = \sin u$

7. $\int \sec^2 u \, du = \tan u$

8. $\int \csc^2 u \, du = -\cot u$

9. $\int \sec u \tan u \, du = \sec u$

10. $\int \csc u \cot u \, du = -\csc u$

11. $\int \sinh u \, du = \cosh u$

12. $\int \cosh u \, du = \sinh u$

13. $\int \operatorname{sech}^2 u \, du = \tanh u$

14. $\int \operatorname{csch}^2 u \, du = -\coth u$

15. $\int \operatorname{sech} u \tanh u \, du = -\operatorname{sech} u$

16. $\int \operatorname{csch} u \coth u \, du = -\operatorname{csch} u$

17. $\int \dfrac{du}{\sqrt{a^2 - u^2}} = \arcsin \dfrac{u}{a}, \quad a > |u|$

18. $\int \dfrac{du}{a^2 + u^2} = \dfrac{1}{a} \arctan \dfrac{u}{a}, \quad a \neq 0$

19. $\int \dfrac{du}{u \sqrt{u^2 - a^2}} = \dfrac{1}{a} \operatorname{arcsec} \dfrac{u}{a}, \quad |u| > a$

20. $\int \dfrac{du}{\sqrt{a^2 + u^2}} = \begin{cases} \operatorname{argsinh} \dfrac{u}{a} \\[2mm] \ln (u + \sqrt{a^2 + u^2}) \end{cases}$

21. $\int \dfrac{du}{a^2 - u^2} = \begin{cases} \dfrac{1}{a} \operatorname{argtanh} \dfrac{u}{a}, \quad |u| < a \quad \text{or} \quad \dfrac{1}{2a} \ln \dfrac{a + u}{a - u}, \quad |u| < a \\[3mm] \dfrac{1}{a} \operatorname{argcoth} \dfrac{u}{a}, \quad |u| > a \quad \text{or} \quad \dfrac{1}{2a} \ln \dfrac{u + a}{u - a}, \quad |u| > a \end{cases}$

22. $\int \dfrac{du}{u \sqrt{a^2 - u^2}} = -\dfrac{1}{a} \operatorname{argsech} \dfrac{u}{a}, \quad \text{or} \quad -\dfrac{1}{a} \ln \left( \dfrac{a + \sqrt{a^2 - u^2}}{u} \right), \quad 0 < u < a$

23. $\displaystyle\int\frac{du}{|u|\sqrt{u^2+a^2}} = -\frac{1}{a}\,\text{argcsch}\,\frac{u}{a}, \quad u\neq 0 \quad\text{or}\quad -\frac{1}{a}\ln\left(\frac{a+\sqrt{u^2+a^2}}{u}\right)$

24. $\displaystyle\int\frac{du}{\sqrt{u^2-a^2}} = \begin{cases}\text{argcosh}\,\dfrac{u}{a}, \\[2mm] \ln|u+\sqrt{u^2-a^2}|,\end{cases} \quad |u|>a>0$

### Algebraic forms

25. $\displaystyle\int\frac{u\,du}{a+bu} = \frac{u}{b} - \frac{a}{b^2}\ln(a+bu)$        26. $\displaystyle\int\frac{du}{u(a+bu)} = \frac{1}{a}\ln\left|\frac{u}{a+bu}\right|$

27. $\displaystyle\int\frac{u\,du}{(a+bu)^2} = \frac{a}{b^2}\left(\frac{1}{a+bu} + \frac{1}{a}\ln|a+bu|\right)$

28. $\displaystyle\int\frac{du}{u(a+bu)^2} = \frac{1}{a(a+bu)} + \frac{1}{a^2}\ln\left|\frac{u}{a+bu}\right|$

29. $\displaystyle\int\frac{du}{u\sqrt{a+bu}} = \frac{1}{\sqrt{a}}\ln\left|\frac{\sqrt{a+bu}-\sqrt{a}}{\sqrt{a+bu}+\sqrt{a}}\right|$

30. $\displaystyle\int u\sqrt{a+bu}\,du = \frac{2(3\,bu-2a)\sqrt{(a+bu)^3}}{15b^2}$

31. $\displaystyle\int\frac{\sqrt{a+bu}}{u}\,du = 2\sqrt{a+bu} + a\int\frac{du}{u\sqrt{a+bu}}$

32. $\displaystyle\int\frac{u\,du}{\sqrt{a+bu}} = \frac{2(bu-2a)}{3b^2}\sqrt{a+bu}$

33. $\displaystyle\int\sqrt{a^2-u^2}\,du = \frac{u}{2}\sqrt{a^2-u^2} + \frac{a^2}{2}\arcsin\frac{u}{a}, \quad |u|<a$

34. $\displaystyle\int\sqrt{u^2\pm a^2}\,du = \frac{u}{2}\sqrt{u^2\pm a^2} \pm \frac{a^2}{2}\ln|u+\sqrt{u^2\pm a^2}|$

35. $\displaystyle\int\frac{\sqrt{a^2\pm u^2}}{u}\,du = \sqrt{a^2\pm u^2} - a\ln\left|\frac{a+\sqrt{a^2\pm u^2}}{u}\right|$

36. $\displaystyle\int\frac{\sqrt{u^2-a^2}}{u}\,du = \sqrt{u^2-a^2} - a\arccos\frac{a}{u}, \quad 0<a<|u|$

### Trigonometric forms

37. $\displaystyle\int\tan u\,du = -\ln|\cos u|$        38. $\displaystyle\int\sec u\,du = \ln|\sec u + \tan u|$

39. $\displaystyle\int\sin^2 u\,du = \tfrac{1}{2}u - \tfrac{1}{4}\sin 2u$

40. $\displaystyle\int\sin^n u\,du = -\frac{\sin^{n-1}u\,\cos u}{n} + \frac{n-1}{n}\int\sin^{n-2}u\,du$

41. $\displaystyle\int \cos^n u \, du = \frac{\cos^{n-1} u \sin u}{n} + \frac{n-1}{n} \int \cos^{n-2} u \, du$

42. $\displaystyle\int \frac{du}{\sin^n u} = -\frac{\cos u}{(n-1)\sin^{n-1} u} + \frac{n-2}{n-1} \int \frac{du}{\sin^{n-2} u}, \quad n \neq 1$

43. $\displaystyle\int \sin mu \sin nu \, du = \frac{\sin(m-n)u}{2(m-n)} - \frac{\sin(m+n)u}{2(m+n)}, \quad m \neq \pm n$

44. $\displaystyle\int \cos mu \cos nu \, du = \frac{\sin(m-n)u}{2(m-n)} + \frac{\sin(m+n)u}{2(m+n)}, \quad m \neq \pm n$

45. $\displaystyle\int \sin mu \cos nu \, du = -\frac{\cos(m-n)u}{2(m-n)} - \frac{\cos(m+n)u}{2(m+n)}, \quad m \neq \pm n$

46. $\displaystyle\int u^n \sin u \, du = -u^n \cos u + n \int u^{n-1} \cos u \, du$

47. $\displaystyle\int u^n \cos u \, du = u^n \sin u - n \int u^{n-1} \sin u \, du$

48. $\displaystyle\int \arcsin u \, du = u \arcsin u + \sqrt{1-u^2}$

49. $\displaystyle\int \arccos u \, du = u \arccos u - \sqrt{1-u^2}$

50. $\displaystyle\int u \arcsin u \, du = \frac{1}{4}[(2u^2 - 1)\arcsin u + u\sqrt{1-u^2}]$

## Logarithmic and exponential forms

51. $\displaystyle\int \ln |u| \, du = u(\ln |u| - 1)$

52. $\displaystyle\int (\ln |u|)^2 \, du = u(\ln |u|)^2 - 2u \ln |u| + 2u$

53. $\displaystyle\int u^n \ln |u| \, du = \frac{u^{n+1}}{n+1} \ln |u| - \frac{u^{n+1}}{(n+1)^2}, \quad n \neq -1$

54. $\displaystyle\int u^n e^u \, du = u^n e^u - n \int u^{n-1} e^u \, du$

55. $\displaystyle\int e^{au} \sin bu \, du = \frac{e^{au}(a \sin bu - b \cos bu)}{a^2 + b^2}$

56. $\displaystyle\int e^{au} \cos bu \, du = \frac{e^{au}(a \cos bu + b \sin bu)}{a^2 + b^2}$

## Miscellaneous forms

57. $\displaystyle\int (a + bu)^n \, du = \frac{(a + bu)^{n+1}}{b(n+1)}, \quad n \neq -1$

58. $\int \dfrac{du}{u^2(a+bu)} = -(au)^{-1} + ba^{-2}[\log(a+bu) - \log u]$

59. $\int \dfrac{du}{(a+bu)^{1/2}(c+du)^{3/2}} = \dfrac{2}{bc-ad}\left(\dfrac{a+bu}{c+du}\right)^{1/2}, \quad bc - ad \neq 0$

60. $\int \dfrac{du}{u(a+bu^n)} = (an)^{-1}[\log(u^n) - \log(a+bu^n)], \quad n \neq 0$

61. $\int \dfrac{u\, du}{(a+bu)^{1/2}} = \dfrac{2(bu-2a)}{3b^2}(a+bu)^{1/2}$

62. $\int \dfrac{du}{u^2(u^2+a^2)^{1/2}} = -(a^2u)^{-1}(u^2+a^2)^{1/2}$

63. $\int \dfrac{(a^2-u^2)^{1/2}}{u^2}\, du = -u^{-1}(a^2-u^2)^{1/2} - \arcsin\dfrac{u}{a}$

64. $\int \sin^4 u\, du = \dfrac{3u}{8} - \dfrac{\sin 2u}{4} + \dfrac{\sin 4u}{32}$

65. $\int \cot^3 u\, du = -\tfrac{1}{2}\cot^2 u - \log \sin u$

# ANSWERS TO ODD-NUMBERED PROBLEMS

## CHAPTER 1

### SECTION 1 (page 6)

**1** $x < 5$  **3** $x > -3$

**5** $x < \frac{14}{11}$  **7** $0 < x < \frac{3}{2}$

**9** $-9 < x < 1$  **11** All $x$ not in interval $[0, \frac{20}{3}]$

**13** All $x$ not in interval $[2, \frac{15}{4}]$

**15** All $x$ in interval $(\frac{5}{3}, 4)$

**17** All $x$ not in interval $[2, 3]$

**19** All $x$ in interval $(\frac{1}{3}, 4)$

**21** All $x$ in interval $(\frac{11}{5}, \frac{36}{5})$

**23** All $x$ in $(-\infty, 1), (\frac{3}{2}, 2), (\frac{17}{7}, \infty)$ satisfy both inequalities

**25** All $x$ in $(-\infty, 0)$ and $(\frac{3}{5}, \frac{5}{4})$ satisfy inequalities

**27** $-4 < x < 0$

**31** *Hint:* Multiply by $x$ and transfer all terms to left side, getting $x^2 - 2x + 1 \geq 0$.

**33** *Hint:* Set $x/y = u$ and use the binomial formula for $(a + b)^4$.

### SECTION 2 (pages 9–10)

**1** $-\frac{5}{2}, \frac{3}{2}$  **3** $\frac{1}{2}, \frac{7}{2}$

**5** $\frac{14}{3}, -\frac{16}{3}$  **7** $-\frac{3}{4}, \frac{7}{10}$

**9** $3, -3$  **11** $\frac{15}{2}$

**13** $-2, 6/7$  **15** $-\frac{13}{2}, -\frac{17}{4}$

**17** $-5 < x < 3$  **19** $-4 < x < -1$

**21** $-\frac{13}{5} < x \leq -\frac{11}{5}$  **23** All $x$ not in $[-\frac{11}{2}, -\frac{5}{6}]$

**25** $-\frac{1}{3} \leq x \leq 7$  **27** $|x| < 2$

**29** $(x, y) = (\frac{11}{5}, -\frac{6}{5}), (-1, 2), (1, -2), (-\frac{11}{5}, \frac{6}{5})$

### SECTION 3 (pages 13–14)

**1** $M = 14$  **3** $M = 196$

**5** $M = 10$  **7** $M = 98.5$

**9** $M = 10$  **11** None

**13** One value is 7/2  **15** One value is 28/3

**17** One value is 7/2

### SECTION 4 (page 18)

**1** $S = [-1, 7]$  **3** $S = \{0, 1, 2, 5, 6, 8, 9\}$  **5** $S = \{6\}$

### SECTION 5 (pages 23–24)

**1** $f(-4) = 17, f(-3) = 10, f(-2) = 5, f(-1) = 2, f(0) = 1,$ $f(1) = 2, f(2) = 5, f(3) = 10$

**3** $f(-3) = -2, \quad f(-2) = -4, \quad f(-1) = -4, \quad f(0) = -2,$ $f(1) = 2, f(2) = 8, f(3) = 16, f(a + 2) = a^2 + 7a + 8$

**5** $f(-4) = 2, f(-3) = \frac{7}{3}, f(-2) = 4, f(-1) = -1, f(0) = \frac{2}{3},$ $f(1) = 1, \quad f(-1000) = \frac{2998}{1997}, \quad f(1000) = \frac{3002}{2003}, \quad f[f(x)] =$ $\frac{13x + 12}{12x + 13},$ No

**7** $x^8 - 8x^6 + 20x^4 - 16x^2 + 2$

**9** $f(-1000) = \frac{-2000}{1,000,001}, f(-3) = \frac{-3}{5}, f(-2) = \frac{-4}{5},$ $f(-1) = -1, f(0) = 0, f(1) = 1, f(2) = \frac{4}{5},$ $f(3) = \frac{3}{5}, f(1000) = \frac{2000}{1,000,001}$

**11**

| $x$ | $-1$ | $0$ | $1$ | $2$ | $3$ |
|---|---|---|---|---|---|
| $f(x)$ | $0$ | $1$ | $\sqrt{2}$ | $\sqrt{3}$ | $2$ |

, domain: $x \geq -1$

**13**

| $x$ | $-4$ | $-3$ | $-2$ | $-1$ | $0$ | $1$ | $2$ |
|---|---|---|---|---|---|---|---|
| $f(x)$ | $0$ | $1$ | $2$ | $3$ | $2$ | $1$ | $0$ |

**15** Domain: all $x$ except $x = 1$

**17** Domain: $-1 - \sqrt{3} \le x \le -1 + \sqrt{3}$

**19** Domain: all $x \ne 2, 6, -2$; $f(x) = G(x) = x + 8$ if $x \ne -2, 2, 6$

**21** $4x + 2h + 1$      **23** $\dfrac{-1}{x(x+h)}$

**25** $\dfrac{1}{\sqrt{x+h} + \sqrt{x}}$      **29** $f[g[h(x)]] = 18x^2 + 12x + 5$

**31** $V = \frac{4}{3}\pi^{5/2} r^3$

## SECTION 6 (page 31)

| | Intercepts | Domain | Range | Asymptote |
|---|---|---|---|---|
| **1** | $(0, 0)$ | $x \in R^1$ | $y \ge 0$ | none |
| **3** | $(0, 0)$ | $x \le 0$ | $y \in R^1$ | none |
| **5** | $(0, 3)$ $(\pm\sqrt{3}, 0)$ | $x \in R^1$ | $y \le 3$ | none |
| **7** | none | $x \ne 0$ | $y > 0$ | $x = 0, y = 0$ |
| **9** | $(\pm\sqrt{3}, 0)$ | $|x| \ge \sqrt{3}$ | $y \in R^1$ | no vert. or horiz. |
| **11** | $(0, \pm\sqrt{2})$ $(\pm\sqrt{3}, 0)$ | $|x| \le \sqrt{3}$ | $|y| \le \sqrt{2}$ | none |
| **13** | none | $x > 0$ | $y \ne 0$ | $x = 0, y = 0$ |
| **15** | $(0, \pm 2\sqrt{3})$ $(\pm 2\sqrt{3}, 0)$ | $|x| \le 4$ | $|y| \le 4$ | none |
| **17** | $(0, \pm 2)$ | $x > -1$ | $y \ne 0$ | $x = -1, y = 0$ |
| **19** | none | $x \ne 0$ | $|y| > 2$ | $x = 0,$ $y = 2, -2$ |
| **21** | $(0, \pm 2)$ | $x \in R^1$ | $|y| \le 2,$ $y \ne 0$ | $y = 0$ |
| **23** | $(0, \frac{4}{3})$ | $x \ne 1, 3$ | $y \le -4$ or $y > 0$ | $x = 1, 3,$ $y = 0$ |
| **25** | $(0, 0)$ | $x \in R^1$ | $y \ne \pm 2$ | $x = 0,$ $y = 2, -2$ |
| **27** | $(0, 0)$ | $-1 < x \le 0,$ $x > 1$ | $y \in R^1$ | $y = 0,$ $x = 1, -1$ |
| **29** | $(0, 0)$ | $x \in R^1$ | $y \in R^1$ | none |
| **31** | $(-2, 0)$ | $-2 \le x < -1,$ $x > 1$ | $y \in R^1$ | $x = +1, -1$ $y = 0$ |
| **33** | $(-\frac{3}{5}, 0)$ $(0, \frac{2}{5})$ | $x \ne 1, 3$ | $y \ge -\frac{1}{5}$ or $y \le -5$ | $x = 1, 3,$ $y = 0$ |

**35** Range: all of $R^1$; intercepts: $(0, 3), (0, -1), (3, 0), (1, 0)$

**37** Domain: all of $R^1$; range all of $R^1$; intercepts: $(0, \pm\sqrt{3})$, $(-3, 0), (-1, 0)$

## SECTION 7 (pages 34–35)

**1** $|AB| = 2\sqrt{2}, |BC| = 3\sqrt{5}, |AC| = \sqrt{41}$

**3** $|AB| = \sqrt{26}, |BC| = 5, |AC| = \sqrt{13}$

**5** $\bar{x} = \frac{9}{2}, \bar{y} = \frac{1}{2}$      **7** $(\frac{19}{4}, 2), (\frac{5}{2}, 3), (\frac{1}{4}, 4)$

**9** $(-16, -1)$      **11** $\frac{1}{2}\sqrt{149}, \frac{1}{2}\sqrt{389}, \frac{1}{2}\sqrt{338}$

**13** $|AB| = |BC| = \sqrt{41}$

**15** $|AB| = \sqrt{13}, |BC| = \sqrt{65}, |AC| = \sqrt{52}; |AC|^2 + |AB|^2 = |BC|^2; \therefore$ right triangle

**19** $(\frac{3}{2}, 2)$

**23** $M_{AB} = (\frac{1}{2}(x_1 + x_2), \frac{1}{2}(y_1 + y_2)), M_{BC} = (\frac{1}{2}(x_2 + x_3), \frac{1}{2}(y_2 + y_3)), \quad |M_{AB}M_{BC}| = \frac{1}{2}\sqrt{(x_3 - x_1)^2 + (y_3 - y_1)^2} = \frac{1}{2}|AC|$. Other sides similar.

**25** $x^2 + y^2 - 2x - 2y - 2 = 0$; circle, center $(1, 1)$, radius 2

**27** $(y + 1)^2 = 8(x - 2)$

**29** $(x - 5)^2 + (y + 1)^2 = 4$, circle, center $(5, -1)$, radius 2

## SECTION 8 (pages 40–41)

**1** Parallel      **3** Perpendicular

**5** Parallel      **7** Yes

**9** No      **11** $3x - 4y - 23 = 0$

**13** Parallelogram      **15** Trapezoid

**17** None of these      **19** None of these

**21** Rhombus      **23** None of these

**25** Completely general since axes may always be chosen so that origin is at $A$ and the base is along the $x$ axis

**27** Completely general for reasons similar to answer to Problem 25. Point of bisection is

$$\left(\frac{a + b + e}{4}, \frac{c + f}{4}\right)$$

**29** Coordinates are: $(a, \frac{1}{2}a\sqrt{3}), (\frac{1}{2}a, a\sqrt{3}), (-\frac{1}{2}a, a\sqrt{3})$ $(-a, \frac{1}{2}a\sqrt{3})$

## SECTION 9 (page 45)

**1** $2x - y + 6 = 0$      **3** $2x + 3y - 11 = 0$

**5** $x - 2y - 8 = 0$      **7** $y = -7$

**9** $y = -2$      **11** $y + 5 = 0$

**13** $4x - 2y - 9 = 0$      **15** $y + 3x = 0$

**17** $y + 5 = 0$      **19** $m = \frac{2}{3}, b = -\frac{7}{3}$

**21** $3x - 7y - 15 = 0$      **23** $4x + 5y - 29 = 0$

**25** $8x - 3y - 38 = 0$

**27** $15x - y + 78 = 0$

**29** $7x - y - 15 = 0$

**35** $\left(\dfrac{a+b}{3}, \dfrac{c}{3}\right)$

**37** $\left(b, \dfrac{ab - b^2}{c}\right)$

**11** $\pm 1, \pm 5$

**15** $5$

**17** $50/53$

**23**

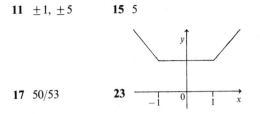

**25** Intercepts: $(0, 2)$, $(\sqrt[3]{-2}, 0)$; domain: all of $R^1$; range: all of $R^1$; no asymptotes

**27** Intercept: $(0, 0)$; domain: $0 \le x < \infty$; range: all of $R^1$; no asymptotes

**29** (a) $(7/2, 7/2)$; (b) $3x + 5y = 35$

**31** All points on the line $6x - 4y + 5 = 0$

**33** $2x - 3y = 7$     **35** $k = 1/4$

## CHAPTER 1 REVIEW PROBLEMS (page 46)

**1** $x < 1/3$

**3** All $x$ not in the interval $[1, 2]$

**5** $x < -11$

**7** All $x$ not in the interval $[5/4, 7/2]$

**9** $x$ in the interval $\left(\dfrac{1 - \sqrt{5}}{2}, \dfrac{1 + \sqrt{5}}{2}\right)$

## CHAPTER 2

## SECTION 1 (page 54)

**1** $-4$     **3** $\frac{32}{5}$     **5** $3$

**7** $\frac{1}{4}$     **9** $-\frac{2}{3}$     **11** $6$

**13** $-1/2$; $\lim\limits_{x \to 2} f(x) = -1/5$

**15** $15$     **17** $8$     **19** $\sqrt{5}/20$

**21** $\frac{4}{3}$     **23** $1/\sqrt{b}$     **25** $-6144$

**27** $1/(2\sqrt[4]{8})$     **29** Limit is 1

**31** Graph indicates limit must be 1

**33** $f(\pm 0.1) = \pm 0.04996$;  $f(\pm 0.01) = \pm 0.005$;  $f(\pm 0.001) = \pm 0.0005$; $f(\pm 0.0001) = \pm 0.00005$; $\lim\limits_{x \to 0} f(x) = 0$

## SECTION 3 (page 61)

**1** $6x$     **3** $4x - 3$     **5** $-4x^3$

**7** $\dfrac{-1}{(x-1)^2}$     **9** $\dfrac{-3}{x^4}$     **11** $\dfrac{1}{(1-x)^2}$

**13** $\dfrac{4x}{(x^2+1)^2}$     **15** $\dfrac{-1}{2x^{3/2}}$     **17** $\frac{3}{2}\sqrt{x}$

**19** $\dfrac{x-2}{2(x-1)^{3/2}}$     **21** $\dfrac{-x}{\sqrt{4-x^2}}$     **23** $\dfrac{3x^2-6}{(x^2+x+2)^2}$

**25** $\dfrac{2x+1}{2\sqrt{x^2+x+1}}$     **27** $-3/(x+1)^4$     **29** $-4/x^5$

**31** $4/(3x+4)^2$

**33** $f'(x) = 1$ for $x > 5$; $f'(x) = -1$ for $x < 5$

## SECTION 2 (pages 57–58)

**1** $24$     **3** $6$     **5** $4a$

**7** $-4$     **9** $-\frac{1}{9}$     **11** $-2x$

**13** $29$     **15** $0$     **17** $\frac{3}{2}\sqrt{x}$

**19** $2c + 3$     **21** $-3x^2/(x^3+1)^2$     **23** $1 + \dfrac{1}{2\sqrt{2}}$

**25** $2/(x+1)^2$     **27** $-3/16$     **29** $\dfrac{y^2+4y+1}{2(y+2)^2}$

## SECTION 4 (pages 65–66)

**1** Tangent: $3x - y - 2 = 0$; normal: $x + 3y - 4 = 0$

**3** Tangent: $4x - y - 5 = 0$; normal: $x + 4y - 14 = 0$

**5** Tangent: $9x + y - 17 = 0$; normal: $x - 9y - 11 = 0$

**7** Tangent: $x - 2y + 2 = 0$; normal: $2x + y - 6 = 0$

**9** Tangent: $x + 16y - 14 = 0$; normal: $32x - 2y - 191 = 0$

**11** Tangent: $y = 4$; normal: $x = 1$

**13** Decreasing if $x \le 2$; increasing if $x \ge 2$; rel. min. at $x = 2$

**15** Increasing if $x \le -1$, $x \ge 2$; decreasing if $-1 \le x \le 2$; rel. max. at $x = -1$; rel. min. at $x = 2$

**17** Decreasing if $x \le 0$, $x \ge 2$; increasing if $0 \le x \le 2$; rel. min., $x = 0$; rel. max., $x = 2$

**19** Decreasing if $x \le -1$, $1 \le x \le 2$; increasing if $-1 \le x \le 1$, $x \ge 2$; rel. min., $x = -1$, 2; rel. max. $x = 1$

**21** $\Delta y = -0.09$, $\dfrac{\Delta y}{\Delta x} = -0.9$

**23** $\Delta y = \dfrac{0.000901}{1.0001}$, $\dfrac{\Delta y}{\Delta x} = \dfrac{0.0901}{1.0001}$

**25** $\dfrac{4x}{(x^2+1)^2}$, at $x = 1$, value is 1

**27** $-3/25$    **29** $5/2$

## SECTION 5 (pages 70–71)

**1** $v = 4$, moving right    **3** $v = 14$, right

**5** $v = 26$, moving right    **7** $-4/25$, left

**9** $v = 2t - 4$; $a = 2$; $v = 0$ at $t = 2$

**11** $v = t^2 - 4t + 3$, $a = 2t - 4$; $v = 0$ at $t = 1, 3$

**13** $v = 2bt + c$; $a = 2b$; $v = 0$ if $t = -c/2b$

**15** $v = (t^2 + 4t + 1)/(t+2)^2$; $a = 6/(t+2)^3$; $v$ never zero

**17** $v = 2t + 3$, $a = 2$; left if $t < -\frac{3}{2}$, right if $t > -\frac{3}{2}$; $a = 2$

**19** $v = 2t - 4$, $a = 2$; moving left if $t < 2$, right if $t > 2$

**21** $v = 6(t^2 - t - 2)$, $a = 6(2t - 1)$; moving left if $-1 < t < 2$, right if $t < -1$ or $t > 2$

**23** $v = \dfrac{(1-t)(1+t)}{(1+t^2)^2}$, $a = -2t(3-t^2)(1+t^2)^{-3}$; moving left if $t < -1$ or $t > 1$, right if $-1 < t < 1$

**25** $v = -2/(t-1)^2$; $a = 4/(t-1)^3$; moving left for all $t \ne 1$

**27** $v = 4bt^3 + c$, $a = 12bt^2$

**29** Max. height is $\frac{245}{4}$ meters, reached when $t = \frac{3}{2}$ sec. The stone hits the ground when $t = 5$ sec.

**31** $g = 1.6$ m/sec$^2$

## SECTION 6 (pages 74–75)

**1** $\delta = 0.0005$    **3** $\delta = 0.005$

**5** $\delta = 0.029701$    **7** $\delta = 0.0398$

**9** $\delta = \frac{84}{121}$    **11** $\delta = \frac{1}{101}$

**13** $\delta = \frac{200}{2601}$    **15** $\dfrac{\sqrt{101} - 10}{10}$

**17** $\delta = 0.05$, although it is not the largest possible $\delta$

**19** Since $f(x) - f(7) = 0$ for all $x$, any $\delta > 0$ is valid for all $\varepsilon > 0$

**21** $\delta = \varepsilon$

**23** $\delta = \varepsilon_1(\varepsilon_1^2 - \sqrt[3]{81}\varepsilon_1 + \sqrt[3]{243})$ where $\varepsilon_1$ is the smaller of $\varepsilon$ and $\sqrt[3]{3}$

**25** $\delta = \dfrac{\varepsilon_1}{1 + \varepsilon_1}$ where $\varepsilon_1$ is the smaller of $\varepsilon$ and 1

## SECTION 7 (pages 82–83)

**1** 9    **3** $-\frac{1}{2}$    **5** 12    **7** $\frac{3}{4}$    **9** $\frac{3}{2}$

**11** 2    **13** $\sqrt{3}$    **15** $1/(2\sqrt{x})$    **17** $-\frac{1}{2}$    **19** 80

**21** No limit

## SECTION 8 (pages 88–89)

**1** 0    **3** 0    **5** 0

**7** No limit    **9** 1    **11** 4

**13** Both limits 0. Yes    **15** Continuous, $-4 < x < 7$

**17** Discontinuous at $x = 2$    **19** Continuous, $0 < x < 2$

**21** Continuous, $-\infty < x < \infty$    **23** Continuous

**25** Discontinuous at $x = 2$    **27** Continuous

**29** Continuous    **31** Continuous for all $x$

## SECTION 9 (pages 93–94)

**1** $\frac{2}{3}$    **3** 0    **5** $-1$    **7** $\infty$

**9** $\infty$    **11** $+\infty$    **13** $+\infty$    **15** $+\infty$

**17** Discontinuous at $x = -3, 3$

**19** Discontinuous at $x = 2$

**21** 0    **23** No limit

**27** $\displaystyle \lim_{x \to \infty} \frac{P(x)}{Q(x)} = \begin{cases} 0 & \text{if } m > n \\ a_n/b_n & \text{if } m = n \\ \infty & \text{if } m < n \end{cases}$

**29** If for some $i$ and $j$ it is true that $d_i = c_j$, then limit is as in Problem 27. If $d_i \ne c_j$ for all $i$, $j$, then limit is $\infty$.

## CHAPTER 2 REVIEW PROBLEMS (pages 94–95)

**1** $-3/2$    **3** $1/42$    **5** $4x - 1$

**7** $3x^2 - 4x$    **9** $-3/(x-1)^2$    **11** $-2x^{-3} + 2$

**13** $4x^3 + 2x$   **15** $x + y + 3 = 0$   **17** $2x - 3y + 5 = 0$

**19** $v = 0$ when $t = -1$; $a$ never zero

**21** $v = 0$ when $t = \frac{1}{3}(-2 \pm \sqrt{7})$; $a = 0$ when $t = -2/3$

**23** $v$ and $a$ never zero

**25** 9.8 sec   **27** $-3/2$   **29** $\sqrt{6}$

**31** 3   **33** 2   **35** $-1 \leq x \leq 8$

**37** $-1 \leq x \leq 5$

**39** all $x$ except $x = -2, 3$

# CHAPTER 3

## SECTION 1 (pages 103–104)

**1** $2x + 3$   **3** $14x^{13} + 20x + 7$

**5** $-2x^{-3} + 20x^{-6} - 24x^{-9}$

**7** $3x^2 + 4x - 3 + 2x^{-2} - 8x^{-3}$

**9** $2x + 2 + 2x^{-3}$   **11** $-\frac{21}{4}x^{-8}$

**13** $9x^2 + 14x + 2$

**15** $(3x^2 + 12x - 2)(x^2 + 3x - 5) + (x^3 + 6x^2 - 2x + 1)(2x + 3) =$
$5x^4 + 36x^3 + 33x^2 - 70x + 13$

**17** $2 - 4x^{-2}$   **19** $3x^{-2} - 8x^{-3} + 6x^{-4}$

**21** $\dfrac{2}{(x+1)^2}$   **23** $\dfrac{13}{(2x+3)^2}$

**25** $(1 - x^2)/(1 + x^2)^2$   **27** $-\dfrac{x^2 + 2x}{(x^2 + 2x + 2)^2}$

**29** $\dfrac{4x^2 + 12x + 1}{(2x+3)^2}$

**31** a) $4(x + 1)(x^2 + 2x - 1)$
   b) $2(x^3 + 7x^2 - 8x - 6)(3x^2 + 14x - 8)$
   c) $2(x^7 - 2x + 3x^{-2})(7x^6 - 2 - 6x^{-3})$

**35** $[Q(x)]^{-2}[(n - m)a_n b_m x^{m+n-1} + \{a_{n-1}b_m(n - m - 1)$
$+ a_n b_{m-1}(n - m + 1)\}x^{m+n-2} + \cdots + a_1 b_0 - a_0 b_1]$

## SECTION 2 (pages 110–111)

**1** $18(2x + 3)^8$   **3** $-20(2 - 5x)^3$

**5** $-8(2x + 1)^{-5}$

**7** $4(x^3 + 2x - 3 + x^{-2})^3(3x^2 + 2 - 2x^{-3})$

**9** $2x(x^2 + 1)(x^2 - 2)(5x^4 - 3x^2 - 2)$

**11** $\dfrac{(x^3 + 2x - 6)^6}{x^2}(23x^6 + 39x^4 - 32x^3 + 14x^2 - 12x - 6)$

**13** $\dfrac{2x(x^2 + 1)^2(x^2 + 4)}{(x^2 + 2)^3}$

**15** $-\dfrac{(x^2 + 2)(2x^5 + 12x^3 - 7x^2 - 6)}{x^2(x^2 + x^{-1})^4}$

**17** $\frac{5}{3}x^{2/3} + \frac{8}{3}x^{1/3} + x^{-4/3}$   **19** $-\frac{2}{3}x^{-5/3} - \frac{4}{3}x^{-7/3} - \frac{8}{7}x^{-3/7}$

**21** $x^{1/3} + x^{-1/2} + 2x^{-2}$   **23** $\frac{20}{3}(2x + 3)^{7/3}$

**25** $3(x + 1)(x^2 + 2x + 3)^{1/2}$

**27** $-(2x^2 - 2x + 1)(2x^3 - 3x^2 + 3x - 1)^{-4/3}$

**29** $2(x^2 + 1)(x^3 + 3x + 2)^{-1/3}$

**31** $\frac{3}{10}(2x + 1)(x^2 + x - 3)^{1/2}$

**33** $(2x + 3)^3(3x - 2)^{4/3}(38x + 5)$

**35** $(2x - 1)^{3/2}(7x - 3)^{-4/7}(41x - 18)$

**37** 0   **39** 3725/6

**41** $-2t^{-2}(2 + 6t)^{-2/3}(1 + 2t)$

**43** $\frac{7}{2}(y^2 + 3y + 4)^{-3/2}$

**45** Tangent: $5x - y - 9 = 0$; normal: $x + 5y - 7 = 0$

**47** Tangent: $7x + 3y - 64 = 0$; normal: $3x - 7y + 72 = 0$

**51** $f'(x) = g'(u + v) \cdot (u' + v')$; $F'(x) = G'(u \cdot v)(uv' + vu')$;

$$H'(x) = h'\left(\frac{u}{v}\right) \cdot \left(\frac{vu' - uv'}{v^2}\right)$$

**53** a) $(x^2 + 1)^{1/2}(2x - 5)^{-2/3}(x^2 + 4) \cdot [(x^2 + 1)(4x)(2x - 5)$
$+ (x^2 + 1)(\frac{2}{3})(x^2 + 4) + 3x(2x - 5)(x^2 + 4)]$
   b) $(x - 1)^2(3 - 2x)(x^2 - 5)^{1/3}$

$$\times \left[ 3(3 - 2x)(x^2 - 5) - 4(x - 1)(x^2 - 5) \right.$$
$$\left. + \frac{8x}{3}(x - 1)(3 - 2x) \right]$$

**55** $\frac{1}{3}x^{-2/3}$

## SECTION 3 (pages 114–115)

**1** $y' = x/2y$   **3** $y' = -x^3/2y^3$

**5** $y' = (2x - 3y)/(3x - 2y)$    **7** $y' = -(x^2 + 2y)/(2x + 5y^2)$

**9** $y' = (6x^2 - 6xy + 2y^2)/(3x^2 - 4xy + 3y^2)$

**11** $y' = (6x^5 + 6x^2y - y^7)/(-2x^3 + 7xy^6)$

**13** $y' = (x - y)/x; \ y = \frac{1}{2}x - \frac{3}{2}x^{-1}$

**15** $y' = -4x^{-2} = -y/x; \ y = 4x^{-1}$

**17** $y' = -x^{-1/3}y^{1/3}; \ y = \pm(a^{2/3} - x^{2/3})^{3/2}$

**19** $y' = -1/y; \ y = \pm(8 - 2x)^{1/2}$

**21** $y' = 2x/y; \ y = \pm(2x^2 - 1)^{1/2}$

**23** $y' = (4x - 3y)/(3x + 8y); \ y = (-3x \pm \sqrt{41x^2 - 80})/8$

**25** $y' = -\frac{1}{2}; \ y = -(2x)^{1/2}$    **27** $y' = 3; \ y = -(10 - x^2)^{1/2}$

**29** $y' = \frac{1}{6}; \ y = \frac{3}{4}x - \frac{1}{4}(16 - 7x^2)^{1/2}$

**31** $y' = 3y/(2y - 3x)$

**35** $x + 2y = 6$          **37** $10x + y = 14$

## SECTION 5 (pages 121–122)

**1** $5$                **3** $\frac{2}{5}$              **5** $\frac{1}{8}$

**7** $1$                **9** $0$               **11** $\frac{1}{8}$

**13** $1$               **15** $1$               **17** $-1$

**19** $3\sec^2 3x$                   **21** $2s\sec(s^2)\tan(s^2)$

**23** $-\csc^3 \frac{1}{3}y \cot \frac{1}{3}y$            **25** $-2\cot^2 2x + 2$

**27** $\sec^2 2s/\sqrt{\tan 2s}$          **29** $x \tan\frac{1}{2}x(x\sec^2\frac{1}{2}x$
$+ 2\tan\frac{1}{2}x)$

**31** $2\sec 2x\tan^3 2x$          **33** $2(x\cos 2x - \sin 2x)/x^3$

**35** $-2\cos 3x(1 + x^2)^{-2}[x\cos 3x + 3(1 + x^2)(\sin 3x)]$

**37** $-6\cos 3x\csc^3 2x\,(\sin 3x + \cos 3x\cot 2x)$

**39** $-2\cot ax[(1 + x^2)a\csc^2 ax + x\cot ax]/(1 + x^2)^2$

**41** $\sin 4x(1 + \sin^2 2x)^{-1/2}$    **43** $-\sin x\cos(\cos x)$

**45** $-[x(1 + \cos^2 3x) + (1 + x^2)\sin 6x] \div$
$((1 + x^2)^{3/2}(1 + \cos^2 3x)^{2/3})$

**51** $x - \sqrt{2}y = \frac{1}{4}\pi - 1$        **53** $2x - y = \frac{1}{2}\pi - 2$

**55** $\sqrt{3}x - 4y = \frac{1}{6}\pi\sqrt{3} - 1$

## CHAPTER 3 REVIEW PROBLEMS (pages 122–123)

**1** $3x^2 - 4x + 7$

**3** $(2x + 2)(x^4 + 7x - 5) + (x^2 + 2x - 1)(4x^3 + 7)$

**5** $(-x^4 - 2x^3 + 5x^2 + 10x + 7)/(x^3 + 2x + 5)^2$

**7** $21(x^3 + 12x + 5)^6(x^2 + 4)$

**9** $(x + 6)^{-1}(x^2 - 2x + 1)(3x^2 + 1) + (x + 6)^{-1}(x^3 + x + 5) \cdot$
$(2x - 2) - (x + 6)^{-2}(x^2 - 2x + 1)(x^3 + x + 5)$

**11** $\frac{3}{2}x^{-1/2} - \frac{2}{3}x^{-2/3}$        **13** $\frac{5}{3}(x^2 + 2x - 1)^{2/3}(2x + 2)$

**15** $-3(x^2 + 2x^{1/3} - 4)^{-4}(2x + \frac{2}{3}x^{-2/3})$

**17** $(s - 1)^{-1/2}(s + 1)^{-3/2}$        **19** $-6(3t + 4)^{-2/3}(3t - 2)^{-4/3}$

**21** $(5y^3 - y^{2/3})^{-2}[(5y^3 - y^{2/3})(3y^2 - 6y - 5)$
$-(y^3 - 3y^2 - 5y + 2)(15y^2 - \frac{2}{3}y^{-1/3})]$

**23** $f'(x) = g'(3u - 2v)\left(3\dfrac{du}{dx} - 2\dfrac{dv}{dx}\right)$

**25** $x^2/2y^2$

**27** $-2(x + 3xy^3 - y)/(9x^2y^2 - 2x)$

**29** $(\frac{1}{3}x^{-2/3} + 1)/(y^{-2/3} + 1)$

**31** $-(y\tan(xy) + xy^2\sec^2(xy) + 2xy^2) \div$
$(x\tan(xy) + x^2y\sec^2(xy) + 2x^2y)$

**33** For $x > 2, \dfrac{dy}{dx} = 2/3$; for $x < 2, \dfrac{dy}{dx} = -2/3$;

$\dfrac{dy}{dx}$ does not exist at $x = 2$

**35** $2\sec^2 2x + \frac{1}{3}\sin(\frac{1}{3}x)$      **37** $-\frac{8}{3}\sin(2t)\cos^{1/3}(2t)$

**39** $4[\sec^2(2x)\tan(2x) - \csc^2(4x)]$

**41** $24[\sec^4(3x)\tan(3x)] - \frac{140}{3}\tan^{1/3}(5x)\sec^2(5x)$

**43** $-8\sin(2x)\cos^3(2x)\sin^2(3x)$
$+ 6\cos^4(2x)\sin(3x)\cos(3x)$

**45** tan line: $y = 1$; norm. line: $x = 0$

**47** tan line: $8x - y + 5 = 0$;
norm line $x + 8y + 25 = 0$

**49** tan line: $34x - y + \frac{17}{18}\pi + 2\sqrt{3} = 0$
norm line: $x + 34y = 68\sqrt{3} + \pi/204$

**51** $-6\cos 3x\sin[2\sin(3x)]$

# CHAPTER 4

## SECTION 1 (pages 129–130)

**1** Max. at $(1, 1), (3, 1)$; min. at $(0, 0), (2, 0)$

**9** $(-1, -2)$          **11** Max., $(1, 2)$; min., $(3, -6)$

**13** Max., $(2, 2); f'(2) = 0$   **15** Max. at $(0, 1)$

**17** No

## SECTION 2 (pages 134–135)

**1** $x_0 = 1$     **3** $x_0 = \frac{11}{3}$     **5** $x_0 = -2 + \sqrt{6}$

**7** $x_0 = \pm\frac{1}{2}\sqrt{2}$     **9** $x_0 = 0, \frac{1}{2}, 1$     **11** $x_0 = 1$

**13** Zeros at $x = -2, 1, 4$; $f'(x) = 0$ at $x = 1 + \sqrt{3}, 1 - \sqrt{3}$

**15** The equation $\frac{5}{2} = -5/(3x+2)^2$ has no solution

**17** There is no number $x_0$. $f$ becomes infinite at $x = 1$; Mean value Theorem not applicable

**21** Mean Value Theorem applicable as $f$ and $f'$ are continuous for $0 < x < 2$

## SECTION 3 (pages 139–140)

**1** Decreasing for $x \leq -\frac{3}{2}$; increasing for $x \geq -\frac{3}{2}$; rel. min. at $x = -\frac{3}{2}$

**3** Increasing for $x \leq \frac{1}{4}$; decreasing for $x \geq \frac{1}{4}$; rel. max. at $x = \frac{1}{4}$

**5** Increasing for $x \leq -2$, $x \geq 1$; decreasing for $-2 \leq x \leq 1$; rel. max. at $x = -2$, rel. min. at $x = 1$

**7** Increasing for $x \leq \dfrac{-2 - \sqrt{13}}{3}$;

decreasing for $\dfrac{-2 - \sqrt{13}}{3} \leq x \leq \dfrac{-2 + \sqrt{13}}{3}$;

increasing for $x \geq \dfrac{-2 + \sqrt{13}}{3}$;

rel. max. at $x = \dfrac{-2 - \sqrt{13}}{3}$; rel. min. at $x = \dfrac{-2 + \sqrt{13}}{3}$

**9** Increasing for all $x$; $f'(x) = 0$ if $x = -2$

**11** Decreasing for $x \leq \frac{1}{3}$, $x \geq 1$; increasing for $\frac{1}{3} \leq x \leq 1$; $x = \frac{1}{3}$, rel. min.; $x = 1$, rel. max.

**13** Decreasing, $x \leq -1$; increasing, $-1 \leq x \leq 0$; decreasing, $0 \leq x \leq 2$; increasing $x \geq 2$; rel. min. at $x = -1, 2$; rel. max. at $x = 0$

**15** Decreasing, $x \leq -\frac{3}{2}$; increasing, $x \geq -\frac{3}{2}$; rel. min. $x = -\frac{3}{2}$

**17** Increasing for $x \leq -1$, $x \geq 1$; decreasing for $-1 \leq x < 0$, $0 < x \leq 1$; rel. max. at $x = -1$; rel. min. at $x = 1$; $x = 0$ is a vertical asymptote

**19** Rel. max. at $x = 0$, $\pm 2\pi$, $\pm 4\pi$, ..., $\pm 2n\pi$, ...; rel. min. at $x = \pm\pi$, $\pm 3\pi$, ..., $\pm(2n+1)\pi$, ...; decreasing, $0 \leq x \leq \pi$, $2\pi \leq x \leq 3\pi$, ..., $2n\pi \leq x \leq (2n+1)\pi$, ...; $-2\pi \leq x \leq -\pi$, $-4\pi \leq x \leq -3\pi$, ..., $-2n\pi \leq x \leq -(2n-1)\pi$, ...; increasing, $\pi \leq x \leq 2\pi$, $3\pi \leq x \leq 4\pi$, ..., $(2n-1)\pi \leq x \leq 2n\pi$, ...; $-\pi \leq x \leq 0$, $-3\pi \leq x \leq -2\pi$, ..., $-(2n+1)\pi \leq x \leq -2n\pi$, ...

**21** rel. max. at: $\dfrac{\pi}{4}$, $\pi + \dfrac{\pi}{4}$, ..., $n\pi + \dfrac{\pi}{4}$, ...; $\dfrac{-3\pi}{4}$, $-\pi - \dfrac{3\pi}{4}$, ..., $-n\pi - \dfrac{3\pi}{4}$, ...; rel. min. at $\dfrac{3\pi}{4}$, $\pi + \dfrac{3\pi}{4}$, ..., $n\pi + \dfrac{3\pi}{4}$, ...; $-\dfrac{\pi}{4}$, $-\pi - \dfrac{\pi}{4}$, ..., $-n\pi - \dfrac{\pi}{4}$, ...; increasing, $-\dfrac{\pi}{4} \leq x \leq \dfrac{\pi}{4}$, $\dfrac{3\pi}{4} \leq x \leq \pi + \dfrac{\pi}{4}$, ..., $n\pi + \dfrac{3\pi}{4} \leq x \leq (n+1)\pi + \dfrac{\pi}{4}$, ...; $-\pi - \dfrac{\pi}{4} \leq x \leq -\dfrac{3\pi}{4}$, $2\pi - \dfrac{\pi}{4} \leq x \leq -\pi - \dfrac{3\pi}{4}$, ..., $-n\pi - \dfrac{\pi}{4} \leq x \leq -(n-1)\pi - \dfrac{3\pi}{4}$, ...; decreasing, $\dfrac{\pi}{4} \leq x \leq \dfrac{3\pi}{4}$, $\pi + \dfrac{\pi}{4} \leq x \leq \pi + \dfrac{3\pi}{4}$, ..., $n\pi + \dfrac{\pi}{4} \leq x \leq n\pi + \dfrac{3\pi}{4}$, ...; $-\dfrac{3\pi}{4} \leq x \leq -\dfrac{\pi}{4}$, $-\pi - \dfrac{3\pi}{4} \leq x \leq -\pi - \dfrac{\pi}{4}$, ..., $-n\pi - \dfrac{3\pi}{4} \leq x \leq -n\pi - \dfrac{\pi}{4}$, ...

**23** Increasing for $x < -1$; increasing for $x > -1$; $x = -1$ a vertical asymptote; $y = 1$, a horizontal asymptote

**25** Decreasing for $-1 < x \leq 0$; increasing for $x \geq 0$; rel. min. at $x = 0$; $x = -1$ is vertical asymptote

**27** Increasing for $x \leq -2$; decreasing for $-2 \leq x \leq 0$; increasing for $x \geq 0$; rel. max. at $x = -2$; rel. min. at $x = 0$; vertical tangent at $x = -3$

**29** Increasing if $x \leq 2$; decreasing if $2 \leq x \leq 3$; rel. max. at $x = 2$

**31** Decreasing for $-\sqrt{2} \leq x \leq -1$, $1 \leq x \leq \sqrt{2}$; increasing for $-1 \leq x \leq 1$; rel. min., $x = -1$; rel. max., $x = 1$

**33** Increasing for all $x$; $x = 2$, a vertical asymptote

**35** Decreasing for $x < -1$; increasing for $x > -1$; rel. min. at $x = -1$

**37** Rel. max., $x = \pi/4$, rel. min., $x = 5\pi/4$; increasing for $0 \leq x \leq \dfrac{\pi}{4}$, $\dfrac{5\pi}{4} \leq x \leq 2\pi$; decreasing for $\dfrac{\pi}{4} \leq x \leq 5\pi/4$

**39** No rel. max. or min.; increasing $0 \leq x \leq 2\pi$

**41** Rel. max. at $x = \pi/6$, rel. min. at $x = 5\pi/6$; increasing, $0 \leq x \leq \dfrac{\pi}{6}$, $\dfrac{5\pi}{6} \leq x \leq 2\pi$; decreasing, $\dfrac{\pi}{6} \leq x \leq 5\pi/6$

**43** No. rel. max. or min.; increasing $0 \leq x \leq 2\pi$

**45** Let $x_1$, $x_2$ be solutions of $\tan x = 3$ such that $0 < x_1 < \dfrac{\pi}{2}$, $\pi < x_2 < \dfrac{3\pi}{2}$. Then there is a rel. max. at $x = x_1$, rel. min. at $x = x_2$; increasing, $0 \leq x \leq x_1$, $x_2 \leq x \leq 2\pi$; decreasing, $x_1 \leq x \leq x_2$. Repeats in each interval of length $2\pi$.

**47** Rel. max. at $x = \frac{1}{3}\pi + 2n\pi$; rel. min. at $x = -\frac{1}{6}\pi + 2n\pi$, $n = 0$, $\pm 1$, $\pm 2$, ...

**49** $f'' > 0$ for $\frac{1}{6}\pi < x < \frac{1}{2}\pi$

## SECTION 4 (pages 146–147)

**1** Minimum at $x = \frac{3}{2}$, concave upward for all $x$

**3** Rel. max. at $x = -3$, rel. min. at $x = 3$; inflection point at $x = 0$; concave downward, $x \le 0$; concave upward, $x \ge 0$

**5** $x = -3$, rel. min.; $x = -2$, inflection point; $x = 0$, inflection point; $x \le -2$, concave upward; $-2 \le x \le 0$, concave downward; $x \ge 0$, concave upward

**7** $x = -1$, 1, rel. minima; concave upward for all $x$; $x = 0$, vertical asymptote

**9** $x = -1$, rel. min.; $x = 1$, rel. max.; points of inflection $x = 0$, $+\sqrt{3}$, $-\sqrt{3}$; concave downward, $x < -\sqrt{3}$, $0 < x < \sqrt{3}$; concave upward, $-\sqrt{3} < x < 0$ and $x > \sqrt{3}$

**11** Rel. max., $x = -1$; rel. min., $x = 2$; inflection point, $x = \frac{1}{2}$; concave downward, $x \le \frac{1}{2}$; concave upward, $x \ge \frac{1}{2}$

**13** Rel. max., $x = \frac{2}{3}$; rel. min., $x = 2$; inflection point, $x = \frac{4}{3}$; concave downward, $x \le \frac{4}{3}$; concave upward, $x \ge \frac{4}{3}$

**15** Inflection point $x = \frac{1}{3}$; concave downward, $x \le \frac{1}{3}$; concave upward, $x \ge \frac{1}{3}$

**17** Rel. min., $x = -1$; inflection points $x = 0$, 2; concave upward $x \le 0$, $x \ge 2$; concave downward, $0 \le x \le 2$; horizontal tangent at $x = 2$

**19** Rel. minima $x = 0$, $(-15 - \sqrt{33})/8$; rel. max., $x = (-15 + \sqrt{33})/8$; concave upward $x \le -2$, $x \ge -\frac{1}{2}$; concave downward, $-2 \le x \le -\frac{1}{2}$; points of inflection $x = -2$, $-\frac{1}{2}$

**21** $x = -2$, rel. min.; $x = 2$, rel. max.; concave downward for $x \le -2\sqrt{3}$ and $0 \le x \le 2\sqrt{3}$; concave upward for $-2\sqrt{3} \le x \le 0$ and $x \ge 2\sqrt{3}$; $x = -2\sqrt{3}$, 0, $2\sqrt{3}$, inflection points

**23** $x = -2$, rel. min.; $x = 2$, rel. max.; concave upward, $-2\sqrt{2} \le x \le 0$; concave downward, $0 \le x \le 2\sqrt{2}$; $x = 0$, inflection point

**25** $x = -4$, rel. max.; $x = 0$, rel. min.; concave downward $-5 \le x \le -4 + \frac{2}{3}\sqrt{6}$; concave upward, $x \ge -4 + \frac{2}{3}\sqrt{6}$; inflection point at $x = -4 + \frac{2}{3}\sqrt{6}$

**27** $x = 2$, rel. max.; concave downward for $x < -2$, $-2 < x \le 2 - \sqrt{6}$; concave upward for $2 - \sqrt{6} \le x \le 0$; concave downward for $0 \le x \le 2 + \sqrt{6}$; concave upward for $x \ge 2 + \sqrt{6}$; inflection points at $x = 2 - \sqrt{6}$, $2 + \sqrt{6}$. Note that the curve is concave upward on one side of $x = 0$ and concave downward on the other. There is a vertical asymptote at $x = -2$.

**29** No rel. max. or rel. min. Concave upward for $-\pi \le x \le 0$; concave downward for $0 \le x \le \pi$; $x = 0$ is point of inflection

**31** Rel. max. at $x = \frac{\pi}{8}, \frac{5\pi}{8}, -\frac{3\pi}{8}, -\frac{7\pi}{8}$; rel. min. at $x = \frac{3\pi}{8}, \frac{7\pi}{8}, -\frac{\pi}{8}, -\frac{5\pi}{8}$; concave upward: $\frac{\pi}{4} \le x \le \frac{\pi}{2}$, $\frac{3\pi}{4} \le x \le \pi$, $-\frac{\pi}{4} \le x \le 0$, $-\frac{3\pi}{4} \le x \le -\frac{\pi}{2}$; concave downward: $0 \le x \le \frac{\pi}{4}$, $\frac{\pi}{2} \le x \le \frac{3\pi}{4}$, $-\frac{\pi}{2} \le x \le -\frac{\pi}{4}$, $-\pi \le x \le -\frac{3\pi}{4}$; points of inflection: $x = 0$, $\pm\frac{\pi}{4}$, $\pm\frac{\pi}{2}$, $\pm\frac{3\pi}{4}$

**33** No rel. max. or rel. min.; concave upward for $0 \le x < \pi/2$; concave downward for $-\frac{\pi}{2} < x \le 0$; $x = 0$ is point of inflection

**35** Rel. max., $x = \frac{\pi}{3}$; rel. min., $x = -2\pi/3$; concave upward for $-\pi \le x \le -\frac{\pi}{6}$, $\frac{5\pi}{6} \le x \le \pi$; concave downward for $-\frac{\pi}{6} \le x \le \frac{5\pi}{6}$; $x = -\frac{\pi}{6}, \frac{5\pi}{6}$ are points of inflection

**37** Rel. max. at $x = 4$, rel. min. at $x = 0$

## SECTION 5 (pages 148–149)

**1** Min. at $(-1, -5)$; max. at $(3, 11)$

**3** Max. at $(1, 3)$; min. at $(-2, -18)$

**5** Minima $(-5, -1)$ or $(1, -1)$; max., $(6, 274)$

**7** Min. at $(1, 0)$; max. at $(5, 576)$

**9** Min. at $(-\frac{1}{2}, -1)$; max. at $(1, \frac{1}{2})$

**11** Min. at $\left(-\frac{\pi}{3}, -\sqrt{3}\right)$; max. at $\left(\frac{\pi}{3}, \sqrt{3}\right)$

**13** Min. at $\left(-\frac{\pi}{3}, -2\right)$; max. at $\left(\frac{\pi}{4}, \frac{1}{2}(\sqrt{6} - \sqrt{2})\right)$

**17** No maximum or minimum. [For interval $-3 \le x \le 3$, max. is $(3, 11)$, min. is $(-3, -49)$.]

**19** No maximum or minimum

**21** Max. at $(2, \frac{1}{3})$; no minimum

**23** No maximum or minimum

**25** $f'(x) = 0 \Leftrightarrow x = (-a \pm \sqrt{a^2 - 3b})/3$; if $a^2 < 3b$, then $f'(x) \ne 0$ for all $x$ and max. and min. must be at endpoints

## SECTION 6 (pages 154–155)

**1** Length = width = 30 meters; length = width = $\frac{1}{4}L$

**3** $h = \frac{2}{3}\sqrt{3}R$; $r = \frac{1}{3}\sqrt{6}R$ **5** 10, $-10$

**7** Radius of semicircle = height of rectangle = $12/(4 + \pi)$ meters

**9** Width = $(2 + 2\sqrt{5})$ cm; height = $(3 + 3\sqrt{5})$ cm

**11** $r = \sqrt[6]{(360)^2/2\pi^2}$ cm; $h = \sqrt[3]{720/\pi}$ cm; $r = \sqrt[6]{9V^2/2\pi^2}$ cm; $h = \sqrt[3]{6V/\pi}$ cm

**13** Land at $\frac{5}{3}\sqrt{3}$ km down beach; walk $6 - \frac{5}{3}\sqrt{3}$ km; time $\approx 3.66$ kr

**15** $(1, 2)$

**17** (a) base = $13/2$; height = $30/13$; (b) base = $H/2$; height = $h/2$

**19** Each number is 10

**21** (a) $r = h = \sqrt[3]{16/\pi}$ cm; (b) $r = h = \sqrt[3]{V/\pi}$ cm

**23** From $(5, 0)$ to $(0, 10)$

**25** Radius = $\frac{1}{2}R\sqrt{2}$; height = $R\sqrt{2}$

**27** (a) radius of circle = $\dfrac{L\sqrt{3}}{18 + 2\pi\sqrt{3}}$, length of side of triangle $= \dfrac{3L}{9 + \pi\sqrt{3}}$; (b) circle only, $r = L/2\pi$

**29** Height = $4R/3$; radius of base = $2\sqrt{2}R/3$

**31** $r = (9V^2/2\pi^2)^{1/6}$; $h = (6V/\pi)^{1/3}$

**33** $r = (3V/20\pi)^{1/3}$; $h = 6(3V/20\pi)^{1/3}$

**35** $4(1 + \sqrt[3]{4})^{3/2}$ meters          **37** 3500

## SECTION 7 (pages 160–161)

**1** $df(x, h) = (4x^3 + 6x - 2)h$

**3** $df(x, h) = \dfrac{(4x^5 + 2x)h}{\sqrt{x^4 + 1}}$

**5** $df(x, h) = \dfrac{(8x^2 + 3x + 4)h}{3(x^2 + 2)^{1/2}(2x + 1)^{2/3}}$

**7** $df(x, h) = (3\cos 3x)h$

**9** $df(x, h) = (\frac{1}{2}\sec^2 \frac{1}{2}x + 2\tan 2x \sec 2x)h$

**11** $df = 0.03$; $\Delta f = 0.0301$

**13** $df = 0.18$; $\Delta f = 0.180901$

**15** $df = -0.01250$; $\Delta f = -0.01220$

**17** $df = -0.05$; $\Delta f = -0.04654$

**19** $df = 0.695$; $\Delta f = 0.455$

**21** 8.0625          **23** 1.02000

**25** 1.98750          **27** 0.049

**29** $1\frac{2}{3}\%$ (approximately)

**31** $6a^2t$ cm$^3$ of paint

## SECTION 8 (page 164)

**1** $15x(x^3 - 3x^2 + 2)^4(x - 2)dx$

**3** $-5(x^3 + 4)^{-6}(3x^2)dx$

**5** $(5x^2 + 6x)(2x + 3)^{-1/2}dx$

**7** $2(x + 1)(2x - 1)^2(5x + 2)dx$

**9** $2(1 - x^2)(x^2 + 1)^{-2}dx$

**11** $\frac{2}{3}x^{-1/3}(x + 1)^{-5/3}dx$

**13** $-(x + 1)^{-1/2}(x - 1)^{-3/2}dx$

**15** $2x(\sin 2x + x \cos 2x)dx$

**17** $\tan \frac{1}{2}x(\tan \frac{1}{2}x + x \sec^2 \frac{1}{2}x)dx$

**19** $(4x + y + 2)/(-x + 2y + 3)$

**21** $-y^{1/2}/x^{1/2}$

**23** $(6x^2 - y^2 + 2)/(2xy + 3y^2 + 1)$

**25** $(2xy \sin x \cos x - y^2 \cos^2 x + x \cos y)/(x \cos^2 x - 2x^2 \sin y)$

**27** $\dfrac{3(x + 1)(t - 1)}{2(x^2 + 2x + 5)^{1/4}(t^2 - 2t + 1)^{1/2}}$

**29** $\dfrac{(2x + 3)(3r^2 - 8)3t^2}{2\sqrt{t^3 + 5}}$

**31** $\dfrac{-(3x^2 + 2y)}{(2x - 3y^2)}(3t^2 - 2)$

**33** $\dfrac{-16y^2(x + 1)(t + 1)^3}{z(z - 2)}$

**35** $\dfrac{dy}{dx} = \dfrac{3x^2 + 7}{16y^3 + 6}$; $\dfrac{d^2y}{dx^2} = \dfrac{6x(16y^3 + 6)^2 - (3x^2 + 7)^2 \cdot 48y^2}{(16y^3 + 6)^3}$

**37** $\dfrac{dx}{dt} =$

$\dfrac{(9yt^2)(3y^2 + 3x) - (2x - 2t - 6t^2)(-3y^2 + 4xy + 3t^3)}{(2x + 2t + 3y)(-3y^2 + 4xy + 3t^3) - (3y^2 + 3x)(6x^2 + 2y^2 + 1)}$

## SECTION 9 (page 169)

**1** 3.606          **3** 8.185

**5** 2.190          **7** 1.581

**9** $-1.24$          **13** 1.30, 0.67, $-2.32$

**15** 0.82, 1.12

## SECTION 10 (pages 172–173)

**1** $(2/\pi)$m/min          **3** $5\sqrt{193}$ km/hr

**5** 11 cm$^2$/min          **7** $(32/\sqrt{89})$m/sec

**9** $(2/15\pi)^{1/3}$ m/min

**11** $(24/\sqrt{61})$ m/sec

**13** 1

**15** $\dfrac{dz}{dt} = \pm 7/6$

**17** $-3$ m/sec

**19** 0.3 m/min

**21** $(4/135)$ m/min

**23** $(0.5 + 27\pi h^2/8)$ m$^3$/min

**25** $(60/13)$ m/min

**27** $(4/75\pi)$ m/min

**29** $(7/\sqrt{173})$ m/sec

**31** $\left| -\dfrac{8hdgt_0}{(h + 2gt_0^2)^2} \right|$

## Section 11 (pages 177–178)

**1** $F(x) = \frac{1}{5}x^5 - \frac{1}{2}x^4 + \frac{7}{2}x^2 - 6x + \frac{24}{5}$

**3** $F(x) = \frac{1}{8}(x+2)^8 + 1$

**5** $F(x) = (2x+1)^3(x+2)^2$

**7** $F(x) = \frac{5}{6}x^{6/5} + \frac{18}{5}x^{5/3} + \frac{1}{3}x^3 + 3$

**9** $F(x) = \frac{4}{5}x^{5/4} + \sec x + 1 - \sqrt{2} - \frac{4}{5}(\pi/4)^{5/4}$

**11** $F(x) = x^2 \cos x + 3$

**13** $F(x) = \frac{1}{2}\sin 2x + \frac{3}{4}\sqrt{3}$

**15** $F(x) = \frac{1}{3}\sin^3 x + \frac{11}{12}\sqrt{2}$

**17** $s = \frac{1}{3}t^3 + \frac{1}{2}t^2 - 2t - \frac{1}{6}$

**19** $s = 6t - t^2 - t^3$

**21** $s = \frac{1}{6}\cos^3 2t + \frac{11}{6}$

**23** Rises for 8 sec; stays in air for 16 sec

**25** $y = x^3 - \frac{5}{2}x^2 + 2x + 6$

**27** $y = -6\cos x + 3$

**29** $y = \frac{27}{4}x^4 - x^3 + 2x^2 + 6x - \frac{7}{4}$

**35** $y = x^3 - \frac{3}{2}x^2 + 7x + 6$

## Chapter 4 Review Problems (pages 178–179)

**1** Rel. max. at $(1, 4)$

**3** Rel. max. when $x = -\frac{4}{3} - \frac{1}{3}\sqrt{22}$; rel. min. when $x = -\frac{4}{3} + \frac{1}{3}\sqrt{22}$

**5** Rel. max. when $x = 1$; rel. min. when $x = -1, 2$

**7** Rel. max. when $x = -3\pi/4, \pi/4$; rel. min. when $x = -\pi/4, 3\pi/4$

**9** Rel. min. when $x = 0$

**11** Rel. max., $(0, 5)$; no points of inflection; concave downward, $-\infty < x < \infty$

**13** Rel. max. when $x = 0$; rel. min. when $x = 3$; point of inflection at $x = 1/6$; concave downward $-\infty < x < \frac{1}{6}$; concave upward $\frac{1}{6} < x < +\infty$

**15** Rel. max. when $x = -1$; rel. min. when $x = -2, 3$; points of inflection when $x = \pm\sqrt{7/3}$; concave upward, $-\infty < x < -\sqrt{7/3}$, $\sqrt{7/3} < x < +\infty$; concave downward, $-\sqrt{7/3} < x < \sqrt{7/3}$

**17** Rel. max. when $x = \pi/6, 5\pi/6, 3\pi/2$; rel. min. when $x = \pi/2, 7\pi/6, 11\pi/6$; points of inflection at $n\pi/3$, $n = 1, 2, 3, 4, 5$; concave downward, $0 < x < \dfrac{\pi}{3}$, $\dfrac{2\pi}{3} < x < \pi$, $\dfrac{4\pi}{3} < x < \dfrac{5\pi}{3}$; concave upward, $\dfrac{\pi}{3} < x < \dfrac{2\pi}{3}$, $\pi < x < \dfrac{4\pi}{3}$, $\dfrac{5\pi}{3} < x < 2\pi$

**19** No rel. max. or min.; no points of inflection; concave downward, $1 < x < +\infty$; concave upward $-\infty < x < -1$

**21** Max. 72 at $x = 4$; min. $-3$ at $x = -1$

**23** Max. 8 at $x = 2$; min. $-1$ at $x = -1$ and 1

**25** Max. 37 at $x = 4$; min. 7 at $x = -1$

**27** Max. 5 at $x = -1$; min. $-5/4$ at $x = 3/2$

**29** Max. 281 at $x = -3$; min. $-10 - 64\sqrt{2}$ at $x = \sqrt{2}$

**31** If $x$, $y$ are numbers, then $3x + y$ may become indefinitely large or small. The values $x = \pm\sqrt{17}$, $y = \pm 51\sqrt{17}$ are rel. max. and min.

**33** $(1, -1)$

**35** $\dfrac{5x^2 + 24x + 39}{(x^2 + 7x + 9)^2}\, dx$

**37** $4\sec^2 2x \tan 2x\, dx$

**39** $-\dfrac{2y^2 + 8z^3}{3y^2 + 4yz}$

**41** $-2.224$

**45** 3.6 m/sec approx.

**47** $y = -2 + 2x - \frac{3}{2}x^2 + \frac{1}{3}x^3$

**49** $y = -\frac{1}{2}\cos 2x + \sin x + 1 - \frac{1}{2}\sqrt{2}$

**51** $y = \frac{1}{3}\sin 3x + \frac{1}{2}\tan 2x + \frac{8}{3} - \frac{1}{2}\sqrt{3}$

# CHAPTER 5

## Section 1 (pages 186–187)

**1** 78

**3** 12

**5** $\frac{137}{60}$

**7** 2046

**11** 3/8

**13** $3[\frac{1}{72} + \frac{1}{71} - \frac{3}{2}]$

**15** 12/25

**19** $A_3^L = 1\frac{7}{8}$; $A_3^U = 2\frac{1}{8}$

**21** $b = 3$

## Section 2 (pages 192–193)

**1** $\underline{S}(\Delta) = 9.006$, $\bar{S}(\Delta) = 12.444$

**3** $\underline{S}(\Delta) = 0.803$; $\bar{S}(\Delta) = 0.873$

**5** $\underline{S}(\Delta) = 0.1588$; $\bar{S}(\Delta) = 0.2702$

**7** $\underline{S}(\Delta) = 0.0322$; $\overline{S}(\Delta) = 2.1710$

**9** $\underline{S}(\Delta) = 1.000$; $\overline{S}(\Delta) = 7.000$

**35** Approx. $= 0.692^-$

**37** Approx. $= 2.221$

## SECTION 3 (pages 195–196)

**1** Sum $= 2.000$; exact area $= 2$

**3** Sum $= 2.700$; exact area $= \frac{8}{3}$

**5** $\frac{1}{3}$    **7** $\frac{4}{3}$    **9** $\frac{13}{6}$    **11** $\frac{13}{6}$

**13** $2$    **15** $1 + \frac{1}{2}\sqrt{3}$    **17** $2.330$    **19** $0.786$

## SECTION 4 (pages 199–200)

**1** Smallest value $= 6$; largest value $= 24$

**3** Smallest value $= -3$; largest value $= 24$

**5** Smallest value $= \frac{3}{4}$; largest value $= \frac{4}{5}$

**7** Smallest value $= 0$; largest value $= \frac{64}{17}$

**9** Smallest value $= 3\sqrt{3}$; largest value $= 3\sqrt{6}$

**11** Smallest value $= 0$; largest value $= 4$

**13** Smallest value $= 0$; largest value $= 5\sqrt{5}$

**15** Smallest value $= 0$; largest value $= \pi/4$

**17** Smallest value $= 0$; largest value $= \pi/8$

**23** Any interval not containing as an interior point or endpoint: $x = \pm\frac{\pi}{2}, \pm\frac{3\pi}{2}, \pm\frac{5\pi}{2}, \ldots$

**25** Any interval not containing as an interior point or endpoint: $x = \pm\frac{\pi}{2}, \pm\frac{3\pi}{2}, \pm\frac{5\pi}{2}, \ldots$

**27** $a \le x \le b$, $a$, $b$ any numbers

**29** Any interval not containing $x = 0$ as an interior point or endpoint

## SECTION 5 (pages 205–206)

**1** $-1\frac{1}{6}$    **3** $-3\frac{11}{15}$    **5** $\frac{38}{5}\sqrt{3} - \frac{26}{15}$

**7** $118\frac{2}{15}$    **9** $\frac{100}{3}$    **11** $\frac{446}{135}$

**13** $x^3/3$    **15** $\sqrt{3}/4$    **17** $1/3$

**19** Approx. $= 2.330$; exact $= 2\frac{1}{3}$

**21** Approx. $= 0.497$; exact $= 0.5$

**23** Approx. $= 10$; exact $= 10$

**25** Approx. $=$ exact $= (A/2)(b^2 - a^2) + B(b - a)$

**27** Approx. $= 10.156$; exact $= 10\frac{1}{6}$

**29** Approx. $= 1.0065$; exact $= 1$

**31** Approx. $= 0$; exact $= 0$

**33** Approx. $= 0.7930^-$

## SECTION 6 (pages 211–212)

**1** $f(2) = 9$    **3** Theorem does not apply

**5** Theorem does not apply

**7** $f(\frac{1}{4}) = \frac{1}{2}$, function is continuous

**9** $\xi = \frac{7}{2}$    **11** $\xi = (3 + \sqrt{129})/3$

**13** $\xi = (3 \pm \sqrt{57})/12$    **15** $\xi = 0, \pm 1, \pm\sqrt{3}$

**17** $\sqrt{14} - \sqrt{6}$    **19** $\frac{2}{3}$

**21** $1/4$    **23** $2\sqrt{2} - \frac{4}{3}\sqrt{3}$

**25** $1/50$    **27** $4$

**29** Integral is zero    **31** $\int_c^d f(t)\,dt$

## SECTION 7 (pages 217–218)

**1** $\frac{1}{18}(3x - 2)^6 + C$    **3** $\frac{1}{3}(2 - x)^{-3} + C$

**5** $\frac{1}{5}(2y + 1)^{5/2} + C$    **7** $\frac{3}{40}(2x^2 + 3)^{10/3} + C$

**9** $-\frac{3}{4}(3 - 2x^2)^{1/3} + C$    **11** $-\frac{1}{6}(3x^2 + 2)^{-1} + C$

**13** $\frac{2}{9}(x^3 + 1)^{3/2} + C$    **15** $\frac{1}{2}(x^3 + 3x^2 + 1)^{2/3} + C$

**17** $-\frac{1}{6}\cos 6x + C$    **19** $-\frac{3}{2}\cot((2x + 1)/3) + C$

**21** $\frac{1}{6}\sec^2 3x + C$    **23** $\frac{1}{40}(x^2 + 1)^4(4x^2 - 1) + C$

**25** $\frac{5}{612}(x^3 + 1)^{12/5}(12x^3 - 5) + C$

**27** $-\frac{1}{2}\cot(2x - 6) - x + C$

**29** $-\cos x(1 - \frac{2}{3}\cos^2 x + \frac{1}{5}\cos^4 x) + C$

**31** $-\frac{1}{2}\cos(x^2) + C$    **33** $\frac{1}{4}\sec^4 x + C$

**35** $\frac{1}{6}\tan^6 x + C$    **37** $-\frac{1}{6}\tan^3(3 - 2x) + C$

**39** $\frac{1}{2}x - \frac{1}{12}\sin 6x + C$    **41** $33/5$

**43** $\frac{1}{4}$    **45** $\frac{1}{4}$

**47** $\frac{1}{2}$    **49** $\frac{2}{3}(\sqrt{15} - 1)$

**51** $\frac{3}{5}[(\sqrt[3]{2} + 2)^5 - 243]$

**53** $0$    **55** $2\sqrt{3}/45$

## CHAPTER 5 REVIEW PROBLEMS (pages 218–220)

**1** $40$    **3** $-19{,}674$

**5** $\frac{1}{3}(\frac{1}{10} + \frac{1}{11} + \frac{1}{12} - \frac{1}{201} - \frac{1}{202} - \frac{1}{203})$

**9** $\overline{S}(\Delta) = \frac{1}{12}(6 + 2\sqrt{2} + \frac{1}{2}\sqrt{3})$; $\underline{S}(\Delta) = \frac{1}{12}(2\sqrt{2} + \sqrt{3} + \frac{1}{2})$

**11** $\overline{S}(\Delta) = \frac{1}{36}\sqrt{3}$; $\underline{S}(\Delta) = -\frac{1}{12}(1 + \frac{4}{3}\sqrt{3})$

**17** Upper bound $= \frac{4}{3}$; lower bound $= \frac{1}{6}$

**19** Upper bound $= \frac{21}{12}\pi\sqrt{3}$; lower bound $= -7\pi/4$

**21** Upper bound $= 32\sqrt{3}/3$; lower bound $= 8\sqrt{15}/15$

**23** None

**25** Upper bound $= \pi\sqrt{2}$; lower bound $= \pi/2$

**27** $\frac{1}{2}(\sqrt[3]{36} - \sqrt[3]{4})$

**29** $\frac{2}{5}[(12)^{5/4} - 5^{5/4}]$

**31** 1/4          **33** 1/4

**37** $\xi = 0.845$          **39** $-\frac{1}{24}\left(2 + \frac{3}{t^4}\right)^2 + C$

**41** $-\frac{1}{8}\cos^4(x^2 + 2x - 6) + C$

**43** $\frac{1}{6}\tan(2x^3) + \frac{1}{18}\tan^3(2x^3) + C$

**45** $-\frac{1}{2}\csc(x^2 + 1) + C$     **47** $\frac{3}{10}\sec^{5/3}(2x) + C$

**49** $-\frac{1}{3}\csc(x^3) + C$

# CHAPTER 6

## SECTION 1 (pages 227–228)

**1** $\frac{74}{3}$          **3** 45          **5** 6

**7** $\frac{3}{2}$          **9** $\frac{4}{15}$          **11** $\frac{4}{3}(2\sqrt{2} - 1)$

**13** $\frac{10}{3}$          **15** 72          **17** $\frac{5}{12}$

**19** $10\frac{2}{3}$          **21** $7\frac{7}{48}$          **23** $\frac{3}{2}$

**25** a) $\frac{5}{4} - \frac{3}{4}\sqrt[3]{4}$   b) $\frac{3}{2} - \frac{3}{4}\sqrt[3]{4}$

**27** $1 + \frac{\pi^2}{8} - \frac{\pi}{2}$     **29** $\frac{\pi^2}{16} + \frac{\pi}{2} - 1$

**31** $\frac{1}{2} - \frac{\pi}{8}$          **33** 2          **35** 19/2

**39** $2\sqrt{a} - 2$; $A \to +\infty$ as $a \to +\infty$

## SECTION 2 (page 232)

**1** $160\pi$

**3** $\frac{\pi}{54}$; $R = \{(x, y): 0 \le x \le \frac{1}{2}, 0 \le y \le \frac{1}{3}(1 - 2x)\}$

**5** $\frac{4\pi}{3}a^3$

**7** $\frac{512}{15}\pi$; $R = \{(x, y): -2 \le x \le 2, 0 \le y \le 4 - x^2\}$

**9** $\frac{\pi}{3}(3\sqrt{3} - \pi)$; $R = \left\{(x, y): 0 \le x \le \frac{\pi}{3}, 0 \le y \le \tan x\right\}$

**11** $\frac{\pi^2}{6} + \frac{\pi}{8}\sqrt{3}$; $R = \left\{(x, y); 0 \le x \le \frac{\pi}{3}, 0 \le y \le \cos x\right\}$

**13** $R = \{(x, y): 0 \le x \le 4, \sqrt{x} \le y \le 2\}$; $8\pi$

**15** $\frac{16\pi}{3}(2\sqrt{2} - 1)$

**17** $16\pi$; $R = \{(x, y): y^2 \le x \le 8 - y^2, 0 \le y \le 2\}$

**19** $\frac{5\pi}{14}$; $R = \{(x, y): 0 \le x \le 1, x^3 \le y \le \sqrt{x}\}$

**21** $\frac{45}{4}\pi$; $R = \{(x, y): \frac{1}{2}(y^2 - 3y) \le x \le y - 2, 1 \le y \le 4\}$

**23** $\frac{3\pi}{7}(3^{7/3} - 2^{7/3})$     **25** $\frac{1}{3}\pi a^2 h$

**27** $\frac{\pi^2}{24} + \frac{\pi}{4}(1 - \frac{1}{2}\sqrt{3})$     **29** $3\pi/10$

## SECTION 3 (pages 235–236)

**1** $32\pi/5$

**3** $\pi/7$; $R = \{(x, y): 0 \le x \le 1, 0 \le y \le x^3\}$

**5** $\sqrt{3} - 1$; $R = \{(x, y): \frac{1}{4} \le x \le \frac{1}{3}, 0 \le y \le \cos \pi x/x\}$

**7** $\frac{28\pi}{3}\sqrt{2}$; $R = \{(x, y): 0 \le x \le \sqrt{y^2 + 4}, 0 \le y \le 2\}$

**9** $256\pi/3$

**11** $2\pi/5$; $R = \{(x, y): 0 \le x \le 1, x^3 \le y \le \sqrt{x}\}$

**13** $\sqrt{3} - 1 - \frac{\pi}{12}$; $R = \left\{(x, y): 0 \le x \le \frac{\tan^2(2\pi y)}{y}, \frac{1}{8} \le y \le \frac{1}{6}\right\}$

**15** $2\pi/35$; $R = \{(x, y): 0 \le x \le 1, x^3 \le y \le x^2\}$

**17** Suppose $m < n$: $\frac{2(n - m)\pi}{(2n + 1)(2m + 1)}$;

$R = \{(x, y): 0 \le x \le 1, x^n \le y \le x^m\}$

**19** $4\pi/3$          **21** $128\pi/3$     **23** $103\pi/6$

**25** $11\pi/4$          **27** $95\pi/6$

**29** $V = \pi \displaystyle\int_a^b \{[f(x) + c]^2 - c^2\}\, dx$

**31** $V = 2\pi \displaystyle\int_{\alpha}^{\beta} (y + \gamma)g(y)\, dy$

**9** $\dfrac{6\pi}{5}(\sqrt{2} + 1)$     **11** $\pi$     **13** $8\pi\sqrt{2}/15$

**15** $\dfrac{25\pi}{2}\sqrt{5}\left(\dfrac{610}{9} - \dfrac{375}{7} + 15\right) - \dfrac{25\pi}{2}\left(-\dfrac{2}{9} - \dfrac{3}{7} + \dfrac{3}{5}\right)$

**17** $\dfrac{\pi}{6}(5\sqrt{5} - 1)$

## SECTION 4 (page 241)

**1** $14/3$     **3** $\frac{1}{27}\{(40)^{3/2} - [9(2)^{2/3} + 4]^{3/2}\}$

**5** $\frac{85}{2} + \frac{45}{256}$     **7** $\sqrt{3}$     **9** $33/16$

**11** $\frac{211}{4} + \frac{19}{3240}$

**13** $\frac{1}{8}(3^8 - 2^8) + \dfrac{1}{(24)(64)} - \dfrac{1}{(24)(3^6)}$

**15** $s = \displaystyle\int_1^4 \sqrt{1 + \dfrac{1}{4x}}\, dx$

**17** $s = \displaystyle\int_1^2 \sqrt{1 + 16y^2(y^2 - 1)^2}\, dy$

**19** $s = \displaystyle\int_0^{\pi/2} \sqrt{1 + 4\sin^2 2x}\, dx$

**21** $s = \displaystyle\int_0^{\pi/4} \sqrt{1 + \tfrac{1}{4}\sec^4 \tfrac{1}{2}y}\, dy$

**23** $3.61$     **25** $2.04$     **27** $1.47$

**29** $4.32$     **31** $8/\sqrt{3}$

**33** Let $b$ be any positive number and choose $a = 3b$

**35** Let $c$ be the point where the jump occurs. Find the length from $a$ to $c$; then find the length from $c$ to $b$. Add the results.

## CHAPTER 6 REVIEW PROBLEMS (pages 246–247)

**1** $9/2$     **3** $\frac{13}{2} - 2\sqrt{3}$     **5** $8 - 4\sqrt{2}$

**7** $1/2$     **9** $25$     **11** $22\pi/5$

**13** $26\pi/77$     **15** $81\pi/2$     **17** $\dfrac{\pi^2}{12} - \dfrac{\pi\sqrt{3}}{16}$

**19** (a) $\pi \displaystyle\int_a^b (f_2(x) - f_1(x))(f_2(x) + f_1(x) + 2k)\, dx$

$\equiv \pi \displaystyle\int_a^b [f_2^2(x) - f_1^2(x) + 2k(f_2(x) - f_1(x))]\, dx$

(b) $236\pi/15$

**21** $1009/16$     **23** $\frac{1}{27}(31)^{3/2} - \frac{8}{27}$

**25** $\dfrac{25\pi}{14}\left(\dfrac{2}{7}(17)^{7/2} - \dfrac{4}{5}(17)^{5/2} + \dfrac{2}{3}(17)^{3/2} - \dfrac{2}{7}(2)^{7/2} + \dfrac{4}{5}(2)^{5/2}\right.$

$\left. - \dfrac{2}{3}(2)^{3/2}\right)$

**27** $\pi ab(b + 2)\sqrt{1 + a^2}$

## SECTION 5 (page 246)

**1** $\dfrac{\pi}{9}(17\sqrt{17} - 1)$     **3** $\dfrac{\pi}{6}(13\sqrt{13} - 5\sqrt{5})$

**5** $13\pi/3$     **7** $1179\pi/256$

# CHAPTER 7

## SECTION 1 (page 255)

**1** $\ln 4 = 1.38630$; $\ln 5 = 1.60944$; $\ln 6 = 1.79176$; $\ln 8 = 2.07945$; $\ln 9 = 2.19722$

**3** $\ln 24 = 3.17806$; $\ln 25 = 3.21888$; $\ln 27 = 3.29583$; $\ln 30 = 3.40120$; $\ln 32 = 3.46575$

**5** $\ln \sqrt{30} = 1.70060$; $\ln \sqrt[3]{270} = 1.86614$; $\ln(\sqrt{2/3}) = -0.20273$; $\ln \sqrt[4]{7.2} = 0.49352$

**7** $\frac{43}{36} < \ln 4.5 < \frac{47}{24}$     **9** $\frac{13}{14} < \ln 14 < 13$

**11** $\frac{1}{21} < \ln 1.05 < 0.05$

**13** $-\frac{17}{3} < \ln 0.15 < -0.85$ or $-\frac{391}{120} < \ln 0.15$

**15** $-\frac{1}{155} < \ln 0.995 < -0.005$

**17** $2/(2x + 3)$     **19** $2(x + 1)/(x^2 + 2x + 3)$

**21** $2\cot x$     **23** $-2\tan x$

**25** $4(\ln x)^3/x$     **27** $1/(x \ln x)$

**29** $-x/(1-x^2)$

**31** $\dfrac{2+3x^2}{x(1+x^2)}$

**33** $-\csc x$

**35** 1.60944

**37** 1.60944

**39** $-1.38630$

**41** 3.17805

**43** 6.90776

**47** $y - \ln 0.75 = \frac{1}{12}(x-3)$    **49** $\frac{5}{6}\ln 15 - \frac{1}{2}$

**51** Domain: $(1, +\infty)$; range: $(-\infty, +\infty)$

## SECTION 2 (page 261)

**1** Domain: $0 \le x < +\infty$; range: $-\infty < y < +\infty$

**3** Domain: $-2 \le x < +\infty$; range: $-\infty < y < +\infty$

**5** Domain: $-\infty < x < +\infty$; range: $-\infty < y < +\infty$

**7** Domain: $-\infty < x \le -1,\ 1 \le x < +\infty$; range: $-\infty < y < +\infty$

**9** Domain: $-\infty < x < +\infty$; range: $-\infty < y \le -1,\ 1 \le y < +\infty$

**11** Domain: $-1 \le x < +\infty$; range: $-\infty < y < +\infty$

**13** $g(x) = \dfrac{x-2}{3}$; domain: $-\infty < x < +\infty$

**15** Inverse exists; domain: $-\infty < x < +\infty$

**17** No inverse function

**19** Inverse: $y = -1 + \sqrt{4-x}$; domain: $-\infty < x < 3$

**21** Inverse: $y = \sqrt{\dfrac{1}{x}}$; domain: $1 \le x < +\infty$

**23** Domain of $f$: $-\infty < x \le -2$; inverse: $g(x) = -2 - \sqrt{x+5}$, $x \ge -5$

Domain of $f$: $-2 \le x < \infty$; $g(x) = -2 + \sqrt{5+x}$, $x \ge -5$

**25** Domain of $f$: $-\infty < x \le -\frac{3}{4}$;

inverse $g(x) = \dfrac{-3 - \sqrt{8x-23}}{4}$, $x \ge 23/8$

Domain of $f$: $-\frac{3}{4} \le x < +\infty$; inverse $g(x) = \dfrac{-3 + \sqrt{8x-23}}{4}$, $x \ge 23/8$

**27** Domain of $f$: all $x \ne -1$; inverse $g(x) = \dfrac{x}{1-x}$, $x \ne 1$

**29** Domain of $f$: all $x \ne -2$; inverse $g(x) = \dfrac{1+2x}{2-x}$, $x \ne 2$

**31** If domain is $-\infty < x \le -3$, inverse function has domain: $-\infty < x \le 34$

If domain is $-3 \le x \le 1$, inverse function has domain: $2 \le x \le 34$

If domain is $1 \le x < \infty$, inverse function has domain: $2 \le x < \infty$

**33** Domain: $-\infty < x < +\infty$; inverse function has domain $-\infty < x < +\infty$

**39** Domain of $g$ is $0 \le x < +\infty$, range is $[0, +\infty)$; $g$ is differentiable except at $n = 0, 1, 2, \ldots$

## SECTION 3 (pages 268–269)

**15** (a) 4; (b) 2; (c) $-5$     **17** (a) 0; (b) $-2$; (c) 1

**23** $\dfrac{\ln 8}{\ln 3}; \dfrac{\ln 12}{\ln 7}; \dfrac{\ln 15}{\ln 2}$

**25** $G(\frac{1}{5}) = 2.488$; $G(\frac{1}{6}) = 2.522$; $G(\frac{1}{7}) = 2.547$; $G(\frac{1}{8}) = 2.566$; $G(\frac{1}{20}) = 2.653$; $G(\frac{1}{50}) = 2.691$

**27** Continuous compounding: \$260.203
Actual interest: \$260.202

**29** \$252.12    **35** 4.907    **37** 1.3802

**39** 0.8114    **41** 0

## SECTION 4 (pages 272–273)

**1** $f'(x) = -3e^{-3x}$

**3** $f'(x) = x^2 e^{4x}(4x+3)$

**5** $H'(x) = e^{3x}\left(3\ln|x| + \dfrac{1}{x}\right)$

**7** $g'(x) = (\sec^2 x)e^{\tan x}$

**9** $f'(x) = 3^{5x} 5 \ln 3$

**11** $f'(x) = 2^{-7x}[3x^2 - 7(x^3+3)\ln 2]$

**13** $F'(x) = x^{-1+\sin x}(\sin x + x \cos x \ln x)$

**15** $G'(x) = (G(x)/2\sqrt{x})(2 + \ln x)$

**17** $H'(x) = H(x)[\ln(\ln x) + (1/\ln x)]$

**19** $f'(x) = f(x)[(2/x) + (1/(2x+3)) - (8x/(x^2+1))]$

**21** $g'(x) = \frac{1}{2}g(x)[(1/(x+2)) + (1/(x+3)) - (1/(x+1))]$

**23** $f'(x) = f(x)\left[\dfrac{3}{x} + 2\cot 2x - \dfrac{x}{1+x^2}\right]$

**25** $f'(x) = f(x)\left[\dfrac{1}{3x} + \dfrac{4}{\sin 4x} - \dfrac{14}{2x-1}\right]$

**27** $f'(x) = f(x)\left[2x - \dfrac{x}{x^2+1} - \dfrac{x}{2x^2+3}\right]$

**29** $\frac{1}{6}\ln|\sec(3x^2) + \tan(3x^2)| + C$

**31** $-\ln|\cot x + \csc x| + \cos x + C$

**33** $\frac{1}{2}e^{-1+x^2} + C$

**35** $\dfrac{1}{a}x^a + C$, where $a = -2/\ln 3$

**37** $\dfrac{1}{\ln 2} e^{x \ln 2} + C$  **39** $-\tfrac{1}{4} e^{-\tan 4x} + C$

## SECTION 5 (page 280)

**9** (a) $\pi/3$, (b) $3\pi/4$, (c) $-\pi/4$

**11** (a) $\pi/6$, (b) $\pi/3$, (c) $-\pi/6$

**13** $f'(x) = 2/(1 + 4x^2)$  **15** $\phi'(y) = -1/2\sqrt{y - y^2}$

**17** $y'(x) = \arccos x - (x/\sqrt{1 - x^2})$

**19** $f'(x) = -[3(1 + x^2) + 2x(1 + 9x^2) \ \text{arccot} \ 3x]/((1 + 9x^2)$
$\times (1 + x^2)^2)$

**21** $f'(x) = 1/(1 + x^2)$  **23** $f'(x) = 1/(1 + x^2)$

**25** $f'(x) = \arcsin 2x$  **27** $f'(x) = \sqrt{x^2 - 4}/x$  $(|x| > 2)$

**29** $f'(x) = 3/(5 + 4 \cos x)$

**31** $\dfrac{dy}{dx} = -3e^{-3x}[\arccos 3x + (1 - 9x^2)^{-1/2}]$

**33** $4/(1 + 16x^2) \arctan 4x$

**35** $\varphi'(x) = xe^{-x}[(2 - x) \arcsin x + x(1 - x^2)^{-1/2}]$

**37** $f'(x) = 2e^{2x}[\arccos 4x - 2(1 - 16x^2)^{-1/2}]$

**39** $f'(t) = t^2 \left\{ \arcsin(1 + t) \dfrac{3t \ln |2 + t|}{2 + t} \right.$

$\left. + t \ln |2 + t|(-2t - t^2)^{-1/2} \right\}$

**41** $f'(x) = 0; f(x) = \text{constant}$

**45** $a = -1, b = \pi/2; f(x) = -x + \pi/2$

## SECTION 6 (page 282)

**1** $\tfrac{1}{4} \arcsin 4x + C$  **3** $\text{arcsec } 3x + C$

**5** $(1/\sqrt{7}) \arctan \sqrt{7}x + C$  **7** $\tfrac{1}{2} \text{arcsec } \tfrac{1}{2}x + C$

**9** $x - \tfrac{3}{2} \arctan \tfrac{1}{2}x + C$  **11** $(1/\sqrt{10}) \text{arcsec } (x/\sqrt{5}) + C$

**13** $\arcsin (e^x) + C$  **15** $\arcsin (\ln x) + C$

**17** $\text{arcsec } (2e^x) + C$

**19** $\tfrac{1}{12} \arctan (\tfrac{1}{3}e^x) + C$

**21** $\pi/2$  **23** $\pi/6$

**25** $\pi/2$  **27** $(3\pi - 2)/12\sqrt{2}$

## SECTION 7 (pages 286–287)

**1** Tangent: $x - 2y - 1 + \pi/2 = 0$;
normal: $2x + y - 2 - \pi/4 = 0$

**3** Tangent: $x + 2\sqrt{3}y + \tfrac{4}{3}\pi\sqrt{3} + 2 = 0$;
normal: $2\sqrt{3}x - y - \tfrac{2}{3}\pi + 4\sqrt{3} = 0$

**5** Tangent: $x - ey = 0$; normal: $ex + y - e^2 - 1 = 0$

**7** Tangent: $2x - y - e = 0$; normal: $x + 2y - 3e = 0$

**9** Tangent: $y = 6x$; normal: $6y + x = 0$

**11** Tangent: $2(y - \log 2) = (4 \log 2 + 1)(x - 1)$;
normal: $(y - \log 2)(4 \log 2 + 1) = -2(x - 1)$

**13** Tangent: $2y - 2 \ln 2 = (1 - 2 \ln 2)x$;
normal: $(y - \ln 2)(2 \ln 2 - 1) - 2x = 0$

**15** $(\tfrac{1}{2}\sqrt{2}, \pi/4); \tan \theta = 2\sqrt{2}$

**17** $(\ln 2, 4); \tan \theta = \tfrac{4}{33}$

**19** $(1, \pi/4); \tan \theta = 2$

**21** $(0, 0), \tan \theta = 1; (1, e^{-1}), \tan \theta = e^{-1}$

**23** $(1, \ln 2), \tan \theta = 3$

**25** No rel. max. or rel. min.; decreasing for all $x$; concave upward for all $x$

**27** Rel. max. at $(2, 4e^{-2})$; rel. min. at $(0, 0)$; points of inflection at $x = 2 \pm 2\sqrt{2}$; decreasing for $-\infty < x \le 0$ and $2 \le x < \infty$; increasing for $0 \le x \le 2$; concave upward for $-\infty < x \le 2 - \sqrt{2}$ and $2 + \sqrt{2} \le x < \infty$; concave downward for $2 - \sqrt{2} \le x \le 2 + \sqrt{2}$

**29** Rel. max. at $(1, 2e^{-1/2})$; rel. min. at $(-1, -2e^{-1/2})$; points of inflection at $x = 0, \pm\sqrt{3}$; decreasing for $|x| \ge 1$; increasing for $|x| \le 1$; concave downward for $x \le -\sqrt{3}, 0 \le x \le \sqrt{3}$; concave upward for $-\sqrt{3} \le x \le 0, \sqrt{3} \le x$

**31** No rel. max., min., or inflection points; domain is $x > -1$; increasing for $x > -1$; concave downward for $x > -1$

**33** Rel. min. at $(e^{-1/2}, -e^{-1})$; inflection at $x = e^{-3/2}$; decreasing for $0 < x \le e^{-1/2}$; increasing for $x \ge e^{-1/2}$; concave downward for $0 \le x \le e^{-3/2}$; concave upward for $x \ge e^{-3/2}$

**35** Rel. max. at $\left(\dfrac{\pi}{4}, \dfrac{\sqrt{2}}{2} e^{-\pi/4}\right)$; rel. min. at $\left(\dfrac{5\pi}{4}, -\dfrac{\sqrt{2}}{2} e^{-5\pi/4}\right)$;
inflections at $x = \dfrac{\pi}{2}, \dfrac{3\pi}{2}$; increasing for $-\dfrac{3\pi}{4} \le x \le \dfrac{\pi}{4}$;
decreasing for $\dfrac{\pi}{4} \le x \le \dfrac{5\pi}{4}$; concave upward for
$\dfrac{\pi}{2} \le x \le \dfrac{3\pi}{2}$; concave downward for $-\dfrac{\pi}{2} \le x \le \dfrac{\pi}{2}$; pattern repeats in each interval of length $2\pi$, but function $f$ is not periodic

**37** No rel. max. or rel. min.; decreasing for all $x$; concave upward for all $x$, $-\infty < x < +\infty$

**39** Rel. min. at $(0, 0)$; decreasing for $-\dfrac{\pi}{2} < x \le 0$; increasing for $0 \le x < \dfrac{\pi}{2}$; concave upward for $-\dfrac{\pi}{2} < x < \dfrac{\pi}{2}$

**41** $(52\pi/3)$ km/min  **43** 440 m/sec

**45** $(3/26)$ rad/sec  **47** Width $\ge 5\sqrt{5}$ m

**49** $\frac{1}{3}\pi(6-2\sqrt{6})$ rad for any radius

**51** $\dfrac{ds}{dt} = \dfrac{50t}{\sqrt{5-t^2}}$

At $t=1$, $ds/dt = (50/3)$ m/sec

As $t \to 0^+$, $ds/dt + (50/\sqrt{10})$ m/sec

**53** $(-11\pi/10)$m/hr

**55** $t = \dfrac{1}{c}\ln 2$        **57** $I = 2e^{-Rt/L}$

## CHAPTER 7 REVIEW PROBLEMS (pages 288–289)

**1** $4/(4x+5)$

**3** $f'(x) = \dfrac{2}{(2x+10)\,\ln\,(2x+10)}$;

$f''(x) = -4\dfrac{1+\ln\,(2x+10)}{(2x+10)^2[\ln\,(2x+10)]^2}$

**5** Tangent: $2y + 2\ln 2 = (4\ln 2 + 1)x - 1$;
normal: $(y - \ln 2)(4\ln 2 + 1) = 2(1-x)$

**7** $\frac{1}{6}\ln(3x^2+4) + C$

**9** $\frac{1}{4}\ln 86$

**11** $g(x) = \sqrt{x-1}$, $1 \leq x < +\infty$

**13** Decreasing $-\infty < x \leq -3/2$, $3/2 \leq x < +\infty$;
increasing $-3/2 \leq x \leq 3/2$;

$g_1(x) = \dfrac{1+\sqrt{1-144x^2}}{8x}$, $0 < x \leq \dfrac{1}{12}$;

$g_2(x) = \dfrac{1-\sqrt{1-144x^2}}{8x}$, $-\dfrac{1}{12} \leq x \leq \dfrac{1}{12}$;

$g_3(x) = \dfrac{1+\sqrt{1-144x^2}}{8x}$, $-\dfrac{1}{12} \leq x < 0$

**15** Domain: $-\infty < x < +\infty$; range: $-\infty < y \leq -\sqrt{2}$, $\sqrt{2} \leq y < +\infty$; inverse relation: $x^4 - 2y^2 = 4$; inverse is not a function

**17** $f'(x) = 3x^2$, which shows that $f$ is monotone increasing for all $x$. Thus $f$ has an inverse which is a function. Domain of $f'$: $-\infty < x < +\infty$; domain of $g'$: $0 \leq x < +\infty$

**29** $(6x-2)e^{3x^2-2x+5}$

**31** $e^{4x+1}\left(\dfrac{1}{2+x} + 4\ln|2+x|\right)$

**33** $F' = F\left(\dfrac{3x}{x^2+1} + \dfrac{4}{3(2x+5)} - \dfrac{7x+11}{4(x+1)(x+2)}\right)$

**35** $e^x\left(\cos x \ln|x| + \dfrac{\sin x}{x} + \sin x \ln|x|\right)$

**37** $f'(x) = e^{-1/x}\left(2x+1+\dfrac{1}{x^2}\right)$;

$f''(x) = e^{-1/x}\left(2 + \dfrac{2}{x} + \dfrac{1}{x^2} - \dfrac{2}{x^3} + \dfrac{1}{x^4}\right)$

**39** $-\frac{1}{2}e^{2-x^2} + C$        **41** $\dfrac{\ln 3}{2}3^{(1/2)x} + C$

**43** $-\frac{1}{2}[\ln(\cos x)]^2 + C$        **45** $2x(2x-x^2)^{-1/2}$

**47** $-\dfrac{1}{x\sqrt{1-x^2}} - \dfrac{\arccos x}{x^2}$        **49** $\dfrac{\arcsin x}{x} + \dfrac{\ln x}{\sqrt{1-x^2}}$

**51** $e^{-3\sin x}\left(\dfrac{1}{x} - 3\cos x \ln x\right)$        **53** $\frac{1}{3}\arcsin\left(\dfrac{3x}{4}\right) + C$

**55** $\frac{1}{3}\arcsin\left(\dfrac{3x}{2}\right) + C$        **57** $\frac{1}{2}\arcsin(\frac{1}{2}e^{2x}) + C$

**59** $3y - \pi/2 = \sqrt{3}(4x+1)$        **61** rel. max. at $(-1, e^2)$

**63** 13,236 yrs        **65** 5.8 hrs

---

# CHAPTER 8

## SECTION 1 (pages 297–299)

**1** $-\sqrt{3-2x} + C$        **3** $-\frac{1}{5}\cos 5x + C$

**5** $\frac{1}{2}\tan 2y + C$        **7** $-\frac{1}{2}\ln|\cos 2x| + C$

**9** $2e^{x/2} + C$        **11** $\dfrac{1}{12}\arctan\dfrac{3x}{4} + C$

**13** $-2\cos\sqrt{x} + C$        **15** $\frac{1}{2}e^{x^2+1} + C$

**21** $\frac{1}{3}\sin^3 x + C$        **23** $\frac{1}{4}\tan^2 2x + C$

**25** $\frac{1}{2}\ln|1+(\ln x)^2| + C$        **27** $\frac{1}{2}\ln|\sin 2x| + C$

**29** $\frac{1}{4}(\ln x)^4 + C$        **31** $-(e^x+2)^{-1} + C$

**33** $\frac{1}{14}\sec^7 2x + C$        **35** $\ln|e^x+1| + C$

**37** $y - \ln|1+e^y| + C$        **39** $2(\sqrt{7}-1)$

**41** $\frac{7}{3}$        **43** $\pi/12$

**45** $\frac{1}{4}$        **47** $\frac{3}{2}$

**49** $\dfrac{1}{3}(e^2+1)^{3/2} - \dfrac{2\sqrt{2}}{3}$        **51** $\frac{1}{2}\ln|x^2+2x+2| + C$

**53** $\ln|1+\tan x| + C$

**55** $\frac{1}{2}\ln|\sin 2x| + C$

**57** $\frac{x^2}{2} + 2x - 4\ln|x+1| + C$    **59** $\frac{x}{2} + \frac{1}{4}\ln|2x-1| + C$

**61** $\frac{1}{2\sqrt{3}}\arctan\frac{\sqrt{3}}{2}x + C$      **63** $-\frac{1}{2}(x^2+1)^{-1} + C$

**65** $\frac{1}{2\ln 2}2^{x^2} + C$

**67** $\frac{-x^2}{1} - 5x - 14\ln|3-x| + C$

**69** $x + \ln|x^2+1| + C$      **71** $2x + \ln|x^2+2x-2| + C$

**73** $e^{\tan x} + C$      **75** $\arcsin\left(\frac{e^x}{\sqrt{3}}\right) + C$

**77** $2\sec\sqrt{x} + C$      **79** $-\frac{1}{2}\ln|2+\cos 2x| + C$

**81** $-\sqrt{5-x^2} - 2\arcsin\left(\frac{x}{\sqrt{5}}\right) + C$

**83** $-\frac{1}{2}(2x+\cos x)^{-2} + C$

**85** $-\frac{\sqrt{3}}{6}\arctan\left[(\cos 2x)/\sqrt{3}\right] + C$

**87** $\ln 4$      **89** $0$

**91** $\frac{1}{3}\tan^2\frac{3\pi}{8} - \frac{1}{3}$      **93** $\ln 8$

**95** $1 + \ln(2+e^2) - \ln(1+2e)$

**27** $\frac{1}{10}e^{-x}(3\sin 3x - \cos 3x) + C$

**29** $e^{ax}(a^2+b^2)^{-1}(a\sin bx - b\cos bx) + C$

**31** $\frac{1}{3}\sin^3 2x + C$

**33** $\frac{5x}{16} - \frac{1}{4}\sin 2x + \frac{3}{64}\sin 4x + \frac{1}{48}\sin^3 2x + C$

**35** $-2$

**37** $\frac{1}{32}\left(1 + \frac{4\pi^2}{9} - \ln\left(1 + \frac{4\pi^2}{9}\right)\right) - \sqrt{3}\pi^3/27$

**39** $\frac{4}{25}(1 + e^{3\pi/4})$      **41** $\frac{\pi}{3}\sqrt{3} + \frac{1}{2}\ln\frac{3}{4}$

**43** $-\cos x\left(\frac{\sin^{n-1}x}{n} + \frac{(n-1)\sin^{n-3}x}{n(n-2)}\right.$

$\left. + \frac{(n-1)(n-3)}{n(n-2)(n-4)}\sin^{n-5}x\right)$

$+ \frac{(n-1)(n-3)(n-5)}{n(n-2)(n-4)}\int\sin^{n-6}x\, dx$

**45** $\frac{b}{a+b}(g'(x)f(x) - g(x)f'(x)) + C$

**47** $\int x^5(\log x)^2\, dx = \frac{1}{6}x^6[(\log x)^2 - \frac{1}{3}(\log x) + \frac{1}{18}] + C$

**51** $a_0 = -\frac{2}{\ln 2}\left[1 + \frac{1!}{\ln 2} + \frac{2!}{(\ln 2)^2} + \frac{3!}{(\ln 2)^3} + \cdots\right.$

$\left. + \frac{(n-1)!}{(\ln 2)^{n-1}}\right];$

$a_1 = 1,\ a_2 = 1,\ a_3 = 2!,\ \ldots,\ a_n = (n-1)!$

## SECTION 2 (pages 304–305)

**1** $\left(\frac{x^2}{2}\right)\ln x - \frac{x^2}{4} + C$

**3** $-x^2\cos x + 2x\sin x + 2\cos x + C$

**5** $x(\ln x)^2 - 2x\ln x + 2x + C$

**7** $x\arctan x - \frac{1}{2}\ln|1+x^2| + C$

**9** $\frac{1}{2}x^2\operatorname{arcsec} x - \frac{1}{2}(x^2-1)^{1/2} + C$

**11** $\frac{1}{3}x^3\arcsin x + \frac{1}{3}(1-x^2)^{1/2} - \frac{1}{9}(1-x^2)^{3/2} + C$

**13** $-2x\cot(x/2) + 4\ln|\sin(x/2)| + C$

**15** $3x\tan 3x - \frac{9x^2}{2} + \ln|\cos 3x| + C$

**17** $2x^3\arcsin 2x + \frac{1}{4}\sqrt{1-4x^2} - \frac{1}{12}(1-4x^2)^{3/2} + C$

**19** $(\frac{1}{3}x^3 - x^2 + 2x)\ln x - \frac{1}{9}x^3 + \frac{1}{2}x^2 + 2x + C$

**21** $\sqrt{2x}\cos\sqrt{2x} + \sin\sqrt{2x} + C$

**23** $e^{x^2}(x^2-1) + C$

**25** $\frac{1}{5}e^x(\sin 2x - 2\cos 2x) + C$

## SECTION 3 (page 309)

**1** $\frac{1}{8}(1/3 - \sqrt{3})$      **3** $-2^{-1/4}(\frac{9}{5}) + \frac{8}{5}$

**5** $\sqrt{3}/27$      **7** $\pi/4$

**9** $3\pi/16$      **11** $3/8$

**13** $\frac{2}{3} + \frac{10}{27}\sqrt{3}$

**15** $-2\cos\frac{x}{2} + \frac{4}{3}\cos^3\frac{x}{2} - \frac{2}{5}\cos^5\frac{x}{2} + C$

**17** $\ln|\sin x| - \frac{1}{2}\sin^2 x + C$      **19** $\frac{1}{4}\tan^2 2x + C$

**21** $\frac{1}{2}\tan 2x + \frac{1}{6}\tan^3 2x + C$

**23** $-\frac{1}{7}\csc^7 x + \frac{2}{5}\csc^5 x - \frac{1}{3}\csc^3 x + C$

**25** $-x + \tan x + C$      **27** $x - \frac{1}{3}\cot^3 x + \cot x + C$

**29** $-\frac{1}{2}\cos(x^2) + \frac{1}{6}\cos^3(x^2) + C$

**31** $\frac{1}{3}\sin^3 x + C$      **33** $\frac{1}{3}\tan^3 x + C$

**35** $-\frac{1}{4}\cos^4 x + C$      **37** $\frac{1}{9}\tan^3 3x + C$

**39** $-\frac{1}{4}\tan^2 2x + C$      **43** $(15/154)2^{-3/4}$

## SECTION 4 (pages 312–313)

**1** $\pi/6$            **3** $-\pi/18$

**5** $32/3$            **7** $\pi/18$

**9** $\dfrac{17}{\sqrt{8}} - 6$

**11** $(2/\sqrt{5})\,\text{arcsec}\,(x/\sqrt{5}) + C$

**13** $\frac{1}{3}(9 - x^2)^{3/2} - 9(9 - x^2)^{1/2} + C$

**15** $\frac{1}{12}\sqrt{2x^2 + 7}\,(2x^2 - 14) + C$

**17** $\dfrac{1}{5a^4}(a^2u^2 + b^2)^{5/2} - \dfrac{b^2}{3a^4}(a^2u^2 + b^2)^{3/2} + C$

**19** $\sqrt{x^2 - a^2} - a \cdot \text{arcsec}\,\dfrac{x}{a} + C$

**21** $-(1/a^2 x)\sqrt{x^2 + a^2} + C$

**23** $-a^{-4}\left(\dfrac{\sqrt{a^2 - x^2}}{x} + \dfrac{(a^2 - x^2)^{3/2}}{3x^3}\right) + C$

**25** $9[\frac{1}{63}(9 - x^2)^{7/2} - \frac{1}{5}(9 - x^2)^{5/2}] + C$

**27** $-\frac{1}{28}(9 - 4x^2)^{7/2} + C$

**29** $\sqrt{2}/32$       **31** $3^{-12} - \dfrac{\pi}{6}$

**33** $\frac{2}{9}(1 - 3^{-1/2})$       **35** $\dfrac{25\pi}{16}$

**37** For type (B), $-\dfrac{\pi}{2} < \theta < \dfrac{\pi}{2}$;

for type (C), $-\pi \le \theta < -\dfrac{\pi}{2}$ or $0 \le \theta < \dfrac{\pi}{2}$

**39** $-\dfrac{1}{3}\dfrac{(1 - x^2)^{3/2}}{x^3} + C$

**41** $n$ is a nonnegative even integer

## SECTION 5 (pages 314–315)

**1** $\arctan(x + 2) + C$

**3** $x - \ln|x^2 + 2x + 5| - \dfrac{3}{2}\arctan\dfrac{(x + 1)}{2} + C$

**5** $\text{arcsec}\,(x + 2) + C$

**7** $\ln|x - 1 + \sqrt{5 - 2x + x^2}| + C$

**9** $\dfrac{1}{2}\ln|x^2 + 2x + 5| + \arctan\dfrac{x + 1}{2} + C$

**11** $\frac{1}{3}x^3 - \frac{3}{2}x^2 + 8x - (21/2)\ln|x^2 + 3x + 1|$

$-\dfrac{47}{2\sqrt{5}}\ln\left|\dfrac{\sqrt{5} + 2x + 3}{\sqrt{5} - 2x - 3}\right| + C$

**13** $3\sqrt{x^2 + 2x} + \ln|x + 1 + \sqrt{x^2 + 2x}| + C$

**15** $\dfrac{11x - 19}{4(x^2 - 2x + 3)} + \dfrac{11\sqrt{2}}{8}\arctan\dfrac{x - 1}{\sqrt{2}} + C$

**17** $\dfrac{8x - 17}{23(2x^2 + 3x + 4)} + \dfrac{16}{(23)^{3/2}}\arctan\dfrac{4x + 3}{\sqrt{23}} + C$

**19** $\dfrac{9x - 13}{\sqrt{x^2 - 2x + 2}} + C$     **21** $\dfrac{5x - 3}{4\sqrt{x^2 + 2x - 3}}$

**23** $(2 - x - x^2)^{-3/2} + \dfrac{8}{81}\left[\dfrac{x + \frac{1}{2}}{(2 - x - x^2)^{1/2}} + \dfrac{(x + \frac{1}{2})^3}{3(2 - x - x^2)^{3/2}}\right]$

**25** $\frac{2}{7}\sqrt{7}(\arctan\sqrt{7} - \arctan\frac{3}{7}\sqrt{7})$    **27** $\arctan(3) - (\pi/4)$

**29** $\frac{3}{40} + \frac{5}{16}\arctan(\frac{1}{2})$       **31** $\pi/8$

**33** $\arcsin\dfrac{2}{\sqrt{6}} - \arcsin\dfrac{1}{\sqrt{6}}$

**35** Integrand is infinite at $x = \pm 1$. Integrable over any interval not containing $x = 1$, $x = -1$

## SECTION 6 (pages 322–323)

**1** $\dfrac{x^2}{2} + 5x + 14\ln|x - 2| + C$

**3** $\dfrac{x^2}{2} - 4x + 13\ln|x + 2| - \ln|x + 1| + C$

**5** $x + 11\ln|x - 2| - 6\ln|x - 1| + C$

**7** $x + (x - 2)^{-1} + 2\ln|x - 2| + C$

**9** $2\ln|x - 1| + \dfrac{1}{x - 1} + \ln|x + 1| + C$

**11** $\frac{1}{2}x^2 + 6x + 27\ln|x - 3| - 27(x - 3)^{-1} + C$

**13** $-\frac{8}{3}\ln|x + 2| + \frac{5}{2}\ln|x + 1| + \frac{1}{6}\ln|x - 1| + C$

**15** $2\ln|x + 3| - \frac{1}{3}\ln|x + 2| + \frac{1}{3}\ln|x - 1| + C$

**17** $\frac{1}{2}\ln|x + 1| + \frac{1}{4}\ln|x^2 + 1| + \frac{5}{2}\arctan x + C$

**19** $\frac{1}{9}\ln|x + 1| - \frac{4}{3}(x + 1)^{-1} - \frac{1}{9}\ln|x - 2| + C$

**21** $-\frac{1}{9}\ln|x + 2| + \frac{1}{9}\ln|x - 1| - \frac{5}{3}(x - 1)^{-1} + C$

**23** $-\frac{3}{4}\ln|x + 1| + \frac{7}{4}\ln|x - 1| + \frac{1}{2}(x - 1)^{-1} - \frac{1}{2}(x - 1)^{-2} + C$

**25** $\ln|x^2 + 4| + \dfrac{11}{6}\arctan\dfrac{x}{2} - \dfrac{1}{2}\ln|x^2 + 1| - \dfrac{2}{3}\arctan x + C$

**27** $\frac{1}{4}\ln|x - 1| - \frac{1}{8}\ln|x + 1| + \frac{7}{8}\ln|x - 3| - \ln|x - 2| + C$

**29**   $-\dfrac{4}{25}\ln|x-1|-\dfrac{2}{5}(x-1)^{-1}+\dfrac{2}{25}\ln|x^2+4|+$

    $\dfrac{19}{50}\arctan\dfrac{x}{2}+C$

**31**   $\frac{1}{9}(x+1)^{-1}-\frac{5}{27}\ln|x+1|+\frac{32}{27}\ln|x-2|-\frac{5}{9}(x-2)^{-1}+C$

**33**   $2\ln|x^2+4|+\dfrac{5}{4}\arctan\dfrac{x}{2}+\dfrac{1}{2}(16-11x)(x^2+4)^{-1}+C$

**35**   $\dfrac{14}{27}\ln|x-3|+\dfrac{13}{54}\ln|x^2+3x+9|+\dfrac{7}{3\sqrt{27}}\arctan\dfrac{2x+3}{\sqrt{27}}$

**37**   $\dfrac{51}{256}\arctan\dfrac{x}{2}-\dfrac{77x}{128(x^2+4)}+\dfrac{17x}{16(x^2+4)^2}+C$

## SECTION 7 (pages 325–326)

**1**   $\frac{4}{3}(x+2)^{3/2}-2(x+2)^{1/2}+C$

**3**   $\frac{2}{5}(x+1)^{5/2}-\frac{2}{3}(x+1)^{3/2}+C$

**5**   $\frac{1}{12}(3x-1)^{4/3}-\frac{5}{3}(3x-1)^{1/3}+C$

**7**   $\frac{3}{64}(2x+1)^{8/3}-\frac{3}{20}(2x+1)^{5/3}+\frac{3}{16}(2x+1)^{2/3}+C$

**9**   $2(x+4)^{1/2}+4\ln|-2+\sqrt{x+4}|-2\ln|x|+C$

**11**   $x+6\sqrt{x}+6\ln|\sqrt{x}-1|+C$

**13**   $\frac{1}{3}(x^2-4)^{3/2}+4(x^2-4)^{1/2}+C$

**15**   $\frac{1}{5}(x^2+4)^{5/2}-2(x^2+4)^{3/2}+8(x^2+4)^{1/2}+C$

**17**   $-5(16-x^2)^{3/2}+\frac{1}{5}(16-x^2)^{5/2}+C$

**19**   $\dfrac{1}{a}\ln|a-\sqrt{a^2-x^2}|-\dfrac{1}{a}\ln|x|+C$

**21**   $-\frac{1}{10}(2x+1)^{-5/2}+\frac{1}{14}(2x+1)^{-7/2}+C$

**23**   $\frac{3}{2}(x+5)^{2/3}+27(x+5)^{-1/3}-\frac{63}{4}(x+5)^{-4/3}+C$

**25**   $\sqrt{a^2-x^2}+a\ln|a-\sqrt{a^2-x^2}|-a\ln|x|+C$

**27**   $\sqrt{x^2-a^2}-a\arctan(\sqrt{x^2-a^2}/a)+C$

**29**   $3x^{1/3}-6x^{1/6}+6\ln|1+x^{1/6}|+C$

**31**   $\dfrac{1}{32}\ln\left|\dfrac{\sqrt{x^2+4}+2}{\sqrt{x^2+4}-2}\right|-\dfrac{\sqrt{x^2+4}}{8x^2}+C$

**33**   $\frac{1}{6}\ln|x|-\frac{1}{6}\ln|\sqrt{x^2+9}+3|-\frac{1}{2}x^{-2}\sqrt{x^2+9}+C$

**37**   $-\dfrac{1}{4}\cot^2\left(\dfrac{x}{2}\right)-\dfrac{1}{2}\ln\left|\tan\dfrac{x}{2}\right|+C$

**39**   $\dfrac{1}{2}\ln\left|\dfrac{1+\tan\dfrac{\theta}{2}}{1-\tan\dfrac{\theta}{2}}\right|-\dfrac{\tan\dfrac{\theta}{2}}{\left(1+\tan\dfrac{\theta}{2}\right)^2}+C$

**41**   $\frac{1}{3}\ln|4\tan(\frac{1}{2}\theta)+2|-\frac{1}{3}\ln|4\tan(\frac{1}{2}\theta)+8|+C$

**43**   $\dfrac{2}{5}\sqrt{5}\arctan\dfrac{3\sqrt{5}}{5}\left(\tan\dfrac{1}{2}\theta-\dfrac{2}{3}\right)+C$

## CHAPTER 8 REVIEW PROBLEMS (pages 326–327)

**1**   $\frac{1}{2}(4x+1)^{1/2}+C$

**3**   $\frac{1}{9}(2x+5)^{9/2}+C$

**5**   $\frac{1}{4}e^{x^4+3}+C$

**7**   $\frac{1}{2}\ln(e^{2x}+4)$

**9**   $(x^2+4x+2)^{1/2}+C$

**11**   $\frac{1}{2}(x^2-1)\ln(1+x)+\frac{1}{4}(1+x)(3-x)+C$

**13**   $\frac{1}{2}x^2\arcsin(x^2)+\frac{1}{2}\sqrt{1-x^4}+C$

**15**   $\frac{1}{2}x^2\arctan(x^2)-\frac{1}{4}\ln(1+x^4)+C$

**17**   $\frac{1}{2}x^2\tan(x^2)+\frac{1}{2}\ln|\cos(x^2)|+C$

**19**   $\frac{1}{32}(\frac{1}{3}\cos^3 4x-\cos 4x)+C$

**21**   $\frac{1}{6}\sec^3 2x-\frac{1}{2}\sec 2x+C$

**23**   $\tan^{1/3} 3x+\frac{1}{7}\tan^{7/3} 3x+C$

**25**   $-\frac{1}{2}x^2\cos(x^2)+\frac{1}{6}x^2\cos^3(x^2)+\frac{1}{3}\sin(x^2)+\frac{1}{18}\sin^3(x^2)$
     $+C$

**27**   $9\ln 3-4\ln 2-5/2$

**29**   $\dfrac{1}{20}\arctan\dfrac{5x}{4}+C$

**31**   $\frac{1}{6}\sqrt{3}\arcsin\frac{2}{5}\sqrt{15}x+C$

**33**   $\dfrac{1}{8}x\sqrt{9+4x^2}-\dfrac{9}{16}\ln\left|\dfrac{1}{3}\sqrt{9+4x^2}+\dfrac{2x}{3}\right|+C$

**35**   $\dfrac{1}{11}\sqrt{11}\arctan\dfrac{3}{\sqrt{11}}\left(x+\dfrac{1}{3}\right)+C$

**37**   $\arcsin(\frac{1}{2}\ln x)+C$

**39**   $\frac{1}{3}x^3+2x^2+12x+33\ln|x-3|+C$

**41**   $x+\frac{10}{3}\ln|x-2|-\frac{1}{3}\ln|x+1|+C$

**43**   $\dfrac{1}{9}\ln|x-1|+\dfrac{8}{9}\ln|x+2|+\dfrac{4}{3(x+2)}+C$

**45**   $\frac{2}{3}\ln|x+2|+\frac{1}{3}\ln|x-1|+C$

**47**   $\frac{1}{4}\ln|x-1|-\frac{1}{8}\ln(x^2+2x+5)+\frac{1}{4}\arctan\frac{1}{2}(x+1)+C$

**49**   $\frac{1}{12}\ln|x-1|-\frac{1}{3}\ln|x+2|+\frac{1}{4}\ln|x+3|+C$

**51**   $\frac{4}{27}(3x+4)^{3/2}-\frac{10}{9}(3x+4)^{1/2}+C$

**53**   $\frac{1}{5}(x^2-9)^{5/2}+6(x^2-9)^{3/2}+81(x^2-9)^{1/2}+C$

**55**   $\dfrac{1}{\sqrt{13}}\ln\dfrac{\tan\dfrac{1}{2}x-\dfrac{3}{2}+\dfrac{1}{2}\sqrt{13}}{\tan\dfrac{1}{2}x-\dfrac{3}{2}-\dfrac{1}{2}\sqrt{13}}+C$

# CHAPTER 9

## SECTION 1 (pages 332–333)

**1** $(\frac{13}{4}, 0)$ **3** $(\frac{2}{5}, 0)$ **5** $(\frac{11}{9}, 0)$

**7** $(\frac{7}{8}, \frac{11}{8})$ **9** $(\frac{23}{17}, \frac{35}{17})$ **11** $(-\frac{22}{15}, \frac{4}{15})$

**13** $(-\frac{15}{17}, \frac{30}{17})$

**15** With one side on the $x$ axis from $(-1, 0)$ to $(1, 0)$, $\bar{x} = 0$, $\bar{y} = 1.3764$

**17** With axes on outer sides of figure, $\bar{x} = 2$, $\bar{y} = 1.5$

**19** Origin at point of tangency of circle and rectangle;

$$\bar{x} = \frac{4\pi - 5}{2\pi + 5}, \bar{y} = \frac{-5}{4\pi + 10}$$

**23** With the base on the $x$ axis from

$$\left(-\frac{b}{2}, 0\right) \text{ to } \left(\frac{b}{2}, 0\right), \bar{x} = 0, \bar{y} = \frac{3}{10}\sqrt{4a^2 - b^2}$$

**19** $\bar{x} = \dfrac{2\pi - 3\sqrt{3}}{2}, \bar{y} = \dfrac{3\sqrt{3}}{8}$

**21** $\bar{x} = \dfrac{24 \ln 2 - 14}{24 \ln 2 - 15}$,

$\bar{y} = \dfrac{4[(\ln 2)^2 - 2 \ln 2 + (23/24)]}{8 \ln 2 - 5}$

**23** $\bar{x} = \dfrac{2 - e^{1/2}}{e^{1/2} - 1}, \bar{y} = \dfrac{1}{4}\dfrac{e - 1}{e^{1/2} - 1}$

**25** $\bar{x} = \dfrac{(\ln 2)^2 - 2 \ln 2 - 1}{2 \ln 2 + 1}, \bar{y} = \dfrac{8 \ln 2 + 9}{8 \ln 2 + 4}$

**27** $\bar{x} = \dfrac{\pi}{2}, \bar{y} = -15/17$

## SECTION 2 (pages 338–339)

**1** $(\bar{x}, \bar{y}) = \left(\dfrac{18}{5}, \dfrac{3\sqrt{6}}{4}\right)$; $R = \{(x, y): 0 \le x \le 6, 0 \le y \le \sqrt{4x}\}$

**3** $(\bar{x}, \bar{y}) = \left(\dfrac{\pi}{2}, \dfrac{\pi}{8}\right)$; $R = \{(x, y): 0 \le x \le \pi, 0 \le y \le \sin x\}$

**5** $(\bar{x}, \bar{y}) = \left(0, \dfrac{4a}{3\pi}\right)$;

$$R = \{(x, y): -a \le x \le a, 0 \le y \le \sqrt{a^2 - x^2}\}$$

**7** $\bar{x} = \dfrac{67}{18(8\sqrt{2} - 7)}, \bar{y} = \dfrac{2(72\sqrt{2} - 53)}{15(8\sqrt{2} - 7)}$

**9** $\bar{x} = e/(e - 1), \bar{y} = e(e + 1)/4$

**11** $\bar{x} = \dfrac{\pi}{4} - \dfrac{1}{2}\ln 2, \bar{y} = 2/3$

**13** $\bar{x} = 8/15, \bar{y} = 8/21$

**15** $\bar{x} = \dfrac{3e^2 - 1}{2(e^2 - 1)}, \bar{y} = \dfrac{1}{4}e^2(e^2 + 1)$

**17** $\bar{x} = \dfrac{1}{3}\left(\dfrac{a^2 + d^2 - b^2 + ad}{a + d - b}\right)$,

$\bar{y} = \dfrac{1}{2}c\left(\dfrac{a + 2d - 2b}{a + d - b}\right)$

## SECTION 3 (pages 340–341)

**1** $\bar{x} = \dfrac{3h}{4}$ **3** $\bar{x} = \dfrac{3a}{8}$

**5** $\bar{x} = \dfrac{256 \ln 2 - 150}{64 \ln 2 - 33}$ **7** $\bar{x} = \dfrac{\pi}{2}$

**9** $\bar{x} = \frac{5}{9}$ **11** $\bar{x} = \frac{25}{32}$

**13** $\bar{x} = \dfrac{\pi - \dfrac{\pi^2}{8} - 2 \ln 2}{4 - \pi}$ **15** $\bar{y} = 910/363$

**17** $\bar{y} = 3$ **19** $\bar{y} = \dfrac{1}{4}\pi - \dfrac{1}{\pi}$

**21** $\bar{x} = \frac{9}{8}$cm, $\bar{y} = \bar{z} = 0$ **23** $\bar{x} = 5/21, \bar{y} = 10/21$

**25** $\bar{x} = 1/2$

## SECTION 4 (page 345)

**1** $\bar{x} = \frac{263}{165}, \bar{y} = \frac{393}{352}$ **3** $\bar{x} = \bar{y} = \dfrac{2a}{5}$

**5** $\bar{x} = \dfrac{17\sqrt{17} - 1}{3[4\sqrt{17} + \ln(4 + \sqrt{17})]}$,

$\bar{y} = \dfrac{132\sqrt{17} - \ln(4 + \sqrt{17})}{16[4\sqrt{17} + \ln(4 + \sqrt{17})]}$

$$7 \begin{cases} \bar{y} = \dfrac{A}{B}, \\[2mm] \text{where } B = \sqrt{1+e^4} - \sqrt{1+e^2} - \ln\left(\dfrac{1+\sqrt{1+e^4}}{e^2}\right) \\[4mm] \qquad\qquad + \ln\dfrac{1+\sqrt{1+e^2}}{e} \\[4mm] \text{and} \quad A = \tfrac{1}{2}[e^2\sqrt{1+e^4} + \ln(e^2 + \sqrt{1+e^4}) \\[1mm] \qquad\qquad - e\sqrt{1+e^2} - \ln(e + \sqrt{1+e^2})] \end{cases}$$

**9** $\bar{x} = \dfrac{2(25\sqrt{5}+1)}{5(5\sqrt{5}-1)}$ **11** $\bar{x} = \dfrac{169\sqrt{13}-31}{130\sqrt{13}-10}$

**13** $\bar{x} = \dfrac{38}{10 + 9\sqrt{5}\,\arcsin\sqrt{5}/3}$

**15** $\bar{y} = \dfrac{A}{B}$, where $A = \displaystyle\int_a^b x\,f(x)\sqrt{1+(dy/dx)^2}\,dx$

and $\qquad B = \displaystyle\int_a^b x\sqrt{1+(dy/dx)^2}\,dx$

**17** $\bar{y} = \dfrac{1217}{63(15 + \ln 16)}$ **19** $\bar{y} = 723/26$

**21** $\bar{y} = \dfrac{A}{B}$, where $A = \dfrac{2284}{45} + \left(\dfrac{2}{3}\ln 2 - 6\ln 3\right)$

and $\qquad B = \dfrac{85}{4} - 9\ln 3 - (\ln 3)^2 + 4(\ln 2) + (\ln 2)^2$

## SECTION 5 (pages 346−347)

**1** $2\pi$ **3** $\dfrac{\pi^2}{2}$ **5** $\dfrac{242\pi}{5}$

**7** $\dfrac{8\pi}{105}$ **9** $V = 2a^2b\pi^2,\ S = 4ab\pi^2$ **11** $\pi r^2 h$

## SECTION 6 (page 352)

**1** 1 **3** 2 **5** Divergent

**7** $1/(1-p),\ p<1$ **9** Divergent **11** 2

**13** $\ln(\tfrac{4}{3} + \tfrac{1}{3}\sqrt{7})$ **15** $\pi/8$ **17** Divergent

**19** $-\dfrac{1}{6}\ln 7 + \dfrac{1}{\sqrt{3}}\left(\dfrac{\pi}{2} - \arctan\dfrac{5}{\sqrt{3}}\right)$

**21** 1 **23** 9 **25** 0

**27** $\tfrac{1}{2}$ **29** Divergent **31** 4

**33** $2(2\sqrt{2}-1)$ **35** $\pi$ **37** Divergent

**39** $2\sqrt{2}$ **41** $\dfrac{4\pi\sqrt{3}}{9}$ **43** $\pi$

**45** $\tfrac{1}{2}$

## SECTION 7 (pages 356−357)

**1** $\dfrac{165}{4}$ **3** $\tfrac{1}{3}(10^{3/2} - 2^{3/2})$

**5** $7/192$ **7** $\tfrac{1}{2}(e^6 - 1)$

**9** $2/3$ **11** 0

**13** $\tfrac{1}{2}(e^{-1} - e^{-9})$ **15** 180 ergs

**17** (a) 40 ergs (b) 120 ergs **19** (a) 2700 ergs (b) 8100 ergs

**21** 400,000 kg-m

**23** $6562.5w$ kg-m, where $w$ = weight of one m$^3$ of water

**25** $\dfrac{40}{3}w$ kg-m, where $w$ = weight of one m$^3$ of water

**27** $4w$ kg-m, where $w$ = weight of one m$^3$ of water

**29** $\dfrac{7}{5}$ **31** $E/24$

## SECTION 8 (pages 361−362)

Throughout $w = 1000$ kg, the weight of a cubic meter of water.

**1** $10{,}000w$ **3** $2000w$

**5** $90{,}000w$ **7** $14{,}000w$

**9** $(170\sqrt{0.96} + 25\pi - 36\arccos 0.2 - 50\arcsin 0.2]w$

**11** $62.5\sqrt{2}w$ **13** 4800 kg

**15** $96w$ **17** $226.5w$

**19** $(80\pi + \tfrac{128}{3})w$ **21** $11{,}250\sqrt{3}w$

**23** $\tfrac{1}{3}(169{,}600\sqrt{3})w$

**27** (a) $10 \times 4^{5/7}$, (b) $1500(1 - 4^{-2/7})$ joules

## SECTION 9 (pages 368−369)

**1** Approx.: 21.1; exact: 21; $E_T \leq 1/8$

**3** Approx.: 0.696; exact: $\ln 2$; $E_T \leq 1/150$

**5** Approx.: 0.880; exact: $\ln(1+\sqrt{2})$; $E_T \leq 1/150$ with $M = 2$

**7** Approx.: 0.392; exact: $\pi/8$; $E_T \leq 1/30$

**9** Approx.: $-0.31$; exact: $\tfrac{1}{2}\ln(5/9)$; $E_T \leq 0.0672$

**11** $26/3$ **13** $38/3$

**15** Approx.: 0.6933; exact: $\ln 2$; $E_S \leq 3/32$

**17** Approx.: 0.8814; exact: $\ln(1+\sqrt{2})$; $E_S \le 105/256$

**19** Approx.: 0.6045; exact: $\pi\sqrt{3}/9$; $E_S \le 16.03$

**21** Approx.: 0.00; exact: 0; $E_S \le 0.0747$

**23** Approx.: 0.8358; $E_S \le 24.61$

**25** Approx.: 1.0948; $E_S \le 79.80$

**27** Approx.: 0.4298; $E_S \le 11.45$

**29** $x^4 + 1$ has the factors $(x^2 + x\sqrt{2} + 1)(x^2 - x\sqrt{2} + 1)$

## SECTION 10 (pages 375–376)

**1** Range is the set of all positive integers

**3** Range is the set of all positive integers

**5** Range is all $X$ such that $0 \le X \le 1$

**7** 72 cents, assuming a person may win more than one prize

**13** $\dfrac{1}{2} - \dfrac{1}{\pi}\arctan 3$

**17** (a) 0; (b) $\frac{1}{9}$; (c) $\frac{1}{3}$; (d) $\frac{7}{16}$

## CHAPTER 9 REVIEW PROBLEMS (pages 376–377)

**1** $\bar{x} = \frac{9}{4}$; $\bar{y} = \frac{4}{3}$

**3** $\bar{x} = 6/11$; $\bar{y} = 6/11$

**5** With coordinate axes along bottom and left sides of shape in Fig. 42, $\bar{x} = 121/30$; $\bar{y} = 49/30$

**7** $\bar{x} = 0$; $\bar{y} = (\pi + 2)\sqrt{2}/16$

**9** $\bar{x} = (e-1)/e^2$; $\bar{y} = \dfrac{e+1}{4e^2}$

**11** $\bar{x} = 2/5$; $\bar{y} = 1$

**13** $a = 10/3$

**15** $\bar{x} = \dfrac{3\pi^2 - 4\pi - 8}{8(\pi - 2)}$

**17** $\bar{x} = 6\ln(3/2)$

**19** $\bar{x} = A/B$ where $A = \dfrac{9 - 2\sqrt{3}}{36}\pi - \dfrac{5\pi^2}{288} + \dfrac{1}{2}\ln(2/3)$, $B = 1 - \dfrac{1}{3}\sqrt{3} - \dfrac{\pi}{12}$

**21** $\bar{x} = 16/5$; $\bar{y} = 16/5$     **23** Divergent

**25** 0     **27** Divergent

**29** 1/2     **31** Divergent

**33** $\frac{16}{3}(10)^4\pi$     **35** 36

**37** 0.40; $E_T < 0.02$     **39** 3.66; $E_T < 2.82$

**41** 25.17; $E_T < 3.38$     **43** 1.15; $E_S < 0.36$

**45** 9.33; $E_S = 0$     **47** 1.002; $E_S < 0.60$

**49** 0.8826; $E_M < 0.19$     **51** 1.0065; $E_M < 0.041$

# CHAPTER 10

## SECTION 1 (pages 385–386)

**1** $\frac{3}{4}$     **3** $\frac{3}{2}$     **5** $\frac{2}{3}$     **7** 0

**9** 3     **11** 4     **13** $-\infty$     **15** $\frac{2}{3}$

**17** $\infty$     **19** 0     **21** $-1$     **23** 0

**25** 0     **27** 2     **29** 1     **31** 0

**33** $-\frac{2}{3}$     **35** 1     **37** 1     **39** $e$

**41** 1     **43** $+\infty$     **45** 0     **47** no limit

**49** 1     **51** 0     **53** (c) 0

**25** Yes. The limit is 1.

**27** Yes. Limit is 1.

**29** Yes. Limit is 0.

## SECTION 2 (pages 391–391)

**1** 0     **3** $-3/2$     **5** $+\infty$     **7** 0

**9** 2     **11** 0     **13** 0     **15** $1/(3\sqrt{2})$

**17** 0     **19** $\infty$

**23** *Hint:* $n = 1 \cdot 2 \cdot \dfrac{3}{2} \cdot \dfrac{4}{3} \cdots \dfrac{(n-1)}{(n-2)} \dfrac{n}{(n-1)}$

and $\dfrac{n}{3^n} = \dfrac{1}{3} \cdot \dfrac{2}{3} \cdot \dfrac{\frac{3}{2}}{3} \cdot \dfrac{\frac{4}{3}}{3} \cdots \dfrac{n/(n-1)}{3}$

## SECTION 3 (pages 396–397)

**1** $\dfrac{71}{99}$     **3** $\dfrac{13}{999}$     **5** $\dfrac{14}{99,000}$

**7** $\dfrac{36,140}{9999}$     **9** $5/11$     **11** $7/15$

**13** $15/21$     **15** $-8/3$     **17** $16/3$

**19** $\frac{1}{2} + \frac{4}{3} + \frac{9}{4} + \frac{16}{5} + \frac{25}{6} +$; $\lim\limits_{n\to\infty} u_n = \infty \ne 0$

**21** $\frac{1}{2} + \frac{3}{11} + \frac{3}{11} + \frac{11}{37} + \frac{9}{28} +$; $\lim\limits_{n\to\infty} u_n = \frac{1}{2} \ne 0$

**23** $\dfrac{4}{\ln 2} + \dfrac{8}{\ln 3} + \dfrac{14}{\ln 4} + \dfrac{22}{\ln 5} + \dfrac{32}{\ln 6} +$; $\lim\limits_{n\to\infty} u_n = \infty \ne 0$

**25** $2.72 + 17.5 + 77.75 + 296.15 + 1038.2 +$; $\lim\limits_{n\to\infty} u_n = \infty \ne 0$

**27** $-\frac{1}{5} + \frac{2}{7} - \frac{1}{3} + \frac{4}{11} - \frac{5}{13}$; $\lim\limits_{n \to \infty} u_n = \frac{1}{2} \neq 0$

**29** $\sum \dfrac{n-2}{n^3} = \sum \dfrac{1}{n^2} - 2\sum \dfrac{1}{n^3}$

**31** $\sum \dfrac{3n^2 - 2n + 4}{n^4} = 3\sum \dfrac{1}{n^2} - 2\sum \dfrac{1}{n^3} + 4\sum \dfrac{1}{n^4}$

**33** $\sum \dfrac{2(n-1)^2}{3n^5} = \dfrac{2}{3}\sum \dfrac{1}{n^3} - \dfrac{4}{3}\sum \dfrac{1}{n^4} + \dfrac{2}{3}\sum \dfrac{1}{n^5}$

## SECTION 4 (page 406)

| | | |
|---|---|---|
| **1** Convergent | **3** Convergent | **5** Divergent |
| **7** Convergent | **9** Divergent | **11** Divergent |
| **13** Convergent | **15** Convergent | **17** Convergent |
| **19** Convergent | **21** Convergent | **23** Divergent |
| **25** Convergent | **27** Convergent | **29** Convergent |
| **31** Convergent | **33** Convergent | **35** Convergent |
| **37** Convergent | **41** $q > 1$ | |

## SECTION 5 (pages 415–416)

**1** Divergent
**3** Divergent
**5** Absolutely convergent
**7** Conditionally convergent
**9** Absolutely convergent
**11** Absolutely convergent
**13** Divergent
**15** Absolutely convergent
**17** Absolutely convergent
**19** Conditionally convergent
**21** Absolutely convergent
**23** Divergent
**25** Conditionally convergent
**27** Conditionally convergent
**29** Absolutely convergent
**31** Absolutely convergent
**33** Divergent
**35** Divergent
**37** Absolutely convergent
**39** Absolutely convergent
**41** Conditionally convergent
**43** Divergent
**45** Absolutely convergent

**47** Converges for all $p$
**57** 0.7930
**59** 0
**61** $-0.8333$

## SECTION 6 (pages 420–421)

**1** Converges for $-1 < x < 1$
**3** Converges for $-\frac{1}{2} < x < \frac{1}{2}$
**5** Converges for $-1 < x < 1$
**7** Converges for $-1 \leq x \leq 3$
**9** Converges for $-3 \leq x < -1$
**11** Converges for $-\infty < x < \infty$
**13** Converges for $-\frac{2}{3} < x \leq \frac{2}{3}$
**15** Converges for $0 < x < 2$
**17** Converges for $-7 \leq x \leq -1$
**19** Converges for $-6 < x < 10$
**21** Converges for $-\frac{4}{3} < x < \frac{4}{3}$
**23** Converges for $-1 \leq x < 1$
**25** Converges for $-1 < x \leq 1$
**27** Converges for $-\frac{3}{2} \leq x \leq \frac{3}{2}$
**29** Converges for $-\infty < x < \infty$
**31** Converges for $-1 < x \leq 1$
**33** Converges for $-1 < x < 1$
**35** Converges for $-3/2 < x < 3/2$
**37** Converges for $-\infty < x < \infty$
**39** Converges only for $x = 0$
**41** Converges for $-2 < x < 0$
**43** (a) Converges for $-1 < x < 1$
**45** Converges for $-1 < x < 1$

## SECTION 7 (pages 425–426)

**1** $\displaystyle\sum_{n=0}^{\infty} \frac{x^n}{n!}$

**3** $\displaystyle\sum_{n=1}^{\infty} \frac{(-1)^{n-1} x^n}{n}$

**5** $\displaystyle\sum_{n=0}^{\infty} (n+1)x^n$

**7** $1 + \displaystyle\sum_{n=1}^{\infty} \frac{\left(\frac{1}{2}\right)\left(\frac{-1}{2}\right)\left(\frac{-3}{2}\right)\cdots\left(-n+\frac{1}{2}\right)x^n}{n!}$

**9** $\ln 3 + \displaystyle\sum_{n=1}^{\infty} \frac{(-1)^{n-1}(x-3)^n}{3^n \cdot n}$

**11** $\dfrac{1}{2} - \dfrac{\sqrt{3}}{2}\left(x - \dfrac{\pi}{3}\right) - \dfrac{1}{2}\dfrac{(x-\pi/3)^2}{2!} + \dfrac{\sqrt{3}}{2}\dfrac{(x-\pi/3)^3}{3!} + \cdots$

**13** $2 + \dfrac{1}{4}(x-4) + 2\displaystyle\sum_{n=2}^{\infty}\dfrac{(-1)^{n-1}\cdot 1\cdot 3\cdots(2n-3)(x-4)^n}{2^n n! 4^n}$

**15** $\cos\left(\tfrac{1}{2}\right) - x\sin\left(\tfrac{1}{2}\right) - \dfrac{x^2}{2!}\cos\left(\tfrac{1}{2}\right) + \dfrac{x^3}{3!}\sin\left(\tfrac{1}{2}\right) + \cdots$

**17** $1 - x^2 + \dfrac{x^4}{2!}$

**19** $1 - x^2 + x^4$

**21** $1 + \dfrac{x}{2} + \dfrac{x^2}{8}$

**23** $\ln 3 + \dfrac{2}{3}(x-1) - \dfrac{2}{9}(x-1)^2 + \dfrac{8}{81}(x-1)^3 - \dfrac{4}{81}(x-1)^4$

**25** $1 - (x - \pi/2)^2$

**27** $1 + x - \dfrac{x^3}{3} - \dfrac{x^4}{6}$

**29** $x + \dfrac{x^3}{2\cdot 3} + \dfrac{3x^5}{2\cdot 4\cdot 5}$

**31** $\dfrac{x^2}{2} + \dfrac{x^4}{12} + \dfrac{x^6}{45}$

**33** $2 - 2\sqrt{3}\left(x - \dfrac{\pi}{6}\right) + 7(x - \pi/6)^2 - \dfrac{23\sqrt{3}(x-\pi/6)^3}{3}$

$\quad + \dfrac{305(x-\pi/6)^4}{12}$

**35** $2 + 2\sqrt{3}\left(x - \dfrac{\pi}{3}\right) + 7\left(x - \dfrac{\pi}{3}\right)^2 + \dfrac{23\sqrt{3}(x-\pi/3)^3}{3} + \cdots$

**37** $x + \dfrac{x^3}{3} + \dfrac{2}{15}x^5 + \dfrac{17}{315}x^7 + \cdots$

**39** (c) Sum of Taylor's series is identically 0

## SECTION 8 (page 432)

**1** $1 - \dfrac{(b-\pi/2)^2}{2} + R_3;\ R_3 = \sin(\zeta)\dfrac{(b-\pi/4)^4}{24}$

**3** $3 + \dfrac{1}{6}(b-3) - \dfrac{1}{216}(b-3)^2 + R_2;\ R_2 = \dfrac{1}{16}(\zeta)^{-5/2}(b-3)^3$

**5** $1 + \dfrac{1}{e}(b-e) - \dfrac{1}{2e^2}(b-e)^2 + \dfrac{1}{3e^3}(b-e)^3 + R_3;$

$\quad R_3 = \dfrac{-(b-e)^4}{4\zeta^4}$

**7** $e^{24} + 2e^{24}(b-12) + 2e^{24}(b-12)^2 + \dfrac{4}{3}e^{24}(b-12)^3 + R_3;$
$\quad R_3 = \dfrac{2}{3}e^{2\zeta}(b-12)^4$

| | | |
|---|---|---|
| **9** 0.81873 | **11** 1.22140 | **13** 0.87758 |
| **15** 0.1823 | **17** 0.36788 | **19** 1.01943 |
| **21** 0.96905 | **23** 1.97435 | **25** 0.95635 |
| **27** −0.22314 | **29** 0.017452 | **31** 0.99619 |

**33** $|R_4| = |R_5| < .011$

**35** $\arctan x = x - \dfrac{x^3}{3} + \dfrac{x^5}{5} - \dfrac{x^7}{7} + \cdots$

## SECTION 9 (pages 439–440)

**1** $\displaystyle\sum_{n=0}^{\infty}\dfrac{(-1)^n x^{2n}}{(2n+1)!}$     **3** $\displaystyle\sum_{n=0}^{\infty}\dfrac{x^n}{(n+1)!}$

**5** $2\displaystyle\sum_{n=0}^{\infty}\dfrac{x^{2n+1}}{2n+1}$     **7** $\dfrac{1}{2}\displaystyle\sum_{n=0}^{\infty}(-1)^n(n+1)(n+2)x^n$

**9** $\displaystyle\sum_{n=1}^{\infty}\dfrac{x^n}{n}$     **11** $2\displaystyle\sum_{n=0}^{\infty}(-1)^n\dfrac{x^{2n+2}}{2n+2}$

**13** $(1-x^2)^{1/2} = 1 + \displaystyle\sum_{n=1}^{\infty}\dfrac{\tfrac{1}{2}(\tfrac{1}{2}-1)\cdots(\tfrac{1}{2}-n+1)}{n!}(-1)^n x^{2n}$  and

$\quad \arcsin x = x + \displaystyle\sum_{n=1}^{\infty}\dfrac{1\cdot 3\cdot 5\cdots(2n-1)}{2\cdot 4\cdot 6\cdots(2n)}\dfrac{x^{2n+1}}{2n+1}$ for $|x| < 1$

**17** 48     **19** 16     **21** $\ln 4$

**23** $\dfrac{3}{4}\ln 3 - \dfrac{1}{2}$     **25** $\dfrac{1}{4}\ln 2$     **27** $1 + \dfrac{(1-x)}{x}\ln|1-x|$

**29** $\displaystyle\sum_{n=0}^{\infty}\dfrac{(n+2)(n+1)x^n}{n!} = (2 + 4x + x^2)e^x$

**33** $x > 0$ and all $p$

**35** Converges for $x < -1$ or $x > 1$. The series can be differentiated or integrated term by term for $|x| > 1$.

## SECTION 10 (pages 444–445)

**1** $1 - \dfrac{3}{2}x + \dfrac{1\cdot 3\cdot 5}{2^2\cdot 2!}x^2 - \dfrac{1\cdot 3\cdot 5\cdot 7}{2^3\cdot 3!}x^3 + \dfrac{1\cdot 3\cdot 5\cdot 7\cdot 9}{2^4\cdot 4!}x^4 \cdots$

**3** $1 - \dfrac{2}{3}x^2 + \dfrac{2\cdot 5}{3^2\cdot 2!}x^4 - \dfrac{2\cdot 5\cdot 8}{3^3\cdot 3!}x^6 + \dfrac{2\cdot 5\cdot 8\cdot 11}{3^4\cdot 4!}x^8 - \cdots$

**5** $1 + 7x^3 + 21x^6 + 35x^9 + 35x^{12} + 21x^{15} + 7x^{18} + x^{21}$

**7** $\dfrac{1}{3^3} - \dfrac{3}{3^4}x^{1/2} + \dfrac{3\cdot 4}{3^5\cdot 2!}x - \dfrac{3\cdot 4\cdot 5}{3^6\cdot 3!}x^{3/2} + \dfrac{3\cdot 4\cdot 5\cdot 6}{3^7\cdot 4!}x^2 - \cdots$

**9** $1 - \dfrac{1}{3}x + \dfrac{2}{9}x^2 - \dfrac{14}{81}x^3 + \cdots$

**11** 3.31

**13** $0.31028;\ 0 < R < \dfrac{1}{6894720}$

**15** $0.74682;\ 0 < R < \dfrac{1}{685440}$

**17** $0.94608;\ 0 < R < \dfrac{1}{9(9!)}$

**19** $0.49258;\ 0 > R > -0.00000023$

**21** $0.33764;\ 0 < R < 12(10^{-7})$

**23** 0.40547

# CHAPTER 10 REVIEW PROBLEMS (pages 445–446)

**1** 4     **3** 0     **5** 0

**7** 0     **9** $\dfrac{\sqrt{2}}{2}$     **11** 0

**13** $+\infty$     **15** 3/11     **17** 6/21

**19** Convergent

**21** Convergent

**23** Divergent

**25** Divergent

**27** Divergent

**29** Divergent

**31** Convergent

**33** Absolutely convergent

**35** Divergent

**37** Absolutely convergent

**39** Absolutely convergent

**41** Absolutely convergent

**43** Absolutely convergent

**45** Divergent

**47** Divergent

**49** Conditionally convergent

**51** $(\frac{1}{3}, 1)$

**53** $(-\infty, \infty)$

**55** $(-\sqrt{2}, \sqrt{2})$

**57** $(-\infty, -1) \cup [1, \infty)$

**59** $(-e, e)$

**61** $e^2(1 + (x-2) + \frac{1}{2}(x-2)^2 + \frac{1}{6}(x-2)^3 + \frac{1}{24}(x-2)^4)$

**63** $-\frac{1}{3} - \frac{1}{9}(x+4) - \frac{1}{27}(x+4)^2$

**65** $1 - \frac{1}{2}x + \frac{3}{8}x^2 - \frac{5}{16}x^3$

**67** 0.38942     **69** $\frac{2}{3}$     **71** 5 ln 5

**73** 38     **75** 0.10013

# CHAPTER 11

## SECTION 1 (pages 454–455)

**1** $\sqrt{5}$     **3** $\frac{15}{4}\sqrt{2}$

**5** $\frac{16}{13}\sqrt{13}$     **7** 2

**9** 1/5     **11** 0

**13** 2     **15** $\sqrt{61}/61$

**17** $9/\sqrt{5}, 9/2$     **19** $11/\sqrt{5}, 11$

**21** $(16/3, 0), (-4/3, 0)$

**23** $(64, -44), (4, -4)$

**25** $x - 2y + 1 \pm 3\sqrt{5} = 0$

**27** $(1, 1)$

**29** $x - 3y + 6 = 0; 3x + y = 0$

**31** $\tan \phi = 11/10$

**33** $\tan \phi = 1/12$

**35** $\tan \phi = -5/12$

**37** $\tan \phi = -3/2$

**39** $\tan \angle CAB = 8; \tan \angle ABC = 2/3; \tan \angle BCA = 2$

**41** $\tan \angle CAB = 1; \tan \angle ABC = 2; \tan \angle BCA = 3$

**45** $3x + 4y - 5 = 0; x = -1$

**47** $(\sqrt{3} - 1)x + (\sqrt{3} + 1)y + 3\sqrt{3} + 1 = 0; (\sqrt{3} + 1)x + (\sqrt{3} - 1)y + 3\sqrt{3} - 1 = 0$

**49** $-6$; no     **51** 8/15

## SECTION 2 (pages 460–461)

**1** Focus: $(2, 0)$; directrix, $x = -2$

**3** Focus: $(-3, 0)$; directrix, $x = 3$

**5** $y^2 = -16x$

**7** $x^2 = -8y$

**9** $2y^2 = 9x$

**11** Vertex: $(1, 2)$; focus: $(1, 2\frac{1}{4})$; directrix: $y = 1\frac{3}{4}$; axis: $x = 1$

**13** Vertex: $(\frac{3}{8}, \frac{9}{16})$; focus: $(\frac{3}{8}, \frac{1}{2})$; directrix: $y = 5/8$; axis: $x = 3/8$

**15** Vertex: $(\frac{3}{2}, -11/8)$; focus: $(\frac{3}{2}, -15/8)$; directrix: $y = -7/8$; axis: $x = 3/2$

**17** Vertex: $(0, 0)$; focus: $(0, -3)$; directrix: $y = 3$; axis: $x = 0$

**19** Vertex: $(0, 0)$; focus: $(-\frac{3}{4}, 0)$; directrix: $x = 3/4$; axis: $y = 0$

**21** Vertex: $(-4, -3)$; focus: $(-9/2, -3)$; directrix: $x = -7/2$; axis: $x = -4$

**23** Vertex: $(-\frac{1}{2}, \frac{1}{4})$; focus: $(-\frac{1}{2}, 0)$; directrix: $y = \frac{1}{2}$; axis: $x = -\frac{1}{2}$

**25** $y^2 = 12(x - 3)$

**27** $y^2 = -8(x - 4)$

**29** $(x + 1)^2 = -8y$

**33** $(2 \pm 6\sqrt{3})x - y - 5 = 0$

**35** $3x - 2y + 3 = 0$

**37** $3x - 2y - 4 = 0$

**39** $8x - y + 9 = 0$

**67** $(y - y_1) = \dfrac{a^2 y_1}{b^2 x_1}(x - x_1)$

**69** Vertices of square $= (\pm\frac{12}{5}, \pm\frac{12}{5})$; perimeter $= \frac{96}{5}$; area $= (\frac{24}{5})^2$

**71** $\sqrt{2}a; \sqrt{2}b$

## SECTION 3 (pages 468–469)

**1** $x^2 + y^2 + 6x - 8y = 0$

**3** $x^2 + y^2 - 4x - 6y + 9 = 0$

**5** $x^2 + y^2 - 12x - 8y + 44 = 0$

**7** $x^2 + y^2 - 6x - 8y + 9 = 0$

**9** $x^2 + y^2 - 4x - 2y - 5 = 0$

**11** $x^2 + y^2 - 4x - 6y - 3 = 0$

**13** $x^2 + y^2 + 12x - 12y + 36 = 0$

**15** Circle; center $(-3, 4)$, radius 5

**17** Point $(2, -1)$

**19** Circle; center $(-\frac{3}{2}, \frac{5}{2})$, radius 3

**21** Circle; center $(-\frac{3}{4}, -\frac{5}{4})$, radius $\frac{3}{4}\sqrt{2}$

**23** $2x + y - 5 = 0$

**25** $9x - 8y - 26 = 0$

**27** $x + 2y + 4 = 0$

**29** $5x + 8y - 38 = 0$

**31** $3x + 5y - 13 = 0, 5x - 3y + 35 = 0$

**33** No tangent lines as $P(3, 0)$ is inside circle

**37** $13; 5; e = 12/13$; foci: $(\pm 12, 0)$; vertices: $(\pm 13, 0)$

**39** $5; 4; e = 3/5$; foci: $(0, \pm 3)$; vertices: $(0, \pm 5)$

**41** $2; \sqrt{3}; e = 1/2$; foci: $(\pm 1, 0)$; vertices: $(\pm 2, 0)$

**43** $\sqrt{5}; \sqrt{3}; e = \sqrt{2/5}$; foci: $(0, \pm\sqrt{2})$; vertices: $(0, \pm\sqrt{5})$

**45** $\dfrac{\sqrt{17}}{2}; \dfrac{\sqrt{17}}{5}; e = 1/\sqrt{5}$; foci: $\left(0, \pm\dfrac{1}{2}\dfrac{\sqrt{17}}{5}\right)$; vertices: $\left(0, \pm\dfrac{\sqrt{17}}{2}\right)$

**47** $25x^2 + 16y^2 = 400$

**49** $20x^2 + 36y^2 = 720$

**51** $16x^2 + 9y^2 = 225$

**53** $45x^2 + 16y^2 = 1189$

**55** $7x^2 + 16y^2 = 508$

**57** $16x^2 + 25y^2 = 1600$

**59** $8x^2 + 9y^2 = 648$

**61** $3x^2 + 4y^2 - 20x + 28 = 0$; major axis: $y = 0$; minor axis: $x = 10/3$

**63** $3\sqrt{2}y - x + 9 = 0$

**65** $9\sqrt{5}x - 12y - 10\sqrt{5} = 0$

## SECTION 4 (pages 475–476)

| | $a$ | $b$ | $c$ | $e$ | foci | vertices |
|---|---|---|---|---|---|---|
| **1** | 3 | 2 | $\sqrt{13}$ | $\sqrt{13}/3$ | $(\pm\sqrt{13}, 0)$ | $(\pm 3, 0)$ |
| **3** | 3 | 2 | $\sqrt{13}$ | $\sqrt{13}/3$ | $(0, \pm\sqrt{13})$ | $(0, \pm 3)$ |
| **5** | 3 | 4 | 5 | $5/3$ | $(\pm 5, 0)$ | $(\pm 3, 0)$ |
| **7** | 5 | 12 | 13 | $13/5$ | $(0, \pm 13)$ | $(0, \pm 5)$ |
| **9** | $\sqrt{3}$ | $\sqrt{2}$ | $\sqrt{5}$ | $\sqrt{5/3}$ | $(\pm\sqrt{5}, 0)$ | $(\pm\sqrt{3}, 0)$ |
| **11** | $\sqrt{2}$ | $\sqrt{5}$ | $\sqrt{7}$ | $\sqrt{7/2}$ | $(\pm\sqrt{7}, 0)$ | $(\pm\sqrt{2}, 0)$ |
| **13** | $1/\sqrt{3}$ | $1/\sqrt{2}$ | $\sqrt{5/6}$ | $\sqrt{5/2}$ | $(\pm\sqrt{5/6}, 0)$ | $(\pm 1/\sqrt{3}, 0)$ |
| **15** | $3\sqrt{3}$ | 3 | 6 | $2/\sqrt{3}$ | $(\pm 6, 0)$ | $(\pm 3\sqrt{3}, 0)$ |
| **17** | $a$ | $a$ | $\sqrt{2}a$ | $\sqrt{2}$ | $(0, \pm\sqrt{2}a)$ | $(0, \pm a)$ |
| **19** | $\sqrt{\frac{1}{11}}$ | $\sqrt{\frac{1}{7}}$ | $3\sqrt{\frac{2}{77}}$ | $3\sqrt{\frac{2}{7}}$ | $(\pm 3\sqrt{\frac{2}{77}}, 0)$ | $(\pm\sqrt{\frac{1}{11}}, 0)$ |

**21** $24x^2 - 25y^2 = 600$

**23** $9x^2 - 7y^2 + 343 = 0$

**25** $4x^2 - y^2 = 32$

**27** $y^2 - x^2 = 5$

**29** $4y^2 - 5x^2 = 19$

**31** $\dfrac{x^2}{4} - \dfrac{y^2}{16} = 1$

**33** $\dfrac{x^2}{9} - \dfrac{y^2}{64} = 1$; or $\dfrac{y^2}{64} - \dfrac{x^2}{9} = 1$

**35** $\dfrac{x^2}{5} - y^2 = 1$; or $y^2 - \dfrac{x^2}{5} = 1$

**37** $36x^2 - 748y^2 + 1683 = 0$

**39** $60(y - 6)^2 - 4(x - 2)^2 = 135$

**41** $\left(\dfrac{45 + 9\sqrt{10}}{4\sqrt{10} - 5}, \dfrac{90}{5 - 4\sqrt{10}}\right)$

**43** $16x^2 - 9y^2 = 144$

**45** $\dfrac{x^2}{20 - 8\sqrt{5}} - \dfrac{y^2}{8\sqrt{5}} = 1$

**47** $60x^2 - 4y^2 = 15$

## SECTION 5 (pages 479–480)

**1** $(7, -1)$, $(1, 3)$, $(-3, -\frac{5}{2})$, $(-2, -6)$

**3** $3x' - 7y' = 0$, $2x' - 5y' = 0$

**5** Ellipse; center $(-1, 2)$; vertices $(-1, 2 \pm 5)$; foci $(-1, 2 \pm 3)$; $e = \frac{3}{5}$; directrices $y = 2 \pm \frac{25}{3}$

**7** Hyperbola: center $(-1, 2)$; vertices $(-1, 2 \pm 3)$; foci $(-1, 2 \pm \sqrt{13})$; $e = \sqrt{13}/3$; directrices $y = 2 \pm 9/\sqrt{13}$; asymptotes $(y - 2) = \pm\frac{3}{2}(x + 1)$

**9** Ellipse; center $(-1, 2)$; vertices $(-1 \pm 3, 2)$; foci $(-1 \pm \sqrt{5}, 2)$; $e = \sqrt{5}/3$; directrices $x = -1 \pm (9/\sqrt{5})$

**11** Ellipse; center $(1, -1)$, vertices $(1, -1 \pm \frac{3}{2})$; foci $(1, -1 \pm(\sqrt{5}/2))$; $e = \sqrt{5}/3$, directrices $y = -1 \pm (9/(2\sqrt{5}))$

**13** Ellipse; center $(2, -1)$; vertices $(2 \pm 2, -1)$; $e = \frac{1}{2}$; foci $(2 \pm 1, -1)$; directrices $x = 2 \pm 4$

**15** Hyperbola; center $(-1, -2)$; vertices $(-1 \pm 2, -2)$; $e = \sqrt{10}/2$; foci $(-1 \pm(\sqrt{10}/2), -2)$; directrices, $x = -1 \pm (2/5)\sqrt{10}$; asymptotes: $(y + 2) = \pm\dfrac{\sqrt{6}}{2}(x + 1)$

**17** Two intersecting lines: $y - 2 = \pm(2/\sqrt{3})(x + 1)$

**19** $16x^2 + 7y^2 - 64x - 14y - 41 = 0$

**21** $x^2 - 5y^2 - 2x + 10y - 20 = 0$

**23** $y^2 - 6x + 2y + 22 = 0$

**25** Ellipse: $4x^2 + 3y^2 - 16x - 12y + 16 = 0$

**27** Circle: $x^2 + y^2 + 12x - 10y + 29 = 0$

## SECTION 6 (page 486)

**1** $5(x')^2 = 9$; two lines $x + 2y = \pm 3$

**3** $(x')^2 - (y')^2 = 10$; hyperbola; vertices $(3, 1)$, $(-3, -1)$

**5** $x'^2 = -4\sqrt{2}y'$; parabola; focus $(1, -1)$; directrix $x - y + 2 = 0$

**7** $17x'^2 + 4y'^2 = 884$; ellipse; vertices $(\mp 3\sqrt{17}, \pm 2\sqrt{17})$

**9** $y'^2(1 + \sqrt{5}) - x'^2(-1 + \sqrt{5}) = 20$; hyperbola; vertices $(\pm(2\sqrt{5})^{1/2}, \mp(3\sqrt{5} - 5)^{1/2})$

**11** $(\sqrt{13} - 3)x'^2 - (\sqrt{13} + 3)y'^2 = 10$; hyperbola; vertices $\left(\pm\dfrac{(\sqrt{13} + 3)}{2}\left(\dfrac{5\sqrt{13}}{13}\right)^{1/2}, \left(\dfrac{5\sqrt{13}}{13}\right)^{1/2}\right)$

**13** $(y' - \frac{7}{5})^2 = 3(x' + 2/5)$; parabola

**15** $9(x')^2 - (y')^2 = 1$; hyperbola

**17** $(y')^2 = 2(x')$; parabola

**19** Hyperbola: $6(x'')^2 - 9(y'')^2 = 1$; vertices: $\left(\dfrac{-35 \pm \sqrt{30}}{30}, \dfrac{20 \pm \sqrt{30}}{15}\right)$

**21** Straight line: $(5y' - 7)^2 = 0$; $4x - 3y + 7 = 0$

**23** Two lines: $y' = \frac{7}{3}$, $y' = -1$ or $2x - \sqrt{5}y + 7 = 0$, $2x - \sqrt{5}y - 3 = 0$

**25** Straight line: $x' = \sqrt{2}$ or $x + y = 2$

**27** Hyperbola: $8(y'')^2 - 9(x'')^2 = 22$; center: $(-\frac{2}{9}, -\frac{1}{3})$; vertices: $\left(\dfrac{-20 \pm \sqrt{220}}{90}, \dfrac{-10 \pm \sqrt{220}}{30}\right)$

**29** Hyperbola: $2(x')^2 - (y')^2 = 8$

**31** Hyperbola: $9(x')^2 - 4(y')^2 = 4$

## CHAPTER 11 REVIEW PROBLEMS (page 487)

**1** 1    **3** 0    **5** $\dfrac{3\sqrt{2}}{2}$

**7** $\frac{1}{8}$    **9** 6/17    **11** $\infty$

**13** Parabola; focus: $(0, -2)$; vertex: $(0, 0)$; directrix: $y = 2$; axis: $x = 0$

**15** Parabola; focus: $(5, 0)$; vertex: $(2, 0)$; directrix: $x = -1$; axis: $y = 0$

**17** Circle; center: $(-3/2, 2)$; radius: $5/2$

**19** Circle; center: $(0, 0)$; radius: $2\sqrt{6}$

**21** Ellipse; major axis length: 1; minor: 2/3; foci: $\left(\pm\dfrac{\sqrt{5}}{6}, 0\right)$; vertices: $\left(\pm\dfrac{1}{2}, 0\right)$; $e = \sqrt{5}/3$.

**23** Ellipse; major axis length: $\dfrac{4\sqrt{11}}{\sqrt{3}}$; minor: $\dfrac{4\sqrt{11}}{\sqrt{7}}$; foci: $(\pm 4\sqrt{11}/\sqrt{21}, 0)$; vertices: $(\pm 2\sqrt{11}/\sqrt{3}, 0)$; $e = 2/\sqrt{7}$

**25** Hyperbola; length of transverse axis: 2; conjugate: 6; foci: $(-1 \pm \sqrt{10}, 0)$; vertices: $(-2, 0)$, $(0, 0)$; $e = \sqrt{10}$; asymptotes: $y = 3x + 3$; $y = -3x - 3$

**27** Hyperbola; length of transverse axis: 8; conjugate: 6; foci: $(-4, 3)$, $(6, 3)$; vertices: $(-3, 3)$, $(5, 3)$; $e = \frac{5}{4}$; asymptotes: $y = \pm\frac{3}{4}(x - 1) + 3$

**29** Circle; center: $(2, -3/2)$; radius: $\sqrt{61}/2$.

**31** Hyperbola; length of transverse axis: 2; conjugate: 2; foci: $(0, \pm\sqrt{2})$; vertices: $(0, \pm 1)$; $e = \sqrt{2}$; asymptotes: $y = \pm x$

**33** Ellipse; major axis length: 10; minor: 8; foci: $(-1, 3)$, $(5, 3)$; vertices: $(-3, 3)$, $(7, 3)$; $e = \frac{3}{5}$

**35** Parabola; focus: $(-\frac{1}{2}, \frac{19}{8})$; vertex: $(-\frac{1}{2}, \frac{5}{2})$; directrix: $y = \frac{21}{8}$; axis: $x = -\frac{1}{2}$

**37** Ellipse; major axis length: $2\sqrt{14}$; minor: $\sqrt{14}$; foci: $(3 \pm \sqrt{\frac{21}{2}}, 0)$; vertices: $(3 \pm \sqrt{14}, 0)$; $e = \sqrt{3}/2$

**39** Hyperbola    **41** Ellipse

**43** Hyperbola    **45** Ellipse

**47** Parabola    **49** Ellipse

# CHAPTER 12

## SECTION 1 (pages 494–495)

**1** $x = 3 - (7t/\sqrt{218})$, $y = -7 + (13t/\sqrt{218})$

**3** $x = 4 - (5t/\sqrt{29})$, $y = -2 + (2t/\sqrt{29})$

**5** $x = 3 + (3t/\sqrt{13})$, $y = 1 + (2t/\sqrt{13})$

**7** $5x + 2y = 0$       **9** $xy = 1$

**11** $4(x+1)^2 + (y-2)^2 = 4$

**13** $(x/2)^{2/3} + (y/2)^{2/3} = 1$

**15** $x^2 - 2xy + y^2 - 7x + 6y + 11 = 0$

**17** $x^2 + y^2 = 25$       **19** $xy = 1$

**21** $x^2 - y^2 = 4$       **23** $x = y + 8$

**25** $y = 2x^2 + 6$       **27** $y = \ln x \quad (x > 0)$

**31**

| $t$ | 2 | 4 | 6 | 8 | 10 | $T = 60$ sec |
|---|---|---|---|---|---|---|
| $x$ | 800 | 1600 | 2400 | 3200 | 4000 | $R = 24{,}000$ m |
| $y$ | 580 | 1120 | 1620 | 2080 | 2500 | 0 |

**33** $x = a\theta - b\sin\theta$; $y = a - b\cos\theta$

## SECTION 2 (pages 502–503)

**1** $\dfrac{dy}{dx} = -\dfrac{5}{4}$; $\dfrac{d^2y}{dx^2} = 0$

**3** $\dfrac{dy}{dx} = -\dfrac{3}{2}t - 4 - t^{-1}$; $\dfrac{d^2y}{dx^2} = \dfrac{3}{4t} - \dfrac{1}{2t^3}$

**5** $\dfrac{dy}{dx} = -3te^{2t}$; $\dfrac{d^2y}{dx^2} = \dfrac{3}{2}e^{4t}(1+2t)$

**7** $\dfrac{dy}{dx} = -\cos t$; $\dfrac{d^2y}{dx^2} = -\dfrac{1}{4}$

**9** $\dfrac{dy}{dx} = -\tan\theta$; $\dfrac{d^2y}{dx^2} = \dfrac{1}{3a\sin\theta\cos^4\theta}$

**11** $\dfrac{dy}{dx} = 2t$; $\dfrac{d^2y}{dx^2} = \dfrac{2}{3t^2+1}$; $\dfrac{d^3y}{dx^3} = \dfrac{-12t}{(3t^2+1)^3}$; $\dfrac{d^4y}{dx^4} = 12\dfrac{(15t^2-1)}{(3t^2+1)^5}$

**13** Tangent: $7x + 2y - 11 = 0$; normal: $2x - 7y - 94 = 0$

**15** Tangent: $x + 2y - 5 = 0$; normal: $4x - 2y - 5 = 0$

**17** Tangent: $x + ey - 2 = 0$; normal: $e^2x - ey + 2 = 0$

**19** Tangent: $x + y - \sqrt{2} = 0$; normal: $x - y = 0$

**21** Tangent: $y = \frac{3}{2}x - \frac{1}{2}$; normal: $y = -\frac{2}{3}x + \frac{5}{3}$

**23** Tangent: $y = 8\sqrt{\pi}x - 8\pi + 4\sqrt{\pi} + 7$; normal: $y = -\dfrac{1}{8\sqrt{\pi}}x + 4\sqrt{\pi} + \dfrac{57}{8}$

**25** Tangent: $y = 1$; normal: $x = 3$

**27** $x$ increasing on $\left(-\dfrac{\pi}{2}, \dfrac{\pi}{2}\right)$; $y$ increasing on $\left(-\dfrac{\pi}{2}, 0\right)$; $y$ decreasing on $\left(0, \dfrac{\pi}{2}\right)$; $(x^2+4)y = 16$

**29** $x$ increasing on $\pi \leqslant \theta \leqslant 2\pi$; $x$ decreasing on $0 \leqslant \theta \leqslant \pi$; $y$ increasing on $0 \leqslant \theta \leqslant \dfrac{\pi}{4}$, $\dfrac{3\pi}{4} \leqslant \theta \leqslant \dfrac{5\pi}{4}$, $\dfrac{7\pi}{4} \leqslant \theta \leqslant 2\pi$; $y$ decreasing on $\dfrac{\pi}{4} \leqslant \theta \leqslant \dfrac{3\pi}{4}$, $\dfrac{5\pi}{4} \leqslant \theta \leqslant \dfrac{7\pi}{4}$; $y^2 = 4x^2 - 4x^4$

**31** $x$ increasing for $-\infty < t \leqslant 0$, $\frac{4}{3} \leqslant t < \infty$; $x$ decreasing for $0 \leqslant t \leqslant \frac{4}{3}$; $y$ increasing for $-\infty < t \leqslant \frac{2}{3}$, $2 \leqslant t < \infty$; $y$ decreasing for $\frac{2}{3} \leqslant t \leqslant 2$

**33** 2.2516       **35** 1.5426       **37** 1.0568

**39** $s = \displaystyle\int_1^5 \sqrt{9t^4 + 16t^2 - 4t + 5}\, dt$

**41** $s = \displaystyle\int_0^\pi \sqrt{1 + t^2}\, dt$

**43** $s = \displaystyle\int_2^\pi \sqrt{\dfrac{1}{t^2} + \sin^2 t}\, dt$

**45** 10   **47** $4\sqrt{13}$   **49** $\frac{2}{27}(13)^{3/2}$   **55** $24\pi^2$   **57** $\frac{2}{3}\pi[\sqrt{8} - (1 - e^{-16})^{3/2}]$

## SECTION 3 (page 507)

**1** $\kappa(x) = \dfrac{4}{(1+16x^2)^{3/2}}$, $R(x) = \dfrac{(1+16x^2)^{3/2}}{4}$

**3** $\kappa(x) = \dfrac{6}{(4x+9)^{3/2}}$, $R(x) = \left|\dfrac{1}{6}(4x+9)^{3/2}\right|$

**5** $\kappa(x) = \dfrac{-\sin x}{(1+\cos^2 x)^{3/2}}$, $R(x) = |\csc x|(1+\cos^2 x)^{3/2}$

**7** $\kappa(x) = \dfrac{ab^2 e^{bx}}{\{1 + a^2b^2e^{2bx}\}^{3/2}}$, $R(x) = \dfrac{(1+a^2b^2e^{2bx})^{3/2}}{|ab^2e^{bx}|}$

**9** $\kappa(x) = x(2-x^2)^{-3/2}$, $R(x) = \dfrac{1}{|\kappa(x)|}$

**11** $\kappa(x) = -a^4b(a^4 - c^2x^2)^{-3/2}$ with $c^2 = a^2 - b^2$

**13** $\kappa(y) = -2\csc^2 y \cot y(1 + \csc^4 y)^{-3/2}$, $R(y) = |\frac{1}{2}\sin^2 y \tan y|(1 + \csc^4 y)^{3/2}$

**15** $\kappa(y) = -\dfrac{1}{a}$, $R(y) = a \quad (a > 0)$

**17** $\kappa(t) = |t|^{-1}(1+t^2)^{-3/2}$, $R(t) = \dfrac{1}{|\kappa(t)|}$

**19** $\kappa(t) = \dfrac{1}{\sqrt{2e^t}}$, $R(t) = |\sqrt{2e^t}|$

**21** $\kappa(t) = -1/3|a \sin \theta \cos \theta|$, $R(t) = 3|a \sin \theta \cos \theta|$

**23** $-0.0533$  **25** $-\dfrac{1}{\sqrt{2\pi}}$  **27** $0.3041$

**29** $0.0135$  **31** $0.61898$

**33** $\kappa(x) = \dfrac{1}{(1+x^2)^{3/2}}$, max. at $x = 0$

**35** $\kappa(x) = 2x^3(x^4+1)^{-3/2}$, max. at $x = 1$; min. at $x = -1$

**39** $\kappa = -\dfrac{4}{7\sqrt{7}}$, $x_c = \dfrac{\pi}{3} - \dfrac{7\sqrt{3}}{4}$, $y_c = -3$

**41** $\kappa = \dfrac{1}{2\sqrt{2}}$, $x_c = -2$, $y_c = 3$

**43** $x_c = \dfrac{1}{a}(\sec^3 t)(a^2 + b^2)$, $y_c = -\dfrac{1}{b}(\tan^3 t) \cdot (a^2 + b^2)$

## SECTION 4 (pages 513–514)

**1** $(2\sqrt{3}, 2)$, $\left(-\dfrac{3\sqrt{2}}{2}, \dfrac{3\sqrt{2}}{2}\right)$, $(-2, 0)$, $(1, 0)$, $(2, 0)$

**3** $(-1, 0)$, $(\sqrt{3}, -1)$, $(2, -2\sqrt{3})$, $\left(\dfrac{3\sqrt{2}}{2}, -\dfrac{3\sqrt{2}}{2}\right)$, $(0, 0)$

**5** $\left(3\sqrt{2}, \dfrac{\pi}{4}\right)$, $\left(4, \dfrac{\pi}{2}\right)$, $\left(2, \dfrac{2\pi}{3}\right)$, $\left(1, \dfrac{-\pi}{2}\right)$, $(2, 0)$

**7** $\left(2\sqrt{2}, \dfrac{3\pi}{4}\right)$, $\left(3\sqrt{2}, \dfrac{-\pi}{4}\right)$, $\left(2, \dfrac{5\pi}{6}\right)$, $\left(4, \dfrac{\pi}{6}\right)$, $\left(4, \dfrac{\pi}{3}\right)$

**9** $r = 5$ is circle of radius 5, center at pole; $\theta = \pi/3$ is straight line through pole, slope $= \sqrt{3}$; curves $r = $ constant and $\theta = $ constant intersect at right angles.

**11** $\sqrt{13}$

**13** $\bar{r} = \dfrac{3}{2}\sqrt{5 + \sqrt{6} + \sqrt{2}}$, $\tan \bar{\theta} = \dfrac{\sqrt{2} + 1}{\sqrt{6} + 1}$

**15** Symmetric with respect to the $x$ axis

**17** Symmetric with respect to the $y$ axis

**19** None

**21** Symmetric with respect to the $x$ axis

**23** Symmetric with respect to the $y$ axis

**25** Symmetric with respect to the $x$ axis

**27** Symmetric with respect to the $y$ axis

**29** Symmetric with respect to the $x$ axis, $y$ axis, and pole

**31** Symmetric with respect to the $x$ axis, $y$ axis, and pole

**33** Symmetric with respect to the $x$ axis

**35** Symmetric with respect to the $x$ axis

**37** Symmetric with respect to the $x$ axis, $y$ axis, pole

**39** Symmetric with respect to the $x$ axis, $y$ axis, and pole

**41** Symmetric with respect to the $y$ axis

**43** Symmetric with respect to the $x$ axis

**45** Symmetric with respect to the $y$ axis

**49** None  **51** None

**53** $(0, 0)$, $\left(\dfrac{1}{2}\sqrt{3}, \dfrac{\pi}{3}\right)$, $\left(\dfrac{1}{2}\sqrt{3}, \dfrac{2}{3}\pi\right)$

**55** $(0, 0)$, $(1, 0)$

**57** $\left(\dfrac{12}{5}, \pm\arccos(2/5)\right)$

**59** $\left(1 + \dfrac{1}{2}\sqrt{2}, \dfrac{\pi}{4}\right)$, $\left(1 - \dfrac{1}{2}\sqrt{2}, \dfrac{5\pi}{4}\right)$

## SECTION 5 (pages 520–521)

**1** $r \cos \theta = -2$

**3** $\theta = \pi/4$

**5** $r(2 \cos \theta + \sqrt{2} \sin \theta) = 4$

**7** $r^2 \sin 2\theta = 8$

**9** $r + 2 \cos \theta - 4 \sin \theta = 0$

**11** $r + 2 \cos \theta + 6 \sin \theta = 0$

**13** $x^2 + y^2 = 49$

**15** $x^2 + y^2 - 3x = 0$

**17** $x = 5$

**19** $x + y = 2\sqrt{2}$

**21** $x^2 - y^2 = 4$

**23** $(x^2 + y^2)^2 = 2xy$

**25** $x^2 = 2y$

**27** $3x^2 - y^2 + 8x + 4 = 0$

**29** $(x^2 + y^2)^2 = a(3x^2 y - y^3)$

**31** $(x^2 + y^2 + 2y)^2 = x^2 + y^2$

**33** $r \cos(\theta - \pi/3) = 2$

**35** $r \sin \theta = -1$

**37** $r \cos \theta = 3$

**39** $r^2 - 6r \cos(\theta - \pi/6) + 5 = 0$

**41** $r = 8 \cos \theta$

**43** $r^2 - 8r \cos(\theta - \pi/3) - 49 + 14(\sqrt{2} + \sqrt{6}) = 0$

**45** $r = \dfrac{10}{3}\sqrt{3} \cos\left(\theta - \dfrac{\pi}{3}\right)$

**47** $r = \dfrac{10}{1 - 2 \cos \theta}$

**49** $r = \dfrac{1}{2 + \sin \theta}$

**51** $r = \dfrac{6}{1 - 2\cos\left(\theta + \frac{4}{3}\pi\right)}$

**53** $d = |p - r_1\cos(\theta_1 - \alpha)|$

**55** $e = \frac{1}{2}$; $r\cos\theta = -9$

**57** $e = 1$; $r\sin\theta = -2$

**59** $e = 2$; $r\cos(\theta - \pi/3) = -\frac{5}{2}$

**61** $e = \frac{1}{6}$; $r\cos(\theta + \pi/2) = -4$

**63** $e = \frac{1}{4}$; $r\cos(\theta - \pi) = -3$

**65** $e = \frac{1}{6}$; $r\cos(\theta - \pi/2) = -3$

**67** $r = \dfrac{ep}{1 + e\sin\theta}$ or $r = \dfrac{-ep}{1 - e\sin\theta}$

**69** Equation in rectangular coordinates: If $e = 1$, $x^2 = -2py$.

**73** $(x^2 + y^2)^3 - (x^2 + y^2)^2 + 8x^2y^2 = 0$; degree 6

## SECTION 7 (pages 528–529)

**1** $\dfrac{2\pi^3}{3}$   **3** $\frac{1}{4}(e^{3\pi} - 1)$   **5** $\dfrac{\pi^5}{10}$

**7** $\dfrac{1}{\pi}$   **9** $\frac{17}{8}\pi$   **11** $\pi$

**13** 2   **15** $6\pi$

**17** $\dfrac{\pi}{4}$ if $n$ is odd, $\dfrac{\pi}{2}$ if $n$ is even.

**19** $\frac{3}{2}\pi$   **21** 4   **23** $\dfrac{10\pi}{3} + \dfrac{7\sqrt{3}}{2}$

**25** $\dfrac{9\sqrt{3}}{2} - 2\pi$   **27** $\dfrac{\pi}{3} + 1 - \sqrt{3}$   **29** 1

**31** $\pi + 4\sqrt{3}$   **33** $\frac{1}{2}a^2$

## SECTION 6 (pages 525–526)

**1** $\dfrac{ds}{d\theta} = |2a|$, $\cot\psi = \cot\theta$

**3** $\dfrac{ds}{d\theta} = \left|6\sin\dfrac{\theta}{2}\right|$, $\cot\psi = \cot\dfrac{\theta}{2}$

**5** $\dfrac{ds}{d\theta} = 5\sqrt{1 + \theta^2}$, $\cot\psi = \dfrac{1}{\theta}$

**7** $\dfrac{ds}{d\theta} = \sqrt{13 + 12\cos\theta}$, $\cot\psi = -2\sin\theta/(3 + 2\cos\theta)$

**9** $\dfrac{ds}{d\theta} = e^{5\theta}\sqrt{26}$, $\cot\psi = 5$

**11** $\dfrac{ds}{d\theta} = \dfrac{2}{(2 - \sin\theta)^2}\sqrt{5 - 4\sin\theta}$, $\cot\psi = \dfrac{\cos\theta}{2 - \sin\theta}$

**13** $\dfrac{ds}{d\theta} = \dfrac{2\sqrt{5 + 4\cos\theta}}{(1 + 2\cos\theta)^2}$, $\cot\psi = \dfrac{2\sin\theta}{1 + 2\cos\theta}$

**15** $\dfrac{1}{\sqrt{3}}$   **17** 2   **19** $\dfrac{1}{\ln 2}$

**21** $s = \frac{2}{3}\sqrt{10}(e^9 - 1)$

**23** $6\sqrt{2}$

**27** $\dfrac{3\sqrt{2}}{4}(1 - \cos\theta)^{-1/2}$

**29** $x = -\dfrac{\sqrt{2}}{2}$

**31** Normal to $r = f(\theta)$ at $(r_1, \theta_1)$; $r\sin(\theta - \alpha) = g$ where $\alpha = \theta_1 + \psi_1 - \dfrac{\pi}{2}$ and $g = r_1\cos\psi_1$

## CHAPTER 12 REVIEW PROBLEMS (pages 529–530)

**1** $y^2 = 1 - x^2$   **3** $y = \dfrac{1}{x^2}$

**5** $y = x^{2/3}$   **7** $y = \dfrac{1}{x - 1}$

**9** $y = 2 - x + (2 - x)^3$

**11** $y = 2 - x$   **13** $y = \dfrac{2x^2}{2x^2 - 2x + 1}$

**15** $\dfrac{dy}{dx} = 4t^{-1}$; $\dfrac{d^2y}{dx^2} = -\dfrac{4}{3}t^{-4}$

**17** $\dfrac{dy}{dx} = 2t^3 - t^4$; $\dfrac{d^2y}{dx^2} = t^5(2t - 3)$

**19** $\dfrac{dy}{dx} = \dfrac{e^t - e^{-t}}{e^t + e^{-t}}$; $\dfrac{d^2y}{dx^2} = \dfrac{8}{(e^t + e^{-t})^3}$

**21** $\dfrac{dy}{dx} = \dfrac{\sec^2\theta}{2\sec 2\theta\tan 2\theta}$;

$\dfrac{d^2y}{dx^2} = \dfrac{\tan 2\theta\sec^2\theta\tan\theta - \sec^2\theta\tan^2 2\theta - \sec^2\theta\sec^2 2\theta}{2\sec^2 2\theta\tan^3 2\theta}$

**23** Tangent: $y = \dfrac{4}{e}x - 5$; normal: $y = -\dfrac{e}{4}x + \left(\dfrac{e^2}{4} - 1\right)$

**25** Tangent: $y = 1$; normal: $x = 0$

**27** Tangent: $y = (2\cos 2)x + \sin 2 - 2\cos 2\ln 2 + 3$;  normal:

$y = -\dfrac{\sec 2}{2}x + \sin 2 + \dfrac{\sec 2}{2}\ln 2 + 3$

**29** 1.15   **31** 16

**33** $\frac{4}{3}[6 + \sqrt{3}\ln(2 + \sqrt{3})]$

**35** $\dfrac{\pi}{2}[\pi(1 + \pi^2)^{1/2} + \ln\{\pi + (1 + \pi^2)^{1/2}\}]$

**37** Radius: $\frac{125}{16}$

**39** Radius: $\frac{1}{8}(17)^{3/2}$

**41** Radius: $6\sqrt{3/2}$

**43** Radius: $\frac{1}{6}(9 + 4t^2)^{3/2}$

**45** Radius: $3|2 \sin \theta \cos \theta|$

**47** Radius: $2(\sin^4 \theta + \cos^4 \theta)^{3/2}$

**61** $e = 2$; directrix perpendicular to the polar axis, passing through $(4, \pi)$

**63** $e = 6$; directrix parallel to the polar axis, passing through $(1, \frac{3}{2}\pi)$

**65** $e = 3/2$; directrix: $r \cos(\theta + \pi/2) = -2$

**67** 0.54    **69** 16    **71** $\frac{3}{2}$    **73** $1380\pi$

# CHAPTER 13

## SECTION 1 (pages 536–537)

**1** $|AB| = \sqrt{30}$; $|BC| = \sqrt{19}$; $|AC| = \sqrt{21}$; scalene

**3** $|AB| = \sqrt{17}$; $|AC| = \sqrt{18}$; $|BC| = \sqrt{35}$; right triangle

**5** $|AB| = \sqrt{21}$; $|AC| = \sqrt{51}$; $|BC| = \sqrt{126}$

**7** $(-4, \frac{3}{2}, \frac{9}{2})$

**9** $(\frac{9}{4}, -\frac{3}{2}, \frac{7}{4}), (\frac{7}{2}, 3, \frac{7}{2}), (\frac{19}{4}, \frac{15}{2}, \frac{21}{4})$

**11** $\frac{1}{2}\sqrt{61}, \frac{1}{2}\sqrt{61}, \frac{1}{2}\sqrt{10}$

**13** $\frac{1}{2}\sqrt{74}, \frac{1}{2}\sqrt{74}, \frac{1}{2}\sqrt{26}$

**15** $(1, 4, -2)$    **17** $(0, -1, 0)$    **19** $(9, 4, -4)$

**21** $P_2(1, -2, 3)$; $Q(1, 0, 3)$

**23** Not on a line

**25** Not on a line

**27** Block extending indefinitely in $x$ and $z$ directions, bounded by planes $y = -2$ and $y = 5$

**29** Interior of sphere of radius 1

**33** $(-0.241, 1.414, 0.585)$

**35** $2x + 4y - 8z + 3 = 0$; plane

**37** $\frac{x^2}{4} + \frac{y^2}{3} + \frac{z^2}{3} = 1$; an ellipsoid

## SECTION 2 (page 541)

**1** Dir. nos.; 1, 4, $-6$; dir. cosines: $\dfrac{1}{\sqrt{53}}, \dfrac{4}{\sqrt{53}}, \dfrac{-6}{\sqrt{53}}$

**3** Dir. nos.: 2, $-6$, $-3$; dir. cosines: $\dfrac{2}{7}, \dfrac{-6}{7}, \dfrac{-3}{7}$

**5** $(3, 5, 2)$    **7** $(0, 4, 2)$

**9** On line    **11** Not on line

**13** Parallel    **15** Not parallel

**17** Parallel

**19** Perpendicular

**21** Not perpendicular

**29** $\dfrac{13}{\sqrt{415}}$

**31** $\dfrac{\sqrt{2}}{2}$

**33** $(0, 0, 0), (2, 0, 0), (1, \sqrt{3}, 0), \left(1, \dfrac{\sqrt{3}}{3}, \dfrac{2\sqrt{6}}{3}\right)$

**35** $(\frac{3}{2}, \frac{3}{2}, \frac{5}{2}), (\frac{5}{2}, \frac{1}{2}, \frac{3}{2}), (4, 1, 4), (3, 2, 5)$; parallelogram

**37** $\left(\dfrac{2}{\sqrt{14}}, \dfrac{1}{\sqrt{14}}, \dfrac{3}{\sqrt{14}}\right), \left(\dfrac{-2}{\sqrt{14}}, \dfrac{-1}{\sqrt{14}}, \dfrac{3}{\sqrt{14}}\right)$

## SECTION 3 (pages 546–548)

**1** $x = 1 + t, y = 3 - 4t, z = 2 + 2t$

**3** $x = 4 - t, y = -2 + 4t, z = -t$

**5** $\dfrac{x-1}{2} = \dfrac{y}{1} = \dfrac{z+1}{-3}$

**7** $\dfrac{x-4}{2} = \dfrac{y}{-1} = \dfrac{z}{-3}$

**9** $\dfrac{x-3}{2} = \dfrac{y+1}{0} = \dfrac{z+2}{0}$

**11** Perpendicular

**13** Not perpendicular

**15** $\dfrac{x-4}{1} = \dfrac{y}{-2} = \dfrac{z-2}{0}$;    $\dfrac{x-3}{0} = \dfrac{y-1}{1} = \dfrac{z-4}{-2}$;    $\dfrac{x-2}{1}$ $= \dfrac{y-5}{-3} = \dfrac{z}{2}$

**17** $\dfrac{x-2}{a} = \dfrac{y+1}{b} = \dfrac{z-5}{-2a+3b}$

**19** $\left(0, \dfrac{-5}{2}, \dfrac{-5}{2}\right), \left(\dfrac{-5}{3}, 0, \dfrac{10}{3}\right), \left(\dfrac{-5}{7}, \dfrac{-10}{7}, 0\right)$

**21** $x = 3 - t,\ y = 1 + 3t,\ z = 5 + t$

**23** $A'B':\ \dfrac{x-2}{5} = \dfrac{y}{0} = \dfrac{x-6}{2};\quad A'C':\ \dfrac{x-2}{3} = \dfrac{y}{0} = \dfrac{z-6}{1};$

$\ \ B'C': \dfrac{x+3}{8} = \dfrac{y}{0} = \dfrac{z-4}{3}$

**25** $(1.414, 3.295, 1.013)$

**27** $3x + y - 4z + 1 = 0$

**29** $3y - 2z - 4 = 0$

**31** $2x - z = 1$

**33** $9x + y - 5z = 16$

**35** $2x + 3y - 4z + 11 = 0$

**37** $3x - 2z = 10$

**39** $\dfrac{x+2}{2} = \dfrac{y-3}{3} = \dfrac{z-1}{1}$

**41** $\dfrac{x+1}{1} = \dfrac{y}{0} = \dfrac{z+2}{2}$

**43** $3x + 2y - z = 0$

**45** $x - 2y - 3z + 5 = 0$

**47** $\dfrac{x-2}{3} = \dfrac{y+1}{-2} = \dfrac{z-3}{4}$

**49** $\dfrac{x-1}{2} = \dfrac{y+2}{-1} = \dfrac{z}{4}$

**51** $7x - 3y - 5z + 18 = 0$

**53** $2x - y - z + 3 = 0$

**55** $x + 5y + z + 3 = 0$

**57** $6x + 5y + 8z - 20 = 0$

**59** $x - 5y + 2z = 0$

**SECTION 4 (pages 551–552)**

**1** $8/21$

**3** $\sqrt{14/17}$

**5** $x = 11 - 5t,\ y = -19 + 8t,\ z = t$

**7** $x = -1 - 2t,\ y = t,\ z = 3$

**9** $(3, 2, -1)$

**11** $\left(-\dfrac{3}{5}, 0, -\dfrac{6}{5}\right)$

**13** $1$

**15** $8/\sqrt{29}$

**17** $10x - 17y + z + 25 = 0$

**19** $14x + 8y - 13z + 15 = 0$

**21** $y + x = 1$

**23** $x - 2y + 3z - 2 = 0$

**25** $\left(\frac{91}{57}, \frac{31}{57}, \frac{25}{57}\right)$          **27** No intersection

**29** $x = 3 + 4t,\ y = -1 + 5t,\ z = 2 - t$

**31** $x = 29t,\ y = 2 + t,\ z = 4 - 22t$

**33** $x = -2t,\ y = 7 - \frac{10}{9}t,\ z = -1 + t$

**35** $5x + 2y + 3z - 3 = 0$

**37** $\sqrt{14};\ 2\sqrt{14}/\sqrt{15}$

**SECTION 5 (pages 555–556)**

**1** $x^2 + y^2 + z^2 - 2x - 8y + 4z + 12 = 0$

**3** $x^2 + y^2 + z^2 - 2y - 8z - 19 = 0$

**5** Sphere: $C(-1, 0, 2),\ r = 2$

**7** No graph

**9** Sphere: $C(3, -2, -1),\ r = 2$

**11** Sphere: $(4, 0, -4),\ r = 4$

**13** Sphere: $(0, 0, 1),\ r = 1$

**15** $x^2 + y^2 + z^2 + 5x + 7y - 12z + 41 = 0$

**17** Plane      **19** Plane

**21** Parabolic cylinder

**23** Elliptic cylinder

**25** Circular cylinder

**27** Circular cylinder

**29** Circle, $C(0, 0, 3),\ r = 4$

**31** No intersection

**SECTION 6 (page 559)**

**1** (a) $\left(3\sqrt{2}, \dfrac{\pi}{4}, 7\right)$

(b) $(4\sqrt{5}, \arctan 2, 2)$

(c) $\left(\sqrt{13}, \arctan\left(\dfrac{-3}{2}\right), 1\right)$

**3** (a) $\left(2\sqrt{3}, \dfrac{\pi}{4}, \arccos\dfrac{1}{\sqrt{3}}\right)$

(b) $\left(2\sqrt{3}, \dfrac{-\pi}{4}, \arccos\dfrac{1}{\sqrt{3}}\right)$

(c) $\left(2\sqrt{2}, \dfrac{2\pi}{3}, \dfrac{\pi}{4}\right)$

**5** (a) $\left(4, \dfrac{\pi}{3}, 0\right)$  (b) $\left(1, \dfrac{2\pi}{3}, -\sqrt{3}\right)$  (c) $\left(\dfrac{7}{2}, \dfrac{\pi}{2}, \dfrac{7}{2}\sqrt{3}\right)$

**7** $r^2 + z^2 = 9$     **9** $r^2 = 4z$

**11** $r^2 = z^2$       **13** $r^2 \cos 2\theta = 4$

**15** $r = 4 \sin \theta$     **17** $\rho = 4 \cos \phi$

**19** $\phi = \dfrac{\pi}{4}$     **21** $\rho = \dfrac{\mp 2}{(1 \pm \cos \phi)}$

**23** Sphere; center at $(0, 0, 0)$ and radius 3

**25** Plane; perpendicular to $xy$ plane

**27** Right circular cylinder; center of $xy$ circle at $(0, 0)$ and radius 6

**29** Plane $z = 1$

**31** Paraboloid opening down with maximum at $(0, 0, 1)$

**33** Hyperbolic cylinder $z = 2/x$ ($y$ is unrestricted)

**35** Circular cylinder; center in $xy$ plane at $(0, 1)$; radius 1

**37** Right circular nappe opening above the $xy$ plane

## CHAPTER 13 REVIEW PROBLEMS (pages 559–560)

**1** Do not lie on the same line

**3** Do lie on the same line

**5** $\dfrac{\sqrt{2}}{2}$

**7** $\dfrac{4}{\sqrt{14}}$

**9** $\dfrac{x - 3}{1} = \dfrac{y - 1}{2} = \dfrac{z - 5}{8}$

**11** $(2, 4, 2)$

**13** $x = 1 + (\sqrt{2} - 1)t$; $y = 2 - 4t$; $z = 3 - 3t$

**15** $x = 1 + t$; $y = -2 - 2t$; $z = 3 + 3t$

**17** $\dfrac{x - 1}{2} = -y = \dfrac{z - 2}{3}$

**19** $2x + y + 4z - 4 = 0$

**21** $\dfrac{\sqrt{14}}{7}$

**23** $10x - 7y - z - 11 = 0$

**25** $C(-4, 2, 0)$; $\rho = \sqrt{19}$

**27** $C(-6, 0, 0)$; $\rho = 6$

**29** $C\left(\dfrac{1}{8}, -\dfrac{1}{4}, -\dfrac{1}{2}\right)$; $\rho = \dfrac{\sqrt{149}}{8} \approx 1.5258$

**31** $(x - 2)^2 + (y - 2)^2 + (z - 2)^2 = 14$

**33** Parabolic cylinder

**35** Parabolic cylinder

**37** The plane $x = 2$

**39** $2xy = z$ (hyperboloid)

**41** Sphere; $C(0, 0, -8)$; $\rho = 8$

**43** $z = -3\dfrac{x}{y} + 3$

**45** The $z$ axis

**47** Right circular cylinder; in $xy$ plane a circle centered at $(0, 2)$, radius 2

**49** $z\sqrt{x^2 + y^2} = 3$

**53** $\rho = 1$

**55** $r^2 = 3z$

**57** $z = \tan \theta$

**59** $1/5 = \cos^2 \phi$

**61** $\rho = -4 \cos \phi$

**63** $z = r^2(1 - 2 \sin^2 \theta)$

# CHAPTER 14

## SECTION 1 (pages 569–570)

**1** $\mathbf{v} = -2\mathbf{i} + 7\mathbf{j}$     **3** $\mathbf{v} = -5\mathbf{i} + 4\mathbf{j}$

**5** $\mathbf{v} = -12\mathbf{i} - 4\mathbf{j}$     **7** $\mathbf{u} = -\frac{5}{13}\mathbf{i} - \frac{12}{13}\mathbf{j}$

**9** $\mathbf{u} = \dfrac{-2}{\sqrt{29}}\mathbf{i} + \dfrac{5}{\sqrt{29}}\mathbf{j}$     **11** $B(4, 6)$

**13** $A(4, 7)$     **15** $A(-3, \frac{1}{2})$, $B(-5, \frac{7}{2})$

**17** $\pm 2\mathbf{i} + 2\mathbf{j}$     **19** $-\mathbf{i} + \mathbf{j}$

**21** (a) $(3, -1)$;  (b) $(1, -5)$;  (c) $(2, -17)$

**23** (a) $(1, -1)$;  (b) $(-1, 1)$;  (c) $(-4, 4)$

**25** (a) $(16, 10)$;  (b) $(20, 2)$;  (c) $(62, 2)$

**27** 5     **29** $2\sqrt{10}$

**31** $2\pi\sqrt{5}$     **33** $10\sqrt{2}$

**35** 1     **37** 6

**39** Insufficient information

**41** $\sqrt{38}$     **43** $\sqrt{5}$

## SECTION 2 (pages 574–575)

**1** $|\mathbf{v}| = 5$, $|\mathbf{w}| = 5$, $\cos \theta = -\frac{7}{25}$, proj. $\mathbf{v}$ on $\mathbf{w} = -\frac{7}{5}$

**3** $|\mathbf{v}| = 5$, $|\mathbf{w}| = 13$, $\cos \theta = -\frac{33}{65}$, proj. $\mathbf{v}$ on $\mathbf{w} = -\frac{33}{13}$

**5** $|\mathbf{v}| = \sqrt{13}$, $|\mathbf{w}| = \sqrt{13}$, $\cos \theta = 0$, proj. $\mathbf{v}$ on $\mathbf{w} = 0$

**7** $-1$      **9** $\dfrac{6}{\sqrt{17}}$      **11** $0$

**13** $\cos \theta = \dfrac{+1}{\sqrt{5}}$, $\cos \alpha = \dfrac{1}{\sqrt{5}}$

**15** $\cos \theta = \dfrac{2}{\sqrt{13}}$, $\cos \alpha = \dfrac{11}{\sqrt{130}}$

**17** $\cos \theta = -1$, $\cos \alpha = 1$

**19** $a = \frac{3}{2}$

**21** Impossible

**23** $a = \dfrac{(-240 + \sqrt{((240)^2 + 69 \cdot 407)}}{407}$

**27** Orthogonal

**29** Orthogonal

**33** $|BC| = 2\sqrt{10}$; proj. of $A\vec{B}$ on $B\vec{C} = -\dfrac{8}{\sqrt{10}}$; proj. of $A\vec{C}$ on

$B\vec{C} = \dfrac{12}{\sqrt{10}}$

**35** Proj. of $A\vec{B}$ on $C\vec{B} = \frac{19}{6}$; proj. of $A\vec{C}$ on $C\vec{B} = -\frac{35}{6}$

**37** $\frac{3}{4}\mathbf{v}[A\vec{B}] + \frac{1}{2}\mathbf{v}[B\vec{C}]$

**39** $-\frac{1}{2}\mathbf{v}[A\vec{B}]$

**43** $\mathbf{v}[O\vec{P}] = \frac{1}{3}[\mathbf{a} + \mathbf{b} + \mathbf{c}]$

## SECTION 3 (pages (579–580)

**1** $-\mathbf{i} + 4\mathbf{j} - 6\mathbf{k}$      **3** $5\mathbf{i} - \mathbf{j} + 3\mathbf{k}$

**5** $-2\mathbf{i} - 3\mathbf{k}$      **7** $\mathbf{j}$

**9** $\mathbf{i} + \mathbf{j} + 2\mathbf{k}$      **11** $\dfrac{1}{\sqrt{29}}(3\mathbf{i} + 2\mathbf{j} - 4\mathbf{k})$

**13** $\dfrac{1}{\sqrt{21}}(2\mathbf{i} - 4\mathbf{j} - \mathbf{k})$      **15** $\dfrac{1}{\sqrt{10}}(\mathbf{i} - 3\mathbf{j})$

**17** $\dfrac{1}{\sqrt{3}}(\mathbf{i} + \mathbf{j} + \mathbf{k})$

**19** $B: (3, 3, -4)$

**21** $A: (-1, -2, 0)$

**23** $A: (\frac{3}{2}, 0, 3)$, $B: (\frac{5}{2}, -2, 5)$

**25** $A: (\frac{7}{4}, -\frac{3}{4}, \frac{7}{2})$, $B: (\frac{3}{4}, \frac{1}{4}, \frac{3}{2})$

**27** $4\mathbf{i} - 5\mathbf{j} + \mathbf{k}$

## SECTION 4 (pages 583–584)

**1** $\dfrac{16}{7\sqrt{6}}$

**3** $\dfrac{1}{\sqrt{3}}$

**5** $\dfrac{-11}{29}$

**7** $-\frac{2}{7}$

**9** $-23/\sqrt{50}$

**11** $96$

**13** $35$

**15** $\frac{1}{7}(2\mathbf{i} - 6\mathbf{j} + 3\mathbf{k})$

**17** $\dfrac{1}{\sqrt{62}}(3\mathbf{i} - 2\mathbf{j} + 7\mathbf{k})$

**19** $k = 2$, $h = \frac{2}{5}$

**21** $k = 3$, $h = \frac{42}{145}$

**25** $3g + 5h = 0$

**27** $4g - 9h = 0$

**29** $h = 1$, $g = -1$

**39** $\cos \theta = \frac{1}{3}$

## SECTION 5 (pages 589–590)

**1** $-2$

**3** $-2$

**5** $0$

**7** $0$

**9** $-65$

**11** $7(\sqrt{2} - \pi)$

**13** $2\mathbf{i} - 3\mathbf{j} - 7\mathbf{k}$

**15** $-4\mathbf{i} - 2\mathbf{j} - 5\mathbf{k}$

**17** $20\mathbf{i} + 30\mathbf{j} - 16\mathbf{k}$

**19** $5(\mathbf{i} - \mathbf{j} + \mathbf{k})$

**21** $4\mathbf{i} - 2\mathbf{j} + \mathbf{k}$

**23** $-5\mathbf{i} + 35\mathbf{k}$

**25** $A = \dfrac{3\sqrt{22}}{2}$; $5x + 13y + 2z = 37$

**27** $A = \dfrac{\sqrt{21}}{2}$; $4x - 2y + z - 11 = 0$

**29** $A = \frac{1}{2}$; $x = 0$

**31** $A = \frac{1}{2}\sqrt{237}$; $8x - 13y - 2z = 0$

**33** $A = \dfrac{\sqrt{13}}{2}$; $-2y + 3z - 5 = 0$

**35** $A = \sqrt{141}$; $10x + 8y + 20z - 40 = 0$

**37** $\dfrac{107}{\sqrt{1038}}$          **39** $\dfrac{3}{\sqrt{5}}$

**41** $\dfrac{x+1}{2} = \dfrac{y-3}{-11} = \dfrac{z-2}{-7}$

**43** $\dfrac{x-1}{1} = \dfrac{y+2}{-1} = \dfrac{z-3}{1}$

**45** $\dfrac{x-3}{-2} = \dfrac{y}{1} = \dfrac{z-1}{3}$

**47** $\dfrac{x+2}{2} = \dfrac{y-1}{11} = \dfrac{z+1}{8}$

**49** $\dfrac{x+1}{-2} = \dfrac{y-1}{2} = \dfrac{z-1}{0}$

**51** $4x - y - 3z + 5 = 0$

**53** $14x - 17y - 4z - 19 = 0$

**55** $x + 5y - 3z + 1 = 0$

**57** $7x - 8y - 5z + 7 = 0$

**59** $22x - 7y - 13z - 21 = 0$

**61** $\dfrac{x-3}{2} = \dfrac{y-3}{1} = \dfrac{z+1}{1}$

**63** Yes

## SECTION 6 (pages 593–594)

**1** $V = 20$          **3** Plane: $x + 2y - 5 = 0$

**7** $\mathbf{i} + 5\mathbf{j} - 2\mathbf{k}$          **9** $8\mathbf{i} + 10\mathbf{j} - 14\mathbf{k}$

**13** $\dfrac{x-2}{160} = \dfrac{y+1}{-45} = \dfrac{z-3}{37}$          **17** 6

**19** $\frac{11}{3}$          **21** $-16\mathbf{v} + 12\mathbf{w}$

**23** $[(\mathbf{t} \times \mathbf{u}) \cdot \mathbf{w}]\mathbf{v} - [(\mathbf{t} \times \mathbf{u}) \cdot \mathbf{v}]\mathbf{w}$

**25** $\mathbf{v} = |\mathbf{a}|^2 p\mathbf{a} - |\mathbf{a}|^{-2}(\mathbf{a} \times \mathbf{b})$

## SECTION 7 (page 597)

**1** Linearly independent

**3** Linearly dependent

**5** Linearly dependent

**7** Linearly independent

**9** Linearly independent

**11** $\mathbf{r} = \mathbf{v} + 2\mathbf{w}$

**13** $\mathbf{r} = 4\mathbf{u} - 6\mathbf{v} - \mathbf{w}$

**15** $-3\mathbf{u} + 2\mathbf{v} - \mathbf{w}$

In problems 19, 2!, 23 only one of many possible answers is given.

**19** $\mathbf{w} = \mathbf{k}$          **21** $\mathbf{w} = \mathbf{i}$          **23** $\mathbf{w} = \mathbf{j}$

**27** $\dfrac{x-2}{3} = \dfrac{y-1}{-2} = \dfrac{z+3}{7}$

## SECTION 8 (page 601)

**1** $\mathbf{f}'(t) = 2t\mathbf{i} + (3t^2 - 3)\mathbf{j}$; $\mathbf{f}''(t) = 2\mathbf{i} + 6t\mathbf{j}$

**3** $\mathbf{f}'(t) = 3 \sec^2 3t\mathbf{i} - \pi \sin \pi t\mathbf{j}$; $\mathbf{f}''(t) = 18 \sec^2 3t \tan 3t\mathbf{i} - \pi^2 \cos \pi t\mathbf{j}$

**5** $\mathbf{f}'(t) = 2e^{2t}\mathbf{i} - 2e^{-2t}\mathbf{j}$; $\mathbf{f}''(t) = 4e^{2t}\mathbf{i} + 4e^{-2t}\mathbf{j}$

**7** $\dfrac{2t(t^2 + 1)}{(t^2 + 2)^3} + 4t$

**9** $\dfrac{2 \sin 4t - 3 \sin 6t}{2(\sin^2 2t + \cos^2 3t)^{1/2}}$

**11** 0          **13** $\dfrac{3t^2 + 40}{t^6}$

**15** $x(t) = 1 + 6t$; $y(t) = 2 + t$

**17** $x(t) = t$; $y(t) = 1 + t$

**19** $x(t) = t$; $y(t) = 1 + t$

**21** $x(t) = 1 + 2t$; $y(t) = 2 + 2t$

**23** $x(t) = 1 + t$; $y(t) = -1 - t$

**25** $x(t) = \sqrt{3} + \dfrac{2}{\sqrt{3}}t$; $y(t) = 8 + 12t$

**31** Circle radius 1, center at origin, $x^2 + y^2 = 1$

## SECTION 9 (pages 603–604)

**1** $\mathbf{v}(t) = 2t\mathbf{i} - 3\mathbf{j}$; $|\mathbf{v}(t)| = s'(t) = \sqrt{4t^2 + 9}$;
  $\mathbf{a}(t) = 2\mathbf{i}$; $|\mathbf{a}(t)| = 2$; $s''(t) = \dfrac{4t}{\sqrt{4t^2 + 9}}$

**3** $\mathbf{v}(t) = 3\mathbf{i} + t^{-2}\mathbf{j}$; $|\mathbf{v}(t)| = s'(t) = \sqrt{9 + t^{-4}}$;
  $\mathbf{a}(t) = -2t^{-3}\mathbf{j}$; $|\mathbf{a}(t)| = |2t^{-3}|$; $s''(t) = \dfrac{-2}{t^3\sqrt{9t^4 + 1}}$

**5** $\mathbf{v}(t) = \tan t\mathbf{i} + \mathbf{j}$; $|\mathbf{v}(t)| = s'(t) = |\sec t|$;
  $\mathbf{a}(t) = \sec^2 t\mathbf{i}$; $|\mathbf{a}(t)| = \sec^2 t$; $s''(t) = \sec t \tan t$

**7** $\mathbf{v}(t) = 2t\mathbf{i} + (3 - 3t^2)\mathbf{j}$; $|\mathbf{v}(t)| = s'(t) = \sqrt{9t^4 - 16t^2 + 9}$;
  $\mathbf{a}(t) = 2\mathbf{i} - 6t\mathbf{j}$; $|\mathbf{a}(t)| = \sqrt{4 + 36t^2} = 2\sqrt{1 + 9t^2}$;
  $s''(t) = \dfrac{18t^3 - 16t}{\sqrt{9t^4 - 16t^2 + 9}}$

**9** $\mathbf{v}(t) = (e^t + te^t)\mathbf{i} + (1 - e^t)\mathbf{j}$;
$|\mathbf{v}(t)| = s'(t) = \sqrt{e^{2t}(2 + 2t + t^2) + 1 - 2e^t}$;
$\mathbf{a}(t) = (2 + t)e^t\mathbf{i} - e^t\mathbf{j}$; $|\mathbf{a}(t)| = e^t\sqrt{5 + 4t + t^2}$;
$$s''(t) = \frac{e^{2t}(3 + 3t + t^2) - e^t}{\sqrt{e^{2t}(2 + 2t + t^2) + 1 - 2e^t}}$$

**11** $\mathbf{v}(t) = 2e^t\mathbf{i} - 3e^{-t}\mathbf{j}$; $|\mathbf{v}(t)| = s'(t) = \sqrt{4e^{2t} + 9e^{-2t}}$;
$\mathbf{a}(t) = 2e^t\mathbf{i} + 3e^{-t}\mathbf{j}$; $|\mathbf{a}(t)| = \sqrt{4e^{2t} + 9e^{-2t}}$;
$$s''(t) = \frac{4e^{2t} - 9e^{-2t}}{\sqrt{4e^{2t} + 9e^{-2t}}}$$

**13** $\mathbf{v}(t) = \mathbf{i} + 2t\mathbf{j}$; $|\mathbf{v}(t)| = s'(t) = \sqrt{1 + 4t^2}$;
$$\mathbf{a}(t) = 2\mathbf{j}; \ |\mathbf{a}(t)| = 2; \ s''(t) = \frac{4t}{\sqrt{1 + 4t^2}}$$

**15** $\mathbf{v}(t) = 2t\mathbf{i} + (6t - 3t^2)\mathbf{j}$;
$|\mathbf{v}(t)| = s'(t) = \sqrt{40t^2 - 36t^3 + 9t^4}$;
$\mathbf{a}(t) = 2\mathbf{i} + 6(1 - t)\mathbf{j}$; $|\mathbf{a}(t)| = 2\sqrt{9t^2 - 18t + 10}$;
$$s''(t) = \frac{2(20t - 27t^2 + 9t^3)}{\sqrt{40t^2 - 36t^3 + 9t^4}}$$

**17** $\mathbf{v}(1) = \mathbf{i}$; $\mathbf{a}(1) = 2\mathbf{i} + 2\mathbf{j}$; $s'(1) = 1$; $s''(1) = 2$

**19** $\mathbf{v}\left(\dfrac{\pi}{6}\right) = 2\mathbf{i} + \dfrac{8}{3}\mathbf{j}$; $\mathbf{a} = \dfrac{10}{\sqrt{3}}\mathbf{i} + \dfrac{16}{3\sqrt{3}}\mathbf{j}$;
$s' = \dfrac{10}{3}$; $s'' = \dfrac{154}{15\sqrt{3}}$

**21** $\mathbf{v} = \dfrac{1}{3}\mathbf{i} - \dfrac{3}{4}\mathbf{j}$; $\mathbf{a} = -\dfrac{1}{9}\mathbf{i} + \dfrac{3}{4}\mathbf{j}$;
$s' = \dfrac{\sqrt{97}}{12}$; $s'' = \dfrac{-259}{36\sqrt{97}}$

**23** $\mathbf{v} = \mathbf{i} + \mathbf{j}$; $\mathbf{a} = \mathbf{i} + 2\mathbf{j}$; $s' = \sqrt{2}$; $s'' = \dfrac{3}{\sqrt{2}}$

**25** $\mathbf{v} = \mathbf{i} - \dfrac{1}{e}\mathbf{j}$; $\mathbf{a} = -\mathbf{i} + \dfrac{1}{e}\mathbf{j}$; $s' = \sqrt{1 + \dfrac{1}{e^2}}$;
$s'' = -\dfrac{1 + 1/e^2}{\sqrt{1 + 1/e^2}}$

**29** $\mathbf{T} = \dfrac{4e^{2t}\mathbf{i} - 3e^{-2t}\mathbf{j}}{(16e^{4t} + 9e^{-4t})^{1/2}}$

## SECTION 10 (pages 608–609)

**1** $y = -4.9t^2 + 150t + 80$; highest point $= 1228$ m
**3** $x = t^2 + 12t + 1$; $v = 2t + 12$
**5** $x = \frac{1}{2}(5e^{2t} - 3)$; $v = 5e^{2t}$
**7** $x = x_0 + (1/k)\log(1 + v_0kt)$; $v = v_0/(1 + v_0kt)$
**9** Distance $= 112.5 \times 10^4\sqrt{3}/g$, $t = 1500\sqrt{3}/g$; $g = 9.78$
**11** $T = (375 + \sqrt{(375)^2 + 12{,}000\,g})/g$; $g = 9.78$
**13** $A = 2(\frac{5}{2})^t$, $A(3) = \frac{125}{4}$

**15** $t = 1690\left(\dfrac{\ln 10}{\ln 2}\right)$ years

**17** $t = \dfrac{\ln 100}{\ln 2}$ weeks

**19** $y = 160 + 240e^{-1/2}$
**21** $(5600 \log 3)/\log 2$ years
**23** $72(A - 8B)$
**25** $y = 5.5e^{12} - 2.5g$

## SECTION 11 (pages 613–614)

**1** $\mathbf{f}' = 2t\mathbf{i} + 2t\mathbf{j} - 3\mathbf{k}$; $\mathbf{f}'' = 2\mathbf{i} + 2\mathbf{j}$
**3** $\mathbf{f}' = -2(\sin 2t)\mathbf{i} + 2(\cos 2t)\mathbf{j} + 2\mathbf{k}$
$\mathbf{f}'' = -4(\cos 2t)\mathbf{i} - 4(\sin 2t)\mathbf{j}$

**5** $\mathbf{f}' = \dfrac{2t}{(t^2 + 1)^2}\mathbf{i} - \dfrac{2t}{(t^2 + 1)^2}\mathbf{j} + 2t\mathbf{k}$
$\mathbf{f}'' = \dfrac{-6t^2 + 2}{(t^2 + 1)^3}\mathbf{i} - \dfrac{-6t^2 + 2}{(t^2 + 1)^3}\mathbf{j} + 2\mathbf{k}$

**7** $\mathbf{f}' = -(\sin t)\mathbf{i} + (\sec^2 t)\mathbf{j} + (\cos t)\mathbf{k}$
$\mathbf{f}'' = -(\cos t)\mathbf{i} + 2(\sec^2 t \tan t)\mathbf{j} - (\sin t)\mathbf{k}$

**9** $\mathbf{f}' = 2t\mathbf{i} - \sin t\mathbf{j} + e^{-t}(1 - t)\mathbf{k}$
$\mathbf{f}'' = 2\mathbf{i} - \cos t\mathbf{j} + e^{-t}(t - 2)\mathbf{k}$

**11** $\mathbf{f}'(t) = (4t^3 - 2t)\mathbf{i} + (3t^2 - 6t^{-4})\mathbf{j} + \dfrac{2t}{t^2 + 1}\mathbf{k}$;
$\mathbf{f}''(t) = (12t^2 - 2)\mathbf{i} + (6t + 24t^{-5})\mathbf{j} + \dfrac{2 - 2t^2}{(t^2 + 1)^2}\mathbf{k}$

**13** Domain: $[0, \infty)$; continuous on $[0, \infty)$
**15** Domain: $(-\infty, 0) \cup (0, \infty)$; continuous on
$(-\infty, 0) \cup (0, \infty)$; if $\dfrac{\sin t}{t}$ is defined to be 1 at $t = 0$, then
domain is $(-\infty, \infty)$ and $\mathbf{f}$ is continuous on $(-\infty, \infty)$
**17** Domain: $(-\infty, 0) \cup (0, \infty)$; continuous on
$(-\infty, 0) \cup (0, \infty)$.
**19** $f'(t) = (t^2 + 2t + 2)e^t + 6t + 3$
**21** $\mathbf{f}'(t) = (2t^2 - 288t^{71})\mathbf{i} + (72(t^{73} + 2t^{71}) - t)\mathbf{j} + (-2t^3 - 4t + 4)\mathbf{k}$
**23** $f'(t) = e^t \cos t \tan t - e^t \sin t \tan t + e^t \cos t \sec^2 t$
**25** 14

**27** $\dfrac{\pi}{8}\sqrt{\pi^2 + 8} + \ln(\pi + \sqrt{\pi^2 + 8}) - \dfrac{3}{2}\ln 2$

**29** $\dfrac{\sqrt{2}}{2} + \dfrac{1}{2}\ln(1 + \sqrt{2})$

**31** $22\pi$

**33** $\mathbf{v}(t) = (\sin t + t \cos t)\mathbf{i} + (\cos t - t \sin t)\mathbf{j} + \mathbf{k}$;
$s'(t) = \sqrt{t^2 + 2}$; $\mathbf{a}(t) = (2\cos t - t \sin t)\mathbf{i} - (2\sin t + t \cos t)\mathbf{j}$

## SECTION 12 (page 619)

**1** $(3t^2\mathbf{i} - \mathbf{j} + 2\mathbf{k})/\sqrt{5 + 9t^4}$

**3** $(3\mathbf{i} + e^t\mathbf{k})/\sqrt{9 + e^{2t}}$

**5** $(\mathbf{i} + 2\mathbf{j} + 3\mathbf{k})/\sqrt{14}$

**7** $(-e^{-2t}\mathbf{i} + e^{2t}\mathbf{j} + t\mathbf{k})/\sqrt{e^{-4t} + e^{4t} + t^2}$

**9** $\dfrac{(\cos t + \sin t)\mathbf{i} + e^t(2\cos t - \sin t)\mathbf{j} - e^{-2t}\mathbf{k}}{[(1 + 2\cos t \sin t) + e^{2t}(2\cos t - \sin t)^2 + e^{-4t}]^{1/2}}$

**11** $\mathbf{T} = (1/\sqrt{6})(\mathbf{i} - \mathbf{j} + 2\mathbf{k})$; $\mathbf{B} = \mathbf{N} = $ not defined; $\kappa = 0$; $(x - 4)/1 = y/(-1) = (z - 10)/2$; osculating plane not defined.

**13** $\mathbf{T} = (1/\sqrt{3})(\mathbf{i} + \mathbf{j} + \mathbf{k})$; $\mathbf{B} = (1/\sqrt{6})(-\mathbf{i} - \mathbf{j} + 2\mathbf{k})$; $\mathbf{N} = (1/\sqrt{2})(-\mathbf{i} + \mathbf{j})$; $\kappa = \sqrt{2}/3$; $(x - 1)/1 = y/1 = (z - 1)/1$; $x + y - 2z + 1 = 0$

**15** $\mathbf{T} = \dfrac{1}{\sqrt{26}}\mathbf{i} + \dfrac{3}{\sqrt{26}}\mathbf{j} + \dfrac{4}{\sqrt{26}}\mathbf{k}$; $\mathbf{B} = \mathbf{N} = \mathbf{O}$; $\kappa = 0$; osculating plane not defined.

**17** $\mathbf{N}(t) = \dfrac{1}{\sqrt{2}}[(-\sin t - \cos t)\mathbf{i} + (\cos t - \sin t)\mathbf{j}]$;

$\kappa(t) = \dfrac{\sqrt{2}}{3}e^{-t}$

**19** $\mathbf{v} = \mathbf{i} + 2\mathbf{j} + \sqrt{3}\mathbf{k}$; $\mathbf{a} = 2\sqrt{3}\mathbf{j} + 4\mathbf{k}$; $s'\left(\dfrac{\pi}{3}\right) = 2\sqrt{2}$; $R = 4$;

$\mathbf{T}\left(\dfrac{\pi}{3}\right) = \dfrac{1}{\sqrt{2}}\mathbf{v}$; $\mathbf{N} = -\dfrac{1}{2}(\sqrt{3}\mathbf{i} - \mathbf{k})$; $a_T = 2\sqrt{6}$; $a_N = 2$

**21** $\mathbf{v} = \mathbf{i} + \mathbf{k}$; $\mathbf{a} = 2\mathbf{j}$; $s' = \sqrt{2}$; $\mathbf{T} = \dfrac{1}{\sqrt{2}}(\mathbf{i} + \mathbf{k})$; $\mathbf{N} = \mathbf{j}$; $R = 1$;

$a_T = 0$; $a_N = 2$

## CHAPTER 14 REVIEW PROBLEMS (pages 619–621)

**1** $-2\mathbf{i} + 4\mathbf{j}$          **3** $-3\mathbf{i} - 3\mathbf{j}$

**5** $-8\mathbf{i} - 5\mathbf{j}$

**7** (a) $(\pi + 2)\mathbf{i} + 3\mathbf{j}$;
(b) $(\pi - 2)\mathbf{i} - 3\mathbf{j}$;
(c) $(2\pi - 6)\mathbf{i} - 9\mathbf{j}$

**11** $-\sqrt{29}$          **13** $\dfrac{-6}{\sqrt{2}}$

**15** $(2 + \sqrt{2})\sqrt{5}$     **17** $\sqrt{34}$

**21** $-\dfrac{\sqrt{2}}{10}$          **23** $5\sqrt{2}$

**25** (a) $\dfrac{2}{\sqrt{14}}\mathbf{i} - \dfrac{1}{\sqrt{14}}\mathbf{j} - \dfrac{3}{\sqrt{14}}\mathbf{k}$;

(b) $\dfrac{1}{\sqrt{30}}\mathbf{i} + \dfrac{5}{\sqrt{30}}\mathbf{j} - \dfrac{2}{\sqrt{30}}\mathbf{k}$

**27** $B = (-1, 6, 1)$

**29** $A = (-2, 12, 4)$

**31** $-13$

**33** $5\mathbf{i} + 14\mathbf{j} - 16\mathbf{k}$

**35** $6\mathbf{i} - 11\mathbf{j} - 47\mathbf{k}$

**37** 2.1077 radians

**39** $\cos\theta = 0.3311$

**41** $\dfrac{-13}{\sqrt{17}}$

**43** (a) $\dfrac{3}{\sqrt{14}}\mathbf{i} - \dfrac{2}{\sqrt{14}}\mathbf{j} + \dfrac{1}{\sqrt{14}}\mathbf{k}$;

(b) $\dfrac{-2}{\sqrt{17}}\mathbf{i} + \dfrac{3}{\sqrt{17}}\mathbf{j} + \dfrac{2}{\sqrt{17}}\mathbf{k}$

**45** $\alpha = 13$; $\beta = 21$

**47** (a) $\dfrac{-7}{\sqrt{138}}\mathbf{i} - \dfrac{8}{\sqrt{138}}\mathbf{j} + \dfrac{5}{\sqrt{138}}\mathbf{k}$;

(b) $\dfrac{7}{\sqrt{138}}\mathbf{i} + \dfrac{8}{\sqrt{138}}\mathbf{j} - \dfrac{5}{\sqrt{138}}\mathbf{k}$

**49** Area $= \dfrac{\sqrt{51}}{2}$; $5x + 5y - z - 29 = 0$

**51** $\dfrac{x - 3}{1} = \dfrac{y - 1}{2} = \dfrac{z - 5}{8}$

**53** $\dfrac{x - 2}{0} = \dfrac{y - 2}{3} = \dfrac{z}{-2}$

**55** $\mathbf{f}'(t) = (3t^2 + 1)\mathbf{i} + 2t\mathbf{j}$;
$\mathbf{f}''(t) = 6t\mathbf{i} + 2\mathbf{j}$

**57** $\mathbf{f}'(t) = -\sin t\mathbf{i} + (\sin t + t\cos t)\mathbf{j}$;
$\mathbf{f}''(t) = -\cos t\mathbf{i} + (2\cos t - t\sin t)\mathbf{j}$

**59** $x = t$, $y = 1 + t$, $z = 1 + t$

**61** $x = 4 + 6t$, $y = 2 + 4t$, $z = 1 + t$

**63** $\mathbf{v}(t) = 3\mathbf{i} + 3t^2\mathbf{j} + 4t^3\mathbf{k}$;
$|\mathbf{v}(t)| = s'(t) = \sqrt{16t^6 + 9t^4 + 9}$;
$\mathbf{a}(t) = 6t\mathbf{j} + 12t^2\mathbf{k}$; $|\mathbf{a}(t)| = 6\sqrt{t^2 + 4t^4}$;
$s''(t) = \dfrac{6(8t^5 + 3t^3)}{\sqrt{16t^6 + 9t^4 + 9}}$

**65** $\mathbf{v}(t) = -\sin t\mathbf{i} + 2\cos t\mathbf{j}$;
$|\mathbf{v}(t)| = s'(t) = \sqrt{1 + 3\cos^2 t}$;
$\mathbf{a}(t) = -\cos t\mathbf{i} - 2\sin t\mathbf{j}$; $|\mathbf{a}(t)| = \sqrt{1 + 3\sin^2 t}$;
$s''(t) = \dfrac{-\cos t \sin t}{\sqrt{1 + 3\cos^2 t}}$

**67** 53, 711 years

**69** $\ln(1 + \sqrt{2})$

**71** $16 - \sqrt{52} \approx 8.789$

# CHAPTER 15

## SECTION 1 (pages 629–630)

**1** Level curves are lines and the graph is a plane through three points $(-2, 0, 0)$, $(0, -3, 0)$ and $(0, 0, 6)$

**3** The level curves are hyperbolas $xy = c$ and the graph is the hyperbolic paraboloid symmetric to the plane $x = y$

**5** The level curve $c = x(x + y)^{-1}$ is a line if $x + y \neq 0$

**7** The level curves are hyperbolas $xy = \ln(c)$

**9** The level curves are lines $y = \pm(\sqrt{1 - c^{-1}})x$

**11** The level curves are the exponential curves $y = e^x - c$

**13** The level curves are $y = \arcsin(x - c)$ for $|x - c| \leq 1$

**15** $f_x = 2xy^4 + 2y$; $f_y = 4x^2y^3 + 2x$

**17** $f_x = 4x(x^2 - y^3)$; $f_y = -6y^2(x^2 - y^3)$

**19** $f_x = 3x^2 + 6xy - 3y^2$; $f_y = 3y^2 + 3x^2 - 6xy$

**21** $f_x = (y^3 - x^2y)/(x^2 + y^2)^2$; $f_y = (x^3 - xy^2)/(x^2 + y^2)^2$

**23** $f_x = x/(x^2 + 3y^2)$; $f_y = 3y/(x^2 + 3y^2)$

**25** $f_x = [(1 + y)^2 - x^2]^{-1/2}$;
$f_y = -x[(1 + y)((1 + y^2) - x^2)^{1/2}]^{-1}$

**27** $f_x = -e^y \sin(xe^y)$; $f_y = -xe^y \sin(xe^y)$

**29** $f_u(3, 3) = e^5(\sec 1)[3 + (\tan 1)/3]$;
$f_v(3, 3) = e^5(\sec 1)[3 - (\tan 1)/3]$

**31** $f_t(0, 1) = 0$; $f_u(0, 1) = -2$

**33** $f_s(3, 3) = 27(1 + \ln 3)$; $f_t(3, 3) = 27(1 + \ln 3)$

**35** $f_x = yz(y^2 + z^2 - x^2)(x^2 + y^2 + z^2)^{-2}$;
$f_y = xz(x^2 + z^2 - y^2)(x^2 + y^2 + z^2)^{-2}$;
$f_z = xy(x^2 + y^2 - z^2)(x^2 + y^2 + z^2)^{-2}$

**37** $f_x = 2x + 2$; $f_y = 3y^2 - 3$; $f_z = 2z + 4$

**39** $f_x = 2x + yz \sin(xyz)$; $f_y = 2y + xz \sin(xyz)$;
$f_z = 2z + xy \sin(xyz)$

**41** $w_x = (y^2 - x^2)[x(x^2 + y^2)]^{-1}$; $w_y = (x^2 - y^2)[y(x^2 + y^2)]^{-1}$

**43** $w_x = -(yx^{-2})e^{\sin(yx^{-1})}\cos(yx^{-1})$;
$w_y = x^{-1}\cos(yx^{-1})e^{\sin(yx^{-1})}$

**45** $z = \ln y - \ln x$, $z_x = -1/x$, $z_y = 1/y$, $xz_x + yz_y = 0$

**47** $xq_x + yq_y + zq_z = \sum_{k=1}^{n} a_k[(2n - 2k) + (k + 1) + (k - 1)]$
$\times x^{2n-2k}y^{k+1}z^{k-1} = 2nq$

**49** (i) $\lim_{x \to 0} g(x, x^2) = \lim_{x \to 0} [x/(1 + x^2)] = 0$;
(ii) $\lim_{x \to 0} g(x, x) = \lim_{x \to 0} [x^2/(2x^2)] = 1/2$
Also $\lim_{x \to 0} g(x, mx) = m/(1 + m^2)$ which depends on $m$

**51** For $\varepsilon > 0$, let $\delta = \sqrt{\varepsilon}$. Then for $|(x, y)| = \sqrt{x^2 + y^2} < \delta$,
$|f(x, y) - 0| \leq |x^2| < \varepsilon$, as required

**53** Since $z = xy$ is continuous and $\sin(z)$ is continuous, the composition $\sin(xy)$ is also continuous. $\lim_{x \to 0} g(\pi/(2y), y)$
$= \sin(\pi/2) = 1 \neq 0$. So, $g$ is not continuous at $(0, 0)$

## SECTION 2 (pages 632–633)

**1** $\dfrac{\partial w}{\partial x} = \dfrac{1 - 6x}{12w}$; $\dfrac{\partial w}{\partial y} = -\dfrac{(1 + 4y)}{12w}$

**3** $\dfrac{\partial w}{\partial x} = -\dfrac{x - y + w}{x + w}$; $\dfrac{\partial w}{\partial y} = \dfrac{x - 3y}{x + w}$

**5** $\dfrac{\partial w}{\partial r} = -\dfrac{2r \cos rw - w(r^2 + s^2) \sin rw}{1 + r(r^2 + s^2) \sin rw}$
$\dfrac{\partial w}{\partial s} = \dfrac{2s \cos rw}{1 + r(r^2 + s^2) \sin rw}$

**7** $\dfrac{\partial w}{\partial x} = \dfrac{y \cos 2xw(w \sin xy + \cos xy) - 2w \sin xy \sin 2xw}{\sin xy(2x \sin 2xw - xy \cos 2xw)}$
$\dfrac{\partial w}{\partial y} = \dfrac{\cos 2xw(w \sin xy + \cos xy)}{\sin xy(2 \sin 2xw - y \cos 2xw)}$

**9** $\dfrac{\partial w}{\partial x} = -\dfrac{z(y + 2x + w)}{xz - yz - 3w^2}$; $\dfrac{\partial w}{\partial y} = \dfrac{z(w - x - z)}{xz - yz - 3w^2}$
$\dfrac{\partial w}{\partial z} = \dfrac{yw - xy - x^2 - xw - 2yz}{xz - yz - 3w^2}$

**11** $\dfrac{\partial w}{\partial x} = \dfrac{ye^{x(y-w)} + w^2}{y^2e^{w(y-x)} - xw - 1}$; $\dfrac{\partial w}{\partial y} = \dfrac{xe^{y(x-w)} - yw - 1}{y^2 - (1 + xw)e^{w(x-y)}}$

**13** $\dfrac{\partial w}{\partial x} = -w^4(4xw^3 - y \sin w)^{-1}$

**15** $\dfrac{\partial w}{\partial x} = (w - 2y\sqrt{xw})(-x + 8\sqrt{xw})^{-1}$

**17** $\dfrac{\partial w}{\partial z} = -\dfrac{e^z}{2y(x + yw)}$, $\dfrac{\partial w}{\partial x} = \dfrac{-1}{y}$

**19** $\left(\dfrac{\partial z}{\partial x}\right)\left(\dfrac{\partial x}{\partial y}\right)\left(\dfrac{\partial y}{\partial z}\right) = \left(\dfrac{x}{z}\right)\left(\dfrac{-y}{x}\right)\left(\dfrac{z}{y}\right) = -1$

**21** $\dfrac{\partial x}{\partial \rho} = \sin \phi \cos \theta$; $\dfrac{\partial x}{\partial \phi} = \rho \cos \phi \cos \theta$; $\dfrac{\partial x}{\partial \theta} = -\rho \sin \phi \cos \theta$
$\dfrac{\partial y}{\partial \rho} = \sin \phi \sin \theta$; $\dfrac{\partial y}{\partial \phi} = \rho \cos \phi \sin \theta$; $\dfrac{\partial y}{\partial \theta} = \rho \sin \phi \cos \theta$
$\dfrac{\partial z}{\partial \rho} = \cos \phi$; $\dfrac{\partial z}{\partial \phi} = -\rho \sin \phi$; $\dfrac{\partial z}{\partial \theta} = 0$

**23** If $A = (a_{ij})$, then $(u, v, w)^t = A(x, y, z)^t$ and $(x, y, z)^t$
$= A^{-1}(u, v, w)$
$\dfrac{\partial v}{\partial y} = a_{22}$ and $\dfrac{\partial y}{\partial v} = b_{22}$ if $A^{-1} = (b_{ij})$

## SECTION 3 (pages 637–638)

**1** $\dfrac{\partial z}{\partial s} = 10s; \dfrac{\partial z}{\partial t} = 10t$

**3** $\dfrac{\partial z}{\partial s} = 4s(s^2 + t^2); \dfrac{\partial z}{\partial t} = 4t(t^2 + s^2)$

**5** $\dfrac{\partial z}{\partial s} = \dfrac{y(2y - x)}{(x^2 + y^2)^{3/2}}; \dfrac{\partial z}{\partial t} = \dfrac{-y(2x + y)}{(x^2 + y^2)^{3/2}}$

**7** $\dfrac{\partial z}{\partial s} = y \sin t + (t \cos s)(x + 3y^2);$

$\dfrac{\partial z}{\partial t} = ys \cos t + (x + 3y^2) \sin s$

**9** $\dfrac{\partial z}{\partial u} = -(ye^{-x} + e^{-y}) + 2(u + 2v)(e^{-x} + xe^{-y});$

$\dfrac{\partial z}{\partial v} = -(ye^{-x} + e^{-y}) + 4(u + 2v)(e^{-x} + xe^{-y})$

**11** $\dfrac{\partial z}{\partial u} = v^2\{2x \cos(xy) - x^2 y \sin(xy)\} - x^3 \sin(xy)$

$\times \{2u(v + 3u)^{102} + 306u^2(v + 3u)^{101}\};$

$\dfrac{\partial z}{\partial v} = 2uv\{2x \cos(xy) - x^2 y \sin(xy)\} - x^3 \sin(xy)$

$\times \{102u^2(v + 3u)^{101}\}$

**13** $\dfrac{\partial w}{\partial s} = 10x + 13y - 4z; \dfrac{\partial w}{\partial t} = -5x + y + 2z$

**15** $\dfrac{\partial w}{\partial r} = 4(2r^3 + 3r^2 - 6r + 1)$

**17** $\dfrac{\partial w}{\partial r} = 2r(3u^2 + 4u - 3); \dfrac{\partial w}{\partial s} = -2s(3u^2 + 4u - 3);$

$\dfrac{\partial w}{\partial t} = 2t(3u^2 + 4u - 3)$

**19** $\dfrac{\partial z}{\partial r} = 0; \dfrac{\partial z}{\partial \theta} = -4$

**21** $\dfrac{1}{8}\sqrt{2}\pi(4 + \sqrt{2}) = \dfrac{dw}{dt}$

**23** $\dfrac{\partial z}{\partial r} = 0; \dfrac{\partial z}{\partial \theta} = \dfrac{1}{2}$

**25** $\dfrac{\partial w}{\partial s} = 0; \dfrac{\partial w}{\partial t} = -16$

**27** $\dfrac{dw}{dt} = -4$ at $t = 0$

**31** $g(t) = f(u, v, w), u = tx, v = ty,$
$w = tz;$ then $g'(t) = f_u u_t + f_v v_t + f_w w_t \rightarrow g'(t) = xf_u + yf_v + zf_w$
as required

## SECTION 4 (pages 640–641)

**1** $\dfrac{290\pi}{\sqrt{13}}$ cm$^2$/sec

**3** $\dfrac{dT}{dt} = \dfrac{250{,}000}{k}$

**5** $\dfrac{40\pi\sqrt{3} - 63}{252}$

**7** $-\frac{1}{15}(1 + 2\sqrt{10\pi})$

**9** (a) $\dfrac{\partial z}{\partial x} = f'(y/x)(-y/x^2); \dfrac{\partial z}{\partial y} = f'(y/x)(1/x)$

**11** (a) $\dfrac{\partial z}{\partial r} = \dfrac{\partial z}{\partial x} \cos \theta + \dfrac{\partial z}{\partial y} \sin \theta;$

$\dfrac{\partial z}{\partial \theta} = \dfrac{\partial z}{\partial x}(-r \sin \theta) + \dfrac{\partial z}{\partial y}(r \cos \theta)$

**13** $u = f(v, w), v = x + at, w = y + bt \rightarrow u_x = f_v, u_y = f_w,$ and
$u_t = f_v v_t + f_w w_t = af_v + bf_w = au_x + bu_y$

**15** $xf_x + yf_y = x[2x + y \ln(y/x) - y]$
$+ y[-2y + x \ln(y/x) + x] = 2f$

**17** $1 + 2\sqrt{3} - \dfrac{6 + 3\sqrt{3}}{2\sqrt{12 - 3\sqrt{3}}}$

**19** $x\dfrac{\partial z}{\partial x} - y\dfrac{\partial z}{\partial y} = xf'(xy)y - yf'(xy)x = 0$

**21** $h'(t) = -13/2$ cm/sec; $V'(t) = 22$ cm$^3$/sec

## SECTION 5 (pages 646–647)

**1** $d_\theta f = 6 \cos \theta + 8 \sin \theta$

**3** $d_\theta f = 6 \cos \theta - 3 \sin \theta$

**5** $d_\theta f(0, \pi/4) = -3 \cos \theta + 2 \sin \theta$

**7** $d_\theta f(4, 3) = (-3 \cos \theta + 4 \sin \theta)/25$

**9** $d_\theta f(0, \pi/3) = (\cos \theta - \sqrt{3} \sin \theta)/2$

**11** $d_\theta f = 2 \cos \theta + \sin \theta;$ max. if $\cos \theta = 2/\sqrt{5}, \sin \theta = 1/\sqrt{5}$

**13** $d_\theta f = (\cos \theta + \sqrt{3} \sin \theta)/2;$
max. if $\sin \theta = \sqrt{3}/2, \cos \theta = 1/2$

**15** $d_\theta f = (17 \cos \theta - 3 \sin \theta);$
max. if $\sin \theta = -3/\sqrt{292}, \cos \theta = 17/\sqrt{292}$

**17** $d_\theta f = (2 \cos \theta + \sin \theta)/4;$
max. if $\sin \theta = 1/\sqrt{5}, \cos \theta = 2/\sqrt{5}$

**19** $D_\mathbf{a} f = -15\lambda + 6\mu + (6 - 2e^3)\nu,$ where $\mathbf{a} = \lambda\mathbf{i} + \mu\mathbf{j} + \nu\mathbf{k}$

**21** $D_\mathbf{a} f = (-3/2)\lambda + (3/2)\mu + (5/2)\nu,$ where $\mathbf{a} = \lambda\mathbf{i} + \mu\mathbf{j} + \nu\mathbf{k}$

**23** $D_\mathbf{a} f = \dfrac{3\sqrt{2}}{16}\lambda + \dfrac{3\sqrt{2}}{16}\mu - 6\sqrt{2}\nu,$ where $\mathbf{a} = \lambda\mathbf{i} + \mu\mathbf{j} + \nu\mathbf{k}$

**25** $D_\mathbf{a} f(1, 2, 0) = 6(3 - 2e^2)/7$

**27** $D_\mathbf{a} f(0, 1, 1) = 2/\sqrt{26}$

**29** $D_\mathbf{a} f(2, -2) = \sqrt{26}/26$

**31** $D_\mathbf{a} f(1, 4, \pi/4) = -19\sqrt{22}/88$

**33** $\nabla f = 55\mathbf{i} - 42\mathbf{j}$

**35** $\nabla f = (2 \cos 2 - \cos 1)\mathbf{i} + (\cos 1 + 2 \cos 2)\mathbf{k}$

**37** $D_{\mathbf{a}}f(1, 0, 2) = 13/\sqrt{17}$; max. value $= \sqrt{29}$

**39** $D_{\mathbf{a}}f(0, 1, 0) = 5/\sqrt{26}$; max. value $= \sqrt{5}$

**41** $D_{\mathbf{a}}f(x, y, z) = \lim\limits_{h \to 0} \dfrac{1}{h}[f(x + \lambda h, y + \lambda h, z + vh) - f(x, y, z)]$

$\qquad = \lim\limits_{h \to 0} \dfrac{1}{\lambda h}[f(x + \lambda h, y + \mu h, z + vh)$
$\qquad\qquad - f(x, y + \mu h, z + vh)] \cdot \lambda$
$\qquad\quad + \lim\limits_{h \to 0} \dfrac{1}{\mu h}[f(x, y + \mu h, z + vh)$
$\qquad\qquad - f(x, y, z + vh)] \cdot \mu$
$\qquad\quad + \lim\limits_{h \to 0} \dfrac{1}{vh}[f(x, y, z + vh) - f(x, y, z)]$
$\qquad = \lambda f_x(x, y, z) + \mu f_y(x, y, z) + v f_z(x, y, z)$

**43** $d_\theta T(4, 3) = 400 \cos\theta - 300 \sin\theta$; $\tan\theta = 4/3$; slope of curve $= 4/3$

**45** $P_1(0, 0, 0)$; $P_2(2/3, 0, 2/3)$; $P_3(\sqrt{3}/3, ((2/\sqrt{3}) - 1)^{1/2}/2, 1/2)$;
$\quad P_4(\sqrt{3}/3, -((2/\sqrt{3}) - 1)^{1/2}, 1/2)$

**47** $\nabla(F(g)) = (F(g))_x\mathbf{i} + (F(g))_y\mathbf{j} + (F(g))_z\mathbf{k}$
$\qquad = F'(u)g_x\mathbf{i} + F'(u)g_y\mathbf{j} + F'(u)g_z\mathbf{k}$
$\qquad = F'(u)\nabla g$; $u = g(x, y, z)$

**49** Let $F = f_1\mathbf{i} + f_2\mathbf{j} + f_3\mathbf{k}$ and $G = g_1\mathbf{i} + g_2\mathbf{j} + g_3\mathbf{k}$ and verify the required equality

**21** Let $(x_1, y_1, z_1)$ be on $ax^2 + by^2 + cz^2 = 1$. Then
$\nabla f(x_1, y_1, z_1) \cdot [(x - x_1)\mathbf{i} + (y - y_1)\mathbf{j} + (z - z_1)\mathbf{k}] = 0 \to$
$2ax_1(x - x_1) + 2by_1(y - y_1) + 2cz_1(z - z_1) = 0 \to$
$axx_1 + byy_1 + czz_1 = ax_1^2 + by_1^2 + cz_1^2 = 1$

**23** Proof is the same as in problem 21

**25** Normal line at $(x_0, y_0, z_0)$ is $\dfrac{x - x_0}{6x_0} = \dfrac{y - y_0}{6y_0} = \dfrac{z - z_0}{2z_0}$.
If $x = 0$, $y = 0$, then $z = 2z_0/3$; $(0, 0, 2z_0/3) =$ point of intersection

**27** $\dfrac{x - 4}{4} = \dfrac{y + 2}{3} = \dfrac{z - 20}{20}$

**29** $\dfrac{x - 4}{3} = \dfrac{y + 3}{4} = \dfrac{z - 16}{24}$

**31** $\dfrac{x - 1}{-2} = \dfrac{y - 2}{2} = \dfrac{z - 2}{-1}$

**33** The normals to the surfaces at $(2, 2, 0)$ are $4\mathbf{i} + 4\mathbf{j}$ and $2\mathbf{i} + 2\mathbf{j}$ that are parallel. Hence the surfaces are tangent to each other.

**35** By problem 23, the tangent plane at $(x_0, y_0, z_0)$ has equation $xx_0 + yy_0 - zz_0 = 1$. The point $(0, 0, 1/z_0)$ is on the plane. The intersection of $x^2 + y^2 - z^2 = 1$ with the plane $z = 1/z_0$ is the circle $x^2 + y^2 = 1 + z_0^{-1}$ with points on both sides of the tangent plane.

**37** $P(-\frac{1}{4}, -\frac{5}{4}, -\frac{25}{8})$

## SECTION 6 (pages 653–654)

**1** $4x - 4y - z - 6 = 0$; $\dfrac{x - 2}{4} = \dfrac{y + 1}{-4} = \dfrac{z - 6}{-1}$

**3** $x - 2y + z - 2 = 0$; $\dfrac{x - 2}{-1} = \dfrac{y + 1}{2} = \dfrac{z + 2}{-1}$

**5** $ex - z = 0$; $\dfrac{x - 1}{e} = \dfrac{y - \pi/2}{0} = \dfrac{z - e}{-1}$

**7** $3x - 4y + 25z = 25(\ln 5 - 1)$; $\dfrac{x + 3}{3} = \dfrac{y - 4}{-4} = \dfrac{z - \ln 5}{25}$

**9** $2x + 2y + 3z - 3 = 0$; $\dfrac{x - 2}{2} = \dfrac{y - 1}{2} = \dfrac{z + 1}{3}$

**11** $4x - 3z - 25 = 0$; $\dfrac{x - 4}{4} = \dfrac{y + 2}{0} = \dfrac{z + 3}{-3}$

**13** $3x + 6y + 2z - 36 = 0$; $\dfrac{x - 4}{3} = \dfrac{y - 1}{6} = \dfrac{z - 9}{2}$

**15** $4x - y + 2z = 0$; $\dfrac{x - 0}{1} = \dfrac{y - 4}{-1/4} = \dfrac{z - 2}{1/2}$

**17** $x + 11y - 7z = 30$; $\dfrac{x - 1}{1} = \dfrac{y - 2}{11} = \dfrac{z + 1}{-7}$

**19** $6x - 6y - z = 0$; $\dfrac{x - 3}{6} = \dfrac{y - 3}{-6} = \dfrac{z - 0}{-1}$

## SECTION 7 (pages 658–659)

**1** $df = -0.17$; $\Delta f = -0.1689$

**3** $df = \dfrac{\pi}{2}(\pi - 5)$; $\Delta f = -\sqrt{3} + \sin\dfrac{13\pi^2}{2}$

**5** $df = -0.45$; $\Delta f = -0.456035$

**7** $df = 0.05$; $\Delta f = 0.0499$

**9** $df = -0.08$; $\Delta f = -0.080384$

**11** $dz = (-5y^4)\,dx + (5y^4 - 20xy^3)\,dy$

**13** $dz = (x\,dx + y\,dy)(x^2 + y^2)^{-1/2}$

**15** $d\theta = (-y\,dx + x\,dy)(x^2 + y^2)^{-1}$

**17** $dw = (x\,dx + y\,dy + z\,dz)(x^2 + y^2 + z^2)^{-1}$

**19** $dw = (2xe^{xy} + yx^2e^{xy})\,dx + (x^3e^{xy} - 2y^{-3})\,dy + (3z^2)\,dz$

**21** $100\left(\dfrac{df}{f}\right) = 100\left(\dfrac{mh}{x} + \dfrac{nk}{y} + \dfrac{pl}{z}\right)$

**23** 12.98      **25** 0.802 degrees

**27** $dV = 10.56\pi$, $dA = 2.208\pi$      **29** 0.0504 m/sec

**31** 0.111427

**33** $[12t^5 + 15t^4 + 48t^3 - 15t^2 + 12t - 35]\,dt$

**35** $(12r^3 - 4rs^2 + 24rs^{-2})h + (-4r^2s + 12s^3 - 24r^2s^{-3})k$

**37** $G_1 = h,\ G_2 = 2k + 6h$

**39** $T \approx \dfrac{6.28}{7} + 0.00185 = 0.89899$

## SECTION 8 (pages 664–665)

**1** $\dfrac{dy}{dx} = \dfrac{2x + 3y + 2}{-3x + 8y + 6}$

**3** $\dfrac{dy}{dx} = -\dfrac{2x + (1 + x^2 + y^2)ye^{xy}}{2y + (1 + x^2 + y^2)xe^{xy}}$

**5** $\dfrac{dy}{dx} = -\dfrac{y}{x}$

**7** $\dfrac{dy}{dx} = \dfrac{y - 3x(x^2 + y^2)^{3/2}}{x + 3y(x^2 + y^2)^{3/2}}$

**9** $\dfrac{dy}{dx} = [(2y)^{-1/2} + (3y)^{-2/3}]^{-1}$

**11** $\dfrac{\partial w}{\partial y} = \dfrac{3xw - 2y}{2w - 3xy}$

**13** $\dfrac{\partial w}{\partial y} = -\dfrac{xe^{xy} + we^{yw} + xw}{ye^{yw} - xe^{xw} + xy}$

**15** $\dfrac{\partial w}{\partial y} = \dfrac{y(w^2 + x^2)(w^2 + x^2 - 2y^2)}{w(x^2 - y^2)(2w^2 + x^2 - y^2)}$

**17** $\dfrac{\partial w}{\partial x} = -\dfrac{2w^3 + 3x^2y^2}{6xw^2 - 4yw + 4}$

**19** $\dfrac{\partial w}{\partial y} = \dfrac{x \sin(xy)}{1 - x \cos(wx)}$

**21** $\dfrac{\partial w}{\partial z} = -\dfrac{x^2 zw(y^2 + w^2)}{x^2 y^2 z^2 + 2x^2 z^2 w^2 - 2y^4 w^2 + x^6 - 3y^3 w}$

**23** $\dfrac{dz}{dx} = -\dfrac{3x + 2y}{3z + 4y};\ \dfrac{dy}{dx} = \dfrac{4x - 2z}{4y + 3z}$

**25** $\dfrac{dz}{dx} = \dfrac{z(x^2 - y^2)}{x(y^2 + z^2)};\ \dfrac{dy}{dx} = -\dfrac{y(x^2 + z^2)}{x(y^2 + z^2)}$

**27** $\dfrac{\partial u}{\partial x} = \dfrac{u}{2(u^2 + v^2)};\ \dfrac{\partial u}{\partial y} = \dfrac{v}{2(u^2 + v^2)};$
$\dfrac{\partial v}{\partial x} = \dfrac{-v}{2(u^2 + v^2)};\ \dfrac{\partial v}{\partial y} = \dfrac{u}{2(u^2 + v^2)}$

**29** $\dfrac{\partial u}{\partial x} = \dfrac{2xv}{v + u};\ \dfrac{\partial u}{\partial y} = \dfrac{1}{2(v + u)};\ \dfrac{\partial v}{\partial x} = \dfrac{2xu}{(u + v)};\ \dfrac{\partial v}{\partial y} = \dfrac{-1}{2(u + v)}$

**31** $\dfrac{\partial u}{\partial x} = 0;\ \dfrac{\partial u}{\partial y} = \dfrac{y}{u};\ \dfrac{\partial v}{\partial x} = -\dfrac{x}{v};\ \dfrac{\partial v}{\partial y} = 0$

**33** $\dfrac{\partial u}{\partial x} = \dfrac{g_v}{f_u g_v - f_v g_u};\ \dfrac{\partial u}{\partial y} = \dfrac{-f_v}{f_u g_v - f_v g_u};\ \dfrac{\partial v}{\partial x} = \dfrac{-g_u}{f_u g_v - f_v g_u};$
$\dfrac{\partial v}{\partial y} = \dfrac{-f_u}{f_u g_v - f_v g_u}$

**35** $\left(\dfrac{-1}{R^2}\right)\dfrac{\partial R}{\partial R_j} = \dfrac{\partial}{\partial R_j}\left(\dfrac{1}{R}\right) = \dfrac{\partial}{\partial R_j}\left(\sum_{i=1}^{n}\dfrac{1}{R_i}\right)$
$= \dfrac{-1}{R_j^2} \to \dfrac{\partial R}{\partial R_j} = \left(\dfrac{R}{R_j}\right)^2,\quad j = 1, 2, \ldots, n$

**37** Solve the following system for $du_i$ and then read off the derivatives $du_i/dx_j$;
$$\sum_{i=1}^{k}\left(\dfrac{\partial(F_p)}{\partial u_i}(du_i)\right) = -\sum_{j=1}^{n}\left[\dfrac{\partial(F_p)}{\partial x_j}(dx_j)\right],\qquad p = 1, 2, \ldots, k$$

**39** $\dfrac{\partial v}{\partial y} = \dfrac{96yuw^2 - 16yuw - 12yw^2}{-96uvw^2 + 16uw + 12w^2}$

## SECTION 9 (pages 669–670)

**1** $f_{xy} = 14x - 4y = f_{yx}$

**3** $f_{xy} = -18x^2y + 6xy^2 - 3y^2 = f_{yx}$

**5** $e^{xy}(\sin x + xy \sin x + x \cos x + \cos y + xy \cos y - y \sin y)$

**7** $f_{xy} = f_{yx} = -e^{x/y}\cos(xy)\left[\dfrac{-1}{y^2} - \dfrac{x}{y^3} - xy\right]$
$\qquad - e^{x/y}\sin(xy) + e^x \cos y$

**9** $\dfrac{-xy}{(x^2 + y^2 - z^2)^{3/2}}$

**11** $u_{xy} = u_{yx} = -2xy/(x^2 + y^2 + z^2)^2;$
$u_{xz} = u_{zx} = -2xz/(x^2 + y^2 + z^2)^2$

**13** $u_{xy} = u_{yx} = -3z;\ u_{xz} = u_{zx} = -3y$

**15** $u_{xy} = u_{yx} = e^{xy}[xy(x^2 + z^2) + z^2](x^2 + z^2)^{-3/2}$
$u_{xz} = u_{zx} = ze^{xy}[3x - y(x^2 + z^2)](x^2 + z^2)^{-5/2}$

**17** $u_{xy} = u_{yx} = [x^2 + y^2 + 2]^{-2}[-2y - 2x(z - 1)]$
$\qquad \times 8xy(x + yz - y)(x^2 + y^2 + 2)^{-3}$
$u_{xz} = -2xy(x^2 + y^2 + 2)^{-2} = u_{zx}$

**19** $u_{xy} = -\sin y = u_{yx};\ u_{xz} = \cos x = u_{zx};\ u_{yz} = \sec^2 z = u_{zy}$

**21** $\dfrac{\partial^4 u}{\partial x^4} = (x + 4)e^x \cos y;\ \dfrac{\partial^4 u}{\partial x^2 \partial y^2} = -(x + 2)e^x \cos y;$
$\dfrac{\partial^4 u}{\partial y^4} = xe^x \cos y \to$ the result

**23** $\dfrac{\partial^2 z}{\partial r^2} = 2 \cos 2s$

**25** $12r^2 - 24rs - 12s^2$

**27** $z_{xx} = 2(x - y)^{41}(990x^2 + 818y^2 + 1804xy);$
$z_x = (x + y)(x - y)^{42}(45x + 41y)$
$z_{yy} = 2(x - y)^{41}(990x^2 + 818y^2 - 1804xy);$
$z_y = -(x + y)(x - y)^{42}(41x + 45y)$
$z_{rr} = z_{xx}\sin^2 r - 2z_{xy}(rs \sin r \cos rs)$
$\qquad - z_x \cos r + z_{yy}s^2 \cos^2(rs) - s^2 z_y \cos(rs)$

**29** $z_{rr} = 2 \sin(s^2)$

**31** $z_{rr} = \tfrac{1}{2}e^{-s}(e^{r+s} + e^{-(r+s)})$

**33** $u_{yy} = F_{yy} + 2F_{yz}z_y + F_{zz}(z_y)^2 + F_z z_{yy}$

**35** $u_{xx} = f''(r) + g''(s)$, $u_{yy} = 4f''(r) + 4g''(s)$ where
$r = x + 2y$, $s = x - 2y \rightarrow u_{xx} - \frac{1}{4}u_{yy} = 0$

**37** $u_{rr} = F_{xx}\cos^2\theta + 2F_{xy}\sin\theta\cos\theta + F_{yy}\sin^2\theta$

**39** $\dfrac{d^2y}{dx^2} = -[F_{xx}F_y^2 - 2F_{xy}F_xF_y + F_{yy}F_x^2](F_y)^{-3}$

## SECTION 10 (pages 675–676)

**1** $10 + 13(x-2) + 4(y-1) + 6(x-2)^2 + 2(x-2)(y-1)$
$+ 2(y-1)^2 + (x-2)^3 + (x-2)(y-1)^2$

**3** $x + y - \frac{1}{6}(x+y)^3$

**5** $1 + (x+y) + \frac{1}{2}(x+y)^2 + \frac{1}{6}(x+y)^3 + \cdots$

**7** $1 - \dfrac{x^2}{2!} - \dfrac{y^2}{2!} + \dfrac{x^4}{4!} + \dfrac{x^2y^2}{2!2!} + \dfrac{y^4}{4!} + \cdots$

**9** $3 + 4(x-1) + 2(y-1) + z + (x-1)^2$
$+ z^2 + 2(x-1)(y-1) + (y-1)z$

**11** $\theta''(0) = 2\lambda^2 + 8\lambda\mu + 2\mu^2$. No

**13** $\displaystyle\sum_{i=0}^{k}\sum_{j=0}^{i}\dfrac{i!k!}{i!j!(k-i)!(i-j)!}A^jB^{i-j}C^{k-i}$

**17** $f(x_1, x_2, \ldots, x_k) = f(a_1, a_2, \ldots, a_k)$
$+ \displaystyle\sum_{1 \le r_1 + \cdots + r_k \le p} \dfrac{\partial^{r_1 + \cdots + r_k}f(a_1, \ldots, a_k)}{\partial x_1^{r_1}, \partial x_2^{r_2}, \ldots, \partial x_k^{r_k}}$
$\times \dfrac{(x_1 - a_1)^{r_1}(x_2 - a_2)^{r_2} \cdots (x_k - a_k)^{r_k}}{r_1!r_2! \cdots r_k!} + R_p$
where $R_p = \displaystyle\sum_{r_1 + \cdots + r_k = p+1} \dfrac{\partial^{r_1 + \cdots + r_k}f}{\partial x_1^{r_1}\partial x_2^{r_2} \cdots \partial x_k^{r_k}}$
$\times \dfrac{(x_1 - a_1)^{r_1}(x_2 - a_2)^{r_2} \cdots (x_k - a_k)^{r_k}}{r_1!r_2! \cdots r_k!}$

## SECTION 11 (pages 684–685)

**1** Rel. min. at $(2, -1)$

**3** Rel. min. at $(1, -2)$

**5** Rel. min. $(\frac{27}{2}, 5)$; saddle point $(\frac{3}{2}, 1)$

**7** No. rel. max. or rel. min.; $(1, \frac{2}{3})$, $(-1, -\frac{4}{3})$, saddle points

**9** Rel. max. at $\left(\dfrac{\pi}{3} \pm 2n\pi, \dfrac{\pi}{3} \pm 2m\pi\right)$;
rel. min. at $\left(-\dfrac{\pi}{3} \pm 2n\pi, -\dfrac{\pi}{3} \pm 2m\pi\right)$;
test fails for $(\pi \pm 2n\pi, \pi \pm 2m\pi)$

**11** Test fails but $(0, 0)$ is rel. max.

**13** Test fails: critical points on line $(x, n\pi)$

**15** $(1, -1)$ and $(-1, 1)$ are saddle points; no relative extremum exists

**17** Rel. min. at $(1, 1)$; $f(1, 1) = 3$

**19** For each integer $n$, $(n\pi, (-1)^{n+1})$ is a saddle point; no relative extremum

**21** Rel. min. at $(0, \pm 3\sqrt{2}/2)$; $(0, 0)$ is a saddle point

**23** Rel. and absolute max. at $(\pi/2, \pi/2)$

**25** Rel. max. at $(\pi/3, \pi/3)$; saddle point at $(\pi, \pi)$

**27** $(\pi/2, \pi/2)$ yields a saddle point

**29** Critical point, $(-\frac{3}{2}, -\frac{1}{2}, -\frac{1}{4})$

**31** $\dot{x} = -\frac{9}{31}$, $y = \frac{55}{62}$, $z = \frac{18}{31}$, $t = -\frac{41}{31}$

**33** $17/\sqrt{10}$    **35** $\frac{1}{2}\sqrt{10}$

**37** If $\alpha = xyz$, then $g(x, y) = x + y + \alpha/(xy)$ has a relative minimum at $(\alpha^{1/3}, \alpha^{1/3})$ for $x > 0$ and $y > 0$. Since $g(\alpha^{1/3}, \alpha^{1/3}) = 3\alpha^{1/3} \le x + y + z$, it follows that $(xyz)^{1/3} \le (x + y + z)/3$

**39** Cube with edge $= 2\sqrt{6}/3$

**41** $h = 3\sqrt{3}$, $l = 3$

**43** $(1, 0, 0)$, $(0, 1, 0)$

## SECTION 12 (pages 691–692)

**1** $\frac{8}{7}$    **3** $\dfrac{d^2}{a^2 + b^2 + c^2}$

**5** Min. $\frac{134}{75}$ at $(\frac{16}{15}, \frac{1}{3}, -\frac{11}{15})$

**7** $x = \pm\frac{1}{2}(1 - \sqrt{5})\sqrt{50 - 10\sqrt{5}}$, $y = \mp\sqrt{50 - 10\sqrt{5}}$

**9** Rel. max. at $\left(\dfrac{1}{\sqrt{2}}, \dfrac{1}{\sqrt{2}}\right)$ and $\left(\dfrac{-1}{\sqrt{2}}, \dfrac{-1}{\sqrt{2}}\right)$; $f_{max} = \dfrac{1}{2}$
Rel. min. at $\left(\dfrac{-1}{\sqrt{2}}, \dfrac{1}{\sqrt{2}}\right)$ and $\left(\dfrac{1}{\sqrt{2}}, \dfrac{-1}{\sqrt{2}}\right)$; $f_{min} = \dfrac{-1}{2}$

**11** Rel. max. at $(3^{1/3}, 3^{1/3}, 3^{1/3})$; $f_{min} = 3$

**13** Rel. min. at $(1, 1)$ and $(-1, -1)$; $f_{min} = 2$

**15** $f_{max} = \sqrt{238}$ at $(\sqrt{\frac{17}{14}}, 2\sqrt{\frac{17}{14}}, 3\sqrt{\frac{17}{14}})$

**17** Rel. min. at $(2, 1, 0)$ and $(0, -1, 2)$, $f_{min} = 5$

**19** $2, 2, 1$

**21** $h = 2\sqrt{5}$; $H = \dfrac{V}{25\pi} - \dfrac{4\sqrt{5}}{3}$

**23** $\dfrac{17}{23}$ occurs when $x = \dfrac{10}{23}$, $y = \dfrac{1}{23}$, $z = \dfrac{-1}{23}$, $t = \dfrac{17}{23}$

**25** Min. $= 132/13$

**27** Critical points $(1, 1, 1)$ and two complex numbers

**29** Dimensions: $\dfrac{2a}{\sqrt{3}}, \dfrac{2b}{\sqrt{3}}, \dfrac{2c}{\sqrt{3}}$

**31** $(1, -1, -1)$, $(1, 1, 1)$, $(-1, 1, 1)$, $(-1, -1, -1)$

**33** $f_{max} = \left[\displaystyle\sum_{i=1}^{k} b_i^2\right]^{1/2}$

## SECTION 13 (pages 697–698)

**1** Exact: $(x^4/4) + x^3y + (y^4/4) + C$

**3** Exact: $2xy - \ln x + \ln y + C$

**5** Not exact

**7** Exact: $e^{x^2} \sin y + C$

**9** Exact: $\arctan (y/x) + C$

**11** Not exact      **13** Not exact

**15** Exact: $x \sin (xy) - 3xy^2 + C$

**17** Exact: $xy^3 - \ln (xy) + C$

**19** Exact: $y^x + C$

**21** Exact: $x^2y + xz^2 + zy^2 + C$

**23** Not exact

**29** $f(x, y) = y/x + 2y - 2y \ln y + C$

**31** $f(x, y) = -e^{-x/y} - \frac{1}{2}y^2 + C, I = y^{-2}$

**33** $f(x, y) = \dfrac{y}{x} - \dfrac{1}{2}xe^{2x} + \dfrac{1}{4}e^{2x} + C, I = x^{-2}$

## CHAPTER 15 REVIEW PROBLEMS (pages 698–700)

**1** The level curves are circles: $x^2 + y^2 = c^2$

**3** The level curves are parabolas: $x^2 = -2(y - c)$

**5** The level curves are logarithmic curves: $y = c - \ln x, \ x > 0$

**7** The level curves are the lines: $y = \dfrac{c^2 - 1}{c^2 + 1}x$

**9** The level curves are inverse sine curves:
$y = -\arcsin (x - c)$

**11** $f_x = 2x + 3y, f_y = 3x + 4y^3, f_{xx} = 2, f_{xy} = 3$

**13** $f_x = \dfrac{1}{1 + (x + 2y)^2}, f_y = \dfrac{2}{1 + (x + 2y)^2}$

**15** $f_x = \left(\dfrac{-2y}{x^3}\right) e^{\sin (y/x^2)} \cos (y/x^2),$

$f_y = \left(\dfrac{1}{x^2}\right) e^{\sin (y/x^2)} \cos (y/x^2)$

**17** If $u = x^2 + y^2 + z^4$, then $f_z = \dfrac{-2z^3 \sin (\sqrt{u})}{\sqrt{u}},$

$f_{xx} = (-\cos \sqrt{u})[x^2 u^{-1}] + (\sin \sqrt{u})x^2 u^{-3/2}$
$- (\sin \sqrt{u})u^{-1/2}$

$f_{xy} = -xyu^{-1} \cos \sqrt{u} + xyu^{-3/2} \sin \sqrt{u}$

**21** $z_s = (2x + 3y)t, z_t = (2x + 3y)s - 3x \sin t$

**23** $z_s = \dfrac{4xst}{1 + (x^2 + y^2)^2} + \dfrac{4yt}{1 + (x^2 + y^2)^2},$

$z_t = \dfrac{2xs^2}{1 + (x^2 + y^2)^2} + \dfrac{2y(2s + 3)}{1 + (x^2 + y^2)^2}$

**25** $z_s = \dfrac{e^y}{s} + xte^y, z_t = \dfrac{e^y}{t} + xse^y$

**27** $\dfrac{-640}{\sqrt{41}} \approx -99.95$ km/hr

**29** $D_a f(0, \pi) = 0$      **31** $D_a f(2, 1, 1) = 0$

**33** Tangent plane: $2x + y - z = 1$;

normal line: $\dfrac{x - 2}{2} = \dfrac{y - 0}{1} = \dfrac{z - 1}{-1}$

**35** Tangent plane: $-2x + 16y - 17z = 34 - \dfrac{17}{2} \ln (17)$;

normal line: $\dfrac{x + 1}{-2} = \dfrac{y - 2}{16} = \dfrac{z - (\ln (17))/2}{-17}$

**37** Tangent plane: $-32x + 48y - z = 128$;

normal line: $\dfrac{x + 2}{-32} = \dfrac{y - 2}{48} = \dfrac{z - 32}{-1}$

**39** $dz = 2x^3(x^4 + y^6)^{-1/2} dx + 3y^5(x^4 + y^6)^{-1/2} dy$

**41** $dw = (x + y)^{-1} dx + (x + y)^{-1} dy - \tan z \, dz$

**43** 4.996

**45** 2.22; max. error = 0.3148

**47** $\dfrac{dy}{dx} = -\dfrac{2x - y^2(1 + x^2 + y^4)e^{-xy^2}}{4y^3 - 2xy(1 + x^2 + y^4)e^{-xy^2}}$

**49** $\dfrac{dy}{dx} = -\dfrac{ye^{-xy} + \cos x}{xe^{-xy} + \sin y}$

**51** $\dfrac{\partial w}{\partial y} = -\dfrac{2ywe^{x^2 + y^2} - xw}{e^{x^2 + y^2} - xy}$

**53** $\dfrac{\partial w}{\partial y} = \dfrac{1 + x + 2y}{1 + 2w}$

**55** $\dfrac{\partial w}{\partial x} = \dfrac{2y(xy - zw) - yzw \sin (xyzw)}{xyz \sin (xyzw) + 2z(xy - zw)}$

**57** $f_{xy} = 7 = f_{yx}$

**59** If $w = x^2 + y^2 + z^3$, then $u_{xx} = 2w^{-1} - 4x^2w^{-2}$,
$u_{xy} = -4xyw^{-2} = u_{yx}, u_{zx} = -6xz^2w^{-2}$

**61** $z_{rr} = z_{xx}x_r^2 + 2z_{xy}x_ry_r + z_{yy}y_r^2 + z_xx_{rr} + z_yy_{rr}$, where
$x_r = -\sin r, x_{rr} = -\cos r, y_r = 3 \cos (rs), y_{rr} = -s^2 \sin (rs)$,
$z_x = 2(x - y)(3x + 4y)^2 + 6(x - y)^2(3x + 4y)$,
$z_y = -2(x - y)(3x + 4y)^2 + 8(x - y)^2(3x + 4y)$,
$z_{xx} = 2(3x + 4y)^2 + 24(x - y)(3x + 4y) + 18(x - y)^2$,
$z_{xy} = -2(3x + 4y)^2 + 4(x - y)(3x + 4y) + 24(x - y)^2$,
$z_{yy} = 2(3x + 4y)^2 - 32(x - y)(3x + 4y) + 32(x - y)^2$

**63** $z_{rr} = z_{xx}x_r^2 + 2z_{xy}x_ry_r + z_{yy}y_r^2 + z_xx_{rr} + z_yy_{rr}$, where
$x_r = 2s, x_{rr} = 0, y_r = 2(r + s), y_{rr} = 2, z_x = -2xye^{-(x^2 + y^2)}$,
$z_y = -2y^2e^{-(x^2 + y^2)}, z_{xx} = (-2y + 4x^2y)e^{-(x^2 + y^2)}$,
$z_{xy} = (-2x + 4xy^2)e^{-(x^2 + y^2)}, z_{yy} = (-4y + 4y^3)e^{-(x^2 + y^2)}$

**65** $f(1, 0) = 1$ is a rel. min.

**67** Saddle points: $(0, 0), (0, 1), (0, -1), (1, 0), (-1, 0)$;

Rel. min.: $\left(\dfrac{1}{2}, \dfrac{-1}{2}\right), \left(\dfrac{-1}{2}, \dfrac{1}{2}\right)$;

Rel. max.: $\left(\dfrac{1}{2}, \dfrac{1}{2}\right), \left(\dfrac{-1}{2}, \dfrac{-1}{2}\right)$

**69** No extrema

**71** Saddle point: $(0, 0)$; No extrema

**73** Exact: $f(x, y) = xy + \cos(xy) + C$

**75** Not exact

**77** Exact: $f(x, y) = (-\cos x + x \ln y) + C$

**79** Exact: $f(x, y) = \sin x + y \cos x + C$

---

# CHAPTER 16

---

## SECTION 1 (page 707)

**1** 0.9690          **3** 3.5355          **5** 3.8140

**7** 690          **9** 3.2148          **11** 0.182

## SECTION 2 (page 709)

**1** Smallest $= 0$; largest $= 80$

**3** Smallest $= 48$; largest $= 1680$

**5** Smallest $= -27\pi\sqrt{2}$; largest $= 27\pi\sqrt{2}$

**7** Smallest $= 9$; largest $= 9\sqrt{10}$

## SECTION 3 (pages 716–717)

**1** $\frac{423}{4}$          **3** $\frac{1}{6}(e^3 - 1)(e^2 - 1)$

**5** $-1498\frac{3}{14}$          **7** $-\frac{18}{7}\sqrt{3} + \frac{16}{21}\sqrt{2} + \frac{559}{8}$

**9** 0          **11** $\frac{1}{4}(2\cos 4 - \cos 8 - 1)$

**13** $2e - \frac{e^4}{2}$          **15** $\frac{9\sqrt{3}}{4} + \frac{\pi^2}{18}\sin 3$

**17** $\frac{\pi}{24} - \frac{1}{2}\ln 2 + \ln 3 - \ln(1 + \sqrt{2}) + \frac{1}{2}(\sqrt{2} - 1)$

**19** $40\frac{88}{105}$          **21** $\frac{\pi}{4}(\sqrt{3} - 1)$

**23** $\frac{2\pi - 3\sqrt{3}}{48}$          **25** $2e^{\sqrt{2}/2} - \sqrt{2}e$

**27** 0.0          **29** $\frac{2\sqrt{2} - 1}{3}$

**31** $\frac{1}{6}(1 - \cos 1)$          **33** $\frac{1}{2}(\sqrt{2} - 1)$

**35** $\frac{2}{\pi}(1 - \cos 1)$          **37** $\frac{1}{30}(4\sqrt{2} - 1)$

**39** $\frac{5}{3}$          **41** $12\pi$

**43** $\frac{a^3}{90}$          **45** 18.9          **47** $4\pi$

## SECTION 4 (page 720)

**1** $\frac{5}{12}$          **3** $\frac{2}{3} + 4\ln\left(\frac{3}{2}\right)$

**5** $34\frac{2}{3}$          **7** $\frac{1}{2}\left(5 - 12\ln\left(\frac{3}{2}\right)\right)$

**9** $(2048)\pi$          **11** $1707/140$

**13** 18          **15** $\frac{ac}{3}(a + b)$

**17** $3a \arctan(ab) - \frac{3}{2b}\ln(1 + a^2b^2)$

**19** $\frac{1}{\pi} + \frac{1}{6}$          **21** $\frac{1}{2}(e - 1)$

## SECTION 5 (pages 725–726)

**1** $\frac{4\pi}{3}$          **3** $2\pi$

**5** $\frac{64\pi}{3} - \frac{256}{9}$          **7** $\sqrt{2} - 1$

**9** $\frac{8\pi}{5}$          **11** $\frac{1}{6}(1 - \cos 1)$

**13** $2\pi(\sin 1 - \cos 1)$

**15** $\frac{64}{3}$          **17** $\frac{1}{6}(8\pi + 12\sqrt{3})$

**19** $\frac{1}{4}(8 + \pi)$          **21** $4\pi$

**23** $\frac{32\pi}{3}(8 - 3\sqrt{3})$          **25** $\frac{64}{9}$

**27** $\frac{9\pi}{2}$          **29** $4\pi\sqrt{3}$

**31** $\frac{\pi}{4}$          **33** $\frac{1}{9}(3\pi - 4)a^3$

**35** $\frac{1}{3}(\sqrt{2} + \ln(1 + \sqrt{2}))a^3$

**37** $\frac{2}{9}\pi c^3$; $\frac{2}{3}\phi c^3$

**41** $\frac{\sqrt{3}\pi}{18}$

## SECTION 6 (pages 733–734)

**1** $I = \dfrac{\rho a^4}{3}$; $R = \dfrac{a}{\sqrt{3}}$

**3** $I = \dfrac{3}{35}\rho$; $R = \dfrac{3}{\sqrt{35}}$

**5** $I = \dfrac{4ka^5}{15}$; $R = a\sqrt{2/5}$

**7** $I = \rho(\pi^2 - 4)$; $R = \sqrt{(\pi^2 - 4)/2}$

**9** $\frac{423}{28}\rho$     **11** $\dfrac{3k}{56}$

**13** $ka^5\dfrac{\pi}{5}$     **15** $\dfrac{\rho\pi a^4}{16}$

**17** $I = \rho\left(\dfrac{\pi}{8} - \dfrac{1}{5}\right)$

**19** $\bar{x} = \frac{235}{112}$, $\bar{y} = \frac{25}{32}$

**21** $\bar{x} = \frac{5}{9}$, $\bar{y} = \frac{4}{7}$

**23** $\bar{x} = \bar{y} = \dfrac{5a}{8}$

**25** $\bar{x} = \frac{5}{3}$, $\bar{y} = 0$

**27** $\bar{x} = -\dfrac{2}{8\sqrt{3} - 1}$, $\bar{y} = 0$

**33** $\dfrac{\pi\delta(r_2^4 - r_1^4)}{2}$; $R^4 = r_2^4 - r_1^4$

## SECTION 7 (pages 738–739)

**1** $\frac{2}{15}(1 + 9\sqrt{3} - 8\sqrt{2})$

**3** $8a^2$     **5** $2\pi\sqrt{2}$

**7** $9\sqrt{2}$     **9** $\dfrac{\pi\sqrt{2}}{2}$

**11** $\frac{1}{9}(160 - 24\pi)$

**13** $2a^2\dfrac{\pi - 2}{3}$     **15** $16$

**17** $\dfrac{a^2}{3}(\pi + 6\sqrt{3} - 12)$

**19** $\frac{1}{12}(\sqrt{5} + 1 + 3\ln(2 + \sqrt{5}))$

**21** $\dfrac{\pi(17\sqrt{17} - 1)}{6}$

**23** $2\sqrt{2}\pi$

## SECTION 8 (pages 740–741)

**1** $\dfrac{4\pi ab^2}{3}$     **3** $\dfrac{64\pi}{15}$

**5** $\dfrac{2\pi a^3}{21}$     **7** $\dfrac{\pi}{7}$

**9** $\pi(e - 2)$     **11** $\dfrac{8\pi}{15}$

**13** $\dfrac{4\pi}{9}$     **15** $\dfrac{269\pi}{24}$

**17** $V(S) = \dfrac{256\pi\sqrt{2}}{105}$; $\bar{x} = \dfrac{256\sqrt{2}}{105}\pi$, $\bar{y} = 0$

**19** $\bar{y}$ is $\frac{2}{3}$ of the way from $(b, c)$ to opposite side. Same for distances from other vertices.

**21** (b) $V = \pi ac(d + \frac{1}{3}c)$; $\bar{y} = d + \frac{1}{3}c$

## SECTION 9 (pages 746–747)

**1** $\frac{1}{8}$     **3** $\frac{109}{1008}$

**5** $\frac{1}{3}(16 - 3\pi)$     **7** $\frac{1}{2}$

**9** $2\pi r^2$     **11** $0$

**13** $\frac{1}{4}\pi a^4$     **15** $\frac{128}{45}a^5$; $(8, 9, 1)$

**17** $\frac{18}{385}$     **19** $\frac{608}{15}$

**21** $\left(\dfrac{128\sqrt{2}}{15}\right)a^3$

**23** $\displaystyle\int_0^1\int_y^1\int_0^{x-y} x\, dz\, dx\, dy$; $\displaystyle\int_0^1\int_0^x\int_0^{x-z} x\, dy\, dz\, dx$

**25** $\displaystyle\int_0^1\int_{y^2}^{\sqrt{y}}\int_0^{y+z} xy\, dx\, dz\, dy$; $\displaystyle\int_0^1\int_0^{1-x}\int_{y^2}^{\sqrt{y}} xy\, dz\, dy\, dx$

**27** $\displaystyle\int_0^4\int_{-\sqrt{4-x}}^{+\sqrt{4-x}}\int_0^{y+2} (xyz)\, dz\, dy\, dx$;

$\displaystyle\int_{-2}^2\int_0^{y+2}\int_0^{\sqrt{4-y^2}} (xyz)\, dx\, dz\, dy$

**29** $\displaystyle\int_0^1\int_0^{1-y}\int_0^{x+y} (x-y)z\, dz\, dx\, dy$;

$\displaystyle\int_0^1\int_0^y\int_{z-y}^{1-y} (x-y)z\, dx\, dz\, dy$

**31**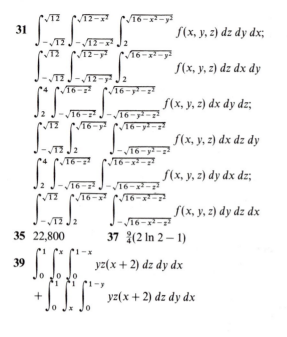

**35** 22,800        **37** $\frac{9}{4}(2 \ln 2 - 1)$

**39** $\int_0^1 \int_0^x \int_0^{1-x} yz(x+2)\, dz\, dy\, dx$

$+ \int_0^1 \int_x^1 \int_0^{1-y} yz(x+2)\, dz\, dy\, dx$

**11** $\dfrac{4ka^6}{9}$        **13** $\dfrac{4\pi\delta abc(a^2 + b^2)}{15}$

**15** $\dfrac{4\pi\delta}{15}((c^2 - a^2)^{3/2}(2c^2 + 3a^2) - (b^2 - a^2)^{3/2}(2b^2 + 3a^2))$

**17** $\bar{x} = \bar{z} = \dfrac{2a}{5};\ \bar{y} = \dfrac{a}{5}$

**19** $\bar{y} = \bar{z} = 0;\ \bar{x} = \frac{8}{7}$

**21** $\bar{x} = \bar{y} = 0;\ \bar{z} = \frac{1}{3}$

**23** $\bar{y} = \bar{z} = 0;\ \bar{x} = 2a$

**25** $\bar{x} = \bar{y} = 0;\ \bar{z} = \dfrac{3a(2 + \sqrt{2})}{16}$

**27** $\bar{x} = \bar{y} = 0;\ \bar{z} = \dfrac{9a}{7}$

**29** $\bar{x} = \bar{y} = 0;\ \bar{z} = \dfrac{4(1591 - 720\sqrt{3})}{7(391 - 192\sqrt{3})}$

**31** $\bar{y} = \bar{z} = 0;\ \bar{x} = \dfrac{a\sqrt{2}}{2}$

## SECTION 10 (page 752)

**1** $\pi\delta a^3 \dfrac{2 - \sqrt{2}}{3}$        **3** $\pi k(b^4 - a^4)$

**5** $12\pi ka^5 \dfrac{\sqrt{3}}{5}$        **7** $128ka^5 \dfrac{15\pi - 26}{225}$

**9** $\dfrac{7\pi ka^4}{6}$        **11** $\dfrac{7\pi\delta a^3}{6}$

**13** $2\delta a^3 \dfrac{3\pi + 20 - 16\sqrt{2}}{9}$        **15** $\dfrac{\pi ka^4}{4}$

**17** $\dfrac{\pi}{2}$        **19** $\dfrac{5\pi a^3}{64}$

**21** $\pi(e^{16} - 1)$        **23** $\dfrac{\pi}{9}(2 - \sqrt{2})$

**25** $\dfrac{4\pi}{3}(\cos 1 - \cos 27)$

## SECTION 11 (page 755)

**1** $\dfrac{2\delta a^5}{3}$        **3** $\dfrac{ka^6}{90}$

**5** $32\delta\left(\dfrac{4}{9} - \dfrac{1}{5} + \dfrac{1}{21}\right) = \dfrac{2944\delta}{315}$

**7** $\dfrac{5\pi\delta}{16}$        **9** $\dfrac{4\pi k}{9}(b^6 - a^6)$

## CHAPTER 16 REVIEW PROBLEMS (pages 755–757)

**1** $\frac{22}{3}$        **3** $\frac{9}{20}$

**5** $\frac{1}{7}$        **7** $\frac{1}{3}(\cos(\frac{1}{2}) - \cos 1)$

**9** $\dfrac{e - 1}{12}$        **11** $\frac{21}{2}$

**13** $(e^2 - e)(1 - e^{-2})$        **15** 2

**17** $\frac{1}{2}(e - 2)$        **19** 9

**21** $2(e^{b/2} - e^{-(a/2)}) - \frac{1}{3}(b^3 + a^3)$ where $(-a, a^2)$ and $(b, b^2)$ are the intersections of the curves $y = x^2$ and $y^2 = e^x$.

**23** $\pi \sin 1$

**25** $\pi(1 - e^{-1})$        **27** $\frac{1}{4}(1 - \cos 1)$

**29** 2        **31** $\dfrac{\pi}{2}$

**33** $I = \frac{27}{20},\ R^2 = \frac{6}{5}$        **35** $\dfrac{20k\pi}{3}$

**37** $I_z = \dfrac{1}{12}(8\pi^3 - 7\pi);\ M = 2\pi;\ R = \left(\dfrac{8\pi^2 - 7}{24}\right)^{1/2}$

**39** $\bar{x} = -\frac{1}{2};\ \bar{y} = \frac{82}{45}$

**41** $\bar{x} = \dfrac{\pi\sqrt{2} - 4}{4(\sqrt{2} - 1)};\ \bar{y} = \dfrac{1}{4(\sqrt{2} - 1)}$

**43** $\bar{x} = \bar{y} = \frac{1}{5}$        **45** $2\sqrt{3}\pi$

**47** (a) $7\pi$;   (b) $10\pi$;   (c) $3\sqrt{2}\dfrac{\pi}{2}$

**49** $\int_0^a \int_0^y \int_0^z f(x)\,dx\,dz\,dy = \int_0^a \int_y^a \int_x^y f(x)\,dz\,dx\,dy$

$$= \int_0^a \int_y^a (y-x)f(x)\,dx\,dy$$

$$= \int_0^a \int_x^a (y-x)f(x)\,dy\,dx$$

$$= \frac{1}{2}\int_0^a (a-x)^2 f(x)\,dx$$

**51** $\frac{1}{8}$

**53** $-\frac{729}{4}$

**55** $\frac{\pi}{6}(8-3\sqrt{3})$

**57** $\frac{\pi}{12}$

**59** $\frac{\pi}{3}(e-1)(2-\sqrt{2})$

**61** $\frac{4}{3}\pi(b^3-a^3)$

**63** $(\bar{x}, \bar{y}, \bar{z}) = (\frac{1}{3}, 1, \frac{1}{3})$

**65** $(\bar{x}, \bar{y}, \bar{z}) = (\frac{2}{5}, \frac{6}{5}, \frac{1}{5})$

**67** $(\bar{x}, \bar{y}, \bar{z}) = (\frac{7}{12}, \frac{34}{45}, \frac{43}{90})$

# CHAPTER 17

## SECTION 2 (pages 765–767)

**1** $\frac{7}{2}$

**3** $\frac{231}{64}$

**5** $\pi + 8$

**7** $4 + 2\ln(\frac{5}{4})$

**9** $\arcsin(\frac{4}{5})$

**11** $\dfrac{508\sqrt{3}}{11}$

**13** $\frac{64}{105}$

**15** $\dfrac{8}{9^4}\left[(37)^{3/2}\left[\dfrac{(37)^3}{9} - \dfrac{3}{7}(37)^2 + \dfrac{3}{5}(37) - \dfrac{1}{3}\right]\right.$

$\left. - (10)^{3/2}\left[\dfrac{(10)^3}{9} - \dfrac{3}{7}(10)^2 + \dfrac{3}{5}(10) - \dfrac{1}{3}\right]\right]$

**17** $\dfrac{\sqrt{61}}{7}(e^7 - 1)$

**19** $0$

**21** $-\frac{17}{2}$

**23** $\frac{2}{3}\pi a^2 (a^2 + b^2)^{1/2}(3a^2 + 4b^2\pi^2)$

**25** $3691/30$    **27** $-3/2$    **29** $8$    **31** $\frac{1}{2}\ln(17)$

**33** $2\pi$

## SECTION 3 (pages 772–773)

**1** $\frac{101}{3}$

**3** $-e + \cos 1$

**5** $-\frac{7}{5}$

**7** $\sin 3 - \cos 2 + e^{-6}\cos 6$

**9** $-9$

**11** $e^{-6}\cos 1 + \sin 2 - \cos 3$

**13** $2\ln 3 + 8\ln 2 + 84$

**15** $0$

**17** $108\pi^3$

**19** $\dfrac{\partial}{\partial y}(\Sigma \alpha_i P_i) = \Sigma \alpha_i \left(\dfrac{\partial P_i}{\partial y}\right) = \Sigma \beta_i \dfrac{\partial Q_i}{\partial x} = \dfrac{\partial}{\partial x}(\Sigma \beta_i Q_i)$

The condition is not necessary.

**23** Work $= \dfrac{1}{2}\ln\left(\dfrac{a^2 + b^2}{c^2 + d^2}\right)$ is independent of the path, provided $(0, 0)$ is not on the path.

**25** Exact differential $\rightarrow$ work $= (d^2 + e^2 + f^2)^{-1/2}$
$\qquad\qquad\qquad\qquad\qquad - (a^2 + b^2 + c^2)^{-1/2}$

**27** $\dfrac{k}{2}\ln(1 + 4\pi^2)$

## SECTION 4 (pages 779–780)

**1** $\displaystyle\int_0^1 \int_0^1 (Q_x - P_y)\,dx\,dy = 2 = \int_\Gamma P\,dx + Q\,dy$

**3** $\displaystyle\int_2^3 \int_1^2 (Q_x - P_y)\,dx\,dy = -\frac{27}{2} = \int_\Gamma (xy\,dx - 2xy\,dy)$

**5** $\displaystyle\int_0^1 \int_0^x (Q_x - P_y)\,dx\,dy = \frac{1}{4} = \int_\Gamma P\,dx + Q\,dy$

**7** Each integral is 0; $V$ is exact.

**9** Exact differential; 0

**11** $108$    **13** $3$    **15** $-\dfrac{19,702}{105}$    **17** $-3\pi$

**19** $\displaystyle\oint_{\partial G} (0\,dx + x\,dy) = \int_G \int (1-0)\,dA = $ Area. To make the equation valid, join the boundary curves with slits and orient the boundaries positively, to define $\partial G$.

**21** $\frac{7}{2}$    **23** $\frac{9}{2}$    **25** $\pi ab$    **27** $27\pi$

**29** $-\frac{988}{5}$    **31** $0$    **33** $\frac{3}{4}\pi a^4$    **35** $2\pi$

**37** $0$

**39**

$$-\oint_{\partial G} v(f_y\,dx - f_x\,dy) = -\int_G\int [(-vf_x)_x - (vf_y)_y]\,dA$$

$$= \int_G\int (v_x f_x + v_y f_y)\,dy\,dx$$

$$+ \int_G\int v(f_{xx} + f_{yy})\,dA$$

$$= \int_G\int (v_x f_x + v_y f_y)\,dy\,dx$$

## SECTION 5 (page 789)

**1** $\dfrac{\sqrt{3}}{6}$     **3** $\dfrac{15\pi\sqrt{2}}{4}$

**5** $2\pi$     **7** $\dfrac{1}{3}\left[\left(1 + \dfrac{\pi^2}{4}\right)^{3/2} - 1\right]$

**9** $\dfrac{\pi}{4}(15\sqrt{2} + 17)$

**11** $\dfrac{1024\delta}{45}$     **13** $\dfrac{(2 + \sqrt{3})\delta}{6}$

**15** $\bar{z} = \dfrac{2 + \sqrt{2}}{4},\ \bar{x} = \bar{y} = 0$

**17** $\bar{x} = \dfrac{4a}{3(\pi - 2)},\ \bar{y} = \bar{z} = 0$

**19** Case (i) $4\pi a^2\dfrac{\delta}{c}$; Case (ii) $4\pi\delta a$

**21** $2\pi\delta$

## SECTION 6 (page 798)

**1** 0     **3** $\dfrac{15\pi}{2}$

**5** $\dfrac{4}{15}$     **7** $\dfrac{64}{3}$

In problems 9 through 13, we give the value of $\displaystyle\int_{\partial S} \mathbf{v}\cdot d\mathbf{r}$.

**9** 0     **11** $2\left(\dfrac{\pi^2}{4} - 1\right)$     **13** 0     **15** 0

## SECTION 7 (pages 805–806)

**1** $a_{11} + a_{22} + a_{33}$

**3** $2x + 2y + 2z$

**5** $e^{xz}(2z\cos(yz) - x)$

**7** 0     **9** 0

**11** $\frac{1}{8}$     **13** $\frac{4}{3}\pi$

**15** $12\pi\sqrt{3}$     **17** 0

**19** $2\pi$     **21** 0

**23** 0     **25** $5\pi$

## CHAPTER 17 REVIEW PROBLEMS (pages 806–807)

**1** 11     **3** $\frac{211}{3}$

**5** $\frac{67}{65}$     **7** $\sqrt{2}\dfrac{\pi}{3}(3 + \pi^2)$

**9** 0     **11** 11

**13** $e^5 - 1$     **15** 42

**17** (a) $c$: Triangle with vertices $(0, 0)$, $(1, 0)$, $(1, 1)$ oriented clockwise; $\frac{5}{6}$;

   (c) $c$: Triangle with vertices $(0, 0, 0)$, $(1, 0, 0)$ and $(1, 0, 1)$ in that order; $\frac{3}{2}$;

   (d) $c$: Triangle with vertices $(0, 0, 0)$, $(1, 0, 0)$, $(1, 1, 0)$; $\frac{1}{3}$

**19** $\frac{9}{2}$     **21** 30

**23** 0     **25** 0

**27** $3\pi$     **29** $\frac{2}{3}(2\sqrt{2} - 1)$

**31** $\frac{4}{3}$     **33** $\pi[3\sqrt{2} - \ln(1 + \sqrt{2})]$

**35** $\sqrt{2}\pi$     **37** $\dfrac{4\pi}{3}$

**39** 0     **41** $-8\pi$

**43** 5     **45** $e^{x+y} - e^{x-y}$

**47** $12x^2 - 4x$     **49** $-\frac{1}{24}$

**51** 2     **53** 0

# CHAPTER 18

## SECTION 1 (page 813)

**1** $(3 + \sin x)' = \cos x$; linear; order 1

**3** $y' = 5e^x + 14e^{2x}$, $y'' = 5e^x + 28e^{2x}$, $y'' - 3y' + 2y = 0$; linear; 2nd order

**5** $y' = 3x^2$, $y'' = 6x$, $y''' = 6 \to x^3 y''' + x^2 y'' - 3xy' - 3y = 0$; linear; 3rd order

**7** $y' = (2xy)^{-1/2} \to \sqrt{2xy}\,y' = 1$; nonlinear; 1st order

**9** $y = Ce^{2x}$

**11** $y = \sec x + 2x + C$

**13** $y = 3x + \dfrac{e^x}{5}(\sin 2x - 2\cos 2x) + C$

**15** $y = \frac{1}{8}(1+x)^8 - \frac{2}{7}(1+x)^7 + \frac{1}{6}(1+x) + C$

**17** $y = -\sin x + C_1 x + C_2$

**19** $y = \displaystyle\sum_{k=0}^{\infty} \dfrac{x^{2k+2}}{(k!)(2k+1)(2k+2)} + \frac{1}{2}x^3 + C_1 x + C_2$

**21** $y = -\dfrac{3}{2} + \dfrac{C}{2}e^{2x}$

## SECTION 2 (pages 817–818)

**1** $\tan y + x^{-1} = C$

**3** $x = -y^{-1} - 2y + C$

**5** $-x^{-1} = 2e^{-y} - y + C$

**7** $\ln|y| = \frac{1}{2}x^2 + C$

**9** $e^x + e^{-2y} = C$

**11** $(x^2+1)(y^2+1) = C$

**13** $y^2 = 2\sin(C - x^4)$

**15** $\dfrac{y^3}{3}\ln|y| - \dfrac{y^3}{9} = \dfrac{x^5}{5}\ln|x| - \dfrac{x^5}{25} + C$

**17** Homogeneous; degree 2

**19** Homogeneous; degree 1

**21** Homogeneous; degree 2

**23** Homogeneous; degree $-1$

**25** Not homogeneous

**27** $2\arctan\left(\dfrac{y}{x}\right) - \dfrac{1}{2}\ln(x^2+y^2) = C$

**29** $(y - 2x)^2(x+y) = C$

**31** $(x-y)(3x+y) = C(x+y)$

**33** $\ln|x| + \arcsin\left(\dfrac{y}{|x|}\right) = C$

**35** $x\ln|x| + \sqrt{x^2+y^2} = Cx$

**37** $y = Cx$

**39** $y = C\csc\left(\dfrac{x}{y}\right), \quad y \ne 0$

**41** Let $y = vx$,

$$\dfrac{dy}{dx} = v + \dfrac{x\,dv}{dx} \to \dfrac{dy}{dx} = v + \dfrac{x\,dv}{dx} = f(x) + g(vx)$$

$$= f(x) + g(x) \to y = \int (f(x) + g(x))\,dx + C$$

## SECTION 3 (page 821)

**1** $\dfrac{x^2}{2} + 2xy - 2x - \dfrac{y^2}{2} + 3y = C$

**3** $-x^3 - \dfrac{y^3}{3} + xy + y^2 + 2x = C$

**5** $\dfrac{x^4}{2} + 3x^2y^2 - 2xy^3 - y^4 + 2x^2 + 3xy - y^2 = C$

**7** $x\arcsin y + 2\sin y = C$

**9** $(x-y)^2(x^2+y^2) = C$

**11** $\frac{1}{2}x^2\ln|y| + xy\ln|x| = C$

**13** $\left(\sin y + \ln(xy) + \dfrac{1}{x}\right)dx + \left(x\cos y + \dfrac{x}{y}\right)dy$
$= d(x\sin y + (x\ln x - x) + x\ln y + \ln x) = 0$
$x\sin y + (x\ln x - x) + x\ln y + \ln x = C$

**15** $y^3 + \arctan xy = C$

**17** $\dfrac{x\,dy - y\,dx}{x^2} = d\left(\dfrac{y}{x}\right) = (y+y^3)\,dy;$

$\dfrac{y}{x} = \dfrac{1}{2}y^2 + \dfrac{1}{4}y^4 + C$

**19** $\dfrac{x\,dy - y\,dx}{x^2} = -xe^{2x}\,dx;\ \dfrac{y}{x} = -\dfrac{1}{2}xe^{2x} + \dfrac{1}{4}e^{2x} + C$

**21** $x\sin x\,dx + 2\sec^2 y\,dy = 0;\ -x\cos x + \sin x + 2\tan y = C$

**23** $\dfrac{x\,dx + y\,dy}{x^2+y^2} + \dfrac{y\,dx - x\,dy}{x^2+y^2} = 0;$

$\dfrac{1}{2}\ln(x^2+y^2) + \arctan\dfrac{y}{x} = C$

**25** $\dfrac{1}{|x|}\,dx = \dfrac{y\,dx - x\,dy}{\sqrt{x^2-y^2}(|x|)} = d\arcsin\left(\dfrac{y}{|x|}\right);$

$\ln|x| = \arcsin\left(\dfrac{y}{|x|}\right) + C$

**27** $(PI)\,dx + (QI)\,dy$ exact $\to (PI)_y - (QI)_x$

$= 0 \to P_y - Q_x = \dfrac{QI_x}{I} - \dfrac{PI_y}{I} \to$

$P_y - Q_x = Q(\log|I|)_x - P(\log|I|)_y$

## SECTION 4 (page 824)

**1** $y = \dfrac{x^{n+1}}{n} + Cx$

**3** $y = e^x(x^3 + 2x^2 - 3x + C)$

**5** $y = (2x)^{-1}(e^{2x} + 2\cos x + C)$

**7** $y = x^2 - 1 + C\sqrt{x^2-1}$

**9** $y = (2x-3)^{-3/2}\dfrac{4x^3 - 9x^2 + 6C}{6}$

**11** $(y \sin x)' = 1;\ y = (\csc x)(x + C)$

**13** $(y \sec x)' = \tan x;\ y = \cos x(C + \ln |\sec x|)$

**15** $y = \dfrac{(1 - x)x + C}{1 + x}$

**17** $y = (\sin 2x)^{-1/2}(\cos 2x + C)$

**19** $y = \dfrac{1}{3}(a - x)(a + x)^2 + \dfrac{C(a - x)}{a + x}$

## SECTION 5 (pages 828–830)

**1** $xy = C$        **3** $x^2 - y^2 = C, \dfrac{y}{x} > 1$

**5** $y = \dfrac{k}{2}(e^{C \pm x/k} + e^{-C \pm x/k})$

**7** $\dfrac{(x)\,dy}{dx} + e^x - xe^x - y = 0$

**9** $(2xy + 1)y = x(xy + 2)\dfrac{dy}{dx}$

**11** $y\left(\dfrac{dy}{dx}\right)^2 + 2x\left(\dfrac{dy}{dx}\right) - y = 0$

**13** $y^2 = Cx^3$      **15** $x^2 + y^2 = Cy$

**17** $P(t) = P(0) + \dfrac{\beta}{\alpha}(e^{\alpha t} - 1)$

**19** $B(t) = 100e^{2t - (\cos \alpha t/\alpha)}$

**21** $T(t) = (2kt + (T(0)^2)^{1/2};\ T(0) = $ thickness at $t = 0$

**23** $s = \dfrac{g}{k}t - \dfrac{g}{k^2}(1 - e^{-kt})$

     ($s = $ distance fallen, $k = $ const in resistance law)

**25** $s = \dfrac{1}{k}\ln \dfrac{1}{2}(e^{akt} + e^{-akt}), a = \sqrt{\dfrac{g}{k}}$

**27** $v_e = 1.866$ km/sec

**29** $v_e = 618.264$ km/sec

**31** $x = ab(1 - e^{(b-a)kt})(a - be^{(b-a)kt})^{-1} \to a$ as $t \to \infty$

## SECTION 6 (page 835)

**1** $y = C_1 e^{2x} + C_2 e^{-2x}$

**3** $y = C_1 e^x + C_2 e^{3x}$

**5** $y = e^{-x}(C_1 \cos x + C_2 \sin x)$

**7** $y = e^x[C_1 \cos(x\sqrt{2}) + C_2 \sin(x\sqrt{2})]$

**9** $y = e^{-3x}(C_1 \cos x + C_2 \sin x)$

**11** $y = A \cos(kx) + B \sin(kx)$

**13** $y = e^{-x}(A \cos 2x + B \sin 2x)$

**15** $y = e^x(A \cos \sqrt{2}x) + B \sin(\sqrt{2}x)$

**17** $y = (A + Bx)e^x$

**19** $y = (A + Bx)e^{-x}$

**21** $y = e^x\left(A \cos\left(\sqrt{3}\dfrac{x}{2}\right) + B \sin\left(\sqrt{3}\dfrac{x}{2}\right)\right)$

**23** $y = (A + Bx)e^{-\sqrt{2}x}$

**25** $y = Ae^{(2 + \sqrt{6})x} + Be^{(2 - \sqrt{6})x}$

**27** $y = -e^{-3x}$

**29** $y = -e^{3x} + 5xe^{3x}$

**31** $y = 6(e^{x/2} - e^{2x/3})$

**33** $y = \dfrac{2(1 - e^2)}{3 + e^4}(e^x - e^{-3x})$

**35** $y = -\dfrac{1}{2}(e^{5(x-3)} + e^{-5(x-3)})$

**37** $y = 2e^{-2x}(\cos x + \sin x)$

## SECTION 7 (page 841)

**1** $y = Ae^{3x} + Be^{-x} - 14 + 12x - 9x^2$

**3** $y = x + \dfrac{1}{2} + e^{-x} + e^{-x}(C_1 \cos x + C_2 \sin x)$

**5** $y = \dfrac{1}{4}x^2 + \dfrac{1}{2}x + \dfrac{3}{8} - \dfrac{1}{16}e^{-2x} + (C_1 x + C_2)e^{2x}$

**7** $y = e^x(\dfrac{1}{2}x^2 + C_1 x + C_2)$

**9** $y = -\dfrac{1}{2}\cos x + C_1 e^x + C_2 e^{-x}$

**11** $y = -\dfrac{1}{5}\cos 2x - \dfrac{1}{10}\sin 2x + e^{-x}(C_1 \cos x + C_2 \sin x)$

**13** $y = -\dfrac{1}{2}x \cos x + C_1 \cos x + C_2 \sin x$

**15** $y = Ae^{-2x} + B + \dfrac{e^x}{3}$

**17** $y = A \cos 2x + B \sin 2x + 3e^x - 2x$

**19** $y = Ae^{2x} + Be^{-2x} + \dfrac{1}{4}xe^{2x} - \dfrac{1}{2}$

**21** $y = Ae^x + Be^{-x} - 5 + \cos 2x$

**23** $y = A \cos 2x + B \sin 2x + \dfrac{1}{2}(1 + \sin 2x)$

**25** $y = \dfrac{1}{12}(e^{-x} - 4e^{2x} + 3e^{3x})$

**27** $y = x - \sin x$

**29** $y = \dfrac{x^2}{4}$

**31** $y = e^{2x} - \dfrac{1}{2}e^{-2x} - \dfrac{1}{2} + 2x$

**33** $y = -2e^{-2x}(\cos x + 2 \sin x) + 2$

**35** $y = (\sin x)(x + B)$, for all constants $B$

## SECTION 8 (pages 845–846)

**1** $y = a_0 \displaystyle\sum_{k=0}^{\infty} \dfrac{x^{2k}}{(2k)!} + a_1 \sum_{k=0}^{\infty} \dfrac{x^{2k+1}}{(2k+1)!}$

**3** $y = a_0 \displaystyle\sum_{k=0}^{\infty} (-1)^k \dfrac{x^{2k}}{1 \cdot 3 \cdots (2k-1)} + a_1 \sum_{k=0}^{\infty} (-1)^k \dfrac{x^{2k+1}}{2^k \cdot k!}$

**5**   $y = a_0 \displaystyle\sum_{k=0}^{\infty} \dfrac{(-2x^2)^k}{1 \cdot 3 \cdots (2k-1)} + a_1 \displaystyle\sum_{k=0}^{\infty} (-1)^k \dfrac{x^{2k+1}}{k!}$

**7**   $y = a_0 \left[ 1 + \displaystyle\sum_{m=1}^{\infty} \dfrac{1 \cdot 4 \cdot 7 \cdots (3m-2)}{(3m)!} x^{2m-2} \right]$

      $+ a_1 \left[ 1 + \displaystyle\sum_{m=1}^{\infty} \dfrac{2 \cdot 5 \cdots (3m-1)}{(3m+1)!} x^{2m-1} \right]$

**9**   $y = 1 + \displaystyle\sum_{m=1}^{\infty} \dfrac{x^{2m}}{2^m (m!)}$

**11**   $y = a_0 \left[ 1 + \displaystyle\sum_{k=1}^{\infty} \dfrac{1 \cdot 4 \cdots (3k-2)}{(3k)!} x^{3k} \right]$

      $+ a_1 \left[ x + \displaystyle\sum_{k=1}^{\infty} \dfrac{2 \cdot 5 \cdots (3k-2)}{(3k+1)!} x^{3k+1} \right]$

**13**   $y = a_0 \left( 1 + \dfrac{x^4}{3 \cdot 4} + \dfrac{x^8}{3 \cdot 4 \cdot 7 \cdot 8} + \dfrac{x^{12}}{3 \cdot 4 \cdot 7 \cdot 8 \cdot 11 \cdot 12} + \cdots \right)$

      $+ a_1 \left( x + \dfrac{x^5}{4 \cdot 5} + \dfrac{x^9}{4 \cdot 5 \cdot 8 \cdot 9} + \dfrac{x^{13}}{4 \cdot 5 \cdot 8 \cdot 9 \cdot 12 \cdot 13} + \cdots \right)$

**15**   $|x| < 2$;   $y = a_0 + a_1 x + \dfrac{a_0 + a_1}{2}(e^x - 1 - x)$

**17**   $y = 1 + x^2 + \tfrac{5}{24} x^4 + \cdots$

**19**   $y = x - \tfrac{1}{2} x^3 + \tfrac{7}{40} x^5 + \cdots$

**21**   $y = 1 + x + x^2 + \tfrac{4}{3} x^3 + \tfrac{7}{6} x^4 + \cdots$

**23**   $y = x - \tfrac{1}{6} x^3 - \tfrac{1}{12} x^4 + \cdots$

**25**   $y_1 = 1 + 2x + 2x^2 + \cdots, \quad y_2 = 1 - x^2 + \cdots$

**27**   $y = x - \tfrac{1}{6} x^3 + \tfrac{1}{24} x^5 + \cdots$

## CHAPTER 18 REVIEW PROBLEMS (pages 846–847)

**1**   $y = ce^{3x}$

**3**   $y = c + \displaystyle\int e^{-x^2} \, dx = c + \displaystyle\sum_{n=0}^{\infty} (-1)^n \dfrac{x^{2n+1}}{(n!)(2n+1)}$

**5**   $y = c + \dfrac{3}{2} x^2 + \dfrac{1}{2} \displaystyle\int (1 + \cos 2x) x^{-2} \, dx$

      $= c + \dfrac{3}{2} x^2 - \dfrac{1}{2x} + \dfrac{1}{2} \displaystyle\sum_{n=0}^{\infty} (-1)^n 4^n \dfrac{x^{2n-1}}{(2n)!(2n-1)}$

**7**   $y = (e^{-x^2} - c)^{-1/2}$

**9**   $y = c(1 + 4e^{-x^2})^{-1/8}$

**11**   $y = c(1 + e^{-3x})^{-1/3}$

**13**   $y = \tan(c - \arctan x)$

**15**   $\dfrac{2y}{x} = \sin\left(\dfrac{2y}{x}\right) + 4 \ln |x| + c$

**17**   Exact; $x^2 + 2xy - y^2 = c$

**19**   Exact; $2x(1 + y^2) + y^2(1 + x^2) = c$

**21**   Exact; $y^2 + 2xy^3 = e^{xy} + c$

**23**   $y = (x + c) \csc x$

**25**   First order linear; $y = (1 + x^2)[x - \arctan x + c]$

**27**   First order linear;
      $y = \tfrac{1}{3} \tan x + \tfrac{2}{3} \tan x \sec^2 x + \sec^3 x$

**29**   First order linear; $y = \tfrac{1}{2}(x^2 - 1 + 2e^{1-x^2})$

**31**   $M(9) = 600e^{1/2\pi}$

**33**   $y^2 = 8x + c$

**35**   $y^2 = x^2(1 + cx^2)^{-1}$

**37**   $y = e^{3x} + Be^{-2x}$

**39**   $y = (3x - 2)e^{x/2}$

**41**   $y = e^{2x}(A \cos(\sqrt{3}x) + B \sin(\sqrt{3}x))$

**43**   $y = 2 \cos x$

**45**   $y = A \cos x + B \sin x - \tfrac{1}{2} x \sin x$

**47**   $y = e^{-2x}(A \cos x + B \sin x) - 8 + 10x + \tfrac{1}{2} e^{3x}$

**49**   $y = Ae^{-x} + Bxe^{-x} + \tfrac{1}{2} x^2 e^{-x}$

**51**   $y = Ae^x + Be^{-2x} + \tfrac{1}{3} xe^x + \tfrac{1}{4} e^{2x}$

# APPENDIX 1

## SECTION 1 (page App. 5)

**5**  $\sin 3\theta = 3 \sin \theta - 4 \sin^3 \theta$;
$\cos 3\theta = 4 \cos^3 \theta - 3 \cos \theta$

**7**  $\sin (\theta/2) = \pm \sqrt{(1 - \cos \theta)/2}$;
$\cos (\theta/2) = \pm \sqrt{(1 + \cos \theta)/2}$

**9**  $\sin 15° = \frac{1}{2}\sqrt{2 - \sqrt{3}}$; $\cos 15° = \frac{1}{2}\sqrt{2 + \sqrt{3}}$;
$\sin 22\frac{1}{2}° = \frac{1}{2}\sqrt{\sqrt{2} - 1}$; $\cos 22\frac{1}{2}° = \frac{1}{2}\sqrt{\sqrt{2} + 1}$;
$\tan 75° = 2 + \sqrt{3}$; $\sec 75° = 2/\sqrt{2 - \sqrt{3}}$

**11**  $\sin \left(\dfrac{3\pi}{2} - \theta\right) = -\cos \theta$; $\cos \left(\dfrac{3\theta}{2} - \theta\right) = -\sin \theta$;
$\sin \left(\dfrac{3\pi}{2} + \theta\right) = -\cos \theta$; $\cos \left(\dfrac{3\pi}{2} + \theta\right) = \sin \theta$;
$\sin (2\pi - \theta) = -\sin \theta$; $\cos (2\pi - \theta) = \cos \theta$

**17**  $\sin \left(\dfrac{\pi}{6} - \theta\right) = \frac{1}{2} \cos \theta - \frac{1}{2}\sqrt{3} \sin \theta = \cos \left(\dfrac{\pi}{3} + \theta\right)$;
$\cos \left(\dfrac{5\pi}{6} - \theta\right) = -\frac{1}{2}\sqrt{3} \cos \theta + \frac{1}{2} \sin \theta$

## SECTION 2 (page App. 9)

**1**  Amp $= 2$; per $= 4\pi$     **3**  Amp $= 4$; per $= 4$

**5**  Amp $= 3$; per $3\pi$     **7**  Amp $= 2$; per $= \pi$

**9**  Amp $= 2$; per $= \pi$     **11**  Amp $= 4$; per $= \pi$

**13**  Amp $= 3$; per $= 4\pi$     **15**  Amp $= 1$; per $= \pi$

**17**  Per $= \pi/2$     **19**  Amp $= \sqrt{2}$; per $= 2\pi$

**21**  Per $= 4\pi$     **23**  Amp $= 4$; per $= \pi$

# APPENDIX 2

## SECTION 1 (page App. 13)

**3** $x^2 + y^2 - 2x + 4y - 20 = 0$; $C(1, -2)$; $r = 5$

**5** $x^2 + y^2 + 3x + 2y - 3 = 0$; $C(-\frac{3}{2}, -1)$; $r = 5/2$

**7** $7x^2 + 7y^2 - 19x + 15y - 58 = 0$

**9** $x^2 + y^2 + 2y - 19 = 0$

**11** $x^2 + (y - k)^2 = 16$

**13** $(x - h)^2 + y^2 = h^2$

**15** $(x - h)^2 + (y - 3h)^2 = \frac{1}{10}(10h + 4)^2$

**19** $x = \frac{1}{10}(1 \pm \sqrt{39})$; $y = \frac{1}{10}(9 \pm \sqrt{39})$

**21** Do not intersect

**23** Circle; $3x^2 + 3y^2 - 26x - 4y + 55 = 0$; $C(\frac{13}{3}, \frac{2}{3})$; $r = 2\sqrt{2}/3$

**25** Point of intersection: $(1, -2)$

## SECTION 2 (pages App. 16–17)

**1** $x - 2y - 3 = 0$       **3** $3x + y - 10 = 0$

**5** $x + y + 2 = 0$; $2x - 3y + 9 = 0$

**7** $2x + (-5 \pm \sqrt{29})y - 27 \pm 5\sqrt{29} = 0$

**9** $(1 \pm \sqrt{37})x + 6y + 19 \pm \sqrt{37} = 0$

**11** $x - 3y - 4 = 0$, $x - 11y - 12 = 0$

**13** No tangents

**17** 30 meters

**19** $y^2 + 18x + 9$, parabola; $V(-\frac{1}{2}, 0)$, $F(4, 0)$, dir. $x = -5$

## SECTION 3 (pages App. 19–20)

**1** $x^2 + 3y^2 = 24$

**3** $7x^2 + 3y^2 = 84$.

**5** $144x^2 + 95y^2 = 4655$

**13** Tangents through $Q$: $y = 3$, $4x - y - 5 = 0$
Tangents through $R$: $3x \pm 2\sqrt{6}y - 15 = 0$

## SECTION 4 (pages App. 23–24)

|   | $a$ | $b$ | $c$ | $e$ | foci | vertices | directrices | asymptotes |
|---|---|---|---|---|---|---|---|---|
| **1** | 3 | 2 | $\sqrt{13}$ | $\dfrac{\sqrt{13}}{3}$ | $(\pm\sqrt{13}, 0)$ | $(\pm 3, 0)$ | $x = \pm\dfrac{9}{\sqrt{13}}$ | $y = \pm\dfrac{2x}{3}$ |
| **3** | 2 | 1 | $\sqrt{5}$ | $\dfrac{\sqrt{5}}{2}$ | $(\pm\sqrt{5}, 0)$ | $(\pm 2, 0)$ | $x = \pm\dfrac{4}{\sqrt{5}}$ | $y = \pm\dfrac{x}{2}$ |
| **5** | 1 | 1 | $\sqrt{2}$ | $\sqrt{2}$ | $(\pm\sqrt{2}, 0)$ | $(\pm 1, 0)$ | $x = \pm\dfrac{1}{\sqrt{2}}$ | $y = \pm x$ |
| **7** | 5 | $\sqrt{8}$ | $\sqrt{33}$ | $\dfrac{\sqrt{33}}{5}$ | $(0, \pm\sqrt{33})$ | $(0, \pm 5)$ | $y = \pm\dfrac{25}{\sqrt{33}}$ | $y = \pm\dfrac{5x}{\sqrt{8}}$ |

**9** $4x^2 - y^2 = 144$   **11** $63x^2 - 81y^2 = 448$   **17** $y^2 - x^2 = 2$   **21** $8x^2 - y^2 - 2x - 55 = 0$

**13** $400x^2 - 225y^2 = 5184$   **15** $9x^2 - 4y^2 = 189$   **23** $R$: $11y \pm \sqrt{22}x \mp 4\sqrt{22} = 0$

## SECTION 5 (page App. 29)

**1** Ellipsoid       **3** Ellipsoid (oblate)

**5** Elliptic hyperboloid of two sheets

**7** Elliptic hyperboloid of one sheet

**9** Elliptic paraboloid; axis; $x$ axis

**11** Circular paraboloid; axis; $y$ axis

**13** Circular cone about $y$ axis

**15** Elliptic hyperboloid of one sheet

**17** Hyperbolic paraboloid

**19** a) $x + y = 0$, $z = 0$; $x - y = 1$, $z = x + y$
b) $x + y = 0$, $z = 0$; $x - y = a$, $z = a(x + y)$

## APPENDIX 3

### SECTION 1 (page App. 33)

**19** $\cosh x = \frac{5}{3}$, $\sinh x = -\frac{4}{3}$, $\coth x = -\frac{5}{4}$,
$\operatorname{sech} x = \frac{3}{5}$, $\operatorname{csch} x = -\frac{3}{4}$

**21** $\sinh x = -\frac{1}{2}$, $\cosh x = \frac{\sqrt{5}}{2}$, $\tanh x = \frac{-1}{\sqrt{5}}$,

$\coth x = -\sqrt{5}$, $\operatorname{sech} x = \frac{2}{\sqrt{5}}$

**23** $\sinh x = -\sqrt{3}$, $\tanh x = \frac{-\sqrt{3}}{2}$, $\coth x = \frac{-2}{\sqrt{3}}$,

$\operatorname{sech} x = \frac{1}{2}$, $\operatorname{csch} x = \frac{-1}{\sqrt{3}}$

**25** Minimum at $(0, 1)$, concave upward for all $x$

**27** No max or min; concave downward for $x < 0$; concave upward for $x > 0$; $x = 0$ and $y = \pm 1$ are asymptotes

**29** No max or min; concave downward for $x > 1$; concave upward for $x < 1$

### SECTION 2 (page App. 36–37)

**1** $\ln\left(\frac{1}{2} + \frac{\sqrt{5}}{2}\right)$

**3** $\ln(2 + \sqrt{3})$

**5** $\ln 3$

**7** No max or min; point of inflection at $(0, 0)$; concave upward for $x \leqslant 0$; concave downward for $x \geqslant 0$

**9** No max or min; $(0, 0)$ is point of inflection; concave upward for $0 \leqslant x < 2$; concave downward for $-2 < x \leqslant 0$

**11** No rel max or min; point of inflection at $x = \sqrt{2}/8$; concave downward for $\sqrt{2}/8 \leqq x < \frac{1}{4}$, concave upward for $0 < x \leqslant \sqrt{2}/8$

**13** $f'(x) = 2/\sqrt{4x^2 + 1}$

**15** $g'(x) = -\frac{1}{x^2}\operatorname{argtanh} x^2 + \frac{2}{1 - x^4}$

**17** $G'(x) = \sec x$

**19** $f'(x) = \frac{4}{\sqrt{4x^2 + 1}}\operatorname{argsinh} 2x$

**21** $\ln(5 + 2\sqrt{6}) - \ln(2 + \sqrt{3})$

**23** $\ln(\sqrt{2} - 1) - \ln(\sqrt{5} - 2)$

**25** $\ln(5 + \sqrt{21}) - \ln(3 + \sqrt{5})$

## APPENDIX 4

### SECTION 1 (pages App. 46–47)

**3** $-2$          **5** $24$          **9** $(1, -2, -1)$

**7** $8$          **11** $\left(-\frac{73}{19}, \frac{21}{19}, -\frac{43}{19}\right)$

# INDEX

# A SHORT TABLE OF INTEGRALS

The constant of integration is omitted.

---

## Trigonometric forms

45. $\displaystyle \int \sin mu \cos nu \, du = -\frac{\cos(m-n)u}{2(m-n)} - \frac{\cos(m+n)u}{2(m+n)}, \quad m \neq \pm n$

46. $\displaystyle \int u^n \sin u \, du = -u^n \cos u + n \int u^{n-1} \cos u \, du$

47. $\displaystyle \int u^n \cos u \, du = u^n \sin u - n \int u^{n-1} \sin u \, du$

48. $\displaystyle \int \arcsin u \, du = u \arcsin u + \sqrt{1-u^2}$

49. $\displaystyle \int \arccos u \, du = u \arccos u - \sqrt{1-u^2}$

50. $\displaystyle \int u \arcsin u \, du = \frac{1}{4}[(2u^2-1)\arcsin u + u\sqrt{1-u^2}]$

---

## Logarithmic and exponential forms

51. $\displaystyle \int \ln |u| \, du = u(\ln |u| - 1)$

52. $\displaystyle \int (\ln |u|)^2 \, du = u(\ln |u|)^2 - 2u \ln |u| + 2u$

53. $\displaystyle \int u^n \ln |u| \, du = \frac{u^{n+1}}{n+1} \ln |u| - \frac{u^{n+1}}{(n+1)^2}, \quad n \neq -1$

54. $\displaystyle \int u^n e^u \, du = u^n e^u - n \int u^{n-1} e^u \, du$

55. $\displaystyle \int e^{au} \sin bu \, du = \frac{e^{au}(a \sin bu - b \cos bu)}{a^2 + b^2}$

56. $\displaystyle \int e^{au} \cos bu \, du = \frac{e^{au}(a \cos bu + b \sin bu)}{a^2 + b^2}$

---

## Miscellaneous forms

57. $\displaystyle \int (a+bu)^n \, du = \frac{(a+bu)^{n+1}}{b(n+1)}, \quad n \neq -1$

58. $\displaystyle \int \frac{du}{u^2(a+bu)} = -(au)^{-1} + ba^{-2}[\log(a+bu) - \log u]$

59. $\displaystyle \int \frac{du}{(a+bu)^{1/2}(c+du)^{3/2}} = \frac{2}{bc-ad}\left(\frac{a+bu}{c+du}\right)^{1/2}, \quad bc-ad \neq 0$

60. $\displaystyle \int \frac{du}{u(a+bu^n)} = (an)^{-1}[\log(u^n) - \log(a+bu^n)], \quad n \neq 0$